Environmental Science

Books in the Brooks/Cole Biology Series

Biology: The Unity and Diversity of Life, Ninth, Starr / Taggart

Biology: Concepts and Applications, Fourth, Starr

Laboratory Manual for Biology, Perry / Morton / Perry

Human Biology, Fourth, Starr / McMillan

Perspectives in Human Biology, Knapp

Human Physiology, Fourth, Sherwood

Psychobiology: The Neuron and Behavior, Hoyenga / Hoyenga

Introduction to Cell and Molecular Biliogy, Wolfe

Introduction to Microbiology, Second, Ingraham / Ingraham

Genetics: The Continuity of Life, Fairbanks

Human Heredity, Fifth, Cummings

Introduction to Biotechnology, Second, Daly / Wilson

Plant Biology, Rost et al.

Plant Physiology, Fourth, Salisbury / Ross

Plant Physiology Laboratory Manual, Ross

Essentials of Ecology, Miller

General Ecology, Second, Krohne

Living in the Environment, Twelfth, Miller

Environmental Science, Ninth, Miller

Sustaining the Earth, Fifth, Miller

Environment: Problems and Solutions, Miller

Environmental Science, Fifth, Chiras

Oceanography: An Invitation to Marine Science, Third, Garrison

Essentials of Oceanography, Second, Garrison

Introduction to Ocean Sciences, Segar

Oceanography: An Introduction, Fifth, Ingmanson / Wallace

Marine Life and the Sea, Milne

For more information about this or any other
Brooks / Cole products, contact:
BROOKS / COLE
511 Forest Lodge Road
Pacific Grove, CA 93950 USA
www.brookscole.com
1-800-423-0563 (Thomson Learning Academic
Resource Center)

Environmental Science

Working with the Earth

NINTH EDITION

G. TYLER MILLER, JR.

President, Earth Education and Research

Adjunct Professor of Human Ecology
St. Andrews Presbyterian College

BROOKS/COLE
THOMSON LEARNING™

Australia • Canada • Mexico • Singapore • Spain • United Kingdom • United States

Publisher: *Jack Carey*
Assistant Editor: *Suzannah Alexander*
Editorial Assistant: *Karoliina Tuovinen*
Production Management: *Electronic Publishing Services Inc., NYC*
Marketing Assistant: *Maureen Griffin*
Advertising Project Manager: *Linda Yip*
Print/Media Buyer: *Vena Dyer*
Photo Researchr: *Linda L. Rill*
Copy Editor: *Electronic Publishing Services Inc., NYC*
Cover Photo: © Corbis Stock Market/ZEFA Germany/2001
Interior Illustration: *Electronic Publishing Services Inc., NYC; Precision Graphics; Sarah Woodward; Darwin and Vally Hennings; Tasa Graphic Arts, Inc.; Alexander Teshin Associates; John and Judith Waller; Raychel Ciemma; and Victor Royer*
Typesetting: *Electronic Publishing Services Inc., NYC*
Printing and Binding: *Transcontinental Printing/Interglobe*

Title Page Photograph: *Crater Lake, Oregon (Jack Carey)*
Part Opening Photographs
Part I: *Composite satellite view of Earth. © Tom Van Sant/ The GeoSphere Project*
Part II: *Endangered green sea turtle. © David B. Fleetham*
Part III: *Area of forest in Czechoslovakia killed by acid deposition and other air pollutants. Silvestris Fotoservice/NHPA*
Part IV: *Highly endangered Florida panther. George Gentry/Fish & Wildlife Service*
Part V: *Sunset. NOAA Corps Collection, photo by Commander John Bortniak, NOAA Corps. (ret.)*

Printed in Canada

1 2 3 4 5 6 7 06 05 04 03 02

For more information about our products, contact us at:
Thomson Learning Academic Resource Center
1-800-423-0563
For permission to use material from this text, contact us by:
Phone: 1-800-730-2214
Fax: 1-800-730-2215
Web: http://www.thomsonrights.com

Library of Congress Cataloging-in-Publication Data

Miller, G. Tyler (George Tyler),
Environmental science : working with the Earth/ G. Tyler Miller. -- 9th ed.
p. cm.
Includes index.
ISBN 0-534-38987-2
1. Environmental sciences 2. Human ecology.
3. Environmental protection. I. Title
GE105.M544 2002
333.7'2—dc21 2001043186

Brooks/Cole-Thomson Learning
511 Forest Lodge Road
Pacific Grove, CA 93950
USA

Asia
Thomson Learning
60 Albert Street, #15-01
Albert Complex
Singapore 189969

Australia
Nelson Thomson Learning
102 Dodds Street
South Melbourne, Victoria 3205
Australia

Canada
Nelson Thomson Learning
1120 Birchmount Road
Toronto, Ontario M1K 5G4
Canada

Europe/Middle East/Africa
Thomson Learning
Berkshire House
168-173 High Holborn
London WC1V7AA
United Kingdom

Latin America
Thomson Learning
Seneca, 53
Colonia Polanco
11560 Mexico D.F.
Mexico

Spain
Paraninfo Thomson Learning
Calle/Magallanes, 25
28015 Madrid, Spain

PREFACE

For Instructors and Students

How Did I Become Involved with Environmental Problems? In 1966, I heard a scientist give a lecture on the problems of overpopulation and pollution. Afterward I went to him and said, "If even a fraction of what you have said is true, I will feel ethically obligated to give up my research on the corrosion of metals and devote the rest of my life to research and education on environmental problems and solutions. Frankly, I don't want to believe a word you have said, and I'm going into the literature to try to prove that your statements are either untrue or grossly distorted."

After 6 months of study I was convinced of the seriousness of these problems. Since then, I have been studying, teaching, and writing about them. This book summarizes what I have learned in more than three decades of trying to understand environmental principles, problems, connections, and solutions.

What Is My Philosophy of Education? In our lifelong pursuit of knowledge, I believe we should do three things:

- *Question everything and everybody*, as any good scientist does.
- *Develop a list of principles, concepts, and rules to serve as guidelines in making decisions*, and continually evaluate and modify this list on the basis of experience. This is based on my belief that the key goal of education is to learn how to sift through mountains of facts and ideas to find the few that are most useful and worth knowing. We need to be *wisdom seekers*, not information vessels. This takes a firm commitment to learning how to think logically and critically. This book is full of facts and numbers, but they are useful only to the extent that they lead to an understanding of key ideas, scientific laws, concepts, principles, and connections.
- *Interact with what you read as a way to sharpen your critical thinking skills.* I do this by marking key sentences and paragraphs with a highlighter or pen. I put an asterisk in the margin next to something I think is important and double asterisks next to something that I think is especially important. I write comments in the margins, such as *Beautiful*, *Confusing*, *Misleading*, or *Wrong*. I fold down the top corner of pages with highlighted passages and the top and bottom corners of especially important pages. This way, I can flip through a book and quickly review the key passages. I urge you to interact in such ways with this book.

Redefining Environmental Science for the 21st Century This is a *science-based* book designed for introductory courses on environmental science. It treats environmental science as an *interdisciplinary* study, combining ideas and information from *natural sciences* (such as biology, chemistry, and geology) and *social sciences* (such as economics, politics, and ethics) to present a general idea of how nature works and how things are interconnected. It is a study of *connections in nature.*

This ninth edition is the most significant revision of this publication since its first edition. This new edition redefines the course by emphasizing the following major shifts in environmental education and environmental policy that have taken place over the past 25 years and that will accelerate in this century.

- ***Increased emphasis on science-based approaches to understanding and solving environmental problems.*** Since its first edition, this book has led the way in using scientific laws, principles, models, and concepts to help us **(1)** understand environmental and resource problems and their possible solutions and **(2)** see how these concepts, problems, and solutions are connected. The first edition had four chapters on basic scientific concepts when other books had a single chapter. In this ninth edition, eight chapters (two more than in the last edition) and 230 pages are devoted to the treatment of scientific principles and concepts—far more than in any other introductory environmental science text of this size. This emphasis on basic science will become increasingly important throughout this century. I have introduced only the concepts and principles necessary for understanding the material in the book, and I have tried to present them simply but accurately.
- ***Increased emphasis on solutions.*** The emphasis in this century is on finding and implementing scientific, technological, economic, and political solutions to environmental problems. This text has stressed solutions as a major theme for many years. In this new edition, 141 pages are devoted to presenting and eval-

uating solutions to environmental problems—far more than in any other introductory environmental science textbook of this size.

- ***Increased emphasis on prevention and precaution.*** Since its first edition this book has categorized proposed solutions to environmental problems as either **(1)** *input* (prevention) solutions such as pollution prevention and waste reduction or **(2)** *output* (cleanup) solutions such as pollution control and waste management. Both approaches are needed, but so far most emphasis has been on output or management solutions. There is a growing awareness of the need to put more emphasis on input or prevention approaches. This edition increases this emphasis based on using the *precautionary principle* as a guideline for solutions to urgent environmental problems when there is insufficient scientific information. This shift in environmental thinking will accelerate during this century.

- ***Increased emphasis on decentralized micropower.*** I highlight the shift from large centralized sources of electricity (mostly coal and nuclear plants) to a dispersed array of smaller micropower plants, including gas turbines, solar cell arrays, and fuel cells. This shift, discussed in Section 20-9, p. 538, is under way and will accelerate during this century.

- ***Greater integration of economics and environment.*** I emphasize the increased use of emissions trading, environmental accounting, full-cost pricing, and evolving eras of environmental management in the greening of businesses. This trend, discussed in Chapter 2 (pp. 28-29), is under way and will increase rapidly as businesses learn that improving environmental quality is one of the greatest investment opportunities of this century.

To help ensure that the material is accurate and up to date, I have consulted more than 10,000 research sources in the professional literature and about the same number of internet sites. I have also benefited from the more than 250 experts and teachers (see list on pp. x–xii) who have provided detailed reviews of this and my other two books in this field.

How Have I Attempted to Eliminate Bias? There are at least two sides to all controversial environmental issues. The challenge for an author is to give a fair and balanced view of opposing viewpoints without injecting personal bias. This allows students to make up their own minds about important issues. Studying a subject as important as environmental science and ending up with no conclusions, opinions, and beliefs means that both the teacher and student have failed. However, such conclusions should be based on using critical thinking to evaluate opposing ideas.

The easiest way to avoid bias is not to discuss controversial environmental issues. I consider this to be a disservice to students, who will have to deal with the many important, complex, and controversial environmental issues we all face.

In this new edition, the publisher and I enlisted the aid of 25 new reviewers, charged with trying to improve this book and to focus especially on detecting and eliminating any hint of bias.

A few examples of my efforts to give a balanced presentation of opposing viewpoints are **(1)** the pros and cons of reducing birth rates (p. 255), **(2)** the Pro / Con box on oil development in the Arctic National Wildlife Refuge (p. 500), **(3)** Section 13-3 (pp. 307–313), on global warming, **(4)** pros and cons of pesticides (pp. 419–421), and **(5)** 22 Pro / Con summary diagrams (such as Figures 15-10, p. 385, Figure 19-17, p. 501, and Figure 20-28, p. 536).

However, bias can be subtle, and I invite instructors and students to write me and point out any remaining bias.

What Are Some Key Features of This Book? This book is *science based, solution oriented,* and *flexible.* The book is divided into five major parts (see Brief Contents, p. xiii). After the introductory chapters in Part I and the scientific principles and concepts chapters in Part II have been covered, the rest of the book can be used in almost any order. In addition, most chapters and many sections within these chapters can be moved around or omitted to accommodate courses with different lengths and emphases.

Each chapter begins with a brief *case study* designed to capture interest and set the stage for the material that follows. In addition to these 20 case studies, 55 other case studies are found throughout the book (some in special boxes and others within the text); they provide a more in-depth look at specific environmental problems and their possible solutions. Eleven *Guest Essays* present an individual researcher's or activist's point of view, which is then evaluated through Critical Thinking questions.

Other special boxes found in the text include **(1)** *Pro/Con boxes* that present both sides of controversial environmental issues, **(2)** *Connections boxes* that show connections in nature and between environmental concepts, problems, and solutions, **(3)** *Solutions boxes* that summarize a variety of solutions to environmental problems proposed by various analysts, **(4)** *Spotlight boxes* that highlight and give insights into key environmental problems and concepts, and **(5)** *Individuals Matter boxes* that describe what people have done to help solve environmental problems. To encourage critical thinking and integrate it throughout the book, all boxes (except Individuals Matter) end with Critical Thinking questions.

This book is an integrated study of environmental problems, connections, and solutions. The eight integrated themes in this book are **(1)** *biodiversity and natural resources (ecosystem services)*, **(2)** *sustainability*, **(3)** *connections in nature*, **(4)** *pollution prevention*, **(5)** *population and exponential growth*, **(6)** *energy and energy efficiency*, **(7)** *solutions to environmental problems*, and **(8)** *the importance of individuals working together to bring about environmental change.*

I hope you will start by looking at the brief table of contents (p. xiii) to get an overview of this book. Then I suggest that you look at the Concepts and Connections diagram inside the back cover, which shows the major components and relationships found in environmental science. In effect, it is a map of the book.

This book has 459 illustrations, 189 of them new to this edition. These illustrations are designed to present complex ideas in understandable ways and to relate learning to the real world.

I have not cited specific sources of information. This is rarely done for an introductory-level text in any field, and it would interrupt the flow of the material. Instead, on the website material for each chapter you will find **(1)** readings, **(2)** internet site references, and **(3)** references to complete articles that can be accessed online on the *InfoTrac* supplement available free to qualified users of this book. These sources **(1)** back up most of the content of this book and **(2)** serve as springboards to further information and ideas. Placing these references on the website also allows me to update them regularly.

Instructors wanting a longer and more comprehensive book covering this material with a different emphasis and organization can use my book *Living in the Environment*, 12th edition (758 pages, Brooks/Cole, 2002). Those wanting a shorter, more integrated, and less expensive book can use *Sustaining the Earth: An Integrated Approach*, 5th edition (385 pages, Brooks/Cole, 2002). An even shorter book, *Essentials of Ecology* (260 pages, Brooks/Cole, 2002), is available for instructors who want to concentrate primarily on basic ecological principles.

What Are the Major Changes in the Ninth Edition? *This edition represents the most significant revision since the first edition came out.* Detailed changes by chapter are listed in the annotated material in the insert provided with the instructor's version of this book and on this book's website. Major changes include the following:

CONTENT

- Updated and revised material throughout the book.
- 189 new figures.
- Two new chapters on *Community Ecology* (Chapter 7) and *Geology* (Chapter 9).
- Expanded coverage of aquatic biodiversity in Chapters 6, 17, and 18.
- Addition of 167 new topics. See the insert provided with the instructor's version of this book and this book's website for a detailed list of these topics listed by chapter. Examples include the following:
    - Effects of globalization (p. 17).
    - Ecological footprints of countries (Figure 1-8, p. 10).
    - Ecological and economic services provided by terrestrial and aquatic systems. Examples include Figures 6-31 (p. 145), 6-41 (p. 154), 16-3 (p. 399), and 17-8 (p. 435).
    - Diagrams summarizing the harmful effects of human activities on natural systems. Examples include Figures 6-17 (p. 133), 6-21 (p. 137), 6-26 (p. 142), and 6-27 (p. 142).
    - Precautionary principle (pp. 37, 178, 232).
    - Genetically modified food (pp. 405–407, Figure 16-10, p. 406, and Figure 16-11, p. 407).
    - Drip irrigation (Solutions, p. 341).
    - Using the internet to save energy and reduce global warming (p. 530).
    - Micropower electricity production (pp. 538–539).
    - Selling services instead of goods (pp. 374–377).
    - Brownfields (p. 390).
    - Sustainable timber certification (Solutions, p. 440).
    - Biodiversity hot spots (Figure 17-21, p. 454 and Figure 18-12, p. 478).
    - Integrated coastal management (pp. 455-456).
    - Smart urban growth (pp. 271–272 and Figure 11-27, p. 274).
- 25 new reviewers to improve content and balance. See the insert provided with the instructor's version of this book and this book's website for a detailed list by chapter of how balance on controversial issues has been achieved.
- 23 Pro/Con summary diagrams summarize a large amount of complex information in an easy-to-understand manner. Examples include Figures 15-8 (p. 383), 15-10 (p. 385), 15-11 (p. 385), 16-11 (p. 407), 16-22 (p. 418), 19-23 (p. 506), and 20-12 (p. 523).
- Two new Guest Essays (see pp. 118 and 531).

LEARNING AIDS

- Greater use of numbered and bulleted lists to make the book simpler and help students comprehend and review key material.

- Use of questions as titles for all subsections to spark interest and serve as a built-in list of learning objectives for readers.
- Cross-references by page number to link concepts and material throughout the book. This **(1)** emphasizes the basic ecological concept that everything is connected and **(2)** serves as a textbook version of interconnected web links.

In-Text Study Aids Each chapter begins with a few general questions to reveal how it is organized and what students will be learning. When a new term is introduced and defined, it is printed in boldface type. A glossary of all key terms is located at the end of the book.

Questions are used as titles for all subsections so that readers know the focus of the material that follows. In effect, this is a built-in set of learning objectives.

Each chapter ends with **(1)** a set of Review Questions covering *all* of the material in the chapter as a study guide for students and **(2)** a set of questions to encourage students to think critically and apply what they have learned to their lives. The Critical Thinking questions are followed by several projects that individuals or groups can carry out.

Internet and Online Study Aids Qualified users of this textbook have free access to the *Brooks/Cole Biology and Environmental Science Resource Center*. The online resource material for this book can be accessed by logging on at

http://www.brookscole.com/product/0534389872s

At this website you will find the following material for each chapter:

- "Flash Cards," which allow you to test your mastery of the Terms and Concepts to Remember for each chapter.
- "Tutorial Quizzes," which provide a multiple-choice practice quiz.
- "Student Guide to InfoTrac," which will lead you to Critical Thinking Projects that use InfoTrac College Edition as a research tool.
- "References," which lists the major books and articles consulted in writing this chapter.
- A brief "What You Can Do" list addressing key environmental problems.
- "Hypercontents," which takes you to an extensive list of websites with news, research, and images related to individual sections of the chapter.

Qualified adopters of this textbook also have free access to *WebTutor Toolbox on WebCT* at

http://e.thomsonlearning.com

It provides access to a full array of study tools, including flashcards (with audio), practice quizzes, online tutorials, and Web links.

Students using *new* copies of this textbook also have free and unlimited access to *InfoTrac College Edition*. This fully searchable online library gives users access to complete environmental articles from several hundred periodicals dating back over the past four years. I have put two practice exercises at the end of each chapter to help students learn how to navigate this valuable source of information.

Other learning tools include:

- *Essential Study Skills for Science Students by Daniel D. Chiras*. This book includes **(1)** chapters on developing good study habits, **(2)** sharpening memory, **(3)** getting the most out of lectures, labs, and reading assignments, **(4)** improving test-taking abilities, and **(5)** becoming a critical thinker. Your instructor can have this book bundled free with your textbook.
- *Laboratory Manual by C. Lee Rocket and Kenneth J. Van Dellen*. This manual includes a variety of laboratory exercises, workbook exercises, and projects that require a minimum use of sophisticated equipment.

Supplementary Materials for Instructors The following supplementary materials are available to instructors adopting this book:

- *Multimedia Presentation Manager.* This CD-ROM, which is free to adopters, places all of Miller's diagrams in PowerPoint slides. There is a PowerPoint file for each of the Miller textbook chapters. The CD allows you to: **(1)** move PowerPoint slides from our PowerPoint file to your PowerPoint file by simply cutting and pasting the slide, **(2)** select a portion of a diagram and increase its size to full screen, **(3)** modify or remove figure labels and, **(4)** easily convert the slides to HTML and post the PowerPoint slides to the web if you are using one of the Miller texts as a required text for your adoption. The CD allows you to link to some CNN video clips that come free with the adoption. You may also save all the PowerPoint slides as jpegs if you prefer this format. All illustrations are presented as high-resolution artwork and images.
- *Transparency Masters and Acetates.* Includes **(1)** 100 color acetates of line art and **(2)** nearly 600 black and

white master sheets of key diagrams for making overhead transparencies. Free to adopters.

- *Case Studies in Environmental Science, Second Edition.* This book presents 12 case studies offering a range of views on key environmental issues. The case studies focus on eight geographic regions in the United States and Canada.

- *CNN™ Today Videos.* These videos, updated annually, contain short clips of news stories about environmental news. Adopters can receive one video free each year for three years.

- Two videos, **(1)** *In the Shadow of the Shuttle: Protecting Endangered Species* and **(2)** *Costa Rica: Science in the Rainforest*, are available to adopters.

- *Instructor's Manual with Test Items.* Free to adopters.

- *ExamView™.* Allows you to **(1)** easily create and customize tests, **(2)** see them on the screen exactly as they will print, and **(3)** print them out.

Help Me Improve This Book Let me know how you think this book can be improved; if you find any errors, bias, or confusing explanations please send them to Jack Carey, Biology Publisher, Brooks/Cole Publishing Company, 10 Davis Drive, Belmont, CA 94002 (e-mail: jack.carey@brookscole.com). He will forward them to me. Most errors can be corrected in subsequent printings of this edition rather than waiting for a new edition.

Annenberg/CPB Television Course This textbook is being offered as part of the Annenberg/CPB Project television series *Race to Save the Planet*, a 10-part public broadcasting series and a college-level telecourse examining the major environmental questions facing the world today. The series takes into account the wide spectrum of opinion about what constitutes an environmental problem and discusses the controversies about appropriate remedial measures. It analyzes problems and emphasizes the successful search for solutions. The course develops a number of key themes that cut across a broad range of environmental issues, including sustainability, the interconnections of the economy and the ecosystem, short-term versus long-term gains, and the trade-offs involved in balancing problems and solutions. A study guide and a faculty guide, both available from Brooks/Cole Publishing Company, integrate the telecourse and this text.

For further information about available television course licenses and duplication licenses, contact PBS Adult Learning Service, 1320 Braddock Place, Alexandria, VA 22314-1698 (1-800-ALS-ALS-8).

For information about purchasing videocassettes and print material, contact the Annenberg/CPB Collection, P.O. Box 2284, South Burlington, VT 05407-2284 (1-800-LEARNER).

Acknowledgments I wish to thank the many students and teachers who responded so favorably to the 8 previous editions of *Environmental Science*, the 12 editions of *Living in the Environment*, and the 5 editions of *Sustaining the Earth* and who corrected errors and offered many helpful suggestions for improvement. I am also deeply indebted to the reviewers, who pointed out errors and suggested many important improvements in this book. Any errors and deficiencies left are mine.

The members of the talented production team, listed on the copyright page, have made vital contributions as well. My thanks also go to **(1)** copyeditor Carol Anne Peschke, **(2)** production editors Scott Hitchcock and Hal Humphrey, **(3)** Eileen Mitchell, Michael Gutch, and other members of the staff of Electronic Publishing Services who have improved the design of this edition and made my life much easier, **(4)** photo researcher Linda L. Rill, **(5)** Brooks/Cole's hard-working sales staff, and **(6)** Pat Waldo, Chris Evers, and their talented colleagues who develop multimedia and the website material associated with this book.

I also thank **(1)** Paul M. Rich for his help with some of the chapters on basic ecology, **(2)** C. Lee Rockett and Kenneth J. Van Dellen for developing the *Laboratory Manual* to accompany this book, **(3)** Jane Heinze-Fry for her work on concept mapping, *Environmental Articles, Critical Thinking and the Environment: A Beginner's Guide*, and the *Internet Booklet*, **(4)** Richard K. Clements for his excellent work on the *Instructor's Manual*, and **(5)** the people who have translated this book into five different languages for use throughout much of the world.

My deepest thanks go to Jack Carey, biology publisher at Brooks/Cole, for his encouragement, help, 36 years of friendship, and superb reviewing system. It helps immensely to work with the best and most experienced editor in college textbook publishing.

I dedicate this book to the earth and to Kathleen Paul Miller, my wife and research assistant.

G. Tyler Miller, Jr.

Guest Essayists and Reviewers

Guest Essayists The following are authors of Guest Essays: **(1) Lester R. Brown**, chair of the board, Worldwatch Institute; **(2) Robert D. Bullard**, professor of sociology and director of the Environmental Justice Resource Center at Clark Atlanta University; **(3) Lois Marie Gibbs**, director, Center for Health, Environment, and Justice; **(4) Garrett Hardin**, professor emeritus of human ecology, University of California, Santa Barbara; **(5) Paul G. Hawken**, environmental author and business leader; **(6) Jane Heinze-Fry**, author, teacher, and consultant in environmental education; **(7) Amory B. Lovins**, energy policy consultant and director of research, Rocky Mountain Institute; **(8) Peter Montague**, director, Environmental Research Foundation; **(9) Norman Myers**, tropical ecologist and consultant in environment and development; **(10) David W. Orr**, professor of environmental studies, Oberlin College; and **(11) Nancy Wicks**, ecopioneer and director of Round Mountain Organics.

Cumulative Reviewers Barbara J. Abraham, Hampton College; Donald D. Adams, State University of New York at Plattsburgh; Larry G. Allen, California State University, Northridge; Susan Allen-Gil, Ithaca College; James R. Anderson, U.S. Geological Survey; Mark W. Anderson, University of Maine; Kenneth B. Armitage, University of Kansas; Samuel Arthur, Bowling Green State University; Gary J. Atchison, Iowa State University; Marvin W. Baker, Jr., University of Oklahoma; Virgil R. Baker, Arizona State University; Ian G. Barbour, Carleton College; Albert J. Beck, California State University, Chico; W. Behan, Northern Arizona University; Keith L. Bildstein, Winthrop College; Jeff Bland, University of Puget Sound; Roger G. Bland, Central Michigan University; Grady Blount II, Texas A&M University, Corpus Christi; Georg Borgstrom, Michigan State University; Arthur C. Borror, University of New Hampshire; John H. Bounds, Sam Houston State University; Leon F. Bouvier, Population Reference Bureau; Daniel J. Bovin, Université Laval; Michael F. Brewer, Resources for the Future, Inc.; Mark M. Brinson, East Carolina University; Dale Brown, University of Hartford; Patrick E. Brunelle, Contra Costa College; Terrence J. Burgess, Saddleback College North; David Byman, Pennsylvania State University, Worthington–Scranton; Lynton K. Caldwell, Indiana University; Faith Thompson Campbell, Natural Resources Defense Council, Inc.; Ray Canterbery, Florida State University; Ted J. Case, University of San Diego; Ann Causey, Auburn University; Richard A. Cellarius, Evergreen State University; William U. Chandler, Worldwatch Institute; F. Christman, University of North Carolina, Chapel Hill; Preston Cloud, University of California, Santa Barbara; Bernard C. Cohen, University of Pittsburgh; Richard A. Cooley, University of California, Santa Cruz; Dennis J. Corrigan; George Cox, San Diego State University; John D. Cunningham, Keene State College; Herman E. Daly, University of Maryland; Raymond F. Dasmann, University of California, Santa Cruz; Kingsley Davis, Hoover Institution; Edward E. DeMartini, University of California, Santa Barbara; Charles E. DePoe, Northeast Louisiana University; Thomas R. Detwyler, University of Wisconsin; Peter H. Diage, University of California, Riverside; Lon D. Drake, University of Iowa; David DuBose, Shasta College; Dietrich Earnhart, University of Kansas; T. Edmonson, University of Washington; Thomas Eisner, Cornell University; Michael Esler, Southern Illinois University; David E. Fairbrothers, Rutgers University; Paul P. Feeny, Cornell University; Richard S. Feldman, Marist College; Nancy Field, Bellevue Community College; Allan Fitzsimmons, University of Kentucky; Andrew J. Friedland, Dartmouth College; Kenneth O. Fulgham, Humboldt State University; Lowell L. Getz, University of Illinois at Urbana–Champaign; Frederick F. Gilbert, Washington State University; Jay Glassman, Los Angeles Valley College; Harold Goetz, North Dakota State University; Jeffery J. Gordon, Bowling Green State University; Eville Gorham, University of Minnesota; Michael Gough, Resources for the Future; Ernest M. Gould, Jr., Harvard University; Peter Green, Golden West College; Katharine B. Gregg, West Virginia Wesleyan College; Paul K. Grogger, University of Colorado at Colorado Springs; L. Guernsey, Indiana State University; Ralph Guzman, University of California, Santa Cruz; Raymond Hames, University of Nebraska, Lincoln; Raymond E. Hampton, Central Michigan University; Ted L. Hanes, California State University, Fullerton; William S. Hardenbergh, Southern Illinois University at Carbondale; John P. Harley, Eastern Kentucky University; Neil A. Harriman, University of Wisconsin, Oshkosh; Grant A. Harris, Washington State University; Harry S. Hass, San Jose City College; Arthur N. Haupt, Population Reference Bureau; Denis A. Hayes, environmental consultant; Stephen Heard, University of Iowa; Gene Heinze-Fry, Department of Utilities, State of Massachusetts; Jane Heinze-Fry, environmental educator; John G. Hewston, Humboldt State University; David L. Hicks, Whitworth College; Kenneth M. Hinkel, University of Cincinnati; Eric Hirst, Oak Ridge National Laboratory; Doug Hix, University of Hartford; S. Holling, University of British Columbia; Donald Holtgrieve, California State University, Hayward; Michael H. Horn, California State University, Fullerton; Mark A. Hornberger, Bloomsberg University; Marilyn Houck, Pennsylvania State University; Richard D. Houk, Winthrop College; Robert J. Huggett, College of William and Mary; Donald Huisingh, North Carolina State

University; Marlene K. Hutt, IBM; David R. Inglis, University of Massachusetts; Robert Janiskee, University of South Carolina; Hugo H. John, University of Connecticut; Brian A. Johnson, University of Pennsylvania, Bloomsburg; David I. Johnson, Michigan State University; Mark Jonasson, Crafton Hills College; Agnes Kadar, Nassau Community College; Thomas L. Keefe, Eastern Kentucky University; Nathan Keyfitz, Harvard University; David Kidd, University of New Mexico; Pamela S. Kimbrough; Jesse Klingebiel, Kent School; Edward J. Kormondy, University of Hawaii–Hilo/West Oahu College; John V. Krutilla, Resources for the Future, Inc.; Judith Kunofsky, Sierra Club; E. Kurtz; Theodore Kury, State University of New York, Buffalo; Steve Ladochy, University of Winnipeg; Mark B. Lapping, Kansas State University; Tom Leege, Idaho Department of Fish and Game; William S. Lindsay, Monterey Peninsula College; E. S. Lindstrom, Pennsylvania State University; M. Lippiman, New York University Medical Center; Valerie A. Liston, University of Minnesota; Dennis Livingston, Rensselaer Polytechnic Institute; James P. Lodge, air pollution consultant; Raymond C. Loehr, University of Texas at Austin; Ruth Logan, Santa Monica City College; Robert D. Loring, DePauw University; Paul F. Love, Angelo State University; Thomas Lovering, University of California, Santa Barbara; Amory B. Lovins, Rocky Mountain Institute; Hunter Lovins, Rocky Mountain Institute; Gene A. Lucas, Drake University; Claudia Luke; David Lynn; Timothy F. Lyon, Ball State University; Stephen Malcolm, Western Michigan University; Melvin G. Marcus, Arizona State University; Gordon E. Matzke, Oregon State University; Parker Mauldin, Rockefeller Foundation; Marie McClune, The Agnes Irwin School (Rosemont, Pennsylvania); Theodore R. McDowell, California State University; Vincent E. McKelvey, U.S. Geological Survey; Robert T. McMaster, Smith College; John G. Merriam, Bowling Green State University; A. Steven Messenger, Northern Illinois University; John Meyers, Middlesex Community College; Raymond W. Miller, Utah State University; Arthur B. Millman, University of Massachusetts, Boston; Fred Montague, University of Utah; Rolf Monteen, California Polytechnic State University; Ralph Morris, Brock University, St. Catherine's, Ontario, Canada; Angela Morrow, Auburn University; William W. Murdoch, University of California, Santa Barbara; Norman Myers, environmental consultant; Brian C. Myres, Cypress College; A. Neale, Illinois State University; Duane Nellis, Kansas State University; Jan Newhouse, University of Hawaii, Manoa; Jim Norwine, Texas A&M University, Kingsville; John E. Oliver, Indiana State University; Carol Page, copyeditor; Eric Pallant, Allegheny College; Charles F. Park, Stanford University; Richard J. Pedersen, U.S. Department of Agriculture, Forest Service; David Pelliam, Bureau of Land Management, U.S. Department of Interior; Rodney Peterson, Colorado State University; Julie Phillips, De Anza College; William S. Pierce, Case Western Reserve University; David Pimentel, Cornell University; Peter Pizor, Northwest Community College; Mark D. Plunkett, Bellevue Community College; Grace L. Powell, University of Akron; James H. Price, Oklahoma College; Marian E. Reeve, Merritt College; Carl H. Reidel, University of Vermont; Charles C. Reith, Tulane University; Roger Revelle, California State University, San Diego; L. Reynolds, University of Central Arkansas; Ronald R. Rhein, Kutztown University of Pennsylvania; Charles Rhyne, Jackson State University; Robert A. Richardson, University of Wisconsin; Benjamin F. Richason III, St. Cloud State University; Ronald Robberecht, University of Idaho; William Van B. Robertson, School of Medicine, Stanford University; C. Lee Rockett, Bowling Green State University; Terry D. Roelofs, Humboldt State University; Christopher Rose, California Polytechnic State University; Richard G. Rose, West Valley College; Stephen T. Ross, University of Southern Mississippi; Robert E. Roth, The Ohio State University; Arthur N. Samel, Bowling Green State University; Floyd Sanford, Coe College; David Satterthwaite, I.E.E.D., London; Stephen W. Sawyer, University of Maryland; Arnold Schecter, State University of New York, Syracuse; Frank Schiavo, San Jose State University; William H. Schlesinger, Ecological Society of America; Stephen H. Schneider, National Center for Atmospheric Research; Clarence A. Schoenfeld, University of Wisconsin, Madison; Henry A. Schroeder, Dartmouth Medical School; Lauren A. Schroeder, Youngstown State University; Norman B. Schwartz, University of Delaware; George Sessions, Sierra College; David J. Severn, Clement Associates; Paul Shepard, Pitzer College and Claremont Graduate School; Michael P. Shields, Southern Illinois University at Carbondale; Kenneth Shiovitz; F. Siewert, Ball State University; E. K. Silbergold, Environmental Defense Fund; Joseph L. Simon, University of South Florida; William E. Sloey, University of Wisconsin, Oshkosh; Robert L. Smith, West Virginia University; Val Smith, University of Kansas; Howard M. Smolkin, U.S. Environmental Protection Agency; Patricia M. Sparks, Glassboro State College; John E. Stanley, University of Virginia; Mel Stanley, California State Polytechnic University, Pomona; Norman R. Stewart, University of Wisconsin, Milwaukee; Frank E. Studnicka, University of Wisconsin, Platteville; Chris Tarp, Contra Costa College; Roger E. Thibault, Bowling Green State University; William L. Thomas, California State University, Hayward; Shari Turney, copyeditor; John D. Usis, Youngstown State University; Tinco E. A. van Hylckama, Texas Tech University; Robert R. Van Kirk, Humboldt State University; Donald E. Van Meter, Ball State University; Gary Varner, Texas A&M University; John D. Vitek, Oklahoma State University; Harry A. Wagner, Victoria College; Lee B. Waian, Saddleback College; Warren C. Walker, Stephen F. Austin State University; Thomas D. Warner, South Dakota State University; Kenneth E. F. Watt, University of California,

Davis; Alvin M. Weinberg, Institute of Energy Analysis, Oak Ridge Associated Universities; Brian Weiss; Margery Weitkamp, James Monroe High School (Granada Hills, California); Anthony Weston, SUNY at Stony Brook; Raymond White, San Francisco City College; Douglas Wickum, University of Wisconsin, Stout; Charles G. Wilber, Colorado State University; Nancy Lee Wilkinson, San Francisco State University; John C. Williams, College of San Mateo; Ray Williams, Rio Hondo College; Roberta Williams, University of Nevada, Las Vegas; Samuel J. Williamson, New York University; Ted L. Willrich, Oregon State University; James Winsor, Pennsylvania State University; Fred Witzig, University of Minnesota at Duluth; George M. Woodwell, Woods Hole Research Center; Robert Yoerg, Belmont Hills Hospital; Hideo Yonenaka, San Francisco State University; Malcolm J. Zwolinski, University of Arizona.

Brief Contents

Detailed Contents

Hawaiian monk seal's mouth caught in plastic

Rocky Mountain Institute, Colorado

National Park Service

Glacier National Park

Couresy of Dennis Gonsalves and Stephen Ferreira/ Cornell University

Tree plantation

U.S. Department of Agriculture

Fast growing Kenaf for making paper, Texas

U.S. Department of Agriculture

Boll weevil

U.S. Fish and Wildlife Service

Wildlife refuge

National Archives/EPA Documerica

Monoculture cropland, California

Arctic National Wildlife Refuge

Caribou herd, Arctic National Wildlife Refuge, Alaska

NOAA Corps Collection/Photographer: Commander John Bortniak, NOAA Corps. (ret.)

Coral reef sanctuary, Tortugas Marine Ecological Reserve, Florida Keys

John Todd at solar sewage plant, Rhode Island

Biosphere 2, Arizona

Appendixes A1

Glossary G1

Index I1

Part I

HUMANS AND SUSTAINABILITY: AN OVERVIEW

The environmental crisis is an outward manifestation of a crisis of mind and spirit. There could be no greater misconception of its meaning than to believe it is concerned only with endangered wildlife, human-made ugliness,

and pollution. These are part of it, but more importantly, the crisis is concerned with the kind of creatures we are and what we must become in order to survive.

LYNTON K. CALDWELL

1 ENVIRONMENTAL PROBLEMS, THEIR CAUSES, AND SUSTAINABILITY

Living in an Exponential Age

Once there were two kings from Babylon who enjoyed playing chess, with the winner claiming a prize from the loser. After one match, the winning king asked the loser to pay him by placing one grain of wheat on the first square of the chessboard, two on the second, four on the third, and so on. The number of grains was to double each time until all 64 squares were filled.

The losing king, thinking he was getting off easy, agreed with delight. It was the biggest mistake he ever made. He bankrupted his kingdom and still could not produce the incredibly large number of grains of wheat he had promised. In fact, it is probably more than all the wheat that has ever been harvested!

This is an example of **exponential growth**, in which a quantity increases by a fixed percentage of the whole in a given time. As the losing king learned, exponential growth is deceptive. It starts off slowly, but after only a few doublings it grows to enormous numbers because each doubling is more than the total of all earlier growth. If plotted on a graph, continuing exponential growth eventually yields a graph shaped somewhat like the letter *J* (Figure 1-1).

Here is another example. Fold a piece of paper in half to double its thickness. If you could do this 42 times, the stack would reach from the earth to the moon, 386,400 kilometers (240,000 miles) away. If you could double it 50 times, the folded paper would almost reach the sun, 149 million kilometers (93 million miles) away!

Six important environmental issues are **(1)** *population growth*, **(2)** *increasing resource use*, **(3)** *global climate change*, **(4)** *premature extinction of plants and animals*, **(5)** *pollution*, and **(6)** *poverty*. All these issues are interconnected and are growing exponentially.

For example, between 1950 and 2001, world population increased from 2.5 billion to 6.1 billion. Unless death rates rise sharply, it may reach 8 billion by 2028, 9 billion by 2050, and 10–14 billion by 2100 (Figure 1-1). Global economic output, much of it environmentally damaging, is a rough measure of resource use. It has increased fivefold since 1950.

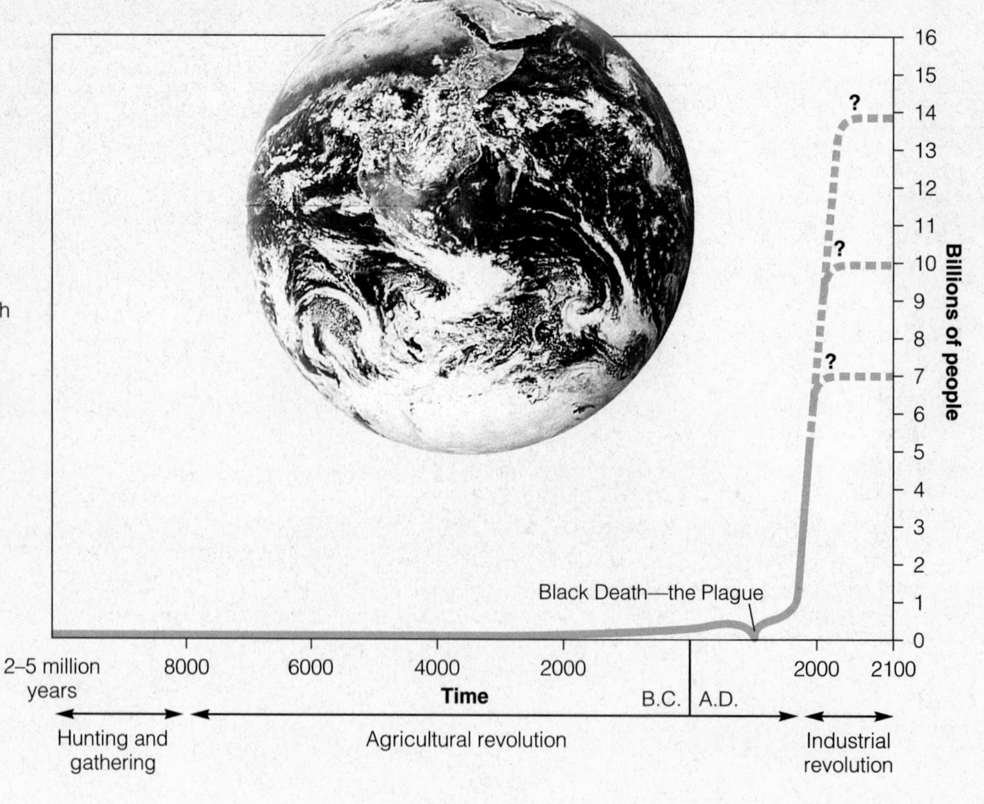

Figure 1-1 The *J*-shaped curve of past exponential world population growth, with projections beyond 2100. Notice that exponential growth starts off slowly, but as time passes the curve becomes increasingly steep. The current world population of 6.1 billion people is projected to reach 7–14 billion sometime during this century. (This figure is not to scale.) (Data from World Bank and United Nations; photo courtesy of NASA)

Alone in space, alone in its life-supporting systems, powered by inconceivable energies, mediating them to us through the most delicate adjustments, wayward, unlikely, unpredictable, but nourishing, enlivening, and enriching in the largest degree—is this not a precious home for all of us? Is it not worth our love?

BARBARA WARD AND RENÉ DUBOS

This chapter is an overview of **(1)** environmental problems, **(2)** their causes, **(3)** controversy over their seriousness, and **(4)** ways we can live more sustainably. It discusses these questions:

- What are natural resources, and why are they important? What is an environmentally sustainable society?
- How fast is the human population increasing?
- What is the difference between economic growth, economic development, and environmentally sustainable economic development?
- What are the earth's main types of resources? How can they be depleted or degraded?
- What are the principal types of pollution? How can pollution be reduced and prevented?
- What are the basic causes of today's environmental problems? How are these causes connected?
- What major effects have hunter-gatherer societies, agricultural societies, and industrialized societies had on the environment?
- Is our current course sustainable? How can we live more sustainably?

1-1 LIVING MORE SUSTAINABLY

What Is the Difference Between Environment, Ecology, and Environmental Science? **Environment** is everything that affects a living organism (any unique form of life). **Ecology** is a biological science that studies the relationships between living organisms and their environment.

This textbook is an introduction to **environmental science**. It is an interdisciplinary science that uses concepts and information from *natural sciences* such as ecology, biology, chemistry, and geology and *social sciences* such as economics, politics, and ethics to help us understand **(1)** how the earth works, **(2)** how we are affecting the earth's life-support systems (environment) for us and other forms of life, and **(3)** how to deal with the environmental problems we face. Many different groups of people are concerned about environmental issues (Spotlight, right).

SPOTLIGHT

Cast of Players in the Environmental Drama

The cast of major characters you will encounter in this book includes the following:

- **Ecologists**, who are biological scientists studying relationships between living organisms and their environment.
- **Environmental scientists**, who use information from the physical sciences and social sciences to **(1)** understand how the earth works, **(2)** learn how humans interact with the earth, and **(3)** develop solutions to environmental problems.
- **Conservation biologists**, who in the 1970s created a multidisciplinary science to **(1)** investigate human impacts on the diversity of life found on the earth (biodiversity) and **(2)** develop practical plans for preserving such biodiversity.
- **Environmentalists**, who **(1)** are concerned about the impact of people on environmental quality and **(2)** believe that some human actions are degrading parts of the earth's life-support systems for humans and many other forms of life. Some of their beliefs and proposals for dealing with environmental problems are based on scientific information and concepts and some are based on their social and ethical environmental beliefs (environmental worldviews). Environmentalists are a broad group of people from different economic groups (rich, middle class, poor) and with different political persuasions (ranging from conservative to liberal).
- **Preservationists**, concerned primarily with setting aside or protecting undisturbed natural areas from harmful human activities.
- **Conservationists**, concerned with using natural areas and wildlife in ways that sustain them for current and future generations of humans and other forms of life.
- **Restorationists**, devoted to the partial or complete restoration of natural areas that have been degraded by human activities.

Many people consider themselves members of several of these groups.

Critical Thinking

Which, if any, of these groups do you most identify with? Why?

What Keeps Us Alive? Our existence, lifestyles, and economies depend completely on the sun and the earth, a blue and white island in the black void of space (Figure 1-1). To economists *capital* is wealth used to sustain a business and to generate more wealth. By analogy, we can think of **(1)** energy from the sun as **solar capital** and **(2)** the planet's air, water, soil, wildlife, minerals, and natural purification, recycling, and pest control processes as **natural resources** or **natural capital** (Guest Essay, p. 6). **Solar energy** is defined broadly to include direct sunlight and indirect forms of solar energy such as **(1)** wind power, **(2)** hydropower (energy from flowing water), and **(3)** biomass (direct solar energy converted to chemical energy stored in biological sources of energy such as wood).

What Is an Environmentally Sustainable Society? To survive and maintain good health, all forms of life must have enough food, clean air, clean water, and shelter to meet their *basic needs*. Additional needs for humans include respectable and safe work, health care, recreation, cultural opportunities, education, and freedom from physical danger. Can you add any other basic needs that you have?

An **environmentally sustainable society** satisfies the basic needs of its people without depleting or degrading its natural resources and thereby preventing current and future generations of humans and other species from meeting their basic needs. *Living sustainably* means **(1)** living off the natural income replenished by soils, plants, air, and water and **(2)** not depleting the natural capital that supplies this income (Guest Essay, p. 6).

For example, imagine that you inherit $1 million. Invest this capital at 10% interest per year and you will have a sustainable annual income of $100,000 without depleting your capital. If you spend $200,000 a year, your $1 million will be gone early in the 7th year; even if you spend only $110,000 a year, you will be bankrupt early in the 18th year.

The lesson here is a very old one: *Protect your capital.* Deplete your capital, and you move from a sustainable to an unsustainable lifestyle.

The same lesson applies to the earth's natural capital. Environmentalists and many leading scientists believe that we are living unsustainably by depleting and degrading the earth's natural capital at an accelerating rate as our population (Figure 1-1) and demands on the earth's resources and life-sustaining processes increase exponentially.

Other analysts do not believe that we are living unsustainably. They contend **(1)** that environmentalists have exaggerated the seriousness of population and environmental problems and **(2)** that any population, resource, and environmental problems we face can be overcome by human ingenuity and technological advances.

1-2 POPULATION GROWTH, ECONOMIC GROWTH, ENVIRONMENTALLY SUSTAINABLE DEVELOPMENT, AND GLOBALIZATION

How Rapidly Is the Human Population Growing? The increasing size of the human population is an example of exponential growth (Figures 1-1 and 1-2 and Spotlight, right). The main reason for the rapid growth of the earth's human population over the past 100 years has been a much greater drop in death rates (mostly because of increases in food supply and better health and sanitation) than in birth rates.

One measure of population growth is **doubling time**: the number of years it takes for a population growing at a specified rate to double its size. A quick way to calculate doubling time is to use the **rule of 70**: 70/percentage growth rate = doubling time (a formula derived from the basic mathematics of exponential growth). For example, in 2001 the world's population grew by 1.33%. If that yearly rate continues, the earth's population will double in about 53 years (70/1.33 = 52.6, or about 53 years).

Studies by researchers at Conservation International suggest that roughly *73% of the earth's habitable land surface (that which is not bare rock, ice, or drifting sand) has been partially or heavily disturbed by human activities* (Figure 1-3). What will happen to the earth's remaining wildlife habitat and species if the human population increases from 6.1 billion to 8 billion between 2001 and 2028 and perhaps to 9 billion by 2050?

What Is Economic Growth? Almost all countries seek **economic growth**: an increase in their capacity to provide people with goods and services. This increase is accomplished by population growth (more consumers and producers), more consumption per person, or both.

Economic growth usually is measured by an increase in several indicators:

World Population Reached	
1 billion in 1804	
2 billion in 1927	(123 years later)
3 billion in 1960	(33 years later)
4 billion in 1974	(14 years later)
5 billion in 1987	(13 years later)
6 billion in 1999	(12 years later)
World Population May Reach	
7 billion in 2013	(14 years later)
8 billion in 2028	(15 years later)
9 billion in 2050	(22 years later)

Figure 1-2 World population milestones. (Data from United Nations Population Division, *World Population Prospects, 1998*)

- **Gross national product (GNP)**: the market value in current dollars of all goods and services produced *within* and *outside* a country by the country's businesses during a year.
- **Gross domestic product (GDP)**: the

SPOTLIGHT

Current Exponential Growth of the Human Population

The world's population is growing exponentially at a rate of about 1.33% per year. The relentless ticking of this population clock means that in 2001 the world's population of 6.1 billion grew by 81 million people (6.1 billion × 0.0133 = 81 million), an average increase of 222,000 people a day, or 9,250 an hour.

At this 1.33% annual rate of exponential growth, it takes only about

- 4 days to add the number of Americans killed in all U.S. wars
- 2 months to add as many people as live in the Los Angeles basin
- 1.6 years to add the 129 million people killed in all wars fought in the past 200 years
- 3.5 years to add 285 million people (the population of the United States in 2001)
- 16 years to add 1.27 billion people (the population of China, the world's most populous country, in 2001)

How much is a billion? If you could live for a billion minutes, you would be 1,902 years old. To travel 1 billion kilometers (0.6 billion miles), you would have to circle the earth about 25,000 times.

Critical Thinking

Some economists argue that population growth is good because it provides more workers, consumers, and problem solvers to keep the global economy growing. Environmentalists argue that population growth threatens economies and the earth's life-support systems through increased pollution and environmental degradation. What is your position? Why?

market value in current dollars of all goods and services produced *within* a country during a year.

- **Gross world product (GWP)**: the market value in current dollars of all goods and services produced in the world each year.
- **Per capita GNP**: the GNP divided by the total population. It gives the average slice of the economic pie per person.

Economic development is the improvement of living standards by economic growth. The United Nations

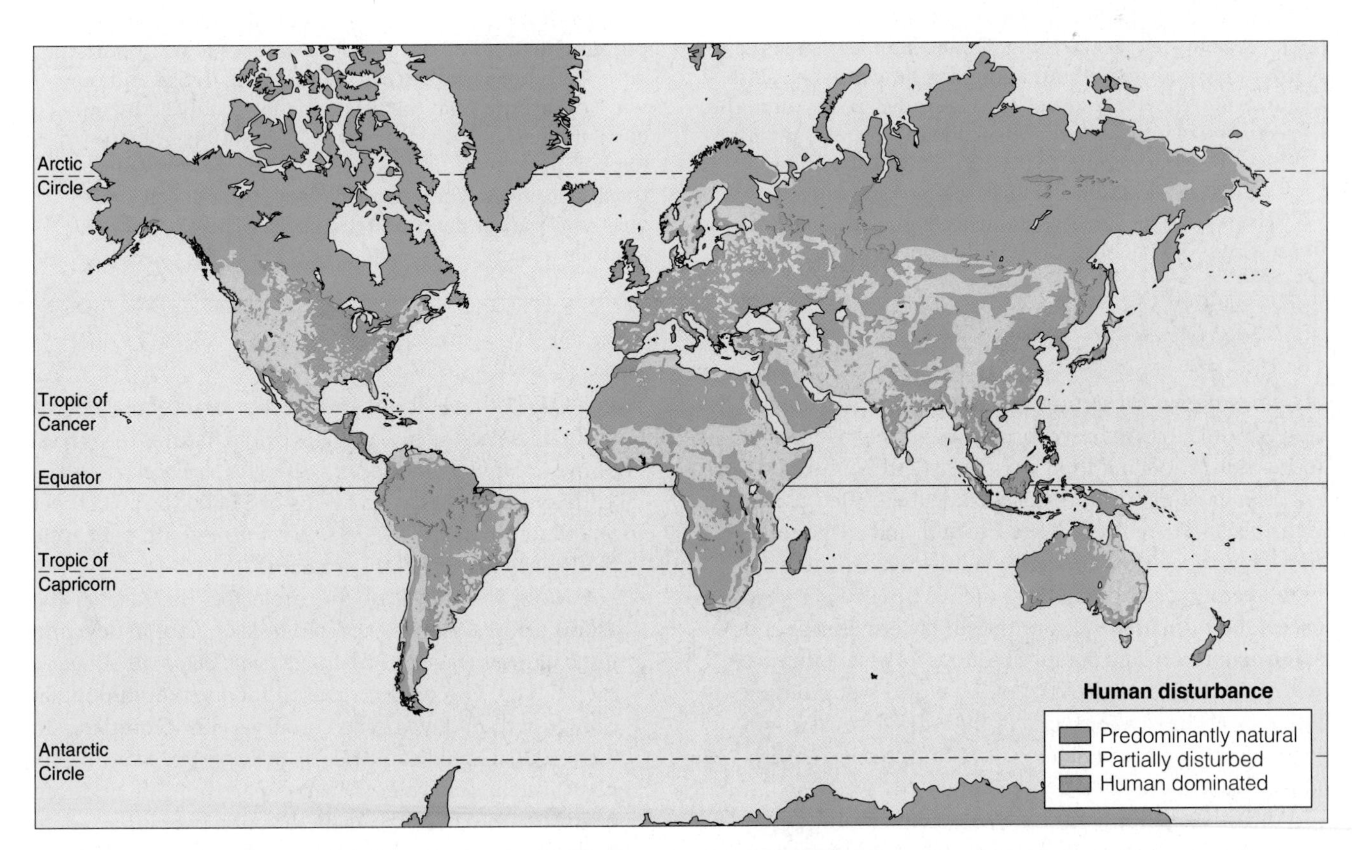

Figure 1-3 Human disturbance of the earth's land area. (Data from Lee Hannah and David Lohse, *1993 Annual Report*, Washington, D.C.: Conservation International)

Natural Capital

GUEST ESSAY

Paul G. Hawken

Paul G. Hawken understands both business and ecology. In addition to founding Smith & Hawken, a retail company known for its environmental initiatives, he has written seven widely acclaimed books, including Growing a Business *(1987),* The Ecology of Commerce *(1993),* Factor 10, The Next Industrial Revolution *(1998, with Amory and Hunter Lovins), and* Natural Capitalism *(1999, with Amory and Hunter Lovins). He produced and hosted* Growing a Business, *a series for public television shown nationwide on 210 stations and now shown in 115 countries.* The Ecology of Commerce *was hailed as the best business book of 1993 and one of the most important books of the 20th century. In 1987,* Inc. *magazine named him one of the 12 best entrepreneurs of the 1980s, and in 1995 he was named by the* Utne Reader *as one of the 100 visionaries who could change our lives.*

Great ideas, in hindsight, seem obvious. The concept of natural capital is such an idea. *Natural capital* is the myriad necessary and valuable resources and ecological processes that we rely on to produce our food, products, and services.

The concept of natural capital is not a new one. Economists have long noted that natural capital is a factor in industrial production, but a marginal factor.

A new view is emerging: Our economic systems cannot long endure without taking the flow of renewable and nonrenewable resources [Figure 1-9, p. 10] through economies into account. The value of natural capital is becoming paramount to the success of all business.

This revision of neoclassical economics, yet to be accepted by most mainstream academicians, provides business and public policy with a powerful new tool for the continued prosperity of business and the preservation and restoration of the earth's living natural systems.

Most Americans are filled with cornucopian fantasies of technological prowess, where human ingenuity bypasses natural limits and creates unimagined abundance. Optimism easily intertwines with the belief that nanotechnology, biotechnology, computers, and technologies yet to developed will eliminate hunger, disease, and want.

Dreams of alleviating human suffering are worthy. However, they usually overlook the absolute necessity of fertile soil, ocean fisheries, a stable climate, biological diversity, and pure water, all of which we are degrading and none of which can be created by any human-made technology known or imagined.

In our pursuit of dominance over the natural world, we have not taken into account the basic principle that industrialism, for all its sophistication, is enormously inefficient with respect to resource use, energy use, and waste production. It is difficult for neoclassical economists, whose hypotheses and theories originated in a time of resource abundance, to understand that the very success of *linear industrial systems* based on increasing economic growth by increasing the rate of flow of materials and energy through economic systems has laid the groundwork for the next stage in economic evolution.

This shift is profoundly biological. It involves incorporating the cycling of material resources that supports natural systems into our ways of making things and our ways of dealing with the waste matter produced by our current linear industrial systems. This shift is going to happen because *cyclical industrial systems* work better than linear ones. They close the loop and reincorporate wastes as part of the production cycle. There are no landfills in a cyclical society.

classifies the world's countries as economically developed or developing based primarily on their degree of industrialization and their per capita GNP (Figure 1-4).

The **developed countries** include the United States, Canada, Japan, Australia, New Zealand, and all the countries of Europe. Most are highly industrialized and have high average per capita GNPs (above $10,000 per year, except for industrialized countries in eastern Europe and some in northern and southern Europe). These countries, with 1.2 billion people (20% of the world's population in 2001), **(1)** have about 85% of the world's wealth and income, **(2)** use about 88% of its natural resources, and **(3)** generate about 75% of its pollution and waste.

All other nations are classified as **developing countries**, most of them in Africa, Asia, and Latin America. Their 4.9 billion people (80% of the world's population in 2001) **(1)** have about 15% of the wealth and income and **(2)** use about 12% of the world's natural resources. Some are *middle-income, moderately developed countries* with average per capita GNPs of $1,000 to $10,000 per year and others are *low-income countries* with per capita GNPs less than $1,000 per year (Figure 1-4).

More than 95% of the projected increase in the world's population is expected to take place in developing countries (Figure 1-5), *where 1 million people are added every 5 days.* The primary reason for such rapid population growth in developing countries (1.6% compared to 0.1% in developed countries) is the *large percentage of people who are under age 15* (33% compared to 18% in developed countries in 2001). As these young people move into their prime reproductive years over the next several decades, they will fuel rapid population growth.

If there is so much inefficiency in our current system, why isn't it more apparent? The inefficiencies are masked by a financial system in which money, prices, and markets give us inaccurate information. Markets are not giving us correct information about how much our suburbs, cars, and plastic drinking water bottles truly cost.

Instead, we are getting warning signals from the beleaguered airsheds and watersheds, the overworked and eroded soils, the life-degrading inner cities and rural counties, the breakdown of stability worldwide, and the conflicts based on resource shortages. These feedbacks from nature are providing the information that our prices should give us but don't.

Prices don't give us good information for a simple reason: improper accounting. Natural capital has never been placed on the balance sheets of companies or the countries of the world. To paraphrase G. K. Chesterton, it could fairly be said that capitalism might be a good idea, but we have not tried it yet. Capitalism cannot be fully attained or practiced until we have an accurate balance sheet, as any accounting student will tell us.

As it stands, our economic system is based on accounting principles that would bankrupt a company. When natural capital is placed on the balance sheet, not as a free resource of infinite supply but as an integral and valuable part of the production process, everything changes. The nearly obsessive pursuit of improvement in *human productivity* becomes balanced by the need for improved *resource productivity*. Using more and more resources to make fewer people more productive flies in the face of what we need to improve our society and the environment. After all, it is people we have more of, not natural resources, so it is people we must use to reduce the flow of matter and energy resources through economies and the resulting pollution and loss of natural capital. Moving from linear industrial systems to cyclical ones that mimic nature accomplishes this.

Many people sincerely believe that an economic system based on the integrity of natural systems is unworkable. To answer that concern, we may want to reverse the question and ask, "How have we created an economic system that tells us it is cheaper to destroy natural capital than to maintain it?" We know this is not the way to take care of our cars, houses, and bridges, but somehow we have managed to overlook a pricing system that discounts the future and sells off the past. Or to put it another way, "How did we create an economic system that confuses capital with income?"

Can we devise and implement a more rational economic system? I think so. It is right before us. It requires no new theories, only common sense. It is based on the simple but powerful proposition that *all capital must be valued.*

There may be no *right* way to value a forest or a river, but there is a *wrong* way, which is to give it little or no value. If we have doubts about how to value a 500-year-old tree, we need only ask how much it would cost to make a new one from scratch. Or a new river. Or a new atmosphere.

Our goal should be to estimate and integrate the worth of living systems into every aspect of our culture and commerce so that human systems mimic natural systems. Only if we do this can our cultures reflect growth and harmony rather than damage and discord.

Critical Thinking

If you were in charge of the world's economy, what are the three most important things you would do? Compare your answers with those of other members of your class.

What Is Environmentally Sustainable Economic Development? Some analysts have called for a shift from an emphasis on traditional economic development fueled by economic growth of essentially any type to an emphasis on **environmentally sustainable economic development**. This type of development uses **(1)** economic rewards (government subsidies and tax breaks) to *encourage* environmentally beneficial and sustainable forms of economic growth and **(2)** economic penalties (government taxes and regulations) to *discourage* environmentally harmful and unsustainable forms of economic growth. Using environmentally sustainable economic development as the basis for developing more environmentally sustainable societies (p. 4) requires that governments, businesses, and individuals integrate social, economic, and environmental goals and policies in their decision making (Figure 1-6).

What Are the Environmental Effects of Poverty? According to the United Nations, about 1.2 billion people—one person in five—are hungry or malnourished (Figure 1-7) and lack access to clean drinking water, decent housing, and adequate health care. One of every three people lacks enough fuel to keep warm and to cook food and does not have access to electricity. About two-thirds of humanity lacks sanitary toilets, and one of every four adults (1.3 billion people) cannot read or write.

Daily life is a harsh struggle for the estimated one of every two people on the earth who try to survive on an income of $1–3 per day. Many poor parents have many children as a form of economic security. Their children

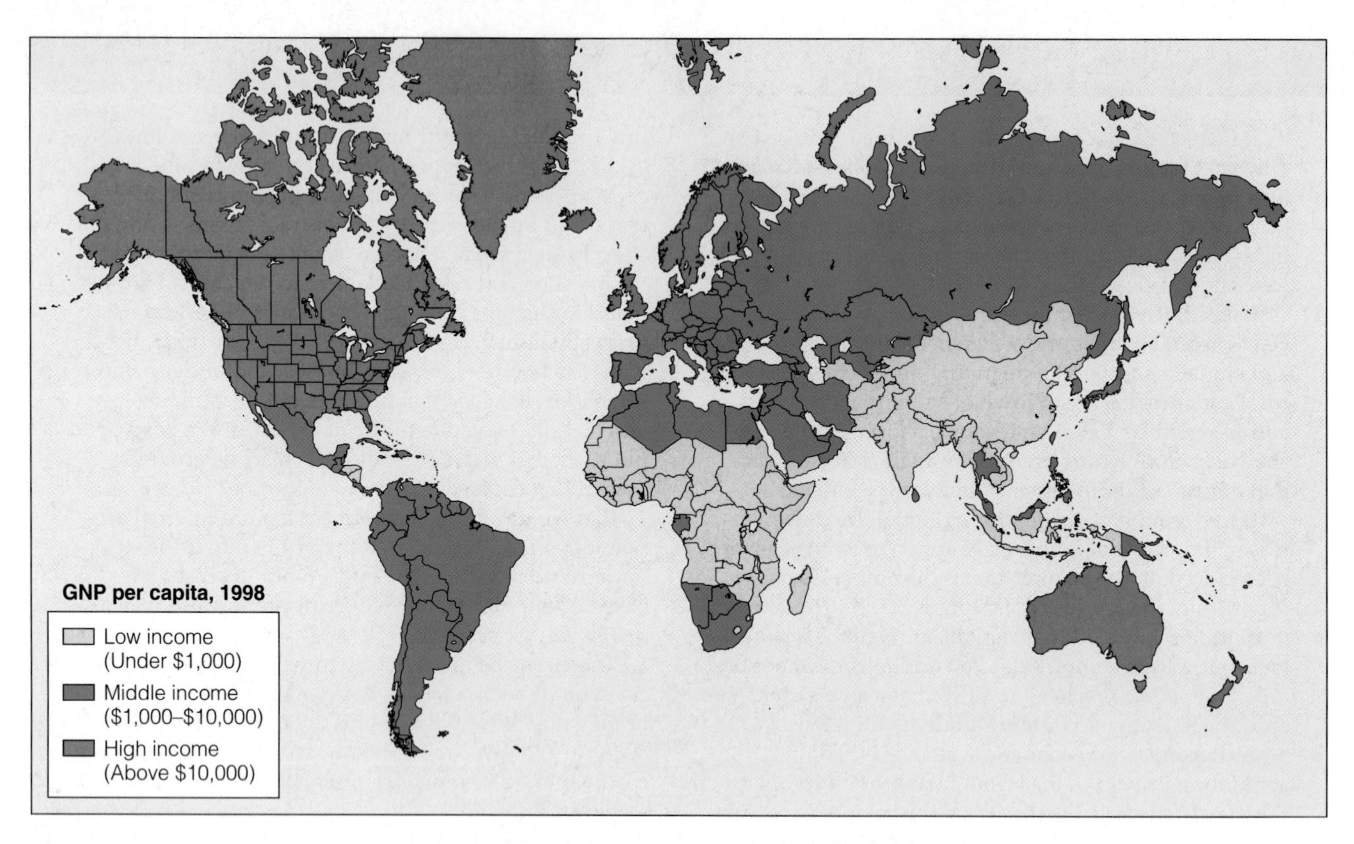

Figure 1-4 Degree of economic development as measured by per capita GNP in 1998. (Data from United Nations and the World Bank)

(1) help them grow food, **(2)** gather fuel (mostly wood and dung), **(3)** haul drinking water, **(4)** tend livestock, **(5)** work, **(6)** beg in the streets, and **(7)** help them survive in their old age (typically their 50s or 60s).

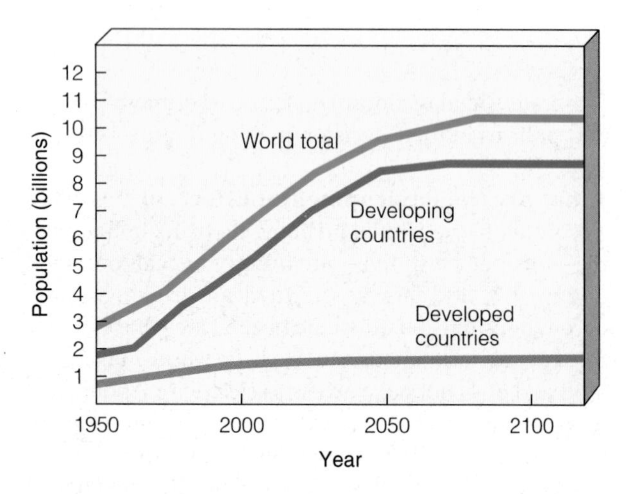

Figure 1-5 Past and projected population size for developed countries, developing countries, and the world, 1950–2120. More than 95% of the addition of 3.6 billion people between 1990 and 2030 is projected to occur in developing countries. (Data from United Nations)

Poverty is related to environmental quality and people's quality of life because poor people

- May deplete and degrade local forests, soil, grasslands, wildlife, and water supplies for short-term survival even though they know it may lead to disaster in the long run. They do not have the luxury of worrying about long-term supplies of natural resources when their daily life is focused on getting enough food and water to survive.
- Often have to live in areas with the highest levels of air and water pollution and with the greatest risk of natural disasters such as floods, earthquakes, hurricanes, and volcanic eruptions.
- Spend an average of **(1)** 4–6 hours *per day* searching for and carrying fuelwood and **(2)** 4–6 hours *per week* drawing and carrying water.
- Must take jobs (if they can find them) that subject them to unhealthy and unsafe working conditions at very low pay.

According to the World Health Organization (WHO), each year, at least 10 million of the desperately poor die prematurely of **(1)** malnutrition (lack of protein and other nutrients needed for good health), **(2)** increased susceptibility to infectious diseases

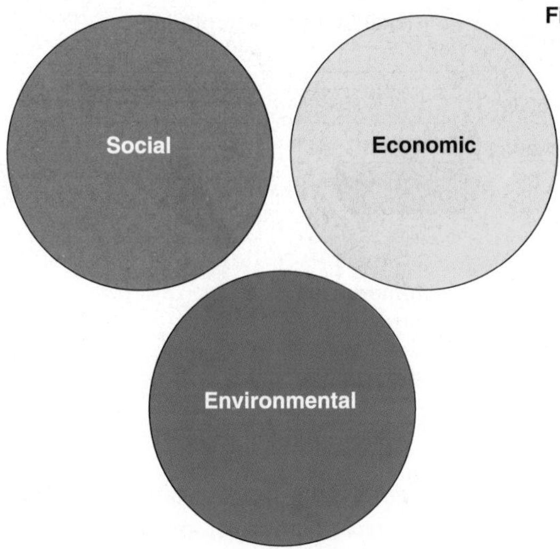

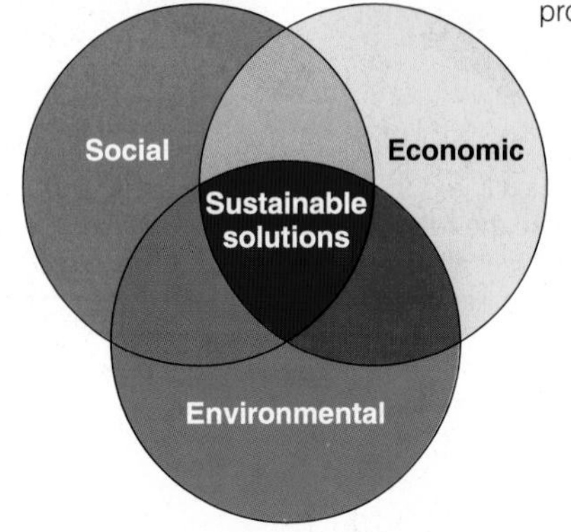

Figure 1-6 Types of decision making in traditional and sustainable societies. The traditional decision making in most societies involves treating social, economic, and environmental issues separately (left). Environmentally sustainable development calls for integrating social, economic, and environmental issues and concepts to find *sustainable solutions* to problems (right).

because of their weakened condition from malnutrition, and **(3)** infectious diseases caused by drinking contaminated water. *This premature death of at least 27,400 human beings per day is equivalent to 69 jumbo jet planes, each carrying 400 passengers, accidentally crashing every day with no survivors.* Half of those dying are children under age 5 (Figure 1-7).

What Is Globalization? One of the major trends since 1950 and especially since 1970 is globalization. **Globalization** is the process of global social, economic, and environmental change that leads to an increasingly integrated world.

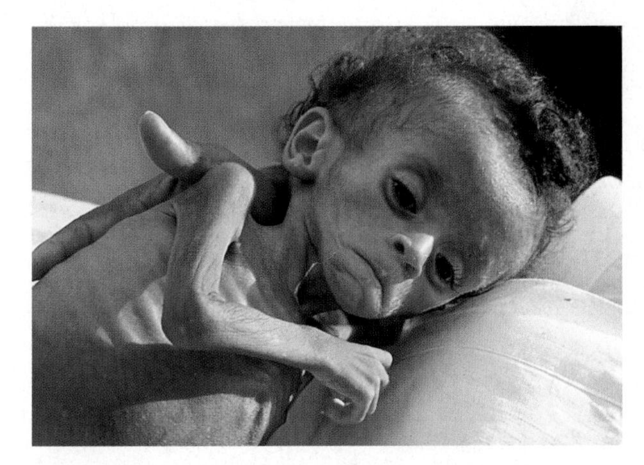

Figure 1-7 One in every three children under age 5, such as this Brazilian child, suffers from malnutrition. According to the World Health Organization, *each day* at least 13,700 children die prematurely from malnutrition and infectious diseases, most from drinking contaminated water and a weakened condition from malnutrition—an average of 10 preventable deaths each minute. Some analysts put the estimated death toll at almost twice this number. (John Bryson/Photo Researchers, Inc.)

Here are a few indicators of globalization:

- Between 1950 and 2001, the global economy grew from $6.7 trillion to $43 trillion.
- Between 1950 and 2001, international trade of goods increased from 5% to 14% of the gross world product.
- Between 1970 and 2001, the number of transnational corporations operating in three or more countries grew from 7,000 to about 54,000.
- By 2001, roughly 1 in every 34 people in the world had internet access on a global basis, and this figure is growing rapidly.
- Since 1950, there has been a huge increase in the number of species and infectious disease organisms (microbes) transported across international borders by trade and travel.
- Since 1950, long-lived pollutants such as DDT, PCBs, radioactive particles, and acidic chemicals have been transferred across the globe by wind, rainfall patterns, ocean currents, and rivers. On an even larger scale, nations now face the global threats of **(1)** widespread ocean pollution, **(2)** depletion of ozone in the upper atmosphere (stratosphere) that keeps much of the sun's harmful ultraviolet radiation from reaching the earth's surface, and **(3)** global and regional climate change caused by chemicals released into the environment by human activities.

Largely as a result of economic growth, population growth, and globalization, the environmental impact or *ecological footprint* of each person in developed countries is large compared with that in developing countries (Figure 1-8).

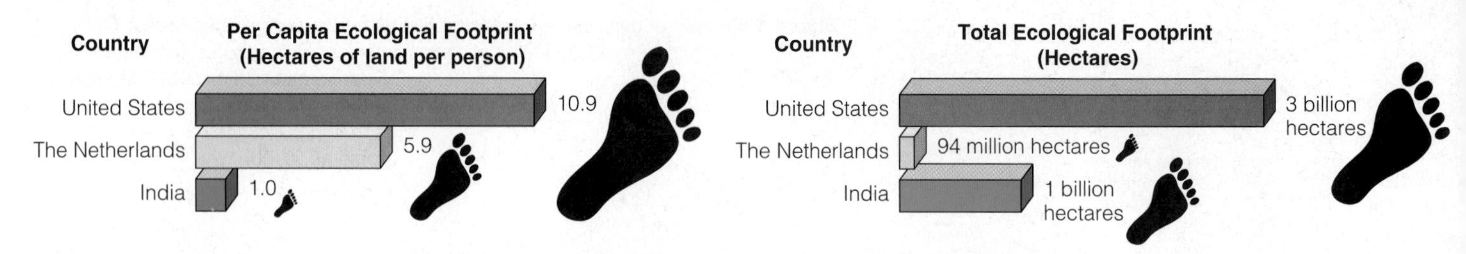

Figure 1-8 Ecological footprints of the United States, the Netherlands, and India. An **ecological footprint** is the amount of land needed to produce the resources needed by an average person in a country. It would take the land area of about three earths if all the world's 6.1 billion people consumed the same amount of resources as is consumed by the 285 million people in the United States. (Data from William Rees and Mathis Wackernagel)

1-3 RESOURCES

What Is a Resource? From a human standpoint, a **resource** is anything obtained from the environment to meet human needs and wants. Examples include food, water, shelter, manufactured goods, transportation, communication, and recreation.

Some resources, such as solar energy, fresh air, wind, fresh surface water, fertile soil, and wild edible plants, are directly available for use. Other resources, such as petroleum (oil), iron, groundwater (water found underground), and modern crops, are not directly available. They become useful to us only with some effort and technological ingenuity. For example, petroleum was a mysterious fluid until we learned how to find, extract, and convert (refine) it into gasoline, heating oil, and other products that could be sold at affordable prices.

On our short human time scale, we classify the material resources we get from the environment as **(1)** *perpetual*, **(2)** *renewable*, or **(3)** *nonrenewable* (Figure 1-9; also see the orange boxes in the "Concepts and Connections" diagram inside the back cover).

What Are Perpetual and Renewable Resources? Solar energy is called a **perpetual resource** because on a human time scale it is renewed continuously. It is expected to last at least 6 billion years as the sun completes its life cycle.

On a human time scale, a **renewable resource** can be replenished fairly rapidly (hours to several decades) through natural processes as long as it is not used up faster than it is replaced. These resources are endlessly renewable but only at the rate at which nature provides them. Examples are **(1)** forests, **(2)** grasslands, **(3)** wild animals, **(4)** fresh water, **(5)** fresh air, and **(6)** fertile soil.

However, renewable resources can be depleted or degraded. The highest rate at which a renewable resource can be used *indefinitely* without reducing its available supply is called **sustainable yield**.

If we exceed a resource's natural replacement rate, the available supply begins to shrink, a process known as **environmental degradation**. Examples of such degradation include **(1)** urbanization of productive land, **(2)** waterlogging and salt buildup in soil, **(3)** excessive topsoil erosion, **(4)** deforestation, **(5)** groundwater depletion, **(6)** overgrazing of grasslands by livestock, **(7)** reduction in the earth's forms of wildlife (biodiversity) by elimination of habitats and species, and **(8)** pollution.

Such forms of environmental degradation can change usable, renewable resources into nonrenewable or unusable resources. A major cause of environmental degradation of renewable resources is a phenomenon known as the *tragedy of the commons* (Connections, right).

Figure 1-9 Major types of material resources. This scheme is not fixed; renewable resources can become nonrenewable if used for a prolonged period at a faster rate than they are renewed by natural processes.

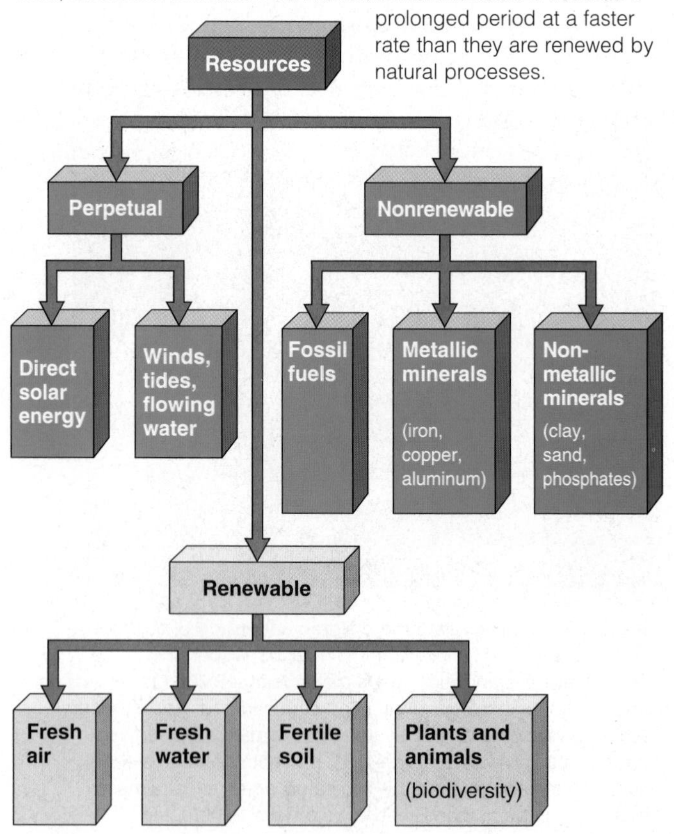

Free-Access Resources and the Tragedy of the Commons

One cause of environmental degradation is the overuse of **common-property** or **free-access resources.** Such resources are owned by no one (or jointly by everyone in a country or area) but are available to all users at little or no charge.

Examples include **(1)** clean air, **(2)** the open ocean and its fish, **(3)** migratory birds, **(4)** wildlife species, **(5)** publicly owned lands (such as national forests, national parks, and wildlife refuges), **(6)** gases of the lower atmosphere, and **(7)** space.

In 1968, biologist Garrett Hardin (Guest Essay, p. 256) called the degradation of renewable free-access resources the **tragedy of the commons**. It happens because each user reasons, "If I do not use this resource, someone else will. The little bit I use or pollute is not enough to matter, and such resources are renewable."

With only a few users, this logic works. However, the cumulative effect of many people trying to exploit a free-access resource eventually exhausts or ruins it. Then no one can benefit from it, and therein lies the tragedy.

Two solutions to this problem are to

- *Use free-access resources at rates well below their estimated sustainable yields or overload limits by reducing population, regulating access, or both.* This prevention approach is rarely used because **(1)** it entails establishing and enforcing regulations that restrict resource use or population growth, and **(2)** it is difficult and expensive to determine the sustainable yield of a forest, grassland, or animal population because such yields vary with weather, climate, and unpredictable biological factors.
- *Convert free-access resources to private ownership.* The reasoning is that owners of land or some other resource have a strong incentive to protect their investment. However, this approach is not practical for global common resources (such as the atmosphere, the open ocean, most wildlife species, and migratory birds) that cannot be divided up and converted to private property.

Experience shows that there is another possibility. Just because a resource is easily available to a community does not always mean that people have free and unregulated access to that resource. There are many examples in which communities have established a set of rules and traditions to regulate and share their access to a common-property resource such as fisheries, grazing lands, and forests.

Critical Thinking

Give three examples of how you cause environmental degradation as a result of the tragedy of the commons. How should we deal with this problem? Explain.

What Are Nonrenewable Resources? Resources that exist in a fixed quantity or stock in the earth's crust are called **nonrenewable resources**. On a time scale of millions to billions of years, geological processes can renew such resources. However, on the much shorter human time scale of hundreds to thousands of years, these resources can be depleted much faster than they are formed.

These exhaustible resources include **(1)** *energy resources* (such as coal, oil, and natural gas, which cannot be recycled), **(2)** *metallic mineral resources* (such as iron, copper, and aluminum, which can be recycled), and **(3)** *nonmetallic mineral resources* (such as salt, clay, sand, and phosphates, which usually are difficult or too costly to recycle).

We know how to find and extract more than 100 nonrenewable mineral resources from the earth's crust. We convert these raw materials into many everyday items and then we discard, reuse, or recycle them.

Figure 1-10 shows the production and depletion cycle of a nonrenewable energy or mineral resource. We never completely exhaust a nonrenewable mineral resource. However, such a resource becomes *economically depleted* when the costs of extracting and using what is left exceed its economic value. At that point, we have six choices: **(1)** try to find more, **(2)** recycle or reuse existing supplies (except for nonrenewable energy resources, which cannot be recycled or reused), **(3)** waste less, **(4)** use less, **(5)** try to develop a substitute, or **(6)** wait millions of years for more to be produced.

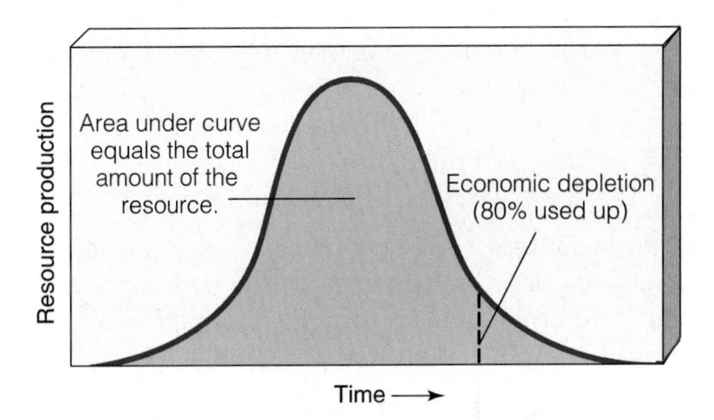

Figure 1-10 Full production and exhaustion cycle of a nonrenewable resource such as copper, iron, oil, or coal. Usually, a nonrenewable resource is considered *economically depleted* when 80% of its total supply has been extracted and used. Normally, it costs too much to extract and process the remaining 20%.

Some nonrenewable material resources, such as copper and aluminum, can be recycled or reused to extend supplies. **Recycling** involves collecting and reprocessing a resource into new products. For example, glass bottles can be crushed and melted to make new bottles or other glass items. **Reuse** involves using a resource over and over in the same form. For example, glass bottles can be collected, washed, and refilled many times.

Recycling nonrenewable metallic resources takes much less energy, water, and other resources and produces much less pollution and environmental degradation than exploiting virgin metallic resources. Reusing such resources takes even less energy and other resources and produces less pollution and environmental degradation than recycling.

Nonrenewable energy resources, such as coal, oil, and natural gas, cannot be recycled or reused. Once burned, the useful energy in these fossil fuels is gone, leaving behind waste heat and polluting exhaust gases.

1-4 POLLUTION

What Is Pollution, and Where Do Pollutants Come From? Any addition to air, water, soil, or food that threatens the health, survival, or activities of humans or other living organisms is called **pollution**. Pollutants can enter the environment **(1)** naturally (for example, from volcanic eruptions) or **(2)** through human (anthropogenic) activities (for example, from burning coal). Most pollution from human activities occurs in or near urban and industrial areas, where pollutants are concentrated. Industrialized agriculture also is a major source of pollution.

Some pollutants contaminate the areas where they are produced; others are carried by wind or flowing water to other areas. Pollution does not respect local, state, or national boundaries.

There are two types of pollutant sources:

- **Point sources**, where pollutants come from single, identifiable sources. Examples are the **(1)** smokestack of a coal-burning power plant, **(2)** drainpipe of a factory, or **(3)** exhaust pipe of an automobile.
- **Nonpoint sources**, where pollutants come from dispersed (and often difficult to identify) sources. Examples are **(1)** runoff of fertilizers and pesticides (from farmlands, golf courses, and suburban lawns and gardens) into streams and lakes and **(2)** pesticides sprayed into the air or blown by the wind into the atmosphere.

It is much easier and cheaper to identify and control pollution from point sources than from widely dispersed nonpoint sources.

What Types of Harm Do Pollutants Cause? Unwanted effects of pollutants include the following:

- Disruption of life-support systems for humans and other species
- Damage to wildlife, human health, and property
- Nuisances such as noise and unpleasant smells, tastes, and sights

Solutions: What Can We Do About Pollution? There are two basic approaches to dealing with pollution: **(1)** prevent it from reaching the environment or **(2)** clean it up if it does. **Pollution prevention** or **input pollution control** reduces or eliminates the production of pollutants, often by using less harmful chemicals or processes. We can prevent (or at least reduce) pollution by following the four *R*s of resource use: *refuse (do not use), reduce (use less), reuse,* and *recycle.*

Pollution cleanup or **output pollution control** involves cleaning up pollutants after they have been produced. Environmentalists have identified three problems with relying primarily on pollution cleanup:

- *It is only a temporary bandage as long as population and consumption levels grow without corresponding improvements in pollution control technology.* For example, adding catalytic converters to car exhaust systems has reduced air pollution. However, increases in the number of cars and in the total distance each travels have reduced the effectiveness of this cleanup approach.
- *It often removes a pollutant from one part of the environment only to cause pollution in another.* For example, we can collect garbage, but the garbage is then **(1)** burned (perhaps causing air pollution and leaving a toxic ash that must be put somewhere), **(2)** dumped into streams, lakes, and oceans (perhaps causing water pollution), or **(3)** buried (perhaps causing soil and groundwater pollution).
- *Once pollutants have entered and become dispersed into the environment at harmful levels, it usually costs too much to reduce them to acceptable levels.*

Both pollution prevention and pollution cleanup are needed. However, environmentalists and some economists urge us to emphasize prevention because it works better and is cheaper than cleanup. As Benjamin Franklin observed long ago, "An ounce of prevention is worth a pound of cure." An increasing number of businesses have found that *pollution prevention pays.*

Governments can encourage both pollution prevention and pollution cleanup by

- Using incentives such as various subsidies and tax write-offs
- Using regulations and taxes

Most analysts believe that a combination of both approaches is best because excessive regulation and too much taxation can cause a political backlash. Achieving the right balance is difficult.

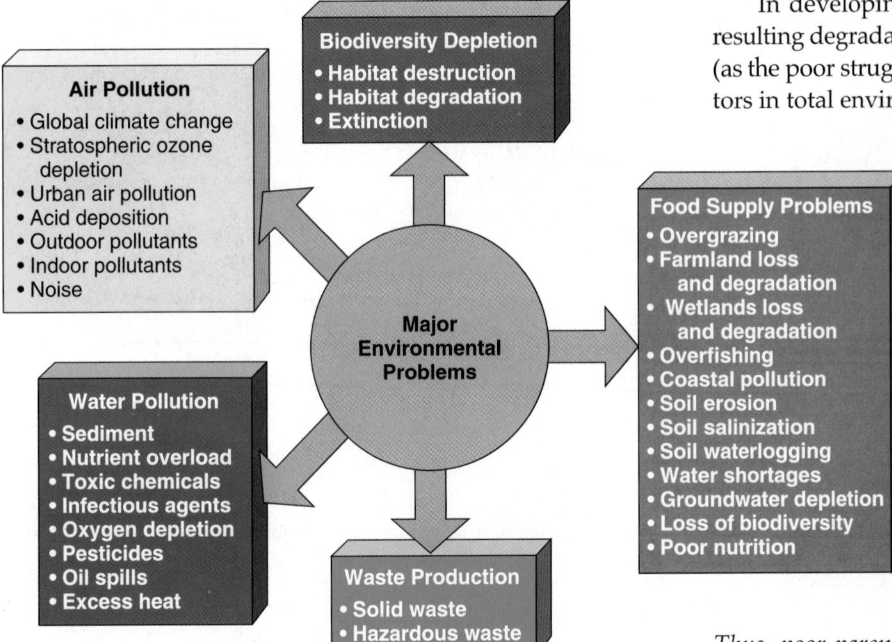

Figure 1-11 Major environmental and resource problems.

1-5 ENVIRONMENTAL AND RESOURCE PROBLEMS: CAUSES AND CONNECTIONS

What Are Key Environmental Problems and Their Basic Causes? We face a number of interconnected environmental and resource problems (Figure 1-11). The first step in dealing with these problems is to identify their underlying causes (Figure 1-12).

How Are Environmental Problems and Their Causes Connected? Once we have identified environmental problems and their root causes, the next step is to understand how they are connected to one another. The three-factor model in Figure 1-13 is a good starting point.

According to this simple model, the environmental impact (**I**) of population on a given area depends on three key factors: **(1)** the number of people (**P**), **(2)** average resource use per person (affluence, **A**), and **(3)** the environmental effects of the technologies (**T**) used to provide and consume each unit of resource.

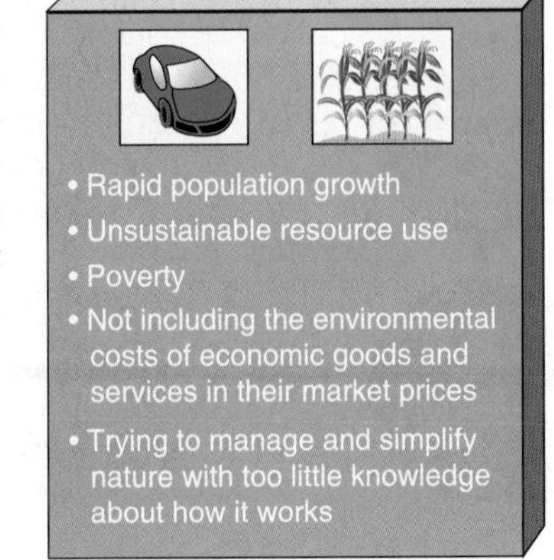

• Rapid population growth
• Unsustainable resource use
• Poverty
• Not including the environmental costs of economic goods and services in their market prices
• Trying to manage and simplify nature with too little knowledge about how it works

Figure 1-12 Environmentalists have identified five basic causes of the environmental problems we face.

In developing countries, population size and the resulting degradation of potentially renewable resources (as the poor struggle to stay alive) tend to be the key factors in total environmental impact (Figure 1-13, top). In such countries per capita resource use is low.

In developed countries, high rates of per capita resource use and the resulting high levels of pollution and environmental degradation per person usually are the key factors determining overall environmental impact (Figure 1-13, bottom). For example, it is estimated that the average U.S. citizen consumes 35 times as much as the average citizen of India and 100 times as much as the average person in the world's poorest countries. *Thus, poor parents in a developing country would need 70–200 children to have the same lifetime resource consumption as 2 children in a typical U.S. family.*

Some forms of technology, such as polluting factories and motor vehicles and energy-wasting devices, increase environmental impact by raising the T factor in the equation. Other technologies, such as pollution control, solar cells, and energy-saving devices, lower environmental impact by decreasing the T factor in the equation. In other words, some forms of technology are *environmentally harmful* and some are *environmentally beneficial.*

The three-factor model in Figure 1-13 can help us understand how key environmental problems and some of their causes are connected. However, these problems involve a number of poorly understood interactions between many more factors than those in this simplified model, as outlined in Figure 1-14. A more detailed model is given inside the back cover.

1-6 CULTURAL CHANGES AND SUSTAINABILITY

What Major Human Cultural Changes Have Taken Place? Evidence from fossils and studies of ancient cultures suggests that the current form of our species, *Homo sapiens sapiens*, has walked the earth for only about 60,000 years (some recent evidence suggests 90,000–176,000 years), an instant in the planet's estimated 4.6-billion-year existence.

Until about 12,000 years ago, we were mostly hunter–gatherers who typically moved as needed to find enough food for survival. Since then, there have been three major cultural changes: **(1)** the *agricultural*

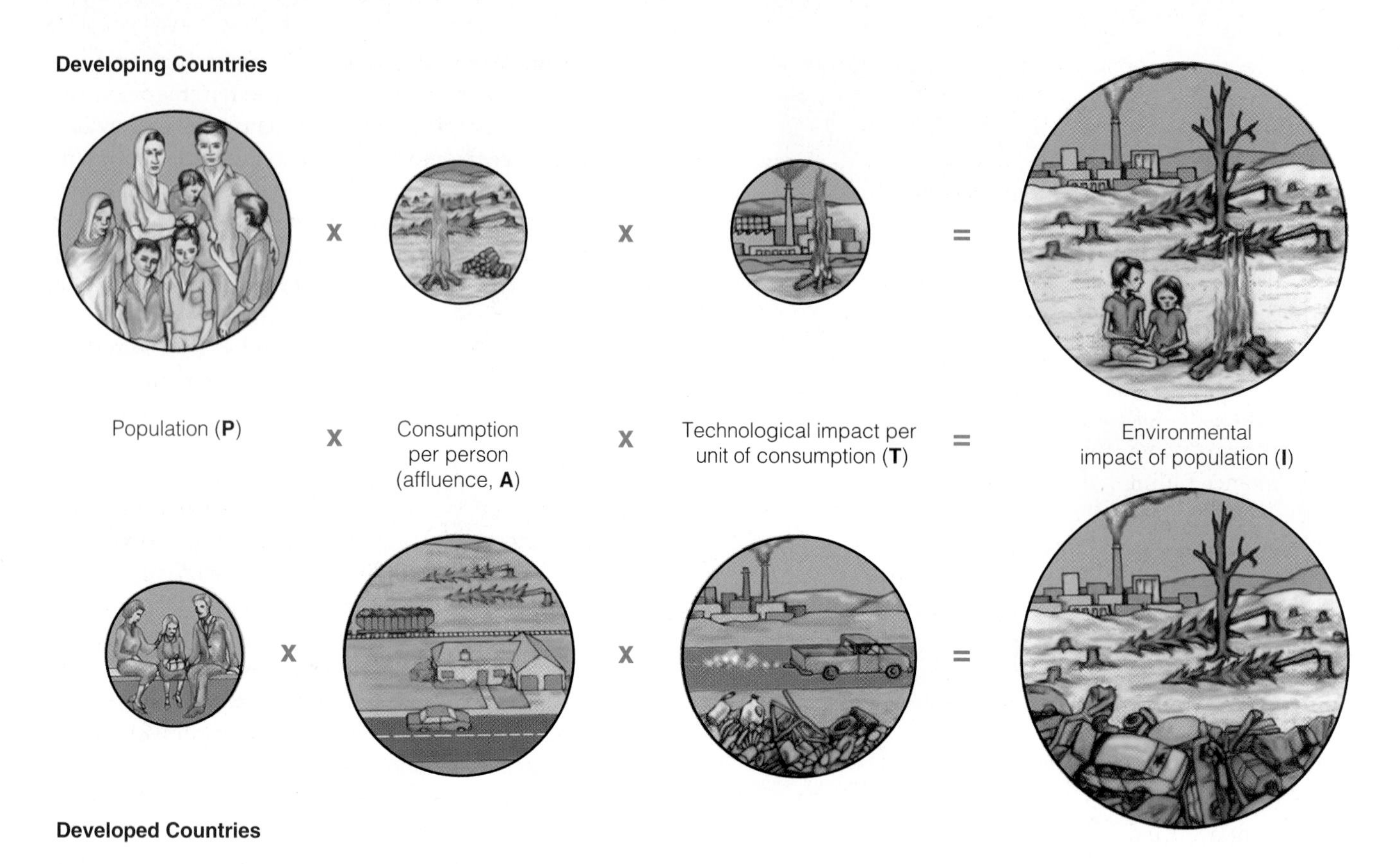

Figure 1-13 Simplified model of how three factors—population, affluence, and technology—affect the environmental impact of population in developing countries (top) and developed countries (bottom).

revolution (which began 10,000–12,000 years ago), **(2)** the *industrial revolution* (which began about 275 years ago), and **(3)** the *information and globalization revolution* (which began about 50 years ago).

These major cultural changes have

- Given us much more energy and new technologies with which to alter and control more of the planet to meet our basic needs and increasing wants
- Allowed expansion of the human population, mostly because of increased food supplies and longer life spans
- Increased our environmental impact because of increased resource use, pollution, and environmental degradation

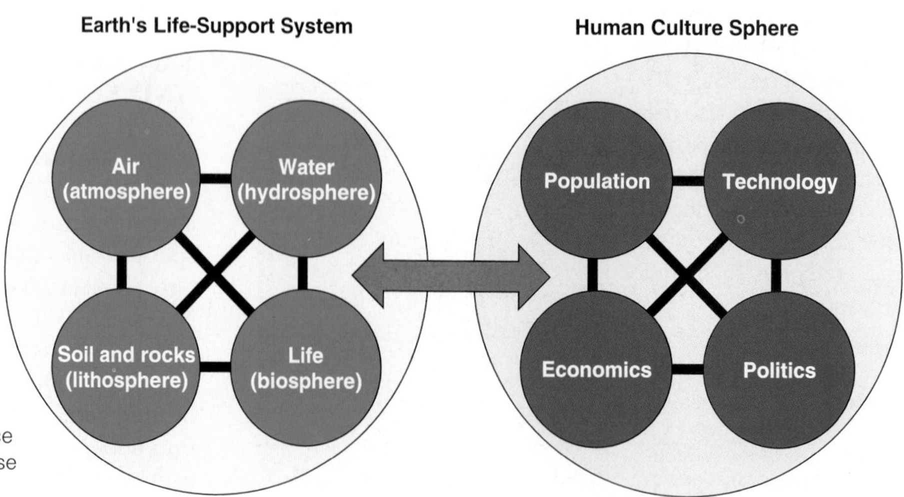

Figure 1-14 Major components and interactions within and between the earth's life-support system and the human sociocultural system (culturesphere). The goal of environmental science is to learn as much as possible about these complex interactions.

How Did Ancient Hunting-and-Gathering Societies Affect the Environment? During most of our 60,000-year existence, we were **hunter–gatherers** who survived by collecting edible wild plant parts, hunting, fishing, and scavenging meat from animals killed by other predators. Our hunter–gatherer ancestors typically lived in small bands (of fewer than 50 people) who worked together to get enough food to survive. Many groups were nomadic, picking up their few possessions and moving seasonally from place to place to find enough food.

The earliest hunter–gatherers (and those still living this way today) survived through expert knowledge and understanding of their natural surroundings. They discovered **(1)** which plants and animals could be eaten and used as medicines, **(2)** where to find water, **(3)** how plant availability changed throughout the year, and **(4)** how some game animals migrated to get enough food. Because of high infant mortality and an estimated average life span of 30–40 years, hunter–gatherer populations grew very slowly.

Advanced hunter–gatherers had a greater impact on their environment than did early hunter–gatherers. They **(1)** used more advanced tools and fire to convert forests into grasslands, **(2)** contributed to the extinction of some large animals (including the mastodon, saber-toothed tiger, giant sloth, cave bear, mammoth, and giant bison), and **(3)** altered the distribution of plants (and animals feeding on such plants) as they carried seeds and plants to new areas.

Early and advanced hunter–gatherers exploited their environment to survive. However, their environmental impact usually was limited and local because of **(1)** their small population sizes, **(2)** low resource use per person, **(3)** migration, which allowed natural processes to repair most of the damage they caused, and **(4)** lack of technology that could have expanded their impact.

How Has the Agricultural Revolution Affected the Environment? Some 10,000–12,000 years ago, a cultural shift known as the **agricultural revolution** began in several regions of the world. It involved a gradual move from usually nomadic hunting-and-gathering groups to settled agricultural communities in which people domesticated wild animals and cultivated wild plants.

Plant cultivation probably developed in many areas, especially in the tropical forests of Southeast Asia, northeast Africa, and Mexico. People discovered how to grow various wild food plants from roots or tubers (fleshy underground stems). To prepare the land for planting, they cleared small patches of tropical forests by cutting down trees and other vegetation and then burning the underbrush (Figure 1-15). The ashes fertilized the nutrient-poor soils in this **slash-and-burn cultivation**.

Early growers also used various forms of **shifting cultivation** (Figure 1-15), primarily in tropical regions.

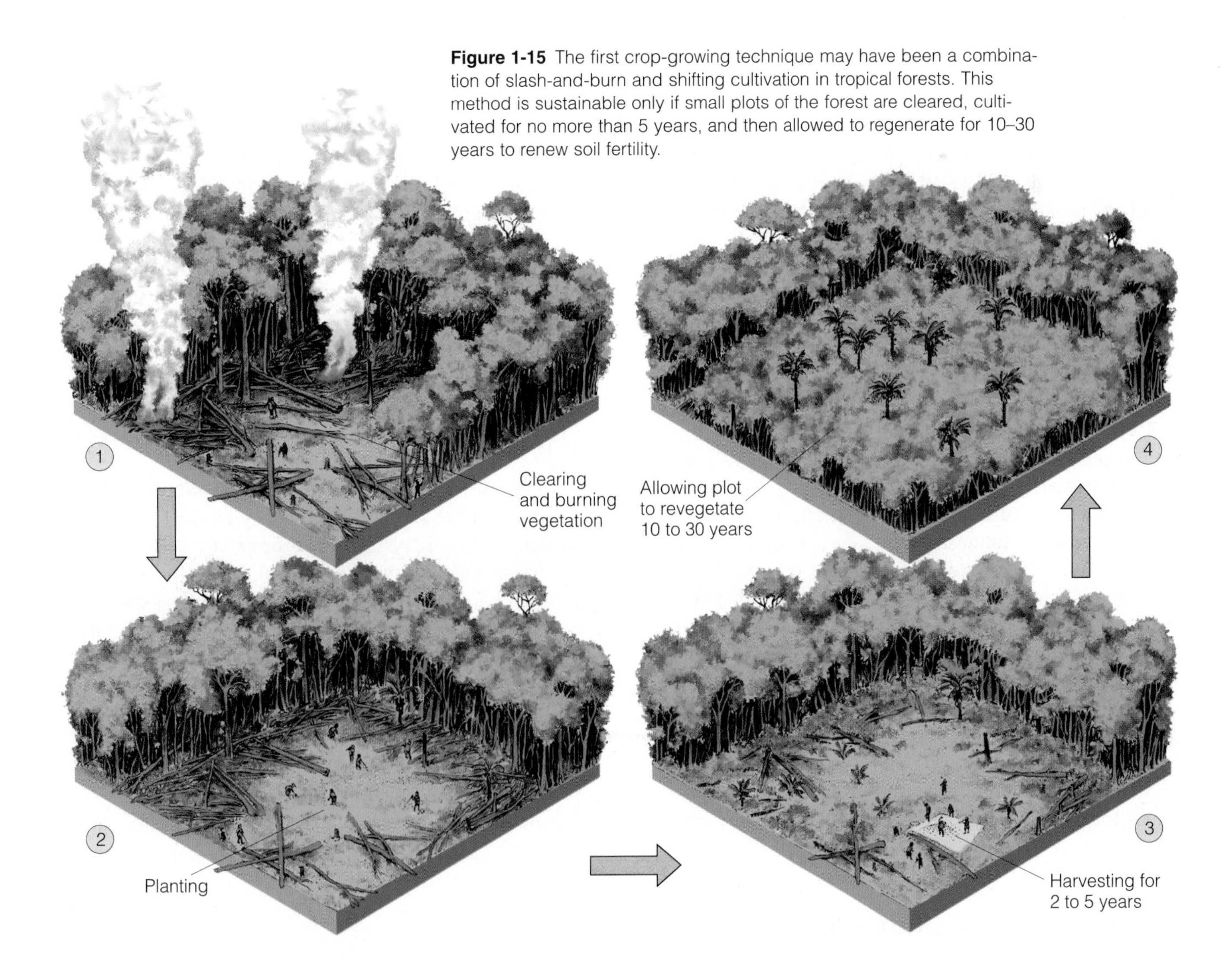

Figure 1-15 The first crop-growing technique may have been a combination of slash-and-burn and shifting cultivation in tropical forests. This method is sustainable only if small plots of the forest are cleared, cultivated for no more than 5 years, and then allowed to regenerate for 10–30 years to renew soil fertility.

CONNECTIONS

Consequences of the Agricultural Revolution

Here are some of the beneficial and harmful effects of the agricultural revolution:

- *Using domesticated animals to plow fields, haul loads, and perform other tasks increased the ability to expand agriculture and support more people.*
- *People* **(1)** *cut down vast forests to supply wood for fuel and building materials,* **(2)** *plowed up large expanses of grassland to grow crops, and* **(3)** *built irrigation systems to transfer water from one place to another.* Such extensive land clearing degraded or destroyed the habitats of many wild plants and animals, causing or hastening their extinction.
- *Soil erosion, salt buildup in irrigated soils, and overgrazing of grasslands by huge herds of livestock helped turn fertile land into desert; topsoil washed into streams, lakes, and irrigation canals.* This environmental degradation was a factor in the downfall of many early civilizations in the Middle East, North Africa, and the Mediterranean.
- *People began accumulating material goods.* Nomadic hunter–gatherers could not carry many possessions in their travels, but farmers living in one place could acquire as much as they could afford.
- *Farmers could grow more than enough food for their families.* They could store the excess for a "rainy day" or use it to barter for other goods and services.
- *Urbanization—the formation of villages, towns, and cities—became practical.* Some villages grew into towns and cities, which served as centers for trade, government, and religion. Towns and cities concentrated sewage and other wastes, polluted the air and water, and greatly increased the spread of diseases.
- *Increased production and use of material goods created growing volumes of waste.*
- *Conflict between societies became more common as ownership of land and water rights became a crucial economic issue.* Armies and their leaders rose to power and conquered large areas of land and water supplies. These rulers forced powerless people (slaves and landless peasants) to do the hard, disagreeable work of producing food and constructing irrigation systems, temples, and walled fortresses.
- *The survival of wild plants and animals, once vital to humanity, became less important.* Wild animals, which competed with livestock for grass and fed on crops, became enemies to be killed or driven from their habitats. Wild plants invading cropfields became weeds to be eliminated.

Critical Thinking

Would we be better off if agricultural practices had never been developed and we were still hunters and gatherers? Explain.

After a plot had been used for several years, the soil would be depleted of nutrients or reinvaded by the forest. Then the growers cleared a new plot. They learned that each abandoned patch normally had to be left fallow (unplanted) for 10–30 years before the soil became fertile enough to grow crops again. While patches were regenerating, growers used them for tree crops, medicines, fuelwood, and other purposes. In this manner, early growers practiced *sustainable cultivation*.

These early farmers had fairly little impact on the environment because **(1)** their dependence mostly on human muscle power and crude stone or stick tools meant that they could cultivate only small plots, **(2)** their population size and density were low, and **(3)** normally there was enough land to move to other areas and leave abandoned plots unplanted for the several decades needed to restore soil fertility. The gradual shift from hunting and gathering to farming had several significant effects (Connections, above).

How Has the Industrial Revolution Affected the Environment? The next cultural shift, the **industrial revolution**, began in England in the mid-1700s and spread to the United States in the 1800s (Case Study, p. 18). It led to a rapid expansion in the production, trade, and distribution of material goods.

The industrial revolution represented a shift from dependence on **(1)** *renewable* wood (with supplies dwindling in some areas because of unsustainable cutting) and flowing water to **(2)** dependence on machines running on *nonrenewable* fossil fuels (first coal and later oil and natural gas). This led to a switch from small-scale, localized production of handmade goods to large-scale production of machine-made goods in centralized factories in rapidly growing industrial cities.

Factory towns grew into cities as rural people came to the factories for work. There they worked long hours under noisy, dirty, and hazardous conditions. Other workers toiled in dangerous coal mines. In these early industrial cities, coal smoke belching out of chimneys was so heavy that many people died prematurely of lung ailments. Ash and soot covered everything, and on some days the smoke was so thick that it blotted out the sun.

Fossil-fuel-powered farm machinery, commercial fertilizers, and new plant-breeding techniques increased per acre crop yields. This helped protect biodiversity by reducing the need to expand the area of cropland to grow food. Because fewer farmers were needed, more people migrated to cities. With a larger and more reliable food supply and longer life spans, the size of the human population began the sharp increase that continues today (Figures 1-1 and 1-2).

After World War I (1914–18), more efficient machines and mass production techniques were developed. These technologies became the basis of today's advanced industrial societies in places such as the United States, Canada, Japan, Australia, and western Europe. Advanced industrial societies have provided numerous benefits along with environmental problems (Connections, right, and Case Study, p. 18).

How Might the Information and Globalization Revolution Affect the Environment? We are in the midst of a new cultural shift, the **information and globalization revolution**, in which new technologies such as the telephone, radio, television, computers, the internet, automated databases, and remote sensing satellites are enabling people to have increasingly rapid access to much more information on a global scale. It is estimated that scientific information now doubles about every 12 years, and general information doubles about every 2.5 years. The World Wide Web contains hundreds of millions of electronic pages and grows by roughly a million electronic pages per day.

This new cultural revolution can have beneficial and harmful environmental effects. On the *positive side*, today's information technologies

- Help us understand more about how the earth, economies, and other complex systems work and how such systems might be affected by our actions
- Allow us to respond to environmental problems more effectively and rapidly
- Allow us to use remote sensing satellites to survey resources and monitor changes in the world's forests, grasslands, oceans, rivers, polar regions, cities, and other systems
- Enable us to develop sophisticated computer models and computer-generated maps of the earth's environmental systems
- Can reduce pollution and environmental degradation by substituting data for materials and energy and communication for transportation

On the *negative side*, information technologies

- Provide an overload of information
- Cause confusion, distraction, and a sense of hopelessness as we try to identify useful environmental information and ideas in a rapidly growing sea of information
- Increase environmental degradation and decrease cultural diversity as a globalized economy spreads over most of the earth and homogenizes the world's cultures

CONNECTIONS

Consequences of Advanced Industrial Societies

The *good news* is that advanced industrial societies provide a variety of benefits to most people living in them, including

- Mass production of many useful and affordable products
- A sharp increase in agricultural productivity
- Lower infant mortality and longer life expectancy because of better sanitation, hygiene, nutrition, and medical care
- A decrease in the rate of population growth
- Better health, birth control methods, and education
- Methods for controlling pollution
- Greater average income and old-age security

However, the *bad news* is the resource and environmental problems we face today (Figure 1-11, p. 13), mostly because of the rise of advanced industrial societies.

Critical Thinking

1. On balance, do you believe that the advantages of the industrial revolution have outweighed its disadvantages? Explain.

2. What three major things would you do to reduce the harmful environmental impacts of advanced industrial societies?

1-7 IS OUR PRESENT COURSE SUSTAINABLE?

Are Things Getting Better or Worse? Experts disagree about **(1)** how serious our population and environmental problems are and **(2)** what we should do about them. Some analysts believe that human ingenuity and technological advances will allow us to **(1)** clean up pollution to acceptable levels, **(2)** find substitutes for any resources that become scarce, and **(3)** keep expanding the earth's ability to support more humans, as we have done in the past. They accuse most scientists and environmentalists of **(1)** exaggerating the seriousness of the problems we face and **(2)** failing to appreciate the progress we have made in improving quality of life and protecting the environment.

On the other hand, environmentalists and many leading scientists contend that we are disrupting the earth's life-support system for us and other forms of life

CASE STUDY

Eras of Environmental History in the United States

The environmental history of the United States can be divided into four eras:

- *Tribal,* when North America was occupied by 5–10 million tribal people (now called Native Americans) for at least 10,000 years before European settlers began arriving in the early 1600s. Although there were exceptions, many Native American cultures had a deep respect for the land and its animals and did not believe in land ownership.

- *Frontier* (1607–1890), when European colonists began settling North America. Faced with a continent containing seemingly inexhaustible forest and wildlife resources and rich soils, the early colonists developed a **frontier environmental worldview**. They viewed most of the continent as a wilderness to be conquered by clearing and planting and with vast resources to be used.

- *Early conservation* (1832–1870), during which some people became alarmed at the scope of resource depletion and degradation in the United States. They urged that part of the unspoiled wilderness on public lands owned jointly by all people (but managed by the government) be protected as a legacy to future generations.

- *Increasing role of the federal government and private citizens in resource conservation, public health, and environmental protection* (1870–present). Appendix 2 summarizes some of the major events during this period.

Critical Thinking

Use the library or the internet to add major environmental events to the latest time line in Appendix 2 that have occurred since 2001.

at an accelerating rate and that this is leading to serious environmental and economic harm (Guest Essay, p. 20). They are encouraged by the progress we have made. However, they point out how much more we must do to help make the earth more sustainable for present and future human generations and for other species that support us and other forms of life.

On November 18, 1992, some 1,680 of the world's senior scientists from 70 countries, including 102 of the 196 living scientists who are Nobel laureates, signed and sent an urgent warning to government leaders of all nations. According to this warning,

> *Our massive tampering with the world's interdependent web of life—coupled with the environmental damage inflicted by deforestation, species loss, and climate change—could trigger widespread adverse effects, including unpredictable collapses of critical biological systems whose interactions and dynamics we only imperfectly understand. . . . No more than one or a few decades remain before the chance to avert the threats we now confront will be lost and the prospects for humanity immeasurably diminished.*

Also in 1992, the prestigious U.S. National Academy of Sciences and the Royal Society of London issued a joint report, their first ever, which began,

> *If current predictions of population growth prove accurate and patterns of human activity on the planet remain unchanged, science and technology may not be able to prevent either irreversible degradation of the environment or continued poverty for much of the world. . . . Sustainable development can be achieved, but only if irreversible degradation of the environment can be halted in time.*

These two major warnings are not the views of a small number of scientists but the consensus of the mainstream scientific community, consisting of most of the world's key researchers on environmental problems.

Leading environmentalists call for us to launch an *environmental or sustainability revolution* to take place over the next 50 years (Guest Essay, p. 20). This would involve shifting our efforts from

- Pollution cleanup to pollution prevention (clean production)
- Waste disposal (mostly burial and burning) to waste prevention and reduction
- Protecting species to protecting the places where they live (habitats)
- Environmental degradation to environmental restoration
- Increased resource use to more efficient (less wasteful) resource use
- Population growth to population stabilization

The Solutions box (right) lists some guidelines various analysts have suggested for living more sustainably by working with the earth.

This chapter has presented an overview of the problems most environmentalists and many of the world's most prominent scientists believe we face and

Some Guidelines for Working with the Earth

SOLUTIONS

- Leave the earth as good as or better than we found it.
- Take no more than we need.
- Try not to harm life, air, water, or soil.
- Sustain the variety of the earth's life-forms (biodiversity).
- Help maintain the earth's capacity for self-repair and adaptation.
- Do not use potentially renewable resources (soil, water, forests, grasslands, and wildlife) faster than they are replenished.
- Do not waste resources.
- Do not release pollutants into the environment faster than the earth's natural processes can dilute or recycle them.
- Emphasize pollution prevention and waste reduction.
- Slow the rate of population growth.
- Have the market prices of all goods and services include all of their harmful environmental costs.
- Reduce poverty.

See various chapters on the website for this book for specific ways you can work with the earth by trying to implement such guidelines.

Critical Thinking

Which of these guidelines do you agree with and which do you disagree with? Why? Can you add any other guidelines?

their root causes. It has also summarized the controversy over how serious environmental problems are and the cultural changes that have influenced how we treat the earth. The rest of this book presents a more detailed analysis of these problems, the controversies they have created, and solutions proposed by analysts.

Try not to be overwhelmed or immobilized by the *bad environmental news* because there is also some *great environmental news*. We are learning a great deal about how nature works and sustains itself, and we have numerous scientific, technological, and economic solutions available to deal with the environmental problems we face, as you will learn in this book.

The challenge is to make creative use of our economic and political systems to implement such solutions. One key is to recognize that most economic and political change comes about as a result of individual actions and individuals acting together to bring about change. Studies by social scientists suggest that it takes only about 5–10% of the population of a country or of the world to bring about major social change. Anthropologist Margaret Mead summarized our potential for change: "Never doubt that a small group of thoughtful, committed citizens can change the world. Indeed, it is the only thing that ever has."

We live in exciting times during what might be called a *hinge of cultural history*. Indeed, if I had to pick a time to live, it would be the next 50 years as we face the challenge of developing more environmentally sustainable societies.

What's the use of a house if you don't have a decent planet to put it on?

Henry David Thoreau

REVIEW QUESTIONS

1. Define the boldfaced terms in this chapter.
2. What is *exponential growth*? What is the connection between exponential growth and environmental problems?
3. Distinguish between *environment*, *ecology*, and *environmental science*. Distinguish between *ecologists*, *environmental scientists*, *conservation biologists*, *environmentalists*, *preservationists*, and *conservationists*.
4. Distinguish between *solar capital* and *natural capital* (*natural resources*). Explain the importance of natural capital to the environment, the economy of the country where you live, and your lifestyle.
5. What is an *environmentally sustainable society*? Distinguish between living on principal and living on interest and relate this to the sustainability of **(a)** the earth's life-support system and **(b)** your lifestyle.
6. What are *doubling time* and the *rule of 70*? Use the rule of 70 to calculate how many years it would take for the population of a country to double if it was growing at 2% per year.
7. Define *economic growth*, *gross national product*, *gross domestic product*, *per capita GNP*, and *economic development*. Distinguish between *developed countries* and *developing countries*.
8. What is *environmentally sustainable economic development*? How does it differ from traditional economic growth and economic development?
9. Why does it make sense for a poor family to have a large number of children?
10. List four ways in which poverty is related to environmental quality, people's quality of life, and premature deaths of poor people.
11. Define *globalization* and list four indicators of this phenomenon. What is the *ecological footprint per person*, and what useful information does it give us?

GUEST ESSAY

Launching the Environmental Revolution

Lester R. Brown

Lester R. Brown is chair of the board of the Worldwatch Institute, a private nonprofit research institute he founded in 1974 that is devoted to analyzing global environmental issues. Under his leadership, the institute publishes the annual State of the World Report, *considered by many environmentalists and world leaders to be the best source of information about key environmental issues. It also publishes monographs on specific topics,* World Watch *magazine, and a series of* Environmental Alert *books. He is author of 19 books, recipient of the MacArthur Foundation Genius Award, and winner of the United Nations' 1989 environment prize. The* Washington Post *described him as "one of the world's most influential thinkers."*

Two challenges facing us in this new century are that the human population is 4 times as large as it was a century ago and the world economy is 17 times as large. This growth has allowed advances in living standards that our ancestors could not have imagined, but it has also undermined natural systems in ways they could not have feared. The policy decisions we make in the years immediately ahead will determine whether our children live in a world of development or decline.

There is no precedent for the rapid and substantial change we need to make. Building an environmentally sustainable future depends on **(1)** a restructuring of the global economy so that it does not destroy or degrade its natural support systems, **(2)** major shifts in human reproductive behavior, and **(3)** dramatic changes in values and lifestyles all within a few decades. If this *environmental* or *sustainability revolution* succeeds, it will rank as one of the great economic and social transformations in human history.

The two overriding challenges facing our global civilization as the new century begins are to stabilize population and stabilize climate. The exciting thing about the population and climate challenges is that we already have the knowledge and technologies to succeed at both.

The key to stabilizing world population is for national governments to formulate strategies for stabilizing population humanely rather than waiting for nature to intervene with its inhumane methods based on rising death rates, as in Africa. Once these strategies are developed, it is in the interest of the international community to support the stabilization effort.

Stabilizing climate means shifting away from a *fossil fuel–* or *carbon-based economy* to a *solar–hydrogen economy*. This new energy system taps a mix of renewable sources of energy from the sun, such as direct sunlight, hydropower, wind power, and wood. Electricity produced by wind turbines and photovoltaic (solar) cells can be used to produce clean-burning hydrogen gas from water to meet most of our energy needs. Making this shift is the greatest investment opportunity in history.

Caught up in the excitement of unprecedented economic growth and the information economy, many people have lost sight of the deterioration of the earth's environmental systems and resources. Whereas economic indicators such as investment, production, and trade are usually positive, the key environmental indicators are increasingly negative. Forests are shrinking,

12. From a human standpoint, what is a *resource?* Distinguish between *perpetual resources, renewable resources,* and *nonrenewable resources* and give an example of each.

13. What are *sustainable yield* and *environmental degradation*? Give four examples of environmental degradation.

14. Define and give three examples of *common-property resources*. What is the *tragedy of the commons*? Give three examples of this tragedy on a global scale and explain how your lifestyle contributes to these examples. List three ways to deal with the tragedy of the commons.

15. List three types of nonrenewable resources. Distinguish between a *physically depleted resource* and an *economically depleted resource*. Distinguish between *reuse* and *recycling*. Draw a depletion curve for a nonrenewable resource and explain how recycling and reuse affect depletion time.

16. What is *pollution*? Distinguish between *point sources* and *nonpoint sources* of pollution. List three types of harm caused by pollution.

17. Distinguish between *pollution prevention (input pollution control)* and *pollution cleanup (output pollution control)*. What are three problems with relying primarily on pollution cleanup? Why is pollution prevention better than pollution control? List two ways that governments can encourage both pollution prevention and pollution cleanup.

18. According to environmentalists, what are five basic causes of the environmental problems we face?

19. Describe a simple model of relationships between population, resource use, environmental degradation, and overall environmental impact. How do these factors differ in developed and developing countries?

20. Distinguish between *hunter–gatherer societies, agricultural societies, industrial societies,* and *information and globalization societies*. Summarize the major beneficial and harmful environmental effects of each of these societies.

21. Define *slash-and-burn cultivation* and *shifting cultivation* and list the benefits and limitations of these types of cultivation.

water tables are falling, soils are eroding, plant and animal species are disappearing, wetlands are disappearing, fisheries are collapsing, rangelands are deteriorating, rivers are running dry, coral reefs are dying, temperatures are rising, glaciers are melting, and there are more destructive storms and more hungry people in the world today than ever before.

The global economy as now structured cannot continue to expand much longer if the natural systems on which it depends continue to deteriorate at the current rate. In effect, we are unravelling parts of the earth's ecological safety net that supports all life.

On the economic front, the signs are equally ominous: Soil erosion, deforestation, and overgrazing are adversely affecting farming, forestry, and livestock productivity and slowing overall economic growth in agriculturally based economies. The decline in living standards that was once predicted by some ecologists from the combination of continuing rapid population growth and spreading environmental degradation has become a reality for one-sixth of humanity.

Suppose we used a more comprehensive system of national economic accounting that incorporated losses of natural capital [Guest Essay, p. 6] such as topsoil and forests, the destruction of productive grasslands, the extinction of plant and animal species, and the health costs of air and water pollution and increased ultraviolet radiation. Such an analysis might well show that most of humanity has suffered a decline in living conditions since the 1980s.

The key environmental limits we face are fresh water, forests, rangelands, ocean fisheries, biological diversity, and the global atmosphere. Our numbers expand, but earth's natural systems do not. Will we recognize the world's natural limits and adjust our economies accordingly, or will we expand our ecological footprint [Figure 1-8, p. 10] until it is too late? Nature has no reset button.

Some *good news* is that there is a growing worldwide recognition outside the environmental community that the economy we now have cannot take us where we want to go. Three decades ago, only environmental activists were speaking out on the need for change. Now the ranks of activists have broadened to include CEOs of major corporations, government ministers, prominent scientists, and intelligence agencies.

The goal is to develop a new type of economy. It is a solar-powered, bicycle- and rail-centered, reuse and recycle economy that uses energy, water, land, and materials much more efficiently and wisely than we do today. In addition to helping sustain the earth's life-support systems, such an economy can lead to greater economic security, healthier lifestyles, and a worldwide improvement in the human condition. This challenge to humanity rivals any in our history.

Critical Thinking

1. Do you agree with the author that we need to bring about an environmental or sustainability revolution within a few decades? Explain.

2. Do you believe that this can be done by making minor adjustments in the global economy or by restructuring the global economy to put less strain on the earth's natural systems? Explain.

22. List three reasons why some analysts believe that the environmental problems we face are not serious.

23. Summarize the 1992 position statement of about 1,700 leading scientists about the state of the environment.

24. List five major changes that environmentalists believe should take place over the next 50 years as part of an *environmental* or *sustainability revolution*. What are two important keys to bringing about such changes?

25. List 12 guidelines that environmentalists have suggested for working with the earth.

CRITICAL THINKING

1. Do you believe that the society you live in is on an unsustainable path? Explain. Do you believe that it is possible for the society you live in to become a more sustainable society within the next 50 years? Explain.

2. Do you favor instituting policies designed to reduce population growth and stabilize **(a)** the size of the world's population as soon as possible and **(b)** the size of the population of the country where you live as soon as possible? Explain. If you agree that population stabilization is desirable, what three major policies do you believe should be implemented to accomplish this goal?

3. Explain why you agree or disagree with the following propositions:

a. The economic growth from high levels of resource use in developed countries provides money for more financial aid to developing countries for reducing pollution, environmental degradation, and poverty.

b. Stabilizing population is not desirable because without more consumers, economic growth would stop.

c. The world will never run out of renewable resources and most currently used nonrenewable resources because technological innovations will produce substitutes, reduce resource waste, or allow use of lower grades of scarce nonrenewable resources.

4. List three forms of economic growth that you believe are environmentally unsustainable and three forms that you believe are environmentally sustainable.

5. When you read that at least 27,400 human beings die prematurely each day (19 per minute) from preventable malnutrition and infectious disease, do you **(a)** doubt whether it is true, **(b)** not want to think about it, **(c)** feel hopeless, **(d)** feel sad, **(e)** feel guilty, or **(f)** want to do something about this problem?

6. How do you feel when you read that **(1)** the average American consumes about 50 times more resources than the average Chinese citizen, **(2)** human activities lead to the premature extinction of at least 10 species per day, **(3)** humans have disturbed about 73% of the earth's habitable land (Figure 1-3, p. 5), and **(4)** human activities are projected to make the earth's climate warmer:
(a) skeptical about their accuracy, **(b)** indifferent, **(c)** sad, **(d)** helpless, **(e)** guilty, **(f)** concerned, or **(g)** outraged? Which of these feelings help perpetuate such problems and which can help alleviate these problems?

7. Do you agree or disagree with the five basic causes of environmental problems listed in Figure 1-12, p. 13? Explain. List any other root causes you believe should be added.

8. Do you believe that your current lifestyle is sustainable? If the answer is yes, explain why and include the impact of the world's other 6.1 billion people on your ability to sustain your current lifestyle. If your answer is no, explain why and list five things you could do now to make your lifestyle more sustainable. Which of these things do you actually plan to do?

PROJECTS

1. What are the major resource and environmental problems in **(a)** the city, town, or rural area where you live and **(b)** the state where you live? Which of these problems affect you directly?

2. Write two-page scenarios describing what your life and that of any children you choose to have might be like 50 years from now if **(a)** we continue on our present path and **(b)** we shift to more sustainable societies throughout most of the world.

3. Make a list of the resources you truly need. Then make another list of the resources you use each day only because you want them. Finally, make a third list of resources you want and hope to use in the future. Compare your lists with those compiled by other members of your class, and relate the overall result to the tragedy of the commons.

4. Make a concept map of this chapter's major ideas using the section heads and subheads and the key terms (in boldface type). Look on the website for this book for information about making concept maps.

INTERNET STUDY RESOURCES AND RESOURCES FOR FURTHER READING AND RESEARCH

The website for this book contains helpful study aids and many ideas for further reading and research. Log on to

http://www.brookscole.com/product/0534389872s

and click on the Chapter-by-Chapter area. Choose Chapter 1 and select a resource:

- "Flash Cards" allows you to test your mastery of the Terms and Concepts to Remember for this chapter.
- "Tutorial Quizzes" provides a multiple-choice practice quiz.
- "Student Guide to InfoTrac" will lead you to Critical Thinking Projects that use InfoTrac College Edition as a research tool.
- "References" lists the major books and articles consulted in writing this chapter.
- "Hypercontents" takes you to an extensive list of sites with news, research, and images related to individual sections of the chapter.

INFOTRAC COLLEGE EDITION

Improve your skills with InfoTrac College Edition, a searchable online database of articles from more than 700 periodicals. Log on to

www.infotrac-college.com

or access InfoTrac through the website for this book. Try to find the following articles:

1. Keeping Score. *The Ecologist*, April 2001 v31 i3 p44. According to the 2001 Environmental Sustainability Index (ESI), the "most 'eco-friendly' nations were the world's most industrialised." *The Ecologist* and Friends of the Earth "reformed and recalculated the methodology by which the ESI was produced." The resulting maps and tables reveal something quite different, and the article explains why. *Hint:* Enter the search term "keeping score" using the keyword "sustainability."

2. Sustainable Development: Concepts and Rankings. Mozaffar Qizilbash. *Journal of Development Studies*, Feb 2001 v37 i3 p134. Do countries that do well in terms of well-being perform badly on environmental concerns? This article discusses the two points of view regarding this question and suggests that the answer is "yes." *Hint:* Enter the search term "concept rankings" using the keyword "sustainability."

2 ENVIRONMENTAL ECONOMICS, POLITICS, AND WORLDVIEWS

Biosphere 2: A Lesson in Humility

In 1991, eight scientists (four men and four women) were sealed into Biosphere 2, a $200-million facility designed to be a self-sustaining life-support system (Figure 2-1).

The project, financed with private capital, was designed to **(1)** provide information and experience in designing self-sustaining stations in space or on the moon or other planets and **(2)** increase our understanding of the earth's life-support system: Biosphere 1.

The 1.3-hectare (3.2-acre) closed and sealed system was built in the desert near Tucson, Arizona. It had a variety of natural living systems, each built from scratch. They included a tropical rain forest, lakes, a desert, streams, freshwater and saltwater wetlands, and a mini-ocean with a coral reef.

The system was designed to mimic the earth's natural chemical recycling systems. Water evaporated from its oceans and other aquatic systems condensed to provide rainfall over the tropical rain forest. This water then trickled through soil filters into the marshes and the ocean to provide fresh water for the crew and other living organisms before being evaporated again.

The facility was stocked with more than 4,000 species of organisms selected to maintain life-support functions. Sunlight and external natural gas-powered generators provided energy.

The Biospherians were supposed to be isolated for 2 years and to **(1)** raise their own food, **(2)** breathe air recirculated by plants, and **(3)** drink water cleansed by natural recycling processes. From the beginning they encountered numerous unexpected problems.

The life-support system began unraveling. Large amounts of oxygen disappeared mysteriously. Additional oxygen had to be pumped in from the outside to keep the Biospherians from suffocating.

The nitrogen and carbon recycling systems also failed to function properly. Carbon dioxide skyrocketed to levels that threatened to poison the humans and spurred the growth of weedy vines that choked out food crops. Plant nutrients leached from the soil and polluted the water systems.

Disruption of the facility's chemical recycling systems and leaks in the seals from the outside disrupted populations of the system's life-forms. Tropical birds disappeared after the first freeze. An Arizona ant species got into the enclosure, proliferated, and killed off most of the system's introduced insect species. The facility was then overrun with cockroaches and katydids. All together, 19 of the Biosphere's 25 small animal species became extinct. Before the 2-year period was up, all plant-pollinating insects became extinct, thereby dooming to extinction most of the plant species.

Despite many problems, all the facility's waste and wastewater were recycled, and the Biospherians were able to produce 80% of their food supply.

Scientists Joel Cohen and David Tilman, who evaluated the project, concluded, "No one yet knows how to engineer systems that provide humans with life-supporting services that natural ecosystems provide for free." In other words, an expenditure of $200 million failed to maintain a life-support system for eight people. The earth—Biosphere 1—does this every day for 6.1 billion people and millions of other species at no cost. If we had to pay for these services at the same annual cost of $12.5 million per person in Biosphere 2, the total bill for the earth's 6.1 billion people would be 1,900 times the annual gross world product.

This world's largest ecological laboratory is now used by Columbia University's Lamont-Doherty Earth Observatory to carry out climate and ecological research.

Figure 2-1 Biosphere 2, constructed near Tucson, Arizona, was designed to be a self-sustaining life-support system for eight people sealed into the facility in 1991. The experiment failed because of a breakdown in its nutrient cycling systems. (Stone/Russell Kaye)

The main ingredients of an environmental ethic are caring about the planet and all of its inhabitants, allowing unselfishness to control the immediate self-interest that harms others, and living each day so as to leave the lightest possible footprints on the planet.

ROBERT CAHN

This chapter addresses the following questions:

- What are economic goods and resources, and how are they provided?
- How can we monitor economic and environmental progress?
- How can economics be used to help control pollution and manage resources?
- How does poverty reduce environmental quality? How can we reduce poverty?
- How can we shift to more environmentally sustainable economies over the next few decades?
- How is environmental policy made in the United States?
- What are guidelines for making environmental policy? How can people affect such decisions?
- What human-centered environmental worldviews guide most industrial societies?
- What are some life-centered and earth-centered environmental worldviews?
- What ethical guidelines might be used to help us work with the earth?
- How can we live more sustainably?

2-1 ECONOMIC SYSTEMS AND ENVIRONMENTAL PROBLEMS

What Supports and Drives Economies? An **economy** is a system of production, distribution, and consumption of goods and services that satisfy people's wants or needs. In an economy, individuals, businesses, and governments make **economic decisions** about **(1)** what goods and services to produce, **(2)** how to produce them, **(3)** how much to produce, and **(4)** how to distribute them.

The kinds of resources (or capital) that produce goods and services in an economy are called **economic resources**. They fall into four groups:

- **Natural resources** or **natural capital**: goods and services produced by the earth's natural processes, which support all economies and all life (Guest Essay, p. 6). These include **(1)** the planet's air, water, and land, **(2)** chemicals (nutrients) in the soil and water used to support all life, **(3)** chemicals (minerals) found in the earth's crust that can be extracted and used to manufacture useful items, **(4)** wild and domesticated plants and animals (biodiversity), and **(5)** nature's dilution, waste disposal, pest control, and recycling services.
- **Human resources**: people's physical and mental talents that provide labor, innovation, culture, and organization.
- **Financial resources**: cash, investments, and monetary institutions used to support the use of natural resources and human resources to provide goods and services.
- **Manufactured resources**: items made from natural resources with the help of human and financial resources. This type of capital includes tools, machinery, equipment, factories, and transportation and distribution facilities used to provide goods and services.

What Are the Major Types of Economic Systems? There are two major types of economic systems: command and market. In a **pure command economic system**, the government makes all *economic decisions* about **(1)** what and how much goods and services are produced, **(2)** how they are produced, and **(3)** for whom they are produced.

Market economic systems can be divided into two types: pure free market and capitalist market. In a **pure free-market economic system**, which so far exists only in theory,

- All economic decisions are made in *markets*, in which buyers (demanders) and sellers (suppliers) of economic goods freely interact without any government or interference.
- All buying and selling are based on *pure competition*, in which no seller or buyer can control or manipulate the market.
- All sellers and buyers have full access to the market and enough information about the beneficial and harmful aspects of economic goods to make informed decisions.
- Prices reflect all harmful costs to society and the environment (full-cost pricing).

The **capitalist market economic systems** found in the real world are designed to subvert many of the theoretical conditions of a truly free market. Here are the basic operating rules of the world's capitalist market economies.

- Drive out all competition and gain monopolistic control of market prices on a global scale.

- Lobby for unrestricted *global free trade* that allows anything to be manufactured anywhere in the world and sold anywhere else.
- Lobby for **(1)** government subsidies, tax breaks, or regulations that give a company's products a market advantage over their competitors and **(2)** governments to bail them out if they make bad investments.
- Withhold information about dangers posed by products and deny consumers access to information that would allow them to make informed choices.
- Maximize profits by passing harmful costs resulting from production and sale of goods and services on to the public, the environment, and in some cases future generations.
- A business has no legal obligation **(1)** to a particular area or nation, **(2)** to supply any particular good or service, or **(3)** to provide jobs, safe workplaces, or environmental protection.
- A company's primary obligation is to produce the highest profit for the owners or stockholders whose financial capital the company is using to do business.

Stockholders and owners of corporations do not have to follow these rules. For example, many corporations **(1)** voluntarily contribute to the well-being of their communities and their nation and **(2)** feel they have a responsibility to provide jobs and safe workplaces and to protect or improve environmental quality. However, CEOs who spend too much money on such things can be fired and legally sued by their stockholders for failure to maximize profits.

Why Have Governments Intervened in Market Economic Systems? Governments have intervened in economies to stop or slow down the accomplishment of some of the goals of capitalist market economies by trying to

- Level the economic playing field by preventing a single seller or buyer (monopoly) or a single group of sellers or buyers (oligopoly) from dominating the market and thus controlling supply or demand and price. This is becoming increasingly difficult because the current global capitalist market economy is dominated by a small number of financial speculators and large transnational corporations with enough financial and political clout to manipulate prices.
- Help ensure economic stability by trying to control boom-and-bust cycles that occur in any type of market system. Again, the power of transnational corporations is making it more difficult for governments to carry out this role.
- Provide basic services such as national security, education, and health care.
- Provide an economic safety net for people who because of health, age, and other factors cannot work and meet their basic needs.
- Protect people from fraud, trespass, theft, and bodily harm.
- Protect the health and safety of workers and consumers.
- Help compensate owners for large-scale destruction of their assets by floods, earthquakes, hurricanes, and other natural disasters.
- Protect common-property resources such as the atmosphere and open oceans (Connections, p. 11).
- Prevent or reduce pollution and depletion of natural resources.
- Manage public land resources.

How Do Conventional and Ecological Economists Differ in Their View of Market-Based Economic Systems? *Conventional economists* often depict a market-based economic system as a circular flow of economic goods and money between households and businesses operating essentially independently of the earth's life-support systems or natural resources (Figure 2-2). They **(1)** assume that the economy is the total system and nature's economy is a subsystem and **(2)** consider the earth's natural resources important but not vital because of our ability to find substitutes for scarce resources and ecosystem services. Consequently, they do not believe that depletion and degradation of natural resources will limit future economic growth.

Ecological economists consider all economies as human subsystems that are supported by the earth's natural resources (Guest Essay, p. 6), most of which have no substitutes (Figure 2-3). Ecological economists distinguish between **(1)** environmentally unsustainable conventional economic growth and **(2)** environmentally sustainable economic development (p. 7 and Figure 2-4).

2-2 MONITORING ECONOMIC AND ENVIRONMENTAL PROGRESS

Are GNP and GDP Useful Measures of Quality of Life and Environmental Degradation? GNP, *gross domestic product* (GDP), and *per capita GNP and GDP* indicators (p. 4) provide a standardized method for comparing the economic outputs of nations. The

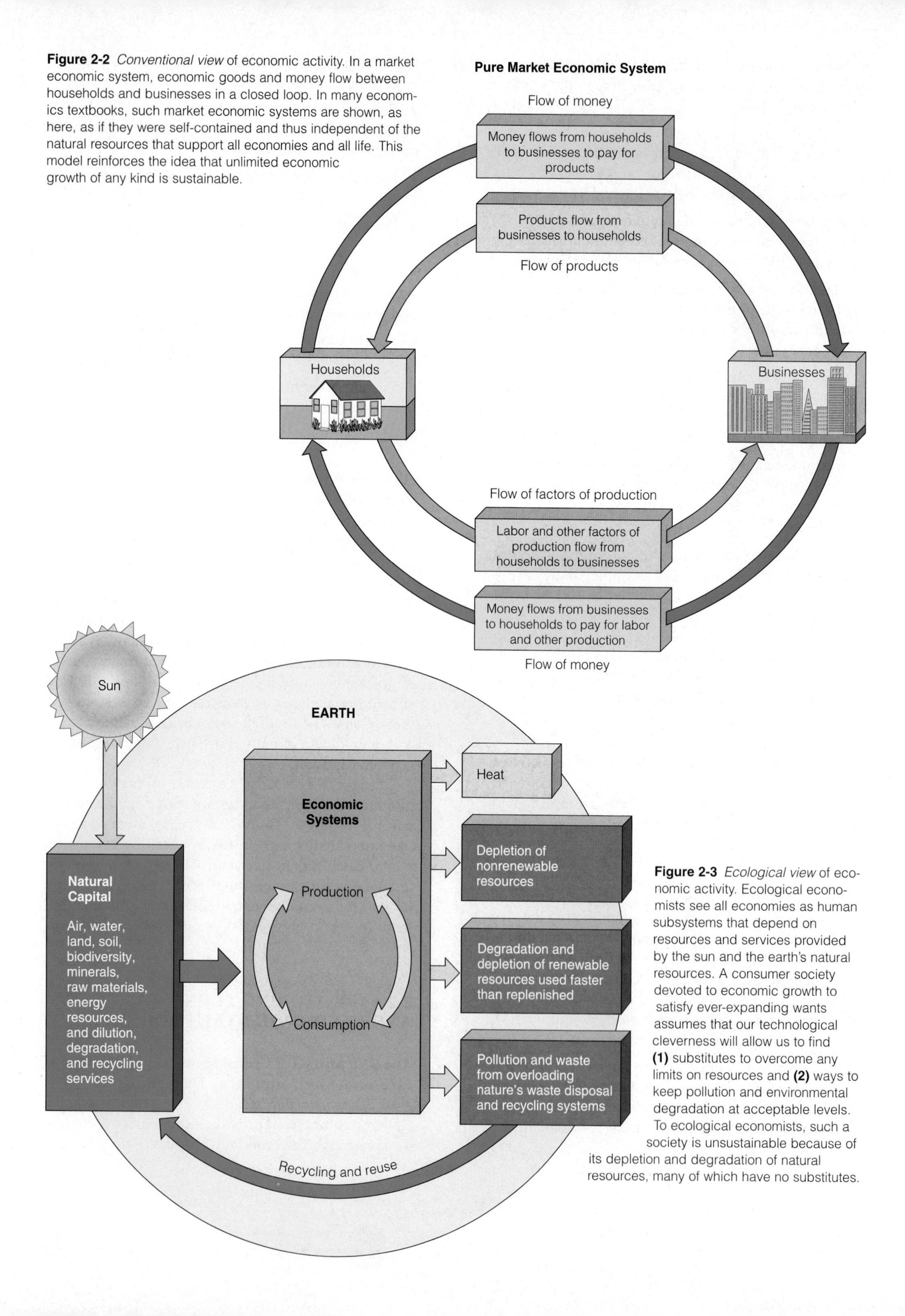

Figure 2-2 *Conventional view* of economic activity. In a market economic system, economic goods and money flow between households and businesses in a closed loop. In many economics textbooks, such market economic systems are shown, as here, as if they were self-contained and thus independent of the natural resources that support all economies and all life. This model reinforces the idea that unlimited economic growth of any kind is sustainable.

Figure 2-3 *Ecological view* of economic activity. Ecological economists see all economies as human subsystems that depend on resources and services provided by the sun and the earth's natural resources. A consumer society devoted to economic growth to satisfy ever-expanding wants assumes that our technological cleverness will allow us to find **(1)** substitutes to overcome any limits on resources and **(2)** ways to keep pollution and environmental degradation at acceptable levels. To ecological economists, such a society is unsustainable because of its depletion and degradation of natural resources, many of which have no substitutes.

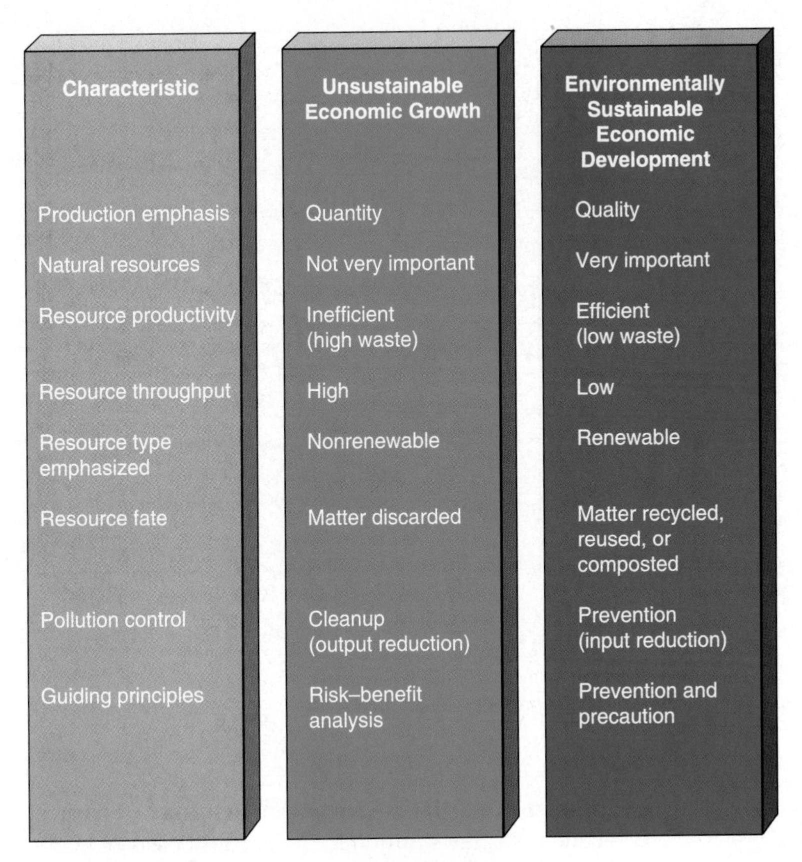

Characteristic
Unsustainable Economic Growth
Environmentally Sustainable Economic Development
Production emphasis
Quantity
Quality
Natural resources
Not very important
Very important
Resource productivity
Inefficient (high waste)
Efficient (low waste)
Resource throughput
High
Low
Resource type emphasized
Nonrenewable
Renewable
Resource fate
Matter discarded
Matter recycled, reused, or composted
Pollution control
Cleanup (output reduction)
Prevention (input reduction)
Guiding principles
Risk–benefit analysis
Prevention and precaution

Figure 2-4 Comparison of unsustainable economic growth and environmentally sustainable economic development.

economists who developed these indicators many decades ago never intended them to be used as measures of economic health, environmental health, or human well-being. However, most governments and business leaders use them for such purposes.

There are several reasons why GNP and GDP indicators are poor measures of economic health, environmental health, and human well being.

- *GNP and GDP hide the harmful environmental and social effects of producing goods and services.* Pollution, crime, sickness, and death are counted as positive gains in the GDP or GNP. Thus, the GNP increases for a country with rising cancer, crime, and divorce rates and increasingly polluted air and water because money is spent to deal with these problems. In other words, these harmful effects are counted as benefits that are added to these indicators instead of as costs that should be subtracted.

- *GNP and GDP do not include the depletion and degradation of natural resources or assets on which all economies depend.* A country can be headed toward ecological bankruptcy, exhausting its mineral resources, eroding its soils, cutting down its forests, destroying its wetlands and estuaries, and depleting its wildlife and fisheries. At the same time, it can have a rapidly rising GNP and GDP, at least for a while, until its environmental debts come due when the country no longer has its environmental assets and the income from these assets.

- *GNP and GDP do not include many beneficial transactions that meet basic needs in which no money changes hands.* Examples are the **(1)** labor we put into volunteer work, **(2)** health care and child care we give loved ones, **(3)** food we grow for ourselves, and **(4)** cooking, cleaning, and repairs we do for ourselves.

- *GNP and GDP tell us nothing about income distribution and economic justice.* They do not reveal how resources, income, or the harmful effects of economic growth (pollution, waste dumps, and land degradation) are distributed among the people in a country. The United Nations Children's Fund (UNICEF) suggests that countries should be ranked not by average per capita GNP or GDP but by average or median income of the poorest 40% of their people.

Solutions: Can Environmental Accounting Help? Ecological economists believe that GNP and GDP indicators should be supplemented with widely used and publicized *environmental indicators* that give a more realistic picture of environmental health, human welfare, and economic health.

Basically, such indicators would **(1)** subtract from the GDP and GNP things that lead to a lower quality of life and depletion of natural resources and **(2)** add to the GNP and GDP things that enhance environmental quality and human well-being but are currently being left out.

One such indicator is the *genuine progress indicator (GPI).* This indicator, developed by Redefining Progress, evaluates economic output by **(1)** subtracting expenses that do not improve environmental quality and human well-being from the GDP and **(2)** adding services that improve environmental quality and human well-being not currently included in the GDP. Figure 2-5 compares the total and per capita values of the GDP and GPI for the United States between 1950 and 1998 and shows a steady decline in both GPI indicators since 1980.

This and other environmental indicators are far from perfect and include many crude estimates. However, without such indicators, we do not **(1)** know much about what is happening to people, the environment, and the planet's natural resource base or **(2)** have an effective way to measure what policies work. In effect, we are trying to guide national and global economies through treacherous economic and environmental waters at ever-increasing speeds using faulty radar.

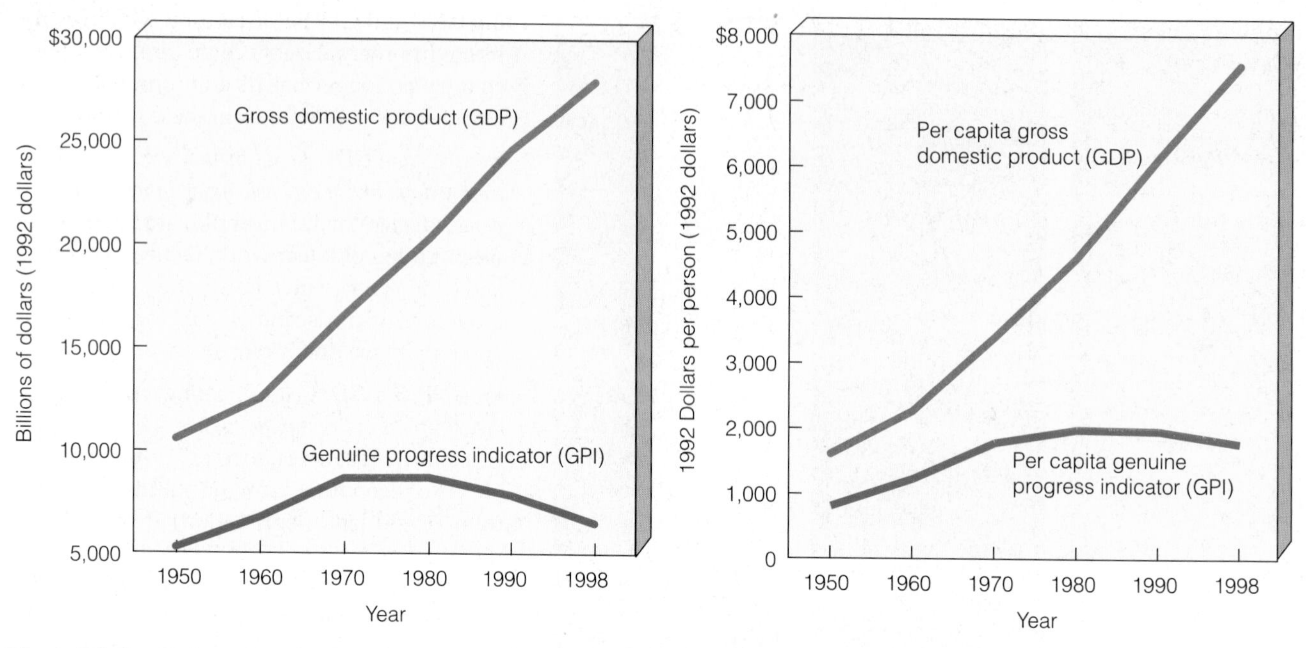

Figure 2-5 Comparison of the gross domestic product (GDP) and genuine progress indicator (GPI, left) and the per capita values for these indicators (right) in the United States between 1950 and 1998. (Data from Clifford Cobb, Mary Sue Goodman, and Mathis Wackernagel)

2-3 SOLUTIONS: USING ECONOMICS TO IMPROVE ENVIRONMENTAL QUALITY

What Are Internal and External Costs? All economic goods and services have both internal and external costs. For example, the price a consumer pays for a car reflects the costs of the factory, raw materials, labor, marketing, and shipping, as well as a markup to allow the car company and its dealers some profits. After a car is purchased, the buyer must pay for gasoline, maintenance, and repair. All these direct and indirect costs, which are paid for by the seller and the buyer of an economic good, are called **internal costs**.

Making, distributing, and using any economic good or service also involve external costs or benefits not included in the market price. For example, if a car dealer builds an aesthetically pleasing showroom, it is an **external benefit** to people who enjoy the sight at no cost.

On the other hand, extracting and processing raw materials to make and propel cars **(1)** deplete nonrenewable energy and mineral resources, **(2)** produce solid and hazardous wastes, **(3)** disturb land, **(4)** pollute the air and water, **(5)** contribute to global climate change, and **(6)** reduce biodiversity. These harmful effects are **external costs** passed on to the public, the environment, and in some cases future generations.

Because these harmful costs are not included in the market price, people do not connect them with car ownership. Still, everyone pays these hidden costs sooner or later, in the form of **(1)** poorer health, **(2)** higher costs for health care and health insurance, and **(3)** higher taxes for pollution control.

Should We Shift to Full-Cost Pricing? For most economists, the solution to the harmful costs of goods and services is to include such costs in the market prices of goods and services. *Internalizing external costs* will not occur unless it is required by government regulation. As long as businesses receive subsidies and tax breaks for extracting and using virgin resources and are not taxed for the pollutants they produce, few will volunteer to reduce short-term profits by becoming more environmentally responsible.

Assume you own a company and believe it's wrong to subject your workers to hazardous conditions and pollute the environment beyond what natural processes can handle. Suppose you voluntarily improve safety conditions for your workers and install pollution controls, but your competitors do not. Then your product will cost more to produce than similar products produced by your competitors, and you will be at a competitive disadvantage. Your profits will decline and you may eventually go bankrupt and have to lay off your employees.

One way to deal with the problem of harmful external costs is for the government to **(1)** levy taxes, **(2)** pass laws and develop regulations, **(3)** provide subsidies, or **(4)** use other strategies that encourage or force producers to include all or most of these costs in the market prices of their economic goods and services. Then the market price would be the **full cost** of these goods and services: internal costs plus short- and long-term external costs. Internalizing the external costs of pollution and environmental degradation would make **(1)** preventing pollution more profitable than cleaning it up and **(2)** waste reduction, recycling, and reuse more profitable than burying or burning most of the waste we produce.

The *bad news* is that when external costs are internalized, the market prices for most goods and services would rise. However, the *good news* is that total price people pay would be about the same because the hidden external costs related to each product would be included in its market price. In addition, full-cost pricing provides consumers with information needed to make informed economic decisions about the effects of their lifestyles on the planet's life-support systems and on human health.

However, as external costs are internalized, economists and environmentalists warn that governments must **(1)** reduce income, payroll, and other taxes and **(2)** withdraw subsidies formerly used to hide and pay for these external costs. Otherwise, consumers will face higher market prices without tax relief—a politically unacceptable policy guaranteed to fail.

Some more *good news* is that some goods and services would cost less because internalizing external costs encourages producers to **(1)** find ways to cut costs by inventing more resource-efficient and less-polluting methods of production and **(2)** offer more environmentally beneficial (or *green*) products. Jobs would be lost in environmentally harmful businesses, but at least as many and probably more jobs would be created in environmentally beneficial businesses. If a shift to full-cost pricing took place over several decades, most current environmentally harmful businesses would have time to transform themselves into profitable environmentally beneficial businesses.

Full-cost pricing seems to make a lot of sense. Why is it not used more widely? There are several reasons:

- Many producers of harmful and wasteful goods would have to charge so much that they could not stay in business.
- Huge government subsidies distort the marketplace and hide many of the harmful environmental and social costs of producing and using some goods and services.
- Producers would have to give up government subsidies and tax breaks that have helped hide the harmful external costs of their goods and services and distort the marketplace. Studies by Norman Myers (Guest Essay, p. 118) and other analysts conservatively estimate that governments around the globe spend about $2 trillion per year to help subsidize **(1)** habitat-destroying road transportation ($780 billion), **(2)** unsustainable agriculture ($510 billion), **(3)** use of nonrenewable fossil fuels and nuclear energy that discourage energy conservation and development of less harmful renewable energy alternatives ($510 billion), **(4)** water projects and groundwater depletion that discourage water conservation ($230 billion), **(5)** unsustainable forestry ($92 billion), and **(6)** overfishing ($25 billion). Economists estimate that eliminating these environmentally harmful (perverse) subsidies and replacing them with environmentally beneficial subsidies probably would cost taxpayers about one-sixth as much.
- The prices of harmful but desirable goods and services would rise.
- It is hard to put a price tag on many harmful environmental and health costs.
- Many business interests, political leaders, and consumers are unaware that they are paying these costs in ways not connected to the market prices of goods and services.

Despite the difficulties, proponents believe that full-cost pricing for harmful environmental and health effects should be phased in over the next 20 years. They argue that doing the best we can to estimate and internalize current external costs is far better than continuing the current pricing system, which gives too little or misleading information about the environmental and health effects of goods and services.

How Much Are We Prepared to Pay to Control Pollution? The cleanup cost for removing specific pollutant gases or wastewater being discharged into the environment rises with each additional unit of pollutant that is removed. As Figure 2-6 shows, a large percentage of the pollutants emitted by a smokestack or wastewater discharge can be removed fairly cheaply.

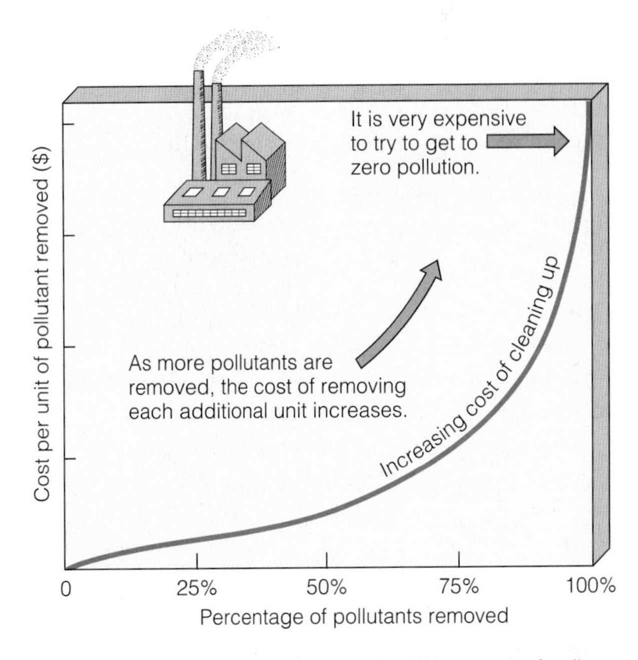

Figure 2-6 The cost of removing each additional unit of pollution rises. Cleaning up a certain amount of pollution is affordable, but at some point the cost of pollution control is greater than the harmful costs of the pollution to society.

However, as more and more pollutants are removed, the cost of removing each additional unit of pollution rises sharply. Beyond a certain point, the cleanup costs exceed the harmful costs of pollution. The *breakeven point* is the level of pollution control at which the harmful costs to society and the costs of cleanup are equal. Pollution control beyond this point costs more than it is worth, and not controlling pollution to this level causes pollution that we can afford to avoid.

To find the breakeven point, economists plot two curves: **(1)** a curve of the estimated economic costs of cleaning up pollution and **(2)** a curve of the estimated harmful external costs of pollution to society. They then add the two curves together to get a third curve showing the total costs. The lowest point on this third curve is the breakeven point, or the *optimum level of pollution* (Figure 2-7).

On a graph, determining the optimum value of pollution looks neat and simple, but there are problems with this approach.

- Ecological economists, health scientists, and business leaders often disagree in their estimates of the harmful costs of pollution, and such costs are difficult to determine.
- Some critics raise environmental justice questions about who benefits and who suffers from allowing optimum levels of pollution. The levels may be optimum for the entire country or an area. However, this is not the case for the people who live near or downwind or downriver from a polluting power plant or factory and are exposed to high levels of pollution.
- Assigning monetary values to things such as lost lives, forests, wetlands, and ecological services is difficult and controversial. On the other hand, assigning little or no value to such things means that they will not be counted in determining optimum pollution levels.

Some analysts point out that we should not be unduly influenced by curves showing the sharply increasing costs of improving pollution control (Figure 2-6). The reason is that stricter regulations and higher pollution control costs can stimulate business to **(1)** find cheaper ways to control pollution or **(2)** redesign manufacturing processes to sharply reduce or eliminate pollution.

For example, in the 1970s the U.S. chemical industry predicted that controlling benzene emissions would cost $350,000 per plant. Shortly after these predictions were made, however, the companies developed a process that substituted other chemicals for benzene and almost eliminated control costs.

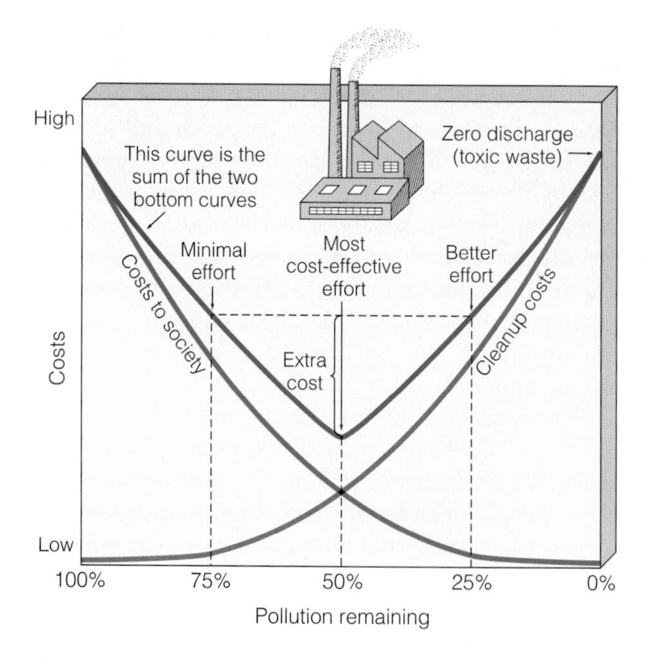

Figure 2-7 Finding the optimum level of pollution. This graph shows the optimum level at 50%, but the actual level varies depending on the pollutant.

What Is Benefit–Cost Analysis, and How Can It Be Improved? A widely used tool for making economic decisions about how to control pollution and manage resources is **benefit–cost analysis.** It involves comparing the estimated short-term and long-term economic benefits (gains) and costs (losses) for various courses of action. Examples are implementing a particular pollution control regulation or deciding whether a dam should be built, a wetland filled, or a forest cleared. Benefit–cost analyses can be useful guides and can indicate the cheapest way to go, but they can also have some serious limitations (Spotlight, right).

To minimize possible abuses, environmentalists and economists advocate the following guidelines for all benefit–cost analyses: **(1)** Use uniform standards, **(2)** state all assumptions, **(3)** evaluate the reliability of all data inputs as high, medium, or low, **(4)** make projections using low, medium, and high discount rates, **(5)** show the estimated range of costs and benefits based on various sets of assumptions, **(6)** estimate the short- and long-term benefits and costs to all affected population groups, and **(7)** evaluate and compare all alternative courses of action.

Should We Rely Mostly on Regulations or Market Forces? Most economists agree that government intervention in the marketplace is needed to control or prevent pollution and reduce resource waste. Such government action can take the form of regulation, the use of market forces, or some combination of these approaches. Each of the various regulatory and market approaches to reducing pollution and resource waste has advantages and disadvantages (Table 2-1).

SPOTLIGHT

Some Problems with Benefit–Cost Analysis

There are several controversies about benefit–cost analysis. One involves the **discount rate**, an estimate of a resource's future economic value compared to its present value. *The size of the discount rate (usually given as a percentage) is a primary factor affecting the outcome of any benefit–cost analysis.*

At a zero discount rate, for example, a stand of redwood trees worth $1 million today will still be worth $1 million 50 years from now.

However, most businesses, the U.S. Office of Management and Budget, and the World Bank typically use a 10% annual discount rate to evaluate how resources should be used. At this rate, the stand of redwood trees will be worth only $10,000 in 50 years. With this discount rate, it makes sense from an economic standpoint to cut these trees down as quickly as possible and invest the money in something else.

Proponents of high (5–10%) discount rates argue that **(1)** inflation will reduce the value of their future earnings, **(2)** innovation or changes in consumer preferences could make a product or service obsolete, and **(3)** without a high discount rate, they can make more money by investing their capital in some other venture.

Critics point out that high discount rates encourage such rapid exploitation of resources for immediate payoffs that sustainable use of most renewable natural resources is virtually impossible. These critics believe that **(1)** a 0% or even a negative discount rate should be used to protect unique and scarce resources and **(2)** discount rates of 1–3% would make it profitable to use other nonrenewable and renewable resources sustainably or slowly.

Another problem with benefit–cost analysis is determining *who benefits and who is harmed*. In the United States, an estimated 100,000 people die prematurely each year because of exposure to hazardous chemicals and other safety hazards at work, and an additional 400,000 are seriously injured by such exposure. In many other countries (especially developing countries), the situation is much worse. Is this a necessary or an unnecessary cost of doing business?

Another limitation of benefit–cost analysis is that *many things we value cannot easily be reduced to dollars and cents*. We can put estimated price tags on human life, good health, clean air and water, pollution and accidents that are prevented, wilderness, an endangered species, and various forms of natural capital (Guest Essay, p. 6). However, the dollar values different people assign to such things vary widely because of different assumptions, discount rates, and value judgments. This leads to a wide range of projected benefits and costs.

Because estimates of benefits and costs are so variable, *figures can be weighted easily to achieve the outcome desired by proponents or opponents of a proposed project or environmental regulation*. For example, one industry-sponsored benefit–cost study estimated that compliance with a standard to protect U.S. workers from vinyl chloride would cost $65–90 billion; in fact, less than $1 billion was actually needed to comply with the standard.

Critics also point out that benefit–cost analyses

- Typically fail to evaluate alternative ways to **(1)** control or prevent a particular form of pollution, **(2)** manage a certain resource, or **(3)** find alternative uses or a substitute for a resource.
- Are too imprecise to use as the *primary way* to make decisions about environmental regulations and resources. To these critics, using this tool is somewhat like trying to detect a car speeding at 160 kilometers per hour (100 miles per hour) with a radar device so unreliable that at best it can tell us only that the car's speed is somewhere between 8 kph (5 mph) and 8,000 mph (5,000 mph).

Critical Thinking

Do you believe that benefit–cost analysis should be used as the primary way to evaluate **(a)** whether any new environmental law or regulation should be put into effect and **(b)** whether any existing environmental law or regulation should be weakened or strengthened? Explain. What should be the role of benefit–cost analysis? Why?

How Can Economic Incentives Be Used to Improve Environmental Quality and Reduce Resource Waste? *Market forces* can help improve environmental quality and reduce resource waste, mostly by encouraging the internalization of external costs by using **(1)** *economic incentives* (rewards) or **(2)** *economic disincentives* (punishments). This is based on a fundamental principle of the marketplace in today's capitalist economic systems: *What we reward (mostly by government subsidies and tax breaks) we tend to get more of, and what we discourage (mostly by regulations and taxes) we tend to get less of.*

One way to put this principle into practice is to **(1)** *phase in government subsidies and tax breaks that encourage environmentally beneficial behavior and* **(2)** *phase out government subsidies and tax breaks that encourage environmentally harmful behavior*. The difficulty is that doing this involves political decisions that often are opposed successfully by powerful economic interests.

Table 2-1 Economic Solutions to Pollution and Resource Waste

Solution	Internalizes External Costs	Innovation	International Competitiveness	Administrative Costs	Increases Government Revenue
Regulation	Partially	Can encourage	Decreased*	High	No
Subsidies	No	Can encourage	Increased	Low	No
Withdrawing harmful subsidies	Yes	Can encourage	Decreased*	Low	Yes
Tradable rights	Yes	Encourages	Decreased*	Low	Yes
Green taxes	Yes	Encourages	Decreased*	Low	Yes
User fees	Yes	Can encourage	Decreased*	Low	Yes
Pollution-prevention bonds	Yes	Encourages	Decreased*	Low	No

*Unless more cost-effective and productive technologies are developed.

How Can Economic Disincentives Be Used to Improve Environmental Quality and Reduce Resource Waste? Economic *disincentive* approaches include

- *Using green taxes or effluent fees* to help internalize many of the harmful environmental costs of production and consumption. Taxes can be levied on each unit of **(1)** pollution discharged into the air or water, **(2)** hazardous or nuclear waste produced, **(3)** virgin resources used, and **(4)** fossil fuel used. Economists point out that such *green taxes* work if **(1)** they reduce or replace income, payroll, or other taxes and **(2)** the poor and middle class are given a safety net to reduce the regressive nature of consumption taxes on essentials such as food, fuel, and housing. In other words, to be politically acceptable environmental taxes must be seen as a *tax-shifting* instead of a *tax-burden* approach (Solutions, right).
- *Charging user fees* that cover all or most costs for activities such as **(1)** extracting lumber and minerals from public lands, **(2)** using water provided by government-financed projects, and **(3)** using public lands for livestock grazing.

What Are the Pros and Cons of Using Tradable Pollution and Resource-Use Rights? Another market approach is for the government to *grant tradable pollution and resource-use rights*. For example, a total limit on emissions of a pollutant or use of a resource such as a fishery could be set. Then permits would be used to allocate or auction the total among manufacturers or users.

Permit holders not using their entire allocation could **(1)** use it as a credit against future expansion, **(2)** use it in another part of their operation, or **(3)** sell it to other companies. Tradable rights could also be established among countries to help **(1)** preserve biodiversity and **(2)** reduce emissions of pollutants with harmful regional (or global) effects.

Some environmentalists and economists support tradable pollution rights as an improvement over the current regulatory approach. Other environmentalists oppose allowing companies to buy and trade rights to pollute because it **(1)** allows the wealthiest companies to continue polluting (thereby excluding smaller companies from the market), **(2)** tends to concentrate pollutants at the dirtiest plants (thereby jeopardizing the health of people downwind or downstream from the plant), **(3)** creates an incentive for fraud because most pollution control regulations are based on self-reporting of pollution outputs, and government monitoring and enforcement of such outputs are inadequate, and **(4)** has no built-in incentives to reduce overall pollution unless the system requires an annual decrease in the total pollution allowed (which is rarely the case).

How Has Environmental Management Changed? Figure 2-8 shows the evolution of several phases of environmental management, with the long-term goal of achieving more environmentally sustainable economies and societies.

The period between 1970 and 1985 can be viewed as the *resistance-to-change management era.* In the 1970s, **(1)** many companies and government environmental regulators developed an adversarial approach in which companies resented and actively resisted environmental regulations, and **(2)** government regulators thought they had to prescribe ways for reluctant companies to clean up their pollution emissions. At this stage, most companies dealt with environmental regulations by **(1)** hiring outside environmental consultants, who usu-

Shifting the Tax Burden from Wages and Profits to Pollution and Waste

SOLUTIONS

An important goal of a more sustainable economic system would be to tax labor, income, and wealth *less* and tax throughput of matter and energy resources *more*. The tax structure in most countries encourages businesses to increase productivity by substituting manufactured capital (machines) and throughput (energy and materials) for workers (human capital).

To many analysts such a tax system is backwards. It **(1)** discourages what we want more of (jobs, income, and profit-driven innovation), **(2)** encourages what we want less of (pollution and resource depletion), and **(3)** leads to tremendous waste of natural and human capital. Such tax systems remain primarily because of political resistance to change and the political power of businesses that profit from them.

Shifting more of the tax burden from wages and profits to pollution and waste **(1)** decreases depletion and degradation of natural capital, **(2)** improves environmental quality by making polluters and resource depleters and degraders pay for the harm they cause (full-cost pricing), **(3)** encourages pollution prevention, waste reduction, and creativity in solving environmental problems (to avoid paying pollution taxes and increase profits), **(4)** rewards recycling and reuse instead of use of nonrenewable virgin resources, **(5)** uses the marketplace, rather than regulations, to help protect the environment, **(6)** reduces unemployment, and **(7)** allows cuts in highly resented and often regressive income, payroll, and sales taxes.

Such a tax shift would have to be phased in over a 15- to 20-year period. This would allow businesses to plan for the future and depreciate existing capital investments over their useful lives.

Because such consumption taxes place a larger burden than income taxes on the poor, governments would need to provide an economic safety net for poor and lower middle-class people.

Trial versions of such tax shifts are taking place in Germany, Sweden, Great Britain, Norway, Denmark, Italy, France, and the Netherlands. If Europe and other countries continue moving toward this type of tax shift over the next two decades, their labor costs will be lowered and their resource productivity will increase. To remain competitive in the global marketplace, the United States and other countries may be forced to follow.

Critical Thinking

What are disadvantages of making such a tax shift for you and for the country where you live? Do you favor making such a shift in what is taxed? Explain.

ally favored end-of-pipe pollution control solutions, **(2)** using lawyers to oppose or find legal loopholes in the regulations, and **(3)** lobbying elected officials to have environmental laws and regulations overthrown or weakened.

By 1985 most company managers accepted environmental regulations and continued to rely mostly on pollution control. However, they placed little emphasis on trying to find innovative solutions to pollution and resource waste problems because they believed them to be too costly.

In 1990s, a growing number of company managers began to realize that environmental improvement is an economic and competitive opportunity instead of a cost to be resisted. This was the beginning of the *innovation management era* that environmental and business visionaries project will go through several phases over the next 40–50 years (Figure 2-8, right).

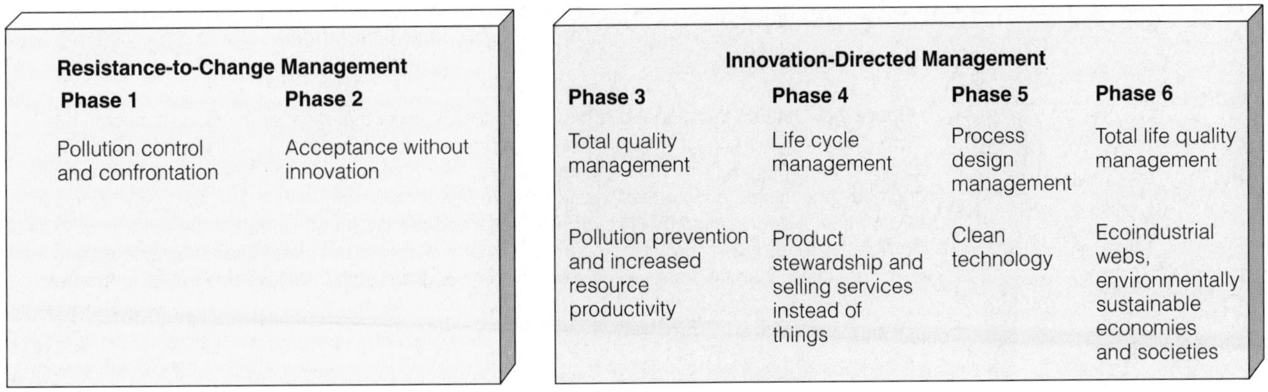

Resistance-to-Change Management	
Phase 1	**Phase 2**
Pollution control and confrontation	Acceptance without innovation

Innovation-Directed Management			
Phase 3	**Phase 4**	**Phase 5**	**Phase 6**
Total quality management	Life cycle management	Process design management	Total life quality management
Pollution prevention and increased resource productivity	Product stewardship and selling services instead of things	Clean technology	Ecoindustrial webs, environmentally sustainable economies and societies

Figure 2-8 Evolving eras of environmental management.

2-4 REDUCING POVERTY TO IMPROVE ENVIRONMENTAL QUALITY AND HUMAN WELL-BEING

What Is the Relationship Between Poverty and Environmental Problems? Poverty usually is defined as the inability to meet one's basic economic needs. Currently, an estimated 1.4 billion people in developing countries—roughly one of every four people on the planet—have an annual income of less than $370 per year. This income of roughly $1 per day is the World Bank's definition of poverty. According to a 2000 World Bank study, half of humanity is trying to live on less than $2 a day.

Poverty has a number of harmful health and environmental effects. It **(1)** causes premature deaths and preventable health problems, **(2)** tends to increase birth rates because the poor have more children to help them grow food or work and to take care of them in their old age, and **(3)** often pushes poor people to use potentially renewable resources unsustainably to survive.

Has Economic Growth Reduced Poverty? Most economists believe that a growing economy is the best way to help the poor because such growth **(1)** creates more jobs, **(2)** enables more of the increased wealth to reach workers, and **(3)** provides greater tax revenues that can be used to help the poor help themselves.

However, since 1960 most of the benefits of global economic growth as measured by income have flowed up to the rich rather than down to the poor. This has made the top one-fifth of the world's people much richer and the bottom one-fifth poorer in terms of total real income and real per capita GNP (adjusted for inflation; Figure 2-9). Since 1980, growth of this wealth gap has accelerated.

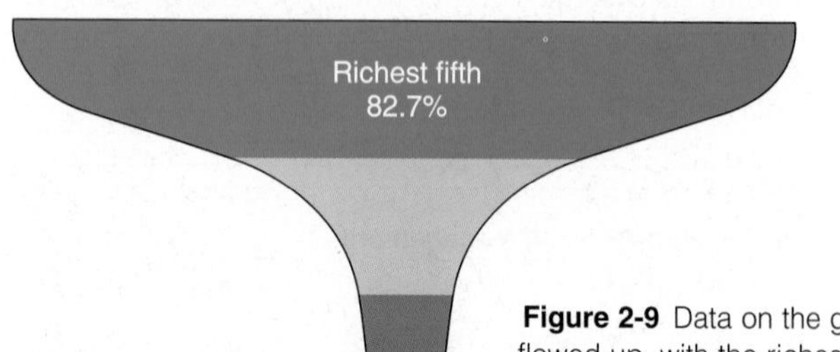

Figure 2-9 Data on the global distribution of income show that most of the world's income has flowed up, with the richest 20% of the world's population receiving more of the world's income than all of the remaining 80% in 1991 (the last year for which such data are available). This upward flow of global income has accelerated since 1960 and especially since 1980. This trend can increase environmental degradation by **(1)** increasing average per capita consumption by the richest 20% of the population and **(2)** causing the poorest 20% of the world's people to use renewable resources faster than they are replenished to survive. (Data from UN Development Programme)

How Can We Reduce Poverty? Analysts point out that reducing poverty requires the governments of most developing countries to make policy changes. Two examples are **(1)** shifting more of the national budget to help the rural and urban poor work their way out of poverty and **(2)** giving villages, villagers, and the urban poor title to common lands and to crops and trees they plant on them.

Analysts also urge developed countries and the wealthy in developing countries to help reduce poverty. Several ways to do this have been suggested.

- *Forgive* **(1)** *at least 60% of the $2.6 trillion debt (up from $0.25 trillion in 1970) that developing countries owe to developed countries and international lending agencies and* **(2)** *all of the $213 billion debt of the most heavily indebted poor countries.* Currently, developing countries pay almost $300 billion per year in interest to developed countries to service this debt. In 2000, the U.S. government announced that it would cancel all the debt owed by the world's poorest countries if the money is spent on basic human needs.
- *Increase nonmilitary government and private aid to developing countries from developed countries, with the aid going directly to the poor to help them become more self-reliant.*
- *Encourage banks and other organizations to make small loans to poor people wanting to increase their income* (Solutions, right).
- *Require international lending agencies to use standard environmental and social impact analysis to evaluate any proposed development project.* Some analysts believe that no project should be supported unless **(1)** its net environmental impact (using full-cost accounting) is favorable, **(2)** most of its benefits go to the poorest 40% of the people affected, and **(3)** the local people it affects are involved in planning and executing the project.
- *Carefully monitor all projects and halt funding if environmental safeguards are not followed.*
- *Establish policies that encourage both developed countries and developing countries to slow population growth and stabilize their populations.*

Microloans to the Poor

SOLUTIONS

Most of the world's poor desperately want to earn more, become more self-reliant, and have a better life. However, they have **(1)** no credit record and **(2)** few if any assets to use for collateral to secure a loan to buy seeds and fertilizer for farming or tools and materials for a small business.

During the last 25 years an innovative tool called *microlending* has increasingly helped deal with this problem. For example, since economist Muhammad Yunus started it in 1976, the Grameen (Village) Bank in Bangladesh has provided more than $2 billion in microloans (varying from $50 to $500) to 2.3 million mostly poor, rural, and landless women in 40,000 villages. About 94% of the loans are to women who start their own small businesses as sewers, weavers, bookbinders, peanut fryers, or vendors.

To stimulate repayment and provide support, the Grameen Bank organizes microborrowers into five-member "solidarity" groups. If one member of the group misses a weekly payment or defaults on the loan, the other members of the group must make the payments.

The Grameen Bank's experience has shown that microlending is both successful and profitable. For example, **(1)** less than 3% of microloan repayments to the Grameen Bank are late, and **(2)** the repayment rate on its loans is an astounding 95%, compared with a repayment rate of conventional loans by commercial banks of only 30%.

About half of Grameen's borrowers move above the poverty line within 5 years, and domestic violence, divorce, and birth rates are lower among borrowers. Microloans to the poor by the Grameen Bank are being used to develop day-care centers, health clinics, reforestation projects, drinking water supply projects, literacy programs, and group insurance programs and to bring solar and wind micropower systems to rural villages.

Grameen's model has inspired the development of microcredit projects in more than 52 countries that have reached 36 million people (including dependents). In 1997, some 2,500 representatives of microlending organizations from 113 countries met at a Microcredit Summit in Washington, D.C., and adopted a goal of reaching 100 million of the world's poorest people by 2005.

Critical Thinking

Why do you think there has been little use of microloans by international development and lending agencies such as the World Bank and the International Monetary Fund? How might this situation be changed?

According to the United Nations Development Programme (UNDP), it will cost about $40 billion a year to provide universal access to basic services such as education, health, nutrition, family planning, reproductive health, safe water, and sanitation. The UNDP notes that this is less than 0.1% of total annual world income.

2-5 MAKING THE TRANSITION TO ENVIRONMENTALLY SUSTAINABLE ECONOMIES

How Can We Make Working with the Earth Profitable? Greening Business Paul Hawken (Guest Essay, p. 6) and several other business leaders and economists have laid out the following principles for making the transition from environmentally unsustainable to more environmentally sustainable economies over the next several decades:

- *Reward (subsidize) environmentally beneficial behavior.*
- *Discourage (tax and do not subsidize) environmentally harmful behavior.*
- *Use environmental and social indicators to measure progress toward environmental and economic sustainability and human well-being* (Figure 2-5).
- *Use full-cost pricing to include the external costs of goods and services in their market prices.*
- *Replace taxes on wages and profits with taxes on matter and energy throughput* (Solutions, p. 33).
- *Use low discount rates for evaluating future worth of irreplaceable or vulnerable resources* (Spotlight, p. 31).
- *Do not use renewable resources (soil, water, forests, grasslands, and wildlife) faster than they can be replenished.*
- *Do not use nonrenewable resources faster than substitutes can be developed and phased in.*
- *Do not release pollutants into the environment faster than the earth's natural processes can dilute or assimilate them.*
- *Reduce poverty.*
- *Slow population growth.*

Paul Hawken's simple golden rule for such an economy is, *"Leave the world better than you found it, take no more than you need, try not to harm life or the environment, and make amends if you do."*

Can We Make the Transition to a More Environmentally Sustainable Economy? Even if people believe that an environmentally sustainable economy is desirable, is it possible to make such a drastic change in the way we think and act? Some environmentalists, economists, and business leaders say that it is not only possible but imperative and that it can be done over the next 40–50 years.

According to Paul Hawken, this new approach to economic thinking and actions recognizes that most business leaders are not evil, earth-degrading ogres. Instead, they are trapped in a system that by design rewards them (with the highest profits and salaries and best chances for promotion) for maximizing short-term profits for owners and investors, regardless of the harmful short- and long-term environmental and social impacts.

Hawken argues that environmentally sustainable economies throughout the world would free business leaders, workers, and investors from this ethical dilemma. In such economies, they would be financially compensated and respected for **(1)** doing socially and ecologically responsible work, **(2)** improving environmental quality, and **(3)** still making hefty profits for owners and stockholders. Making this shift should also create jobs (Connections, right).

The problem in making this shift is not economics but politics. It involves the difficult task of convincing more business leaders, elected officials, and voters to begin changing current government systems of economic rewards and penalties.

Here are three pieces of *great news*:

- Making the shift to environmentally sustainable economies could be an extremely profitable enterprise that will **(1)** create many jobs, **(2)** greatly improve environmental quality, and **(3)** sharply reduce poverty. According to Lester R. Brown (Guest Essay, p. 20) and Christopher Flavin, "Converting the economy of the 21st century into one that is environmentally sustainable represents the greatest investment opportunity in history."
- We already have most of the technologies needed to implement this economic shift.
- Governments would not have to spend more money. Such a shift would be revenue neutral if governments **(1)** shifted environmentally harmful subsidies to environmentally beneficial enterprises and **(2)** taxed pollution and resource waste instead of wages and income (Solutions, p. 33).

Forward-looking investors, corporate executives, and political leaders recognize that *the environmental revolution is also an economic revolution.* This involves recognizing that economic and ecological systems are interdependent (Figure 2-3).

CONNECTIONS

Jobs and the Environment

Environmental protection is a major growth industry that creates new jobs. According to Worldwatch Institute estimates, annual sales of global ecotechnology industries are $600 billion—on a par with the global car industry—and these industries employ about 11 million people. In 2000, the environmental industry in the United States employed nearly 1.4 million people and generated annual revenues of more than $185 billion.

Studies by the EPA show that environmental laws create far more jobs than have been lost. Indeed, the U.S. Clean Air and Clean Water Acts have created more than 300,000 jobs in pollution control.

A congressional study concluded that investing $115 billion per year in solar energy and in improving energy efficiency in the United States would **(1)** eliminate about 1 million jobs in oil, gas, coal, and electricity production but **(2)** create 2 million other new jobs. Investing the money saved by reducing energy waste could create another 2 million jobs.

Critical Thinking

What major things (if any) have the national and local governments of the country where you live done to stimulate the growth of environmental jobs? What major things (if any) have these governments done to discourage the growth of environmental jobs?

2-6 POLITICS AND ENVIRONMENTAL POLICY

How Does Social Change Occur in Representative Democracies? **Politics** is the process by which individuals and groups try to influence or control the policies and actions of governments at the local, state, national, or international levels. Politics is concerned with **(1)** who has power over the distribution of resources and **(2)** who gets what, when, and how.

Democracy is government by the people through elected officials and representatives. In a *constitutional democracy*, a constitution **(1)** provides the basis of government authority, **(2)** limits government power by mandating free elections, and **(3)** guarantees freely expressed public opinion.

Political institutions in constitutional democracies are designed to allow gradual change to ensure economic and political stability. In the United States, for example, rapid and destabilizing change is curbed by the system of checks and balances that distributes power between the three branches of government—*legislative, executive,* and *judicial*—and between federal, state, and local governments.

In passing laws, developing budgets, and formulating regulations, elected and appointed government officials must deal with pressure from many competing *special-interest groups*. Each group advocates passing laws or establishing regulations favorable to its cause and weakening or repealing laws and regulations unfavorable to its position.

Some special-interest groups (such as corporations) are *profit-making organizations,* and others are *nonprofit, nongovernment organizations (NGOs)*. Examples of NGOs are **(1)** educational institutions, **(2)** labor unions, and **(3)** mainstream and grassroots environmental organizations.

What Principles Can Be Used as Guidelines in Making Environmental Policy Decisions? Analysts have suggested that legislators and individuals evaluating existing or proposed environmental policy should to be guided by the following principles:

- *The humility principle*: Recognize and accept that we have a limited capacity to manage nature because our understanding of nature and of the consequences of our actions will always be limited.
- *The reversibility principle*: Try not to do something that cannot be reversed later if the decision turns out to be wrong. For example, most biologists believe that the current large-scale destruction and degradation of forests, wetlands, wild species, and other components of the earth's biodiversity is unwise because much of it could be irreversible on a human time scale.
- *The precautionary principle*: When there is much evidence that an activity raises threats of harm to human health or the environment, we should take precautionary measures to prevent or reduce harm even if some of the cause-and-effect relationships are not fully established scientifically. In such cases, it is better to be safe than sorry.
- *The prevention principle*: Whenever possible, make decisions that help prevent a problem from occurring or becoming worse.
- *The integrative principle*: Make decisions that involve integrated solutions to environmental and other problems.
- *The environmental justice principle*: Establish environmental policy so that no group of people bears a disproportionate share of the harmful environmental risks from industrial, municipal, and commercial operations or from the execution of environmental laws, regulations, and policies. The EPA defines **environmental justice** as "the fair treatment and meaningful involvement of all people regardless of race, color, national origin, or income with respect to the development, implementation, and enforcement of environmental laws, regulations, and policies."

How Can Individuals Affect Environmental Policy? A major theme of this book is that *individuals matter*. History shows that significant change usually comes from the *bottom up* when individuals join with others to bring about change. Without grassroots political action by millions of individual citizens and organized groups, the air you breathe and the water you drink today would be much more polluted, and much more of the earth's biodiversity would have disappeared.

Individuals can influence and change government policies in constitutional democracies by

- Running for office (especially local offices)
- Trying to get appointed to local planning and zoning boards and environmental commissions that study and make recommendations to elected officials
- Appearing at hearings before local planning and zoning boards, environmental commissions, and public meetings of elected officials to make their views known
- Voting for candidates and ballot measures
- Contributing money and time to candidates seeking office
- Writing, faxing, e-mailing, calling, or meeting with elected representatives, asking them to **(1)** pass or oppose certain laws, **(2)** establish certain policies, and **(3)** fund various programs
- Forming or joining nongovernment organizations (NGOs) that lobby elected and regulatory officials to support a particular position
- Using education and persuasion to convert elected officials and other citizens to a particular position
- Exposing fraud, waste, and illegal activities in government (whistleblowing)

What Is Environmental Leadership? It is important to distinguish between leaders, rulers, and managers.

- A *leader* is a person other people follow voluntarily because of his or her vision, credibility, or charisma.
- A *ruler* is a person who has enough power to make people follow against their will.

- A *manager* is a person who knows how to organize things, get things done, pay attention to detail, and delegate responsibility. Some leaders and rulers can also be good managers and vice versa.

Individuals can provide leadership on environmental (or other) issues by

- *Leading by example,* using one's lifestyle to show others that change is possible and beneficial.

- *Working within existing economic and political systems to bring about environmental improvement.* People can influence political elites by campaigning and voting for candidates and by communicating with elected officials. They can also work within the system by choosing environmental careers (Individuals Matter, right). See the website material for this chapter for information on how to communicate with elected officials.

- *Proposing and working for better solutions to environmental problems.* Leadership is more than being against something; it also involves **(1)** coming up with better ways to accomplish various goals and **(2)** getting people to work together to achieve such goals.

INDIVIDUALS MATTER

Environmental Careers

In the United States (and in other developed countries), economists claim that the *green job market* is one of the fastest-growing segments of the economy.

Many employers are actively seeking environmentally educated graduates. They are especially interested in people with **(1)** scientific and engineering backgrounds and **(2)** double majors (business and ecology, for example) or double minors.

Environmental career opportunities exist in a large number of fields: environmental engineering (currently the fastest-growing job market), sustainable forestry and range management, parks and recreation, air and water quality control, solid-waste and hazardous-waste management, recycling, urban and rural land-use planning, computer modeling, ecological restoration, and soil, water, fishery, and wildlife conservation and management.

Environmental careers can also be found in education, environmental planning, environmental management, environmental health, toxicology, geology, ecology, conservation biology, chemistry, climatology, population dynamics and regulation (demography), law, risk analysis, risk management, accounting, environmental journalism, design and architecture, energy conservation and analysis, renewable-energy technologies, hydrology, consulting, public relations, activism and lobbying, economics, diplomacy, development and marketing, publishing (environmental magazines and books), and law enforcement (pollution detection and enforcement teams).

Critical Thinking

Have you considered an environmental career? Why or why not?

2-7 CASE STUDY: ENVIRONMENTAL POLICY IN THE UNITED STATES

How Is Environmental Policy Made in the United States? The major function of the federal government in the United States is to develop and implement *policy* for dealing with various issues. This policy typically is composed of various **(1)** *laws* passed by the legislative branch, **(2)** *regulations* instituted by the executive branch to put laws into effect, and **(3)** *funding* to implement and enforce the laws and regulations. Figure 2-10 is a greatly simplified overview of how individuals and lobbyists for and against a particular environmental law interact with the three branches of government in the United States.

There are several steps in establishing federal environmental policy (or any other policy) in the United States:

- Persuade lawmakers that an environmental problem exists and that the government has a responsibility to address it.

- Try to influence how laws are written and to pass laws to deal with the problem. Most environmental bills are evaluated by as many as 10 committees in the House of Representatives and the Senate. Effective proposals often are weakened by this fragmentation and by lobbying from groups opposing the law. Since the 1970s a number of environmental laws have been passed in the United States (see list on website for this chapter and Solutions, p. 40).

- Appropriate enough funds to implement and enforce each law. Indeed, developing and adopting a budget is the most important and controversial thing the executive and legislative branches do. Developing a budget involves answering two key questions: **(1)** What programs will be funded, and **(2)** how much money will be used to address each problem?

- Have the appropriate government department or agency draw up regulations for implementing each law. Groups try to influence how the regulations are

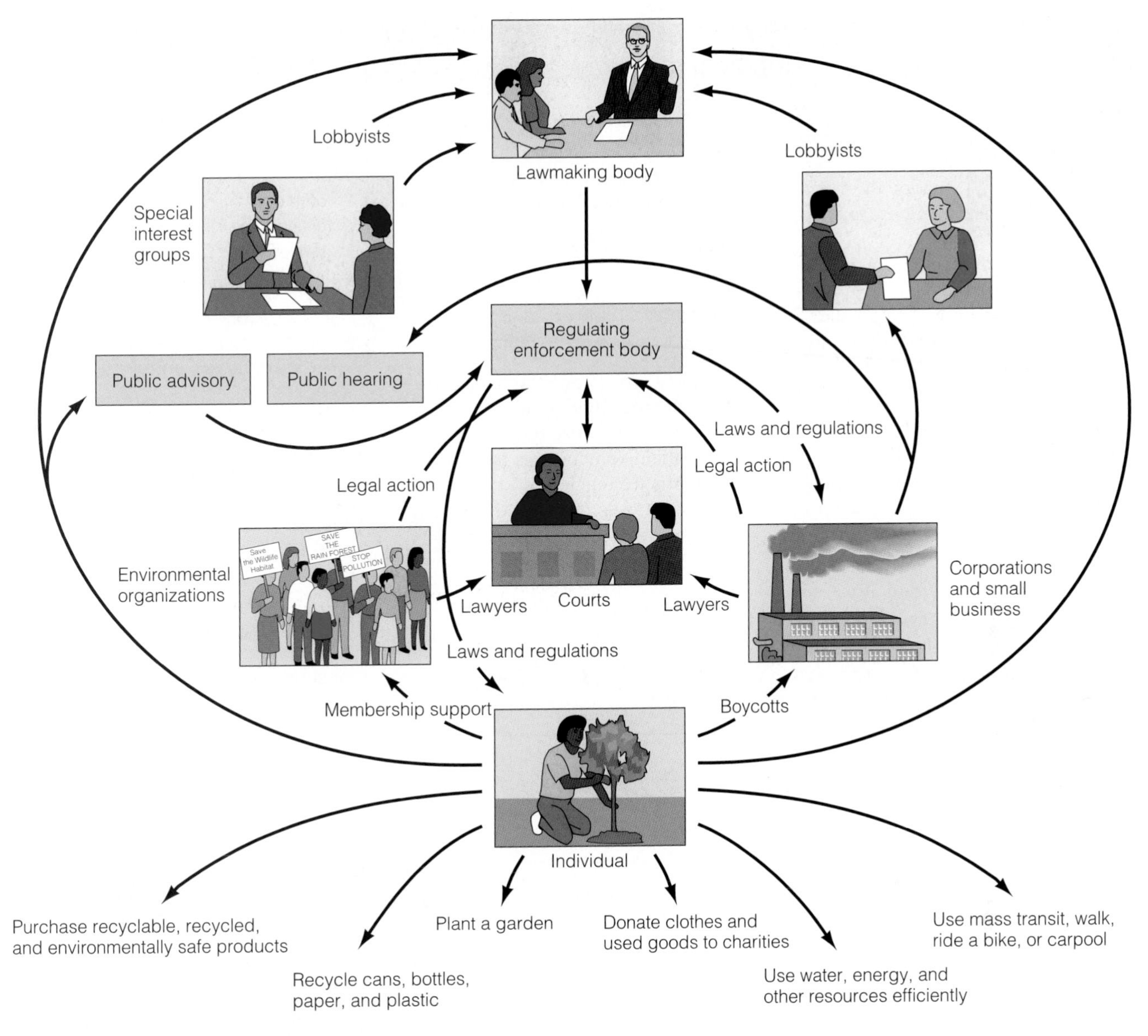

Figure 2-10 Greatly simplified overview of how individuals and lobbyists for and against a particular environmental law interact with the legislative, executive, and judicial branches of government in the United States. The bottom of this diagram also shows some ways in which individuals can bring about environmental change through their own lifestyles.

written and enforced and sometimes challenge the final regulations in court.

- Implement and enforce the approved regulations. Proponents or affected groups may take the agency to court for failing to implement and enforce the regulations or for enforcing them too rigidly.

According to social scientists, the development of public policy in democracies often goes through a *policy life cycle* consisting of four stages: **(1)** recognition, **(2)** formulation, **(3)** implementation, and **(4)** control. Figure 2-11 shows the general position of several major environmental problems in the policy life cycle in the United States and most other developed countries.

What Are the Roles of Mainstream Environmental Groups? In the United States, more than 8 million citizens belong to at least 30,000 NGOs dealing with environmental issues at the international, national, state, and local levels.

Some of these environmental organizations are multi-million-dollar *mainstream* groups, led by chief executive officers and staffed by expert lawyers, scientists, and economists. Mainstream environmental

Types of Environmental Laws in the United States

SOLUTIONS

Environmentalists and their supporters have persuaded the U.S. Congress to enact a number of important federal environmental and resource protection laws, as discussed throughout this text and listed on the website for this chapter. These laws use the following approaches:

- *Setting standards for pollution levels or limiting emissions or effluents for various classes of pollutants* (Federal Water Pollution Control Act and Clean Air Acts)
- *Screening new substances for safety before they are widely used* (Toxic Substances Control Act)
- *Requiring comprehensive evaluation of the environmental impact of an activity before it is undertaken by a federal agency* (National Environmental Policy Act)
- *Setting aside or protecting various ecosystems, resources, and species from harm* (Wilderness Act and Endangered Species Act)
- *Encouraging resource conservation* (Resource Conservation and Recovery Act and National Energy Act)

Some environmental laws contain glowing rhetoric about goals but little guidance about how to meet them, leaving this task to regulatory agencies and the courts. In other cases, the laws or presidential executive orders specify one or more of the following general principles for setting regulations:

- *No unreasonable risk:* food regulations in the Food, Drug, and Cosmetic Act
- *No risk:* the zero-discharge goals of the Safe Drinking Water and Clean Water Acts
- *Standards based on best available technology:* the Clean Air, Clean Water, and Safe Drinking Water Acts
- *Risk–benefit analysis*: pesticide regulation
- *Benefit–cost analysis* (p. 31): the Toxic Substances Control Act

Critical Thinking

Pick a U.S. environmental law listed on the website for this chapter (or a law in the country where you live). Use the library or the internet to evaluate the law's major strengths and weaknesses. Decide whether the law should be weakened, strengthened, or abolished and explain why. List the three most important ways you believe the law should be changed.

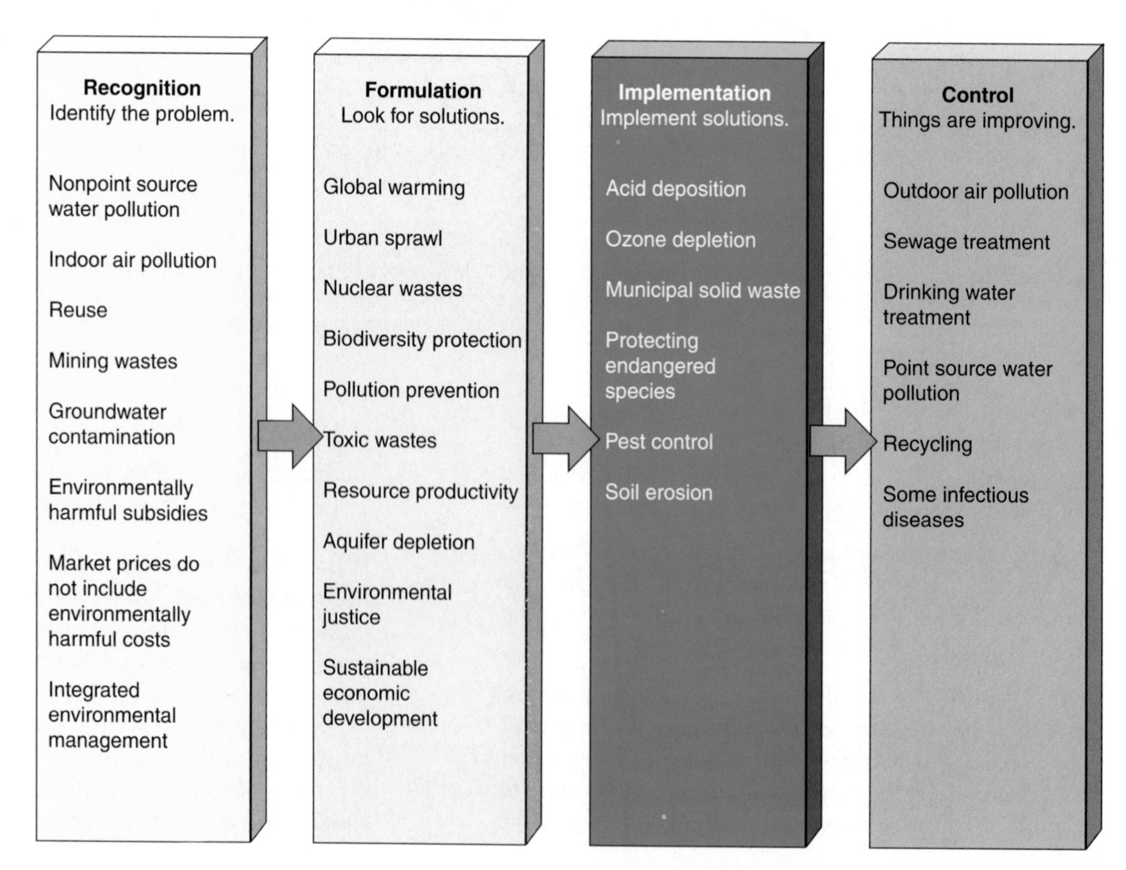

Figure 2-11 General position of several major environmental problems in the *policy life cycle* in most developed countries.

groups are active primarily at the national level and to a lesser extent at the state level; sometimes they form coalitions to work together on issues. Some mainstream organizations such as the Sierra Club funnel substantial funds to local activists and projects.

Mainstream groups work within the political system. Many of these NGOs have been major forces in persuading Congress to pass and strengthen environmental laws (see list on website for this chapter) and fighting off attempts to weaken or repeal such laws. However, these groups must guard against **(1)** being subverted by the political system they work to improve and **(2)** losing touch with ordinary people and nature in the insulated atmosphere of national and state capitals.

Instead of acting as adversaries, some industries and environmental groups have worked together to find solutions to environmental problems. For example, the Environmental Defense Fund has worked with **(1)** McDonald's to redesign its packaging system to eliminate its polyethylene foam clamshell hamburger containers, **(2)** General Motors to remove high-pollution cars from the road, and **(3)** a number of businesses to promote the use of recycled paper.

Some environmental groups have also shifted some of their resources away from demonstrating and litigating to publicizing research on innovative solutions to environmental problems. For example, to promote the use of chlorine-free paper, Greenpeace Germany **(1)** printed a magazine using such paper and **(2)** encouraged readers to demand that magazine publishers switch to chlorine-free paper. Shortly thereafter, several major magazines made such a shift.

There are also more than 3,000 international NGOs working on environmental issues. These international networks **(1)** monitor the environmental activities of governments, corporations, and international agencies and **(2)** push for improved environmental performance by such organizations and by governments.

What Are the Roles of Grassroots Environmental Groups? The base of the environmental movement in the United States and throughout the world consists of thousands of grassroots citizens' groups organized to improve environmental quality, often at the local level. According to political analyst Konrad von Moltke, "There isn't a government in the world that would have done anything for the environment if it weren't for the citizen groups."

These groups carry out a number of environmental roles such as

- Preventing environmental harm to their members and their local communities by opposing projects such as landfills, waste incinerators, nuclear waste dumps, clear-cutting of forests, and harmful development projects
- Getting government officials to take action when they have been victims of environmental harm or environmental injustice because of the unequal distribution of environmental risks
- Voicing the concerns of people who often are not heard in state and national policy debates on environmental policy
- Forming land trusts and other local organizations to **(1)** save wetlands, forests, farmland, and ranchland from development, **(2)** restore degraded rivers and wetlands, and **(3)** convert abandoned urban lots into community gardens and parks
- Forming coalitions of workers and environmentalists to improve worker safety and health
- Setting up internet service providers and networks to **(1)** improve environmental education and health, **(2)** provide individuals and NGOs with information on toxic releases and other environmentally harmful activities by local industries, and **(3)** publicize successful environmental projects
- Working to make colleges and public schools more environmentally sustainable (Individuals Matter, p. 42)

How Successful Have Environmental Groups Been? During the past 30 years, a variety of groups have **(1)** raised understanding of environmental issues by the general public and some business leaders, **(2)** gained public support for an array of environmental and resource use laws in the United States (see list on website for this chapter) and other developed countries, and **(3)** helped individuals deal with a number of local environmental problems.

Polls show that three-fourths of U.S. citizens are strong supporters of environmental laws and regulations and do not want them weakened. However, polls also show that less than 10% of the U.S. public views the environment as one of the nation's most pressing problems. As a result, environmental concerns increasingly do not get transferred to the ballot box. As one political scientist put it, "Environmental concerns are like the Everglades, a mile wide but only a few inches deep."

What Are the Goals of the Anti-Environmental Movement in the United States? Despite general public approval, there is strong opposition to many environmental proposals by

- Leaders of some corporations and people in positions of economic and political power who see environmental laws and regulations as threats to their wealth and power
- Citizens who see environmental laws and regulations as threats to their private property rights and jobs

Environmental Action on Campuses

Since 1988, there has been a boom in environmental awareness on college campuses and in some public schools across the United States. Much of this momentum began in 1989, when the Student Environmental Action Coalition (SEAC), then at the University of North Carolina at Chapel Hill (UNC), held the first national student environmental conference on the UNC campus.

SEAC groups are active on 700 campuses, and the National Wildlife Federation's Campus Ecology Program (launched in 1989) has groups on about 600 campuses.* Most student environmental groups work with members of the faculty and administration to bring about environmental improvements on their own campuses and in their local communities.

Many of these groups focus on making *environmental audits* of their own campuses or schools. Then they use the data gathered to propose changes that will make their campuses or schools more ecologically sustainable, usually saving them money in the process.**

Such audits have resulted in numerous improvements. For example, Morris A. Pierce, a graduate student at the University of Rochester in New York, developed an energy management plan adopted by that school's board of trustees. Under this plan, a capital investment of $33 million is projected to save the university $60 million over 20 years. Students have also induced almost 80% of universities and colleges in the United States to develop recycling programs.

At Bowdoin College in Maine, chemistry professor Dana Mayo and student Caroline Foote developed the concept of *microscale experiments*, in which smaller amounts of chemicals are used. This has reduced toxic wastes and saved the chemistry department more than $34,000. Today more than 50% of all undergraduates in chemistry in the United States use such microscale techniques, as do universities in a growing number of other countries. Students at Oberlin College in Ohio helped design a sustainable environmental studies building (Figure 20-19, p. 528).

A 1997 report by the National Wildlife Foundation's Campus Ecology Program found that 23 student-researched and student-motivated projects had saved the participating universities and colleges $16.3 million. According to this study, implementing similar programs in the nation's 3,700 universities and colleges could **(1)** help improve environmental quality and environmental education and **(2)** lead to a savings of more than $12.6 billion. Such student-spurred environmental activities and research studies are spreading to universities in at least 42 other countries.

*These efforts are described in the book *Ecodemia: Campus Environmental Stewardship at the Turn of the 21st Century* (Washington, D.C.: National Wildlife Federation, 1995) and in the *Campus Environmental Yearbook*, published annually by the National Wildlife Federation.

**Details for conducting such audits are found in April Smith and the Student Environmental Action Coalition, *Campus Ecology: A Guide to Assessing Environmental Quality and Creating Strategies for Change* (Los Angeles, Calif.: Living Planet Press, 1993), and Jane Heinze-Fry, *Green Lives, Green Campuses*, available free on the website for this textbook.

- People who disagree with the basic beliefs behind some environmental worldviews (p. 45)
- Some state and local government officials who are tired of having to implement federal environmental laws and regulations without federal funding (unfunded mandates) or who disagree with certain federal environmental regulations

Since 1980, businesses, individuals, and grassroots groups in the United States have mounted a strong campaign to **(1)** weaken or repeal existing environmental laws, **(2)** change the way in which public lands are used (p. 433), and **(3)** destroy the reputation and effectiveness of the environmental movement. Some of the tactics used by this anti-environmental movement are listed on the website for this book. Because of the anti-environmental movement, since 1980 mainstream environmental groups in the United States have spent most of their time and money trying to prevent existing environmental laws from being weakened or repealed.

2-8 GLOBAL ENVIRONMENTAL POLICY

Should We Expand the Concept of National and Global Security? Countries are legitimately concerned with *military security* and *economic security*. However, environmentalists point out that **(1)** all economies are supported by natural resources (Figure 2-3) and **(2)** many environmental problems do not recognize political boundaries. Thus, military and economic security also depend on national and global environmental security.

Proponents call for all countries to make environmental security a major focus of diplomacy and government policy at all levels. They propose that national governments have a council of advisers made up of highly qualified experts in environmental, economic, and military security. Any major decision would integrate all three security concerns.

What Progress Has Been Made in Developing International Environmental Cooperation and Policy? Since the 1972 UN Conference on the Human Environment in Stockholm, Sweden, some progress has been made in addressing environmental issues at the global level. Today, 115 nations have environmental protection agencies, and nearly 240 international environmental treaties and agreements between various countries have been signed. They address issues such as endangered species, ozone depletion, ocean pollution, climate change, ozone layer depletion, biodiversity, and hazardous waste export. The 1972 conference also created the UN Environment Programme (UNEP) to negotiate environmental treaties and to help monitor and implement them.

In June 1992, the second UN Conference on the Human Environment, known as the *Rio Earth Summit*, was held in Rio de Janeiro, Brazil. More than 100 heads of state, thousands of officials, and more than 1,400 accredited NGOs from 178 nations met to develop plans to address environmental issues.

The major official results included **(1)** an *Earth Charter*, a nonbinding statement of broad principles for guiding environmental policy that commits countries that sign it to pursue sustainable development and work toward eradicating poverty, **(2)** *Agenda 21*, a nonbinding detailed action plan to guide countries toward sustainable development and protection of the global environment during the 21st century, **(3)** a *forestry agreement* that is a broad, nonbinding statement of principles of forest management and protection, **(4)** a *convention on climate change* that requires countries to use their best efforts to reduce their emissions of greenhouse gases, **(5)** a *convention on protecting biodiversity* that calls for countries to develop strategies for the conservation and sustainable use of biological diversity, and **(6)** the *UN Commission on Sustainable Development*, composed of high-level government representatives charged with carrying out and overseeing the implementation of these agreements.

Most environmentalists were disappointed because these accomplishments consisted of nonbinding agreements without sufficient incentives or funding for their implementation. According to analyses by the United Nations, by 2000 there was little improvement in the major environmental problems discussed at the Rio summit.

However, there is hope for greater progress in the slowly moving arena of international cooperation because

- The conference gave the world a forum for discussing and seeking solutions to environmental problems. This led to general agreement on some key principles, which with enough political pressure from citizens and NGOs could be implemented or improved.
- Paralleling the official meeting was a Global Forum that brought together 20,000 concerned citizens and activists from more than 1,400 NGOs in 178 countries. These people outnumbered the conference's official representatives by at least two to one. These NGOs **(1)** worked behind the scenes to influence official policy, **(2)** formulated their own agendas and treaties for environmental sustainability, **(3)** learned from one another, and **(4)** developed a series of new global networks, alliances, and projects. In the long run, these newly formed networks and alliances may play the greatest role in helping **(1)** monitor, support, and implement the commitments and plans developed by the 1992 conference and **(2)** set the agenda for the third UN Conference on the Human Environment in 2002.

2-9 HUMAN-CENTERED ENVIRONMENTAL WORLDVIEWS

What Is an Environmental Worldview? There are conflicting views about how serious our environmental problems are and what we should do about them. These conflicts arise mostly out of differing **environmental worldviews**: **(1)** how people think the world works, **(2)** what they think their role in the world should be, and **(3)** what they believe is right and wrong environmental behavior (**environmental ethics**).

People with widely differing environmental worldviews can take the same data, be logically consistent, and arrive at quite different conclusions because they start with different assumptions and values.

There are many different types of environmental worldviews, as summarized in Figure 2-12. Most can be divided into two groups according to whether they are *individual centered* (atomistic) or *earth centered* (holistic). Atomistic environmental worldviews tend to be *human centered* (anthropocentric) or *life centered* (biocentric, with the primary focus on individual species or individual organisms). Holistic or ecocentric environmental worldviews are centered on the sustaining the earth's natural systems (ecosystems), life-forms (biodiversity), and life-support systems (biosphere).

What Are the Major Human-Centered Environmental Worldviews? Most people in today's industrial consumer societies have a **planetary management worldview**, which has become increasingly common during the past 50 years. According to this human-centered environmental worldview, human beings, as the planet's most important and dominant species, can and should manage the planet mostly for their own benefit. Other species and parts of nature are seen as having only *instrumental value* based on how useful they are to us.

Figure 2-12 General types of environmental worldviews. (Diagram developed by Jane Heinze-Fry)

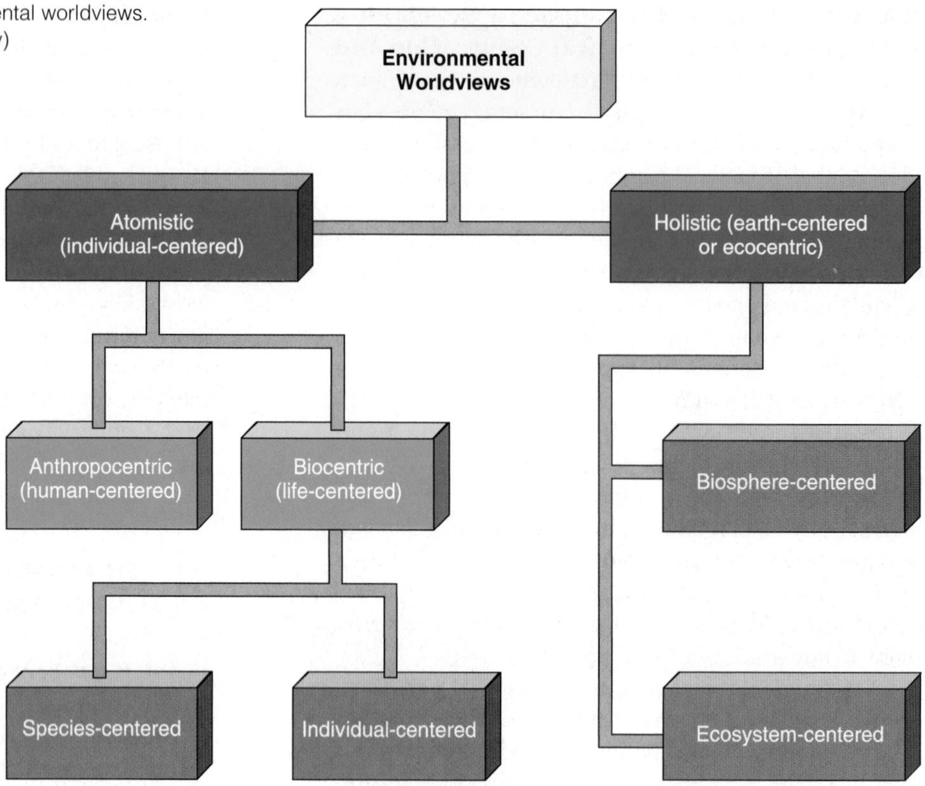

Environmental Worldviews
Atomistic (individual-centered)
Holistic (earth-centered or ecocentric)
Anthropocentric (human-centered)
Biocentric (life-centered)
Biosphere-centered
Species-centered
Individual-centered
Ecosystem-centered

The basic environmental beliefs of this worldview include the following:

- ***We are the planet's most important species, and we are apart from and in charge of the rest of nature.*** This idea sometimes surfaces when people talk about "our" planet, "our" earth, or "saving the earth."

- ***There is always more.*** The earth has an essentially unlimited supply of resources for use by us through science and technology. If we deplete a resource, we will find substitutes. To deal with pollutants, we can invent technology to clean them up, dump them into space, or move into space ourselves. If we extinguish other species, we can use genetic engineering to create new and better ones.

- ***All economic growth is good, and the potential for global economic growth is essentially limitless.***

- ***Our success depends on how well we can understand, control, and manage the earth's life-support systems for our benefit.***

All or most aspects of this worldview are widely supported because it is said to be the primary driving force behind the major improvements in the human condition since the beginning of the industrial revolution.

There are several variations of this environmental worldview.

- The *no-problem school.* There are no environmental, population, or resource problems that we cannot solve with more economic growth and development, better management, and better technology.

- The *free-market school.* The best way to manage the planet for human benefit is through a free-market global economy with minimal government interference and regulations. Free-market advocates would convert all public property resources to private property resources and let the global marketplace, governed by pure free-market competition (p. 24), decide essentially everything.

- The *responsible planetary management school.* We have serious environmental problems, but we can sustain our species with a mixture of market-based competition, better technology, and some government intervention that **(1)** promotes environmentally sustainable forms of economic development, **(2)** protects environmental quality and private property rights, and **(3)** protects and manages public and common property resources. People holding this view follow the pragmatic principle of *enlightened self-interest*: Better earth care is better self-care.

- The *spaceship-earth school.* Earth is seen as a spaceship: a complex machine that we can understand, dominate, change, and manage to prevent environmental overload and provide a good life for everyone. This view developed as a result of photographs taken from space showing the earth as a finite planet or island "floating" in space (Figure 1-1, p. 2). This powerful image led many people to see that the earth is our only home and that we had better treat it right.

- The *stewardship school.* We have an ethical responsibility to be caring and responsible managers, or stewards, of the earth. According to this view, we can and should make the world a better place for our species and other species through love, care, knowledge, and technology.

2-10 LIFE-CENTERED ENVIRONMENTAL WORLDVIEWS

Can We Manage the Planet? Some people believe that any human-centered worldview will eventually fail because it wrongly assumes that we now have (or can gain) enough knowledge to become effective managers or stewards of the earth.

These people argue that the unregulated global free-market approach will not work because it **(1)** is based on increased losses of natural capital (Guest Essay, p. 6, and Figure 2-3) that support all life and economies and **(2)** focuses on short-term economic benefits regardless of the harmful long-term environmental and social consequences.

The image of the earth as an island or spaceship in space has played an important role in raising global environmental awareness. However, critics argue that thinking of the earth as a spaceship that we should and can manage is an oversimplified and misleading way to view an incredibly complex and ever-changing planet, as the failure of Biosphere 2 demonstrated (p. 23). As biologist David Ehrenfeld puts it, "In no important instance have we been able to demonstrate comprehensive successful management of the world, nor do we understand it well enough to manage it even in theory."

Even if we had enough knowledge and wisdom to manage spaceship earth, some critics see this approach as requiring us to give up individual freedom to survive. Life on spaceship earth under a comprehensive system of planetary management or world government might be very much like the regimented life of astronauts in their capsule. The astronauts have almost no individual freedom, with essentially all their actions dictated by a central command (ground control).

What Are Some Major Biocentric and Ecocentric Worldviews? Critics of human-centered environmental worldviews believe that such worldviews should be expanded to recognize the *inherent value* of all forms of life. According to this view, all species are part of a community of living things and have just as much right to exist as humans do. In other words, each species has *intrinsic value* unrelated to its potential or actual use to us.

Most people with a life-centered (biocentric) worldview believe that we have an ethical responsibility to not cause the premature extinction of a species because of our activities.

Others believe that we must go beyond this biocentric worldview, which focuses mostly on species. They believe that we have an ethical responsibility not to degrade the earth's natural systems (ecosystems), lifeforms (biodiversity), and life-support systems (biosphere) for this and future generations of humans (Connections, p. 46). In other words, they have an *earth-centered*, or *ecocentric*, environmental worldview, devoted to preserving the earth's biodiversity and the functioning of its life-support systems for all forms of life.

According to the ecocentric worldview, we are part of, not apart from, the community of life and the ecological processes that sustain all life. Aldo Leopold (Individuals Matter, p. 47) summed up this idea in 1948: "All ethics rest upon a single premise: that the individual is a member of a community of interdependent parts."

There are many life-centered and earth-centered environmental worldviews, and several of them overlap in some of their beliefs. One ecocentric environmental worldview is the **environmental wisdom worldview**. It is based on the following major beliefs, which are the opposite of those making up the planetary management worldview:

- ***We are part of nature, and nature does not exist just for us.*** We need the earth, but the earth does not need us. Each species has an inherent right to exist.
- ***There is not always more.*** The earth's resources are limited, should not be wasted, and should be used efficiently and sustainably for us and all other species.
- ***Some forms of technology and economic growth are environmentally beneficial and should be encouraged, but some are environmentally harmful and should be discouraged.***
- ***Our success depends on* (1) *learning how the earth sustains itself and adapts to ever-changing environmental conditions and* (2) *integrating such scientific lessons from nature (environmental wisdom) into the ways we think and act.***

Others say we do not need to be biocentrists or ecocentrists to value life or the earth. They point out that human-centered stewardship and planetary management environmental worldviews also call for us to value individuals, species, and the earth's life-support systems as part of our responsibility as earth's caretakers.

2-11 SOLUTIONS: LIVING MORE SUSTAINABLY

Solutions: What Are Some Ethical Guidelines for Living More Sustainably? Ethicists and philosophers have developed ethical guidelines for living more sustainably on the earth (Solutions, p. 19).

Biosphere and Ecosystems

- We should try to understand and work with the rest of nature to help sustain the natural resources,

Why Should We Care About Future Generations?

CONNECTIONS

Some people believe that our only ethical obligation is to the present human generation. They ask, "What has the future done for me?" or believe that we cannot know enough about the condition of the earth for future generations to be concerned about it.

According to biologist David W. Ehrenfeld, caring about future generations enough not to degrade the earth's life-support systems is important because it gives future generations options for dealing with the problems they will face. He points out that if our ancestors had left for us the ecological degradation we appear to be leaving our descendants, our options for enjoyment—perhaps even for survival—would be quite limited.

And in response to the question, "What can future generations do for us?" Ehrenfeld gives the following answer: "They give us a reason for treating our ecological home respectfully, so that our lives as well as theirs will be enriched."

According to this view, as we use the earth's natural resources we **(1)** are borrowing from the earth and from future generations and **(2)** have an ethical responsibility to leave the earth as good as or better than it is now for future generations. In thinking about our responsibility toward future generations, some analysts believe that we should consider the wisdom given to us in the 18th century by the Iroquois Confederation of Native Americans: *In our every deliberation, we must consider the impact of our decisions on the next seven generations.*

Critical Thinking

What obligations, if any, concerning the environment do you have to future generations? To how many future generations do you have responsibilities? Be honest about your feelings.

biodiversity, and adaptability of the earth's life-support systems.

- When we alter nature to meet our needs or wants, we should carefully evaluate our proposed actions and choose methods that do the least possible short- and long-term environmental harm.

Species and Cultures

- We should strive not to cause the premature extinction of any wild species.
- The best ways to protect species and individuals of species are to protect the places where they live and to help restore places we have degraded.
- No human culture should become extinct because of our actions.

Individual Responsibility

- We should not inflict unnecessary suffering or pain on any animal we raise or hunt for food or use for scientific or other purposes.
- We should leave the earth as good as or better than we found it.
- We should use no more of the earth's resources than we need.
- We should work with the earth to help heal ecological wounds we have inflicted.

Solutions: How Can We Implement Earth Education? Most environmentalists believe that learning how to live more sustainably takes a foundation of environmental or earth education. According to its proponents, the most important goals of such an education are to

- *Develop respect or reverence for all life.*
- *Understand as much as we can about how the earth works and sustains itself and use such knowledge to guide our lives, communities, and societies.*
- *Use critical thinking skills to become wisdom seekers instead of vessels of information.*
- *Understand and evaluate one's worldview and see this as a lifelong process* (Individuals Matter, p. 48).
- *Learn how to evaluate the beneficial and harmful consequences of one's lifestyle and profession on the earth, today and in the future.*
- *Foster a desire to make the world a better place and act on this desire.* As David Orr (Guest Essay, p. 272) puts it, education should help students "make the leap from 'I know' to 'I care' to 'I'll do something.'"

According to environmental educator Mitchell Thomashow, four basic questions should be at the heart of environmental education:

- Where do the things I consume come from?
- What do I know about the place where I live?
- How am I connected to the earth and other living things?
- What are my purpose and responsibility as a human being?

Aldo Leopold and His Land Ethic

INDIVIDUALS MATTER

Aldo Leopold (Figure 2-13) is best known for being a strong proponent of *land ethics*, a philosophy in which humans as part of nature have an ethical responsibility to preserve wild nature.

After earning a master's degree in forestry from Yale University, he joined the U.S. Forest Service. He became **(1)** alarmed by overgrazing and land deterioration on public lands where he worked and **(2)** convinced that the United States was losing too much of its mostly untouched wilderness lands.

In 1933, Leopold became a professor of game management at the University of Wisconsin and founded the profession of game management. In 1935 he was one of the founders of the Wilderness Society.

As years passed, he developed a deep understanding and appreciation for wildlife and urged that nature should be included in our ethical concerns. Through his writings and teachings he became one of the founders of the *conservation* and *environmental movements* of the 20th century.

Leopold died in 1948 while fighting a brush fire at a neighbor's farm. His weekends of planting, hiking, and observing nature at his own nearby farm provided material he used to write his most famous book, *A Sand County Almanac*, published posthumously in 1949. Since then more than 2 million copies of this important book have been sold.

The following quotes from his writings reflect Leopold's land ethic and form the basis of many of the beliefs of the modern *environmental wisdom worldview* (p. 44):

- *That land is a community is the basic concept of ecology, but that land is to be loved and respected is an extension of ethics.*
- *The land ethic... changes the role of* Homo sapiens *from conqueror of the land-community to plain member and citizen of it.*
- *We abuse land because we regard it as a commodity belonging to us. When we see land as a community to which we belong, we may begin to use it with love and respect.*
- *A thing is right when it tends to preserve the integrity, stability, and beauty of the biotic community. It is wrong when it tends otherwise.*

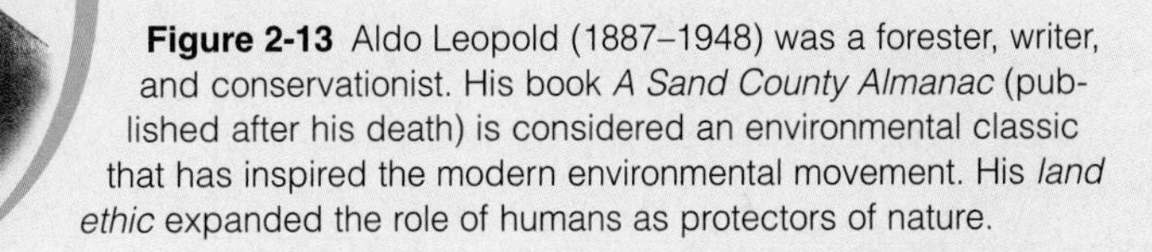

Figure 2-13 Aldo Leopold (1887–1948) was a forester, writer, and conservationist. His book *A Sand County Almanac* (published after his death) is considered an environmental classic that has inspired the modern environmental movement. His *land ethic* expanded the role of humans as protectors of nature.

How we answer these questions determines our *ecological identity*.

In addition to formal education, some analysts believe that we need to experience nature directly to help us learn to walk more lightly on the earth (Connections, p. 49).

How Can We Live More Simply? Many analysts urge us to *learn how to live more simply*. Although seeking happiness through the pursuit of material things is considered folly by almost every major religion and philosophy, it is preached incessantly by modern advertising.

Some affluent people in developed countries are adopting a lifestyle of *voluntary simplicity*, doing and enjoying more with less by learning to live more simply. Voluntary simplicity is based on Mahatma Gandhi's *principle of enoughness*: "The earth provides enough to satisfy every person's need but not every person's greed. . . . When we take more than we need, we are simply taking from each other, borrowing from the future, or destroying the environment and other species."

Implementing this principle means asking oneself, "How much is enough?" This is not an easy thing to do because people in affluent societies are conditioned to want more and more, and they often think of such wants as vital needs.

Voluntary simplicity begins by asking a series of questions before buying anything: **(1)** Do I really need this? **(2)** Can I buy it secondhand (reuse)? **(3)** Can I borrow, rent, lease, or share it? **(4)** Can I build it myself?

The decision to buy something triggers another set of questions: **(1)** Is the product produced in an environmentally sustainable manner? **(2)** Did the workers producing it get fair wages for their work, and did they

Mindquake: Evaluating One's Environmental Worldview

Questioning and perhaps changing one's environmental worldview can be difficult and threatening. It can set off a cultural *mindquake* that involves examining many of one's most basic beliefs. However, once people change their worldviews, it no longer makes sense for them to do things in the old ways. If enough people do this, then tremendous cultural change, once considered impossible, can take place rapidly.

Most environmentalists urge us to think about what our basic environmental beliefs are and why we have them. They believe that evaluating our beliefs and being open to the possibility of changing them should be one of our most important lifelong activities.

As this book emphasizes, most environmental issues are filled with controversy and uncertainty. A clearly right or wrong path is not easy to discover and usually is strongly influenced by one's environmental worldview. As philosopher Georg Hegel pointed out nearly two centuries ago, tragedy is not the conflict between right and wrong but the conflict between right and right.

Critical Thinking

What are the basic beliefs of your environmental worldview?

have safe and healthful working conditions? **(3)** Is it designed to last as long as possible? **(4)** Is it easy to repair, upgrade, reuse, and recycle?

Voluntary simplicity by those who have more than they need should not be confused with the *forced simplicity* of the poor, who do not have enough to meet their basic needs for food, clothing, shelter, clean water and air, and good health.

After a lifetime of studying the growth and decline of the world's human civilizations, historian Arnold Toynbee summarized the true measure of a civilization's growth in what he called the *law of progressive simplification*: "True growth occurs as civilizations transfer an increasing proportion of energy and attention from the material side of life to the nonmaterial side and thereby develop their culture, capacity for compassion, sense of community, and strength of democracy."

How Can We Move Beyond Blame, Guilt, and Denial to Responsibility? According to many psychologists, when we first encounter an environmental problem, our initial response often is to find someone or something to blame: greedy industrialists, uncaring politicians, environmentalists, and people with misguided worldviews. It is the fault of such villains, and we are the victims.

This can lead to despair, denial, and inaction because we feel powerless to stop or influence these forces. There are also many complex and interconnected environmental problems and conflicting views about their seriousness and possible solutions. As a result, we feel overwhelmed and wonder whether there is any way out—an emotion leading to denial and inaction.

Upon closer examination we may realize that we all make some direct or indirect contributions to the environmental problems we face. As Pogo said, "We have met the enemy and it is us." We do not want to feel guilty or bad about all of the things we are not doing, so we avoid thinking about them—another path to denial and inaction.

How do we move beyond blame, fear, and guilt to more responsible environmental actions in our daily lives? Analysts have suggested several ways to do this.

- Recognize and avoid common mental traps that lead to denial, indifference, and inaction. These traps include **(1)** *gloom-and-doom pessimism* (it's hopeless), **(2)** *blind technological optimism* (science and technofixes will save us), **(3)** *fatalism* (we have no control over our actions and the future), **(4)** *extrapolation to infinity* (if I cannot change the entire world quickly, I will not try to change any of it), **(5)** *paralysis by analysis* (searching for the perfect worldview, philosophy, solutions, and scientific information before doing anything), and **(6)** *faith in simple, easy answers*.
- Understand that no one can even come close to doing all of the things people suggest (or that we know we should be doing) to work with the earth. Try to determine what things you do have the greatest environmental impact and concentrate on them. Focus on things you feel strongly about and that you can do something about. Rejoice in the good things you have done and continually expand your efforts make the earth a better place.
- Keep your empowering feelings of hope slightly ahead of your immobilizing feelings of despair. In working with the earth we should be guided by historian Arnold Toynbee's observation, "If you make the world ever so little better, you will have done splendidly, and your life will have been worthwhile," and by George Bernard Shaw's reminder that "indifference is the essence of inhumanity."
- Do not use guilt and fear to motivate other people to work with the earth and other people. We need to nurture, reassure, understand, and care for one another.

Learning from the Earth

Formal environmental education is important, but many earth thinkers believe that it is not enough. They urge us to take the time to escape the cultural and technological body armor we use to insulate ourselves from nature and to experience nature directly.

They suggest that we reenchant our senses and kindle a sense of awe, wonder, and humility by standing under the stars, sitting in a forest, taking in the majesty and power of an ocean, or experiencing a stream, lake, or other part of nature.

We might pick up a handful of soil and try to sense the teeming microscopic life in it that keeps us alive. We might look at a tree, mountain, rock, or bee and try to sense how they are a part of us and we a part of them as interdependent participants in the earth's life-sustaining recycling processes.

Earth thinker Michael J. Cohen suggests that we recognize who we really are by saying,

> *I am a desire for water, air, food, love, warmth, beauty, freedom, sensations, life, community, place, and spirit in the natural world. . . . I have two mothers: my human mother and my planet mother, Earth. The planet is my womb of life.*

Many psychologists believe that consciously or unconsciously we spend much of our lives in a search for roots: something to anchor us in a bewildering and frightening sea of change. As philosopher Simone Weil observed, "To be rooted is perhaps the most important and least recognized need of the human soul."

Earth philosophers say that to be rooted, each of us needs to find a *sense of place*: a stream, a mountain, a yard, a neighborhood lot, or any piece of the earth we feel at one with as a place we know, experience emotionally, and love. It can be a place where we live or a place we occasionally visit and experience in our inner being. When we become part of a place, it becomes a part of us. Then we are driven to defend it from harm and to help heal its wounds.

To many earth thinkers, emotionally experiencing our connectedness with the earth leads us to recognize that the healing of the earth and the healing of the human spirit are one and the same. They call for us to discover and tap into what Aldo Leopold calls "the green fire that burns in our hearts" and use this as a force for respecting and working with the earth and with one another.

Critical Thinking

Some analysts believe that learning environmental wisdom by experiencing the earth and forming an emotional bond with its life-forms and processes is unscientific, mystical poppycock based on a romanticized view of nature. They believe that better scientific understanding of how the earth works and improved technology are the only ways to achieve sustainability. Do you agree or disagree? Explain.

- Recognize that there is no single correct or best solution to the environmental problems we face. Indeed, one of nature's most important lessons is that preserving diversity or a rainbow of flexible and adaptable solutions to the problems we face is the best way to adapt to earth's largely unpredictable, ever-changing conditions.

- Have fun and take time to enjoy life. Every day we should laugh and enjoy nature, beauty, friendship, and love.

What Are the Major Components of the Environmental Revolution? The *environmental revolution* that many environmentalists call for us to bring about would have several components:

- An *efficiency revolution* that involves not wasting matter and energy resources.

- A *pollution-prevention revolution* that reduces pollution and environmental degradation by **(1)** reducing the waste of matter and energy resources, **(2)** keeping highly toxic substances from being released into the environment by recycling or reusing them within industrial processes, and **(3)** trying to find less polluting substitutes for such substances or not producing them at all.

- A *sufficiency revolution*. This involves trying to meet the basic needs of all people on the planet and asking how many material things we really need to have a decent and meaningful life.

- A *demographic revolution* based on bringing the size and growth rate of the human population into balance with the earth's ability to support humans and other species without serious environmental degradation.

- An *economic revolution* in which we use economic systems to reward environmentally beneficial behavior and discourage environmentally harmful behavior (p. 35).

Opponents of such a cultural change like to paint environmentalists as messengers of gloom, doom, and hopelessness. However, *the real message of environmentalism is not gloom and doom, fear, and catastrophe but hope and a positive vision of the future.* This is an exciting message of challenge and adventure as we struggle to find better and more responsible ways to live on this planet.

Envision the earth's life-sustaining processes as a beautiful and diverse web of interrelationships—a kaleidoscope of patterns, rhythms, and connections whose very complexity and multitude of possibilities remind us that cooperation, sharing, honesty, humility, and love should be the guidelines for our behavior toward one another and the earth.

When there is no dream, the people perish.
PROVERBS 29:18

REVIEW QUESTIONS

1. Describe the Biosphere 2 project and the lessons it taught us.

2. Distinguish between *natural, human, financial,* and *manufactured resources* used in an economic system.

3. Distinguish between *pure command, pure free-market,* and *capitalist market* systems. List 10 reasons why governments intervene in economic systems.

4. Explain how conventional economists and ecological economists differ in their view of market-based economic systems. Distinguish between *economic growth* and *environmentally sustainable economic development.* What are three characteristics of environmentally sustainable economic development?

5. List four reasons why GNP and GDP are not useful measures of economic health, environmental health, or human well-being. Describe an environmental indicator that could be used to provide such information.

6. Distinguish between *internal costs* and *external costs* and give an example of each. What is *full-cost pricing,* and what are the pros and cons of using this approach to internalize external environmental costs? List six reasons why full-cost pricing has not been widely used.

7. How do economists determine the *optimum level of pollution* for a particular chemical? What are the pros and cons of using this approach?

8. What is *benefit–cost analysis,* and what are the pros and cons of using this tool to evaluate alternative courses of action?

9. Describe two types of *economic incentives* (rewards) that can be used to improve environmental quality and reduce resource waste and list the pros and cons of each type. Describe two types of *economic disincentives* (punishments) that can be used to improve environmental quality and reduce resource waste and list the pros and cons of each type.

10. What are the pros and cons of using tradable pollution and resource-use rights to reduce pollution and resource waste?

11. List six phases or eras in the evolution of environmental management.

12. What is *poverty,* and what are its harmful health and environmental effects?

13. List two ways in which the governments of developing countries can reduce poverty. List six ways in which governments of developed countries can help reduce poverty. What are *microloans,* and how are they being used to reduce poverty?

14. List 11 principles for shifting to environmentally sustainable economies over the next several decades. List three great pieces of news about making the shift to environmentally sustainable economies over the next several decades and explain why the environmental revolution is also an economic revolution.

15. What is *politics*? What is a *democracy*?

16. List six principles that can be used as guidelines in making environmental policy decisions. What is *environmental justice*?

17. List nine ways in which individuals can influence the environmental policies of local, state, and federal governments.

18. Distinguish between *leaders, rulers,* and *managers.* Describe four types of environmental leadership.

19. Describe the five steps used to develop federal environmental policy in the United States. Describe the four phases of a *policy life cycle.*

20. Describe the key roles of mainstream and grassroots environmental groups. Describe some of the environmentally beneficial activities that have been carried out by high school and college students.

21. Describe three major accomplishments of environmental groups over the past 30 years. Describe the *anti-environmental movement.*

22. Distinguish between environmental, economic, and military security and explain the importance of making environmental security a key priority of governments.

23. What is an *environmental worldview*? What are the four major beliefs of the *planetary management environmental worldview*? Describe the **(a)** *no problem,* **(b)** *free-market,* **(c)** *responsible planetary management,* **(d)** *spaceship earth,* and **(e)** *stewardship* variations of the planetary management environmental worldview.

24. Explain why many environmentalists believe that planetary management worldviews will not work.

25. Give two reasons for caring about future generations.

26. What are the four major beliefs of the *environmental wisdom worldview*?

27. List nine ethical guidelines for living more sustainably.

28. List six guidelines of earth education and four questions that should be answered to determine one's *ecological identity*. Explain why some people believe that it is important to learn more about the earth from direct experience.

29. What is *voluntary simplicity*?

30. List six ways to move beyond blame, guilt, and denial and become more environmentally responsible. Describe six mental traps that can lead to denial, indifference, and inaction about environmental (and other) problems.

31. What are five major components of an environmental revolution that most environmentalists believe we should bring about over the next 50 years?

CRITICAL THINKING

1. Some analysts argue that the problems with Biosphere 2 resulted mostly from inadequate design and that a better team of scientists and engineers could make it work. Explain why you agree or disagree with this view.

2. The primary goal of current economic systems is to maximize economic growth by producing and consuming more and more economic goods. Do you agree with that goal? Explain. What are the alternatives?

3. Suppose that over the next 20 years the current harmful environmental and health costs of goods and services are internalized so that their market prices reflect their total costs. What harmful and beneficial effects might this have on **(a)** your lifestyle and **(b)** any child you might have?

4. Do you believe that we should establish optimum levels or zero-discharge levels for toxic chemicals we release into the environment? Explain.

5. Do you agree or disagree with the proposals various analysts have made for sharply reducing poverty as discussed on p. 34? Explain.

6. Do you agree or disagree with the guidelines for an environmentally sustainable economy listed on p. 35? Explain.

7. What are the greatest strengths and weaknesses of the system of government in your country with respect to **(a)** protecting the environment and **(b)** ensuring environmental justice for all? What three major changes, if any, would you make in this system?

8. Suppose that a presidential candidate ran on a platform calling for the federal government to phase in a tax on gasoline so that, over 5–10 years, the price of gasoline would rise to $5–7 a gallon (as is the case in Japan and most western European nations). The candidate argues that this tax increase is necessary to encourage oil and gasoline conservation, reduce air pollution, and enhance future economic, environmental, and military security. The candidate also says that the tax revenue should be used to **(1)** reduce income taxes on the poor and middle class by an amount roughly equal to the increase in gasoline taxes and **(2)** reduce taxes on wages and profits (Solutions, p. 33). Would you vote for this candidate who wants to triple the price of gasoline? Explain.

9. This chapter has summarized a number of different environmental worldviews. Go through these worldviews and find the beliefs you agree with to describe your environmental worldview. Which of your beliefs were added or modified as a result of taking this course?

10. Explain why you agree or disagree with the following ideas: **(a)** Everyone has the right to have as many children as they want, **(b)** each member of the human species has a right to pollute and to use as many resources as they want, and **(c)** individuals should have the right to do anything they want with land they own. Relate your answers to the beliefs of your environmental worldview as summarized in your answer to question 9.

PROJECTS

1. List all the goods you use, and then identify those that meet your basic needs and those that satisfy your wants. Identify any economic wants **(a)** you would be willing to give up, **(b)** you believe you should give up but are unwilling to give up, and **(c)** you hope to give up in the future. Relate the results of this analysis to your personal impact on the environment. Compare your results with those of your classmates.

2. What student environmental groups (if any) are active at your school? How many people actively participate in these groups? What environmentally beneficial things have they done? What actions (if any) taken by such groups do you disagree with? Why?

3. Does your school's curriculum provide *all* graduates with the basic elements of ecological literacy? To what extent are the funds in its financial endowments invested in enterprises that are working to develop or encourage environmental sustainability? Over the past 20 years, what important roles have its graduates played in making the world a better and more sustainable place to live? Using such information, rate your school on a 1–10 scale in terms of its contributions to environmental awareness and sustainability. Develop a detailed plan illustrating how your school could become better at achieving such goals and present this information to school officials, alumni, parents, and financial backers.

4. If you knew you were going to die and had an opportunity to address everyone in the world for 5 minutes, what would you say? Write out your 5-minute speech and compare it with those of other members of your class.

5. Make a concept map of this chapter's major ideas, using the section heads and subheads and the key terms (in boldface type). Look on the website for this book for information about making concept maps.

INTERNET STUDY RESOURCES AND RESOURCES FOR FURTHER READING AND RESEARCH

The website for this book contains helpful study aids and many ideas for further reading and research. Log on to

http://www.brookscole.com/product/0534389872s

and click on the Chapter-by-Chapter area. Choose Chapter 2 and select a resource:

- "Flash Cards" allows you to test your mastery of the Terms and Concepts to Remember for this chapter.
- "Tutorial Quizzes" provides a multiple-choice practice quiz.
- "Student Guide to InfoTrac" will lead you to Critical Thinking Projects that use InfoTrac College Edition as a research tool.
- "References" lists the major books and articles consulted in writing this chapter.
- "Hypercontents" takes you to an extensive list of sites with news, research, and images related to individual sections of the chapter.

INFOTRAC COLLEGE EDITION

Improve your skills with InfoTrac College Edition, a searchable online database of articles from more than 700 periodicals. Log on to

http://www.infotrac-college.com

or access InfoTrac through the website for this book. Try to find the following articles:

1. Environmental Taxation: A New Tool for Local Planning? Tony Jackson. *Regional Studies*, Feb 2001 v35 i1 p80. "Research provides increasing evidence favoring the use of economic instruments to deliver policies promoting sustainable development on a discretionary basis at local levels. There are substantiated clear economic as well as environmental gains from the application of such tools either as alternatives or supplements to traditional command-and-control environmental regulations." *Hint*: Enter the search term "environmental taxation" using the keywords "using regulations."

2. People, Nature, and Ethics. Paul Wapner. *Current History*, Nov 2000 v99 i640 p355. "Environmental abuse is not only about how humans treat the nonhuman world but also about how they treat each other. *Hint*: Enter the search term "nature ethics" using the keywords "environmental ethics."

PART II

SCIENTIFIC PRINCIPLES AND CONCEPTS

Animal and vegetable life is too complicated a problem for human intelligence to solve, and we can never know how wide a circle of disturbance we produce in the harmonies of nature when we throw the smallest pebble into the ocean of organic life.

GEORGE PERKINS MARSH

3 SCIENCE, SYSTEMS, MATTER, AND ENERGY

Saving Energy, Money, and Jobs in Osage, Iowa

Osage, Iowa (population about 4,000), has become the energy-efficiency capital of the United States. Wes Birdsall began its transformation in 1974. As general manager of Osage Municipal Gas and Electric Company, he started urging the townspeople to save energy and reduce their natural gas and electric bills. The utility would also save money by not having to add new electrical generating facilities.

Birdsall started his crusade by telling homeowners about the importance of insulating walls and ceilings and of plugging leaky windows and doors. He also advised people to **(1)** replace their incandescent light bulbs with more efficient fluorescent bulbs and **(2)** turn down the temperature on water heaters and wrap them with insulation. These suggestions were economic boons to the local hardware and lighting stores. The utility company even gave away free water heater blankets. Birdsall also suggested saving water and fuel by installing low-flow showerheads.

Birdsall then stepped up his efforts by offering to give every building in town a free thermogram, an infrared scan that shows where heat escapes (Figure 3-1). When people could see the energy (and their money) hemorrhaging out of their buildings, they took action to plug the leaks, again helping the local economy. Birdsall then stepped up his campaign even more, announcing that no new houses could be hooked up to the company's natural gas line unless they met minimum energy-efficiency standards.

Since 1974, the town has reduced its natural gas consumption by 45%, no mean feat in a locale with frigid winter temperatures. In addition, the utility company saved enough money to accumulate a cash surplus and cut inflation-adjusted electricity rates by a third (which attracted two new factories to the area).

Each household saves more than $1,000 per year; this money supports jobs, and most of it circulates in the local economy. Before the town's energy-efficiency revolution, about $1.2 million a year left town to buy energy. What are your local utility companies and community doing to improve energy efficiency and stimulate the local economy?

Figure 3-1 An infrared photo showing heat loss (red, white, and yellow colors) around the windows, doors, roofs, and foundations of houses and stores in Plymouth, Michigan. Wes Birdsall provided similar thermograms of houses in Osage, Iowa. The average house in the United States has heat leaks and air infiltration equivalent to leaving a window wide open during the heating season. Because of poor design, most office buildings and houses in this country waste about half the energy used to heat and cool them. Americans pay about $300 billion a year for this wasted heat—about equal to the entire annual military budget. (VANSCAN® Continuous Mobile Thermogram by Daedalus Enterprises, Inc.)

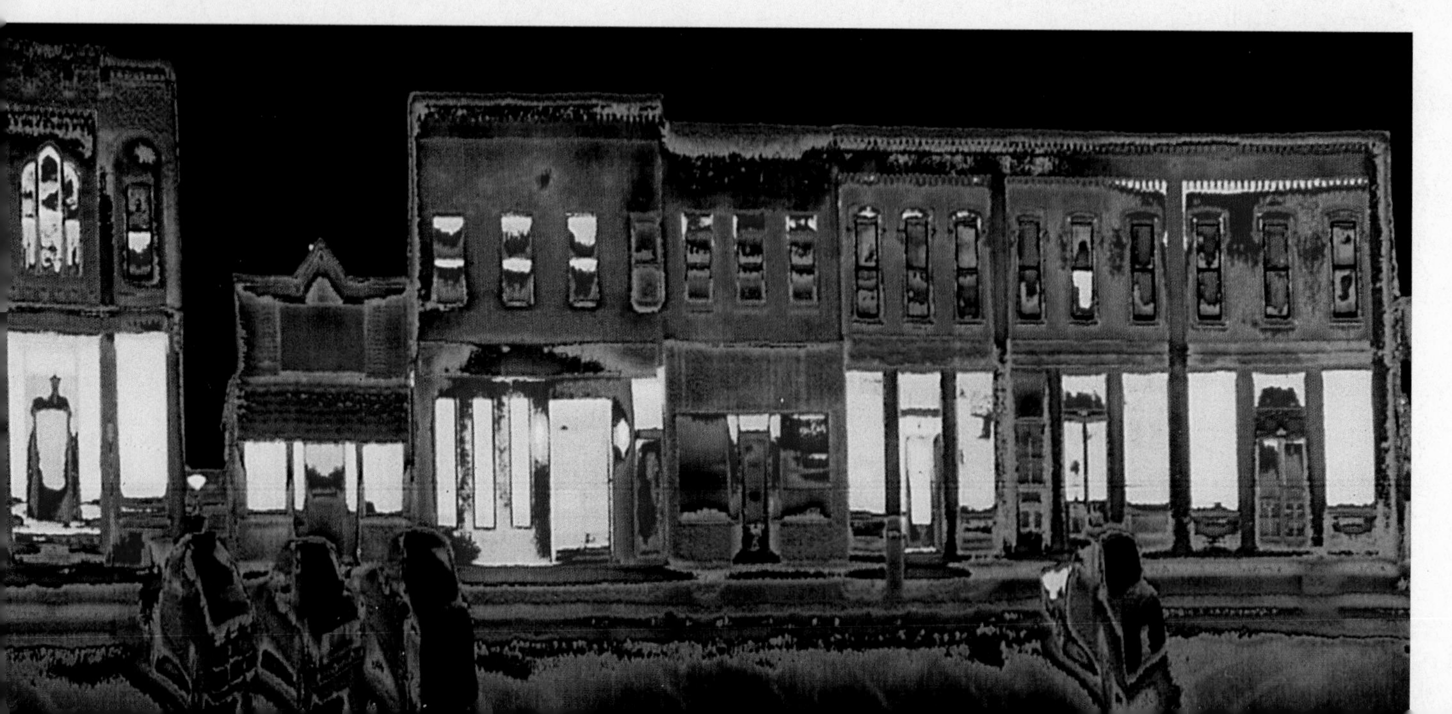

Science is an adventure of the human spirit. It is essentially an artistic enterprise, stimulated largely by curiosity, served largely by disciplined imagination, and based largely on faith in the reasonableness, order, and beauty of the universe.

WARREN WEAVER

This chapter addresses the following questions:

- What is science, and what do scientists do? What is critical thinking? What are some limitations of environmental science?
- What are major components and behaviors of complex systems?
- What are the basic forms of matter? What is matter made of? What makes matter useful to us as a resource?
- What are the major forms of energy? What makes energy useful to us as a resource?
- What are physical and chemical changes? What scientific law governs changes of matter from one physical or chemical form to another?
- What three main types of nuclear changes can matter undergo?
- How can exposure to radioactivity affect human health?
- What two scientific laws govern changes of energy from one form to another?
- How are the scientific laws governing changes of matter and energy from one form to another related to resource use and environmental disruption?

3-1 SCIENCE, ENVIRONMENTAL SCIENCE, AND CRITICAL THINKING

What Is Science and What Do Scientists Do?

Science is a pursuit of knowledge about how the world works. It is an attempt to discover order in nature and use that knowledge to make predictions about what is likely to happen in nature. Figure 3-2 and the Guest Essay on p. 58 summarize the systematic version of the critical thinking process used by scientists.

The first thing scientists must do is ask a question or identify a problem to be investigated. Then scientists working on this problem collect **scientific data**, or facts, by making observations and measurements. Repeated observations and measurements must confirm the resulting scientific data or facts, ideally by several different investigators.

The primary goal of science is not facts themselves but a new idea, principle, or model that **(1)** connects and explains certain facts and **(2)** leads to useful predictions about what is likely to happen in nature. Scientists working on a particular problem try to come up with a variety of possible or tentative explanations, or **scientific hypotheses**, of what they (or other scientists) observe in nature.

To be accepted, a scientific hypothesis must **(1)** explain scientific data and phenomena and **(2)** make predictions that can be tested by further experiments. One method scientists use to test a hypothesis is to develop a **model**, an approximate representation or simulation of a system being studied, as discussed on p. 57.

If many experiments by different scientists support a particular hypothesis, it becomes a **scientific theory.** It is an idea, principle, or model that **(1)** usually ties together and explains many facts that previously appeared to be unrelated and **(2)** is supported by a great deal of evidence. To scientists, theories are not to be taken lightly. They are ideas or principles with a high degree of certainty because they are supported by extensive evidence.

Nonscientists often use the word *theory* incorrectly when they mean to refer to a *scientific hypothesis,* a tentative explanation that needs further evaluation. The statement, "Oh, that's just a theory," made in everyday conversation, implies a lack of knowledge and careful testing—the opposite of the scientific meaning of the word.

Another important result of science is a **scientific** or **natural law**: a description of what we find happening in nature over and over in the same way, without

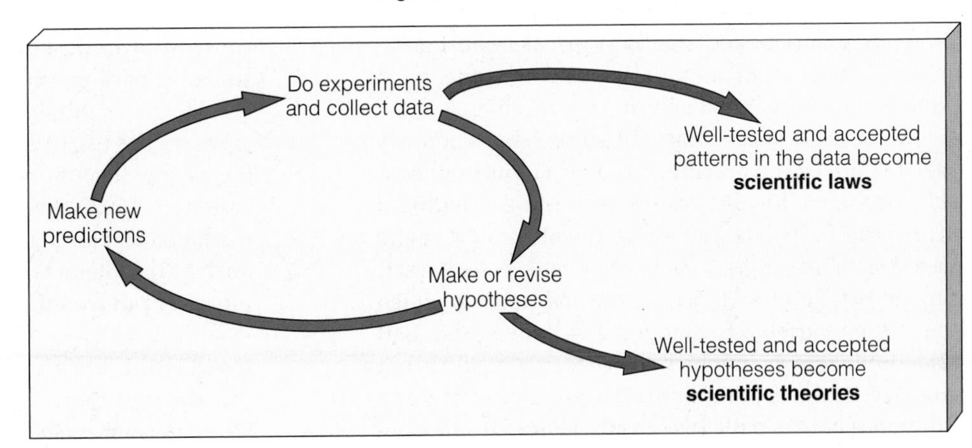

Figure 3-2 What scientists do.

known exception. For example, after making thousands of observations and measurements over many decades, scientists discovered what is called the *second law of energy* or *thermodynamics*. One simple way of stating this law is that heat always flows spontaneously from hot to cold—something you learned the first time you touched a hot object. *Scientific laws* describe what we find happening in nature in the same way, whereas *scientific theories* are widely accepted explanations of data and laws.

How Do Scientists Learn About Nature? We often hear about *the* scientific method. In reality, there are many **scientific methods**: ways scientists gather data and formulate and test scientific hypotheses, models, theories, and laws (Figure 3-2). Instead of being a recipe, each scientific method involves trying to answer a set of questions:

- What is the question about nature to be answered?
- What relevant facts and data are already known?
- What new data (observations and measurements) should be collected, and how should this be done?
- After the data are collected, can they be used to formulate a scientific law or model?
- How can a hypothesis be invented that explains the data and predicts new facts? Is this the simplest and only reasonable hypothesis?
- What new experiments can be done to test the hypothesis (and modify it if necessary) so it can become a scientific theory?

New discoveries happen in many ways. Some follow this sequence: data → law → hypothesis → theory. At other times, scientists simply follow a hunch, bias, or belief and then do experiments to test the idea or hypothesis. Some discoveries occur when an experiment gives totally unexpected results and the scientist investigates to find out what happened.

According to physicist Albert Einstein, "There is no completely logical way to a new scientific idea." Intuition, imagination, and creativity are as important in science as they are in poetry, art, music, and other great adventures of the human spirit.

Most processes or parts of nature that scientists seek to understand are influenced by a number of *variables* or *factors*. Ideally, scientists conduct a *controlled experiment* to isolate and study the effect of a single variable. This *single-variable analysis* is done by setting up two groups: **(1)** an *experimental group*, in which the chosen variable is changed in a known way, and **(2)** a *control group*, in which the chosen variable is not changed. If the experiment is designed properly, any difference between the two groups should result from a variable that was changed in the experimental group (Connections, above).

CONNECTIONS

What's Harming the Robins?

Suppose a scientist observes an abnormality in the growth of robin embryos in a certain area. She knows that the area has been sprayed with a pesticide and suspects that the chemical may be causing the abnormalities she has observed.

To test this hypothesis, the scientist carries out a *controlled experiment*. She maintains two groups of robin embryos of the same age in the laboratory. Each group is exposed to exactly the same conditions of light, temperature, food supply, and so on, except that the embryos in the experimental group are exposed to a known amount of the pesticide in question.

The embryos in both groups are then examined over an identical period of time for the abnormality. If there is a significantly larger number of the abnormalities in the experimental group than in the control group, then the results support the hypothesis that the pesticide is the culprit.

To be sure there were no errors in the procedure, the experiment should be repeated several times by the original researcher, and ideally by one or more other scientists.

Critical Thinking

Can you find flaws in this experiment that might lead you to question the scientist's conclusions? (*Hint*: What other factors in nature—not the laboratory—and in the embryos themselves could possibly explain the results?)

A basic problem is that many of the questions investigated by environmental scientists involve a huge number of interacting variables. In such cases, it is often difficult or impossible to carry out meaningful controlled experiments.

However, this limitation is being overcome in some cases by the use of *multivariable analysis*, made possible by the development of computing and communication technologies. This way to do science uses mathematical models run on high-speed computers to analyze the interactions of many variables without having carry out traditional controlled experiments.

How Valid Are the Results of Science? Scientists can do two major things: **(1)** disprove things and **(2)** establish that a particular model, theory, or law has a very high degree of probability or certainty of being true. However, like scholars in any field, scientists cannot prove that their theories, models, and laws are *absolutely* true.

When people say that something has or has not been "scientifically proven," they can mislead us by falsely implying that science yields absolute proof or certainty. Although it may be extremely low, there is always some degree of uncertainty involved in any scientific theory, model, or law.

In addition, scientists (like other people) have personal biases and beliefs that can consciously or unconsciously affect their objectivity. The goal of the rigorous scientific process is to reduce the degree of uncertainty and lack of objectivity as much as possible. Because of the high standards of evidence required, science is the best way we have to get reliable knowledge about how nature works.

How Does Frontier Science Differ from Consensus Science? News reports often focus on **(1)** new scientific "breakthroughs" and **(2)** disputes between scientists over the validity of preliminary (untested) data, hypotheses, and models (which are by definition tentative). These preliminary results, called **frontier science**, are controversial because they have not been widely tested and accepted. At this preliminary frontier stage, it is normal and healthy for reputable scientists in a field to state contradictory opinions about **(1)** meaning and accuracy of scientific data and **(2)** the validity of various hypotheses.

By contrast, **consensus science** consists of data, theories, and laws that are widely accepted by scientists considered experts in the field involved. This aspect of science is very reliable but is rarely considered newsworthy. One way to find out what scientists generally agree on is to seek out reports by scientific bodies such as the U.S. National Academy of Sciences and the British Royal Society that attempt to summarize consensus among experts in key areas of science.

The history of science shows that occasionally the scientific consensus about an idea can be modified or overturned by new information or better ideas. However, until such an event occurs, current scientific consensus is our most useful guideline.

What Are Some Limitations of Environmental Science? There is controversy over some of the knowledge provided by environmental science, for much of it falls into the realm of frontier science. At the preliminary frontier stage, it is normal and healthy for reputable scientists in a field to state contradictory opinions about the **(1)** meaning and accuracy of scientific data and **(2)** validity of various hypotheses.

One problem involves *arguments over the validity of data*. For example, there is no way to measure accurately how many **(1)** metric tons of soil are eroded worldwide, **(2)** hectares of tropical forest are cut, **(3)** species become extinct, or **(4)** metric tons of certain pollutants are emitted into the atmosphere or bodies of water each year. However, scientists can use sampling and statistical techniques and mathematical models to make estimates of such data.

The point environmental scientists want to make is that the trends in these phenomena are significant enough to be evaluated and addressed. Such environmental data should not be dismissed because they are "only estimates" (which are all we can ever have). However, this does not relieve investigators of the responsibility to get the best estimates possible and point out that they *are* estimates.

Another limitation is that *most environmental problems involve so many variables and such complex interactions that we often do not have enough information or sufficiently sophisticated mathematical models to aid in understanding them very well.* However, at some point we must use critical thinking skills (Guest Essay, p. 58) to

- Evaluate the available (but always limited) scientific information about a particular environmental problem.
- List the possible solutions (including doing nothing).
- Predict positive and negative consequences of each alternative solution and the probability that each consequence will occur.
- Evaluate the alternatives and choose the best solution.

3-2 MODELS AND BEHAVIOR OF SYSTEMS

What Is a System, and What Are Its Major Components? A **system** is a set of components that **(1)** function and interact in some regular and theoretically predictable manner and **(2)** can be isolated for the purposes of observation and study. The environment consists of a vast number of interacting systems involving living and nonliving things.

Most *systems* have the following key components:

- **Inputs** of things such as matter, energy, or information into the system.
- **Flows** or **throughputs** of matter, energy, or information within the system at certain rates.
- **Stores** or **storage areas** within a system where energy, matter, or information can accumulate for various lengths of time before being released. For example, your body stores various chemicals with different residence times, and water vapor typically

Critical Thinking and Environmental Studies

GUEST ESSAY

Jane Heinze-Fry

Jane Heinze-Fry has a Ph.D. in environmental education and teaches environmental science and biology at Emerson College in Boston, Massachusetts. She is author of Critical Thinking and Environmental Studies: A Beginner's Guide *(with G. Tyler Miller as coauthor). Previously, she taught and directed environmental studies at Sweet Briar College in Virginia. She also taught biology to students at the junior high, high school, and college levels. Her interdisciplinary orientation is reflected in her concept maps, including the one inside the back cover and those on the website for this book.*

Learning how to think critically is essential in helping you evaluate the validity and usefulness of what you **(1)** read in newspapers, magazines, and books (such as this textbook), **(2)** hear in lectures and speeches, and **(3)** see and hear on the news and in advertisements.

Learners engaged in critical thinking try to

- Connect new knowledge to prior knowledge and experience
- Evaluate the validity of claims made by people
- Relate what they have learned to their own life experiences
- Understand and evaluate their environmental worldviews
- Take and defend positions on issues
- Develop and implement strategies for dealing with problems

Whenever we are faced with new information, we need to evaluate it by using critical thinking. Do we believe the information or not, and why? Do the claims seem reasonable or exaggerated? Here are some rules for evaluating scientific evidence and claims:

1. Gather all the information you can.
2. Understand the definitions of all key terms and concepts.
3. Question how the information (data) was obtained.
   - Were the studies well designed and carried out?
   - Was there an experimental group and a control group? Were the control and experimental groups treated identically except for the variable changed in the experimental group?
   - Did the investigators repeat their experiments several times and get essentially the same results?
   - Did one or more other investigators verify the results?
4. Question the conclusions derived from the data.
   - Do the data support the claims, conclusions, and predictions?
   - Are there other, more reasonable interpretations?

remains in the lower atmosphere for about 10 days before it is replaced.

- **Outputs** of certain forms of matter, energy, or information that flow out of the system into *sinks* in the environment (such as the atmosphere, bodies of water, underground water, soil, and land surfaces).

Why Are Models of Complex Systems Useful? Over time, people have learned the value of using models as approximate representations or simulations of real systems to **(1)** find out how systems work and **(2)** evaluate which ideas or hypotheses work.

Some of the most powerful and useful technologies invented by humans are mathematical models, which are used to supplement our mental models. *Mathematical models* consist of one or more equations used to **(1)** describe the behavior of a system and **(2)** make predictions about the behavior of a system.

Making a mathematical model usually requires going many times through three steps: **(1)** Make a guess and write down some equations, **(2)** compute the predictions implied by the equations, and **(3)** compare the predictions with observations, the predictions of mental models, existing experimental data, and scientific hypotheses, laws, and theories.

Mathematical models are important because they can give us improved perceptions and predictions, especially in situations where our mental models are weak. Research has shown that mental models tend to be especially unreliable when **(1)** there are many interacting variables, **(2)** we attempt to extrapolate from too few experiences to a general case, **(3)** consequences follow actions only after long delays, **(4)** consequences of actions lead to other consequences, **(5)** responses vary from one time to the next, and **(6)** controlled experiments (Connections, p. 56) are impossible, too slow, or too expensive to conduct.

Most of our effects on the environment involve these characteristics that limit the usefulness of mental models. By using well-designed mathematical models of some aspect of the environment, we can analyze the interactions of multiple variables and gain the equivalent of hundreds or thousands of years of experience in a few weeks or months.

After building and testing a mathematical model, scientists use it to predict what is *likely* to happen under a

- Are the conclusions based on the results of original research by experts in the field involved, or are they drawn by reporters or scientists in other fields?
- Are the conclusions based on stories or reports of isolated events (*anecdotal information*) or on careful analysis of a large number of related observations?

5. Try to determine the assumptions and biases of the investigators and then question them.

- Do the investigators have a monetary or political advantage in the outcome of the investigation or issue involved?
- Would investigators with different basic assumptions or worldviews take the same data and come to different conclusions?

6. Are the data, claims, and conclusions based on the tentative results of *frontier science* or the more reliable and widely accepted results of *consensus science*?

7. Based on these steps, take a position by either rejecting or conditionally accepting the claims.

Other ways to improve your critical thinking skills involve using

- *Thinking strategies* such as constructing models, brainstorming, creating alternative solutions, and visualizing future possibilities
- *Attitude and value strategies* such as reflecting on the effects of your lifestyle on the environment and understanding and evaluating your environmental worldview
- *Action strategies* such as evaluating alternative solutions, creating plans of action, and developing strategies for implementing action plans

In the environmental course you are taking, there are many opportunities to develop your critical thinking skills. Your textbook offers Critical Thinking questions at the ends of chapters and in most boxes. If your course uses the supplement *Critical Thinking and Environmental Studies: A Beginner's Guide*, you will learn the critical thinking strategies mentioned here.

Critical Thinking

1. Can you come up with an example in which critical thinking has helped you make a major change in one or more of your beliefs or helped you make an important personal decision? Can you think of a decision that may have come out better if you had used critical thinking skills such as those discussed in this essay?

2. Rote learning often involves the "memorize and spit back" strategy. Meaningful learning (including critical thinking) goes far beyond memorization and requires us to evaluate the validity of what we learn. Currently, about what percentage of your learning involves rote learning and what percentage involves critical thinking as discussed in this essay?

variety of conditions. In effect, they use mathematical models to answer *if–then* questions: "*If* we do such and such, *then* what is likely to happen now and in the future?"

Despite its usefulness, a mathematical model is nothing more than a set of hypotheses or assumptions about how we think a certain system works. Such models (like all other models) are no better than **(1)** the assumptions built into them and **(2)** the data fed into them to make projections about the behavior of complex systems.

How Do Feedback Loops Affect Systems? In making and using mathematical and other models, it is important to know how systems operate and change. Systems undergo change as a result of feedback loops. A **feedback loop** occurs when one change leads to some other change, which either reinforces or slows the original change.

Feedback loops occur when an output of matter, energy, or information is fed back into the system as an input that changes the system. For example, recycling aluminum cans involves melting aluminum and feeding it back into an economic system to make new aluminum products. This feedback loop of matter reduces the **(1)** need to find, extract, and process virgin aluminum ore and **(2)** flow of waste matter (discarded aluminum cans) into the environment.

Feedback loops can be either positive or negative. In a **positive feedback loop** a change in a certain direction provides information that causes a system to change further in the same direction. Often such loops, called *vicious circles* or *runaway cycles*, destabilize a system.

An example involves projected global warming of the atmosphere. Such warming appears to play a role in shrinking the reflective white Arctic ice pack and exposing more open ocean. Because the open ocean absorbs much more heat than ice pack, as the ice pack diminishes the oceans will warm more rapidly. This can in turn melt more ice and lead to even faster warming of the ocean's surface in a positive feedback loop.

Another example involves depositing money in a bank at compound interest. In this case, the interest increases the balance, which through a positive feedback loop leads to more interest.

In a **negative feedback loop**, one change leads to a lessening of that change. For example, to survive, you must maintain your body temperature within a certain range, regardless of whether the temperature outside is steamy or freezing. This phenomenon is called **homeostasis**: the maintenance of favorable internal conditions despite fluctuations in external conditions. Normally negative feedback is desirable because it helps stabilize a system.

Most systems contain one or a series of *coupled positive and negative feedback loops*. A negative (or corrective) feedback loop coupled to a positive feedback loop can dampen or even halt a positive feedback loop of runaway growth.

For example, the temperature-regulating system of your body involves coupled negative and positive feedback loops (Figure 3-3). Normally a negative feedback loop regulates your body temperature. However, if your body temperature exceeds 42°C (108°F), your built-in negative feedback temperature control system breaks down as your body produces more heat than your sweat-dampened skin can get rid of. Then a positive feedback loop caused by overloading the system (Figure 3-3) overwhelms the negative or corrective feedback loop. These conditions produce a net gain in body heat, which produces even more body heat, and so on, until you die from heatstroke.

How Do Time Delays Affect Complex Systems? Complex systems often show **time delays** between the input of a stimulus and the response to it. A long time delay can mean that corrective action comes too late. For example, a smoker exposed to cancer-causing chemicals in cigarette smoke may not get lung cancer for 20–30 years.

Time delays allow a problem to build up slowly until it reaches a *threshold level* and causes a fundamental shift in the behavior of a system. Examples in which prolonged delays dampen the negative feedback mechanisms that might slow, prevent, or halt environmental problems are **(1)** population growth, **(2)** leaks from toxic waste dumps, **(3)** depletion of ozone in the stratosphere (the second layer of the earth's atmosphere), **(4)** global warming and climate change from carbon dioxide and other chemicals we add to the atmosphere, and **(5)** degradation of forests from prolonged exposure to air pollutants.

What Is Synergy, and How Can It Affect Complex Systems? In arithmetic, 1 plus 1 always equals 2. However, in some of the complex systems found in nature, 1 plus 1 may add up to more than 2 because of synergistic interactions. A **synergistic interaction** occurs when two or more processes interact so that the combined effect is greater than the sum of their separate effects.

Synergy can result when two people work together to accomplish a task. For example, suppose you and I need to move a 140-kilogram (300-pound) tree that has fallen across the road. By ourselves, each

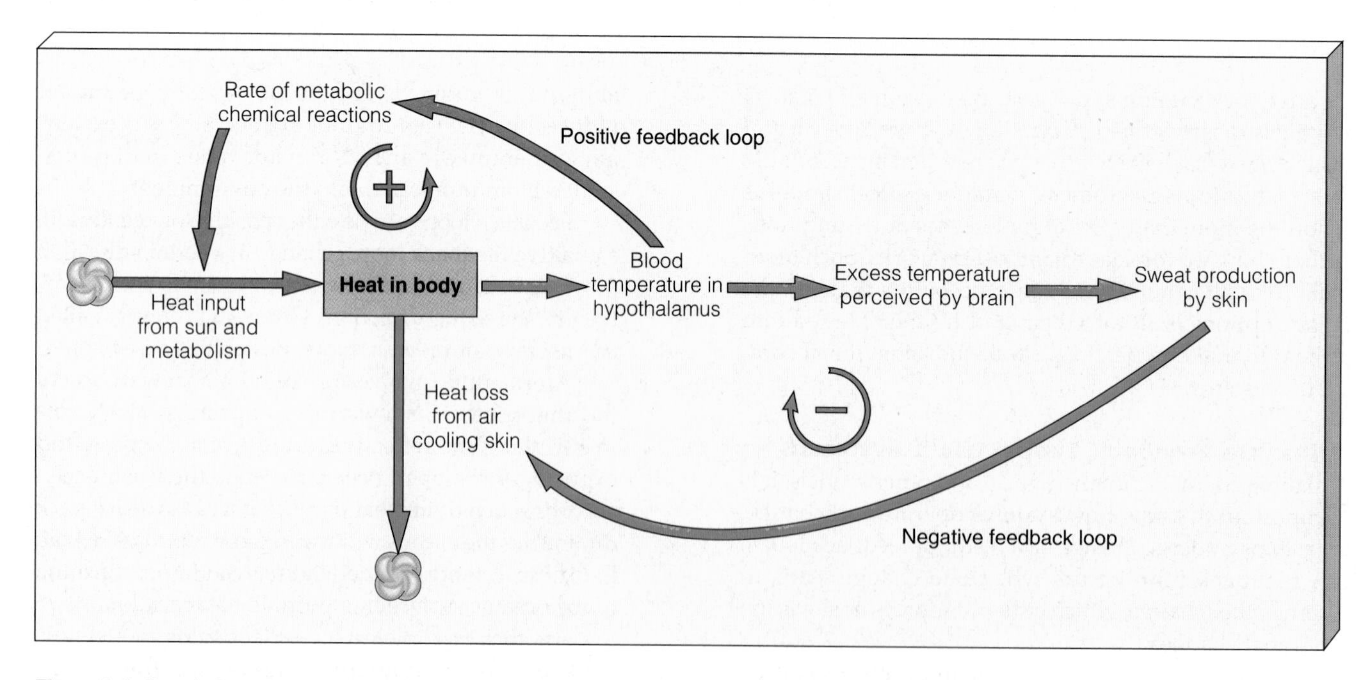

Figure 3-3 Coupled negative and positive feedback loops involved in temperature control of the human body. Homeostasis works in a limited range only: Above a certain body temperature, body metabolic rates get out of control and generate large amounts of heat. This positive (runaway) feedback loop generates more heat than the negative feedback loop can get rid of, and body temperature increases out of control, resulting in death.

of us can lift only, say, 45 kilograms (100 pounds). However, if we cooperate and use our muscles properly, together we can move the tree out of the way. That's using synergy to solve a problem. Research in the social sciences suggests that most political changes or changes in cultural beliefs are brought about by only about 5% (and rarely more than 10%) of a population working together (synergizing) and expanding their efforts to influence other people.

By identifying potentially harmful synergistic interactions and the leverage points that can activate them, we may be able to **(1)** counter harmful synergisms, **(2)** promote beneficial ones, and **(3)** improve the quality of life on the earth.

What Is the Law of Conservation of Problems? Biologist Eric Davidson has proposed a law of technodynamics that he calls the *law of conservation of problems.* According to this law, the technological solution of one problem usually creates one or more new unanticipated problems. The reason is that nature's systems are connected, so solving a problem by changing one part of a system can affect other parts or other connected systems in unpredictable and often undesirable ways.

For example, modern technology provides cheap chemical fertilizers that can increase crop productivity and replace some of the plant nutrients lost by erosion and poor farming practices. However, the widespread use of such fertilizers creates the new problem of pollution of streams, lakes, and underground water supplies with excess inputs of plant nutrients. In addition, the new water pollution problem is not a local problem confined to a farmer's field. Instead, it becomes a regional problem that affects people who live downstream or who extract and drink polluted groundwater.

3-3 MATTER: FORMS, STRUCTURE, AND QUALITY

What Are Nature's Building Blocks? **Matter** is anything that has mass (the amount of material in an object) and takes up space. Matter includes the solids, liquids, and gases around us and within us. Matter is found in two *chemical forms*:

- **Elements:** the distinctive building blocks of matter that make up every material substance
- **Compounds:** two or more different elements held together in fixed proportions by attractive forces called *chemical bonds*

Various elements, compounds, or both can be found together in **mixtures**.

All matter is built from the 115 known chemical elements (92 of them occur naturally and the other 23 have been synthesized in laboratories). To simplify things, chemists represent each element by a one- or two-letter symbol; hydrogen (H), carbon (C), oxygen (O), nitrogen (N), phosphorus (P), sulfur (S), chlorine (Cl), fluorine (F), bromine (Br), sodium (Na), calcium (Ca), lead (Pb), mercury (Hg), arsenic (As), and uranium (U) are but a few. These are the elements discussed in this book.

If you had a supermicroscope capable of looking at individual elements and compounds, you could see that they are made up of three types of building blocks:

- **Atoms:** the smallest unit of matter that are unique to a particular element
- **Ions:** electrically charged atoms or combinations of atoms
- **Molecules**: combinations of two or more atoms or ions of the same or different elements held together by chemical bonds

Some elements are found in nature as molecules. Examples are nitrogen and oxygen, which together make up about 99% of the volume of air you breathe. Two atoms of nitrogen (N) combine to form a gaseous molecule, with the shorthand formula N_2 (read as "N-two"). The subscript after the element's symbol indicates the number of atoms of that element in a molecule. Similarly, most of the oxygen gas in the atmosphere exists as O_2 (read as "O-two") molecules. A small amount of oxygen, found mostly in the second layer of the atmosphere (stratosphere), exists as O_3 (read as "O-three") molecules, a gaseous form of oxygen called *ozone.*

What Are Atoms Made Of? If you increased the magnification of your supermicroscope, you would find that each different type of atom contains a certain number of *subatomic particles*. The main building blocks of an atom are **(1)** positively charged **protons** (*p*), **(2)** uncharged **neutrons** (*n*), and **(3)** negatively charged **electrons** (*e*).

Each atom consists of **(1)** an extremely small center, or **nucleus**, containing protons and neutrons, and **(2)** one or more electrons in rapid motion somewhere outside the nucleus. Atoms are incredibly small. For example, more than 3 million hydrogen atoms could sit side by side on the period at the end of this sentence.

Each atom has an equal number of positively charged protons (inside its nucleus) and negatively charged electrons (outside its nucleus). Because these electrical charges cancel one another, *the atom as a whole has no net electrical charge.*

Each element has its own specific **atomic number**, equal to the number of protons in the nucleus of each of its atoms. The simplest element, hydrogen (H), has only 1 proton in its nucleus, so its atomic number is 1. Carbon (C), with 6 protons, has an atomic number of

6, whereas uranium (U), a much larger atom, has 92 protons and an atomic number of 92.

Because atoms are electrically neutral, the atomic number of an atom tells us the number of positively charged protons in its nucleus and the equal number of negatively charged electrons outside its nucleus. For example, an atom of uranium with an atomic number of 92 has 92 protons in its nucleus and 92 electrons outside, and thus no net electrical charge.

Because electrons have so little mass compared with the mass of a proton or a neutron, *most of an atom's mass is concentrated in its nucleus*. The mass of an atom is described in terms of its **mass number**: the total number of neutrons and protons in its nucleus. For example, **(1)** a hydrogen atom with 1 proton and no neutrons in its nucleus has a mass number of 1, and **(2)** an atom of uranium with 92 protons and 143 neutrons in its nucleus has a mass number of 235 (92 + 143 = 235).

All atoms of an element have the same number of protons in their nuclei. However, they may have different numbers of uncharged neutrons in their nuclei, and thus may have different mass numbers. Various forms of an element having the same atomic number but a different mass number are called **isotopes** of that element. Isotopes are identified by attaching their mass numbers to the name or symbol of the element. For example, hydrogen has three isotopes: hydrogen-1 (H-1), hydrogen-2 (H-2, common name *deuterium*), and hydrogen-3 (H-3, common name *tritium*). A natural sample of an element contains a mixture of its isotopes in a fixed proportion or percentage abundance by weight (Figure 3-4).

What Are Ions? Atoms of some elements can lose or gain one or more electrons to form **ions**: atoms or groups of atoms with one or more net positive (+) or negative (−) electrical charges. For example, an atom of sodium (Na, atomic number 11) with 11 positively charged protons and 11 negatively charged electrons can lose one of its electrons. It then becomes a sodium ion with a positive charge of 1 (Na^+) because it now has 11 positive charges (protons) but only 10 negative charges (electrons). An atom of chlorine (Cl, with an atomic number of 17) can gain an electron and become a chlorine ion with a negative charge of 1 (Cl^-) because it then has 17 positively charged protons and 18 negatively charged electrons.

The number of positive or negative charges on an ion is shown as a superscript after the symbol for an atom or a group of atoms. Examples of other positive ions are hydrogen ions (H^+), calcium ions (Ca^{2+}) and ammonium ions (NH_4^+); other common negative ions are nitrate ions (NO_3^-), sulfate ions (SO_4^{2-}), and phosphate ions (PO_4^{3-}). These are the ions discussed in this book.

The amount of a substance in a unit volume of air, water, or other medium is called its **concentration.** The concentration of *hydrogen ions* (H^+) in a water solution is a measure of its acidity or alkalinity. **pH** is a measure of the concentration of H^+ in a water solution. On a *pH scale* of 0 to 14, *acids* have a pH less than 7, *bases* have a pH greater than 7, and a *neutral solution* has a pH of 7 (Figure 3-5).

What Holds the Atoms and Ions in Compounds Together? Most matter exists as compounds. Chemists use a shorthand **chemical formula** to show the number of atoms (or ions) of each type in a compound. The formula **(1)** contains the symbols for each of the elements present and **(2)** uses subscripts to represent the number of atoms or ions of each element in

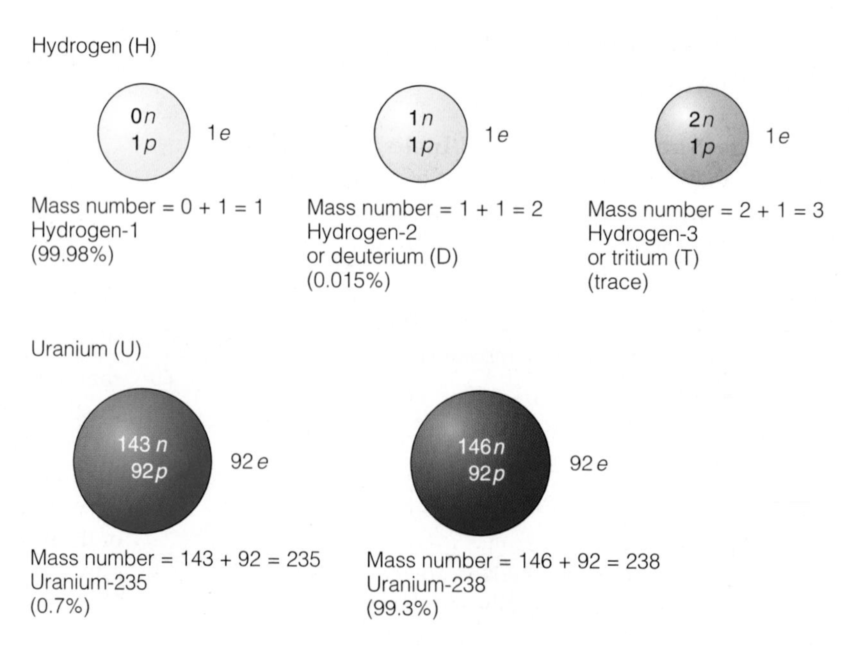

Figure 3-4 Isotopes of hydrogen and uranium. All isotopes of hydrogen have an atomic number of 1 because each has one proton in its nucleus; similarly, all uranium isotopes have an atomic number of 92. However, each isotope of these elements has a different mass number because its nucleus contains a different number of neutrons. Figures in parentheses indicate the percentage abundance by weight of each isotope in a natural sample of the element.

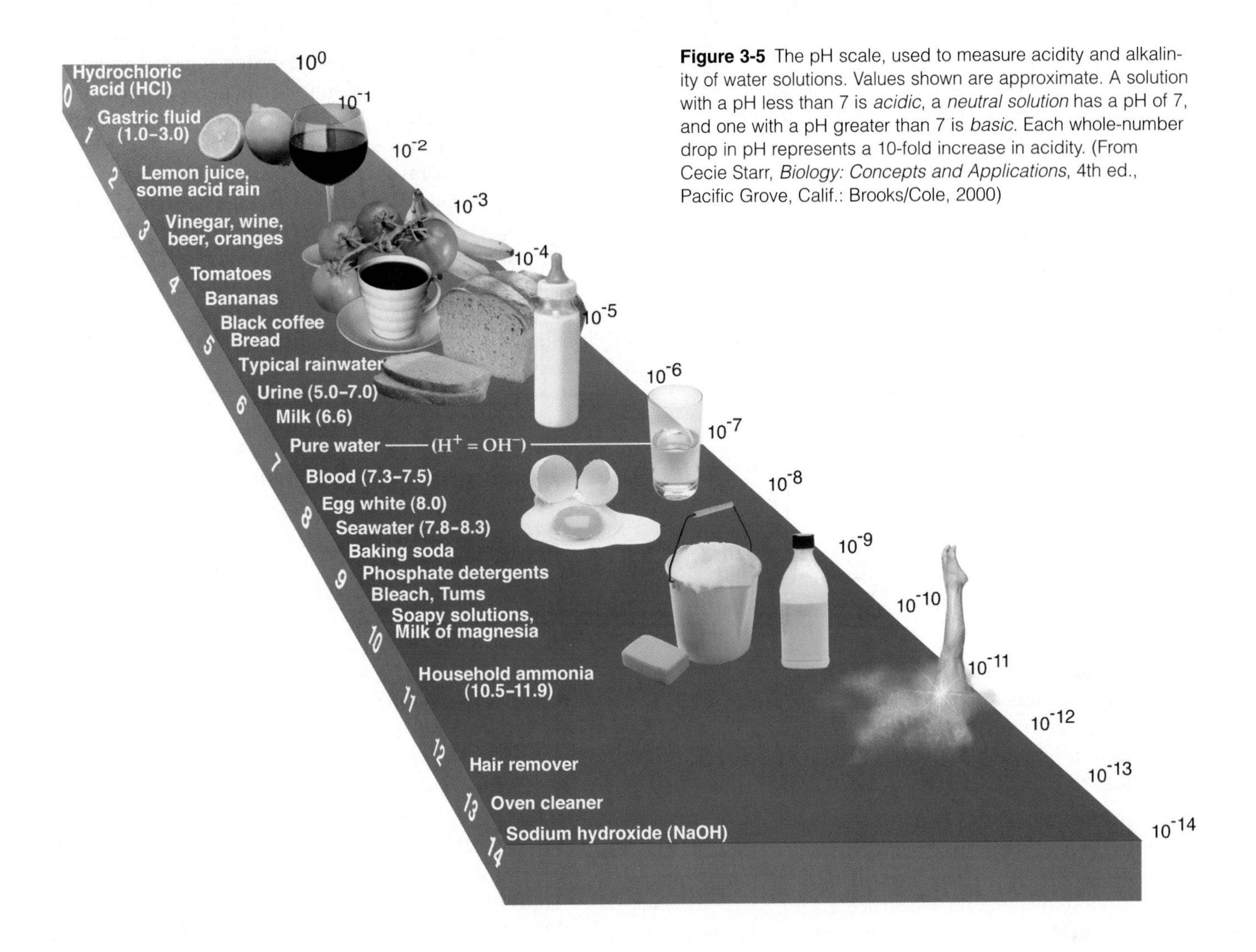

Figure 3-5 The pH scale, used to measure acidity and alkalinity of water solutions. Values shown are approximate. A solution with a pH less than 7 is *acidic*, a *neutral solution* has a pH of 7, and one with a pH greater than 7 is *basic*. Each whole-number drop in pH represents a 10-fold increase in acidity. (From Cecie Starr, *Biology: Concepts and Applications*, 4th ed., Pacific Grove, Calif.: Brooks/Cole, 2000)

the compound's basic structural unit. Compounds made up of oppositely charged ions are called *ionic compounds*. Those made up of molecules of uncharged atoms are called *covalent* or *molecular compounds*.

Sodium chloride (table salt), an *ionic compound*, is represented by NaCl. It consists of a network of oppositely charged *ions* (Na^+ and Cl^-) held together by the forces of attraction between opposite charges. The strong forces of attraction between such oppositely charged ions are called *ionic bonds*.

Water, a *covalent* or *molecular compound*, consists of molecules made up of uncharged atoms of hydrogen (H) and oxygen (O). Each water molecule consists of two hydrogen atoms chemically bonded to an oxygen atom, yielding H_2O (read as "H-two-O") molecules. The bonds between the atoms in such molecules are called *covalent bonds* and are formed when the atoms in the molecule share one or more pairs of their electrons.

What Are Organic and Inorganic Compounds?

Table sugar, vitamins, plastics, aspirin, penicillin, and many other important materials have one thing in common: They are **organic compounds**, containing carbon atoms combined with each other and with atoms of one or more other elements such as hydrogen, oxygen, nitrogen, sulfur, phosphorus, chlorine, and fluorine. Almost all organic compounds are molecular compounds held together by covalent bonds. Organic compounds can be either natural or synthetic (such as plastics and many drugs made by humans).

The millions of known organic (carbon-based) compounds include the following:

- *Hydrocarbons*: compounds of carbon and hydrogen atoms. An example is methane (CH_4), the main component of natural gas.
- *Chlorinated hydrocarbons*: compounds of carbon, hydrogen, and chlorine atoms. Examples are **(1)** the insecticide DDT ($C_{14}H_9Cl_5$, an insecticide) and **(2)** toxic polychlorinated biphenyls or PCBs (such as $C_{12}H_5Cl_5$), oily compounds used as insulating materials in electric transformers.
- *Chlorofluorocarbons* (CFCs): compounds of carbon, chlorine, and fluorine atoms. An example is Freon-12

(CCl_2F_2), until recently widely used as **(1)** a coolant in refrigerators and air conditioners, **(2)** an aerosol propellant, and **(3)** a foaming agent for making some plastics.

- *Simple carbohydrates* (simple sugars): certain types of compounds of carbon, hydrogen, and oxygen atoms. An example is glucose ($C_6H_{12}O_6$), which most plants and animals break down in their cells to obtain energy.

Larger and more complex organic compounds, called *polymers*, consist of a number of basic structural or molecular units (*monomers*) linked by chemical bonds, somewhat like cars linked in a freight train. The three major types of organic polymers are

- *Complex carbohydrates*, which are made by linking a number of simple carbohydrate molecules such as glucose ($C_6H_{12}O_6$). Examples are the complex starches in rice and potatoes and cellulose found in the walls around plant cells. Simple and complex carbohydrates are broken down in cells to supply energy.
- *Proteins*, which are produced in cells by the linking of different sequences of about 20 different monomers known as *alpha-amino acids*. From only about 20 alpha-amino acid molecules, the earth's life-forms can make tens of millions of different protein molecules.
- *Nucleic acids*, such as DNA and RNA, which are made by linking hundreds to thousands of five different types of monomers, called *nucleotides*. DNA molecules contain the hereditary instructions for assembling **(1)** new cells and **(2)** the proteins each cell needs to survive and reproduce.

Genes consist of specific sequences of nucleotides in a DNA molecule. Each gene carries codes (each consisting of three nucleotides) needed to make various proteins. These coded units of genetic information about specific traits are passed on from parents to offspring during reproduction.

Chromosomes are combinations of genes that make up a single DNA molecule, together with a number of proteins. Each chromosome typically contains thousands of genes. Genetic information coded in your chromosomal DNA is what makes you different from an oak leaf, an alligator, or a flea and from your parents. The relationships of genetic material to cells are depicted in Figure 3-6.

All other compounds are called **inorganic compounds.** Such compounds do not have carbon–carbon or carbon–hydrogen covalent bonds. Some of the inorganic compounds discussed in this book are sodium chloride ($NaCl$), water (H_2O), nitrous oxide (N_2O), nitric oxide (NO), carbon monoxide (CO), carbon dioxide (CO_2), nitrogen dioxide (NO_2), sulfur dioxide (SO_2), ammonia (NH_3), hydrogen sulfide (H_2S), sulfuric acid (H_2SO_4), and nitric acid (HNO_3). These are the major compounds discussed in this book.

What Are Matter Quality and Material Efficiency? **Matter quality** is a measure of how useful a form of matter is to us as a resource, based on its availability and concentration. **High-quality matter** **(1)** is concentrated, **(2)** usually is found near the earth's surface, and **(3)** has great potential for use as a matter resource. **Low-quality matter** **(1)** is dilute, **(2)** often is deep underground or dispersed in the ocean or the atmosphere, and **(3)** usually has little potential for use as a matter resource (Figure 3-7).

An aluminum can is a more concentrated, higher-quality form of aluminum than aluminum ore containing the same amount of aluminum. That's why it takes

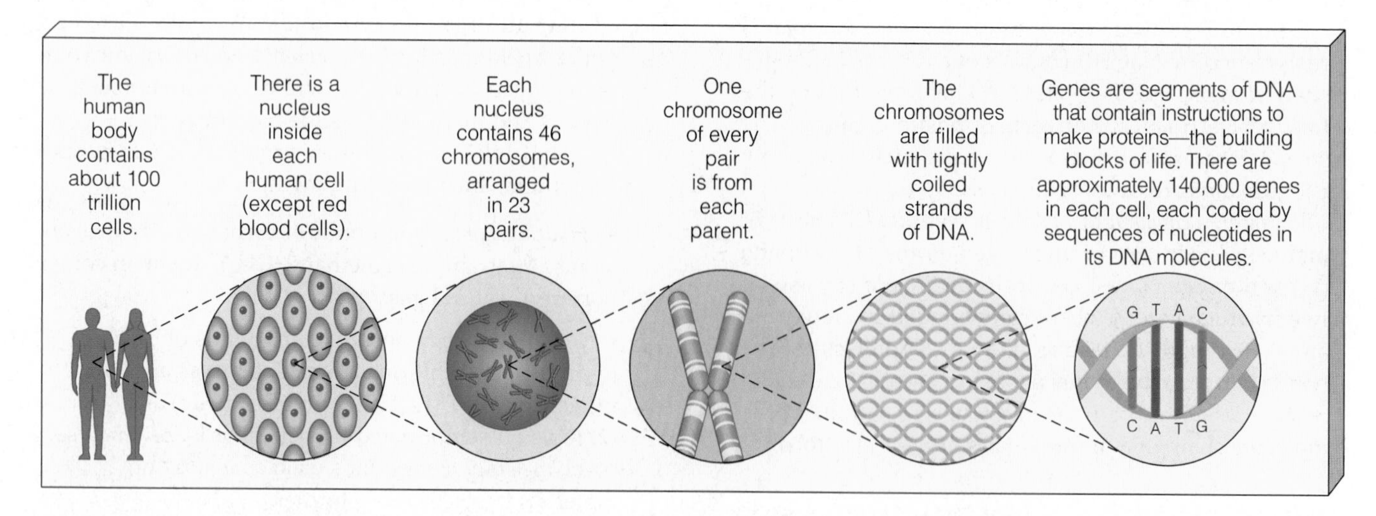

The human body contains about 100 trillion cells.
There is a nucleus inside each human cell (except red blood cells).
Each nucleus contains 46 chromosomes, arranged in 23 pairs.
One chromosome of every pair is from each parent.
The chromosomes are filled with tightly coiled strands of DNA.
Genes are segments of DNA that contain instructions to make proteins—the building blocks of life. There are approximately 140,000 genes in each cell, each coded by sequences of nucleotides in its DNA molecules.
G T A C
C A T G

Figure 3-6 Relationships between cells, nuclei, chromosomes, DNA, and genes.

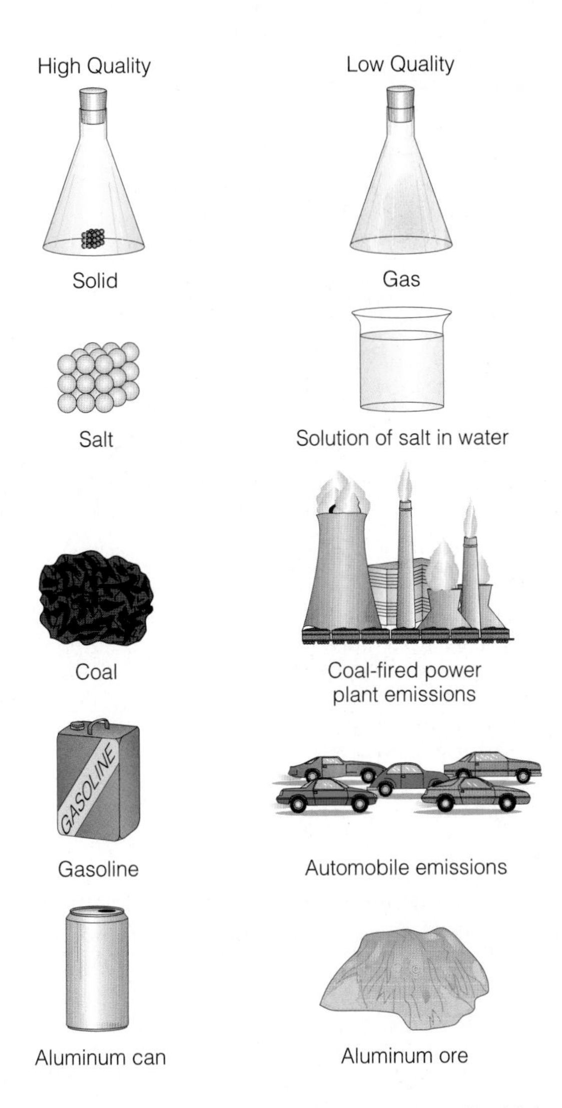

Figure 3-7 Examples of differences in matter quality. High-quality matter (left-hand column) is fairly easy to extract and is concentrated; low-quality matter (right-hand column) is more difficult to extract and is more dispersed than high-quality matter.

less energy, water, and money to recycle an aluminum can than to make a new can from aluminum ore.

Material efficiency or **resource productivity** is the total amount of material needed to produce each unit of goods or services. Although resource productivity has been improving, only about 2–6% of the matter resources flowing through the economies of developed countries ends up providing useful goods and services. Because of such waste, business expert Paul Hawken (Guest Essay, p. 6) and physicist Amory Lovins (Guest Essay, p. 517) contend that resource productivity in developed countries could be improved by 75–90% within two decades using existing technologies, as discussed in more detail in the Solutions box on p. 374.

3-4 ENERGY: FORMS AND QUALITY

What Are Different Forms of Energy? Energy is the capacity to do work and transfer heat. Work is performed when an object—be it a grain of sand, this book, or a giant boulder—is moved over some distance. Work, or matter movement, also is needed **(1)** to boil water or **(2)** to burn natural gas to heat a house or cook food. Energy is also the heat that flows automatically from a hot object to a cold object when they come in contact.

Scientists classify energy as either kinetic or potential. **Kinetic energy** is the energy that matter has because of its mass and its speed or velocity. It is energy in action or motion. Examples are **(1)** wind (a moving mass of air), **(2)** flowing streams, **(3)** heat flowing from a body at a high temperature to one at a lower temperature, **(4)** electricity (flowing electrons), **(5) electromagnetic radiation** (Figure 3-8), **(6) heat** (the total kinetic energy of all the moving atoms, ions, or molecules within a given substance, excluding the overall motion of the whole object), and **(7) temperature** (the average speed of motion of the atoms, ions, or molecules in a sample of matter at a given moment).

Potential energy is stored energy that is potentially available for use. Examples are **(1)** a rock held in your hand, **(2)** an unlit stick of dynamite, **(3)** still water behind a dam, **(4)** the chemical energy stored in gasoline molecules, and **(5)** the nuclear energy stored in the nuclei of atoms.

Potential energy can be changed to kinetic energy. When you drop a rock, its potential energy changes into kinetic energy. When you burn gasoline in a car engine, the potential energy stored in the chemical bonds of its molecules changes into heat, light, and mechanical (kinetic) energy that propels the car.

What Is Energy Quality? **Energy quality** is a measure of an energy source's ability to do useful work (Figure 3-9). **High-quality energy** is organized or concentrated and can perform much useful work. Examples are **(1)** electricity, **(2)** the chemical energy stored in coal and gasoline, **(3)** concentrated sunlight, and **(4)** nuclei of uranium-235 used as fuel in nuclear power plants.

By contrast, **low-quality energy** is disorganized or dispersed and has little ability to do useful work. An example is heat dispersed in the moving molecules of a large amount of matter (such as the atmosphere or a large body of water) so that its temperature is low. For example, the total amount of heat stored in the Atlantic Ocean is greater than the amount of high-quality chemical energy stored in all the oil deposits of Saudi

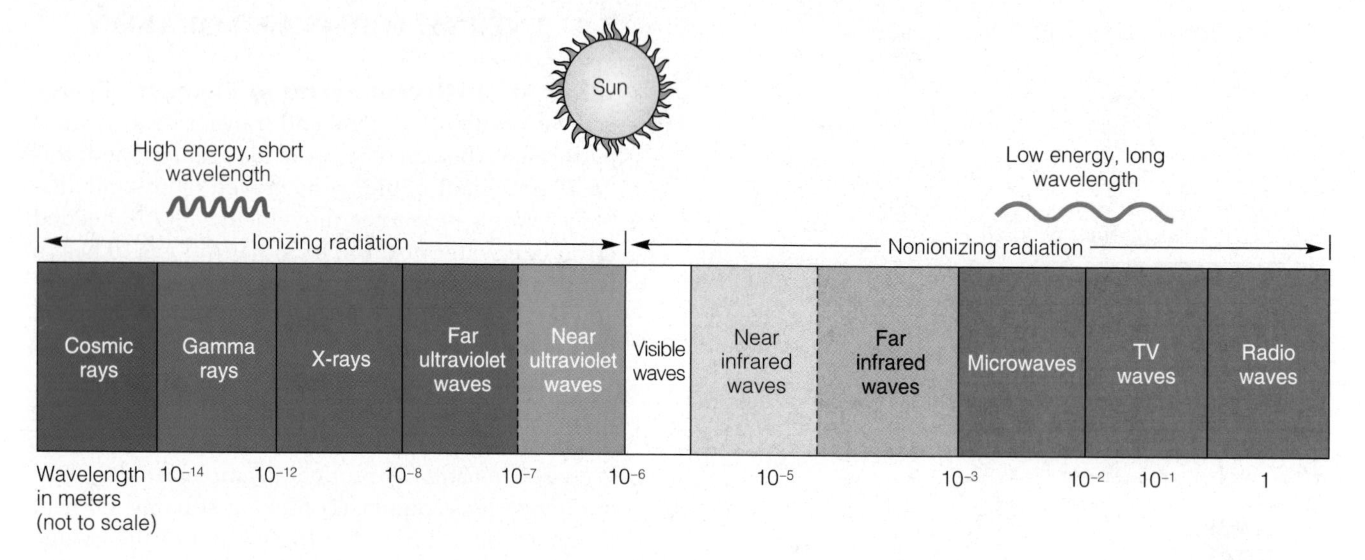

Figure 3-8 The *electromagnetic spectrum*: the range of electromagnetic waves, which differ in wavelength (distance between successive peaks or troughs) and energy content. Cosmic rays, gamma rays, X-rays, and ultraviolet radiation are called *ionizing radiation* because they have enough kinetic energy to knock electrons from atoms and change them to positively charged ions. The resulting highly reactive electrons and ions can **(1)** disrupt living cells, **(2)** interfere with body processes, and **(3)** cause many types of sickness, including various cancers. The other forms of electromagnetic radiation (right side) do not contain enough kinetic energy to form ions and are called *nonionizing radiation.*

Arabia. Yet the ocean's heat is so widely dispersed that it cannot be used to move things or to heat things to high temperatures. It makes sense to match the quality of an energy source with the quality of energy needed to perform a particular task (Figure 3-9) because doing so saves energy and usually money (unless government subsidies or taxes have distorted the energy marketplace).

3-5 PHYSICAL AND CHEMICAL CHANGES AND THE LAW OF CONSERVATION OF MATTER

What Is the Difference Between a Physical and a Chemical Change? A **physical change** involves no change in chemical composition. Cutting a piece of aluminum foil into small pieces is one example. Changing a substance from one physical state to another is a second example. When solid water (ice) is melted or liquid water is boiled, none of the H_2O molecules involved are altered; instead, the molecules are organized in different spatial (physical) patterns.

In a **chemical change** or **chemical reaction**, the chemical compositions of the elements or compounds are altered. Chemists use shorthand chemical equations to represent what happens in a chemical reaction. For example, when coal burns completely, the solid carbon (C) it contains combines with oxygen gas (O_2) from the atmosphere to form the gaseous compound carbon dioxide (CO_2):

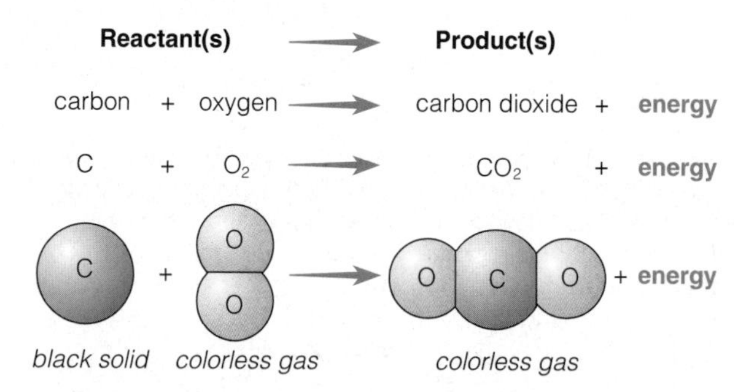

Energy is given off in this reaction, making coal a useful fuel. The reaction also shows how the complete burning of coal (or any of the carbon-containing compounds in wood, natural gas, oil, and gasoline) gives off carbon dioxide gas, which is a key gas that can lead to increased warming of the lower atmosphere (troposphere).

The Law of Conservation of Matter: Why Is There No "Away"? People commonly talk about consuming or using up material resources, but the truth is that *we do not consume matter*. Instead, we only use some of the earth's resources for a while. We take materials from the earth, carry them to another part of the globe, and process them into prod-

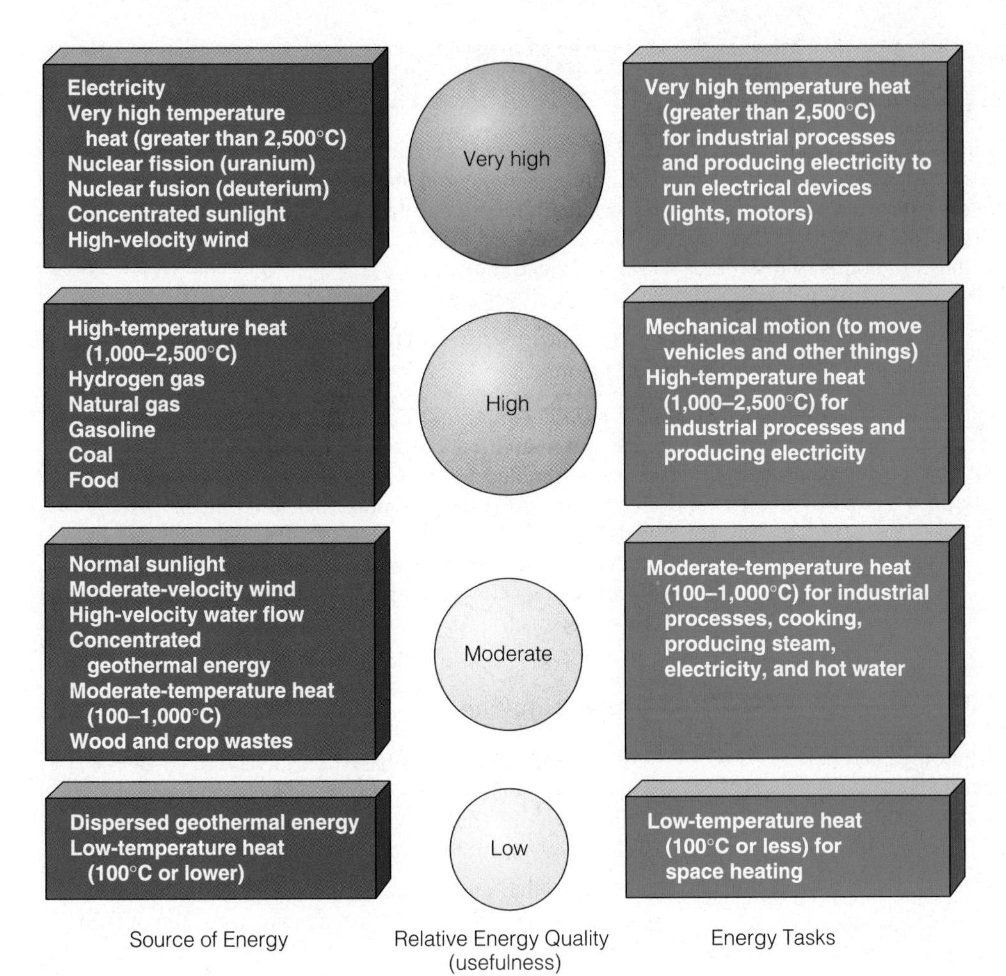

Figure 3-9 Categories of the quality of different sources of energy. *High-quality energy* is concentrated and has great ability to perform useful work. *Low-quality energy* is dispersed and has little ability to do useful work. To avoid unnecessary energy waste, it is best to match the quality of an energy source with the quality of energy needed to perform a task.

ucts that are used and then discarded, burned, buried, reused, or recycled.

We may change various elements and compounds from one physical or chemical form to another, but in no physical and chemical change can we create or destroy any of the atoms involved. All we can do is rearrange them into different spatial patterns (physical changes) or different combinations (chemical changes). The italicized statement, based on many thousands of measurements, is known as the **law of conservation of matter**. In describing chemical reactions, chemists use a shorthand system to make sure atoms are neither created nor destroyed, as required by the law of conservation of matter (Spotlight, p. 68).

The law of conservation of matter means that there is no "away" in "to throw away." *Everything we think we have thrown away is still here with us in one form or another.* We can collect dust and soot from the smokestacks of industrial plants, but these solid wastes must then be put somewhere. We can remove substances from polluted water at a sewage treatment plant, but the gooey sludge must be **(1)** burned (producing some air pollution), **(2)** buried (possibly contaminating underground water supplies), or **(3)** cleaned up and applied to the land as fertilizer (dangerous if the sludge contains nondegradable toxic metals such as lead and mercury). Banning use of the pesticide DDT in the United States but still selling it abroad means that it can return to the United States as **(1)** DDT residues in imported coffee, fruit, and other foods, or **(2)** fallout from air masses moved long distances by winds.

How Harmful Are Pollutants? We can make the environment cleaner and convert some potentially harmful chemicals into less harmful physical or chemical forms. However, *the law of conservation of matter means that we will always face the problem of what to do with some quantity of wastes and pollutants.* By placing much greater emphasis on pollution prevention, waste reduction, and more efficient resource use, we can greatly reduce the amount of wastes and pollution we add to the environment.

Thus, regardless of what we do we will always have certain pollutants that can cause harm to humans or other forms of life. Three factors that determine how severe the harmful effects of a pollutant are its

- *Chemical nature.*
- *Concentration*, which is sometimes expressed in *parts per million (ppm)*, with 1 ppm corresponding to 1 part pollutant per 1 million parts of the gas, liquid, or solid mixture in which the pollutant is found. Smaller concentration units are parts per billion (ppb) and parts per trillion (ppt). The concentration of a

SPOTLIGHT

Keeping Track of Atoms

In keeping with the law of conservation of matter, each side of a chemical equation must have the same number of atoms of each element involved. When this is the case, the equation is said to be *balanced*. The equation for the burning of carbon ($C + O_2 \rightarrow CO_2$) is balanced because there is one atom of carbon and two atoms of oxygen on both sides of the equation.

Consider the following chemical reaction: When electricity is passed through water (H_2O), the latter can be broken down into hydrogen (H_2) and oxygen (O_2), as represented by the following equation:

H_2O	$\rightarrow$	H_2	+	O_2
2 H atoms 1 O atom		2 H atoms		2 O atoms

This equation is unbalanced because there is one atom of oxygen on the left but two atoms on the right.

We cannot change the subscripts of any of the formulas to balance this equation because then we would be changing the arrangements of the atoms involved. Instead, we could use different numbers of the *molecules* involved to balance the equation. For example, we could use two water molecules:

$2\,H_2O$	$\rightarrow$	H_2	+	O_2
4 H atoms 2 O atom		2 H atoms		2 O atoms

This equation is still unbalanced because even though the numbers of oxygen atoms on both sides are now equal, the numbers of hydrogen atoms are not.

We can correct this by: having the reaction produce two hydrogen molecules.

$2\,H_2O$	$\rightarrow$	$2\,H_2$	+	O_2
4 H atoms 2 O atoms		4 H atoms		2 O atoms

Now the equation is balanced, and the law of conservation of matter has not been violated. We see that for every two molecules of water through which we pass electricity, two hydrogen molecules and one oxygen molecule are produced.

Try to balance the chemical equation for the reaction of nitrogen gas (N_2) with hydrogen gas (H_2) to form ammonia gas (NH_3).

Critical Thinking

1. Balancing equations is based on the law of conservation of matter. Do you believe that this is an ironclad law of nature or one that through new scientific discoveries could be overthrown? Explain.

2. Imagine that you have the power to revoke the law of conservation of matter. List three major ways this would affect your life.

pollutant can be reduced by dumping it into the air or a large volume of water, but there are limits to the effectiveness of this dilution approach.

- **Persistence**, or how long it stays in the air, water, soil, or body.

Pollutants can be classified into three categories based on their persistence as

- **Degradable**, or **nonpersistent, pollutants** that are broken down completely or reduced to acceptable levels by natural physical, chemical, and biological processes. Complex chemical pollutants broken down (metabolized) into simpler chemicals by living organisms (usually specialized bacteria) are called **biodegradable pollutants**. Human sewage in a river, for example, is biodegraded fairly quickly by bacteria if the sewage is not added faster than it can be broken down.
- **Slowly degradable**, or **persistent**, **pollutants** that take decades or longer to degrade. Examples include the insecticide DDT and most plastics.
- **Nondegradable pollutants** that cannot be broken down by natural processes. Examples include the toxic elements lead, mercury, and arsenic.

3-6 NUCLEAR CHANGES

What Is Natural Radioactivity? In addition to physical and chemical changes, matter can undergo a third type of change known as a **nuclear change**. This occurs when nuclei of certain isotopes spontaneously change or are made to change into one or more different isotopes. Three types of nuclear change are **(1)** natural radioactive decay, **(2)** nuclear fission, and **(3)** nuclear fusion.

Natural radioactive decay is a nuclear change in which unstable isotopes spontaneously emit fast-moving chunks of matter (called particles), high-energy radiation, or both at a fixed rate. The unstable isotopes are called **radioactive isotopes** or **radioisotopes**. Radioactive decay into various isotopes continues until the original isotope is changed into a stable isotope that is not radioactive.

Radiation emitted by radioisotopes is damaging ionizing radiation (Figure 3-8). The most common form of ionizing energy released from radioisotopes is **gamma rays**, a form of high-energy electromagnetic radiation. High-speed ionizing particles emitted from the nuclei of radioactive isotopes are most commonly of two types: **(1) alpha particles** (fast-moving, positively charged chunks of matter that consist of two protons and two neutrons) and **(2) beta particles** (high-speed electrons).

Each type of radioisotope spontaneously decays at a characteristic rate into a different isotope. This rate of decay can be expressed in terms of **half-life**: the time needed for *one-half* of the nuclei in a radioisotope to decay and emit their radiation to form a different isotope. The decay continues, often producing a series of different radioisotopes, until a nonradioactive isotope is formed. Each radioisotope has a characteristic half-life, which may range from a few millionths of a second to several billion years (Table 3-1).

An isotope's half-life cannot be changed by temperature, pressure, chemical reactions, or any other known factor. Half-life can be used to estimate how long a sample of a radioisotope must be stored in a safe container before it decays to what is considered a safe level. A general rule is that such decay takes about 10 half-lives. Thus, people must be protected from radioactive waste containing iodine-131 (which concentrates in the thyroid gland and has a half-life of 8 days) for 80 days (10 × 8 days). Plutonium-239 (which has a half-life of 24,000 years, is produced in nuclear reactors, and is used as the explosive in some nuclear weapons) can cause lung cancer when its particles are inhaled in minute amounts. Thus, it must be stored safely for 240,000 years (10 × 24,000 years)—about four times longer than the latest version of our species has existed.

How Much Ionizing Radiation Are We Exposed To? Each year people are exposed to some ionizing radiation (Figure 3-8) from natural or background sources and from human activities (Figure 3-10).

Table 3-1 Half-Lives of Selected Radioisotopes

Isotope	Radiation Half-Life	Emitted
Potassium-42	12.4 hours	Alpha, beta
Iodine-131	8 days	Beta, gamma
Cobalt-60	5.27 years	Beta, gamma
Hydrogen-3 (tritium)	12.5 years	Beta
Strontium-90	28 years	Beta
Carbon-14	5,370 years	Beta
Plutonium-239	24,000 years	Alpha, gamma
Uranium-235	710 million years	Alpha, gamma
Uranium-238	4.5 billion years	Alpha, gamma

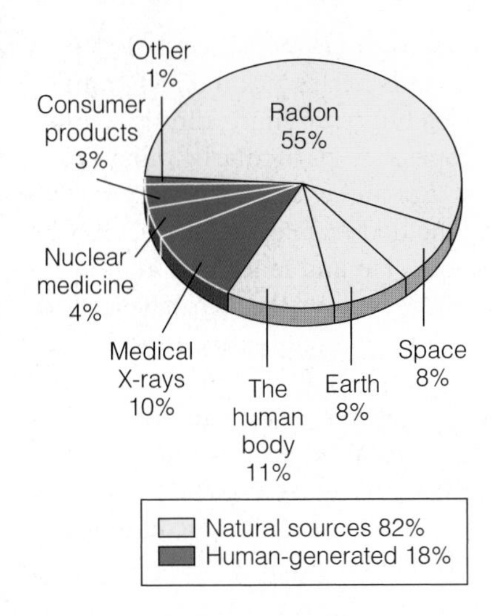

Figure 3-10 Natural and human sources of the average annual dosage of ionizing radiation received by people in the United States. Most studies indicate that there is no safe dosage of ionizing radiation. (Data from National Council on Radiation Protection and Measurements)

Sources of natural ionizing radiation include cosmic rays from outer space, soil, rocks, air, water, and food.

Nuclear power plants provide very low exposure if they are operating properly. However, a serious nuclear accident, such as the one that occurred in 1986 at the Chernobyl nuclear power plant in Ukraine (p. 490), can release large quantities of radioactive materials that can kill and harm people and make large areas uninhabitable.

Most human-caused exposure to ionizing radiation comes from medical X-rays and from diagnostic tests and treatments using radioactive isotopes. The federal government estimates that one-third of the X-rays taken each year in the United States are unnecessary.

What Are the Effects of Ionizing Radiation? Ionizing radiation can cause harm by **(1)** penetrating a human cell, **(2)** knocking loose (ionizing) one or more electrons from a cellular chemical, and **(3)** altering molecules needed for normal cellular functioning. Exposure to ionizing radiation can damage cells in two ways:

- *Genetic damage* from mutations in DNA molecules that alter genes and chromosomes (Figure 3-6). If the mutation is harmful, genetic defects can become apparent in the next generation of offspring or several generations later.
- *Somatic damage* to tissues, which causes harm during the victim's lifetime. Examples include burns, miscarriages, eye cataracts, and certain cancers.

The effects of ionizing radiation vary with the **(1)** type and penetrating power, **(2)** source (outside or inside the body), and **(3)** half-life of the radioisotope (Table 3-1). Alpha particles lack the penetrating power of beta particles but have more energy. Thus, alpha-emitting isotopes are particularly dangerous when breathed in or ingested with food or water. Alpha particles outside the body can cause skin cancer but cannot penetrate the skin and reach vital organs. Because beta particles can penetrate the skin, a beta emitter outside the body can damage internal organs.

According to the U.S. National Academy of Sciences, exposure over an average lifetime to average levels of ionizing radiation from natural and human sources (Figure 3-10) causes about 1% of all fatal cancers and 5–6% of all normally encountered genetic defects in the U.S. population.

Is Nonionizing Electromagnetic Radiation Harmful? The simple answer is that we don't know. *Electromagnetic fields (EMFs)* are low-energy, nonionizing forms of electromagnetic radiation (Figure 3-8) given off when an electric current passes through a wire or a motor.

Sources of these weak electrical and magnetic fields include **(1)** overhead power lines and **(2)** household electrical appliances (such as microwave ovens, hair dryers, electric blankets, waterbed heaters, electric razors, computer monitors, and TV sets).

Since the late 1960s there has been growing public concern and controversy over the possibility that EMFs could have harmful health effects on humans. Numerous studies have suggested that prolonged exposure to EMFs could lead to increased risk from **(1)** some cancers (including childhood leukemia, brain tumors, and breast cancer), **(2)** miscarriages, **(3)** birth defects, and **(4)** Alzheimer's disease. However, more recent studies show an insignificant correlation between EMFs and such health effects. It may be decades before these contradictory results and evaluations are resolved.

What Is Nuclear Fission? Splitting Nuclei **Nuclear fission** is a nuclear change in which nuclei of certain isotopes with large mass numbers (such as uranium-235) are split apart into lighter nuclei when struck by neutrons; each fission releases two or three more neutrons and energy (Figure 3-11). Each of these neutrons, in turn, can cause an additional fission. For these multiple fissions to take place, enough fissionable nuclei must be present to provide the **critical mass** needed for efficient capture of these neutrons.

Multiple fissions within a critical mass form a **chain reaction**, which releases an enormous amount of energy (Figure 3-12). Living cells can be damaged by the ionizing radiation released by the radioactive lighter nuclei and by high-speed neutrons produced by nuclear fission.

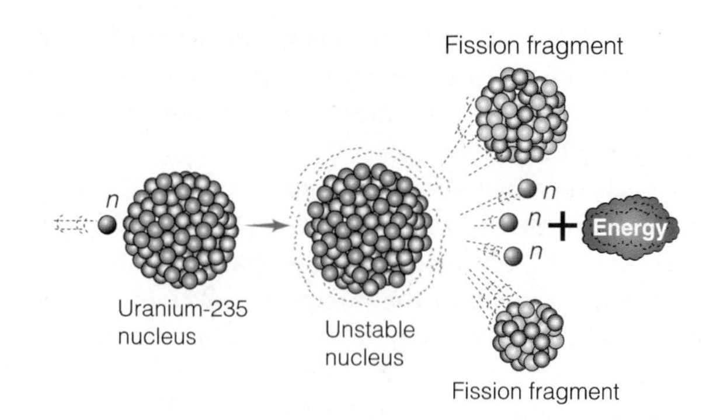

Figure 3-11 Fission of a uranium-235 nucleus by a neutron (*n*).

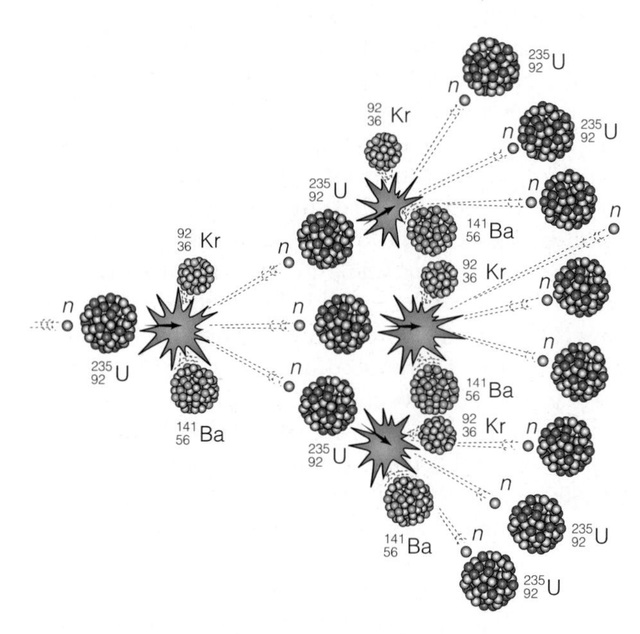

Figure 3-12 A nuclear chain reaction initiated by one neutron triggering fission in a single uranium-235 nucleus. This figure illustrates only a few of the trillions of fissions caused when a single uranium-235 nucleus is split within a critical mass of uranium-235 nuclei. The elements krypton (Kr) and barium (Ba), shown here as fission fragments, are only two of many possibilities.

In an atomic bomb, an enormous amount of energy is released in a fraction of a second in an uncontrolled nuclear fission chain reaction. This reaction is initiated by an explosive charge, which **(1)** suddenly pushes two masses of fissionable fuel together, and **(2)** causes the fuel to reach the critical mass needed for a chain reaction.

In the reactor of a nuclear power plant, the rate at which the nuclear fission chain reaction takes place is controlled so that under normal operation only one of every two or three neutrons released is used to split another nucleus. In conventional nuclear fission reactors, the splitting of uranium-235 nuclei releases heat, which produces high-pressure steam to spin turbines and thus generate electricity.

What Is Nuclear Fusion? Forcing Nuclei to Combine **Nuclear fusion** is a nuclear change in which two isotopes of light elements, such as hydrogen, are forced together at extremely high temperatures until they fuse to form a heavier nucleus, releasing energy in the process. Temperatures of at least 100 million °C are needed to force the positively charged nuclei (which strongly repel one another) to fuse.

Nuclear fusion is much more difficult to initiate than nuclear fission, but once started it releases far more energy per unit of fuel than does fission. Fusion of hydrogen nuclei to form helium nuclei is the source of energy in the sun and other stars.

After World War II, the principle of *uncontrolled nuclear fusion* was used to develop extremely powerful hydrogen, or thermonuclear, weapons. These weapons use the D–T fusion reaction, in which a hydrogen-2, or deuterium (D), nucleus and a hydrogen-3 (tritium, T) nucleus are fused to form a larger, helium-4 nucleus, a neutron, and energy (Figure 3-13).

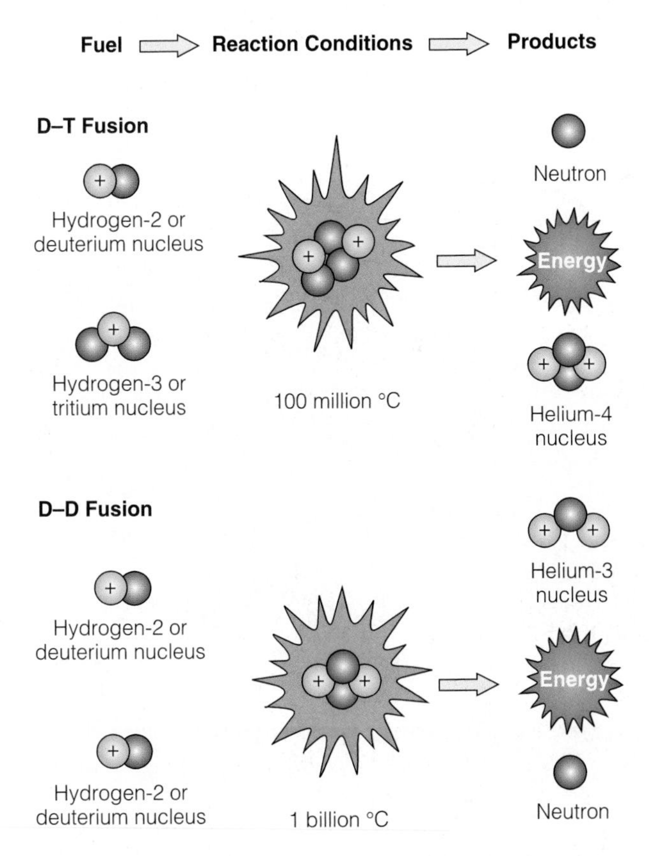

Figure 3-13 The deuterium–tritium (D–T) and deuterium–deuterium (D–D) nuclear fusion reactions, which take place at extremely high temperatures.

Scientists have also tried to develop *controlled nuclear fusion*, in which the D–T reaction is used to produce heat that can be converted into electricity. Despite more than 50 years of research, this process is still in the laboratory stage. Even if it becomes technologically and economically feasible, many energy experts don't expect it to be a practical source of energy until perhaps 2100, if then.

3-7 THE TWO IRONCLAD LAWS OF ENERGY

What Is the First Law of Energy? You Cannot Get Something for Nothing Scientists have observed energy being changed from one form to another in millions of physical and chemical changes, but they have never been able to detect the creation or destruction of any energy (except in nuclear changes). The results of their experiments have been summarized in the **law of conservation of energy**, also known as the **first law of energy** or the **first law of thermodynamics**: *In all physical and chemical changes, energy is neither created nor destroyed, but it may be converted from one form to another.*

This scientific law tells us that when one form of energy is converted to another form in any physical or chemical change, *energy input always equals energy output*. No matter how hard we try or how clever we are, we cannot get more energy out of a system than we put in; in other words, *we cannot get something for nothing in terms of energy quantity*.

What Is the Second Law of Energy? You Cannot Even Break Even Because the first law of energy states that energy can be neither created nor destroyed, it is tempting to think that there will always be enough energy. Yet if we fill a car's tank with gasoline and drive around or use a flashlight battery until it is dead, something has been lost. If it is not energy, what is it? The answer is *energy quality* (Figure 3-9), the amount of energy available that can perform useful work.

Countless experiments have shown that when energy is changed from one form to another, a decrease in energy quality always occurs. The results of these experiments have been summarized in what is called the **second law of energy** or the **second law of thermodynamics**: *When energy is changed from one form to another, some of the useful energy is always degraded to lower-quality, more dispersed, less useful* *energy*. This degraded energy usually takes the form of heat given off at a low temperature to the surroundings (environment). There it is dispersed by the random motion of air or water molecules and becomes even more disorderly and less useful. Another way to state the second law of energy is that *heat always flows spontaneously from hot (high-quality energy) to cold (lower-quality energy)*.

Basically, this law says that in any energy conversion, we always end up with *less* usable energy than we started with. So not only can we not get something for nothing in terms of energy quantity, *we cannot even break even in terms of energy quality because energy always goes from a more useful to a less useful form*. No one has ever found a violation of this fundamental scientific law (see quote at the end of this chapter).

Here are three examples of the second energy law in action:

- When a car is driven, only about 10% of the high-quality chemical energy available in its gasoline fuel is converted into mechanical energy (to propel the vehicle) and electrical energy (to run its electrical systems). The remaining 90% is degraded to low-quality heat that is released into the environment and eventually lost into space.
- When electrical energy flows through filament wires in an incandescent light bulb, it is changed into about 5% useful light and 95% low-quality heat that flows into the environment. In other words, this so-called *light bulb* is really a *heat bulb*.
- In living systems, solar energy is converted into chemical energy (food molecules) and then into mechanical energy (moving, thinking, and living). During each of these conversions high-quality energy is degraded and flows into the environment as low-quality heat (Figure 3-14).

The second law of energy also means that *we can never recycle or reuse high-quality energy to perform useful work*. Once the concentrated energy in a serving of food, a liter of gasoline, a lump of coal, or a chunk of uranium is released, it is degraded to low-quality heat that is dispersed into the environment. We can heat air or water at a low temperature and upgrade it to high-quality energy, but the second law of energy tells us that it will take more high-quality energy to do this than we get in return.

Energy efficiency or **energy productivity** is a measure of how much useful work is accomplished by a particular input of energy into a system. As with material efficiency (p. 65), there is plenty of room for

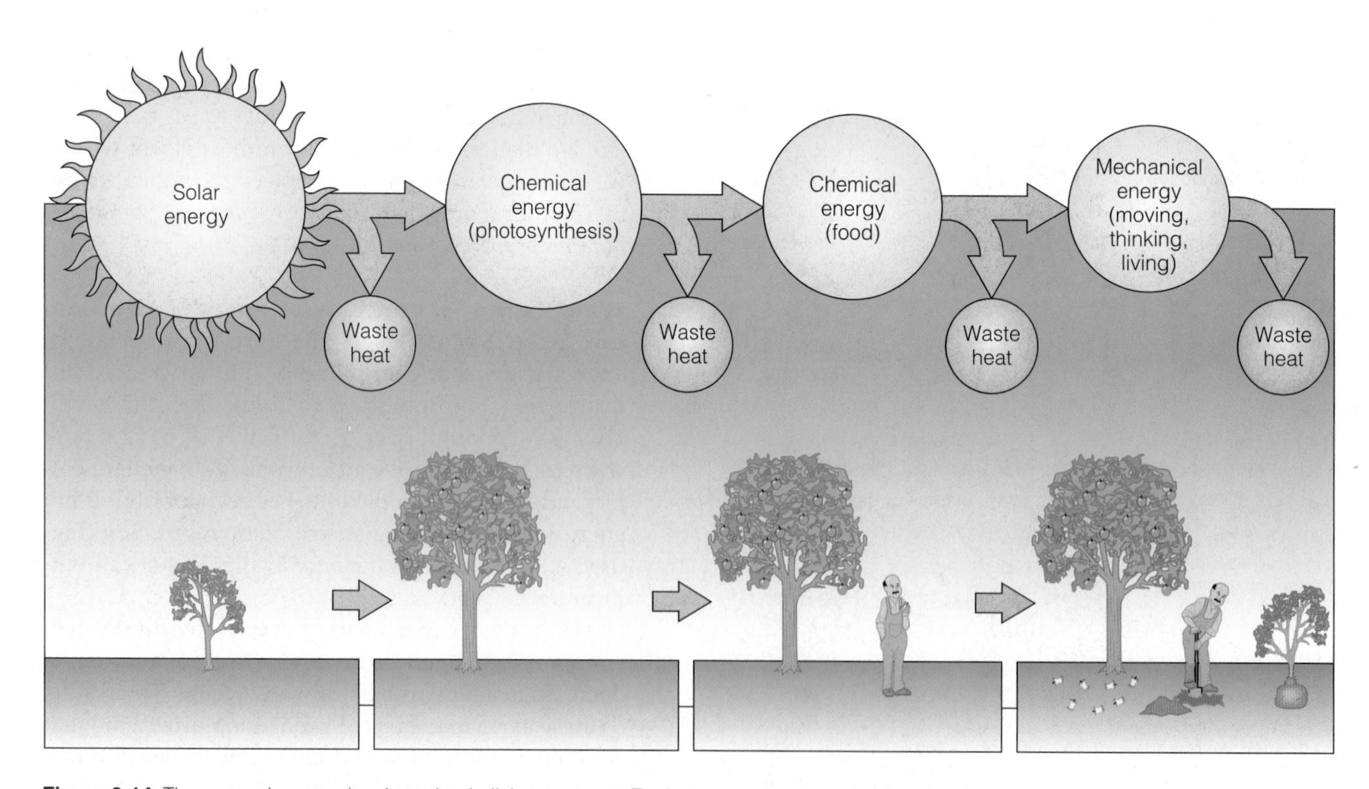

Figure 3-14 The second energy law in action in living systems. Each time energy is changed from one form to another, some of the initial input of high-quality energy is degraded, usually to low-quality heat that disperses into the environment.

improvement. For example, scientists estimate that only about 16% of the energy used in the United States ends up performing useful work. The remaining 84% is either unavoidably wasted because of the second law of energy (41%) or unnecessarily wasted (43%).

Connections: How Does the Second Energy Law Affect Life? To form and maintain the highly ordered arrangement of molecules and the organized biochemical processes in your body, you must continually get and use high-quality matter and energy resources from your surroundings. As you use these resources, you add low-quality heat and low-quality waste matter to your surroundings. Your body continuously gives off heat roughly equal to that of a 100-watt incandescent light bulb, which is why a closed room full of people gets warm. You also continuously break down solid, large molecules (such as glucose) into smaller molecules of carbon dioxide gas and water vapor, which are dispersed in the atmosphere.

Planting, growing, processing, and cooking food all use high-quality energy and matter resources that add low-quality heat and waste materials to the environment. In addition, enormous amounts of low-quality heat and waste matter are added to the environment when concentrated deposits of minerals and fuels are extracted from the earth's crust, processed, and used.

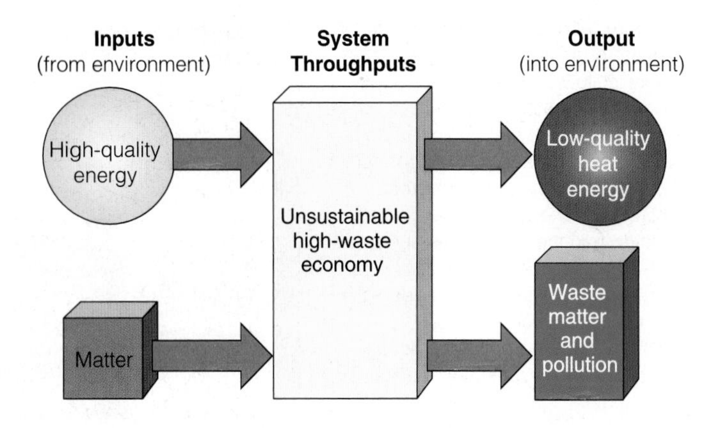

Figure 3-15 The high-throughput economy found in most developed countries is based on maximizing the rates of energy and matter flow. This process rapidly converts high-quality matter and energy resources into waste, pollution, and low-quality heat.

3-8 CONNECTIONS: MATTER AND ENERGY LAWS AND ENVIRONMENTAL PROBLEMS

What Is a High-Throughput Economy? As a result of the law of conservation of matter and the second law of energy, individual resource use automatically adds some waste heat and waste matter to the environment. Most of today's advanced industrialized countries have a **high-throughput (high-waste) economy** that attempts to sustain ever-increasing economic growth by increasing the flow of matter and energy resources through their economic systems (Figure 3-15). These resources flow through their economies into planetary *sinks* (air, water, soil, organisms), where pollutants and wastes end up and can accumulate to harmful levels.

What happens if more and more people continue to use and waste more and more energy and matter resources at an increasing rate? The scientific laws of matter and energy discussed in this chapter tell us that eventually this will exceed the capacity of the environment to **(1)** dilute and degrade waste matter and **(2)** absorb waste heat.

What Is a Matter-Recycling Economy? A stopgap solution to this problem is to convert a high-throughput economy to a **matter-recycling economy**. The goal of such a conversion is to allow economic growth to continue without depleting matter resources or producing excessive pollution and environmental degradation.

Even though recycling matter saves energy, the two laws of energy tell us that *recycling matter resources always* **(1)** *requires using high-quality energy (which cannot be recycled) and* **(2)** *adds waste heat to the environment.*

Changing to a matter-recycling economy is an important way to buy some time. However, it does not allow more and more people to use more and more resources indefinitely, even if all of them were somehow perfectly recycled.

What Is a Low-Throughput Economy? Learning from Nature The three scientific laws governing matter and energy changes suggest that the best long-term solution to our environmental and resource problems is to shift from an economy based on maximizing matter and energy flow (throughput) to a more sustainable **low-throughput (low-waste) economy,** as summarized in Figure 3-16.

The next four chapters are devoted to **(1)** applying the three basic scientific laws of matter and energy to living systems and **(2)** looking at some *biological principles* that can teach us how to live more sustainably by working with nature.

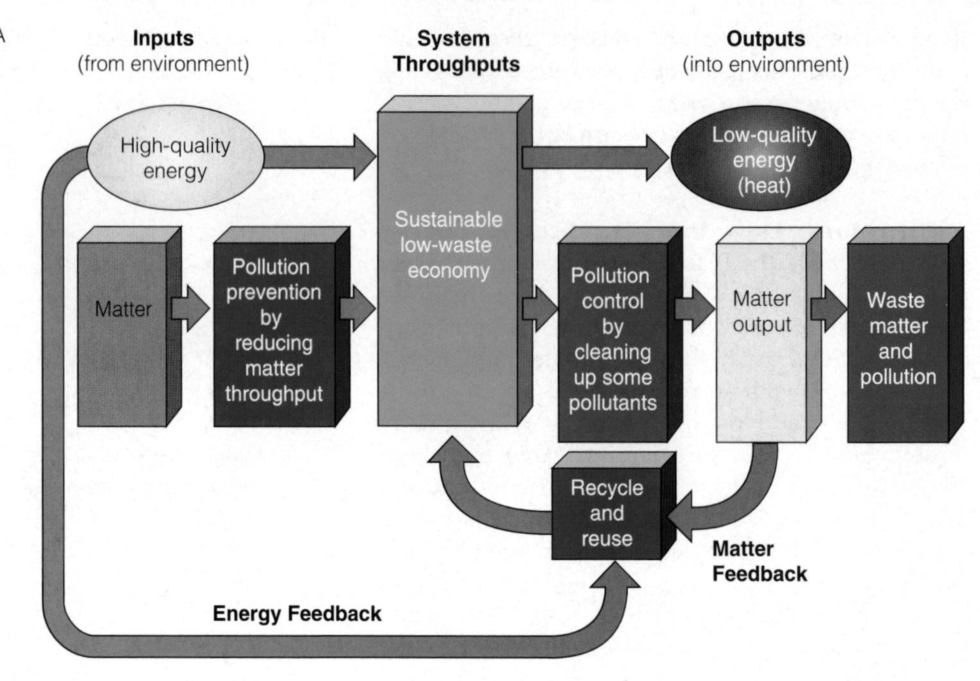

Figure 3-16 Lessons from nature. A low-throughput economy, based on energy flow and matter recycling, works with nature to reduce throughput and waste. This is done by **(1)** reusing and recycling most nonrenewable matter resources, **(2)** using renewable resources no faster than they are replenished, **(3)** using matter and energy resources efficiently, **(4)** reducing unnecessary consumption, **(5)** emphasizing pollution prevention and waste reduction, and **(6)** controlling population growth.

The second law of thermodynamics holds, I think, the supreme position among laws of nature. . . . If your theory is found to be against the second law of thermodynamics, I can give you no hope.

ARTHUR S. EDDINGTON

REVIEW QUESTIONS

1. Define the boldfaced terms in this chapter.

2. Describe how the community of Osage, Iowa, saved money and created jobs by saving energy.

3. Define *science* and explain how it works. Distinguish between *scientific data, scientific hypothesis, scientific model, scientific theory, scientific law,* and *scientific methods.* Explain why a scientific theory should be taken seriously.

4. What are *scientific methods,* and what five questions do scientists try to answer in studying some aspect of nature? What is a *controlled experiment*? What is *multivariable analysis*?

5. What does it mean to say that something has been scientifically proven? If scientists cannot establish absolute proof, what do they establish?

6. Distinguish between *frontier science* and *consensus science.*

7. What are two major limitations of environmental science?

8. What is a *mathematical model,* and how is such a model made? Why are mathematical models important?

9. What is a *system*? Distinguish between the *inputs, flows* or *throughputs, stores,* and *outputs* of a system.

10. What is a *feedback loop*? Distinguish between a *positive feedback loop* and a *negative feedback loop* and give an example of each. What is *homeostasis,* and how is it maintained in a system? Define and give an example of a *time delay* in a system. Define *synergy* and give an example of how it can change a system.

11. Distinguish between *matter, elements, compounds,* and *mixtures.*

12. Distinguish between *atoms, ions,* and *molecules* and give an example of each. What three major types of subatomic particles are found in atoms? Which two of these particles are found in the nucleus, and which is found outside the nucleus?

13. Distinguish between *atomic number* and *mass number.* What is an *isotope* of an atom?

14. What is the *concentration* of a chemical? What is *pH*?

15. What is a *chemical formula*? Distinguish between *ionic compounds* and *covalent compounds* and give the names and chemical formulas for an example of each of these types of compounds.

16. Distinguish between *organic compounds* and *inorganic compounds* and give an example of each type. Distinguish between *hydrocarbons, chlorinated hydrocarbons, chlorofluorocarbons, simple carbohydrates, polymers, complex carbohydrates, proteins, nucleic acids,* and *nucleotides.*

17. Distinguish between *genes* and *chromosomes.*

18. Distinguish between *high-quality matter* and *low-quality matter* and give an example of each. What is material efficiency?

19. What is *energy*? Distinguish between *kinetic energy* and *potential energy* and give an example of each. List three types of electromagnetic radiation. Distinguish between *ionizing radiation* and *nonionizing radiation.*

20. Distinguish between *high-quality energy* and *low-quality energy* and give an example of each. What is energy efficiency?

21. Distinguish between a *physical change* and a *chemical change* and give an example of each.

22. What is the *law of conservation of matter*? What is its environmental significance? Explain why there is no "away" as a repository for pollution. What is a *balanced chemical equation*, and how is it related to the law of conservation of matter?

23. What three factors determine the harm caused by a pollutant? Distinguish between concentrations of *parts per million, parts per billion,* and *parts per trillion*. What is the *persistence* of a pollutant? Distinguish between *degradable (nonpersistent), biodegradable, slowly degradable (persistent),* and *nondegradable pollutants* and give an example of each type.

24. What is a *nuclear change*? Distinguish between *natural radioactive decay, radioisotopes, gamma rays, alpha particles,* and *beta particles*. What is the *half-life* of a radioactive isotope? For how many half-lives should radioactive material be stored safely before it decays to an acceptable level of radioactivity? What are the major sources and effects of human exposure to ionizing radiation?

25. Distinguish between *nuclear fission* and *nuclear fusion*. Distinguish between *critical mass* and a *nuclear chain reaction*.

26. Distinguish between the *first law of energy (thermodynamics)* and the *second law of energy (thermodynamics)* and give an example of each law in action. Use the second law of thermodynamics to explain why energy cannot be recycled.

27. Distinguish between a *high-throughput (high-waste) economy*, a *low-throughput (low-waste) economy*, and a *matter-recycling economy*. Use the law of conservation of matter and the first and second laws of thermodynamics to explain the need to shift from a high-throughput economy to a matter-recycling economy and eventually to a low-throughput economy.

CRITICAL THINKING

1. Respond to the following statements:

   a. It has never been scientifically proven that anyone has ever died from smoking cigarettes.

   b. The greenhouse theory—that certain gases (such as water vapor and carbon dioxide) warm the atmosphere—is not a reliable idea because it is only a scientific theory.

2. See whether you can find an advertisement or an article describing some aspect of science in which **(a)** the concept of scientific proof is misused, **(b)** the term *theory* is used when it should have been *hypothesis*, and **(c)** a consensus scientific finding is dismissed or downplayed because it is "only a theory."

3. How does a scientific law (such as the law of conservation of matter) differ from a societal law (such as one imposing maximum speed limits for vehicles)? Can each be broken?

4. Explain why we do not really consume anything and why we can never really throw matter away.

5. A tree grows and increases its mass. Explain why this is not a violation of the law of conservation of matter.

6. If there is no "away," why isn't the world filled with waste matter?

7. Someone wants you to invest money in an automobile engine that will produce more energy than the energy in the fuel (such as gasoline or electricity) you use to run the motor. What is your response? Explain.

8. Use the second energy law to explain why a barrel of oil can be used only once as a fuel.

9. **(a)** Use the law of conservation of matter to explain why a matter-recycling economy will sooner or later be necessary. **(b)** Use the first and second laws of energy to explain why, in the long run, we will need a low-throughput economy, not just a matter-recycling economy.

10. **(a)** Imagine that you have the power to violate the law of conservation of energy (the first energy law) for 1 day. What are the three most important things you would do with this power? **(b)** Repeat this process, imagining that you have the power to violate the second law of energy for 1 day.

PROJECTS

1. If you have the use of a sensitive balance, try to demonstrate the law of conservation of mass in a physical change. Weigh a container with a lid (a glass jar will do), add an ice cube and weigh it again, and then allow the ice to melt and weigh it again.

2. Use the library or internet to find examples of various perpetual motion machines and inventions that allegedly violate the two laws of energy (thermodynamics) by producing more high-quality energy than the high-quality energy needed to make them run. What has happened to these schemes and machines (many of them developed by scam artists to attract money from investors)?

3. Make a concept map of this chapter's major ideas using the section heads and subheads and the key terms (in boldface). Look on the website for this book for information about making concept maps.

INTERNET STUDY RESOURCES AND RESOURCES FOR FURTHER READING AND RESEARCH

The website for this book contains helpful study aids and many ideas for further reading and research. Log on to

http://www.brookscole.com/product/0534389872s

and click on the Chapter-by-Chapter area. Choose Chapter 3 and select a resource:

- "Flash Cards" allows you to test your mastery of the Terms and Concepts to Remember for this chapter.

- "Tutorial Quizzes" provides a multiple-choice practice quiz.
- "Student Guide to InfoTrac" will lead you to Critical Thinking Projects that use InfoTrac College Edition as a research tool.
- "References" lists the major books and articles consulted in writing this chapter.
- "Hypercontents" takes you to an extensive list of sites with news, research, and images related to individual sections of the chapter.

INFOTRAC COLLEGE EDITION 

Improve your skills with InfoTrac College Edition, a searchable online database of articles from more than 700 periodicals. Log on to

http://www.infotrac-college.com

or access InfoTrac through the website for this book. Try to find the following articles:

1. Breaking the Law (perpetual motion). Peter Weiss. *Science News*, Oct 7, 2000 v158 i15 p234. We have long understood that perpetual motion machines violate the laws of thermodynamics. Now some scientists think that a certain class of perpetual motion machines may be possible, at least on a microscopic level. *Hint*: Enter the search term "perpetual motion" using the keywords "energy laws."

2. Fusion in Miniature (nuclear fusion) (Brief Article). Charles W. Petit. *U.S. News & World Report*, April 5, 1999 v126 i13 p58(1). Government researchers announced that they have produced "tabletop" fusion using laser power. They did it in an ordinary laboratory with equipment available to almost any decent team of physicists. It is not cold fusion, another small-scale method that caused a sensation 10 years ago until most scientists concluded that it was bogus. Instead, this is hot fusion, the same nuclear process that fuels the sun and hydrogen bombs. *Hint*: Enter the search term "nuclear fusion" using the keywords "energy laws."

4 ECOSYSTEMS: COMPONENTS, ENERGY FLOW, AND MATTER CYCLING

Have You Thanked the Insects Today?

Insects have a bad reputation. We classify many insect species as *pests* because they **(1)** compete with us for food, **(2)** spread human diseases (such as malaria), and **(3)** invade our lawns, gardens, and houses. Some people have "bugitis," fear all insects, and think that the only good bug is a dead bug. However, this view fails to recognize the vital roles insects play in helping sustain life on the earth.

A large proportion of the earth's plant species (including many trees) depend on insects to pollinate their flowers (Figure 4-1, left). In turn, we and other land-dwelling animals depend on plants for food, either by eating them or by consuming animals that eat them. If there were no pollinating insects, there would be very few fruits and vegetables for us and plant-eating animals to eat.

Insects, such as the praying mantis (Figure 4-1, right), that eat other insects help control the populations of at least half the species of insects we call pests. This free pest control service is an important part of nature's services that help sustain us.

Suppose all insects disappeared today. Within a year most of the earth's animals would become extinct because of the disappearance of so much plant life. The earth would be covered with rotting vegetation and animal carcasses being decomposed by unimaginably huge hordes of bacteria and fungi.

Fortunately, this is not a realistic scenario because insects, which have been around for at least 400 million years, are phenomenally successful life-forms. They were the first animals to invade the land and, later, the air. Today they are by far the planet's most diverse, abundant, and successful animals.

Insects **(1)** can rapidly evolve new genetic traits, such as resistance to pesticides, **(2)** have an exceptional ability to evolve into new species when faced with new environmental conditions, and **(3)** are extremely resistant to extinction.

This is another example of the law of conservation of problems (p. 61). Applying chemical pesticides to protect crops from pests can help grow more food. However, the pesticides can also **(1)** contaminate the air and water far from where they are applied, **(2)** harm beneficial insects that help protect crops from other insects, **(3)** threaten the health of wildlife and people, and **(4)** accelerate the natural ability of rapidly reproducing insect pests to develop genetic resistance (immunity) to such pesticides. Thus, in the long run pesticide technology can backfire and become less effective in solving the problem of crop losses from insect pests.

The environmental lesson is that although insects can thrive without newcomers such as us, we and most other land organisms would perish quickly without them. Learning about the roles insects play in nature requires us to understand how insects and other organisms living in a biological *community* (such as a forest or pond) interact with one another and with the nonliving environment. *Ecology* is the science that studies such relationships and interactions in nature, as discussed in this chapter and the four chapters that follow.

Figure 4-1 Insects play important roles in helping sustain life on earth. The bright green caterpillar moth feeding on pollen in a crocus (left) and other insects pollinate flowering plants that serve as food for many plant eaters. The praying mantis eating a monarch butterfly (right) and many other insect species help control the populations of at least half of the insect species we classify as pests.

Students who can begin early in their lives to think of things as connected, even if they revise their views every year, have begun the life of learning.

MARK VAN DOREN

This chapter addresses the following questions:

- What is ecology?
- What basic processes keep us and other organisms alive?
- What are the major components of an ecosystem?
- What happens to energy in an ecosystem?
- What happens to matter in an ecosystem?
- How do scientists study ecosystems?
- What are ecosystem services, and how do they affect the sustainability of the earth's life-support systems?

4-1 THE NATURE OF ECOLOGY

What Is Ecology? **Ecology** (from the Greek words *oikos*, "house" or "place to live," and *logos*, "study of") is the study of how organisms interact with one another and with their nonliving environment. In effect it is a study of *connections in nature*.

What Are Organisms? Ecologists focus on trying to understand the interactions between organisms, populations, communities, ecosystems, and the biosphere (Figure 4-2).

An **organism** is any form of life. The **cell** is the basic unit of life in organisms. Organisms may consist of a single cell (bacteria, for instance) or many cells.

On the basis of their cell structure, organisms can be classified as either *eukaryotic* or *prokaryotic*. Each cell of a **eukaryotic** organism **(1)** is surrounded by a membrane, **(2)** has a distinct *nucleus* (a membrane-bounded structure containing genetic material in the form of DNA), and **(3)** has several other internal parts called *organelles* (Figure 4-3a). All organisms except bacteria are eukaryotic.

The cell of a **prokaryotic** organism is surrounded by a membrane, but inside the cell there is no distinct nucleus or other internal parts enclosed by membranes (Figure 4-3b). All bacteria are single-celled prokaryotic organisms. Although most familiar organisms are eukaryotic, they could not exist without hordes of prokaryotic organisms (bacteria; Connections, p. 80). These bacteria are examples of *microorganisms*, so small that they can be seen only with the aid of a microscope.

What Are Species? Organisms can be classified into **species**, or groups of organisms that resemble one another in appearance, behavior, chemistry, and genetic makeup. Species differ in how they produce offspring. **Asexual reproduction** is common in species such as bacteria with only one cell, which divides to produce two identical cells that are clones or replicas of the original cell.

Sexual reproduction occurs in organisms that produce offspring by combining sex cells or gametes (such as ovum and sperm) from both parents. This produces offspring that have combinations of genetic traits from each parent. Sexual reproduction usually gives the offspring a greater chance of survival under changing environmental conditions than the genetic clones produced by asexual reproduction. Organisms that reproduce sexually are classified as the members of the same species if, under natural conditions, they can **(1)** actually or potentially breed with one another and **(2)** produce live, fertile offspring.

We do not know how many species exist on the earth. Estimates range from 5 million to 100 million. Most are insects (p. 77) and microorganisms too small to be seen with the naked eye (Connections, p. 80). Excluding hordes of bacterial species, a best guess of the number of species is about 10–14 million.

So far biologists have identified and named about 1.8 million species, not including bacteria. Biologists know a fair amount about roughly one-third of the known species but understand the detailed roles and interactions of only a few. Each year researchers identify only about 10,000 new species, so we have a long way to go to identify even a fraction of the estimated number of species.

What Is a Population? A **population** consists of a group of interacting individuals of the same species that occupy a specific area at the same time (Figure 4-4). Examples are all **(1)** sunfish in a pond, **(2)** white oak trees in a forest, and **(3)** people in a country. In most natural populations, individuals vary slightly in their genetic makeup, which is why they do not all look or behave exactly alike—a phenomenon called **genetic diversity** (Figure 4-5). In response to changes in environmental conditions, populations change in **(1)** size, **(2)** age distribution (number of individuals in each age group), **(3)** density (number of individuals per unit of space), and **(4)** genetic composition.

The place where a population (or an individual organism) normally lives is its **habitat**. It may be as large as an ocean or prairie or as small as the underside of a rotting log or the intestine of a termite.

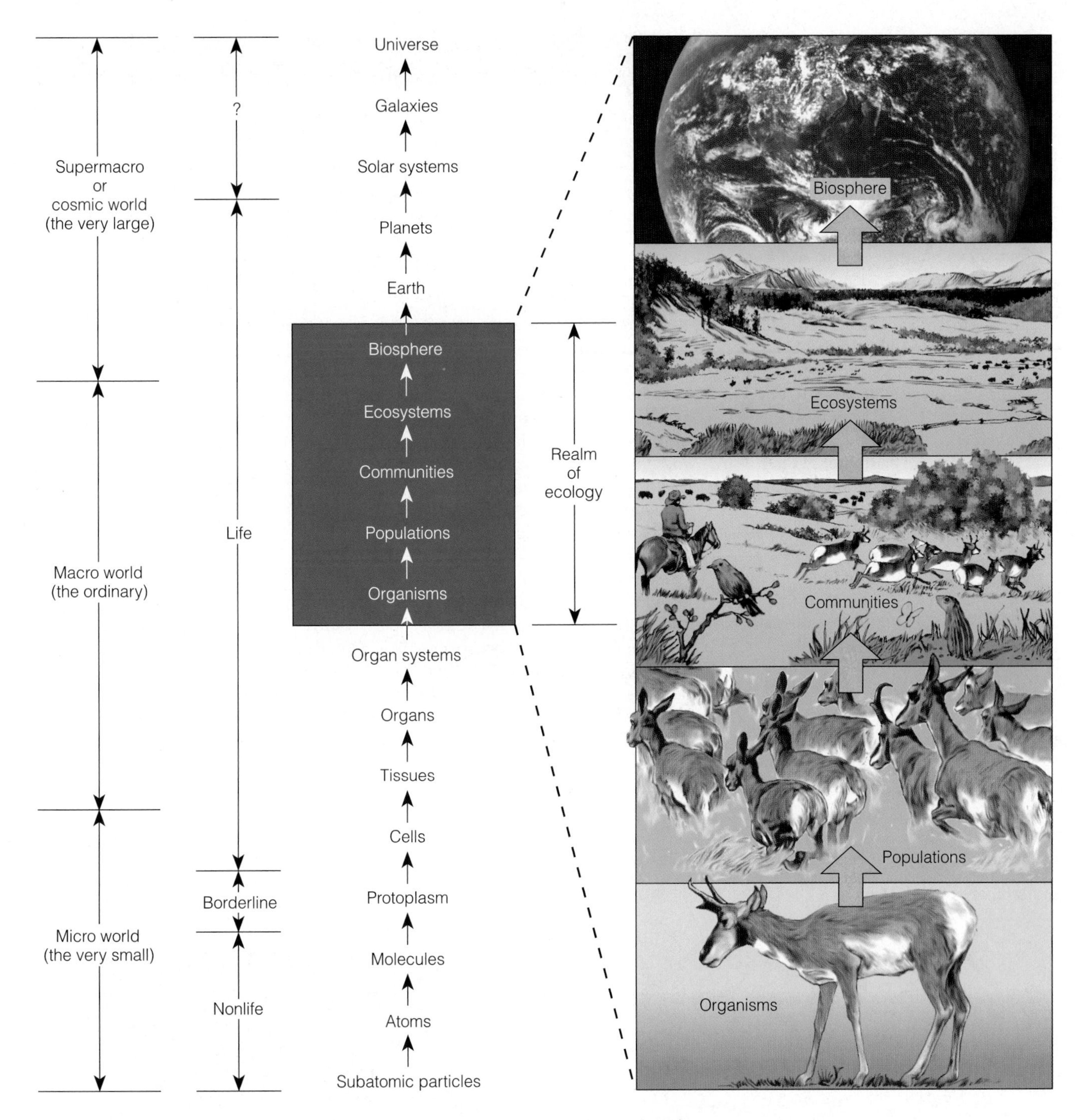

Figure 4-2 Levels of organization of matter in nature. Notice the five levels that ecology focuses on.

What Are Communities, Ecosystems, and the Biosphere? Populations of the different species occupying a particular place make up a **community**, or **biological community.** It is a complex interacting network of plants, animals, and microorganisms.

An **ecosystem** is a community of different species interacting with one another and with their nonliving environment of matter and energy. Ecosystems can range in size from a puddle of water to a stream, a patch of woods, an entire forest, or a desert. Ecosystems can be natural or artificial (human-created). Examples of human-created ecosystems are cropfields, farm ponds, and reservoirs. All the earth's ecosystems together make up what we call the **biosphere**.

Microbes: The Invisible Rulers of the Earth

CONNECTIONS

They are everywhere, and there are trillions of them. Billions are found inside your body, on your body, in a handful of soil, and in a cup of river water.

These mostly invisible rulers of the earth are *microbes*, a catchall term for many thousands of species of bacteria, protozoa, fungi, and yeasts, most of which are too small to be seen with the naked eye.

Most microbes do not get the respect they deserve. Most of us think of them as threats to our health in the form of **(1)** infectious bacteria or "germs," **(2)** fungi that cause athlete's foot and other skin diseases, and **(3)** protozoa that cause often fatal diseases such as malaria. However, these potentially harmful microbes are in the minority.

Most of the earth's hordes of microbes not only are harmless but also make the rest of life possible. Some of them play a vital role in producing foods such as bread, cheese, yogurt, vinegar, tofu, soy sauce, beer, and wine. Others provide us with food by converting nitrogen gas in the atmosphere into forms that plants can take up from the soil as nutrients.

Bacteria and fungi in the soil decompose organic wastes into nutrients that can be taken up by plants. Bacteria in your intestinal tract break down the food you eat. Some microbes in your nose prevent harmful bacteria from reaching your lungs.

Other microbes have been the source of disease-fighting antibiotics, including penicillin, erythromycin, and streptomycin.

Another vital ecological service provided by some microbes is the control of some plant diseases and populations of insect species that attack food crops. Enlisting some of these microbes for pest control can reduce the use of potentially harmful chemical pesticides.

In addition, bioengineering is being used to develop microbes than can **(1)** extract metals from ores, **(2)** break down various pollutants, and **(3)** help clean up toxic waste sites.

Harvard biologist Edward O. Wilson, who has developed many important ecological theories and is one of the world's experts on ants, says that if he were starting over he would study microbes.

Critical Thinking

1. A bumper sticker reads, "Have You Thanked Microbes Today?" Give reasons for doing so and explain why microbes are the real rulers of the earth.

2. What are some potentially harmful effects of **(a)** using genetic engineering to design microbes to break down oil and toxic chemicals and **(b)** overusing antibacterial soaps, sprays, and antibiotics?

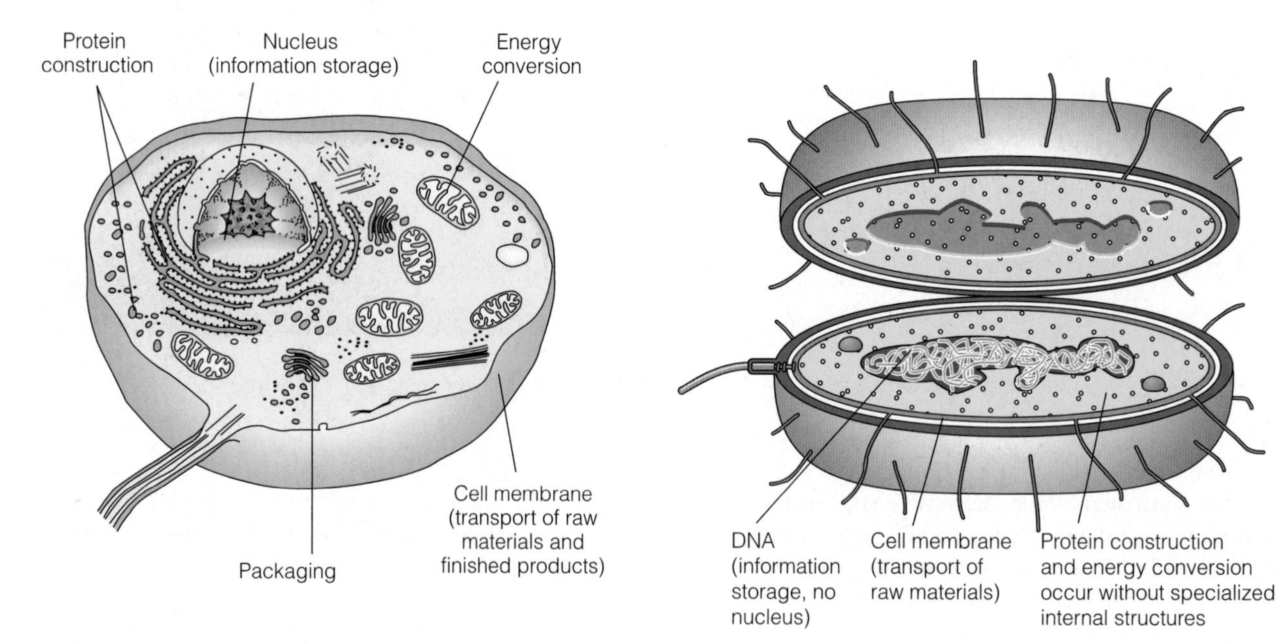

Figure 4-3 **(a)** Generalized structure of a eukaryotic cell. The parts and internal structure of cells in various types of organisms such as plants and animals differ somewhat from this generalized model. **(b)** Generalized structure of a prokaryotic cell. Note that prokaryotic cells lack a distinct nucleus. (Adapted from Cecie Starr and Ralph Taggart, *Biology: The Unity and Diversity of Life*, 8th ed., Belmont, Calif.: Wadsworth, 1998)

Figure 4-4 A population of monarch butterflies. The geographic distribution of this butterfly coincides with that of the milkweed plant, on which monarch larvae and caterpillars feed.

4-2 THE EARTH'S LIFE-SUPPORT SYSTEMS

What Are the Major Parts of the Earth's Life-Support Systems? We can think of the earth as being made up of several spherical layers (Figure 4-6). The **atmosphere** is a thin envelope of air around the planet. Its inner layer, the **troposphere**, extends only about 17 kilometers (11 miles) above sea level but contains most of the planet's air, mostly nitrogen (78%) and oxygen (21%). The next layer, stretching 17–48 kilometers (11–30 miles) above the earth's surface, is the **stratosphere**. Its lower portion contains enough ozone (O_3) to filter out most of the sun's harmful ultraviolet radiation, thus allowing life to exist on land and in the surface layers of bodies of water.

The **hydrosphere** consists of the earth's **(1)** liquid water (both surface and underground), **(2)** ice (polar ice, icebergs, and ice in frozen soil layers, or permafrost), and **(3)** water vapor in the atmosphere. The **lithosphere** is the earth's crust and upper mantle. The lithosphere's crust contains nonrenewable fossil fuels and minerals we use as well as renewable soil chemicals (nutrients) needed for plant life.

The **biosphere** is the portion of the earth in which living (biotic) organisms exist and interact with one another and with their nonliving (abiotic) environment. The biosphere includes most of the hydrosphere and parts of the lower atmosphere and upper lithosphere.

Figure 4-5 The genetic diversity among individuals of one species of Caribbean snail is reflected in the variations in shell color and banding patterns.

It reaches from the deepest ocean floor, 20 kilometers (12 miles) below sea level, to the tops of the highest mountains. If the earth were an apple, the biosphere would be no thicker than the apple's skin. *The goal of ecology is to understand the interactions in this thin, life-supporting global skin of air, water, soil, and organisms.*

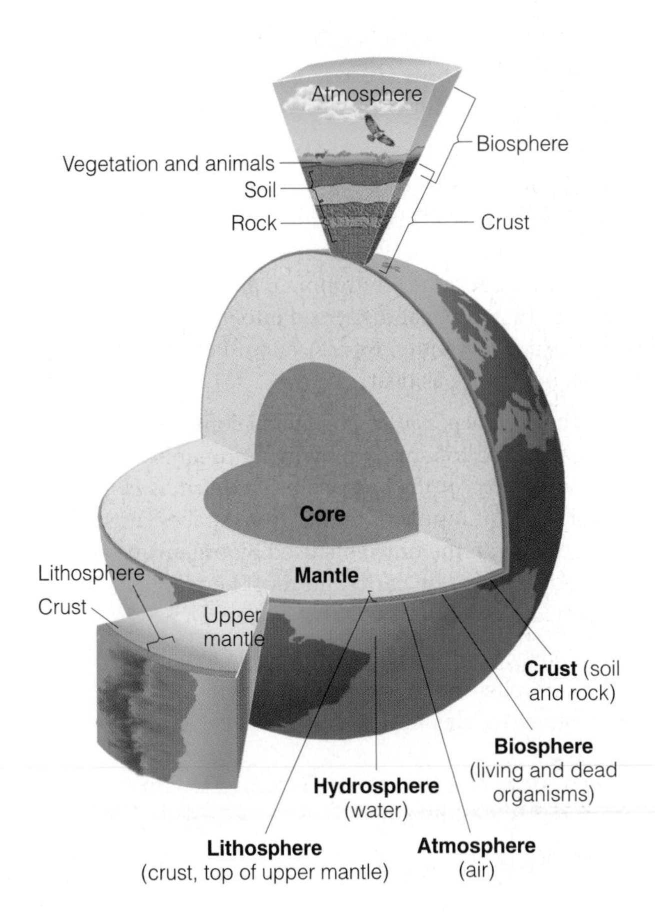

Figure 4-6 The general structure of the earth.

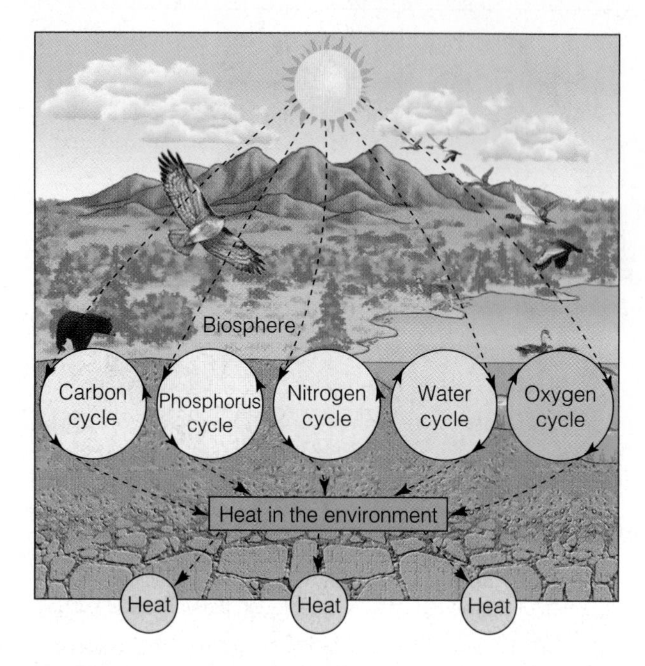

Figure 4-7 Life on the earth depends on **(1)** the *one-way flow of energy* (dashed lines) from the sun through the biosphere, **(2)** the *cycling of crucial elements* (solid lines around circles), and **(3)** *gravity*, which keeps atmospheric gases from escaping into space and draws chemicals downward in the matter cycles. This simplified model depicts only a few of the many cycling elements.

What Sustains Life on Earth? Life on the earth depends on three interconnected factors (Figure 4-7):

- The *one-way flow of high-quality energy* (Figure 3-9, p. 67) from the sun, **(1)** through materials and living things in their feeding interactions, **(2)** into the environment as low-quality energy (mostly heat dispersed into air or water molecules at a low temperature), and **(3)** eventually back into space as heat.
- The *cycling of matter* (the atoms, ions, or molecules needed for survival by living organisms) through parts of the biosphere. The earth is closed to significant inputs of matter from space. Thus, essentially all the nutrients used by organisms are already present on earth and must be recycled again and again for life to continue.
- *Gravity*, which **(1)** allows the planet to hold onto its atmosphere and **(2)** causes the downward movement of chemicals in the matter cycles.

How Does the Sun Help Sustain Life on Earth? The sun is a middle-aged star whose energy

- Lights and warms the planet
- Supports *photosynthesis*, the process used by green plants and some bacteria to make compounds such as carbohydrates that keep them alive and that feed most other organisms
- Powers the cycling of matter
- Drives the climate and weather systems that distribute heat and fresh water over the earth's surface

What Happens to Solar Energy Reaching the Earth? Because the earth is a tiny sphere in the vastness of space, it receives only about one-billionth of the sun's output of energy. Much of this energy is either reflected away or absorbed by chemicals in the atmosphere (Figure 4-8).

Most of what reaches the troposphere is **(1)** visible light, **(2)** infrared radiation (heat), and **(3)** the small amount of ultraviolet radiation that is not absorbed by ozone in the stratosphere (Figure 3-8, p. 66). This incoming energy **(1)** warms the troposphere and land, **(2)** evaporates water and cycles it through the biosphere, and **(3)** generates winds. A tiny fraction is captured by green plants, algae, and bacteria to fuel photosynthesis and make the organic compounds most forms of life need to survive.

Most unreflected solar radiation is degraded into infrared radiation (which we experience as heat) as it interacts with the earth. Greenhouse gases (such as water vapor, carbon dioxide, methane, nitrous oxide, and ozone) in the atmosphere reduce this

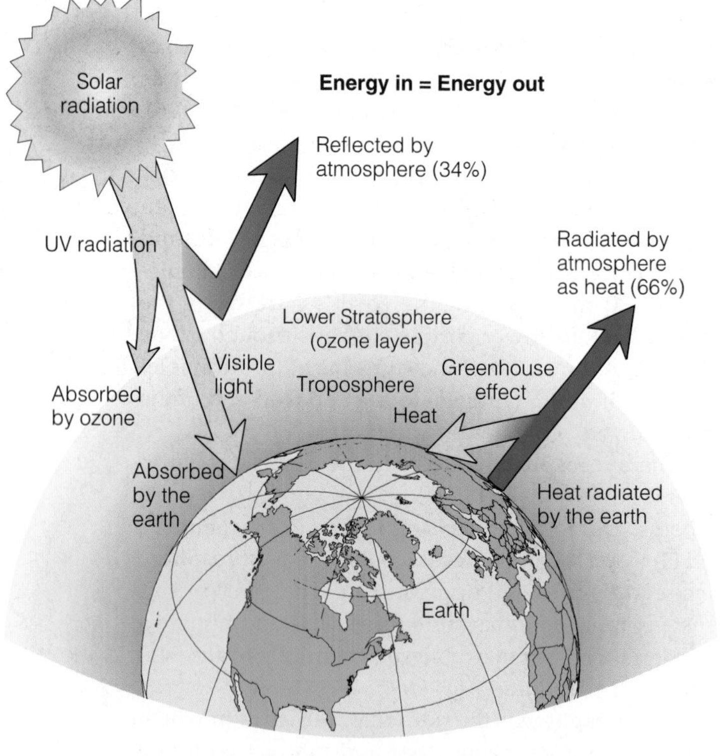

Figure 4-8 The flow of energy to and from the earth.

flow of heat back into space. This helps warm the earth by acting somewhat like the glass in a greenhouse or the windows in a closed car, which allow a buildup of heat. Without this **natural greenhouse effect** (Figure 4-8), the earth would be too cold for life as we know it to exist.

4-3 ECOSYSTEM CONCEPTS AND COMPONENTS

What Are Biomes and Aquatic Life Systems? Viewed from outer space, the earth resembles an enormous jigsaw puzzle consisting of large masses of land and vast expanses of ocean (p. 1 and Figure 1-1, p. 2). Biologists have classified the terrestrial (land) portion of the biosphere into **biomes** ("BY-ohms"). They are large regions such as forests, deserts, and grasslands characterized by **(1)** a distinct climate and **(2)** specific life-forms (especially vegetation) adapted to them (Figure 4-9).

Climate—long-term patterns of weather—is the main factor determining what type of life, especially what plants, will thrive in a given land area. Each biome consists of a patchwork of many different ecosystems whose communities have adapted to differences in climate, soil, and other factors throughout the biome.

Marine and freshwater portions of the biosphere can be divided into **aquatic life zones**, each containing numerous ecosystems. Aquatic life zones are the aquatic equivalent of biomes. Examples include **(1)** *freshwater life zones* (such as lakes and streams) and **(2)** *ocean or marine life zones* (such as estuaries, coastlines, coral reefs, and the deep ocean). The earth's major land biomes and aquatic life zones are discussed in more detail in Chapter 6.

What Are the Major Components of Ecosystems? The biosphere and its ecosystems can be separated into two parts: **(1) abiotic**, or nonliving, components (water, air, nutrients, and solar energy) and **(2) biotic**, or living, components (plants, animals, and microorganisms, sometimes called *biota*). Figures 4-10 and 4-11 are greatly simplified diagrams of some of the

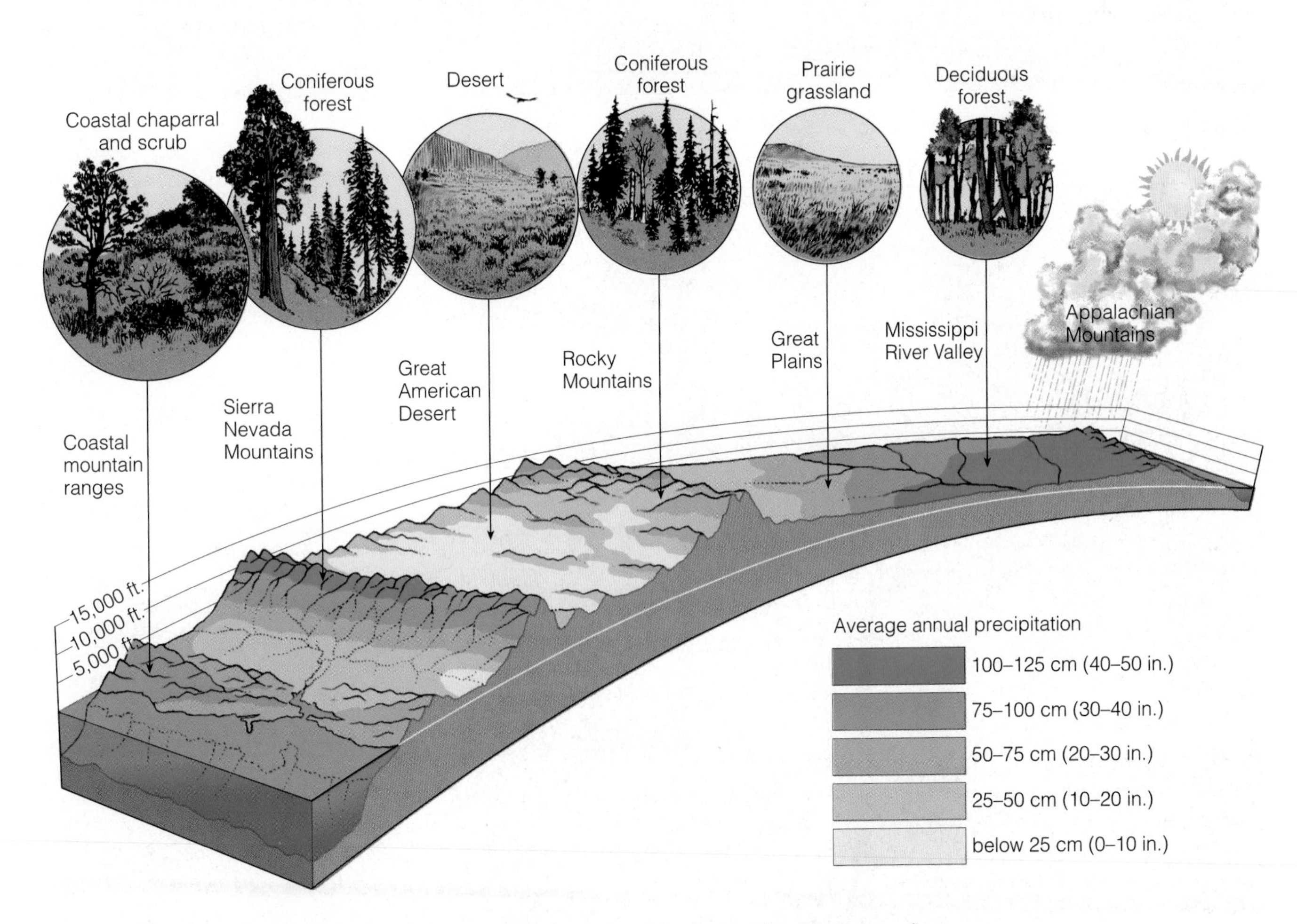

Figure 4-9 Major biomes found along the 39th parallel across the United States. The differences reflect changes in climate, mainly differences in average annual precipitation and temperature (not shown).

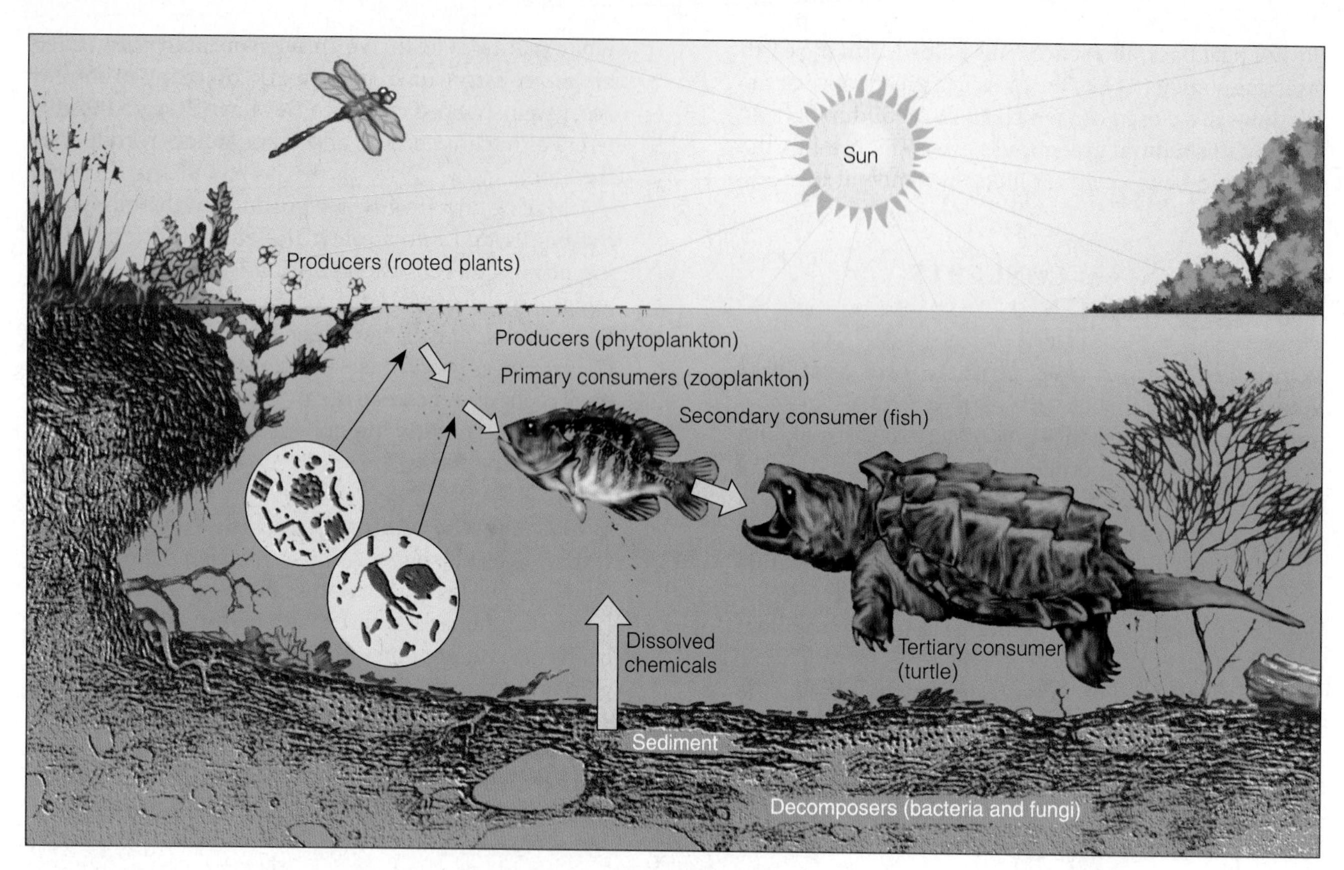

Figure 4-10 Major components of a freshwater ecosystem.

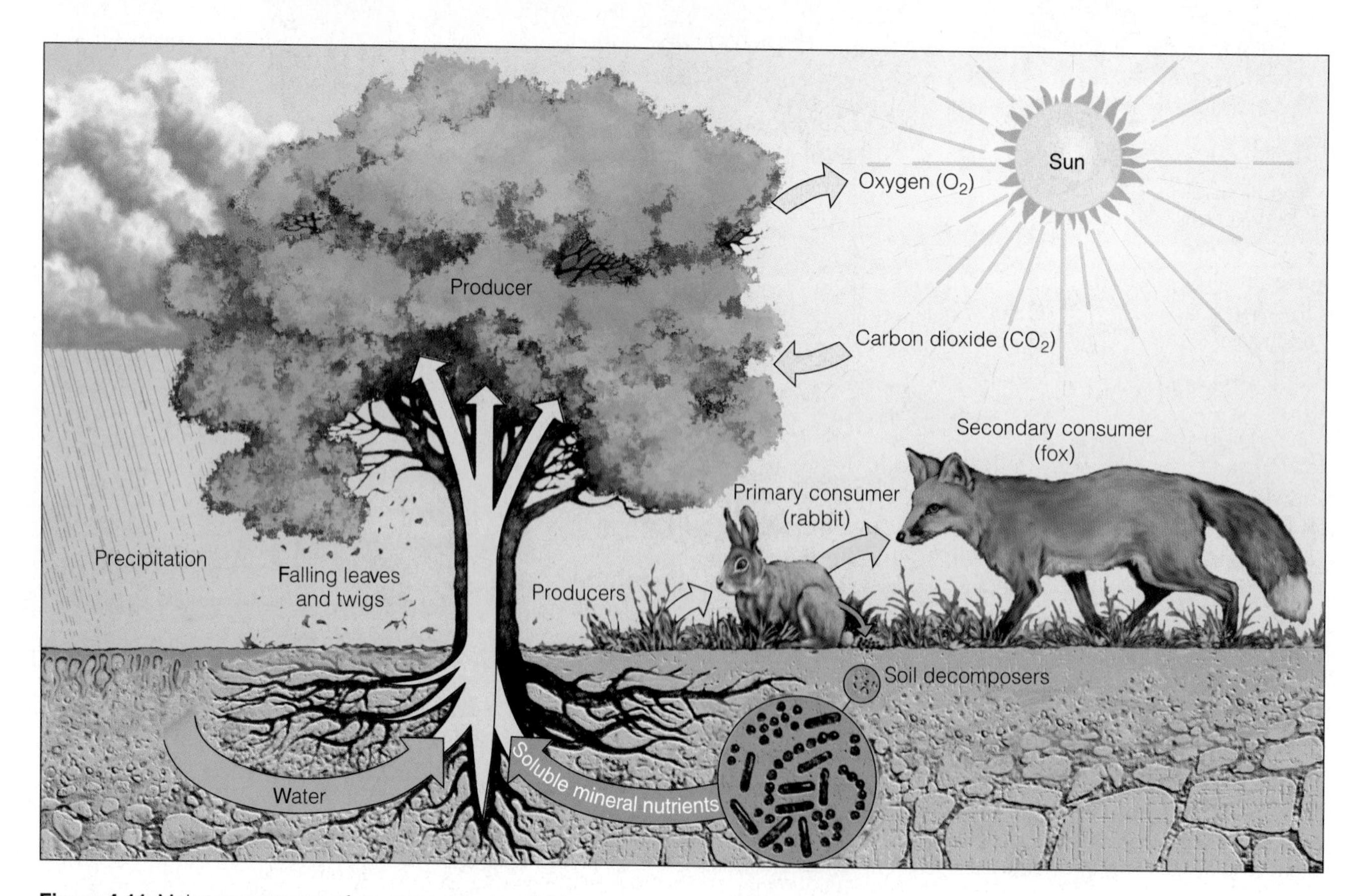

Figure 4-11 Major components of an ecosystem in a field.

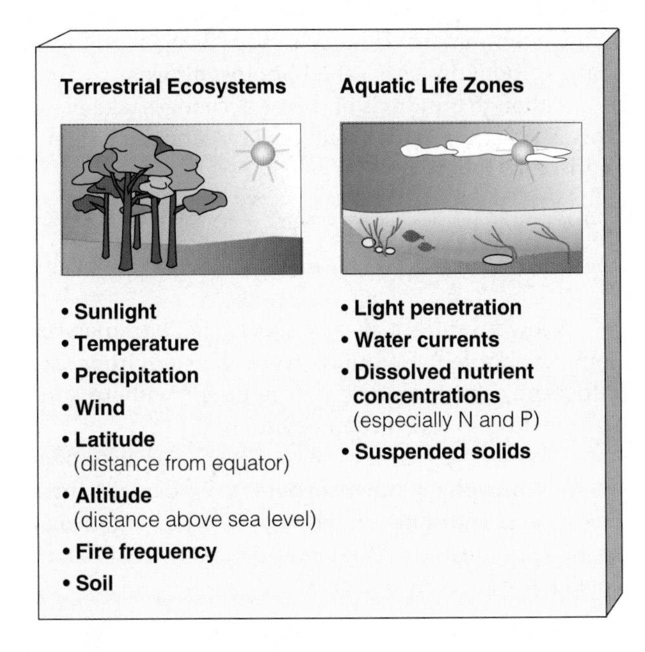

Terrestrial Ecosystems
Aquatic Life Zones
• Sunlight
• Temperature
• Precipitation
• Wind
• Latitude
(distance from equator)
• Altitude
(distance above sea level)
• Fire frequency
• Soil
• Light penetration
• Water currents
• Dissolved nutrient concentrations
(especially N and P)
• Suspended solids

Figure 4-12 Key physical and chemical or abiotic factors affecting terrestrial ecosystems (left) and aquatic life zones (right).

biotic and abiotic components in a freshwater aquatic ecosystem and a terrestrial ecosystem.

What Are the Major Nonliving Components of Ecosystems? The nonliving, or abiotic, components of an ecosystem are the physical and chemical factors that influence living organisms in land (terrestrial) ecosystems and aquatic life zones (Figure 4-12).

Different species thrive under different physical conditions. Some need bright sunlight, and others thrive better in shade. Some need a hot environment and others a cool or cold one. Some do best under wet conditions and others under dry conditions.

Each population in an ecosystem has a **range of tolerance** to variations in its physical and chemical environment (Figure 4-13). Individuals within a population may also have slightly different tolerance ranges for temperature or other factors because of small differences in genetic makeup, health, and age. Thus, although a trout population may do best within a narrow band of temperatures (*optimum level or range*), a few individuals can survive above and below that band. As Figure 4-13 shows, tolerance has its limits, beyond which none of the trout can survive.

These observations are summarized in the **law of tolerance**: *The existence, abundance, and distribution of a species in an ecosystem are determined by whether the levels of one or more physical or chemical factors fall within the range tolerated by that species.* In other words, there are minimum and maximum limits for physical conditions (such as temperature) and concentrations of chemical substances, called **tolerance limits**, beyond which no members of a particular species can survive.

A species may have a wide range of tolerance to some factors and a narrow range of tolerance to others.

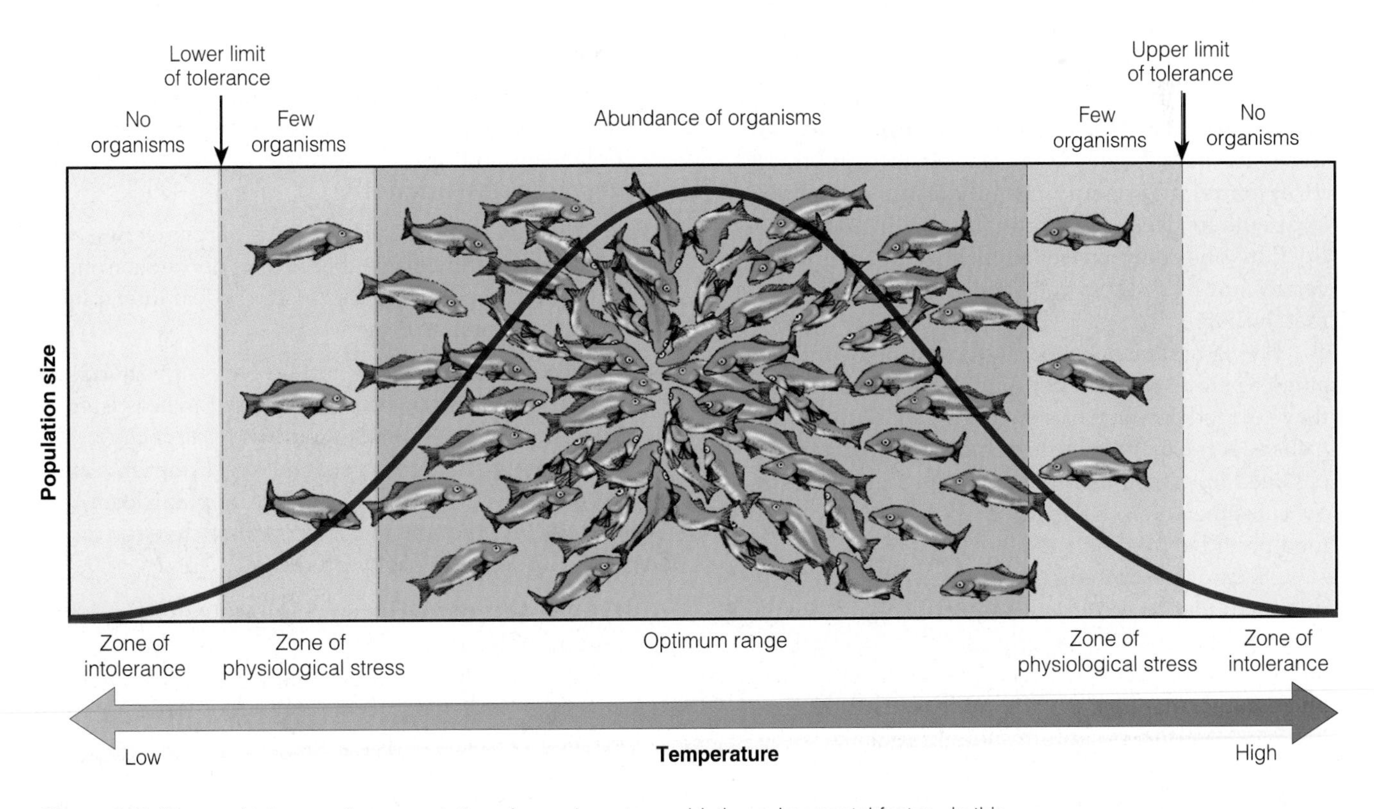

Figure 4-13 Range of tolerance for a population of organisms to an abiotic environmental factor—in this case, temperature.

Most organisms are least tolerant during juvenile or reproductive stages of their life cycles. *Highly tolerant species can live in a variety of habitats with widely different conditions.*

A variety of factors can affect the number of organisms in a population. However, sometimes one factor, known as a **limiting factor**, is more important in regulating population growth than other factors. This ecological principle, related to the law of tolerance, is called the **limiting factor principle**: *Too much or too little of any abiotic factor can limit or prevent growth of a population, even if all other factors are at or near the optimum range of tolerance.*

On land, precipitation often is the limiting factor. Lack of water in a desert limits plant growth. Soil nutrients also can act as a limiting factor on land. Suppose a farmer plants corn in phosphorus-poor soil. Even if water, nitrogen, potassium, and other nutrients are at optimum levels, the corn will stop growing when it uses up the available phosphorus.

Too much of an abiotic factor can also be limiting. For example, too much water or too much fertilizer can kill plants, a common mistake of many beginning gardeners.

Important limiting factors for aquatic ecosystems include **(1)** temperature, **(2)** sunlight, **(3) dissolved oxygen content** (the amount of oxygen gas dissolved in a given volume of water at a particular temperature and pressure), and **(4)** nutrient availability. Another limiting factor in aquatic ecosystems is **salinity** (the amounts of various inorganic minerals or salts dissolved in a given volume of water).

What Are the Major Living Components of Ecosystems? Living organisms capture and transform matter and energy from their environment to supply their needs for survival, growth, and reproduction. The complete set of chemical reactions that carries out this role in cells and organisms is called **metabolism**.

Living organisms in ecosystems usually are classified as either *producers* or *consumers*, based on how they get food. **Producers**, sometimes called **autotrophs** (self-feeders), make their own food from compounds obtained from their environment. All other organisms are consumers, which depend directly or indirectly on food provided by producers.

On land, most producers are green plants. In freshwater and marine ecosystems, algae and plants are the major producers near shorelines. In open water the dominant producers are *phytoplankton* (most of them microscopic) that float or drift in the water.

Most producers capture sunlight to make carbohydrates (such as glucose, $C_6H_{12}O_6$) and other complex organic compounds from inorganic (abiotic) nutrients in the environment. This process for using sunlight to make carbohydrates is called **photosynthesis**.

Although hundreds of chemical changes take place during photosynthesis, the overall reaction can be summarized as follows:

carbon dioxide + water + **solar energy** ⟶ glucose + oxygen

$$6\,CO_2 + 6\,H_2O + \textbf{solar energy} \longrightarrow C_6H_{12}O_6 + 6\,O_2$$

A few producers, mostly specialized bacteria, can convert simple compounds from their environment into more complex nutrient compounds without sunlight, a process called **chemosynthesis**.

All other organisms in an ecosystem are **consumers** or **heterotrophs** ("other-feeders"), which get their energy and nutrients by feeding on other organisms or their remains. Based on their primary source of food, consumers are classified as

- **Herbivores** (plant eaters) or **primary consumers** feeding directly on producers.
- **Carnivores** (meat eaters) feeding on other consumers. Those feeding only on primary consumers are called **secondary consumers**, and those feeding on other carnivores are called **tertiary (higher-level) consumers**.
- **Omnivores** (such as pigs, rats, foxes, bears, cockroaches, and humans) that eat plants and animals.
- **Scavengers** (such as vultures, flies, hyenas, and some species of sharks and ants) that feed on dead organisms.
- **Detritivores** (detritus feeders and decomposers) feeding on **detritus** ("di-TRI-tus"), or parts of dead organisms and cast-off fragments and wastes of living organisms (Figure 4-14).
- **Detritus feeders** (such as crabs, carpenter ants, termites, and earthworms) that extract nutrients from partly decomposed organic matter in leaf litter, plant debris, and animal dung.
- **Decomposers** (mostly certain types of bacteria and fungi) that recycle organic matter in ecosystems. They do this by **(1)** breaking down (*biodegrading*) dead organic material (detritus) to get nutrients and **(2)** releasing the resulting simpler inorganic compounds into the soil and water, where they can be taken up as nutrients by producers.

Figures 4-10 and 4-11 show various types of producers and consumers.

Both producers and consumers use the chemical energy stored in glucose and other organic compounds to fuel their life processes. In most cells, this energy is released by **aerobic respiration**, which uses oxygen to convert organic nutrients back into carbon

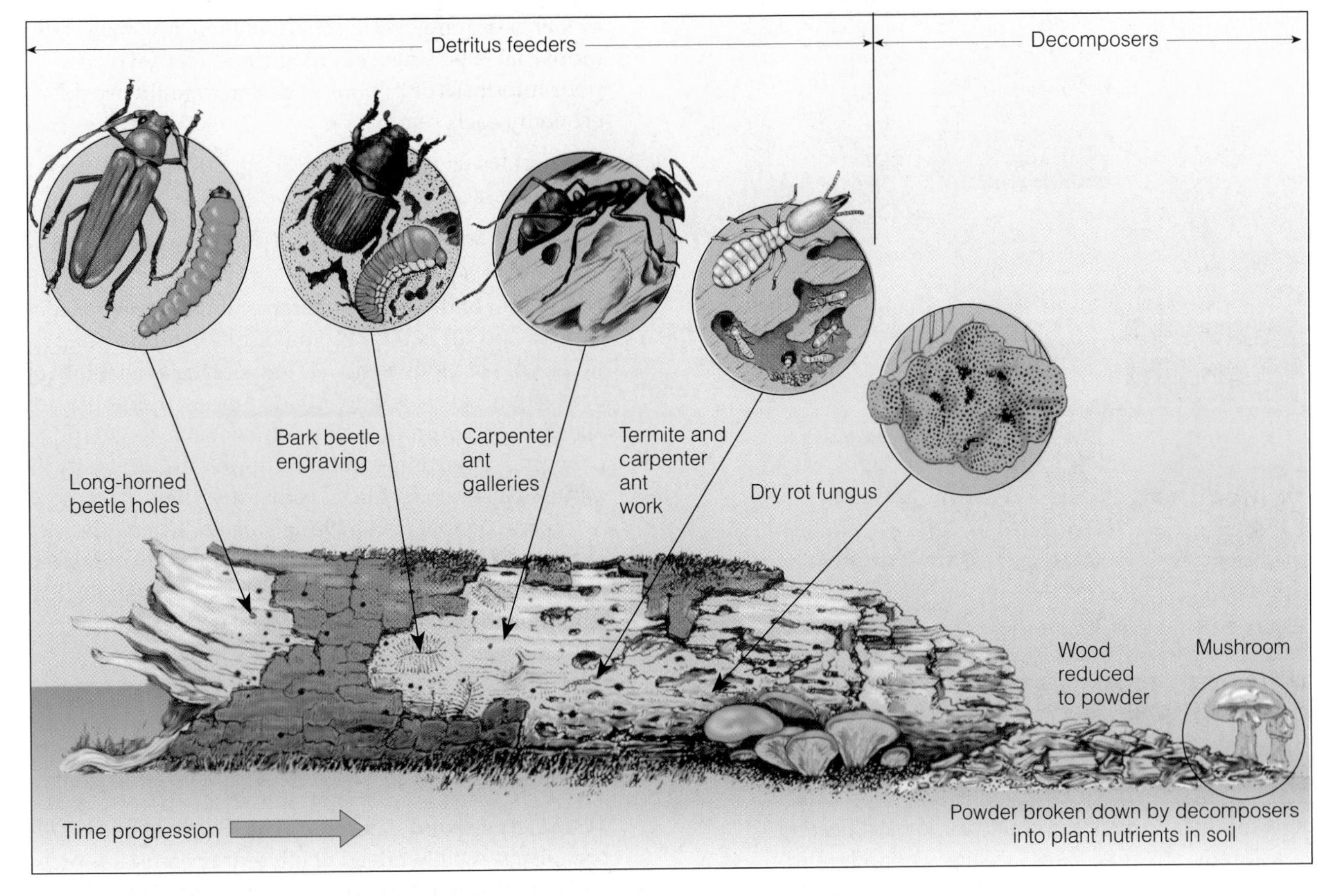

Figure 4-14 Some detritivores, called *detritus feeders*, directly consume tiny fragments of this log. Other detritivores, called *decomposers* (mostly fungi and bacteria), digest complex organic chemicals in fragments of the log into simpler inorganic nutrients. These nutrients can be used again by producers if they are not washed away or otherwise removed from the system.

dioxide and water. The net effect of the hundreds of steps in this complex process is represented by the following reaction:

glucose + oxygen → carbon dioxide + water + **energy**

$$C_6H_{12}O_6 + 6\,O_2 \longrightarrow 6\,CO_2 + 6\,H_2O + \textbf{energy}$$

Although the detailed steps differ, the net chemical change for aerobic respiration is the opposite of that for photosynthesis.

Some decomposers get the energy they need by breaking down glucose (or other organic compounds) in the absence of oxygen. This form of cellular respiration is called **anaerobic respiration** or **fermentation**. Instead of carbon dioxide and water, the end products of this process are compounds such as **(1)** methane gas (CH_4, the main component of natural gas), **(2)** ethyl alcohol (C_2H_6O), **(3)** acetic acid ($C_2H_4O_2$, the key component of vinegar), and **(4)** hydrogen sulfide (H_2S, when sulfur compounds are broken down).

The survival of any individual organism depends on the *flow of matter and energy* through its body. However, an ecosystem as a whole survives primarily through a combination of *matter recycling* (rather than one-way flow) and *one-way energy flow* (Figure 4-15).

Decomposers complete the cycle of matter by breaking down detritus into inorganic nutrients that are used by producers. Without decomposers, **(1)** the entire world would be knee-deep in plant litter, dead animal bodies, animal wastes, and garbage and **(2)** most life as we know it would no longer exist.

What Is Biodiversity, and Why Is It Important?

One important renewable resource is **biological diversity** or **biodiversity**: the different life-forms and life-sustaining processes that can best survive the variety of conditions currently found on the earth. Kinds of biodiversity include

- **Genetic diversity** (variety in the genetic makeup among individuals within a species, Figure 4-5)
- **Species diversity** (variety among the species or distinct types of living organisms found in different habitats of the planet; Figure 4-16)

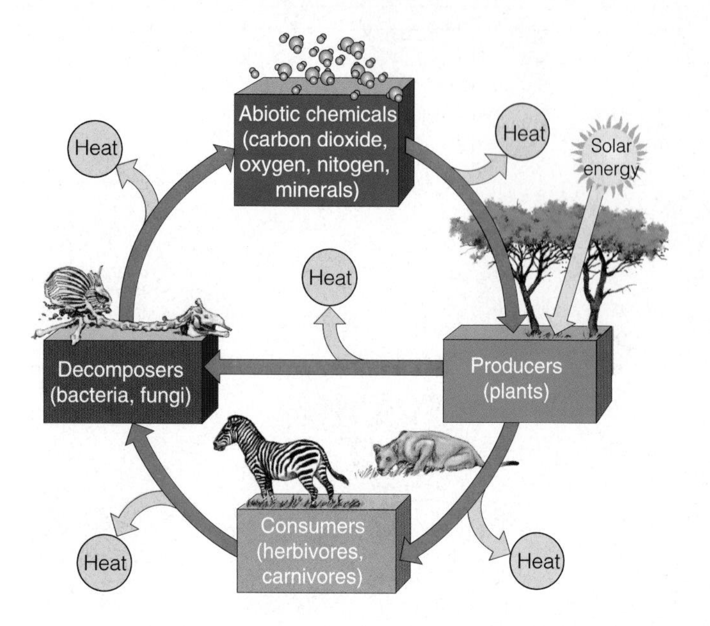

Figure 4-15 The main structural components (energy, chemicals, and organisms) of an ecosystem are linked by matter recycling and the flow of energy from the sun, through organisms, and then into the environment as low-quality heat.

- **Ecological diversity** (variety of forests, deserts, grasslands, streams, lakes, oceans, wetlands, and other biological communities, Figure 4-9)
- **Functional diversity** (biological and chemical processes or functions such as energy flow and matter cycling needed for the survival of species and biological communities, Figure 4-15)

This rich variety of genes, species, biological communities, and life-sustaining biological and chemical processes

- Gives us food, wood, fibers, energy, raw materials, industrial chemicals, and medicines, all of which pour hundreds of billions of dollars into the world economy each year
- Provides us with free recycling, purification, and natural pest control services

Every species here today **(1)** contains genetic information that represents thousands to millions of years of adaptation to the earth's changing environmental conditions and **(2)** is the raw material for future adaptations. Loss of biodiversity **(1)** reduces the availability of ecosystem services and **(2)** decreases the ability of species, communities, and ecosystems to adapt to changing environmental conditions. Biodiversity is nature's insurance policy against disasters.

Some people also include *human cultural diversity* as part of the earth's biodiversity. The variety of human cultures represents numerous social and technological solutions to changing environmental conditions.

4-4 CONNECTIONS: FOOD WEBS AND ENERGY FLOW IN ECOSYSTEMS

What Are Food Chains and Food Webs? All organisms, whether dead or alive, are potential sources of food for other organisms. A caterpillar eats a leaf, a robin eats the caterpillar, and a hawk eats the robin. Decomposers consume the leaf, caterpillar, robin, and hawk after they die. As a result, *there is little matter waste in natural ecosystems.*

The sequence of organisms, each of which is a source of food for the next, is called a **food chain**. It determines how energy and nutrients move from one organism to another through an ecosystem (Figure 4-17).

Figure 4-16 Two species found in tropical forests are part of the earth's biodiversity. On the right is the world's largest flower, the flesh flower (*Rafflesia arnoldi*), growing in a tropical rain forest in Sumatra. The flower of this leafless plant can be as large as 1 meter (4.3 feet) in diameter and weigh 7 kilograms (15 pounds). The plant gives off a smell like rotting meat, presumably to attract flies and beetles that pollinate its flower. After blossoming once a year for a few weeks, the flower dissolves into a slimy black mass. On the left is a cotton top tamarin.

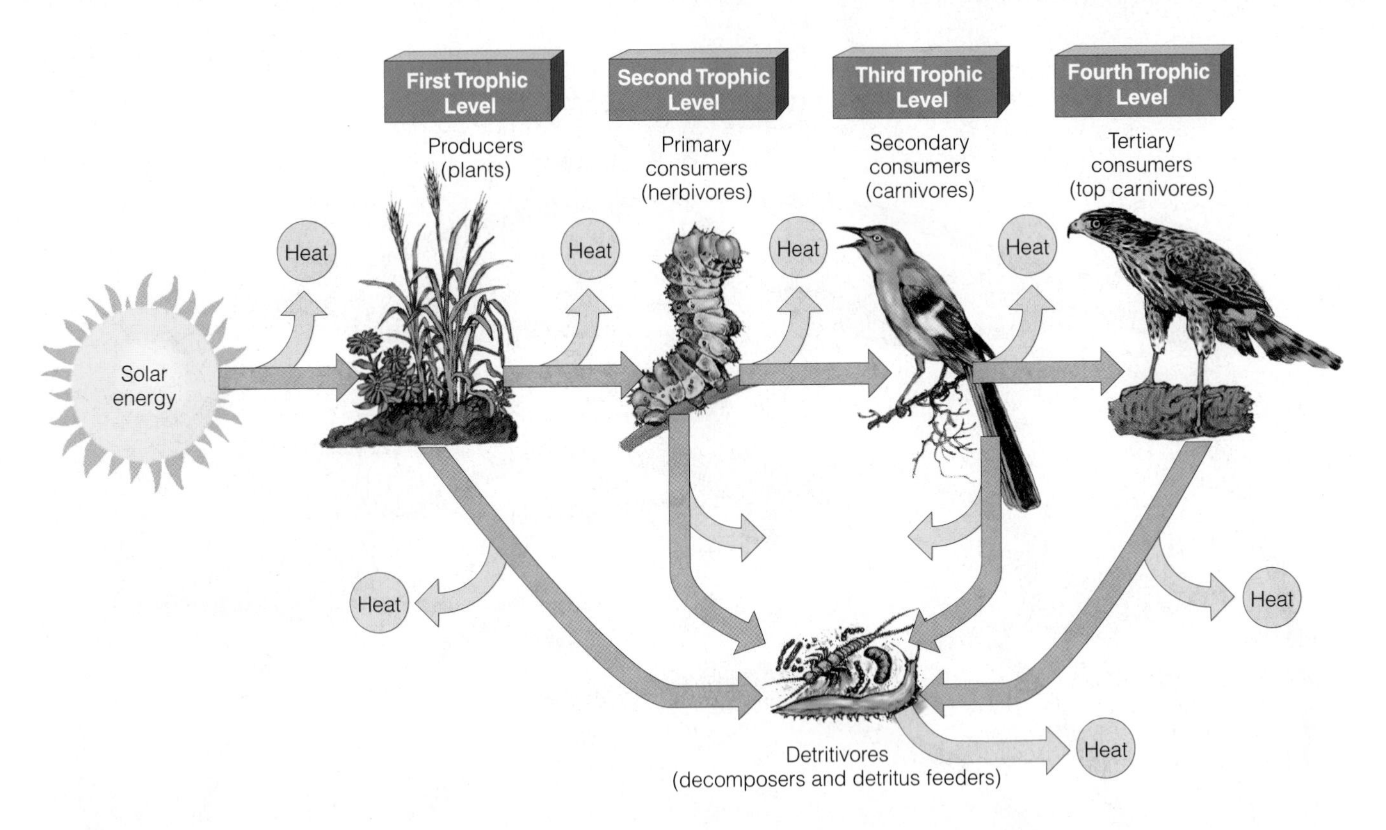

Figure 4-17 Model of a food chain. The arrows show how chemical energy in food flows through various *trophic levels* or energy transfers. Most of the energy flowing through the chain is degraded to heat, in accordance with the second law of energy. Food chains rarely have more than four trophic levels. Can you explain why?

Ecologists assign each organism in an ecosystem to a *feeding level*, or **trophic level** (from the Greek word *trophos*, "nourishment"), depending on whether it is a producer or a consumer and on what it eats or decomposes. Producers belong to the first trophic level, primary consumers to the second trophic level, secondary consumers to the third, and so on. Detritivores and decomposers process detritus from all trophic levels.

Real ecosystems are more complex than this. Most consumers feed on more than one type of organism, and most organisms are eaten by more than one type of consumer. Because most species participate in several different food chains, the organisms in most ecosystems form a complex network of interconnected food chains called a **food web** (Figure 4-18). Trophic levels can be assigned in food webs just as in food chains.

How Can We Represent the Energy Flow in an Ecosystem? Pyramids of Energy Flow Each trophic level in a food chain or web contains a certain amount of **biomass**, the dry weight of all organic matter contained in its organisms. In a food chain or web, chemical energy stored in biomass is transferred from one trophic level to another. With each transfer some usable energy is degraded and lost to the environment as low-quality heat. Thus, **(1)** only a small portion of what is eaten and digested is actually converted into an organism's bodily material or biomass, and **(2)** the amount of usable energy available to each successive trophic level declines.

The percentage of usable energy transferred as biomass from one trophic level to the next is called **ecological efficiency**. It ranges from 5% to 20% (that is, a loss of 80–95%) depending on the types of species and the ecosystem involved, but 10% is typical.

Assuming 10% ecological efficiency (90% loss) at each trophic transfer, if green plants in an area manage to capture 10,000 units of energy from the sun, then only about 1,000 units of energy will be available to support herbivores and only about 100 units to support carnivores.

The more trophic levels or steps in a food chain or web, the greater the cumulative loss of usable energy as energy flows through the various trophic levels. The **pyramid of energy flow*** in Figure 4-19 illustrates this energy loss for a simple food chain, assuming a 90%

*Because such pyramids represent energy flows, not energy storage, they should not be called pyramids of energy (a common error in some biology and environmental science textbooks).

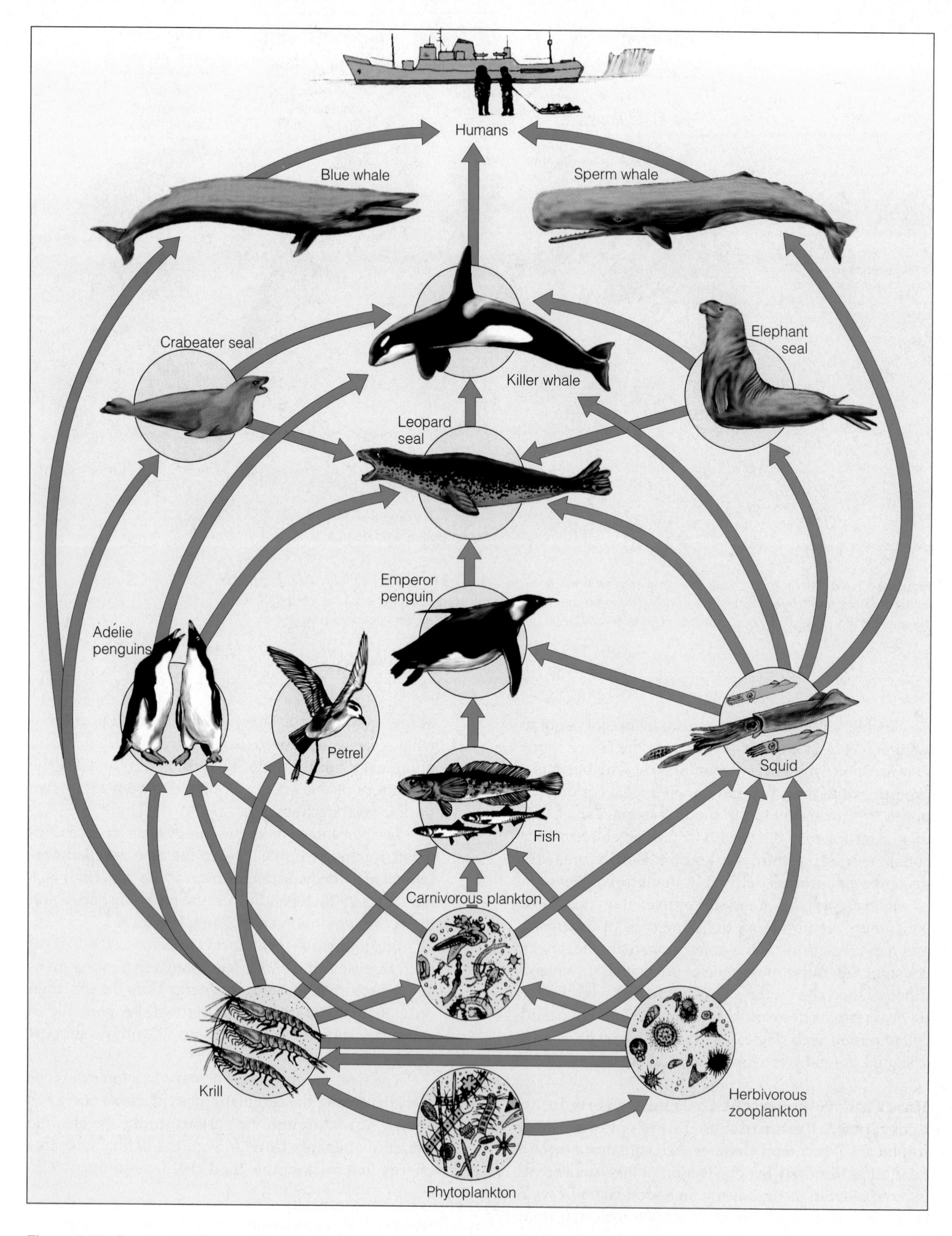

Figure 4-18 Greatly simplified food web in the Antarctic. Many more participants in the web, including an array of decomposer organisms, are not depicted here.

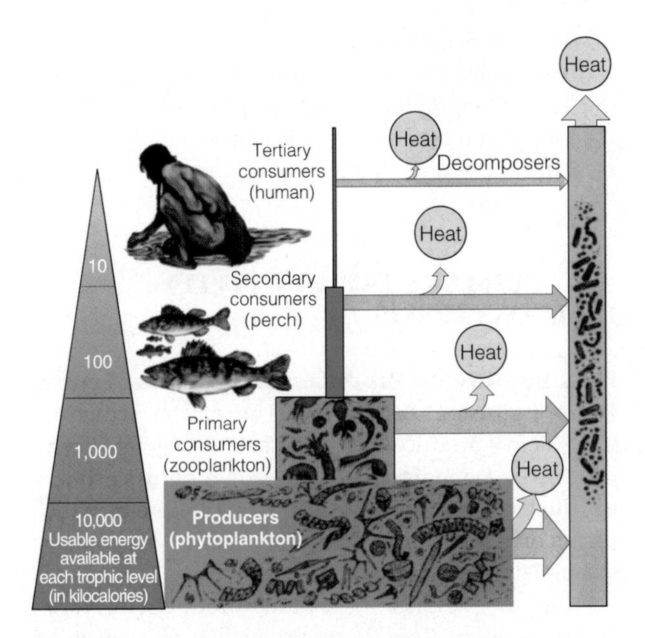

Figure 4-19 Generalized *pyramid of energy flow* showing the decrease in usable energy available at each succeeding trophic level in a food chain or web. In nature, ecological efficiency varies from 5% to 20%, with 10% efficiency being common. This model assumes a 10% ecological efficiency (90% loss in usable energy to the environment, in the form of low-quality heat) with each transfer from one trophic level to another.

energy loss with each transfer. Figure 4-20 shows the pyramid of energy flow during 1 year for an aquatic ecosystem in Silver Springs, Florida. Pyramids of energy flow *always* have an upright pyramidal shape because of the automatic degradation of energy quality required by the second law of energy.

Energy flow pyramids explain why the earth can support more people if they eat at lower trophic levels by consuming grains, vegetables, and fruits directly (for example, grain → human) rather than passing such crops through another trophic level and eating grain eaters (grain → steer → human).

The large energy loss between successive trophic levels also explains why food chains and webs rarely have more than four or five trophic levels. In most cases, too little energy is left after four or five transfers to support organisms feeding at these high trophic levels. This explains why **(1)** there are so few top carnivores such as eagles, hawks, tigers, and white sharks, **(2)** such species usually are the first to suffer when the ecosystems that support them are disrupted, and **(3)** these species are so vulnerable to extinction.

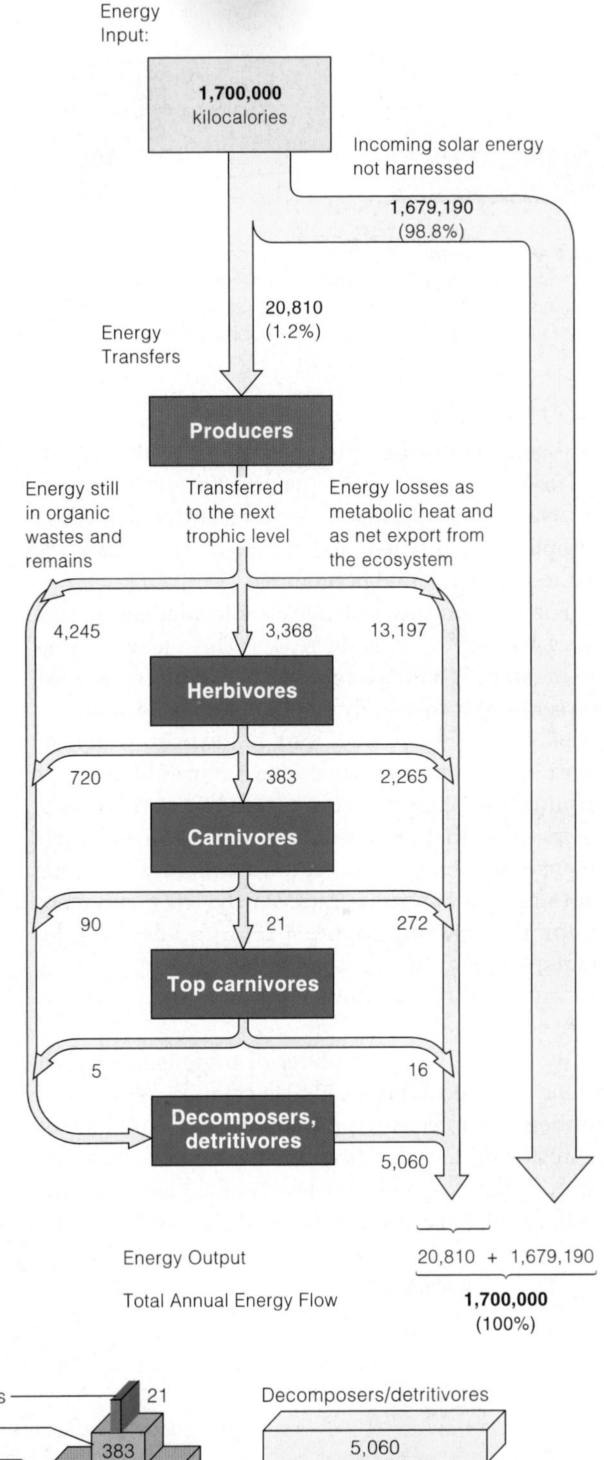

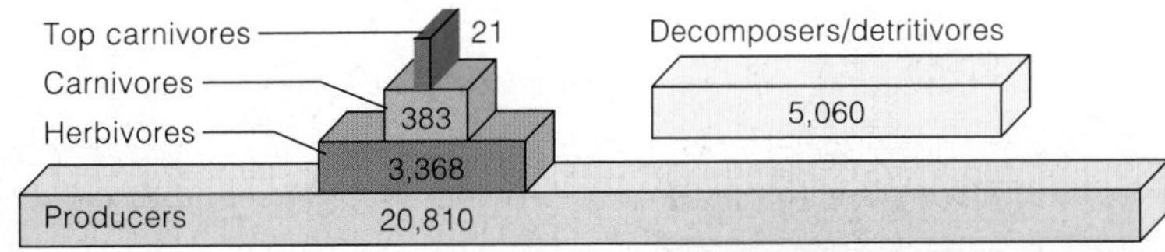

Figure 4-20 Annual pyramid of energy flow (in kilocalories per square meter per year) for an aquatic ecosystem in Silver Springs, Florida. The pyramid is constructed by using the data on energy flow through this ecosystem shown in the top drawing. (From Cecie Starr, *Biology: Concepts and Applications*, 4th ed., Pacific Grove, Calif.: Brooks/Cole, 2000)

How Can We Represent Biomass Storage in an Ecosystem? Pyramids of Biomass and Numbers Ecologists use a **pyramid of biomass** (Figure 4-21) to represent the storage of biomass at various trophic levels in an

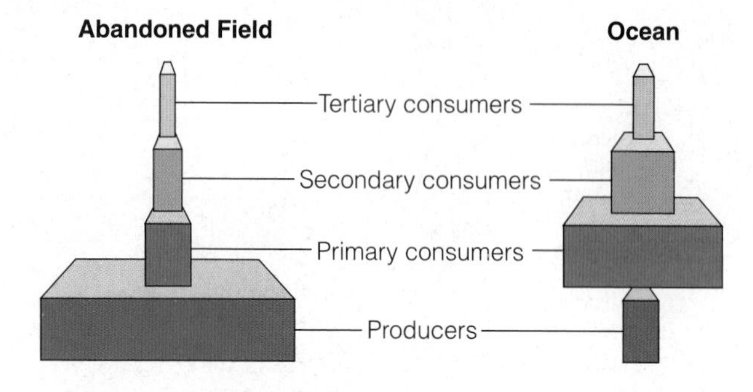

Figure 4-21 Generalized graphs of *biomass of organisms* in the various trophic levels for two ecosystems. The size of each tier in this conceptual model represents the dry weight per square meter of all organisms at that trophic level.

ecosystem. They estimate biomass by harvesting organisms from random patches or narrow strips in an ecosystem. The sample organisms are then sorted according to trophic levels, dried, and weighed. These data are used to plot a pyramid of biomass.

For most land ecosystems, the total biomass at each successive trophic level decreases. This yields a pyramid of biomass with a large base of producers, topped by a series of increasingly smaller biomasses at higher trophic levels (Figure 4-21, left). In the open waters of aquatic ecosystems, however, the biomass of primary consumers (zooplankton) can exceed that of producers. The reason is that the producers are microscopic phytoplankton that grow and reproduce rapidly, not large plants that grow and reproduce slowly. The graph is not an upright pyramid (Figure 4-21, right) because the zooplankton eat the phytoplankton almost as fast as they are produced so that the producer population is never very large.

By estimating the number of organisms at each trophic level, ecologists can also create a **pyramid of numbers** for an ecosystem (Figure 4-22). Numbers of organisms for grasslands and many other ecosystems taper off from the producer level to the higher trophic levels, forming an upright pyramid (Figure 4-22, left).

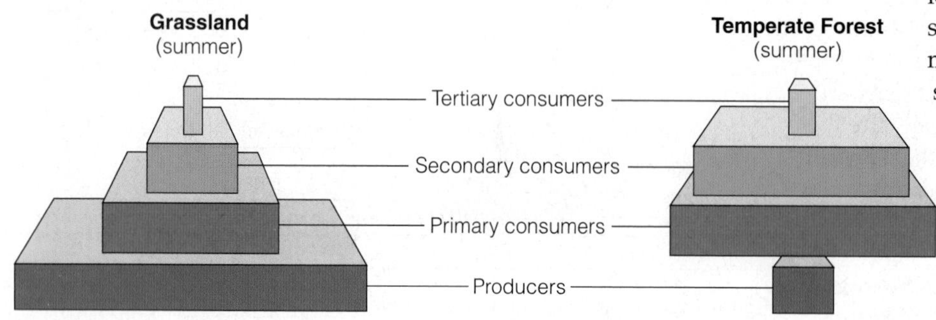

Figure 4-22 Generalized graphs of *numbers of organisms* in the various trophic levels for two ecosystems.

For other ecosystems, however, the graph can take a different shape. For example, a temperate forest (Figure 4-22, right) has a few large producers (the trees) that support a much larger number of small primary consumers (insects) that feed on the trees.

4-5 PRIMARY PRODUCTIVITY OF ECOSYSTEMS

How Rapidly Do Producers in Different Ecosystems Produce Biomass? The *rate* at which an ecosystem's producers convert solar energy into chemical energy as biomass is the ecosystem's **gross primary productivity (GPP)**. In effect, it is the rate at which plants or other producers use photosynthesis to make more plant material (biomass).

Figure 4-23 shows how this productivity varies across the earth. This figure shows that gross primary productivity generally is greatest in **(1)** the shallow waters near continents, **(2)** along coral reefs where abundant light, heat, and nutrients stimulate the growth of algae, and **(3)** where upwelling currents bring nitrogen and phosphorus from the ocean bottom to the surface. The lowest gross primary productivity is in **(1)** deserts and other arid regions because of their low precipitation and high temperatures and **(2)** the open ocean because of a lack of nutrients and sunlight except near the surface.

To stay alive, grow, and reproduce, an ecosystem's producers must use some of the total biomass they produce for their own respiration. Only what is left, called **net primary productivity (NPP)**, is available for use as food by other organisms (consumers) in an ecosystem:

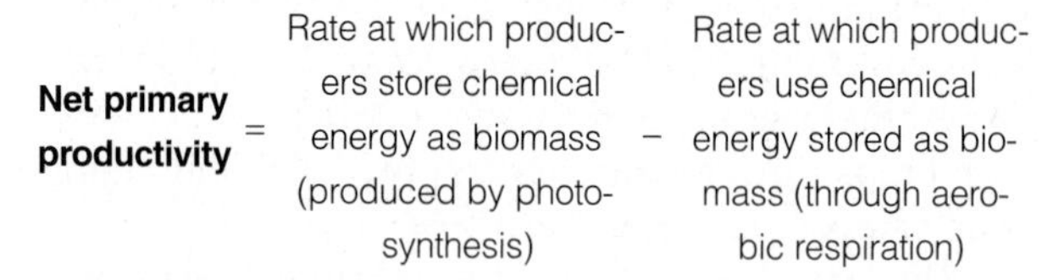

Net primary productivity is the *rate* at which energy for use by consumers is stored in new biomass (cells, leaves, roots, and stems). It is measured in units of the energy or biomass available to consumers in a specified area over a given time. It is typically measured in **(1)** kilocalories per square meter per year (kcal/m^2/yr) or **(2)** grams of biomass created per square meter per year (g/m^2/yr).

Various ecosystems and life zones differ in their net primary productivity (Figure 4-24). The most productive are

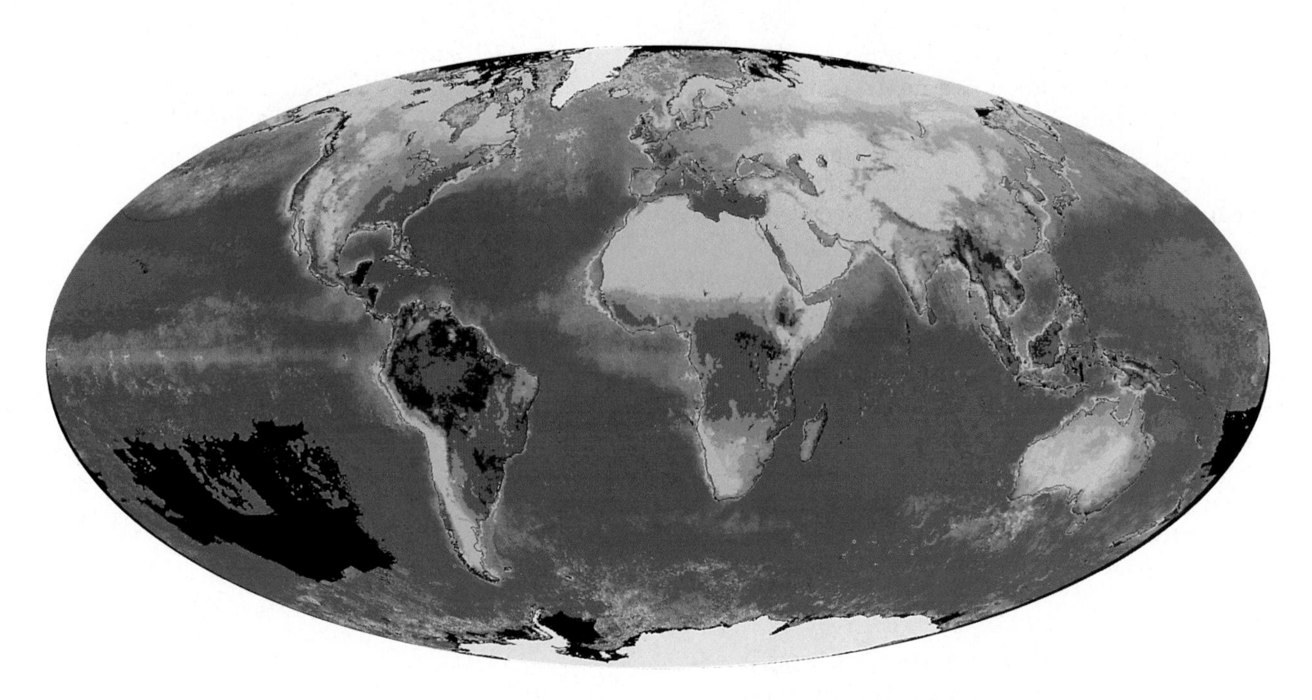

Figure 4-23 Three years of satellite data on the earth's gross primary productivity. Rain forests and other highly productive areas appear as dark green and deserts as yellow. The concentration of phytoplankton, a primary indicator of ocean productivity, ranges from red (highest) to orange, yellow, green, and blue (lowest). (Gene Carl Feldman, Compton J. Tucker-NASA/Goddard Space Flight Center)

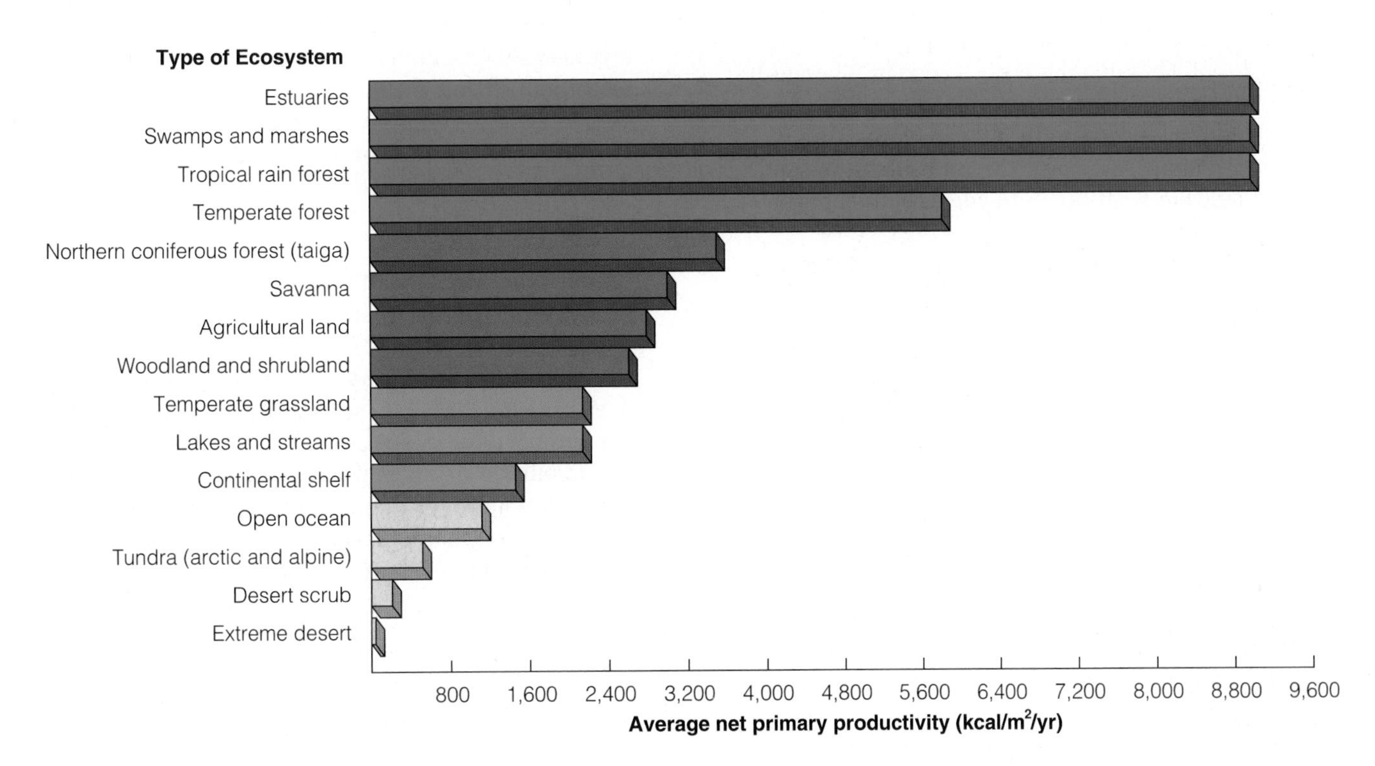

Figure 4-24 Estimated annual average *net primary productivity* (NPP) per unit of area in major life zones and ecosystems, expressed as kilocalories of energy produced per square meter per year (kcal/m^2/yr). (Data from R. H. Whittaker, *Communities and Ecosystems*, 2nd ed., New York: Macmillan, 1975)

(1) estuaries, **(2)** swamps and marshes, and **(3)** tropical rain forests. The least productive are **(1)** open ocean, **(2)** tundra (arctic and alpine grasslands), and **(3)** desert. Despite its low net primary productivity, there is so much open ocean that it produces more of the earth's net primary productivity per year than any of the other ecosystems and life zones shown in Figure 4-24.

Agricultural land is a highly modified and managed ecosystem. The goal is to increase the net primary productivity and biomass of selected crop plants by adding water (irrigation) and nutrients (fertilizers). Nitrogen as nitrate (NO_3^-) and phosphorus as phosphate (PO_4^{3-}) are the most common nutrients in fertilizers because they are most often the nutrients limiting crop growth. Despite such inputs, the net primary productivity of agricultural land is not particularly high compared with that of other ecosystems (Figure 4-24).

How Does the World's Net Rate of Biomass Production Limit the Populations of Consumer Species? Ultimately, the planet's net primary productivity limits the number of consumers (including humans) that can survive on the earth. In other words, *the earth's net primary productivity is the upper limit determining the planet's carrying capacity for all consumer species.*

It is tempting to conclude from Figure 4-24 that a good way to feed the world's hungry millions would be to harvest plants in estuaries, swamps, and marshes. Ecologists point out that this is not a good idea because **(1)** most plants in estuaries, swamps, and marshes cannot be eaten by people and **(2)** these plants are vital food sources (and spawning areas) for fish, shrimp, and other aquatic life-forms that provide us and other consumers with protein.

We might also conclude from Figure 4-24 that we could grow more food for human consumption by clearing tropical forests and planting food crops. According to most ecologists, this is also a bad idea. The basic problem is that in tropical forests most of the nutrients needed to grow food crops are stored in the vegetation rather than in the soil. When the trees are removed, the nutrient-poor soils are rapidly depleted of their nutrients by frequent rains and growing crops. Crops can be grown only for a short time without massive and expensive applications of commercial fertilizers.

Because the earth's vast open oceans provide the largest percentage of the earth's net primary productivity, why not harvest its primary producers (floating and drifting phytoplankton) to help feed the rapidly growing human population? The problem is that harvesting the widely dispersed, tiny floating producers in the open ocean would **(1)** take much more fossil fuel and other types of energy than the food energy we would get and **(2)** disrupt the food webs of the open ocean (Figure 4-18) that provide us and other consumer organisms with important sources of energy and protein from fish and shellfish.

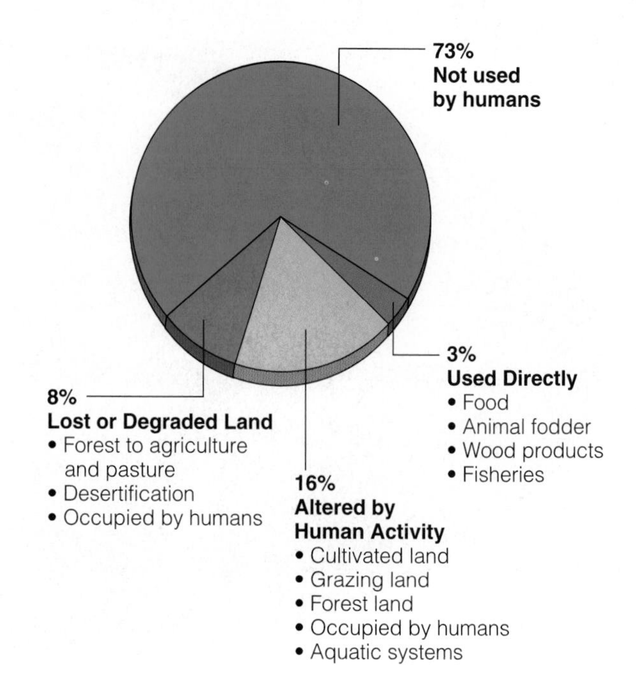

Figure 4-25 Human use of the biomass produced by photosynthesis. Humans destroy, alter, and directly use about **(1)** 27% of the earth's total net primary productivity and **(2)** 40% of the net primary productivity of the earth's terrestrial ecosystems. (Data from Peter Vitousek)

How Much of the World's Net Rate of Biomass Production Do We Use? Peter Vitousek and other ecologists estimate that humans now use, waste, or destroy about **(1)** 27% of the earth's total potential net primary productivity and **(2)** 40% of the net primary productivity of the planet's terrestrial ecosystems (Figure 4-25).

This is the main reason why we are crowding out or eliminating the habitats and food supplies of a growing number of other species. What might happen to us and to other consumer species if **(1)** the human population doubles over the next 40–50 years and **(2)** per capita consumption of resources such as food, timber, and grassland rises sharply?

4-6 CONNECTIONS: MATTER CYCLING IN ECOSYSTEMS

What Are Biogeochemical Cycles? A **nutrient** is any atom, ion, or molecule an organism needs to live, grow, or reproduce. Some elements (such as carbon, oxygen, hydrogen, nitrogen, phosphorus, sulfur, and calcium) are needed in fairly large amounts, whereas others (such as sodium, zinc, copper, and iodine) are needed in small or even trace amounts.

These nutrient atoms, ions, and molecules are continuously cycled from the nonliving environment (air, water, soil, and rock) to living organisms (biota) and then back again in what are called **nutrient cycles**, or **biogeochemical cycles** (literally, life–earth–chemical cycles). These cycles, driven directly or indirectly by incoming solar energy and gravity, include the carbon, oxygen, nitrogen, phosphorus, and hydrologic (water) cycles (Figure 4-7).

The earth's chemical cycles also connect past, present, and future forms of life. Some of the carbon atoms in your skin may once have been part of a leaf, a dinosaur's skin, or a layer of limestone rock. Your grandmother, Plato, or a hunter–gatherer who lived 25,000 years ago may have inhaled some of the oxygen molecules you just inhaled.

How Is Water Cycled in the Biosphere? The **hydrologic cycle**, or **water cycle**, which collects, purifies, and distributes the earth's fixed supply of water, is shown in simplified form in Figure 4-26.

The main processes in this water recycling and purifying cycle are **(1)** *evaporation* (conversion of water into water vapor), **(2)** *transpiration* (evaporation from leaves of water extracted from soil by roots and transported throughout the plant), **(3)** *condensation* (conversion of water vapor into droplets of liquid water), **(4)** *precipitation* (rain, sleet, hail, and snow), **(5)** *infiltration* (movement of water into soil), **(6)** *percolation* (downward flow of water through soil and permeable rock formations to groundwater storage areas called aquifers), and **(7)** *runoff* (downslope surface movement back to the sea to resume the cycle).

The water cycle is powered by energy from the sun and by gravity. Incoming solar energy evaporates water from oceans, streams, lakes, soil, and vegetation. About 84% of water vapor in the atmosphere comes from the oceans, and the rest comes from land.

Winds and air masses transport water vapor over various parts of the earth's surface, often over long distances. Falling temperatures cause the water vapor to condense into tiny droplets that form clouds or fog. For precipitation to occur, air must contain **condensation nuclei**: tiny particles on which droplets of water vapor can collect. Sources of such particles include **(1)** volcanic ash, **(2)** soil dust, **(3)** smoke, **(4)** sea salts, and **(5)** particulate matter emitted by factories, coal-burning power plants, and motor vehicles. The temperature at which condensation occurs is called the **dew point**.

Some of the fresh water returning to the earth's surface as precipitation becomes locked in glaciers. Most of the precipitation falling on terrestrial ecosystems becomes *surface runoff* flowing into streams and lakes, which eventually carry water back to the oceans, where it can be evaporated to cycle again.

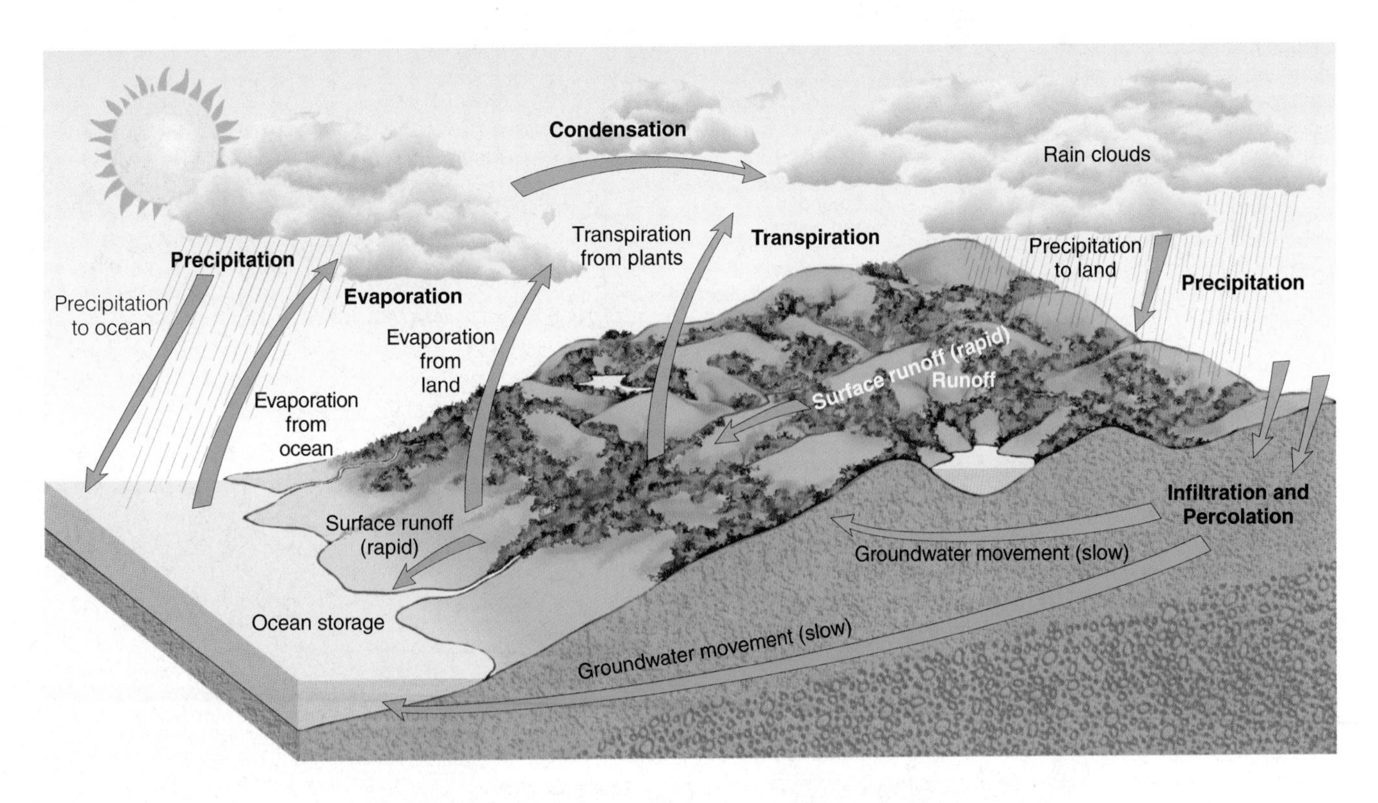

Figure 4-26 Simplified model of the hydrologic cycle.

Besides replenishing streams and lakes, surface runoff also causes soil erosion, which moves soil and weathered rock fragments from one place to another. Water is thus the primary sculptor of the earth's landscape. Because water dissolves many nutrient compounds, it is a major medium for transporting nutrients within and between ecosystems.

Some of the water returning to the land **(1)** soaks into (infiltrates) the soil and porous rock and **(2)** percolates downward, dissolving minerals from porous rocks on the way. This water is stored as *groundwater* in the pores and cracks of rocks. Where the pores are joined, a network of water channels allows water to flow through the porous rock. Such water-laden rock is called an *aquifer*, and the level of the earth's land crust to which it is filled is called the *water table*. This underground water flows slowly downhill through rock pores and seeps out into streams and lakes or comes out in springs. Eventually, this water evaporates or reaches the sea to continue the cycle.

Throughout the hydrologic cycle, many natural processes act to purify water. Evaporation and subsequent precipitation act as a natural distillation process that removes impurities dissolved in water. Water flowing above ground through streams and lakes and below ground in aquifers is naturally filtered and purified by chemical and biological processes. Thus, the hydrologic cycle also can be viewed a cycle of natural renewal of water quality.

How Are Human Activities Affecting the Water Cycle? During the past 100 years, we have been intervening in the earth's current water cycle in several ways:

- Withdrawing large quantities of fresh water from streams, lakes, and underground sources. In heavily populated or heavily irrigated areas, withdrawals have led to groundwater depletion or intrusion of ocean salt water into underground water supplies.
- Clearing vegetation from land for agriculture, mining, road and building construction, and other activities. This **(1)** increases runoff, **(2)** reduces infiltration that recharges groundwater supplies, **(3)** increases the risk of flooding, and **(4)** accelerates soil erosion and landslides.
- Modifying water quality by **(1)** adding nutrients (such as phosphates and nitrates found in fertilizers) and other pollutants and **(2)** changing ecological processes that purify water naturally. Chapter 14

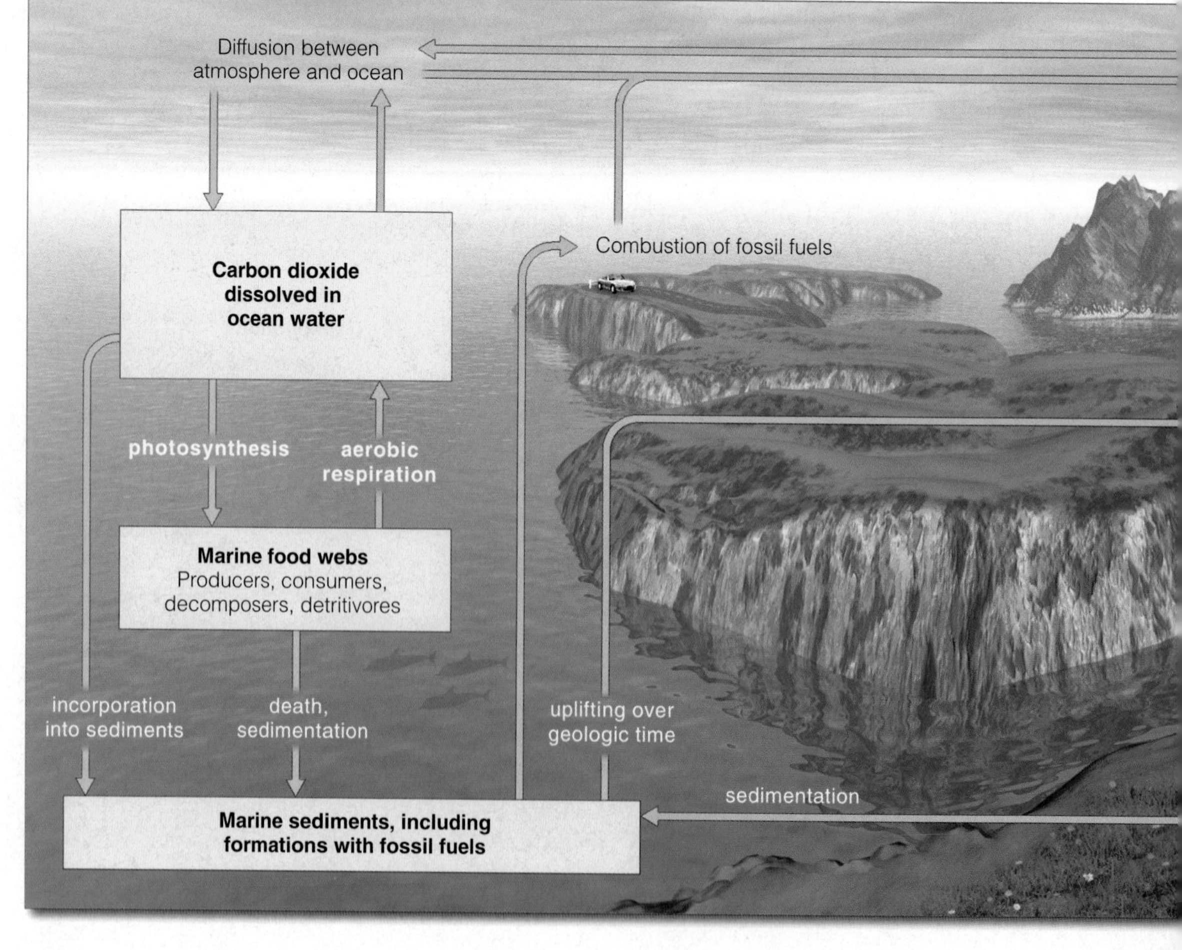

Figure 4-27 Simplified model of the global carbon cycle. The left portion shows the movement of carbon through marine systems, and the right portion shows its movement through terrestrial ecosystems. Carbon reservoirs are shown as boxes; processes that change one form of carbon to another are shown in unboxed print. (From Cecie Starr, *Biology: Concepts and Applications*, 4th ed., Pacific Grove, Calif.: Brooks/Cole, 2000)

(p. 327) examines issues of water resource use and water quality.

How Is Carbon Cycled in the Biosphere? Carbon is essential to life as we know it. It is the basic building block of the carbohydrates, fats, proteins, DNA, and other organic compounds necessary for life.

The **carbon cycle** (Figure 4-27) is based on carbon dioxide gas, which makes up 0.036% of the volume of the troposphere and is also dissolved in water. Carbon dioxide is a key component of nature's thermostat. If the carbon cycle removes too much CO_2 from the atmosphere, the atmosphere will cool; if the cycle generates too much, the atmosphere will get warmer. Thus, even slight changes in the carbon cycle can affect climate and ultimately the types of life that can exist on various parts of the planet.

Terrestrial producers remove CO_2 from the atmosphere and aquatic producers remove it from the water. They then use photosynthesis to convert CO_2 into complex carbohydrates such as glucose ($C_6H_{12}O_6$).

The cells in oxygen-consuming producers, consumers, and decomposers then carry out aerobic respiration. This breaks down glucose and other complex organic compounds and converts the carbon back to CO_2 in the atmosphere or water for reuse by producers. This linkage between *photosynthesis* in producers and *aerobic respiration* in producers, consumers, and decomposers circulates carbon in the biosphere and is a major part of the global carbon cycle. Oxygen and hydrogen, the other elements in carbohydrates, cycle almost in step with carbon.

Over millions of years, buried deposits of dead plant matter and bacteria are compressed between layers of sediment, where they form carbon-containing *fossil fuels* such as coal and oil (Figure 4-27). This carbon is not released to the atmosphere as CO_2 for recycling until **(1)** these fuels are extracted and burned or **(2)** long-term geological processes expose these deposits to air. In only a few hundred years, we have extracted and burned fossil fuels that took millions of years to form. This is why fossil fuels are nonrenewable resources on a human time scale.

Rocks such as limestone ($CaCO_3$) deposited as sediments on the ocean floor and on continents are the largest storage reservoir for the earth's carbon. This carbon reenters the cycle very slowly, when some of the sediments dissolve and form dissolved CO_2 gas that can enter the atmosphere. Geologic processes can also bring bottom sediments to the surface, exposing the

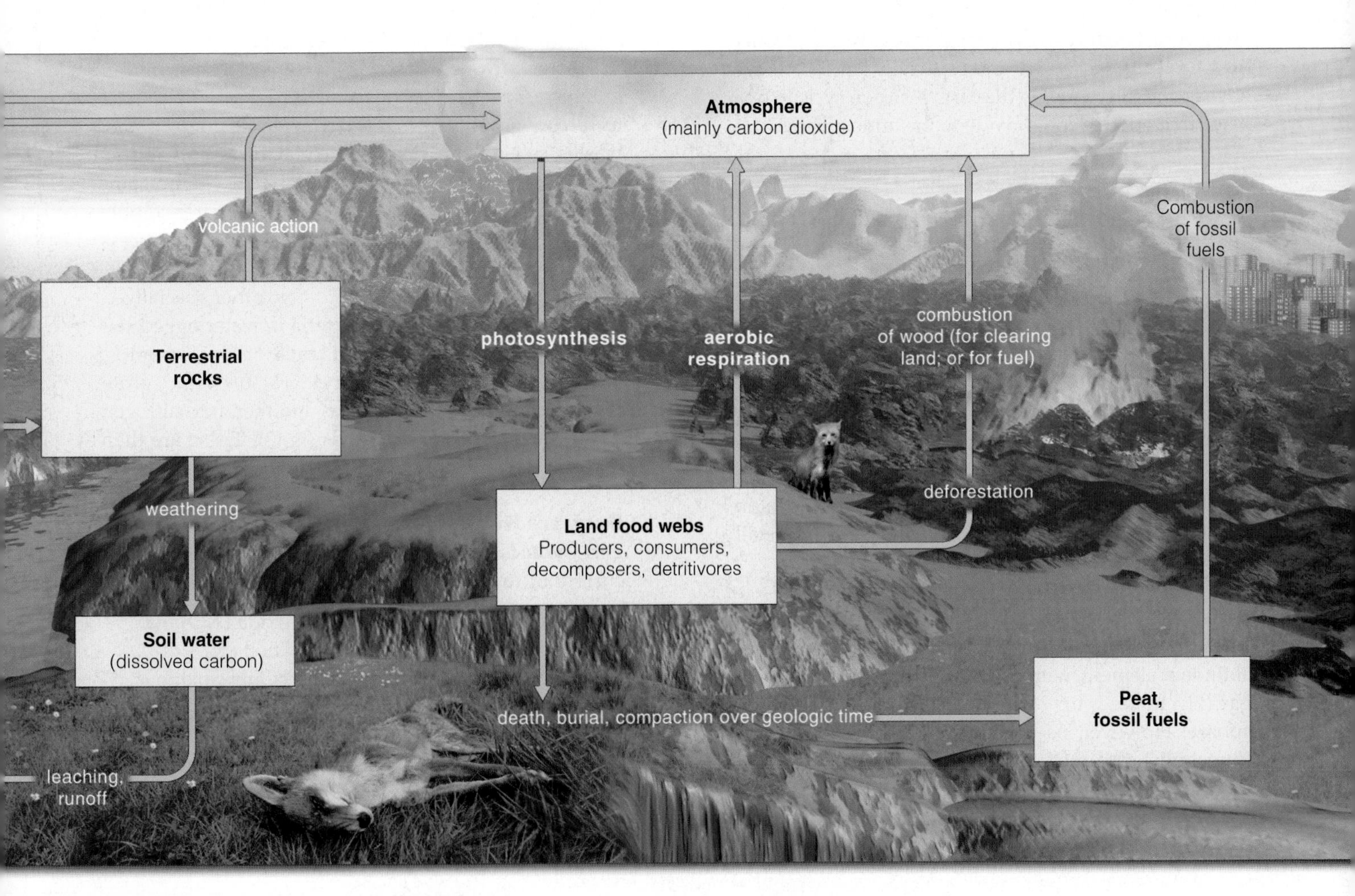

carbonate rock to chemical attack by oxygen and converting it to carbon dioxide gas. Carbon dioxide also is released into the atmosphere when acidic rain falls on and dissolves exposed limestone rock.

The oceans are the second largest storage reservoir in the carbon cycle. Oceans also play a major role in regulating the level of carbon dioxide in the atmosphere. Some carbon dioxide gas, which is readily soluble in water, **(1)** stays dissolved in the sea, **(2)** some is removed by photosynthesizing producers, and **(3)** some reacts with seawater to form carbonate ions (CO_3^{2-}) and bicarbonate ions (HCO_3^-). As water warms, more dissolved CO_2 returns to the atmosphere, just as more carbon dioxide fizzes out of a carbonated beverage when it warms.

In marine ecosystems, some organisms take up dissolved CO_2 molecules, carbonate ions, or bicarbonate ions from ocean water. These ions can then react with calcium ions (Ca^{2+}) in seawater to form slightly soluble carbonate compounds such as calcium carbonate ($CaCO_3$) to build the shells and skeletons of marine organisms. When these organisms die, tiny particles of their shells and bone drift slowly to the ocean depths and are buried for eons (as long as 400 million years) in deep bottom sediments (Figure 4-27). There, under immense pressure, they are converted into limestone rock.

How Are Human Activities Affecting the Carbon Cycle? Since 1800 and especially since 1950, we have been intervening in the earth's carbon cycle in two ways that add carbon dioxide to the atmosphere:

- Clearing trees and other plants that absorb CO_2 through photosynthesis
- Adding large amounts of CO_2 by burning fossil fuels and wood

Computer models of the earth's climate systems suggest that increased concentrations of atmospheric CO_2 and other gases we are adding to the atmosphere could enhance the planet's *natural greenhouse effect* that helps warm the lower atmosphere (troposphere) and the earth's surface (Figure 4-8). The resulting *global warming* could **(1)** disrupt global food production and wildlife habitats and **(2)** raise the average sea level in various parts of the world, as discussed in more detail in Section 13-4 (p. 311).

How Is Nitrogen Cycled in the Biosphere? Bacteria in Action Nitrogen is the atmosphere's most abundant element, with chemically unreactive nitrogen gas (N_2) making up 78% of the volume of the troposphere. However, N_2 cannot be absorbed and used (metabolized) directly as a nutrient by multicellular plants or animals.

Thus, nitrogen must be "fixed" or combined with hydrogen or oxygen to provide compounds that plants can use. Fortunately, **(1)** atmospheric electrical discharges in the form of lightning (which causes nitrogen and oxygen in the atmosphere to react and produce oxides of nitrogen, such as $N_2 + O_2 \rightarrow 2NO$) and **(2)** certain bacteria in the soil and aquatic systems convert nitrogen gas into compounds that can enter food webs as part of the **nitrogen cycle** (Figure 4-28).

In the first step in the nitrogen cycle, called *nitrogen fixation*, specialized bacteria convert gaseous nitrogen (N_2) to ammonia (NH_3) that can be used by plants by the reaction $N_2 + 3H_2 \rightarrow 2NH_3$. This is done mostly by **(1)** cyanobacteria in soil and water and **(2)** *Rhizobium* bacteria living in small nodules (swellings) on the root systems of a wide variety of plant species, including soybeans and alfalfa.

In a two-step process called *nitrification*, most of the ammonia in soil is converted by specialized aerobic bacteria to nitrite ions (NO_2^-), which are toxic to plants, and then to nitrate ions (NO_3^-), which are easily taken up by plants as a nutrient.

In a process called *assimilation*, plant roots then absorb inorganic ammonia, ammonium ions, and nitrate ions formed by nitrogen fixation and nitrification in soil water. They use these ions to make nitrogen-containing organic molecules such as DNA, amino acids, and proteins. Animals in turn get their nitrogen by eating plants or plant-eating animals.

After nitrogen has served its purpose in living organisms, vast armies of specialized decomposer bacteria convert the nitrogen-rich organic compounds, wastes, cast-off particles, and dead bodies of organisms into **(1)** simpler nitrogen-containing inorganic compounds such as ammonia (NH_3) and **(2)** water-soluble salts containing ammonium ions (NH_4^+). This process is known as *ammonification*.

In a process called *denitrification*, other specialized bacteria (mostly anaerobic bacteria in waterlogged soil or in the bottom sediments of lakes, oceans, swamps, and bogs) then convert NH_3 and NH_4^+ back into nitrite (NO_2^-) and nitrate (NO_3^-) ions and then into nitrogen gas (N_2) and nitrous oxide gas (N_2O). These are then released to the atmosphere to begin the cycle again.

How Are Human Activities Affecting the Nitrogen Cycle? Major human interventions in the earth's current nitrogen cycle over the past 100 years include

- Adding large amounts of nitric oxide (NO) into the atmosphere when we burn any fuel ($N_2 + O_2 \rightarrow 2NO$). In the atmosphere, this nitric oxide combines with oxygen to form nitrogen dioxide gas (NO_2), which can react with water vapor to form nitric acid (HNO_3). Droplets of HNO_3 dissolved in rain or snow are components of *acid deposition*, commonly called *acid rain*. Nitric acid, along with other air pollutants, can **(1)** damage and weaken trees,

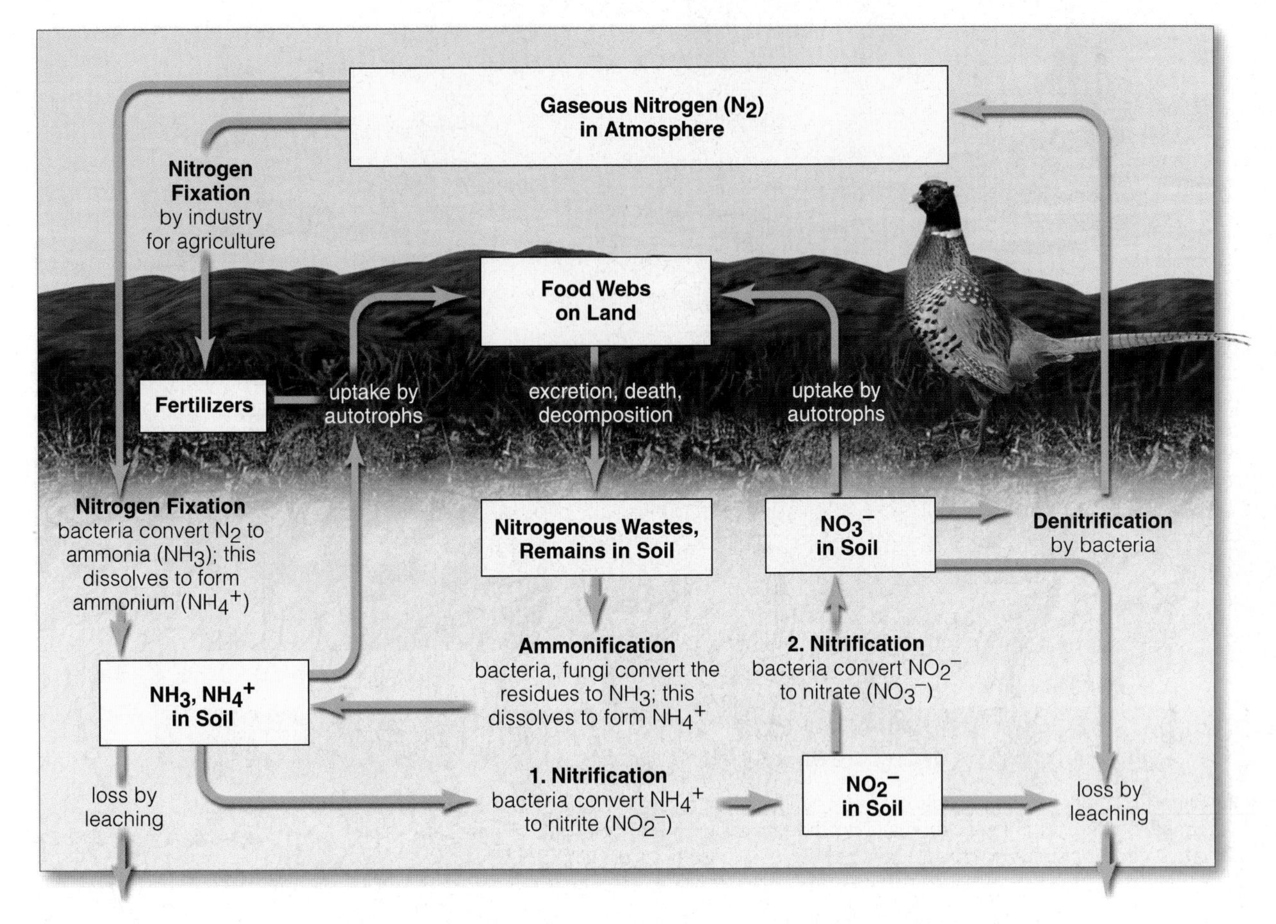

Figure 4-28 Greatly simplified model of the nitrogen cycle in a terrestrial ecosystem. Nitrogen reservoirs are shown as boxes; processes changing one form of nitrogen to another are shown in unboxed print. (Adapted from Cecie Starr and Ralph Taggart, *Biology: The Unity and Diversity of Life*, 9th ed., Pacific Grove, Calif.: Brooks/Cole, 2001)

(2) upset aquatic ecosystems, **(3)** corrode metals, and **(4)** damage marble, stone, and other building materials, as discussed in Section 12-6 (p. 291).

- Adding nitrous oxide (N_2O) to the atmosphere through the action of anaerobic bacteria on livestock wastes and commercial inorganic fertilizers applied to the soil. When N_2O reaches the stratosphere, it can **(1)** help warm the atmosphere by enhancing the natural greenhouse effect and **(2)** contribute to depletion of the earth's ozone shield, which filters out harmful ultraviolet radiation from the sun.

- Removing nitrogen from topsoil when we **(1)** harvest nitrogen-rich crops, **(2)** irrigate crops, and **(3)** burn or clear grasslands and forests before planting crops.

- Adding nitrogen compounds to aquatic ecosystems in agricultural runoff and discharge of municipal sewage. This excess of plant nutrients stimulates rapid growth of photosynthesizing algae and other aquatic plants. The subsequent breakdown of dead algae by aerobic decomposers can **(1)** deplete the water of dissolved oxygen and **(2)** disrupt aquatic ecosystems by killing some types of fish and other oxygen-using (aerobic) organisms, as discussed in Section 14-6 (p. 345).

- Accelerating the deposition of acidic nitrogen compounds (such as NO_2 and HNO_3) from the atmosphere onto terrestrial ecosystems (Section 12-4, p. 284). This excessive input of nitrogen can stimulate the growth of weedy plant species, which can outgrow and perhaps eliminate other plant species that cannot take up nitrogen as efficiently.

How Is Phosphorus Cycled in the Biosphere? Phosphorus circulates through water, the earth's crust, and living organisms in the **phosphorus cycle** (Figure 4-29). Bacteria are less important here than in the nitrogen cycle. Very little phosphorus circulates in the atmosphere because at the earth's normal temperatures and pressures, phosphorus and its compounds are not gases. Phosphorus is found in the atmosphere only as small particles of dust. In contrast to the carbon cycle, the phosphorus cycle is slow, and on a short human

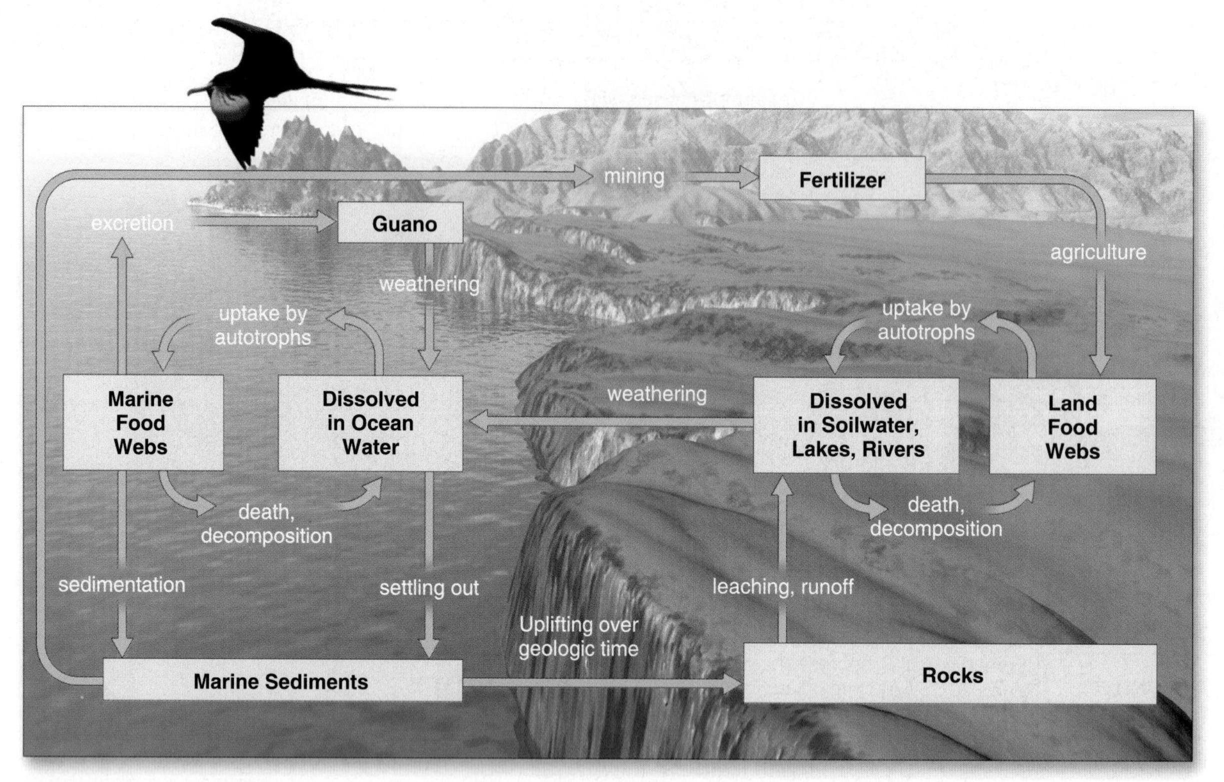

Figure 4-29 Simplified model of the phosphorus cycle. Phosphorus reservoirs are shown as boxes; processes that change one form of phosphorus to another are shown in unboxed print. (From Cecie Starr, *Biology: Concepts and Applications*, 4th ed., Pacific Grove, Calif.: Brooks/Cole, 2000)

time scale much phosphorus flows one way from the land to the oceans.

Phosphorus typically is found as phosphate salts containing phosphate ions (PO_4^{3-}) in terrestrial rock formations and ocean bottom sediments. Because most soils contain little phosphate, it is often the *limiting factor* for plant growth on land unless phosphorus (as phosphate salts mined from the earth) is applied to the soil as a fertilizer. Phosphorus also limits the growth of producer populations in many freshwater streams and lakes because phosphate salts are only slightly soluble in water. This explains why adding phosphate compounds to lakes greatly increases their biological productivity.

How Are Human Activities Affecting the Phosphorus Cycle? During the past 100 years humans have been intervening in the earth's phosphorus cycle by

- Mining large quantities of phosphate rock for use in commercial inorganic fertilizers and detergents.
- Reducing the available phosphate in tropical forests by removing trees. When such forests are cut and burned, most remaining phosphorus and other soil nutrients are washed away by heavy rains, and the land becomes unproductive.
- Adding excess phosphate to aquatic ecosystems in **(1)** runoff of animal wastes from livestock feedlots, **(2)** runoff of commercial phosphate fertilizers from cropland, and **(3)** discharge of municipal sewage. Too much of this nutrient causes explosive growth of cyanobacteria, algae, and aquatic plants. When these plants die and are decomposed they use up dissolved oxygen and disrupt aquatic ecosystems.

How Is Sulfur Cycled in the Biosphere? Sulfur circulates through the biosphere in the **sulfur cycle**, which is a *gaseous cycle* (Figure 4-30). Much of the earth's sulfur is stored underground in rocks and minerals, including sulfate (SO_4^{2-}) salts buried deep under ocean sediments.

Sulfur also enters the atmosphere from several natural sources. Hydrogen sulfide (H_2S) is a colorless, highly poisonous gas with a rotten-egg smell. It is released from active volcanoes and by the breakdown of organic matter in swamps, bogs, and tidal flats caused by decomposers that do not use oxygen (anaerobic decomposers). Sulfur dioxide (SO_2), a colorless, suffocating gas, also comes from volcanoes. Particles of sulfate (SO_4^{2-}) salts, such as ammonium sulfate, enter the atmosphere from sea spray.

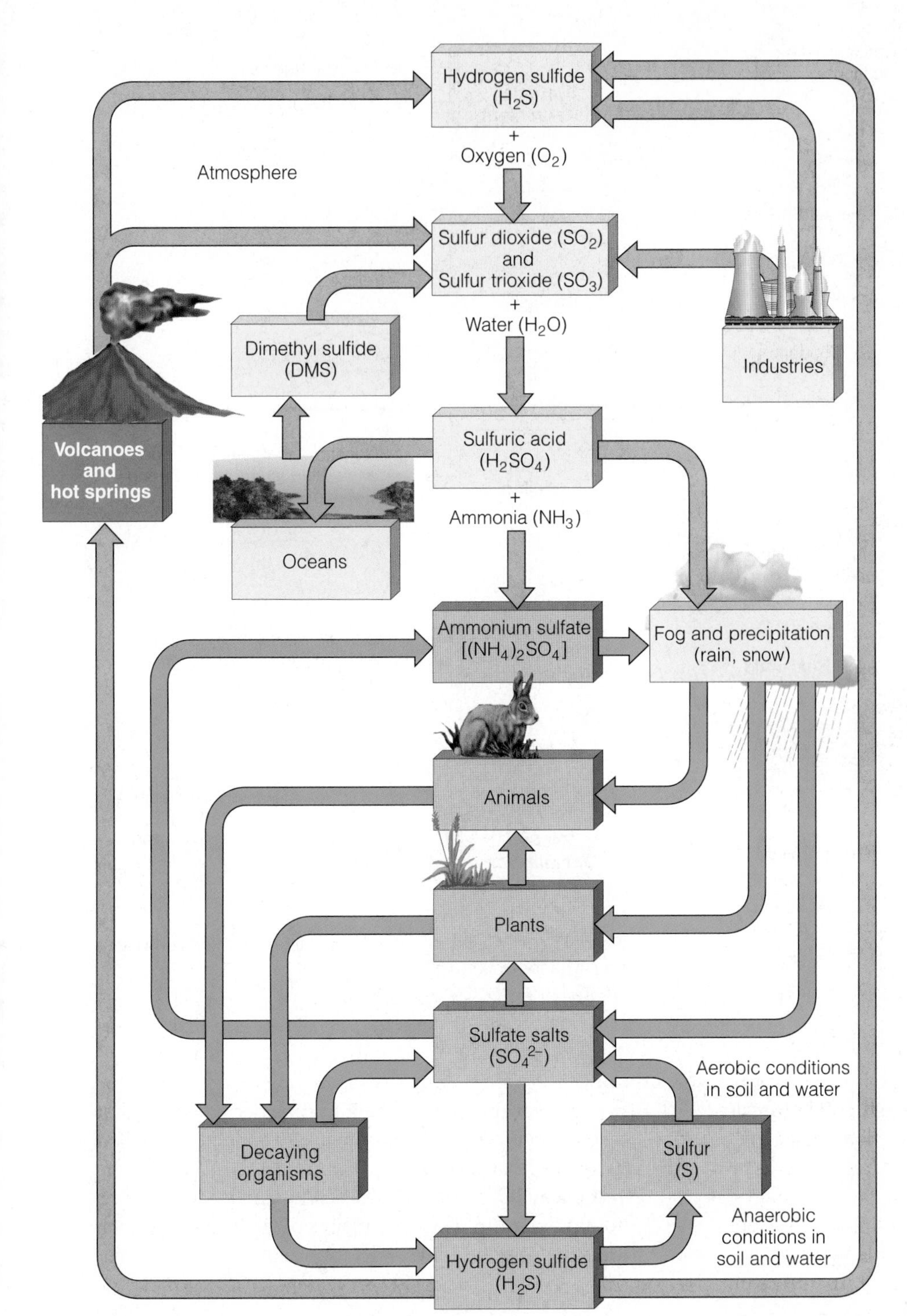

Figure 4-30 Simplified model of the sulfur cycle. Green shows the movement of sulfur compounds in living organisms, blue in aquatic systems, and orange in the atmosphere.

Certain marine algae produce large amounts of volatile dimethyl sulfide, or DMS (CH_3SCH_3). Tiny droplets of DMS serve as nuclei for the condensation of water into droplets found in clouds. Thus, changes in DMS emissions can affect cloud cover and climate.

In the atmosphere, sulfur dioxide reacts with oxygen to produce sulfur trioxide gas (SO_3). Some of the sulfur trioxide then reacts with water droplets in the atmosphere to produce tiny droplets of sulfuric acid (H_2SO_4). Sulfur dioxide also reacts with other chemicals in the atmosphere such as ammonia to produce tiny particles of sulfate salts. These droplets and particles fall to the earth as components of *acid deposition,* which along with other air pollutants can harm trees and aquatic life, as discussed in Section 12-4, p. 283.

How Are Human Activities Affecting the Sulfur Cycle? Over the past 100 years we have been intervening in the atmospheric phase of the earth's current sulfur cycle by

- Burning sulfur-containing coal to produce electric power, producing about two-thirds of the human inputs of sulfur dioxide
- Refining sulfur-containing petroleum to make gasoline, heating oil, and other useful products
- Using smelting to convert sulfur compounds of metallic minerals into free metals such as copper, lead, and zinc

4-7 HOW DO ECOLOGISTS LEARN ABOUT ECOSYSTEMS?

What Is Field Research? Increasingly, ecologists are using new technologies to collect field data. These include **(1)** *remote sensing* from aircraft and satellites and **(2)** *geographic information systems* (*GISs*), in which information gathered from broad geographic regions is stored in a spatial databases (Figure 4-31). Then computers and GIS software can analyze and manipulate the data and combine them with ground and other data to produce computerized maps of various ecological

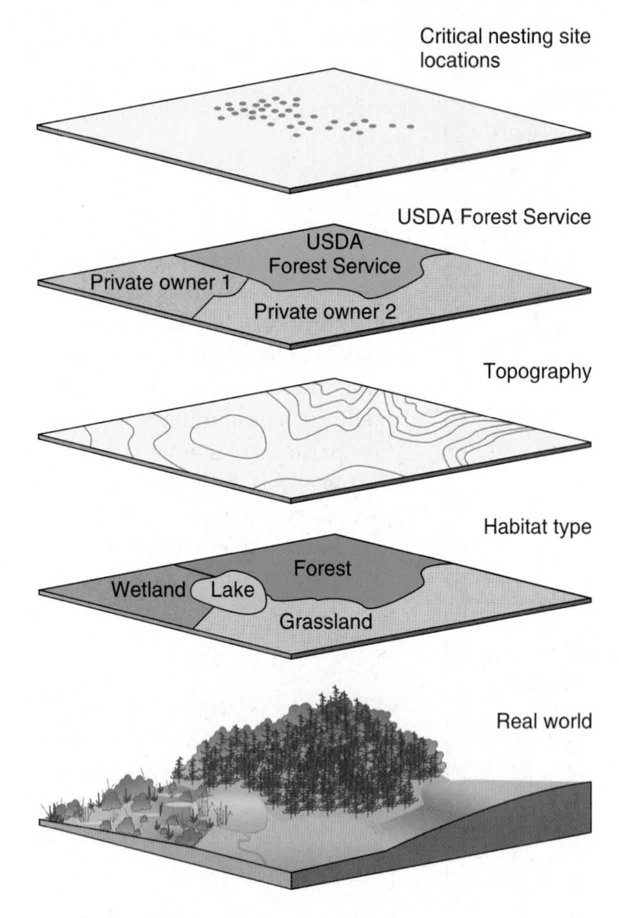

Figure 4-31 Geographic information systems (GISs) provide the computer technology for organizing, storing, and analyzing complex data collected over broad geographic areas. GISs enable scientists to overlay many layers of data (such as soils, topography, distribution of endangered populations, and land protection status).

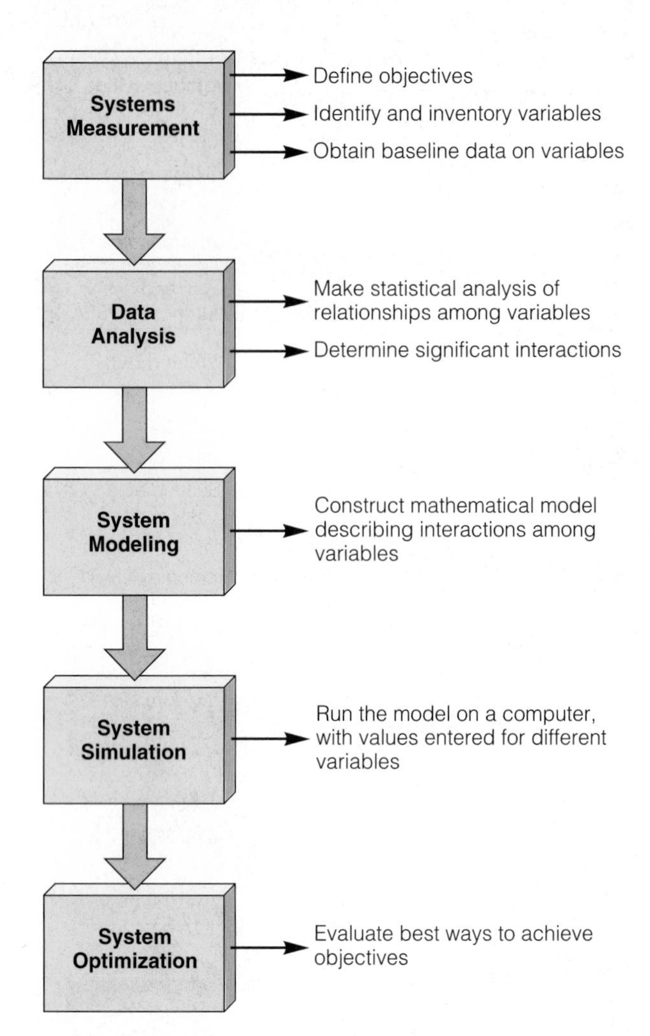

Figure 4-32 Major stages of systems analysis. (Modified data from Charles Southwick)

and environmental data on items such as **(1)** forest cover, **(2)** water resources, **(3)** air pollution emissions, and **(4)** changes in global sea temperatures.

What Is Laboratory Research? In the past 50 years, ecologists have increasingly supplemented field research by using *laboratory research* to set up, observe, and make measurements of model ecosystems and populations under laboratory conditions. Such simplified systems have been set up in containers such as **(1)** culture tubes, **(2)** bottles, **(3)** aquarium tanks, and **(4)** greenhouses and in indoor and outdoor chambers where temperature, light, CO_2, humidity, and other variables can be controlled carefully.

In such systems, it is easier for scientists to carry out controlled experiments. In addition, such laboratory experiments often are quicker and cheaper than similar experiments in the field.

However, it is important to consider whether what scientists observe and measure in a simplified, controlled system under laboratory conditions takes place in the same way in the more complex and dynamic conditions found in nature. Thus, the results of laboratory research must be coupled with and supported by field research.

What Is Systems Analysis? Since the late 1960s ecologists have made increasing use of *systems analysis* to develop mathematical and other models that simulate ecosystems. Computer simulation of such models can help us understand large and very complex systems (such as rivers, oceans, forests, grasslands, cities, and climate) that cannot be adequately studied and modeled in field and laboratory research. Figure 4-32 outlines the major stages of systems analysis.

Researchers can change values of the variables in their computer models to **(1)** project possible changes in environmental conditions, **(2)** help anticipate environmental surprises, and **(3)** analyze the effects of various alternative solutions to environmental problems.

However, simulations and predictions made using ecosystem models are no better than the data and assumptions used to develop the models. Thus, careful field and laboratory ecological research must be used to provide the baseline data and determine the causal relationships between key variables needed to develop and test ecosystem models.

4-8 ECOSYSTEM SERVICES AND SUSTAINABILITY

What Are Ecosystem Services? We depend on nature for food, air, water, and almost everything else we use. Ecosystems provide us and other species with a number of **ecosystem services** (Figure 4-33). Without these services performed by diverse communities of species, we would be starving, gasping for breath, and drowning in our own wastes.

It is very expensive to **(1)** build sewage treatment plants to replace free water purification services provided by wetlands, **(2)** rely mostly on pesticides rather than natural biological controls to control crop and forest pests, **(3)** try to save species whose premature extinction could have been prevented, and **(4)** try to restore ecosystems that we have degraded. In addition, these costly replacements are rarely as effective as the free ecological services nature provides.

What Are the Two Basic Principles of Ecosystem Sustainability? In this chapter we have seen that almost all natural ecosystems and the biosphere itself achieve *sustainability* by

- Using renewable solar energy as their energy source
- Recycling the chemical nutrients its organisms need for survival, growth, and reproduction

These two principles for sustainability arise from the **(1)** structure and function of natural ecosystems (Figures 4-7 and 4-15), **(2)** law of conservation of matter (p. 66), and **(3)** two laws of energy (p. 71). Thus, the results of basic research in both the physical and biological sciences provide us with the same guidelines or lessons from nature on how we can live more sustainably on the earth, as summarized in Figure 3-16, p. 74.

In this chapter we have learned about **(1)** how the earth's life-support systems work, **(2)** the living and nonliving components of ecosystems, **(3)** how energy flows through ecosystems, **(4)** how rapidly biomass is produced in different ecosystems, **(5)** how matter cycles in ecosystems, **(6)** how ecologists learn about ecosystems, and **(7)** how ecosystem services sustain life. In Chapter 5 we will learn about how life developed on the earth and reached its current level of biodiversity.

All things come from earth, and to earth they all return.

MENANDER (342–290 B.C.)

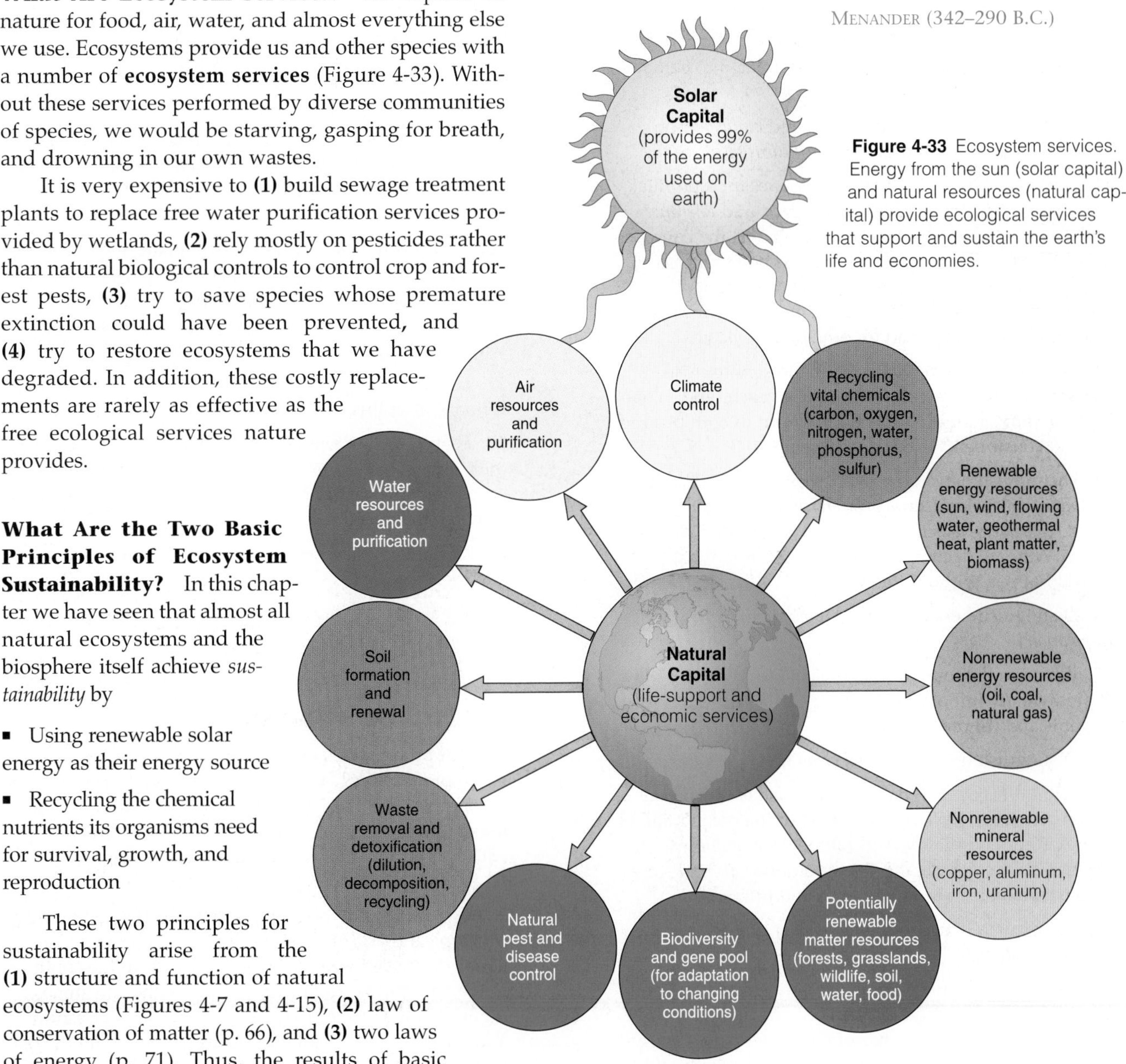

Figure 4-33 Ecosystem services. Energy from the sun (solar capital) and natural resources (natural capital) provide ecological services that support and sustain the earth's life and economies.

REVIEW QUESTIONS

1. Define the boldfaced terms in this chapter.

2. Why are insects important for many forms of life and for you and your lifestyle?

3. What is *ecology*? What five levels of the organization of matter are the main focus of ecology?

4. Distinguish between *organism, eukaryotic organism, prokaryotic organism, species, population, genetic diversity, habitat, community, ecosystem,* and *biosphere.*

5. Explain why microbes (microorganisms) are so important.

6. Distinguish between the *atmosphere, troposphere, stratosphere, hydrosphere, lithosphere,* and *biosphere.*

7. What three processes sustain life on earth?

8. How does the sun help sustain life on the earth? How is this related to the earth's natural greenhouse effect?

9. What are *biomes,* and how are they related to climate? What are *aquatic life zones*?

10. Distinguish between the *abiotic* and *biotic* components of ecosystems and give three examples of each.

11. Distinguish between *range of tolerance* for a population in an ecosystem, the *law of tolerance,* and *tolerance limits.* How does each of these factors affect the composition (structure) of ecosystems? What is a *limiting factor,* and how do such factors affect the composition of ecosystems? What are two important limiting factors for terrestrial ecosystems and for aquatic ecosystems?

12. Distinguish between *producers* and *consumers* in ecosystems and give three examples of each type. What is *photosynthesis,* and why is it important to both producers and consumers? What is *chemosynthesis*?

13. Distinguish between *primary consumers (herbivores), secondary consumers (carnivores), tertiary consumers, omnivores, scavengers, detritivores, detritus feeders,* and *decomposers.* Why are decomposers important, and what would happen without them?

14. Distinguish between *aerobic respiration* and *anaerobic respiration.*

15. What are the four components of biodiversity? Why is biodiversity important to **(a)** the earth's life-support systems and **(b)** the economy?

16. Distinguish between a *food chain* and a *food web.*

17. What is *biomass*? What is the *pyramid of energy flow* for an ecosystem? What is the effect of the second law of energy (thermodynamics) on **(a)** the flow of energy through an ecosystem and **(b)** the amount of food energy available to top carnivores and humans?

18. Distinguish between the *pyramid of biomass* and the *pyramid of numbers.*

19. Distinguish between *gross primary productivity* and *net primary productivity.* Explain how net primary productivity affects the number of consumers in an ecosystem and on the earth. List two of the most productive ecosystems or aquatic life zones and two of the least productive ecosystems or aquatic life zones. Use the concept of net primary productivity to explain why harvesting plants from estuaries, clearing tropical forests to grow crops, and harvesting the primary producers in oceans to feed the human population are not good ideas.

20. About what percentages of total potential net primary productivity of **(a)** the entire earth and **(b)** the earth's terrestrial ecosystems are used, wasted, or destroyed by humans?

21. What is a *biogeochemical cycle*? How do such cycles connect past, present, and future forms of life?

22. Describe the *water cycle.* Distinguish between *absolute humidity* and *relative humidity* and between *condensation nuclei* and *dew point.* What is *groundwater*? What is an *aquifer*?

23. List three human activities that alter the water cycle.

24. Describe the *carbon cycle* and list two human activities that alter this cycle.

25. Describe the *nitrogen cycle.* Distinguish between *nitrogen fixation, nitrification, assimilation, ammonification,* and *denitrification.* List five ways in which humans alter this cycle.

26. Describe the *phosphorus cycle.* Explain why the level of phosphorus in soil often limits plant growth on land and why phosphorus also limits the growth of producers in many freshwater streams and lakes. List three ways in which humans alter this cycle.

27. Describe the *sulfur cycle* and list three ways in which humans alter this cycle.

28. Distinguish between *field research, laboratory research,* and *systems analysis* as methods for learning about ecosystems. What are *geographic information systems,* and how are they used to learn about ecosystems?

29. Define *ecosystem services* and list nine examples of such services.

30. What are two basic principles of ecosystem sustainability?

CRITICAL THINKING

1. (a) A bumper sticker asks, "Have you thanked a green plant today?" Give two reasons for appreciating a green plant. **(b)** Trace the sources of the materials that make up the bumper sticker and decide whether the sticker itself is a sound application of the slogan. **(c)** Explain how decomposers help keep you alive.

2. (a) How would you set up a self-sustaining aquarium for tropical fish? **(b)** Suppose you have a balanced aquarium sealed with a clear glass top. Can life continue in the aquarium indefinitely as long as the sun shines regularly on it? **(c)** A friend cleans out your aquarium and removes all the soil and plants, leaving only the fish and water. What will happen?

3. Using the second law of energy, explain why there is such a sharp decrease in usable energy as energy flows through a food chain or web. Doesn't an energy loss at each step violate the first law of energy? Explain.

4. Using the second law of energy, explain why many poor people in developing countries live on a mostly vegetarian diet.

5. Why do farmers not need to apply carbon to grow their crops but often need to add fertilizer containing nitrogen and phosphorus?

6. Carbon dioxide (CO_2) in the atmosphere fluctuates significantly on a daily and seasonal basis. Why are CO_2 levels higher during the day than at night?

7. Why could the total amount of animal flesh on the earth never exceed the total amount of plant flesh, even if all animals were vegetarians?

8. Which causes a larger loss of energy from an ecosystem: a herbivore eating a plant or a carnivore eating an animal? Explain.

9. Why are there more mice than lions in an African ecosystem supporting both types of animals?

10. What would happen to an ecosystem if **(a)** all its decomposers and detritus feeders were eliminated or **(b)** all its producers were eliminated?

PROJECTS

1. Visit several types of nearby aquatic life zones and terrestrial ecosystems. For each site, try to determine **(a)** the major producers, consumers, detritivores, and decomposers and **(b)** the shapes of the pyramids of energy flow, biomass, and numbers.

2. Make a concept map of this chapter's major ideas using the section heads and subheads and the key terms (in boldface). Look on the website for this book for information about making concept maps.

INTERNET STUDY RESOURCES AND RESOURCES FOR FURTHER READING AND RESEARCH

The website for this book contains helpful study aids and many ideas for further reading and research. Log on to

http://www.brookscole.com/product/0534389872s

and click on the Chapter-by-Chapter area. Choose Chapter 4 and select a resource:

- "Flash Cards" allows you to test your mastery of the Terms and Concepts to Remember for this chapter.
- "Tutorial Quizzes" provides a multiple-choice practice quiz.
- "Student Guide to InfoTrac" will lead you to Critical Thinking Projects that use InfoTrac College Edition as a research tool.
- "References" lists the major books and articles consulted in writing this chapter.
- "Hypercontents" takes you to an extensive list of sites with news, research, and images related to individual sections of the chapter.

INFOTRAC COLLEGE EDITION

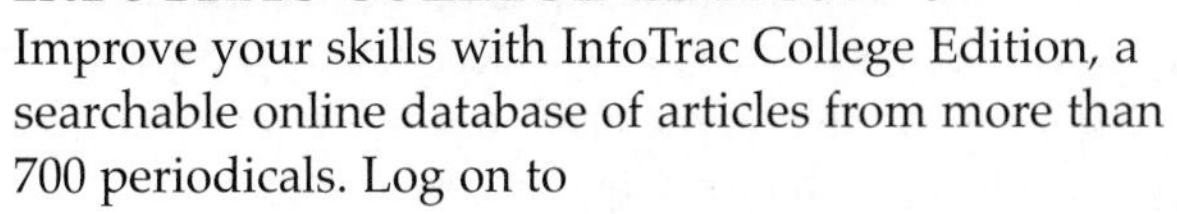

Improve your skills with InfoTrac College Edition, a searchable online database of articles from more than 700 periodicals. Log on to

http://www.infotrac-college.com

or access InfoTrac through the website for this book. Try to find the following articles:

1. Nutrient Limitation of Decomposition in Hawaiian Forests. Sarah E. Hobbie and Peter M. Vitousek. *Ecology*, July 2000 v81 i7 p1867. The researchers examined whether the same nutrients limit decomposition and aboveground net primary production (ANPP). *Hint*: Enter the search term "nutrient limitation" using the keyword "Hawaii."

2. Seed Size, Nitrogen Supply, and Growth Rate Affect Tree Seedling Survival in Deep Shade. Michael B. Walters and Peter B. Reich. *Ecology*, July 2000 v81 i7 p1887. Species differences in seedling survival in deeply shaded understories (i.e., shade tolerance) may depend on both seed size and growth rates. *Hint*: Enter the search term "nitrogen supply" using the keyword "Midwest."

5 EVOLUTION AND BIODIVERSITY: ORIGINS, NICHES, AND ADAPTATION

Earth: The Just-Right, Resilient Planet

Life on the earth as we know it (Figure 5-1) needs a certain temperature range: Venus is much too hot and Mars is much too cold, but the earth is *just right.* (Otherwise, you would not be reading these words.)

Life as we know it depends on the liquid water that dominates the earth's surface. Again, temperature is crucial; life on the earth needs average temperatures between the freezing and boiling points of water, between 0°C and 100°C (32°F and 212°F) at the earth's range of atmospheric pressures.

The earth's orbit is the right distance from the sun to provide these conditions. If the earth were much closer, it would be too hot—like Venus—for water vapor to condense to form rain. If it were much farther away, its surface would be so cold—like Mars—that its water would exist only as ice. The earth also spins; if it did not, the side facing the sun would be too hot and the other side too cold for water-based life to exist.

The earth is also the right size; that is, it has enough gravitational mass to keep its iron–nickel core molten and to keep the gaseous molecules in its atmosphere from flying off into space. (A much smaller earth would be unable to hold onto an atmosphere consisting of such light molecules as N_2, O_2, CO_2, and H_2O.)

The slow transfer of its internal heat (geothermal energy) to the surface also helps keep the planet at the right temperature for life. And thanks to the development of photosynthesizing bacteria more than 2 billion years ago, an ozone sunscreen protects us and many other forms of life from an overdose of ultraviolet radiation.

On a time scale of millions of years, the earth is enormously resilient and adaptive. During the 3.7 billion years since life arose, the average surface temperature of the earth has remained within the narrow range of 10–20°C (50–68°F), even with a 30–40% increase in the sun's energy output. In short, the earth is just right for life as we know it.

We can summarize the 3.7-billion-year biological history of the earth in one sentence. *Organisms convert solar energy to food, chemicals cycle, and a variety of species with different biological roles (niches) have evolved in response to changing environmental conditions.*

Each species here today represents a long chain of evolution, and each of these species plays a unique ecological role in the earth's communities and ecosystems.

This chapter is devoted to helping us understand how the earth's species evolved and the nature of their niches or biological roles. This information is important for helping us **(1)** understand the effects of human actions on wild species and **(2)** protect species—including the human species—from premature extinction.

Figure 5-1 The earth is a blue and white planet in the black void of space. Currently, it has the right physical and chemical conditions to allow the development of life as we know it today. (NASA)

There is a grandeur to this view of life . . . that, whilst this planet has gone cycling on . . . endless forms most beautiful and most wonderful have been, and are being, evolved.

CHARLES DARWIN

This chapter addresses the following questions:

- How do scientists account for the emergence of life on the earth?
- What is evolution, and how has it led to the current diversity of organisms on the earth?
- How does evolution affect the way organisms fit into their environment?
- What is an ecological niche, and how does it relate to adaptation to changing environmental conditions?
- How do extinction of species and formation of new species affect biodiversity?

5-1 ORIGINS OF LIFE

How Did Life Emerge on the Earth? How did a barren planet become a living jewel in the vastness of space (Figure 5-1)? How did life on the earth evolve to its present incredible diversity of species, living in an interlocking network of matter cycles, energy flows, and species interactions? We do not know the full answer to these questions, but a growing body of evidence suggests what might have happened.

Evidence about the earth's early history comes from chemical analysis and measurements of radioactive elements in primitive rocks and fossils. Chemists have also conducted laboratory experiments showing how simple inorganic compounds in the earth's early atmosphere might have reacted to produce organic molecules such as amino acids, simple sugars, and other building-block molecules for large biopolymer molecules (such as proteins, complex carbohydrates, RNA, and DNA) needed for life.

From this diverse evidence scientists have hypothesized that life on the earth developed in two phases over the past 4.7–4.8 billion years (Figure 5-2):

- *Chemical evolution* of the organic molecules, biopolymers, and systems of chemical reactions needed to form the first protocells (taking about 1 billion years)
- *Biological evolution* from single-celled prokaryotic bacteria (Figure 4-3b, p. 80), to single-celled eukaryotic creatures (Figure 4-3a, p. 80), and then to multicellular organisms (taking about 3.7–3.8 billion years) (Figure 5-3).

How Do We Know What Organisms Lived in the Past? Most of what we know of the earth's life history comes from **fossils**: mineralized or petrified replicas of skeletons, bones, teeth, shells, leaves, and seeds or impressions of such items. Such fossils **(1)** give us physical evidence of organisms that lived long ago and **(2)** show us what their internal structures looked like.

Despite its importance, the fossil record is uneven and incomplete. Some life-forms left no fossils, some fossils have decomposed, and others are yet to be found. So far we have found fossils representing only about 1% of the species believed to have ever lived.

Other sources of information include **(1)** chemical and radioactive dating of fossils, **(2)** nearby ancient rocks, **(3)** material in cores drilled out of buried ice, and **(4)** the DNA of organisms alive today.

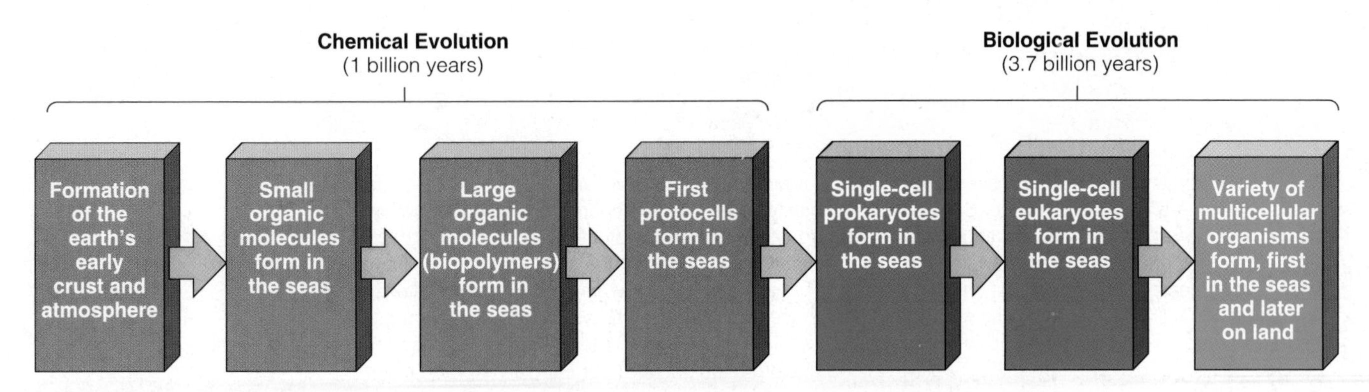

Figure 5-2 Summary of the hypothesized chemical and biological evolution of life on the earth. This drawing is not to scale. Note that the time span for biological evolution is almost four times longer than that for chemical evolution.

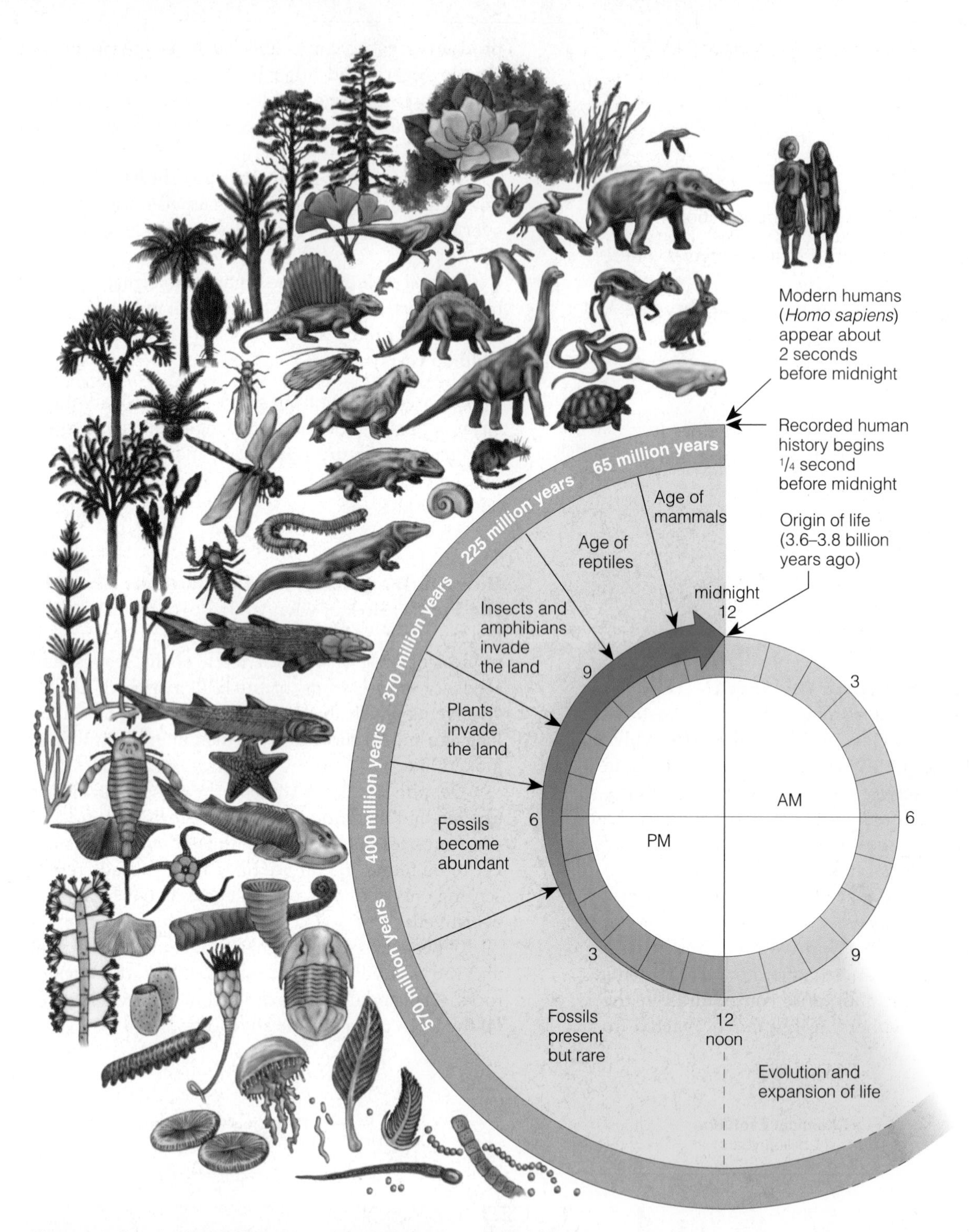

Figure 5-3 Greatly simplified overview of the biological evolution of life on the earth, which was preceded by about 1 billion years of chemical evolution. Evidence indicates that the early span of biological evolution on the earth, between about 3.7 billion and 1 billion years ago, was dominated by microorganisms (mostly bacteria and, later, protists) that lived in water. Plants and animals evolved first in the seas. Recent fossil and DNA evidence suggests that plants began moving onto land as early as 1 billion years ago, with more intense invasion about 400 million years ago. Humans arrived on the scene only a very short time ago. If we compress the earth's roughly 3.7-billion-year history of biological evolution to a 24-hour time scale, the first human species (*Homo habilis*) appeared about 47–94 seconds before midnight, and our species (*Homo sapiens sapiens*) appeared about 2 seconds before midnight. Agriculture began only 0.25 second before midnight, and the industrial revolution has been around for only 0.007 second. (Adapted from George Gaylord Simpson and William S. Beck, *Life: An Introduction to Biology*, 2d ed., New York: Harcourt Brace Jovanovich, 1965)

5-2 EVOLUTION AND ADAPTATION

What Is Evolution? According to scientific evidence, the major driving force of adaptation to changes in environmental conditions is **biological evolution**, or **evolution**: the change in a population's genetic makeup (gene pool) through successive generations. Note that *populations*, not individuals, evolve by becoming genetically different.

According to the **theory of evolution**, all species descended from earlier, ancestral species. This widely accepted scientific theory explains how life has changed over the past 3.7 billion years and why life is so diverse today.

Biologists use the term ***microevolution*** to describe the small genetic changes that occur in a population. The term ***macroevolution*** is used to describe long-term, large-scale evolutionary changes through which new species are formed from ancestral species and other species are lost through extinction.

How Does Microevolution Work? The first step in evolution is the development of *genetic variability* in a population. Recall that **(1)** genetic information in *chromosomes* is contained in various sequences of chemical units (called *nucleotides*) in DNA molecules, and **(2)** genes found in chromosomes are segments of DNA that are coded for certain traits that can be passed on to offspring (Figure 3-6, p. 64).

A population's **gene pool** is the set of all genes in the individuals of the population of a species. *Microevolution* is a change in a population's gene pool over time.

Although members of a population generally have the same number and kinds of genes, a particular gene may have two or more different molecular forms, called **alleles**. Sexual reproduction leads to a random shuffling or recombination of alleles. As a result, each individual in a population has a different combination of alleles.

Microevolution works through a combination of four processes that change the genetic composition of a population:

- **Mutation**, which involves random changes in the structure or number of DNA molecules in a cell and is the ultimate source of genetic variability in a population
- **Natural selection**, which occurs when some individuals of a population have genetically based traits that cause them to survive and produce more offspring than other individuals
- **Gene flow**, which involves movement of genes between populations and can lead to changes in the genetic composition of local populations
- **Genetic drift**, which involves change in the genetic composition of a population by chance and is especially important for small populations

What Is the Role of Mutation in Microevolution? Genetic variability in a population originates through **mutations**: random changes in the structure or number of DNA molecules in a cell. Mutations can occur in two ways:

- Exposure of DNA to external agents such as radioactivity, X-rays, and natural and human-made chemicals (called *mutagens*)
- Random mistakes that sometimes occur in coded genetic instructions when DNA molecules are copied (each time a cell divides and whenever an organism reproduces)

Mutations can occur in any cells, but only those in reproductive cells are passed on to offspring.

Some mutations are harmless, but most are harmful and alter traits so that an individual cannot survive (lethal mutations). Every so often, a mutation is beneficial. The result is new genetic traits that give their bearer and its offspring better chances for survival and reproduction, either under existing environmental conditions or when such conditions change.

It is important to understand that mutations are **(1)** random and unpredictable, **(2)** the only source of totally new genetic raw material (alleles), and **(3)** rare events. Once created by mutation, new alleles can be shuffled together or recombined *randomly* to create new combinations of genes in populations of sexually reproducing species.

What Role Does Natural Selection Play in Microevolution? The process of **natural selection** occurs when some individuals of a population have genetically based traits that increase their chances of survival and their ability to produce offspring. This idea was developed in 1846 by Charles Darwin and published in 1859 in his book *On the Origin of Species by Means of Natural Selection*. Darwin recognized that three conditions are necessary for evolution of a population by natural selection to occur:

- There must be natural *variability* for a trait in a population.
- The trait must be *heritable*, meaning that it must have a genetic basis such that it can be passed from one generation to another.
- The trait must somehow lead to **differential reproduction**, meaning that it must enable individuals with the trait to leave more offspring than other members of the population.

Natural selection causes any allele or set of alleles that result in a beneficial trait to become more common in succeeding generations and other alleles to become less common. A heritable trait that enables organisms to better survive and reproduce under a given set of environmental conditions is called an **adaptation**, or **adaptive trait**.

When faced with a change in environmental conditions, a population of a species can **(1)** adapt to the new conditions through natural selection, **(2)** migrate (if possible) to an area with more favorable conditions, or **(3)** become extinct.

It is important to understand that environmental conditions do not create favorable heritable characteristics. Instead, natural selection favors some individuals over others by acting on inherited genetic variations (alleles) already present in the gene pool of a population.

The process of microevolution can be summarized as follows: *Genes mutate, individuals are selected, and populations evolve.* The genetic characteristics of populations of a species also can be changed through *artificial selection* (Spotlight, p. 112).

What Is an Example of Microevolution by Natural Selection? One of the best-documented examples of microevolution by natural selection involves camouflage coloration in the peppered moth, which is found in England (Figure 5-4).

Natural selection in the peppered moth occurred because **(1)** there were two color forms (*variability*), **(2)** color form was genetically based (*heritability*), and **(3)** survival and reproduction by one of the color forms were greater (*differential reproduction*). First an environmental change in the form of soot caused a change in the background color of tree trunks. This environmental change then allowed bird predators to find and eat the moths with the coloration that no longer blended in with the background.

What Are Three Types of Natural Selection? Biologists recognize three types of natural selection (Figure 5-5):

- *Directional natural selection* (Figure 5-5, left), in which changing environmental conditions cause allele frequencies to shift so that individuals with traits at one end of the normal range become more common than midrange forms. Examples of this "it pays to be different" type of natural selection are **(1)** the changes in the varieties of peppered moths (Figure 5-4) and **(2)** the evolution of genetic resistance to pesticides among insects and to antibiotics among disease-carrying bacteria. This type of natural selection is most common during periods of environmental change or when members of a population migrate to a new habitat with different environmental conditions.

Figure 5-4 Two varieties of peppered moths found in England illustrate one kind of adaptation: camouflage. Before the industrial revolution in the mid-1800s, the speckled light-gray form of this moth was prevalent. When these night-flying moths rested on light-gray lichens on tree trunks during the day, their color camouflaged them from their predators (top). A dark-gray form also existed but was quite rare. During the industrial revolution, soot and other pollutants from factory smokestacks began killing lichens and darkening tree trunks. As a result, the dark form of moth became the common one, especially near industrial cities. In this new environment, the dark form of moth blended in with the blackened trees, whereas the light form of moth was highly visible to predators (bottom). Through natural selection, the dark form began to survive and reproduce at a greater rate than its light-colored kin. (Both varieties appear in each image. Can you spot them?)

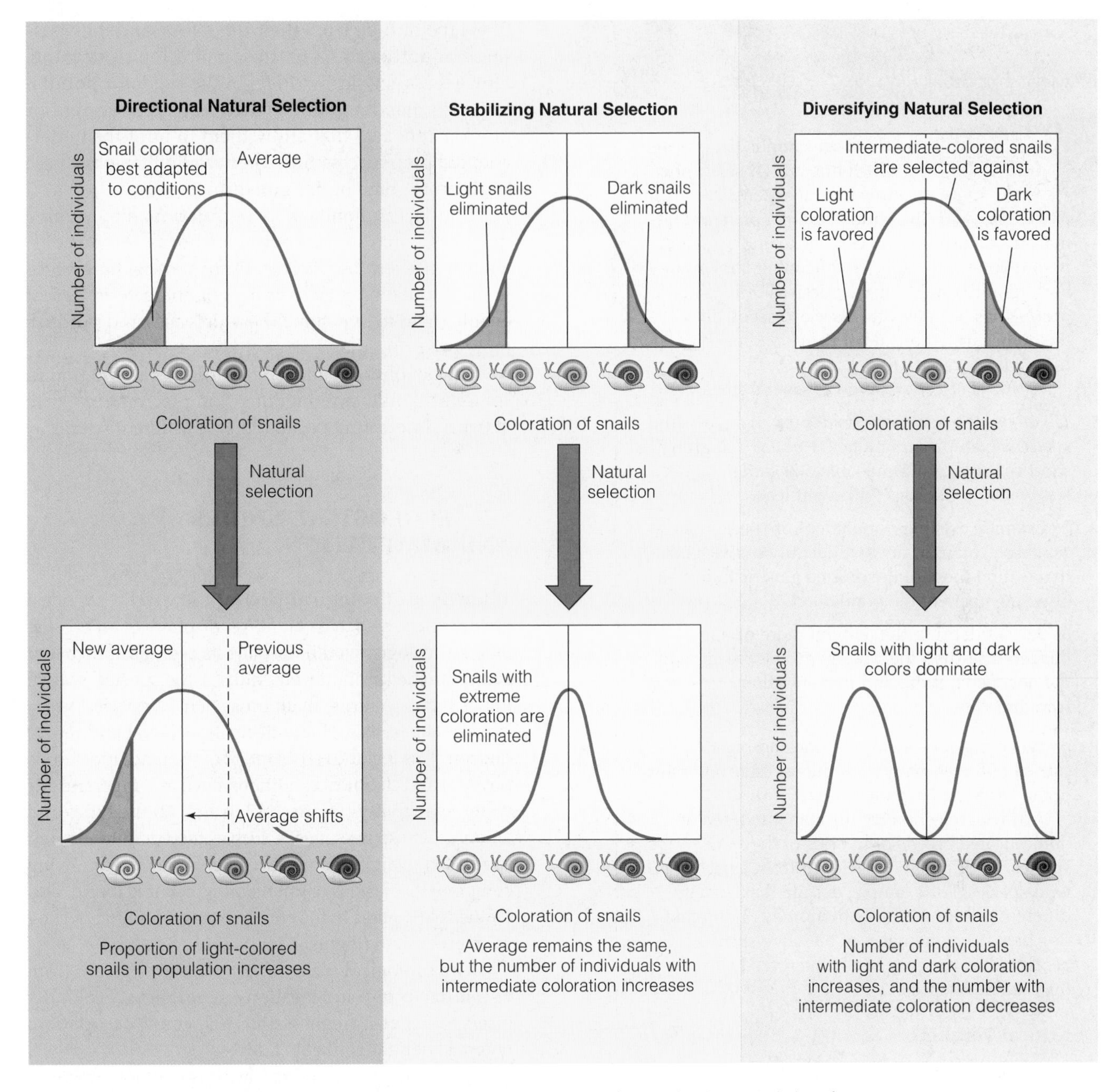

Figure 5-5 Three ways in which natural selection can occur, using the trait of coloration in a population of snails. In *directional natural selection*, changing environmental conditions select organisms with alleles that deviate from the norm so that their offspring (lighter-colored snails) make up a larger proportion of the population. In *stabilizing selection*, environmental factors eliminate fringe individuals (light- and dark-colored snails) and increase the number of individuals with average genetic makeup (intermediate-colored snails). In *diversifying natural selection*, environmental factors favor individuals with uncommon traits (light- and dark-colored snails) and greatly reduce those with average traits (intermediate-colored snails).

- *Stabilizing natural selection,* which tends to eliminate individuals on both ends of the genetic spectrum and favor individuals with an average genetic makeup (Figure 5-5, center). This "it pays to be average" type of natural selection occurs when **(1)** an environment changes little and **(2)** most members of the population are well adapted to that environment.
- *Diversifying natural selection,* which occurs when environmental conditions favor individuals at both extremes of the genetic spectrum and eliminate or sharply reduce numbers of individuals with normal or intermediate genetic traits (Figure 5-5, right). In this "it does not pay to be normal" type of natural selection, a population is split into two groups.

What Is Artificial Selection?

SPOTLIGHT

Artificial selection is the process by which humans **(1)** select one or more desirable genetic traits in the population of a plant or animal and **(2)** use *selective breeding* to end up with populations of the species containing large numbers of individuals with the desired traits. This process has been used to develop essentially all the domesticated breeds of plants and animals from wild populations.

Artificial selection involves several steps:

1. Choose the genetic traits desired in a particular species. Examples might be **(1)** wheat that grows short so it will not topple over, **(2)** cattle that need less water, or **(3)** dogs with short legs.

2. Examine existing populations of species (such as wheat, cattle, or dogs) to identify individuals that exhibit more of the desired genetic trait than other members of the population.

3. Select offspring that exhibit more of the desired trait than their parents to be breeders for the next generation, and prevent other offspring from breeding.

When this process of selection and breeding is carried out over many generations, more and more of the offspring have the selected or desired traits.

Artificial selection results in many different domesticated breeds or hybrids of the same species, all originally developed from a particular wild species. For example, despite their widely different genetic traits, the hundreds of different dog breeds are members of the same species because they can potentially interbreed and produce fertile offspring.

Critical Thinking

How does artificial selection differ from natural selection?

What Is Coevolution? Some biologists have proposed that *interactions between species* also can result in microevolution in each of their populations. According to this hypothesis, when populations of two different species interact over a long time, changes in the gene pool of one species can lead to changes in the gene pool of the other species. This process is called **coevolution**.

Suppose that certain individuals in a population of carnivores (such as owls) become better at hunting prey (such as mice). Because of genetic variation, certain individuals of the prey have traits that allow them to escape or hide from their predators, and they pass these adaptive traits on to some of their offspring. However, a few individuals in the predator population also may have traits (such as better eyesight or quicker reflexes) that allow them to hunt the better-adapted prey successfully. They would then pass these traits on to some of their offspring.

Similarly, plants in a population may evolve defenses, such as camouflage, thorns, or poisons, against efficient herbivores. In turn, some herbivores in the population may have genetic characteristics that enable them to overcome these defenses and produce more offspring than those without such traits.

In coevolution, adaptation follows adaptation in something like an ongoing, long-term arms race between interacting populations of different species.

5-3 ECOLOGICAL NICHES AND ADAPTATION

What Is an Ecological Niche? If asked what role a certain species such as an alligator plays in an ecosystem, an ecologist would describe its **ecological niche**, or simply **niche** (pronounced "nitch"), the species' way of life or functional role in an ecosystem. A species' niche involves everything that affects its survival and reproduction. This includes **(1)** its range of tolerance for various physical and chemical conditions, such as temperature or water availability (Figure 4-13, p. 85), **(2)** the types and amounts of resources it uses, such as food or nutrients and space, **(3)** how it interacts with other living and nonliving components of the ecosystems in which it is found, and **(4)** the role it plays in the energy flow and matter cycling in an ecosystem (Figure 4-15, p. 88).

The ecological niche of a species is different from its **habitat** or physical location where it lives. Ecologists often say that a niche is like a species' occupation, whereas habitat is like its address.

Understanding a species' niche is important because it **(1)** can help us prevent it from becoming prematurely extinct and **(2)** can be useful in helping us assess the environmental changes we make in terrestrial and aquatic systems. For example, how will the niches of various species be changed by **(1)** clearing a forest, **(2)** plowing up a grassland, **(3)** filling in a wetland, or **(4)** dumping pollutants into a lake or stream?

What Is the Difference Between a Species' Fundamental Niche and Its Realized Niche? A species' **fundamental niche** is the full potential range of physical, chemical, and biological conditions and resources it could theoretically use if there were no direct competition from other species. However, in a particular ecosystem, species often compete with one

Cockroaches: Nature's Ultimate Survivors

SPOTLIGHT

Cockroaches, the bugs many people love to hate, **(1)** have been around for about 350 million years and **(2)** are one of the great success stories of evolution. The major reason they are so successful is that they are *generalists*.

The earth's 4,000 cockroach species can **(1)** eat almost anything (including algae, dead insects, fingernail clippings, salts in tennis shoes, electrical cords, glue, paper, and soap) and **(2)** live and breed almost anywhere except in polar regions.

Some species can **(1)** go for months without food, **(2)** survive for a month on a drop of water from a dishrag, and **(3)** withstand huge doses of radiation. One species can survive being frozen for 48 hours.

They usually can evade their predators and a human foot in hot pursuit because **(1)** the antennae of most cockroach species can detect minute movements of air, **(2)** they have vibration sensors in their knee joints, and **(3)** their response times are lightning-fast (faster than you can blink). Some even have wings.

They also have high reproductive rates. In only a year, a single Asian cockroach (especially prevalent in Florida) and its young can add about 10 million new cockroaches to the world. Their high reproductive rate also helps them quickly develop genetic resistance to almost any poison we throw at them.

Most cockroaches sample food before it enters their mouths and learn to shun foul-tasting poisons. They also clean up after themselves by eating their own dead and, if food is scarce enough, their living.

Only about 25 species of cockroach live in homes. However, such species can **(1)** carry viruses and bacteria that cause diseases such as hepatitis, polio, typhoid fever, plague, and salmonella and **(2)** cause people to have allergic reactions ranging from watery eyes to severe wheezing. Indeed, about 60% of the 12 million Americans suffering from asthma are allergic to dead or live cockroaches.

Critical Thinking

If you could, would you exterminate all cockroach species? What might be some ecological consequences of doing this?

another for one or more of the same resources. This means that the niches of competing species overlap.

To survive and avoid competition for the same resources, a species usually occupies only part of its fundamental niche in a particular community or ecosystem—what ecologists call its **realized niche**. By analogy, you may be capable of being president of a particular company (your *fundamental professional niche*), but competition from others may mean that you may become only a vice president (your *realized professional niche*).

Is It Better to Be a Generalist or a Specialist Species? Broad and Narrow Niches The niches of species can be used to broadly classify them as *generalists* or *specialists*. **Generalist species** have broad niches (Figure 5-6, right curve). They can **(1)** live in many different places, **(2)** eat a variety of foods, and **(3)** tolerate a wide range of environmental conditions. Flies, cockroaches (Spotlight, above), mice, rats, white-tailed deer, raccoons, coyotes, copperheads, channel catfish, and humans are generalist species.

Specialist species have narrow niches (Figure 5-6, left curve). They may be able to **(1)** live in only one type of habitat, **(2)** use only one or a few types of food, or **(3)** tolerate only a narrow range of climatic and other environmental conditions. This makes them more prone

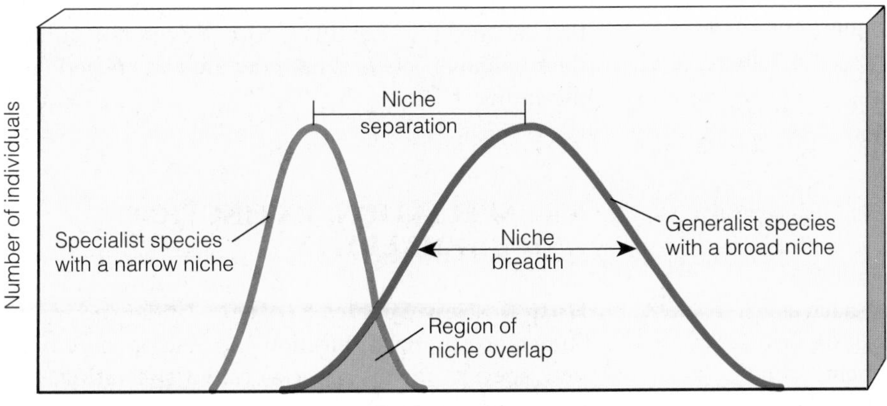

Figure 5-6 Overlap of the niches of two different species: a specialist and a generalist. In the overlap area the two species compete for one or more of the same resources. As a result, each species can occupy only a part of its fundamental niche and thus occupy its realized niche. Generalist species have a broad niche (right), and specialist species have a narrow niche (left).

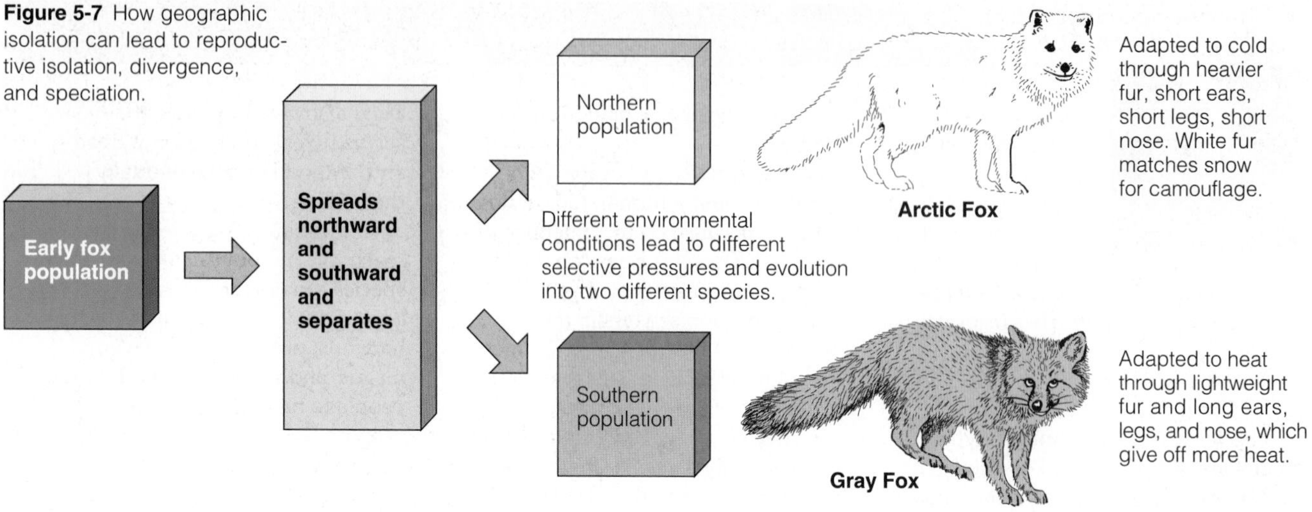

Figure 5-7 How geographic isolation can lead to reproductive isolation, divergence, and speciation.

Early fox population

Spreads northward and southward and separates

Northern population

Southern population

Different environmental conditions lead to different selective pressures and evolution into two different species.

Arctic Fox

Adapted to cold through heavier fur, short ears, short legs, short nose. White fur matches snow for camouflage.

Gray Fox

Adapted to heat through lightweight fur and long ears, legs, and nose, which give off more heat.

to extinction when environmental conditions change. Examples of specialists are **(1)** *tiger salamanders*, which can breed only in fishless ponds so their larvae will not be eaten, **(2)** *red-cockaded woodpeckers*, which carve nest-holes almost exclusively in old (at least 75 years) longleaf pines, **(3)** *spotted owls*, which need old-growth forests in the Pacific Northwest for food and shelter, and **(4)** China's highly endangered giant pandas, which feed almost exclusively on various types of bamboo.

Is it better to be a generalist than a specialist? It depends. When environmental conditions are fairly constant, as in a tropical rain forest, specialists have an advantage because they have fewer competitors. However, under rapidly changing environmental conditions, the generalist usually is better off than the specialist.

What Limits Adaptation? Shouldn't evolution lead to perfectly adapted organisms? Shouldn't adaptations to new environmental conditions allow **(1)** our skin to become more resistant to the harmful effects of ultraviolet radiation, **(2)** our lungs to cope with air pollutants, and **(3)** our livers to become better at detoxifying pollutants? The answer to these questions is *no* because of the following limits to adaptations in nature:

- *A change in environmental conditions can lead to adaptation only for traits already present in the gene pool of a population.*
- *Even if a beneficial heritable trait is present in a population, that population's ability to adapt can be limited by its reproductive capacity.* Populations of genetically diverse species that reproduce quickly—such as weeds, mosquitoes, rats, bacteria, or cockroaches—often can adapt to a change in environmental conditions in a short time. In contrast, populations of species such as elephants, tigers, sharks, and humans, which cannot produce large numbers of offspring rapidly, take a long time (typically thousands or even millions of years) to adapt through natural selection.
- *Even if a favorable genetic trait is present in a population, most of the population would have to die or become sterile so that individuals with the trait could predominate and pass the trait on.* This is hardly a desirable solution to the environmental problems the human species faces.

What Are Two Common Misconceptions About Evolution? Two common misconceptions about evolution are as follows:

- "Survival of the fittest" means "survival of the strongest." To biologists, *fitness* is a measure of reproductive success, not strength. Thus, the fittest individuals are those that leave the most descendants.
- Evolution involves some grand plan of nature in which species become progressively more perfect. From a scientific standpoint, there is no plan or goal of perfection in the evolutionary process. However, some people (creationists) believe that there is a conflict between the scientific theory of evolution and their religious beliefs about how life was created on the earth.

5-4 SPECIATION, EXTINCTION, AND BIODIVERSITY

How Do New Species Evolve? Under certain circumstances natural selection can lead to an entirely new species. In this process, called **speciation**, two species arise from one.

The most common mechanism of speciation (especially among animals) takes place in two phases: geographic isolation and reproductive isolation. **Geographic isolation** occurs when two populations of a species or two groups of the same population become physically separated for fairly long periods into areas with different environmental conditions. For example, part of a population may migrate in search of food and then begin living in another area with different environmental conditions (Figure 5-7). Populations also may become separated **(1)** by a physical barrier (such as a mountain range, stream, lake, or road), **(2)** by a change such as a volcanic eruption or earthquake, or **(3)** when a few individuals are carried to a new area by wind, water, or people.

The second phase of speciation is **reproductive isolation**. It occurs when mutation and natural selection operate independently in two geographically isolated populations and change the allele frequencies in different ways. If this process, called *divergence*, continues long enough, members of the geographically and reproductively isolated populations may become so different in genetic makeup that **(1)** they cannot interbreed or **(2)** if they do, they cannot produce live, fertile offspring. Then one species has become two, and *speciation* has occurred through *divergent evolution*.

In a few rapidly reproducing organisms this type of speciation may occur within hundreds of years. However, with most species such speciation takes from tens of thousands to millions of years. Given this time scale, it is difficult to observe and document the appearance of a new species. As a result, there are many controversial hypotheses about the details of speciation.

How Do Species Become Extinct? After evolution, the second process affecting the number and types of species on the earth is **extinction**. When environmental

Figure 5-8 Continental drift, the extremely slow movement of continents over millions of years on several gigantic plates (discussed in more detail on pp. 195–197). This process plays a role in the extinction of species and the rise of new species. Populations are geographically and eventually reproductively isolated as land masses float apart and new coastal regions are created. Rock and fossil evidence indicates that about 200–250 million years ago all of the earth's present-day continents (bottom left) were locked together in a supercontinent called Pangaea (top left). About 180 million years ago, Pangaea began splitting apart as the earth's huge plates separated and eventually resulted in today's continents (bottom right).

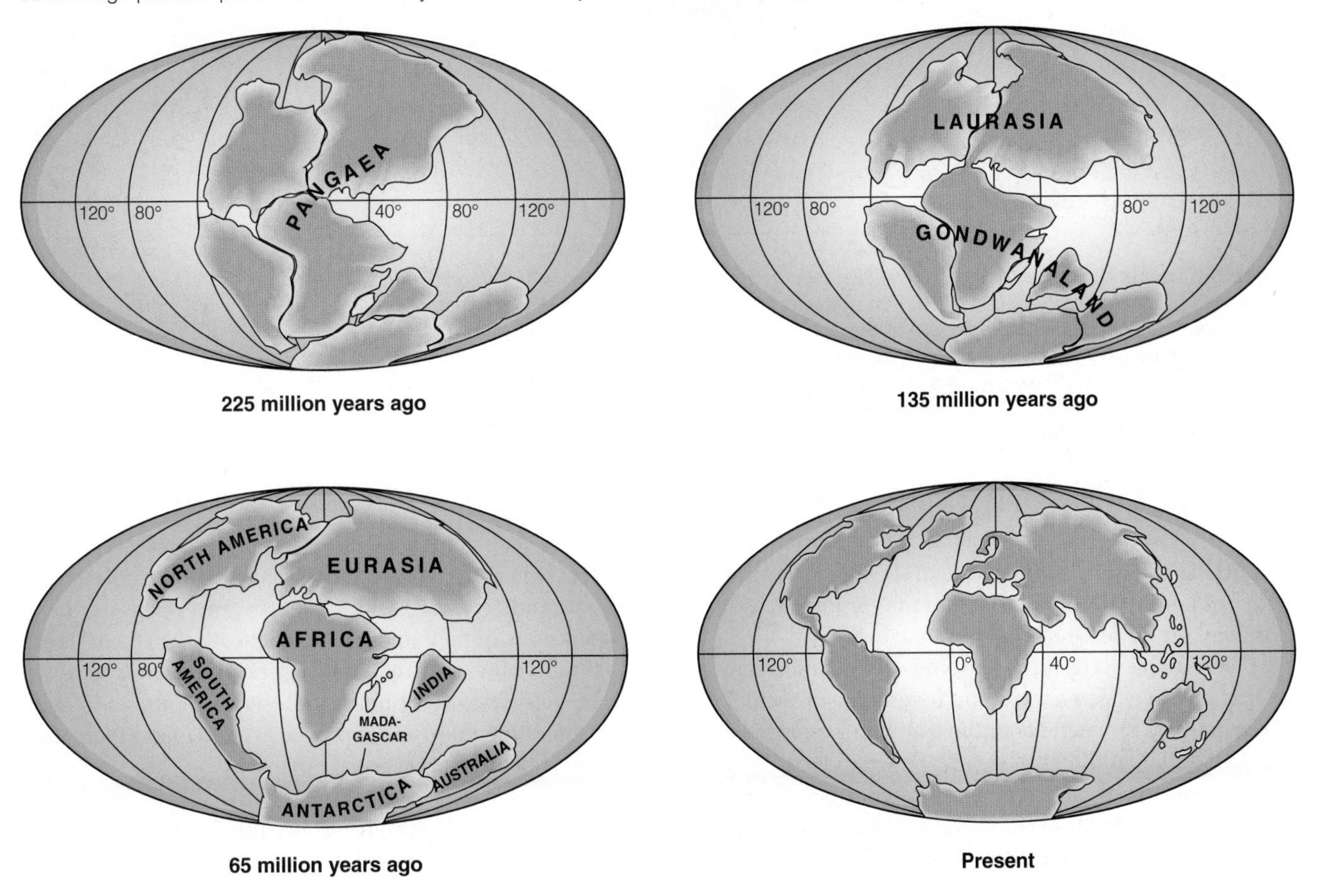

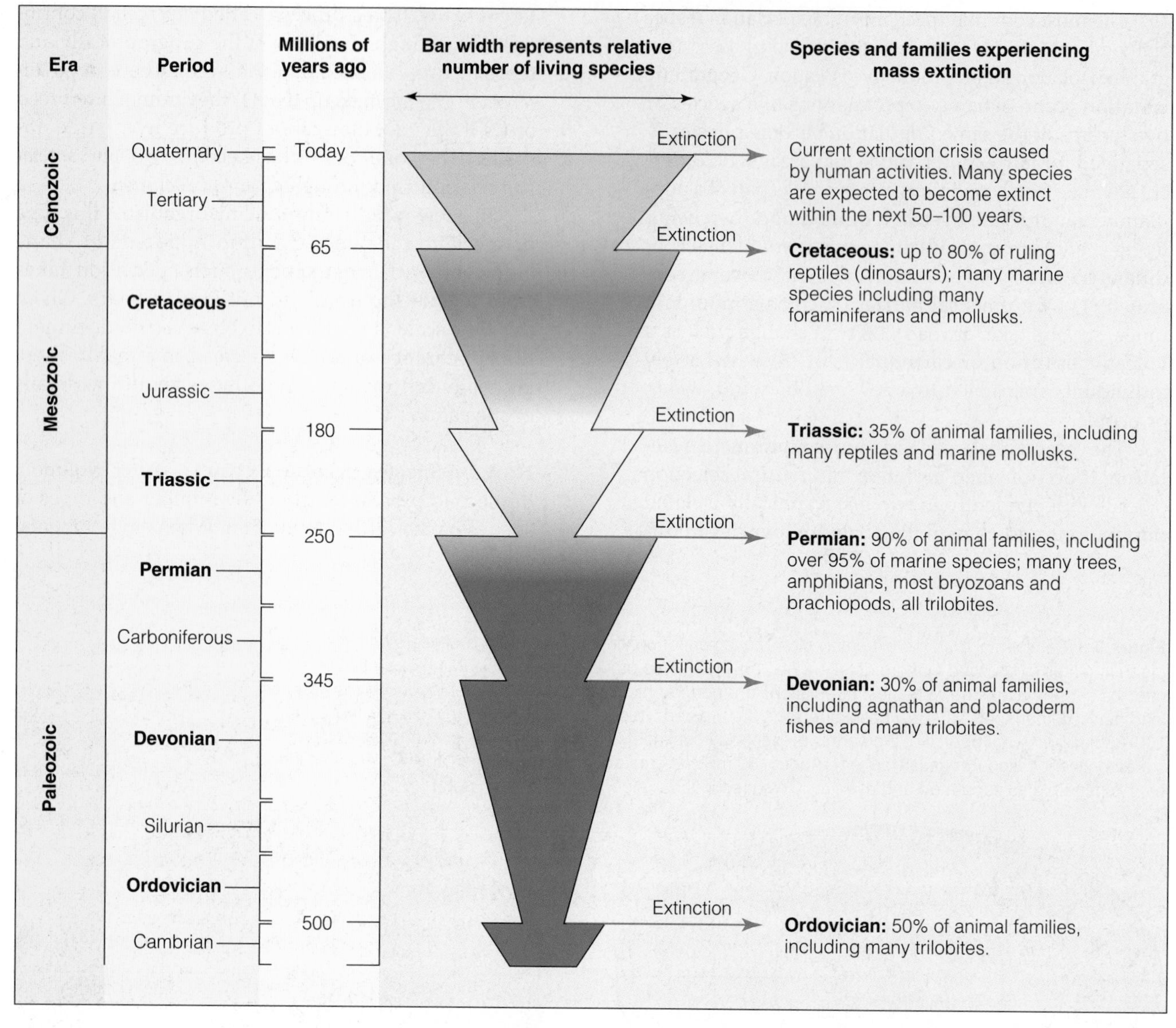

Figure 5-9 Fossils and radioactive dating indicate that five major *mass extinctions* (indicated by arrows) have taken place over the past 500 million years. Mass extinctions leave large numbers of organism roles (niches) unoccupied and create new ones. As a result, each mass extinction has been followed by periods of recovery (represented by the wedge shapes) called *adaptive radiations.* During these periods, which last over 10 million years or more, new species evolve to fill new or vacated ecological roles (niches). Many experts believe that we are now in the midst of a sixth mass extinction, caused primarily by human activities.

conditions change, a species must **(1)** evolve (become better adapted), **(2)** move to a more favorable area (if possible), or **(3)** cease to exist (become extinct).

The earth's long-term patterns of speciation and extinction have been affected by several major factors: **(1)** large-scale movements of the continents (continental drift) over millions of years (Figure 5-8, p. 115), **(2)** gradual climate changes caused by continental drift and slight shifts in the earth's orbit around the sun, and **(3)** rapid climate change caused by catastrophic events (such as large volcanic eruptions, huge meteorites and asteroids crashing into the earth, and release of large amounts of methane trapped beneath the ocean floor). Some of these events create dust clouds that shut down or sharply reduce photosynthesis long enough to eliminate huge numbers of producers and, soon thereafter, the consumers that fed on them.

Extinction is the ultimate fate of all species, just as death is for all individual organisms. Biologists estimate that 99.9% of all the species that have ever existed

Figure 5-10 Adaptive radiation of mammals began in the first 10–12 million years of the Cenozoic era (which began about 65 million years ago) and continues today. This evolution of a large number of new species is thought to have resulted when huge numbers of new and vacated ecological niches became available after the mass extinction of dinosaurs near the end of the Mesozoic era. (Used by permission from Cecie Starr and Ralph Taggart, *Biology: The Unity and Diversity of Life*, 8th ed., Belmont, Calif.: Wadsworth, 1998)

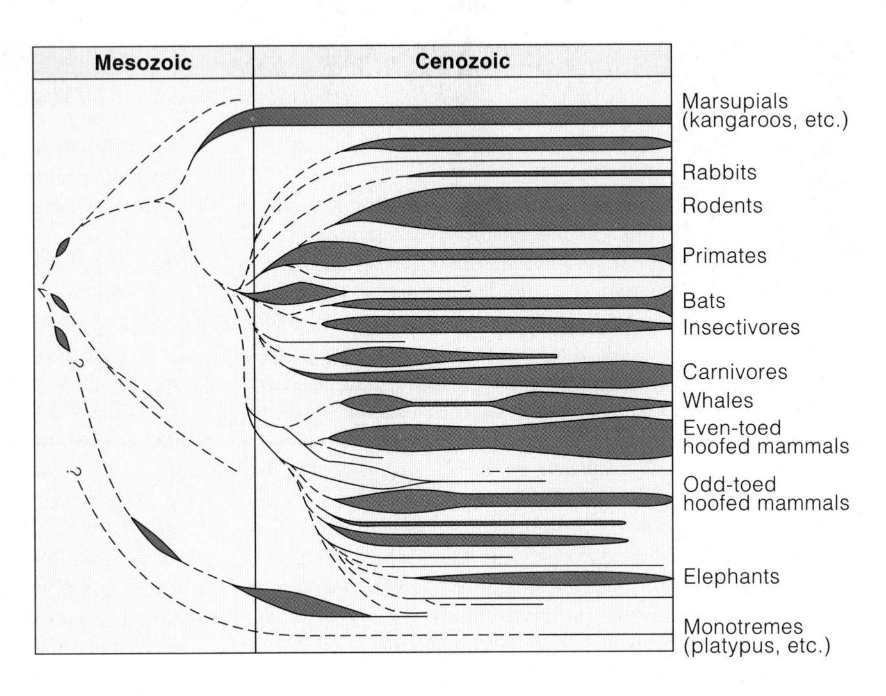

are now extinct. As local environmental conditions change, a certain number of species disappear at a low rate, called **background extinction.** In contrast, **mass extinction** is an abrupt rise in extinction rates above the background level. It is a catastrophic, widespread (often global) event in which large groups of existing species (perhaps 25–90%) are wiped out. Fossil and geological evidence indicates that the earth's species have experienced five great mass extinctions (20–60 million years apart) during the past 500 million years (Figure 5-9).

A crisis for one species is an opportunity for another. The existence of millions of species today means that speciation, on average, has kept ahead of extinction. Evidence shows that the earth's mass extinctions have been followed by periods of recovery called **adaptive radiations** in which numerous new species evolve to fill new or vacated ecological roles or niches in changed environments (Figure 5-10). Fossil records suggest that it takes 10 million years or more for adaptive radiations to rebuild biological diversity after a mass extinction.

How Do Speciation and Extinction Affect Biodiversity? Speciation minus extinction equals *biodiversity,* the planet's genetic raw material for future evolution in response to changing environmental conditions. In this long-term give-and-take between extinction and speciation, mass extinctions temporarily reduce biodiversity. However, they also create evolutionary opportunities for surviving species to undergo adaptive radiations to fill unoccupied and new biological roles or niches (Figure 5-10).

Although extinction is a natural process, there is much evidence that humans have become a major force in the premature extinction of species. Biologist Stuart Primm estimates that during the 20th century extinction rates increased by 100–1,000 times the natural background rate. As human population and resource consumption increase over the next 50 years, we are expected to take over more of the earth's **(1)** surface (Figure 1-3, p. 5) and **(2)** net primary productivity (Figure 4-24, p. 93). This may cause the premature extinction of up to a quarter of the earth's current species and constitute a *sixth mass extinction* (Guest Essay, p. 118).

On our short time scale, such catastrophic losses cannot be recouped by formation of new species; it took tens of millions of years after each of the earth's five great mass extinctions for life to recover to the previous level of biodiversity (Figure 5-9). Genetic engineering cannot stop this loss of biodiversity because genetic engineers do not create new genes. Rather, they transfer existing genes or gene fragments from one organism to another and thus rely on natural biodiversity for their raw material.

Nothing in biology makes sense except in the light of evolution.

THEODOSIUS DOBZHANSKY

REVIEW QUESTIONS

1. Define the boldfaced terms in this chapter.
2. Describe the conditions that make life on the earth just right for life as we know it.
3. Distinguish between *chemical evolution* and *biological evolution.*
4. What are *fossils,* and how do they help us formulate ideas about how life developed on the earth?
5. Distinguish between *biological evolution, the theory of evolution, microevolution,* and *macroevolution.*
6. Use the concepts of *genes, gene pool, alleles, mutations, gene flow,* and *genetic drift* to explain how microevolution takes place.

GUEST ESSAY

The Future of Evolution*

Norman Myers

Norman Myers is a tropical ecologist and international consultant in environment and development, with emphasis on conservation of wildlife species and tropical forests. He is one of the world's leading environmental experts. His research and consulting have taken him to 80 countries. He has served as a consultant for many development agencies and research organizations, including the U.S. National Academy of Sciences, the World Bank, the Organization for Economic Cooperation and Development, various UN agencies, and the World Resources Institute. Among his many publications (see Further Readings on the website for this book) are The Primary Source: Tropical Forests and Our Future *(1992),* The Gaia Atlas of Planet Management *(1993),* Scarcity or Abundance: A Debate on Environment *(1994),* Ultimate Security: The Environmental Basis of Political Security *(1996), and* Perverse Subsidies *(2001).*

Human activities have brought the earth to a biotic crisis. Many biologists have commented that this crisis will result in the loss of large numbers of species, possibly 25–50% within the lifetimes of students reading this book. However, surprisingly few biologists have recognized that in the longer term these extinctions will impoverish evolution's course for several million years.

So the future of evolution should be regarded as one of the most challenging issues humankind has ever encountered. After all, we are effectively conducting a planet-scale experiment, with little clue as to how it might turn out, except that it will prove irreversible and will severely reduce human well-being. We could get by without half of all mammals and other vertebrates, but if we lost half of all insects with their pollinating functions [p. 77], let alone their many other services, we would be in trouble in the first crop-growing season.

In addition, the mass extinction under way is the biggest of our environmental problems in terms of the duration of its impact and the numbers of people to be affected. All our other problems are potentially reversible. If we wanted to clean up acid deposition [p. 284], we could do it within a few decades. We could push back the deserts, restore topsoil, and allow the ozone layer to be repaired within a century or so. We could probably restore climate stability in the wake of global warming within a thousand years. But once a species is gone, it is gone for good.

Of course, in the long run evolution will generate replacement species with numbers and variety to match today's [Figures 5-9 and 5-10]. But that is likely to take millions of years. We are witnessing the gross reduction if not the elimination of entire sectors of biomes, notably tropical forests, coral reefs, and wetlands, all of which may have served as centers of new speciation in the prehistoric past.

Suppose, as has happened after the five mass extinctions of the prehistoric past [Figure 5-9], that the bounce-back period lasts at least 5 million years. This would be 20 times longer than humans have been a species. Suppose that the average number of people on the earth during that period is 2.5 billion people, as opposed to the 6.1 billion today. Then the total number of people affected by what we do (or don't do) to protect the biosphere during the next few decades will be about 500

*For scientific details, see Norman Myers, "The Biodiversity Crisis and the Future of Evolution," *The Environmentalist*, 16 (1996): 37–47.

7. Define *natural selection, differential reproduction,* and *adaptation* and explain their roles in microevolution.

8. Give an example of microevolution by natural selection. Describe three types of natural selection.

9. What is *coevolution,* and what is its importance?

10. What is the *ecological niche* of a species? Why is it important to understand the niches of species? What is the difference between a species' *habitat* and its *niche*? What is the difference between a species' *fundamental niche* and its *realized niche*?

11. Distinguish between the niches of *specialist species* and *generalist species*. Explain why cockroaches have been such successful species.

12. List three factors that limit adaptation.

13. What are two common misconceptions about evolution?

14. What is *speciation*? Distinguish between *geographic isolation and reproductive isolation* and explain how they can lead to speciation through divergent evolution.

15. What is *extinction*? List three factors that have affected the earth's long-term patterns of speciation and extinction. Distinguish between *background extinction* and *mass extinction*. What is believed to have caused the five great mass extinctions in the past?

16. What is an *adaptive radiation*? How can such a radiation lead to recovery after a mass extinction?

17. What evidence suggests that human activities may be bringing about a sixth mass extinction?

18. Explain how speciation and extinction result in the planet's biodiversity.

CRITICAL THINKING

1. How would you respond to someone who tells you that **(a)** he or she does not believe in biological evolution because it is "just a theory" and **(b)** we should not

trillion, in contrast with the 50 billion people who have ever existed. Even 1 trillion is a big number: Figure out how long a period of time in years is made up of 1 trillion seconds.

In short, we are engaged in by far the biggest decision ever made by one human community on behalf of future human communities. Yet the issue is almost entirely disregarded, whether scientifically, ethically, or otherwise. How often do you hear leading scientists even mention the issue?

Despite our gross ignorance of what lies ahead, we can venture a few hypotheses:

- *A temporary outburst of speciation.* As large numbers of niches are vacated, there could be an outburst of speciation, although not nearly enough to match the extinction spasm.
- *A proliferation of opportunistic species* such as cockroaches [Spotlight, p. 113], rats, flies, and others that prosper when new niches open up. This proliferation will be enhanced by the likely elimination of species that naturally control opportunistic species.
- *An end to large vertebrates.*
- *An end to speciation of large vertebrates.* Even if larger vertebrates were to survive the extinction spasm ahead, our largest protected areas will prove far too small for further speciation of elephants, rhinoceroses, apes, bears, and the bigger cats, among other large vertebrates.

What does all this imply for our conservation efforts? By far the predominant strategy of conservationists is to save as many species as possible. But we now need to safeguard evolutionary processes as well. A prime goal is to look out especially for endemic species (found only in a particular place) or species confined to small habitats. Examples include the California condor, the black-footed ferret, the giant panda, and the gorilla.

However, the fossil record shows that endemic species often turn out to be evolutionary dead ends: Generally they do not throw off new species. So should we shift our conservation priority from endemic species to broader-ranging species in the hope that they have more genetic variability and thus more of a diversified resource stock on which natural selection can work its creative impact?

Similarly, should we devote more attention to protecting the evolutionary powerhouses such as the forests, coral reefs, and wetlands of the tropics? All these are in dire trouble and may be all but eliminated within just a few decades. Do they deserve preferential treatment ahead of, say, temperate-zone woodlands, grasslands, and boreal forests with their lower species diversity, ecological complexity, and evolutionary potential?

If within your lifetime we allow the current biotic crisis to proceed unchecked (which is what the recent record suggests), it is possible that your children and grandchildren will ask you a key question: "When the evolutionary debacle was becoming all too plain at the start of the new millennium, what did you do to help ward off disaster?" I hope you will engage yourself in dealing with this crucial issue.

Critical Thinking

1. Do you agree or disagree with the thesis of this essay? Explain.

2. If you agree, list three things you could do to help prevent the outcomes described in this essay.

worry about air pollution because through natural selection the human species will develop lungs that can detoxify pollutants.

2. How would you respond to someone who says that because extinction is a natural process, we should not worry about the loss of biodiversity?

3. Why is the realized niche of a species narrower, or more specialized, than its fundamental niche?

4. As well as you can, describe the major differences between the ecological niches of humans and cockroaches. Are these two species in competition? If so, how do they manage to coexist?

5. In what ways do humans occupy generalist niches and in what ways do they occupy specialist niches?

6. By analogy, use the concepts of generalist and specialist to evaluate the roles of humans in today's societies. In general, the role of college and graduate education is to create specialists in a particular field or a narrow portion of a field. What are the pros and cons of relying mainly on this approach? Is there a need for more generalists? Or are they people who may know a lot about many things (and about connections between things) but not enough about anything in particular? Can they serve a useful role and make a satisfactory living in today's increasingly specialized societies? Explain your answers.

PROJECTS

1. An important adaptation of humans is a strong opposable thumb, which allows us to grip and manipulate things with our hands. As a demonstration of the importance of this trait, fold each of your thumbs into the palm of its hand and then tape them securely in that position for an entire day. After the demonstration, make a list of the things you could not do without the use of your thumbs.

2. Visit a local forest, pond, or lake, choose a particular organism, and then use the library or the internet to describe as much as you can about its niche, including what species it depends on and what species help support its survival. Predict what might happen if your selected species disappeared from the local environment.

3. Make a concept map of this chapter's major ideas using the section heads and subheads and the key terms (in boldface). Look on the website for this book for information about making concept maps.

INTERNET STUDY RESOURCES AND RESOURCES FOR FURTHER READING AND RESEARCH

The website for this book contains helpful study aids and many ideas for further reading and research. Log on to

http://www.brookscole.com/product/0534389872s

and click on the Chapter-by-Chapter area. Choose Chapter 5 and select a resource:

- "Flash Cards" allows you to test your mastery of the Terms and Concepts to Remember for this chapter.
- "Tutorial Quizzes" provides a multiple-choice practice quiz.
- "Student Guide to InfoTrac" will lead you to Critical Thinking Projects that use InfoTrac College Edition as a research tool.
- "References" lists the major books and articles consulted in writing this chapter.
- "Hypercontents" takes you to an extensive list of sites with news, research, and images related to individual sections of the chapter.

INFOTRAC COLLEGE EDITION

Improve your skills with InfoTrac College Edition, a searchable online database of articles from more than 700 periodicals. Log on to

http://www.infotrac-college.com

or access InfoTrac through the website for this book. Try to find the following articles:

1. Great Balls of Fire: What Caused Earth's Biggest Extinction? The Answer May Lie in Tiny Buckyballs Filled with Extraterrestrial Gas (Science) (asteroid or comet impact is thought to have caused the Permian extinction) (Brief Article). Frederic Golden. *Time,* March 5, 2001 v157 i9 p59. *Hint*: Enter the search term "buckyballs" using the keyword "extinction."

2. Flying Apart: Mating Behavior and Speciation. Christine R. B. Boake. *BioScience,* June 2000 v50 i6 p501. If sexual selection drives speciation, it should be possible to find traits that are both sexually selected and involved in behavioral isolation. *Hint*: Enter the search term "speciation" using the keyword "evolution."

6 CLIMATE, TERRESTRIAL BIODIVERSITY, AND AQUATIC BIODIVERSITY

Connections: Blowing in the Wind

One of the things that connects all life on the earth is *wind*, a vital part of the planet's circulatory system. Without wind, the tropics would be unbearably hot, and most of the rest of the planet would freeze.

Winds also transport nutrients from one place to another. Dust rich in phosphates blows across the Atlantic from the Sahara Desert in Africa (Figure 6-1). This helps to replenish rain forest soils in Brazil and build up agricultural soils in the Bahamas. Iron-rich dust blowing from China's Gobi Desert falls into the Pacific Ocean between Hawaii and Alaska and stimulates the growth of phytoplankton, the minute producers that support ocean food webs. That is the *good news*.

The *bad news* is that wind also transports harmful substances. Particles of reddish-brown soil and pesticides banned in the United States are blown from Africa's deserts and eroding farmlands into the sky over Florida. This **(1)** makes it difficult for the state to meet federal air pollution standards and **(2)** is suspected to be a factor in degrading or killing coral reefs in the Florida Keys and the Caribbean.

Pollution and dust from rapidly industrializing China and central Asia blow across the Pacific Ocean and degrade air quality over the western United States, especially Washington and Oregon. Studies show that Asian pollution contributes as much as 10% to West Coast smog, and this threat is expected to increase.

There's *mixed news* as well. Particles from volcanic eruptions ride the winds, circle the globe, and change the earth's climate for a while. Emissions from the 1991 eruption of Mount Pinatubo in the Philippines cooled the earth slightly for 3 years, temporarily masking signs of global warming. On the other hand, volcanic ash, like the blowing desert dust, adds valuable trace minerals to the soil where it settles.

The lesson, once again, is that *there is no "away."* One reason is that wind is part of the planet's circulatory system for heat, moisture, plant nutrients, and long-lived pollutants we put into the air. Movement of soil particles from one place to another by wind and water is a natural phenomenon, but when we disturb the soil and leave it unprotected we hasten the process.

Wind is also an important factor in climate through its influence on global air circulation patterns. Climate, in turn, is crucial for determining what kinds of plant and animal life are found in the major biomes and aquatic life zones of the biosphere, as we shall see in this chapter.

Figure 6-1 Some of the dust shown here blowing from Africa's Sahara Desert can end up as **(1)** soil nutrients in Amazonian rain forests and **(2)** particles of toxic air pollutants in Florida and the Caribbean. (NOAA/USGS/NMD Eros Data Center)

To do science is to search for repeated patterns, not simply to accumulate facts, and to do the science of geographical ecology is to search for patterns of plant and animal life that can be put on a map.

ROBERT H. MACARTHUR

This chapter addresses the following broad questions about geographic patterns of ecology:

- What key factors determine the earth's weather and climate?
- How does climate determine where the earth's major biomes are found?
- What are the major types of desert and grassland biomes, and how do human activities affect them?
- What are the major types of forest biomes, and how do human activities affect them?
- What are the major types of saltwater life zones, and how do human activities affect them?
- What are the major types of freshwater life zones, and how do human activities affect them?

6-1 WEATHER AND CLIMATE: A BRIEF INTRODUCTION

What Is Weather? At every moment at any spot on the earth, the *troposphere* (the inner layer of the atmosphere containing most of the earth's air) has a particular set of physical properties. Examples are **(1)** temperature, **(2)** pressure, **(3)** humidity, **(4)** precipitation, **(5)** sunshine, **(6)** cloud cover, and **(7)** wind direction and speed. These short-term properties of the troposphere at a particular place and time are **weather**.

Meteorologists use weather balloons, aircraft, ships, radar, satellites, and other devices to obtain data on variables such as atmospheric pressures, precipitation, temperatures, wind speeds, and locations of air masses and fronts. These data are fed into computer models to draw weather maps for each of seven levels of the troposphere, ranging from the ground to 19 kilometers (12 miles) up. Computer models use the map data to forecast the weather in each box of a seven-layer grid for the next 12 hours. Other computer models project the weather for the next several days by calculating the probabilities that air masses, winds, and other factors will move and change in certain ways.

What Is Climate? **Climate** is a region's general pattern of atmospheric or weather conditions over a long period. *Average temperature* and *average precipitation* are the two main factors determining a region's climate (Figure 6-2). Figure 6-3 is a generalized map of the earth's major climate zones.

How Does Global Air Circulation Affect Regional Climates? The temperature and precipitation patterns that lead to different climates (Figure 6-3) are caused primarily by the way air circulates over the earth's surface. The following factors determine global air circulation patterns:

- *Uneven heating of the earth's surface* because air is heated much more at the equator (where the sun's rays

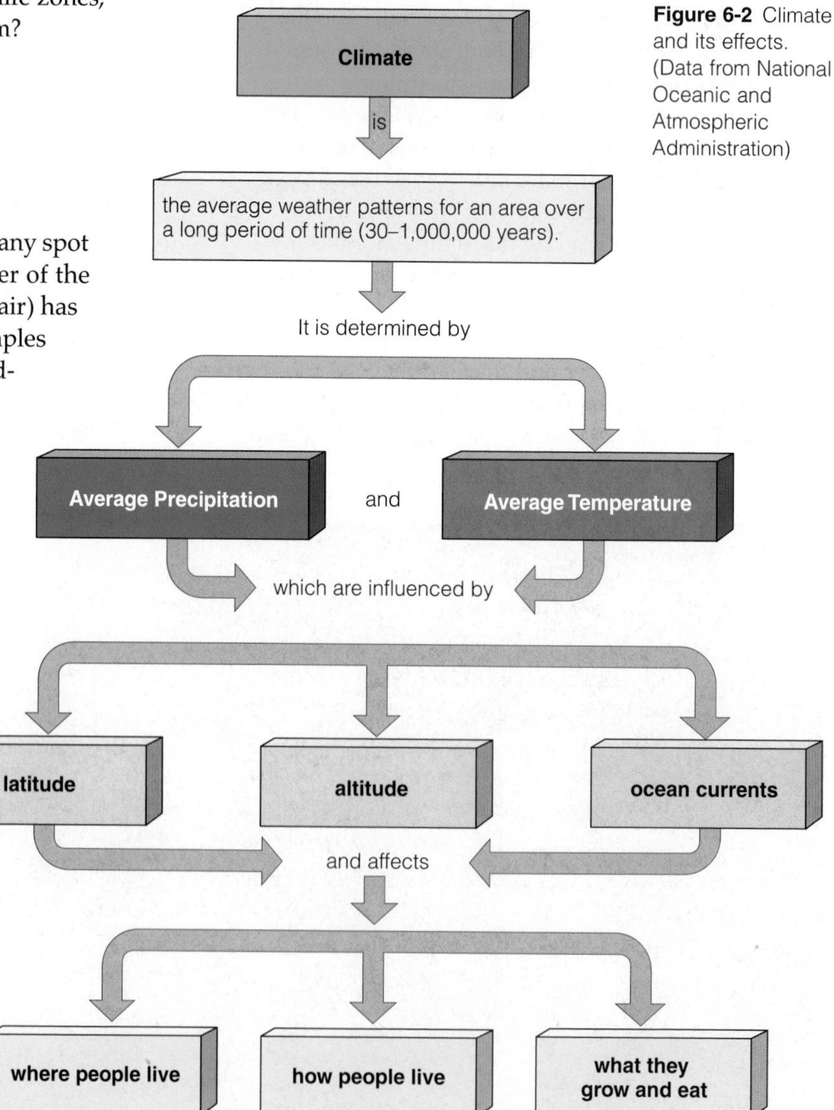

Figure 6-2 Climate and its effects. (Data from National Oceanic and Atmospheric Administration)

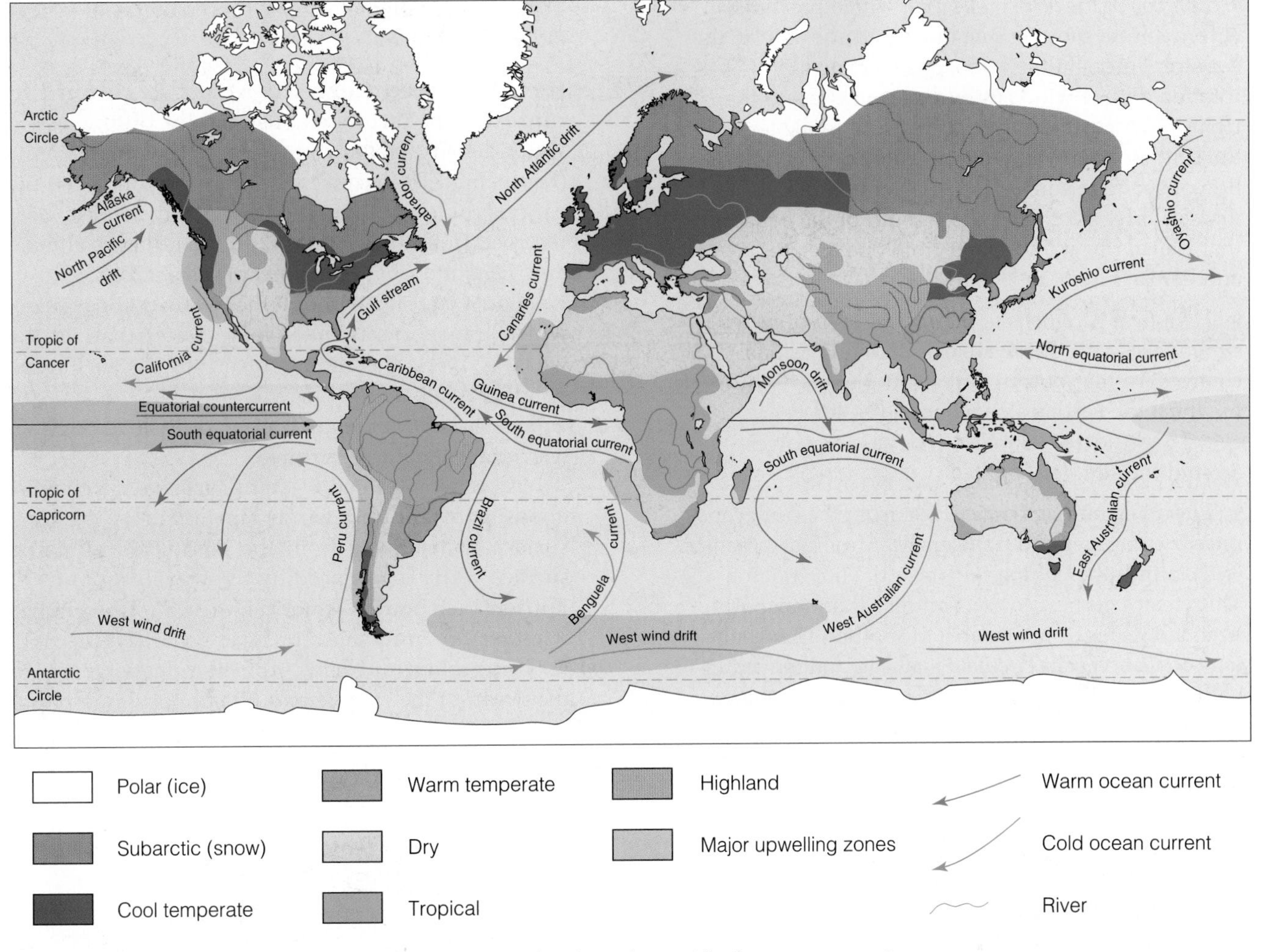

Figure 6-3 Generalized map of global climate zones, showing the major contributing ocean currents and drifts.

strike directly throughout the year) than at the poles (where sunlight strikes at an angle and thus is spread out over a much greater area). These differences in incoming solar energy help explain why **(1)** tropical regions near the equator are hot, **(2)** polar regions are cold, and **(3)** temperate regions in between generally have intermediate average temperatures (Figure 6-3).

- *Seasonal changes in temperature and precipitation* because the earth's axis (an imaginary line connecting the north and south poles) is tilted. As a result, various regions are tipped toward or away from the sun as the earth makes its annual revolution around the sun (Figure 6-4). This creates opposite seasons in the northern and southern hemispheres.

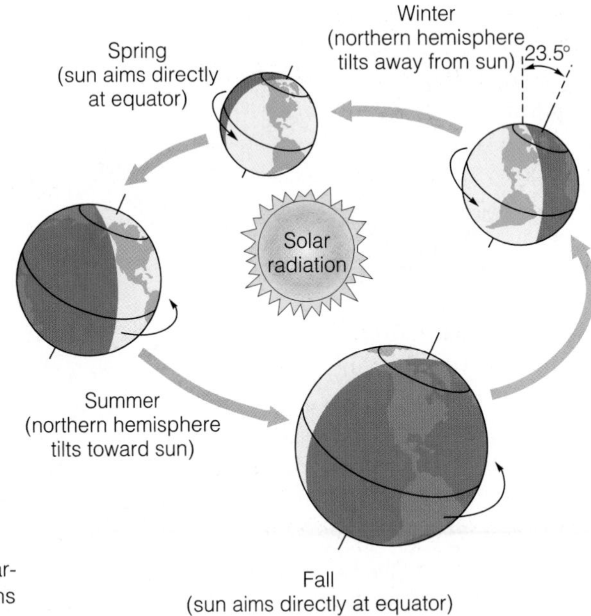

Figure 6-4 The effects of the earth's tilted axis on climate. As the planet makes its annual revolution around the sun on an axis tilted about 23.5°, various regions are tipped toward or away from the sun. The resulting variations in the amount of solar energy reaching the earth create the seasons.

- *Rotation of the earth on its axis*, which prevents air currents from moving due north and south from the equator. Forces in the atmosphere created by this rotation deflect winds (moving air masses) to the right in the northern hemisphere and to the left in the southern hemisphere. This results in the formation of six huge convection cells of swirling air masses, three north and three south of the equator, that transfer heat and water from one area to another (Figure 6-5).
- *Long-term variations in the amount of solar energy striking the earth*. These are caused by occasional changes in solar output and slight planetary shifts in which the earth's axis wobbles (22,000-year cycle) and tilts (44,000-year cycle) as it revolves around the sun.
- *Properties of air and water*. Heat from the sun evaporates ocean water and transfers heat from the oceans to the atmosphere, especially near the hot equator. This creates cyclical convection cells that transport heat and water from one area to another. The resulting convection cells circulate air, heat, and moisture both vertically and horizontally in the troposphere.

This leads to different climates and patterns of vegetation (Figure 6-6).

How Do Ocean Currents Affect Regional Climates? The factors just listed, plus differences in water density, create warm and cold ocean currents (Figure 6-3). These currents, driven by winds and the earth's rotation (Figure 6-5), **(1)** redistribute heat received from the sun and **(2)** thus influence climate and vegetation, especially near coastal areas.

For example, without the warm Gulf Stream, which transports 25 times more water than all the world's rivers, the climate of northwestern Europe would be subarctic. Currents also help mix ocean waters and distribute nutrients and dissolved oxygen needed by aquatic organisms.

Along some steep western coasts of continents, almost constant tradewinds blow offshore, pushing surface water away from the land. This outgoing surface water is replaced by an **upwelling** of cold, nutrient-rich bottom water (Figure 6-7). Upwellings, whether far from shore or near shore (Figure 6-3), **(1)** bring plant nutrients from the deeper parts of the ocean to the surface and **(2)** support large popu-

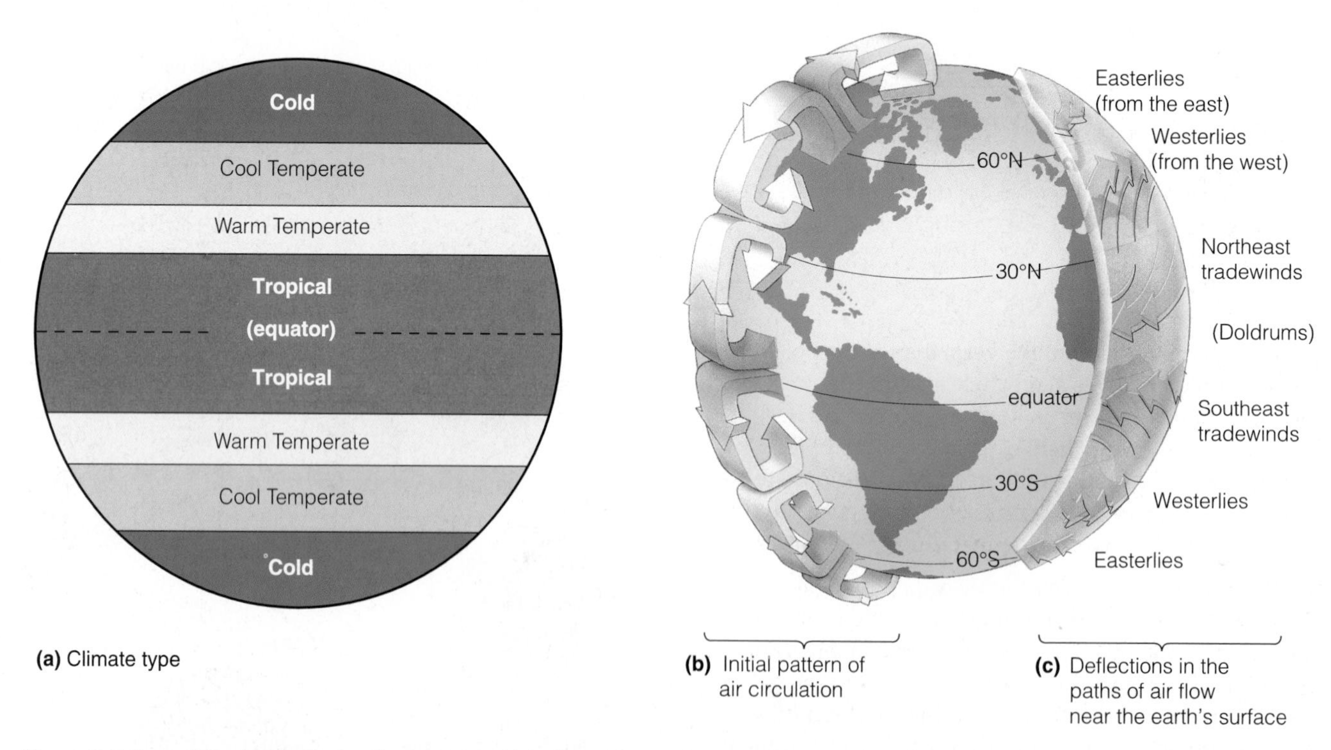

Figure 6-5 Formation of prevailing surface winds, which disrupt the general flow of air from the equator to the poles and back to the equator. As the earth rotates, its surface turns faster beneath air masses at the equator and slower beneath those at the poles. This deflects air masses moving north and south to the west or east, creating six huge convection cells in which air swirls upward and then descends toward the earth's surface at different latitudes. The direction of air movement in these cells sets up belts of prevailing winds that distribute air and moisture over the earth's surface. These winds affect the general types of climate found in different areas and drive the circulation of ocean currents. (Used by permission from Cecie Starr, *Biology: Concepts and Applications*, 4th ed., Pacific Grove, Calif.: Brooks/Cole, 2000)

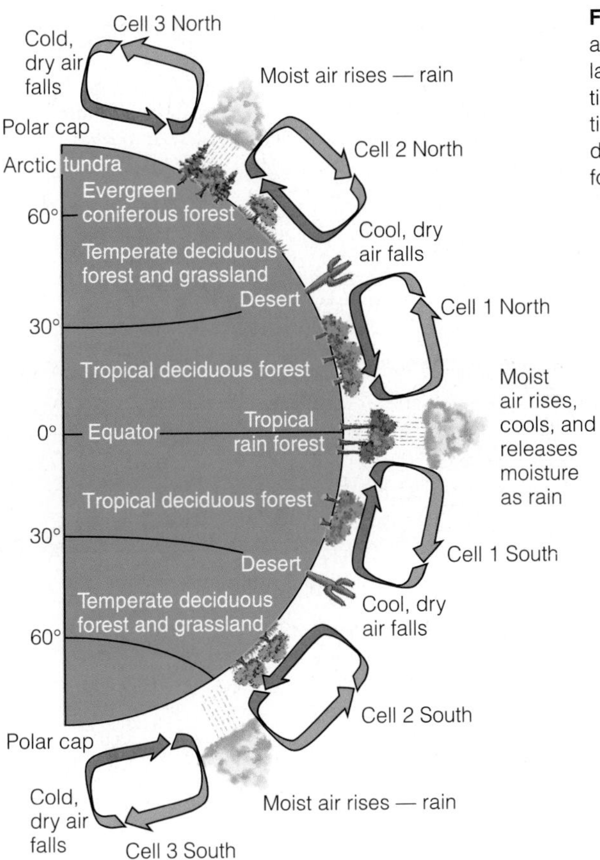

Figure 6-6 Model of global air circulation and biomes. Heat and moisture are distributed over the earth's surface by vertical currents that form six large convection cells (called Hadley cells) at different latitudes. The direction of air flow and the ascent and descent of air masses in these convection cells determine the earth's general climatic zones. The resulting uneven distribution of heat and moisture over the planet's surface leads to the forests, grasslands, and deserts that make up the earth's biomes.

lations of phytoplankton, zooplankton, fish, and fish-eating seabirds.

What Is the El Niño–Southern Oscillation? Every few years in the Pacific Ocean, normal coastal upwelling (Figure 6-8, left) is affected by changes in climate patterns called the *El Niño–Southern Oscillation* or *ENSO* (Figure 6-8, right). In an ENSO, often called *El Niño,* **(1)** the prevailing westerly winds weaken or cease, **(2)** surface water along the South and North American coasts becomes warmer, and **(3)** the normal upwelling of cold, nutrient-rich water is suppressed, which reduces primary productivity and causes a sharp decline in the populations of some fish species.

A strong ENSO can trigger extreme weather changes over at least two-thirds of the globe, especially in lands along the Pacific and Indian Oceans (Figure 6-9). Figure 6-10 shows the occurrence of ENSOs between 1950 and 2000. Some models project that if the earth's atmosphere continues to warm (Section 13-2, p. 303), El Niño–like weather may become the norm and cause major ecological and socioeconomic problems.

What Is La Niña? Sometimes an El Niño is followed by its cooling counterpart, *La Niña,* as occurred from July 1998 into 1999 (Figure 6-10). Typically a La Niña means **(1)** more Atlantic Ocean hurricanes, **(2)** colder winters in Canada and the Northeast, **(3)** warmer and drier winters in the southeastern and southwestern United States, **(4)** wetter winters in the Pacific Northwest, **(5)** torrential rains in Southeast Asia, **(6)** lower wheat yields in Argentina, and **(7)** more wildfires in Florida.

How Does the Chemical Makeup of the Atmosphere Lead to the Greenhouse Effect? Small amounts of certain gases play a key role in determining

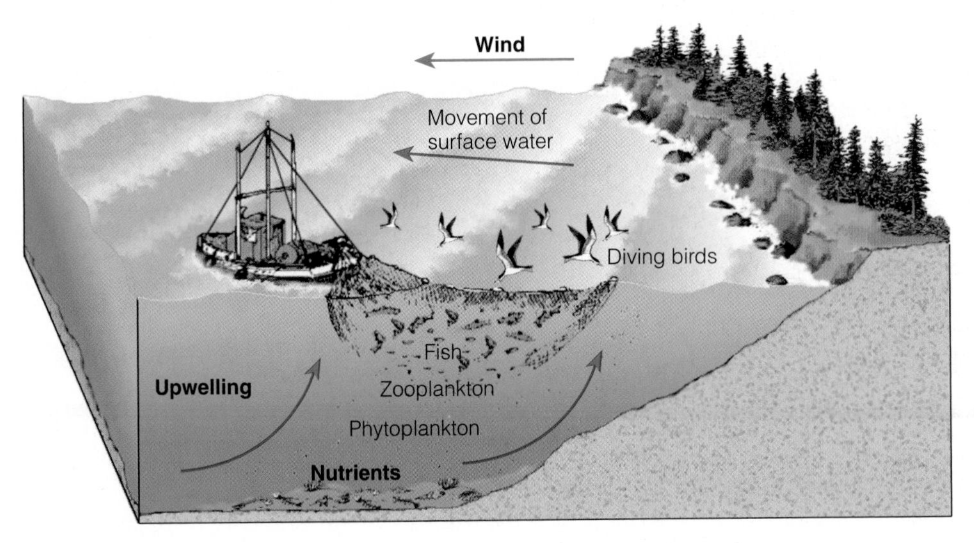

Figure 6-7 A *shore upwelling* (shown here) occurs when deep, cool, nutrient-rich waters are drawn up to replace surface water moved away from a steep coast by wind-driven currents. Such areas support large populations of phytoplankton, zooplankton, fish, and fish-eating birds. *Equatorial upwellings* occur in the open sea near the equator when northward and southward currents interact to push deep waters and their nutrients to the surface, thus greatly increasing primary productivity in such areas.

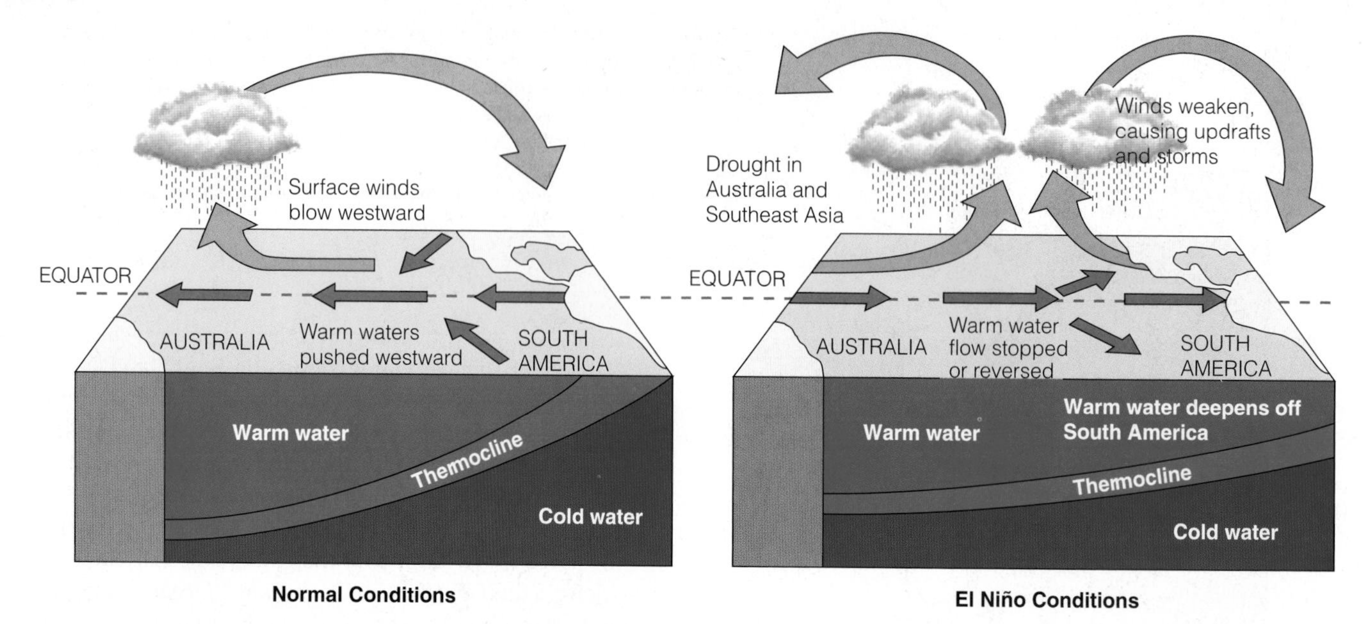

Figure 6-8 Normal surface winds blowing westward cause shore upwellings of cold, nutrient-rich bottom water in the tropical Pacific Ocean near the coast of Peru (left). The warm and cold water are separated by a zone of gradual temperature change called the *thermocline*. Every few years a climate shift known as the El Niño–Southern Oscillation (ENSO) disrupts this pattern. Westward surface winds weaken, which depresses the coastal upwellings and warms the surface waters off South America (right). When an ENSO lasts 12 months or longer, it severely disrupts populations of plankton, fish, and seabirds in upwelling areas and can trigger extreme weather changes over much of the globe (Figure 6-9).

the earth's average temperatures and thus its climates. These gases include **(1)** water vapor (H_2O), **(2)** carbon dioxide (CO_2), **(3)** methane (CH_4), **(4)** nitrous oxide (N_2O), **(5)** synthetic chlorofluorocarbons (CFCs), **(6)** synthetic perfluorocarbons (PFCs), and **(7)** synthetic trifluoromethyl sulfur pentafluoride (SF_5CF_3).

Together, these gases are known as **greenhouse gases.** They allow light, infrared radiation, and some ultraviolet (UV) radiation from the sun (Figure 3-8, p. 66) to pass through the troposphere. The earth's surface absorbs much of this solar energy and degrades it to longer-wavelength, infrared radiation (heat), which then rises into the troposphere (Figure 4-8, p. 82).

Some of this heat escapes into space, and some is absorbed by molecules of greenhouse gases and emitted into the troposphere as even longer-wavelength infrared radiation, which warms the air. This natural warming effect of the troposphere is called the **greenhouse effect*** (Figure 6-11).

The primary greenhouse gas in the atmosphere is *water vapor*. However, because its concentration in the

*It is scientifically incorrect to say that that the earth's atmosphere *traps* heat or *reradiates* heat it has absorbed back toward the earth's surface. Molecules of greenhouse gases absorb various wavelengths of infrared radiation and transform them into infrared radiation with different (longer) wavelengths. Because the originally absorbed wavelengths of infrared radiation no longer exist, it is incorrect to say that they have been trapped or reradiated.

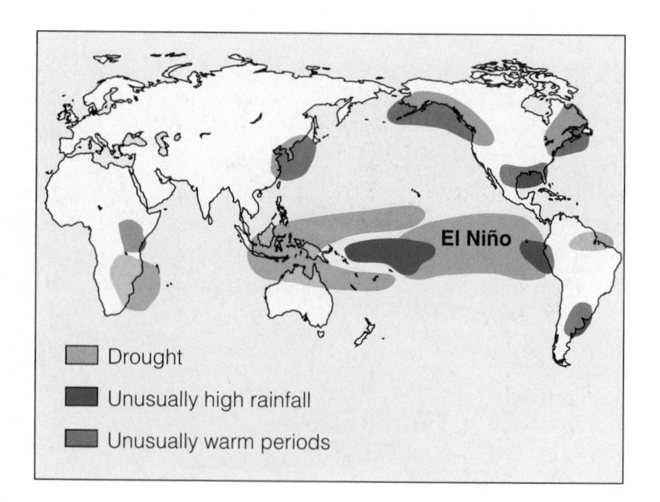

Figure 6-9 Typical global climatic effects of an El Niño–Southern Oscillation. During the 1997–98 ENSO, huge waves battered the California coast, and torrential rains caused widespread flooding and mudslides. In Peru, floods and mudslides killed hundreds of people, left about 250,000 people homeless, and ruined crops. Drought in Brazil, Indonesia, and Australia led to massive wildfires in tinder-dry forests. India and parts of Africa also experienced severe drought. A catastrophic ice storm hit Canada and the northeastern United States, but the southeastern United States had fewer hurricanes. (Data from United Nations Food and Agriculture Organization)

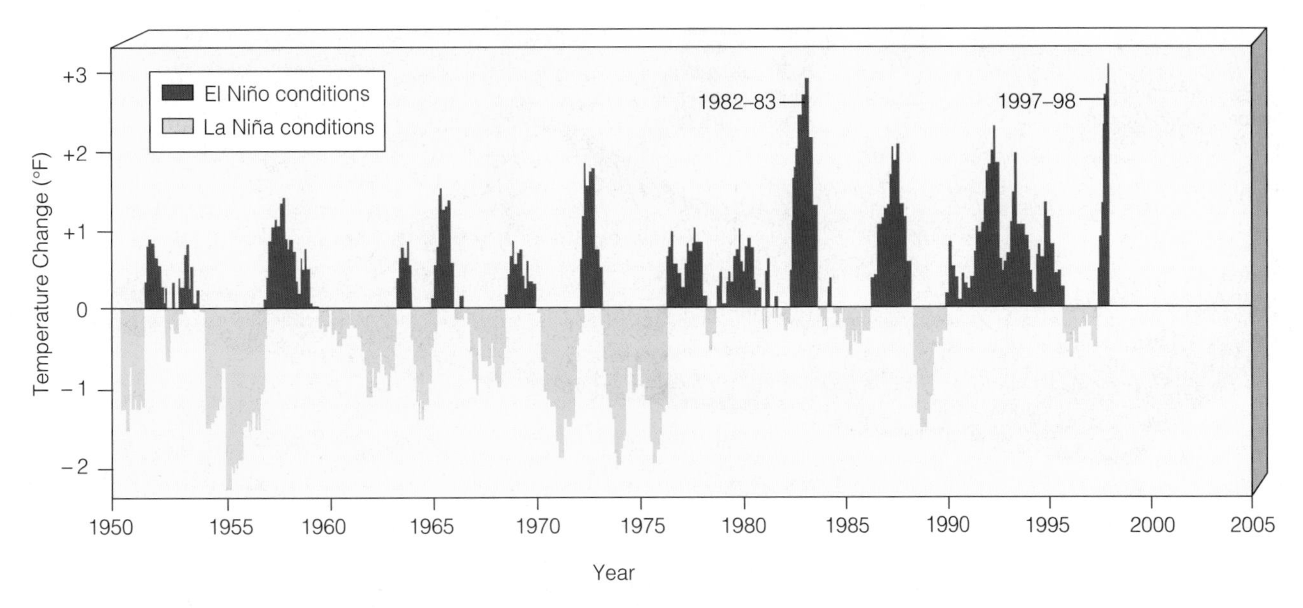

Figure 6-10 El Niño and La Niña conditions between 1950 and 2000. (Data from U.S. National Weather Service)

troposphere is high (1–5%), inputs of water vapor from human activities have little effect on this chemical's greenhouse effects. By contrast, the concentration of carbon dioxide in the atmosphere is so small (0.036%) that the fairly large input of CO_2 from human activities can significantly increase the amount of heat in the lower atmosphere.

The basic principle behind the greenhouse effect is well established. Indeed, without its current greenhouse gases (especially water vapor), the earth would be a cold and mostly lifeless planet.

We and other species currently benefit from a comfortable level of greenhouse gases that typically undergo only minor, slow fluctuations over hundreds to thousands of years. However, mathematical models of the earth's climate indicate that natural or human-induced global warming taking place over a few decades could be disastrous for human

(a) Rays of sunlight penetrate the lower atmosphere and warm the earth's surface.

(b) The earth's surface absorbs much of the incoming solar radiation and degrades it to longer-wavelength infrared radiation (heat), which rises into the lower atmosphere. Some of this heat escapes into space and some is absorbed by molecules of greenhouse gases and emitted as infrared radiation, which warms the lower atmosphere.

(c) As concentrations of greenhouse gases rise, their molecules absorb and emit more infrared radiation, which adds more heat to the lower atmosphere.

Figure 6-11 The *greenhouse effect.* Without the atmospheric warming provided by this natural effect, the earth would be a cold and mostly lifeless planet. According to the widely accepted greenhouse theory, when concentrations of greenhouse gases in the atmosphere rise, the average temperature of the troposphere also rises. (Modified by permission from Cecie Starr, *Biology: Concepts and Principles*, 4th ed., Pacific Grove, Calif.: Brooks/Cole, 2000)

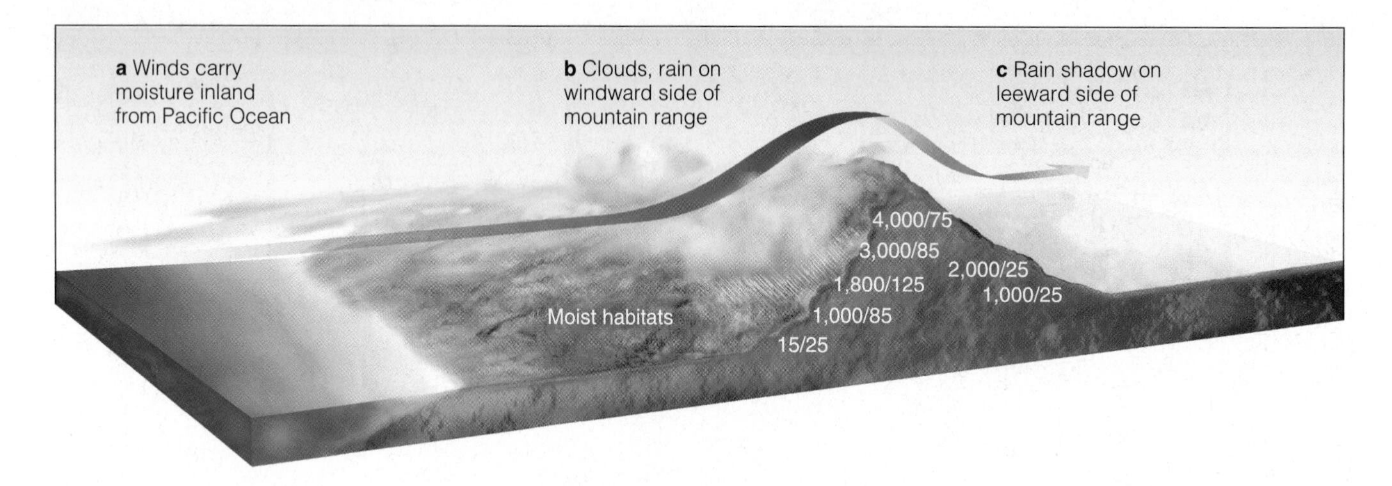

Figure 6-12 The *rain shadow effect* is a reduction of rainfall on the side of high mountains facing away from prevailing surface winds. It occurs when warm, moist air in prevailing onshore winds loses most of its moisture as rain and snow on the windward (wind-facing) slopes of a mountain range. This leads to semiarid and arid conditions on the leeward side of the mountain range and the land beyond. The Mojave Desert, east of the Sierra Nevada in California, is produced by this effect. Numbers on the right of each pair represent average annual precipitation (in centimeters) in California's Sierra Nevada, and numbers on the left signify elevation (in meters). (Used by permission from Cecie Starr, *Biology: Concepts and Principles*, 4th ed., Pacific Grove, Calif.: Brooks/Cole, 2000)

societies and many forms of life, as discussed in Section 13-4, p. 311.

How Does the Chemical Makeup of the Atmosphere Create the Ozone Layer? In a band of the stratosphere 16–26 kilometers (11–16 miles) above the earth's surface, oxygen (O_2) is continuously converted to ozone (O_3) and back to oxygen by a sequence of reactions initiated by UV radiation from the sun ($3O_2 + UV \rightleftharpoons 2O_3$). The result is a thin veil of protective ozone at very low concentrations (up to 12 parts per million).

Normally, the average levels of ozone in this life-saving layer do not change much because the rate of ozone destruction is equal to its rate of formation. This stratospheric ozone prevents at least 95% of the sun's harmful UV radiation from reaching the earth's surface (Figure 4-8, p. 82).

This ozone also creates warm layers of air that prevent churning gases in the troposphere from entering the stratosphere. This *thermal cap* is important in determining the average temperature of the troposphere and thus the earth's current climates. There is much evidence that chemicals added to the atmosphere by human activities are decreasing levels of protective ozone in the stratosphere, as discussed in more detail in Section 13-6 (p. 318).

How Do Topography and Other Features of the Earth's Surface Create Microclimates? Various topographic features of the earth's surface create local climatic conditions, or **microclimates**, that differ from the general climate of a region. For example, mountains interrupt the flow of prevailing surface winds and the movement of storms. When moist air blowing inland from an ocean reaches a mountain range, it cools as it is forced to rise and expand. This causes the air to lose most of its moisture as rain and snow on the windward (wind-facing) slopes. As the drier air mass flows down the leeward (away from the wind) slopes, it draws moisture out of the plants and soil over which it passes. The lower precipitation and the resulting semiarid or arid conditions on the leeward side of high mountains are called the **rain shadow effect** (Figure 6-12).

Cities also create distinct microclimates. Bricks, concrete, asphalt, and other building materials absorb and hold heat, and buildings block wind flow. Motor vehicles and the climate control systems of buildings release large quantities of heat and pollutants. As a result, cities tend to have **(1)** more haze and smog, **(2)** higher temperatures, and **(3)** lower wind speeds than the surrounding countryside.

6-2 BIOMES: CLIMATE AND LIFE ON LAND

Why Are There Different Organisms in Different Places? Why is one area of the earth's land surface a desert, another a grassland, and another a forest? Why are there different types of deserts, grasslands, and forests?

The general answer to these questions is differences in *climate* (Figure 6-3), caused mostly by differ-

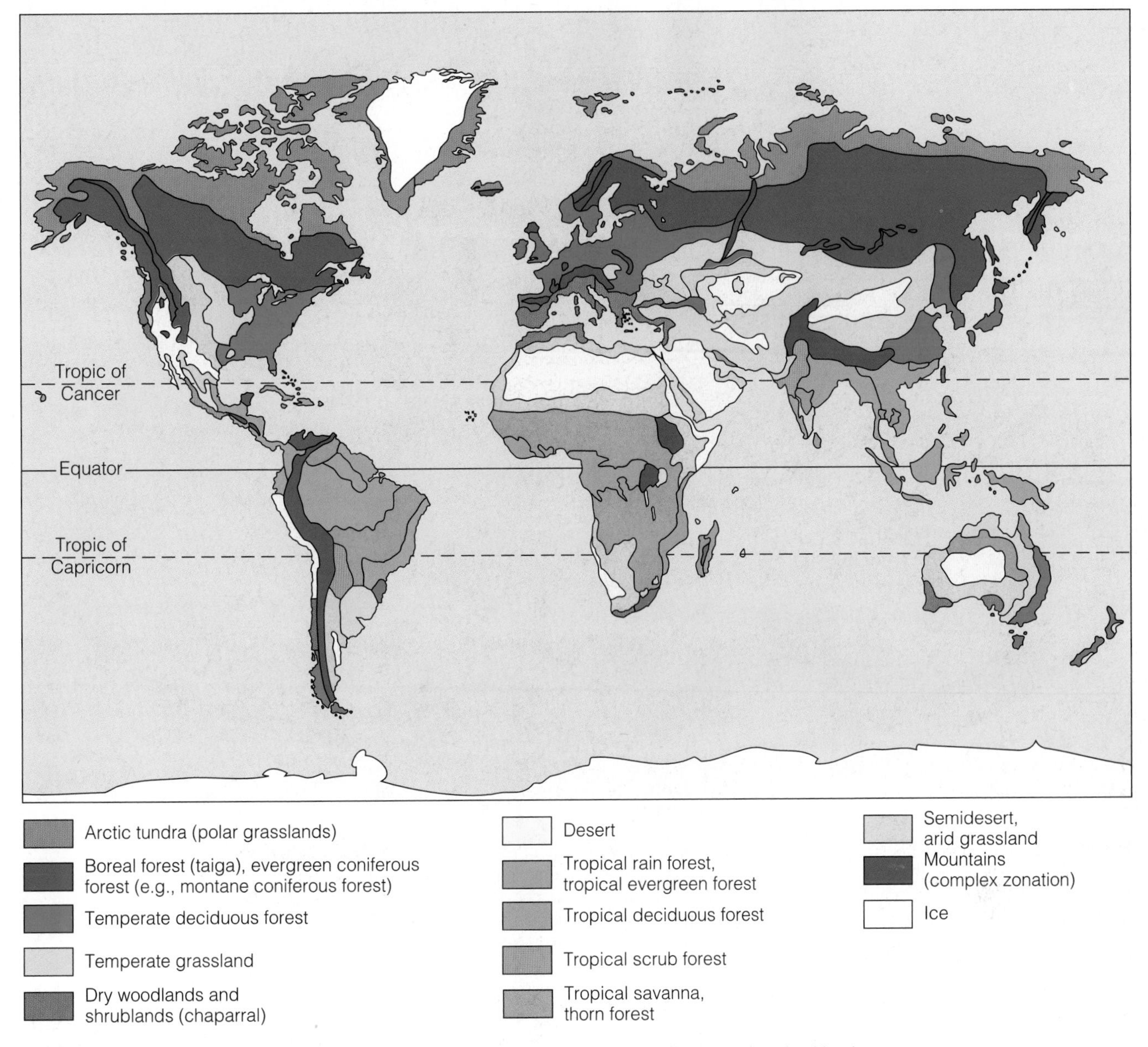

Figure 6-13 The earth's major biomes—the main types of natural vegetation in different undisturbed land areas—result primarily from differences in climate (Figure 6-3). Each biome contains many ecosystems whose communities have adapted to differences in climate, soil, and other environmental factors. In some areas, people have altered these biomes by removing or changing much of this natural vegetation for farming, livestock grazing, lumber and fuelwood, mining, and construction (Figure 1-3, p. 5).

ences in average temperature and precipitation caused by global air circulation (Figure 6-6). Different climates promote different communities of organisms.

Figure 6-13 and the photo on p. 1 show global distributions of *biomes*: terrestrial regions with characteristic types of natural, undisturbed ecological communities adapted to the climate of the region. By comparing Figure 6-13 with Figure 6-3, you can see how the world's major biomes vary with climate. Figure 4-9 (p. 83) shows major biomes in the United States as one moves through different climates along the 39th parallel. Taken together, average annual precipitation and temperature (along with soil type) are the most important factors in producing tropical, temperate, or polar deserts, grasslands, and forests (Figure 6-14).

On maps such as the one in Figure 6-13, biomes are presented as having sharp boundaries and as being covered with the same general type of vegetation. In reality, most biomes do not have sharp boundaries and blend into one another in transitional zones or *ecotones*. Also, the types and numbers of plants in a biome vary from one location to another because of variations in **(1)** local climate (microclimates), **(2)** soil types, and **(3)** natural and human-caused disturbances (Figure 1-3, p. 5). As a result,

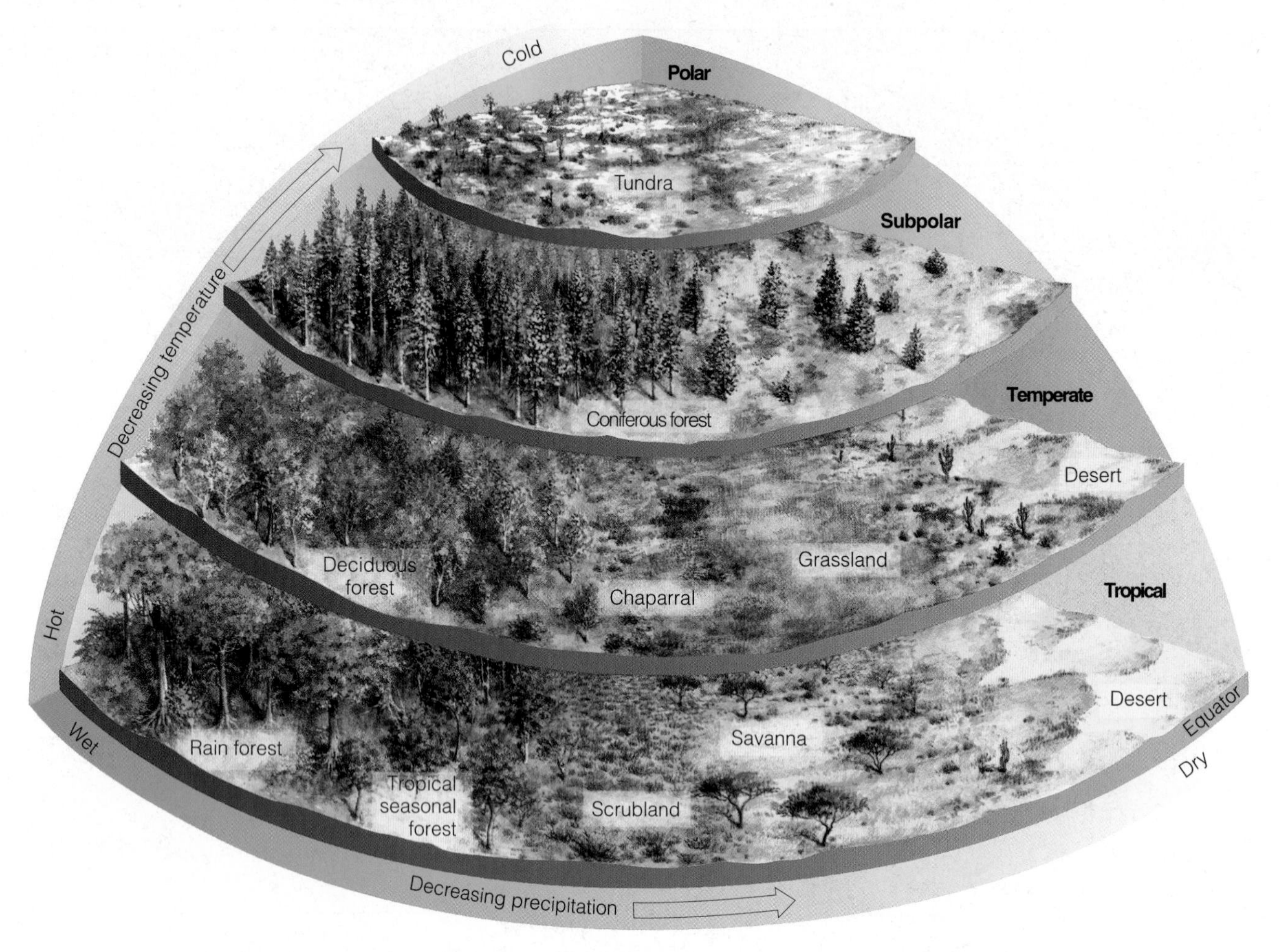

Figure 6-14 Average precipitation and average temperature, acting together as limiting factors over a period of 30 or more years, determine the type of desert, grassland, or forest biome in a particular area. Although the actual situation is much more complex, this simplified diagram explains how climate determines the types and amounts of natural vegetation found in an area left undisturbed by human activities. (Used by permission of Macmillan Publishing Company, from Derek Elsom, *The Earth*, New York: Macmillan, 1992. Copyright © 1992 by Marshall Editions Developments Limited)

biomes are not uniform. They consist of a *mosaic of patches*, all with somewhat different biological communities but with similarities unique to the biome.

Climate and vegetation vary with **latitude** (distance from the equator) and **altitude** (elevation above sea level). If you travel from the equator toward either pole, you will generally encounter colder climates and zones of vegetation adapted to those climates (Figure 6-15, right). Similarly, as elevation above sea level increases, climate becomes colder. Thus, if you climb a tall mountain from its base to its summit, you can observe changes in plant life similar to those you would encounter in traveling from the equator to the earth's poles (Figure 6-15, left).

Why Do Plant Sizes, Shapes, and Survival Strategies Differ? Arctic soils are wet and nutrient rich. So why are there no trees in the Arctic, and why are the plants there so close to the ground? Why are there no leaves on desert plants such as cacti? Why do trees in most forests found in both the warm tropics and in cold areas such as Canada and Sweden keep their leaves year-round, whereas most trees in temperate forests lose their leaves in winter?

The answers to these questions involve evolutionary responses of plants to different climates.

- Plants exposed to cold air year-round or during winter have traits that keep them from losing too much heat and water. For example, trees or tall plants cannot survive in the cold, windy arctic grasslands (tundra) because they would lose too much of their heat for survival.

- Desert plants exposed to the sun all day long must **(1)** be able to lose enough heat so that they do not overheat and die and **(2)** conserve enough water for survival. **Succulent** (fleshy) **plants**, such as the saguaro ("sah-WAH-row") cactus, survive in dry climates by

Figure 6-15 Generalized effects of latitude (right) and altitude (left) on climate and biomes. Parallel changes in vegetation type occur when we travel from the equator to the poles or from lowlands to mountaintops.

(1) having no leaves, **(2)** storing water and synthesizing food in their expandable, fleshy tissue, and **(3)** reducing water loss by opening their pores (stomata) to take up carbon dioxide (CO_2) only at night.

- Trees of wet tropical rain forests tend to be **broadleaf evergreen plants**, which keep most of their broad leaves year-round. The large surface area of the leaves allows them to **(1)** collect ample sunlight for photosynthesis and **(2)** radiate heat during hot weather.
- In a climate with a cold (and sometimes dry) winter, keeping such leaves would cause plants to lose too much heat and water for survival. In such climates, **broadleaf deciduous plants**, such as oak and maple trees, survive drought and cold by shedding their leaves and becoming dormant during such periods.
- If we move further north to areas such as Canada and Sweden, where summers are cool and short, this strategy is less successful. Instead evolution has favored **coniferous** (cone-bearing) **evergreen plants** (such as spruces, pines, and firs). These plants keep some of their narrow, pointed leaves (needles) all year. The waxy coating, shape, and clustering of conifer needles slow down heat loss and evaporation during the long, cold winter. Additionally, by keeping their leaves all winter, such trees are ready to take advantage of the brief summer without having to take time to grow new needles.

6-3 DESERT AND GRASSLAND BIOMES

What Are the Major Types of Deserts? A **desert** is an area where evaporation exceeds precipitation. Precipitation typically is **(1)** less than 25 centimeters (10 inches) a year and **(2)** often scattered unevenly throughout the year. Deserts have sparse, widely spaced, mostly low vegetation.

A combination of low rainfall and different average temperatures creates tropical, temperate, and cold deserts (Figure 6-14).

- In *tropical deserts*, such as the southern Sahara in Africa, **(1)** temperatures usually are high year-round and **(2)** there is little rain, which typically falls during only 1 or 2 months of the year. These driest places on earth typically have few plants and a hard, wind-blown surface strewn with rocks and some sand.
- In *temperate deserts*, such as the Mojave in southern California (Figure 6-12), **(1)** daytime temperatures are high in summer and low in winter and **(2)** there is more precipitation than in tropical deserts. The vegetation is sparse, consisting mostly of widely dispersed, drought-resistant shrubs and cacti or other succulents, and animals are adapted to the lack of water and temperature variations (Figure 6-16).
- In *cold deserts*, such as the Gobi Desert in China, **(1)** winters are cold, **(2)** summers are warm or hot, and **(3)** precipitation is low.

In the semiarid zones between deserts and grasslands, we find *semidesert*. This biome is dominated by thorn trees and shrubs adapted to a long dry spells followed by brief, sometimes heavy rains.

How Do Desert Plants and Animals Survive? Adaptations for survival in the desert have two themes: "Beat the heat" and "Every drop of water counts."

Figure 6-16 Some components and interactions in a *temperate desert biome.* When these organisms die, decomposers break down their organic matter into minerals used by plants. Transfers of matter and energy between producers, primary consumers (herbivores), and secondary (or higher-level) consumers (carnivores) are indicated by colored arrows. Organisms are not drawn to scale.

Desert plants are characterized by various combinations of the following adaptations:

- Having wax-coated leaves that minimize transpiration (evergreens such as the creosote bush).
- Using deep roots to tap into groundwater.
- Using widely spread, shallow roots to collect water after brief showers and store it in their spongy tissues (prickly pear, Figure 6-16, and saguaro cacti).
- Dropping their leaves to survive in a dormant state during long drying spells (mesquite and creosote plants).
- Becoming dormant during dry periods (mosses and lichens).
- Storing much of their biomass in seeds during dry periods and remaining inactive (sometimes for years) until they receive enough water to germinate (annual wildflowers and grasses). Shortly after a rain the seeds **(1)** germinate, **(2)** grow, **(3)** carpet such deserts with a dazzling array of colorful flowers, **(4)** produce new seed, and **(5)** die, all in only a few weeks.

Most desert animals are small and beat the heat and reduce water loss by evaporative cooling by

- Hiding in cool burrows or rocky crevices by day and coming out at night or in the early morning.
- Having physical adaptations for conserving water (Spotlight, p. 136). Insects and reptiles **(1)** have thick outer coverings to minimize water loss through evaporation and **(2)** reduce water loss by excreting dry feces and a dried concentrate of urine.
- Getting their water from dew or from the food they eat (many spiders and insects). Arabian oryxes survive by licking the dew that accumulates at night on rocks and on one another's hair.
- Becoming dormant during periods of extreme heat or drought.

Figure 6-17 shows major human impacts on deserts. Deserts take a long time to recover from disturbances because of their **(1)** slow plant growth, **(2)** low species diversity, **(3)** slow nutrient cycling (because of little bacterial activity in their soils), and **(4)** water shortages. Desert vegetation destroyed by livestock overgrazing and off-road vehicles may take decades to grow back.

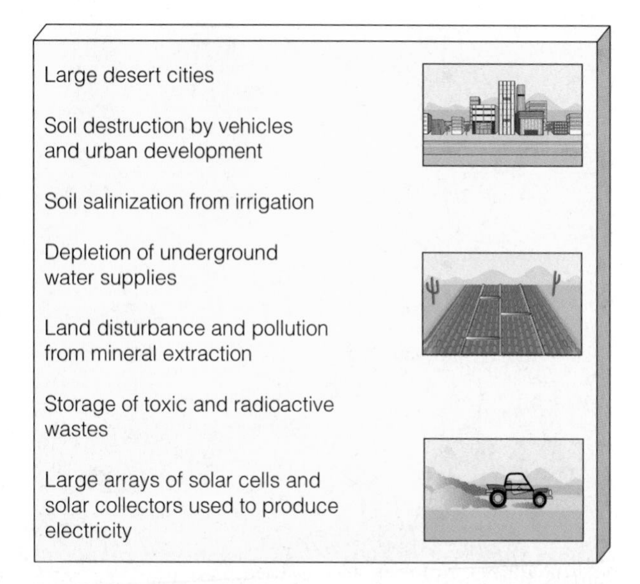

Large desert cities
Soil destruction by vehicles and urban development
Soil salinization from irrigation
Depletion of underground water supplies
Land disturbance and pollution from mineral extraction
Storage of toxic and radioactive wastes
Large arrays of solar cells and solar collectors used to produce electricity

Figure 6-17 Major human impacts on deserts.

What Are the Major Types of Grasslands? **Grasslands** are regions with enough average annual precipitation to allow grass (and in some areas, a few trees) to prosper but with precipitation so erratic that drought and fire prevent large stands of trees from growing. Most grasslands are found in the interiors of continents (Figure 6-13).

Grasslands persist because of a combination of **(1)** seasonal drought, **(2)** grazing by large herbivores, and **(3)** occasional fires, all of which keep large numbers of shrubs and trees from invading and becoming established. If not overgrazed by large herbivores, grasses in these biomes are renewable resources because these plants grow out from the bottom. This allows their stems to grow again after being nibbled off by grazing animals.

The three main types of grasslands—tropical, temperate, and polar (tundra)—result from combinations of low average precipitation and various average temperatures (Figure 6-14). *Tropical grasslands* are found in areas with **(1)** high average temperatures, **(2)** low to moderate precipitation, and **(3)** a prolonged dry season.

One type of tropical grassland, called a *savanna*, usually has **(1)** warm temperatures year-round, **(2)** two prolonged dry seasons, and **(3)** abundant rain the rest of the year. African tropical savannas contain enormous herds of *grazing* (grass- and herb-eating) and *browsing* (twig- and leaf-nibbling) hoofed animals, including wildebeests, gazelles, zebras, giraffes, and antelopes (Figure 6-18). These and other large herbivores have evolved specialized eating habits that minimize competition between species for vegetation. For example, **(1)** giraffes eat leaves and shoots from the tops of trees, **(2)** elephants eat leaves and branches further down, **(3)** Thompson's gazelles and wildebeests prefer short grass, and **(4)** zebras graze on longer grass and stems.

In *temperate grasslands*, **(1)** winters are bitterly cold, **(2)** summers are hot and dry, and **(3)** annual precipitation is fairly sparse and falls unevenly through the year. Drought, occasional fires, and intense grazing inhibit the growth of trees and bushes, except along rivers. Because the aboveground parts of most of the

Figure 6-18 Some of the grazing animals found in different parts of the African savanna. These species share vegetation resources by having different feeding niches.

grasses die and decompose each year, organic matter accumulates to produce a deep, fertile soil. This soil is held in place by a thick network of intertwined roots of drought-tolerant perennial grasses unless the topsoil is plowed up and allowed to blow away by prolonged exposure to high winds found in these biomes.

Types of temperate grasslands are the **(1)** *tall-grass prairies* (Figure 6-19) and *short-grass prairies* of the

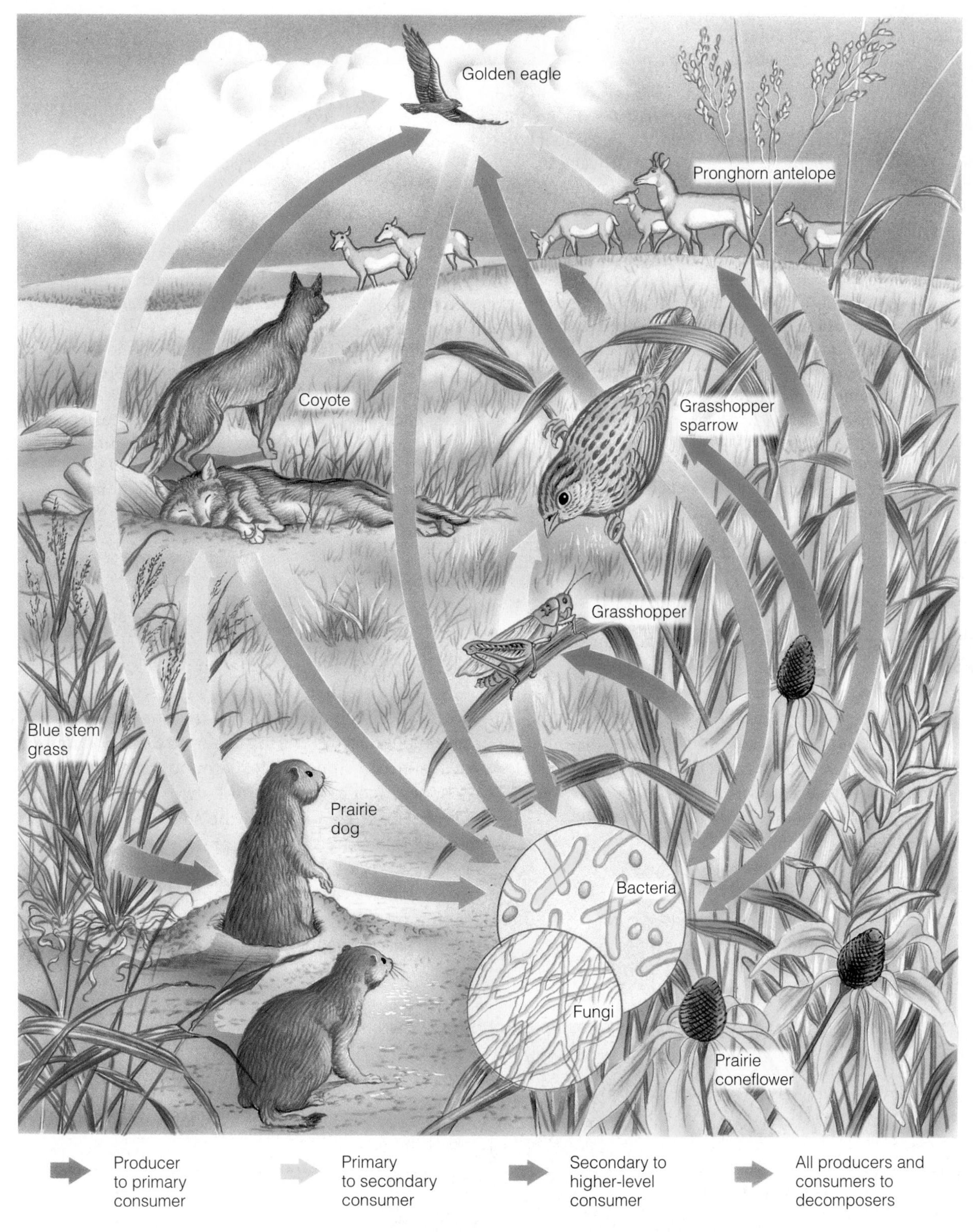

Figure 6-19 Some components and interactions in a *temperate tall-grass prairie ecosystem* in North America. When these organisms die, decomposers break down their organic matter into minerals used by plants. Transfers of matter and energy between producers, primary consumers (herbivores), and secondary (or higher-level) consumers (carnivores) are indicated by colored arrows. Organisms are not drawn to scale.

SPOTLIGHT

The Kangaroo Rat: Water Miser and Keystone Species

The kangaroo rat (Figure 6-16) is a remarkable mammal superbly adapted for conserving water in its desert environment.

As the desert's chief seed eater, it is also a *keystone species* that helps support other desert species and helps keep desert shrubland from becoming grassland.

This rodent comes out of its burrow only at night, when the air is cool and water evaporation has slowed. It seeks dry seeds that it quickly stuffs into its cheek pouches.

After a night of foraging it empties its cache of seeds into its cool burrow, where they soak up water exhaled in the rodent's breath. When the rodent eats these seeds, it gets this water back.

The kangaroo rat does not drink water; its water comes from the recycled moisture in the seeds and from water produced when sugars in the seeds undergo aerobic respiration during digestion.

Some of the water vapor in the rat's breath also condenses on the cool inside surface of its nose. This condensed water then diffuses back to its body.

Kangaroo rats have no sweat glands, so they do not lose water by perspiration. In addition, they save water by excreting hard, dry feces and thick, nearly solid urine produced by their extremely efficient kidneys.

Critical Thinking

Water is scarce in much of the southwestern United States, where the kangaroo rat lives. However, this area has one of the highest rates of human population growth. As this happens, what ecological lesson can we learn from the kangaroo rat about how to survive in this area (and other water-short areas throughout the world)?

midwestern and western United States and Canada, **(2)** South American *pampas,* **(3)** African *veldt,* and **(4)** *steppes* of central Europe and Asia. Here winds blow almost continuously, and evaporation is rapid, often leading to fires in the summer and fall.

Because of their thick and fertile soils, temperate grasslands are widely used to grow crops (Figure 6-20). However, plowing breaks up the complex soil structure and leaves it vulnerable to erosion by wind and water.

Polar grasslands, or *arctic tundra,* occur just south of the arctic polar ice cap (Figure 6-13). During most of the year these treeless plains are **(1)** bitterly cold, **(2)** swept by frigid winds, and **(3)** covered with ice and snow. Winters are long and dark, and the scant precipitation falls mostly as snow.

This biome is carpeted with a thick, spongy mat of low-growing plants, primarily grasses, mosses, and dwarf woody shrubs. Most of the annual growth of these plants occurs during the 6- to 8-week summer, when sunlight shines almost around the clock.

One effect of the extreme cold is **permafrost**, a perennially frozen layer of the soil that forms when the water there freezes. In summer, water near the surface thaws, but the permafrost soil layer below stays frozen and prevents liquid water at the surface from seeping into the ground. Thus, during the brief summer the soil above the permafrost layer remains water-

Figure 6-20 Replacement of a temperate grassland with a monoculture crop in California. When the tangled root network of natural grasses is removed, the fertile topsoil is subject to severe wind erosion unless it is covered with some type of vegetation. (National Archives/EPA Documerica)

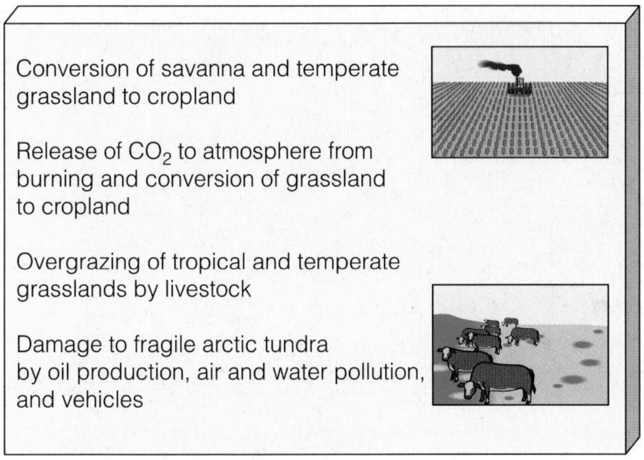

Conversion of savanna and temperate grassland to cropland

Release of CO_2 to atmosphere from burning and conversion of grassland to cropland

Overgrazing of tropical and temperate grasslands by livestock

Damage to fragile arctic tundra by oil production, air and water pollution, and vehicles

Figure 6-21 Major human impacts on grasslands.

logged, forming a large number of shallow lakes, marshes, bogs, ponds, and other seasonal wetlands. Hordes of mosquitoes, blackflies, and other insects thrive in these shallow surface pools. They feed large colonies of migratory birds (especially waterfowl) that return from the south to nest and breed in the bogs and ponds.

Another type of tundra, called *alpine tundra*, occurs above the limit of tree growth but below the permanent snow line on high mountains (Figure 6-15, left). The vegetation there is similar to that found in arctic tundra, but it gets more sunlight than arctic vegetation and has no permafrost layer.

Figure 6-21 lists major human impacts on grasslands.

6-4 FOREST AND MOUNTAIN BIOMES

What Are the Major Types of Forests? Undisturbed areas with moderate to high average annual precipitation tend to be covered with **forest**, which contains various species of trees and smaller forms of vegetation. The three main types of forest—*tropical*, *temperate*, and *boreal* (polar)—result from combinations of this precipitation level and various average temperatures (Figure 6-14).

Tropical rain forests are a type of broadleaf evergreen forest (Figure 6-22) found near the equator (Figure 6-13), where hot, moisture-laden air rises and dumps its moisture (Figure 6-6). These forests have **(1)** a warm annual mean temperature (which varies little daily or seasonally), **(2)** high humidity, and **(3)** heavy rainfall almost daily.

Tropical rain forests have incredible biological diversity. These diverse forms of life occupy a variety of specialized niches in distinct layers, based mostly on their need for sunlight (Figure 6-23). The stratification of specialized plant and animal niches in various layers of a tropical rain forest enables coexistence of a great variety of species (biodiversity). Although tropical rain forests cover only about 2% of the earth's land surface, they are habitats for 50–80% of the earth's terrestrial species.

Dropped leaves, fallen trees, and dead animals decompose quickly because of the warm, moist conditions and hordes of decomposers. This rapid recycling of scarce soil nutrients is why there is usually little litter on the ground. Instead of being stored in the soil, most minerals released by decomposition are taken up quickly by plants. Thus, most of a tropical rain forest's nutrients are stored in the biomass of its living organisms.

Because of the dense vegetation, little wind blows in tropical rain forests, eliminating the possibility of wind pollination. Many of the plants have evolved elaborate flowers (Figure 4-16, right, p. 88) that attract particular insects, birds, or bats as pollinators.

Moving a little farther from the equator (Figure 6-13), we find *tropical deciduous forests* (sometimes called *tropical monsoon forests* or *tropical seasonal forests*). These forests **(1)** are warm year-round and **(2)** get most of their plentiful rainfall during a wet (monsoon) season that is followed by a long dry season. They contain a mixture of **(1) deciduous trees** (which lose their leaves to survive the dry season) and **(2)** drought-tolerant **evergreen trees** (which retain most of their leaves year-round). Where the dry season is especially long, we find *tropical scrub forests* (Figure 6-13) containing mostly small deciduous trees and shrubs.

Temperate deciduous forests (Figure 6-24, p. 140) grow in areas with moderate average temperatures that change significantly with the season. These areas have **(1)** long, warm summers, **(2)** cold but not too severe winters, and **(3)** abundant precipitation, often spread fairly evenly throughout the year.

This biome is dominated by a few species of broadleaf deciduous trees such as oak, hickory, maple, poplar, and sycamore. They survive cold winters by dropping their leaves in the fall and becoming dormant. Each spring they grow new leaves that change in the fall into an array of reds and golds before dropping. Because of the fairly low rate of decomposition, these forests accumulate a thick layer of slowly decaying leaf litter that is a storehouse of nutrients.

Evergreen coniferous forests, also called *boreal forests* and *taigas* (pronounced "TIE-guhs"), are found just south of the arctic tundra in northern regions across North America, Asia, and Europe (Figure 6-13). In this subarctic climate, winters are long, dry, and extremely cold; in the northernmost taiga, sunlight is available only 6–8 hours a day. Summers are short, with mild to

Figure 6-22 Some components and interactions in a *tropical rain forest ecosystem.* When these organisms die, decomposers break down their organic matter into minerals used by plants. Transfers of matter and energy between producers, primary consumers (herbivores), and secondary (or higher-level) consumers (carnivores) are indicated by colored arrows. Organisms are not drawn to scale.

Figure 6-23 Stratification of specialized plant and animal niches in various layers of a *tropical rain forest*. The presence of these specialized niches enables species to avoid or minimize competition for resources and results in the coexistence of a great variety of species (biodiversity).

warm temperatures, and the sun typically shines 19 hours a day.

Most boreal forests are dominated by a few species of coniferous (cone-bearing) evergreen trees such as spruce, fir, cedar, hemlock, and pine that keep some of their narrow pointed leaves (needles) all year long. The small, needle-shaped, waxy-coated leaves of these trees **(1)** can withstand the intense cold and drought of winter when snow blankets the ground and **(2)** are ready to take advantage of the brief summers in these areas without having to take time to grow new needles. Plant diversity is low in these forests because few species can survive the winters when soil moisture is frozen.

Beneath the stands of trees, there is a deep layer of partially decomposed conifer needles and leaf litter. Decomposition is slow because of the **(1)** low temperatures, **(2)** waxy coating of conifer needles, and **(3)** high soil acidity. As the conifer needles decompose, they make the thin, nutrient-poor soil acidic and prevent most other plants (except certain shrubs) from growing on the forest floor.

These biomes contain a variety of wildlife (Figure 6-25). During the brief summer the soil becomes waterlogged, forming acidic bogs, or *muskegs*, in low-lying areas of these forests. Warblers and other insect-eating birds feed on hordes of flies, mosquitoes, and caterpillars.

Coastal coniferous forests or *temperate rain forests* are found in scattered coastal temperate areas with ample rainfall or moisture from dense ocean fogs. Along the coast of North America, from Canada to northern California, undisturbed areas of biomes are dominated by

Figure 6-24 Some components and interactions in a *temperate deciduous forest ecosystem.* When these organisms die, decomposers break down their organic matter into minerals used by plants. Transfers of matter and energy between producers, primary consumers (herbivores), and secondary (or higher-level) consumers (carnivores) are indicated by colored arrows. Organisms are not drawn to scale.

Figure 6-25 Some components and interactions in an *evergreen coniferous (boreal or taiga) forest ecosystem.* When these organisms die, decomposers break down their organic matter into minerals used by plants. Transfers of matter and energy between producers, primary consumers (herbivores), and secondary (or higher-level) consumers (carnivores) are indicated by colored arrows. Organisms are not drawn to scale.

www.brookscole.com/product/0534389872s

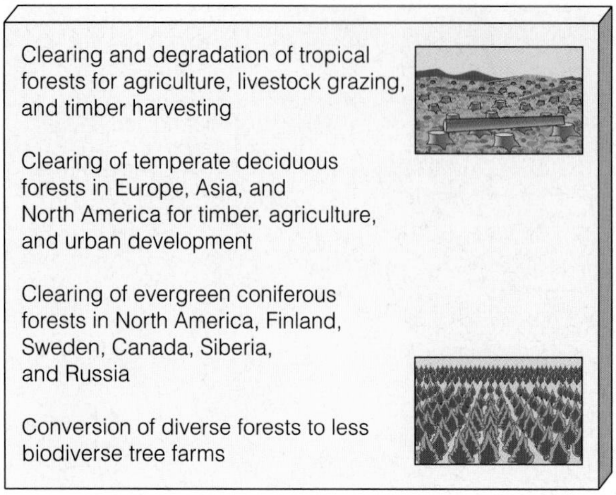

Clearing and degradation of tropical forests for agriculture, livestock grazing, and timber harvesting

Clearing of temperate deciduous forests in Europe, Asia, and North America for timber, agriculture, and urban development

Clearing of evergreen coniferous forests in North America, Finland, Sweden, Canada, Siberia, and Russia

Conversion of diverse forests to less biodiverse tree farms

Figure 6-26 Major human impacts on forests.

dense stands of large conifers such as Sitka spruce, Douglas fir, and redwoods.

Figure 6-26 lists major human impacts on forests.

Why Are Mountains Ecologically Important? Some of the world's most spectacular and important environments are mountains, which make up about 20% the earth's land surface. Mountains are places where dramatic changes in altitude, climate, soil, and vegetation take place over a very short distance (Figure 6-15). Because of the steep slopes, mountain soils are especially prone to erosion when the vegetation holding them in place is removed by natural disturbances (such as landslides and avalanches) or human activities (such as timber cutting and agriculture).

Many freestanding mountains are *islands of biodiversity* surrounded by a sea of lower-elevation landscapes transformed by human activities. Mountains play a number of important ecological roles by

- Containing the majority of the world's forests, which are habitats much of the world's terrestrial biodiversity
- Often containing endemic species found nowhere else on earth
- Serving as sanctuaries for animal species driven from lowland areas
- Helping regulate the earth's climate when mountaintops covered with ice and snow reflect solar radiation back into space
- Affecting sea levels as a result of decreases or increases in glacial ice, most of which is locked up in Antarctica, the most mountainous of all continents
- Playing a critical role in the hydrologic cycle (Figure 4-26, p. 95) by gradually releasing melting ice, snow, and water stored in the soils and vegetation of mountainsides to small streams

Landless poor migrating uphill to survive

Timber extraction

Mineral resource extraction

Hydroelectric dams and reservoirs

Increasing tourism (such as hiking and skiing)

Air pollution from industrial and urban centers

Increased ultraviolet radiation from ozone depletion

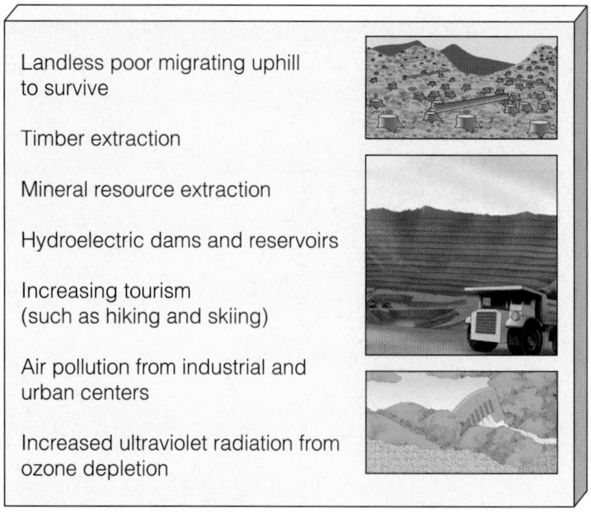

Figure 6-27 Major human impacts on mountains.

Despite their importance, the fate of mountain ecosystems has not been a high priority of governments or many environmental organizations. Mountain ecosystems are coming under increasing pressure from several human activities (Figure 6-27).

6-5 AQUATIC ENVIRONMENTS: TYPES AND CHARACTERISTICS

What Are the Two Major Types of Aquatic Life Zones? The aquatic equivalents of biomes are called *aquatic life zones*. The major types of organisms found in aquatic environments are determined by the water's *salinity* (the amounts of various salts such as sodium chloride [NaCl] dissolved in a given volume of water). As a result, aquatic life zones are divided into two major types: **(1)** *saltwater* or *marine* (particularly estuaries, coastlines, coral reefs, coastal marshes, mangrove swamps, and oceans) **(2)** *freshwater* (particularly lakes and ponds, streams and rivers, and inland wetlands). Figure 6-28 shows the distribution of the world's major oceans, lakes, rivers, coral reefs, and mangroves.

What Are the Main Kinds of Organisms in Aquatic Life Zones? Saltwater and freshwater life zones contain several major types of organisms:

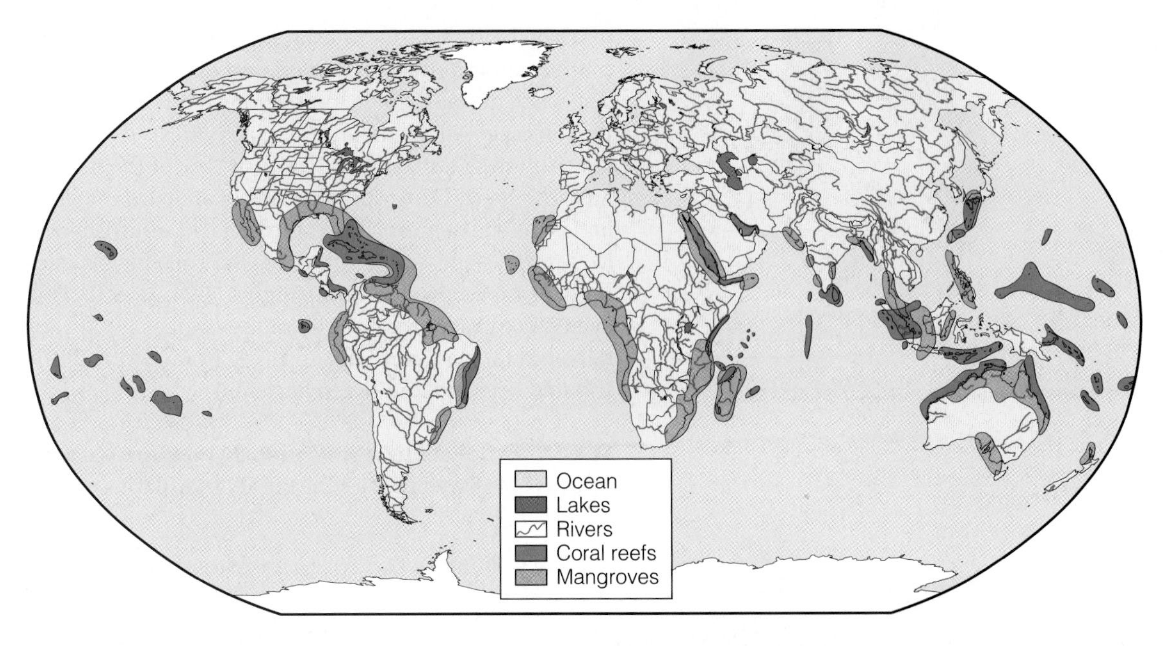

Figure 6-28 Distribution of the world's major saltwater oceans, coral reefs, mangroves, and freshwater lakes and rivers.

(1) weakly swimming, free-floating **plankton** (consisting mostly of *phytoplankton*, or *plant plankton*, and *zooplankton*, or *animal plankton*), **(2)** strongly swimming consumers (**nekton**) such as fish, turtles, and whales, **(3)** bottom-dwellers (**benthos**) such as barnacles and oysters that anchor themselves to one spot, worms that burrow into the sand or mud, and lobsters and crabs that walk about on the bottom, and **(4) decomposers** (mostly bacteria) that break down the organic compounds in the dead bodies and wastes of aquatic organisms into simple nutrient compounds for use by producers.

What Factors Limit Life at Different Depths in Aquatic Life Zones? Most aquatic life zones can be divided into three layers: **(1)** surface, **(2)** middle, and **(3)** bottom. Important environmental factors determining the types and numbers of organisms found in these layers are **(1)** *temperature*, **(2)** *access to sunlight for photosynthesis*, **(3)** *dissolved oxygen content*, and **(4)** *availability of nutrients* such as carbon (as dissolved CO_2 gas), nitrogen (as NO_3^-), and phosphorus (mostly as PO_4^{3-}) for producers.

Photosynthesis is confined mostly to the upper layer, or **euphotic zone**, of deep aquatic systems through which sunlight can penetrate. The depth of the euphotic zone in oceans and deep lakes can be reduced by excessive algal growth (algal blooms) that make water cloudy.

In shallow waters in streams, ponds, and oceans, there are usually ample supplies of nutrients for primary producers. By contrast, in the open ocean nitrates, phosphates, iron, and other nutrients often are in short supply and limit net primary productivity (Figure 4-24, p. 93). However, net primary productivity is much higher in parts of the open ocean where upwellings (Figure 6-3, p. 123, and Figure 6-7, p. 125) bring such nutrients from the ocean bottom to the surface for use by producers.

The creatures that live on the bottoms of the deep ocean and deep lakes depend on animal and plant plankton that die and fall into deep waters. Because this food is limited, deep-dwelling species tend to be slow to reproduce, which makes then especially vulnerable to fishing pressure.

6-6 SALTWATER LIFE ZONES

Why Are the Oceans Important? A more accurate name for Earth would be *Ocean* because saltwater oceans cover about 71% of the planet's surface (Figure 6-29). The world's oceans (1) have an average depth of about 3 kilometers (2 miles), (2) contain a variety of species (Figure 6-30), and (3) provide many important ecological and economic services (Figure 6-31).

What Is the Coastal Zone? Oceans have two major life zones: the *coastal zone* and the *open sea* (Figure 6-32, p. 146). The **coastal zone** is the warm, nutrient-rich, shallow water that extends from the high-tide

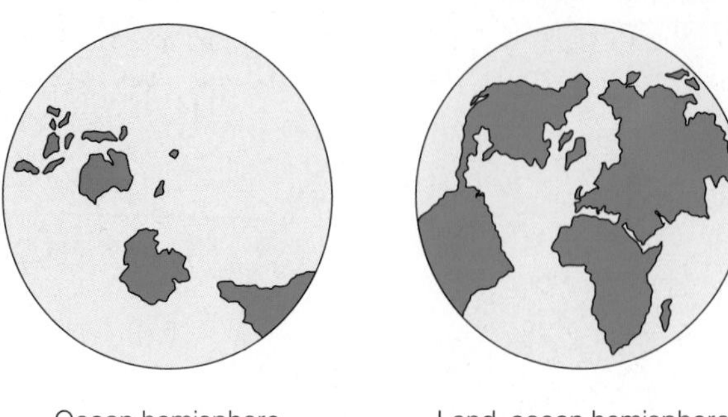

Figure 6-29 The ocean planet. The salty oceans cover about 71% of the earth's surface. About 97% of the earth's water is in the interconnected oceans, which cover 90% of the planet's mostly ocean southern hemisphere (left) and 50% of its land–ocean northern hemisphere (right).

mark on land to the gently sloping, shallow edge of the *continental shelf* (the submerged part of the continents). This zone has numerous interactions with the land and thus is easily affected by human activities.

Although it makes up less than 10% of the ocean's area, the coastal zone contains 90% of all marine species and is the site of most large commercial marine fisheries. Most ecosystems found in the coastal zone have a very high primary productivity (Figure 4-24, p. 93). This occurs because of the zone's ample supplies of **(1)** sunlight and **(2)** plant nutrients (flowing from land and distributed by wind and ocean currents).

What Are Estuaries, Coastal Wetlands, and Mangrove Swamps? One highly productive area in

Figure 6-30 Marine biodiversity. Some marine inhabitants.

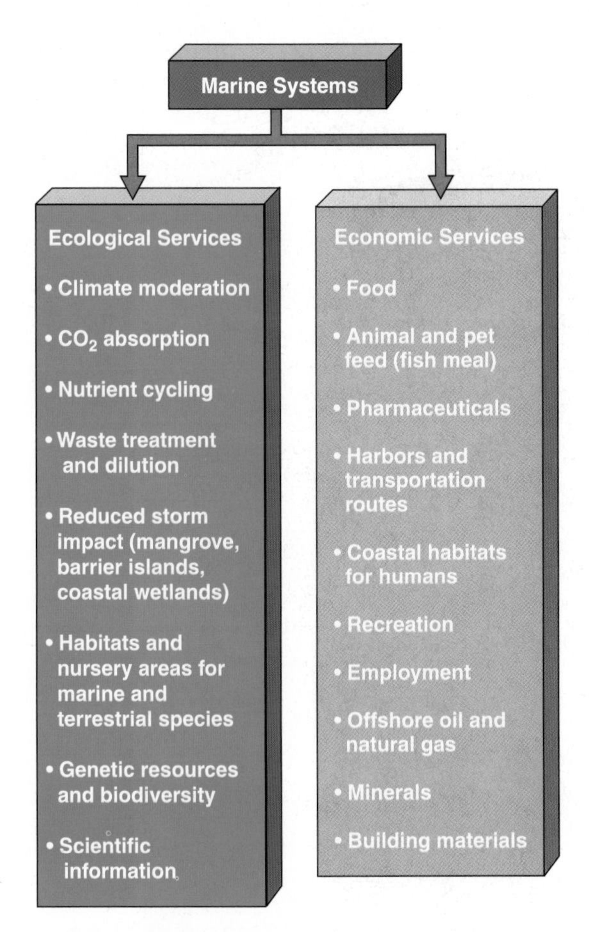

Marine Systems
Ecological Services
• Climate moderation
• CO_2 absorption
• Nutrient cycling
• Waste treatment and dilution
• Reduced storm impact (mangrove, barrier islands, coastal wetlands)
• Habitats and nursery areas for marine and terrestrial species
• Genetic resources and biodiversity
• Scientific information
Economic Services
• Food
• Animal and pet feed (fish meal)
• Pharmaceuticals
• Harbors and transportation routes
• Coastal habitats for humans
• Recreation
• Employment
• Offshore oil and natural gas
• Minerals
• Building materials

Figure 6-31 Major ecological and economic services provided by marine systems.

the coastal zone is an **estuary**, a partially enclosed area of coastal water where seawater mixes with fresh water and nutrients from rivers, streams, and runoff from land (Figure 6-33). Estuaries and their associated **coastal wetlands** (land areas covered with water all or part of the year) include **(1)** river mouths, **(2)** inlets, **(3)** bays, **(4)** sounds, **(5)** mangrove forest swamps in tropical waters (Figure 6-28), and **(6)** salt marshes in temperate zones (Figure 6-34, p. 148).

The constant water movement in estuaries and coastal wetlands stirs up the nutrient-rich silt, making it available to producers. This explains why these aquatic systems are some of the earth's most productive ecosystems (Figure 4-24, p. 93) and provide other important ecological and economic services (Figure 6-31).

What Are Rocky and Sandy Shores? The area of shoreline between low and high tides is called the **intertidal zone**. Organisms living in this stressful zone must be able to avoid being **(1)** swept away or crushed by waves, **(2)** immersed during high tides, and **(3)** left high and dry (and much hotter) at low tides. They must also cope with changing levels of salinity when heavy rains dilute salt water. To deal with such stresses, most intertidal organisms hold onto something, dig in, or hide in protective shells.

Some coasts have steep *rocky shores* pounded by waves. The numerous pools and other niches in the rocks in the intertidal zone of rocky shores contain a great variety of species (Figure 6-35, top, p. 149).

Other coasts have gently sloping *barrier beaches*, or *sandy shores*, with niches for different marine organisms, including crabs, lugworms, clams, ghost shrimp, sand dollars, and flounder (Figure 6-35, bottom, p. 149). Most of them are hidden from view and survive by burrowing, digging, and tunneling in the sand.

One or more rows of natural sand dunes on undisturbed barrier beaches (with the sand held in place by the roots of grasses) serve as the first line of defense against the ravages of the sea (Figure 6-36, p. 150). However, when coastal developers remove the protective dunes or build behind the first set of dunes, storms can flood and even sweep away seaside buildings and severely erode the sandy beaches.

What Are Barrier Islands? **Barrier islands** are long, thin, low offshore islands of sediment that generally run parallel to the shore. They are found along some coasts such as most of North America's Atlantic and Gulf coasts. These islands help protect the mainland, estuaries, and coastal wetlands by dispersing the energy of approaching storm waves.

Their low-lying beaches are constantly shifting, with gentle waves building them up and storms flattening and eroding them. Currents running parallel to the beaches constantly take sand from one area and deposit it in another. Sooner or later, many of the structures humans build on low-lying barrier islands (Figure 6-37, p. 150), such as Atlantic City, New Jersey, and Miami Beach, Florida, are damaged or destroyed by flooding, severe beach erosion, or major storms (including hurricanes).

What Are Coral Reefs? In the shallow coastal zones of warm tropical and subtropical oceans we often find **coral reefs** (Figure 6-28). These beautiful natural wonders are among the world's oldest, most diverse and productive ecosystems and are homes for about one-fourth of all marine species (Figure 6-38, p. 151).

Coral reefs are formed by massive colonies of tiny animals called *polyps* that are close relatives of jellyfish. They slowly build reefs by secreting a protective crust of limestone (calcium carbonate) around their soft bodies. When they die, their empty

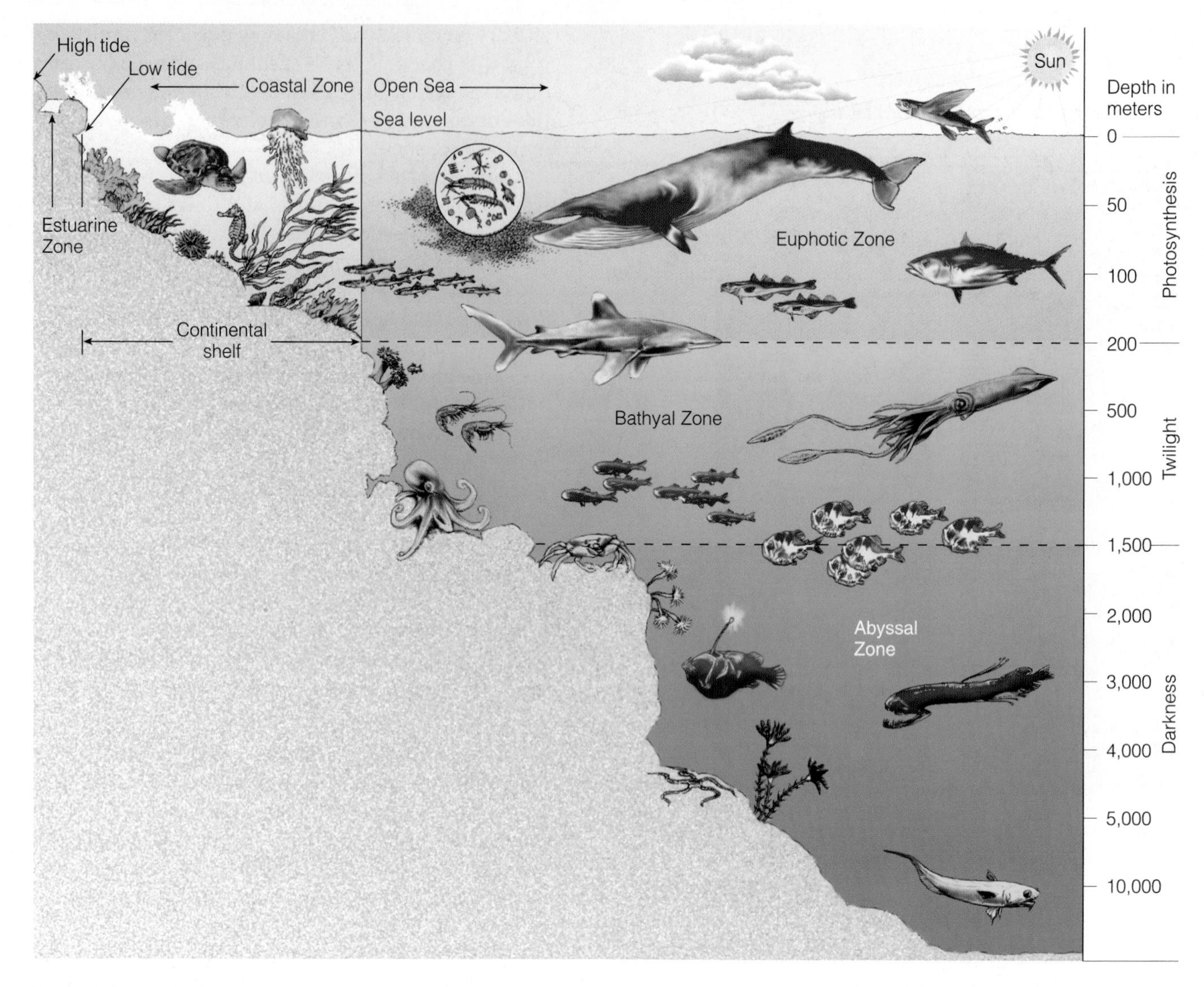

Figure 6-32 Major life zones in an ocean.

crusts or outer skeletons remain as a platform for more reef growth. The result is an elaborate network of crevices, ledges, and holes that serve as calcium carbonate "condominiums" for a variety of marine animals.

Coral reefs are a joint venture between the polyps and tiny single-celled algae called *zooxanthellae* ("zoh-ZAN-thel-ee") that live in the tissues of the polyps. The algae provide the polyps with color, food, and oxygen through photosynthesis. The polyps in turn provide a well-protected home for the algae.

Although coral reefs cover less than 2% of the ocean floor (Figure 6-28) , they provide a number of important ecological and economic services valued conservatively at $375 billion a year. These services include the following:

- Removing some of carbon dioxide from the atmosphere as part of the carbon cycle (when coral polyps form limestone shells).
- Acting as natural barriers that **(1)** help protect 15% of the world's coastlines from storm damage, erosion, and flooding and **(2)** allow the ocean to replenish beaches with sand.
- Supporting at least one-fourth of all identified marine species and 65% of marine fish species.
- Producing roughly 10% of the global fish catch and 25% of the catch in developing countries.
- Providing fish and shellfish, jobs, and building materials for some of the world's poorest countries.
- Supporting fishing and tourism industries worth billions of dollars each year. For example,

Figure 6-33 View of an estuary taken from space. The photo shows the sediment plume at the mouth of Madagascar's Betsiboka River as it flows through the estuary and into the Mozambique Channel. Because of its topography, heavy rainfall, and the clearing of forests for agriculture, Madagascar is the world's most eroded country. (NASA)

Caribbean reefs alone bring in $140 billion a year in tourism revenue.

- Giving us an underwater world to study and enjoy.

What Biological Zones Are Found in the Open Sea? The sharp increase in water depth at the edge of the continental shelf separates the coastal zone from the vast volume of the ocean called the **open sea**. Based primarily on the penetration of sunlight, it is divided into three vertical zones (Figure 6-32):

- *Euphotic zone*: the lighted upper zone where **(1)** photosynthesis occurs mostly by phytoplankton, **(2)** nutrient levels are low (except around upwellings, Figure 6-7), and **(3)** levels of dissolved oxygen are high. Fast-swimming predatory fish such as swordfish, sharks, and bluefin tuna are found in this zone.
- *Bathyal zone*: dimly lit middle zone that does not contain photosynthesizing producers because of a lack of sunlight. It contains various types of zooplankton and smaller fish, many of which migrate to feed on the surface at night.
- *Abyssal zone*: dark lower zone that is **(1)** very cold, **(2)** has little dissolved oxygen, and **(3)** has enough nutrients on the ocean floor to support about 98% of the 250,000 known marine species.

Average primary productivity and net primary productivity per unit of area are quite low in the open sea (Figure 4-24, p. 93) except at occasional equatorial upwellings, where currents bring up nutrients from the ocean bottom. However, because the open sea covers so much of the earth's surface (Figure 6-29), it makes the largest contribution to the earth's overall net primary productivity.

What Impacts Do Human Activities Have on Marine Systems? In their desire to live near the coast, people are destroying or degrading the resources that make coastal areas so enjoyable and economically and ecologically valuable. Currently, about 40% of the world's population live along coasts or within 100 kilometers (62 miles) of a coast. Some 13 of the world's 19 meagacities with populations of 10 million or more people are in coastal zones. By 2030, more than 6.3 billion people—more than the current global population—are expected to live in or near coastal areas.

Here are some major impacts of human activities on marine systems:

- Salt marshes, mangrove forests, and sea-grass meadows—the sea's three great nurseries—are being lost and degraded at a staggering rate to make way for real estate developments, marinas, golf courses, and shrimp farms.
- Scientists estimate that half the world's original coastal wetlands and 53% those in the lower 48 U.S. states (91% in California) have disappeared since 1800, mostly by being filled in for agriculture and coastal development. According to scientists, by 2080 half of the world's remaining coastal wetlands are likely to be lost to agriculture, urban development, and rising sea levels from climate change.

Figure 6-34 Some components and interactions in a *salt marsh ecosystem* in a temperate area such as the United States. When these organisms die, decomposers break down their organic matter into minerals used by plants. Transfers of matter and energy between consumers (herbivores) and secondary (or higher-level) consumers (carnivores) are indicated by colored arrows. Organisms are not drawn to scale.

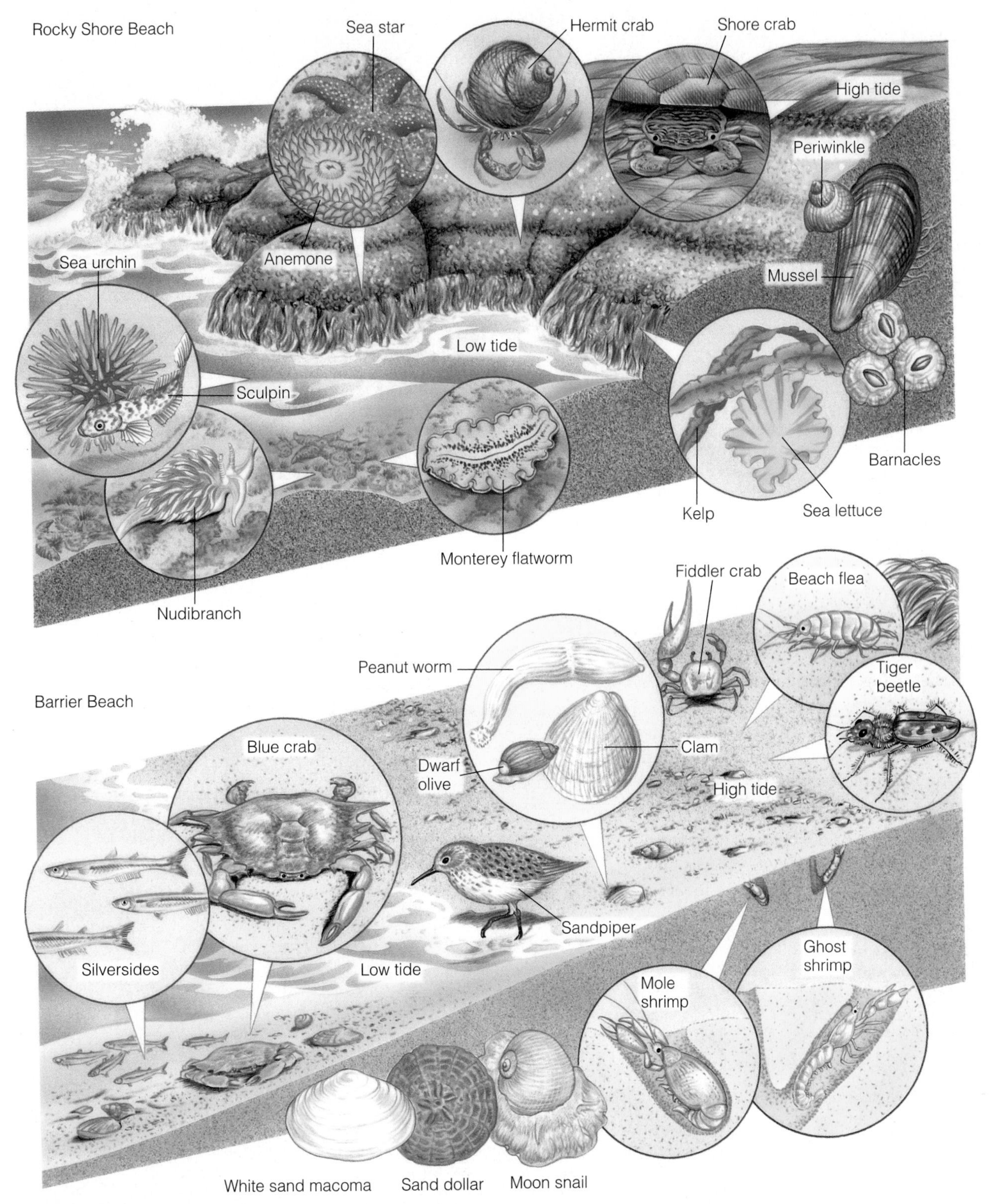

Figure 6-35 Living between the tides. Some organisms with specialized niches found in various zones on rocky shore beaches (top) and barrier or sandy beaches (bottom). Organisms are not drawn to scale.

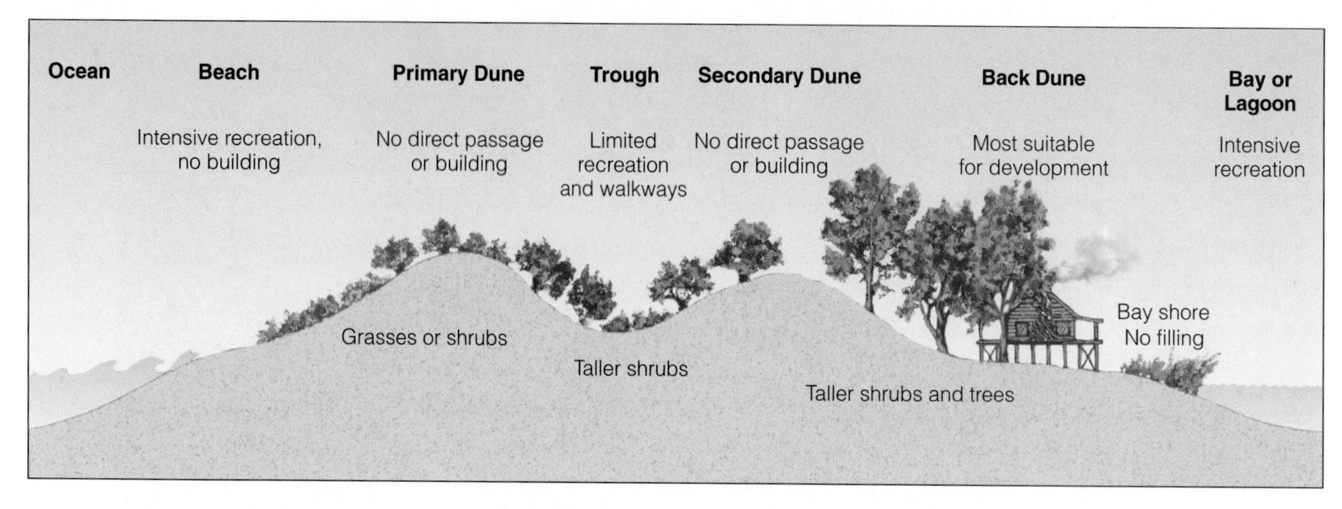

Figure 6-36 Primary and secondary dunes on gently sloping sandy beaches play an important role in protecting the land from erosion by the sea. The roots of various grasses that colonize the dunes help hold the sand in place. Ideally, construction and development should be allowed only behind the second strip of dunes, and walkways to the beach should be built over the dunes to keep them intact. This not only helps preserve barrier beaches but also helps protect human structures from being damaged and washed away by wind, high tides, beach erosion, and flooding from storm surges. This type of protection is rare because the short-term economic value of oceanfront land is considered to be much higher than its long-term ecological and economic values.

- At least half of the world's original mangrove forests (Figure 6-28) have disappeared, mostly because of clearing for coastal development, rice fields, and aquaculture shrimp farms. Much of the remaining mangroves are threatened. The clearing of mangroves also causes erosion that kills coral reefs by smothering the corals in plumes of soil.
- According to a World Wildlife Fund study, almost 70% of the world's beaches are eroding rapidly because of coastal developments and a rising sea level (caused mostly by global warming).
- Ocean bottom habitats are being degraded and destroyed by dredging operations and trawler boats, which drag huge nets weighted down with chains over ocean bottoms to harvest bottom fish and shellfish (Figure 6-39).
- According to a 2000 report from the Global Coral Reef Monitoring Network, about 27% of the world's coral reefs have been severely damaged (up from 10% in 1992), and 11% have been destroyed. Another 70% could be gone by 2050.

Figure 6-37 A developed barrier island. To keep up with shifting sands, taxpayers spend millions of dollars to pump sand onto the beaches and rebuild natural sand dunes. Barrier islands lack effective protection against flooding and damage from severe storms. If global warming raises average sea levels, as projected, most of these valuable pieces of real estate will be under water.

Figure 6-38 Some components and interactions in a *coral reef ecosystem*. When these organisms die, decomposers break down their organic matter into minerals used by plants. Transfers of matter and energy between producers, primary consumers (herbivores), and secondary (or higher-level) consumers (carnivores) are indicated by colored arrows. Organisms are not drawn to scale.

Figure 6-39 Area of ocean bottom before (left) and after (right) a trawler net scraped it like a gigantic plow. These ocean floor communities can take decades or centuries to recover. According to marine scientist Elliot Norse, "Bottom trawling is probably the largest human-caused disturbance to the biosphere." Trawler fishers disagree and claim that ocean bottom life recovers after trawling. (Peter J. Auster, National Undersea Research Center)

6-7 FRESHWATER LIFE ZONES

What Are Freshwater Life Zones? **Freshwater life zones** occur where water with a dissolved salt concentration of less than 1% by volume accumulates on or flows through the surfaces of terrestrial biomes. Examples are **(1)** *standing* (lentic) bodies of fresh water such as lakes, ponds, and inland wetlands and **(2)** *flowing* (lotic) systems such as streams and rivers. Although freshwater systems cover less than 1% of the earth's surface (Figure 6-28), they contain a variety of species (Figure 6-40) and provide a number of important ecological and economic services (Figure 6-41).

What Life Zones Are Found in Freshwater Lakes? **Lakes** are large natural bodies of standing fresh water formed when precipitation, runoff, or groundwater seepage fills depressions in the earth's surface. Causes of such depressions include **(1)** glaciation (the Great Lakes of North America), **(2)** crustal displacement (Lake Nyasa in East Africa), and **(3)** volcanic activity (Crater Lake in Oregon). Lakes are fed by rainfall, melting snow, and streams that drain the surrounding watershed.

Lakes normally consist of four distinct zones that are defined by their depth and distance from shore (Figure 6-42):

- *Littoral zone* ("LIT-tore-el"), consisting of the shallow, sunlit waters near the shore to the depth at which rooted plants stop growing. It has a high biological diversity.

- *Limnetic zone* ("limb-NET-ic"), which is the open, sunlit water surface layer away from the shore that extends to the depth penetrated by sunlight. As the main photosynthetic body of a lake, it produces the food and oxygen that support most of the lake's consumers.

- *Profundal zone* ("pro-FUN-dahl") which is the deep, open water where it is too dark for photosynthesis. Without sunlight and plants, oxygen levels are low. It is inhabited by fish adapted to its cooler, darker water, and most of their food is produced in the limnetic and littoral zones.

- *Benthic zone* ("BEN-thick"), at the bottom of a lake. It is inhabited mostly by organisms that tolerate cool temperatures and low oxygen levels.

How Do Plant Nutrients Affect Lakes? Ecologists classify lakes according to their nutrient content and primary productivity. A newly formed lake generally has a small supply of plant nutrients and is called an **oligotrophic** (poorly nourished) **lake** (Figure 6-43, top). This type of lake is often deep, with steep banks. Because of its low net primary productivity, such a lake

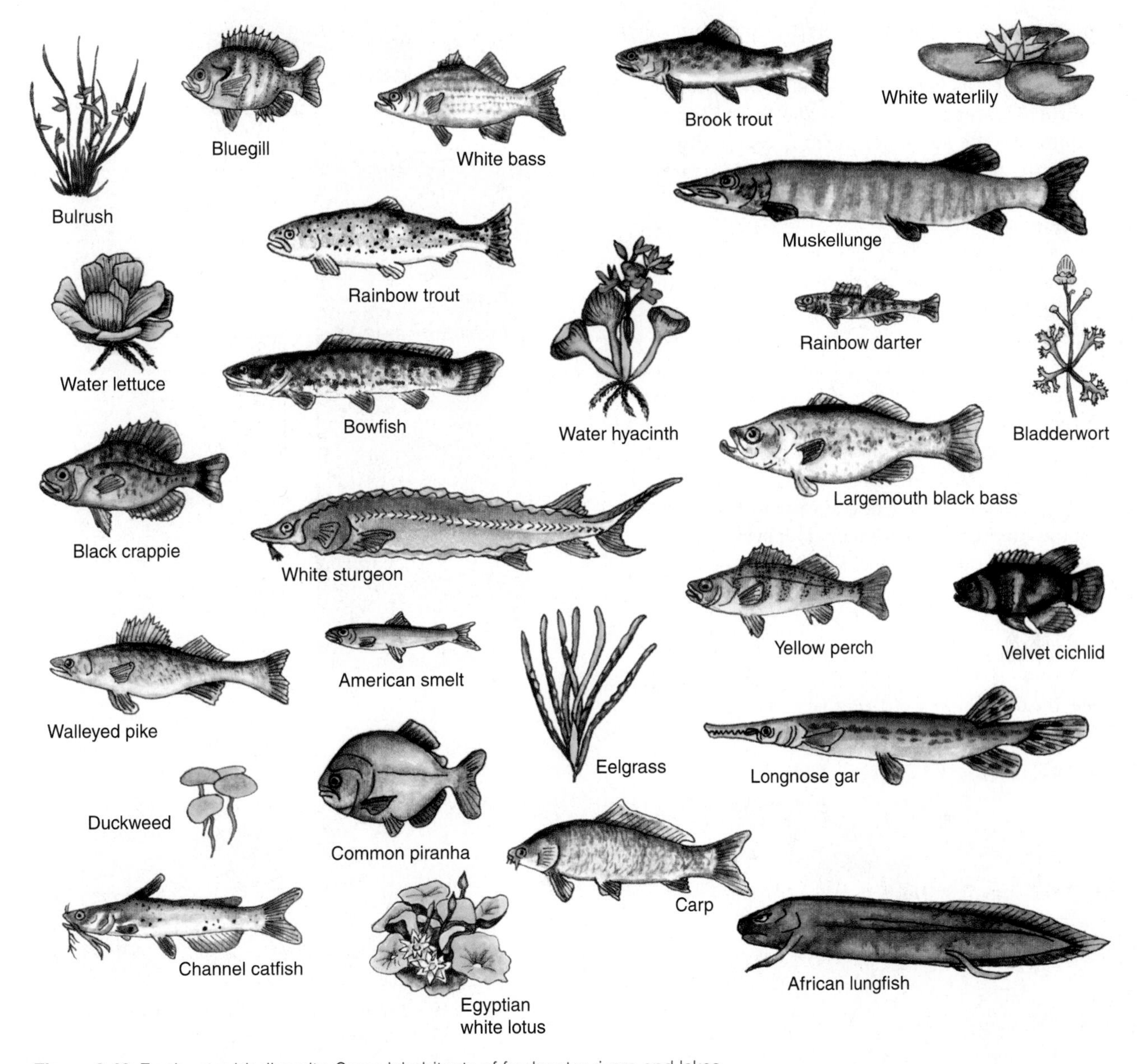

Figure 6-40 Freshwater biodiversity. Some inhabitants of freshwater rivers and lakes.

usually has **(1)** crystal-clear blue or green water and **(2)** small populations of phytoplankton and fish (such as smallmouth bass and trout).

Over time, sediment washes into an oligotrophic lake and plants grow and decompose to form bottom sediments. A lake with a large or excessive supply of nutrients (mostly nitrates and phosphates) needed by producers is called a **eutrophic** (well-nourished) **lake** (Figure 6-43, bottom). Such lakes typically are shallow and have murky brown or green water with poor visibility. Because of their high levels of nutrients, these lakes have a high net primary productivity. Human inputs of nutrients from the atmosphere and from nearby urban and agricultural areas can accelerate the eutrophication of lakes, a process called *cultural eutrophication*. Many lakes fall somewhere between the two extremes of nutrient enrichment and are called **mesotrophic lakes**.

What Are the Major Characteristics of Freshwater Streams and Rivers? Precipitation that does not sink into the ground or evaporate is **surface water**. It becomes **runoff** when it flows into streams. The land area that delivers runoff, sediment, and dissolved substances to a stream is called a **watershed**, or **drainage basin**. Small streams join to form rivers, and rivers flow downhill to the ocean (Figures 6-28 and 6-44) as part of the hydrologic cycle (Figure 4-26, p. 95).

In many areas, streams begin in mountainous or hilly areas that collect and release water falling to the earth's surface as rain or snow. The downward flow of surface water and groundwater from mountain highlands to the sea takes place in three different aquatic life zones with different environmental conditions: the *source zone, transition zone,* and *floodplain zone* (Figure 6-44).

As streams flow downhill, they become powerful shapers of land. Over millions of years the friction of moving water levels mountains and cuts deep canyons, and the rock and soil the water removes are deposited as sediment in low-lying areas.

Streams are fairly open ecosystems that receive many of their nutrients from bordering land ecosystems. Such nutrient inputs come from falling leaves, animal feces, insects, and other forms of biomass washed into streams during heavy rainstorms or by melting snow. To protect a stream or river system from excessive inputs of nutrients and pollutants, one must protect its watershed, the land around it.

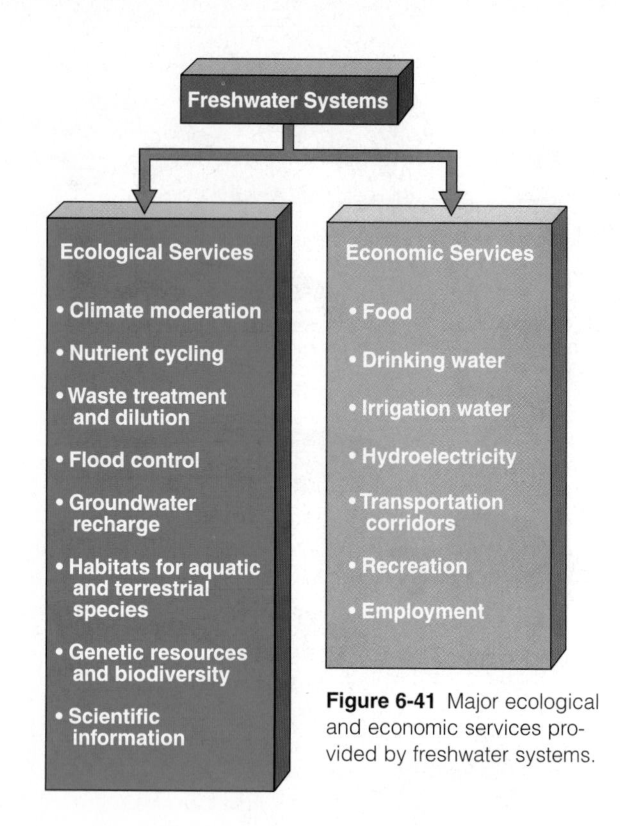

Figure 6-41 Major ecological and economic services provided by freshwater systems.

Figure 6-42 The distinct zones of life in a fairly deep temperate-zone lake.

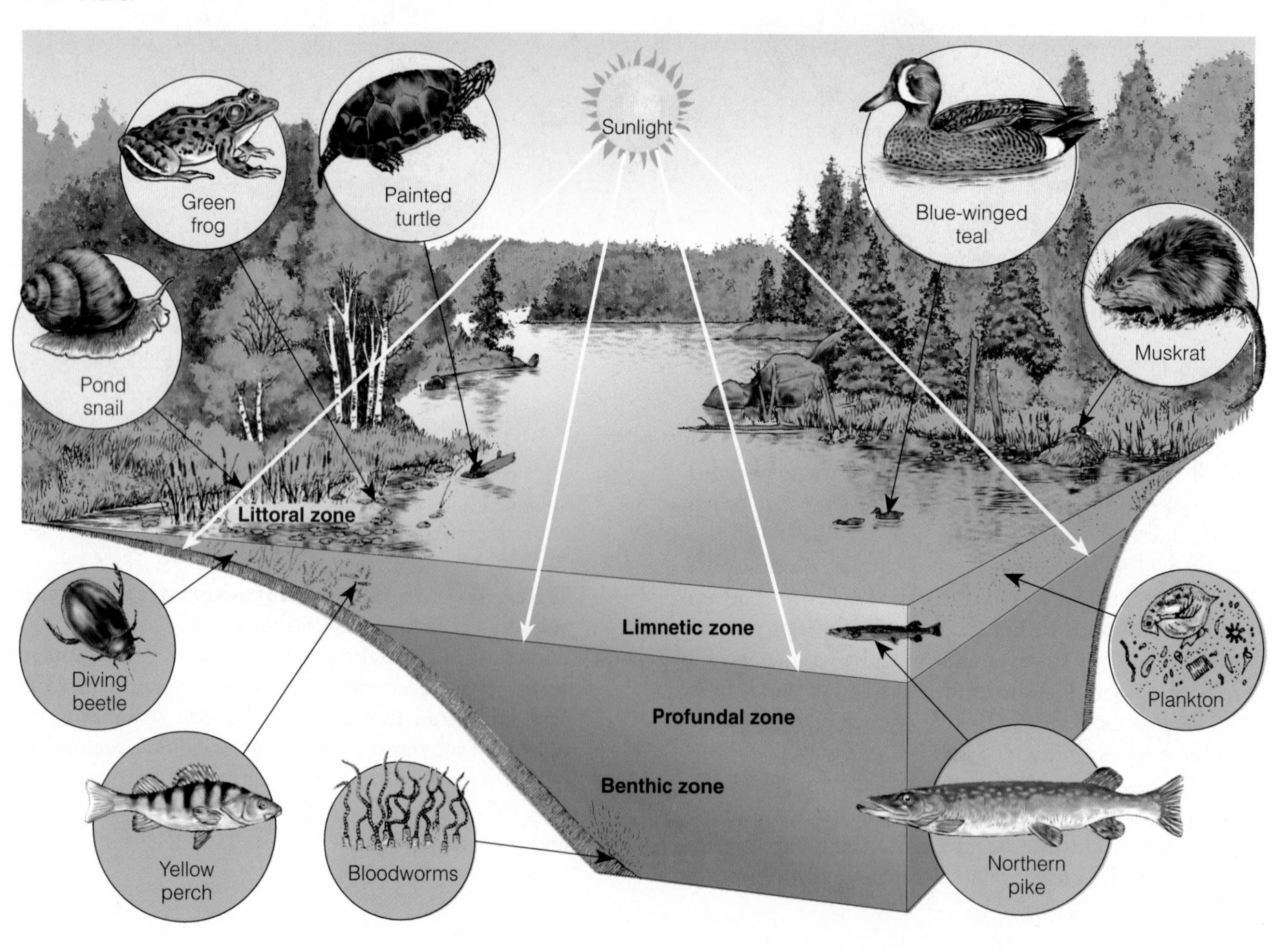

What Are Freshwater Inland Wetlands? Inland **wetlands** are lands covered with fresh water all or part of the time (excluding lakes, reservoirs, and streams) and located away from coastal areas. They include **(1)** *marshes* with few trees, **(2)** *swamps* dominated by trees and shrubs, **(3)** *prairie potholes,* which are depressions carved out by glaciers, **(4)** *floodplains,* which receive excess water during heavy rains and help reduce the impact of floods, and **(5)** *wet arctic tundra* in summer. Some wetlands are huge; others are small.

Some of these wetlands are covered with water year-round. Others, called *seasonal wetlands,* usually are underwater or soggy for only a short time each year. They include prairie potholes, floodplain wetlands, and bottomland hardwood swamps. Some stay dry for years before being covered with water again. In such cases, scientists must use the composition of the soil or the presence of certain plants (such as cattails, bulrushes, or red maples) to determine that a particular area is really a wetland.

What Impacts Do Human Activities Have on Freshwater Systems? Here are some major impacts of human activities on freshwater systems:

- According to a 2000 study by the World Resources Institute, almost 60% of the world's 237 large rivers are strongly or moderately fragmented by dams, diversions, or canals. This alters and destroys wildlife habitats along rivers and in coastal deltas and estuaries by reducing water flow.

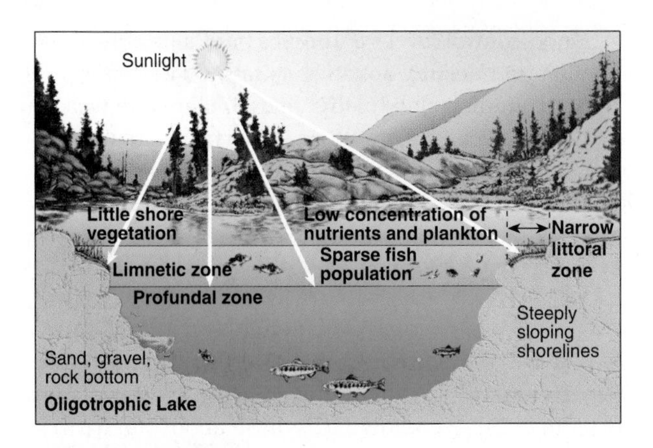

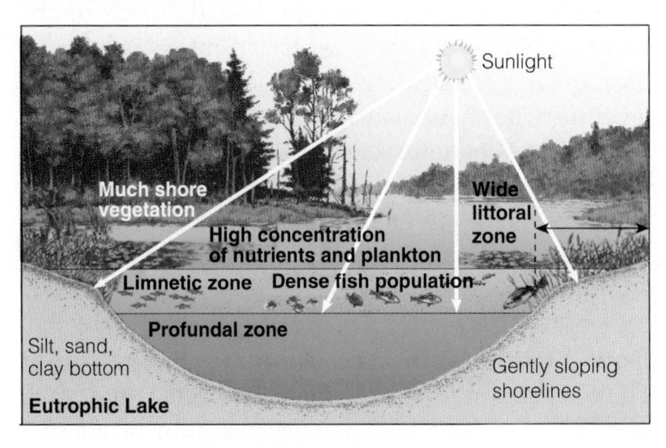

Figure 6-43 An oligotrophic, or nutrient-poor, lake (top) and a eutrophic, or nutrient-rich, lake (bottom). Mesotrophic lakes fall between these two extremes of nutrient enrichment. Nutrient inputs from human activities can accelerate eutrophication.

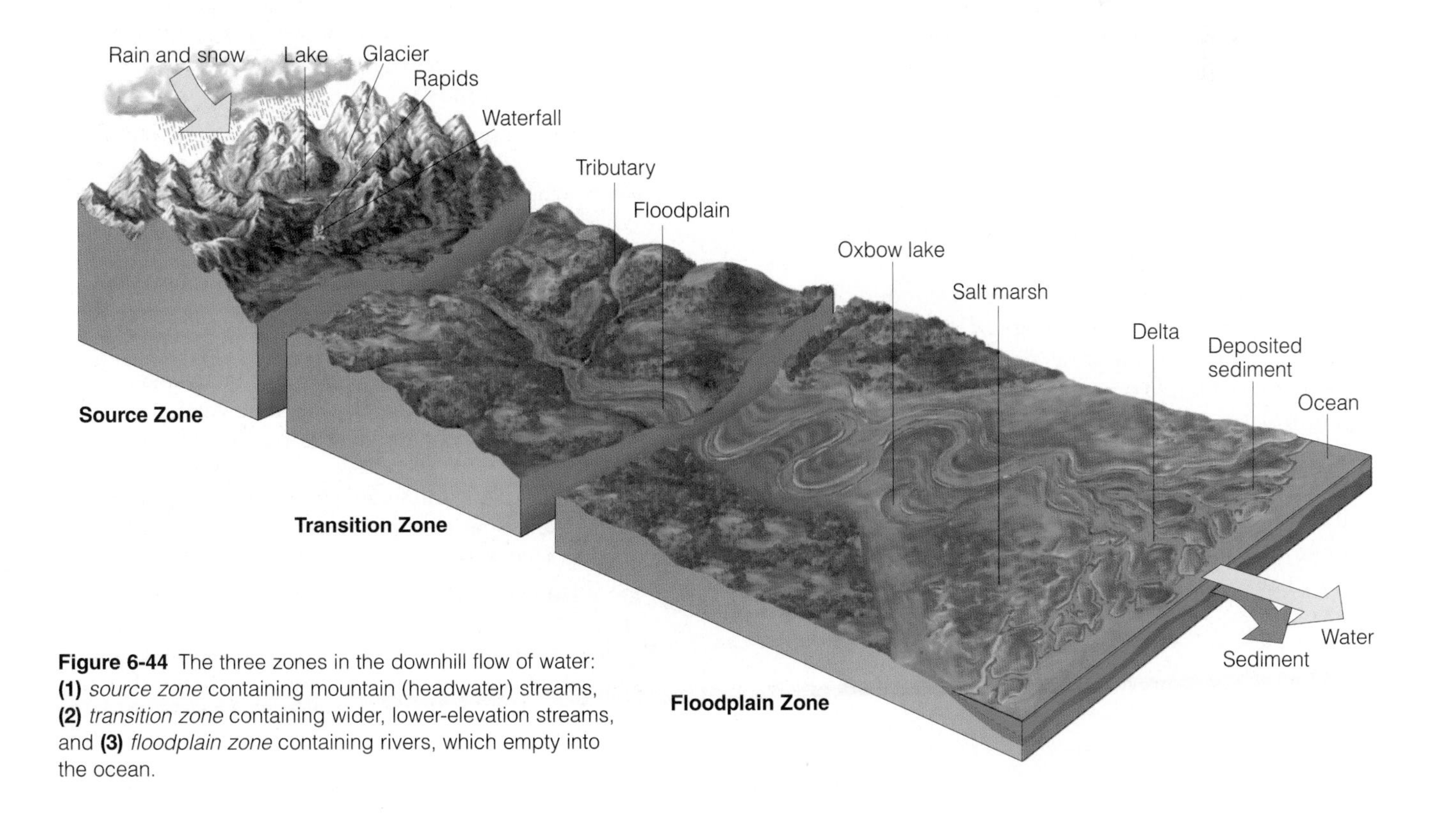

Figure 6-44 The three zones in the downhill flow of water: **(1)** *source zone* containing mountain (headwater) streams, **(2)** *transition zone* containing wider, lower-elevation streams, and **(3)** *floodplain zone* containing rivers, which empty into the ocean.

- Flood control levees and dikes built along rivers **(1)** alter and destroy aquatic habitats, **(2)** disconnect rivers from their floodplains, and **(3)** eliminate wetlands and backwaters that are important spawning grounds for fish.

6-8 SUSTAINABILITY OF AQUATIC LIFE ZONES

What Do We Know About the Earth's Aquatic Biodiversity? Although we live on a water planet (Figure 6-29), we know fairly little about the earth's aquatic systems. For example, less than 5% of the earth's global ocean from which scientists believe all life arose has been explored and mapped with the same level of detail as the surface of the moon and Mars. According to aquatic scientists, the scientific investigation of poorly understood marine and freshwater aquatic systems is a greatly underfunded *research frontier* whose study could result in immense ecological and economic benefits.

How Sustainable Are Aquatic Ecosystems? The *bad news* in terms of human impacts is that each stream, river, and lake reflects the sum of all that occurs in their watersheds. Also, many of the nutrients, wastes, and pollutants produced by human activities end up in the ocean. Recent research shows that many of the chemicals reaching aquatic systems come from the atmosphere.

The *good news* is that aquatic life zones are constantly renewed because **(1)** water is purified by natural hydrologic processes (Figure 4-26, p. 95), **(2)** nutrients cycle in and out, and **(3)** populations of biological organisms can be replenished, given sufficient opportunity and time. However, these life-sustaining processes work only if they are not **(1)** overloaded with pollutants and excessive nutrients and **(2)** overfished.

Scientists have made a start in understanding important aspects of the ecology of the world's terrestrial and aquatic systems, as discussed in this chapter. One of the main lessons from such research is that in nature *everything is connected*. However, there is an urgent need for greatly increased research on how the world's terrestrial and aquatic systems work. With such information we will have a clearer picture of **(1)** the impacts of our activities on the earth's biodiversity and **(2)** what we can do to live more sustainably.

All at last returns to the sea—to Oceanus, the ocean river, like the ever-flowing stream of time, the beginning and the end.

RACHEL CARSON

REVIEW QUESTIONS

1. Define the boldfaced terms in this chapter.

2. How does wind affect climate, global distribution of nutrients, and pollution?

3. Distinguish between *weather* and *climate*. What two main factors determine a region's climate?

4. What five factors affect global air circulation? How does each factor affect global air circulation? What causes opposite seasons in the northern and southern hemispheres?

5. How do ocean currents affect regional climates? What would happen to the climate of northwestern Europe if the Gulf Stream did not exist? What is an *upwelling*, and why are upwellings important to life?

6. What is *the El Niño–Southern Oscillation (ENSO)*? How does it affect ocean life and weather in various parts of the world? What is *La Niña*, and how does it affect weather in various parts of the world?

7. What are *greenhouse gases*? What are the two primary greenhouse gases? What is the *greenhouse effect*, and how does it affect the earth's climate and life?

8. What is the *ozone layer*? How does it help protect life on the earth and affect the earth's climate?

9. What are *microclimates*? What is the *rain shadow effect*, and how can it affect the microclimate on each side of high mountains? How can cities affect microclimates?

10. What is a *biome*? What two factors determine whether an area of the earth's surface is a tropical, temperate, or polar desert, grassland, or forest? How do climate and vegetation vary with latitude and altitude?

11. What are *succulent plants*, where are they found, and how do they survive? What are *broadleaf evergreen plants*, where are they found, and how do they survive? What are *broadleaf deciduous plants*, where are they found, and how do they survive? What are *coniferous evergreen plants*, where are they found, and how do they survive?

12. What is a *desert*? What are the three major types of desert, and how do they differ in climate and biological makeup? How do desert plants and animals survive heat and a lack of water? Describe how the kangaroo rat conserves water. Why are desert ecosystems vulnerable to disruption, and what seven types of human activities have harmful impacts on deserts?

13. What is a *grassland*? What are the three major types of grassland, and how do they differ in climate and biological makeup? Why are grasslands vulnerable to disruption? List four types of human activities that have harmful impacts on grasslands.

14. What is a *forest*? What are the three major types of forest, and how do they differ in climate and biological makeup? Distinguish between *tropical rain forests*, *tropical deciduous forests*, and *temperate rain forests*. List three types of human activities that have harmful impacts on forests.

15. Why are mountains ecologically important, and what factors make them vulnerable to ecological disruption? List seven types of human activities that have harmful impacts on mountains.

16. What are the two major types of aquatic life zones? What two factors determine the major types of organisms found in aquatic systems?

17. List the major ecological and economic services provided by the world's marine systems.

18. List four major factors determining the types and numbers of organisms found in the surface, middle, and bottom layers of aquatic systems.

19. What is the *coastal zone*? Distinguish between *estuaries, coastal wetlands,* and *mangroves*. Explain why estuaries and coastal wetlands have such high net primary productivity.

20. What is the *intertidal zone*? Distinguish between *rocky shores* and *sandy beaches* and describe the major types of aquatic life found in each. Why is it important to preserve the dunes on barrier beaches?

21. What are *barrier islands*? Why are they so attractive for human development, and why are human structures built there so vulnerable to destruction?

22. What are *coral reefs*, and how are they formed? List seven ecological and economic services provided by coral reefs.

23. What is the *open sea*, and what are its three major zones?

24. List six major harmful human impacts on coastal zones.

25. What is a *freshwater life zone*, and what are the two major types of such zones? List the major ecological and economic services provided by freshwater systems.

26. What is a *lake*? Distinguish between the *littoral, limnetic, profundal,* and *benthic zones* of a lake.

27. What are the three types of lakes, based on their nutrient content and primary productivity?

28. Distinguish between *surface water, runoff,* and a *watershed*. What are the three zones of a river as it flows from mountain highlands to the sea?

29. What are *freshwater inland wetlands*? List five examples of such wetlands.

30. List two major harmful human impacts on freshwater systems.

31. Summarize *bad* and *good* news about the sustainability of aquatic systems.

32. Explain why a study of terrestrial and aquatic systems reinforces the basic ecological principle that *everything is connected*.

CRITICAL THINKING

1. List a limiting factor for each of the following ecosystems: **(a)** a desert, **(b)** arctic tundra, **(c)** the floor of a tropical rain forest, and **(d)** a temperate deciduous forest.

2. Why do deserts and arctic tundra support a much smaller biomass of animals than do tropical forests?

3. Why do most animals in a tropical rain forest live in its trees?

4. What biomes are best suited for **(a)** raising crops and **(b)** grazing livestock?

5. How would you respond to someone who proposes that we use the deep portions of the world's oceans to deposit our radioactive and other hazardous wastes because the deep oceans are vast and are located far away from human habitats? Give reasons for your response.

6. Developers want to drain a large area of inland wetlands in your community and build a large housing development. List **(a)** the main arguments the developers would use to support this project and **(b)** the main arguments ecologists would use in opposing this project. If you were an elected city official, would you vote for or against this project? Can you come up with a compromise plan?

7. You are a defense attorney arguing in court for sparing an undeveloped old-growth tropical rain forest and a coral reef from severe degradation or destruction by development. Write your closing statement for the defense of each of these ecosystems. If the judge decides you can save only one of the ecosystems, which one would you choose, and why?

PROJECTS

1. How has the climate changed in the area where you live during the past 50 years? Investigate the beneficial and harmful effects of these changes. How have these changes benefited or harmed you personally?

2. What type of biome do you live in? What effects have human activities over the past 50 years had on the characteristic vegetation and animal life normally found in the biome you live in? How is your own lifestyle affecting this biome?

3. If possible, visit a nearby lake, pond, or reservoir. Would you classify it as oligotrophic, mesotrophic, or eutrophic? What are the primary factors contributing to its nutrient enrichment? Which of these factors are related to human activities?

4. Make a concept map of this chapter's major ideas, using the section heads and subheads and the key terms

(in boldface). Look on the website for this book for information about making concept maps.

INTERNET STUDY RESOURCES AND RESOURCES FOR FURTHER READING AND RESEARCH

The website for this book contains helpful study aids and many ideas for further reading and research. Log on to

http://www.brookscole.com/product/0534389872s

and click on the Chapter-by-Chapter area. Choose Chapter 6 and select a resource:

- "Flash Cards" allows you to test your mastery of the Terms and Concepts to Remember for this chapter.
- "Tutorial Quizzes" provides a multiple-choice practice quiz.
- "Student Guide to InfoTrac" will lead you to Critical Thinking Projects that use InfoTrac College Edition as a research tool.
- "References" lists the major books and articles consulted in writing this chapter.
- "Hypercontents" takes you to an extensive list of sites with news, research, and images related to individual sections of the chapter.

INFOTRAC COLLEGE EDITION

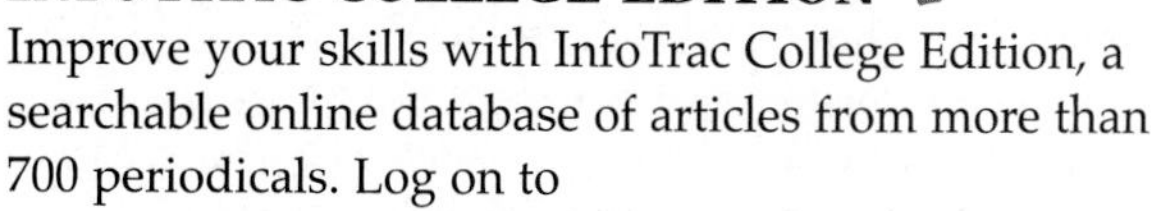

Improve your skills with InfoTrac College Edition, a searchable online database of articles from more than 700 periodicals. Log on to

http://www.infotrac-college.com

or access InfoTrac through the website for this book. Try to find the following articles:

1. Is El Niño Now a Man-Made Phenomenon? (Special Issue on Climate Change). Alain-Claude Galtie. *The Ecologist,* March–April 1999 v29 i2 p64(4). The El Niño phenomenon has become more frequent and severe in recent years. Scientists believe that El Niño and other related natural disasters are the direct result of global warming and massive destruction of tropical rain forests. *Hint*: Enter the search term "el nino" using the keyword "climate."

2. Habitats (low desert plant life) (Brief Article). *Natural History,* March 2000 v109 i2 p16. This is a brief discussion of the plants in various parts of the low desert environment. *Hint*: Enter the search term "low desert" using the keywords "southwest" and "west."

7 COMMUNITY ECOLOGY: STRUCTURE, SPECIES INTERACTIONS, SUCCESSION, AND SUSTAINABILITY

Flying Foxes: Keystone Species in Tropical Forests

The durian (Figure 7-1, bottom right) is one of the most prized fruits growing in Southeast Asian tropical forests. The odor of this football-sized fruit is so strong that it is illegal to have them on trains and in many hotel rooms in Southeast Asia. However, its custardlike flesh has been described as "exquisite," "sensual," "intoxicating," and "the world's finest fruit."

Durian fruits come from a wild tree that grows in the tropical rain forest. The tree depends on nectar- and pollen-feeding flying foxes (Figure 7-1, top) to pollinate the flowers that hang high in the durian trees (Figure 7-1, bottom left). Pollination by flying foxes is an example of *mutualism*: an interaction between two species in which both species benefit. Hundreds of tropical plant species are entirely dependent on various flying fox species for pollination and seed dispersal.

Many species of flying foxes are now listed as *endangered*, and most populations are much smaller than historic numbers. One reason for these population declines is *deforestation*. Another is *hunting* of the bats for their meat, which is sold in China and other parts of Asia. The bats also are viewed as pests and are killed to keep them from eating commercially grown fruits (even though these fruits are picked green). Flying foxes are easy to hunt because they tend to congregate in large numbers when they feed or sleep.

According to ecologists, flying foxes are *keystone species* in tropical forest ecosystems. They are important for **(1)** the plant species they pollinate, **(2)** the plant seeds they disperse in their droppings, and **(3)** the many other species that depend on them. A study in Samoa found that 80–100% of the seeds landing on the ground during the dry season were deposited by flying foxes. They are also critical in regenerating deforested areas because other tropical forest animals are reluctant to venture into the open.

Rain forest ecologists are concerned that the decline of flying fox populations could lead to a cascade of linked extinctions. Their decline also has important economic effects. Studies have also shown that flying foxes are economically important for many products including **(1)** fruits (such as the durian and wild bananas), **(2)** other foods, **(3)** medicine, **(4)** timber (such as ebony and mahogany), **(5)** fibers, **(6)** dyes, **(7)** medicines, **(8)** animal fodder, and **(9)** fuel.

The story of flying foxes and durians illustrates the unique role (niche) of each species in a community or ecosystem and shows how interactions between species can affect ecosystem structure and function. This chapter deals with these topics and the subjects of *community structure, ecological succession,* and *ecosystem sustainability*.

Figure 7-1 Flying foxes (top) are bats that play key ecological roles in tropical rain forests in Southeast Asia by pollinating (left) and spreading the seeds of durian trees, which produce durians, a highly prized tropical fruit (right).

What is this balance of nature that ecologists talk about?
STUART L. PIMM

This chapter addresses the following questions:

- What determines the number of species in a community?
- How can we classify species according to their roles?
- How do species interact with one another?
- How do communities and ecosystems change as environmental conditions change?
- Does high species diversity increase the stability of ecosystems?

7-1 COMMUNITY STRUCTURE: APPEARANCE AND SPECIES DIVERSITY

What Is Community Structure? One property of a community or ecosystem is the *structure* or *spatial distribution* of its individuals and populations. Ecologists usually describe the structure of a community or ecosystem in terms of four characteristics:

- *Physical appearance*: relative sizes, stratification, and distribution of its populations and species
- *Species diversity or richness*: the number of different species
- *Species abundance*: the number of individuals of each species
- *Niche structure*: the number of ecological niches (Section 5-3, p. 112), how they resemble or differ from each other, and how they interact (species interactions)

How Do Communities Differ in Physical Appearance and Population Distribution? The types, relative sizes, and stratification of plants and animals vary in different terrestrial communities and biomes (Figure 7-2). There are also marked differences in the physical structures of different types aquatic life zones such as **(1)** oceans (Figure 6-32, p. 146), **(2)** rocky shore and sandy beaches (Figure 6-35, p. 149), **(3)** lakes (Figure 6-42, p. 154), **(4)** river systems (Figure 6-44, p. 155), and **(5)** inland wetlands. The distribution of populations and species in a terrestrial or aquatic community can be vertical as well as horizontal, as shown in Figure 6-23 (p. 139) for a tropical rain forest and in Figure 6-32 (p. 146) for an ocean.

The physical structure within a particular type of community or ecosystem also can vary. A close look at most large terrestrial communities, ecosystems, and biomes reveals that they usually consist of a mosaic of *vegetation patches* of differing size.

Differences in the physical structure and physical properties (such as sunlight, temperature, wind, and humidity) at boundaries and in transition zones between two ecosystems (*ecotones*) are called **edge effects**. For example, the edge area between a forest and an open field may **(1)** be sunnier, warmer, and drier than the forest interior and **(2)** have a different combination of species than the forest and field interiors.

Popular wild game animals, such as pheasants and white-tailed deer, often are more plentiful in edges and ecotones between forests and fields. Wild game managers sometimes create patches and edges to increase populations of such species for sport hunters.

However, the edge effects associated with habitat fragmentation can reduce the overall biodiversity of ecosystems. This occurs because the increased edge from habitat fragmentation **(1)** makes many species more vulnerable to stresses such as predators and fire and **(2)** creates barriers that can prevent some species from colonizing new areas and finding food and mates.

Conservation biologists urge that overall biodiversity be protected by **(1)** preserving

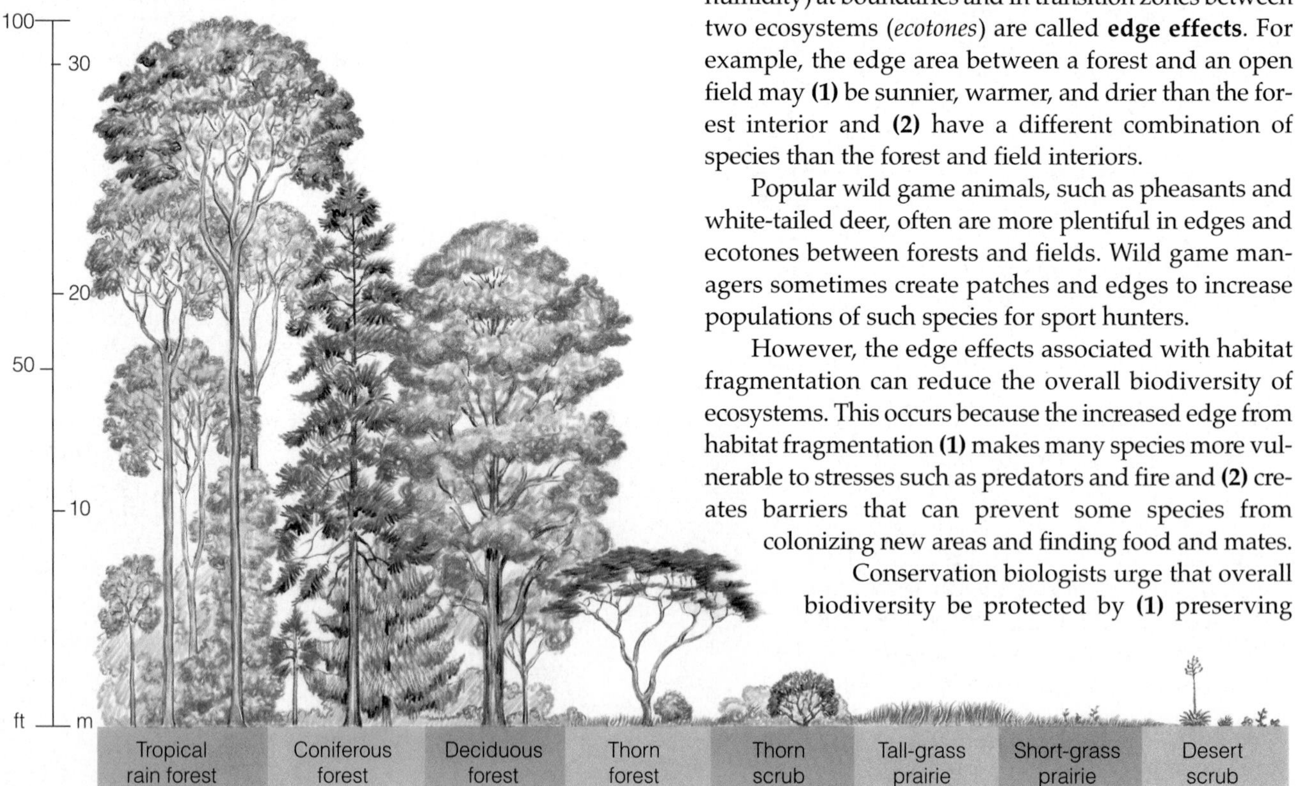

Figure 7-2 Generalized types, relative sizes, and stratification of plant species in various terrestrial communities or ecosystems.

large areas of habitat and **(2)** using migration corridors to link smaller habitat patches, as discussed in Section 17-7, p. 449.

Where Is Most of the World's Biodiversity Found? Studies indicate that the most species-rich environments are **(1)** tropical forests (Figure 6-23, p. 138), **(2)** coral reefs (Figure 6-38, p. 151), **(3)** the deep sea, and **(4)** large tropical lakes.

Other places with an abundance of different species are **(1)** tropical dry habitats (deserts, shrublands, and grasslands) and **(2)** temperate shrublands with a Mediterranean climate (such as South Africa, parts of southern California, Chile, southwestern Australia, and the countries of the Mediterranean Basin).

Communities such as tropical rain forests or coral reefs with large numbers of different species (high species diversity) generally have only a few members of each species (low species abundance).

Field investigations by ecologists have found that three major factors affect species diversity: **(1)** *latitude* (distance from the equator in terrestrial communities (Figure 7-3), **(2)** *depth* in aquatic systems, and **(3)** *pollution* in aquatic systems (Figure 7-4).

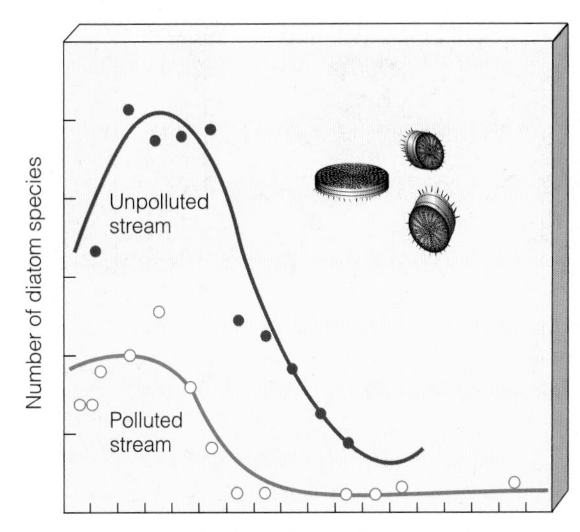

Figure 7-4 Changes in the diversity and abundance of diatom species in an unpolluted stream and a polluted river. Note that both species diversity and species abundance decrease with pollution.

What Determines the Number of Species on Islands? Two factors affecting the species diversity found in an isolated ecosystem such as an island are its *size* and *degree of isolation*. In the 1960s, biologists Robert MacArthur and Edward O. Wilson began studying communities on islands to discover why large islands tend to have more species of a certain category (such as insects, birds, or ferns) than do small islands.

To explain these differences in species diversity with island size, MacArthur and Wilson proposed what is called the **species equilibrium model** or the **theory of island biogeography**. According to this model, the number of species found on an island is determined by a balance between two factors: **(1)** the rate at which new species immigrate to the island and **(2)** the rate at which species become extinct on the island. The model predicts that at some point the rates of immigration and extinction will reach an equilibrium point (Figure 7-5a) that determines the island's average number of different species (species diversity).

The model also predicts that immigration and extinction rates (and thus species diversity) are affected by two important features of the island: **(1)** its *size* (Figure 7-5b) and **(2)** its distance from the nearest mainland (Figure 7-5c). According to the model, a small island tends to have a lower species diversity than a large one for two reasons: **(1)** a small island generally has a lower immigration rate because it is a smaller target for potential colonizers, and **(2)** a small island should have a higher extinction rate because it generally has fewer resources and less diverse habitats for colonizing species.

The model also predicts that an island's distance from a

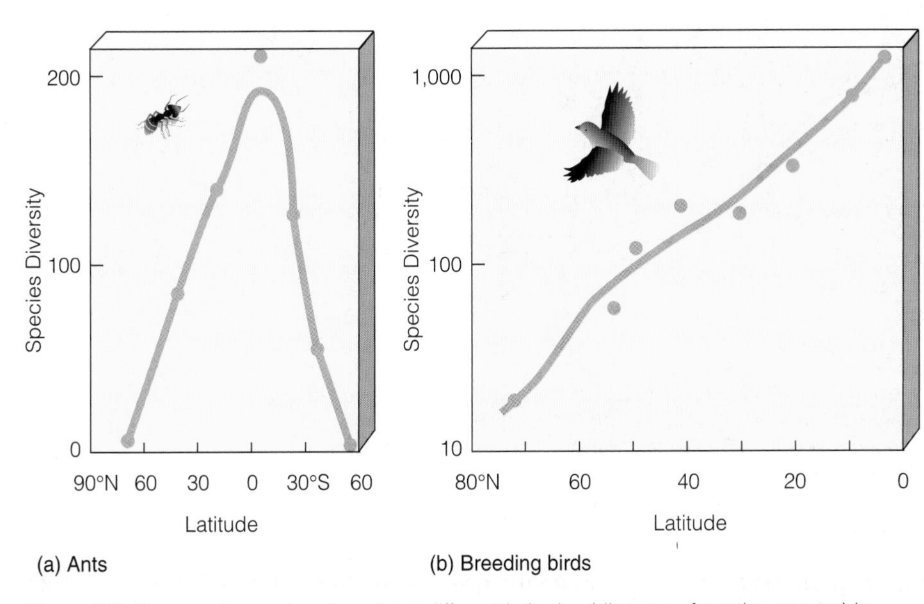

Figure 7-3 Changes in species diversity at different latitudes (distances from the equator) in terrestrial communities for **(a)** ants and **(b)** breeding birds of North and Central America. As a general rule, species diversity steadily declines as we go away from the equator toward either pole. (Modified by permission from Cecie Starr, *Biology: Concepts and Applications*, 4th ed., Pacific Grove, Calif.: Brooks/Cole, 2000)

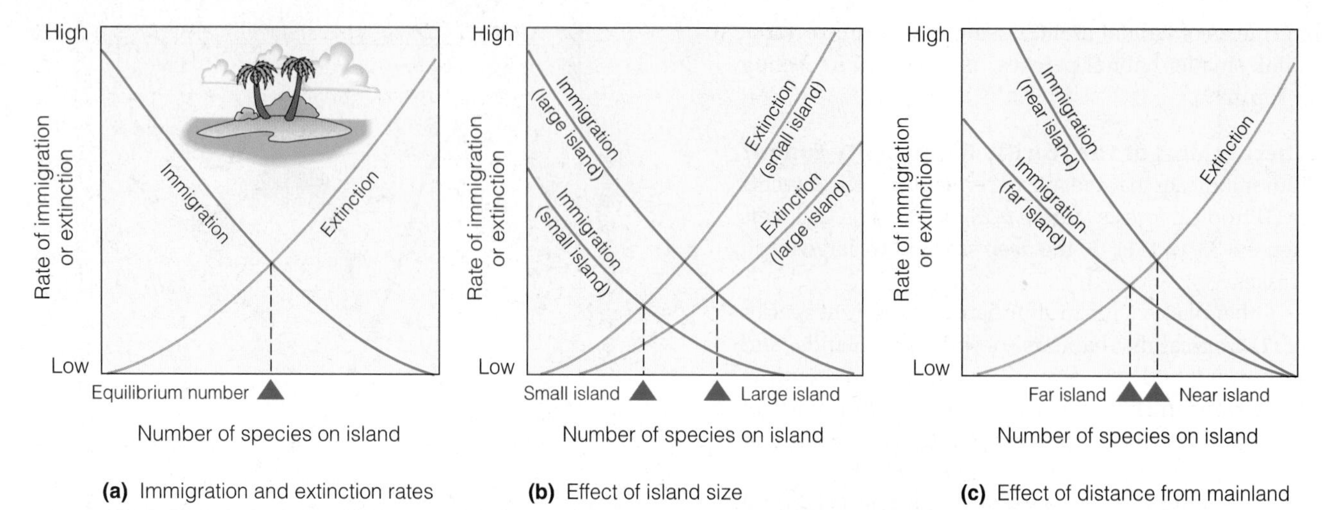

(a) Immigration and extinction rates **(b)** Effect of island size **(c)** Effect of distance from mainland

Figure 7-5 The *species equilibrium model* or *theory of island biogeography*, developed by Robert MacArthur and Edward O. Wilson. **(a)** The equilibrium number of species (blue triangle) on an island is determined by a balance between the immigration rate of new species and the extinction rate of species already on the island. **(b)** With time, large islands have a larger equilibrium number of species than smaller islands because of higher immigration rates and lower extinction rates on large islands. **(c)** Assuming equal extinction rates, an island near a mainland will have a larger equilibrium number of species than a more distant island because the immigration rate to a near island is higher than that to a more distant one.

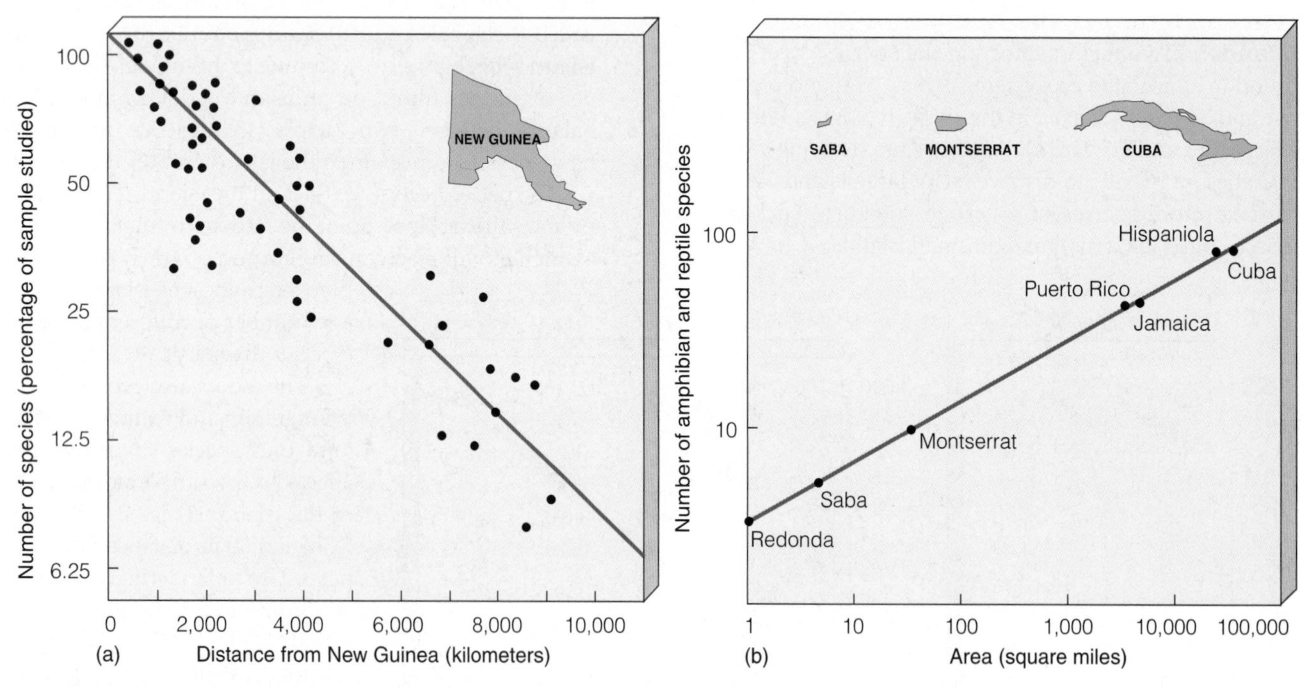

Figure 7-6 Research data supporting the theory of island biogeography: **(a)** The species diversity of birds occupying lowland areas of South Pacific islands decreases with distance from New Guinea, and **(b)** species diversity increases with island size. (Used by permission from Cecie Starr, *Biology: Concepts and Applications*, 4th ed., Pacific Grove, Calif.: Brooks/Cole, 2000)

mainland source of new species is important in determining species diversity. For two islands of about equal size and other factors, the island closest to a mainland source of immigrant species will have the higher immigration rate and thus a higher species diversity (assuming that extinction rates on both islands are about the same).

MacArthur and Wilson's original model or scientific hypothesis has been tested and supported by a series of field experiments (Figure 7-6). As a result, biologists have accepted it well enough to elevate it to the status of an important and useful scientific theory, although it may apply to a limited number of cases over short periods of time. In recent years, it has also been applied to conservation efforts to protect wildlife on land in *habitat islands* such as national parks surrounded by a sea of developed and fragmented land.

7-2 GENERAL TYPES OF SPECIES

What Different Roles Do Various Species Play in Ecosystems? When examining ecosystems, ecologists often apply particular labels—such as *native, nonnative, indicator,* or *keystone*—to various species to clarify their ecological roles or niches (p. 112). Any given species may function as more than one of these four types in a particular ecosystem.

How Can Nonnative Species Cause Problems? Species that normally live and thrive in a particular ecosystem are known as **native species**. Others that migrate into an ecosystem or are deliberately or accidentally introduced into an ecosystem by humans are called **nonnative species**, **exotic species**, or **alien species**. Some of these introduced species (such as crops and game species for sport hunting) are beneficial to humans, but some thrive and crowd out native species.

From a human standpoint, introduction of some nonnative species can become a nightmare. In 1957, wild African bees were imported to Brazil to help increase honey production. Instead, these bees have displaced domestic honeybees and reduced the honey supply.

Since then these nonnative bee species, popularly known as "killer bees," have moved northward into Central America (killing 150 people in Mexico since 1986). They have become established in Texas (one death in 1994), Arizona (one death in 1993), New Mexico, Puerto Rico, and California and are heading north at 240 kilometers (150 miles) per year. They should be stopped eventually by cold winters in the central United States unless they can adapt genetically to cold weather.

Although they are not the killer bees portrayed in some horror movies, these bees are aggressive and unpredictable. They have killed thousands of domesticated animals and an estimated 1,000 people in the western hemisphere. Fortunately, most people not allergic to bee stings can run away. Most people killed by these honeybees have died because they fell down or became trapped and could not flee. The deliberate and accidental introduction of nonnative species is discussed in more detail on pp. 468–471.

What Are Indicator Species? Species that serve as early warnings of damage to a community or an ecosystem are called **indicator species**. Birds are excellent biological indicators because they are found almost everywhere and respond quickly to environmental change. Research indicates that a major factor in the current decline of migratory, insect-eating songbirds in North America is habitat loss or fragmentation. The tropical forests of Latin America and the Caribbean that are winter habitats for such birds are disappearing rapidly. Their summer habitats in North America also are disappearing or are being fragmented into patches that make the birds more vulnerable to attack by predators and parasites.

The presence or absence of trout species in water at temperatures within their range of tolerance (Figure 4-13, p. 85) is an indicator of water quality because trout need clean water with high levels of dissolved oxygen. Some amphibians (frogs, toads, and salamanders), which live part of their lives in water and part on land, are also believed to be indicator species (Connections, p. 164).

What Are Keystone Species? The roles of some species in an ecosystem are much more important than their abundance or biomass suggests. Ecologists call such species **keystone species**, although this designation is controversial.* Such species play pivotal roles in the structure and function of an ecosystem because **(1)** their strong interactions with other species affect the health and survival of these species, and **(2)** they process material out of proportion to their numbers or biomass.

Critical roles of keystone species include **(1)** pollination of flowering plant species by bees, hummingbirds, bats (Figure 7-1), and other species, **(2)** dispersion of seeds by fruit-eating animals such as bats (Figure 7-1), with the undigested seeds being scattered in their feces, **(3)** habitat modification, **(4)** predation by top carnivores to control the populations of various species, **(5)** improving the ability of plant species to obtain soil minerals and water, and **(6)** efficient recycling of animal wastes. Some species play more than one of these roles.

Beneficial *habitat modifications* by keystone species include the following:

- Elephants push over, break, or uproot trees, creating forest openings in the savanna grasslands and woodlands of Africa. This promotes the growth of grasses and other forage plants that benefit smaller grazing species such as antelope and accelerates nutrient cycling rates.
- Bats (p. 159) and birds regenerate deforested areas by depositing plant seeds in their droppings.

*All species play some role in their ecosystems and thus are important. Whereas some scientists consider all species equally important, others consider certain species to be more important than others in helping maintain the structure and function of ecosystems of which they are a part.

Why Are Amphibians Vanishing?

Amphibians (frogs, toads, and salamanders) first appeared about 350 million years ago. These cold-blooded creatures range in size from a frog that can sit on your thumb to a Japanese salamander that is about 1.5 meters (5 feet) long.

Since 1980, populations of hundreds of the world's estimated 5,100 amphibian species (including 2,700 frog and toad species) have been vanishing or declining in almost every part of the world, even in protected wildlife reserves and parks.

In some locales, frog deformities (such as extra legs and missing legs) are occurring in unusually high numbers. This information was discovered and published by schoolchildren in Henderson, Minnesota, who were catching frogs in a farm pond.

According to the World Conservation Union, 25% of all known amphibian species are extinct, endangered, or vulnerable. The Nature Conservancy estimates that 38% of the amphibian species in the United States are endangered.

Frogs are especially vulnerable to environmental disruption. As tadpoles they live in water and eat plants, and as adults they live mostly on land and eat insects (which can expose them to pesticides). Their eggs have no protective shells to block ultraviolet radiation or pollution. As adults, they take in water and air through their thin, permeable skins that can readily absorb pollutants from water, air, or soil.

Because each type of frog lives in a small, specialized habitat, they are sensitive indicators of environmental conditions in a variety of places.

Although no single cause has been identified to explain amphibian declines, scientists have identified a number of contributing factors. They include the following:

- *Habitat loss*, especially because of the **(1)** draining and filling of inland wetlands, **(2)** clear-cutting of forests, and **(3)** fragmentation of habitats into pieces too small to support populations of some amphibians.
- *Prolonged drought*, which dries up breeding pools so that few tadpoles survive. Dehydration can also weaken amphibians, making them more susceptible to fatal viruses, bacteria, fungi, and parasites.
- *Pollution*. Frog eggs, tadpoles, and adults are very sensitive to many pollutants (especially pesticides). Exposure to such pollutants may harm amphibians' immune and endocrine systems and make them more vulnerable to bacterial infections and perhaps to a newly discovered type of skin fungus.
- *Increases in ultraviolet radiation* caused by reductions in stratospheric ozone (pp. 318–324). This radiation can be especially harmful to amphibian embryos in shallow ponds.
- *Warmer weather*, which causes less rain and snow and results in shallower lakes and ponds and exposes the eggs of amphibians to more ultraviolet light. This can make the eggs more susceptible to various water molds and kill the embryos.
- *Increased incidence of parasitism* by a flatworm (trematode), which may account for many frog deformities but not the worldwide decline of amphibians.
- *Overhunting*, especially in Asia and France, where frog legs are a delicacy.
- *Epidemic diseases* such as the chytrid fungus and iridoviruses. Pollution and other factors that weaken the immune systems of amphibians may make them more susceptible to these and other disease organisms.
- *Immigration or introduction of nonnative predators and competitors* (such as fish) *and disease organisms* such as the chytrid (which may have been introduced into Australia by imported tropical fish).

In most cases, the decline or disappearance of amphibian species probably is caused by a combination of such factors. Scientists are concerned about amphibians' decline for three reasons:

- It suggests that the world's environmental health is deteriorating because amphibians generally **(1)** are tough survivors and **(2)** are sensitive bioindicators of changes in environmental conditions such as habitat loss and degradation, pollution, exposure to ultraviolet light, and climate change.
- Adult amphibians play important roles in the world's ecosystems. For example, amphibians eat more insects (including mosquitoes) than do birds. In some habitats, extinction of certain amphibian species could also result in extinction of other species, such as reptiles, birds, aquatic insects, fish, mammals, and other amphibians that feed on them or their larvae.
- From a human perspective amphibians represent a genetic storehouse of pharmaceutical products waiting to be discovered. Hundreds of secretions from amphibian skin have been isolated, and some of these compounds are being used as painkillers and antibiotics and in treating burns and heart attacks.

As possible indicator species, amphibians may be sending us an important message. They do not need us, but we and other species need them.

Critical Thinking

On an evolutionary time scale, all species eventually become extinct. Some people suggest that the widespread disappearance of amphibians is the result of natural responses to changing environmental conditions. Others contend that these losses are caused mostly by human activities and that such declines are a warning of possible danger for our own species and other species. What is your position? Why?

CONNECTIONS

Why Should We Care About Alligators?

The American alligator, North America's largest reptile, has no natural predators except humans. This species, which has been around for about 200 million years, has been able to adapt to numerous changes in the earth's environmental conditions.

This changed when hunters in the United States began killing large numbers of these animals for **(1)** their meat and **(2)** their supple belly skin, used to make shoes, belts, and pocketbooks.

Other people considered alligators to be useless and dangerous and hunted them for sport or out of hatred. Between 1950 and 1960, hunters wiped out 90% of the alligators in Louisiana, and by the 1960s the alligator population in the Florida Everglades also was near extinction.

People who say "So what?" are overlooking the alligator's important ecological role or *niche* in subtropical wetland ecosystems. Alligators dig deep depressions, or gator holes, that **(1)** collect fresh water during dry spells, **(2)** serve as refuges for aquatic life, and **(3)** supply fresh water and food for many animals.

In addition, large alligator nesting mounds provide nesting and feeding sites for herons and egrets. Alligators also eat large numbers of gar (a predatory fish) and thus help maintain populations of game fish such as bass and bream.

As alligators move from gator holes to nesting mounds, they help keep areas of open water free of invading vegetation. Without these ecosystem services, freshwater ponds and shrubs and trees would fill in coastal wetlands in the alligator's habitat, and dozens of species would disappear.

Some ecologists classify the North American alligator as a *keystone species* because of these important ecological roles in helping maintain the structure and function of its natural ecosystems.

In 1967, the U.S. government placed the American alligator on the endangered species list. Protected from hunters, the alligator population made a strong comeback in many areas by 1975—too strong, according to those who find alligators in their backyards and swimming pools and to duck hunters, whose retriever dogs sometimes are eaten by alligators.

In 1977, the U.S. Fish and Wildlife Service reclassified the American alligator from an *endangered* species to a *threatened* species in Florida, Louisiana, and Texas, where 90% of the animals live. In 1987 this reclassification was extended to seven other states.

Alligators now number perhaps 3 million, most in Florida and Louisiana. It is generally illegal to kill members of a threatened species, but limited kills by licensed hunters are allowed in some areas of Florida, Louisiana, and South Carolina to control the population. To biologists, the comeback of the American alligator from near extinction caused by overhunting is an important success story in wildlife conservation.

The increased demand for alligator meat and hides has created a booming business in alligator farms, especially in Florida. Such success reduces the need for illegal hunting of wild alligators.

Critical Thinking

Some homeowners in Florida believe that they should have the right to kill any alligator found on their property. Others argue that this should not be allowed because **(1)** alligators are a threatened species and **(2)** housing developments have invaded the habitats of alligators, not the other way around. What is your opinion on this issue? Explain.

- Sea otters help keep sea urchins from depleting kelp forests in offshore waters from Alaska to southern California. Kelp forests **(1)** provide essential habitats for a variety of species, **(2)** inhibit shore erosion, and **(3)** reduce the impact of storm waves on coastlines.

Top predator keystone species exert a stabilizing effect on their ecosystems by feeding on and regulating the populations of certain species. Examples are the wolf (p. 429), leopard, lion, alligator (Connections, above), sea otter, and great white shark.

The loss of a keystone species can lead to population crashes and extinctions of other species that depend on it for certain services, a ripple or domino effect that spreads throughout an ecosystem. According to biologist Edward O. Wilson, "The loss of a keystone species is like a drill accidentally striking a power line. It causes lights to go out all over."

7-3 SPECIES INTERACTIONS: COMPETITION AND PREDATION

How Do Species Interact? An Overview When different species in an ecosystem have activities or resource needs in common, they may interact with one another. Members of these species may **(1)** be harmed by, **(2)** benefit from, or **(3)** be unaffected by the interaction. There are five basic types of interactions between species: **(1)** *interspecific competition,*

(2) *predation*, **(3)** *parasitism*, **(4)** *mutualism*, and **(5)** *commensalism*.

These interactions tend to regulate the populations of species and can help them survive changes in environmental conditions, as discussed in more detail in Section 8-2, p. 185.

How Do Species Compete for Resources? Intraspecific and Interspecific Competition

Competition between members of the same species for the same resources is called **intraspecific competition**. Intraspecific competition can be intense because members of a particular species compete directly for the same resources.

Competition between members of two or more different species for food, space, or any other limited resource is called **interspecific competition**.

As long as commonly used resources are abundant, different species can share them. This allows each species to come closer to occupying the *fundamental niche* it would occupy if there were no competition from other species.

However, most species face competition from other species for one or more limited resources (such as food, sunlight, water, soil nutrients, space, nesting sites, and good places to hide). Because of such *interspecific competition*, parts of the fundamental niches of different species overlap (Figure 5-6, p. 113). The more the niches of two species overlap, the more they compete with one another. With significant niche overlap, one of the competing species must **(1)** migrate to another area (if possible), **(2)** shift its feeding habits or behavior through natural selection and evolution (Section 5-2, p. 109), **(3)** suffer a sharp population decline, or **(4)** become extinct in that area.

How Have Some Species Reduced or Avoided Competition?

Over a time scale long enough for evolution to occur, some species that compete for the same resources evolve adaptations that reduce or avoid competition or overlap of their fundamental niches (Figure 5-6, p. 113). One way this happens is through **resource partitioning**, the dividing up of scarce resources so that species with similar needs use them at different times, in different ways, or in different places (Figure 7-7). In effect, they evolve traits that allow them to share the wealth.

Each of the competing species occupies a *realized niche* that makes up only part of its *fundamental niche*. The result is that through evolution the fairly broad niches of two competing species (Figure 7-8, top) become more specialized (Figure 7-8, bottom). Resource partitioning through niche specialization is also found in the different layers of tropical rain forests (Figure 6-23, p. 139).

Here are some other examples of resource partitioning. When lions and leopards live in the same area, lions take mostly larger animals as prey, and leopards take smaller ones. Hawks and owls feed on similar prey, but hawks hunt during the day and owls hunt at night. Some bird species feed on the ground, whereas others seek food in trees and shrubs.

Ecologist Robert H. MacArthur studied the feeding habits of five species of warblers (small insect-eating birds) that coexist in the forests of the northeastern United States and in the adjacent area of Canada.

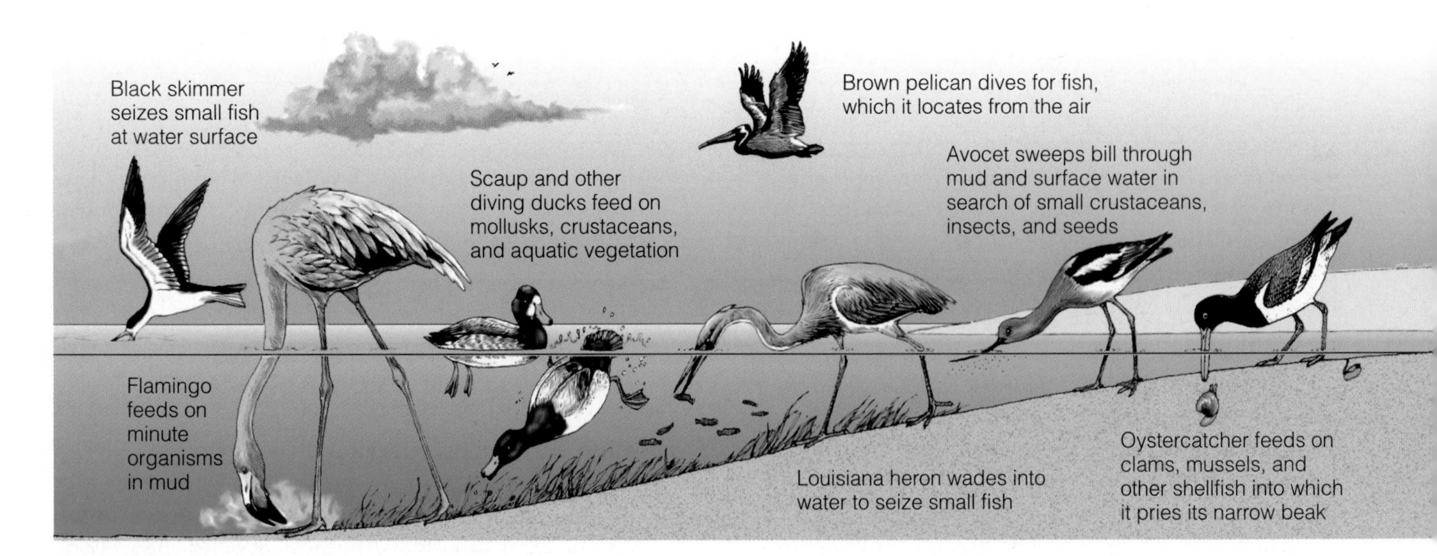

Black skimmer seizes small fish at water surface
Brown pelican dives for fish, which it locates from the air
Avocet sweeps bill through mud and surface water in search of small crustaceans, insects, and seeds
Scaup and other diving ducks feed on mollusks, crustaceans, and aquatic vegetation
Flamingo feeds on minute organisms in mud
Louisiana heron wades into water to seize small fish
Oystercatcher feeds on clams, mussels, and other shellfish into which it pries its narrow beak

Figure 7-7 Specialized feeding niches of various bird species in a coastal wetland. Such resource partitioning reduces competition and allows sharing of limited resources.

Figure 7-8 *Resource partitioning* and *niche specialization* as a result of competition between two species. The top diagram shows the overlapping niches of two competing species. The bottom diagram shows that through evolution the niches of the two species become separated and more specialized (narrower) so that they avoid competing for the same resources.

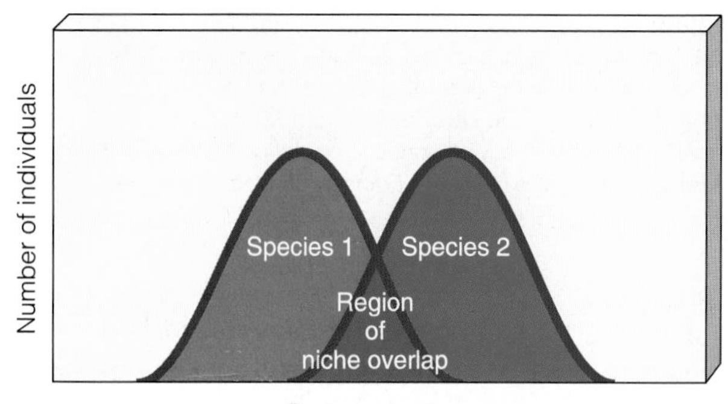
Number of individuals
Species 1
Species 2
Region
of
niche overlap
Resource use

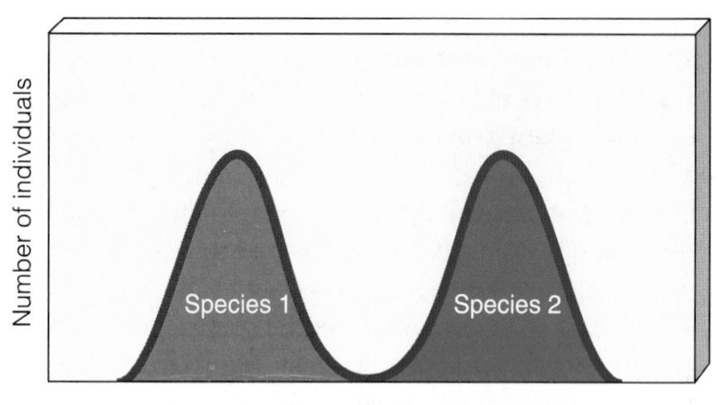
Number of individuals
Species 1
Species 2
Resource use

Although they appear to be competing for the same food resources, MacArthur found that the bird species reduce competition through resource partitioning by spending at least half their time hunting for insects in different parts of trees (Figure 7-9).

How Do Predator and Prey Species Interact? In **predation**, members of one species (the *predator*) feed directly on all or part of a living organism of another species (the *prey*). However, they do not live on or in the prey, and the prey may or may not die from the interaction. In this interaction, the predator benefits and the individual prey is clearly harmed. Together, the two kinds of organisms, such as lions (the predator or hunter) and zebras (the prey or hunted), are said to have a **predator–prey relationship**, as depicted in Figures 4-10 (p. 84), 4-11 (p. 84), 4-17 (p. 89), and 4-18 (p. 90).

At the individual level, members of the prey species are clearly harmed. However, at the population level predation can benefit the prey species because predators such as tigers and some types of sharks often kill the sick, weak, and aged members (Case Study, p. 170). Reducing the prey population **(1)** gives remaining prey greater access to the available food supply and **(2)** can improve the genetic stock of the prey population, which enhances its chances of reproductive success and long-term survival. The effects of predation on populations of predator and prey species are discussed in more detail in Section 8-2, p. 185.

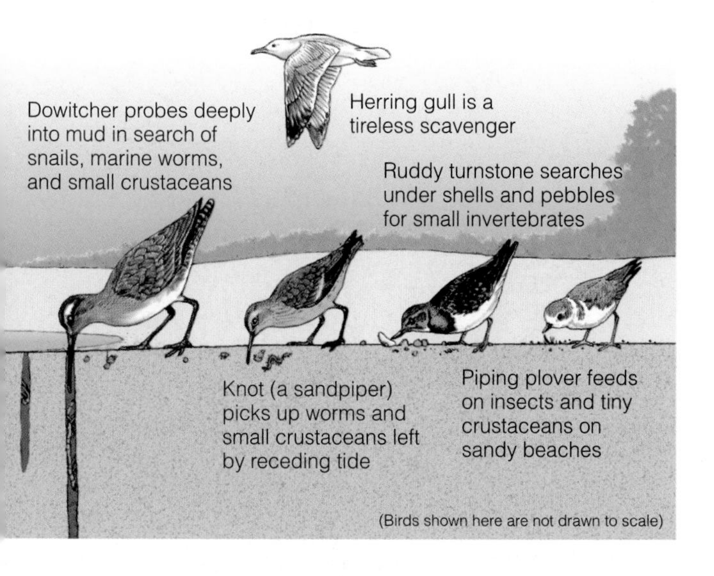
Dowitcher probes deeply into mud in search of snails, marine worms, and small crustaceans
Herring gull is a tireless scavenger
Ruddy turnstone searches under shells and pebbles for small invertebrates
Knot (a sandpiper) picks up worms and small crustaceans left by receding tide
Piping plover feeds on insects and tiny crustaceans on sandy beaches
(Birds shown here are not drawn to scale)

Some people tend to view predators with contempt. When a hawk tries to capture and feed on a rabbit, some tend to root for the rabbit. Yet the hawk (like all predators) is merely trying to get enough food to feed itself and its young; in the process, it is playing an important ecological role in controlling rabbit populations.

How Do Predators Increase Their Chances of Getting a Meal? Predators have a variety of methods that help them capture prey. *Herbivores* can simply walk, swim, or fly to the plants they feed upon.

Carnivores feeding on mobile prey have two main options: *pursuit* and *ambush*. Some, such as the cheetah, catch prey by being able to run fast; others, such as the American bald eagle, fly and have keen eyesight; still others, such as wolves and African lions, cooperate in capturing their prey by hunting in packs.

Other predators have characteristics or strategies that enable them to hide and ambush their prey. Examples include **(1)** praying mantises (Figure 4-1, right, p. 77) sitting in flowers of a similar color and ambushing visiting insects, **(2)** white ermines (a type of weasel) and snowy owls hunting in snow-covered areas, **(3)** the alligator snapping turtle lying camouflaged on its

Figure 7-9 *Resource partitioning* of five species of common insect-eating warblers in spruce forests of Maine. Each species minimizes competition with the others for food by **(1)** spending at least half its feeding time in a distinct portion (shaded areas) of the spruce trees and **(2)** consuming somewhat different insect species. (After R. H. MacArthur, "Population Ecology of Some Warblers in Northeastern Coniferous Forests," *Ecology* 36 (1958): 533–36)

stream-bottom habitat and dangling its worm-shaped tongue to entice fish into its powerful jaws, and **(4)** people camouflaging themselves and using traps to ambush wild game.

How Do Prey Defend Themselves Against or Avoid Predators? Species have various characteristics that enable them to avoid predators. They include **(1)** the ability to run, swim, or fly fast, **(2)** a highly developed sense of sight or smell that alerts them to the presence of predators, **(3)** protective shells (as on armadillos, which roll themselves up into an armor-plated ball, and turtles), **(4)** thick bark (giant sequoia), and **(5)** spines (porcupines) or thorns (cacti and rose bushes). Many lizards have brightly colored tails that break off when they are attacked, often giving them enough time to escape.

Other prey species use *camouflage* by having certain shapes or colors (Figure 7-10a) or the ability to change color (chameleons and cuttlefish). A tree frog may be almost invisible against its background (Figure 7-10b), and an arctic hare in its white winter fur blends into the snow. Some insect species have evolved shapes that look like twigs or bird droppings on leaves.

Chemical warfare is another common strategy. Some prey species discourage predators with chemicals that are **(1)** poisonous (oleander plants), **(2)** irritating (bombardier beetles, Figure 7-10c), **(3)** foul smelling (skunks, skunk cabbages, and stinkbugs), or **(4)** bad tasting (buttercups and monarch butterflies, Figure 7-10d). Scientists have identified more than 10,000 defensive chemicals made by plants, including cocaine, caffeine, nicotine, cyanide, opium, strychnine, peyote, and rotenone (used as an insecticide).

Many bad-tasting, bad-smelling, toxic, or stinging prey species have evolved *warning coloration,* brightly colored advertising that enables experienced predators to recognize and avoid them. Examples are **(1)** brilliantly colored poisonous frogs (Figure 7-10e) and red-, yellow-, and black-striped coral snakes and **(2)** foul-tasting monarch butterflies (Figure 7-10d) and grasshoppers. Other butterfly species, such as the nonpoisonous viceroy (Figure 7-10f), gain some protection by looking and acting like the poisonous monarch (Figure 7-10d), a protective device known as *mimicry.*

Some prey species use behavioral strategies to avoid predation. Some attempt to scare off predators by **(1)** puffing up (blowfish), **(2)** spreading their wings (peacocks), or **(3)** mimicking a predator (Figure 7-10h). To help fool or frighten would-be predators, some moths have wings that look like the eyes of much larger animals (Figure 7-10g). Other prey gain some protection by living in large groups (schools of fish, herds of antelope, flocks of birds).

(a) African stoneplants

(b) Canyon tree frog

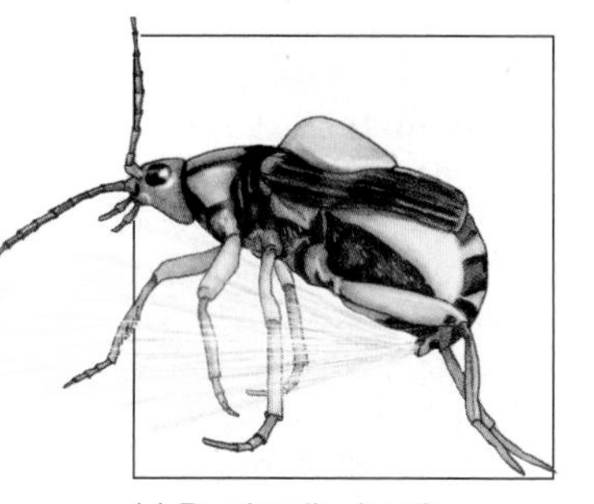

(c) Bombardier beetle

(d) Foul-tasting monarch butterfly

(e) Poison dart frog

(f) Viceroy butterfly mimics monarch butterfly

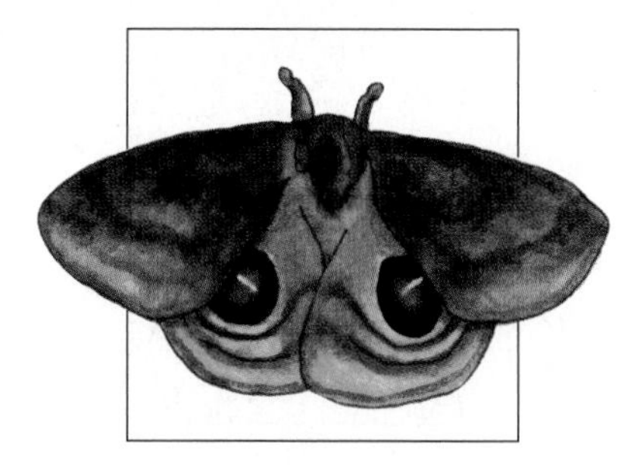

(g) Hind wings of Io moth resemble eyes of a much larger animal.

(h) When touched snake caterpillar changes shape to look like head of snake.

Figure 7-10 Some ways in which prey species avoid their predators by **(1)** *camouflage* (a and b), **(2)** *chemical warfare* (c and e), **(3)** *warning coloration* (d and e), **(4)** *mimicry* (f), **(5)** *deceptive looks* (g), and **(6)** *deceptive behavior* (h).

7-4 SPECIES INTERACTIONS: PARASITISM, MUTUALISM, AND COMMENSALISM

What Are Parasites, and Why Are They Important? **Parasitism** occurs when one species (the *parasite*) feeds on part of another organism (the *host*) by living on or in the host. In this relationship, the parasite benefits and the host is harmed.

Parasitism can be viewed as a special form of predation, but unlike a conventional predator, a parasite **(1)** usually is smaller than its host (prey), **(2)** remains closely associated with, draws nourishment from, and may gradually weaken its host over time, and **(3)** rarely kills its host.

Tapeworms, disease-causing microorganisms (pathogens), and other parasites live *inside* their hosts. Other parasites, such as ticks, fleas, mosquitoes, mistletoe plants, and fungi (that cause diseases such as athlete's foot) attach themselves to the *outside* of their hosts. Some parasites move from one host to another, as fleas and ticks do; others, such as tapeworms, spend their adult lives with a single host.

From the host's point of view parasites are harmful, but parasites play important ecological roles. Collectively, the incredibly complex network of parasitic relationships in an ecosystem acts somewhat like glue that helps hold the species in an ecosystem together. Parasites also promote biodiversity by helping prevent some species from becoming too plentiful and eliminating other species through competition.

How Do Species Interact So That Both Species Benefit? In **mutualism** two species interact in ways that benefit both. Such benefits include **(1)** having pollen and seeds dispersed for reproduction, **(2)** being supplied with food, or **(3)** receiving protection.

The *pollination* relationship between flowering plants and animals such as insects (Figure 4-1, left, p. 77), birds, and bats (p. 159) is one of the most common forms of mutualism. Examples of *nutritional mutualism* include the following:

- *Lichens*, hardy species that can grow on trees or barren rocks, consist of colorful photosynthetic algae and chlorophyll-lacking fungi living together (Figure 12-1, p. 277). The fungi provide a home for the algae, and their bodies collect and hold moisture and mineral nutrients used by both species. The algae, through photosynthesis, provide sugars as food for themselves and the fungi.
- Plants in the legume family support root nodules, where *Rhizobium* bacteria convert atmospheric nitrogen into a form usable by the plants, and the plants provide the bacteria with some simple sugars.

Why Are Sharks Important Species?

CASE STUDY

The world's 350 shark species vary widely in size. The smallest is the dwarf dog shark, about the size of a large goldfish. The largest is the whale shark, the world's largest fish, which can grow to 18 meters (60 feet) long and weigh as much as two full-grown African elephants.

Various shark species, feeding at the top of food webs, cull injured and sick animals from the ocean and thus play an important ecological role. Without such shark species the oceans would be overcrowded with dead and dying fish.

Many people, influenced by movies and popular novels, think of sharks as people-eating monsters. However, the three largest species—the whale shark, basking shark, and megamouth shark—are gentle giants that swim through the water with their mouths open, filtering out and swallowing huge quantities of *plankton* (small free-floating sea creatures).

Every year, members of a few shark species—mostly great white, bull, tiger, gray reef, lemon, and blue—injure about 100 people worldwide and kill between 10 and 15 people. Most attacks are by great white sharks, which feed on sea lions and other marine mammals and sometimes mistake divers and surfers for their usual prey.

For every shark that injures a person, we kill at least 1 million sharks, for a total of about 100 million sharks each year. Sharks are killed mostly for their fins, widely used in Asia as a soup ingredient or pharmaceutical cure-all and worth as much as $563 per kilogram ($256 per pound). In Hong Kong, a single bowl of shark fin soup can sell for as much as $100.

According to a 2001 study by Wild Aid, shark fins sold in restaurants throughout Asia and in Chinese communities in cities such as New York, San Francisco, and London contain dangerously high levels of mercury. Consumption of high levels of mercury is especially dangerous for pregnant women and their babies.

Sharks are also killed for their **(1)** livers, **(2)** meat (especially mako and thresher), **(3)** hides (a source of high-quality leather), and **(4)** jaws (especially great whites, whose jaws are worth thousands of dollars to collectors), or **(5)** just because we fear them. Some sharks (especially blue, mako, and oceanic whitetip) die when they are trapped as bycatch in nets or lines deployed to catch swordfish, tuna, shrimp, and other commercially important species.

Sharks also help save human lives. In addition to providing people with food, they are helping us learn how to fight cancer (which sharks almost never get), bacteria, and viruses. Their highly effective immune system is being studied because it allows wounds to heal without becoming infected.

Sharks have several natural traits that make them prone to population declines from overfishing. They **(1)** have only a few offspring (between 2 and 10) once every year or two, **(2)** take 10–24 years to reach sexual maturity and begin reproducing, and **(3)** have long gestation (pregnancy) periods, up to 24 months for some species.

Sharks are among the most vulnerable and least protected animals on the earth. Eight of the world's shark species, including great whites, sandtigers, and kitefins, are now considered critically endangered, endangered, or vulnerable to extinction. Of the 125 countries that commercially catch more than 100 million sharks per year, only four—Australia, Canada, New Zealand, and the United States—have implemented management plans for shark fisheries, and these plans are hard to enforce.

With more than 400 million years of evolution behind them, sharks have had a long time to get things right. Preserving their evolutionary genetic development begins with the knowledge that sharks do not need us, but we and other species need them.

Critical Thinking

After reading this information, has your attitude toward sharks changed? If so, how has it changed?

- Vast armies of bacteria in the digestive systems of animals break down (digest) their food. The bacteria gain a safe home with a steady food supply; the animal gains more efficient access to a large source of energy.

Examples of mutualistic relationships involving *nutrition* and *protection* are as follows:

- Birds ride on the backs of large animals such as African buffalo, elephants, and rhinoceroses (Figure 7-11a). The birds remove and eat parasites from the animal's body and often make noises warning the animal when predators approach.
- Clownfish species live within sea anemones, whose tentacles sting and paralyze most fish that touch them (Figure 7-11b). The clownfish, which are not harmed by tentacles, gain protection from predators and feed on the detritus left from the meals of the anemones. The sea anemones benefit because the clownfish protect them from some of their predators.
- Minute fungi called mycorrhizae live on the roots of many plants. The fungi get nutrition from a plant's roots and in turn benefit the plant by using their myriad networks of hairlike extensions to

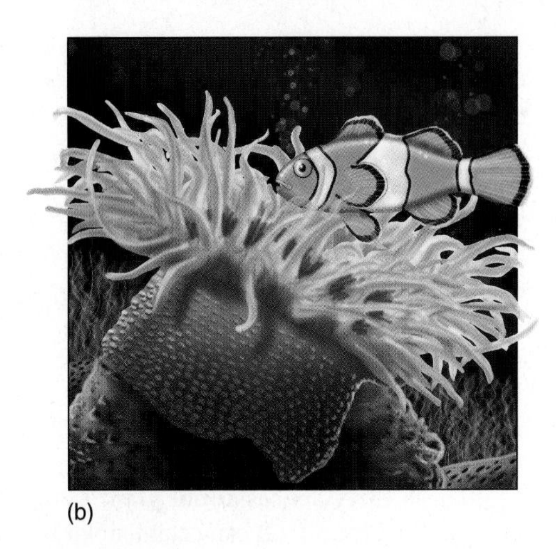

(a) (b)

Figure 7-11 Two examples of *mutualism.* **(a)** Oxpeckers (or tickbirds) feed on the parasitic ticks that infest large, thick-skinned animals such as a black rhinoceros, and **(b)** a clownfish lives among deadly stinging sea anemones.

improve the plant's ability to extract nutrients and water from the soil.

It is tempting to think of mutualism as an example of cooperation between species, but actually it involves each species benefiting by exploiting the other.

How Do Species Interact So That One Benefits but the Other Is Not Harmed? **Commensalism** is an interaction that benefits one species but neither harms nor helps the other species much, if at all. For example, redwood sorrel, a small herb, benefits from growing in the shade of tall redwood trees, with no known negative effects on the redwood trees.

Another example is the commensalistic relationship between various trees and other plants called *epiphytes* (such as some types of orchids and bromeliads) that attach themselves to the trunks or branches of large trees (Figure 7-12) in tropical and subtropical forests. These so-called air plants benefit by **(1)** having a solid base on which to grow and **(2)** living in an elevated spot that gives them better access to sunlight, water from the humid air and rain, and nutrients falling from the tree's upper leaves and limbs.

Figure 7-12 *Commensalism* between a white orchid (an epiphyte or air plant from the tropical forests of Latin America) that roots in the fork of a tree rather than the soil. In this interaction, the epiphytes gain access to water, nutrient debris, and sunlight; the tree apparently remains unharmed unless it contains a large number of epiphytes.

7-5 ECOLOGICAL SUCCESSION: COMMUNITIES IN TRANSITION

How Do Ecosystems Respond to Change? One characteristic of all communities and ecosystems is that their structures change constantly in response to changing environmental conditions. The gradual change in species composition of a given area is called **ecological succession**. During succession some species colonize an area and their populations become more numerous, whereas populations of other species decline and even disappear.

Ecologists recognize two types of ecological succession: *primary* and *secondary*, depending on the conditions present at the beginning of the process:

- **Primary succession** involves the gradual establishment of biotic communities on nearly lifeless ground.
- **Secondary succession**, the more common type of succession, involves the reestablishment of biotic communities in an area where some type of biotic community is already present.

What Is Primary Succession? Establishing Life on Lifeless Ground Primary succession begins with an essentially lifeless area where there is no soil in a terrestrial ecosystem (Figure 7-13) or no bottom sediment in an aquatic ecosystem. Examples include **(1)** bare rock exposed by a retreating glacier or severe soil erosion, **(2)** newly cooled lava, **(3)** an abandoned highway or parking lot, or **(4)** a newly created shallow pond or reservoir.

Before a community of plants (producers), consumers, and decomposers can become established on land, there must be *soil*: a complex mixture of rock particles, decaying organic matter, air, water, and living organisms. Depending mostly on the climate, it takes natural processes several hundred to several thousand years to produce fertile soil.

Soil formation begins when hardy **pioneer species** attach themselves to inhospitable patches of bare rock. Examples are wind-dispersed lichens (Figure 12-1, p. 277) and mosses, which can withstand the lack of moisture and soil nutrients and hot and cold temperature extremes found in such habitats.

As patches of soil build up and spread, eventually the community of lichens and mosses is replaced by a community of **(1)** small perennial grasses (plants that live for more than 2 years without having to reseed) and **(2)** herbs (ferns in tropical areas), whose seeds germinate after being blown in by the wind or carried there in the droppings of birds or on the coats of mammals.

These **early successional plant species (1)** grow close to the ground, **(2)** can establish large populations quickly under harsh conditions, and **(3)** have short lives. Some of their roots penetrate the rock and help break it up into more soil particles, and the decay of their wastes and dead bodies adds more nutrients to the soil.

After hundreds of years the soil may be deep and fertile enough to store enough moisture and nutrients to support the growth of less hardy **midsuccessional plant species** of herbs, grasses, and low shrubs. These, in turn, usually are replaced by trees that need lots of sunlight and are adapted to the area's climate and soil.

As these tree species grow and create shade, they are replaced by **late successional plant species** (mostly trees) that can tolerate shade. Unless fire, flooding, severe erosion, tree cutting, climate change, or other natural or human processes disturb the area, what was once bare rock becomes a complex forest community (Figure 7-13).

What Is Secondary Succession? Secondary succession begins in an area where the natural community of organisms has been disturbed, removed, or destroyed but the soil or bottom sediment remains. Candidates for secondary succession include **(1)** abandoned farmlands, **(2)** burned or cut forests, **(3)** heavily polluted streams, and **(4)** land that has been dammed or flooded. Because some soil or sediment is present, new vegetation usually can begin to grow within a few weeks. Seeds can be present in soils, or they can be carried from nearby plants by wind or by birds and animals.

In the central (Piedmont) region of North Carolina, European settlers cleared the mature native oak and hickory forests and replanted the land with crops. Some of the land was abandoned later because of erosion and loss of soil nutrients. Figure 7-14 shows how such abandoned farmland has undergone secondary succession.

Descriptions of ecological succession usually focus on changes in vegetation. However, these changes in turn affect food and shelter for various types of animals. Thus, as succession proceeds the numbers and types of animals and decomposers also change. Figure 7-15 shows some of the wildlife species likely to be found at various stages of secondary ecological succession in areas with a temperate climate.

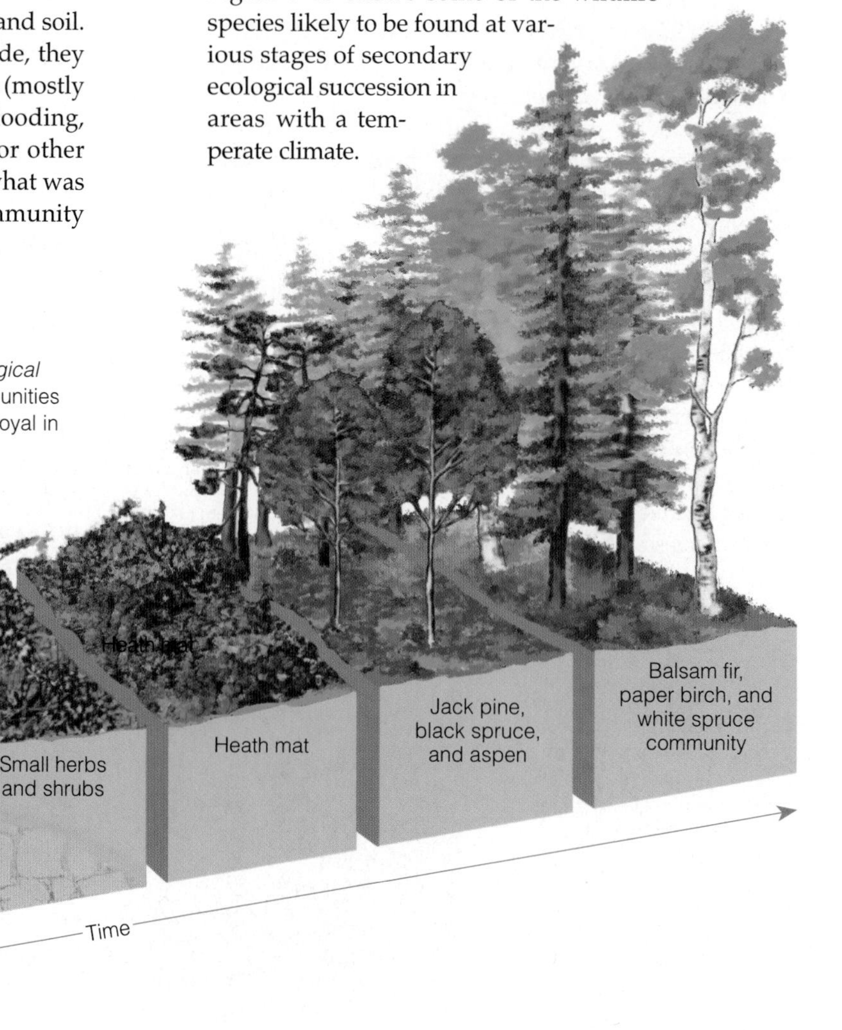

Figure 7-13 Starting from ground zero. *Primary ecological succession* over several hundred years of plant communities on bare rock exposed by a retreating glacier on Isle Royal in northern Lake Superior.

Because primary and secondary succession involve changes in community structure, it is not surprising that the various stages of succession have different patterns of species diversity, trophic structure, niches, nutrient cycling, and energy flow and efficiency, as shown by the work of ecologists such as Eugene Odum (Table 7-1).

How Do Disturbances Affect Succession and Species Diversity? A **disturbance** is a change in environmental conditions that disrupts an ecosystem or community. Such disturbances can be catastrophic or gradual and caused by natural or human-caused changes (Table 7-2). At any time during primary or secondary succession, disturbances such as those listed in Table 7-2 can convert a particular stage of succession to an earlier stage.

Many people think of all environmental disturbances as harmful processes. Large catastrophic disturbances (Table 7-2) can devastate communities and ecosystems. However, many ecologists contend that in the long run some types of disturbances such as fires can be beneficial for the species diversity of some communities and ecosystems. Such disturbances create new conditions that can discourage or eliminate some species but encourage others by releasing nutrients and creating unfilled niches.

According to the *intermediate disturbance hypothesis*, communities that experience fairly frequent but moderate disturbances have the greatest species diversity (Figure 7-16). It is hypothesized that in such communities, moderate disturbances are large enough to create openings for colonizing species in the disturbed areas but mild and infrequent enough to allow the survival of some mature species in undisturbed areas. Some field experiments have supported this hypothesis.

How Predictable Is Succession, and Is Nature in Balance? It is tempting to conclude that ecological succession is an orderly sequence in which each stage leads predictably to the next, more stable stage. According to this classic view, succession proceeds until an area is occupied by a generally predictable and stable type of *climax community* that is dominated by a few long-lived plant species and is in balance with its environment. This equilibrium model of succession is what ecologists meant when they talked about the *balance of nature*.

Over the last several decades many ecologists have changed their views about balance and equilibrium in nature. Under the old *balance-of-nature* view, a large terrestrial community undergoing succession was viewed as eventually being covered with a predictable green blanket of climax vegetation. However, a close look at almost any ecosystem or community

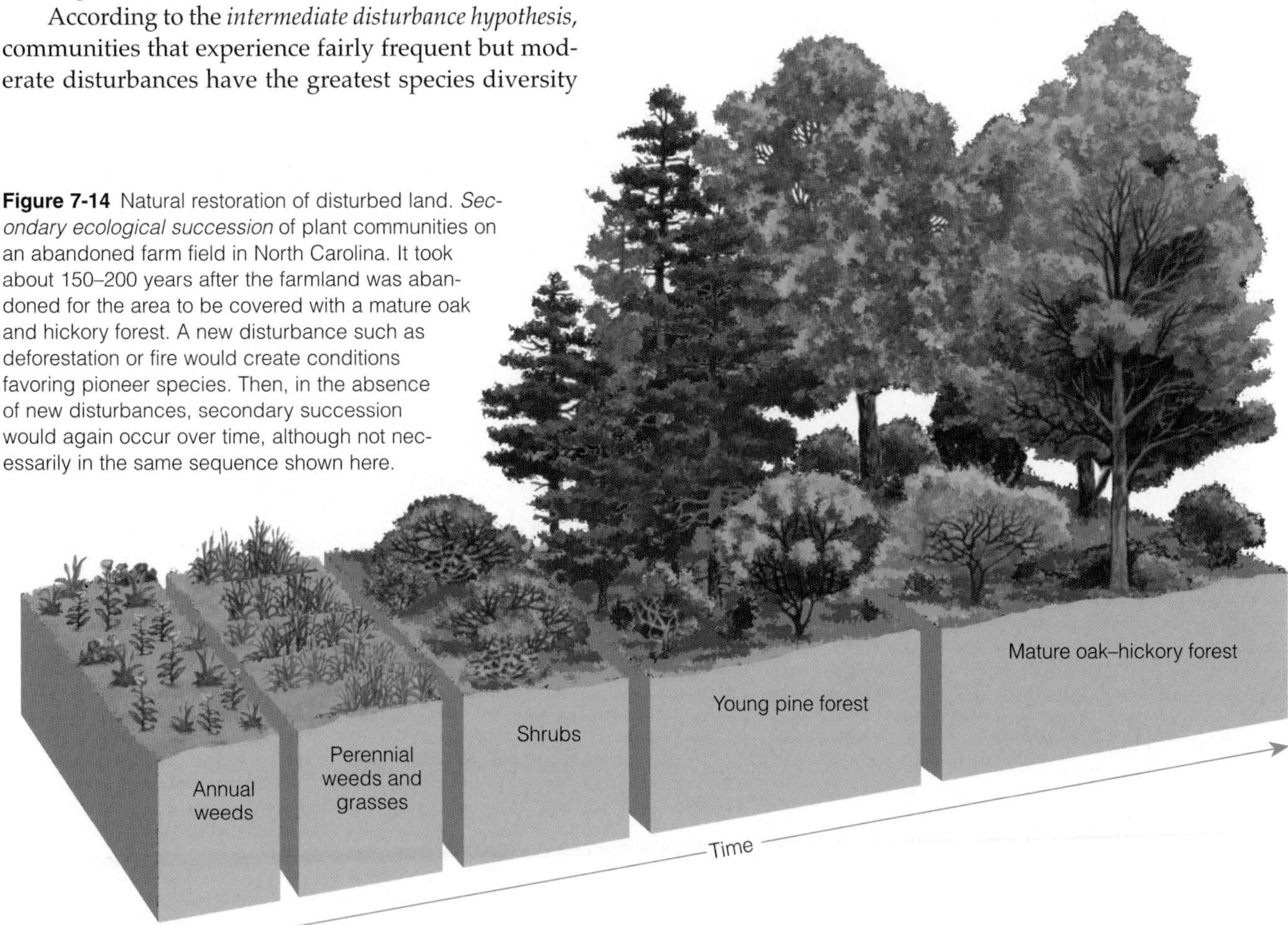

Figure 7-14 Natural restoration of disturbed land. *Secondary ecological succession* of plant communities on an abandoned farm field in North Carolina. It took about 150–200 years after the farmland was abandoned for the area to be covered with a mature oak and hickory forest. A new disturbance such as deforestation or fire would create conditions favoring pioneer species. Then, in the absence of new disturbances, secondary succession would again occur over time, although not necessarily in the same sequence shown here.

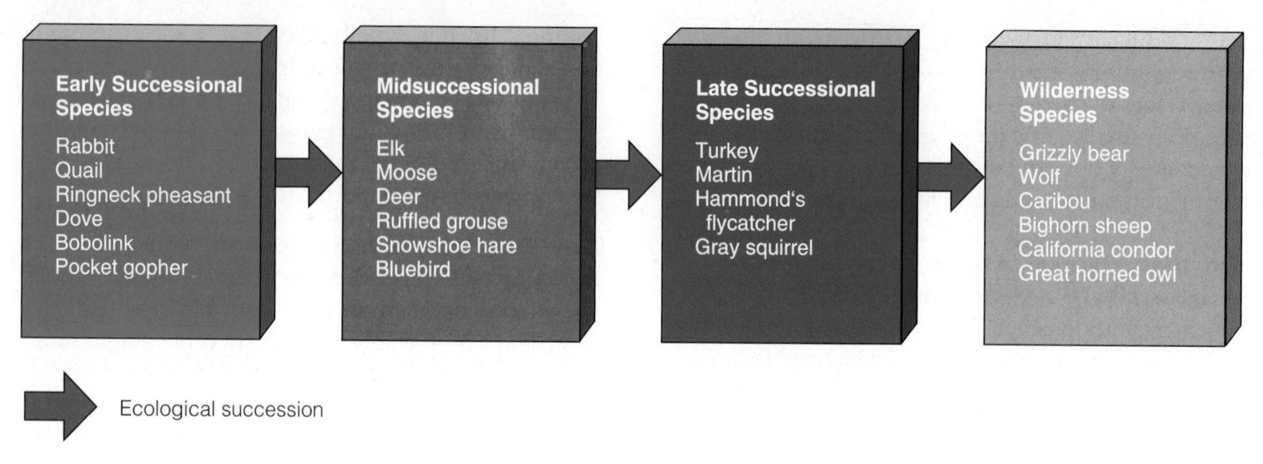

Figure 7-15 Examples of wildlife species typically found at different stages of ecological succession in areas of the United States with a temperate climate.

reveals that it consists of an ever-changing mosaic of vegetation patches at different stages of succession. These patches result from a variety of mostly unpredictable small and medium-sized disturbances (Figure 7-16). This irregular quilt of vegetation increases the diversity of plant and animal life and provides sites where early successional species can gain a foothold.

Such research indicates that *we cannot predict the course of a given succession or view it as preordained progress toward an ideally adapted climax community.* Rather, succession reflects the ongoing struggle by different species for enough light, nutrients, food, and space to **(1)** survive and **(2)** gain reproductive advantages over other species by occupying as much of their fundamental niches as possible.

This change in the way we view what is happening in nature explains why a growing number of ecologists prefer terms such as *biotic change* instead of *succession* (which implies an ordered and predictable sequence of changes). Many ecologists have

Table 7-1 Ecosystem Characteristics at Immature and Mature Stages of Ecological Succession

Characteristic	Immature Ecosystem (Early Successional Stage)	Mature Ecosystem (Late Successional Stage)
Ecosystem Structure		
Plant size	Small	Large
Species diversity	Low	High
Trophic structure	Mostly producers, few decomposers	Mixture of producers, consumers, and decomposers
Ecological niches	Few, mostly generalized	Many, mostly specialized
Community organization (number of interconnecting links)	Low	High
Ecosystem Function		
Biomass	Low	High
Net primary productivity	High	Low
Food chains and webs	Simple, mostly plant → herbivore with few decomposers	Complex, dominated by decomposers
Efficiency of nutrient recycling	Low	High
Efficiency of energy use	Low	High

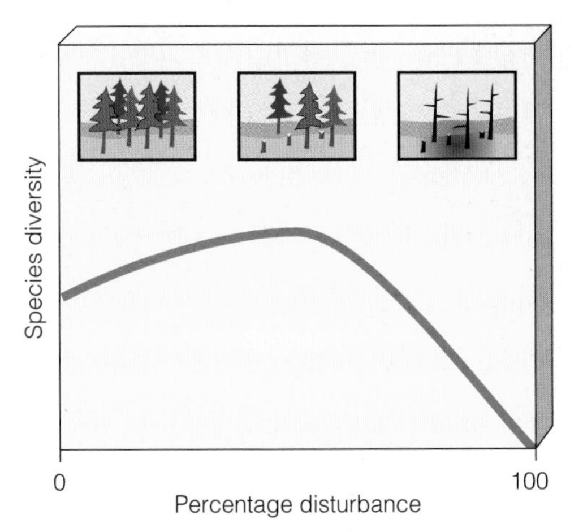

Figure 7-16 According to the *intermediate disturbance hypothesis*, moderate disturbances in communities promote greater species diversity than small or major disturbances.

also replaced the term *climax community* with terms such as *mature community* or a mosaic of *vegetation patches* at different stages of succession. Ecologists do not consider nature to be totally chaotic and unpredictable, but they have lowered their expectations in being able to make accurate predictions about the course of succession.

7-6 ECOLOGICAL STABILITY AND SUSTAINABILITY

What Is Stability? All living systems, from single-celled organisms to the biosphere, contain complex networks of negative and positive feedback loops (p. 59) that interact to provide some degree of stability or sustainability over each system's expected life span.

This stability is maintained only by constant dynamic change in response to changing environmental conditions. For example, in a mature tropical rain forest, some trees die and others take their places. However, unless the forest is cut, burned, or otherwise destroyed you will still recognize it as a tropical rain forest 50 or 100 years from now.

It is useful to distinguish between three aspects of stability or sustainability in living systems:

- **Inertia**, or **persistence**: the ability of a living system to resist being disturbed or altered
- **Constancy:** the ability of a living system such as a population to keep its numbers within the limits imposed by available resources
- **Resilience:** the ability of a living system to bounce back after an external disturbance that is not too drastic

Table 7-2 Changes Affecting Ecosystems

Catastrophic*		Gradual*	
Natural	Drought	Natural	Climatic changes
	Flood		Immigration
	Fire		Adaption and evolution
	Volcanic eruption		Ecological succession
	Earthquake		Disease
	Hurricane or tornado	Human-caused	Salinization and waterlogging of soils from irrigation
	Landslide		Soil compaction
	Change in stream course		Groundwater depletion
	Disease		Water and air pollution
Human-caused	Deforestation		Loss and degradation of wildlife habitat
	Overgrazing		"Pests" and predator elimination
	Plowing		Exotic species introduction
	Erosion		Overhunting and overfishing
	Pesticide application		Toxic contamination
	Fire		Urbanization
	Mining		Excessive tourism
	Toxic contamination		
	Urbanization		
	Water and air pollution		
	Loss and degradation of wildlife habitat		

*Many changes can be either catastrophic or gradual.

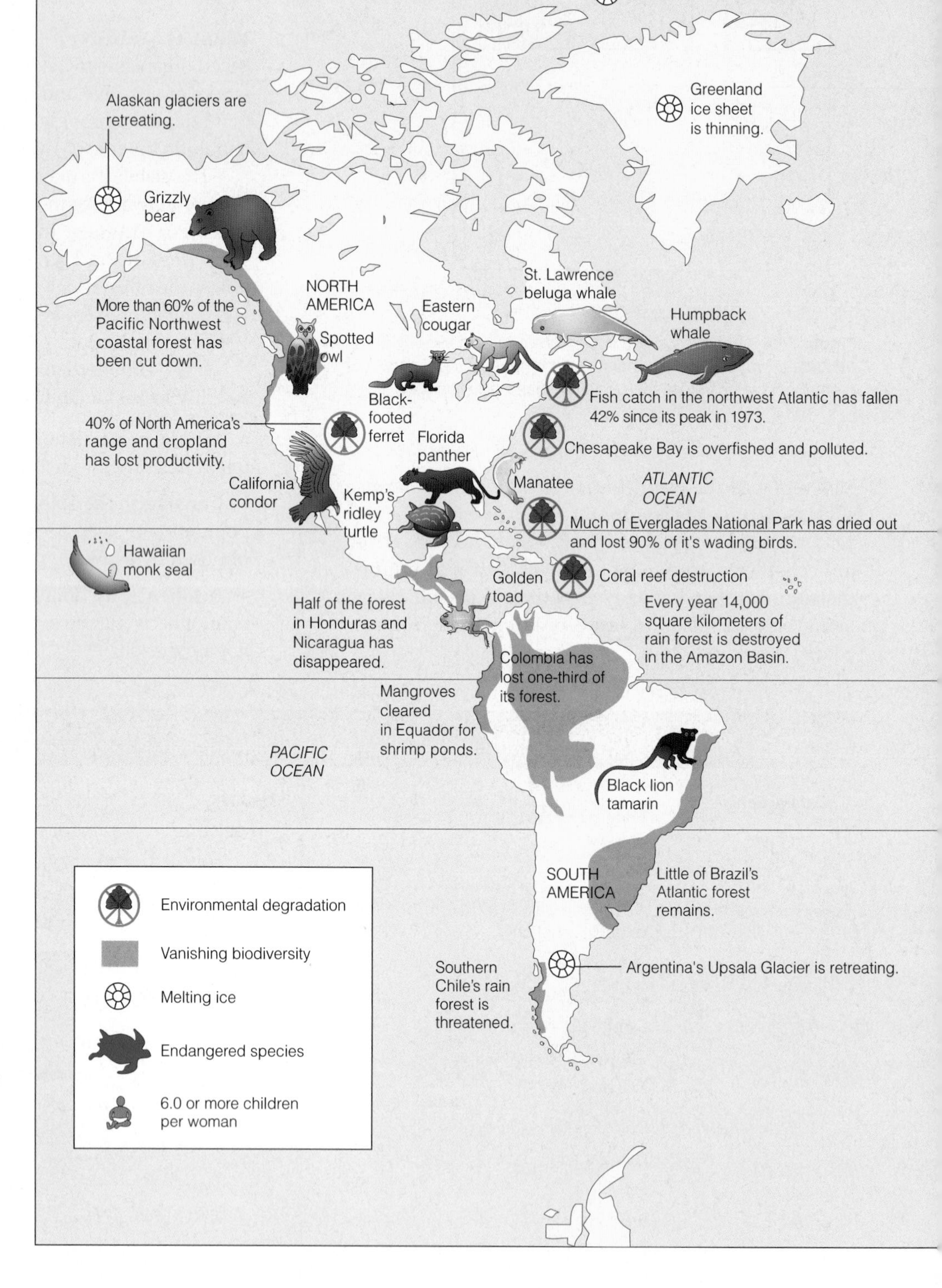

Figure 7-17 Examples of how some of the earth's natural resources are being depleted and degraded at an accelerating rate as a result of the exponential growth of the human population and resource use by humankind. (Data from the World Conservation Union, World Wildlife Fund, Conservation International, United Nations, Population Reference Bureau, U.S. Fish and Wildlife Service, and Daniel Boivin)

Does Species Diversity Increase Ecosystem Stability? In the 1960s, most ecologists believed that the greater the species diversity and the accompanying web of feeding and biotic interactions in an ecosystem, the greater its stability. According to this hypothesis, an ecosystem with a variety of species and feeding paths has more ways to respond to most environmental stresses because it does not have "all its eggs in one basket." However, most recent research indicates that there are exceptions to this intuitively appealing idea.

Because no ecosystem can function without some plants and decomposers, there is a minimum thresh-

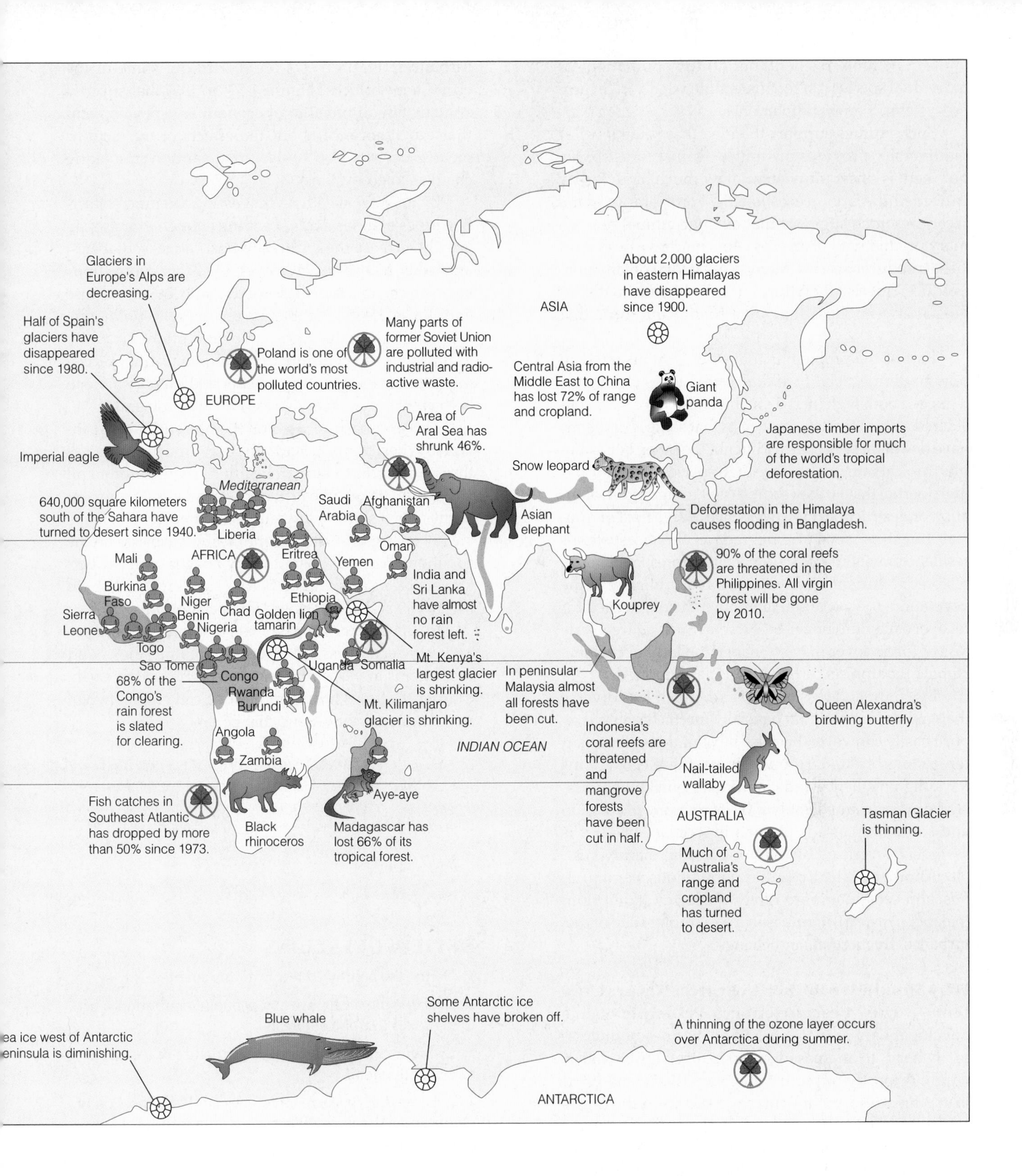

Glaciers in Europe's Alps are decreasing.
Half of Spain's glaciers have disappeared since 1980.
Imperial eagle
Poland is one of the world's most polluted countries.
EUROPE
Many parts of former Soviet Union are polluted with industrial and radio-active waste.
ASIA
About 2,000 glaciers in eastern Himalayas have disappeared since 1900.
Central Asia from the Middle East to China has lost 72% of range and cropland.
Giant panda
Area of Aral Sea has shrunk 46%.
Japanese timber imports are responsible for much of the world's tropical deforestation.
Snow leopard
Mediterranean
640,000 square kilometers south of the Sahara have turned to desert since 1940.
Saudi Arabia
Afghanistan
Asian elephant
Deforestation in the Himalaya causes flooding in Bangladesh.
Liberia
AFRICA
Eritrea
Oman
Mali
Yemen
India and Sri Lanka have almost no rain forest left.
90% of the coral reefs are threatened in the Philippines. All virgin forest will be gone by 2010.
Burkina Faso
Niger
Benin
Chad
Ethiopia
Kouprey
Sierra Leone
Nigeria
Golden lion tamarin
Togo
Sao Tome
Uganda
Somalia
Mt. Kenya's largest glacier is shrinking.
68% of the Congo's rain forest is slated for clearing.
Congo
Rwanda
Burundi
Mt. Kilimanjaro glacier is shrinking.
In peninsular Malaysia almost all forests have been cut.
Queen Alexandra's birdwing butterfly
Angola
Zambia
INDIAN OCEAN
Indonesia's coral reefs are threatened and mangrove forests have been cut in half.
Nail-tailed wallaby
Aye-aye
Fish catches in Southeast Atlantic has dropped by more than 50% since 1973.
Black rhinoceros
Madagascar has lost 66% of its tropical forest.
AUSTRALIA
Tasman Glacier is thinning.
Much of Australia's range and cropland has turned to desert.
Blue whale
Some Antarctic ice shelves have broken off.
A thinning of the ozone layer occurs over Antarctica during summer.
ea ice west of Antarctic eninsula is diminishing.
ANTARCTICA

old of species diversity below which ecosystems cannot function.

Beyond this, it is difficult to know whether simple ecosystems are less stable than complex ones or to identify the threshold below which complex ecosystems fail. In part because some species play redundant roles (niches) in ecosystems, we do not know how many or which species can be eliminated before the entire ecosystem begins to lose stability or collapse.

Recent research by ecologist David Tilman and other researchers indicates that **(1)** ecosystems with more species tend to have higher net primary productivities and can be more resilient, but **(2)** the populations of individual species can fluctuate more widely in diverse ecosystems than in simpler ones.

Such studies support the idea that some level of biodiversity provides insurance against catastrophe, but there is uncertainty about how much biodiversity is needed in various ecosystems. For example, some recent research suggests that average annual net primary productivity of an ecosystem reaches a peak with 10–40 producer species. Many ecosystems contain more producer species than this, but it is difficult to distinguish between those that are essential and those that are not.

Part of the problem is that ecologists disagree on how to define *stability* and *diversity*. Does an ecosystem need both high inertia and high resilience to be considered stable? Evidence suggests that some ecosystems have one of these properties but not the other. For example, tropical rain forests have high species diversity and high inertia; that is, they are resistant to significant alteration or destruction. However, once a large tract of tropical forest is severely degraded, the ecosystem's resilience sometimes is so low that the forest may not be restored. Nutrients (which are stored primarily in the vegetation, not in the soil) and other factors needed for recovery may no longer be present. Such a large-scale loss of forest cover may so change the local or regional climate that forests can no longer be supported.

By contrast, grasslands **(1)** are much less diverse than most forests and **(2)** have low inertia because they burn easily. However, because most of their plant matter is stored in underground roots, these ecosystems have high resilience and can recover quickly. A grassland can be destroyed only if **(1)** its roots are plowed up and something else is planted in its place or **(2)** it is severely overgrazed by livestock or other herbivores.

Another difficulty is that populations, communities, and ecosystems are rarely, if ever, at equilibrium (balance). Instead, nature is in a continuing state of disturbance, fluctuation, and change.

Why Should We Bother to Protect Natural Systems? The Precautionary Principle Some developers argue that if biodiversity does not necessarily lead to increased ecological stability and if nature is mostly unpredictable, then there is no point in trying preserve and manage old-growth forests and other ecosystems. Why not **(1)** cut down diverse old-growth forests, use the timber resources, and replace the forests with simplified tree farms, **(2)** convert the world's grasslands to cropfields (Figure 6-20, p. 136), **(3)** drain and develop inland wetlands, and **(4)** dump our toxic and radioactive wastes into the deep ocean? Why worry about the premature extinction of species?

Ecologists and conservation biologists point out that there is overwhelming evidence that human disturbances (Figure 7-17) are disrupting some of the ecosystem services (Figure 4-33, p. 103) that support and sustain all life and all economies. They contend that our ignorance about the effects of our actions means that we need to use great caution in making changes to ecosystems.

By analogy, we know that eating too much of certain types of foods and not getting enough exercise can greatly increase the chances of heart attacks, diabetes, and other disorders. However, the exact connections between chemicals in these foods and between exercise and these health problems are largely unknown. Instead of using this uncertainty and unpredictability as an excuse to continue overeating and not exercising, the wise course is to eat better and exercise more to help *prevent* potentially serious health problems.

This approach is based on the **precautionary principle**: When there is evidence that an activity raises threats of harm to human health or the environment, we should take precautionary measures to prevent harm even if some of the cause and effect relationships are not fully established scientifically. It is based on the commonsense idea behind many adages such as "Better safe than sorry," Look before you leap," "First, do no harm," and "Slow down for speed bumps."

In this chapter we have seen that interdependence and connectedness of species, communities, and ecosystems are essential features of life on the earth. Chapter 8 shows how populations can change when environmental conditions change.

The old idea of a static landscape, like a single musical chord sounded forever, must be abandoned, for such a landscape never existed except in our imagination. Nature undisturbed by human influence seems more like a symphony whose harmonies arise from variation and change over every interval of time.

DANIEL B. BOTKIN

REVIEW QUESTIONS

1. Define the boldfaced terms in this chapter.

2. Describe the ecological and economic importance of flying foxes in tropical forests.

3. List four characteristics of the structure of a community or ecosystem.

4. Describe the types, relative sizes, and stratification of plants and animals in **(a)** a tropical rain forest, **(b)** an ocean, and **(c)** a lake.

5. Define and give an example of an *edge effect*. What are the advantages and disadvantages of edge effects?

6. What are the three most species-rich environments? How does species diversity vary with **(a)** latitude in terrestrial communities and **(b)** pollution in aquatic systems?

7. What two factors determine the species diversity found on an isolated ecosystem such as an island? What is the *theory of island biogeography*? How do the size of an island and its distance from a mainland affect its species diversity?

8. Distinguish between *native, nonnative, indicator,* and *keystone* species and give an example of each.

9. Why are birds good indicator species? Explain why amphibians are considered to be indicator species and list reasons for declines in their populations.

10. Describe the keystone ecological roles of **(a)** flying foxes, **(b)** alligators, and **(c)** some shark species. What can happen in an ecosystem that loses a keystone species?

11. Distinguish between *intraspecific competition* and *interspecific competition* and give an example of each.

12. What are four options when the niches of two species competing in the same area overlap to a large degree?

13. Define and give two examples of *resource partitioning.* How does it allow species to avoid overlap of their fundamental niches?

14. What is *predation*? Describe the *predator–prey relationship* and give two examples of this type of species interaction.

15. Give two examples of how predators increase their chances of finding prey by **(a)** pursuit and **(b)** ambush.

16. List six ways (adaptations) used by prey to avoid their predators and give an example of each type.

17. Define and give two examples of *parasitism,* and explain how it differs from predation. What is the ecological importance of parasitism?

18. Define and give two examples of **(a)** *mutualism* and **(b)** *commensalism.*

19. Distinguish between *primary succession* and *secondary succession.* Distinguish between *pioneer* (or *early successional*) *species, midsuccessional plant species,* and *late successional plant species.*

20. Give three examples of environmental disturbances and explain how they can affect succession. How can some disturbances be beneficial to ecosystems? What is the *intermediate disturbance hypothesis?* Explain how occasional fires can promote succession and benefit species in some types of ecosystems.

21. Explain why most ecologists contend that **(a)** the details of succession are not predictable and **(b)** there is no *balance of nature.*

22. Distinguish between *inertia, constancy,* and *resilience* and explain how they help maintain stability in an ecosystem.

23. Does high species diversity always increase ecosystem stability? Explain.

24. What is the *precautionary principle,* and why do many scientists believe that it is a useful strategy for dealing with the environmental problems we face?

CRITICAL THINKING

1. Why do deer hunters sometimes plant corn on strips of open land used for firebreaks or telephone poles?

2. Why are more tree species per hectare usually found in tropical forests than in temperate forests?

3. How would you respond to someone who claims that it is not important to protect areas of temperate and polar biomes because most of the world's biodiversity is in the tropics?

4. What two factors determine the number of different species found on an island? Why is the species diversity of a large island usually higher than that on a smaller island?

5. How would you determine whether a particular species found in a given area is a keystone species?

6. How would you reply to someone who argues that **(a)** we should not worry about our effects on natural systems because succession will heal the wounds of human activities and restore the balance of nature, **(b)** if nature is unpredictable, we should not bother to preserve any natural systems, and **(c)** because there is no balance in nature and no stability in species diversity, we should cut down diverse, old-growth forests and replace them with tree farms?

PROJECTS

1. Use the library or internet to find and describe two species not discussed in this textbook that are engaged in **(a)** a commensalistic interaction, **(b)** a mutualistic interaction, and **(c)** a parasite–host relationship.

2. Visit a nearby natural area and identify examples of **(a)** mutualism and **(b)** resource partitioning.

3. Use the library or internet to identify the parasites likely to be found in your body.

4. Visit a nearby land area such as a partially cleared or burned forest or grassland or an abandoned cropfield and record signs of secondary ecological succession. Study the area carefully to see whether you can find patches that are at different stages of succession because of various disturbances.

5. Make a concept map of this chapter's major ideas, using the section heads and subheads and the key terms (in boldface). Look on the website for this book for information about making concept maps.

INTERNET STUDY RESOURCES AND RESOURCES FOR FURTHER READING AND RESEARCH

The website for this book contains helpful study aids and many ideas for further reading and research. Log on to

http://www.brookscole.com/product/0534389872s

and click on the Chapter-by-Chapter area. Choose Chapter 7 and select a resource:

- "Flash Cards" allows you to test your mastery of the Terms and Concepts to Remember for this chapter.
- "Tutorial Quizzes" provides a multiple-choice practice quiz.
- "Student Guide to InfoTrac" will lead you to Critical Thinking Projects that use InfoTrac College Edition as a research tool.

- "References" lists the major books and articles consulted in writing this chapter.
- "Hypercontents" takes you to an extensive list of sites with news, research, and images related to individual sections of the chapter.

INFOTRAC COLLEGE EDITION

Improve your skills with InfoTrac College Edition, a searchable online database of articles from more than 700 periodicals. Log on to

http://www.infotrac-college.com

or access InfoTrac through the website for this book. Try to find the following articles:

1. Species Interactions in Intertidal Food Webs: Prey or Predation Regulation of Intermediate Predators? (Statistical Data Included). Sergio A. Navarrete, Bruce A. Menge, Bryon A. Daley. *Ecology*, August 2000 v81 i8 p2264. This article investigates the relative importance of direct predation and competition by a top predator on an intermediate predator. *Hint*: Enter the search term "intertidal food" using the keyword "Oregon."

2. Complementary Foraging Behaviors Allow Coexistence of Two Consumers. W. G. Wilson, C. W. Osenberg, R. J. Schmitt, R. M. Nisbet. *Ecology*, Oct 1999 v80 i7 p2358. This research presents a mathematical model for an exploitation competition system. *Hint*: Enter the search term "population dynamics" using the keyword "model."

8 POPULATION DYNAMICS, CARRYING CAPACITY, AND CONSERVATION BIOLOGY

Two Islands: Can We Treat This One Better?

Easter Island (Rapa Nui) is a small, isolated island in the great expanse of the South Pacific. It was first colonized by Polynesians about 2,500 years ago.

The civilization they developed was based on the island's towering palm trees, which were used for shelter, tools, fishing boats, fuel, food, rope, and clothing. Using these resources, they developed an impressive civilization and a technology capable of making and moving large stone structures, including their famous statues (Figure 8-1).

The people flourished, with the population peaking at about 10,000 (with estimates ranging from 7,000 to 20,000) by 1400. However, they used up the island's precious trees faster than they were regenerated—an example of the tragedy of the commons (p. 11). Each person who cut a tree reaped immediate personal benefits while helping to doom the civilization in the long run.

Once the trees were gone the islanders could not build canoes for hunting porpoises and catching fish. Without the forest to absorb and slowly release water, **(1)** springs and streams dried up, **(2)** exposed soils eroded, **(3)** crop yields plummeted, and **(4)** famine struck.

The starving people turned to warfare and possibly cannibalism. Both the population and the civilization collapsed. When Dutch explorers first reached the island on Easter Day, 1722, they found only about 2,000 inhabitants, struggling under primitive conditions on a mostly barren island.

Like Easter Island at its peak, the earth is an isolated island (in the vastness of space) with no other suitable planet to migrate to. As on Easter Island, **(1)** our population is growing, and **(2)** we are consuming exhaustible and renewable resources and threatening the existence of many wild species at a rapid pace.

Will the humans on Earth Island re-create the tragedy of Easter Island on a grander scale, or will we learn how to live more sustainably on this planet that is our only home? Answering this question requires that we understand *population dynamics, conservation biology,* and *human impacts on the earth's life-support systems,* as discussed in this chapter.

Figure 8-1 These massive stone figures on Easter Island are the remains of the technology created by an ancient civilization of Polynesians. This civilization collapsed because the people used up the trees (especially large palm trees) that were the basis of their livelihood. More than 200 of these stone statues once stood on huge stone platforms lining the coast. At least 700 additional statues were abandoned in rock quarries or on ancient roads between the quarries and the coast. It is presumed that the islanders moved these large structures by felling large trees and using them to roll and erect the statues.

In looking at nature . . . never forget that every single organic being around us may be said to be striving to increase its numbers.

CHARLES DARWIN, 1859

This chapter addresses the following questions:

- How do populations change in size, density, and makeup in response to environmental stress?
- What is the role of predators in controlling population size?
- What different reproductive patterns do species use to enhance their survival?
- What is conservation biology?
- What impacts do human activities have on populations, communities, and ecosystems?
- How can we live more sustainably?

8-1 POPULATION DYNAMICS AND CARRYING CAPACITY

What Are the Major Characteristics of a Population? Populations are dynamic and change in response to environmental stress or changes in environmental conditions. They change in **(1)** *size* (number of individuals), **(2)** *density* (number of individuals in a certain space), **(3)** *dispersion* (spatial pattern such as clumping, uniform dispersion, or random dispersion, depending mostly on resource availability; Figure 8-2), and **(4)** *age distribution* (proportion of individuals of each age in a population). These changes, called **population dynamics**, occur in response to **(1)** environmental stress (Table 7-2, p. 175) or **(2)** changes in environmental conditions.

What Limits Population Growth? Four variables—*births, deaths, immigration,* and *emigration*—govern changes in population size. A population gains individuals by birth and immigration and loses them by death and emigration:

Population change = (Births + Immigration) – (Deaths + Emigration)

These variables depend on changes in resource availability or on other environmental changes (Figure 8-3). If the number of individuals added from births and immigration equals the number lost to deaths and immigration, then there is **zero population growth.**

Populations vary in their capacity for growth, also known as the **biotic potential** of the population. The **intrinsic rate of increase (r)** is the rate at which a population would grow if it had unlimited resources. Generally, individuals in populations with a high intrinsic rate of increase **(1)** *reproduce early in life,* **(2)** *have short generation times* (the time between successive generations), **(3)** *can reproduce many times* (have a long reproductive life), and **(4)** *have many offspring each time they reproduce.*

No population can grow indefinitely. In the real world, a rapidly growing population reaches some size limit imposed by a shortage of one or more limiting factors, such as light, water, space, or nutrients. *There are always limits to population growth in nature.*

Environmental resistance consists of all the factors acting jointly to limit the growth of a population. The population size of a species in a given place and time is determined by the interplay between its biotic potential and environmental resistance (Figure 8-3).

Together biotic potential and environmental resistance determine the **carrying capacity (K)**, the number of individuals of a given species that can be sustained indefinitely in a given space (area or volume).

The intrinsic rate of increase (r) of many species depends on having a certain minimum population size, called the **minimum viable population (MVP).** If a population declines below the MVP needed to support a breeding population, **(1)** certain individuals may not be able to locate mates, **(2)** genetically related individ-

Clumped (elephants)

Uniform (creosote bush)

Random (dandelions)

Figure 8-2 Generalized *dispersion patterns* for individuals in a population throughout their habitat. The most common pattern is one in which members of a population exist in clumps throughout their habitat (left), mostly because resources usually are found in patches.

uals may interbreed and produce weak or malformed offspring, and **(3)** the genetic diversity may be too low to enable adaptation to new environmental conditions. Then the intrinsic rate of increase falls and extinction is likely.

What Is the Difference Between Exponential and Logistic Population Growth?

A population that has few if any resource limitations grows exponentially. *Exponential growth* starts out slowly and speeds up as the population increases. If the number of individuals is plotted against time, this sequence yields a *J*-shaped exponential growth curve (Figure 8-4a).

Logistic growth involves exponential population growth when the population is small and a steady decrease in population growth with time as the population encounters environmental resistance and approaches the carrying capacity of its environment. After leveling off, a population with this type of growth typically fluctuates slightly above and below the carrying capacity. A plot of the number of individuals against time yields a sigmoid or *S*-shaped logistic growth curve (Figure 8-4b). A classic case of logistic growth involves the increase of the sheep population on the island of Tasmania, south of Australia, in the early 19th century (Figure 8-5).

POPULATION SIZE

0

Growth factors (biotic potential)	Decrease factors (environmental resistance)
Abiotic	**Abiotic**
Favorable light	Too much or too little light
Favorable temperature	Temperature too high or too low
Favorable chemical environment (optimal level of critical nutrients)	Unfavorable chemical environment (too much or too little of critical nutrients)
Biotic	**Biotic**
High reproductive rate	Low reproductive rate
Generalized niche	Specialized niche
Adequate food supply	Inadequate food supply
Suitable habitat	Unsuitable or destroyed habitat
Ability to compete for resources	Too many competitors
Ability to hide from or defend against predators	Insufficient ability to hide from or defend against predators
Ability to resist diseases and parasites	Inability to resist diseases and parasites
Ability to migrate and live in other habitats	Inability to migrate and live in other habitats
Ability to adapt to environmental change	Inability to adapt to environmental change

Figure 8-3 Factors that tend to increase or decrease the size of a population. Whether the size of a population grows, remains stable, or decreases depends on interactions between its growth factors (*biotic potential*) and decrease factors (*environmental resistance*).

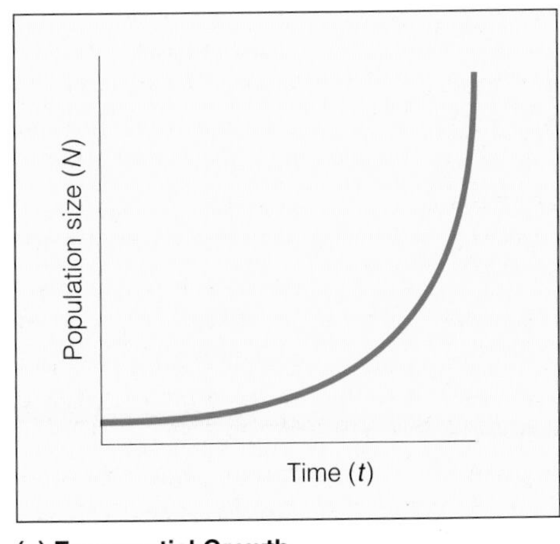

(a) Exponential Growth

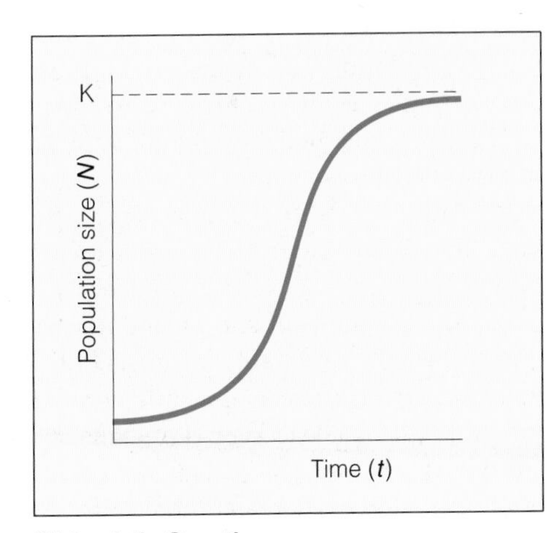

(b) Logistic Growth

Figure 8-4 Theoretical population growth curves. **(a)** *Exponential growth*, in which the population's growth rate increases with time. Exponential growth occurs when resources are not limiting and a population can grow at its *intrinsic rate of increase* (r). Exponential growth of a population cannot continue forever because eventually some factor limits population growth. **(b)** *Logistic growth*, in which the growth rate decreases as the population gets larger. With time, the population size stabilizes at or near the *carrying capacity* (K) of its environment.

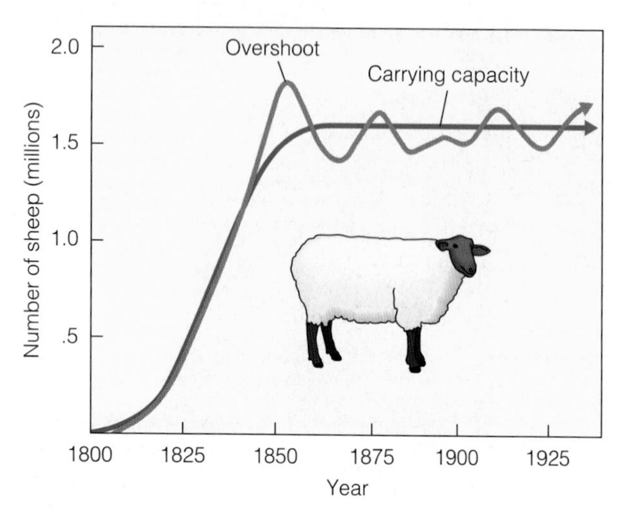

Figure 8-5 *Logistic growth* of a sheep population on the island of Tasmania between 1800 and 1925. After sheep were introduced in 1800 their population grew exponentially because of ample food. By 1855 they overshot the land's carrying capacity. Their numbers then stabilized and fluctuated around a carrying capacity of about 1.6 million sheep.

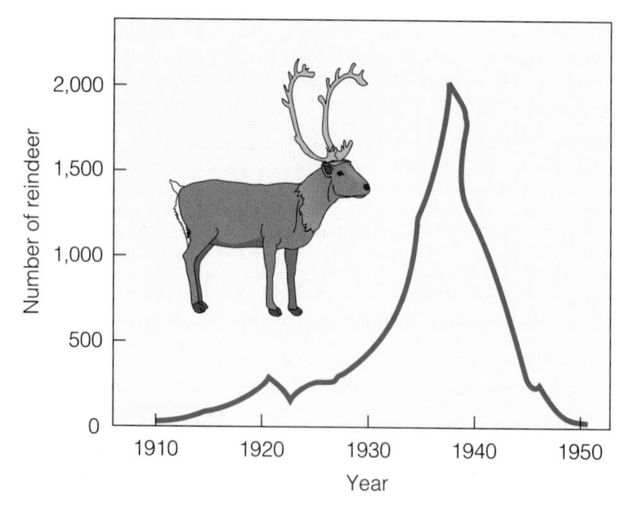

Figure 8-6 *Exponential growth*, overshoot, and population crash of reindeer introduced to a small island off the southwest coast of Alaska. When 26 reindeer (24 of them female) were introduced in 1910, lichens, mosses, and other food sources were plentiful. By 1935 the herd's population had soared to 2,000, overshooting the island's carrying capacity. This led to a population crash, with the herd plummeting to only 8 reindeer by 1950.

What Happens if the Population Size Exceeds the Carrying Capacity? The populations of some species do not make a smooth transition from exponential growth to logistic growth. Instead, such a population uses up its resource base and temporarily *overshoots* or exceeds the carrying capacity of its environment. This overshoot occurs because of a *reproductive time lag*, the period needed for the birth rate to fall and the death rate to rise in response to resource overconsumption.

In such cases the population suffers a *dieback* or *crash* unless the excess individuals switch to new resources or move to an area with more favorable conditions. A classic case of such a population crash occurred when reindeer were introduced in 1910 onto a small island off the southwest coast of Alaska (Figure 8-6).

Humans are not exempt from overshoot and dieback, as shown by the tragedy on Easter Island (p. 181). Ireland also experienced a population crash after a fungus destroyed the potato crop in 1845. About 1 million people died, and 3 million people emigrated to other countries.

Technological, social, and other cultural changes have extended the earth's carrying capacity for the human species. We have increased food production and used large amounts of energy and matter resources to make normally uninhabitable areas of the earth habitable. However, there is growing concern about how long we will be able to keep doing this on a planet with a finite size and resources with an exponentially growing population and per capita resource use.

How Does Population Density Affect Population Growth? *Density-independent population controls* affect a population's size regardless of its population density. Examples include floods, hurricanes, severe drought, unseasonable weather, fire, habitat destruction (such as clearing a forest of its trees or filling in a wetland), and pesticide spraying. For example, a severe freeze in late spring can kill many individuals in a plant population, regardless of its density.

Some limiting factors have a greater effect as a population's density increases. Examples of such *density-dependent population controls* are competition for resources, predation, parasitism, and disease.

Infectious disease is a classic example of density-dependent population control. An example is the *bubonic plague*, which swept through Europe during the 14th century. The bacterium that causes this disease normally lives in rodents. It was transferred to humans by fleas that fed on infected rodents and then bit humans. The disease spread like wildfire through crowded cities, where sanitary conditions were poor and rats were abundant. At least 25 million people in European cities died from the disease.

Health scientists are becoming increasingly alarmed about the possibility of new epidemics of common infectious diseases in crowded urban areas. The primary reason is that many common strains of disease-causing bacteria are becoming genetically resistant to most existing antibiotics (Spotlight, p. 233).

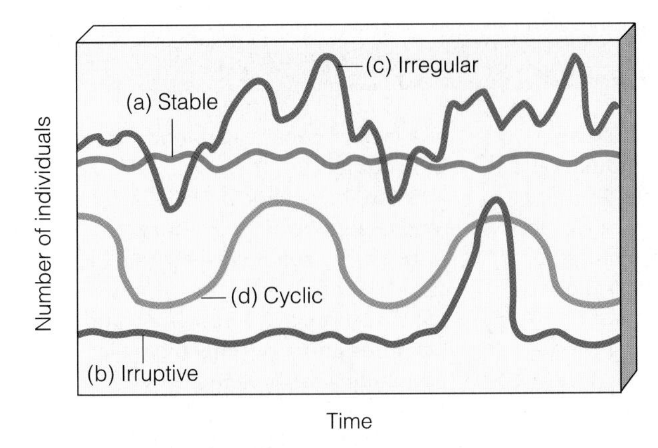

Figure 8-7 General types of simplified *population change curves* found in nature. **(a)** The population size of a species with a fairly *stable* population fluctuates slightly above and below its carrying capacity. **(b)** The populations of some species may occasionally explode, or *irrupt*, to a high peak and then crash to a more stable lower level. **(c)** The population sizes of some species change irregularly for mostly unknown reasons. **(d)** Other species undergo sharp increases in their numbers, followed by crashes over fairly regular time intervals. Predators sometimes are blamed, but the actual causes of such boom–bust cycles are poorly understood.

What Kinds of Population Change Curves Do We Find in Nature? In nature we find that over time species have four general types of *population fluctuations*: stable, irruptive, irregular, and cyclic (Figure 8-7). A species whose population size fluctuates slightly above and below its carrying capacity is said to have a fairly stable population size (Figures 8-5 and 8-7a). Such stability is characteristic of many species found in undisturbed tropical rain forests, where there is little variation in average temperature and rainfall.

Some species, such as the raccoon and feral house mouse, normally have a fairly stable population that may occasionally explode, or *irrupt*, to a high peak, and then crash to a more stable lower level or in some cases to a very low level (Figures 8-6 and 8-7b). The population explosion is caused by some factor that temporarily increases carrying capacity for the population, such as more food or fewer predators.

Some populations exhibit what appears to be irregular *chaotic behavior* in their changes in population size, with no recurring pattern (Figure 8-7c). Some scientists attribute such behavior to chaos in such systems. Other scientists contend that the behavior may be caused by orderly, nonchaotic behavior whose details and interactions are poorly understood.

The fourth type consists of cyclic fluctuations of population size that occur over a regular time period (Figure 8-7d) for poorly understood reasons. Examples are **(1)** lemmings, whose populations rise and fall every 3–4 years and **(2)** grouse, lynx, and snowshoe hares, whose populations generally rise and fall on a 10-year cycle.

8-2 THE ROLE OF PREDATION IN CONTROLLING POPULATION SIZE

Do Predators Control Population Size? The Lynx–Hare Cycle Some species that interact as predator and prey undergo cyclic changes in their numbers, with sharp increases in their numbers followed by seemingly periodic crashes (Figure 8-8). Predator–prey interactions often are blamed, but the actual causes of such *predator–prey cycles* are poorly understood.

For decades, predation has been the explanation for the correlation and time lag between the 10-year population cycles of the snowshoe hare and its predator, the Canadian lynx (Figure 8-8). According to this *top-down control* hypothesis, lynx preying on hares periodically reduce their population. The shortage of hares then reduces the lynx population, which allows the hare population to build up again. At some point the lynx population increases to take

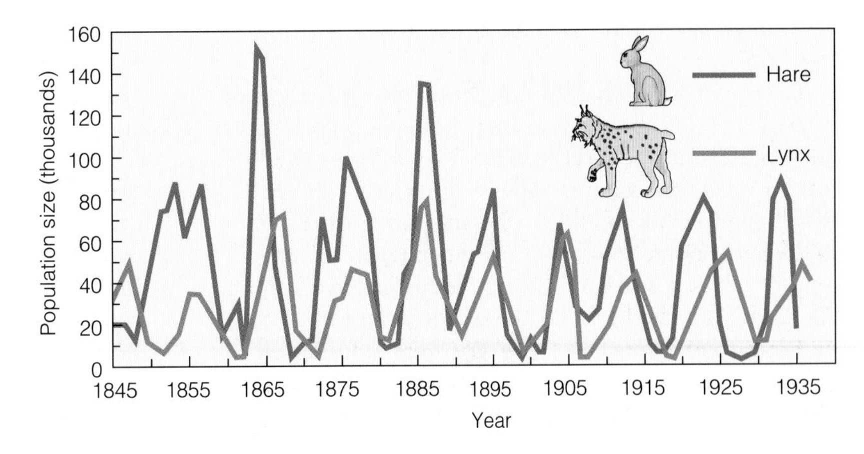

Figure 8-8 Population cycles for the snowshoe hare and Canadian lynx. At one time it was widely believed that these curves provided circumstantial evidence that these predator and prey populations regulated one another. More recent research suggests that the periodic swings in the hare population are caused by a combination of **(1)** predation by lynx and other predators (*top-down population control*) and **(2)** changes in the availability of the food supply for hares, with the rise of hare population helping to determine the lynx population (*bottom-up population control*). (Data from D. A. MacLulich)

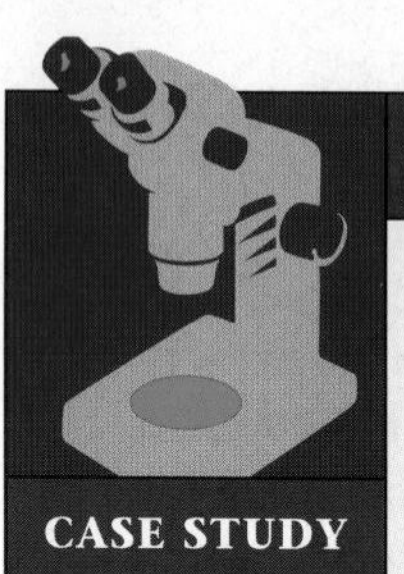

Wolf and Moose Interactions on Isle Royale

CASE STUDY

For decades wildlife biologists have been studying the relationship between the moose and wolf populations on Isle Royale, an island in Lake Superior between Minnesota in the United States and Ontario in Canada.

In the early 1900s, a small herd of moose wandered across the frozen ice of Lake Superior to this island. With an abundance of food, the moose population exploded (Figure 8-9). In 1928, a wildlife biologist visiting the island correctly predicted that the large moose population would soon crash because the moose had stripped the island of most of their preferred food plants.

Sometime during the 1940s timber wolves (probably a single pair) reached Isle Royale by traveling over the ice from the Canadian mainland during winter. They reproduced and slowly grew in numbers. During winter the wolves hunt in packs and concentrate on killing the old, sick, and young moose. The moose is the wolf's largest and most dangerous prey, and these individuals are the easiest to kill without undue risk. Once a target moose is selected, the wolves encircle it and try to get it to run so they can attack it from behind.

Since 1958, wildlife biologists have been tracking the populations of the two species (Figure 8-9). You might think that the wolves would have completely exterminated the moose, but instead the two species have been interacting in what appears to be an oscillating predator–prey cycle (Figure 8-9). If the wolves could drive the moose to extinction they probably would, but the moose are too formidable for this to happen.

Since 1980 the wolf population declined from a high of about 50 and has fluctuated between 12 and 25 individuals. Possible reasons for this decline are **(1)** a canine virus introduced to wolves by dogs and **(2)** a low reproduction rate because of a lack of genetic variability from inbreeding.

With the decline in wolves, the moose population rose sharply until 1995. Then it crashed from a combination of lack of food, poor reproduction, a severe winter, and a tick infestation. By 1999 the wolf population, with plenty of weakened prey, had grown to 25. If their population continues to grow, they may hold the moose numbers in check and allow damaged vegetation to recover and begin a new cycle of interactions.

Critical Thinking

What is the primary ecological lesson to be learned from the moose–wolf interaction on Isle Royale?

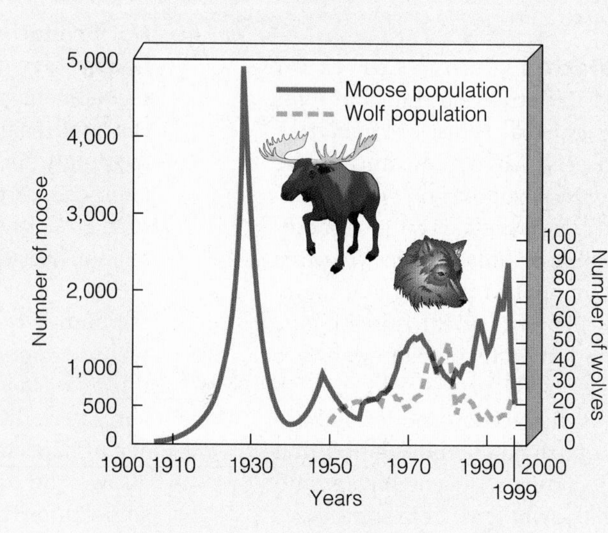

Figure 8-9 Changes in moose and wolf populations on Isle Royale from 1900 to 1999. (Data from Rolf O. Peterson)

advantage of the increased supply of hares, starting the cycle again.

Recent research has cast doubt on this appealing hypothesis because snowshoe hare populations have been found to have similar 10-year boom-and-bust cycles on islands where lynx are absent. Researchers now hypothesize that the periodic crashes in the hare population may occur when **(1)** large numbers of hares consume food plants faster than they can be replenished and **(2)** the quantity and quality of their food decrease. Once the hare population crashes, the plants recover, and the hare population begins rising again in a hare–plant cycle. If this *bottom-up control* hypothesis is correct, instead of lynx controlling hare populations, the changing hare population size may be causing fluctuations in the lynx population.

According to biologists, there are genuine cases of top-down control by predators in a number of ecosystems. Examples are **(1)** wolves controlling deer populations and moose populations (Case Study, above), **(2)** large predatory fish controlling other fish populations in lakes (Connections, right), **(3)** sheep and rabbits controlling plant growth in a pasture, and **(4)** sharks (Connections, p. 170) and alligators (Connections, p. 165) controlling some fish populations.

Can Lake Victoria Be Saved?

Lake Victoria, shared by Kenya, Tanzania, and Uganda in East Africa, is the world's second largest freshwater lake and the source of the Nile River, which drains into the Mediterranean Sea. The lake has been in ecological trouble for more than two decades, and things are getting worse.

Until the early 1980s, Lake Victoria had more than 350 species of fish found nowhere else. About 80% of them were small, algae-eating fishes known as cichlids (pronounced "SIK-lids," Figure 6-40, p. 153), each with a slightly different ecological niche. These fishes were the main source of protein for the more than 30 million people living in the area surrounding the lake and provided a fishing livelihood for many local people.

Currently, only one native minnow species and two introduced fish species dominate the lake. All the remaining fish species are endangered or extinct.

Several factors played a role in this dramatic loss of aquatic biodiversity:

- A large increase in the population of the Nile perch, which was deliberately introduced into the lake in 1960 to stimulate local economies and the fishing industry, despite the protests of some biologists. The population of this large, prolific, and voracious fish exploded; they preyed on the cichlids and by 1985 had wiped out most of these species. The native people who depended on the cichlids for protein cannot afford the perch, and the mechanized fishing industry has put most small-scale fishers and fish vendors out of business. This has increased poverty and protein malnutrition.
- In the 1980s, the lake began experiencing frequent algal blooms because of **(1)** nutrient runoff from surrounding farms, **(2)** deforested land, **(3)** untreated sewage, and **(4)** declines in the populations of the algae-eating cichlids. This **(1)** greatly decreased oxygen levels in the lower depths of the lake and **(2)** drove remaining native cichlids and other fish species to shallower waters, where they were more vulnerable to Nile perch and fishing nets.
- Since 1987, the nutrient-rich lake has been invaded by the water hyacinth (Figure 6-40, p. 153). This rapidly growing plant now carpets large areas of the lake and **(1)** blocks out sunlight, **(2)** deprives fish and plankton of oxygen, **(3)** hinders the movement of small fishing boats, and **(4)** creates stagnant water that is the breeding ground for malaria-spreading mosquitoes and snails that host bilharzia (a human parasite that attacks the liver, lungs, and eyes). Men are vacating villages in search of jobs, often leaving behind women and children who face severe poverty, disease, and protein malnutrition.

Critical Thinking

Congratulations. You have been put in charge of saving Lake Victoria. List the three major components of your strategy.

8-3 REPRODUCTIVE PATTERNS AND SURVIVAL

What Are Opportunist or r-Selected Species?

Each species has a characteristic mode of reproduction. At one extreme are species that reproduce early and put most of their energy into reproduction. They **(1)** have many (usually small) offspring each time they reproduce, **(2)** reach reproductive age rapidly, **(3)** have short generation times, **(4)** give their offspring little or no parental care or protection to help them survive, and **(5)** are short-lived (usually with a life span of less than a year). Species with this reproductive pattern overcome the massive loss of their offspring by producing so many unprotected young that a few will survive to reproduce many offspring to begin the cycle again.

Species with such a capacity for a high intrinsic rate of increase (r) are called **r-selected species** (Figure 8-10, left). Algae, bacteria, rodents, annual plants (such as dandelions), and most insects (p. 77) are examples.

Such species tend to be *opportunists*. They reproduce and spread rapidly when conditions are favorable or when a disturbance (Table 7-2, p. 175) opens up a new habitat or niche for invasion, as in the early stages of ecological succession (Figures 7-13, p. 172, and 7-14, p. 173).

Changed environmental conditions from disturbances can allow opportunist species to gain a foothold. However, once established, their populations may crash because of **(1)** changing or unfavorable environmental conditions or **(2)** invasion by more competitive species. Therefore, most r-selected or opportunist

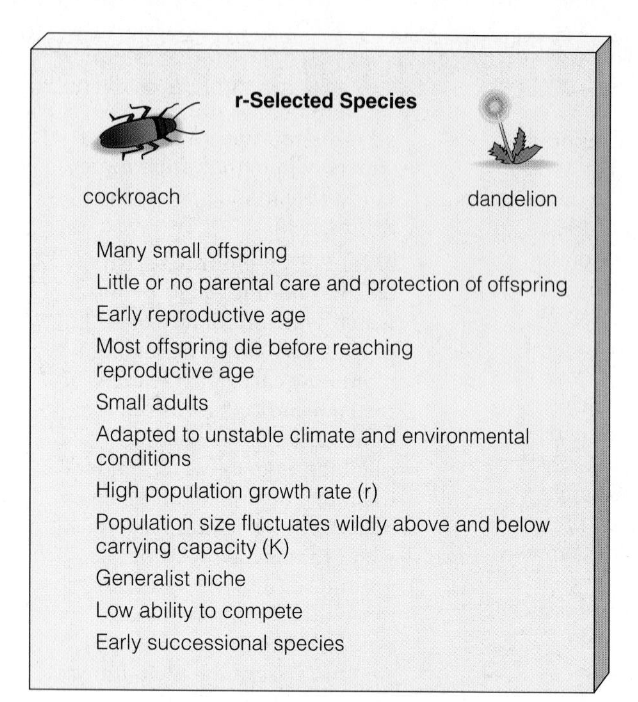

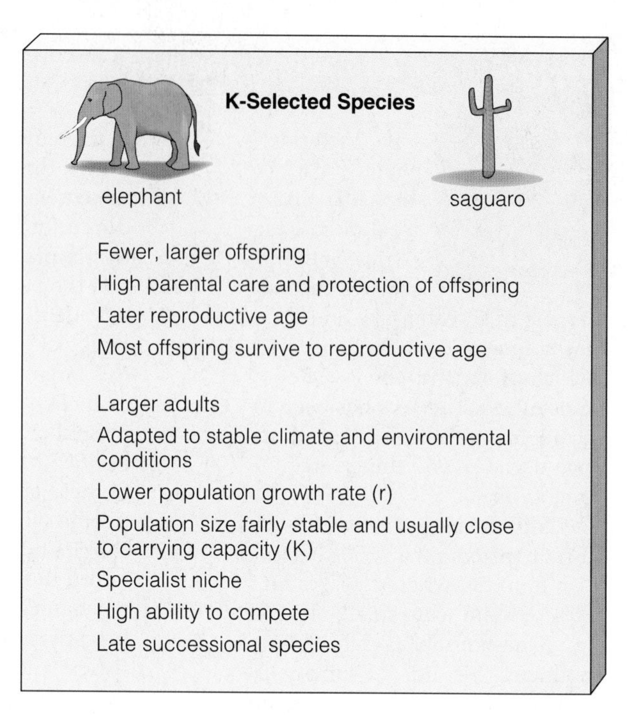

Figure 8-10 Generalized characteristics of r-selected or opportunist species and K-selected or competitor species. Many species have characteristics between these two extremes.

species go through irregular and unstable boom–bust cycles in their population size.

What Are Competitor or K-Selected Species? At the other extreme are *competitor* or **K-selected species.** These species **(1)** put fairly little energy into reproduction, **(2)** tend to reproduce late in life, **(3)** have few offspring with long generation times, and **(4)** put most of their energy into nurturing and protecting their young until they reach reproductive age.

Typically these offspring **(1)** develop inside their mothers (where they are safe), **(2)** are fairly large, **(3)** mature slowly, and **(4)** are cared for and protected by one or both parents until they reach reproductive age. This reproductive pattern results in a few big and strong individuals that can compete for resources and reproduce a few young to begin the cycle again (Figure 8-10, right).

Such species are called K-selected species because they tend to do well in competitive conditions when their population size is near the carrying capacity (K) of their environment. Their populations typically follow a logistic growth curve (Figure 8-4b). Examples are **(1)** most large mammals (such as elephants, whales, and humans), **(2)** birds of prey, and **(3)** large and long-lived plants (such as the saguaro cactus, oak trees, redwood trees, and most tropical rain forest trees). Many K-selected species, especially those with long generation times and low reproductive rates (such as elephants, rhinoceroses, and sharks, Case Study, p. 170), are prone to extinction.

Most competitor or K-selected species thrive best in ecosystems with fairly constant environmental conditions. In contrast, opportunists thrive in habitats that have experienced disturbances (Table 7-2, p. 175) such as a tree falling, a forest fire, or the clearing of a forest or grassland for raising crops.

Many organisms have reproductive patterns between the extremes of r-selected species and K-selected species. In agriculture we raise both r-selected species (crops) and K-selected species (livestock).

The reproductive pattern of a species may give it a temporary advantage, but *the availability of suitable habitat for individuals of a population in a particular area is what determines its ultimate population size.* Regardless of how fast a species can reproduce, there can be no more dandelions than there is dandelion habitat and no more zebras than there is zebra habitat in a particular area.

What Are Survivorship Curves? Individuals of species with different reproductive strategies tend to have different *life expectancies.* One way to represent the age structure of a population is with a **survivorship curve**, which shows the number of survivors of each age group for a particular species. There are three generalized types of survivorship curves: *late loss, early loss,* and *constant loss* (Figure 8-11).

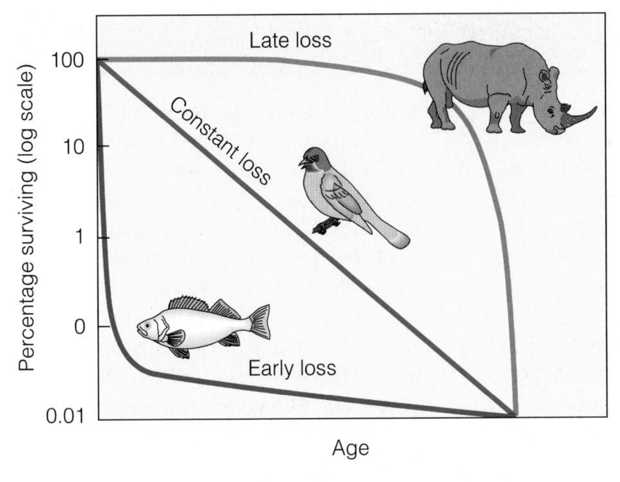

Figure 8-11 Three general *survivorship curves* for populations of different species, obtained by showing the percentages of the members of a population surviving at different ages. For a *late loss* population (such as elephants, rhinoceroses, and humans), there is typically high survivorship to a certain age, then high mortality. A *constant loss* population (such as many songbirds) shows a fairly constant death rate at all ages. For an *early loss* population (such as annual plants and many bony fish species), survivorship is low early in life. These generalized survivorship curves only approximate the behavior of species.

A table of the numbers of individuals at each age from a survivorship curve is called a *life table.* It shows the projected life expectancy and probability of death for individuals at each age. Insurance companies use life tables of human populations in various countries or regions to determine policy costs for their customers. Because life tables show that women in the United States survive an average of 7 years longer than men, a 65-year-old man normally pays more for life insurance than a 65-year-old woman.

8-4 CONSERVATION BIOLOGY: SUSTAINING WILDLIFE POPULATIONS

What Is Conservation Biology? **Conservation biology** is a multidisciplinary science, originated in the 1970s, that uses the best available science to take action to preserve species and ecosystems. Conservation biology seeks answers to three questions:

- *Which species are in danger of extinction?*
- *What is the status of the functioning of ecosystems, and what ecosystem services* (Figure 4-33, p. 103) *of value to humans and other species are we in danger of losing?*
- *What measures can we take to ensure that that ecosystem functions and viable populations of wild species can be sustained?*

These are challenging questions, and the answers require extensive field research and a strong grounding in ecological theory. To understand the status of natural populations, we must **(1)** measure the current population size, **(2)** determine how population size is likely to change with time, and **(3)** determine whether existing populations are likely to be sustainable.

Conservation biology rests on three underlying principles:

- Biodiversity is necessary to all life on earth and should not be reduced by human actions.
- Humans should not cause or hasten the extinction of wildlife populations and species or disrupt vital ecological processes.
- The best way to preserve earth's biodiversity and ecological functions is to protect intact ecosystems that provide sufficient habitat for sustaining natural populations of species.

Conservation biology is based on Aldo Leopold's ethical principle that something is right when it tends to maintain the earth's life-support systems for us and other species and wrong when it does not (Individuals Matter, p. 47).

Chapter 17 (p. 429) examines methods for sustaining the biodiversity of terrestrial and aquatic systems. Chapter 18 (p. 460) examines the issues of premature extinction of terrestrial and aquatic species.

8-5 HUMAN IMPACTS ON ECOSYSTEMS: LEARNING FROM NATURE

How Have Humans Modified Natural Ecosystems? To survive and support growing numbers of people, we have greatly increased the number and area of the earth's natural systems that we have modified, cultivated, built on, or degraded (Figure 1-3, p. 5). We have used technology to alter much of the rest of nature in the following ways:

- *Fragmenting and degrading habitat.*
- *Simplifying natural ecosystems.* When we plow grasslands and clear forests, we often replace their thousands of interrelated plant and animal species with one crop (Figure 6-20, p. 136) or one kind of tree—called *monocultures*—or with buildings, highways, and parking lots. Then we spend a lot of time, energy, and money trying to protect such monocultures from invasion by **(1)** opportunist species of plants (weeds), **(2)** pests (mostly insects, to which a monoculture crop is like an all-you-can-eat

Ecological Surprises

CONNECTIONS

Malaria once infected 9 out of 10 people in North Borneo, now known as Sabah. In 1955 the World Health Organization (WHO) began spraying the island with dieldrin (a DDT relative) to kill malaria-carrying mosquitoes. The program was so successful that the dreaded disease was nearly eliminated.

However, unexpected things began to happen. The dieldrin also killed other insects, including flies and cockroaches living in houses. The islanders applauded this turn of events, but then small lizards that also lived in the houses died after gorging themselves on dieldrin-contaminated insects. Next, cats began dying after feeding on the lizards. Then, in the absence of cats, rats flourished and overran the villages. When the people became threatened by sylvatic plague carried by rat fleas, the WHO parachuted healthy cats onto the island to help control the rats.

Then the villagers' roofs began to fall in. The dieldrin had killed wasps and other insects that fed on a type of caterpillar that either avoided or was not affected by the insecticide. With most of its predators eliminated, the caterpillar population exploded, munching its way through its favorite food: the leaves used in thatched roofs.

Ultimately, this episode ended happily: Both malaria and the unexpected effects of the spraying program were brought under control. Nevertheless, the chain of unforeseen events emphasizes the unpredictability of interfering with an ecosystem. It reminds us that when we intervene in nature, we need to ask, "And then what?"

Critical Thinking

Do you believe that the beneficial effects of spraying pesticides on Sabah outweighed the resulting unexpected and harmful effects? Explain.

restaurant), and **(3)** pathogens (fungi, viruses, or bacteria that harm the plants we want to grow).

- *Strengthening some populations of pest species and disease-causing bacteria by* **(1)** *speeding up natural selection* (Figure 5-5, left, p. 111) *and* **(2)** *causing genetic resistance through overuse of pesticides and antibiotics* (Spotlight, p. 233).

- *Eliminating some predators.* Some ranchers want to eradicate bison or prairie dogs that compete with their sheep or cattle for grass. They also want to eliminate wolves, coyotes, eagles, and other predators that occasionally kill sheep. Big game hunters also push for elimination of predators that prey on game species.

- *Deliberately or accidentally introducing new or nonnative species,* some beneficial (such as most food crops) and some harmful to us and other species.

- *Overharvesting renewable resources.* Ranchers and nomadic herders sometimes allow livestock to overgraze grasslands until erosion converts these ecosystems to less productive semideserts or deserts. Farmers sometimes deplete soil nutrients by excessive crop growing. Fish species are overharvested. Illegal hunting (poaching) endangers wildlife species with economically valuable parts (such as elephant tusks, rhinoceros horns, and tiger skins).

- *Interfering with the normal chemical cycling and energy flows in ecosystems.* Soil nutrients can **(1)** erode from monoculture crop fields, tree farms, construction sites, and other simplified ecosystems and **(2)** overload and disrupt other ecosystems such as lakes and coastal ecosystems (Section 14-6, p. 345). Chemicals such as chlorofluorocarbons (CFCs) released into the atmosphere can increase the amount of harmful ultraviolet energy reaching the earth by reducing ozone levels in the stratosphere (p. 318). Emissions of carbon dioxide and other greenhouse gases—from burning fossil fuels and from clearing and burning forests and grasslands—can trigger global climate change by altering energy flow through the atmosphere (p. 303).

To survive we must exploit and modify parts of nature. However, we are beginning to understand that any human intrusion into nature has multiple effects, most of them unpredictable (Connections, above).

Solutions: What Can We Learn from Nature About Living More Sustainably? Scientific research indicates that living systems have six key features: *interdependence, diversity, resilience, adaptability, unpredictability,* and *limits.* Organisms, populations, and ecosystems are remarkably resilient when exposed to stresses (Table 7-2, p. 175) caused by natural or human-induced changes in environmental conditions. However, scientific research indicates that environmental stresses have harmful effects on organisms, populations, and ecosystems that can affect their environmental health and long-term sustainability (Figure 8-12).

Many biologists believe that the best way for us to live more sustainably is to **(1)** learn about the processes and adaptations by which nature sustains itself (Solutions, right) and **(2)** mimic these lessons from nature.

Principles of Sustainability: Learning from Nature

SOLUTIONS

Here are four basic ecological lessons or principles of sustainability derived from observing how nature works:

- ***Most ecosystems use renewable solar energy as their primary source of energy.*** Thus, a sustainable society would be powered mostly by current sunlight, not ancient sunlight stored as polluting fossil fuels.
- ***Ecosystems replenish nutrients and dispose of wastes by recycling chemicals.*** There is almost no waste in nature because the waste outputs and decomposed remains of one organism are resource inputs for other organisms. Thus, a sustainable society would emphasize **(1)** preventing and reducing waste and **(2)** recycling and reusing resources (Figure 3-16, p. 74).
- ***Biodiversity (1) helps maintain the sustainability and ecological functioning of ecosystems and (2) serves as a source of adaptations to changing environmental conditions.*** Thus, a sustainable society emphasizes conserving biodiversity by protecting ecosystems and preventing the premature extinction of species.
- ***In nature there are always limits to population growth and resource consumption.*** The population size and growth rate of all species are controlled by their interactions with other species and with their nonliving environment, especially resource availability. Therefore, a sustainable society emphasizes controlling human population growth and resource consumption.

Critical Thinking

List two ways in which human activities violate each of these four principles of sustainability. In what ways does your lifestyle violate these principles? Would you be willing to change these practices? What beneficial and harmful effects would such changes have on your lifestyle?

Biologists have used these lessons from nature (Solutions, above) to formulate several principles to guide us in our search for more sustainable lifestyles:

- *Our lives, lifestyles, and economies are totally dependent on the sun and the earth.*
- *Everything is connected to everything else.* The primary goal of ecology is to discover which connections in nature are the strongest, most important, and most vulnerable to disruption.
- *We can never do merely one thing.* Any human intrusion into nature has mostly unpredictable side effects. When we alter nature, we should ask, "And then what?"
- *We should reduce and minimize the damage we do to nature*—a prevention or precautionary strategy (p. 178)—and help heal some of the ecological wounds we have inflicted.
- *We should use care, restraint, humility, and cooperation with nature as we alter the biosphere to meet our needs and wants.*

Using such guidelines, we can create a more ecologically and economically sustainable society that lives within its ecological means by **(1)** taking no more than we need, **(2)** using renewable resources no faster than nature replaces them, **(3)** preserving biodiversity and human cultural diversity, and **(4)** not depleting natural capital (Figure 4-33, p. 103).

We cannot command nature except by obeying her.
SIR FRANCIS BACON

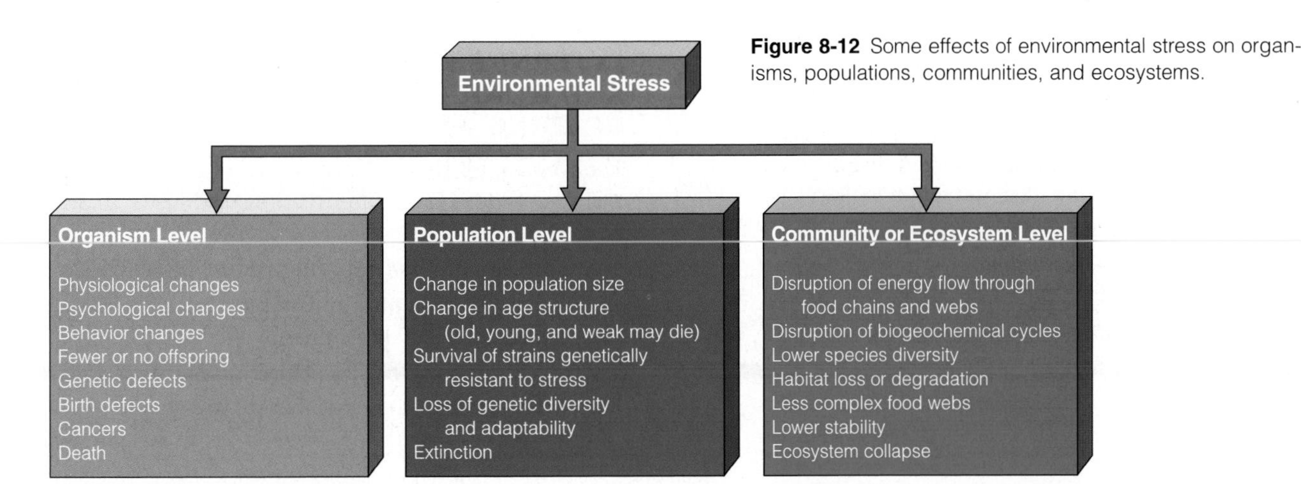

Figure 8-12 Some effects of environmental stress on organisms, populations, communities, and ecosystems.

REVIEW QUESTIONS

1. Define the boldfaced terms in this chapter.

2. Relate the fall of the civilization on Easter Island to the tragedy of the commons.

3. What four factors affect population change?

4. Write an equation showing how population change is related to births, deaths, immigration, and emigration.

5. What is the *biotic potential* of a population? What are four characteristics of a population with a high *intrinsic rate of increase* (r)?

6. What are *environmental resistance* and *carrying capacity*? How do biotic potential and environmental resistance interact to determine carrying capacity? What is a *minimum viable population* size?

7. Distinguish between *exponential* and *logistic growth* of a population and give an example of each type.

8. How can a population overshoot its carrying capacity, and what are the consequences of doing this?

9. Distinguish between *density-dependent* and *density-independent factors* that affect a population's size and give an example of each.

10. Distinguish between stable, irruptive, irregular, and cyclic forms of population change.

11. Distinguish between *top-down control* and *bottom-up control* of a population's size. Use these concepts to describe the effects of the predator–prey interactions between the snowshoe hares and the Canadian lynx on the population of each species.

12. Describe the predator–prey interactions between wolf and moose populations on Isle Royale. Is this a *top-down* or *bottom-up* form of population control?

13. Describe the predator–prey interactions between cichlids and the Nile perch in Africa's Lake Victoria. Is this a *top-down* or *bottom-up* form of population control?

14. List the characteristics of **(a)** *r-selected* or *opportunist species* and **(b)** *K-selected* or *competitor* species and give two examples of each type. Under what environmental conditions are you most likely to find **(a)** r-selected species and **(b)** K-selected species?

15. What is a *survivorship curve*, and how is it used? List three general types of survivorship curves and give an example of an organism characterized by each type.

16. What is *conservation biology*? What three questions does conservation biology try to answer? What are the three underlying principles of conservation biology?

17. List seven potentially harmful ways in which humans modify natural ecosystems.

18. List four *principles of sustainability* derived from observing how natural systems are sustained.

19. List five principles we could use to help us live more sustainably.

CRITICAL THINKING

1. Why do you think the people on Easter Island (p. 181) did not realize that they were consuming the resources that supported them and stop before it was too late? Do you believe that we are facing a similar situation on Earth Island? Explain.

2. Why do **(a)** biotic factors that regulate population growth tend to depend on population density and **(b)** abiotic factors that regulate population growth tend to be independent of population density?

3. Suppose that because of disease or genetic defects from inbreeding, the wolves on Isle Royale die off. Should we **(a)** intervene and import new wolves to help control the moose population or **(b)** let the moose population grow until it exceeds its carrying capacity and suffers another population crash? Explain.

4. Why are pest species likely to be extreme r-selected species? Why are many endangered species likely to be extreme K-selected species?

5. Explain why a simplified ecosystem such as a cornfield usually is much more vulnerable to harm from insects and plant diseases than a more complex, natural ecosystem such as a grassland.

6. Explain why you agree or disagree with the four principles for living more sustainably listed on p. 191.

PROJECTS

1. Use the principles of sustainability derived from the scientific study of how nature sustains itself (Solutions, p. 191) to evaluate the sustainability of the following parts of human systems: **(a)** transportation, **(b)** cities, **(c)** agriculture, **(d)** manufacturing, **(e)** waste disposal, and **(f)** your own lifestyle. Compare your analysis with those made by other members of your class.

2. Use the library and internet to choose one wild plant species and one animal species and analyze the factors that are likely to limit the populations of each species.

3. Make a concept map of this chapter's major ideas, using the section heads and subheads and the key terms (in boldface). Look on the website for this book for information about making concept maps.

INTERNET STUDY RESOURCES AND RESOURCES FOR FURTHER READING AND RESEARCH

The website for this book contains helpful study aids and many ideas for further reading and research. Log on to

http://www.brookscole.com/product/0534389872s

and click on the Chapter-by-Chapter area. Choose Chapter 8 and select a resource:

- "Flash Cards" allows you to test your mastery of the Terms and Concepts to Remember for this chapter.

- "Tutorial Quizzes" provides a multiple-choice practice quiz.
- "Student Guide to InfoTrac" will lead you to Critical Thinking Projects that use InfoTrac College Edition as a research tool.
- "References" lists the major books and articles consulted in writing this chapter.
- "Hypercontents" takes you to an extensive list of sites with news, research, and images related to individual sections of the chapter.

INFOTRAC COLLEGE EDITION

Improve your skills with InfoTrac College Edition, a searchable online database of articles from more than 700 periodicals. Log on to

http://www.infotrac-college.com

or access InfoTrac through the website for this book. Try to find the following articles:

1. Watching Wolves on a Wild Ride: For 25 Winters, Researcher Rolf Peterson Has Tracked the Turbulent Twists and Turns in the Lives of Isle Royale's Top Predators and Prey. Les Line. *National Wildlife,* Dec 2000–Jan 2001 pNA. The article describes the work of a wildlife ecologist at Michigan Technological University who "has led what is believed to be the longest-running study anywhere" of this kind, on an island in Lake Superior. *Hint*: Enter the search term "watching wolves" using the keyword "michigan."

2. How Ecosystems Respond to Stress. David J. Rapport, Walter G. Whitford. *BioScience,* March 1999 v49 i3 p193(1). Ecosystems respond to anthropogenic stresses by transforming from complex to simple systems. This transformation sets the stage for recovery. *Hint*: Enter the search term "ecosystem stress" using the keywords "conservation biology."

9 GEOLOGY: PROCESSES, MINERALS, HAZARDS, AND SOILS

The Mount St. Helens Eruption

For 123 years the volcano at Mount St. Helens, in the Cascade Range near the Washington–Oregon border, had slumbered. On May 18, 1980, it erupted with an enormous explosive force (Figure 9-1), causing the worst volcanic disaster in U.S. history.

Devastation occurred in three semicircular zones reaching out about 44 kilometers (27 miles) to the north of the volcano:

- In the *blast zone* or *tree-removal zone*, extending about 13 kilometers (8 miles), everything was obliterated or carried away.
- In the *tree-down zone*, extending 30 kilometers (19 miles) beyond the blast zone, a wave of ash-laden air blew the trees down like matchsticks.
- In the *seared zone*, 1–2 kilometers (0.6–1.2 miles) further out, trees were left standing but were scorched brown.

The explosion also threw ash more than 7 kilometers (4 miles) up into the atmosphere, high enough to be injected into global atmospheric circulation patterns. Several hours after the blast, ash-darkened sky triggered automatic streetlights during the morning in Yakima and Spokane, Washington. Within two weeks the ash cloud had traveled around the globe and eventually circled the planet several times before settling to the ground.

Fifty-seven people died in the eruption, and several hundred cabins and homes were destroyed or severely damaged. Tens of thousand of hectares of forest were obliterated, along with campgrounds and bridges. An estimated 7,000 big game animals (bear, deer, elk, and mountain lions) died, as did millions of smaller animals and birds and some 11 million fish.

Salmon hatcheries were damaged, and crops (including alfalfa, apples, potatoes, and wheat) were lost. Many people living in the area lost their jobs.

On the positive side, trace elements from ash that were added to the soil may eventually benefit agriculture. Increased tourism to the area to view the destruction also brought new jobs and income.

By 1990, many biologists were surprised at how fast various forms of life had begun colonizing many of the most devastated areas. This rapid recovery has taught biologists important and often surprising lessons about nature's ability to recover from what seems to be devastation.

We live on a dynamic planet. Energy from the sun and from the earth's interior, coupled with the erosive power of flowing water, have created continents, mountains, valleys, plains, and ocean basins in an ongoing process that continues to change the landscape. **Geology** is the science devoted to the study of these dynamic processes. Geologists study and analyze rocks and the features and processes of the earth's interior and surface. Some of these processes lead to geologic hazards such as earthquakes and volcanic eruptions (Figure 9-1), and others produce the mineral resources and soil that support life and economies.

Figure 9-1 Mount St. Helens, a composite volcano in Washington, near the Oregon border, before (top) and shortly after (bottom) its major eruption in May 1980. (U.S. Geological Survey)

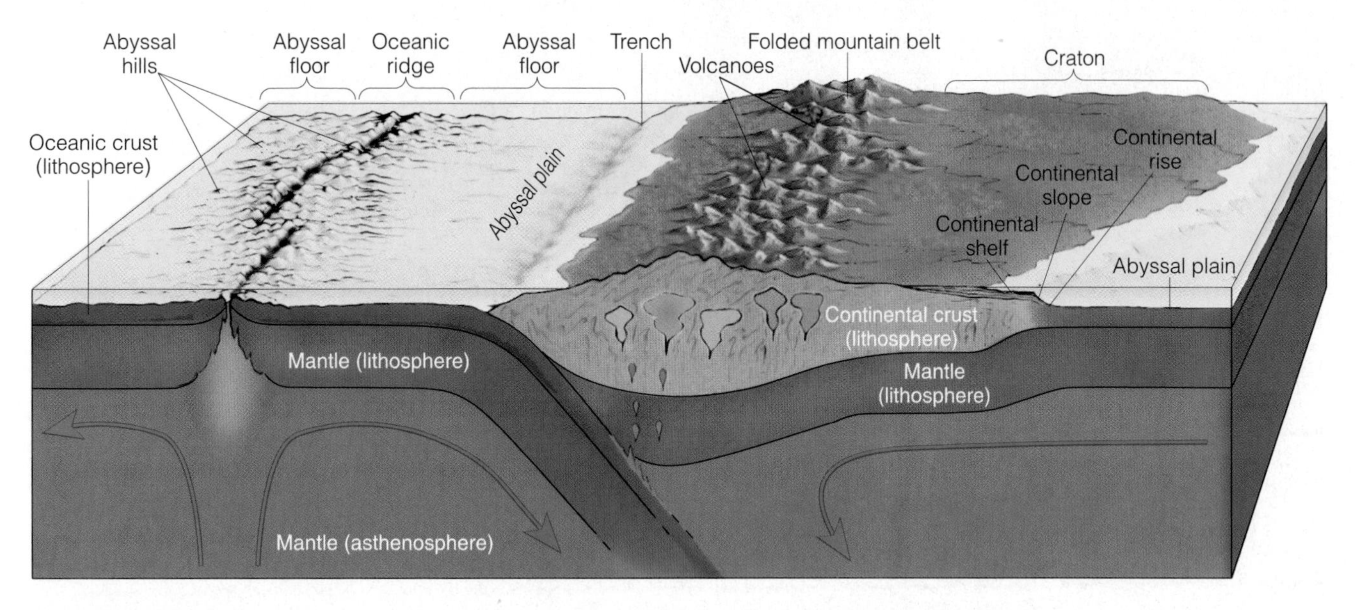

Figure 9-2 Major features of the earth's crust and upper mantle. The *lithosphere*, composed of the crust and outermost mantle, is rigid and brittle. The *asthenosphere*, a zone in the mantle, can be deformed by heat and pressure.

Civilization exists by geological consent, subject to change without notice.

WILL DURANT

This chapter addresses the following questions:

- What major geologic processes occur within the earth and on its surface?
- What are rocks, and how are they recycled by the rock cycle?
- How do we find and extract mineral resources from the earth's crust?
- Will there be enough nonrenewable mineral resources for future generations?
- What are the hazards from earthquakes and volcanic eruptions?
- What are soils, and how are they formed?
- What is soil erosion, and how can it be reduced?

9-1 GEOLOGIC PROCESSES

What Is the Earth's Structure? As the primitive earth cooled over eons, its interior separated into three major concentric zones, which geologists identify as the *core*, the *mantle*, and the *crust* (Figure 4-6, p. 81)

The earth's innermost zone, the **core** (Figure 4-6, p. 81), is very hot and has a solid inner part, surrounded by a liquid core of molten material.

A thick, solid zone called the **mantle** surrounds the earth's core. Most of the mantle is solid rock, but under its rigid outermost part there is a zone of very hot, partly melted rock that flows like soft plastic. This plastic region of the mantle is called the *asthenosphere.*

The outermost and thinnest zone of the earth is called the **crust** (Figure 4-6, p. 81). It consists of **(1)** the *continental crust*, which underlies the continents (including the continental shelves extending into the oceans, Figure 6-32, p. 146), and **(2)** the *oceanic crust*, which underlies the ocean basins and covers 71% of the earth's surface (Figure 9-2).

9-2 INTERNAL AND EXTERNAL EARTH PROCESSES

What Is Plate Tectonics? Internal Geologic Processes We tend to think of the earth's crust, mantle, and core as fairly static and unchanging. However, they are constantly changed by geologic processes taking place within the earth and on the earth's surface, most over thousands to millions of years.

A map of the earth's earthquakes and volcanoes shows that most of these phenomena occur along certain lines or belts on the earth's surface (Figure 9-3a). The areas of the earth outlined by these major belts are called **plates** (Figure 9-3b). They are about 100 kilometers (60 miles) thick and are composed of the crust and the rigid, outermost part of the mantle (above the asthenosphere), a combination called the **lithosphere**.

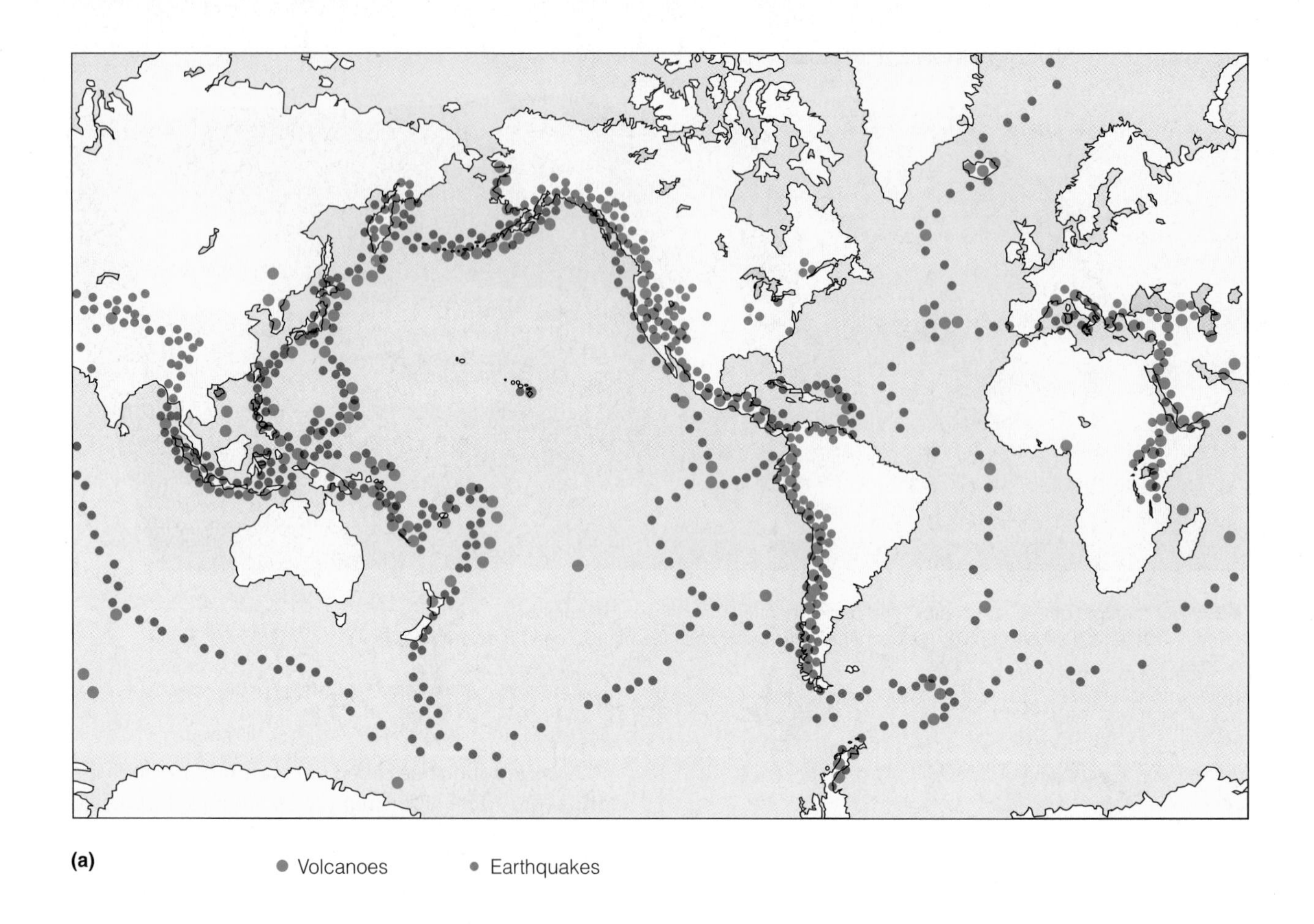

(a) ● Volcanoes ● Earthquakes

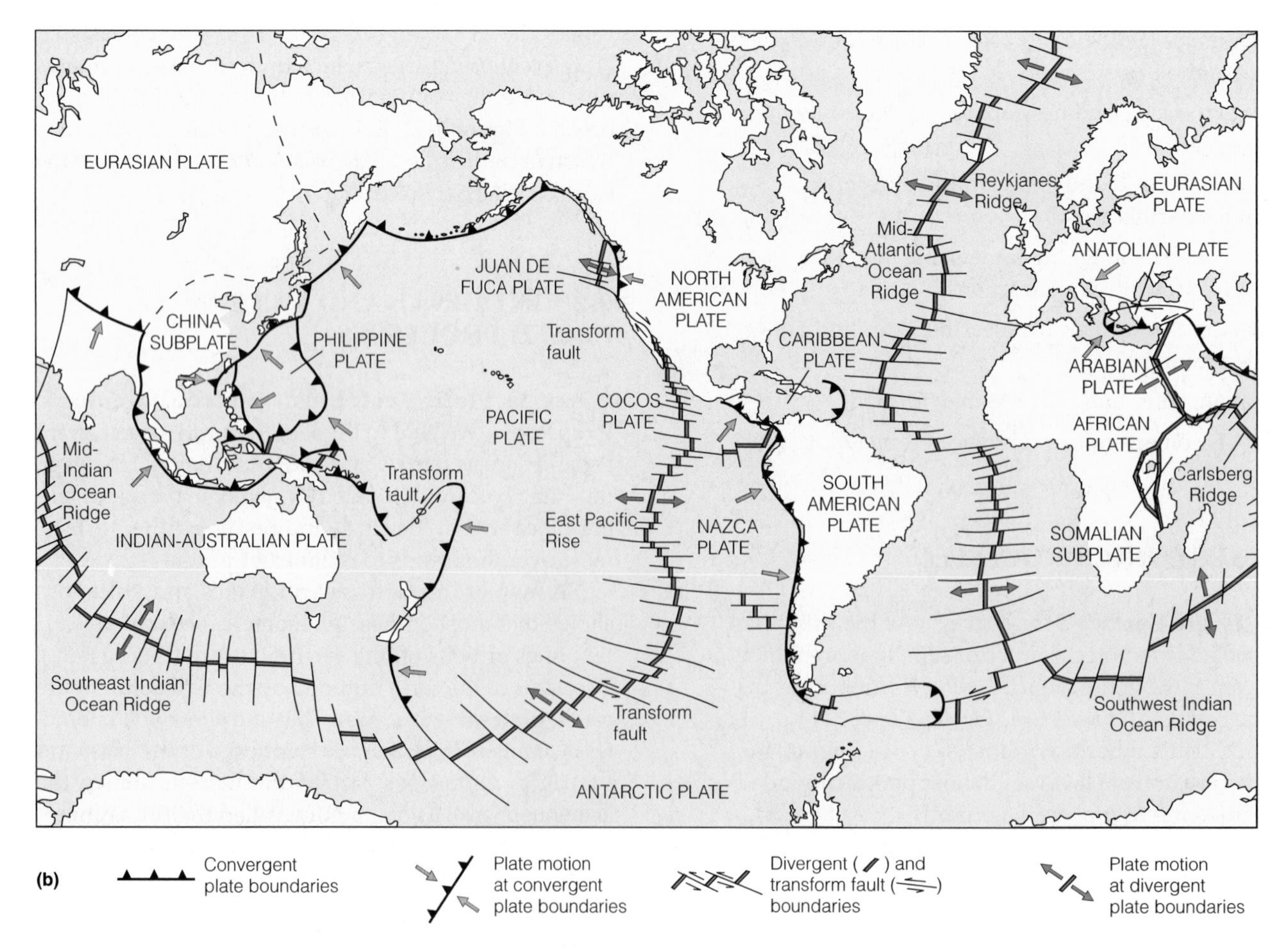

(b) Convergent plate boundaries; Plate motion at convergent plate boundaries; Divergent ( ) and transform fault ( ) boundaries; Plate motion at divergent plate boundaries

Figure 9-3 Earthquake and volcano sites are distributed mostly in bands along the planet's surface **(a)**. These bands correspond to the patterns for the types of lithospheric plate boundaries **(b)** shown in Figure 9-4.

These plates move constantly, supported by the slowly flowing asthenosphere like large pieces of ice floating on the surface of a lake. Some plates move faster than others do, but a typical speed is about the rate at which fingernails grow.

The theory explaining the movements of the plates and the processes that occur at their boundaries is called **plate tectonics**. The concept, which became widely accepted by geologists in the 1960s, was developed from an earlier idea called *continental drift*. Throughout the earth's history, continents have split and joined as plates have drifted thousands of kilometers back and forth across the planet's surface (Figure 5-8, p. 115).

Plate motion produces mountains (including volcanoes), the oceanic ridge system, trenches, and other features of the earth's surface (Figure 9-2). Natural hazards such as volcanoes and earthquakes are likely to be found at plate boundaries (Figure 9-3), and plate movements and interactions concentrate many of the minerals we extract and use.

The theory of plate tectonics also helps explain how certain patterns of biological evolution occurred. By reconstructing the course of continental drift over millions of years (Figure 5-8, p. 115), we can trace how life-forms migrated from one area to another when continents that are now far apart were still joined together. As the continents separated, populations became geographically and reproductively isolated, and speciation occurred (Figure 5-7, p. 114).

What Types of Boundaries Occur Between the Earth's Plates? Lithospheric plates have three types of boundaries (Figure 9-4):

- **Divergent plate boundaries**, where the plates move apart in opposite directions (Figure 9-4, top)
- **Convergent plate boundaries,** where the plates are pushed together by internal forces (Figures 9-3b and 9-4, middle). At most convergent plate boundaries, oceanic lithosphere is carried downward (subducted) under the island arc or the continent at a **subduction zone**. A *trench* ordinarily forms at the boundary between the two converging plates (Figure 9-4, middle). Stresses in the plate undergoing subduction cause earthquakes at convergent plate boundaries.
- **Transform faults**, which occur where plates slide past one another along a fracture (fault) in the lithosphere (Figure 9-4, bottom). Like the other types of plate boundaries, most transform faults are on the ocean floor.

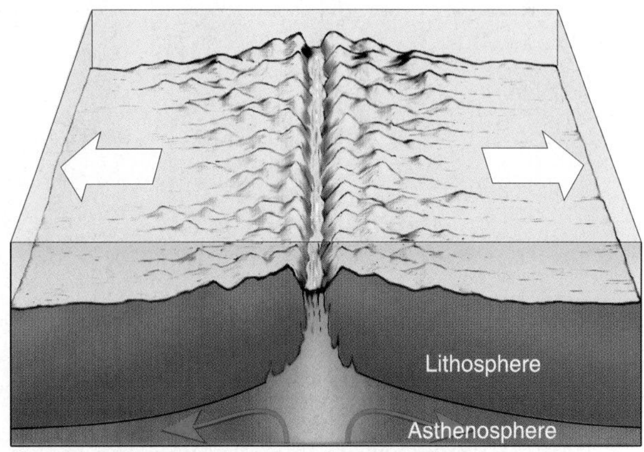

Oceanic ridge at a divergent plate boundary

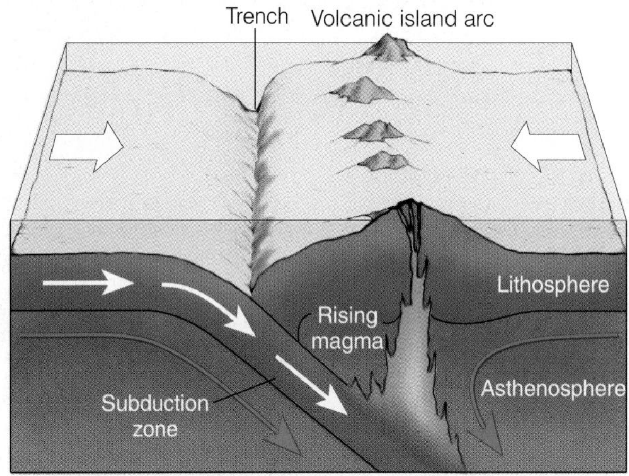

Trench and volcanic island arc at a convergent plate boundary

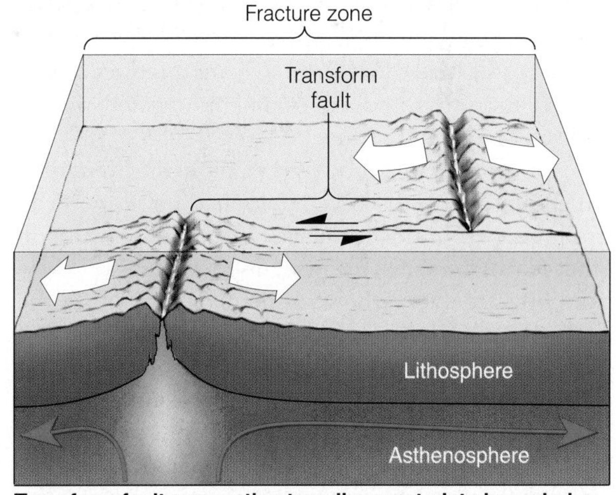

Transform fault connecting two divergent plate boundaries

Figure 9-4 Types of boundaries between the earth's lithospheric plates. All three boundary types occur both in oceans and on continents.

What Geologic Processes Occur on the Earth's Surface? Erosion and Weathering Geological changes based directly or indirectly on energy from the sun and on gravity (rather than on heat in the earth's

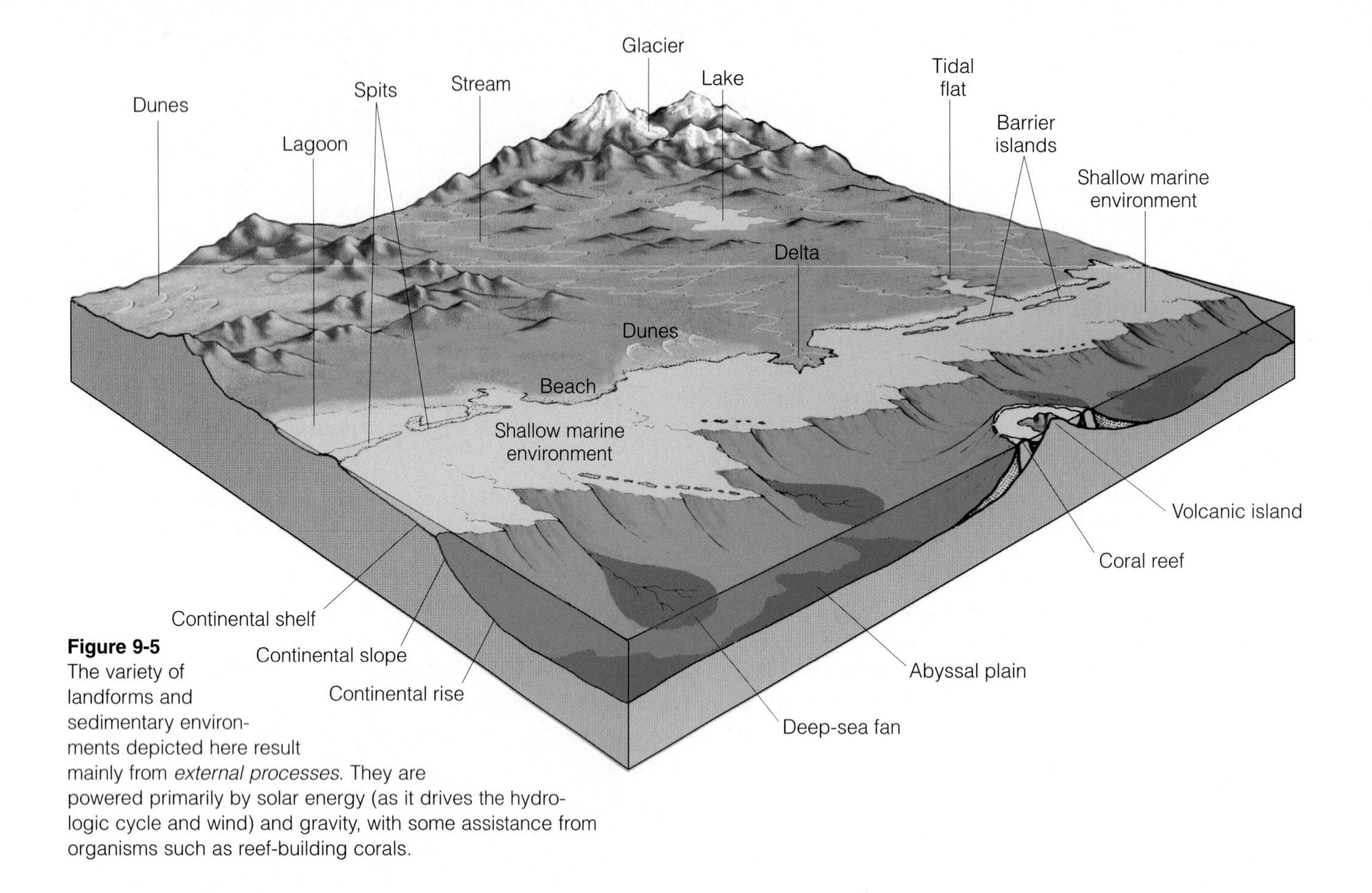

Figure 9-5 The variety of landforms and sedimentary environments depicted here result mainly from *external processes.* They are powered primarily by solar energy (as it drives the hydrologic cycle and wind) and gravity, with some assistance from organisms such as reef-building corals.

interior) are called *external processes*. Whereas internal processes generally build up the earth's surface, external processes tend to wear it down and produce a variety of landforms and environments created by the buildup of eroded sediment (Figure 9-5).

A major external process is **erosion**: the process by which material is dissolved, loosened, or worn away from one part of the earth's surface and deposited in other places. Streams, the most important agent of erosion, operate everywhere on the earth except in the polar regions. They produce valleys and canyons and may form deltas where streams flow into lakes and oceans (Figure 6-44, p. 155). Some erosion is caused when wind blows particles of soil from one area to another (Figure 6-1, p. 121). Human activities, particularly those that destroy vegetation, accelerate erosion, as discussed in Section 9-8, p. 214.

Weathering caused by mechanical or chemical processes usually produces loosened material that can be eroded. In *mechanical weathering*, a large rock mass is broken into smaller fragments of the original material, similar to the results you would get by using a hammer to break a rock into small fragments. The most important agent of mechanical weathering is *frost wedging*, in which water **(1)** collects in pores and cracks of rock, **(2)** expands upon freezing, and **(3)** splits off pieces of the rock.

In *chemical weathering*, a mass of rock is decomposed by one or more chemical reactions. Most chemical weathering involves a reaction of rock material with oxygen, carbon dioxide, and moisture in the atmosphere and the ground.

9-3 MINERALS, ROCKS, AND THE ROCK CYCLE

What Are Minerals and Rocks? The earth's crust, which is still forming in various places, is composed of minerals and rocks. It is the source of almost all the nonrenewable resources we use: fossil fuels, metallic minerals, and nonmetallic minerals (Figure 1-9, p. 10). It is also the source of soil and of the elements that make up our bodies and those of other living organisms.

A **mineral** is an element or inorganic compound that occurs naturally and is solid. Some minerals consist of a single element, such as gold, silver, diamond (carbon), and sulfur. However, most of the more than 2,000 identified minerals occur as inorganic compounds formed by various combinations of elements. Examples are salt, mica, and quartz.

Rock is any material that makes up a large, natural, continuous part of the earth's crust. Some kinds of rock, such as limestone (calcium carbonate, or $CaCO_3$) and quartzite (silicon dioxide, or SiO_2), contain only one mineral, but most rocks consist of two or more minerals.

What Are the Three Major Rock Types? Based on the way it forms, rock is placed in three broad classes:

- **Igneous rock,** which forms below or on the earth's surface when molten rock material (magma) **(1)** wells up from the earth's upper mantle or deep crust, **(2)** cools, and **(3)** hardens into rock. Examples are granite (formed underground) and lava rock (formed above ground when molten lava cools and hardens). Although often covered by sedimentary rocks or soil, igneous rocks form the bulk of the earth's crust. They also are the main source of many nonfuel mineral resources.
- **Sedimentary rock**, formed from sediment when preexisting rocks are **(1)** weathered and eroded into small pieces, **(2)** transported from their sources, and **(3)** deposited in a body of surface water. Examples are **(1)** sandstone and shale formed from pressure created by deposited layers of sediment, **(2)** dolomite and limestone formed from the compacted shells, skeletons, and other remains of dead organisms, and **(3)** lignite and bituminous coal derived from plant remains (Figure 19-16, p. 501).
- **Metamorphic rock,** which is produced when a preexisting rock is subjected to **(1)** high temperatures (which may cause it to melt partially), **(2)** high pressures, **(3)** chemically active fluids, or **(4)** a combination of these agents. Examples are anthracite (a form of coal), slate, and marble.

What Is the Rock Cycle? Rocks are constantly exposed to various physical and chemical conditions that can change them over time. The interaction of processes that change rocks from one type to another is called the **rock cycle** (Figure 9-6).

The slowest of the earth's cyclic processes, the rock cycle recycles material over millions of years. It is responsible for concentrating the planet's nonrenewable mineral resources on which humans depend.

What Are Mineral Resources? A **mineral resource** is a concentration of naturally occurring material in or on the earth's crust that can be extracted and processed into useful materials at an affordable cost. Over millions to billions of years the earth's internal and external geologic processes have produced numerous nonfuel mineral resources and energy resources. Because they take so long to produce, they are classified as *nonrenewable resources*.

We know how to find and extract more than 100 nonrenewable minerals from the earth's crust. They include **(1)** *metallic mineral resources* (iron, copper, aluminum), **(2)** *nonmetallic mineral resources* (salt, gypsum, clay, sand, phosphates, water, and soil), and **(3)** *energy resources* (coal, oil, natural gas, and uranium).

Ore is rock containing enough of one or more metallic minerals to be mined profitably. We convert about 40 metals extracted from ores into many everyday items that we either **(1)** use and discard (Figure 3-15, p. 73) or **(2)** learn to reuse, recycle, or use less wastefully (Figure 3-16, p. 74).

The U.S. Geological Survey (USGS) divides nonrenewable mineral resources into two broad categories: *identified* and *undiscovered* (Figure 9-7). **Identified resources** are **(1)** deposits of a nonrenewable mineral resource that have a *known* location, quantity, and quality or **(2)** deposits based on direct geological evidence and measurements. **Undiscovered resources** are potential supplies of a nonrenewable mineral resource that are assumed to exist on the basis of geologic knowledge

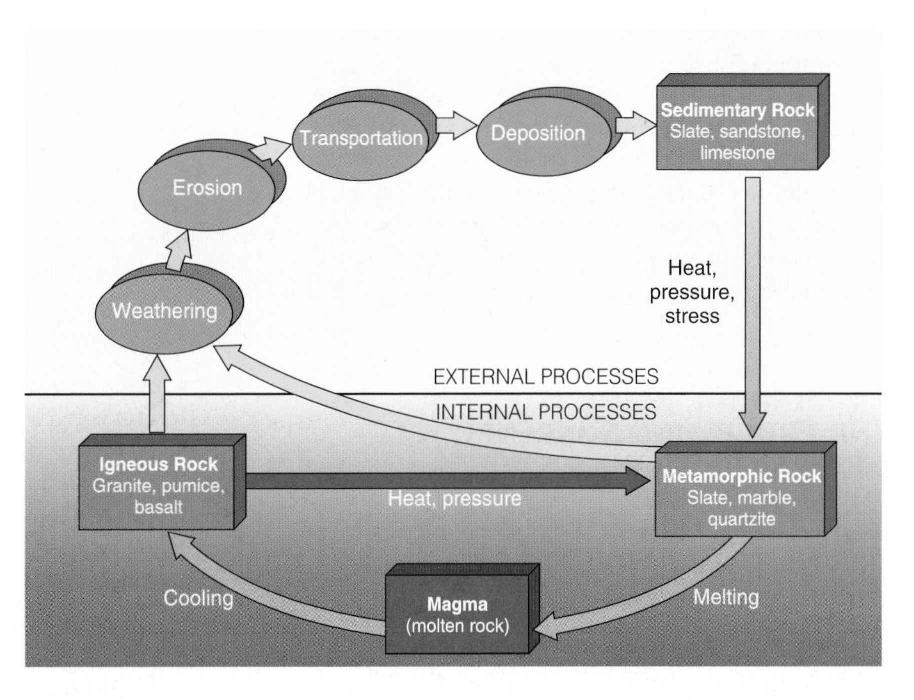

Figure 9-6 The *rock cycle*, the slowest of the earth's cyclic processes. The earth's materials are recycled over millions of years by three processes: melting, erosion, and metamorphism, which produce igneous, sedimentary, and metamorphic rocks. Rock of any of the three classes can be converted to rock of either of the other two classes (or can even be recycled within its own class).

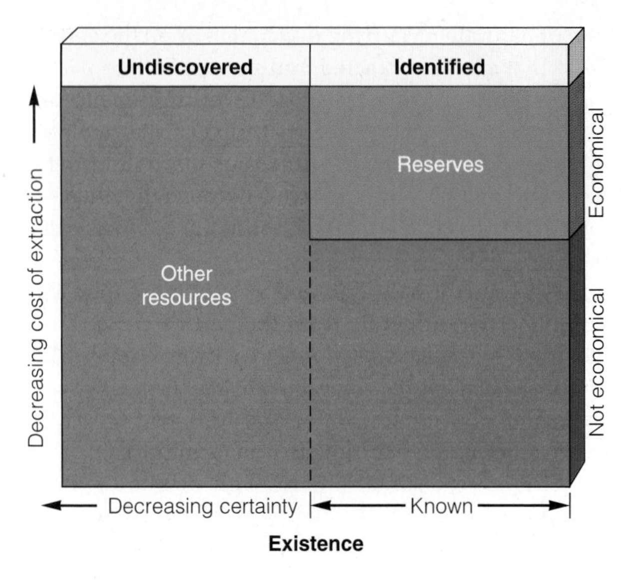

Figure 9-7 General classification of mineral resources. (The area shown for each class does not represent its relative abundance.) In theory, all mineral resources classified as *other resources* could become reserves because of rising mineral prices or improved mineral location and extraction technology. In practice, geologists expect only a fraction of other resources to become reserves.

and theory (although specific locations, quality, and amounts are unknown).

Reserves are identified resources from which a usable nonrenewable mineral can be extracted profitably at current prices. **Other resources** are identified and undiscovered resources not classified as reserves.

Most published estimates of the supply of a given nonrenewable resource refer to *reserves*. Reserves can increase when **(1)** when new deposits are found or **(2)** price increases or improved mining technology make it profitable to extract deposits that previously were too expensive to extract. Theoretically, all of the *other resources* could eventually be converted to reserves, but this is highly unlikely.

9-4 FINDING, REMOVING, AND PROCESSING NONRENEWABLE MINERAL RESOURCES

How Are Buried Mineral Deposits Found? Mining companies use several methods to find promising mineral deposits. They include

- Using aerial photos and satellite images to reveal protruding rock formations (outcrops) associated with certain minerals
- Using planes equipped with **(1)** *radiation measuring equipment* to detect deposits of radioactive metals such as uranium and **(2)** a *magnetometer* to measure changes in the earth's magnetic field caused by magnetic minerals such as iron ore
- Using a *gravimeter* to measure differences in gravity because the density of an ore deposit usually differs from that of the surrounding rock
- Drilling a deep well and extracting core samples
- Putting sensors in existing wells to detect electrical resistance or radioactivity to pinpoint the location of oil and natural gas
- Making *seismic surveys* on land and at sea by detonating explosive charges that send shock waves to get information about the makeup of buried rock layers
- Performing *chemical analysis* of water and plants to detect deposits of underground minerals that have leached into nearby bodies of water or have been absorbed by plant tissues

After suitable mineral deposits are located, several different types of mining techniques are used to remove the deposits, depending on their location and type. Shallow deposits are removed by **surface mining** (Figure 9-8), and deep deposits are removed by **subsurface mining** (Figure 9-9).

In surface mining, mechanized equipment strips away the **overburden** of soil and rock and usually discards it as waste material called **spoil**. In the United States, surface mining extracts about 90% of the nonfuel mineral and rock resources and about 60% of the coal by weight.

The type of surface mining used depends on the resource being sought and on local topography. Methods include the following:

- **Open-pit mining** (Figure 9-8a), in which machines dig holes and remove ores such as iron and copper and sand, gravel, and stone (such as limestone and marble).
- **Dredging** (Figure 9-8b), in which chain buckets and draglines scrape up underwater mineral deposits.
- **Area strip mining** (Figure 9-8c), used where the terrain is fairly flat. An earthmover strips away the overburden, and a power shovel digs a cut to remove the mineral deposit. After the mineral is removed, the trench is filled with overburden and a new cut is made parallel to the previous one. The process is repeated over the entire site. If the land is not restored, area strip mining leaves a wavy series of highly erodible hills of rubble called *spoil banks*.
- **Contour strip mining** (Figure 9-8d), used on hilly or mountainous terrain. A power shovel cuts a series of terraces into the side of a hill. An earthmover removes the overburden, and a power shovel extracts

(a) Open Pit Mine

(b) Dredging

(c) Area Strip Mining

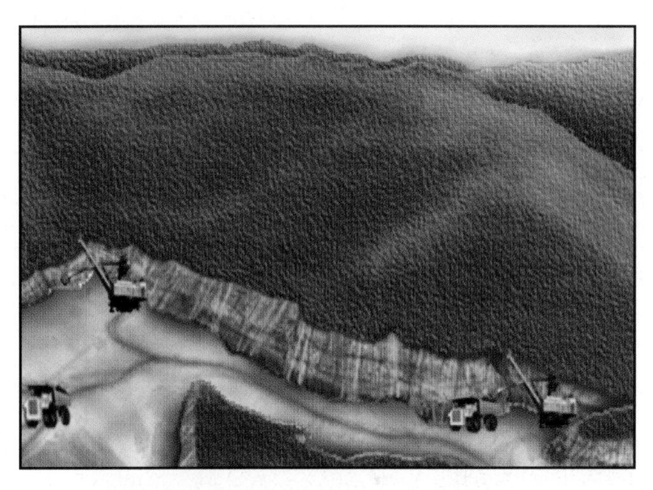

(d) Contour Strip Mining

Figure 9-8 Major mining methods used to extract surface deposits of solid mineral and energy resources.

the coal, with the overburden from each new terrace dumped onto the one below. Unless the land is restored, a wall of dirt is left in front of a highly erodible bank of soil and rock called a *highwall.*

Surface-mined land can be restored (except in arid and semiarid areas), but this is expensive and is not done in many countries. In the United States, the Surface Mining Control and Reclamation Act of 1977 requires mining companies to restore most surface-mined land so that it can be used for the same purpose as it was before it was mined.

The law also levied a tax on mining companies to restore land that was disturbed by surface mining before the law was passed. However, more than 6,000 abandoned coal and metal mines, covering an area of about the size of the state of Virginia, have not been restored. An even larger area of abandoned rock quarries and gravel and sand mines has not been reclaimed.

Subsurface mining (Figure 9-9) is used to remove coal and various metal ores that are too deep to be extracted by surface mining. Miners **(1)** dig a deep vertical shaft, **(2)** blast subsurface tunnels and chambers to get to the deposit, and **(3)** use machinery to remove the ore or coal and transport it to the surface.

Subsurface mining disturbs less than one-tenth as much land as surface mining and usually produces less waste material. However, it leaves much of the resource in the ground and is more dangerous and expensive than surface mining. Hazards include **(1)** collapse of roofs and walls of underground mines, **(2)** explosions of dust and natural gas, and **(3)** lung diseases caused by prolonged inhalation of mining dust.

What Are the Environmental Impacts of Extracting, Processing, and Using Mineral Resources? Mining, processing, and using mineral resources take

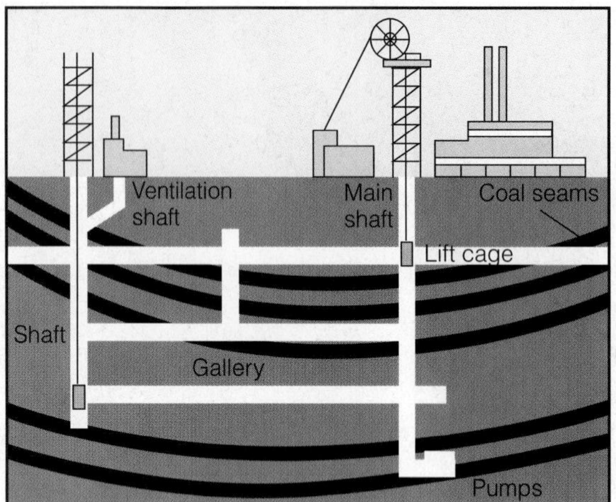

(a) Underground Coal Mine

Figure 9-9 Major mining methods used to extract underground deposits of solid mineral and energy resources. **(a)** Mine shafts and tunnels are dug and blasted out. **(b)** In *room-and-pillar* mining, machinery is used to gouge out coal and load it onto a shuttle car in one operation, and pillars of coal are left to support the mine roof. **(c)** In *longwall coal mining*, movable steel props support the roof, and cutting machines shear off the coal onto a conveyor belt. As the mining proceeds, roof supports are moved forward and the roof behind is allowed to fall (often causing the land above to sink or subside).

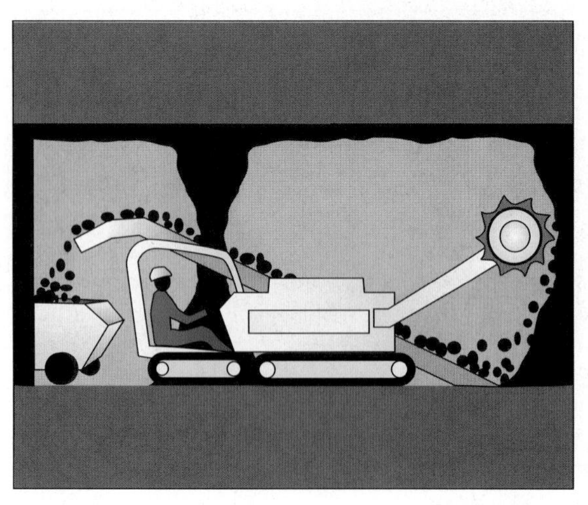

(b) Room-and-Pillar

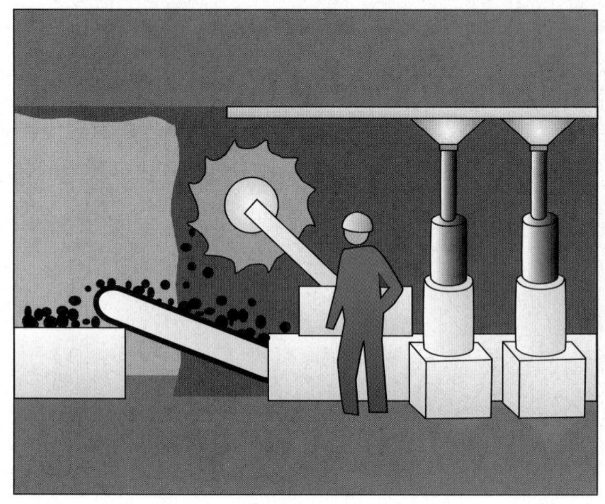

(c) Longwall Mining of Coal

enormous amounts of energy and often cause land disturbance, soil erosion, and air and water pollution (Figure 9-10).

Mining can affect the environment in several ways, including

- Scarring and disruption of the land surface (Figure 9-8).
- Collapse or subsidence of land above underground mines, which can cause **(1)** houses to tilt, **(2)** sewer lines to crack, **(3)** gas mains to break, and **(4)** groundwater systems to be disrupted.
- Wind- or water-caused erosion of toxin-laced mining wastes.
- Acid mine drainage, when rainwater seeping through a mine or mine wastes **(1)** carries sulfuric acid (H_2SO_4, produced when aerobic bacteria act on iron sulfide minerals in spoil) to nearby streams and groundwater (Figure 9-11), **(2)** contaminates water supplies, and **(3)** destroys aquatic life.
- Emissions of toxic chemicals into the atmosphere. In the United States, the mining industry produces more toxic emissions than any other industry (typically accounting for almost half of such emissions). According to Alan Septoff of the Mineral Policy Center, "Mining is the largest toxic polluter in the United States, responsible for polluting 40% of the headwaters of watersheds in the Western states."
- Exposure of wildlife to toxic mining wastes stored in holding ponds and leakage of toxic wastes from such ponds.

Figure 9-12 shows the typical life cycle of a metal resource. Ore extracted from the earth's crust typically has two components: **(1)** the *ore mineral* containing the desired metal and **(2)** waste material called *gangue*. Removing the gangue from ores produces piles of waste called *tailings*. Particles of toxic metals blown or leached from tailings by rainfall can contaminate the air, surface water, and groundwater.

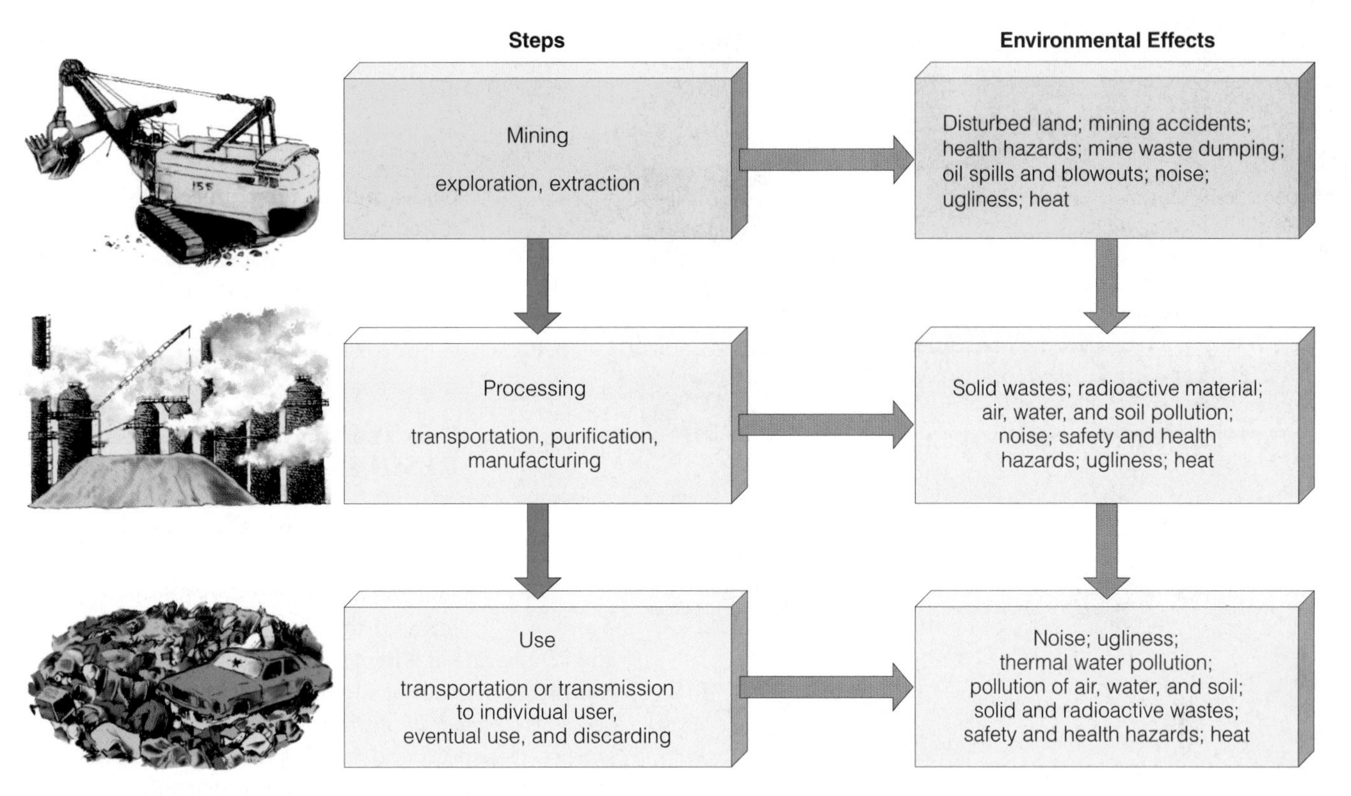

Figure 9-10 Some harmful environmental effects of extracting, processing, and using nonrenewable mineral and energy resources. The energy used to carry out each step causes additional pollution and environmental degradation.

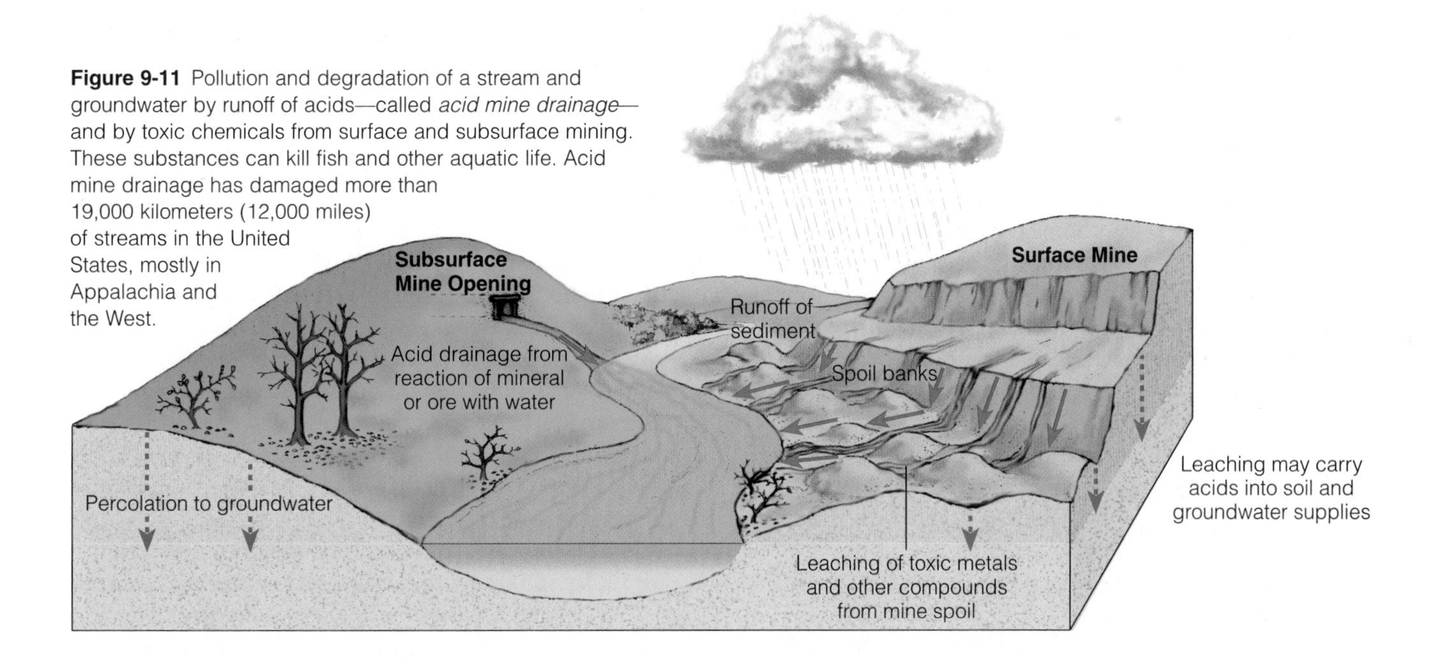

Figure 9-11 Pollution and degradation of a stream and groundwater by runoff of acids—called *acid mine drainage*—and by toxic chemicals from surface and subsurface mining. These substances can kill fish and other aquatic life. Acid mine drainage has damaged more than 19,000 kilometers (12,000 miles) of streams in the United States, mostly in Appalachia and the West.

Most ores consist of one or more compounds of the desired metal. After gangue has been removed, **smelting** is used to separate the metal from the other elements in the ore mineral. Without effective pollution control equipment, smelters emit enormous quantities of air pollutants, which damage vegetation and soils in the surrounding area. It can take decades for vegetation in such areas to be restored by a combination of secondary ecological succession (Figure 7-14, p. 173) and expensive restoration efforts.

Smelters also cause water pollution and produce liquid and solid hazardous wastes that must be

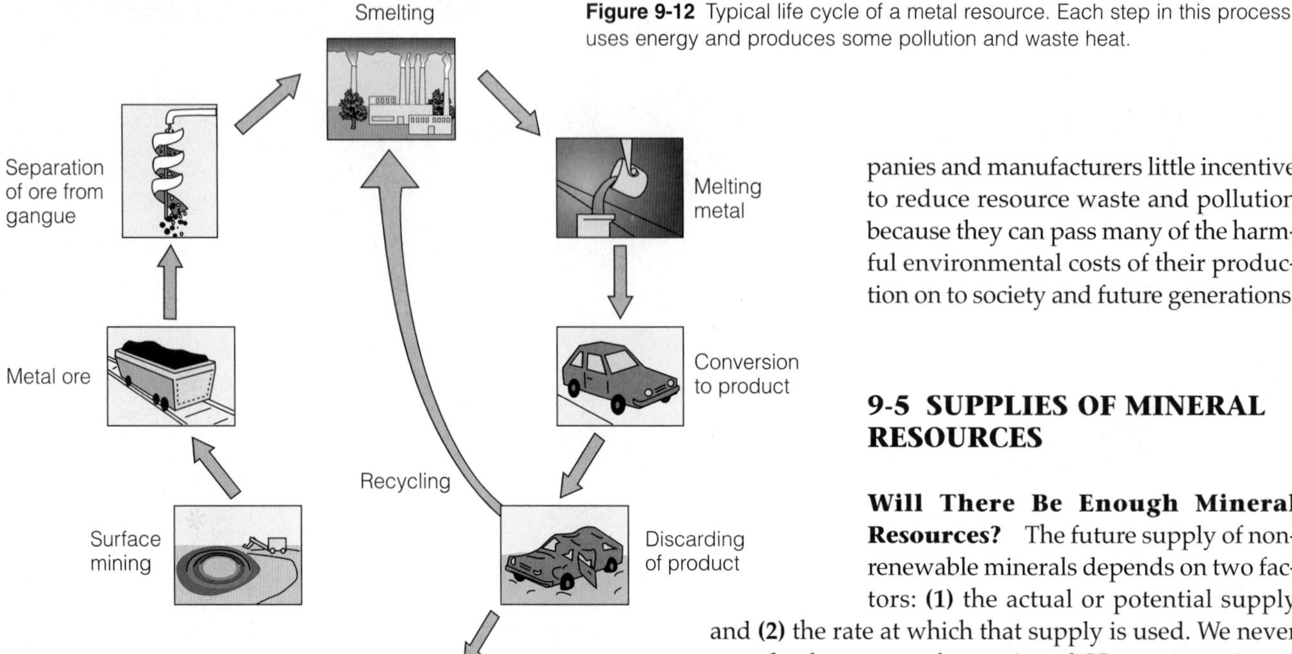

Figure 9-12 Typical life cycle of a metal resource. Each step in this process uses energy and produces some pollution and waste heat.

disposed of safely. Some companies are using improved technology to **(1)** reduce pollution from smelting, **(2)** lower production costs, **(3)** save costly cleanup bills, and **(4)** decrease liability for damages.

Once the pure metal has been produced by smelting, it is usually melted and converted to desired products, which are then used and discarded or recycled (Figure 9-12).

Are There Environmental Limits to Resource Extraction and Use? Some environmentalists and resource experts believe that the greatest danger from the world's continually increasing levels of resource consumption is not the exhaustion of nonrenewable resources but the environmental damage caused by their extraction, processing, and conversion to products (Figure 9-10).

The mineral industry accounts for 5–10% of world energy use, making it a major contributor to air and water pollution and to emissions of greenhouse gases such as carbon dioxide. The environmental impacts from mining an ore are affected by its percentage of metal content, or *grade*. Usually more accessible and higher-grade ores are exploited first. As they are depleted, it takes more money, energy, water, and other materials to exploit lower-grade ores, and land disruption, mining waste, and pollution increase accordingly.

Currently, most of the harmful environmental costs of mining and processing minerals are not included in the prices for processed metals and consumer products produced from such metals. This gives mining companies and manufacturers little incentive to reduce resource waste and pollution because they can pass many of the harmful environmental costs of their production on to society and future generations.

9-5 SUPPLIES OF MINERAL RESOURCES

Will There Be Enough Mineral Resources? The future supply of nonrenewable minerals depends on two factors: **(1)** the actual or potential supply and **(2)** the rate at which that supply is used. We never completely run out of any mineral. However, a mineral becomes *economically depleted* when it costs more to find, extract, transport, and process the remaining deposit than it is worth. At that point, there are five choices: **(1)** Recycle or reuse existing supplies, **(2)** waste less, **(3)** use less, **(4)** find a substitute, or **(5)** do without.

Depletion time is the time it takes to use up a certain proportion (usually 80%) of the reserves of a mineral at a given rate of use (Figure 1-10, p. 11). When experts disagree about depletion times, they are often using different assumptions about supply and rate of use (Figure 9-13).

A traditional measure of the projected availability of nonrenewable resources is the **reserve-to-production ratio**: the number of years that proven reserves of a particular nonrenewable mineral will last at current annual production rates. Reserve estimates are continually changing because **(1)** new deposits often are discovered and **(2)** new mining and processing techniques can allow some of the minerals classified as other resources (Figure 9-7) to be converted to reserves. Under these circumstances, the reserve-to-production ratio is the best available projection of the current estimated supply and its estimated depletion time.

The shortest depletion time assumes no recycling or reuse and no increase in reserves (curve A, Figure 9-13). A longer depletion time assumes that recycling will stretch existing reserves and that better mining technology, higher prices, and new discoveries will increase reserves (curve B, Figure 9-13). An even longer depletion time assumes that new discoveries will further expand reserves and that recycling, reuse, and reduced consumption will extend supplies (curve C, Figure 9-13). Finding a substitute for a resource leads to a new set of depletion curves for the new resource.

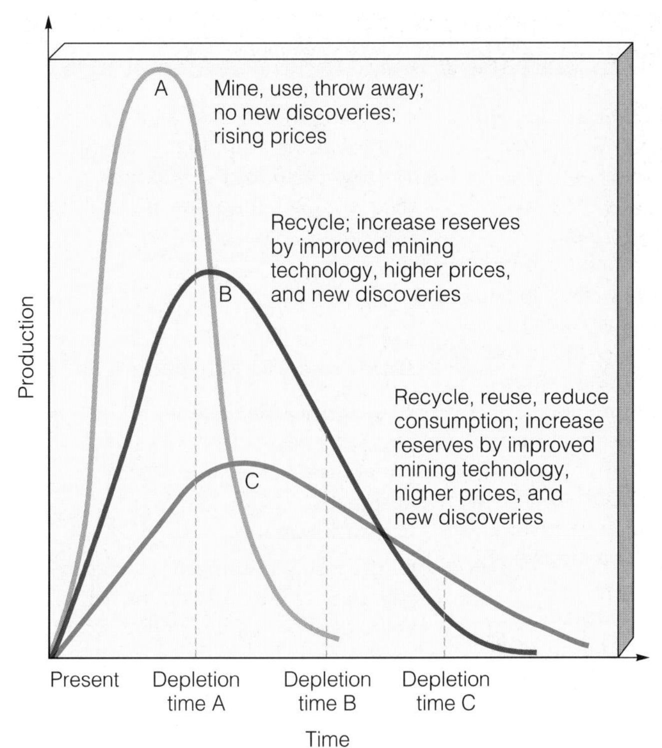

Figure 9-13 Depletion curves for a nonrenewable resource (such as aluminum or copper) using three sets of assumptions. Dashed vertical lines represent times when 80% depletion occurs.

How Does Economics Affect Mineral Resource Supplies? Geologic processes determine the quantity and location of a mineral resource in the earth's crust, but economics determines what part of the known supply is used.

According to standard economic theory, in a competitive free market a plentiful mineral resource is cheap when its supply exceeds demand. However, when a resource becomes scarce its price rises. This can **(1)** encourage exploration for new deposits, **(2)** stimulate development of better mining technology, **(3)** make it profitable to mine lower-grade ores, **(4)** encourage a search for substitutes, and **(5)** promote conservation.

However, according to some economists this theory may no longer apply to most developed countries. One reason is that industry and government in such countries control the supply, demand, and prices of minerals to such a large extent that a truly competitive free market does not exist.

Most mineral prices are low because governments subsidize development of their domestic mineral resources to help promote economic growth and national security. In the United States, for instance, mining companies **(1)** get depletion allowances amounting to 5–22% of their gross income (depending on the mineral) and **(2)** can deduct much of the cost of finding and developing mineral deposits. In addition, hardrock mining companies operating in the United States can buy public land at 1872 prices and pay no royalties to the government on the minerals they extract (Pro/Con, p. 206).

Between 1982 and 2000, U.S. mining companies received more than $6 billion in government subsidies. Critics argue that taxing rather than subsidizing the extraction of nonfuel mineral resources would **(1)** provide governments with revenue, **(2)** create incentives for more efficient resource use, **(3)** promote waste reduction and pollution prevention, and **(4)** encourage recycling and reuse.

Mining company representatives say they need subsidies and low taxes to **(1)** keep the prices of minerals low for consumers and **(2)** encourage them not to move their mining operations to other countries without such taxes and less stringent mining regulations.

Two other economic problems also hinder the development of new supplies of mineral resources:

- Mineral scarcity does not raise the market price of products very much because the cost of mineral resources is only a small part of the final cost of goods.
- Exploring for new mineral resources takes lots of increasingly scarce investment capital and is a risky financial venture. Typically, if geologists identify 10,000 possible deposits of a given resource, only 1,000 sites are worth exploring; only 100 justify drilling, trenching, or tunneling; and only 1 becomes a producing mine or well.

Should More Mining Be Allowed on Public Lands in the United States? About one-third of the land in the United States is public land owned jointly by all U.S. citizens. This land, consisting of national forests, parks, resource lands, and wilderness, is managed by various government agencies under laws passed by Congress. About 72% of this public land is in Alaska, and 22% is in western states (where 60% of all land is public land).

For decades, resource developers, environmentalists, and conservationists have argued over how this land should be used (Section 17-2, p. 432). Mineral resource extractors complain that three-fourths of the country's vast public lands, with many areas containing rich mineral resource deposits, are off limits to mining.

In recent decades, they have stepped up efforts to have Congress **(1)** open up most of these lands to mineral development or **(2)** sell off mineral-rich public lands to private interests. Conservation biologists and

Controversy over the General Mining Law of 1872

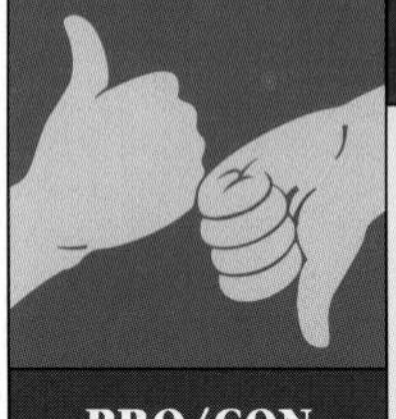

PRO/CON

Some people have gotten rich by using the little-known General Mining Law of 1872. This law was designed to **(1)** encourage mineral exploration and mining of gold, silver, copper, zinc, nickel, uranium, and other *hardrock minerals* on U.S. public lands and **(2)** help develop the then-sparsely populated West.

Under this 1872 law, a person or corporation can assume legal ownership of parcels of land on essentially all U.S. public land except parks and wilderness areas by *patenting* it. This involves **(1)** declaring the belief that the land contains valuable hardrock minerals, **(2)** spending $500 to improve the land for mineral development, **(3)** filing a claim, **(4)** paying an annual fee of $100 to maintain the claim, and, if desired, **(5)** paying the federal government $6–12 per hectare ($2.50–5.00 an acre) for the land. Once purchased, the land can be used, leased, or sold for essentially any purpose.

So far, public lands containing an estimated $240–385 billion (adjusted for inflation) of publicly owned mineral resources have been transferred to private interests at 1872 prices. Domestic and foreign mining companies operating under this law remove mineral resources worth about $2–3 billion per year on once-public land they bought at very low prices. Coal, oil, and natural gas companies pay an 8–16% royalty on the net value of fossil fuels they remove from public lands, but hardrock mining companies pay no royalties on the minerals they extract from such lands.

There is also no provision in the 1872 law requiring mining companies to pay for environmental cleanup of any damage they cause to these lands and nearby water resources. It is estimated that cleanup costs for land and streams damaged by 557,000 abandoned hardrock mines and open pits (mostly in the West) will cost U.S. taxpayers $33–72 billion.

Mining companies defend the 1872 law. They point out that

- They must invest large sums of money (often $100 million or more) to locate and develop an ore site before they make any profits from mining hardrock minerals.
- Their mining operations **(1)** provide high-paying jobs to miners, **(2)** supply vital resources for industry, **(3)** stimulate the national and local economies, **(4)** reduce trade deficits, and **(5)** save American consumers money on products produced from such minerals.
- Less than 0.25% of U.S. public lands have been transferred to private ownership.
- Paying royalties on their profits and requiring them to pay cleanup costs would force them to move their mining operations to other countries.

For decades, environmentalists have been trying, without success, to have this law revised to protect taxpayers and the environment by

- Permanently banning the patenting (sale) of public lands but allowing 20-year leases of designated public land for hardrock mining.
- Requiring mining companies to pay an 8–12% royalty on the *net* value of all minerals removed from public land.
- Making mining companies legally and financially responsible for environmental cleanup and restoration of each site or charging them an additional mining fee to help pay for cleanup costs.

Canada, Australia, South Africa, and other countries that are major extractors of hardrock minerals have laws with such requirements.

Critical Thinking

Do you support or oppose the three major changes environmentalists believe should be made in the U.S. General Mining Law of 1872? Explain.

environmentalists strongly oppose such efforts. They argue that this would **(1)** increase environmental degradation and **(2)** decrease biodiversity (Section 17-2, p. 432).

Can We Get Enough Minerals by Mining Lower-Grade Ores? Some analysts contend that all we need to do to increase supplies of a mineral is to extract lower grades of ore. They point to the development of **(1)** new earth-moving equipment, **(2)** improved techniques for removing impurities, and **(3)** other technological advances in mineral extraction and processing.

In 1900, for instance, the average copper ore mined in the United States was about 5% copper by weight. Today it is 0.5%, and copper costs less (adjusted for inflation). New methods of mineral extraction may allow even lower-grade ores of some metals to be used (Solutions, right).

However, the mining of lower-grade ores can be limited by **(1)** the cost of mining and processing larger volumes of ore, **(2)** the availability of fresh water needed to mine and process some minerals (especially in arid and semiarid areas), and **(3)** the environmental impact of the increased land disruption, waste material, and pollution produced during mining and processing (Figure 9-10).

Can We Find Substitutes for Scarce Nonrenewable Mineral Resources? The Materials Revolution Some analysts believe that even if supplies of key minerals become very expensive or scarce, human ingenuity will find substitutes. They point to the cur-

Mining with Microbes

SOLUTIONS

One way to improve mining technology is to use microorganisms for in-place (*in situ*, pronounced "in-SEE-two") mining. This biological approach to mining would **(1)** remove desired metals from ores while leaving the surrounding environment undisturbed, **(2)** reduce air pollution associated with the smelting of metal ores, and **(3)** reduce water pollution associated with using hazardous chemicals such as cyanides and mercury to extract gold.

Once a commercially viable ore deposit has been identified, wells are drilled into it and the ore is fractured. Then the ore is inoculated with natural or genetically engineered bacteria to extract the desired metal. Next the well is flooded with water, which is pumped to the surface, where the desired metal is removed. Then the water is recycled.

This technique permits economical extraction from low-grade ores, which are increasingly being used as high-grade ores are depleted. Since 1958, the copper industry has been using natural strains of the bacterium *Thiobacillus ferroxidans* to remove copper from low-grade copper ore. Currently, more than 30% of all copper produced worldwide, worth more than $1 billion a year, comes from such *biomining*. If naturally occurring bacteria cannot be found to extract a particular metal, genetic engineering techniques could be used to produce such bacteria.

However, microbiological ore processing is slow. It can take decades to remove the same amount of material that conventional methods can remove within months or years. So far, biological mining methods are economically feasible only with low-grade ore (such as gold and copper) for which conventional techniques are too expensive.

Critical Thinking

If you had a large sum of money to invest, would you invest it in the microbiological mining of aluminum ore? Explain.

rent *materials revolution* in which silicon and new materials, particularly ceramics and plastics, are being developed and used as replacements for metals.

Ceramics have many advantages over conventional metals. They are harder, stronger, lighter, and longer lasting than many metals, and they withstand intense heat and do not corrode. Within a few decades we may have high-temperature ceramic superconductors in which electricity flows without resistance. Such a development may lead to faster computers, more efficient power transmission, and affordable electromagnets for propelling high-speed magnetic levitation trains.

Plastics also have advantages over many metals. High-strength plastics and composite materials strengthened by lightweight carbon and glass fibers are likely to transform the automobile and aerospace industries. They **(1)** cost less to produce than metals because they take less energy, **(2)** do not need painting, and **(3)** can be molded into any shape. New plastics and gels also are being developed to provide superinsulation without taking up much space. One new plastic can withstand extremely high temperatures and is not affected by exposure to the most intense laser beams.

Substitutes undoubtedly can be found for many scarce mineral resources. However, finding substitutes for some key materials may be difficult or impossible. Examples are **(1)** helium, **(2)** phosphorus for phosphate fertilizers, **(3)** manganese for making steel, and **(4)** copper for wiring motors and generators.

In addition, some substitutes are inferior to the minerals they replace. For example, aluminum could replace copper in electrical wiring. However, **(1)** producing aluminum takes much more energy than producing copper and **(2)** aluminum wiring is a greater fire hazard than copper wiring.

9-6 NATURAL HAZARDS: EARTHQUAKES AND VOLCANIC ERUPTIONS

What Are Earthquakes? Stress in the earth's crust can cause solid rock to deform until it suddenly fractures and shifts along the fracture, producing a *fault* (Figure 9-4, bottom) The faulting or a later abrupt movement on an existing fault causes an **earthquake**.

An earthquake has certain features and effects (Figure 9-14). When the stressed parts of the earth suddenly fracture or shift, energy is released as shock waves, which move outward from the earthquake's focus like ripples in a pool of water. The *focus* of an earthquake is the point of initial movement, and the *epicenter* is the point on the surface directly above the focus (Figure 9-14).

One way to measure the severity of an earthquake is by its *magnitude* on a modified version of the Richter scale. The magnitude is a measure of the amount of energy released in the earthquake, as indicated by the amplitude (size) of the vibrations when they reach a recording instrument (seismograph). Using this approach, seismologists rate earthquakes as **(1)** *insignificant* (less than 4.0 on the Richter scale), **(2)** *minor* (4.0–4.9), **(3)** *damaging* (5.0–5.9), **(4)** *destructive* (6.0–6.9),

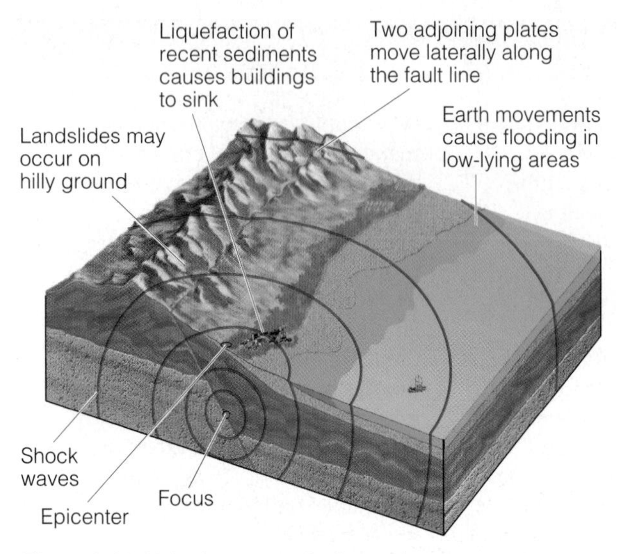

Figure 9-14 Major features and effects of an earthquake.

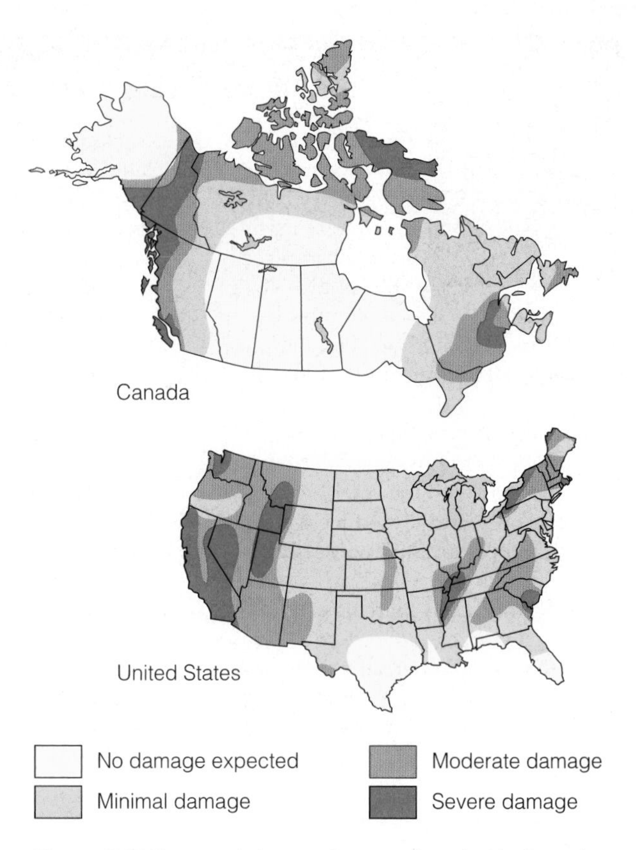

Figure 9-15 Expected damage from earthquakes in Canada and the contiguous United States. This map is based on earthquake records. (Data from U.S. Geological Survey)

(5) *major* (7.0–7.9), and **(6)** *great* (over 8.0). Each unit on the Richter scale represents an amplitude that is 10 times greater than the next smaller unit. Thus, a magnitude 5.0 earthquake is 10 times greater than a magnitude 4.0, and a magnitude 6.0 quake is 100 times greater than a magnitude 4.0 quake. Earthquakes often have *aftershocks* that gradually decrease in frequency over a period of up to several months, and some have *foreshocks* from seconds to weeks before the main shock.

The *primary effects of earthquakes* include shaking and sometimes a permanent vertical or horizontal displacement of the ground. These effects may have serious consequences for people and for buildings, bridges, freeway overpasses, dams, and pipelines.

Secondary effects of earthquakes include rockslides, urban fires, and flooding caused by subsidence (sinking) of land. Coastal areas also can be severely damaged by large, earthquake-generated water waves, called *tsunamis* (misnamed "tidal waves," even though they have nothing to do with tides) that travel as fast as 950 kilometers (590 miles) per hour.

Between 1985 and 1999, nearly 561,000 people died prematurely from natural catastrophes such as floods, earthquakes, volcanic eruptions, and windstorms. About 30%, or 169,000, of these deaths resulted from earthquakes and volcanic eruptions.

Solutions: How Can We Reduce Earthquake Hazards? We can reduce loss of life and property from earthquakes by **(1)** examining historical records and making geologic measurements to locate active fault zones, **(2)** making maps showing high-risk areas (Figure 9-15), **(3)** establishing building codes that regulate the placement and design of buildings in high-risk areas, and **(4)** trying to predict when and where earthquakes will occur.

Engineers know how to make homes, large buildings, bridges, and freeways more earthquake resistant. However, this can be expensive, especially if existing structures must be reinforced.

What Are Volcanoes? An active **volcano** occurs where magma (molten rock) reaches the earth's surface through a central vent or a long crack (fissure; Figure 9-16). Volcanic activity can release **(1)** *ejecta* (debris ranging from large chunks of lava rock to ash that may be glowing hot), **(2)** liquid lava, and **(3)** gases (such as water vapor, carbon dioxide, and sulfur dioxide) into the environment (Figure 9-1).

Volcanic activity is concentrated for the most part in the same areas as seismic activity (Figure 9-3a). Some volcanoes, such as those at Mount St. Helens in Washington (Figure 9-1) and Mount Pinatubo in the Philippines (which erupted in 1991), have a steep, flaring cone shape. They usually erupt explosively and eject large quantities of gases and particulate matter (soot and mineral ash) high into the troposphere.

Most of the particles of soot and ash soon fall back to the earth's surface. However, gases such as sulfur dioxide remain in the atmosphere and are converted to tiny droplets of sulfuric acid, many of which

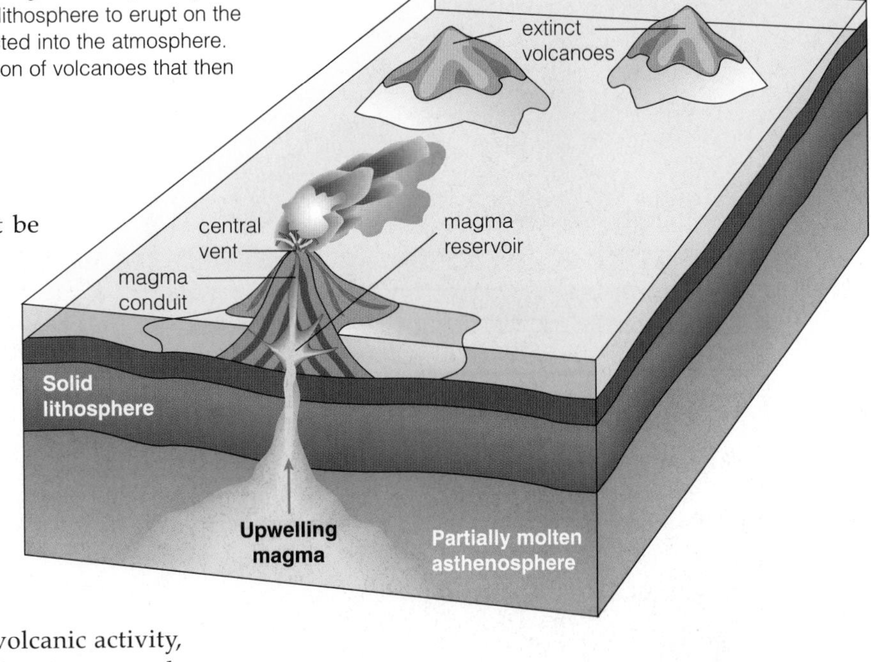

Figure 9-16 A volcano erupts when molten magma in the partially molten asthenosphere rises in a plume through the lithosphere to erupt on the surface as lava that can spill over or be ejected into the atmosphere. Chains of islands can be created by the action of volcanoes that then become inactive.

stay above the clouds and may not be washed out by rain for up to 3 years. These tiny droplets reflect some of the sun's energy and can cool the atmosphere by as much as 0.5°C (1°F) for 1–4 years.

Other volcanic eruptions at divergent boundaries (as in Iceland) and ocean islands (such as the Hawaiian Islands) usually erupt more quietly. They involve primarily lava flows, which can cover roads and villages and ignite brush, trees, and homes.

We tend to think negatively of volcanic activity, but it also provides some benefits. One is outstanding scenery in the form of majestic mountains, some lakes (such as Crater Lake in Oregon), and other landforms. Perhaps the most important benefit of volcanism is the highly fertile soils produced by the weathering of lava.

Solutions: How Can We Reduce Volcano Hazards? We can reduce the loss of human life and sometimes property from volcanic eruptions by **(1)** land-use planning, **(2)** better prediction of volcanic eruptions, and **(3)** effective evacuation plans. The eruptive history of a volcano or volcanic center can provide some indication of where the risks are.

Scientists are also studying phenomena that precede an eruption such as **(1)** tilting or swelling of the cone, **(2)** changes in magnetic and thermal properties of the volcano, **(3)** changes in gas composition, and **(4)** increased seismic activity.

9-7 SOIL RESOURCES: FORMATION AND TYPES

What Major Layers Are Found in Mature Soils? **Soil** is a complex mixture of eroded rock, mineral nutrients, decaying organic matter, water, air, and billions of living organisms, most of them microscopic decomposers (Figure 9-17). Although soil is a renewable resource, it is produced very slowly by the **(1)** weathering of rock, **(2)** deposit of sediments by erosion, and **(3)** decomposition of organic matter in dead organisms.

Mature soils are arranged in a series of zones called **soil horizons**, each with a distinct texture and composition that varies with different types of soils. A cross-sectional view of the horizons in a soil is called a **soil profile**. Most mature soils have at least three of the possible horizons (Figure 9-17).

The top layer, the *surface litter layer*, or *O horizon*, consists mostly of freshly fallen and partially decomposed leaves, twigs, animal waste, fungi, and other organic materials. Normally, it is brown or black. The *topsoil layer*, or *A horizon*, is a porous mixture of partially decomposed organic matter, called **humus**, and some inorganic mineral particles. It is usually darker and looser than deeper layers. The roots of most plants and most of a soil's organic matter are concentrated in these two upper layers. As long as these layers are anchored by vegetation, soil stores water and releases it in a nourishing trickle instead of a devastating flood.

The two top layers of most well-developed soils teem with bacteria, fungi, earthworms, and small insects that interact in complex food webs (Figure 9-18). Bacteria and other decomposer microorganisms found by the billions in every handful of topsoil recycle the nutrients we and other land organisms need (Figure 9-19, p. 212). They break down some complex organic compounds into simpler inorganic compounds soluble in water. Soil moisture carrying these dissolved nutrients is drawn up by the roots of plants and transported through stems and into leaves.

Some organic litter in the two top layers is broken down into a sticky, brown residue of partially decomposed organic material (humus). Because this humus is only slightly soluble in water, most of it stays in the topsoil layer. A fertile soil that produces high crop yields has a thick topsoil layer with lots of

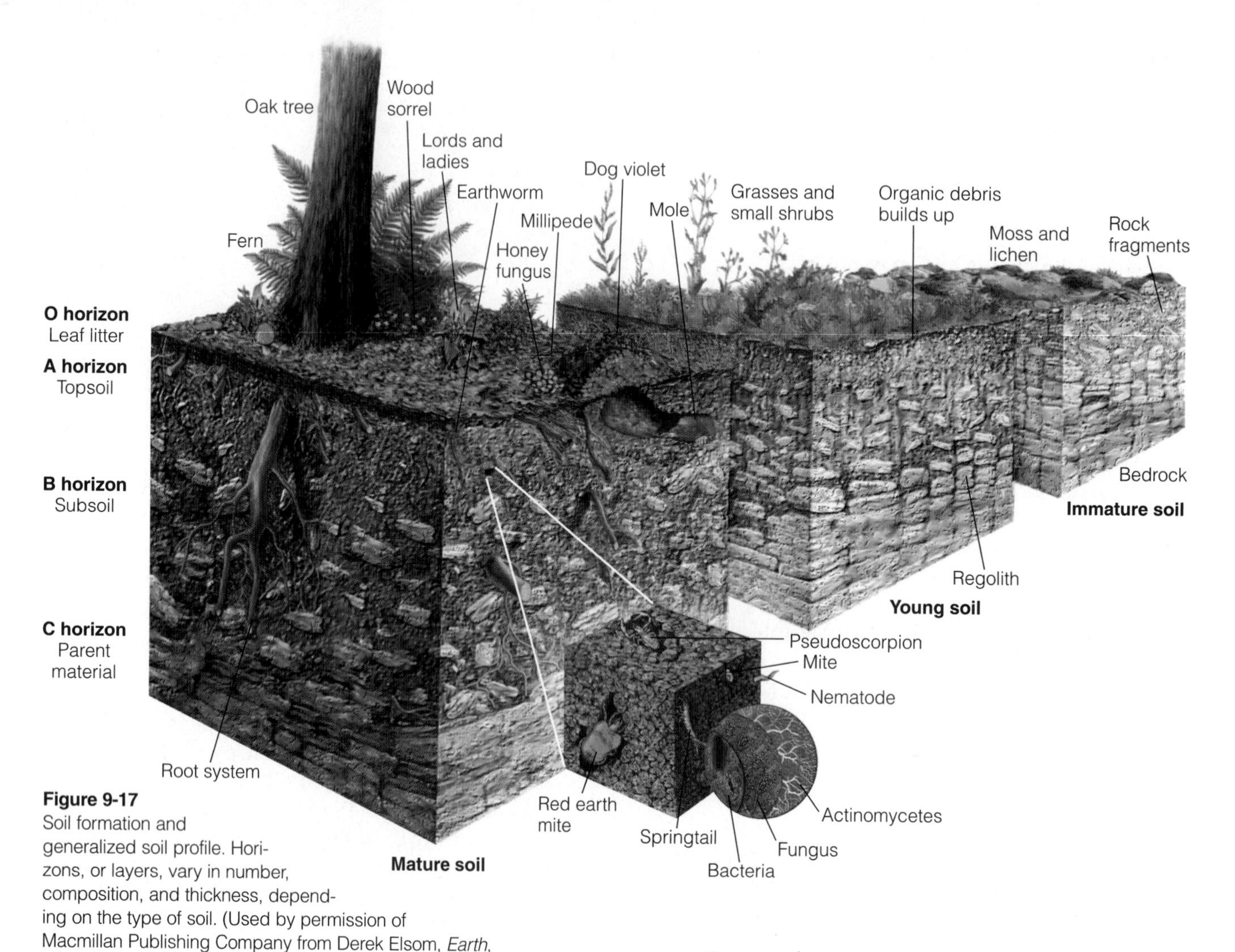

Figure 9-17 Soil formation and generalized soil profile. Horizons, or layers, vary in number, composition, and thickness, depending on the type of soil. (Used by permission of Macmillan Publishing Company from Derek Elsom, *Earth*, New York: Macmillan, 1992. Copyright © 1992 by Marshall Editions Developments Limited)

humus. This helps topsoil hold water and nutrients taken up by plant roots.

The color of its topsoil tells us a lot about how useful a soil is for growing crops. For example, dark-brown or black topsoil is nitrogen-rich and high in organic matter. Gray, bright yellow, or red topsoils are low in organic matter and need nitrogen enrichment to support most crops.

The *B horizon (subsoil)* and the *C horizon (parent material)* contain most of a soil's inorganic matter, mostly broken-down rock consisting of varying mixtures of sand, silt, clay, and gravel. The C horizon lies on a base of unweathered parent rock called *bedrock*.

The spaces, or pores, between the solid organic and inorganic particles in the upper and lower soil layers contain varying amounts of air (mostly nitrogen and oxygen gas) and water. Plant roots need oxygen for cellular respiration.

Some of the precipitation that reaches the soil percolates through the soil layers and occupies many of the soil's open spaces or pores. This downward movement of water through soil is called **infiltration**. As the water seeps down, it dissolves various soil components in upper layers and carries them to lower layers in a process called **leaching.**

Soils develop and mature slowly. It can take 200 to 1,000 years to develop an inch (2.5 centimeters) of topsoil (A horizon). Five important soil types, each with a distinct profile, are shown in Figure 9-20. Most of the world's crops are grown on soils exposed when grasslands (Figure 6-20, p. 136) and deciduous forests are cleared.

How Do Soils Differ in Texture, Porosity, and Acidity? Soils vary in their content of **(1)** *clay* (very fine particles), **(2)** *silt* (fine particles), **(3)** *sand* (medium-size particles), and **(4)** *gravel* (coarse to very coarse particles). The relative amounts of the different sizes and types of mineral particles determine **soil texture**. Soils with roughly equal mixtures of clay, sand, silt, and humus are called **loams**.

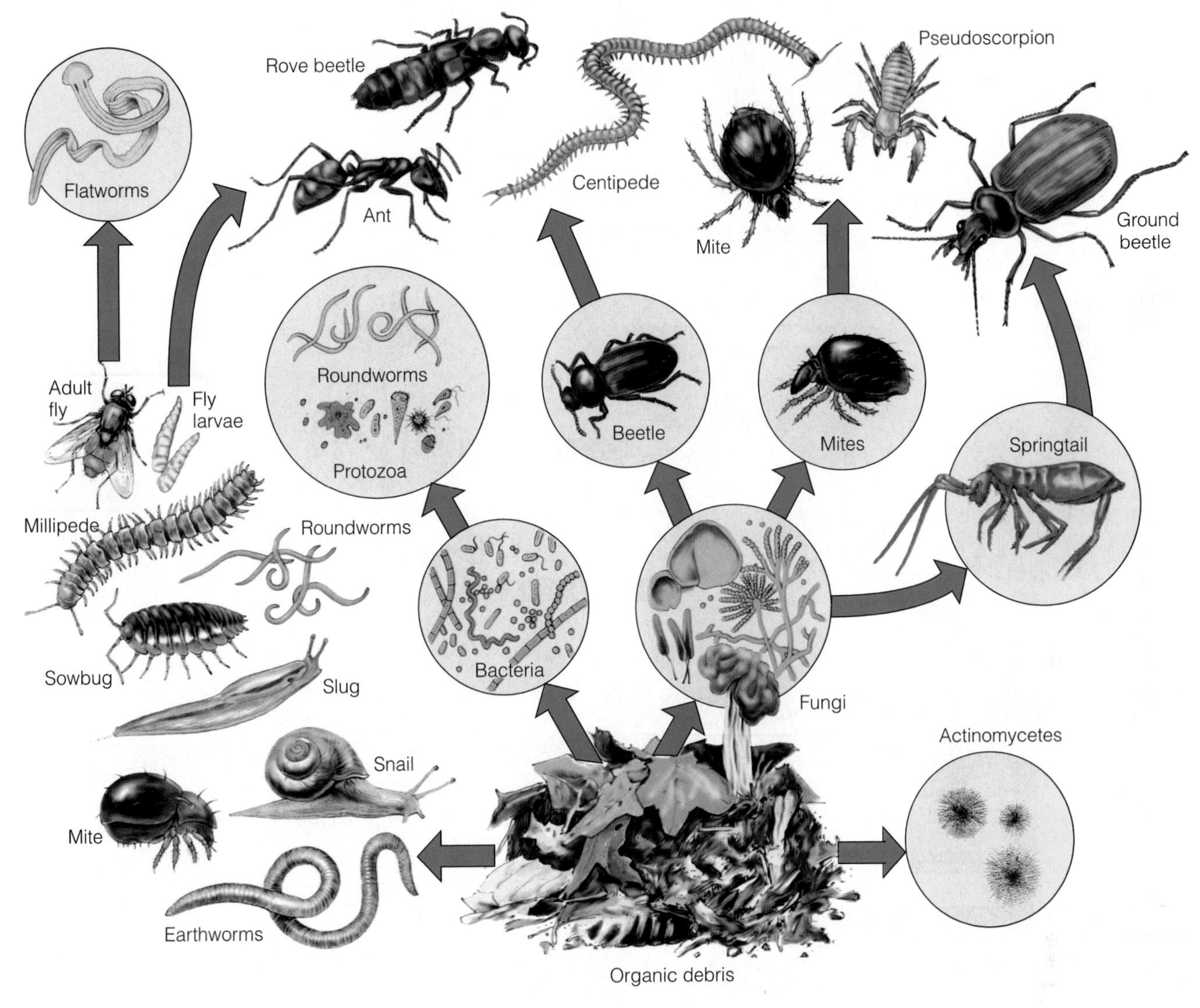

Figure 9-18 Greatly simplified food web of living organisms found in soil.

To get an idea of a soil's texture, take a small amount of topsoil, moisten it, and rub it between your fingers and thumb. A gritty feel means that it contains a lot of sand. A sticky feel means a high clay content, and you should be able to roll it into a clump. Silt-laden soil feels smooth, like flour. A loam topsoil, which is best suited for plant growth, has a texture between these extremes—a crumbly, spongy feeling—with many of its particles clumped loosely together.

Soil texture helps determine **soil porosity**, a measure of the volume of pores or spaces per volume of soil and of the average distances between those spaces. Fine particles are needed for water retention and coarse ones for air spaces. A porous soil has many pores and can hold more water and air than a less porous soil. The average size of the spaces or pores in a soil determines **soil permeability**: the rate at which water and air move from upper to lower soil layers. Soil porosity is also influenced by **soil structure**: the ways in which soil particles are organized and clumped together (Figure 9-21, p. 214). Table 9-1 compares the main physical and chemical properties of sand, clay, silt, and loam soils.

Loams are the best soils for growing most crops because they hold lots of water but not too tightly for plant roots to absorb. Sandy soils are easy to work, but water flows rapidly through them (Figure 9-21, left). They are useful for growing irrigated crops or those with low water needs, such as peanuts and strawberries.

The particles in clay soils are very small and easily compacted. When these soils get wet, they form

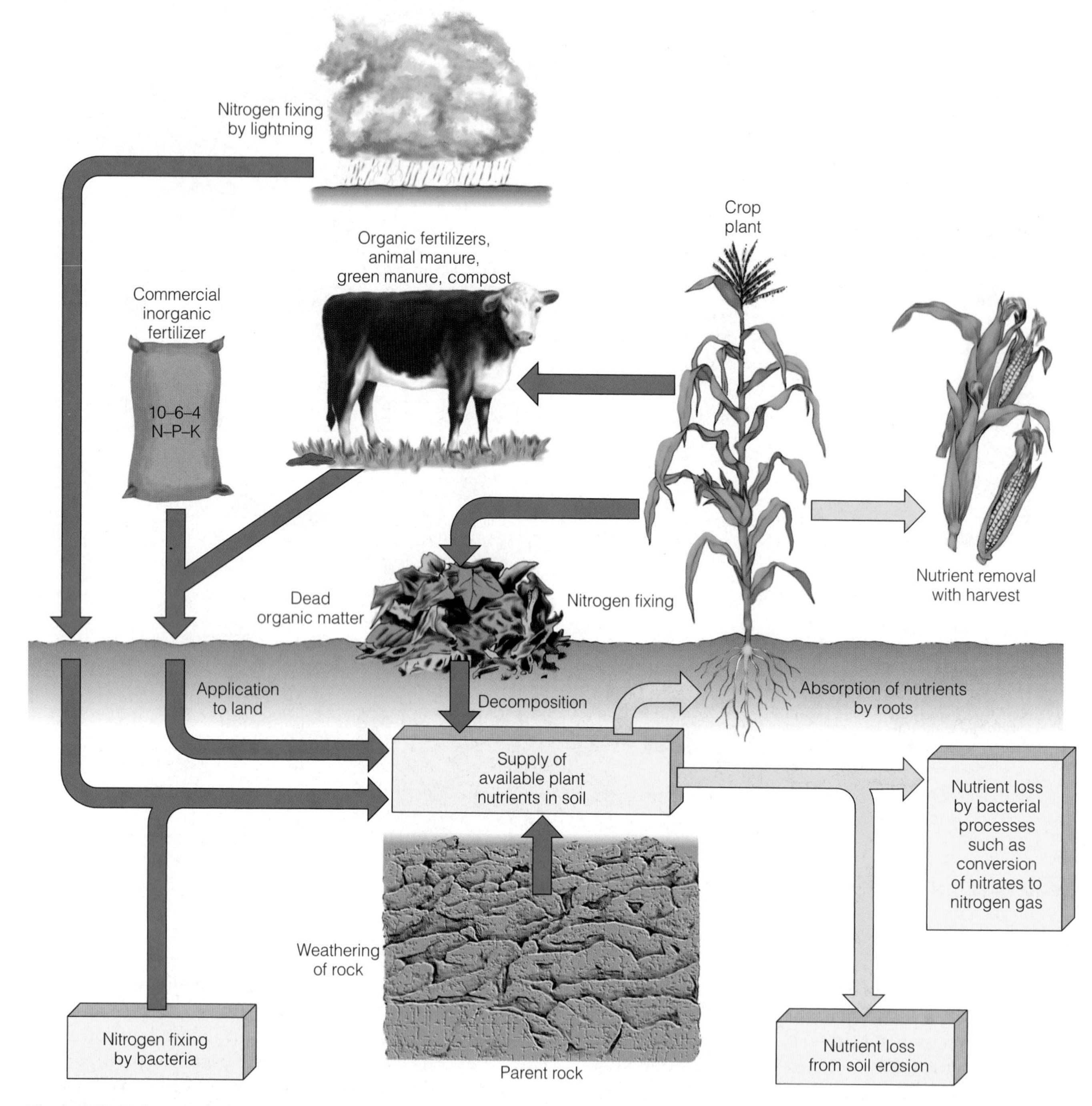

Figure 9-19 Pathways of plant nutrients in soils.

Table 9-1 Properties of Soils with Different Textures

Soil texture	Nutrient-Holding Capacity	Water Infiltration Capacity	Water-Holding Capacity	Aeration	Workability
Clay	Good	Poor	Good	Poor	Poor
Silt	Medium	Medium	Medium	Medium	Medium
Sand	Poor	Good	Poor	Good	Good
Loam	Medium	Medium	Medium	Medium	Medium

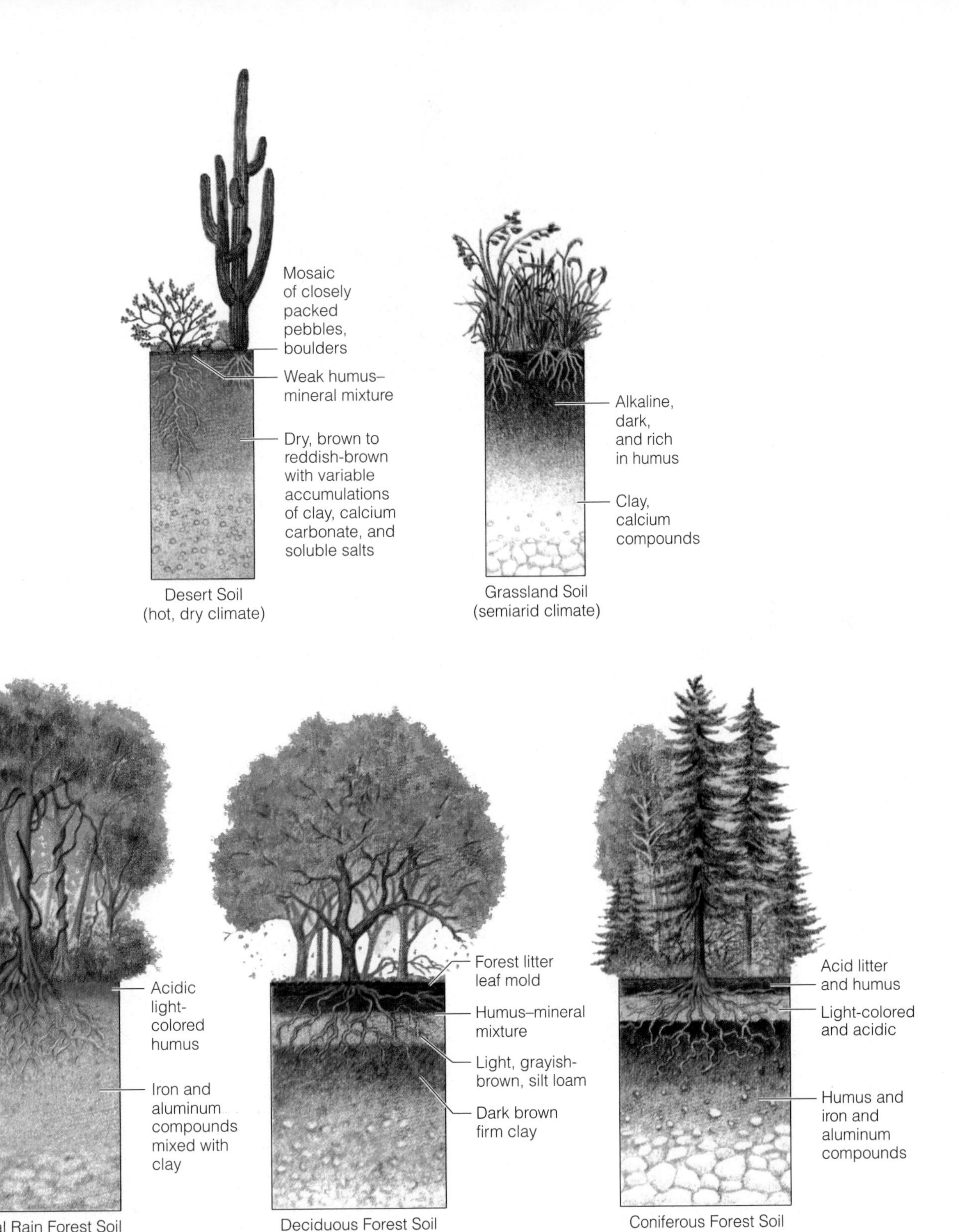

Figure 9-20 Soil profiles of the principal soil types typically found in five different biomes.

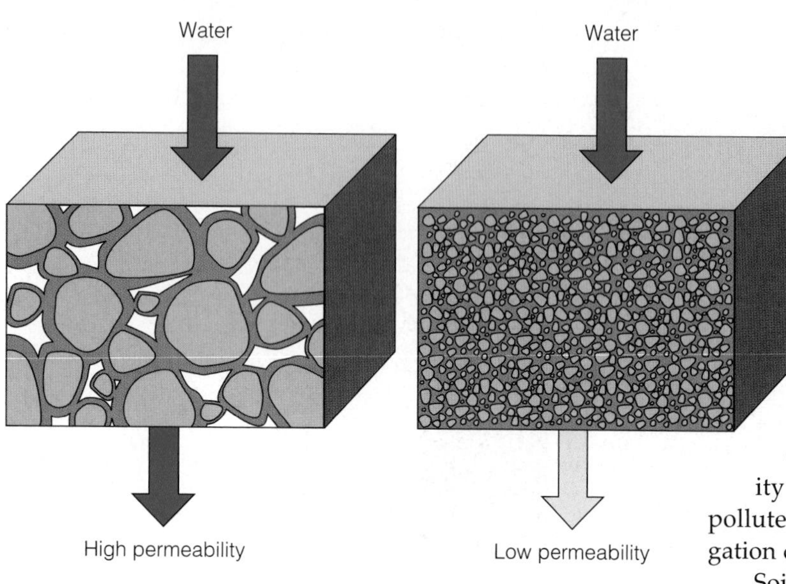

Figure 9-21 The size, shape, and degree of clumping of soil particles determine the number and volume of spaces for air and water within a soil.

large, dense clumps, which is why wet clay can be molded into bricks and pottery. Clay soils are more porous and have a greater water-holding capacity than sandy soils, but the pore spaces are so small that these soils have a low permeability (Figure 9-21, right). Because little water can infiltrate to lower levels, the upper layers can easily become too waterlogged for most crops.

The acidity or alkalinity of a soil, as measured by its pH (Figure 3-5, p. 63), influences the uptake of soil nutrients by plants. When soils are too acidic, the acids can be partially neutralized by an alkaline substance such as lime. Because lime speeds up the decomposition of organic matter in the soil, however, manure or another organic fertilizer should be added to maintain soil fertility.

In dry regions such as much of the western and southwestern United States, rain does not leach away calcium and other alkaline compounds. Therefore, soils in such areas may be too alkaline (pH above 7.5) for some crops. Adding sulfur, which is gradually converted into sulfuric acid by soil bacteria, reduces soil alkalinity.

9-8 SOIL EROSION

What Causes Soil Erosion? **Soil erosion** is the movement of soil components, especially surface litter and topsoil (Figure 9-17), from one place to another. It results in the buildup of sediments and sedimentary rock on land and in bodies of water (Figure 9-5). The two main agents of erosion are flowing water and wind. Some soil erosion is natural, and some is caused by human activities. In undisturbed vegetated ecosystems, the roots of plants help anchor the soil, and usually soil is not lost faster than it forms.

Farming, logging, construction, overgrazing by livestock, off-road vehicles, deliberate burning of vegetation, and other activities that destroy plant cover leave soil vulnerable to erosion. Such human activities can speed up erosion and destroy in a few decades what nature took hundreds to thousands of years to produce. Moving water causes most soil erosion.

The two major harmful effects of soil erosion are **(1)** loss of soil fertility and its ability to hold water and **(2)** runoff of sediment that pollutes water, kills fish and shellfish, and clogs irrigation ditches, boat channels, reservoirs, and lakes.

Soil, especially topsoil, is classified as a renewable resource because natural processes regenerate it. However, in tropical and temperate areas it takes 200–1,000 years (depending on climate and soil type) for 2.54 centimeters (1 inch) of new topsoil to form. If topsoil erodes faster than it forms on a piece of land, the soil becomes a nonrenewable resource.

How Serious Is Global Soil Erosion? Several studies document the seriousness of soil erosion:

- A United Nations (UN) Environment Programme survey found that topsoil is eroding faster than it forms on about one-third of the world's cropland, causing an estimated 85% of the world's land degradation from human activities (Figure 1-3, p. 5).
- A 1992 study by the World Resources Institute and the UN Environment Programme found that soil on an area equal to the size of China and India combined had been seriously eroded since 1945 (Figure 9-22). The study also found that about 15% of land

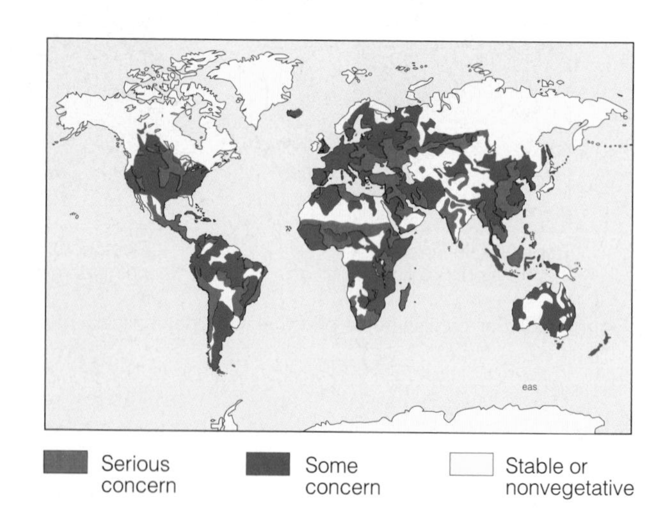

Figure 9-22 Global soil erosion. (Data from UN Environment Programme and the World Resources Institute)

scattered across the globe was too eroded to grow crops anymore because of a combination of **(1)** overgrazing (35%), **(2)** deforestation (30%), and **(3)** unsustainable farming (28%). Two-thirds of these seriously degraded lands are in Asia and Africa.

- According to a 2000 study by the Consultative Group on International Agricultural Research, **(1)** nearly 40% of the world's land (75% in Central America) used for agriculture is seriously degraded by erosion, salt buildup (salinization), and waterlogging, and **(2)** soil degradation has reduced food production on about 16% of the world's cropland.

The situation is worsening as many poor farmers in some developing countries plow up marginal (easily erodible) lands to survive. Soil expert David Pimentel estimates that worldwide soil erosion causes at least $375 billion per year (an average of $42 million per hour) in **(1)** direct damage to agricultural lands and **(2)** indirect damage to waterways, infrastructure, and human health.

How Serious Is Soil Erosion in the United States? According to the National Resources Conservation Service, about one-third of the nation's original prime topsoil has been washed or blown into streams, lakes, and oceans, mostly as a result of overcultivation, overgrazing, and deforestation (Case Study, p. 216).

According to the U.S. Department of Agriculture (USDA), soil on cultivated land in the United States is eroding about 16 times faster than it can form. Erosion rates are even higher in heavily farmed regions. An example is the Great Plains, which has lost one-third or more of its topsoil in the 150 years since it was first plowed. Some of the country's most productive agricultural lands, such as those in Iowa, have lost about half their topsoil.

Because of soil conservation efforts, the USDA estimates that soil erosion in the United States decreased by about 40% between 1985 and 1997. Using these data, USDA researchers estimate that soil erosion cost the United States about $30 billion in 1997, an average loss of $3.4 million per hour.

Critics such as Pierre Crosson say that these estimates of soil erosion and damages from such erosion are **(1)** exaggerated and **(2)** based on inexact models instead of field measurements of soil loss and sedimentation rates in nearby bodies of water.

However, David Pimentel and others point out that current estimates by models and a few on-site measurements do not include all the ecological effects of soil erosion. Such effects include reductions in **(1)** soil depth, **(2)** availability of soil water for crops, and **(3)** soil organic matter and nutrients. When such effects are included, some soil scientists and ecologists estimate that soil erosion causes a 15–30% reduction in U.S. crop productivity.

Inorganic fertilizers can be used to help replace nutrients eroded from cropland soil. However, because fertilizers are not a substitute for fertile soil, there is a limit to the amount of fertilizer that can be applied before crop yields level off and then begin to decline.

What Is Desertification, and How Serious Is This Problem? **Desertification** is a process whereby the productive potential of arid or semiarid land falls by 10% or more because of human activities and climate changes. Desertification can be **(1)** *moderate* (with a 10–25% drop in productivity), **(2)** *severe* (with a 25–50% drop), and **(3)** *very severe* (with a drop of 50% or more, usually creating huge gullies and sand dunes). Desertification is a serious and growing problem in many parts of the world (Figure 9-23).

Desertification is caused mainly by prolonged drought and unsustainable human activities. Human practices that leave topsoil vulnerable to desertification include **(1)** overgrazing on fragile arid and semiarid rangelands, **(2)** deforestation without reforestation, **(3)** surface mining without land reclamation, **(4)** irrigation techniques that lead to increased erosion, **(5)** salt buildup in irrigated soil, **(6)** farming on land with unsuitable terrain or soils, and **(7)** soil compaction by farm machinery and cattle.

Major symptoms of desertification include

- Loss of native vegetation
- Increased erosion of the dry soil (especially by wind; Figure 6-1, p. 121)
- Salt buildup in the soil (salinization)
- Lowering of the water table as wells are dug deeper
- Reduced surface water supply as streams and ponds dry up

The consequences of desertification include **(1)** worsening drought because of lower amounts of water evaporated into the atmosphere to form rain clouds, **(2)** lower

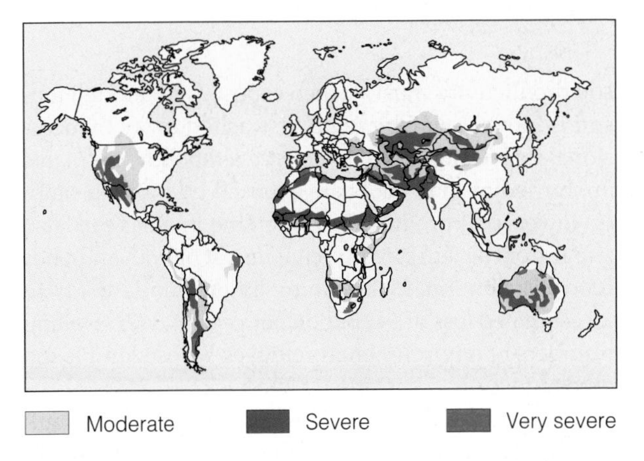

Figure 9-23 Desertification of arid and semiarid lands. (Data from UN Environmental Programme and Harold E. Drengue)

The Dust Bowl

CASE STUDY

In the 1930s, Americans learned a harsh environmental lesson when much of the topsoil in several dry and windy midwestern states was lost through a combination of poor cultivation practices and prolonged drought.

Before settlers began grazing livestock and planting crops there in the 1870s, the deep and tangled root systems of native prairie grasses anchored the fertile topsoil firmly in place (Figure 9-20, top right). Plowing the prairie tore up these roots, and the agricultural crops the settlers planted annually in their place had less extensive root systems.

After each harvest, the land was plowed and left bare for several months, exposing it to high winds. Overgrazing also destroyed large expanses of grass, denuding the ground. The stage was set for severe wind erosion and crop failures; all that was needed was a long drought.

Such a drought occurred between 1926 and 1934. In the 1930s, dust clouds created by hot, dry windstorms darkened the sky at midday in some areas; rabbits and birds choked to death on the dust.

During May 1934, a cloud of topsoil blown off the Great Plains traveled some 2,400 kilometers (1,500 miles) and blanketed most of the eastern United States with dust. Journalists gave the Great Plains a new name: the *Dust Bowl* (Figure 9-24).

During the "dirty thirties," large areas of cropland were stripped of topsoil and severely eroded. Thousands of displaced farm families from Oklahoma, Texas, Kansas, and Colorado migrated to California or to the industrial cities of the Midwest and East. Most found no jobs because the country was in the midst of the Great Depression.

In May 1934, Hugh Bennett of the U.S. Department of Agriculture (USDA) went before a congressional hearing in Washington to plead for new programs to protect the country's topsoil. Lawmakers took action when Great Plains dust began seeping into the hearing room.

In 1935, the United States passed the Soil Erosion Act, which established the Soil Conservation Service (SCS) as part of the USDA. With Bennett as its first head, the SCS (now called the Natural Resources Conservation Service) began promoting sound conservation practices, first in the Great Plains states and later elsewhere. Soil conservation districts were formed throughout the country, and farmers and ranchers were given technical assistance in setting up soil conservation programs.

Climate researchers see signs of a returning Dust Bowl period because of a megadrought that lasts 2 to 4 decades. By examining tree rings, archaeological finds, lake sediments, and sand dunes, scientists have found that prolonged megadroughts generally hit twice a century as part of a complex drought cycle. They also found that smaller 2-year droughts strike every 20 years or so. If the earth warms as projected, the region could become even drier, and farming might have to be abandoned.

Figure 9-24 The Dust Bowl of the Great Plains, where a combination of extreme drought and poor soil conservation practices led to severe wind erosion of topsoil in the 1930s.

Critical Thinking

Do you think Americans learned a lesson about protecting soil as a result of the Dust Bowl in the 1930s? Explain.

soil productivity, mostly when exposed and eroded topsoil is blown away by wind or washed away by occasional rainstorms, **(3)** damaged vegetation when plants are buried or their roots are exposed by moving sand, **(4)** lower food production, **(5)** famine in areas that also suffer from severe poverty, civil unrest or war, and poor food distribution, **(6)** declining living standards, with an estimated loss of $42 billion per year, and **(7)** swelling numbers of environmental refugees whose land is too eroded to grow crops or feed livestock.

An estimated 8.1 million square kilometers (3.1 million square miles)—an area the size of Brazil and 12 times the size of Texas—have become desertified in the past 50 years. According to a 2001 United Nations report on desertification, **(1)** about 40% of the world's land, and 70% of all drylands, is suffering from the effects of desertification, and **(2)** each year about 150,000 square kilometers (58,000 square miles)—an area larger than Greece—becomes desertified. This threatens the livelihoods of at least 1 billion people in 100 countries.

Solutions: How Can We Slow Desertification? The most effective way to slow desertification is to reduce overgrazing, deforestation, and destructive forms of planting, irrigation, and mining. In addition, planting trees and grasses will anchor soil and hold water while slowing desertification and reducing the threat of global warming.

How Do Excess Salts and Water Degrade Soils? The approximately 17% of the world's cropland that is irrigated produces almost 40% of the world's food. Irrigated land can produce crop yields that are two to three times greater than those from rain watering.

However, irrigation also has a downside. Most irrigation water is a dilute solution of various salts, picked up as the water flows over or through soil and rocks. Small quantities of these salts are essential nutrients for plants, but they are toxic in large amounts.

Irrigation water not absorbed into the soil evaporates, leaving behind a thin crust of dissolved salts (such as sodium chloride) in the topsoil. The accumulation of these salts, called **salinization** (Figure 9-25), stunts crop growth, lowers yields, and eventually kills plants and ruins the land. According to a 1995 study, severe salinization has reduced yields on 21% of the world's irrigated cropland, and another 30% has been moderately salinized. The most severe salinization occurs in Asia, especially in China, India, and Pakistan. As irrigation continues throughout heavily farmed parts of the world, salinization will increase and crop yields will fall.

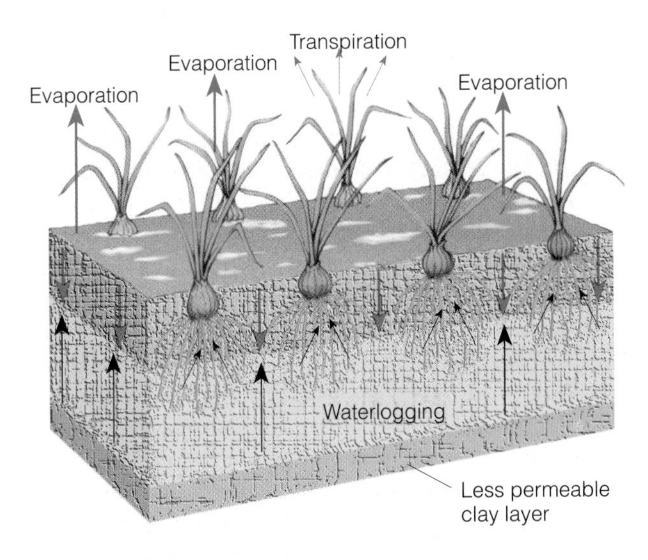

Salinization

1. Irrigation water contains small amounts of dissolved salts.
2. Evaporation and transpiration leave salts behind.
3. Salt builds up in soil.

Waterlogging

1. Precipitation and irrigation water percolate downward.
2. Water table rises.

Figure 9-25 Salinization and waterlogging of soil on irrigated land without adequate drainage lead to decreased crop yields.

In the United States, salinization affects 23% of all irrigated cropland. However, the proportion is much higher in some heavily irrigated western states. This includes 66% of the irrigated land in the lower Colorado Basin and 35% of such land in California.

Precipitation can desalinate soil, but this takes thousands of years in arid and semiarid areas where irrigation is used. Salts can be flushed out of soil by applying much more irrigation water than is needed for crop growth. However, this practice increases pumping and crop production costs and wastes enormous amounts of water.

Heavily salinized soil also can be renewed by **(1)** taking the land out of production for 2 to 5 years, **(2)** installing an underground network of perforated drainage pipes, and **(3)** flushing the soil with large quantities of low-salt water. However, this costly scheme only slows the salt buildup; it does not stop the process. Flushing salts from the soil also makes downstream irrigation water saltier unless the saline water can be drained into evaporation ponds rather than returned to the stream or canal. Such ponds can be toxic to various forms of wildlife, especially migrating birds that rely on wetlands for food.

Methods for reducing the threat of salinization include **(1)** reducing the use of irrigation water (p. 332), **(2)** switching to more salt-tolerant crops such as barley, cotton, sugarbeet, and semidwarf wheat, and **(3)** planting salt-loving plants (halophytes), such as saltbush, to convert heavily salinized cropland to grazing land.

Another problem with irrigation is **waterlogging** (Figure 9-25). Farmers often apply large amounts of irrigation water to leach salts deeper into the soil. Without adequate drainage, however, water accumulates underground and gradually raises the water table. Saline water then envelops the deep roots of plants, lowering their productivity and killing them after prolonged exposure. At least one-tenth of all irrigated land worldwide suffers from waterlogging, and the problem is getting worse.

9-9 SOLUTIONS: SOIL CONSERVATION

How Can Conservation Tillage Reduce Soil Erosion? **Soil conservation** involves reducing soil erosion and restoring soil fertility. For hundreds of years, farmers have used various methods to reduce soil erosion, mostly by keeping the soil covered with vegetation.

In **conventional-tillage farming** the land is plowed and the soil is broken up and smoothed to make a planting surface. In areas such as the midwestern United States, harsh winters prevent plowing just before the spring growing season. Thus, cropfields often are plowed in the fall. This leaves the soil bare during the winter and early spring and makes it vulnerable to erosion, loss of soil organic matter, and leaching of nutrients to streams and groundwater.

To reduce erosion, many U.S. farmers are using **conservation-tillage farming** (either *minimum-tillage* or *no-till farming*). The idea is to disturb the soil as little as possible while planting crops. With *minimum-tillage farming*, special tillers break up and loosen the subsurface soil without turning over the topsoil, previous crop residues, and any cover vegetation. In *no-till farming*, special planting machines inject seeds, fertilizers, and weed-killers (herbicides) into slits made in the unplowed soil.

Besides reducing soil erosion, conservation tillage **(1)** decreases decomposition of soil organic matter, **(2)** saves fuel, **(3)** cuts costs, **(4)** holds more water in the soil, **(5)** keeps the soil from getting packed down, **(6)** allows more crops to be grown during a season (multiple cropping), **(7)** gives yields at least as high as those from conventional tillage, and **(8)** reduces the release of carbon dioxide from the soil to the air, which helps ease global warming.

At first, conservation tillage was thought to increase herbicide use, but a 1990 USDA study of corn production in the United States found no real difference in levels of herbicide use between conventional and conservation-tillage systems. However, no-till cultivation of corn does leave stalks. They can serve as habitats for the corn borer insect pest, which can increase the need for pesticides. Crop residues also are a good home for a fungal disease called wheat scab, which can devastate monoculture wheat crops.

By 2001, conservation tillage was used on about 45% of U.S. cropland and is projected to be used on more than half of it by 2005. In Indiana, the Nature Conservancy is giving farmers money to buy no-till equipment in exchange for a promise to use conservation tillage for at least 3 years. The USDA estimates that using conservation tillage on 80% of U.S. cropland would reduce soil erosion by at least half. So far, the practice is not widely used in other parts of the world.

How Can Terracing, Contour Farming, Strip Cropping, and Alley Cropping Reduce Soil Erosion? **Terracing** can reduce soil erosion on steep slopes, each of which is converted into a series of broad, nearly level terraces that run across the land contour (Figure 9-26a). Terracing retains water for crops at each level and reduces soil erosion by controlling runoff.

In mountainous areas such as the Himalayas and the Andes, farmers traditionally built elaborate systems of terraces to grow crops. Today, however, some of these slopes are being farmed without terraces. This leaves the land too nutrient poor to grow crops or generate new forest after 10–40 years. Although most poor farmers know the risk of not terracing, many have too little time and too few workers to build terraces; they must plant crops or starve. The resultant loss of protective vegetation and topsoil also greatly intensifies flooding and buildup of sediment in streams in the valleys below.

Contour farming can reduce soil erosion by 30–50% on gently sloping land. It involves plowing and planting crops in rows across the sloped contour of the land (Figure 9-26b). Each row planted along the contour of the land acts as a small dam to help hold soil and slow water runoff.

In **strip cropping**, a row crop such as corn alternates in strips with another crop (such as a grass or a grass–legume mixture) that completely covers the soil and thus reduces erosion (Figure 9-26b). The strips of the cover crop **(1)** trap soil that erodes from the row crop, **(2)** catch and reduce water runoff, and **(3)** help prevent the spread of pests and plant diseases. Planting nitrogen-fixing legumes (such as soybeans or alfalfa) in some of the strips helps restore soil fertility.

Alley cropping, or **agroforestry**, also can reduce erosion. It is a form of *intercropping* in which several crops are planted together in strips or alleys between trees and shrubs that can provide fruit or fuelwood (Figure 9-26c). The trees provide shade (which reduces water loss by evaporation) and help to retain and slowly release soil moisture. The tree and shrub trimmings can be used as mulch (green manure) for the crops and as fodder for livestock.

How Can Gully Reclamation, Windbreaks, and Land Classification Reduce Soil Erosion? **Gully reclamation** can restore sloping bare land on which water runoff quickly creates gullies. Ways to do this include **(1)** seeding small gullies with quick-growing plants such as oats, barley, and wheat for the first season, **(2)** building small dams at the bottoms of deep gullies to collect silt and gradually fill in the channels, **(3)** planting fast-growing shrubs, vines, and trees to stabilize the soil, and **(4)** building channels to divert water from the gully and prevent further erosion.

Long rows of trees can be planted as **windbreaks**, or **shelterbelts**, to reduce wind erosion (Figure 9-26d). Windbreaks also **(1)** help retain soil moisture, **(2)** supply some wood for fuel, and **(3)** provide habitats for birds, pest-eating and pollinating insects, and other animals. Many of the windbreaks planted in the upper Great Plains of the United States after the 1930s Dust Bowl disaster (Case Study, p. 216) have been cut down to make way for large irrigation systems and modern farm machinery.

Land classification can be used to identify easily erodible (marginal) land that should be neither planted

(a) Terracing

(b) Contour planting and strip cropping

(c) Alley cropping

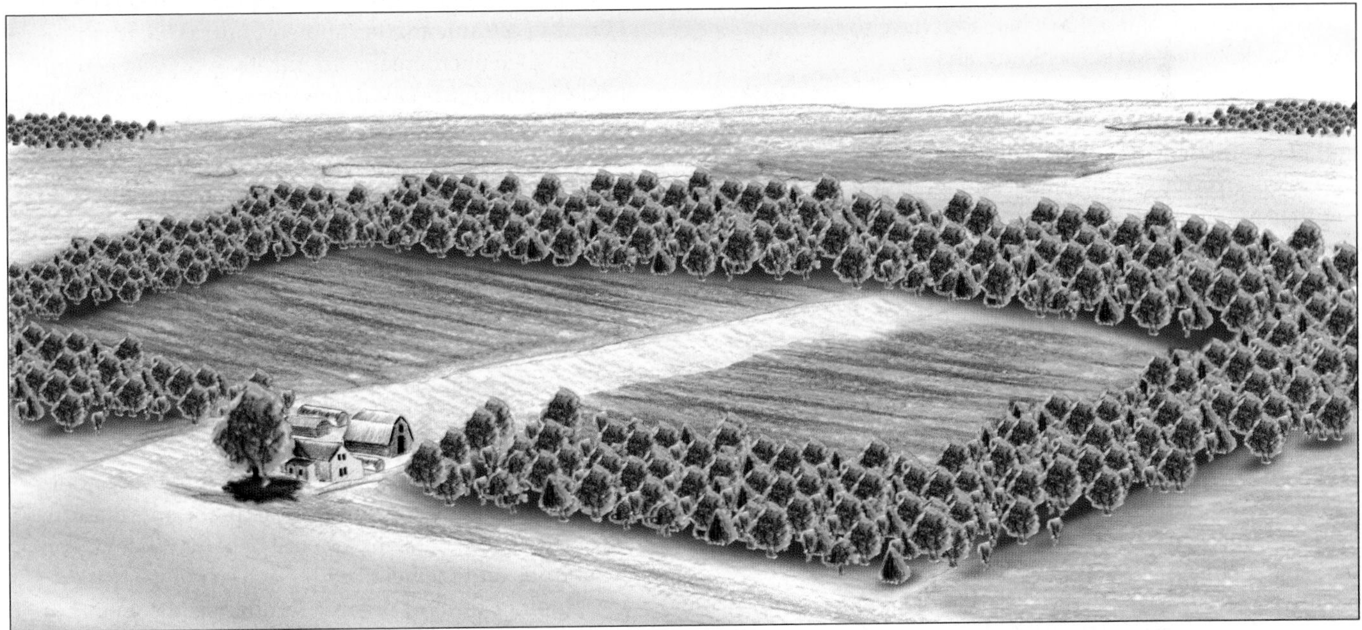
(d) Windbreaks

Figure 9-26 Soil conservation methods: **(a)** terracing, **(b)** contour planting and strip cropping, **(c)** alley cropping, and **(d)** windbreaks.

CASE STUDY

Slowing Soil Erosion in the United States

Of the world's major food-producing countries, only the United States is sharply reducing some of its soil losses through conservation tillage and government-sponsored soil conservation programs.

The 1985 Farm Act established a strategy for reducing soil erosion in the United States. In the first phase of this program, farmers are given a subsidy for taking highly erodible land out of production and replanting it with soil-saving grass or trees for 10–15 years. By early 2001, approximately 13.6 million hectares (33.5 million acres) were in this Conservation Reserve Program (CRP).

The land in such a *conservation reserve* cannot be farmed, grazed, or cut for hay. Farmers who violate their contracts must pay back all subsidies plus interest.

According to the U.S. Department of Agriculture, since 1985 this program has cut soil losses on cropland in the United States by about 65%—a shining example of *good news*—and could eventually cut such losses as much as 80%. In 1996, Congress reauthorized the CRP until 2002.

The second phase of the program required all farmers with highly erodible land to develop government-approved 5-year soil conservation plans for their entire farms by the end of 1990. A third provision of the Farm Act authorizes the government to forgive all or part of farmers' debts to the Farmers Home Administration if they agree not to farm highly erodible cropland or wetlands for 50 years. The farmers must plant trees or grass on this land or restore it to wetland.

The 1985 Farm Act made the United States the first major food-producing country to make soil conservation a national priority. Even though these efforts to slow soil erosion are an important step, effective soil conservation is practiced on only about half of all U.S. agricultural land and on less than half of the country's most erodible cropland.

Critical Thinking

Do you believe that U.S. tax dollars should be used to pay farmers for taking highly erodible land out of production? Explain. What are the alternatives?

in crops nor cleared of vegetation. In the United States, the National Resources Conservation Service has set up a classification system to identify types of land that are suitable or unsuitable for cultivation. Such efforts and recent farm legislation have helped reduce soil erosion in the United States (Case Study, above).

How Can We Maintain and Restore Soil Fertility? Fertilizers partially restore plant nutrients lost by erosion, crop harvesting, and leaching. Farmers can use either **organic fertilizer** from plant and animal materials or **commercial inorganic fertilizer** produced from various minerals.

Three basic types of *organic fertilizer* are animal manure, green manure, and compost. **Animal manure** includes the dung and urine of cattle, horses, poultry, and other farm animals. It **(1)** improves soil structure, **(2)** adds organic nitrogen, and **(3)** stimulates beneficial soil bacteria and fungi.

Despite its effectiveness, the use of animal manure in the United States has decreased. Reasons for this are **(1)** replacement of most mixed animal-raising and crop-farming operations with separate operations for growing crops and raising animals, **(2)** the high costs of transporting animal manure from feedlots near urban areas to distant rural crop-growing areas, and **(3)** replacement of horses and other draft animals that added manure to the soil with tractors and other motorized farm machinery.

Green manure is fresh or growing green vegetation plowed into the soil to increase the organic matter and humus available to the next crop. **Compost** is a sweet-smelling, dark-brown, humuslike material that is rich in organic matter and soil nutrients. It is produced when microorganisms (mostly fungi and aerobic bacteria) in soil break down organic matter such as leaves, food wastes, paper, and wood in the presence of oxygen. Compost is a rich natural fertilizer and soil conditioner that **(1)** aerates soil, **(2)** improves its ability to retain water and nutrients, **(3)** helps prevent erosion, and **(4)** prevents nutrients from being wasted by being dumped in landfills.

Farmers, homeowners, and communities produce compost by piling up alternating layers of **(1)** nitrogen-rich wastes (such as grass clippings, weeds, animal manure, and vegetable kitchen scraps), **(2)** carbon-rich plant wastes (dead leaves, hay, straw, sawdust), and **(3)** topsoil. Compost **(1)** provides a home for microorganisms that help decompose the plant and manure layers and **(2)** reduces the amount of plant wastes taken to landfills and incinerators.

Another form of organic fertilizer is the spores of mushrooms, puffballs, and truffles. Rapidly growing and spreading mycorrhizae fungi in the spores attach to plant roots and help them take in moisture and nutrients from the soil. Unlike typical fertilizers that must be applied every few weeks, one application of mushroom fungi lasts all year and costs just pennies per plant. The fungi also produce a bigger root system that makes plants more disease resistant.

Corn, tobacco, and cotton can deplete the topsoil of nutrients (especially nitrogen) if planted on the same land several years in a row. One way to reduce such losses is **crop rotation.** Farmers plant areas or strips with nutrient-depleting crops one year. In the next year they plant the same areas with legumes (whose root nodules add nitrogen to the soil). In addition to helping restore soil nutrients, this method **(1)** reduces erosion by keeping the soil covered with vegetation and **(2)** helps reduce crop losses to insects by presenting them with a changing target.

Can Inorganic Fertilizers Save the Soil? Today, many farmers (especially in developed countries) rely on *commercial inorganic fertilizers* containing **(1)** nitrogen (as ammonium ions, nitrate ions, or urea), **(2)** phosphorus (as phosphate ions), and **(3)** potassium (as potassium ions). Other plant nutrients may also be present in low or trace amounts.

Inorganic commercial fertilizers are easily transported, stored, and applied. Worldwide, their use increased about 9-fold between 1950 and 1989 but has leveled off since then. Today, the additional food commercial fertilizers help produce feeds one of every three people in the world; without them, world food output would drop an estimated 40%.

Commercial inorganic fertilizers have some disadvantages, however. These include **(1)** not adding humus to the soil, **(2)** reducing the soil's content of organic matter and thus its ability to hold water (unless animal manure and green manure are also added to the soil), **(3)** lowering the oxygen content of soil and keeping fertilizer from being taken up as efficiently, **(4)** typically supplying only 2 or 3 of the 20 or so nutrients needed by plants, **(5)** requiring large amounts of energy for their production, transport, and application, and **(6)** releasing nitrous oxide (N_2O), a greenhouse gas that can enhance global warming, from the soil.

The widespread use of commercial inorganic fertilizers, especially on sloped land near streams and lakes, also causes water pollution as nitrate (NO_3^-) and phosphate (PO_4^{3-}) fertilizer nutrients are washed into nearby bodies of water. The resulting plant nutrient enrichment (cultural eutrophication) causes algae blooms that use up oxygen dissolved in the water, thereby killing fish, as discussed in more detail on p. 348. Rainwater seeping through the soil also can leach nitrates in commercial fertilizers into groundwater. Drinking water drawn from wells containing high levels of nitrate ions can be toxic, especially for infants, and causes bladder cancer.

According to soil scientists, responsibility for reducing soil erosion should not be limited to farmers. Timber cutting, overgrazing, mining, and urban development that are carried out without proper regard for soil conservation also cause soil erosion. See the website material for this chapter for some things you can do to reduce soil erosion.

Below that thin layer comprising the delicate organism known as the soil is a planet as lifeless as the moon.

G. Y. JACKS AND R. O. WHYTE

REVIEW QUESTIONS

1. Define the boldfaced terms in this chapter.

2. Describe the harmful and beneficial effects of the Mount St. Helens volcanic eruption in the United States in 1980.

3. Distinguish between the earth's *core*, *mantle*, and *crust*.

4. What are *tectonic plates*? What is the *lithosphere*? What is the *theory of plate tectonics*, and what is its importance to physical and biological processes on the earth?

5. What are the three different types of boundaries between the earth's lithospheric plates?

6. What is *erosion*, and what are its two major causes?

7. Distinguish between a *mineral* and a *rock*. Distinguish between *igneous*, *sedimentary*, and *metamorphic rock* and give two examples of each type.

8. Describe the *rock cycle* and explain its importance.

9. Distinguish between a *mineral resource* and an *ore*. Distinguish between *identified resources*, *undiscovered resources*, *reserves*, and *other resources*. List two factors that can increase the reserves of a mineral resource.

10. List seven methods mining companies use to find mineral deposits. Distinguish between **(a)** *overburden* and *spoil*, **(b)** *surface mining* and *subsurface mining*, and **(c)** *open-pit mining*, *dredging*, *area strip mining*, and *contour strip mining*.

11. List five major environmental impacts of mining, processing, and using mineral resources. Describe the life cycle of a metal resource. Distinguish between *ore mineral*, *gangue*, and *tailings*. What is *smelting*, and what are its major environmental impacts? What are three environmental limits associated with extracting and using mineral resources?

12. What is *economic depletion* of a mineral resource? When such depletion occurs, what five choices are available? Distinguish between *depletion time* and the *reserve-to-production ratio* for a mineral resource. How can the estimated depletion time for a resource be extended?

13. Explain how a competitive free market should increase supplies of a scarce mineral resource and list three reasons why this idea may not be effective under today's economic conditions.

14. Describe the basic features of the 1872 Mining Law in the United States and list its pros and cons.

15. List the pros and cons of allowing more mineral exploration and mining on public lands in the United States.

16. List the pros and cons of increasing supplies of mineral resources by **(a)** mining lower-grade ores and **(b)** finding substitutes for scarce nonrenewable mineral resources.

17. List the pros and cons of using bacteria to extract metals from ores.

18. What is an *earthquake,* and what are its major harmful effects? List ways to reduce the hazards from earthquakes.

19. What is a *volcanic eruption*? What are some of the hazards and benefits of volcanic eruptions? List ways to reduce the hazards from volcanic eruptions.

20. What is *soil*? Distinguish between a *soil horizon* and a *soil profile.*

21. What is *humus,* and what is its importance? What does the color of topsoil tell you about how useful it is for growing crops?

22. Distinguish between *soil infiltration* and *leaching.* Distinguish between *soil texture, soil porosity,* and *soil permeability.*

23. What is *soil erosion,* and what are its major natural and human-related causes?

24. What are the major harmful effects of soil erosion?

25. How serious is soil erosion **(a)** globally and **(b)** in the United States?

26. Describe the *Dust Bowl* event in the United States.

27. Describe how the U.S. government is reducing soil erosion.

28. What is *desertification*? How serious is this problem, and what are its major causes? How can we slow desertification?

29. Distinguish between *salinization* and *waterlogging* of soils. How serious are these problems? List six ways to reduce the threat of soil salinization.

30. What is *soil conservation*? Distinguish between *conventional-tillage farming* and *conservation-tillage farming.* What are the advantages of conservation-tillage farming?

31. Distinguish between *terracing, contour farming, strip cropping, alley cropping, gully reclamation,* and *windbreaks* as methods for reducing soil erosion.

32. Distinguish between *organic fertilizer* and *commercial inorganic fertilizer* and list the advantages of each approach for maintaining or restoring soil fertility. Distinguish between *animal manure, green manure,* and *compost* as methods for fertilizing soil. What is *crop rotation,* and why is it useful in helping maintain soil fertility?

33. List the advantages and disadvantages of using commercial inorganic fertilizers to maintain and restore soil fertility.

CRITICAL THINKING

1. What might be some beneficial and harmful climatic effects of having all the current continents clustered together in one supercontinent, as was the case in the distant past (Figure 5-8, top left, p. 115)?

2. Use the second law of energy (p. 71) to analyze the scientific and economic feasibility of each of the following processes: **(a)** extracting most minerals dissolved in seawater, **(b)** mining increasingly lower-grade deposits of minerals, **(c)** using inexhaustible solar energy to mine minerals, and **(d)** continuing to mine, use, and recycle minerals at increasing rates.

3. Why does extracting and minerals from low-grade ores cause more environmental damage than extracting them from high-grade ores?

4. List the pros and cons of including the environmental and health costs caused by mining, processing, and producing mineral resources (Figure 9-10) in the prices of metals to manufacturers and in the prices of consumer products. Do you favor including environmental costs in the prices of products? Explain. How would you institute such a policy?

5. In the area where you live, are you more likely to experience an earthquake or a volcanic eruption? What can you do to escape or reduce the harm if such a disaster strikes? What actions can you take when it occurs?

6. How does your lifestyle directly or indirectly contribute to soil erosion?

7. What are the main advantages and disadvantages of using commercial inorganic fertilizers to restore or increase soil fertility? Why should both inorganic and organic fertilizers be used?

PROJECTS

1. Write a brief scenario describing the consequences to us and to other forms of life if the rock cycle stopped functioning.

2. What mineral resources are extracted in your local area? What mining methods are used, and what have been their environmental impacts? How has mining these resources benefited the local economy?

3. Use the library or the internet to find out where earthquakes and volcanic eruptions have occurred during the past 30 years, and then stick small flags on a map of the world or place dots on Figure 9-3a. Compare their locations with the plate boundaries shown in Figure 9-3b.

4. Conduct a survey of soil erosion and soil conservation in and around your community on cropland, construction sites, mining sites, grazing land, and deforested land. Use these data to develop a plan for reducing soil erosion in your community.

5. Make a concept map of this chapter's major ideas using the section heads and subheads and the key terms (in boldface). Look on the website for this book for information about making concept maps.

INTERNET STUDY RESOURCES AND RESOURCES FOR FURTHER READING AND RESEARCH

The website for this book contains helpful study aids and many ideas for further reading and research. Log on to

http://www.brookscole.com/product/0534389872s

and click on the Chapter-by-Chapter area. Choose Chapter 9 and select a resource:

- "Flash Cards" allows you to test your mastery of the Terms and Concepts to Remember for this chapter.
- "Tutorial Quizzes" provides a multiple-choice practice quiz.
- "Student Guide to InfoTrac" will lead you to Critical Thinking Projects that use InfoTrac College Edition as a research tool.
- "References" lists the major books and articles consulted in writing this chapter.
- "Hypercontents" takes you to an extensive list of sites with news, research, and images related to individual sections of the chapter.

INFOTRAC COLLEGE EDITION

Improve your skills with InfoTrac College Edition, a searchable online database of articles from more than 700 periodicals. Log on to

http://www.infotrac-college.com

or access InfoTrac through the website for this book. Try to find the following articles:

1. Waiting for the Second Geological Shoe to Drop (earthquake near Seattle, Washington, may affect still-active volcano Mt. Rainier) (Notebook / The Other Big One) (Brief Article). Michael D. Lemonick. *Time,* March 12, 2001 v157 i10 p18. The Seattle area experienced a bone-rattling earthquake in March, and there is likely to be an eruption of Mount Rainier sometime. An eruption probably would resemble the 1980 eruption of Mount St. Helens, including mud flows. *Hint*: Enter the search term "second shoe" using the keywords "natural hazards."

2. Shifting Sands. Nick Middleton. *Geographical,* April 2000 v72 i4 p24. Desertification has been eclipsed by global warming as the perceived "most pressing environmental problem of today." This article reconsiders the seriousness of desertification. *Hint*: Enter the search term "shifting sands" using the keyword "climate."

10 RISK, TOXICOLOGY, AND HUMAN HEALTH

The Big Killer

What is roughly the diameter of a 30-caliber bullet, can be bought almost anywhere, is highly addictive, and kills about 11,000 people every day, or 460 per hour? It's a cigarette. *Cigarette smoking is the single most preventable major cause of death and suffering among adults.*

The World Health Organization (WHO) estimates that each year tobacco contributes to the premature deaths of at least 4 million people from 25 illnesses including **(1)** heart disease, **(2)** lung cancer, **(3)** other cancers, **(4)** bronchitis, **(5)** emphysema, and **(6)** stroke. The annual death toll from smoking-related diseases is projected to reach 10 million by 2030 (70% of them in developing countries)—an average of about 27,400 preventable deaths per day.

Smoking kills about 431,000 Americans per year, an average of 1,180 deaths per day (Figure 10-1). This death toll is roughly equivalent to three fully loaded jumbo (400-passenger) jets crashing accidentally every day with no survivors. Smoking causes more deaths each year in the United States than do all illegal drugs, alcohol (the second most harmful legal drug after nicotine), automobile accidents, suicide, and homicide combined (Figure 10-1).

According to a 1998 study, secondhand smoke (inhaled by nonsmokers) causes 30,000–60,000 premature deaths per year in the United States. Each year, parental smoking prematurely kills an estimated 6,000 children and causes 5.4 million serious child ailments in the United States.

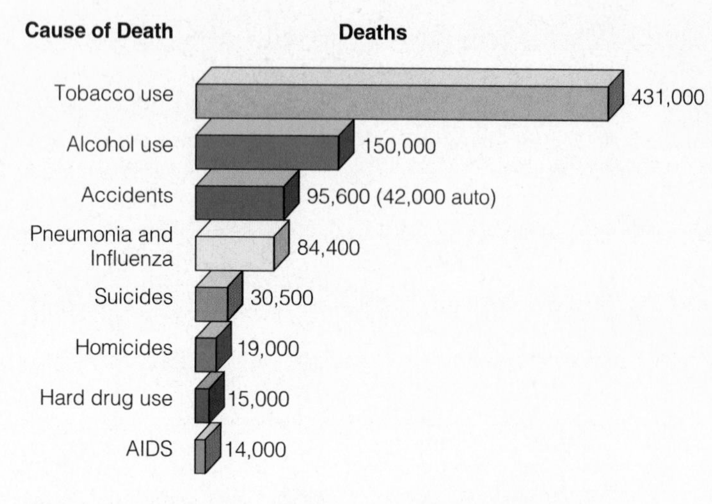

Figure 10-1 Annual deaths in the United States from tobacco use and other causes. Smoking is by far the nation's leading cause of preventable death, causing more premature deaths each year than all the other categories in this figure combined. (Data from National Center for Health Statistics)

The overwhelming consensus in the scientific community is that the nicotine (and probably the acetaldehyde) inhaled in tobacco smoke is highly addictive. Only 1 in 10 people who try to quit smoking succeed, about the same relapse rate as for recovering alcoholics and those addicted to heroin or crack cocaine. A British government study showed that adolescents who smoke more than one cigarette have an 85% chance of becoming smokers. According to a 1999 World Bank study, each day some 80,000–100,000 young people become regular long-term smokers, primarily in developing countries, which contain 70% of the world's smokers.

Worldwide, the cost of treating smoking-related illnesses is estimated at $200 billion a year. In the United States $70–100 billion a year is spent on **(1)** medical bills, **(2)** increased insurance costs, **(3)** disability, **(4)** lost earnings and productivity because of illness, and **(5)** property damage from smoking-caused fires. This is an average of $3–4 per pack of cigarettes sold in the United States.

Many health experts urge that a $2–4 federal tax be added to the price of a pack of cigarettes in the United States. Such a tax would mean that the users of cigarettes (and other tobacco products), not the rest of society, would pay a much greater share of the health, economic, and social costs associated with their smoking: a *user-pays* approach. WHO and the World Bank estimate that a 10% global tax on cigarettes would cause 40 million smokers to quit and would prevent the premature deaths of 10 million people alive today.

Other suggestions for reducing the death toll and health effects of smoking in the United States include **(1)** banning all cigarette advertising, **(2)** prohibiting the sale of cigarettes and other tobacco products to anyone under 21 (with strict penalties for violators), **(3)** banning all cigarette vending machines, **(4)** classifying nicotine as an addictive and dangerous drug (and placing its use in tobacco or other products under the jurisdiction of the Food and Drug Administration), **(5)** eliminating all federal subsidies and tax breaks to U.S. tobacco farmers and tobacco companies, and **(6)** using cigarette tax income to finance an aggressive antitobacco advertising and education program.

The dose makes the poison.
PARACELSUS, 1540

This chapter addresses the following questions:

- What types of hazards do people face?
- What is toxicology, and how do scientists measure toxicity?
- What chemical hazards do people face, and how can they be measured?
- What types of disease (biological hazards) threaten people in developing countries and developed countries?
- How can risks be estimated, managed, and reduced?

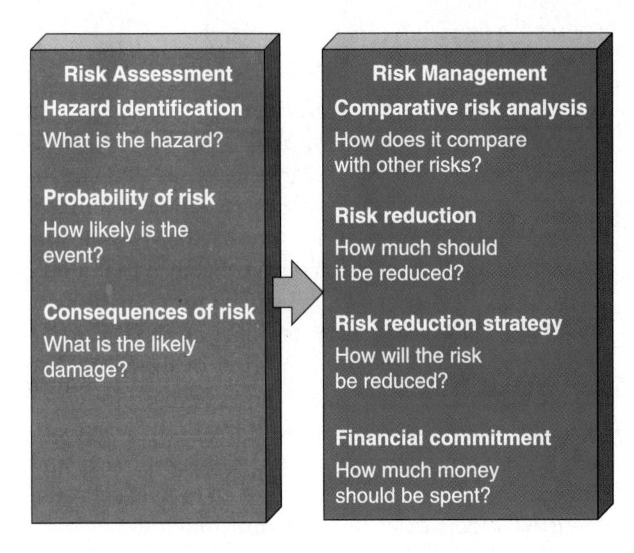

Figure 10-2 Risk assessment and risk management. These are important, difficult, and controversial processes.

10-1 RISK, PROBABILITY, AND HAZARDS

What Is Risk? **Risk** is the possibility of suffering harm from a hazard that can cause injury, disease, economic loss, or environmental damage. Risk is expressed in terms of **probability**: a mathematical statement about how likely it is that some event or effect will occur. In these terms, *risk* is defined as the probability of exposure times the probability of harm (Risk = Exposure × Harm).

Probability often is stated in terms such as "The lifetime probability of developing cancer from exposure to a certain chemical is 1 in 1 million." This means that one of every 1 million people exposed to the chemical at a specified average daily dosage will develop cancer over a typical lifetime (usually considered to be 70 years).

How Are Risks Assessed and Managed? **Risk assessment** involves **(1)** identifying a real or potential hazard ("What is the hazard?"), **(2)** determining the probability of its occurrence ("How likely is the event?"), and **(3)** assessing the severity of its health, environmental, economic, and social impact ("How much damage is it likely to cause?"; Figure 10-2, left).

This is a complex, difficult, and controversial process. For example, assessing the risk of exposure to a toxic chemical involves estimating **(1)** the number of people or other organisms exposed, **(2)** the level and duration of exposure, and **(3)** other possible contributing factors such as age, health, sex, personal habits, and interactions with other chemicals.

After a risk has been assessed, the next step is **risk management**, in which people make decisions about **(1)** how serious it is compared to other risks (*comparative risk analysis*), **(2)** how much (if at all) the risk should be reduced, **(3)** how such risk reduction can be accomplished, and **(4)** how much money should be devoted to reducing the risk to an acceptable level (Figure 10-2, right). This is even more difficult and controversial than risk assessment because of **(1)** a lack of information and **(2)** the economic, health, and political implications of such decisions.

What Are the Major Types of Hazards? The various kinds of hazards we face can be categorized as follows:

- *Cultural hazards* such as unsafe working conditions, smoking (left), poor diet, drugs, drinking, driving, criminal assault, unsafe sex, and poverty.
- *Chemical hazards* from harmful chemicals in the air (Chapter 12), water (Chapter 14), soil, and food. The bodies of most human beings contain small amounts of about 500 synthetic organic chemicals—whose health effects are mostly unknown—that did not exist in 1920.
- *Physical hazards* such as ionizing radiation (p. 69), fire, earthquake (p. 207), volcanic eruption (p. 208), flood (p. 342), tornadoes, and hurricanes.
- *Biological hazards* from pathogens (bacteria, viruses, and parasites), pollen and other allergens, and animals such as bees and poisonous snakes.

According to a 1998 study by Cornell University scientist David Pimentel, environmental factors such as malnutrition, smoking, cooking fires, skin cancer, exposures to pesticides and other hazardous chemicals, and air and water pollution contribute to about 40% of the world's annual deaths.

10-2 TOXICOLOGY

What Determines Whether a Chemical Is Harmful? Dose and Response **Toxicity** is a measure of how harmful a substance is. Whether a chemical (or other agent such as ionizing radiation) is harmful depends on several factors. One is the **dosage**, the amount of a potentially harmful substance that a person has ingested, inhaled, or absorbed through the skin. Whether a chemical is harmful depends on **(1)** the size of the dosage over a certain period of time, **(2)** how often an exposure occurs, **(3)** who is exposed (adult or child, for example), **(4)** how well the body's detoxification systems (liver, lungs, and kidneys) work, and **(5)** genetic makeup that determines an individual's sensitivity to a particular toxin (Figure 10-3).

The harm caused by a substance can also be affected by

- *Solubility*. *Water-soluble toxins* (which are often inorganic compounds) can move throughout the environment and get into water supplies. *Oil- or fat-soluble toxins* (which are usually organic compounds) can accumulate in body tissues and cells.
- *Persistence*. Many chemicals, such as plastics, chlorofluorocarbons (CFCs), chlorinated hydrocarbons, and plastics, are used widely because of their persistence or resistance to breakdown. However, this persistence also means that they can have long-lasting effects on the health of wildlife and people.
- **Bioaccumulation**, in which some molecules are absorbed and stored in specific organs or tissues at levels higher than normally would be expected.

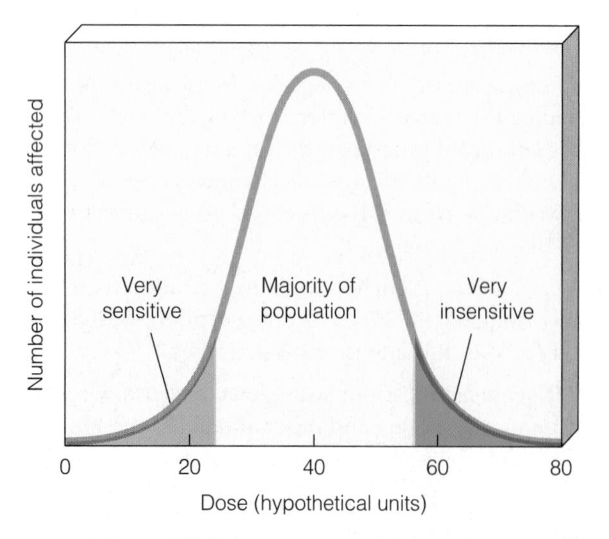

Figure 10-3 Typical variations in sensitivity to a toxic chemical within a population, mostly because of differences in genetic makeup. Some individuals in a population are very sensitive to small does of a toxin (left), and others are very insensitive (right). Most people fall between these two extremes (middle).

- **Biomagnification**, in which the levels of some toxins in the environment are magnified as they pass through food chains and webs (Figure 10-4). Examples of chemicals that can be biomagnified include long-lived, fat-soluble organic compounds such as **(1)** the pesticide DDT, **(2)** PCBs (oily chemicals used in electrical transformers), and **(3)** some radioactive isotopes (such as strontium-90, Table 3-1, p. 69). Stored in body fat, such chemicals can be passed along to offspring during gestation or egg laying and as mothers nurse their young.
- *Chemical interactions* that can decrease or multiply the harmful effects of a toxin. An *antagonistic interaction* can reduce the harmful response. For example, vitamins E and A apparently interact to reduce the body's response to some carcinogens. A *synergistic interaction* (p. 60) multiplies harmful effects. For example, workers exposed to asbestos increase their chances of getting lung cancer 20-fold. However, asbestos workers who also smoke have a 400-fold increase in lung-cancer rates.

The type and amount of health damage that result from exposure to a chemical or other agent are called the **response**. An *acute effect* is an immediate or rapid harmful reaction to an exposure; it can range from dizziness or a rash to death. A *chronic effect* is a permanent or long-lasting consequence (kidney or liver damage, for example) of exposure to a harmful substance.

Should We Be Concerned About Trace Levels of Toxic Chemicals in the Environment and in Our Bodies? The answer is that it depends on the chemical and its concentration. The detection of trace amounts of a chemical in air, water, or food does not necessarily mean that it is there at a level harmful to most people or to wildlife.

A basic concept of toxicology is that any synthetic or natural chemical (even water) can be harmful if ingested in a large enough quantity. Drinking 100 cups of strong coffee one after another would expose most people to a lethal dosage of caffeine. Similarly, downing 100 tablets of aspirin or 1 liter (1.1 quarts) of pure alcohol (ethanol) would kill most people.

The critical question is how much exposure to a particular toxic chemical causes a harmful response. This is the meaning of the quote by German scientist Paracelsus about the dose making the poison (found at the beginning of this chapter).

Most chemicals have some safe or *threshold level* of exposure below which their harmful effects are insignificant because

- The human body has mechanisms for breaking down (usually by enzymes found in the liver), diluting, or excreting small amounts of most toxins to keep them from reaching harmful levels.

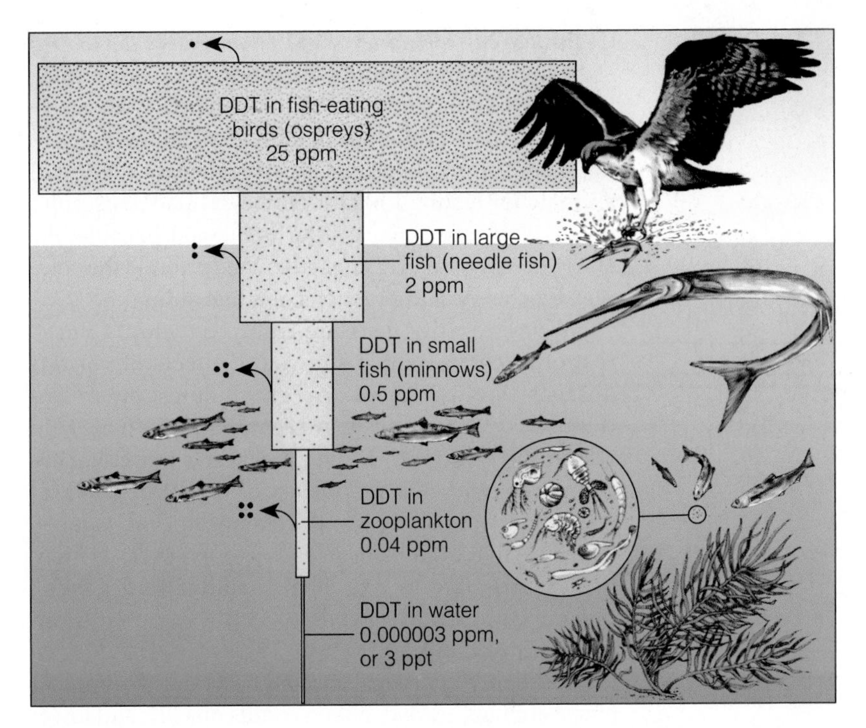

Figure 10-4 *Bioaccumulation* and *biomagnification.* DDT is a fat-soluble chemical that can accumulate in the fatty tissues of animals. In a food chain or food web, the accumulated concentrations of DDT can be biologically magnified in the bodies of animals at each higher trophic level. This diagram shows that the concentration of DDT in the fatty tissues of organisms was biomagnified about 10 million times in this food chain in an estuary near Long Island Sound in New York. If each phytoplankton organism in such a food chain takes up from the water and retains one unit of DDT, a small fish eating thousands of zooplankton (which feed on the phytoplankton) will store thousands of units of DDT in its fatty tissue. Then each large fish that eats 10 of the smaller fish will ingest and store tens of thousands of units, and each bird (or human) that eats several large fish will ingest hundreds of thousands of units. Dots represent DDT, and arrows show small losses of DDT through respiration and excretion.

- Individual cells have enzymes that can repair damage to DNA and protein molecules.
- Cells in some parts of the body (such as the skin and linings of the gastrointestinal tract, lungs, and blood vessels) reproduce fast enough to replace damaged cells. However, such high rates of cell reproduction can sometimes be altered by exposure to ionizing radiation and certain chemicals so that cell growth accelerates and creates a nonmalignant or malignant (cancerous) tumor.

Some people have the mistaken idea that all natural chemicals are safe and all synthetic chemicals are harmful. In fact, many synthetic chemicals are quite safe if used as intended, and many natural chemicals are deadly. For example, the average person is far more likely to be killed by aflatoxin in peanut butter than by lightning or a shark. However, the chance of dying from eating several spoonfuls of peanut butter a day is quite small.

In addition, the ability of chemists to detect increasingly small amounts of potentially toxic chemicals in air, water, and food can give the false impression that dangers from toxic chemicals are increasing. In 1980, chemists could routinely detect concentrations of substances in parts per million (ppm). By 1990, chemists could detect parts per billion (ppb), and today they can detect concentrations of parts per trillion (ppt) and in some cases parts per quadrillion (ppq).

What Is a Poison? Legally, a **poison** is a chemical that has an LD_{50} of 50 milligrams or less per kilogram of body weight. The $\mathbf{LD_{50}}$ is the **median lethal dose**: the amount of a chemical received in one dose that kills exactly 50% of the animals (usually rats and mice) in a test population (usually 60–200 laboratory animals) within a 14-day period (Figure 10-5).

Chemicals vary widely in their toxicity (Table 10-1). Some poisons can cause serious harm or death after a single acute exposure at very low dosages. Others cause such harm only at such huge dosages that it is nearly impossible to get enough into the body. Most chemicals fall between these two extremes.

What Methods Do Scientists Use to Determine Toxicity? Scientists use three methods to determine the level at which a substance poses a health threat:

- *Case reports* (usually made by physicians) provide information about people suffering some adverse health effect or death after exposure to a chemical. Such information often involves accidental poisonings, drug overdoses, homicides, or suicide attempts. Most case reports are not a reliable source for determining toxicity because the actual dosage and the exposed person's health status often are not known. However, such reports can provide clues about environmental hazards and suggest the need for laboratory investigations.
- *Laboratory investigations* (usually on test animals) are used to determine **(1)** toxicity, **(2)** residence time, **(3)** what parts of the body are affected, and **(4)** sometimes how the harm takes place.
- **Epidemiology** ("ep-i-deem-ee-OL-oh-gee") in populations of humans exposed to certain chemicals or diseases is used to find out why some people get sick and others do not.

How Are Laboratory Experiments Used to Determine Toxicity? Acute toxicity and chronic toxicity usually are determined by exposing a population of live laboratory animals (especially mice and rats, which are

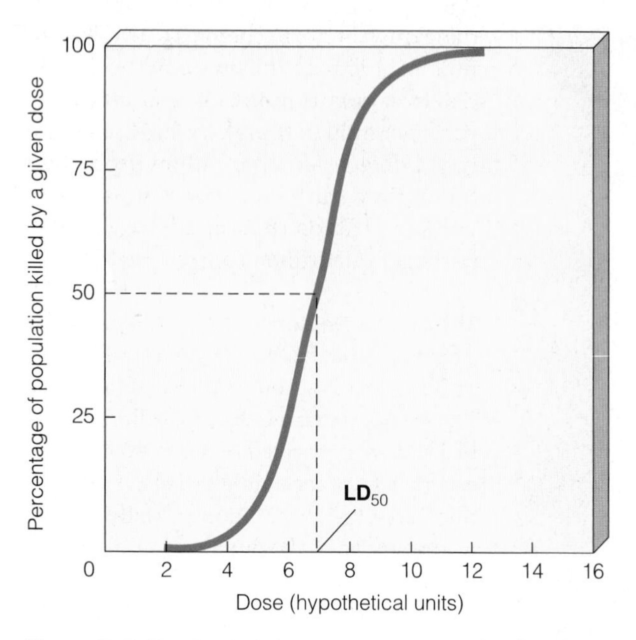

Figure 10-5 Hypothetical dose–response curve showing determination of the LD_{50}, the dosage of a specific chemical that kills 50% of the animals in a test group.

small and prolific and can be housed inexpensively in large numbers) to measured doses of a specific substance under controlled conditions. Animal tests take 2–5 years and cost $200,000 to $2 million per substance tested.

Animal welfare groups want to **(1)** limit or ban use of test animals or **(2)** ensure that experimental animals are treated in the most humane manner possible. More humane methods for carrying out toxicity tests include using **(1)** bacteria, **(2)** cell and tissue cultures, and **(3)** chicken egg membranes. In 1999, scientists developed a cheaper and much more sensitive way to determine toxicity by almost continuous measurement of changes in the electrical properties of individual animal cells.

These alternatives can greatly decrease the use of animals for testing toxicity. However, scientists point out that some animal testing is needed because the alternative methods cannot adequately mimic the complex biochemical interactions of a live animal.

Acute toxicity tests are run to develop a **dose–response curve**, which shows the effects of various dosages of a toxic agent on a group of test organisms (Figure 10-6). Such tests are *controlled experiments* in which the effects of the chemical on a *test group* are compared with the responses of a *control group* of organisms not exposed to the chemical. Care is taken to ensure that organisms in each group are **(1)** as identical as possible in age, health status, and genetic makeup and **(2)** are exposed to the same environmental conditions.

Fairly high dosages are used to reduce the number of test animals needed, obtain results quickly, and lower costs. Otherwise, tests would have to be run on millions of laboratory animals for many years, and manufacturers could not afford to test most chemicals. For the same reasons, the results of high-dose exposures usually are extrapolated to low-dose levels using mathematical models. Then the extrapolated low-dose results on the test organisms are extrapolated to humans to estimate LD_{50} values for acute toxicity (Table 10-1).

According to the *nonthreshold dose–response model* (Figure 10-6, left), any dosage of a toxic chemical or ionizing radiation causes harm that increases with the

Table 10-1 Toxicity Ratings and Average Lethal Doses for Humans

Toxicity Rating	LD_{50} (milligrams per kilogram of body weight)*	Average Lethal Dose†	Examples
Supertoxic	Less than 0.01	Less than 1 drop	Nerve gases, botulism toxin, mushroom toxins, dioxin (TCDD)
Extremely Toxic	Less than 5	Less than 7 drops	Potassium cyanide, heroin, atropine, parathion, nicotine
Very Toxic	5–50	7 drops to 1 teaspoon	Mercury salts, morphine, codeine
Toxic	50–500	1 teaspoon to 1 ounce	Lead salts, DDT, sodium hydroxide, sodium fluoride, sulfuric acid, caffeine, carbon tetrachloride
Moderately Toxic	500–5,000	1 ounce to 1 pint	Methyl (wood) alcohol, ether, phenobarbital, amphetamines (speed), kerosene, aspirin
Slightly toxic	5,000–15,000	1 pint to 1 quart	Ethyl alcohol, Lysol, soaps
Essentially nontoxic	15,000 or greater	More than 1 quart	Water, glycerin, table sugar

*Dosage that kills 50% of individuals exposed
†Amounts of substances that are liquids at room temperature when given to a 70.4-kilogram (155-pound) human

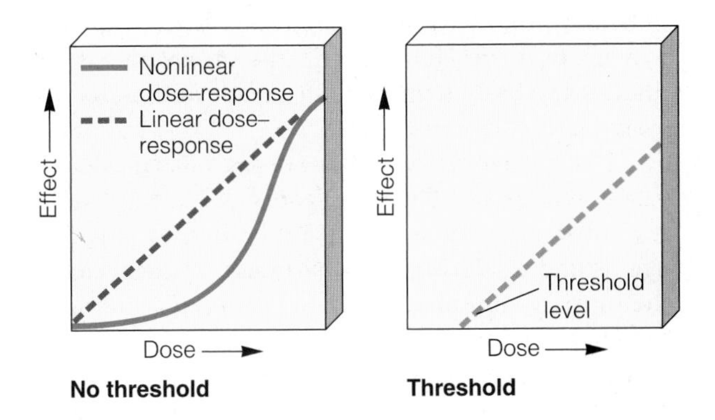

Figure 10-6 Hypothetical dose–response curves. The linear and nonlinear curves in the left graph show that exposure to any dosage of a chemical or ionizing radiation has a harmful effect that increases with the dosage. The curve on the right shows that a harmful effect occurs only when the dosage exceeds a certain *threshold level*. There is much uncertainty about which of these models applies to various harmful agents because of the difficulty in estimating the response to very low dosages.

dosage. Many chemicals that cause birth defects or cancers show this kind of response.

With the *threshold dose–response model* (Figure 10-6, right) a threshold dosage must be reached before any detectable harmful effects occur, presumably because the body can repair the damage caused by low dosages of some substances. It is extremely difficult to establish which of these models applies at low dosages. To be on the safe side, the nonthreshold dose–response model often is assumed.

Some scientists challenge the validity of extrapolating data from test animals to humans because human physiology and metabolism often are different from those of the test animals. Also, different species of test animals can react differently to the same toxin because of differences in body size, physiology, metabolism, and toxin sensitivity (Figure 10-3). Other scientists counter that such tests and models work fairly well (especially for revealing cancer risks) when the correct experimental animal is chosen or when a chemical is toxic to several different test animal species.

How Is Epidemiology Used to Determine Toxicity? In an *epidemiological study*, the health of people exposed to a particular toxic agent or disease organism (the *experimental group*) is compared with the health of another group of statistically similar people not exposed to these conditions (the *control group*). The goal of such studies is to determine whether the statistical association (if any) between a hazard and a health problem is strong, moderate, or weak.

Three major limitations of epidemiology are that **(1)** too few people have been exposed to sufficiently high levels of many toxic agents to allow detection of statistically significant differences, **(2)** conclusively linking an observed effect with exposure to a particular hazard is very difficult because people are exposed to many different toxic agents and disease-causing factors throughout their lives, and **(3)** it cannot be used to evaluate hazards from new technologies, substances, or diseases to which people have not been exposed.

How Valid Are Estimates of Toxicity? As we have seen, all methods for estimating toxicity levels and risks have serious limitations. However, they are all we have. To take this uncertainty into account and minimize harm, standards for allowed exposure to toxic substances and ionizing radiation typically are set at levels 1/100 or even 1/1,000 of the estimated harmful levels.

Despite their many limitations, carefully conducted and evaluated toxicity studies are important sources of information we can use to **(1)** understand dose–response effects and **(2)** estimate and set exposure standards. However, citizens, lawmakers, and regulatory officials must recognize the huge uncertainties and guesswork involved in all such studies.

10-3 CHEMICAL HAZARDS

What Are Toxic and Hazardous Chemicals? **Toxic chemicals** generally are defined as substances that are fatal to more than 50% of test animals (LD_{50}) at given concentrations. **Hazardous chemicals** cause harm by **(1)** being flammable or explosive, **(2)** irritating or damaging the skin or lungs (strong acidic or alkaline substances such as oven cleaners), **(3)** interfering with or preventing oxygen uptake and distribution (asphyxiants such as carbon monoxide and hydrogen sulfide), or **(4)** inducing allergic reactions of the immune system (allergens).

What Are Mutagens? **Mutagens** are agents, such as chemicals and ionizing radiation, that cause random *mutations*, or changes in the DNA molecules found in cells. Mutations in a sperm or egg cell can be passed on to future generations and cause diseases such as **(1)** bipolar disorder, **(2)** cystic fibrosis, **(3)** hemophilia, **(4)** sickle-cell anemia, **(5)** Down's syndrome, and **(6)** some types of cancer. Mutations in other cells are not inherited but may cause harmful effects.

Most mutations are harmless, probably because all organisms have biochemical repair mechanisms that can correct mistakes or changes in the DNA code. In addition, some mutations play a vital role in microevolution (p. 109).

What Are Teratogens? **Teratogens** are chemicals, radiation, or viruses that cause birth defects while the human embryo is growing and developing during pregnancy, especially during the first 3 months. Chemicals known to cause birth defects in laboratory animals include **(1)** PCBs, **(2)** thalidomide, **(3)** steroid hormones, and **(4)** heavy metals such as arsenic, cadmium, lead (p. 386), and mercury.

What Are Carcinogens? According to the WHO, environmental and lifestyle factors play a key role in causing or promoting up to 80% of all cancers. Major sources of carcinogens are **(1)** cigarette smoke (30–40% of cancers), **(2)** diet (20–30%), **(3)** occupational exposure (5–15%), and **(4)** environmental pollutants (1–10%). Inherited genetic factors and certain viruses cause about 10–20% of all cancers.

Carcinogens are chemicals, radiation, or viruses that cause or promote the growth of a malignant (cancerous) tumor, in which certain cells multiply uncontrollably. Many cancerous tumors spread by **metastasis** when malignant cells break off from tumors and travel in body fluids to other parts of the body. There, they start new tumors, making treatment much more difficult.

Because there are more than 100 types of cancer (depending on the types of cells involved), there are many different causes. These include **(1)** genetic predisposition, **(2)** viral infections, and **(3)** exposure to various mutagens and carcinogens.

Typically, 10–40 years may elapse between the initial exposure to a carcinogen and the appearance of detectable symptoms. Partly because of this time lag, many healthy teenagers and young adults have trouble believing that their smoking (p. 224), drinking, eating, and other lifestyle habits today could lead to some form of cancer before they reach age 50.

How Can Chemicals Harm the Immune, Nervous, and Endocrine Systems? Since the 1970s there has been a growing body of research on wildlife and laboratory animals and epidemiological studies of humans indicating that long-term (often low-level) exposure to various toxic chemicals in the environment can disrupt the body's immune, nervous, and endocrine systems.

The *immune system* consists of specialized cells and tissues that protect the body against disease and harmful substances by forming antibodies to invading agents and rendering them harmless. Viruses such as the human immunodeficiency virus (HIV), ionizing radiation, malnutrition, and some synthetic chemicals (including several widely used pesticides) can weaken the human immune system. This can leave the body vulnerable to attacks by allergens, infectious bacteria, viruses, and protozoans.

Synthetic chemicals in the environment threaten the human *nervous system* (brain, spinal cord, and peripheral nerves). Many poisons are *neurotoxins*, which attack nerve cells (neurons). Examples are **(1)** chlorinated hydrocarbons (DDT, PCBs, dioxins), **(2)** organophosphate pesticides (Section 16-7, p. 418), **(3)** formaldehyde, **(4)** various compounds of arsenic, mercury, lead, and cadmium, and **(5)** widely used industrial solvents such as trichloroethylene (TCE), toluene, and xylene.

The *endocrine system* is a complex network of glands and hormones that regulates many of the body's functions. Each type of hormone has a specific molecular shape that allows it to attach only to certain cell receptors (Figure 10-7, left). Once bonded together, the hormone and its receptor molecule move to the cell's nucleus to execute the chemical message carried by the hormone.

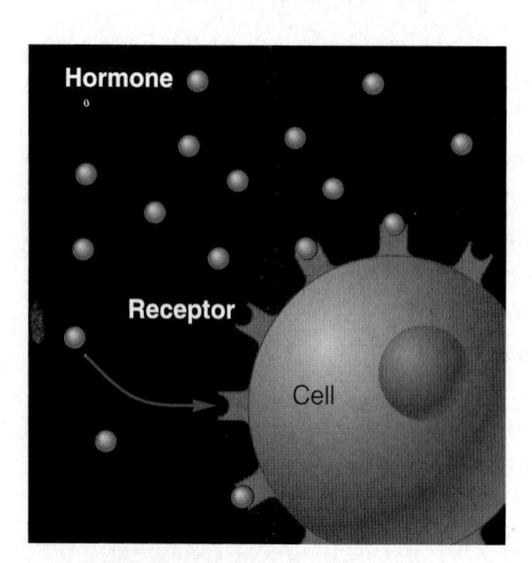

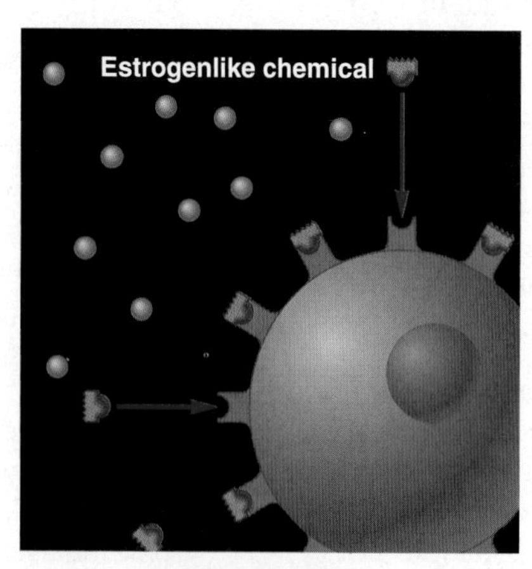

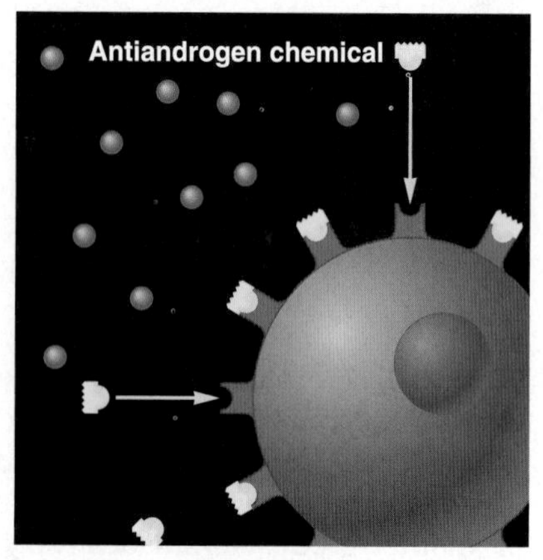

Figure 10-7 Hormones are molecules that act as messengers in the endocrine system to regulate various bodily processes, including reproduction, growth, and development. Each type of hormone has a unique molecular shape that allows it to attach to specially shaped receptors on the surface of or inside cells and transmit its chemical message (left). Molecules of certain pesticides and other molecules have shapes similar to those of natural hormones. Some of these hormone impostors, called *hormone mimics*, disrupt the endocrine system by attaching to estrogen receptor molecules (center). Others, called *hormone blockers*, prevent natural hormones such as androgens (male sex hormones) from attaching to their receptors (right). Some pollutants called *thyroid disrupters* may disrupt hormones released by thyroid glands and cause growth and weight disorders and brain and behavioral disorders.

CONNECTIONS

Are Hormone Disrupters and Mimics a Health Threat?

Over the last 20 years experts from a number of disciplines have been piecing together **(1)** field studies on wildlife, **(2)** studies on laboratory animals, and **(3)** epidemiological studies of human populations. This analysis suggests that a variety of human-made chemicals can act as *hormone* or *endocrine disrupters*, known as *hormonally active agents (HAAs)*. So far more than 60 endocrine disrupters had been identified, and the list of HAAs could reach several hundred.

Some, called *hormone mimics*, are estrogenlike chemicals that disrupt the endocrine system by attaching to estrogen receptor molecules (Figure 10-7, center).

Others, called *hormone blockers*, disrupt the endocrine system by preventing natural hormones such as androgens (male sex hormones) from attaching to their receptors (Figure 10-7, right). There is also growing concern about pollutants that can act as *thyroid disrupters* and cause growth, weight, brain, and behavioral disorders.

Most natural hormones are broken down or excreted. However, many synthetic hormone impostors are stable, fat-soluble compounds whose concentrations can be biomagnified as they move through food chains and webs (Figure 10-4). Thus, they can pose a special threat to humans and other carnivores dining at the top of food webs.

Numerous wildlife and laboratory studies reveal various possible effects of estrogen mimics and hormone blockers (HAAs). Here are a few of many examples:

- Ranch minks fed Lake Michigan fish contaminated with endocrine disrupters such as DDT and PCBs failed to reproduce.
- Exposure to PCBs has reduced penis size in some test animals and in 118 boys born to women who were exposed to a PCB spill in Taiwan in 1979.
- A 1999 study by Michigan State University zoologists found that female rats exposed to PCBs were reluctant to mate, raising the possibility that such contaminants could cause low sex drives in women.
- In 1973, estrogen mimics called PBBs accidentally got into cattle feed in Michigan, and from there into beef. Pregnant women who ate the beef (and whose breast milk had high levels of PBBs) had sons with undersized penises and malformed testicles.
- During the past 50 years there have been dramatic increases in testicular and prostate cancer in humans almost everywhere.
- Average sperm counts among men in the United States and Europe have declined by 50% during the past 60 years.

In 1999, the U.S. National Academy of Sciences released a report based on a 4-year review of the scientific literature on hormone disrupters (HAAs). The panel concluded that far too little is known about the effects of such chemicals to come to a definitive conclusion about their effects on humans.

Scientists on this panel called for greatly increased research to **(1)** verify current frontier science findings and **(2)** determine whether low levels of most hormone-disrupting chemicals in the environment pose a threat to the human population. However, the report also concluded that at present the 75,000 or more industrial chemicals in commercial use cannot be tested to determine whether they are hormone disrupters because the necessary tests do not exist.

If such research (which will take decades) shows that exposure to small amounts of hormone disrupters is harmful to humans and some forms of wildlife, the only reasonable choice may be to prevent such chemicals from reaching the environment. This will be a difficult and controversial economic and political decision.

Some health scientists believe that we should begin sharply reducing the use of potential hormone disrupters now because they meet the two requirements of the *precautionary principle*: great scientific uncertainty and a reasonable suspicion of harm, pp. 178 and 232.

Critical Thinking

1. Do you consider the possible threat from hormone disrupters a problem that could affect you or any child you might have? Explain.

2. Do you believe that the precautionary approach should be used to deal with this problem while more definite research is carried out over the next two decades? Explain. What harmful effects could using this approach have on the economy and on your lifestyle? Do such effects outweigh the risks? Explain.

The endocrine glands release very small amounts of *hormones* into the bloodstream that act as natural chemical messengers to control body functions such as sexual reproduction, growth, development, and behavior in humans and other animals. These naturally occurring hormones have profound effects on the human nervous, reproductive, and immune systems. There is concern that human exposure to low levels of synthetic chemicals, known as *hormonally active agents* (HAAs), can mimic and disrupt the effects of natural hormones (Connections, above).

Why Do We Know So Little About the Harmful Effects of Chemicals? According to risk assessment expert Joseph V. Rodricks, "Toxicologists know a great deal about a few chemicals, a little about many, and next to nothing about most." The U.S. National Academy of Sciences estimates that only about **(1)** 10%

of at least 75,000 chemicals in commercial use have been thoroughly screened for toxicity, and **(2)** 2% have been adequately tested to determine whether they are carcinogens, teratogens, or mutagens. Hardly any of the chemicals in commercial use have been screened for damage to the nervous, endocrine, and immune systems.

Each year about 1,000 new synthetic chemicals are introduced into the marketplace, with little knowledge about their potentially harmful effects. Currently, federal and state governments do not regulate about 99.5% of the commercially used chemicals in the United States. There are three major reasons for this lack of information and regulation.

- Under existing laws most chemicals are considered innocent until proven guilty. No one is required to investigate whether they are harmful.
- There are not enough funds, personnel, facilities, and test animals to provide such information for more than a small fraction of the many chemicals we encounter in our daily lives.
- It is too difficult and expensive to analyze the combined effects of multiple exposures to various chemicals and the possible interactions of such chemicals. For example, just studying the possible different three-chemical interactions of the 500 most widely used industrial chemicals would take 20.7 million experiments—a physical and financial impossibility.

What Is the Precautionary Approach? Because of the difficulty and expense of getting information about the harmful effects of chemicals, an increasing number of scientists and health officials are pushing for much greater emphasis on *pollution prevention*. This strategy greatly reduces the need for statistically uncertain and controversial toxicity studies and exposure standards. It also reduces the risk posed by exposure to potentially hazardous chemicals and products and their possible but poorly understood multiple interactions.

This approach is based on the **precautionary principle**. According to this concept, when there is scientific uncertainty about potentially serious harm from chemicals or technologies, decision makers should act to prevent harm to humans and the environment. It is based on familiar axioms: "Look before you leap," "better safe than sorry," and "an ounce of prevention is worth a pound of cure."

Under this approach, those proposing to introduce a new chemical or technology would bear the burden of establishing its safety. In other words, new chemicals and technologies would be assumed to be guilty until proven innocent. Manufacturers and businesses contend that doing this would make it too expensive and almost impossible to introduce any new chemical or technology.

10-4 BIOLOGICAL HAZARDS: DISEASE IN DEVELOPED AND DEVELOPING COUNTRIES

What Are Nontransmissible Diseases? A **nontransmissible disease** is not caused by living organisms and does not spread from one person to another. Examples are **(1)** cardiovascular (heart and blood vessel) disorders, **(2)** most cancers, **(3)** diabetes, **(4)** asthma, **(5)** emphysema, and **(6)** malnutrition. Such diseases typically have multiple (and often unknown) causes and tend to develop slowly and progressively over time.

What Are Transmissible Diseases? A **transmissible disease** is caused by a living organism (such as a bacterium, virus, protozoa, or parasite) and can be spread from one person to another. These infectious agents are called *pathogens* and are spread by air, water, food, body fluids, some insects, and other nonhuman carriers called *vectors*.

Typically, a *bacterium* is a one-celled microorganism capable of replicating itself by simple cell division. A *virus* is a microscopic, noncellular infectious agent. Its DNA contains instructions for making more viruses, but it has no apparatus to do so. To replicate, a virus must invade a host cell and take over the cell's DNA to create a factory for producing more viruses.

Antibiotics have greatly reduced the incidence of infectious disease caused by bacteria. However, their widespread use and misuse have increased the genetic resistance of many disease-causing bacteria, which can reproduce rapidly (Spotlight, right).

Worldwide, infectious diseases cause about one of every four deaths each year. According to the WHO, the world's seven deadliest infectious diseases are **(1)** *acute respiratory infections*, mostly pneumonia and flu (caused by bacteria and viruses and killing about 3.7 million people per year), **(2)** *acquired immune deficiency syndrome* (AIDS, a viral disease, 3 million), **(3)** *diarrheal diseases* (caused by bacteria and viruses, 2.5 million), **(4)** *tuberculosis* (TB, a bacterial disease, 2 million; Case Study, p. 234), **(5)** *malaria* (caused by parasitic protozoa, 1.5 million, p. 235), **(6)** *measles* (a viral disease, 1 million), and **(7)** *hepatitis B* (a viral disease, 1 million).

As a country industrializes, it usually makes an *epidemiological transition*. The infectious diseases of childhood become less important, and the chronic diseases of adulthood (heart disease and stroke, cancer, and respiratory conditions) become more important in causing mortality. In 1999, for example, infectious and parasitic diseases were responsible for 43% of all deaths in developing countries but only 1% in developed countries.

How Rapidly Are Viral Diseases Spreading? Viral diseases include **(1)** *influenza* or *flu* (transmitted by the bodily fluids or airborne emissions of an infected

SPOTLIGHT

Are We Losing the War Against Infectious Bacteria?

There is growing evidence that we may be losing our war against infectious bacterial diseases because bacteria are among the earth's ultimate survivors. When a colony of bacteria is dosed with an antibiotic such as penicillin, most of the bacteria are killed.

However, a few have mutant genes that make them immune to the drug. Through natural selection (Figure 5-5, left, p. 111), a single mutant can pass such traits on to most of its offspring, which can amount to 16,777,216 in only 24 hours.

Even worse, bacteria can become genetically resistant to antibiotics they have never been exposed to. When a resistant and a nonresistant bacterium touch one another (say, on a hospital bedsheet or in a human stomach), they can exchange a small loop of DNA called a plasmid, thereby transferring genetic resistance from one organism to another.

The incredible genetic adaptability of bacteria is one reason the world faces a potentially serious rise in the incidence of some infectious bacterial diseases once controlled by antibiotics. Other factors also play a key role, including **(1)** spread of bacteria (some beneficial and some harmful) around the globe by human travel and the trade of goods, **(2)** overuse of antibiotics by doctors, often at the insistence of their patients (with a 2000 study by Richard Wenzel and Michael Edward suggesting that at least half of all antibiotics used to treat humans are prescribed unnecessarily), **(3)** failure of many patients to take all of their prescribed antibiotics, which promotes bacterial resistance, **(4)** availability of antibiotics in many countries without prescriptions, **(5)** overuse of pesticides (p. 420), which increases populations of pesticide-resistant insects and other carriers of bacterial diseases, and **(6)** widespread use of antibiotics in the livestock and dairy industries to control disease in livestock animals and to promote animal growth.

The result of these factors acting together is that every major disease-causing bacterium now has strains that resist at least one of the roughly 160 antibiotics we use to treat bacterial infections. In 1998, health officials were alarmed to learn of the existence of a strain of bubonic plague in Madagascar that is resistant to multiple antibiotics.

In 2000, officials at the U.S. Centers for Disease Control and Prevention estimated that about 2.2 million people (most with a weakened immune system) a year get sick, and at least 88,000 die from infectious diseases they pick up in U.S. hospitals, nursing homes, or home health-care settings. Most of these infections are caused by **(1)** contaminated catheters, intravenous lines, and breathing tubes and **(2)** failure of doctors and other health-care personnel to carefully wash their hands with water or alcohol-based antimicrobial hand rubs and frequently change their latex gloves. Patients can reduce such infections by asking any doctor or health-care worker coming into their room, "Did you wash your hands?" or "Did you change your gloves?"

There is growing controversy over the widespread use of antibiotics to increase the growth rate of about 80% of the livestock animals raised each year in the United States, mainly cattle, pigs, and poultry. According to a 2000 study by Margaret Mellon and other researchers, about 84% of all antibiotics in the United States are used as feed additives to boost livestock production. Resistant strains of infectious diseases that develop in livestock animals can spread to humans through contact with infected animals or water and through food chains.

The European Union, the World Health Organization, the American Public Health Association, and the U.S. Centers for Disease Control and Prevention all favor the immediate phaseout of all antibiotics used to promote growth in livestock animals that are the same as or closely related to antibiotics used in humans. Several European countries have imposed such bans, and since 1986 Sweden has banned all use of antibiotics for growth promotion in livestock.

Critical Thinking

1. What role, if any, have you played in the increase in genetic resistance of bacteria to widely used antibiotics? List three ways to reduce this threat.

2. Do you believe that the use of the same antibiotics to treat human illness and to fatten livestock should be banned in the United States (or the country where you live)? Explain. Would you favor using small amounts of such antibiotics to treat disease in livestock?

person), **(2)** *Ebola* (transmitted by the blood or other body fluids of an infected person), **(3)** *rabies* (transmitted by dogs, coyotes, raccoons, skunks, and bats), and **(4)** *AIDS* (transmitted by the blood or other body fluids of an infected person). Viruses, like bacteria, can genetically adapt rapidly to different conditions.

Although health officials worry about the emergence of new viral diseases (such those caused by Ebola viruses), they recognize that the greatest virus health threat to humans is the emergence of new, very virulent stains of influenza. Flu viruses move through the air and are highly contagious. During 1918 and 1919, a flu epidemic infected more than half the world's population and killed 20–30 million people (including about 500,000 in the United States). Today, flu kills about 1 million people per year (20,000 of them in the United States).

The Global Tuberculosis Epidemic

CASE STUDY

Since 1990 one of the world's most underreported stories has been the rapid spread of tuberculosis (TB). According to the World Health Organization, this highly infectious bacterial disease kills about 2 million people and infects about 8 million people per year (Figure 10-8).

The bacterium causing TB infection moves from person to person, mainly in airborne droplets produced by coughing, sneezing, singing, or even talking. About one of every three people in the world is infected with the TB bacillus.

During their lifetime about 5–10% of these people will become sick or infectious with active TB, especially when their immune system is weakened. Left untreated, each person with active TB will infect 10–15 other people. Many infected people do not appear to be sick and about half of them do not know they are infected. As a result, this health problem has been called a *silent global epidemic.*

Major reasons for the recent increase in TB are **(1)** poor TB screening and control programs (especially in developing countries, where about 95% of the new cases occur), **(2)** development of strains of the tuberculosis bacterium that are genetically resistant to almost all effective antibiotics (typically leading to mortality rates of more than 50%), **(3)** population growth and increased urbanization (which increase contacts between people), **(4)** poverty, and **(5)** the spread of AIDS, which greatly weakens the immune system and allows TB bacteria to multiply.

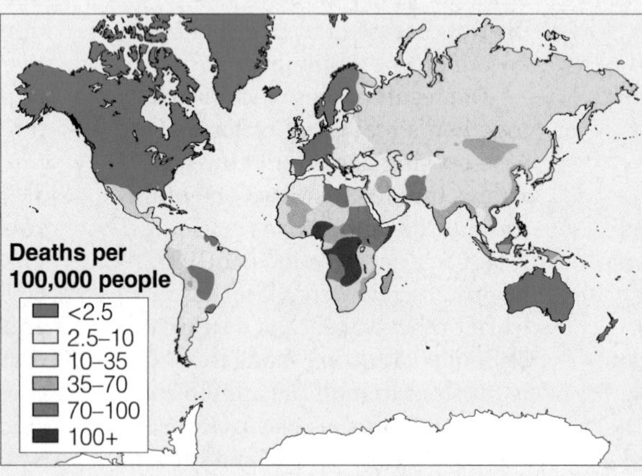

Figure 10-8 The current global tuberculosis epidemic. This easily transmitted disease is spreading rapidly and now kills about 2 million people a year. (Data from World Health Organization)

Slowing the spread of the disease involves early identification and treatment of people with active TB, usually those with a chronic cough. Treatment with a combination of four inexpensive drugs can cure 90% of those with active TB. However, to be successful the drugs must be taken every day for 6 to 8 months. Because the symptoms disappear after a few weeks, many patients think they are cured and stop taking the drugs. This allows the disease to recur in a hard-to-treat form. It then spreads to other people, and drug-resistant strains of TB bacteria develop.

According to the World Health Organization, a worldwide campaign to help control TB would cost about $360 million to help save at least 20 million lives during the next decade.

Critical Thinking

Before you read this report, were you aware of the serious global TB epidemic? Why do you think that this important story has gotten so little media attention compared to other diseases that cause many fewer deaths per year?

Sex can be dangerous to one's health. In the United States, sexually transmitted diseases (STDs) infect about 15.3 million people each year. According to a 1998 report by the American Social Health Association, at least one in every three sexually active people in the United States will contract an STD by age 24. Some of these diseases can cause infertility in men and women. Others can cause warts and genital cancers or, in the case of HIV, death.

There is a growing threat from the spread of *acquired immune deficiency syndrome (AIDS),* which is caused by the human immunodeficiency virus (HIV). The virus itself is not deadly, but it kills immune cells and leaves the body defenseless against infectious bacteria and other viruses. HIV can be transmitted **(1)** during unprotected sexual activity, **(2)** from one intravenous drug user to another through shared needles, **(3)** from an infected mother to an infant before or during birth or through breastfeeding after birth, and **(4)** by exposure to infected blood.

According to the WHO, by the end of 2000 some 58 million people worldwide (two-thirds of them in sub-Saharan Africa and 1.2 million of them children under age 15) were infected with HIV. During 2000, 5.4 million people (80% of them in Africa and Asia) were newly infected with HIV—an average of almost 15,000 new infections per day.

Within about 7–10 years, at least half of those with HIV develop AIDS. This long incubation period means that infected people often spread the virus for several years without knowing that they are infected. So far, there is no cure for AIDS, although drugs may help some infected people live longer (if they can afford the treatment).

According to the United Nations, by the end of 2000 about 21.8 million people (4 million of them children under age 15 and 420,000 people in the United States) had died of AIDS-related diseases. During the last two decades, people in eight African countries have lost more than 10 years of their expected life span. In two African countries, Zimbabwe and Botswana, average life expectancy has dropped by more than 30 years since 1980. Because of AIDS, the populations of 35 African countries are projected to drop by 10% between 2000 and 2015.

Once a viral infection starts, it is much harder to fight than infections by bacteria and protozoans. Only a few antiviral drugs exist because most drugs that will kill a virus also harm the cells of its host. Treating viral infections (such as colds, flu, and most mild coughs and sore throats) with antibiotics is useless and increases genetic resistance in disease-causing bacteria (Spotlight, p. 233).

Medicine's only effective weapons against viruses are *vaccines* that stimulate the body's immune system to produce antibodies to ward off viral infections. Immunization with vaccines has helped reduce the spread of viral diseases such as **(1)** smallpox, **(2)** polio, **(3)** rabies, **(4)** influenza, **(5)** measles, and **(6)** hepatitis B.

Case Study: Malaria, a Protozoal Disease About 40% of the world's people live in tropical and subtropical regions where malaria is present (Figure 10-9). Currently, an estimated 300–500 million people are infected with malaria parasites worldwide, and there are 270–500 million new cases each year.

Malaria's symptoms come and go and include **(1)** fever and chills, **(2)** anemia, **(3)** an enlarged spleen, **(4)** severe abdominal pain and headaches, **(5)** extreme weakness, and **(6)** greater susceptibility to other diseases. The disease kills about 1.5 million people each year, more than half of them children under age 5.

Malaria is caused by four protozoa species of the genus *Plasmodium*. Most cases of the disease are transmitted when an uninfected female of any one of 60 *Anopheles* mosquito species **(1)** bites an infected person, **(2)** ingests blood that contains the parasite, and **(3)** later bites an uninfected person (Figure 10-10). When this happens, *Plasmodium* parasites **(1)** move out of the mosquito and into the human's bloodstream, **(2)** multiply in the liver, and **(3)** enter blood cells to continue multiplying. Malaria also can be transmitted by blood transfusions or by sharing needles.

The malaria cycle repeats itself until immunity develops, treatment is given, or the victim dies. *Over the*

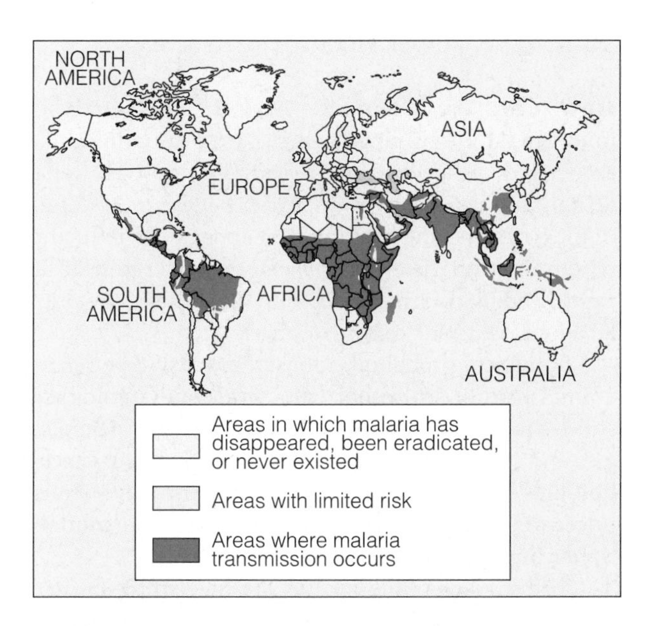

Figure 10-9 Worldwide distribution of malaria. About 40% of the world's current population lives in areas in which malaria is present, with the disease killing at least 1.5 million people a year. If the world becomes warmer, as projected by current climate models, by 2046 malaria could affect 60% of the world's population. (Data from the World Health Organization)

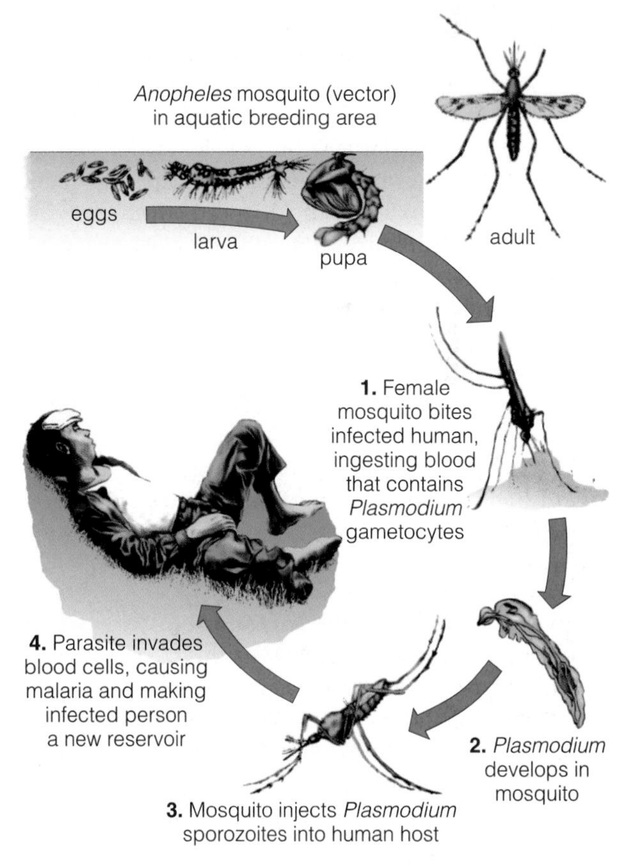

Figure 10-10 The life cycle of malaria. This life cycle of *Plasmodium* circulates from mosquito to human and back to mosquito. Although various mosquito species carry diseases (such as malaria, yellow fever, encephalitis, and dengue fever to humans and heartworm to dogs), mosquitoes also play important ecological roles. Their eggs are a major food source for fish, various insects, and frogs and other amphibians, and adult mosquitoes are an important source of food for bats, spiders, and many insect and bird species.

course of human history, malarial protozoa probably have killed more people than all the wars ever fought.

During the 1950s and 1960s, the spread of malaria was sharply curtailed by **(1)** draining swamplands and marshes, **(2)** spraying breeding areas with insecticides, and **(3)** using drugs to kill the parasites in the bloodstream. Since 1970, however, malaria has come roaring back. Most species of the malaria-carrying *Anopheles* mosquito have become genetically resistant to most insecticides. Worse, the *Plasmodium* parasites have become genetically resistant to common antimalarial drugs. According to the United Nations, since the 1980s death rates from malaria among African children have tripled, mostly because of an increase in drug-resistant strains.

Researchers are working to develop new antimalarial drugs (such as artemisinins derived from the Chinese herbal remedy quihaosu), vaccines, and biological controls for *Anopheles* mosquitoes. However, such approaches are underfunded and have proved more difficult than originally thought. Researchers also are studying the feasibility of altering the genetic makeup of mosquitoes so that they cannot carry and transmit the parasite to humans.

According to health experts, prevention is the best approach to slowing the spread of malaria. Methods include **(1)** increasing water flow in irrigation systems to prevent mosquito larvae from developing (an expensive and wasteful use of water), **(2)** using mosquito nets dipped in a nontoxic insecticide (permethrin) in windows and doors of homes, **(3)** cultivating fish that feed on mosquito larvae (biological control), **(4)** clearing vegetation around houses, **(5)** planting trees that soak up water in low-lying marsh areas where mosquitoes thrive (a method that can degrade or destroy ecologically important wetlands), **(6)** using zinc and vitamin A supplements to boost children's resistance to malaria, and **(7)** greatly increasing public education. In 1998, the World Bank, WHO, and several other organizations launched a Roll Back Malaria program to help reduce the mortality rate.

How Can We Reduce the Incidence of Infectious Diseases? Figure 10-11 lists measures that health scientists and public health officials suggest for preventing or reducing the incidence of infectious diseases that affect humanity (especially in developing countries). They also call for increased emphasis on preventive health care in developing countries (Solutions, right).

The WHO estimates that only 2% of the world's global research and development funds are devoted to infectious diseases in developing countries, even though more people worldwide suffer and die from these diseases than from all others combined. Indeed, according to a 2000 study by the International Federation of Red Cross, an estimated 150 million people have died from tuberculosis, malaria, and AIDS since 1945, compared to 23 million in wars.

Figure 10-11 *Solutions.* Ways to prevent or reduce the incidence of infectious diseases.

10-5 RISK ANALYSIS

How Can We Estimate Risks? Risk analysis involves **(1)** identifying hazards and evaluating their associated risks (*risk assessment,* Figure 10-2, left), **(2)** ranking risks (*comparative risk analysis*), **(3)** determining options and making decisions about reducing or eliminating risks (*risk management,* Figure 10-2, right), and **(4)** informing decision makers and the public about risks (*risk communication*).

Statistical probabilities based on past experience, animal testing and other tests, and epidemiological studies are used to estimate risks from older technologies and chemicals (Section 10-1). To evaluate new technologies and products, risk evaluators use more uncertain statistical probabilities, based on models rather than actual experience.

The left side of Figure 10-12 is an example of *comparative risk analysis,* summarizing the greatest ecological and health risks identified by a panel of scientists acting as advisers to the U.S. Environmental Protection Agency. Note the difference between the comparison of relative risk by scientists (Figure 10-12, left) and the general public (Figure 10-12, right). These differ-

Improving Health Care in Developing Countries

SOLUTIONS

With adequate funding, the health of people in developing countries and the poor in developed countries can be improved dramatically, quickly, and cheaply by providing the following forms of mostly preventive health care:

- Better nutrition, prenatal care, and birth assistance for pregnant women.
- Better nutrition for children.
- Greatly improved postnatal care (including promotion of breast-feeding, except for mothers infected with HIV) to reduce infant mortality. Breast-fed babies get natural immunity to many diseases from antibodies in their mothers' milk.
- Immunization against the world's five largest preventable infectious diseases: tetanus, measles, diphtheria, typhoid fever, and polio. Since 1971 the percentage of children in developing countries immunized against these diseases has increased from 10% to 84%, saving about 10 million lives a year.
- Oral rehydration therapy for victims of diarrheal diseases, which cause about one-fourth of all deaths of children under age 5. A simple solution of boiled water, salt, and sugar or rice, at a cost of only a few cents per person, can prevent death from dehydration.
- Careful and selective use of antibiotics for infections (Spotlight, p. 233).
- Clean drinking water and sanitation facilities for the one-third of the world's population that lacks them. In 2001, the WHO began promoting a simple, do-it-yourself technique of disinfecting water with sunlight in a process called Solar Water Disinfection (SODIS). Using only sunlight, empty plastic soft drink bottles, and a black surface, it costs almost nothing. The process is simple: **(1)** Fill a transparent plastic bottle with contaminated water and **(2)** lay it horizontally on a flat black surface (which absorbs more heat and thus kills more pathogens). The heat and ultraviolet rays of the sun kill most illness-causing microorganisms in polluted water. This method is especially useful in tropical countries where sunlight is intense.

According to the World Health Organization, extending such primary health care to all the world's people would cost an additional $10 billion per year, about 4% of what the world spends every year on cigarettes or devotes every 4 days to military spending. The cost of this primary health care program is about $1 per child.

Critical Thinking

1. Do you believe that developed countries should foot at least half the bill for implementing such proposals? What economic and environmental benefits would this provide for developed countries?

2. How many dollars per year of your taxes would you be willing to spend for such a preventive health program in developing countries?

ences result largely from failure of professional risk evaluators to educate the public about the nature of risks and their relative importance. Some risk experts contend that much of our risk education is based on often misleading media reports on the latest risk scare (based mainly on frontier science, p. 57) that do not put such risks in perspective.

Once a risk assessment has been completed, decision makers much decide what level of risk is acceptable. Figure 10-13 shows four methods used to determine the acceptability of a risk. The most widely used method is *benefit–cost analysis,* which attempts to determine whether the estimated short- and long-term risks or costs of using a particular technology or chemical outweigh its the estimated short- and long-term benefits (p. 30). However, this approach has some limitations (Spotlight, p. 31).

What Are the Greatest Risks People Face? The greatest risks many people face today are rarely dramatic enough to make the daily news. In terms of reduced life span, *the greatest risk by far is poverty* (Figure 10-14, p. 240).

After the health risks associated with poverty, the greatest risks of premature death are mostly the result of voluntary choices people make about their lifestyles (Figures 10-1 and 10-14).

By far the best ways to reduce one's risk of premature death and serious health risks are to **(1)** not smoke, **(2)** avoid excess sunlight (which ages skin and causes skin cancer), **(3)** not drink alcohol or drink only in moderation (no more than two drinks in a single day), **(4)** reduce consumption of foods containing cholesterol and saturated fats, **(5)** eat a variety of fruits and vegetables, **(6)** exercise regularly, **(7)** lose excess weight, and **(8)** for those who can afford a car, drive as safely as possible in a vehicle with the best available safety equipment.

How Can We Estimate Risks for Technological Systems? The more complex a technological system and the more people needed to design and run it, the more difficult it is to estimate the risks. The overall reliability of any technological system (expressed as a percentage) is the product of two factors:

$$\text{System reliability (\%)} = \text{Technology reliability} \times \text{Human reliability}$$

Figure 10-12 *Comparative risk analysis* of the most serious ecological and health problems according to scientists acting as advisers to the U.S. Environmental Protection Agency (left column). Risks in each of these categories are not listed in rank order. The right side of this figure represents polls showing how U.S. citizens rank the ecological and health risks they perceive as being the most serious. Why do you think there is such a great difference between the ranking by risk experts and by the general public? (Data from Science Advisory Board, *Reducing Risks*, Washington, D.C.: Environmental Protection Agency, 1990)

Scientists (Not in rank order in each category)	**Citizens** (In rank order)
High-Risk Health Problems • Indoor air pollution • Outdoor air pollution • Worker exposure to industrial or farm chemicals • Pollutants in drinking water • Pesticide residues on food • Toxic chemicals in consumer products **High-Risk Ecological Problems** • Global climate change • Stratospheric ozone depletion • Wildlife habitat alteration and destruction • Species extinction and loss of biodiversity	**High-Risk Problems** • Hazardous waste sites • Industrial water pollution • Occupational exposure to chemicals • Oil spills • Stratospheric ozone depletion • Nuclear power-plant accidents • Industrial accidents releasing pollutants • Radioactive wastes • Air pollution from factories • Leaking underground tanks
Medium-Risk Ecological Problems • Acid deposition • Pesticides • Airborne toxic chemicals • Toxic chemicals, nutrients, and sediment in surface waters	**Medium-Risk Problems** • Coastal water contamination • Solid waste and litter • Pesticide risks to farm workers • Water pollution from sewage plants
Low-Risk Ecological Problems • Oil spills • Groundwater pollution • Radioactive isotopes • Acid runoff to surface waters • Thermal pollution	**Low-Risk Problems** • Air pollution from vehicles • Pesticide residues in foods • Global climate change • Drinking water contamination

With careful design, quality control, maintenance, and monitoring, a highly complex system such as a nuclear power plant or space shuttle can achieve a high degree of technology reliability. However, human reliability usually is much lower than technology reliability and is almost impossible to predict: To err is human.

Suppose that the technology reliability of a nuclear power plant is 95% (0.95) and that human reliability is 75% (0.75). Then the overall system reliability is 71% ($0.95 \times 0.75 = 0.71 = 71\%$). Even if we could make the technology 100% reliable (1.0), the overall system reliability would still be only 75% ($1.0 \times 0.75 = 0.75 = 75\%$). The crucial dependence of even the most carefully designed systems on unpredictable human reliability helps explain essentially "impossible" tragedies such as the **(1)** Chernobyl nuclear power-plant accident (p. 490) and **(2)** explosion of the space shuttle *Challenger*.

One way to make a system more foolproof or fail-safe is to move more of the potentially fallible elements from the human side to the technical side. However, **(1)** chance events such as a lightning bolt can knock out an automatic control system, **(2)** no machine or computer program can completely replace human judgment, **(3)** the parts in any automated control system are manufactured, assembled, tested, certified, and maintained by fallible human beings, and **(4)** computer software programs used to monitor and control complex systems can also contain human error or can be deliberately modified by computer viruses to malfunction.

What Are the Limitations of Risk Analysis? Here are some of the key questions involved in evaluating risk analysis:

- How reliable are risk assessment data and models?
- Who profits from allowing certain levels of harmful chemicals into the environment, and who suffers? Who decides this?
- Should estimates emphasize short-term risks, or should more weight be put on long-term risks? Who should make this decision?
- Should the primary goal of risk analysis be to **(1)** determine how much risk is acceptable (the current approach) or **(2)** figure out how to do the least damage (a prevention approach)?

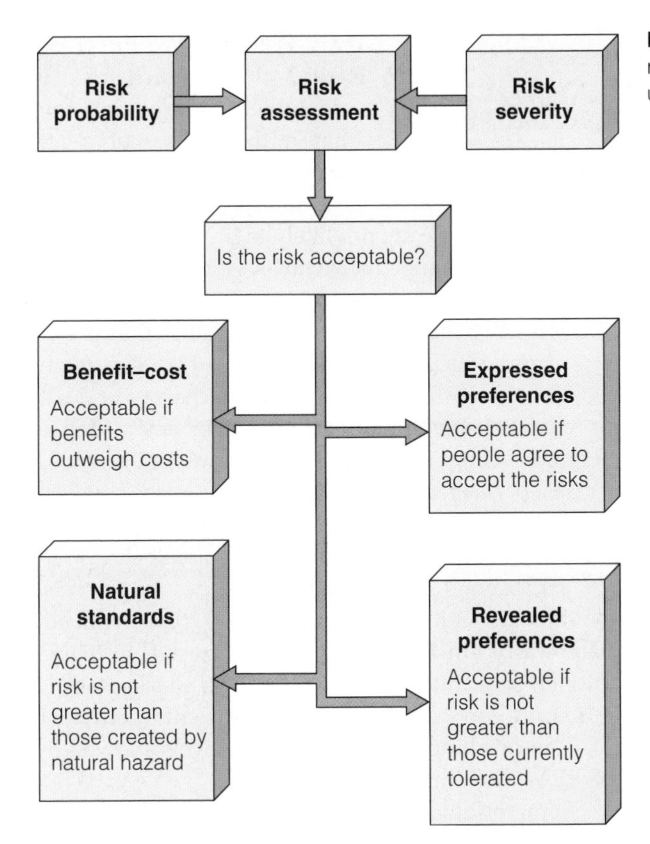

Figure 10-13 Methods for determining the acceptability of a risk after a risk assessment has been made. Benefit–cost analysis is the most widely used method.

- Who should do a particular risk analysis, and who should review the results? A government agency? Independent scientists? The public?
- Should cumulative effects of various risks be considered, or should risks be considered separately, as is usually done? Suppose a pesticide is found to have an annual risk of killing 1 person in 1 million from cancer, the current EPA limit. Cumulatively, however, effects from 40 such pesticides might kill 40, or 400, of every 1 million people. Is this acceptable, and to whom is it acceptable?
- How widespread is each risk?
- Should risk levels be higher for workers (as is almost always the case) than for the general public? What say should workers and their families have in this decision? According to government estimates, workers' exposure to toxic chemicals in the United States causes 50,000–70,000 deaths (at least half from cancer) and 350,000 new cases of illness per year. The situation is much worse in developing countries, with more than 1 million work-related deaths occurring worldwide each year.
- How much risk is acceptable, and to whom is it acceptable? According to the National Academy of Sciences, exposure to toxic chemicals is responsible for 2–4% of the 521,000 cancer deaths in the United States; this amounts to 10,400–20,800 premature cancer deaths per year.

Proponents contend that risk analysis is useful way to **(1)** organize and analyze available scientific information, **(2)** identify significant hazards, **(3)** focus on areas that warrant more research, **(4)** help regulators decide how money for reducing risks should be allocated, and **(5)** stimulate people to make more informed decisions about health and environmental goals and priorities.

However, critics point out that results of risk analysis are very uncertain. For example, a recent study documented the significant uncertainties involved in even simple risk analysis. Eleven European governments established 11 different teams of their best scientists and engineers (including those from private companies) to assess the hazards and risks from a small plant storing only one hazardous chemical (ammonia). The 11 teams, consisting of world-class experts analyzing this very simple system, disagreed with one another on fundamental points and varied in their assessments of the hazards by a factor of 25,000.

Such built-in uncertainty in risk analysis is analogous to a radar device that can detect a car speeding at 160 kilometers (100 miles) per hour but can tell us only that the car is traveling somewhere between 0.16 kilometer (0.1 mile) per hour and 160,000 kilometers (100,000 miles) per hour. Such inherent uncertainty explains why regulators setting human exposure levels for toxic substances usually divide the best results by 100 to 1,000 to provide the public with a margin of safety.

According to critics, the main decision-making tool we should rely on is not to find out how much risk is acceptable, which is mostly a political decision. Instead, it should be to find out the least damaging reasonable alternatives by asking, "Which alternative will bring sufficient benefits and minimize damage to humans and to the earth?" To these critics, the emphasis should be on *alternative assessment*, not *risk assessment* (see Guest Essay on p. 372).

How Should Risks Be Managed? **Risk management** includes the administrative, political, and economic actions taken to decide whether and how to reduce a particular societal risk to a certain level and at what cost.

Risk management involves deciding

- The reliability of the risk analysis for each risk
- Which risks should be given the highest priority
- How much risk is acceptable (Figure 10-13)
- How much it will cost to reduce each risk to an acceptable level

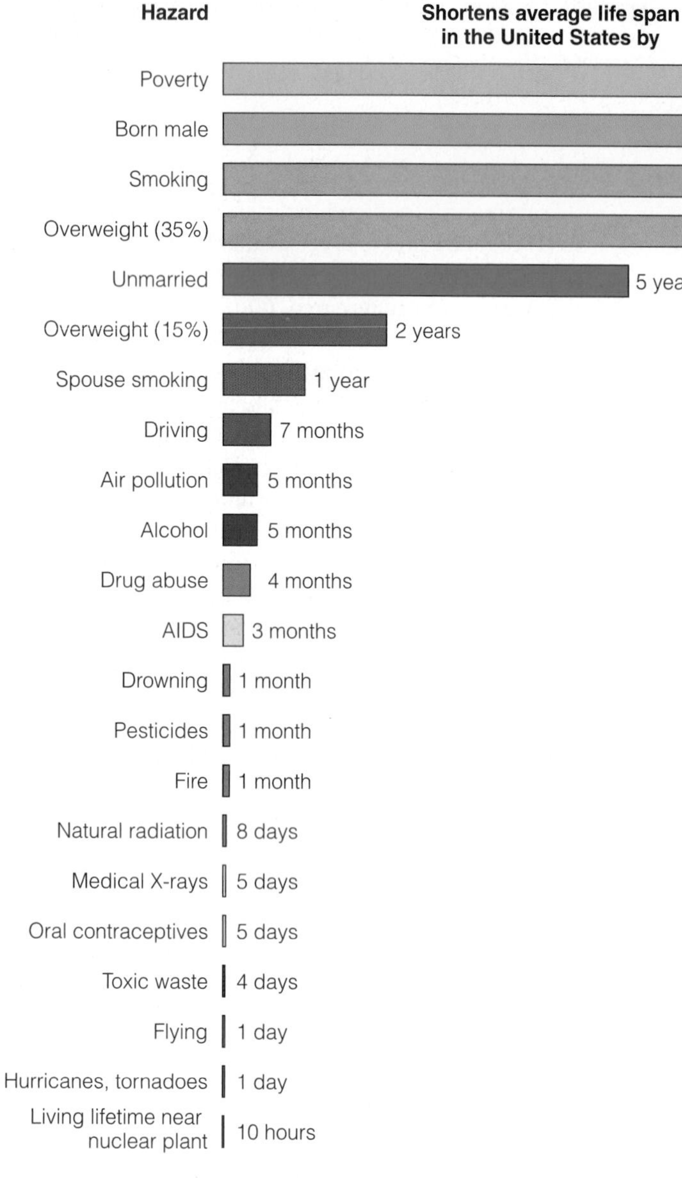

Figure 10-14 Comparison of risks people face, expressed in terms of shorter average life span. After poverty, the greatest risks people face result mostly from voluntary choices they make about their lifestyles. These are only generalized relative estimates. Individual response to some of these risks can vary with factors such as **(1)** genetic variation (Figure 10-3), **(2)** family medical history, **(3)** emotional makeup, **(4)** stress, and **(5)** social ties and support. (Data from Bernard L. Cohen)

- How limited funds should be spent to provide the greatest benefit
- How the risk management plan will be monitored, enforced, and communicated to the public

Each step in this process involves making value judgments and weighing trade-offs to find some reasonable compromise among conflicting political, economic, health, and environmental interests.

How Well Do We Perceive Risks? How much risk is acceptable? Studies indicate that if the chance of death from a chemical or activity is less than 1 in 100,000, most people are not likely to be worried enough to change their ways.

However, most of us do poorly in assessing the relative risks from the hazards that surround us (Figures 10-12 and 10-14). Also, many people deny or shrug off the high-risk chances of death (or injury) from voluntary activities they enjoy, such as **(1)** motorcycling (1 death in 50 participants), **(2)** smoking (1 in 300 participants by age 65 for a pack-a-day smoker), **(3)** hang-gliding (1 in 1,250), and **(4)** driving (1 in 2,500 without a seatbelt and 1 in 5,000 with a seatbelt).

Yet some of these same people may be terrified about the possibility of dying from a **(1)** commercial airplane crash (1 in 4.6 million), **(2)** train crash (1 in 20 million), **(3)** snakebite (1 in 36 million), **(4)** shark attack (1 in 300 million), or **(5)** exposure to trichloroethylene (TCE) in drinking water at the trace levels allowed by the EPA (1 in 2 billion).

Being bombarded with news about people killed or harmed by various hazards distorts our sense of risk. However, *the most important good news each year is that about 99.1% of the people on the earth did not die.* Despite the greatly increased use of synthetic chemicals in food production and processing, the general health and average life expectancy of people in the United States (and most developed countries) have increased during the past 50 years.

Our perceptions of risk and our responses to perceived risks often have little to do with how risky the experts say something is (Figures 10-12 and 10-14). The public generally sees a technology or a product as being riskier than experts do when

- *It is new or complex rather than familiar.* Examples include genetic engineering and genetically modified food (p. 405) or nuclear power, as opposed to large dams or coal-fired power plants (Figure 19-17, p. 501).
- *It is perceived as being mostly involuntary.* Examples include nuclear power plants or food additives, as opposed to driving or smoking.
- *It is viewed as unnecessary rather than as beneficial or necessary.* Examples might include using chlorofluorocarbon (CFC) propellants in aerosol spray cans or using food additives that increase sales appeal, as opposed to cars or aspirin.
- *Its use involves a large, well-publicized death toll from a single catastrophic accident rather than the same or an even larger death toll spread out over a longer time.* Examples might include a severe nuclear power plant acci-

dent (p. 490), an industrial explosion, or an accidental plane crash, as opposed to coal-burning power plants, automobiles, or smoking.

- *Its use involves unfair distribution of the risks.* Citizens are outraged when government officials decide to put a hazardous-waste landfill or incinerator in or near their neighborhood, even when the decision is based on risk analysis. This is usually seen as politics, not science. Residents will not be satisfied by estimates that the lifetime risks of cancer death from the facility are not greater than, say, 1 in 100,000. Living near the facility means that they, not the vast majority people living farther away, have a much higher risk of dying from cancer by having this risk involuntarily imposed on them.
- *The people affected are not involved in the decision-making process from start to finish.*
- *Its use does not involve a sincere search for and evaluation of alternatives.* People who believe that their lives and the lives of their families are being threatened want to know **(1)** what the alternatives are and **(2)** which alternative causes the least harm to them and the earth.

Better education and communication about the nature of risks will help bring the public's perceptions of various risks closer to those of professional risk evaluators. However, such education will not eliminate the emotional, cultural, and ethical factors that decision makers must take into account in determining the acceptability of a particular risk and evaluating the possible alternatives.

The burden of proof imposed on individuals, companies, and institutions should be to show that pollution prevention options have been thoroughly examined, evaluated, and used before lesser options are chosen.

JOEL HIRSCHORN

REVIEW QUESTIONS

1. Define the boldfaced terms in this chapter.

2. What human activity kills the largest number of people each year? List six ways to help reduce the harmful effects of smoking.

3. What are *risk* and *probability*? Distinguish between *risk assessment* and *risk management.*

4. List two **(a)** cultural hazards, **(b)** chemical hazards, **(c)** physical hazards, and **(d)** biological hazards.

5. What is *toxicity*? Distinguish between *dosage* and *response* for a potentially harmful substance. List five factors that determine whether a chemical is harmful. Distinguish between *bioaccumulation* and *biomagnification.* List three mechanisms by which the human body can reduce the harmful effects of most harmful chemicals.

6. What is a *poison*? What is an LD_{50}? List three methods used to determine toxicity and list the limitations of each method. Describe how laboratory tests are used to determine toxicity. What is a *dose–response curve*? Distinguish between a *linear dose–response curve* and a *threshold dose–response curve*.

7. Distinguish between *toxic chemicals* and *hazardous chemicals*. Distinguish between *mutagens, teratogens,* and *carcinogens* and give one example of each.

8. Distinguish between the *immune system, nervous system,* and *endocrine system* and give an example of something that causes harm to each system. What are *hormone disrupters* and *hormone mimics*? List two examples of such chemicals.

9. About what percentage of the 75,000 chemicals in commercial use in the United States have been screened **(a)** to assess toxicity, **(b)** to determine whether they are carcinogens, teratogens, or mutagens, and **(c)** to determine whether they damage the nervous, endocrine, or immune systems?

10. List three reasons for the lack of information about the potentially harmful effects of most chemicals in commercial use. Distinguish between the regulation strategy and the pollution prevention strategy for protecting the public from potentially harmful chemicals. What is the *precautionary principle*? Why is it rarely used?

11. Distinguish between *nontransmissible* and *transmissible diseases* and give two examples of each type. What are the seven deadliest infectious diseases in order of the number of deaths they cause each year?

12. How do infectious bacteria become resistant to antibiotics?

13. What is the best way to treat **(a)** a bacterial disease and **(b)** a viral disease? List two examples of each type of disease.

14. What causes tuberculosis, and how is it transmitted? About how many people died of TB during the past year? List five reasons for the increase in TB infections in recent years. How can the spread of this bacterial infectious disease be slowed?

15. Distinguish between *HIV* and *AIDS*. List four ways in which HIV can be transmitted. About how many people in the world **(a)** are infected with HIV and **(b)** have died of AIDS? List ways to prevent the spread of this viral infectious disease.

16. What causes malaria? About how many people die from malaria each year? List seven ways to help prevent this protozoal infectious disease.

17. List 10 ways to prevent or reduce the incidence of infectious diseases throughout the world. List seven major ways to improve health care in developing countries.

18. What is *risk analysis*? What are the major limitations of risk analysis?

19. List five of the greatest risks people face in terms of reduced life span. List eight ways to reduce your risk of premature death and serious health problems. How can we estimate the risks from technological systems?

20. What is *risk management*? What six questions do risk managers try to answer? About what percentage of the people on the earth die each year? List seven reasons why people often perceive that certain risks are greater than experts say they are.

CRITICAL THINKING

1. Explain why you agree or disagree with the proposals made by health officials for reducing the death toll and other harmful effects of smoking listed on p. 224.

2. Do you think chemicals should be regulated based on their effects on the nervous, immune, and endocrine systems? Explain.

3. Should we have zero pollution levels for all hazardous chemicals? Explain.

4. Evaluate the following statements:

- **a.** We should not get so worked up about exposure to toxic chemicals because almost any chemical can cause some harm at a large enough dosage.
- **b.** We should not worry so much about exposure to toxic chemicals because through genetic adaptation we can develop immunity to such chemicals.
- **c.** We should not worry so much about exposure to toxic chemicals because we can use genetic engineering to reduce or eliminate such problems.

5. What are the five major risks you face from your lifestyle, where you live, and what you do for a living? Which of these risks are voluntary and which are involuntary? List the five most important things you can do to reduce these risks. Which of these things do you actually plan to do?

6. How would you answer each of the questions raised about **(a)** risk analysis on pp. 238–239 and **(b)** risk assessment and risk management on pp. 239–240? Explain each of your answers.

PROJECTS

1. Assume that members of your class (or small manageable groups in your class) have been appointed to a technology benefit–risk assessment board. As a group, decide why you would approve or disapprove of widespread use of each of the following: **(a)** drugs to retard aging, **(b)** electrical or chemical devices that would stimulate the brain to eliminate anxiety, fear, unhappiness, and aggression, and **(c)** genetic engineering to produce people with superior intelligence and strength.

2. Use the library or the internet to find recent articles describing the rise of genetic resistance of disease-causing bacteria to commonly used antibiotics. Evaluate the evidence and claims in these articles.

3. Pick a specific viral disease and use the library or internet to find out about **(a)** how it spreads, **(b)** its effects, **(c)** strategies for controlling its spread, and **(d)** possible treatments.

4. Make a concept map of this chapter's major ideas, using the section heads and subheads and the key terms (in boldface). Look on the website for this book for information about making concept maps.

INTERNET STUDY RESOURCES AND RESOURCES FOR FURTHER READING AND RESEARCH

The website for this book contains helpful study aids and many ideas for further reading and research. Log on to

http://www.brookscole.com/product/0534389872s

and click on the Chapter-by-Chapter area. Choose Chapter 10 and select a resource:

- "Flash Cards" allows you to test your mastery of the Terms and Concepts to Remember for this chapter.
- "Tutorial Quizzes" provides a multiple-choice practice quiz.
- "Student Guide to InfoTrac" will lead you to Critical Thinking Projects that use InfoTrac College Edition as a research tool.
- "References" lists the major books and articles consulted in writing this chapter.
- "Hypercontents" takes you to an extensive list of sites with news, research, and images related to individual sections of the chapter.

INFOTRAC COLLEGE EDITION

Improve your skills with InfoTrac College Edition, a searchable online database of articles from more than 700 periodicals. Log on to

http://www.infotrac-college.com

or access InfoTrac through the website for this book. Try to find the following articles:

1. Is Your Bathtub a Toxic Dump? (disease risk factors higher for children). Andreas Schuld, George Glasser. *Earth Island Journal*, Summer 2001 v16 i2 p26. In 1991, the EPA concluded that the average person can absorb more contaminants through the skin from bathing and showering than from drinking polluted water. Children are most at risk. *Hint*: Enter the search term "toxic dump" using the keywords "chemical hazards."

2. Endocrine Disruptors: Present Issues, Future Directions. David Crews, Emily Willingham, James K. Skipper. *Quarterly Review of Biology*, Sept 2000 v75 i3 p243. "A variety of natural products and synthetic chemicals, known collectively as endocrine-disrupting compounds (EDCs), mimic or interfere with the mechanisms that govern vertebrate reproductive development and function." These compounds may affect present and future generations. *Hint*: Enter the search term "endocrine disrupters" using the keywords "chemical hazards."

Part III

POPULATION, RESOURCES, AND SUSTAINABILITY

I recognize the right and duty of this generation to develop and use our natural resources, but I do not recognize the right to waste them, or to rob by wasteful use, the generations that come after us.

Theodore Roosevelt, 1900

11 THE HUMAN POPULATION: GROWTH AND DISTRIBUTION

Slowing Population Growth in Thailand

Can a country sharply reduce its population growth in only 15 years? Thailand did.

In 1971, Thailand adopted a policy to reduce its population growth. When the program began the country's population was growing at a rate of 3.2% per year, and the average Thai family had 6.4 children.

Fifteen years later, in 1986, the country's population growth rate had been cut in half to 1.6%. By 2001 the rate had fallen to 0.8%, and the average number of children per family was 1.8. Thailand's population is projected to grow from 62 million in 2001 to 72 million by 2025.

There are several reasons for this impressive feat: **(1)** the creativity of the government-supported family-planning program, **(2)** a high literacy rate among women (90%), **(3)** an increasing economic role for women and advances in women's rights, **(4)** better health care for mothers and children, **(5)** the openness of the Thai people to new ideas, **(6)** the willingness of the government to encourage and financially support family planning and to work with the private, nonprofit Population and Community Development Association (PCDA), and **(7)** support of family planning by the country's religious leaders (95% of Thais are Buddhist).

This transition was catalyzed by the charismatic leadership of Mechai Viravidaiya, a public relations genius and former government economist who launched the PCDA in 1974 to help make family planning a national goal. PCDA workers handed out condoms at festivals, movie theaters, and even traffic jams. Between 1971 and 2001, the percentage of married women using modern birth control rose from 15% to 70%—higher than the 60% usage in developed countries and the 51% usage in developing countries.

Mechai helped establish a German-financed revolving loan plan to enable people participating in family-planning programs to install toilets and drinking water systems. Low-rate loans were available to farmers practicing family planning. The government also offers loans to individuals from a fund that increases as their village's level of contraceptive use rises.

All is not completely rosy. Although Thailand has done well in slowing population growth and raising per capita income, it has been less successful in reducing pollution and improving public health. Its capital, Bangkok, remains one of the world's most polluted and congested cities. It is plagued with notoriously high levels of traffic congestion and air pollution (Figure 11-1). The typical motorist in Bangkok spends 44 days per year sitting in traffic, costing $4 billion in lost work time.

Figure 11-1 This policeman and schoolchildren in Bangkok, Thailand, are wearing masks to reduce their intake of air polluted mainly by automobiles. Bangkok is one of the world's most car-clogged cities, with car commutes averaging 3 hours per day. Roughly one of every nine of its residents has a respiratory ailment.

The problems to be faced are vast and complex, but come down to this: 6.1 billion people are breeding exponentially. The process of fulfilling their wants and needs is stripping earth of its biotic capacity to produce life; a climactic burst of consumption by a single species is overwhelming the skies, earth, waters, and fauna.

PAUL HAWKEN

This chapter addresses the following questions:

- How is population size affected by birth, death, fertility, and migration rates?
- How is population size affected by the percentage of males and females at each age level?
- How can population growth be slowed?
- What success have India and China had in slowing population growth?
- How can global population growth be reduced?
- How is the world's population distributed between rural and urban areas, and what factors determine how urban areas develop?
- What are the major resource and environmental problems of urban areas?
- How do transportation systems shape urban areas and growth, and what are the pros and cons of various forms of transportation?
- How can cities be made more sustainable and more desirable places to live?

11-1 FACTORS AFFECTING HUMAN POPULATION SIZE

How Is Population Size Affected by Birth Rates and Death Rates? Populations grow or decline through the interplay of three factors: *births*, *deaths*, and *migration*. **Population change** is calculated by subtracting the number of people leaving a population (through death and emigration) from the number entering it (through birth and immigration) during a specified period of time (usually a year):

$$\text{Population change} = (\text{Births} + \text{Immigration}) - (\text{Deaths} + \text{Emigration})$$

When births plus immigration exceed deaths plus emigration, population increases; when the reverse is true, population declines. When these factors balance out, population size remains stable, a condition known as **zero population growth (ZPG)**.

Instead of using the total numbers of births and deaths per year, demographers use **(1)** the **birth rate**, or **crude birth rate** (the number of live births per 1,000 people in a population in a given year), and **(2)** the **death rate**, or **crude death rate** (the number of deaths per 1,000 people in a population in a given year). Figure 11-2 shows the crude birth and death rates for various groupings of countries in 2001.

Birth rates and death rates are coming down worldwide, but death rates have fallen more sharply than birth rates. As a result, there are more births than deaths; every time your heart beats 26 more babies are added to the world's population. At this rate, we share the earth and its resources with about 222,000 more people each day (95% of them in developing countries).

The rate of the world's annual population change (excluding migration) usually is expressed as a percentage:

$$\text{Annual rate of natural population change (\%)} = \frac{\text{Birth rate} - \text{Death rate}}{1{,}000 \text{ people}} \times 100 = \frac{\text{Birth rate} - \text{Death rate}}{10}$$

Exponential population growth has not disappeared but is occurring at a slower rate. The rate of the world's annual population growth (natural increase) dropped 39% between 1963 and 2001, from 2.2% to 1.33%. This is good news, but during the same period

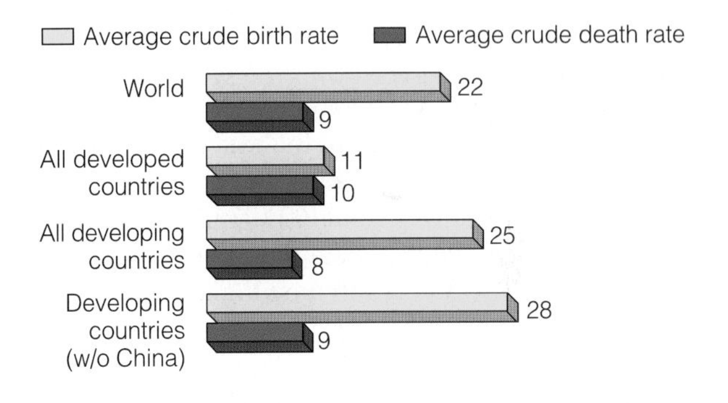

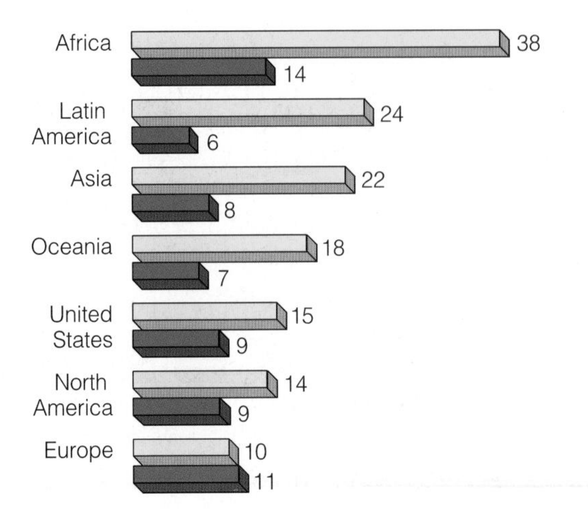

Figure 11-2 Average crude birth and death rates for various groupings of countries in 2001. (Data from Population Reference Bureau)

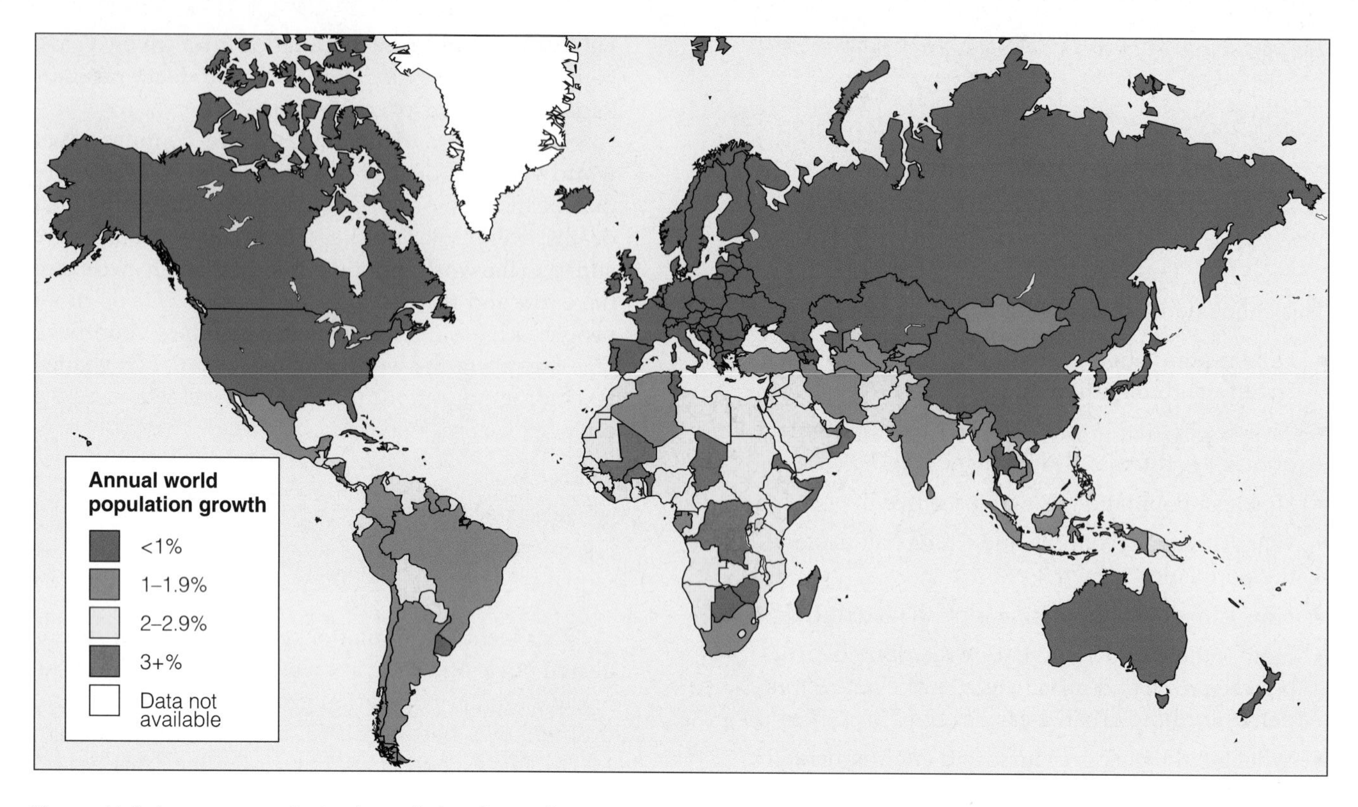

Figure 11-3 Average annual rate of population change (natural increase) in 2001. (Data from Population Reference Bureau)

the population base rose by about 91%, from 3.2 billion to 6.1 billion. This drop in the rate of population increase is somewhat like learning that a truck heading straight at you has slowed from 100 kilometers per hour (kph) to 61 kph while its weight has almost doubled. The world's population is still growing fast enough to double in 53 years.

Figure 11-3 presents the annual rates of population change for major parts of the world in 2001. An annual natural increase rate of 1–3% may seem small, but the current exponential growth rate of 1.33% adds about 81 million people per year. This is roughly equal to adding **(1)** another New York City every month, **(2)** a Germany every year, and **(3)** a United States every 3.5 years. Despite the drop in the rate of population growth, the larger base of population means that 81 million people were added in 2001, compared to only 69 million in 1963, when the world's population growth rate reached its peak. Figure 11-4 shows **(1)** the average annual percentage increase in the world's population and **(2)** the increase in population size, 1950–2001, and projected increases to 2050.

In numbers of people, China (with 1.27 billion in 2001, about one of every five people in the world) and India (with 1 billion) dwarf all other countries. Together they make up 37% of the world's population. The United States, with 285 million people in 2001, has the world's third largest population but only 4.7% of the world's people.

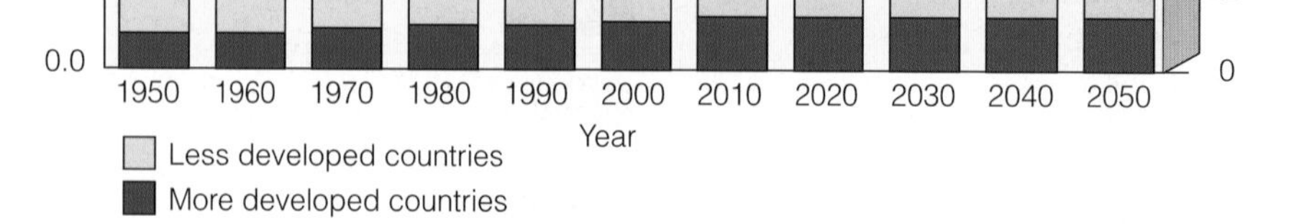

Figure 11-4 Average annual increase in the world's population size, 1950–2001, and projected increase, 2001–2050. (Data from United Nations, *World Population Prospects: The 2000 Revision*, 2001)

How Have Global Fertility Rates Changed? Two types of fertility rates affect a country's population size and growth rate. The first type, **replacement-level fertility**, is the number of children a couple must bear to replace themselves. It is slightly higher than two children per couple (2.1 in developed countries and as high as 2.5 in some developing countries in 2000), mostly because some female children die before reaching their reproductive years.

Does reaching replacement-level fertility mean an immediate halt in population growth (zero population growth)? No, because there are so many future parents already alive. If each of today's couples had an average of 2.1 children and their children also had 2.1 children, the world's population would continue to grow for 50 years or more (assuming that death rates do not rise).

The second type of fertility rate is the **total fertility rate (TFR)**: an estimate of the average number of children a woman will have during her childbearing years if between ages 15 and 49 she bears children at the same rate as women did this year. TFRs have dropped sharply since 1950 (Figure 11-5). Figure 11-6 shows TFRs throughout the world in 2001. In 2001, the average global TFR was 2.8 children per woman. It was 1.6 in developed countries (down from 2.5 in 1950) and 3.2 in developing countries (down from 6.5 in 1950). This drop in the average number of children born to women in developing countries is an impressive decline, but this level of fertility is still far above the replacement level of 2.1.

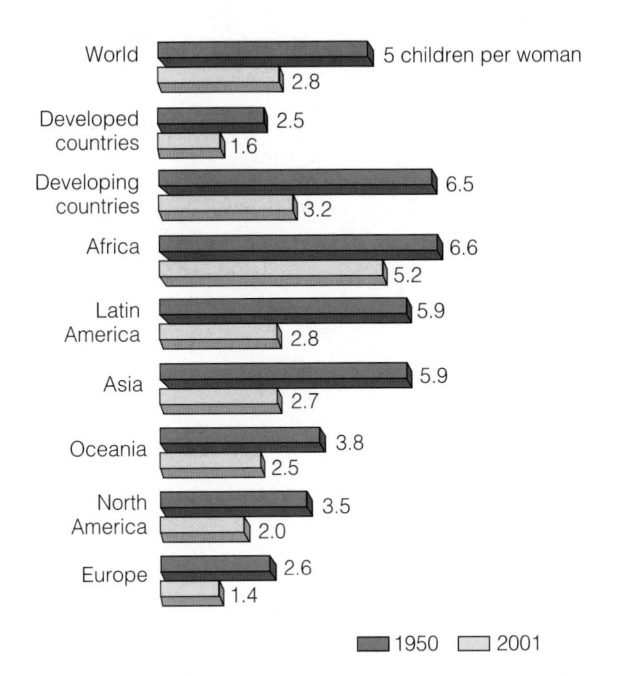

Figure 11-5 Decline in total fertility rates for various groupings of countries, 1950–2001. (Data from United Nations)

United Nations population projections to 2050 vary depending on the world's projected average TFR (Figure 11-7). More than 95% of this growth is projected to take place in developing countries, where acute

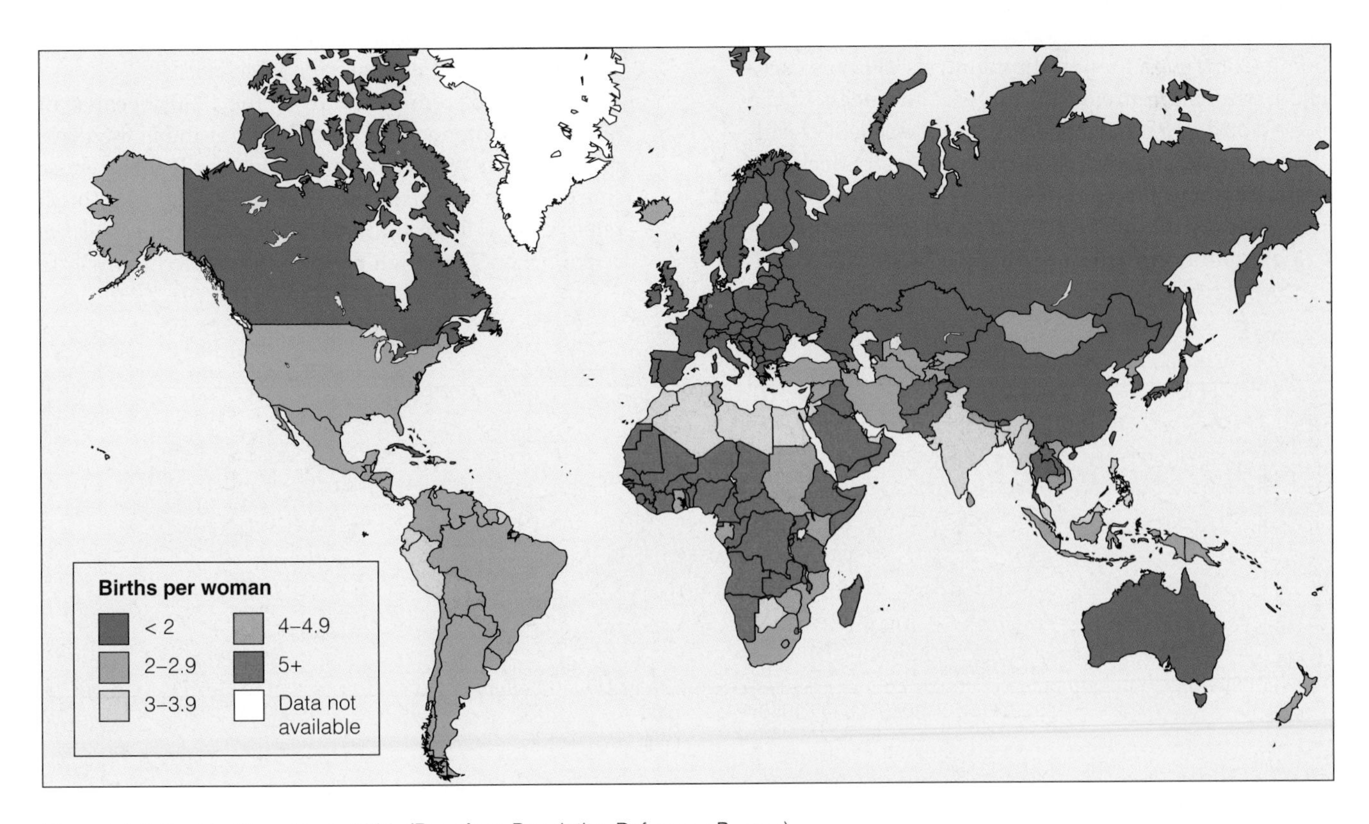

Figure 11-6 Total fertility rates in 2001. (Data from Population Reference Bureau)

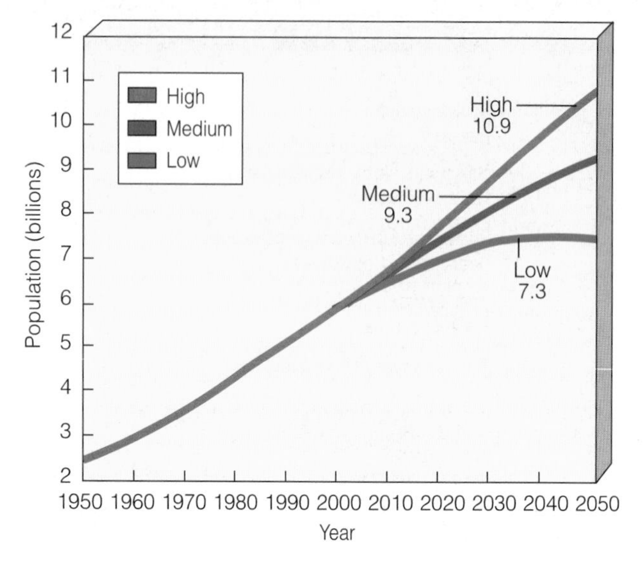

Figure 11-7 United Nations world population projection, assuming that by 2050 the world's total fertility rate is 2.6 (high), 2.1 (medium), or 1.7 (low) children per woman. (Data from United Nations, *World Population Prospects: The 2000 Revision*, 2001)

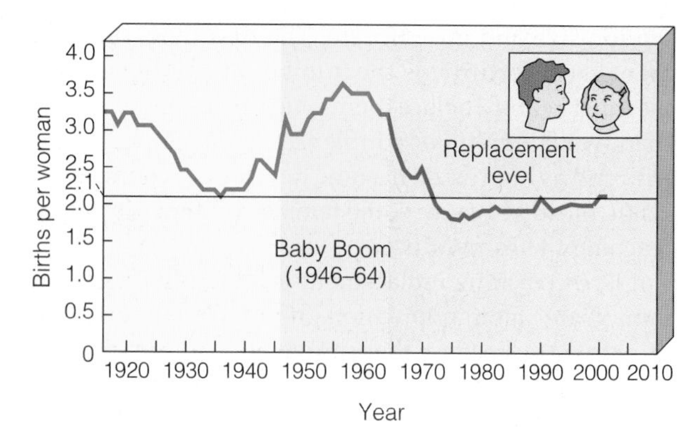

Figure 11-8 Total fertility rates for the United States between 1917 and 2001. (Data from Population Reference Bureau and U.S. Census Bureau)

poverty is a way of life for about 1.5 billion people. Between 2001 and 2050, using medium estimates, the population of developing countries is projected to increase from 5 billion to 8 billion.

Case Study: How Have Fertility Rates Changed in the United States? The population of the United States has grown from 76 million in 1900 to 285 million in 2001, even though the country's TFR has oscillated wildly (Figure 11-8). In 1957, the peak of the post–World War II baby boom, the TFR reached 3.7 children per woman. Since then it has generally declined, remaining at or below replacement level since 1972.

The drop in the TFR has led to a decline in the rate of population growth in the United States. However, the country's population is still growing faster (1% a year) than that of any other developed country and is not even close to zero population growth. In 2001, this growth added about 3 million people: **(1)** 1.7 million more births than deaths (accounting for about 60% of the growth), **(2)** almost 1 million legal immigrants and refugees, and **(3)** an estimated 300,000 illegal immigrants.

Figure 11-9 shows U.S. birth rates between 1910 and 2001. Between 1910 and 1930, birth rates fell sharply as **(1)** the country underwent industrialization and urbanization and **(2)** more women were educated and began working outside the home. This shift from high birth rates to low birth rates during industrialization is called a *demographic transition.*

Birth rates remained low in the 1930s because of the Great Depression and then began rising in the 1940s during World War II. After World War II there was a sharp rise in the birth rate. This period of high birth rates between 1946 and 1964 is known as the *baby-boom period,* when 79 million people were added to the U.S.

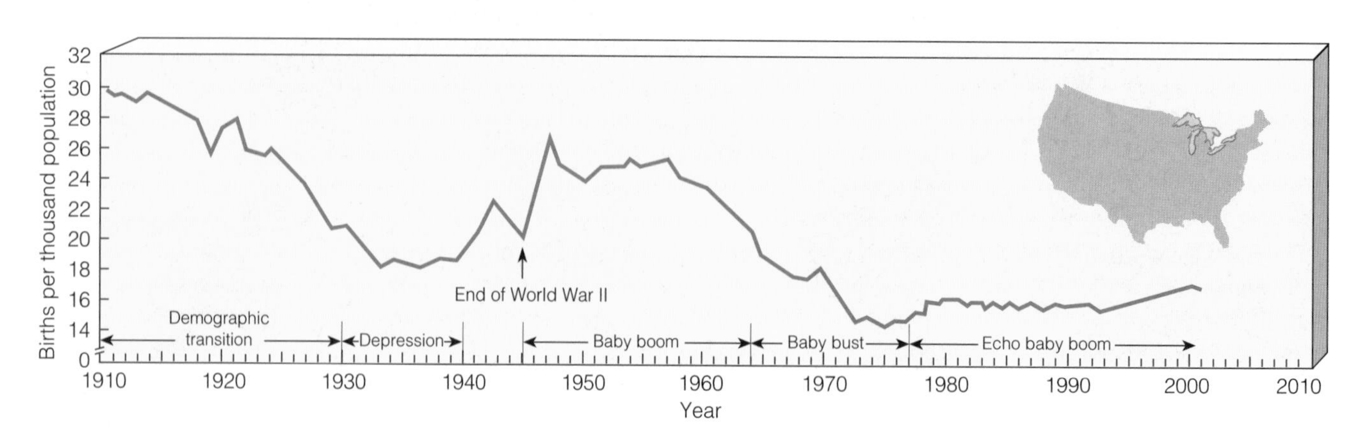

Figure 11-9 Birth rates in the United States from 1910 to 2001. (Data from U.S. Bureau of Census and U.S. Commerce Department)

population. Between 1956 and 1972, birth rates began to decline as more women began working outside the home and the desired family size dropped from 4 to 2 (or no) children.

Between 1977 and 1980, a small *echo boom* in the number of births per year occurred as a large number of people born during the baby boom began having children. Birth rates are projected to rise again between 2001 and 2050. According to U.S. Bureau of Census medium projections, the U.S. population will increase from 285 million to 410 million between 2001 and 2050, with no stabilization on the horizon. Because of a high per capita rate of resource use, each addition to the U.S. population has an enormous environmental impact (Figure 1-8, p. 10).

What Factors Affect Birth Rates and Fertility Rates? Key factors affecting a country's average birth rate and TFR are the following:

- *Importance of children as a part of the labor force.* Rates tend to be higher in developing countries (especially in rural areas, where children begin working to help raise crops at an early age).
- *Urbanization.* People living in urban areas **(1)** usually have better access to family-planning services and **(2)** tend to have fewer children than those living in rural areas, where children are needed to perform essential tasks.
- *Cost of raising and educating children.* Rates tend to be lower in developed countries, where raising children is much more costly because children do not enter the labor force until their late teens or early 20s, or even later.
- *Educational and employment opportunities for women.* Rates tend to be low when women have access to education and paid employment outside the home. In developing countries, women with no education generally have two more children than women with a secondary school education.
- *Infant mortality rate.* In areas with low infant mortality rates, people tend to have fewer children because fewer children die at an early age.
- *Average age at marriage* (or, more precisely, the average age at which women have their first child). Women normally have fewer children when their average age at marriage is 25 or older.
- *Availability of private and public pension systems.* Pensions eliminate the need of parents to have many children to help support them in old age.
- *Availability of legal abortions.* According to the United Nations and the World Bank, an estimated 26 million legal abortions and 20 million illegal (and often unsafe) abortions are performed worldwide each year among the roughly 190 million pregnancies per year.
- *Availability of reliable birth control methods* (Figure 11-10).
- *Religious beliefs, traditions, and cultural norms.* In some countries, these factors favor large families and strongly oppose abortion and some forms of birth control.

What Factors Affect Death Rates? The rapid growth of the world's population over the past 100 years is not the result of a rise in the crude birth rate. Instead, it has been caused largely by a decline in crude death rates, especially in developing countries (Figure 11-11).

More people started living longer (and fewer infants died) because of **(1)** increased food supplies and distribution, **(2)** better nutrition, **(3)** improvements in medical and public health technology (such as immunizations and antibiotics), **(4)** improvements in sanitation and personal hygiene, and **(5)** safer water supplies (which have curtailed the spread of many infectious diseases).

Two useful indicators of overall health of people in a country or region are **(1) life expectancy** (the average number of years a newborn infant can expect to live) and **(2)** the **infant mortality rate** (the number of babies out of every 1,000 born who die before their first birthday).

The *good news* is that global life expectancy at birth **(1)** increased from 48 years to 67 years (75 years in developed countries and 64 years in developing countries) between 1955 and 2001 and **(2)** is projected to reach 73 by 2025. Between 1900 and 2001, life expectancy in the United States increased from 46 to 74 years for men and from 48 to 80 years for women. The *bad news* is that in the world's 38 poorest countries, mainly in Africa, life expectancy is 50 years or less and in many African countries will fall because of increased deaths from AIDS (p. 235).

Because it reflects the general level of nutrition and health care, infant mortality probably is the single most important measure of a society's quality of life. A high infant mortality rate usually indicates **(1)** insufficient food (undernutrition), **(2)** poor nutrition (malnutrition), and **(3)** a high incidence of infectious disease (usually from contaminated drinking water).

Between 1965 and 2001, the world's infant mortality rate dropped from 20 per 1,000 live births to 8 in developed countries and from 118 to 61 in developing countries. This is an impressive achievement, but it still means that at least 8 million infants die of preventable causes during their first year of life—an average of 22,000 mostly unnecessary infant deaths

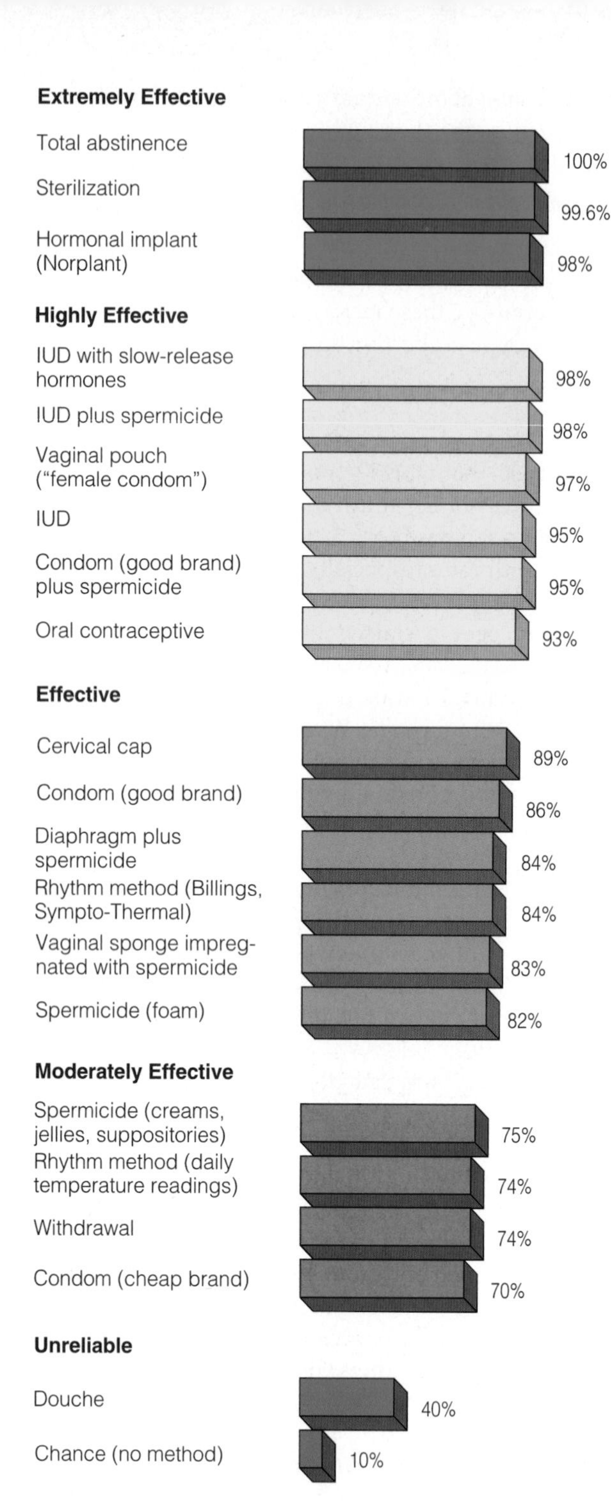

Figure 11-10 Typical effectiveness rates of birth control methods in the United States. Percentages are based on the number of undesired pregnancies per 100 couples using a specific method as their sole form of birth control for a year. For example, the effectiveness rating of 93% for oral contraceptives means that for every 100 women using the pill regularly for 1 year, 7 will get pregnant. Effectiveness rates tend to be lower in developing countries, primarily because of lack of education. (Data from Alan Guttmacher Institute)

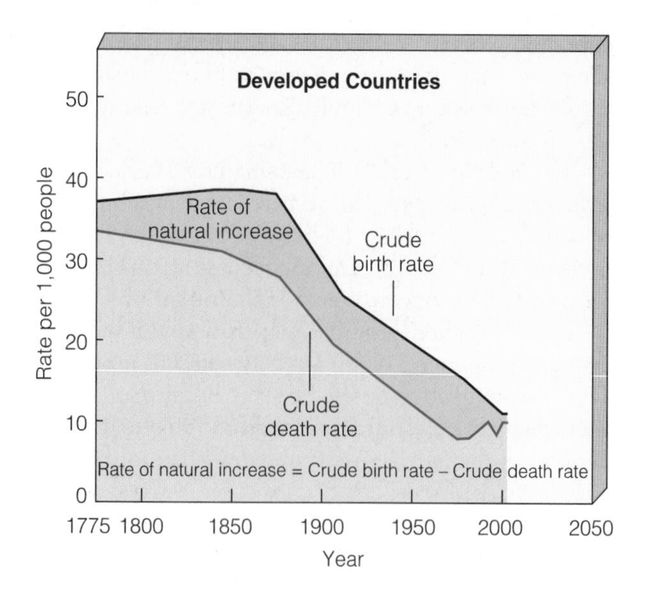

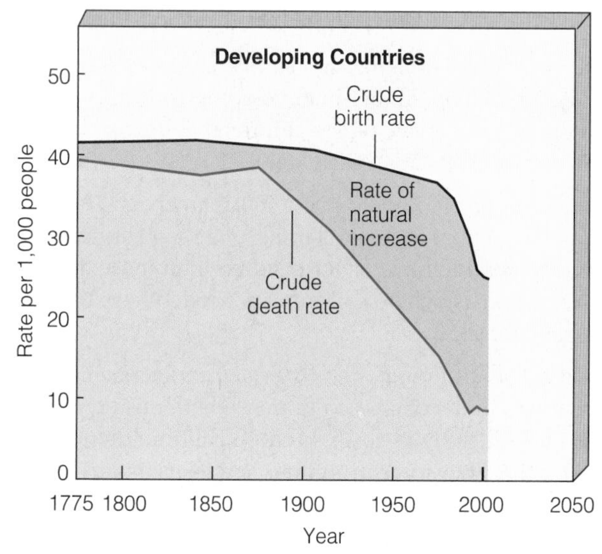

Figure 11-11 Changes in crude birth and death rates for developed and developing countries between 1775 and 2001. (Data from Population Reference Bureau and United Nations)

per day. About 99% of these deaths occur in developing countries.

The U.S. infant mortality rate has declined steadily since 1917. Despite this improvement, 30 countries had lower infant mortality rates than the United States in 2001. Three factors that keep the U.S. infant mortality rate higher than it could be are **(1)** inadequate health care (for poor women during pregnancy and for their babies after birth), **(2)** drug addiction among pregnant women, and **(3)** the high birth rate among teenagers.

The *good news* is that the U.S. birth rate among girls ages 15–19 in 2001 was lower than at any time since 1940. The *bad news* is that the United States has the highest teenage pregnancy rate of any industrialized country. Each year about 872,000 teenage girls become pregnant in the United States (78% of them unplanned). About 253,000 of these girls have abortions, and the rest miscarry. Babies born to teenagers are more likely to have low birth weights, the most important factor in infant deaths.

11-2 POPULATION AGE STRUCTURE

What Are Age Structure Diagrams? As mentioned earlier, even if the replacement-level fertility rate of 2.1 were magically achieved globally tomorrow, the world's population would keep growing for at least another 50 years. The reason for this is a population's **age structure**: the proportion of the population (or of each sex) at each age level.

Demographers typically construct a population age structure diagram by plotting the percentages or numbers of males and females in the total population in each of three age categories: **(1)** *prereproductive* (ages 0–14), **(2)** *reproductive* (ages 15–44), and **(3)** *postreproductive* (ages 45 and up). Figure 11-12 presents generalized age structure diagrams for countries with rapid, slow, zero, and negative population growth rates.

How Does Age Structure Affect Population Growth? Any country with many people below age 15 (represented by a wide base in Figure 11-12, left) has a powerful built-in momentum to increase its population size unless death rates rise sharply. The number of births rises even if women have only one or two children because of the large number of girls who will soon be moving into their reproductive years.

In 2001, 30% of the people on the planet were under 15 years old. These 1.8 billion young people are poised to move into their prime reproductive years. In developing countries the number is even higher: 33%, compared with 18% in developed countries. This powerful force for continued population growth, mostly in developing countries, could be slowed by **(1)** an effective program to reduce birth rates or **(2)** a sharp rise in death rates. Suppose that somehow the world's average TFR fell immediately to the replacement level of 2.1 children. According to the UN Population Fund, even then more than 75% of the world's projected population growth (Figure 11-7) would still occur.

Figure 11-13 shows the age structure in developed and developing countries in 2001. We live in a demographically divided world, as shown by demographic data in the United States, Brazil, and Nigeria (Figure 11-14).

How Can Age Structure Diagrams Be Used to Make Population and Economic Projections? The 79-million-person increase that occurred in the U.S. population between 1946 and 1964, known as the *baby boom* (Figures 11-8 and 11-9), will continue to move up through the country's age structure as the members of this group grow older (Figure 11-15).

Baby boomers now make up nearly half of all adult Americans. As a result, they dominate the population's demand for goods and services and play an

Figure 11-12 Generalized population age structure diagrams for countries with **(1)** rapid (1.5–3%), **(2)** slow (0.3–1.4%), **(3)** zero (0–0.2%), and **(4)** negative population growth rates. (Data from Population Reference Bureau)

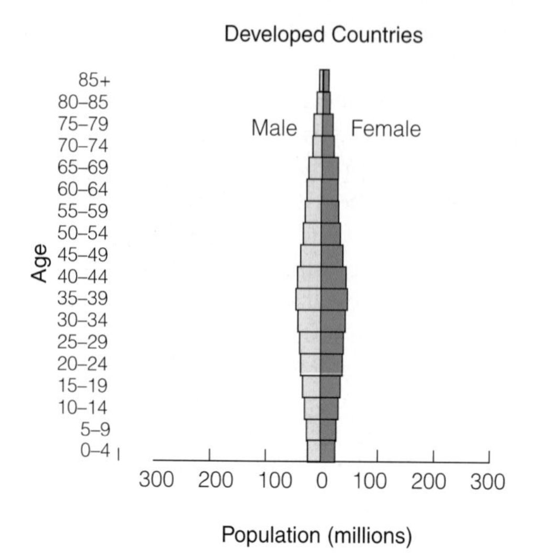

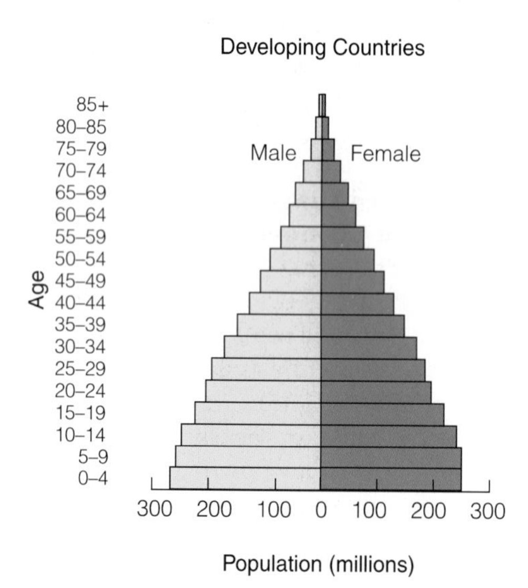

Figure 11-13 Population structure by age and sex in developing countries and developed countries, 2001. In 2001, there were **(1)** 1 billion young people in their prime reproductive years of 15–24 and **(2)** 1.8 billion people under age 15, moving into their reproductive years. (Data from United Nations Population Division and Population Reference Bureau)

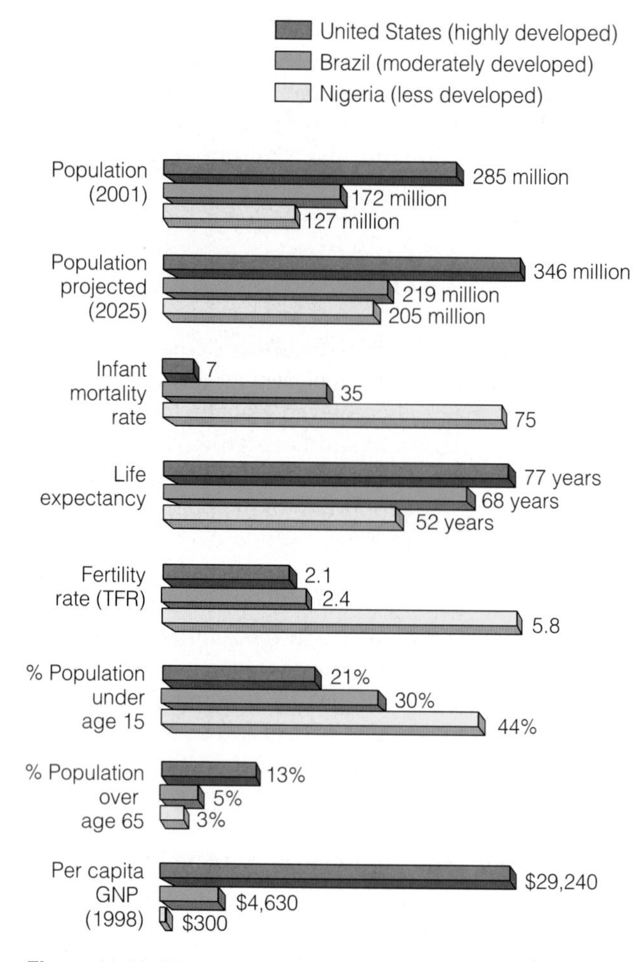

Figure 11-14 Comparison of key demographic indicators in **(1)** highly developed (United States), **(2)** moderately developed (Brazil), and **(3)** less developed (Nigeria) countries in 2001. (Data from Population Reference Bureau)

increasingly important role in deciding who gets elected and what laws are passed. Baby boomers who created the youth market in their teens and 20s are now creating the 50-something market and will soon move on to create a 60-something market.

Much of the economic burden of helping support a large number of retired baby boomers will fall on the *baby-bust generation* (also called *generation X*): the 44 million people born between 1965 and 1976 (when TFRs fell sharply and have remained below 2.1 since 1970; Figure 11-8). Retired baby boomers are beginning to use their political clout to force the smaller number of people in the baby-bust generation to pay higher income, health-care, and Social Security taxes. This could cause resentment and conflicts between the two generations.

The baby-bust generation is being followed by the *echo-boom generation* (Figure 11-9) consisting of about 83 million people born from 1977 to 2001. This largest generation ever is also known as *generation Y* and the *millennials*.

In some respects, the baby-bust generation should have an easier time than the baby-boom generation. There will be **(1)** fewer people competing for educational opportunities, jobs, and services, and **(2)** labor shortages may drive up their wages, at least for jobs requiring education or technical training beyond high school. On the other hand, members of the baby-bust group may find it difficult to get job promotions as they reach middle age because members of the much larger baby boom group will occupy most upper-level positions. Many baby boomers may delay retirement because of **(1)** improved health, **(2)** the need to accumulate adequate retirement funds, or **(3)** extension of the retirement age needed to begin collecting Social Security.

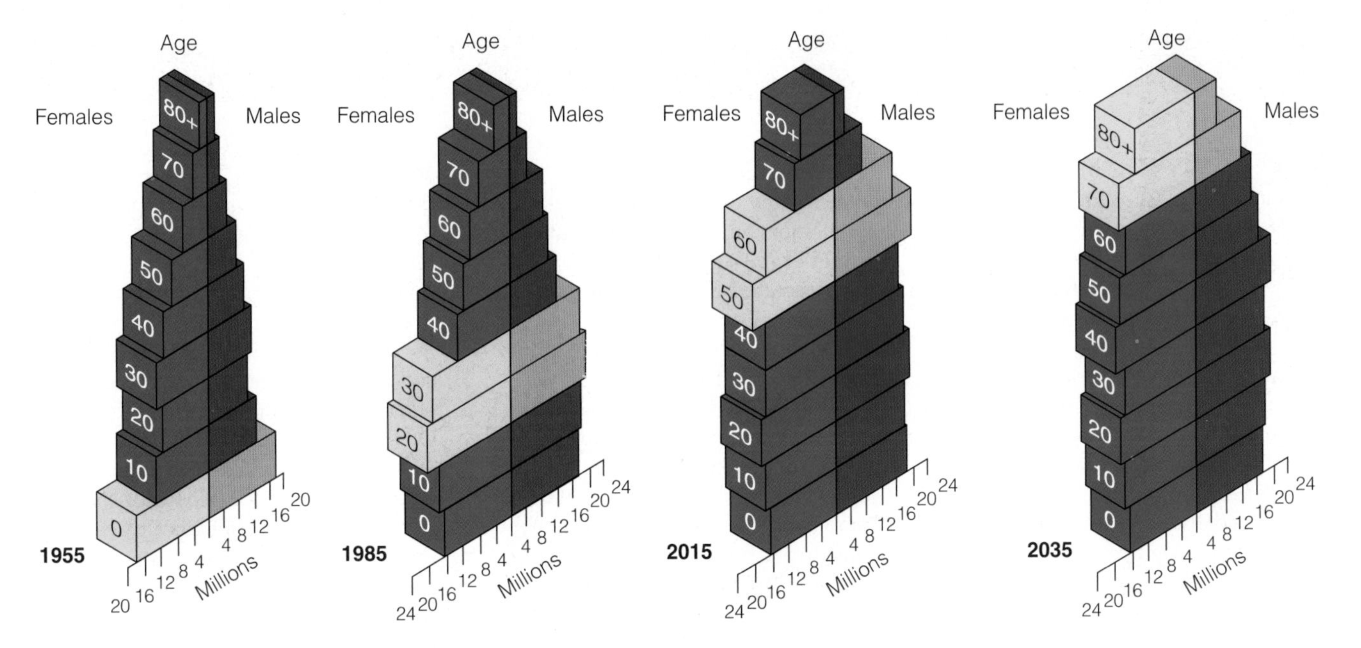

Figure 11-15 Tracking the baby-boom generation in the United States. (Data from Population Reference Bureau and U.S. Census Bureau)

From these few projections, we can see that any booms or busts in the age structure of a population create social and economic changes that ripple through a society for decades.

What Are Some Effects of Population Decline?

The populations of most of the world's countries are projected to grow throughout most of the 21st century. By 2001, however, 35 countries (most of them in Europe) with about 796 million people had roughly stable populations (annual growth rates below 0.3%) or declining populations. In other words, about 13% of humanity in industrialized Europe and Japan has achieved a stable population.

As the age structure of the world's population changes and the percentage of people age 60 or older increases (Figure 11-16), more and more countries will begin experiencing population declines. By 2020 an estimated 1 billion people will be age 60 or older. According to a 2001 UN study, by 2050 the populations of 39 countries (including Japan, Germany, Italy, Hungary, and Ukraine) are projected to be smaller than today.

If population decline is gradual, its negative effects usually can be managed. However, rapid population decline, like rapid population growth, can lead to severe economic and social problems. A country undergoing rapid population decline **(1)** has a sharp rise in the proportion of older people, who consume a large share of medical care, Social Security, and other costly public services funded by working taxpayers, and **(2)** can face labor shortages unless it relies on greatly increased automation or immigration of foreign workers.

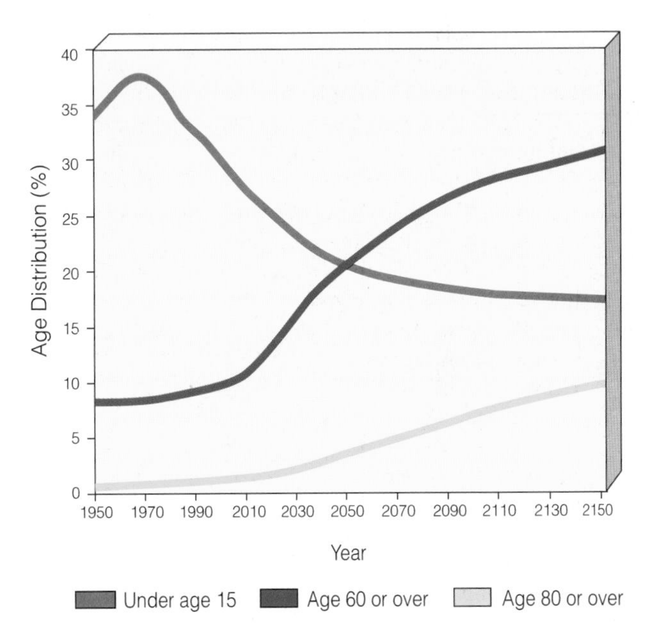

Figure 11-16 *Global aging.* Projected percentage of world population **(1)** under age 15, **(2)** age 60 or over, and **(3)** age 80 or over, 1950–2150, assuming the medium fertility projection shown in Figure 11-7. Between 1998 and 2050 the number of people over age 80 is projected to increase from 66 million to 370 million. The cost of supporting a much larger elderly population will place enormous strains on the world's economy. (Data from the United Nations)

11-3 SOLUTIONS: INFLUENCING POPULATION SIZE

How Is Population Size Affected by Migration?

The population of an area is affected by movement of

Immigration in the United States

CASE STUDY

Between 1820 and 2001, the United States admitted almost twice as many immigrants and refugees as all other countries combined. However, the number of legal immigrants has varied during different periods because of changes in immigration laws and rates of economic growth (Figure 11-17).

In 2001, the United States received about 1 million legal immigrants and refugees and 300,000 illegal immigrants, together accounting for 43% of the country's population growth. The U.S. Census Bureau estimates that there are 9–11 million illegal immigrants in the United States.

Currently, more than 75% of all legal immigrants live in six states: California, Florida, Illinois, New York, New Jersey, and Texas. If illegal immigrants are included, this figure rises to about 90%.

Immigrants place a tax burden on residents of such states. In California, for example, the average household pays an extra $1,200 in taxes per year because of immigrants. However, according to a 1997 study by the National Academy of Sciences, **(1)** the work and taxes paid by immigrants add $1–10 billion per year to the overall U.S. economy, largely because immigration holds down wages (and thus prices) for some jobs, and **(2)** during their lifetimes immigrants pay an average of $80,000 more per person in taxes than they cost in services.

Between 1820 and 1960, most legal immigrants to the United States came from Europe; since then, most have come from Latin America (53%) and Asia (30%). Between 2001 and 2050, the percentage of Latinos in the U.S. population is projected to almost double, from 13% to 24%.

In 1995, the U.S. Commission on Immigration Reform recommended reducing the number of legal immigrants and refugees to about 700,000 per year for a transition period and then to 550,000 a year. Some demographers and environmentalists go further and call for **(1)** lowering the annual ceiling for legal immigrants and refugees into the United States to 300,000–450,000 or **(2)** limiting legal immigration to about 20% of annual population growth. They would accept immigrants only if they can support themselves, arguing that providing immigrants with public services turns the United States into a magnet for the world's poor.

Most of these analysts also support efforts to sharply reduce illegal immigration. However, some are concerned that a crackdown on illegal immigrants can also lead to discrimination against legal immigrants.

Proponents argue that reducing immigration would allow the United States to stabilize its population sooner and help reduce the country's enormous environmental impact. Others oppose reducing current levels of legal immigration, arguing that **(1)** it would diminish the historical role of the United States as a place of opportunity for the world's poor and oppressed, **(2)** immigrants pay taxes and take many menial, low-paying jobs that other Americans shun, **(3)** few immigrants receive public assistance, **(4)** many immigrants open businesses and create jobs, and **(5)** according to the U.S. Census Bureau, after 2020 higher immigration levels will be needed to supply enough workers as baby boomers retire.

Critical Thinking

Should the United States reduce its current level of legal immigration and crack down on illegal immigration? Explain.

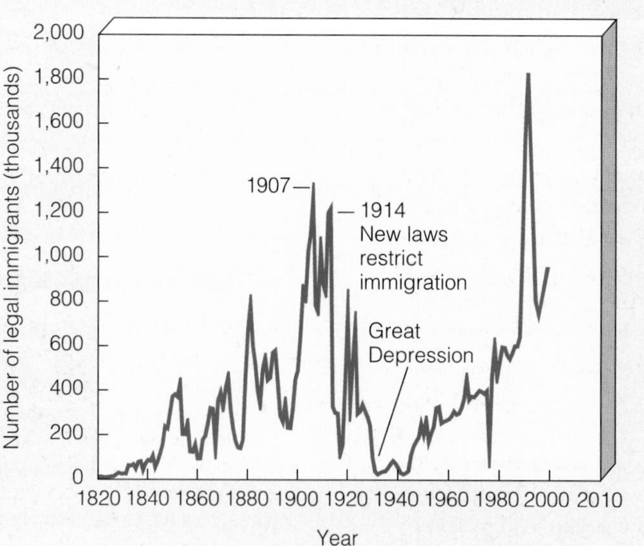

Figure 11-17 Legal immigration to the United States, 1820–2000. The large increase in immigration since 1989 resulted mostly from the Immigration Reform and Control Act of 1986, which granted legal status to illegal immigrants who could show that they had been living in the country for several years. (Data from U.S. Immigration and Naturalization Service)

people into (immigration) and out of (emigration) that area. Most countries influence their rates of population growth to some extent by restricting immigration. Only a few countries—chiefly Canada, Australia, and the United States (Case Study, above)—allow large annual increases in population from immigration.

International migration to developed countries absorbs only about 1% of the annual population growth in developing countries. Thus, population change for most countries is determined mainly by the difference between their birth rates and death rates.

Migration within countries, especially from rural to urban areas, plays an important role in the population dynamics of cities, towns, and rural areas, as discussed on p. 261.

What Are the Pros and Cons of Reducing Births? The projected medium increase of the human population from 6.1 to 9.3 billion or more between 2001 and 2050 raises an important question: *Can the world provide an adequate standard of living for 3.2 billion more people without causing widespread environmental damage?*

There is intense controversy over **(1)** this question, **(2)** whether the earth is already overpopulated, and **(3)** what measures, if any, should be taken to slow population growth. To some the planet is already overpopulated, but others disagree. Some analysts, mostly economists, argue that we should encourage population growth to help stimulate economic growth.

Others believe that asking how many people the world can support is the wrong question, equivalent to asking how many cigarettes one can smoke before getting lung cancer. Instead, they say, we should be asking what the *optimum sustainable population* of the earth might be, based on the planet's *cultural carrying capacity* (Guest Essay, p. 256).

Such an optimum level would allow most people to live in reasonable comfort and freedom without impairing the ability of the planet to sustain future generations. No one knows what this optimum population might be. Some consider it a meaningless concept; some put it at 20 billion, others at 8 billion, and others as low as 2 billion.

Those who do not believe that the earth is overpopulated point out that the average life span of the world's 6.1 billion people is longer today than at any time in the past. They say that the world can support billions more people, and people are the world's most valuable resource for solving the problems we face and stimulating economic growth by becoming consumers.

Some people view any form of population regulation as a violation of their religious beliefs, whereas others see it as an intrusion into their privacy and personal freedom. They believe that all people should be free to have as many children as they want. Some developing countries and some members of minorities in developed countries regard population control as a form of genocide to keep their numbers and power from rising.

Proponents of slowing and eventually stopping population growth point out that we fail to provide the basic necessities for one out of six people on the earth today. If we cannot (or will not) do this now, they ask, how will we be able to do this for the projected 3.2 billion more people added by 2050?

Proponents of slowing population growth contend that if we do not sharply lower birth rates, we are deciding by default to raise death rates for humans and greatly increase environmental harm. In 1992, for example, the U.S. National Academy of Sciences and the Royal Society of London issued the following joint statement: "If current predictions of population growth and patterns of human activity on the planet remain unchanged, science and technology may not be able to prevent either irreversible degradation of the environment or continued poverty for much of the world."

Proponents of this view recognize that population growth is not the only cause of environmental and resource problems. However, they argue that adding several hundred million more people in developed countries and several billion more in developing countries can only intensify existing environmental and social problems.

These analysts believe that people should have the freedom to produce as many children as they want. However, such freedom would apply only if it did not reduce the quality of other people's lives now and in the future, either by impairing the earth's ability to sustain life or by causing social disruption. They point out that limiting the freedom of individuals to do anything they want to protect the more basic freedoms of other individuals is the basis of most laws in modern societies. What is your opinion on this issue?

How Can Economic Development Help Reduce Birth Rates? Demographers have examined the birth and death rates of western European countries that industrialized during the 19th century. From these data they developed a hypothesis of population change known as the **demographic transition**: As countries become industrialized, first their death rates and then their birth rates decline.

According to this hypothesis, the transition takes place in four distinct stages (Figure 11-18, p. 258):

- *Preindustrial stage,* when there is little population growth because harsh living conditions lead to both a high birth rate (to compensate for high infant mortality) and a high death rate.
- *Transitional stage,* when industrialization begins, food production rises, and health care improves. Death rates drop and birth rates remain high, so the population grows rapidly (typically 2.5–3% a year).
- *Industrial stage,* when the birth rate drops and eventually approaches the death rate as industrialization and modernization become widespread. Population growth continues, but at a slower and perhaps fluctuating rate, depending on economic conditions. Most developed countries are now in this third stage (Figure 11-18), and a few developing countries are entering this stage.

GUEST ESSAY

Moral Implications of Cultural Carrying Capacity

Garrett Hardin

As longtime professor of human ecology at the University of California at Santa Barbara, Garrett Hardin made important contributions to relating ethics to biology. He has raised hard ethical questions, sometimes taken unpopular stands, and forced people to think deeply about environmental problems and their possible solutions. He is best known for his 1968 essay "The Tragedy of the Commons," which has had a significant impact on the disciplines of economics and political science and on the management of potentially renewable resources. His 17 books include Filters Against Folly: How to Survive Despite Economists, Ecologists, and the Merely Eloquent, Living Within Limits, *and* The Ostrich Factor: Our Population Myopia *(see Further Readings for this chapter on the website for this book).*

For many years, Angel Island in San Francisco Bay was plagued with too many deer. A few animals transplanted there in the early 1900s lacked predators and rapidly increased to nearly 300 deer—far beyond the carrying capacity of the island. Scrawny, underfed animals tugged at the heartstrings of Californians, who carried extra food for them from the mainland to the island.

Such well-meaning charity worsened the plight of the deer. Excess animals trampled the soil, stripped the bark from small trees, and destroyed seedlings of all kinds. The net effect was to lower the island's carrying capacity, year by year, as the deer continued to multiply in a deteriorating habitat.

State game managers proposed that the excess deer be shot by skilled hunters. "How cruel!" some people protested. Then the managers proposed that coyotes be introduced onto the island. Though not big enough to kill adult deer, coyotes can kill fawns, thereby reducing the size of the herd. However, the Society for the Prevention of Cruelty to Animals was adamantly opposed to this proposal.

In the end, it was agreed that some deer would be transported to other areas suitable for deer. A total of 203 animals were caught and trucked many miles away. From the fate of a sample of animals fitted with radio collars, it was estimated that 85% of the transported deer died within a year (most of them within 2 months) from various causes: predation by coyotes, bobcats, and domestic dogs, shooting by poachers and legal hunters, and being run over by cars.

The net cost (in 1982 dollars) for relocating each animal surviving for a year was $2,876. The state refused to continue financing the program, and no volunteers stepped forward to pay future bills.

Angel Island is a microcosm of the planet as a whole. Organisms reproduce exponentially, but the environment does not increase at all. The moral is a simple ecological commandment: *Thou shalt not transgress the carrying capacity.*

Now let's examine the situation for humans. A competent physicist has placed global human carrying capacity at 50 billion, about eight times the current world population. Before you give in to the temptation to urge women to have more babies, consider what Robert Malthus said nearly 200 years ago: "There should be no more people in a country than could enjoy daily a glass of wine and piece of beef for dinner."

A diet of grain or bread and water is symbolic of minimum living standards; wine and beef are symbolic of higher living standards that make greater demands on the environment. When land that could produce plants for direct human consumption is used to grow grapes for wine or corn for cattle, more energy is expended to feed the human population. Because carrying capacity is defined as the *maximum* number of animals (humans) an area can support, using part of the area to support such cultural luxuries as wine and beef reduces the carrying capacity. This reduced capacity is called the *cultural carrying capacity,* and it is always smaller than simple carrying capacity.

Energy is the common coin of the realm for all competing demands on the environment. Energy saved by giving up a luxury can be used to produce more food staples and support more people. We could increase the simple carrying capacity of the earth by giving up any (or all) of the following "luxuries": street lighting, vacations, private cars, air conditioning, and artistic performances of all sorts. But what we consider luxuries

- *Postindustrial stage,* when the birth rate declines further, equaling the death rate and thus reaching zero population growth. Then the birth rate falls below the death rate, and total population size decreases slowly. Thirty-five countries (most of them in Europe) containing about 13% of the world's population have entered this stage.

In most developing countries today, death rates have fallen much more than birth rates. In other words, these developing countries—mostly in Southeast Asia, Africa, and Latin America—are still in the transitional stage, halfway up the economic ladder, with high population growth rates. Some economists believe that developing countries will make the demographic transition over the next few decades without increased family-planning efforts.

However, despite encouraging declines in fertility (Figure 11-5), some population analysts fear that the still-rapid population growth in many developing countries will outstrip economic growth and overwhelm local life-support systems. This could cause

depends on our values as individuals and societies, and values are largely matters of choice. At one extreme, we could maximize the number of human beings living at the lowest possible level of comfort. Or we could try to optimize the quality of life for a much smaller human population.

The carrying capacity of the earth is a scientific question. It may be possible to support 50 billion people at a bread-and-water level. Is that what we choose? The question, "What is the cultural carrying capacity?" requires that we debate questions of value, about which opinions differ.

An even greater difficulty must be faced. So far, we have been treating carrying capacity as a *global* issue, as if there were some global sovereignty capable of enforcing a solution on all people. But there is no global sovereignty ("one world"), nor is there any prospect of one in the foreseeable future. Thus, we must ask how some 200 nations are to coexist in a finite global environment if different sovereignties adopt different standards of living.

Consider a protected redwood forest that produces neither food for humans nor lumber for houses. Because people must travel many kilometers to visit it, the forest is a net loss in the national energy budget. However, for those fortunate enough to wander through the cathedral-like aisles beneath an evergreen vault, a redwood forest does something precious for the human spirit. But then intrudes an appeal from a distant land, where millions are starving because their population has overshot the carrying capacity; we are asked to save lives by sending food. As long as we have surpluses, we may safely indulge in the pleasures of philanthropy. But after we have run out of our surpluses, then what?

A spokesperson for the needy from that land makes a proposal: "If you would only cut down your redwood forests, you could use the lumber to build houses and then grow potatoes on the land, shipping the food to us. Since we are all passengers together on Spaceship Earth, are you not duty bound to do so? Which is more precious, trees or human beings?"

This last question may sound ethically compelling, but let's look at the consequences of assigning a preemptive and supreme value to human lives. At least 2 billion people in the world are poorer than the 34 million "legally poor" in America, and their numbers are increasing by about 1 million per year. Unless this increase is halted, sharing food and energy on the basis of need would require the sacrifice of one amenity after another in rich countries. The ultimate result of sharing would be complete poverty everywhere on the earth to maintain the earth's simple carrying capacity. Is that the best humanity can do?

To date, there has been overwhelmingly negative reaction to all proposals to make international philanthropy conditional on the cessation of population growth by overpopulated recipient nations. Foreign aid is governed by two apparently inflexible assumptions:

- The right to produce children is a universal, irrevocable right of every nation, no matter how hard it presses against the carrying capacity of its territory.
- When lives are in danger, the moral obligation of rich countries to save human lives is absolute and undeniable.

Considered separately, each of these two well-meaning doctrines might be defensible; taken together, they constitute a fatal recipe. If humanity gives maximum carrying capacity precedence over problems of cultural carrying capacity, the result will be universal poverty and environmental ruin.

Or do you see an escape from this harsh dilemma?

Critical Thinking

1. What population size would allow the world's people to have good quality of life? What do you believe is the cultural carrying capacity of the country where you live? Should your country have a national policy to establish this population size as soon as possible? Explain.

2. Do you support the two principles this essay lists as the basis of foreign aid to needy countries? If not, what changes would you make in the requirements for receiving such aid?

many of these countries to be caught in a *demographic trap*, something that is currently happening in a number of developing countries, especially in Africa.

Analysts also point out that some of the conditions that allowed developed countries to develop are not available to many of today's developing countries. Even with large and growing populations, many developing countries **(1)** do not have enough skilled workers to produce the high-tech products needed to compete in the global economy, **(2)** lack the capital and resources needed for rapid economic development, and **(3)** since 1980 have experienced a drop in economic assistance from developed countries and a rise in their debt to such countries. Indeed, since the mid-1980s, developing countries have paid developed countries $40–50 billion a year (mostly in debt interest) more than they have received from these countries.

How Can Family Planning Help Reduce Birth and Abortion Rates and Save Lives? Family **planning** provides educational and clinical services that help couples choose how many children to have

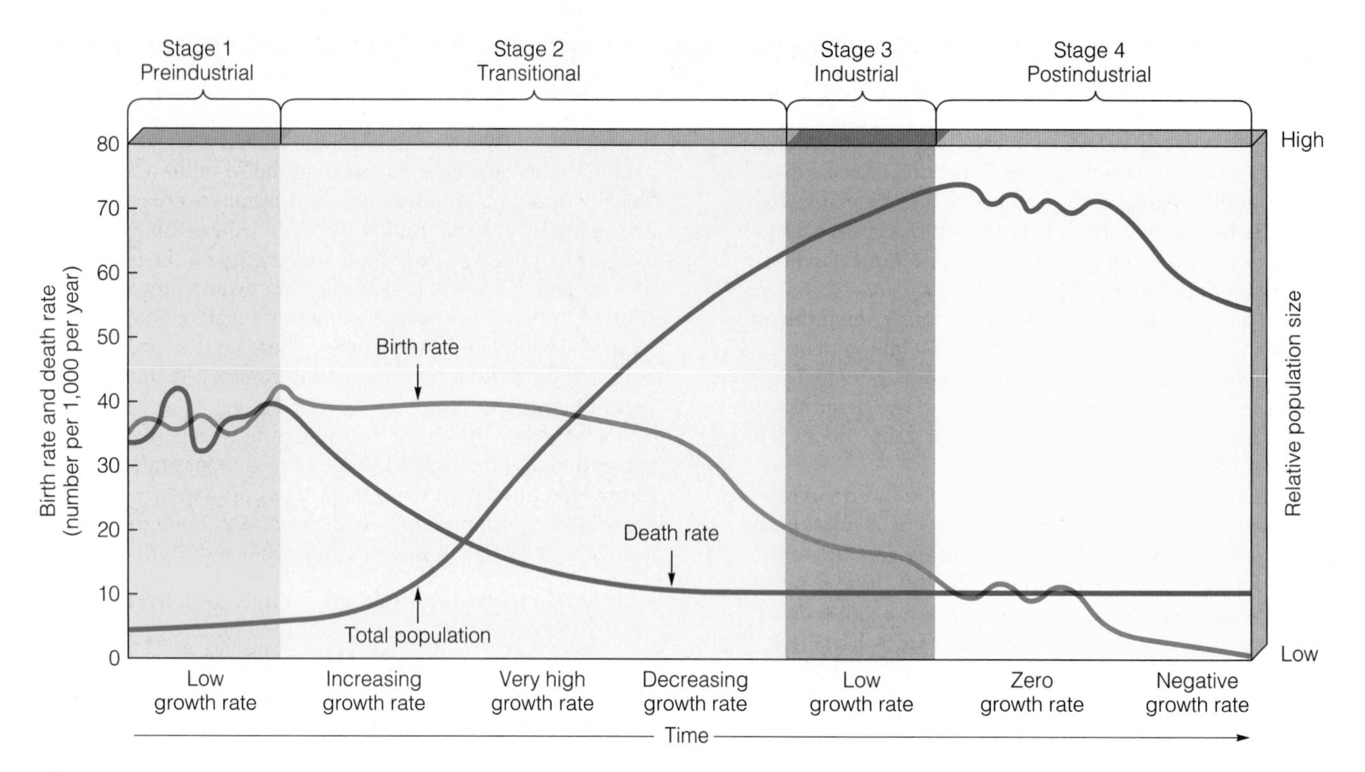

Figure 11-18 Generalized model of the demographic transition.

and when to have them. Such programs vary from culture to culture, but most provide information on **(1)** birth spacing, **(2)** birth control, and **(3)** health care for pregnant women and infants.

Advantages of family planning include the following:

- Helping increase the proportion of married women in developing countries who use modern forms of contraception from 10% of married women of reproductive age in the 1960s to 51% of these women in 2001 (40% if China is excluded).
- Being responsible for at least 55% of the drop in TFRs in developing countries, from 6 in 1960 to 3.2 in 2001 (3.7 if China is excluded).
- Reducing the number of legal and illegal abortions per year.
- Decreasing the risk of death from pregnancy.

Despite some successes, an estimated 150–300 million women in developing countries want to limit the number and determine the spacing of their children but lack access to services. Extending family-planning services to these women and to those who will soon be entering their reproductive years could prevent an estimated 5.8 million births a year and more than 5 million abortions a year.

Other analysts call for

- Expanding existing family-planning programs to include teenagers and sexually active unmarried women, who often are excluded
- Pro-choice and pro-life groups to join forces in greatly reducing unplanned births and abortions, especially among teenagers
- Programs to educate men about the importance of having fewer children and taking more responsibility for raising them
- Increased research on developing new, more effective, and more acceptable birth control methods for men

According to the United Nations, family planning could be provided in developing countries to all couples who want it for about $17 billion a year, the equivalent of less than a week's worth of worldwide military expenditures. If developed countries provided one-third of this $17 billion, each person in the developed countries would spend only $4.80 a year. This would help reduce the world's population by about 2.9 billion people.

How Can Empowering Women Help Reduce Birth Rates? Studies show that women tend to have fewer and healthier children and live longer when they **(1)** have access to education and to paying jobs outside the home and **(2)** live in societies in which their rights are not suppressed.

Women, roughly half of the world's population, **(1)** do almost all of the world's domestic work and child care, **(2)** provide more health care with little or no pay than all the world's organized health services combined, and **(3)** do 60–80% of the work associated with growing food, gathering fuelwood, and hauling water in rural areas of Africa, Latin America, and Asia. As one

pick up dried excrement, a *fecal snow* often falls on parts of the city. This bacteria-laden fallout leads to widespread salmonella and hepatitis infections, especially among children.

Some 3.5 million motor vehicles and 30,000 factories spew pollutants into the atmosphere. Air pollution is intensified because the city lies in a basin surrounded by mountains, and frequent thermal inversions trap pollutants at ground level (Figure 12-7, p. 285, top). Since 1982, the amount of contamination in the city's smog-choked air has more than tripled. Indeed, breathing the city's air is said to be roughly equivalent to smoking three packs of cigarettes a day.

The city's air and water pollution cause an estimated 100,000 premature deaths per year. Writer Carlos Fuentes has nicknamed this megacity "Makesicko City."

The Mexican government is industrializing other parts of the country in an attempt to slow migration to Mexico City. Other efforts include **(1)** banning cars from a 50-block central zone, **(2)** taking taxis built before 1985 off the streets, **(3)** having buses and trucks run only on liquefied petroleum gas (LPG), **(4)** planting 25 million trees, **(5)** buying some land for green space, **(6)** phasing out use of leaded gasoline, **(7)** barring cars without catalytic converters from city streets one day a week, and **(8)** enforcing stricter industrial emission standards.

Some progress has been made, but the city still fails to meet minimum air-quality standards on an average of 300 days a year. Older Volkswagen taxis and minibuses make up less than one-fourth of the city's vehicles but produce nearly half of all emissions.

If the city's population continues to grow as projected, these problems, already at crisis levels, are expected to become even worse. If you were in charge of Mexico City, what would you do?

How Urbanized Is the United States? Between 1800 and 2001, the percentage of the U.S. population living in urban areas increased from 5% to 75%. This rural-to-urban population shift has taken place in four phases.

- *Migration to large central cities.* Currently, **(1)** 75% of Americans live in 271 *metropolitan areas* (cities and towns with at least 50,000 people) and **(2)** nearly half of the country's population lives in consolidated metropolitan areas containing 1 million or more residents (Figure 11-22).
- *Migration from large central cities to suburbs and smaller cities.* Currently, about 51% of the U.S. population lives in suburbs.
- *Migration from the North and East to the South and West.* Since 1980, about 80% of the U.S. population increase has occurred in the South and West, particularly near the coasts. This shift is expected to continue.
- *Migration from urban areas back to rural areas* since the 1970s and especially since 1990.

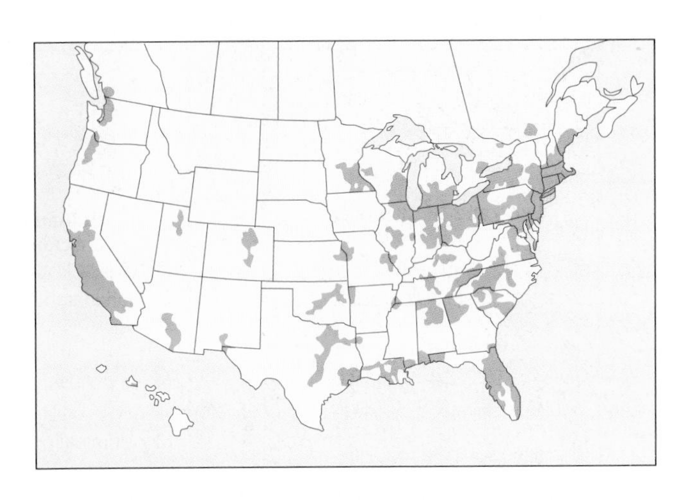

Figure 11-22 Major urban regions in the United States. About 75% of Americans live in urban areas occupying about 3% of the country's land area. Nearly half (48%) of Americans live in *consolidated metropolitan areas* with 1 million or more people. (Data from U.S. Census Bureau)

What Are the Major Urban Problems in the United States? Here is some *good news*.

- Since 1920, many of the worst urban environmental problems in the United have been reduced significantly.
- Most people have better working and housing conditions, and air and water quality have improved.
- Better sanitation, public water supplies, and medical care have slashed death rates and the prevalence of sickness from malnutrition and transmissible diseases such as measles, diphtheria, typhoid fever, pneumonia, and tuberculosis.
- Concentrating most of the population in urban areas has helped protect the country's biodiversity by reducing the destruction and degradation of wildlife habitat.

Here is some *bad news*. A number of cities in the United States (especially older ones) have **(1)** deteriorating services, **(2)** aging infrastructures (streets, schools, bridges, housing, and sewers), **(3)** budget crunches from rising costs as some businesses and people move to the suburbs or rural areas and reduce revenues from property taxes, and **(4)** rising poverty in many central city areas, where unemployment typically is 50% or higher.

Another major problem in the United States is **urban sprawl**: the growth of low-density development

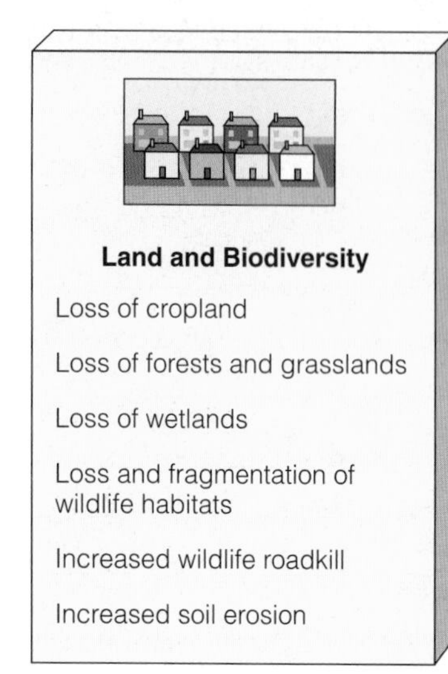

Land and Biodiversity
Loss of cropland
Loss of forests and grasslands
Loss of wetlands
Loss and fragmentation of wildlife habitats
Increased wildlife roadkill
Increased soil erosion

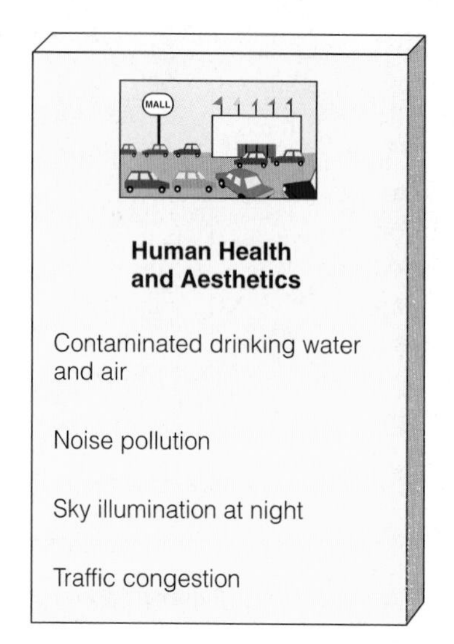

MALL
Human Health and Aesthetics
Contaminated drinking water and air
Noise pollution
Sky illumination at night
Traffic congestion

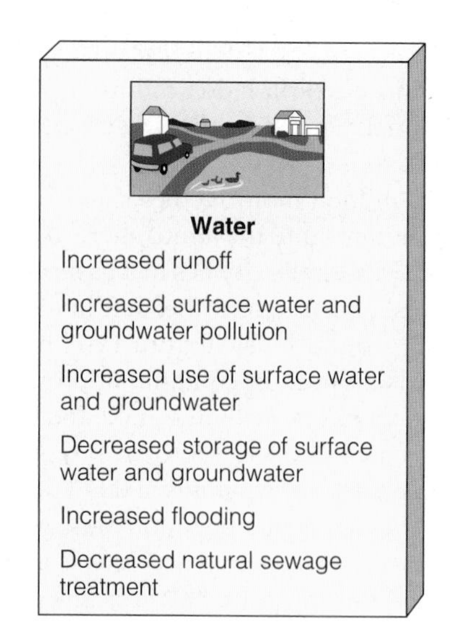

Water
Increased runoff
Increased surface water and groundwater pollution
Increased use of surface water and groundwater
Decreased storage of surface water and groundwater
Increased flooding
Decreased natural sewage treatment

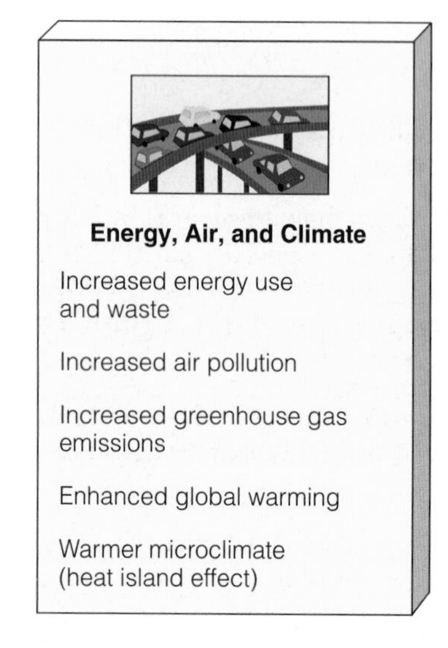

Energy, Air, and Climate
Increased energy use and waste
Increased air pollution
Increased greenhouse gas emissions
Enhanced global warming
Warmer microclimate (heat island effect)

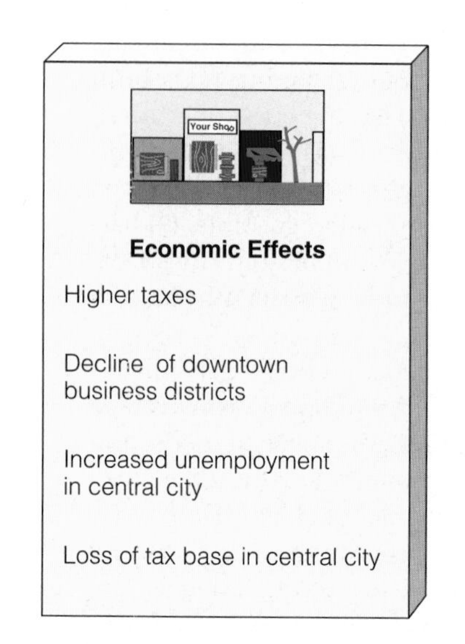

Figure 11-23 Some of the undesirable impacts of urban sprawl and car-dependent development.

on the edges of cities and towns that encourages dependence on cars. Factors promoting urban sprawl in the United States since 1945 include **(1)** ample land for expansion, **(2)** government loan guarantees for new single-family homes for veterans of World War II, **(3)** government and state funding of highways that encourages the development of once-inaccessible outlying tracts of land, **(4)** low-cost gasoline, which encourages automobile use, **(5)** greater availability of mortgages for homes in new subdivisions than in older cities and suburbs, and **(6)** state and local zoning laws that require large residential lots and separation of residential and commercial use of land in new communities. Segregating businesses from homes greatly increases reliance on cars for getting from home to work and shops.

Figure 11-23 shows some of the undesirable consequences of urban sprawl. Some critics say the concern over urban sprawl is overblown because only about 5% of the land in the United States is developed. However, they fail to note that three out of four Americans live in developed urban areas and one out of two reside in sprawling suburban areas.

11-7 URBAN RESOURCE AND ENVIRONMENTAL PROBLEMS

What Are the Economic and Environmental Advantages of Urbanization? Historically, cities have **(1)** been the centers of industry and commerce, **(2)** spurred economic development, and **(3)** nurtured innovations in science and technology. The high density of urban populations has also provided governments with significant cost advantages in delivering goods and services. Urban goods and services with enormous economic value include **(1)** human habitats,

(2) transportation networks, and **(3)** a wide variety of income opportunities.

Urbanization also has some important environmental benefits:

- In many parts of the world, urban populations live longer and have lower infant mortality rates than do rural populations.
- Urban dwellers generally have better access to **(1)** medical care, **(2)** family planning, **(3)** education, **(4)** social services, and **(5)** environmental information than do people in rural areas.
- Urban areas have played a crucial role in reducing fertility and thereby slowing population growth.
- Environmental pressures from population growth are lower because birth rates in urban areas usually are one-fourth to one-third of those in rural areas.
- Recycling is more economically feasible because of the large concentration of recyclable materials.
- Per capita expenditures on environmental protection are higher in urban areas.
- The 46% of the world's people currently living in urban areas occupy only about 4% of the planet's land area (Figure 11-20).
- Concentrating people in urban areas helps preserve biodiversity by reducing the stress on wildlife habitats.

What Are the Environmental Disadvantages of Urbanization? Urbanization also has some harmful environmental effects:

- Although urban dwellers occupy only 4% of the earth's land area, they consume 75% of the earth's resources.
- Large areas of the earth's land area must be disturbed and degraded (Figure 1-3, p. 5, and Figure 1-8, p. 10) to provide urban dwellers with food, water, energy, minerals, and other resources. This decreases and degrades the earth's biodiversity.
- Because of their high resource consumption, urban dwellers produce most of the world's air pollution, water pollution, and solid and hazardous wastes.
- Pollution levels normally are higher in urban areas than in rural areas because large amounts of pollutants are produced in a small area and cannot be as readily dispersed and diluted as those produced in rural areas.
- Most of the world's cities are not self-sustaining systems because of their high resource input and high waste output (Figure 11-24).
- High population densities in urban areas can increase **(1)** the spread of infectious diseases (especially if adequate drinking water and sewage systems are not available), **(2)** physical injuries (mostly from industrial and traffic accidents), **(3)** *crime rates* (Connections, p. 267), and **(4)** excessive noise (Figure 11-25).

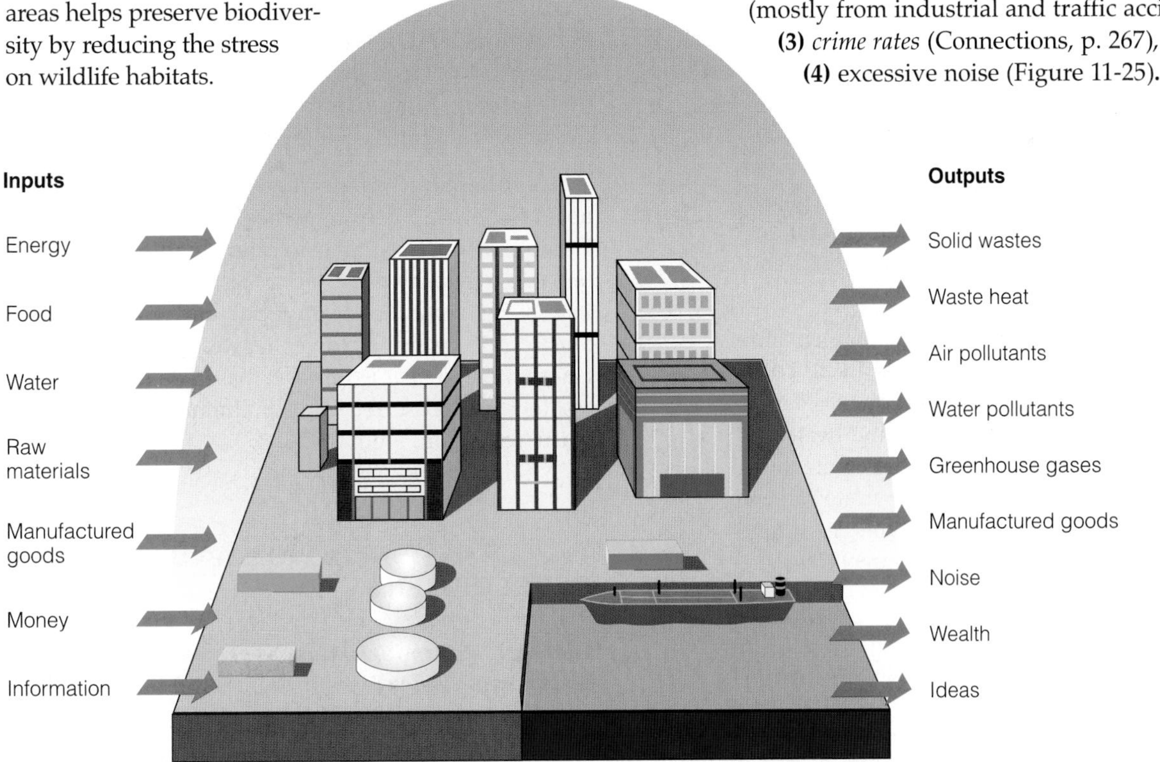

Figure 11-24 Urban areas rarely are sustainable systems. The typical city is an open system that depends on other areas for large inputs of matter and energy resources and for large outputs of waste matter and heat. Large areas of nonurban land must be used to supply urban areas with resources. For example, according to an analysis by Mathis Wackernagel and William Rees, 58 times the land area of London is needed to supply its residents with resources. They estimate that meeting the needs of all the world's people at the same rate of resource use as London would take at least three more earths.

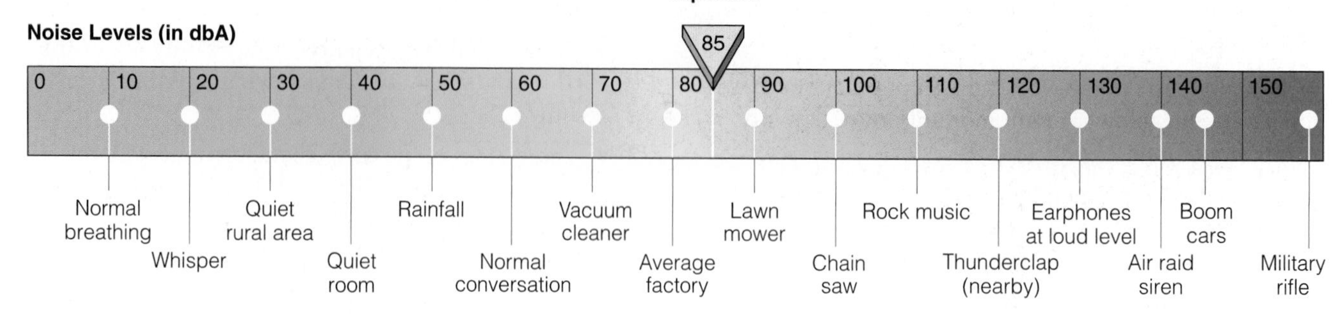

Figure 11-25 Noise levels (in decibel-A [dbA] sound pressure units) of some common sounds. In general, noise levels above 65 dbA are considered unacceptable, and prolonged exposure to levels above 85 dbA can cause permanent hearing damage. You are being exposed to a sound level high enough to cause permanent hearing damage if **(1)** you need to raise your voice to be heard above the racket, **(2)** a noise causes your ears to ring, or **(3)** nearby speech seems muffled. Prolonged exposure to lower noise levels and occasional loud sounds may not damage your hearing but can increase internal stress and cardiovascular disease. Noise pollution can be reduced by **(1)** modifying noisy activities and devices to produce less noise, **(2)** shielding noisy devices or processes, **(3)** shielding workers or other receivers from the noise, **(4)** moving noisy operations or things away from people, and **(5)** using antinoise, a new technology that cancels out one noise with another.

- Most cities have few trees, shrubs, or other plants that **(1)** absorb air pollutants, **(2)** give off oxygen, **(3)** help cool the air, **(4)** reduce soil erosion, **(5)** muffle noise, **(6)** provide wildlife habitats, and **(7)** give aesthetic pleasure. As one observer remarked, "Most cities are places where they cut down the trees and then name the streets after them."

- Most cities produce little of their own food. However, people can grow food by **(1)** planting community gardens in unused lots, **(2)** using window boxes and balcony planters, **(3)** creating gardens or greenhouses on the roofs of apartment buildings and on patios, and **(4)** raising fish in aquaculture tanks and sewage lagoons.

- Cities generally are warmer, rainier, foggier, and cloudier than suburbs and nearby rural areas. The enormous amounts of heat generated by cars, factories, furnaces, lights, air conditioners, and heat-absorbing dark roofs and roads in cities create an *urban heat island* surrounded by cooler suburban and rural areas. As urban areas grow and merge, individual heat islands also merge. This can affect the climate of a large area and keep polluted air from being diluted and cleansed.

- Many cities have water supply problems. As cities grow and their water demands increase, expensive reservoirs and canals must be built and deeper wells drilled. This can deprive rural and wild areas of surface water and deplete groundwater faster than it is replenished.

- Flooding tends to be greater in cities because **(1)** many cities are built on floodplain areas subject to natural flooding, and **(2)** covering land with buildings, asphalt, and concrete causes precipitation to run off quickly and overload storm drains.

- Rural cropland, fertile soil, forests, wetlands, and wildlife habitats are lost as cities expand. Once prime agricultural land or forestland is paved over or built on, it is lost for food production and habitat for most of its former wildlife. In coastal areas, urban growth destroys or pollutes ecologically valuable wetlands.

Some analysts call for developing a more sustainable relationship between cities and the living world. Doing this would entail converting high-waste, unsustainable cities with a *linear metabolism* (based on an ever-increasing resource throughput and waste output; Figure 3-15, p. 73) to lower-waste, more sustainable cities with a *circular metabolism* (based on more efficient resource use, reuse, recycling, pollution prevention, and waste reduction; Figure 3-16, p. 74).

11-8 TRANSPORTATION AND URBAN DEVELOPMENT

How Do Transportation Systems Affect Urban Development? Transportation and land-use decisions are linked together and determine **(1)** where people live, **(2)** how far they must go to get to work and buy food and other necessities, **(3)** how much land is paved over, and **(4)** how much air pollution people are exposed to.

If a city cannot spread outward, it must grow vertically—upward and downward (below ground)—so

CONNECTIONS

How Can Reducing Crime Help the Environment?

Almost everyone favors reducing crime, but few people realize that doing this can also help improve environmental quality. Crimes such as robbery, assault, and shootings

- Drive people out of cities, which are our most energy-efficient living arrangements. Every brick in an abandoned urban building represents an energy waste equivalent to burning a 100-watt light bulb for 12 hours. Each new suburb means replacing farmland or reservoirs of natural biodiversity with dispersed, energy- and resource-wasting roads, houses, and shopping centers.
- Make people less willing to use walking, bicycles, and energy-efficient public transit systems.
- Force many people to use more energy to deter burglars by **(1)** leaving lights, TVs, and radios on and **(2)** clearing away trees and bushes near houses that can reduce solar heat gain in the summer and provide windbreaks in the winter.
- Cause overpackaging of many items to deter shoplifting or poisoning of food or drug items.

Critical Thinking

Can you think of any environmental benefits of certain types of crimes?

that it occupies a small land area with a high population density. Most people living in such *compact cities* walk, ride bicycles, or use energy-efficient mass transit. Many European cities and urban areas such as Hong Kong and Tokyo are compact and tend to be more energy-efficient than the dispersed cities in the United States, Canada, and Australia, where ample land often is available for outward expansion.

A combination of cheap gasoline, plentiful land, and a network of highways produces sprawling, automobile-oriented cities with low population density that have a number of undesirable effects (Figure 11-23). Most people in such urban areas live in single-family houses with unshared walls that lose and gain heat rapidly unless they are well insulated and airtight. Urban sprawl also **(1)** gobbles up unspoiled forests and natural habitats, **(2)** paves over fertile farmland, **(3)** promotes heavy dependence on the automobile, **(4)** degrades watersheds and air, **(5)** wastes resources, and **(6)** promotes age and economic inequalities.

Dispersed car-centered cities use up to 10 times more energy per person for transportation than more compact cities that allow better use of mass transit, bicycles, and walking. In Europe, for example, walking and bicycling are used for 40–50% of all land-based trips and mass transit for 10%. By contrast, in the United States 95% of all trips are by car, 3% by mass transit, and 2% by bicycling and walking. In addition, spread-out cities use more building materials, roads, power lines, and water and sewer lines.

What Are the Pros and Cons of Motor Vehicles? There are two main types of ground transportation: **(1)** *individual* (such as cars, motor scooters, bicycles, and walking) and **(2)** *mass* (mostly buses and rail systems). Only about 10% of the world's people can afford a car. Thus, about 90% of all travel in the world is by foot, bicycle, or motor scooter.

Despite having only 4.7% of the world's people, the United States has 35% of the world's 700 million cars, trucks, and buses. In the United States the car is used for 98% of all urban transportation and 91% of travel to work (with 76% of Americans driving to work alone). Americans drive 2.5 trillion kilometers (1.6 trillion miles) each year—as far as the rest of the world combined. Despite their many advantages, there are a number of drawbacks to relying on motor vehicles as the major form of transportation (Pro/Con, p. 268).

Is It Feasible to Reduce Automobile Use? Two recent estimates by economists put the harmful costs of driving in the United States at roughly $300–350 billion per year. These largely hidden costs include **(1)** deaths and injuries from accidents, **(2)** higher health insurance costs, **(3)** air and water pollution, **(4)** CO_2 emissions that increase global warming, **(5)** the value of time wasted in traffic jams, and **(6)** decreased property values near roads because of noise and congestion. All Americans, whether they drive or not, pay these costs but rarely associate them with driving.

There are additional costs. According to a study by the World Resources Institute, federal, state, and local governments provide automobile subsidies in the United States amounting to $300–600 billion a year (depending on the costs included). This amounts to an average subsidy of $1,400–2,800 per vehicle.

Environmentalists and a number of economists suggest that one way to reduce the harmful effects of automobile use is to make drivers pay directly for most of the full costs of automobile use—a *user-pays* approach. This could be done by **(1)** including the estimated harmful costs of driving as a tax on gasoline, **(2)** phasing out government subsides for motor vehicle owners, and

Good and Bad News About Motor Vehicles

The automobile provides convenience and mobility. To many people, cars are also symbols of power, sex, excitement, social status, and success. Moreover, much of the world's economy is built on producing motor vehicles and supplying roads, services, and repairs for them. In the United States, **(1)** $1 of every $4 spent and one of every six nonfarm jobs is connected to the automobile, and **(2)** five of the seven largest U.S. industrial firms produce either cars or their fuel.

Despite their important benefits, motor vehicles have many destructive effects on people and the environment. Since 1885, when Karl Benz built the first automobile, almost 18 million people have been killed in motor vehicle accidents. According to the World Health Organization and the World Bank, car accidents throughout the world annually **(1)** kill an estimated 885,000 people (an average of 2,400 deaths per day, equal to eight fatal accidental jumbo jet crashes each day), and **(2)** injure or permanently disable another 15 million people.

In the United States alone, 16 million motor vehicle accidents (up from 7 million in 1970) per year **(1)** kill more than 42,000 people and **(2)** injure another 3 million (at least 300,000 of them severely). *More Americans have been killed by cars than have died in all wars in the country's history.*

Motor vehicles also are the largest source of air pollution (including 23% of global CO_2 emissions), laying a haze of smog over the world's cities (p. 280). Transportation is also the fastest-growing source of climate-changing carbon dioxide emissions.

In the United States, motor vehicles produce at least 50% of the air pollution, even though emission standards are as strict as any in the world. Two-thirds of the oil used in the United States and one-third of the world's total oil consumption are devoted to transportation.

By making long commutes and shopping trips possible, automobiles and highways have helped create urban sprawl and reduced use of more efficient forms of transportation. Worldwide, at least a third of urban land is devoted to roads, parking lots, gasoline stations, and other automobile-related uses.

In the United States, more land is devoted to cars than to housing. Half the land in an average U.S. city is used for cars, prompting urban expert Lewis Mumford to suggest that the U.S. national flower should be the concrete cloverleaf.

Roads also have harmful ecological effects, including **(1)** increasing the killing of wild animals by vehicles, **(2)** promoting dispersal of nonnative species, **(3)** blocking movements of some species, and **(4)** dividing populations of various species into smaller, less viable subpopulations.

In 1907, the average speed of horse-drawn vehicles through the borough of Manhattan in New York City was 18.5 kilometers (11.5 miles) per hour; today cars and trucks creep along Manhattan streets at an average speed of 5 kilometers (3 miles) per hour. If current trends continue, U.S. motorists will spend an average of 2 years of their lifetimes in traffic jams.

Even if the money is available, building more roads is not the answer because, as economist Robert Samuelson put it, "Cars expand to fill available concrete." According to transportation expert Michael Replogle, "Adding highway capacity to solve traffic congestion is like buying larger pants to deal with your weight problem."

Critical Thinking

If you own a car (or hope to own one), what conditions (if any) would encourage you to rely less on the automobile and travel to school or work by bicycle, on foot, by mass transit, or by a carpool or vanpool?

(3) using the gasoline tax revenues and savings from reduced motor vehicle subsidies to lower taxes on income and wages (Solutions, p. 33) and to help finance mass transit systems, bike paths, and sidewalks.

If U.S. drivers had to pay the hidden costs of driving directly in the form of a gasoline tax (as is done in most European countries and Japan), the tax on each gallon would be about $5–7. Such a tax would face intense political opposition unless taxpayers faced an equivalent drop in income or other taxes to compensate for increases in gasoline taxes (Solutions, p. 33). Such taxes would also spur the use of more energy-efficient motor vehicles (p. 518).

Other ways to reduce automobile use and congestion are **(1)** charging tolls on roads, tunnels, and bridges (especially during peak traffic times), **(2)** raising parking fees, **(3)** reducing mortgage charges on federally financed loans for homeowners who do not use a car to get to work, and **(4)** encouraging employers to use staggered hours and have more employees use telecommuting to work at home.

Including the hidden costs in the market prices of cars, trucks, and gasoline up front may make economic and environmental sense and has worked in a number of other countries. However, most analysts say that it is not feasible in the United States because

- It faces strong political opposition from the public (mostly because they are unaware of the huge hidden costs they are already paying) and from powerful transportation-related industries (such as oil and tire companies, road builders, carmakers, and many real estate developers). However, taxpayers might accept sharp increases in gasoline taxes if the extra costs were offset by decreases in taxes on wages and income (Solutions, p. 33).

- Fast, efficient, reliable, and affordable mass transit options and bike paths are not widely available as alternatives to automobile travel.
- The dispersed nature of most U.S. urban areas makes people dependent on the car.
- Most people who can afford cars are virtually addicted to them, and most people who cannot afford a car hope to buy one someday.

What Are Alternatives to the Car? Cars are an important form of transportation, but some countries are encouraging alternatives, including

- *Motor scooters,* especially for people in developing countries who cannot afford cars. Most burn a mixture of oil and kerosene in small, inefficient, and noisy engines that emit clouds of air pollutants. However, quiet electric scooters that produce very little pollution except at power plants supplying the electricity for battery recharging could replace them.
- *Bicycles,* which globally outsell cars by more than two to one. They are widely used in countries such as China (50% of urban trips) and the Netherlands (30% of urban trips) but make up only about 1% of urban trips in the United States. Besides being inexpensive to buy and maintain, bicycles **(1)** produce no pollution, **(2)** are rarely a serious danger to pedestrians or cyclists, **(3)** take few resources to make, and **(4)** are the most energy-efficient form of transportation (including walking). Using separate bike paths or lanes running along roads, cyclists can make most trips shorter than 8 kilometers (5 miles) faster than drivers can. Several companies have also developed electric bicycles, powered as needed by a small battery-powered motor, with 1.1 million sold in 1999 (750,000 of them in China). Bicycles powered by small fuel cells may be available within a few years. For longer trips, secure bike parking spaces can be provided at mass transit stations, and buses and trains can be equipped to carry bicycles. Such *bike-and-ride* systems are widely used in Japan, Germany, the Netherlands, and Denmark.
- *Mass transit rail systems within urban areas.* They consist of **(1)** *heavy-rail* systems (subways, elevated railways, and metros operating on exclusive right-of-way tracks) and **(2)** *light-rail* systems (streetcars, trolley cars, and tramways running along tracks that may not be separated from other traffic). Compared with cars, such systems **(1)** are more energy-efficient, **(2)** produce less air pollution, **(3)** cause fewer injuries and deaths, and **(4)** take up less land. However, they are efficient and cost-effective only where many people live along a narrow corridor and can easily reach properly spaced stations (as in Hong Kong, with a density of 300 people per hectare).
- *Rapid rail systems between urban areas.* High-speed trains between cities can greatly reduce the need for travel by car and plane. In western Europe and Japan, *bullet* or supertrains travel on new or upgraded tracks at speeds up to 330 kilometers (200 miles) per hour. Such trains **(1)** are ideal for trips of approximately 200–1,000 kilometers (120–620 miles) and **(2)** are much more energy efficient per rider over the same distance than a commercial airplane or a car carrying only one person. A high-speed train network could replace airplanes, buses, and private cars for most medium-distance travel between major American cities (Figure 11-26). However, these systems **(1)** are expensive to run and maintain, **(2)** must operate along heavily used transportation routes to be profitable, **(3)** cause noise and vibration for nearby residents, and **(4)** can have accidents if not adequately maintained and managed.
- *Buses.* Bus systems **(1)** are more flexible than rail systems because they can run throughout sprawling cities and be rerouted as needed (even overnight) if transportation patterns change, **(2)** use less capital and have lower operating costs than heavy-rail systems, and **(3)** can greatly reduce car use (Solutions, p. 270). However, bus systems **(1)** often cost more to operate than they bring in because they must offer low fares to attract riders and **(2)** often get caught in traffic unless they operate in separate express lanes.

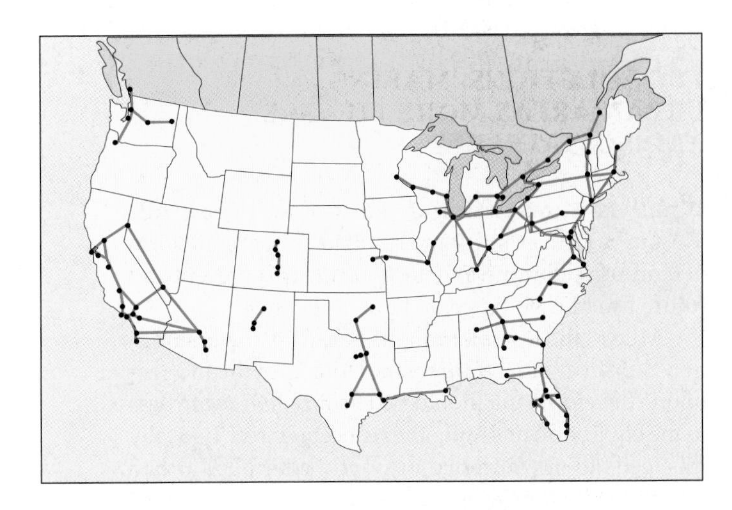

Figure 11-26 Potential routes for high-speed bullet trains in the United States and parts of Canada. Such a system could allow rapid, comfortable, safe, and affordable travel between major cities in a region. It would greatly reduce dependence on cars, buses, and airplanes for trips between these urban areas. (Data from High Speed Rail Association)

Curitiba, Brazil

SOLUTIONS

One of the world's showcase ecocities is Curitiba, Brazil, with a population of 2.2 million. The city is one of Latin America's most livable cities and has a worldwide reputation for its innovative urban planning and environmental protection efforts.

Trees are everywhere in Curitiba because city officials have given neighborhoods more than 1.5 million trees to plant and care for. No tree in the city can be cut down without a permit, and two trees must be planted for each one that is cut down.

The air is clean because **(1)** the city is not built around the car and **(2)** officials have integrated land-use and transportation planning. There are 160 kilometers (100 miles) of bike paths, and more are being built. With the support of shopkeepers, many streets in the downtown shopping district have been converted to pedestrian zones in which no cars are allowed.

Curitiba probably has the world's best bus system. Each weekday a network of clean and efficient buses carries more than 1.9 million passengers—75% of the city's commuters and shoppers—at a low cost (20–40¢ per ride, with unlimited transfers) on express bus lanes. This system radiates out from the center city along main routes like spokes in a bicycle wheel along express lanes dedicated to buses. Since 1974 car traffic has declined by 30% even though the city's population has doubled. Two-thirds of all trips in the city are made by bus. The city's bus stations are linked to the city's network of bike paths. Special buses, taxis, and other services are provided for the disabled.

Only high-rise apartment buildings are allowed near major bus routes. Each building must devote the bottom two floors to stores, which reduces the need for residents to travel. As a result, Curitiba has one of Brazil's lowest outdoor air pollution rates. The system has also reduced traffic congestion and saved energy.

The city recycles roughly 70% of its paper and 60% of its metal, glass, and plastic, which is sorted by households for collection three times a week. Recovered materials are sold mostly to the city's more than 500 major industries. Litter and graffiti are almost nonexistent because of the civic pride of Curitiba's residents.

Instead of being torn down, existing buildings and sites are recycled for new uses. A glue factory was converted to a creativity center where children make handicrafts, which the city's tourist shops sell to help fund social programs. A garbage dump was converted to a botanical garden that houses 220,000 species, and a quarry was filled and converted to the Free University of the Environment.

The city **(1)** bought a plot of land downwind of downtown as an industrial park, **(2)** put in streets, services, housing, and schools, **(3)** ran a special workers' bus line to the area, and **(4)** enacted stiff air and water pollution control laws. This development has attracted national and foreign corporations and 500 nonpolluting industries that provide one-fifth of the city's jobs. Most of these workers can walk or bike to work from their nearby homes.

To give the poor basic technical training skills needed for jobs, the city set up old buses as roving technical training schools. Each bus gives courses, costing the equivalent of two bus tokens (less than a dollar), for 3 months in a particular area and then moves to another area. Other retired buses become **(1)** classrooms, **(2)** health clinics, **(3)** soup kitchens, and **(4)** some of the city's 200 day-care centers, which are open 11 hours a day and are free for low-income parents.

11-9 SOLUTIONS: MAKING URBAN AREAS MORE LIVABLE AND SUSTAINABLE

What Is Conventional Land-Use Planning? Most urban areas and some rural areas use some form of **land-use planning** to determine the best present and future use of each parcel of land in the area.

Much land-use planning is based on the assumption that substantial future population growth and economic development should be encouraged, regardless of the environmental and other consequences. Typically this leads to uncontrolled or poorly controlled urban growth and sprawl (Figure 11-23).

A major reason for this often destructive process is that 90% of the revenue local governments in the United States use to provide schools, police and fire protection, water and sewer systems, and other public services comes from *property taxes* levied on all buildings and property based on their economic value. Thus, local governments often try to raise money by promoting economic growth because they usually cannot raise property taxes enough to meet expanding needs. Typically the long-term result is poorly managed economic growth leading to more environmental degradation.

What Are the Pros and Cons of Using Zoning to Control Land Use? Once a land-use plan is developed, governments control the uses of various parcels of land by legal and economic methods. The most widely used approach is **zoning**, in which various parcels of land are designated for certain uses.

Zoning can be used to control growth and protect areas from certain types of development. For example, cities such as Portland, Oregon, and Curitiba, Brazil (Solutions, above), have used zoning to encourage

The city provides environmental education for adults and children by **(1)** using signs along roadways to present environmental information and **(2)** having all schoolchildren study ecology. Older children are given apprenticeships, job training, and entry-level jobs, often emphasizing environmental skills such as forestry, water pollution control, ecological restoration, and public health.

The city's 700,000 poorest residents can swap sorted trash for locally grown vegetables and fruits or bus tokens and receive free medical, dental, and child care; there are also 40 feeding centers for street children. Poor children get free checkups and regular visits from health workers. Preventive health care is emphasized throughout the city's schools, day care, and teen centers. As a result, the infant mortality rate has fallen by more than 60% since 1977.

In Curitiba, **(1)** 99.5% of households have electricity and drinking water and 98% have trash collection, **(2)** the literacy rate is 95%, **(3)** 83% of the adults have at least a high school education, and **(4)** the average per capita gross domestic product is about $8,000 (compared to $4,630 for Brazil as a whole).

The city has a *build-it-yourself* system that gives each poor family ownership of a plot of land, building materials, two trees, and an hour's consultation with an architect. A public geographic information system (GIS) gives everyone equal access to information about all the land in the city. The government has an array of telephone- and Web-based systems that respond quickly to the problems and inquiries of its citizens.

All of these things have been accomplished despite Curitiba's enormous population growth, from 300,000 in 1950 to 2.2 million in 2001, as rural poor people have flocked to the city. Another million residents are expected by 2020.

Curitiba has slums, shantytowns, and squatter settlements on its outskirts and most of the problems of other cities. However, most of its citizens **(1)** have a sense of vision, solidarity, pride, and hope and **(2)** are committed to making their city even better. Polls show that 99% of the city's inhabitants would not want to live anywhere else.

The entire program is the brainchild of architect and former college teacher Jaime Lerner, an energetic and charismatic leader who has served as the city's mayor three times since the 1970s. Under his leadership, the municipal government has dedicated itself to **(1)** finding solutions to problems that are *simple, innovative, fast, cheap,* and *fun* and **(2)** establishing a government that is honest, accountable, and open to public scrutiny. For example, when people pay their property taxes they get to vote for the improvements they would like to see in their neighborhood.

The government innovates and takes risks with the expectation that mistakes will be made and then quickly detected, diagnosed, and corrected (adaptive management). The World Bank cites Curitiba as an example of what can be done to make cities more sustainable and livable through innovative civic leadership and community effort. Curitiba's well-planned bus network and bicycle paths are being replicated in Bogotá, Columbia; Lima, Peru; and several other Latin American cities.

Critical Thinking

1. Why do you think Curitiba has been so successful in its efforts to become an ecocity, compared with the generally unsustainable cities found in most developed and developing countries?

2. What is the city or area where you live doing to make itself more ecologically and economically sustainable?

high-density development along major mass transit corridors to reduce automobile use and air pollution.

Despite its usefulness, zoning has some drawbacks.

- It can be influenced or modified by developers in ways that cause environmental harm such as destruction of wetlands, prime cropland, forested areas, and open space.
- It often favors high-priced housing and factories, hotels, and other businesses over protecting environmentally sensitive areas because local governments depend on property taxes for revenue.
- Overly strict zoning can discourage innovative approaches to solving urban problems. For example, the pattern in the United States (and in some other countries) has been to prohibit businesses in residential areas, which increases suburban sprawl. There is renewed interest in returning to mixed-use zoning to help reduce urban sprawl, but current zoning laws often prohibit this.

How Is Smart Growth Being Used to Control Growth and Sprawl? There is growing use of the concept of **smart growth** to encourage development that requires less dependence on cars. It recognizes that growth will occur but uses zoning laws (in which various parcels of land are designated for certain uses) and an array of other tools to **(1)** prevent urban sprawl, **(2)** direct growth to certain areas, **(3)** protect ecologically sensitive and important lands and waterways, and **(4)** develop urban areas that are more environmentally sustainable and more enjoyable places to live.

It involves a return to 19th-century town-planning principles such as **(1)** building shopping areas within walking distance of residences, **(2)** encouraging sufficient population density to support mass transit, and

GUEST ESSAY

The Ecological Design Arts

David W. Orr

David W. Orr is professor of environmental studies at Oberlin College and one of the nation's most respected environmental educators. He is author of numerous environmental articles and three books, including Ecological Literacy *and* Earth in Mind. *He is education editor for* Conservation Biology *and a member of the editorial advisory board of* Orion Nature Quarterly. *With help from students, faculty, and townspeople he used ecological design to build an innovative environmental studies building at Oberlin College in Ohio (Figure 20-19, p. 528).*

If *Homo sapiens sapiens* entered its industrial civilization in an intergalactic design competition, it would be tossed out in the qualifying round. It does not fit. It will not last. The scale is wrong, and even its defenders admit that it's not very pretty. The most glaring design failures of industrial and technologically driven societies are the loss of diversity of all kinds, impending climate change, pollution, and soil erosion.

Of course, industrial civilization was not designed at all; it was mostly imposed by single-minded individuals, armed with one doctrine of human progress or another, each requiring a homogenization of nature and society. These individuals for the most part had no knowledge of ecological design arts.

Good ecological design incorporates knowledge about how nature works [Solutions, p.191] into the ways we design, build, and live in just about anything that directly or indirectly uses energy or materials or governs their use.

When human artifacts and systems are well designed, they are in harmony with the ecological patterns in which they are embedded. When poorly designed, they undermine those larger patterns, creating pollution, higher costs, and social stress.

Good ecological design has certain common characteristics, including correct scale, simplicity, efficient use of resources, a close fit between means and ends, durability, redundancy, and resilience. These characteristics often are place-specific, or, in John Todd's words, "elegant solutions predicated on the uniqueness of place."

Good design also solves more than one problem at a time and promotes human competence, efficient and frugal resource use, and sound regional economies. Where good design becomes part of the social fabric at all levels, unanticipated positive side effects multiply. When people fail to design with ecological competence, unwanted side effects and disasters multiply.

The pollution, violence, social decay, and waste all around us indicate that we have designed things badly, for, I think, three primary reasons. *First*, as long as land and energy were cheap and the world was relatively empty, we did not need to master the discipline of good design. The result was **(1)** sprawling cities, **(2)** wasteful economies, **(3)** waste dumped into the environment, **(4)** bigger and less efficient automobiles and buildings, and **(5)** conversion of entire forests into junk mail and Kleenex—all in the name of economic growth and convenience.

Second, design intelligence fails when greed, narrow self-interest, and individualism take over. Good design is a cooperative community process requiring people who share common values and goals that bring them together and hold them together, as has been done in Curitiba, Brazil [Solutions, p. 270], and Chattanooga, Tennessee. Most American cities, with their extremes of poverty and opulence, are products of people who believe they have little in common with one another.

(3) allocating more open space for squares and parks. Existing car-dependent suburbs can be retrofitted with sidewalks, bicycle paths, and bus lanes. Abandoned or existing nearby shopping centers can be rebuilt as town centers where people could work, shop, and live. Figure 11-27 lists smart growth tools used to prevent and control urban growth and sprawl.

Some developers, real estate firms, and business interests strongly oppose many of the measures listed in Figure 11-27, arguing that they **(1)** hinder economic growth, **(2)** restrict what private landowners can do with their land, and **(3)** involve too much regulation by local, state, and federal agencies. However, several studies have shown that most forms of smart growth provide more jobs and spur more economic renewal than conventional economic growth.

Most European countries have been successful in discouraging urban sprawl and encouraging compact cities by **(1)** keeping a tight grip on development at the national level, **(2)** imposing high gasoline taxes to encourage people to live closer to work and shops and to discourage car use, **(3)** using high taxes on heating fuel to encourage living in apartments and small houses, and **(4)** using gasoline and heating fuel tax revenues to develop efficient train and other mass transit systems within and between cities.

Solutions: How Can We Make Cities More Sustainable and More Desirable Places to Live? According to most environmentalists and urban planners, increased urbanization and urban density are better than spreading people out over the countryside, which would destroy more of the planet's biodiversity. To these analysts, the primary problem is not urbanization but our failure to make most cities more sustainable and livable.

Greed, suspicion, and fear undermine good community and good design alike.

Third, poor design results from poorly equipped minds. Only people who understand harmony, patterns, and systems can do good design. Industrial cleverness, on the contrary, is evident mostly in the minutiae of things, not in their totality or their overall harmony. Good design requires a breadth of view that causes people to ask how human artifacts and purposes fit within a particular culture and place. It also requires ecological intelligence, by which I mean an intimate familiarity with how nature works.

An example of good ecological design is found in John Todd's living machines, which are carefully orchestrated ensembles of plants, aquatic animals, technology, solar energy, and high-tech materials to purify wastewater, but without the expense, energy use, and chemical hazards of conventional sewage treatment technology. Todd's living machines resemble greenhouses filled with plants and aquatic animals [p. 362]. Wastewater enters at one end, and purified water leaves at the other. In between, an ensemble of organisms driven by sunlight use and remove nutrients, break down toxins, and incorporate heavy metals in plant tissues.

Ecological design standards also apply to the making of public policy. For example, the Clean Air Act of 1970 required car manufacturers to install catalytic converters to remove air pollutants. Two decades later, emissions per vehicle are down substantially, but because more cars are on the road, air quality is about the same—an example of inadequate ecological design. A sounder design approach to transportation would **(1)** create better access to housing, schools, jobs, stores, and recreation areas, **(2)** build better public transit systems, **(3)** restore and improve railroads, and **(4)** create bike trails and walkways.

An education in the ecological design arts would foster the ability to see things in their ecological context, integrating firsthand experience and practical competence with theoretical knowledge about how nature works. It would equip people to build households, institutions, farms, communities, corporations, and economies that **(1)** do not emit carbon dioxide or other greenhouse gases, **(2)** operate on renewable energy, **(3)** preserve biological diversity, **(4)** recycle material and organic wastes, and **(5)** promote sustainable local and regional economies.

The outline of a curriculum in ecological design arts can be found in recent work in ecological restoration, ecological engineering, solar design, landscape architecture, sustainable agriculture, sustainable forestry, energy efficiency, ecological economics, and least-cost, end-use analysis. A program in ecological design would weave these and similar elements together around actual design objectives that aim to make students smarter about systems and about how specific things and processes fit in their ecological context. With such an education we can develop the habits of mind, analytical skills, and practical competence needed to help sustain the earth for us and other species.

Critical Thinking

1. Does your school offer courses or a curriculum in ecological design? If not, suggest some reasons why it does not.

2. Use Orr's three principles of good ecological design to evaluate how well your campus is designed. Suggest ways to improve its design.

Over the next few decades, they call for us to make new and existing urban areas more self-reliant, sustainable, and enjoyable places to live through good ecological design (Guest Essay, above).

In a more environmentally sustainable city, called an *ecocity* or *green city,* the demands on the earth's resources are reduced by mimicking the circular metabolism of nature (Figure 3-16, p. 74). In such an ecocity, emphasis is placed on **(1)** preventing pollution and reducing waste, **(2)** using energy and matter resources efficiently, **(3)** recycling, reusing, and composting at least 60% of all municipal solid wastes, **(4)** using solar and other locally available renewable energy resources, **(5)** encouraging biodiversity, and **(6)** using solar-powered living machines (p. 362) and wastewater gardens (Figure 14-32, p. 363) to treat sewage.

An ecocity is a people-oriented city, not a car-oriented city. Its residents are able to **(1)** walk or bike to most places, including work, and **(2)** use low-polluting mass transit (Solutions, p. 270).

An ecocity takes advantage of locally available energy sources and requires that all buildings, vehicles, and appliances meet high energy-efficiency standards. Trees and plants adapted to the local climate and soils are planted throughout to **(1)** provide shade and beauty, **(2)** reduce pollution, noise, and soil erosion, and **(3)** supply wildlife habitats. Small organic gardens and a variety of plants adapted to local climate conditions often replace monoculture grass lawns.

Abandoned lots and industrial sites (brownfields) and polluted creeks and rivers are cleaned up and restored. Nearby forests, grasslands, wetlands, and farms are preserved instead of being devoured by urban sprawl. Much of an ecocity's food comes from **(1)** nearby organic farms, **(2)** solar greenhouses,

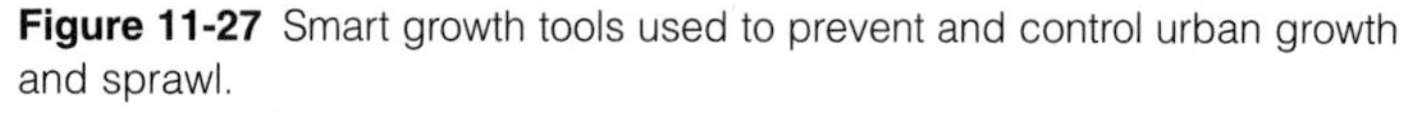

Limits and Regulations
• Limit building permits
• Urban growth boundaries
• Green belts around cities
• Public review of new development
Zoning
• Encourage mixed use
• Concentrate development along mass transportation routes
• Promote high-density cluster housing developments
Planning
• Ecological land-use planning
• Environmental impact analysis
• Integrated regional planning
• State and national planning
Protection
• Preserve existing open space
• Buy new open space
• Buy development rights that prohibit certain types of development on land parcels
Taxes
• Tax land, not buildings
• Tax land on value of actual use (such as forest and agriculture) instead of highest value as developed land
Tax Breaks
• For owners agreeing legally to not allow certain types of development (conservation easements)
• For cleaning up and developing abandoned urban sites (brownfields)
Revitalization and New Growth
• Revitalize existing towns and cities
• Build well-planned new towns

Figure 11-27 Smart growth tools used to prevent and control urban growth and sprawl.

(3) community gardens, and **(4)** small gardens on rooftops, in yards, and in window boxes.

The ecocity is not a futuristic dream. Examples of cities that have attempted to become more environmentally sustainable and livable include **(1)** Curitiba, Brazil (Solutions, p. 270), **(2)** Waitakere City, New Zealand, **(3)** Chattanooga, Tennessee, **(4)** Copenhagen, Denmark, **(5)** Portland, Oregon, and **(6)** Davis, California.

A sustainable world will be powered by the sun; constructed from materials that circulate repeatedly; made mobile by trains, buses, and bicycles; populated at sustainable levels; and centered around just, equitable, and tight-knit communities.

GARY GARDNER

REVIEW QUESTIONS

1. Define the boldfaced terms in this chapter.

2. How did Thailand reduce its birth rate?

3. How is population change calculated? What is *zero population growth*? What are the *crude birth rate* and the *crude death rate*? About how many people are added to the world's population each year and each day? How is the *annual rate of population change* calculated? What three countries have the world's largest populations?

4. Distinguish between *replacement-level fertility* and *total fertility rate*. Explain why replacement-level fertility is higher than 2. Explain why reaching replacement-level fertility does not mean an immediate halt in population growth.

5. How have fertility rates and birth rates changed in the United States since 1910? How rapidly is the U.S. population growing?

6. List 10 factors that affect birth rates and fertility rates.

7. List five reasons why the world's death rate has declined over the past 100 years.

8. Distinguish between *life expectancy* and *infant mortality rate*. Why is infant mortality the best measure of a society's quality of life? List three factors that keep the U.S. infant mortality higher than it could be.

9. What is the *age structure* of a population? Explain why the current age structure of the world's population means that the global population will keep growing for at least another 50 years even if a replacement-level rate of 2.1 is somehow reached globally tomorrow. Draw the general shape of an age structure diagram for a country undergoing **(a)** rapid population growth, **(b)** moderate population growth, and **(c)** slow or zero population growth.

10. What percentage of population is under age 15 in **(a)** the world, **(b)** developed countries, and **(c)** developing countries? Explain how age structure diagrams can be used to make population and economic projections.

11. What are the benefits and potentially harmful effects of rapid population decline?

12. List the major arguments for and against reducing birth rates globally.

13. Describe immigration in the United States in terms of numbers and list the pros and cons of reducing immigration limits.

14. What is the *demographic transition*, and what are its four phases? What factors might keep many developing countries from making the demographic transition?

15. What is *family planning*, and what are the advantages of using this approach to reduce the birth rate?

16. Explain how empowering women can help reduce birth rates.

17. What economic rewards and penalties have some countries used to reduce birth rates? What four conditions increase the success of using such economic rewards and penalties?

18. Briefly describe and compare the success China and India have had in reducing their birth rates. What are the major components of China's population control program?

19. List the eight goals of the current United Nations plan to stabilize the world's population at 7.8 billion by 2050 instead of the projected 9.3 billion.

20. Distinguish between **(a)** an *urban area* and a *rural area* and **(b)** the *degree of urbanization* and *urban growth*.

21. List factors that *push* people and *pull* people to migrate from rural areas to urban areas.

22. What are five major trends in urbanization and urban growth? What percentage of the population lives in urban areas in **(a)** the world, **(b)** developed countries, and **(c)** developing countries?

23. Summarize the major urban problems of Mexico City.

24. Summarize the major problems of the urban poor.

25. What four major shifts in the U.S. population have taken place since 1800? What percentage of the U.S. population lives in **(a)** urban areas and **(b)** suburban areas?

26. List four pieces of *good news* and four pieces of *bad news* about urban problems in the United States.

27. What is *urban sprawl*? List the major harmful effects of urban sprawl.

28. List eight environmental benefits of urbanization. List nine environmentally harmful effects of urbanization. List four ways in which crime decreases environmental quality.

29. How do transportation systems affect urban development? What are two basic types of cities in terms of population density and transportation systems?

30. What are the pros and cons of motor vehicles? List seven ways to reduce dependence on the automobile. List four factors promoting dependence on the automobile in the United States.

31. List the pros and cons of **(a)** motor scooters, **(b)** bicycles, **(c)** mass transit rail systems, and **(d)** buses.

32. Describe the cycle that leads to uncontrolled urban growth and sprawl.

33. What is *zoning*, and what are its pros and cons?

34. What is *smart growth*? List 10 tools cities can use to promote smart growth.

35. Describe the major characteristics of a more sustainable *ecocity* or *green city*. Briefly describe the progress that Curitiba, Brazil, has made in becoming an ecocity.

CRITICAL THINKING

1. Why is it rational for a poor couple in India to have five or six children? What changes might induce such a couple to consider their behavior irrational?

2. List what you consider to be a major local, national, and global environmental problem and describe the role of population growth in this problem.

3. Suppose that all women in the world began bearing children at replacement-level fertility rates of 2.1 children per woman today. Explain why this would not immediately stop global population growth. About how long would it take for population growth to stabilize?

4. Do you believe that the population of **(a)** your own country and **(b)** the area where you live is too high? Explain.

5. Evaluate the claims made by those opposing a reduction in births and those promoting a reduction in births, as discussed on p. 255. Which position do you support, and why?

6. Explain why you agree or disagree with each of the following proposals:

a. The number of legal immigrants and refugees allowed into the United States each year should be reduced sharply.

b. Illegal immigration into the United States should be decreased sharply. If you agree, how would you go about achieving this?

c. Families in the United States should be given financial incentives to have more children to prevent population decline.

d. The United States should adopt an official policy to stabilize its population and reduce unnecessary resource waste and consumption as rapidly as possible.

e. Everyone should have the right to have as many children as they want.

7. Some people have proposed that the earth could solve its population problem by shipping people off to space colonies, each containing about 10,000 people. Assuming that we could build such large-scale, self-sustaining space stations, how many people would have to be shipped off each day to provide living spaces for the 81 million people being added to the earth's population each year? Current space shuttles can handle about 6 to 8 passengers. If this capacity could be increased to 100 passengers per shuttle, how many shuttles would have to be launched per day to offset the 81 million people being added each year? According to your calculations, determine whether this proposal is a logical solution to the earth's population problem.

8. Some people believe that the most important goal is to sharply reduce the rate of population growth in developing countries, where 95% of the world's population growth is expected to take place. Some people in developing countries agree that population growth in these countries can cause local environmental problems. However, they contend that the most serious environmental problem the world faces is disruption of the global life-support system by high levels of resource consumption per person in developed countries, which use 80% of the world's resources. What is your view on this issue? Explain.

9. Do you believe that the United States (or the country where you live) should develop a comprehensive and integrated mass transit system over the next 20 years, including building an efficient rail network for travel

within and between its major cities? How would you pay for such a system?

10. How environmentally sustainable is the area where you live? List five ways to make it more environmentally sustainable.

11. Congratulations. You have been put in charge of the world. List the five most important features of your **(a)** population policy and **(b)** urban policy.

PROJECTS

1. Assume that your entire class (or manageable groups of your class) is charged with coming up with a plan for halving the world's population growth rate within the next 20 years. Develop a detailed plan that would achieve this goal, including any differences between policies in developing countries and developed countries. Justify each part of your plan. Predict what problems you might face in implementing the plan, and devise strategies for dealing with these problems.

2. Prepare an age structure diagram for your community. Use the diagram to project future population growth and economic and social problems.

3. As a class project, **(a)** evaluate land use and land-use planning by your school, **(b)** draw up an improved plan based on ecological principles, and **(c)** submit the plan to school officials.

4. Make a concept map of this chapter's major ideas, using the section heads and subheads and the key terms (in boldface). Look on the website for this book for information about making concept maps.

INTERNET STUDY RESOURCES AND RESOURCES FOR FURTHER READING AND RESEARCH

The website for this book contains helpful study aids and many ideas for further reading and research. Log on to

http://www.brookscole.com/product/0534389872s

and click on the Chapter-by-Chapter area. Choose Chapter 11 and select a resource:

- "Flash Cards" allows you to test your mastery of the Terms and Concepts to Remember for this chapter.
- "Tutorial Quizzes" provides a multiple-choice practice quiz.
- "Student Guide to InfoTrac" will lead you to Critical Thinking Projects that use InfoTrac College Edition as a research tool.
- "References" lists the major books and articles consulted in writing this chapter.
- "Hypercontents" takes you to an extensive list of sites with news, research, and images related to individual sections of the chapter.

INFOTRAC COLLEGE EDITION

Improve your skills with InfoTrac College Edition, a searchable online database of articles from more than 700 periodicals. Log on to

http://www.infotrac-college.com

or access InfoTrac through the website for this book. Try to find the following articles:

1. Four Key Issues: At the Top of the U.S. Agenda Are Economic Growth, Government Entitlements, Immigration, and Education. John C. Soper. *World and I*, May 2001 v16 i5 p32. Four key issues that are likely to dominate the nation over the next few years are **(1)** economic growth, **(2)** government entitlements, especially Social Security, **(3)** immigration, and **(4)** education. *Hint*: Enter the search term "four key issues" using the keyword "immigration."

2. Balancing ACT. Jim Motavalli. *E*, Nov 2000 v11 i6 p26. The article examines the environmental effects of increased population in the United States, including the typical American's environmental impact. *Hint*: Enter the search term "balancing act population" using the keyword "population."

12 AIR AND AIR POLLUTION

When Is a Lichen Like a Canary?

Nineteenth-century coal miners took canaries with them into the mines—not for their songs but for the moment when they stopped singing. Then the miners knew it was time to get out of the mine because the air contained methane, which could ignite and explode.

Today we use sophisticated equipment to monitor air quality, but living things such as lichens (Figure 12-1) still can warn us of bad air. A lichen consists of a fungus and an alga living together, usually in a mutually beneficial (mutualistic) partnership.

These hearty pioneer species are good air pollution detectors because they are always absorbing air as a source of nourishment. Certain lichen species are sensitive to specific air-polluting chemicals. Old man's beard (*Usnea trichodea*) (Figure 12-1, right) and yellow *Evernia* lichens, for example, sicken or die in the presence of too much sulfur dioxide.

Because lichens are widespread, long-lived, and anchored in place, they can also help track pollution to its source. The scientist who discovered sulfur dioxide pollution on Isle Royale in Lake Superior (Case Study, p. 186), where no car or smokestack had ever intruded, used *Evernia* lichens to point the finger northward to coal-burning facilities at Thunder Bay, Canada.

Radioactive particles spewed into the atmosphere by the Chernobyl nuclear power-plant disaster (p. 490) fell to the ground over much of northern Scandinavia and were absorbed by lichens that carpet much of Lapland. The area's Saami people depend on reindeer meat for food, and the reindeer feed on lichens. After Chernobyl more than 70,000 reindeer had to be killed and the meat discarded because it was too radioactive to eat. Scientists helped the Saami identify which of the remaining reindeer to move by analyzing lichens (which absorbed some of the radioactive fallout) to pinpoint the most contaminated areas.

Last but not least, lichens can replace electronic monitoring stations that cost more than $100,000 each. This is not so much a triumph of nature over technology as a partnership between the two, for technicians use highly sophisticated methods to analyze lichens for pollution and measure their rates of photosynthesis.

We all must breathe air from a global atmospheric commons in which air currents and winds can transport some pollutants long distances. Lichens can alert us to the danger, but as with all forms of pollution, the best solution is prevention.

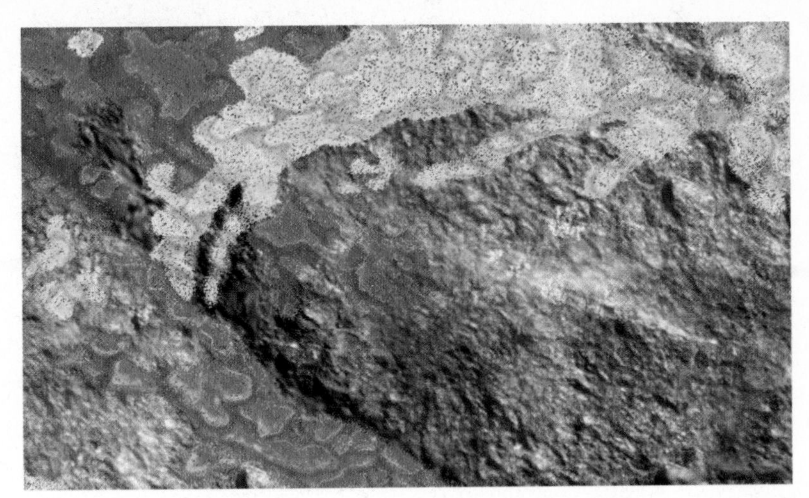

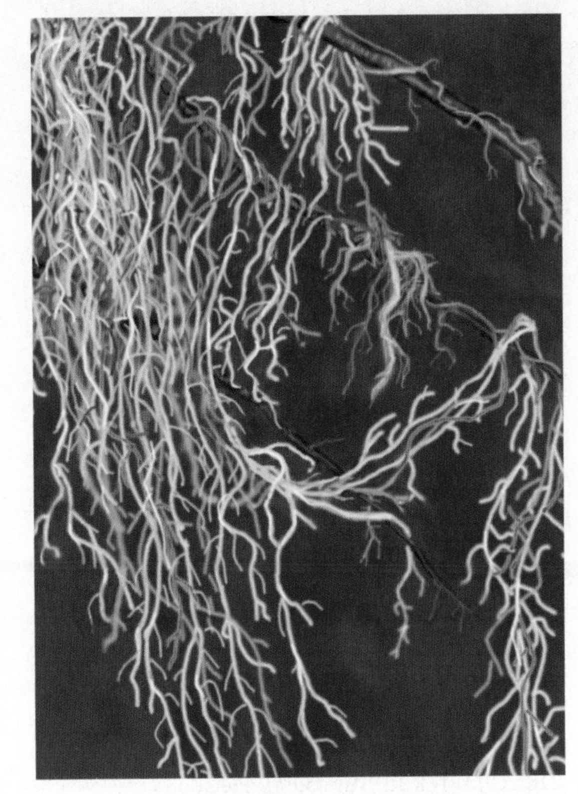

Figure 12-1 Red and yellow crustose lichens growing on slate rock in California (left), and *Usnea trichodea* lichen growing on a branch of a larch tree in Washington (right). The vulnerability of various lichen species to specific air pollutants can help researchers detect levels of these pollutants and track down their sources.

I thought I saw a blue jay this morning. But the smog was so bad that it turned out to be a cardinal holding its breath.

MICHAEL J. COHEN

This chapter addresses the following questions:

- What layers are found in the atmosphere?
- What are the major outdoor air pollutants, and where do they come from?
- What are two types of smog?
- What is acid deposition, and how can it be reduced?
- What are the harmful effects of air pollutants?
- How can we prevent and control air pollution?

12-1 THE ATMOSPHERE

What Is the Troposphere? Weather Breeder We live at the bottom of a sea of air called the **atmosphere**. This sea of life-sustaining gases surrounding the earth is divided into several spherical layers (Figure 12-2). Each layer is characterized by abrupt changes in temperature, the result of differences in the absorption of incoming solar energy.

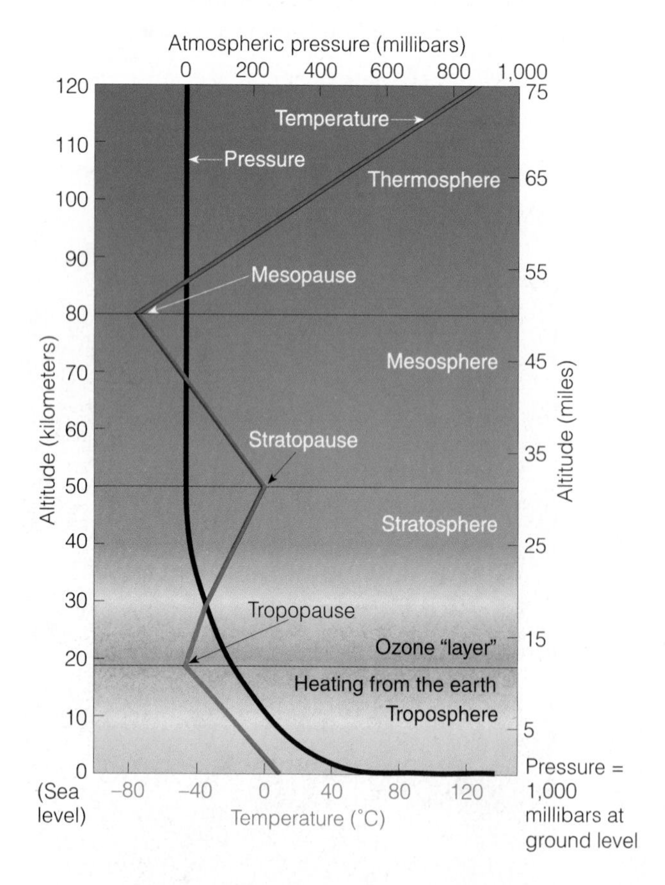

Figure 12-2 The earth's current atmosphere consists of several layers. The average temperature of atmosphere varies with altitude (red line). The average temperature of the atmosphere at the earth's surface is determined by a combination of **(1)** *natural heating* by incoming sunlight and certain greenhouse gases that release absorbed energy as heat into the lower troposphere (the natural *greenhouse effect*, Figure 6-11, p. 127) and **(2)** *natural cooling* by surface evaporation of water and convection processes that transfer heat to higher altitudes and latitudes (Figure 6-6, p. 125). Most UV radiation from the sun is absorbed by ozone (O_3), which is found primarily in the stratosphere in the *ozone layer* 12–26 kilometers (10–16 miles) above sea level.

About 75–80% of the mass of the earth's air is found in the atmosphere's innermost layer, the **troposphere**, which extends only about 17 kilometers (11 miles) above sea level at the equator and about 8 kilometers (5 miles) over the poles. If the earth were the size of an apple, this lower layer containing the air we breathe would be no thicker than the apple's skin. This thin and turbulent layer of rising and falling air currents and winds is the planet's *weather breeder*.

During several billion years of chemical and biological evolution, the composition of the earth's atmosphere has varied. Today, about 99% of the volume of the air you inhale with each breath consists of two gases: nitrogen (78%) and oxygen (21%). The remainder consists of **(1)** water vapor (varying from 0.01% at the frigid poles to 4% in the humid tropics), **(2)** slightly less than 1% argon (Ar), **(3)** 0.037% carbon dioxide (CO_2), and **(4)** trace amounts of several other gases.

What Is the Stratosphere? Earth's Global Sunscreen The atmosphere's second layer is the **stratosphere**, which extends from about 17 to 48 kilometers (11 to 30 miles) above the earth's surface (Figure 12-2). Although the stratosphere contains less matter than the troposphere, its composition is similar, with two notable exceptions: **(1)** Its volume of water vapor is about 1/1,000 as much, and **(2)** its concentration of ozone is much higher (Figure 12-3).

Stratospheric ozone is produced when some of the oxygen molecules there interact with ultraviolet (UV) radiation emitted by the sun. This "global sunscreen" of ozone in the stratosphere keeps about 95% of the sun's harmful UV radiation from reaching the earth's surface. This UV filter **(1)** allows humans and other forms of life to exist on land, **(2)** helps protect humans from sunburn, skin and eye cancer, cataracts, and damage to the immune system, and **(3)** prevents much of the oxygen in the troposphere from being converted to photochemical ozone, a harmful air pollutant (Figure 12-3).

There is much evidence that some human activities are **(1)** *decreasing* the amount beneficial ozone in the stratosphere (Section 13-6, p. 318) and **(2)** *increasing* the amount of harmful ozone in the troposphere.

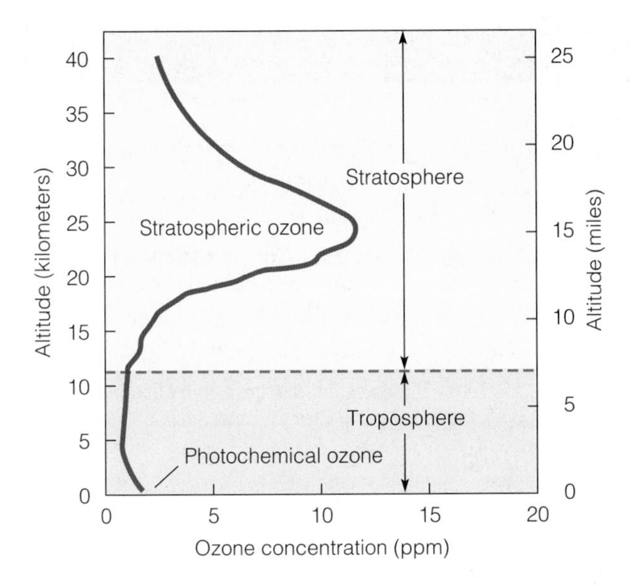

Figure 12-3 Average distribution and concentrations of ozone in the troposphere and stratosphere. *Beneficial ozone* that forms in the stratosphere protects life on earth by filtering out most incoming harmful ultraviolet radiation emitted by the sun. *Harmful* or *photochemical* ozone forms in the troposphere when various air pollutants undergo chemical reactions under the influence of sunlight. Ozone in this portion of the atmosphere near the earth's surface damages plants, lung tissues, and some materials such as rubber.

12-2 OUTDOOR AIR POLLUTION

What Are the Major Types and Sources of Air Pollution? **Air pollution** is the presence of one or more chemicals in the atmosphere in sufficient quantities and duration to cause harm to humans, other forms of life, and materials. Air pollution is not new (Spotlight, p. 283).

Table 12-1 lists the major classes of pollutants commonly found in outdoor (ambient) air. Such air pollutants come from both natural sources and human (anthropogenic) activities. Most natural sources of air pollution are spread out and, except for those from volcanic eruptions and some forest fires, rarely reach harmful levels. Most outdoor pollutants in urban areas enter the atmosphere from the burning of fossil fuels in **(1)** power plants and factories (*stationary sources*) and **(2)** motor vehicles (*mobile sources*).

Scientists distinguish between primary and secondary air pollutants in outdoor air. **Primary pollutants** are those emitted directly into the troposphere in a potentially harmful form. While in the troposphere, some of these primary pollutants may react with one another or with the basic components of air to form new pollutants, called **secondary pollutants** (Figure 12-4).

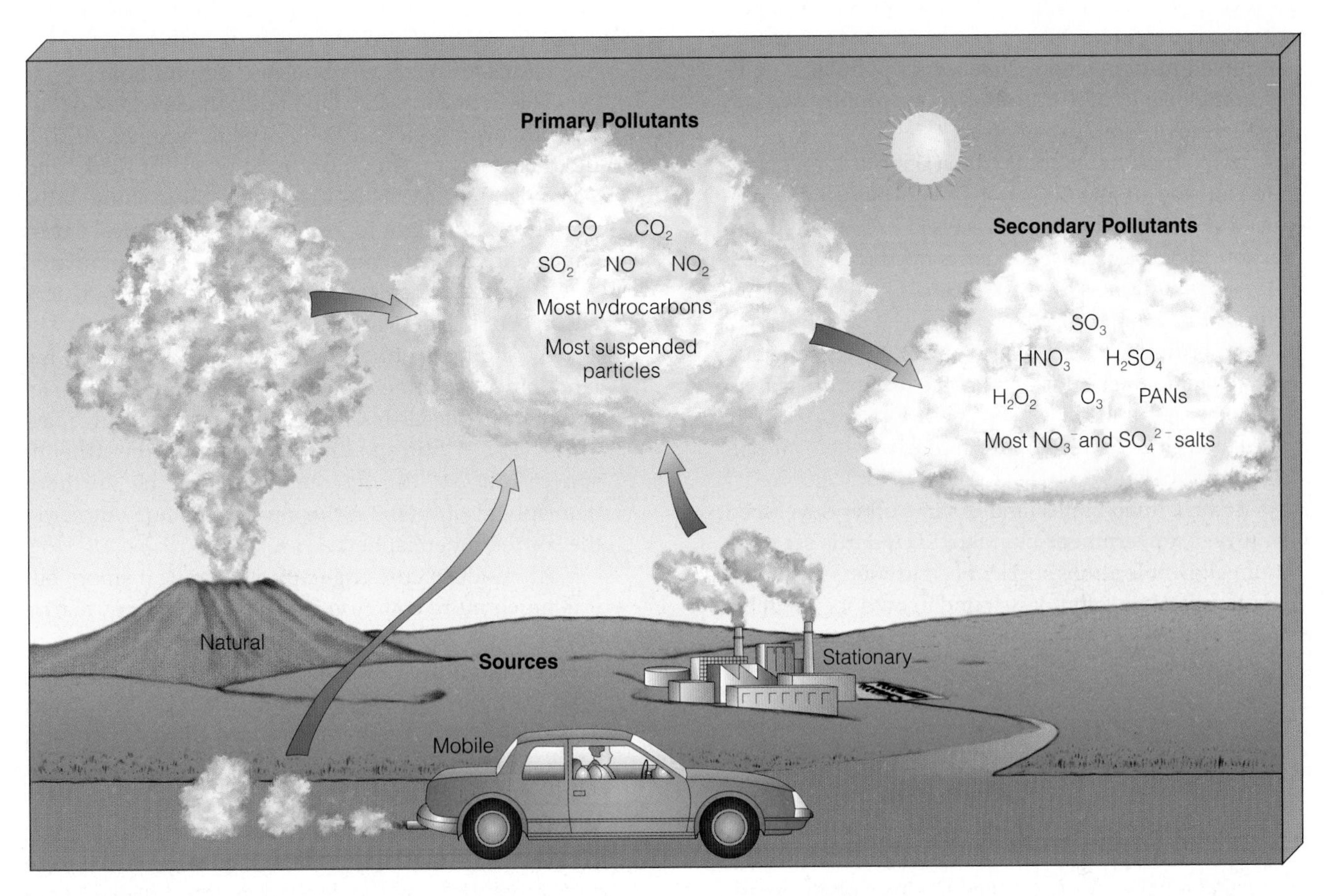

Figure 12-4 Sources and types of air pollutants. Human inputs of air pollutants may come from *mobile sources* (such as cars) and *stationary sources* (such as industrial and power plants). Some *primary air pollutants* may react with one another or with other chemicals in the air to form *secondary air pollutants.*

Table 12-1 Major Classes of Air Pollutants

Class	Examples
Carbon oxides	Carbon monoxide (CO) and carbon dioxide (CO_2)
Sulfur oxides	Sulfur dioxide (SO_2) and sulfur trioxide (SO_3)
Nitrogen oxides	Nitric oxide (NO), nitrogen dioxide (NO_2), nitrous oxide (N_2O) (NO and NO_2 often are lumped together and labeled NO_x)
Volatile organic compounds (VOCs)	Methane (CH_4), propane (C_3H_8), chlorofluorocarbons (CFCs)
Suspended particulate matter (SPM)	Solid particles (dust, soot, asbestos, lead, nitrate, and sulfate salts), liquid droplets (sulfuric acid, PCBs, dioxins, and pesticides)
Photochemical oxidants	Ozone (O_3), peroxyacyl nitrates (PANs), hydrogen peroxide (H_2O_2), aldehydes
Radioactive substances	Radon-222, iodine-131, strontium-90, plutonium-239 (Table 3-2, p. 69)
Hazardous air pollutants (HAPs), which cause health effects such as cancer, birth defects, and nervous system problems	Carbon tetrachloride (CCl_4), methyl chloride (CH_3Cl), chloroform ($CHCl_3$), benzene (C_6H_6), ethylene dibromide ($C_2H_2Br_2$), formaldehyde (CH_2O_2)

With their large concentrations of cars and factories, cities normally have higher air pollution levels than rural areas. However, prevailing winds can spread long-lived primary and secondary air pollutants emitted in urban and industrial areas to the countryside and to other downwind urban areas.

Indoor pollutants come from **(1)** infiltration of polluted outside air and **(2)** various chemicals used or produced inside buildings, as discussed in more detail in Section 12-5. Risk analysis experts rate indoor and outdoor air pollution as high-risk human health problems (Figure 10-12, left, p. 238).

According to the World Health Organization (WHO), more than 1.1 billion people live in urban areas where the air is unhealthy to breathe. Most live in densely populated cities in developing countries where air pollution control laws do not exist or are poorly enforced.

In the United States (and in most other developed countries), government-mandated standards set maximum allowable atmospheric concentrations for six *criteria air pollutants* commonly found in outdoor air (Table 12-2).

12-3 PHOTOCHEMICAL AND INDUSTRIAL SMOG

What Is Photochemical Smog? Brown-Air Smog Any chemical reaction activated by light is called a *photochemical reaction*. Air pollution known as **photochemical smog** is a mixture of primary and secondary pollutants formed under the influence of sunlight (Figure 12-5). The resulting mixture of more than 100 chemicals is dominated by *photochemical ozone*, a highly reactive gas that harms most living organisms (Figure 12-3).

The hotter the day, the higher the levels of ozone and other components of photochemical smog. As traffic increases in the morning, levels of NO_x and unburned hydrocarbons rise and begin reacting in the presence of sunlight to produce photochemical smog. On a sunny day the photochemical smog (dominated by O_3) builds up to peak levels by early afternoon, irritating people's eyes and respiratory tracts.

According to atmospheric chemist Sherwood Rowland, since 1900 the concentration of photochemical ozone near the earth's surface has increased by a factor of **(1)** 5–8 during summer in the northern hemisphere, **(2)** 2–4 during the winter in the northern hemisphere, and **(3)** 2–5 during winter and summer in the southern hemisphere.

All modern cities have photochemical smog, but it is much more common in cities with sunny, warm, dry climates and lots of motor vehicles. Examples of such cities are Los Angeles, California; Denver, Colorado; and Salt Lake City, Utah, in the United States, as well as Sydney, Australia; Mexico City, Mexico; and São Paulo and Buenos Aires in Brazil.

What Is Industrial Smog? Gray-Air Smog Fifty years ago cities such as London, England, and Chicago and Pittsburgh in the United States burned large amounts of coal and heavy oil (which contain sulfur

Table 12-2 Common Criteria Air Pollutants in the United States*

CARBON MONOXIDE (CO)

Description: Colorless, odorless gas that is poisonous to air-breathing animals; forms during the incomplete combustion of carbon-containing fuels ($2C + O_2 \longrightarrow 2CO$).

Major human sources: Cigarette smoking (p. 224), incomplete burning of fossil fuels. About 77% (95% in cities) comes from motor vehicle exhaust.

Health effects: Reacts with hemoglobin in red blood cells and reduces the ability of blood to bring oxygen to body cells and tissues. This impairs perception and thinking; slows reflexes; causes headaches, drowsiness, dizziness, and nausea; can trigger heart attacks and angina; damages the development of fetuses and young children; and aggravates chronic bronchitis, emphysema, and anemia. At high levels it causes collapse, coma, irreversible brain cell damage, and death.

NITROGEN DIOXIDE (NO_2)

Description: Reddish-brown, irritating gas that gives photochemical smog its brownish color; in the atmosphere can be converted to nitric acid (HNO_3), a major component of acid deposition.

Major human sources: Fossil fuel burning in motor vehicles (49%) and power and industrial plants (49%).

Health effects: Lung irritation and damage; aggravates asthma and chronic bronchitis; increases susceptibility to respiratory infections such as the flu and common colds (especially in young children and older adults).

Environmental effects: Reduces visibility; acid deposition of HNO_3 can damage trees, soils, and aquatic life in lakes.

Property damage: HNO_3 can corrode metals and eat away stone on buildings, statues, and monuments; NO_2 can damage fabrics.

SULFUR DIOXIDE (SO_2)

Description: Colorless, irritating; forms mostly from the combustion of sulfur-containing fossil fuels such as coal and oil ($S + O_2 \longrightarrow SO_2$); in the atmosphere can be converted to sulfuric acid (H_2SO_4), a major component of acid deposition.

Major human sources: Coal burning in power plants (88%) and industrial processes (10%).

Health effects: Breathing problems for healthy people; severe restriction of airways in people with asthma; chronic exposure can cause a permanent condition similar to bronchitis. According to the World Health Organization, at least 625 million people are exposed to unsafe levels of sulfur dioxide from fossil fuel burning.

Environmental effects: Reduces visibility; acid deposition of H_2SO_4 can damage trees, soils, and aquatic life in lakes.

Property damage: SO_2 and H_2SO can corrode metals and eat away stone on buildings, statues, and monuments; SO_2 can damage paint, paper, and leather.

SUSPENDED PARTICULATE MATTER (SPM)

Description: Variety of particles and droplets (aerosols) small and light enough to remain suspended in atmosphere for short periods (large particles) to long periods (small particles; Figure 12-6); we see these particles as smoke, dust, and haze.

Major human sources: Burning coal in power and industrial plants (40%), burning diesel and other fuels in vehicles (17%), agriculture (plowing, burning off fields), unpaved roads, construction.

Health effects: Nose and throat irritation, lung damage, and bronchitis; aggravates bronchitis and asthma; causes early death; toxic particulates (such as lead, cadmium, PCBs, and dioxins) can cause mutations, reproductive problems, or cancer.

Environmental effects: Reduces visibility; acid deposition of H_2SO_4 droplets can damage trees, soils, and aquatic life in lakes.

Property damage: Corrodes metal; soils and discolors buildings, clothes, fabrics, and paints.

OZONE (O_3)

Description: Highly reactive, irritating gas with an unpleasant odor that forms in the troposphere as a major component of photochemical smog (Figures 12-3 and 12-5).

Major human sources: Chemical reaction with volatile organic compounds (VOCs, emitted mostly by cars and industries) and nitrogen oxides to form photochemical smog (Figure 12-5).

Health effects: Breathing problems; coughing; eye, nose, and throat irritation; aggravates chronic diseases such as asthma, bronchitis, emphysema, and heart disease; reduces resistance to colds and pneumonia; may speed up lung tissue aging.

Environmental effects: Ozone can damage plants and trees; smog can reduce visibility.

Property damage: Damages rubber, fabrics, and paints.

LEAD

Description: Solid toxic metal and its compounds, emitted into the atmosphere as particulate matter.

Major human sources: Paint (old houses), smelters (metal refineries), lead manufacture, storage batteries, leaded gasoline (being phased out in developed countries).

Health effects: Accumulates in the body; brain and other nervous system damage and mental retardation (especially in children); digestive and other health problems; some lead-containing chemicals cause cancer in test animals.

Environmental effects: Can harm wildlife.

*Data from U.S. Environmental Protection Agency.

impurities) in power plants and factories and for space heating. During winter, people in such cities were exposed to **industrial smog** consisting mostly of **(1)** sulfur dioxide, **(2)** suspended droplets of sulfuric acid (formed from sulfur dioxide, Figure 12-4), and **(3)** a variety of suspended solid particles and droplets (called aerosols; Figure 12-6, p. 284).

Urban industrial smog is rarely a problem today in most developed countries because coal and heavy oil are burned only in large boilers with reasonably good pollution control or with tall smokestacks (which transfer the pollutants to downwind areas). However, industrial smog is a problem in industrialized urban areas of China, India, Ukraine, and some eastern European countries,

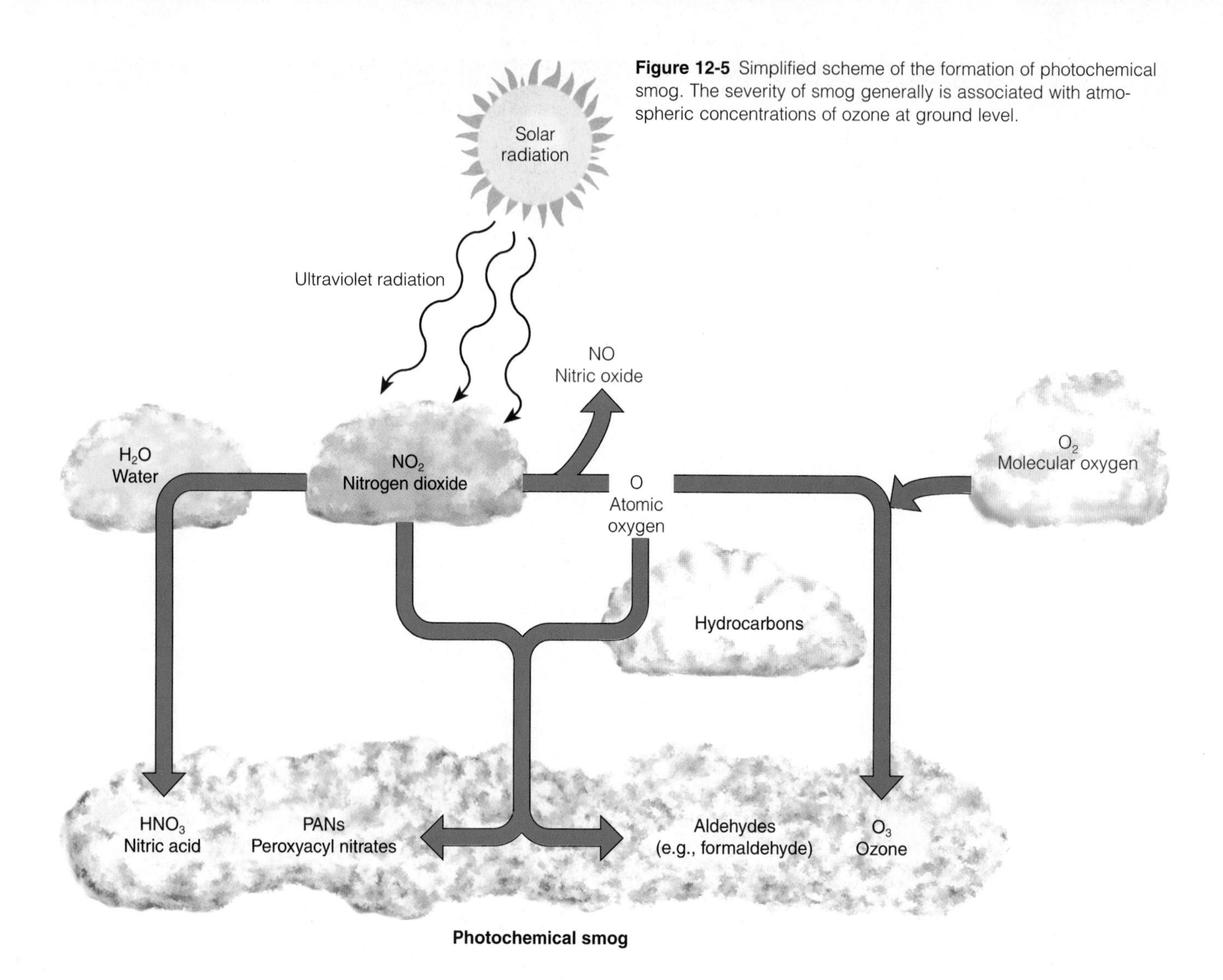

Figure 12-5 Simplified scheme of the formation of photochemical smog. The severity of smog generally is associated with atmospheric concentrations of ozone at ground level.

where large quantities of coal are burned with inadequate pollution controls.

What Factors Influence the Formation of Photochemical and Industrial Smog? The frequency and severity of smog in an area depend on **(1)** local climate and topography, **(2)** population density, **(3)** the amount of industry, and **(4)** the fuels used in industry, heating, and transportation.

Air pollution can be reduced by

- *Rain and snow*, which help cleanse the air of pollutants. This helps explain why cities with dry climates are more prone to photochemical smog than ones with wet climates.
- *Winds*, which **(1)** help sweep pollutants away, **(2)** dilute pollutants by mixing them with cleaner air, and **(3)** bring in fresh air. However, these pollutants are blown somewhere else or are deposited from the sky onto surface waters, soil, and buildings.

Three factors that can increase air pollution are

- *Urban buildings*, which can slow wind speed and reduce dilution and removal of pollutants
- *Hills and mountains*, which tend to reduce the flow of air in valleys below them and allow pollutant levels to build up at ground level
- *High temperatures* that promote the chemical reactions that lead to photochemical smog formation

What Are Temperature Inversions? During daylight the sun warms the air near the earth's surface. Normally, this warm air and most of the pollutants it contains rise to mix with the cooler air above it. This mixing of warm and cold air creates turbulence, which disperses the pollutants.

Under certain atmospheric conditions, however, a layer of warm air can lie atop a layer of cooler air nearer the ground, a situation known as a **temperature inversion**. Because the cooler air is denser than the warmer air above it, the air near the surface does not rise and mix with the air above it. Pollutants can concentrate in this stagnant layer of cool air near the ground.

There are two types of temperature inversions:

Air Pollution in the Past: The Bad Old Days

Modern civilization did not invent air pollution. It probably began when humans discovered fire and used it to burn wood in poorly ventilated caves.

During the Middle Ages a haze of wood smoke hung over densely packed urban areas. The industrial revolution brought even worse air pollution as coal was burned to power factories and heat homes. By the 1850s, London was well known for its "pea soup" fog, consisting of a mixture of coal smoke and fog that blanketed the city. In 1880, a prolonged coal fog killed an estimated 2,200 people. Another in 1911 killed more than 1,100 Londoners. The authors of a report on this disaster coined the word *smog* for the deadly mixture of smoke and fog that enveloped the city.

In 1952, an even worse yellow fog lasted for 5 days and killed 4,000 Londoners, prompting Parliament to pass the Clean Air Act of 1956. Additional air pollution disasters in 1956, 1957, and 1962 killed 2,500 more people. Because of strong air pollution laws, London's air today is much cleaner, and "pea soup" fogs are a thing of the past.

The industrial revolution, powered by coal-burning factories and homes, brought air pollution to the United States. Large industrial cities such as Pittsburgh, Pennsylvania, and St. Louis, Missouri, were known for their smoky air. By the 1940s, the air over some cities was so polluted that people had to use their automobile headlights during the day.

The first documented air pollution disaster in the United States occurred during October 1948, at the town of Donora in Pennsylvania's Monongahela River Valley south of Pittsburgh. Pollutants from the area's industries became trapped in a fog that stagnated over the valley for 5 days. After several days the fog was so dense that people could not see well enough to drive, even at noon with their headlights on. About 7,000 of the town's 14,000 inhabitants became sick, and 22 of them died. This killer fog resulted from a combination of mountainous terrain surrounding the valley and weather conditions that trapped and concentrated deadly pollutants emitted by the community's steel mill, zinc smelter, and sulfuric acid plant (Figure 12-7, top, p. 284).

In 1963, high concentrations of air pollutants accumulated in the air over New York City, killing about 300 people and injuring thousands. Other episodes in New York, Los Angeles, and other large cities in the 1960s led to much stronger air-pollution control programs in the 1970s.

Congress passed the original version of the Clean Air Act in 1963, but it did not have much effect until a stronger version of this law was enacted in 1970. The Clean Air Act of 1970 empowered the federal government to set air-pollution emission standards (with an adequate safety margin) for automobiles and industries that each state was required to enforce. Even stricter emission standards were imposed by amendments to the Clean Air Act in 1977 and 1990. Mostly as a result of these laws and actions by states and local areas, the United States has not had any more Donora or New York City incidents.

Critical Thinking

Explain why you agree or disagree with the statement that air pollution in the United States should not be a major concern because of the significant progress in reducing outdoor air pollution since 1970.

- A **subsidence temperature inversion**, which occurs when a large mass of warm air moves into a region at a high altitude and floats over a mass of colder air near the ground. This keeps the air over a city stagnant and prevents vertical mixing and dispersion of air pollutants. Normally such conditions do not last long, but sometimes warmer air masses can remain over cooler air below for days and allow pollutants to build up to harmful levels.
- A **radiation temperature inversion**, which typically occurs at night as the air near the ground cools faster than the air above it. As the sun rises and warms the earth's surface, a radiation inversion normally disappears by noon and disperses the pollutants built up during the night.

Under certain conditions, radiation or subsidence temperature inversions can last for several days and allow pollutants to build up to dangerous concentrations. Areas with two types of topography and weather conditions are especially susceptible to prolonged temperature inversions (Figure 12-7).

One such area is a town or city located in a valley surrounded by mountains that experiences cloudy and cold weather during part of the year (Figure 12-7, top). In such cases, the surrounding mountains block out the winter sun needed to reverse the nightly radiation temperature inversion. As long as these stagnant conditions persist, concentrations of pollutants in the valley below build up to harmful and even lethal concentrations. This is what happened during the 1948 air pollution disaster in the valley town of Donora, Pennsylvania (Spotlight, above).

A city with several million people and motor vehicles in an area with a **(1)** sunny climate, **(2)** light winds, **(3)** mountains on three sides, and **(4)** the ocean on the other has ideal conditions for photochemical smog worsened by frequent subsidence thermal inversions

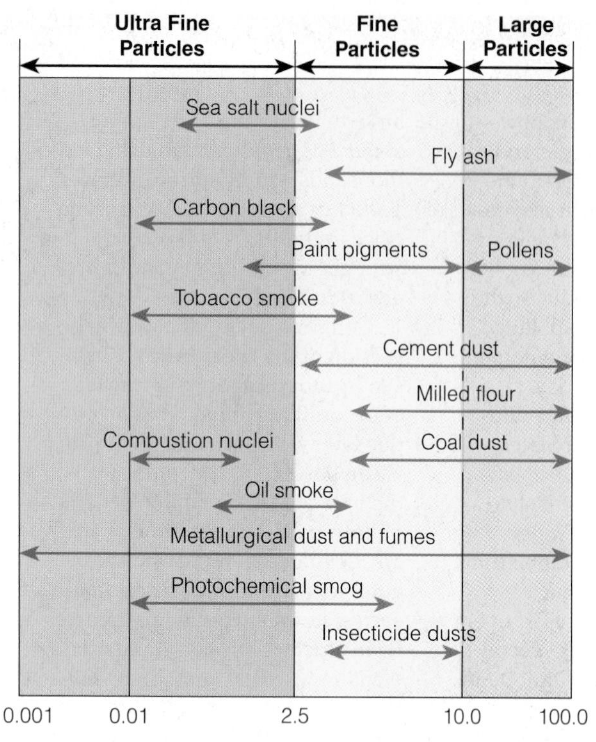

Figure 12-6 Suspended particulate matter consists of particles of solid matter and droplets of liquid that are small and light enough to remain suspended in the atmosphere for short periods (large particles) to long periods (small particles). Suspended particles are found in a wide variety of types and sizes, ranging in diameter from 0.001 micrometer to 100 micrometers (a micrometer, or micron, is one millionth of a meter, or about 0.00004 inches). Since 1987, the EPA has focused on *fine particles* smaller than 10 microns (known as *PM-10*). In 1997, the agency began focusing on reducing emissions of *ultrafine particles* with diameters less than 2.5 microns (known as *PM-2.5*) because these particles are small enough to reach the lower part of human lungs and contribute to respiratory diseases.

(Figure 12-7, bottom). This describes California's heavily populated Los Angeles basin, which has prolonged subsidence temperature inversions at least half of the year, mostly during the warm summer and fall. High-pressure air off the coast of California much of the year creates a descending warm air mass that sits atop an air mass below that is cooled by the nearby ocean. When this subsidence thermal inversion persists throughout the day, the surrounding mountains prevent the polluted surface air from being blown away by sea breezes (Figure 12-7, bottom).

12-4 REGIONAL OUTDOOR AIR POLLUTION FROM ACID DEPOSITION

What Is Acid Deposition? Most coal-burning power plants, ore smelters, and other industrial plants in developed countries use tall smokestacks to emit sulfur dioxide, suspended particles, and nitrogen oxides above the inversion layer (Figure 12-7, top), where mixing, dilution, and removal by wind are more effective. Thus, tall smokestacks reduce *local* air pollution but increase *regional* air pollution downwind.

The primary pollutants, sulfur dioxide and nitrogen oxides, emitted into the atmosphere above the inversion layer are transported as much as 1,000 kilometers (600 miles) by prevailing winds. During their trip, they form secondary pollutants such as **(1)** nitric acid vapor, **(2)** droplets of sulfuric acid, and **(3)** particles of acid-forming sulfate and nitrate salts (Figure 12-4).

These acidic substances remain in the atmosphere for 2–14 days, depending mostly on prevailing winds, precipitation, and other weather patterns. During this period they descend to the earth's surface in two forms: **(1)** *wet deposition* (as acidic rain, snow, fog, and cloud vapor) and **(2)** *dry deposition* (as acidic particles). The resulting mixture is called **acid deposition** (Figure 12-8), sometimes called *acid rain*, and its acidity usually is reported in terms of its pH (Figure 3-5, p. 63). Most dry deposition occurs within about 2–3 days fairly near the emission sources, whereas most wet deposition occurs takes place in 4–14 days in more distant downwind areas (Figure 12-8).

What Areas Are Most Affected by Acid Deposition? Acid deposition is a regional problem in the eastern United States and in other parts of the world (Figure 12-9). Most of these regions are downwind from coal-burning power plants, smelters, or factories or are major urban areas with large numbers of motor vehicles.

In the United States, coal-burning power and industrial plants in the Ohio Valley emit the largest quantities of sulfur dioxide and other acidic pollutants. Mostly as a result of these emissions, typical precipitation in the eastern United States has a pH of 4.2 to 4.7—10 or more times the acidity of natural precipitation with a pH of 5.6. Some mountaintop forests in the eastern United States and east of Los Angeles, California, are bathed in fog and dews as acidic as lemon juice, with a pH of 2.3—about 1,000 times the acidity of normal precipitation.

In some areas, soils contain basic compounds that can neutralize or *buffer* some input of acids. The areas most sensitive to acid deposition are **(1)** those containing thin, acidic soils derived mostly from rock without such natural buffering (Figure 12-9, green and most red areas) and **(2)** those in which the buffering capacity of soils has been depleted by decades of acid deposition (some red areas in Figure 12-9).

Many acid-producing chemicals generated by power plants, factories, smelters, and cars in one country are exported to other countries by prevailing winds. For example, studies show that some acid deposition in

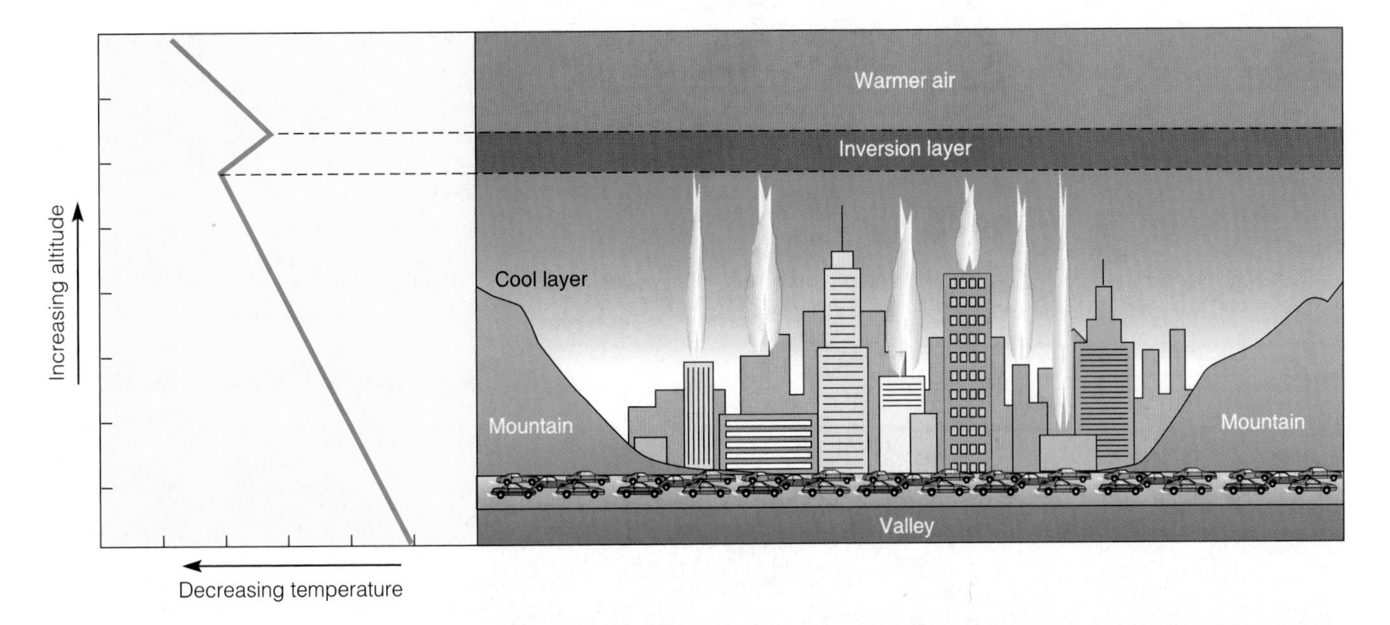

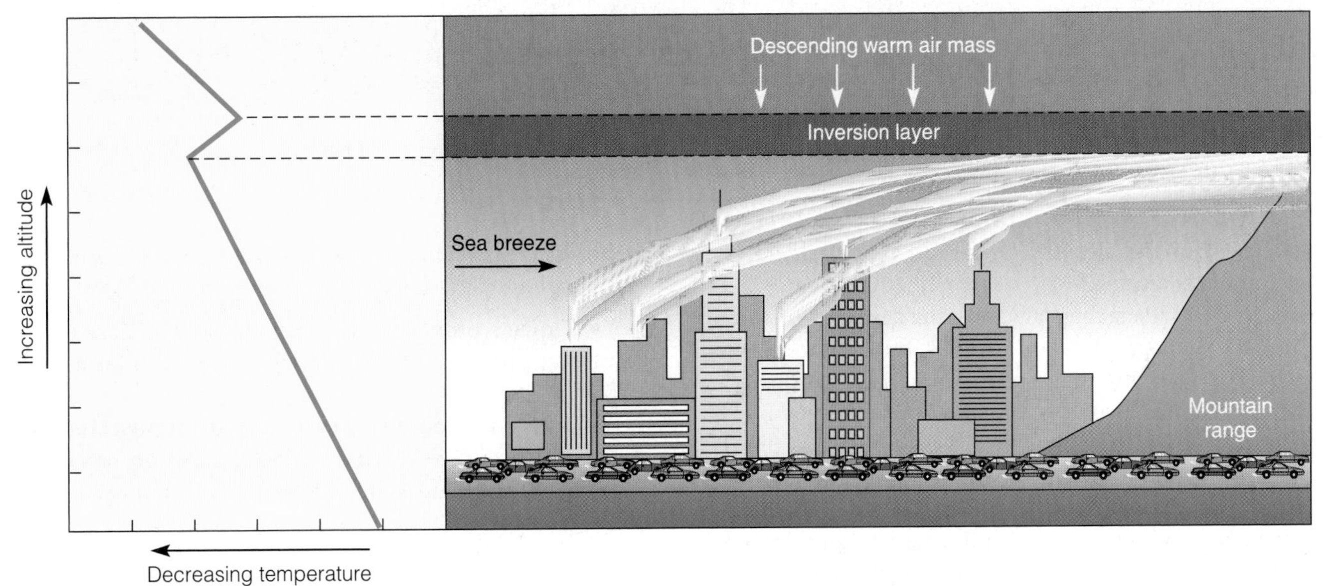

Figure 12-7 Topography and weather conditions that can create more frequent and prolonged *radiation temperature inversions*, in which a layer of warm air sits atop a cooler layer of air near the ground. In such cases, pollutant concentrations in the air near the earth's surface can build up to harmful levels. The top figure shows how air pollutants can build up in the air near the ground in a valley surrounded by mountains. The bottom figure shows how frequent and prolonged radiation temperature inversions can occur in an area (such as Los Angeles, California) with a sunny climate, light winds, mountains on three sides, and the ocean on the other. The layer of descending warm air (bottom) prevents ascending air currents from dispersing and diluting pollutants from the cooler air near the ground. Because of their topography, Los Angeles in the United States and Mexico City in Mexico have frequent thermal inversions, many of them prolonged during the summer.

- Norway, Switzerland, Austria, Sweden, the Netherlands, and Finland is blown to those countries from industrialized areas of western Europe (especially the United Kingdom and Germany) and eastern Europe

- Southeastern Canada (Figure 12-9) can be traced to SO_2 and other emissions from the Ohio Valley of the United States

- The eastern United States has been traced to emissions from two large metal smelters in southeastern Canada

- Japan and North and South Korea comes from China

The worst acid deposition is in Asia, especially in China, which gets 73% of its energy from burning coal. Acid deposition is also a growing problem in

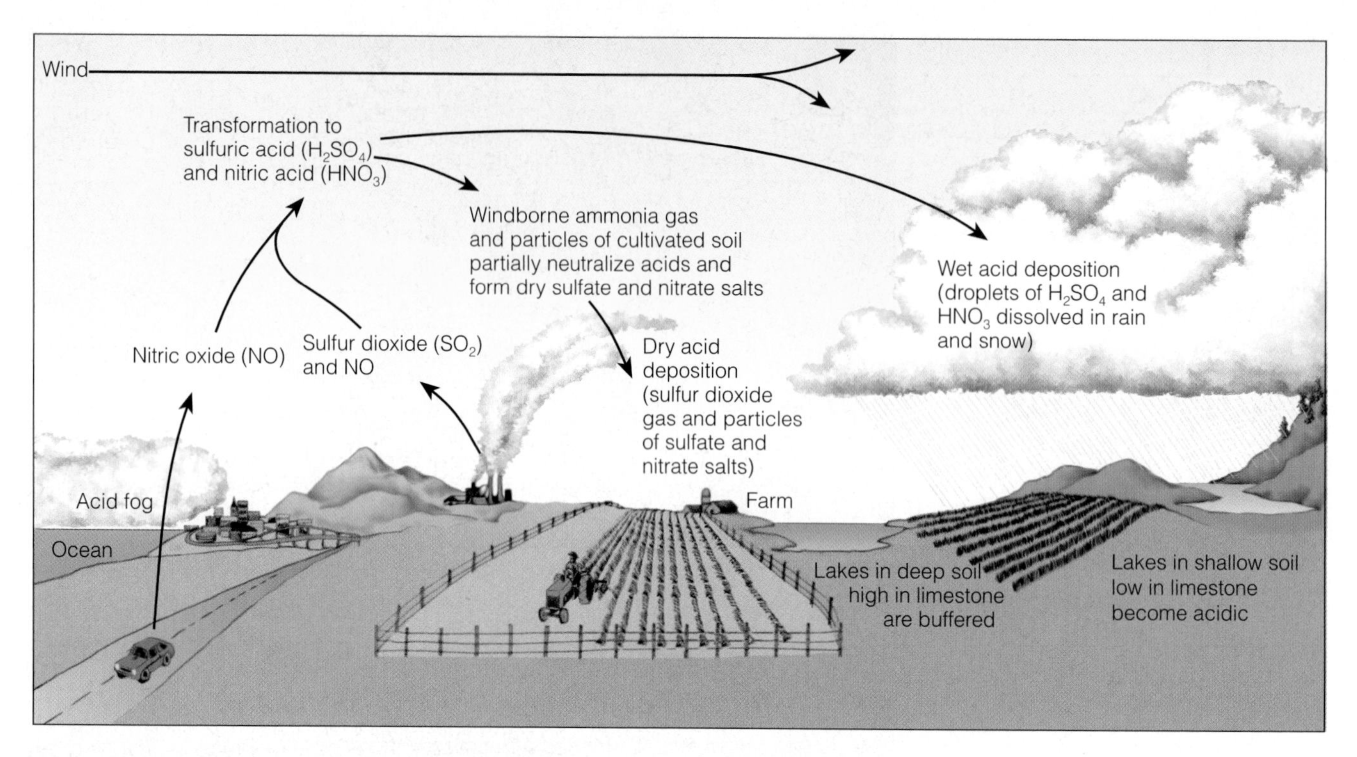

Figure 12-8 Acid deposition, which consists of rain, snow, dust, or gas with a pH lower than 5.6, is commonly called acid rain. Soils and lakes vary in their ability to buffer or remove excess acidity.

eastern Europe, Russia, Nigeria, Venezuela, and Colombia (Figure 12-9).

In 1999, researchers at California's Scripps Institution of Oceanography found a thick, brown haze of air pollution covering much of the Indian Ocean during winter. The haze-covered area is about the size of the continental United States and rises as high as 3,000 meters (10,000 feet). During the late spring and summer, when prevailing winds reverse, some of the haze can be blown back onto the land, where it can combine with monsoon rains and fall as acid deposition.

What Are the Effects of Acid Deposition on Human Health, Materials, and the Economy? Acid deposition

- Contributes to human respiratory diseases such as bronchitis and asthma, as discussed in Section 12-6
- Can leach toxic metals such as lead and copper from water pipes into drinking water
- Damages statues, buildings, metals, and car finishes
- Decreases atmospheric visibility (mostly because of sulfate particles)
- Lowers profits and causes job losses because of lower productivity in fisheries, forests, and farms

What Are the Effects of Acid Deposition on Aquatic Ecosystems? Acid deposition has many harmful ecological effects when the pH of most aquatic

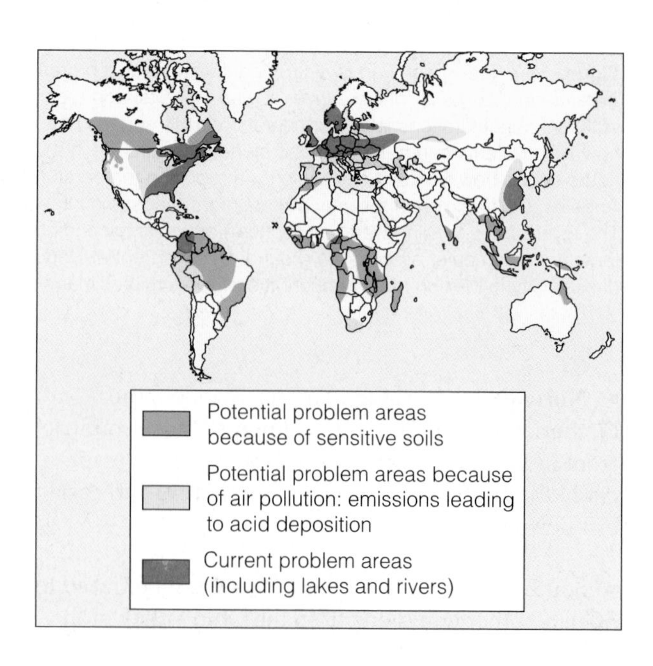

Figure 12-9 Regions where acid deposition is now a problem (red) and regions with the potential to develop this problem (yellow and green). Such regions **(1)** have large inputs of air pollution (mostly from coal-burning power plants, industrial plants, and ore smelters) or **(2)** are sensitive areas with soils and bedrock that cannot neutralize (buffer) inputs of acidic compounds (green areas and most red areas). (Data from World Resources Institute and U.S. Environmental Protection Agency)

systems falls below 6 and especially below 5. These effects include

- Loss of essentially all fish populations below a pH of 4.5 (Figure 12-10).
- Release of aluminum ions (Al^{3+}) attached to minerals in nearby soil into lakes, where they can kill many kinds of fish by stimulating excessive mucus formation. This asphyxiates the fish by clogging their gills.
- Contamination of lakes, streams, and fish in downwind areas by toxic mercury emitted by coal-burning plants. Humans who eat the contaminated fish can suffer permanent kidney failure, tremors, severe brain damage, and death.

Much of the damage to aquatic life in sensitive areas with little buffering capacity (Figure 12-9) is a result of *acid shock*. This is caused by the sudden runoff of large amounts of highly acidic water and aluminum ions into lakes and streams, when snow melts in the spring or after unusually heavy rains.

Because of excess acidity,

- In Norway and Sweden, at least 16,000 lakes contain no fish, and 52,000 more lakes have lost most of their acid-neutralizing capacity.
- In Canada, some 14,000 acidified lakes contain few if any fish, and some fish populations in 150,000 more lakes are declining because of increased acidity.
- In the United States, about 9,000 lakes (most in the Northeast and upper Midwest) are threatened with excess acidity, one-third of them seriously.

What Are the Effects of Acid Deposition on Plants and Soil Chemistry? Acid deposition (often along with other air pollutants such as ozone) can harm forests and crops, especially when the soil pH falls below 5.1. Effects of acid deposition on trees and other plants are caused by chemical interactions in forest and cropland soils (Figure 12-11) and include

- Damaging leaves and needles directly
- Leaching essential plant nutrients such as calcium and magnesium salts from soils, which reduces plant productivity and the ability of the soils to buffer or neutralize acidic inputs
- Releasing aluminum ions (Al^{3+}) attached to insoluble soil compounds, which can hinder uptake and use of soil nutrients and water by plants
- Dissolving insoluble soil compounds and releasing ions of metals such as lead, cadmium, and mercury that can be absorbed by plants and are highly toxic to plants and animals
- Promoting the growth of acid-loving mosses that can kill trees
- Weakening trees and other plants so they become more susceptible to other types of damage such as **(1)** severe cold, **(2)** diseases, **(3)** insect attacks, **(4)** drought, and **(5)** harmful mosses

Mountaintop forests are the terrestrial areas hardest hit by acid deposition because **(1)** they tend to have thin soils without much buffering capacity, and

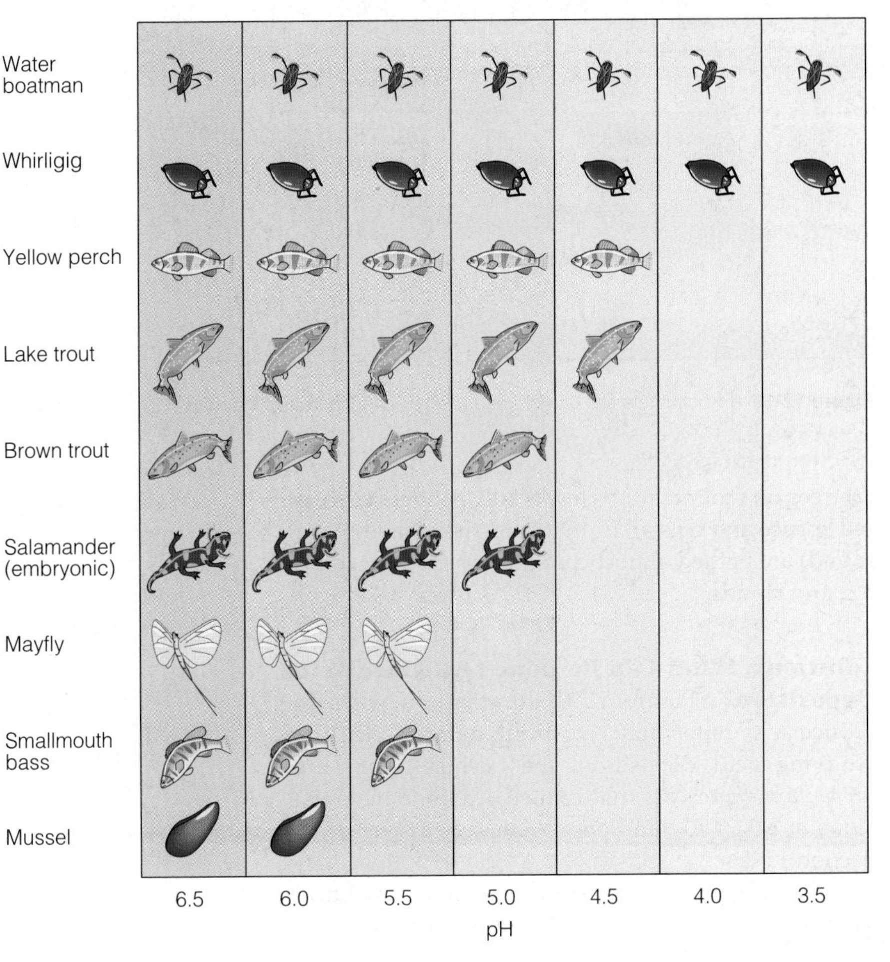

Figure 12-10 Fish and other aquatic organisms vary in their sensitivity to acidity. The figure shows the lowest pH (highest acidity) at which the various species can survive. Note that the greatest effects occur when the pH drops below 5.5.

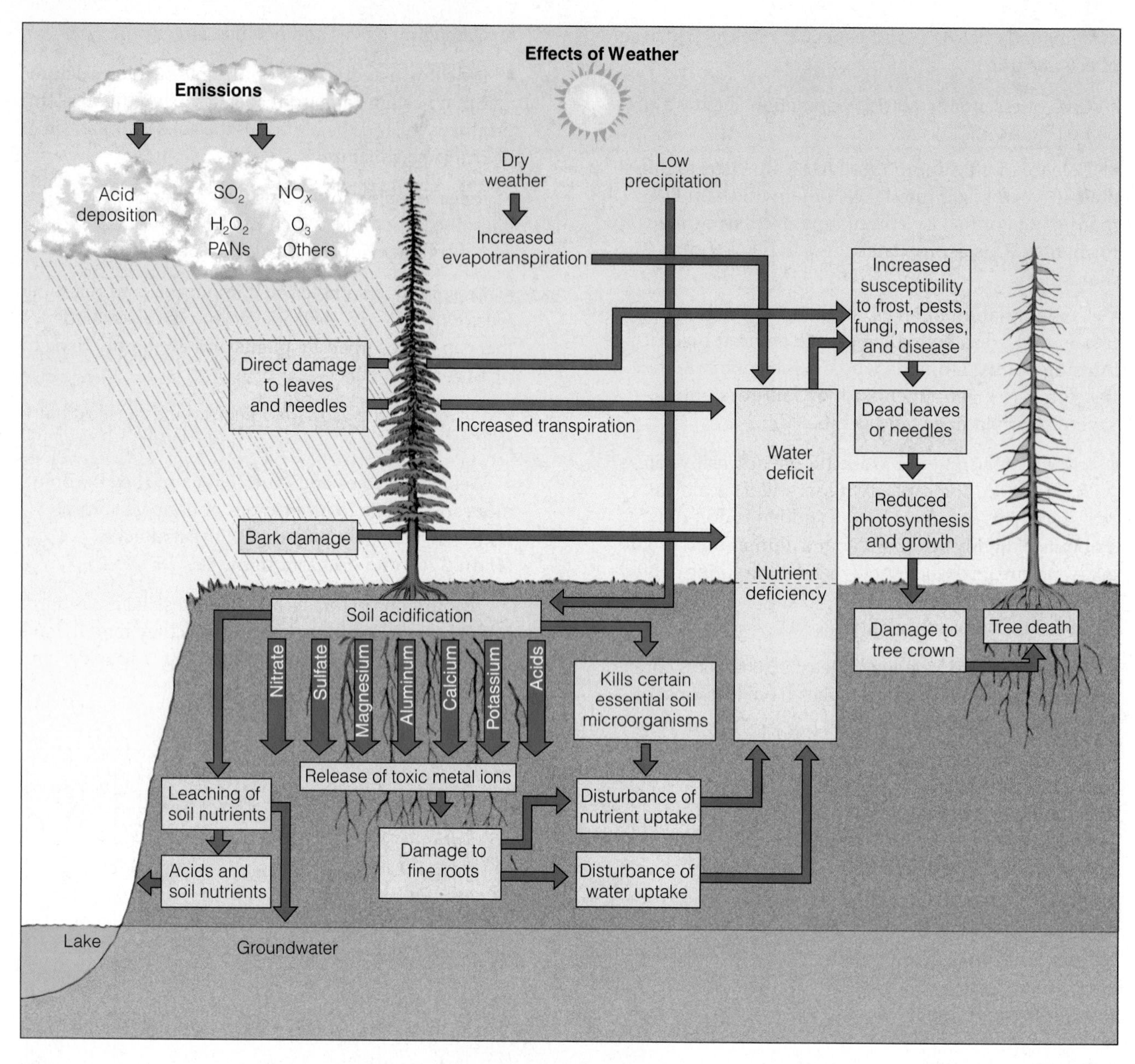

Figure 12-11 Air pollutants are one of several interacting stresses that can damage, weaken, or kill trees.

(2) trees on mountaintops (especially conifers such as red spruce and balsam fir that keep their leaves year-round) are bathed almost continuously in very acidic fog and clouds.

Solutions: What Can Be Done to Reduce Acid Deposition? Figure 12-12 summarizes ways to reduce acid deposition. According to most scientists studying acid deposition, the best solutions are *prevention approaches* that reduce or eliminate emissions of SO_2, NO_x, and particulates, as discussed in Section 12-7.

Controlling acid deposition is a difficult political problem because **(1)** the people and ecosystems it affects often are quite distant from those who cause the problem, **(2)** countries with large supplies of coal (such as China, India, Russia, and the United States) have a strong incentive to use it as a major energy resource, and **(3)** owners of coal-burning power plants say that the costs of adding air-pollution reducing equipment, using low-sulfur coal, or removing sulfur from coal are too high.

Large amounts of limestone or lime can be used to neutralize acidified lakes or surrounding soil, the only cleanup approach now being used. However, there are several problems with liming.

- It is an expensive and temporary remedy that usually must be repeated annually.

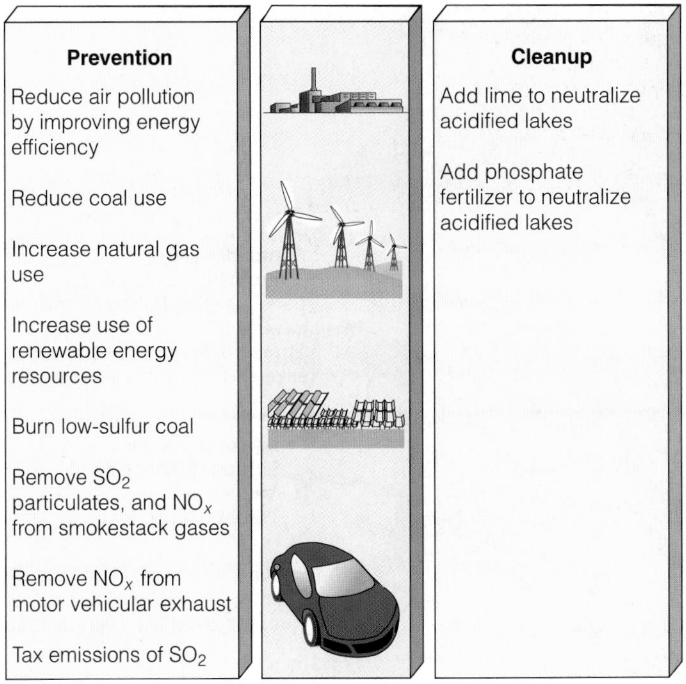

Prevention

Reduce air pollution by improving energy efficiency

Reduce coal use

Increase natural gas use

Increase use of renewable energy resources

Burn low-sulfur coal

Remove SO_2 particulates, and NO_x from smokestack gases

Remove NO_x from motor vehicular exhaust

Tax emissions of SO_2

Cleanup

Add lime to neutralize acidified lakes

Add phosphate fertilizer to neutralize acidified lakes

Figure 12-12 Solutions: methods for reducing acid deposition.

- It can kill some types of plankton and aquatic plants and can harm wetland plants that need acidic water.
- It is difficult to know how much of the lime to put where (in the water or at selected places on the ground).

Recently researchers in England found that adding a small amount of phosphate fertilizer can neutralize excess acidity in a lake. However, the effectiveness of this approach is still being evaluated.

12-5 INDOOR AIR POLLUTION

What Are the Types and Sources of Indoor Air Pollution? If you are reading this book indoors, you may be inhaling more air pollutants with each breath than if you were outside (Figure 12-13). According to U.S. Environmental Protection Agency (EPA) studies, in the United States

- Levels of 11 common pollutants generally are **(1)** two to five times higher inside homes and commercial buildings than outdoors and **(2)** as much as 100 times higher in some cases.
- Levels of fine particles (Figure 12-6), which can contain toxins and metals such as lead and cadmium, can be as much as 60% higher indoors than outdoors.
- Concentrations of several pesticides (such as chlordane), approved for outdoor use only, were 10 times greater inside than outside monitored homes (some coming from pesticide dust tracked in on shoes).
- Pollution levels inside cars in traffic-clogged U.S. urban areas can be up to 18 times higher than those outside the vehicles.

The health risks from exposure to such chemicals are magnified because people typically spend 70–98% of their time indoors or inside vehicles. In 1990, the EPA placed indoor air pollution at the top of the list of 18 sources of cancer risk, and it is rated by risk analysis scientists as a high-risk health problem for humans (Figure 10-12, left, p. 238).

According to the EPA, more than 3,000 cases of cancer per year in the United States may be caused by exposure to indoor air pollutants. At greatest risk are **(1)** smokers, **(2)** infants and children under age 5, **(3)** the old, **(4)** the sick, **(5)** pregnant women, **(6)** people with respiratory or heart problems, and **(7)** factory workers.

Danish and U.S. EPA studies have linked pollutants found in buildings to dizziness, headaches, coughing, sneezing, nausea, burning eyes, chronic fatigue, and flulike symptoms, known as the *sick building syndrome*. New buildings are more commonly "sick" than old ones because of **(1)** reduced air exchange (to save energy) and **(2)** chemicals released from new carpeting and furniture. According to the EPA, at least 17% of the 4 million commercial buildings in the United States are considered "sick" (including EPA headquarters). Indoor air pollution in the United States costs an estimated $100 billion per year in **(1)** absenteeism, **(2)** reduced productivity, and **(3)** health-care costs. Mostly because of differences in genetic makeup, some individuals can be acutely sensitive to one or a number of indoor air pollutants (Figure 10-3, p. 226).

According to the EPA and public health officials, the three most dangerous indoor air pollutants are **(1)** cigarette smoke (p. 224), **(2)** formaldehyde, and **(3)** radioactive radon-222 gas. A 2001 EPA study showed that burning candles with multiple wicks or a number of candles can lead to unsafe indoor levels of particulates and toxic lead from imported candles with lead cores.

Worker exposure to asbestos fibers in mines and in factories making asbestos material also is a serious indoor air pollution problem, especially in developing countries. A number of research studies on laboratory animals have also identified tiny fibers of *fiberglass* as a widespread and potentially potent carcinogen in indoor air.

The chemical that causes most people difficulty is *formaldehyde*, a colorless, extremely irritating gas widely used to manufacture common household materials. As many as 20 million Americans suffer from chronic breathing problems, dizziness, rash, headaches, sore throat, sinus and eye irritation, wheezing, and nausea caused by daily exposure to low levels of formaldehyde

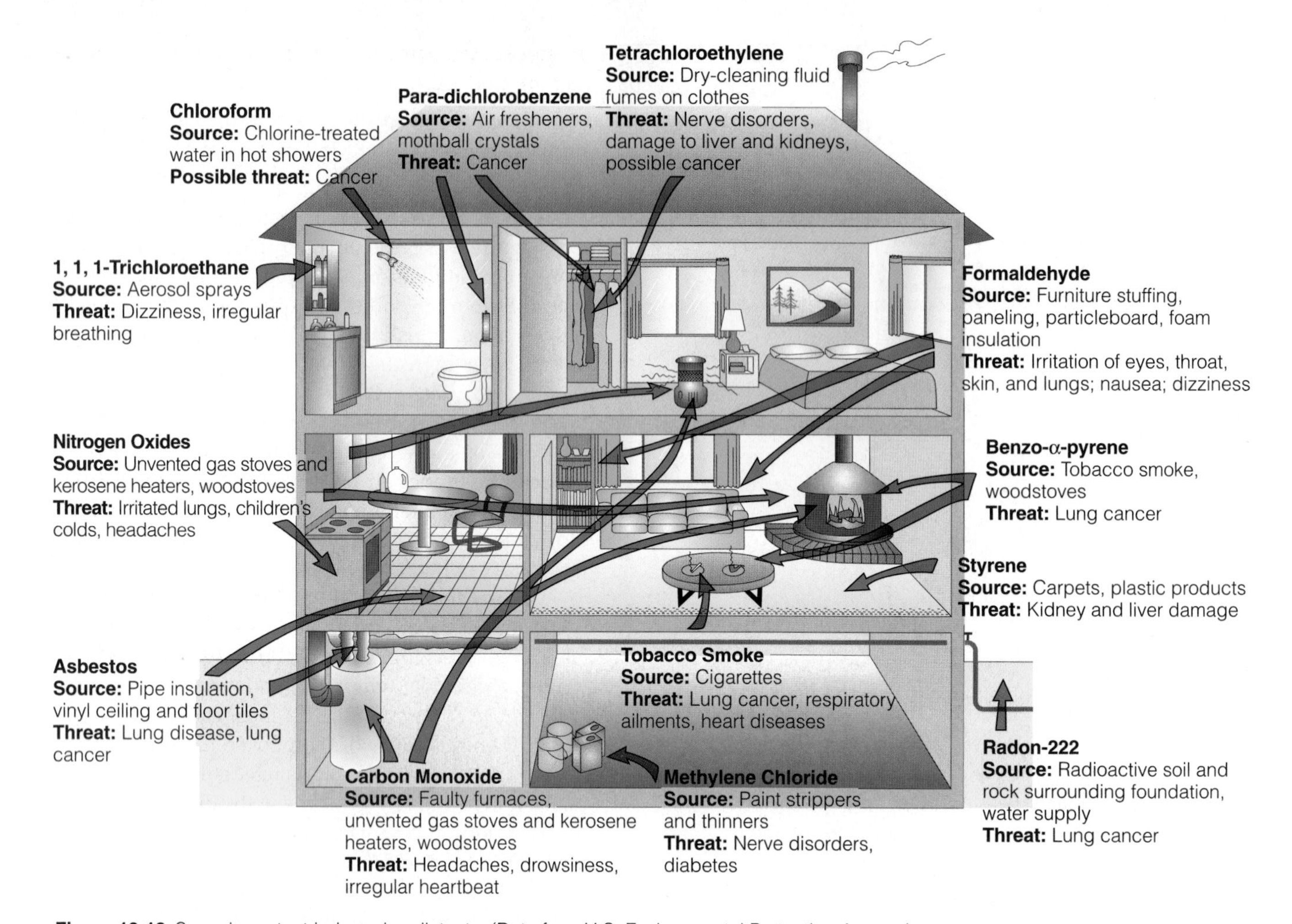

Figure 12-13 Some important indoor air pollutants. (Data from U.S. Environmental Protection Agency)

emitted (outgassed) from common household materials. These include **(1)** building materials (such as plywood, particleboard, paneling, and high-gloss wood used in floors and cabinets), **(2)** furniture, **(3)** drapes, **(4)** upholstery, **(5)** adhesives in carpeting and wallpaper, **(6)** urethane–formaldehyde insulation, **(7)** fingernail hardener, and **(8)** wrinkle-free coating on permanent-press clothing (Figure 12-13). The EPA estimates that as many as 1 of every 5,000 people who live in manufactured homes for more than 10 years will develop cancer from formaldehyde exposure.

In developing countries, the burning of wood, dung, crop residues, and coal in open fires or in unvented or poorly vented stoves for cooking and heating exposes inhabitants (especially women and young children) to very high levels of particulate air pollution.

Case Study: Is Your Home Contaminated with Radon Gas? Radon-222 is naturally occurring radioactive gas that you cannot see, taste, or smell. It is produced by the radioactive decay of uranium-238. Small amounts of uranium-238 are found in most soil and rock, but this isotope is much more concentrated in underground deposits of minerals such as uranium, phosphate, granite, and shale.

When radon gas from such deposits seeps upward through the soil and is released outdoors, it disperses quickly in the atmosphere and decays to harmless levels. However, radon gas can enter buildings above such deposits through **(1)** cracks in foundations and walls, **(2)** openings around sump pumps and drains, and **(3)** hollow concrete blocks (Figure 12-14) and build up to high levels, especially in unventilated lower levels of homes and buildings.

Radon-222 gas quickly decays into solid particles of other radioactive elements that, if inhaled, expose lung tissue to a large amount of ionizing radiation (Figure 3-8, p. 66) from alpha particles. This can damage lung tissue and lead to lung cancer over the course of a 70-year lifetime. Your chances of getting lung cancer from radon depend mostly on **(1)** how much radon is in your home, **(2)** how much time you spend in your home, and **(3)** whether you are a smoker or have ever smoked.

In 1998, the National Academy of Science estimated that prolonged exposure for a lifetime of 70 years to low levels of radon or radon acting together with

Figure 12-14 Sources and paths of entry for indoor radon-222 gas. (Data from U.S. Environmental Protection Agency)

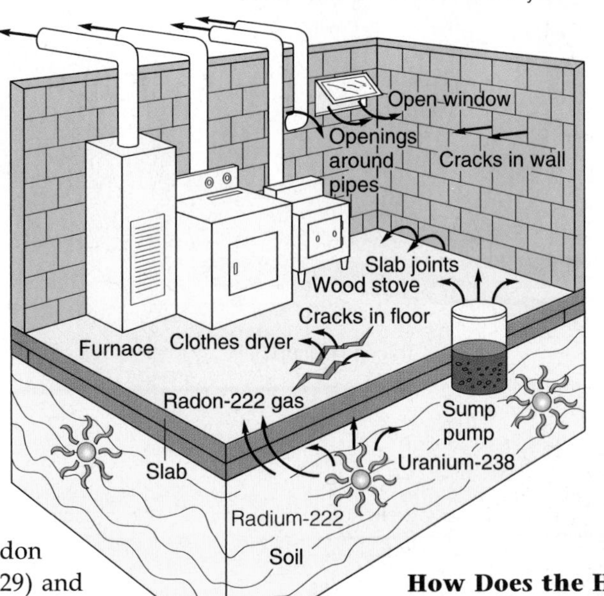

smoking is responsible for 15,000–22,000 (or 12%) of the lung cancer deaths each year in the United States. This makes radon the second leading cause of lung cancer after smoking (p. 224). Most of the deaths are among smokers or former smokers, with about 2,100 to 2,900 among nonsmokers.

These estimates are based on the assumptions that **(1)** there is no safe threshold dose for radon exposure (Figure 10-6, left, p. 229) and that **(2)** the incidence of lung cancer in uranium miners exposed to high levels of radon in mines can be extrapolated to estimate lung cancer deaths for people in homes exposed to much lower levels of radon. Some scientists question these assumptions and say that these estimates are too high. They also point to the contradictory results of several epidemiological studies on the risks of lung cancer from radon exposure.

EPA indoor radon surveys suggest that 4–5 million U.S. homes may have annual radon levels above 4 picocuries* per liter of air and that 50,000–100,000 homes may have levels above 20 picocuries per liter. If the 4 picocuries per liter standard is adopted (as proposed by the EPA), the cost of testing and correcting the problem could run about $50 billion, with a 15–20% reduction in radon-related deaths. Some researchers argue that it makes more sense to spend perhaps only $500 million to find and fix homes and buildings with radon levels above 20 picocuries per liter until more reliable data are available on the threat from exposure to lower levels of radon.

Because radon hot spots can occur almost anywhere, it is impossible to know which buildings have unsafe levels of radon without conducting tests. In 1988, the EPA and the U.S. Surgeon General's Office recommended that everyone living in a detached house, a town house, a mobile home, or on the first three floors of an apartment building test for radon. Ideally, radon levels should be monitored continuously in the main living areas (not basements or crawl spaces) for 2 months to a year. By 2001, only about 6% of U.S. households had conducted radon tests (most lasting only 2 to 7 days and costing $20–100 per home).

If testing reveals an unacceptable level, homeowners can consult the free EPA publication *Radon Reduction Methods* for ways to reduce radon levels and health risks. According to the EPA, radon control could add $350–500 to the cost of a new home, and correcting a radon problem in an existing house could run $800–2,500.

12-6 EFFECTS OF AIR POLLUTION ON LIVING ORGANISMS AND MATERIALS

How Does the Human Respiratory System Help Protect Us from Air Pollution? Table 12-2 (p. 281) listed the major health effects from the six most common (criteria) outdoor air pollutants. Your respiratory system (Figure 12-15) has a number of mechanisms that help protect you from the harmful effects of such air pollutants. They include

- Hairs in your nose, which filter out large particles.
- Sticky mucus in the lining of your upper respiratory tract, which captures smaller (but not the smallest) particles and dissolves some gaseous pollutants.
- Sneezing and coughing, which expel contaminated air and mucus when pollutants irritate your respiratory system.
- Hundreds of thousands of tiny, mucus-coated hairlike structures called *cilia,* which line your upper respiratory tract. They continually wave back and forth and transport mucus and the pollutants they trap to your throat (where they are swallowed or expelled).

Years of smoking and exposure to air pollutants can overload or break down these natural defenses. This can cause or contribute to respiratory diseases such as **(1)** *lung cancer,* **(2)** *asthma* (typically an allergic reaction causing sudden episodes of muscle spasms in the bronchial walls, resulting in acute shortness of breath), **(3)** *chronic bronchitis* (persistent inflammation and damage to the cells lining the bronchi and bronchioles, causing mucus buildup, painful coughing, and shortness of breath), and **(4)** *emphysema* (irreversible damage to air sacs or alveoli leading to abnormal dilation of air spaces, loss of lung elasticity, and acute shortness of breath (Figure 12-16). Older adults, infants, pregnant women, and people with heart disease, asthma, or other respiratory diseases are especially vulnerable to air pollution.

*A *picocurie* is a trillionth of a curie, which is the amount of radioactivity emitted by a gram of radium.

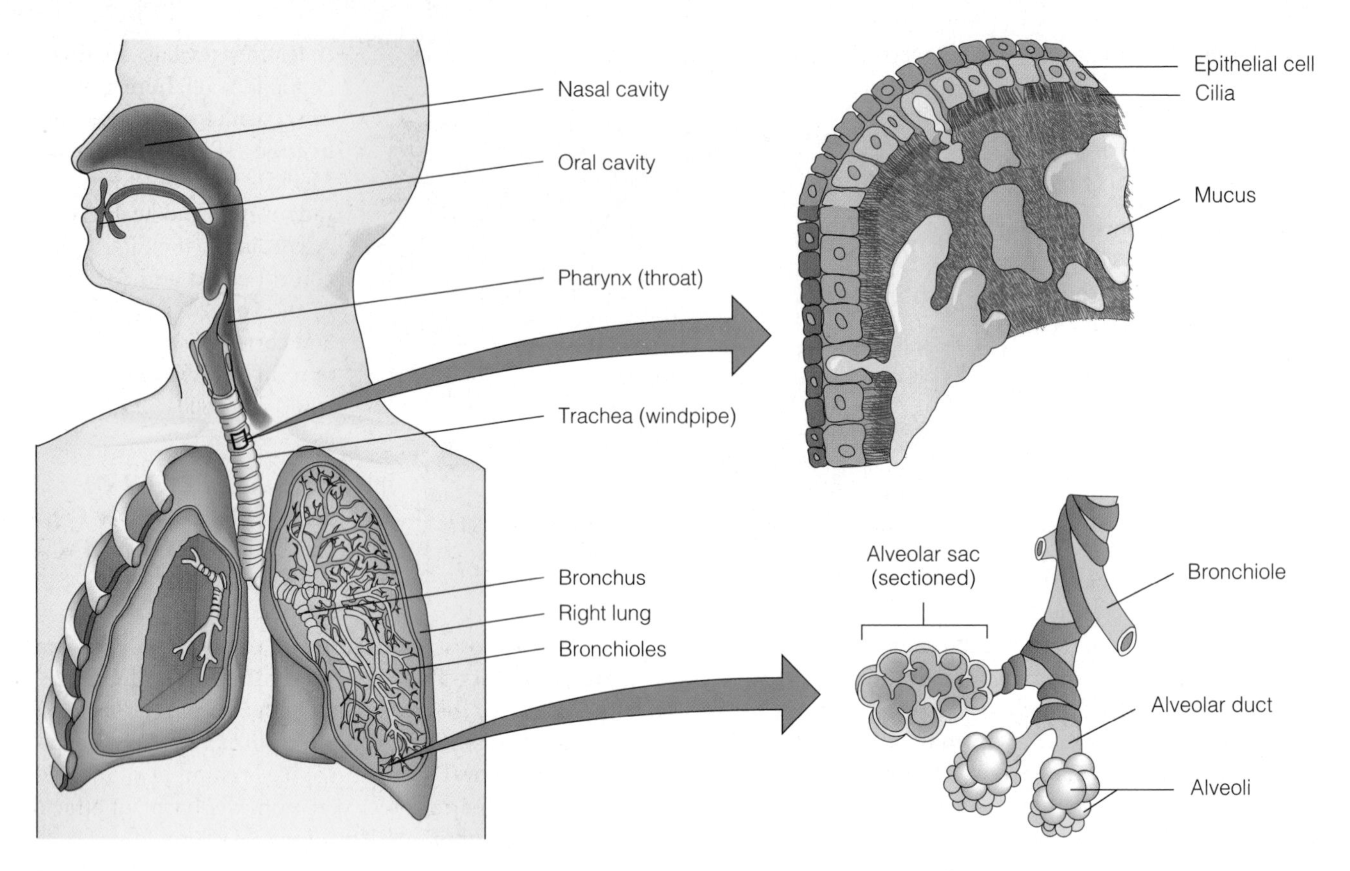

Figure 12-15 Major components of the human respiratory system.

How Many People Die Prematurely from Air Pollution? It is difficult to estimate by risk analysis how many people die prematurely from respiratory or cardiac problems caused or aggravated by air pollution because people are exposed to so many different pollutants over their lifetimes.

In the United States, estimates of annual deaths related to outdoor air pollution range from 65,000 to 200,000 (most from exposure to fine particles; Spotlight, p. 294). If indoor air pollution is included, estimated annual deaths from air pollution in the United States range from 150,000 to 350,000 people—equivalent to 1–2 fully loaded 400-passenger jumbo jets accidentally crashing *each day* with no survivors.

Millions more become ill and lose work time. A 2000 study by state and local air pollution officials estimated that each year more than 125,000 Americans (120,000 of them in urban areas) get cancer from breathing diesel fumes from buses, trucks, and other diesel engines. According to the EPA and the American Lung Association, air pollution in the United States costs at least $150 billion annually in health care and lost work productivity, with $100 billion of that caused by indoor air pollution.

According to a 1999 study by Australia's Commonwealth Science Council, worldwide at least 3 million people (most of them in Asia) die prematurely each year from the effects of air pollution, an average of 8,200 deaths per day. About 2.8 million of these deaths are from *indoor* air pollution, and 200,000 are from *outdoor* pollution. Most people who live in large cities in developing countries breathe air that is the equivalent of smoking 2–3 packs of cigarettes a day.

How Are Plants Damaged by Air Pollutants? The effects of exposure of trees to a combination of air pollutants may not become visible for several decades, when large numbers of trees suddenly begin dying because of **(1)** soil nutrient depletion and **(2)** increased susceptibility to pests, diseases, fungi, and drought. This phenomenon, known as *Waldsterben* (forest death), has turned whole forests of spruce, fir, and beech into stump-studded meadows (see photo on p. 243) and mountainsides. It is estimated that air pollution has been a key factor in reducing the overall productivity of European forests by about 16% and causing damage valued at roughly $30 billion per year.

Forest diebacks have also occurred in the United States. The most seriously affected areas are high-elevation spruce trees that populate the ridges of the Appalachian Mountains from Maine to Georgia, including the Shenandoah and Great Smoky Mountain national parks.

Air pollution, mostly by ozone, also **(1)** threatens some crops (especially corn, wheat, and soybeans) and **(2)** reduces U.S. food production by 5–10%. In the United States, estimates of agricultural losses caused by air pollution (mostly by ozone) range from \$2 to \$6 billion per year, with an estimated \$1 billion of damages in California alone.

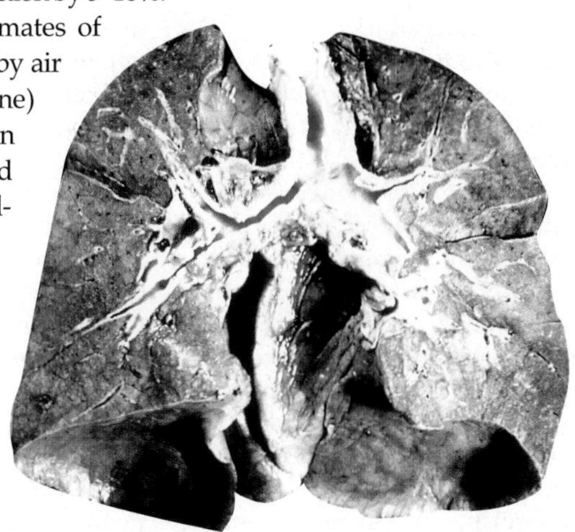

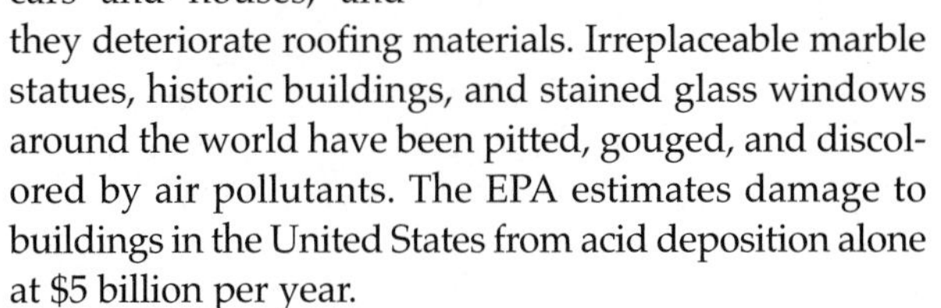

Figure 12-16 Normal human lungs (left) and the lungs of a person who died of emphysema (right). Prolonged smoking and exposure to air pollutants can cause emphysema in anyone, but about 2% of emphysema cases result from a defective gene that reduces the elasticity of the air sacs in the lungs. Anyone with this hereditary condition, for which testing is available, should not smoke and should not live or work in a highly polluted area. (O. Auerbach/Visuals Unlimited)

What Are the Harmful Effects of Air Pollutants on Materials? Each year, air pollutants cause billions of dollars in damage to various materials we use (Table 12-3). The fallout of soot and grit on buildings, cars, and clothing necessitates costly cleaning. Air pollutants break down exterior paint on cars and houses, and they deteriorate roofing materials. Irreplaceable marble statues, historic buildings, and stained glass windows around the world have been pitted, gouged, and discolored by air pollutants. The EPA estimates damage to buildings in the United States from acid deposition alone at \$5 billion per year.

12-7 SOLUTIONS: PREVENTING AND REDUCING AIR POLLUTION

How Have Laws Been Used to Reduce Air Pollution in the United States? The U.S. Congress passed Clean Air Acts in 1970, 1977, and 1990. With these laws the federal government establishes air pollution regulations that are enforced by each state and by major cities.

Congress directed the EPA to establish *national ambient air quality standards (NAAQS)* for six outdoor criteria pollutants (Table 12-2). Each standard specifies the maximum allowable level, averaged over a specific period, for a certain pollutant in outdoor (ambient) air. The EPA has also established national emission standards for more than 100 different toxic air pollutants that are known to cause or suspected of causing cancer or other adverse health effects.

Table 12-3 Harmful Effects of Air Pollution on Materials

Material	Effects	Principal Air Pollutants
Stone and concrete	Surface erosion, discoloration, soiling	Sulfur dioxide, sulfuric acid, nitric acid, particulate matter
Metals	Corrosion, tarnishing, loss of strength	Sulfur dioxide, sulfuric acid, nitric acid, particulate matter, hydrogen sulfide
Ceramics and glass	Surface erosion	Hydrogen fluoride, particulate matter
Paints	Surface erosion, discoloration, soiling	Sulfur dioxide, hydrogen sulfide, ozone, particulate matter
Paper	Embrittlement, discoloration	Sulfur dioxide
Rubber	Cracking, loss of strength	Ozone
Leather	Surface deterioration, loss of strength	Sulfur dioxide
Textiles	Deterioration, fading, soiling	Sulfur dioxide, nitrogen dioxide, ozone, particulate matter

Health Dangers from Fine Particles

SPOTLIGHT

Research indicates that invisible particles—especially *fine particles* with diameters less than 10 microns (PM-10) and *ultrafine particles* with diameters less than 2.5 microns (PM-2.5)—pose a significant health hazard. Such particles are emitted by incinerators, motor vehicles, radial tires, wind erosion, wood-burning fireplaces, and power and industrial plants (Figure 12-6).

Such tiny particles **(1)** are not effectively captured by modern air-pollution control equipment (Figure 12-18, p. 296), **(2)** are small enough to penetrate the respiratory system's natural defenses against air pollution (Figure 12-15), and **(3)** can bring with them droplets or other particles of toxic or cancer-causing pollutants that become attached to their surfaces.

Once they are lodged deep within the lungs, these fine particles cause chronic irritation that can **(1)** trigger asthma attacks, **(2)** aggravate other lung diseases, **(3)** cause lung cancer, and **(4)** interfere with the blood's ability to take in oxygen and release CO_2. This strains the heart, increasing the risk of death from heart disease.

Several recent studies of air pollution in U.S. cities have indicated that fine and ultrafine particles prematurely kill 65,000–200,000 Americans each year. There is no known threshold level below which the harmful effects of fine particles disappear.

Exposure to particulate air pollution is much worse in most developing countries, where urban air quality has generally deteriorated. The World Bank estimates that if particulate levels were reduced globally to WHO guidelines, 300,000–700,000 premature deaths per year could be prevented.

In 1997, the EPA announced stricter emission standards for ultrafine particles with diameters less than 2.5 microns (PM-2.5). The EPA estimates the cost of implementing the standards at $7 billion per year, with the resulting health and other benefits estimated at $120 billion per year.

According to industry officials, the new standard is based on flimsy scientific evidence, and its implementation will cost $200 billion per year. EPA officials say that their review of the scientific evidence—one of the most exhaustive ever undertaken by the agency—supports the need for the new standard for ultrafine particles. Furthermore, a 2000 study by the Health Effects Institute of 90 large American cities confirmed the link between fine and ultrafine particles and higher rates of death and disease.

Critical Thinking

Are you for or against the stricter standard for emissions of ultrafine particles? Explain. Use the library or internet to determine whether stricter standards for ultrafine particles have been implemented.

Here is some *good news.* According to the EPA,

- Between 1970 and 1998, national total emissions of the six criteria pollutants declined 31%, while U.S. population increased 31%, gross domestic product increased 114%, and vehicle miles traveled rose 127%.
- Between 1978 and 1998, mean concentrations of the six criteria air pollutants in the troposphere decreased by **(1)** 97% for lead, **(2)** 60% for carbon monoxide, **(3)** 58% for sulfur dioxide, **(4)** 30% for ground-level ozone, **(5)** 25% for suspended particulate matter (10 micrometers or less in diameter), and **(6)** 2% for nitrogen dioxide.
- The mean estimated human health and environmental benefits from air pollution regulations between 1970 and 1990 amounted to $6.8 trillion, compared to $436 million (in 1990 dollars) spent to implement all federal, state, and local air pollution regulations. Thus, the net economic benefit of the Clean Air Act between 1970 and 1990 was $6.4 trillion. According to the EPA, during this 20-year period, the act prevented an estimated 1.6 million premature deaths and 300 million cases of respiratory disease.

Here is some *bad news.*

- Between 1970 and 1998, emissions of nitrogen oxides (NO_x) increased 11%.
- Despite continued improvements in air quality, in 1999 approximately 62 million people lived in 130 areas with air that did not meet the primary standards for one or more of the six criteria pollutants.
- According to a 2001 study by the American Lung Association, some 141 million Americans—about half of the nation's population—live in communities with dangerously high smog levels.

How Can U.S. Air Pollution Laws Be Improved? The Clean Air Act of 1990 was an important step in the right direction, but many environmentalists point to the following deficiencies in this law:

- *Continuing to rely mostly on pollution cleanup rather than prevention.* In the United States, the air pollutant with the largest drop (97% between 1970 and 1998) in its atmospheric level was lead, which was largely banned in gasoline.

- *Failing to increase fuel efficiency standards for cars and light trucks.* According to environmental scientists, this would **(1)** reduce air pollution more quickly and effectively than any other method, **(2)** reduce CO_2 emissions, **(3)** save energy (Section 20-1, p. 514), and **(4)** save consumers enormous amounts of money.

- *Not adequately regulating emissions from inefficient, two-cycle gasoline engines.* These engines are used in devices such as lawnmowers, leaf blowers, chain saws, and personal marine engines used to power jet skis, outboard motors, and personal watercraft. According to recent studies by the California Air Resources Board, **(1)** a 1-hour ride on a typical jet ski creates more air pollution than the average U.S. car does in a year, **(2)** operating a 100-horsepower marine engine for 7 hours emits more air pollutants than a new car driven 160,000 kilometers (100,000 miles), and **(3)** each year the fuel and oil spilled by the 14 million small marine engines in the United States is 15 times the amount spilled in 1989 by the *Exxon Valdez* oil tanker in Prince William Sound near Alaska.

- *Doing too little to reduce emissions of carbon dioxide and other greenhouse gases* (Section 13-2, p. 303).

Executives of companies affected by such policies strongly oppose such changes in air pollution laws. They claim that implementing such changes would cost too much, harm economic growth, and cost jobs.

Proponents contend that history has shown that almost all industry estimates of implementing various air pollution control standards in the United States were many times the actual cost of implementation. In addition, implementing such standards has helped increase economic growth and create jobs by stimulating companies to develop new technologies for reducing air pollution emissions. Many of these technologies are sold in the international marketplace.

Should We Use the Marketplace to Reduce Pollution? To help reduce SO_2 emissions, the Clean Air Act of 1990 allows an *emissions trading policy,* which enables the 110 most polluting power plants in 21 states (primarily in the Midwest and East) to buy and sell SO_2 pollution rights.

Each year a power plant is given a specific number of pollution credits or rights that allow it to emit a certain amount of SO_2. A utility that emits less SO_2 than its limit receives more pollution credits. It can use these credits **(1)** to avoid reductions in SO_2 emissions from some of its other facilities, **(2)** bank them for future plant expansions, or **(3)** sell them to other utilities, private citizens, or environmental groups. Proponents of this system argue that it allows the marketplace to determine the cheapest, most efficient way to get the job done instead of having the government dictate how to control pollution.

Some environmentalists see this market approach as an improvement over the regulatory approach, as long as it achieves a net reduction in SO_2 pollution. This would be done by limiting the total number of credits and gradually lowering the annual number of credits, something that is not required by the 1990 amendments to the Clean Air Act.

Some environmentalists contend that marketing pollution rights

- Allows utilities with older, dirtier power plants to buy their way out and keep on emitting unacceptable levels of SO_2

- Creates incentives to cheat because air quality regulation is based largely on self-reporting of emissions, and pollution monitoring is incomplete and imprecise

Here is some *good news*. Between 1994 and 1997, the emission trading system helped reduce SO_2 emissions in the United States by 30%. The cost of doing this was less than one-tenth the cost projected by industry because this market-based system motivated companies to reduce emissions in more efficient ways.

In 1997, the EPA proposed a voluntary emissions trading program involving smog-forming nitrogen oxides (NO_x) for 22 eastern states and the District of Columbia. Emissions trading may also be implemented for particulate emissions and volatile organic compounds.

How Can We Reduce Outdoor Air Pollution? Figure 12-17 summarizes ways to reduce emissions of sulfur oxides, nitrogen oxides, and particulate matter from stationary sources (such as electric power plants and industrial plants that burn coal). Until recently, emphasis has been on **(1)** dispersing and diluting the

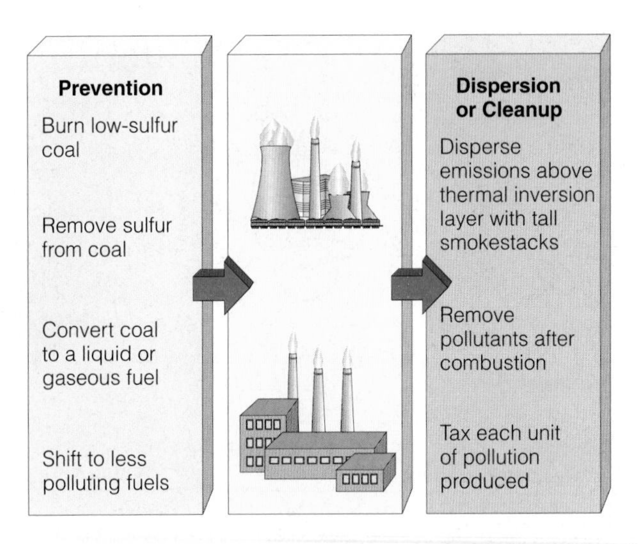

Figure 12-17 *Solutions*: methods for reducing emissions of sulfur oxides, nitrogen oxides, and particulate matter from stationary sources such as coal-burning electric power plants and industrial plants.

pollutants by using tall smokestacks or **(2)** adding equipment that removes some of the particulate pollutants after they are produced (Figure 12-18). However, under the sulfur reduction requirements of the 1990 amendments to the Clean Air Act, more utilities are switching to low-sulfur coal to reduce SO_2 emissions. Environmentalists call for taxes on air pollutant emissions and greater emphasis on prevention.

Figure 12-19 lists ways to reduce emissions from motor vehicles, the primary culprits in producing photochemical smog. Use of alternative vehicle fuels to reduce air pollution is evaluated in Table 12-4.

An important way to make significant reductions in air pollution is to get older, high-polluting vehicles off the road. According to EPA estimates, 10% of the vehicles on the road in the United States emit 50–70% of the pollutants. A problem is that people who cannot afford to buy a newer car often own old cars. One suggestion would be to pay people to take their old cars off the road, which would result in huge savings in health and air-pollution control costs.

Here is some *good news*. Since the 1960s, Tokyo, Japan (the world's most populous city, with a population of about 28 million), has implemented a strict air-pollution control program that has sharply reduced levels of sulfur dioxide, carbon monoxide, and ozone. During the past 30 years outdoor air quality in most western European cities has also improved. The *bad*

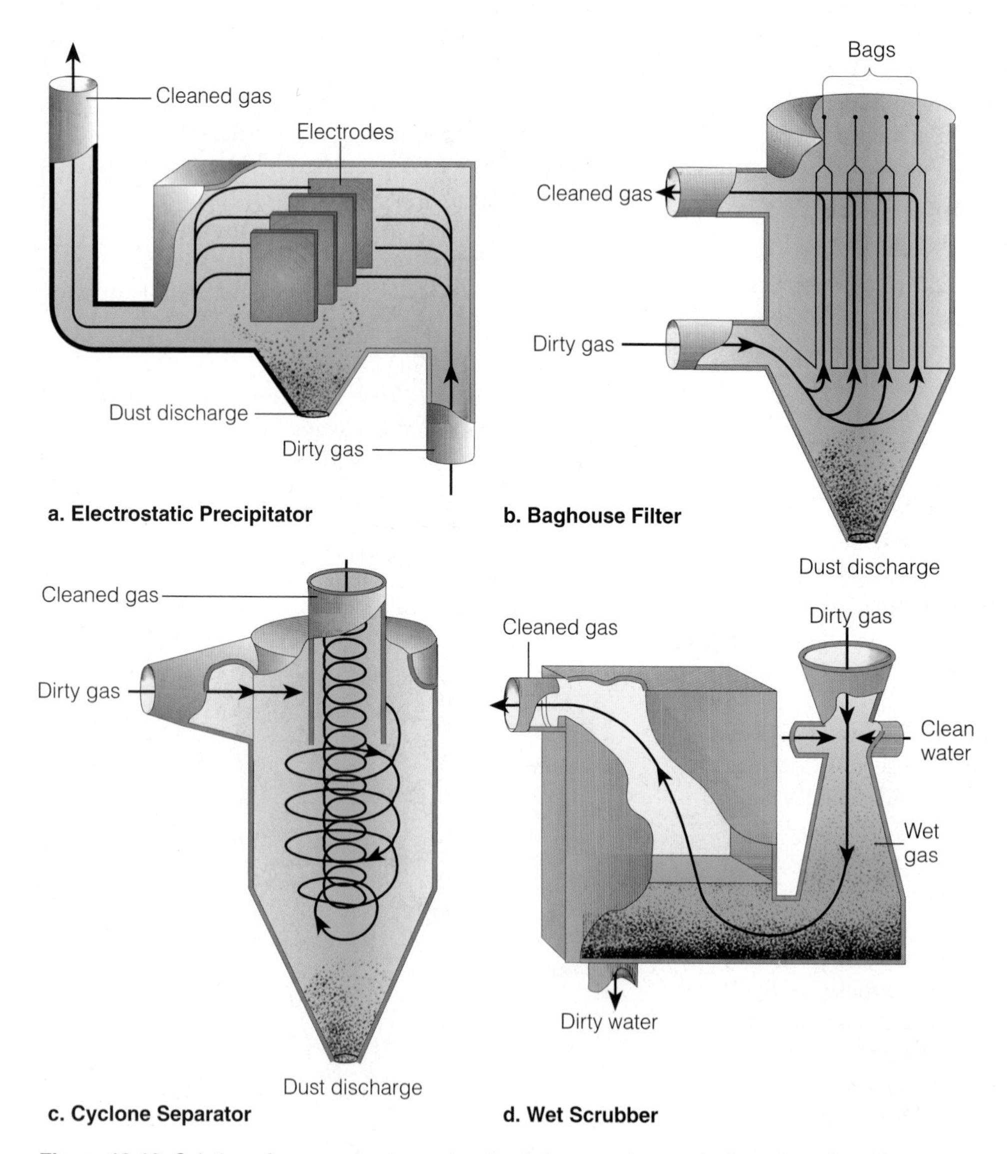

Figure 12-18 *Solutions*: four commonly used methods for removing particulates from the exhaust gases of electric power and industrial plants. Of these, only baghouse filters remove many of the more hazardous fine particles. All these methods produce hazardous materials that must be disposed of safely, and except for cyclone separators, all of them are expensive. The wet scrubber can also reduce sulfur dioxide emissions.

Table 12-4 Evaluation of Alternatives to Gasoline

Advantages	Disadvantages
Compressed Natural Gas	
Fairly abundant, inexpensive domestic and global supplies Low hydrocarbon, CO, and CO_2 emissions Vehicle development advanced; well suited for fleet vehicles Reduced engine maintenance	Large fuel tank needed; one-fourth the range Expensive engine modification needed ($2,000) New filling stations needed Nonrenewable resource
Electricity	
Renewable if not generated from fossil fuels or nuclear power Zero vehicle emissions Electric grid in place Efficient and quiet	Limited range and power Batteries expensive Slow refueling (6–8 hours) Power plant emissions if generated from coal or oil
Reformulated Gasoline (Oxygenated Fuel)	
No new filling stations needed Low to moderate CO emissions reduction No engine modification needed	Nonrenewable resource Dependence on imported oil perpetuated No CO_2 emission reduction Higher cost Groundwater contaminated by leakage and spills (especially by MTBE, a possible human carcinogen) No longer needed because of improved emission control system
A-55 (55% water, 45% naphtha)	
Can be sold in conventional filling station Much lower emissions of nitrogen oxide and particulates than diesel fuel Cannot explode or catch fire Lower cost (25–50%) Naphtha produces 90% less pollution at refineries than gasoline or diesel fuel Low-cost engine modification ($300 for cars, $1,000 for trucks and buses) Modified engine can run A-55, gasoline, or diesel	Not yet widely available Independent tests needed to verify pollution reduction claims Refineries may limit supply or drive up price of less-profitable naphtha Large amounts of water needed to produce
Methanol	
High octane Reduction of CO_2 emissions (total amount depends on method of production) Reduced total air pollution (30–40%)	Large fuel tank needed; one-half the range Corrosive to metal, rubber, plastic Increased emissions of potentially carcinogenic formaldehyde High CO_2 emissions if generated by coal High capital cost to produce Hard to start in cold weather
Ethanol	
High octane Reduction of CO_2 emissions (total amount depends on distillation process and efficiency of crop growing) Reduction of CO emissions Potentially renewable	Large fuel tank needed; lower range Much higher cost Corn supply limited Competition with food growing for cropland Smog formation possible Corrosive Hard to start in cold weather
Solar–Hydrogen	
Renewable if produced using solar energy Lower flammability than gasoline Virtually emission-free No emissions of CO_2 Nontoxic	Nonrenewable if generated by fossil fuels or nuclear power Large fuel tank needed No distribution system in place Engine redesign needed Currently expensive

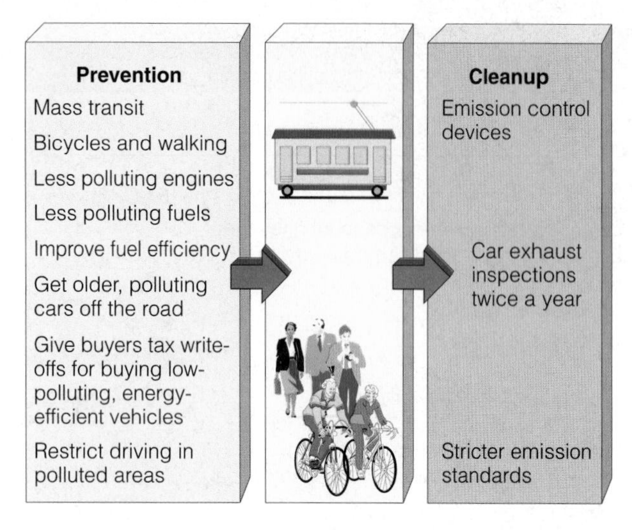

Figure 12-19 Solutions: methods for reducing emissions from motor vehicles.

news is that outdoor air quality has remained about the same or has gotten worse in most rapidly growing urban areas in developing countries.

How Can We Reduce Indoor Air Pollution? In the United States indoor air pollution poses a much greater health risk for many people than outdoor air pollution. Yet the EPA spends about $500 million per year fighting outdoor air pollution and only about $13 million a year on indoor air pollution.

To reduce indoor air pollution, it is not necessary to impose indoor air quality standards and monitor the more than 100 million homes and buildings in the United States. Instead, air pollution experts suggest that indoor air pollution can be reduced by several means (Figure 12-20). Another possibility for cleaner indoor air in high-rise buildings is rooftop greenhouses through which building air can be circulated.

In developing countries, indoor air pollution from open fires and leaky and inefficient stoves that burn wood, charcoal, or coal (and the resulting high levels of respiratory illnesses) could be reduced if governments **(1)** gave people simple stoves that burn biofuels more efficiently (which would also reduce deforestation) and are vented outside or **(2)** provided them with simple solar cookers (Figure 20-16d, p. 526).

How Can We Protect the Atmosphere? An Integrated Approach Environmentalists believe that protecting the atmosphere, and thus the health of people and many other organisms, will take a global approach that integrates many different strategies. Suggestions for doing this over the next 40–50 years include the following:

- *Putting more emphasis on pollution prevention*

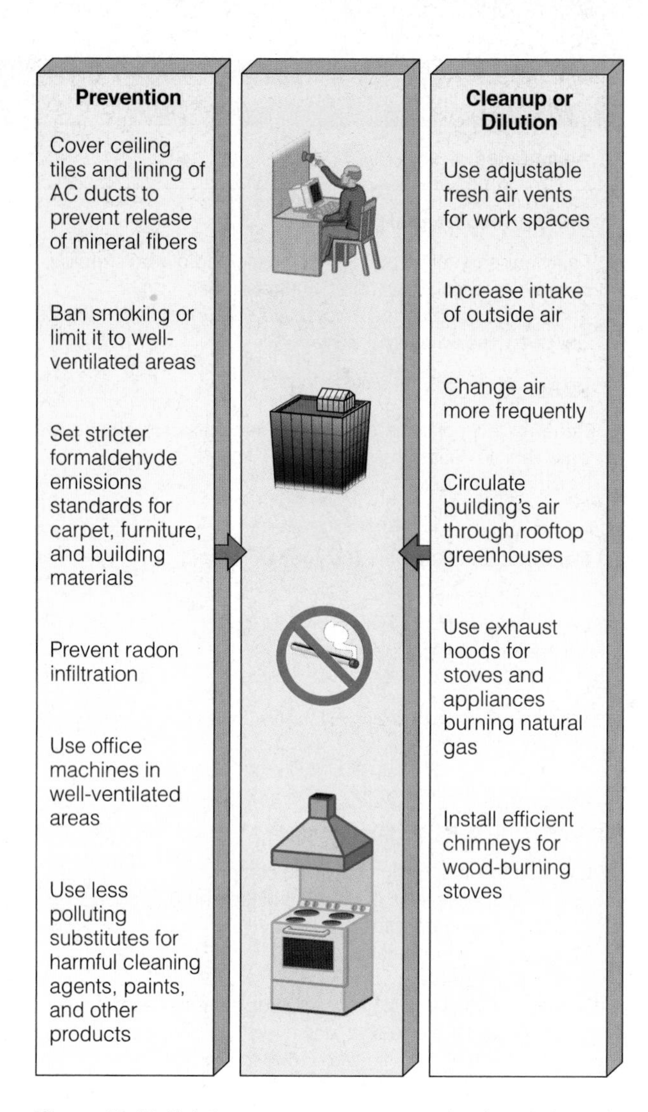

Figure 12-20 Solutions: ways to prevent and reduce indoor air pollution.

- *Improving energy efficiency*
- *Reducing use of fossil fuels (especially coal and oil)*
- *Increasing use of renewable energy*
- *Slowing population growth*
- *Integrating air pollution, water pollution, energy, land-use, population, economic, and trade policies*
- *Regulating air quality for an entire region or airshed*
- *Taxing the production of air pollutants and using the revenue to reduce taxes on income and wealth* (Solutions, p. 000)
- *Distributing cheap and efficient cookstoves and solar cookstoves in developing countries*
- *Transferring the latest energy-efficiency, renewable-energy, pollution prevention, and pollution control technologies to developing countries*

Making such changes will be controversial and expensive. However, proponents argue that not implementing such an integrated approach will cost far more in money, poor human health, premature death, and ecological damage. See the website material for this chapter for ways you can reduce your exposure to indoor and outdoor pollutants.

Turning the corner on air pollution requires moving beyond patchwork, end-of-pipe approaches to confront pollution at its sources. This will mean reorienting energy, transportation, and industrial structures toward prevention.

HILARY F. FRENCH

REVIEW QUESTIONS

1. Define the boldfaced terms in this chapter.

2. How can lichens be used to detect air pollutants?

3. Briefly describe the history of air pollution in Europe and the United States.

4. Distinguish between *atmosphere, troposphere,* and *stratosphere.* What key role does the stratosphere play in maintaining life on the earth?

5. Distinguish between *air pollution, primary air pollutants,* and *secondary air pollutants.* List the major classes of pollutants found in outdoor air. Distinguish between *stationary* and *mobile* sources of pollution for outdoor air. What are the two major sources of indoor air pollution?

6. List the six *criteria* air pollutants regulated in the United States (and in most developed countries). For each of these pollutants, summarize its major human sources and health effects.

7. What is *photochemical smog,* and how does it form? What is *industrial smog,* and how does it form?

8. List major factors that can **(a)** reduce air pollution and **(b)** increase air pollution. What is a *temperature inversion,* and what are its harmful effects? Distinguish between a *subsidence temperature inversion* and a *radiation temperature inversion.* What types of places are most likely to suffer from prolonged inversions of each type?

9. What is *acid deposition,* and what are its major components and causes? Distinguish between *acid deposition, wet deposition,* and *dry deposition.* What areas tend to be affected by acid deposition? What two types of areas are most sensitive to acid deposition?

10. What are the major harmful effects of acid deposition on **(a)** human health, **(b)** materials, **(c)** soils, **(d)** aquatic life, **(e)** trees and other plants, and **(f)** some forms of wildlife?

11. List eight ways to prevent acid deposition. What are the advantages and disadvantages of liming acidified lakes to reduce the effects of acid deposition?

12. How serious is indoor air pollution, and what are some of its sources? What is the *sick building syndrome*? According to the EPA, what are the three most dangerous indoor air pollutants in the United States? What is the most dangerous indoor air pollutant in most developing countries?

13. Summarize the problems of indoor pollution from **(a)** formaldehyde and **(b)** radioactive radon gas.

14. List four defenses your body has against air pollution. Describe the health dangers from inhaling fine particles.

15. About how many people die prematurely each year from exposure to air pollutants in **(a)** the United States and **(b)** the world? What percentage of these deaths occurs in developing countries? What percentage of these deaths is the result of indoor air pollution?

16. What is the Clean Air Act, and how has it helped reduce outdoor air pollution in the United States?

17. Summarize the major *good news* and *bad news* about the effectiveness of the Clean Air Act in reducing outdoor air pollution in the United States. According to environmentalists, what are four weaknesses of the current Clean Air Act in the United States?

18. What is an *emission trading policy,* and what are the pros and cons of using this approach to help reduce air pollution?

19. List the major prevention and cleanup methods for dealing with air pollution from **(a)** emissions of sulfur oxides, nitrogen oxides, and particulate matter from stationary sources, **(b)** automobile emissions, **(c)** indoor air pollution in developed countries, and **(d)** indoor air pollution in developing countries.

20. What are 10 components of an integrated approach for dealing with air pollution?

CRITICAL THINKING

1. Evaluate the pros and cons of the following statement: "Because we have not proven absolutely that anyone has died or suffered serious disease from nitrogen oxides, current federal emission standards for this pollutant should be relaxed."

2. Identify climate and topographic factors in your local community that **(a)** intensify air pollution and **(b)** help reduce air pollution.

3. Should all tall smokestacks be banned? Explain.

4. Why are most severe air pollution episodes associated with subsidence temperature inversions rather than radiation temperature inversions?

5. Should annual government-held auctions of marketable trading permits be used as a primary way to control and reduce air pollution? Explain. What conditions, if any, would you put on this approach?

6. Do you agree or disagree with the four possible weaknesses of the U.S. Clean Air Act listed on pp. 294–295? Defend each of your choices. Can you identify other weaknesses?

7. Congratulations. You have been put in charge of reducing air pollution in the country where you live. List the three most important features of your policy.

PROJECTS

1. Evaluate your exposure to some or all of the indoor air pollutants in Figure 12-13 in your school, workplace, and home. Come up with a plan for reducing your exposure to these pollutants.

2. Have buildings at your school been tested for radon? If so, what were the results? What has been done about areas with unacceptable levels? If this testing has not been done, talk with school officials about having it done.

3. Make a concept map of this chapter's major ideas, using the section heads and subheads and the key terms (in boldface). Look on the website for this book for information about making concept maps.

INTERNET STUDY RESOURCES AND RESOURCES FOR FURTHER READING AND RESEARCH

The website for this book contains helpful study aids and many ideas for further reading and research. Log on to

http://www.brookscole.com/product/0534389872s

and click on the Chapter-by-Chapter area. Choose Chapter 12 and select a resource:

- "Flash Cards" allows you to test your mastery of the Terms and Concepts to Remember for this chapter.
- "Tutorial Quizzes" provides a multiple-choice practice quiz.
- "Student Guide to InfoTrac" will lead you to Critical Thinking Projects that use InfoTrac College Edition as a research tool.
- "References" lists the major books and articles consulted in writing this chapter.
- "Hypercontents" takes you to an extensive list of sites with news, research, and images related to individual sections of the chapter.

INFOTRAC COLLEGE EDITION

Improve your skills with InfoTrac College Edition, a searchable online database of articles from more than 700 periodicals. Log on to

http://www.infotrac-college.com

or access InfoTrac through the website for this book. Try to find the following articles:

1. California Breathing (effect of air pollution on growth of children's lungs) (Brief Article) (Statistical Data Included). Tanya Zimbardo. *Mother Earth News*, April 2001 p20. Air pollution in parts of Los Angeles is causing a reduction in lung growth and capacity in children. *Hint*: Enter the search term "california breathing" using the keyword "california."

2. Why Tiny Particles Pose Big Problems (airborne dust causes health problems). Peter Jaret. *National Wildlife*, Feb–March 2001 pNA. This article explains why particulate matter may cause health problems for humans and other species. *Hint*: Enter the search term "tiny particles" using the keywords "effects of air pollution."

13 CLIMATE CHANGE AND OZONE LOSS

A.D. 2060: Green Times on Planet Earth

Mary Wilkins sat in the living room of the solar-powered and earth-sheltered house (Figure 13-1) she shared with her daughter Jane and her family. It was July 4, 2060: Independence Day.

She got up and walked into the greenhouse that provided much of her home's heat. There she heard the hum of solar-powered pumps trickling water to rows of organically grown vegetables and glanced at the fish in the aquaculture and waste treatment tanks.

Mary returned to the coolness of her earth-sheltered house and began putting the finishing touches on her grandchildren's costumes for this afternoon's pageant in Rachel Carson Park. It would honor earth heroes who began the Age of Ecology in the 20th century and those who continued this tradition in the 21st century.

She was delighted that her 12-year-old grandson Jeffrey had been chosen to play Aldo Leopold (Figure 2-13, p. 47), who in the late 1940s began urging people to work with the earth (Individuals Matter, p. 47). Her pride swelled when her 10-year-old granddaughter Lynn was chosen to play Rachel Carson, who in the 1960s alerted us to threats from increasing exposure to pesticides and other harmful chemicals (Individuals Matter, p. 420). Her neighbor's son, Manuel, had been chosen to play biologist Edward O. Wilson, who in the last third of the 20th century alerted us to the need to preserve the earth's biodiversity.

Even in her most idealistic dreams, she had never guessed that she would see the loss of global biodiversity slowed to a trickle. Most air pollution gradually disappeared when energy from the sun (p. 523), wind (p. 531), and hydrogen (produced by using solar energy to decompose water, p. 534) replaced most use of fossil fuels. Most food was grown by sustainable agriculture (Figure 16-29, p. 425).

Preventing pollution and reducing resource waste had become important, money-saving priorities for businesses and households based on the four *R*s of resource consumption: *reduce, reuse, recycle,* and *refuse*. Walking and bicycling had increased in cities and towns designed as vibrant communities for people instead of cars (p. 272). Low-polluting and safe ecocars (p. 518) got 128 kilometers per liter (300 miles per gallon), and most urban areas had efficient mass transportation.

World population had stabilized at 8 billion in 2028 and then had begun a slow decline. The threat of climate change from atmospheric warming enhanced by human activities lessened as the use of fossil fuels declined. International treaties enacted in the 1990s effectively banned the chemicals that had begun depleting ozone in the stratosphere (p. 318). By 2050, ozone levels in the stratosphere had returned to 1980 levels.

Two hours later, she, her daughter Jane, and her son-in-law Gene watched with pride as 40 beautiful children honored the leaders of the Age of Ecology. At the end, Lynn stepped forward and said, "Today we have honored many earth heroes, but the real heroes are the ordinary people in this audience and around the world who worked to help sustain the earth's life-support systems for us and other species. Thank you, Grandma, Mom, Dad, and everyone here for giving us such a wonderful gift. We promise to devote our lives to leaving the earth even better for our children and grandchildren and all living creatures."

Figure 13-1 An earth-sheltered house in the United States. Solar cells on the roof provide most of the house's electricity. About 13,000 families across the United States have built such houses. Mary Wilkins's fictional house in 2060 could be similar to this one.

We are embarked on the most colossal ecological experiment of all time—doubling the concentration in the atmosphere of an entire planet of one of its most important gases—and we really have little idea of what might happen.

PAUL A. COLINVAUX

This chapter addresses the following questions:

- How has the earth's climate changed in the past?
- How might the earth's climate change in the future?
- What factors can affect changes in the earth's average temperature?
- What are some possible effects of climate change from a warmer earth?
- What can we do to slow or adapt to climate change caused by natural processes, human activities, or both?
- Are human activities depleting ozone in the stratosphere, and why should we care?
- What can we do to slow and eventually reverse ozone depletion in the stratosphere caused by human activities?

13-1 PAST CLIMATE CHANGE AND THE NATURAL GREENHOUSE EFFECT

How Have the Earth's Temperature and Climate Changed in the Past? Climate change is neither new nor unusual. The earth's average surface temperature and climate have been changing throughout the world's 4.7-billion-year history, sometimes gradually (over hundreds to millions of years) and at other times fairly quickly (over a few decades). Figure 13-2 shows how the *estimated* average global temperature of the atmosphere near the earth's surface has changed during four time scales in the past: **(1)** 900,000 years, **(2)** 22,000 years, **(3)** 1,000 years, and **(4)** 140 years.

Over the past 900,000 years the average temperature of the atmosphere near the earth's surface has undergone prolonged periods of *global cooling* and *global warming* (Figure 13-2, top). During each cold period, thick glacial ice covered much of the earth's surface for about 100,000 years. Each of these periods was followed by a warmer interglacial period lasting 10,000–12,500 years, during which most of the ice melted. During the past 10,000 years we have had the good fortune to live in an interglacial period with a fairly stable climate and average global surface temperature (Figure 13-2, top middle).

However, even during this generally stable period, significant changes in regional climates have taken place. For example, about 7,000 years ago most of the current Sahara desert received annual rainfall of more than 20 centimeters (8 inches) per year, but its current annual rainfall is less than 1.3 centimeters (0.5 inch).

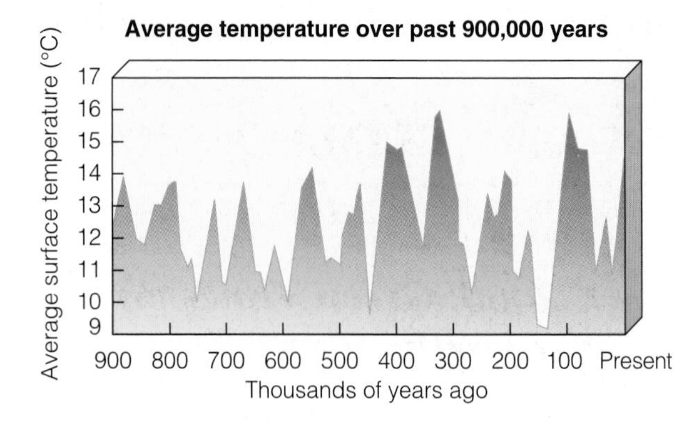

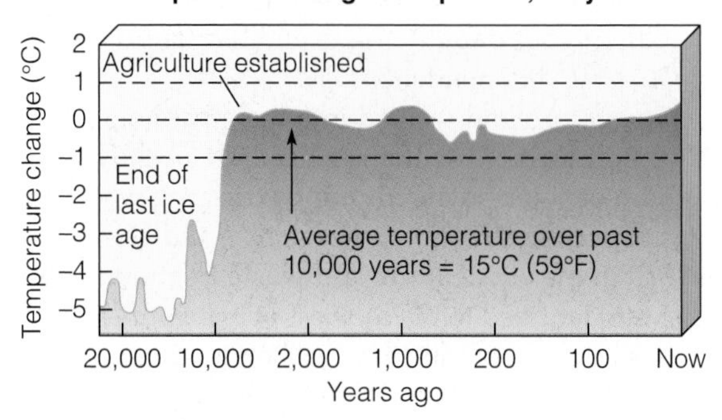

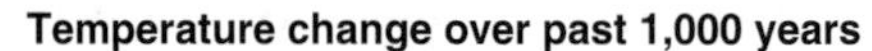

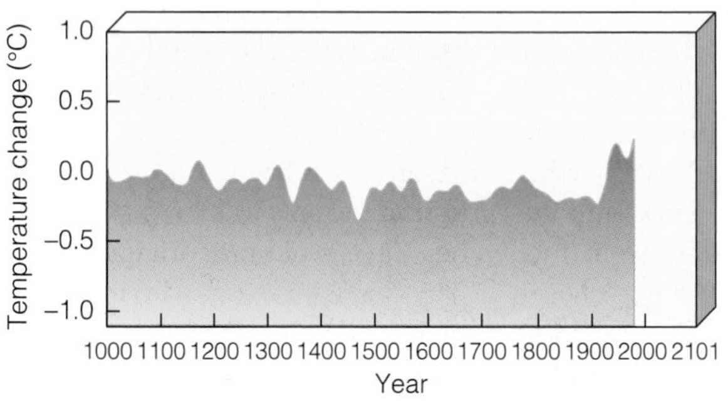

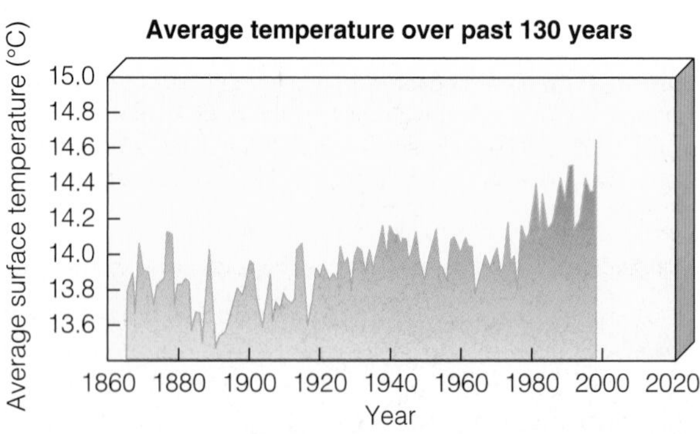

Figure 13-2 Estimated changes in the average global temperature of the atmosphere near the earth's surface over different periods of time. Past temperature changes are estimated by analysis of **(1)** plankton and isotopes in ocean sediments, **(2)** ice cores from ancient glaciers, **(3)** temperature measurements at different depths in boreholes drilled deep into the earth's surface, **(4)** pollen from lake bottoms and bogs, **(5)** tree rings, **(6)** historical records, and **(7)** temperature measurements (since 1860). (Data from Goddard Institute for Space Studies, Intergovernmental Panel on Climate Change, National Academy of Sciences, National Aeronautics and Space Agency, National Center for Atmospheric Research, and National Oceanic and Atmospheric Administration)

What Is the Greenhouse Effect? For the earth and its entire atmosphere to remain at a constant temperature, incoming solar energy must be balanced by an equal amount of outgoing energy (Figure 3-8, p. 66, and Figure 4-8, p. 82). Although the overall average temperature of the atmosphere is constant, the average temperature at various altitudes varies (Figure 12-2, p. 278, red line).

In addition to incoming sunlight, a natural process called the *greenhouse effect* (Figure 6-11, p. 127) warms the earth's lower troposphere and surface. Recall that it occurs because molecules of certain atmospheric gases, called *greenhouse gases,* warm the earth's surface. They do this by absorbing some of the infrared radiation (heat) radiated by the earth's surface. This causes their molecules to vibrate and transform the absorbed energy into longer-wavelength infrared radiation (heat) in the troposphere (Figure 6-11, p. 127).*

Swedish chemist Svante Arrhenius first recognized this natural tropospheric heating effect in 1896. Since then it **(1)** has been confirmed by numerous laboratory experiments and measurements of atmospheric temperatures at different altitudes and **(2)** is one of the most widely accepted theories in the atmospheric sciences.

The two greenhouse gases with the largest concentrations in the atmosphere are **(1)** water vapor, controlled by the hydrologic cycle (Figure 4-26, p. 95), and **(2)** carbon dioxide, controlled by the carbon cycle (Figure 4-27, p. 96). Other greenhouse gases present in lower concentrations include **(1)** methane (CH_4), **(2)** nitrous oxide (N_2O), **(3)** synthetic chlorofluorocarbons (CFCs), **(4)** synthetic perfluorocarbons (PFCs), and **(5)** synthetic trifluoromethyl sulfur pentafluoride (SF_5CF_3). Inputs of these gases into the atmosphere can come from **(1)** natural sources and **(2)** human activities (Table 13-1), except for CFCs, SF_6, and SF_5CF_3, which come only from human sources.

Analysis of gases in bubbles trapped at various depths in ancient glacial ice shows that over the past 160,000 years levels of tropospheric water vapor (the primary greenhouse gas) have remained fairly constant. However, during most of this period, CO_2 levels have fluctuated between 190 and 290 parts per million. These estimated changes in tropospheric CO_2 levels correlate fairly closely with estimated variations in the atmosphere's average global temperature near the earth's surface during the past 160,000 years (Figure 13-3).

*It is scientifically incorrect to say that the earth's atmosphere *traps* heat or *reradiates* heat it has absorbed back toward the earth's surface. Molecules of greenhouse gases absorb various wavelengths of infrared radiation and transform them into infrared radiation with different (longer) wavelengths. Because the originally absorbed wavelengths of infrared radiation no longer exist, it is incorrect to say that they have been trapped or reradiated.

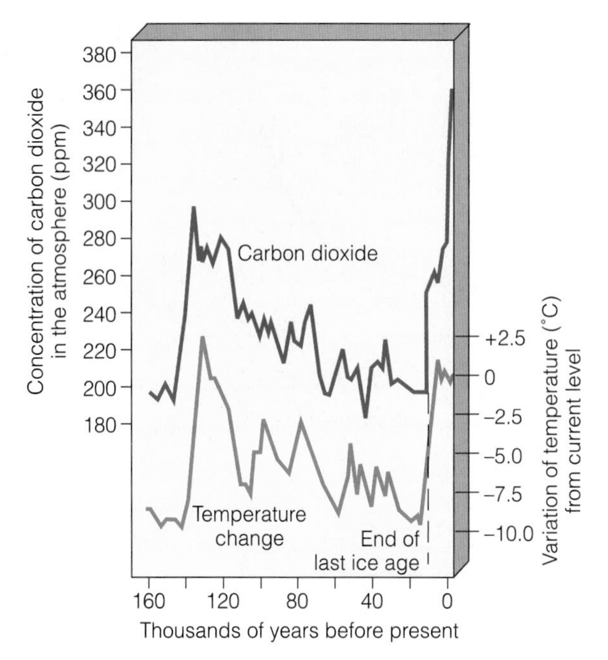

Figure 13-3 Estimated long-term variations in average global temperature of the atmosphere near the earth's surface and average tropospheric carbon dioxide levels over the past 160,000 years. These CO_2 levels were obtained by inserting metal tubes deep into antarctic glaciers, removing the ice, and analyzing bubbles of ancient air trapped in ice at various depths throughout the past. The rough correlation between tropospheric CO_2 levels and temperature shown in these estimates based on ice core data suggests a connection between these two variables, although no definitive causal link has been established. In 1999, the world's deepest ice core sample revealed a similar correlation between air temperatures and the greenhouse gases CO_2 and CH_4 going back for 460,000 years. (Data from Intergovernmental Panel on Climate Change and National Center for Atmospheric Research)

13-2 CLIMATE CHANGE AND HUMAN ACTIVITIES

What Is Global Warming? Since the beginning of the industrial revolution around 1750 (and especially since 1950) there has been a sharp rise in **(1)** the use of fossil fuels, which release large amounts of the greenhouse gases CO_2 and CH_4 into the troposphere, **(2)** deforestation and clearing and burning of grasslands to raise crops, which release CO_2 and N_2O into the atmosphere, and **(3)** cultivation of rice in paddies and use of inorganic fertilizers, which release N_2O into the troposphere.

Figure 13-4 shows that since 1860 there has been a sharp rise in the concentrations of the greenhouse gases CO_2, CH_4, and N_2O, with especially sharp increases since 1950. According to BP Amoco, 54.5% of the world's carbon dioxide emissions in 1999 came from **(1)** the United States (26.5%), **(2)** the European Union (14.5%), and **(3)** China (13.5%). Emissions of CO_2 from U.S. coal-burning power and industrial plants alone exceed the combined CO_2 emissions of 146 nations, which contain 75% of the world's people.

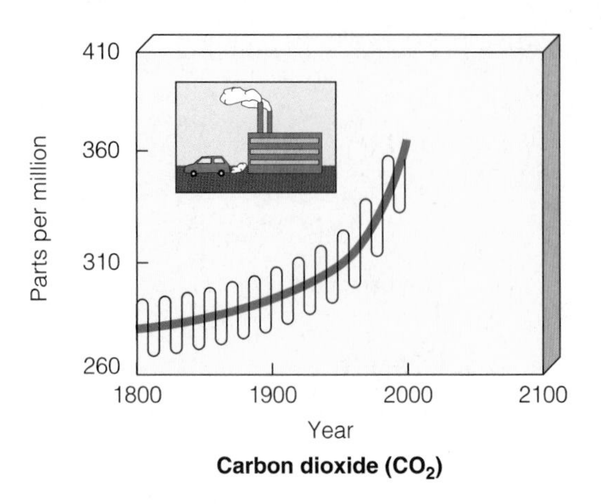

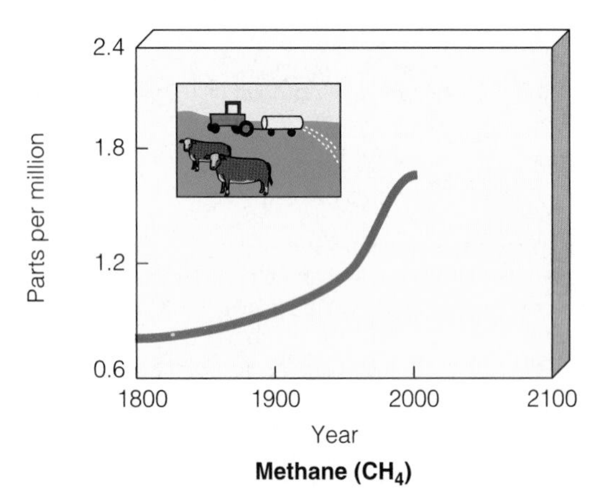

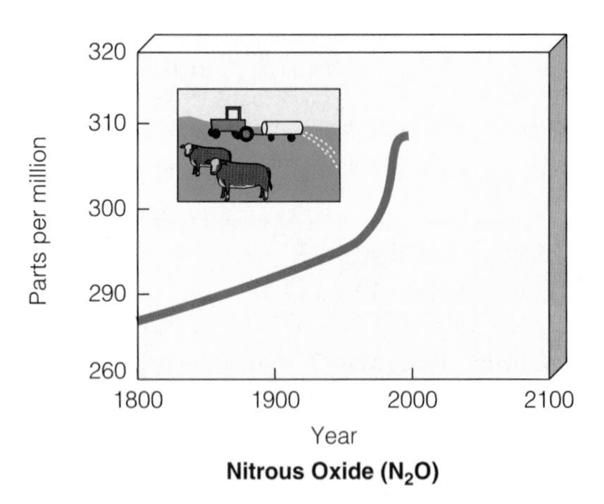

Figure 13-4 Increases in average concentrations of the greenhouse gases carbon dioxide, methane, and nitrous oxide in the troposphere between 1860 and 1999. (Data from Intergovernmental Panel on Climate Change, National Center for Atmospheric Research, and World Resources Institute)

Based on evidence about past changes in atmospheric CO_2 concentrations and atmospheric temperatures (Figure 13-3), climate scientists contend that it is highly likely that increased inputs of CO_2 and other greenhouse gases from human activities (Figure 13-4) will **(1)** enhance the earth's natural greenhouse effect and **(2)** raise the average global temperature of the atmosphere near the earth's surface. This *enhanced greenhouse effect* is called **global warming** and should not be confused with the different problem of ozone depletion (Table 13-2 and Section 13-6, p. 318).

There is much evidence of a correlation between atmospheric concentrations of carbon dioxide and average tropospheric temperatures. Analysis of ice core samples, temperature measurements at different levels in several hundred boreholes in the earth's surface, and atmospheric temperature measurements show that

- The concentration of CO_2 in the troposphere is higher than it has been in the past 420,000 years (and probably in the last 20 million years).
- The 20th century was the hottest century in the past 1,000 years (Figure 13-2, lower middle).
- Since 1861 (when atmospheric temperature measurements began), the average global temperature of the troposphere near the earth's surface has risen $0.6 \pm 0.2°C$ ($1.1 \pm 0.4°F$), with most of this increase taking place since 1946 (Figure 13-2, bottom).
- The 1990s was the warmest decade and 1998 the warmest year since 1861, when atmospheric temperature measurements began.

Other observed signs of a warmer troposphere during recent decades include

- Increased temperatures and melting of ice caps and floating ice at the earth's poles (Connections, p. 308)
- An average global sea-level rise of 10–25 centimeters (4–10 inches) over the past 100 years
- Shrinking of some glaciers on the tops of mountains in the Alps, Andes, Himalayas, and northern Cascades of Washington (Figure 7-17, p. 176)
- Northward migration of some warm-climate fish and trees
- Earlier spring arrival and later autumn frosts in many parts of the world (which affects patterns of crop growth and animal migrations)
- More common occurrences of unusual weather patterns such as warm spells, droughts, and unexpected storms and hurricanes in parts of the world

Scientists project that there will be significant increases in the emissions of CO_2, CH_4, and N_2O during the 21st century (Figure 13-5) and that such increases could enhance the earth's natural greenhouse effect.

Regardless of the cause, significant climate change caused by atmospheric warming or cooling over several decades to a hundred years has important implications for human life, wildlife, and the world's economies. Such rapid climate change can **(1)** affect the availability of water resources by altering rates of evaporation and precipitation, **(2)** shift areas where crops can be grown, **(3)** change average sea levels, and **(4)** alter the structure and location of the world's biomes (Figure 6-13, p. 129), as discussed in more detail in Section 13-4.

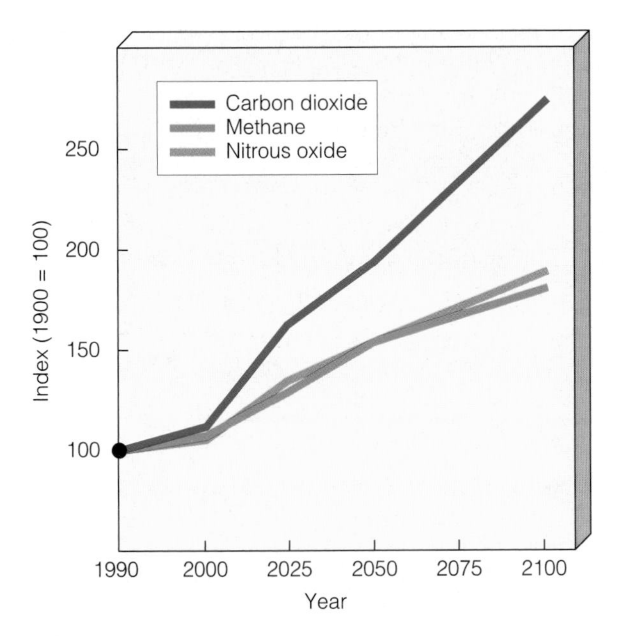

Figure 13-5 Projected emissions of three important greenhouse gases as a result of human activities, 1990–2100. (Data from United Nations Food and Agriculture Organization, 1997)

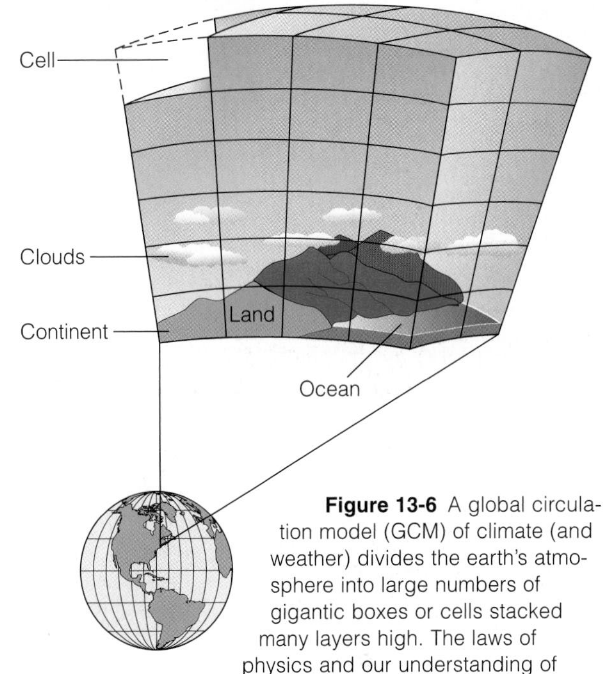

Figure 13-6 A global circulation model (GCM) of climate (and weather) divides the earth's atmosphere into large numbers of gigantic boxes or cells stacked many layers high. The laws of physics and our understanding of global air circulation patterns (Figure 6-6, p. 125) are used to describe numerically **(1)** what happens to major variables affecting climate in each cell and **(2)** how they change from one cell to another.

How Do Scientists Model Climate Changes? Computer Models as Crystal Balls To project the effects of increases in greenhouse gases (Figure 13-5) on average global temperature and the earth's climate, scientists develop *mathematical models* of the global air circulation system (Figure 6-6, p. 125) and run them on supercomputers. These global circulation models (GCMs) simulate the earth's atmosphere mathematically by covering the earth's surface with 9–64 stacked layers of gigantic cells or boxes, each several hundred kilometers on a side and about 3 kilometers (2 miles) high (Figure 13-6). The flow of various types of energy and matter in and out of the box-filled atmosphere is simulated with a set of mathematical equations.

Table 13-1 Major Greenhouse Gases from Human Activities

Greenhouse Gas	Human Sources	Average Time in the Troposphere	Relative Warming Potential (compared to CO_2)
Carbon dioxide (CO_2)	Fossil fuel burning (especially coal), deforestation, and plant burning	50–500 years	1
Methane (CH_4)	Rice paddies, guts of cattle and termites, landfills, coal production, coal seams, and natural gas leaks from oil and gas production and pipelines	9–15 years	24
Nitrous oxide (N_2O)	Fossil fuel burning, fertilizers, livestock wastes, and nylon production	120 years	360
Chlorofluorocarbons (CFCs)	Air conditioners, refrigerators, plastic foams	11–20 years (65–110 years in the stratosphere)	1,500–7,000

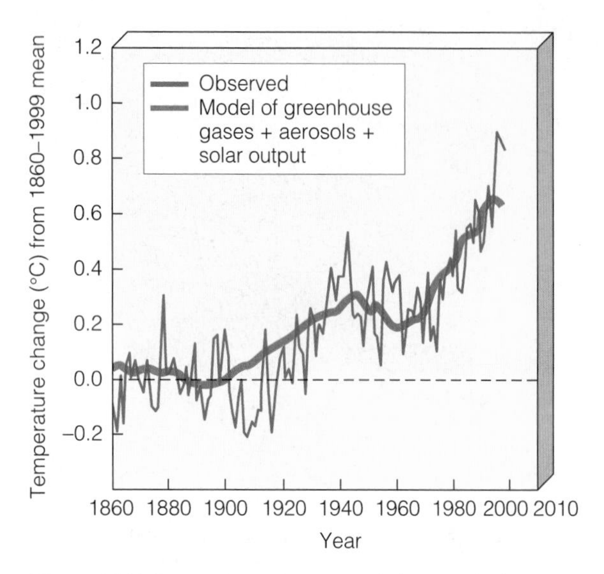

Figure 13-7 Comparison of measured changes in the average atmospheric temperature near the earth's surface (blue line) with those projected by the latest climate models between 1860 and 1999 (red line). (Data from Tom L. Wigley, Pew Center on Global Climate Change)

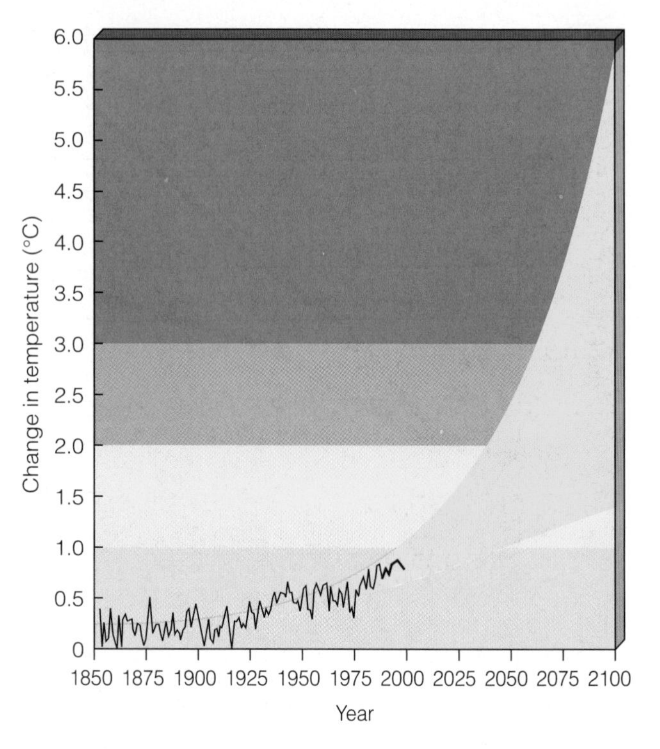

Figure 13-8 Comparison of measured changes in the average temperature of the atmosphere at the earth's surface between 1860 and 2000 and the projected range of temperature increase during the 21st century. Current models indicate that the average global temperature at the earth's surface will rise by 1.4–5.8°C (2.5–10.4°F) sometime during the 21st century. Because of the uncertainty involved in current climate models, the projected warming shown here could be *overestimated* or *underestimated* by a factor of two. (Data from U.S. National Academy of Sciences, National Center for Atmospheric Research, and Intergovernmental Panel on Climate Change)

Greatly improved climate models and measurements have increased our ability to account for past changes (Figure 13-7) and predict future changes in the earth's average surface temperature. However, there are still scientific uncertainties, as discussed in Section 13-3.

What Is the Scientific Consensus About Future Climate Change and Its Effects? The Intergovernmental Panel on Climate Change (IPCC) is a network of about 2,000 of the world's leading climate experts from 70 nations. It was established by the United Nations and the World Meteorological Organization to study climate change. In 1990, 1995, and 2001, the IPCC published major reports evaluating the available evidence about **(1)** past changes in global temperatures (Figure 13-2) and **(2)** climate models projecting future changes in average global temperatures and climate. The U.S. National Academy of Sciences and the American Geophysical Union have also evaluated possible future climate changes.

According to the 2001 IPCC report,

- "There is new and stronger evidence that most of the warming observed over the last 50 years is attributable to human activities."
- There is a 90–95% chance that the earth's mean surface temperature will increase 1.4–5.8°C (2.5–10.4°F) between 2000 and 2100 (Figure 13-8).

According to a 2001 report by the National Academy of Sciences (prepared at the request of President Bush),

- "Greenhouse gases are accumulating in the Earth's atmosphere as a result of human activities, causing surface air temperatures and subsurface ocean temperatures to rise. . . . The changes observed over the last several decades are likely mostly due to human activities. . . . Global warming could have serious adverse societal and ecological impacts by the end of this century."
- Global mean surface air temperatures warmed 0.4–0.8°C (0.7–1.5°F) during the 20th century (Figure 13-7).
- If present trends continue, global average temperatures will rise a further 1.4–5.8°C (2.5–10.4°F) by 2100, with a mid-level projection of 3°C (5.4°F).

At the 1998 World Economic Forum in Davos, Switzerland, the CEOs of the world's 1,000 largest corporations voted climate change as the most critical problem facing humanity. Growing numbers of corporate leaders also recognize that a transition to a greatly increased dependence on energy-efficient and renewable energy sources (Chapter 20, p. 513) could lead to an unprecedented global economic boom.

However, a *very small minority* of atmospheric scientists inside and outside the IPCC contend that

- There is still insufficient knowledge about natural climate variables that could change the assessment (up or down), as discussed in Section 13-3.

Table 13-2 Major Characteristics of Global Warming and Ozone Depletion		
Characteristic	**Global Warming**	**Ozone Depletion**
Region of atmosphere involved	Troposphere.	Stratosphere.
Major substances involved	CO_2, CH_4, N_2O (greenhouse gases).	O_3, O_2, chlorofluorocarbons (CFCs).
Interaction with radiation	Molecules of greenhouse gases absorb infared (IR) radiation from the earth's surface, vibrate, and release longer-wavelength IR radiation (heat) into the lower troposphere. This natural greenhouse effect helps warm the lower troposphere (Figure 6-11, p. 127).	About 95% of incoming ultraviolet (UV) radiation from the sun is absorbed by O_3 molecules in the stratosphere and does not reach the earth's surface (Figure 12-2, p. 278).
Nature of problem	There is a high probability that increasing concentrations of greenhouse gases in the troposphere from burning fossil fuels, deforestation, and agriculture are enhancing the natural greenhouse effect and raising the earth's average surface temperature. (Figure 13-2, bottom, and Figure 13-7)	CFCs and other ozone-depleting chemicals released into the troposphere by human activities have made their way to the stratosphere, where they decrease O_3 concentration. This can allow more harmful UV radiation to reach the earth's surface.
Possible consequences	Changes in climate, agricultural productivity, water supplies, and sea level.	Increased incidence of skin cancer, eye cataracts, and immune system suppression and damage to crops and phytoplankton.
Possible responses	Decrease fossil fuel use and deforestation.	Eliminate CFCs and other ozone-depleting chemicals and find acceptable substitutes.

- Computer models used to predict climate change are improving but still are not very reliable.

IPCC and other climate scientists holding the consensus view

- Agree that we need to **(1)** support greatly increased research on how the climate system works and **(2)** improve climate models to help narrow the range of uncertainty in projected temperature changes (Figure 13-8).
- Point out that the biggest concern is not just a temperature increase but how rapidly it rises. They warn that warming or cooling of the earth's surface by even 1°C (1.8°F) per century **(1)** would be faster than any temperature change occurring in the last 10,000 years, since the beginning of agriculture (Figure 13-2) and **(2)** could seriously disrupt the earth's ecosystems, economies, and human societies (p. 311).

13-3 FACTORS AFFECTING CHANGES IN THE EARTH'S AVERAGE TEMPERATURE

Will the Earth Continue to Get Warmer? The average temperature at the earth's surface has increased significantly during the past 20 years (Figure 13-7). Will this trend continue?

The answer is *that this is highly likely, but we do not know for sure.* Scientists have identified a number of natural and human-influenced factors that might *amplify* (positive feedback) or *dampen* (negative feedback) changes in the earth's average surface temperature. These factors could influence **(1)** how much and how fast temperatures climb or drop and **(2)** what the effects are in various areas.

How Might Changes in Solar Output Affect the Earth's Temperatures? Solar output varies by about 0.1% over the 13-year and 22-year sunspot cycles and over 80-year and other much longer cycles. These up-and-down changes in solar output can temporarily warm or cool the earth and thus affect the projections of climate models.

Atmospheric scientists have not been able to identify a mechanism related to changes in solar output that could account for more than 50% of the atmospheric warming between 1900 and 2000 (Figure 13-7).

How Might Changes in the Earth's Reflectivity Affect Atmospheric Temperatures? Different parts of the earth's surface vary in their **albedo**, or ability to reflect light. White or shiny surfaces such as snow, ice, and sand have a high albedo reflect most of the sunlight that hits them.

Albedo increases when polar ice caps expand during glacial periods and decreases when they melt and expose

CONNECTIONS

Early Warnings from the Earth's Ice

As the atmosphere warms, it causes more convection that transfers heat from equatorial to polar areas (Figure 6-6, p. 125). Thus, temperature increases tend to be greater at the earth's poles than at its middle latitudes. This explains why the earth's frigid poles are regarded as early warning sentinels of global warming.

The sea ice floating in the Arctic is large enough to cover the United States and is highly sensitive to changes in the air above and the ocean below. News from the Arctic is alarming. Between 1968 and 1997, surface temperatures at nine stations north of the arctic circle rose by about 5.5°C (9.9°F)—about 10 times the average global temperature increase during this period (Figure 13-7, p. 306).

The bright color (high albedo) of the floating sea ice in the Arctic Ocean helps cool the earth by reflecting 80% of the sunlight it receives back into space. If this ice melted, the darker Arctic Ocean would become a heat collector and absorb 80% of its input of sunlight. This would drastically affect global climate, especially in the northern hemisphere.

According to a 2001 report by the IPCC, satellite data and measurements by surface boats and submarines indicate that since the 1950s floating sea ice during summer and spring in the Arctic Ocean has **(1)** shrunk in area by 10–15% (an area almost twice the size of Texas) and **(2)** decreased in thickness by 40%.

It is not known whether this shrinkage and thinning of arctic sea ice is the result of natural polar climate fluctuations or global warming caused by increases in greenhouse gases.

Because it is floating, large-scale melting of Arctic Ocean ice will not raise global sea levels (just as an ice cube in a glass of water does not raise the water level when it melts). However, widespread melting would greatly amplify warming of the Arctic region. This could **(1)** reroute warm ocean currents (Figure 6-3, p. 123) and weather patterns further south and **(2)** cause significant cooling in parts of the northern hemisphere, especially in Europe and eastern North America.

Studies also show that as the frozen north warms and thaws, peat buried in the arctic tundra soil would decay and release large amounts of CO_2. According to climate researchers at San Diego State University (California), since 1982 the arctic tundra has warmed so much that it has been giving off more carbon dioxide than it absorbs—a positive feedback loop that could speed up the rate of increase of CO_2 in the atmosphere and amplify global warming.

The news from the Antarctic is also disturbing. The huge antarctic ice cap, which is almost twice the area of Australia or Europe, contains **(1)** 70% of the world's fresh water and **(2)** 90% of the world's highly reflective ice, which helps cool the earth.

Since 1947, the average temperature of the Antarctic Peninsula has risen by about 2.8°C (5°F) in summer and 5.6°C (10°F) in winter. As a result, huge pieces of ice shelf—some as large as the state of Delaware—have begun breaking off (calving) from the peninsula's eastern shore.

The western antarctic ice sheet—the size of Texas and Colorado combined—has been melting slowly for decades. The breakup and eventual collapse of this ice sheet appear to be part of an ongoing natural cycle initiated by the melting of northern hemispheric ice sheets at the end of the last glacial period, when average sea levels began their huge rise (Figure 13-10, p. 310).

However, this breakup could speed up if other natural processes or human activities continue to warm the atmosphere and oceans. A partial melting or breakdown of the western antarctic ice sheet would raise global sea levels by as much as 0.9 meters (3 feet) by 2100. Some scientists are considering the possibility that a much larger area of this sheet could be gone by 2100. If that occurred, the average global sea level would rise by 5–6 meters (15–20 feet) and severely flood many of the world's low-lying coastal cities, wetlands, and islands.

For some scientists, the most disturbing news comes from Greenland, more than 14,500 kilometers (9,000 miles) north of Antarctica. A 2000 study of ice cores found that Greenland's glaciers, which cover an area larger than Mexico, are more likely to melt because they are closer to the equator than the west antarctic ice sheet. If this occurred, as it did in a previous interglacial warm period 110,000–130,000 years ago (Figure 13-10), average sea levels would rise by 4–7 meters (13–23 feet). This influx of fresh water **(1)** would cause flooding in the world's low-lying coastal areas and cities and **(2)** could shut off currents such as the Gulf Stream and North Atlantic, which keep Europe warmer than it would otherwise be.

These changes in ice cover at the earth's poles and Greenland may result from natural climate cycles, human activities, or a combination of both. In any case, they can have serious long-term implications for wildlife and human economies.

One comedian jokes that he plans to buy land in Kansas because it will probably become valuable beachfront property. Another boasts that she is not worried because she lives in a houseboat—the "Noah strategy."

Critical Thinking

What difference might it make in your life and in that of any child you might have if human activities play an important role in continued warming of the earth's polar regions and Greenland? What three important things could you do to help slow down such warming?

less reflective land and ocean surfaces (Connections, left). Satellite measurements of how albedo varies throughout the world and projections of how this might change have been incorporated into current climate models used to make the projections in Figure 13-8.

How Might the Oceans Affect Climate? Currently, the oceans help moderate the earth's average surface temperature by removing about 29% of the excess CO_2 we pump into the atmosphere as part of the global carbon cycle (Figure 4-27, p. 96). We do not know whether the oceans can absorb more CO_2.

The oceans also affect global climate by absorbing heat from the atmosphere and transferring some of it to the deep ocean, where it is stored temporarily. Climate modelers hypothesize that **(1)** this deep-ocean absorption of heat has delayed part of the warming of the earth's atmosphere, and **(2)** some of this stored heat probably will be released back to the atmosphere and could amplify global warming.

There is also concern that global warming could disrupt ocean currents (Figure 6-3, p. 123), which are driven largely by differences in water density and winds. Connected deep and surface currents act like a gigantic conveyor belt to transfer heat from one place to another and store carbon dioxide and heat in the deep sea (Figure 13-9). Scientists are concerned that an influx of fresh water from thawing ice in the Arctic and the Antarctic (Connections, left) might **(1)** slow or disrupt this conveyer belt and **(2)** decrease the amount of heat it brings to the North Atlantic region. If this loop stalls out, evidence from past climate changes indicates that this could trigger atmospheric temperature changes of more than 5°C (9°F) over periods as short as 40 years.

Changes in average sea level affect **(1)** the amount of heat and CO_2 that can be stored in the ocean and **(2)** changes in the earth's biomes (Figure 6-13, p. 129). Figure 13-10 shows estimated changes in the earth's average sea level over the last 250,000 years based on data obtained from cores drilled in the ocean floor. According to the 2001 IPCC report, the rate of sea level rise is now faster than at any other time during the past 1,000 years.

How Do Water Vapor Content and Clouds Affect Climate? Warmer temperatures increase evaporation of surface water and create more clouds. These additional clouds could have **(1)** a warming effect by absorbing and releasing heat into the troposphere or **(2)** a cooling effect by reflecting sunlight back into space.

The net result of these two opposing effects depends on **(1)** whether it is day or night and **(2)** the type (thin or thick) and altitude of clouds. Scientists do not know **(1)** which of these factors might predominate or **(2)** how cloud types and heights might vary in different parts of the world as a result of global warming. What they do know about the effects of clouds has been included in recent climate models, but much uncertainty about these effects remains.

How Might Air Pollution Affect Climate? Global warming (Figure 13-8) might be partially offset by *aerosols* (tiny droplets and solid particles, Figure 12-6, p. 282) of various air pollutants released or formed in the atmosphere by volcanic eruptions and human activities. Climate scientists hypothesize that higher levels of aerosols attract enough water molecules to form condensation nuclei, leading to increased cloud formation.

Some of the resulting clouds have a high albedo and reflect more incoming sunlight back into space during the day. This could help counteract the heating effects of increased greenhouse gases.

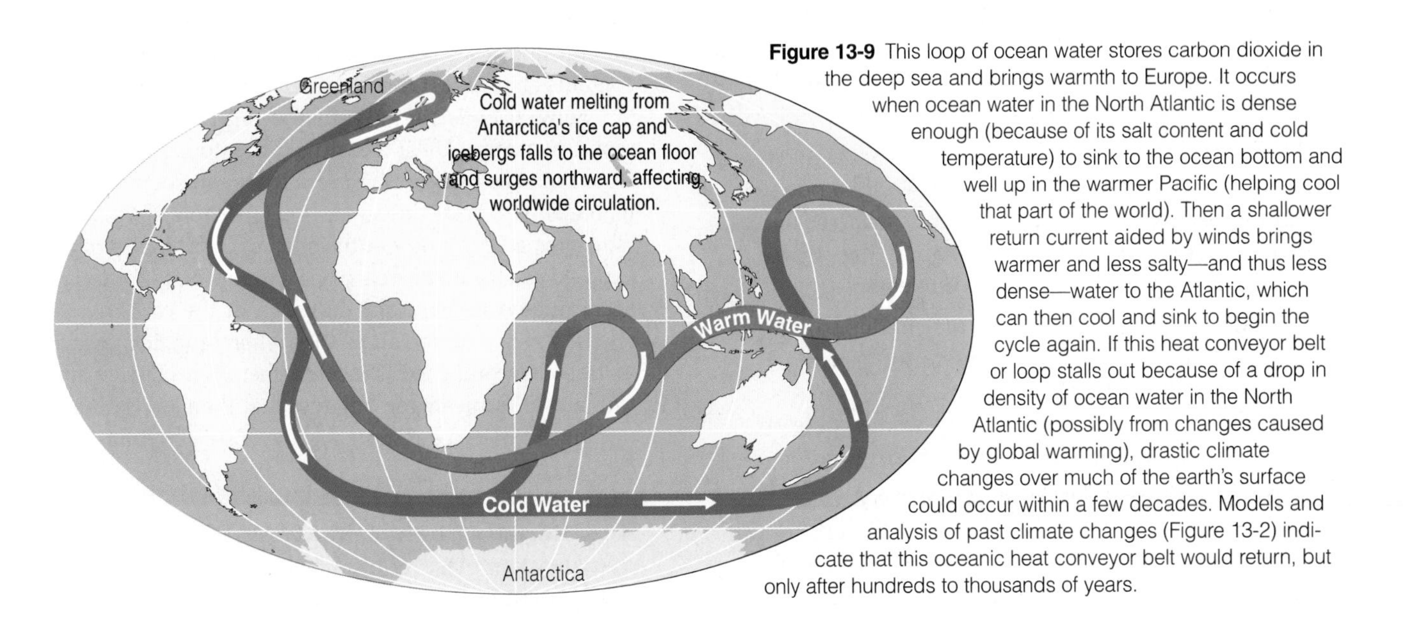

Figure 13-9 This loop of ocean water stores carbon dioxide in the deep sea and brings warmth to Europe. It occurs when ocean water in the North Atlantic is dense enough (because of its salt content and cold temperature) to sink to the ocean bottom and well up in the warmer Pacific (helping cool that part of the world). Then a shallower return current aided by winds brings warmer and less salty—and thus less dense—water to the Atlantic, which can then cool and sink to begin the cycle again. If this heat conveyor belt or loop stalls out because of a drop in density of ocean water in the North Atlantic (possibly from changes caused by global warming), drastic climate changes over much of the earth's surface could occur within a few decades. Models and analysis of past climate changes (Figure 13-2) indicate that this oceanic heat conveyor belt would return, but only after hundreds to thousands of years.

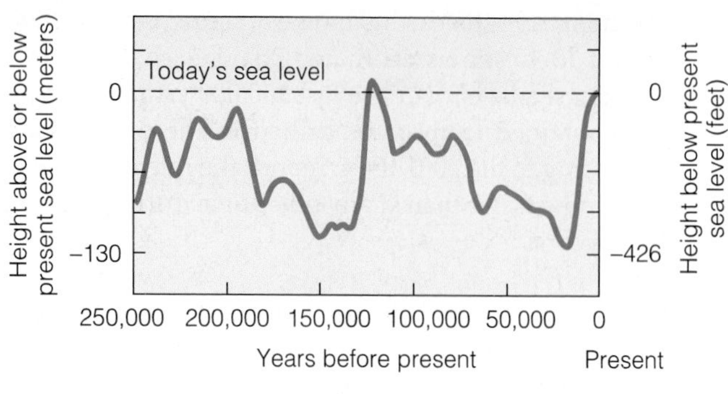

Figure 13-10 Changes in average sea level over the past 250,000 years based on data from cores removed from the ocean floor. The coming and going of glacial periods (ice ages) largely determine the rise and fall of sea level. As glaciers have melted and retreated since the peak of the last glacial period about 18,000 years ago, the earth's average sea level has risen about 125 meters (410 feet) and changed the earth's coastal zones significantly. (Adapted with permission from Tom Garrison, *Oceanography: An Invitation to Marine Science*, 3d ed., Pacific Grove, Calif.: Brooks/Cole, 1998)

Nights would be warmer because the clouds would still be there and prevent some of the heat stored in the earth's surface (land and water) during the day from being radiated into space. These pollutants may explain why most of the recent warming in the northern hemisphere occurs at night.

However, these interactions are complex. Pollutants in the lower troposphere can either warm or cool the air, depending on the reflectivity of the underlying surface. In 2001, Stanford University professor Mark Jacobson published a study indicating that tiny particles of *soot*, produced mainly from burning coal and diesel fuel, may be responsible for 15–30% of the observed global warming over the past 50 years. If this preliminary research is correct, soot would be the second biggest contributor to global warming, just behind the greenhouse gas carbon dioxide.

Climate scientists do not expect air pollutants to counteract global warming very much in the next 50 years because

- Aerosols fall back to the earth or are washed out of the atmosphere within weeks or months, whereas CO_2 and other greenhouse gases remain in the atmosphere for decades to several hundred years.
- Aerosols are major components of acid deposition (Figure 12-8, p. 286), which can slow forest growth and weaken or kill trees (Figure 12-11, p. 288). This reduces the ability of growing trees to absorb some of the CO_2 we are putting into the atmosphere and could accelerate atmospheric warming.
- Aerosol inputs into the atmosphere are being reduced (Section 12-7, p. 293) because they **(1)** kill or sicken large numbers of people each year (p. 292) and **(2)** damage food crops, trees, and other forms of vegetation (Section 12-6, p. 291).

How Might Increased CO_2 Levels Affect Photosynthesis and Methane Emissions? Some studies suggest that more CO_2 in the atmosphere could increase the rate of photosynthesis in areas with adequate amounts of water and other soil nutrients. This would remove more CO_2 from the atmosphere and help slow atmospheric warming. However, according to 2001 studies **(1)** this effect would slow as the plants reach maturity and take up less CO_2, and **(2)** when the plants die and are decomposed or are burned, the carbon they stored is returned to the atmosphere as carbon dioxide.

Other studies suggest that **(1)** this effect varies with different types of plants and in different climate zones, and **(2)** much of any increased plant growth could be offset by plant-eating insects that would breed more rapidly and year-round in warmer temperatures.

Atmospheric warming could be accelerated by increased release of methane (a potent greenhouse gas, Table 13-1) from

- Bogs and other wetlands.
- Icelike compounds called methane hydrates trapped beneath the Arctic permafrost and sediments on the floor of Arctic Ocean. Large amounts of methane would be released into the troposphere if **(1)** the blanket of permafrost in tundra soils melts and the Arctic Ocean warms (Connections, p. 308) or **(2)** methane hydrates on the ocean floor are extracted to increase natural gas supplies (a possibility now under study in the United States).

Can Soils Absorb More of the Excess Carbon Dioxide? When a plant dies and is decomposed, some of its carbon returns to the atmosphere as carbon dioxide and some is released into the soil. Globally, soil contains about five times as much carbon as vegetation.

Some scientists suggest that we may be able to slow global warming by increasing the amount of carbon stored in soils by planting more trees, improving forest management, and conserving soils (Section 9-9, p. 217). However, a 2001 study of the soil around loblolly pines exposed to elevated levels of CO_2 found that **(1)** the soil accumulated carbon, but **(2)** much of the carbon was released back into the air as CO_2 when organic material in the soil decomposed. The researchers, John Lichter and William H. Schlesinger, suggest that their preliminary findings "call into question the role of soils as long-term carbon sinks."

How Rapidly Could Climate Shift? If moderate change takes place gradually over several hundred years, people in areas with unfavorable climate changes may be able to adapt to the new conditions. However, if global temperature change takes place over several decades, we may not have enough time (and money) to **(1)** switch food-growing regions, **(2)** relocate people from low-lying coastal areas, and **(3)** build elaborate systems of dikes and levees to help protect the large portion of the world's population living near coastal areas. Such rapid changes could **(1)** lead to widespread starvation and to social and economic chaos, especially in developing countries, and **(2)** reduce the earth's biodiversity because many species could not move or adapt.

Recent data from analyses of ice cores and deep-sea sediments suggest that **(1)** average temperatures during the warm interglacial period that began about 125,000 years ago (Figure 13-2, top) varied as much as 10°C (18°F) in only a decade or two, and **(2)** such warming and cooling periods each lasted 1,000 years or more. If these findings are correct and also apply to the current interglacial period, fairly small rises in greenhouse gas concentrations could trigger rapid up-and-down shifts in the earth's average surface temperatures.

As a result of uncertainties in climate models and the factors discussed in this section, climate scientists estimate that *atmospheric warming and rises in average sea levels during the next 50–100 years could be half (the best-case scenario) or twice (the worst-case scenario) the current projections shown in Figure 13-8.*

13-4 SOME POSSIBLE EFFECTS OF A WARMER WORLD

Why Should We Worry if the Earth's Temperature Rises a Few Degrees? So what's the big deal? Why should we worry about a possible rise of only a few degrees in the earth's average surface temperature? We often have that much change between May and July, or even between yesterday and today.

The key point is that we are not talking about normal swings in *local weather* but a *global* change in *climate*—weather averaged over decades, centuries, and millennia.

What Are Some Possible Effects of Atmospheric Warming? A warmer global climate could have a number of harmful and beneficial effects (Figure 13-11) depending on where one lives. If the earth's surface warms, climate models project that

- Deserts will expand and droughts will grow more severe, especially in parts of Africa and Asia.
- Warmer soil, especially at high latitudes, will speed up plant decomposition and release more CO_2.

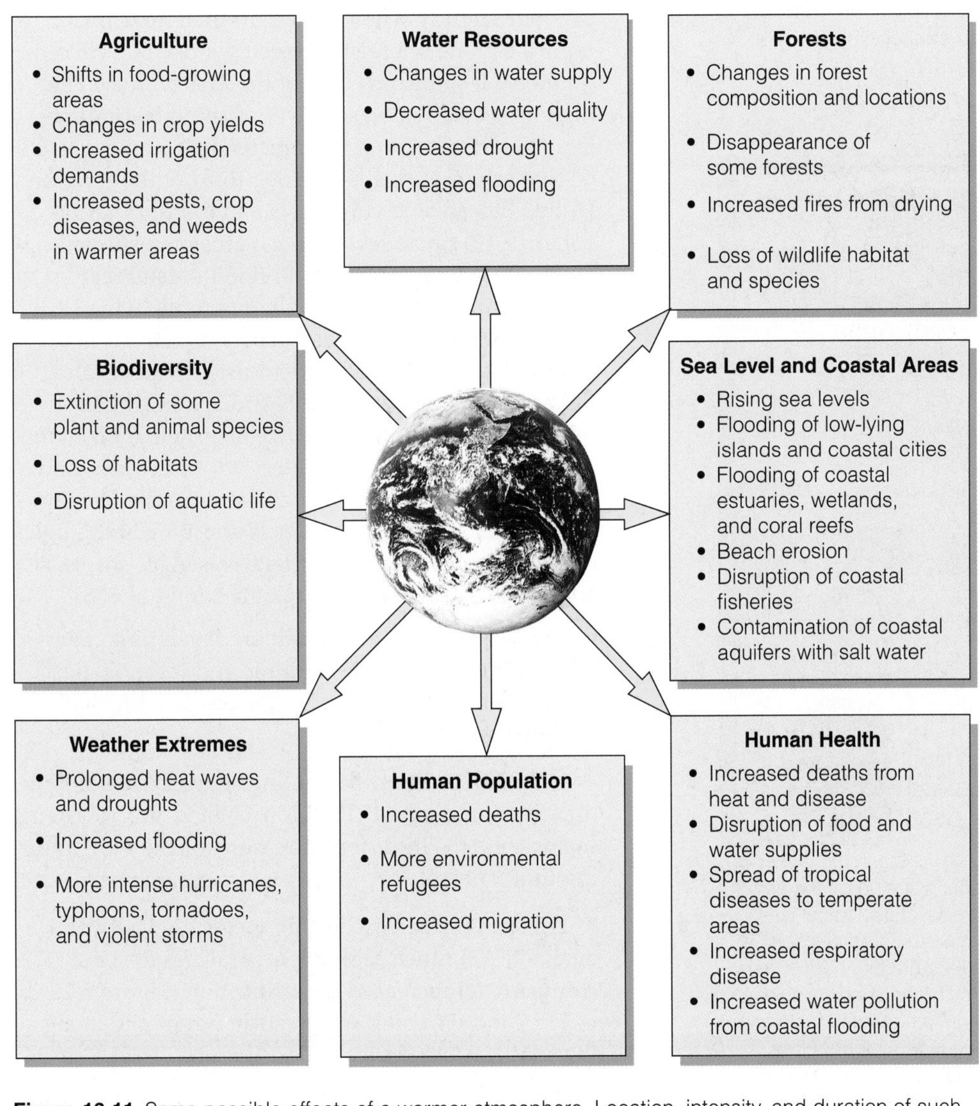

Agriculture
• Shifts in food-growing areas
• Changes in crop yields
• Increased irrigation demands
• Increased pests, crop diseases, and weeds in warmer areas
Water Resources
• Changes in water supply
• Decreased water quality
• Increased drought
• Increased flooding
Forests
• Changes in forest composition and locations
• Disappearance of some forests
• Increased fires from drying
• Loss of wildlife habitat and species
Biodiversity
• Extinction of some plant and animal species
• Loss of habitats
• Disruption of aquatic life
Sea Level and Coastal Areas
• Rising sea levels
• Flooding of low-lying islands and coastal cities
• Flooding of coastal estuaries, wetlands, and coral reefs
• Beach erosion
• Disruption of coastal fisheries
• Contamination of coastal aquifers with salt water
Weather Extremes
• Prolonged heat waves and droughts
• Increased flooding
• More intense hurricanes, typhoons, tornadoes, and violent storms
Human Population
• Increased deaths
• More environmental refugees
• Increased migration
Human Health
• Increased deaths from heat and disease
• Disruption of food and water supplies
• Spread of tropical diseases to temperate areas
• Increased respiratory disease
• Increased water pollution from coastal flooding

Figure 13-11 Some possible effects of a warmer atmosphere. Location, intensity, and duration of such effects will vary in different parts of the world.

■ The largest temperature increases will take place at the earth's poles and probably cause more melting of floating ice and glaciers (Connections, p. 308). This could **(1)** decrease the earth's ability to reflect incoming sunlight (albedo), **(2)** amplify global warming, **(3)** reduce the amount of tundra available to Arctic species and species that spend part of each year in the Arctic, and **(4)** cause declines in populations of some penguin species in the Antarctic.

■ For each 1°C (1.8°F) rise in the earth's average temperature, climate belts in midlatitude regions would shift toward the earth's poles by 100–150 kilometers (60–90 miles) or upward 150 meters (500 feet) in altitude. Such shifts could **(1)** change areas where crops could be grown and **(2)** affect the makeup and location of at least one-third of today's forests.

■ Tree species whose seeds are spread by wind may not be able to migrate fast enough to keep up with climate shifts and would die out (Figure 13-12). In addition, projected climate changes would make many existing forest trees more susceptible to fire and disease.

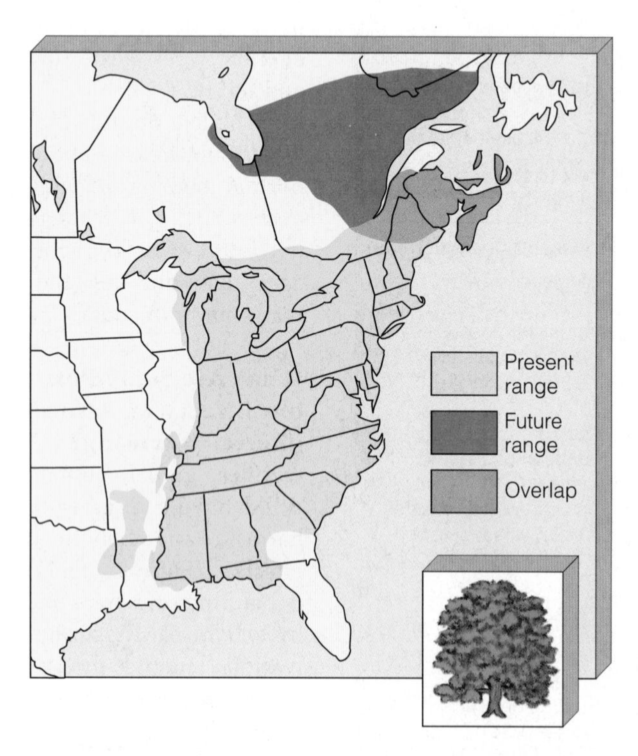

Figure 13-12 Possible effects of global warming on the geographic range of beech trees based on archaeological evidence and computer models. According to one projection, if CO_2 emissions doubled between 1990 and 2050, beech trees (now common throughout the eastern United States) would survive only in a greatly reduced range in northern Maine and southeastern Canada. This is only one of a number of tree species whose geographic ranges could be changed drastically by increased atmospheric warming. (Data from Margaret B. Davis and Catherine Zabinski, University of Minnesota)

■ Other sources of large inputs of CO_2 into the environment include **(1)** forest diebacks from climate change, **(2)** greatly increased wildfires in areas where the climate becomes drier, and **(3)** tree deaths from increased disease and pest populations. This could accelerate atmospheric warming and affect timber production, wildlife habitat, recreational activities, and homes and other human structures.

■ Shifts in regional climate would threaten many parks, wildlife reserves, wilderness areas, wetlands, and coral reefs, wiping out many current efforts to stem the loss of biodiversity (Chapter 17, p. 429).

■ Ocean currents may shift and change (Figure 13-9). This could cause sharp temperature drops in areas such as northern Europe and Japan, where climates are more temperate because of warm-water currents (such as the Gulf Stream and Kuroshio, Figure 6-3, p. 123).

■ Global sea levels will rise, mainly because water expands slightly when heated. In their 2001 IPCC report, climate scientists projected a rise of 9–88 centimeters (4–35 inches) during the 21st century. The high projected rise in sea level of about 88 centimeters (35 inches) would **(1)** threaten half of the world's coastal estuaries, wetlands (one-third of those in the United States), and coral reefs and disrupt marine fisheries, **(2)** cause severe beach erosion (especially along the U.S. East Coast), **(3)** flood coastal regions and put an estimated 200 million people living in 30 of the world's largest coastal cities directly at risk (Figure 13-13), **(4)** flood agricultural lowlands and deltas in parts of Bangladesh, India, and China, where much of the world's rice is grown, **(5)** contaminate freshwater coastal aquifers with saltwater, and **(6)** submerge some low-lying islands in the Pacific and Caribbean (Figure 13-13). If warming at the poles causes increased melting of land-based ice sheets, sea levels will rise much more (Connections, p. 308).

■ The largest burden will fall on developing nations, which do not have the economic and technological resources needed to adapt to the adverse impacts of climate change.

In 2001, a report by the Intergovernmental Panel on Climate Change (IPCC) projected the following major changes in different regions of the world during this century:

■ **Africa: (1)** Decreased grain yields, **(2)** less water available, **(3)** much higher temperatures in some countries, **(4)** increased desertification (Figure 9-23, p. 215), and **(5)** rising sea levels in Egypt and along the southeastern African coast (Figure 13-13).

■ **Asia: (1)** Much higher temperatures in much of the continent, **(2)** increased drought in some areas,

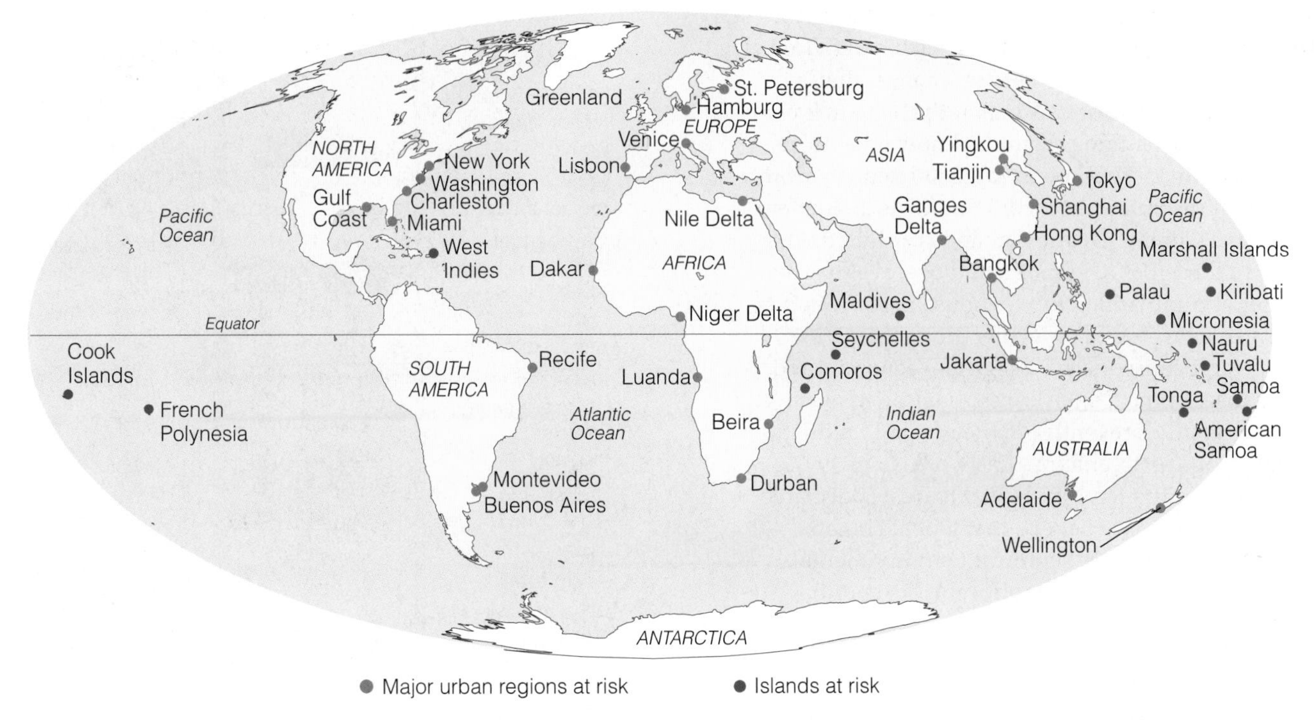

Figure 13-13 The high projected sea-level rise by 2100 could cause severe flooding of many of the world's major cities and low-lying islands. (Data from Intergovernmental Panel on Climate Change, 2000)

(3) increased floods in some areas (Figure 13-13), **(4)** soil degradation, **(5)** decreased food production, and **(6)** rises in sea level and more intense tropical cyclones in coastal areas that will displace tens of millions of people.

- **Europe: (1)** Increased agricultural productivity with moderate temperature increases, **(2)** increased drought in southern Europe, **(3)** increased flooding in some coastal areas (Figure 13-13), and **(4)** disappearance of half of Alpine glaciers.

- **Latin America: (1)** Increased drought in some areas, **(2)** increased floods in some areas (Figure 13-13), and **(3)** lower yields of key crops in many areas (such as Mexico and northeastern Brazil).

- **North America: (1)** Much higher temperatures in Canada, **(2)** increased food production in some areas of the United States from moderate warming, **(3)** decreased crop yields in the U.S. Great Plains and Canada's prairies, **(4)** faster and earlier snowmelt in the Rocky Mountains, which would increase spring flooding, make summers drier, and increase wildfires, and **(5)** increased sea level, coastal erosion, flooding, and greater risk of storm surges, especially in Florida and along the Atlantic coast (Figure 13-13). According to a 2000 report by the U.S. Global Climate Change Research Program, average temperatures in the United States are likely to increase by 3–5°C (5–9°F) by 2100.

- **Polar areas: (1)** Very large temperature increases and **(2)** increased ice melting (Connections, p. 308), which can raise global sea levels and change climate patterns by altering global ocean circulation (Figure 13-9).

- **Small island states:** Greatly increased coastal erosion and disappearance of some low-lying islands from flooding (Figure 13-13) as a result of sea-level rises.

13-5 SOLUTIONS: DEALING WITH THE THREAT OF CLIMATE CHANGE

What Are Our Options? There are four schools of thought concerning global warming.

- *Do nothing.* A dozen or so scientists contend that climate change from human activities is not a threat, and a few popular press commentators and writers even claim that global warming is a hoax.

- *Do more research before acting.* A second group of scientists and economists point to the considerable uncertainty about climate change and its effects. They call for more research before making such far-reaching economic and political decisions as phasing out fossil fuels and sharply reducing deforestation.

- *Act now to reduce the risks from climate change.* A third group of scientists and economists urge us to adopt a *precautionary strategy*. They believe that when dealing with risky and far-reaching environmental problems such as climate change, the safest course is to take informed preventive action *before* there is overwhelming scientific knowledge to justify acting. In 1997, more than 2,500 scientists from a variety of disciplines signed a Scientists' Statement on Global Climate Disruption and concluded, "We endorse those [IPCC] reports and observe that the further accumulation of greenhouse gases commits the earth irreversibly to further global climatic change and consequent ecological, economic, and social disruption. The risks associated with such changes justify preventive action through reductions in emissions of greenhouse gases." Also in 1997, 2,700 economists led by 8 Nobel laureates declared, "As economists, we believe that global climate change carries with it significant environmental, economic, social, and geopolitical risks and that preventive steps are justified."

- *Act now as part of a no-regrets strategy.* Scientists and economists supporting this approach say that we should take the key actions needed to slow atmospheric warming even if it is not a serious threat because such actions lead to other important environmental, health, and economic benefits (Solutions, right). For example, a reduction in the combustion of fossil fuels, especially coal, will lead to sharp reductions in air pollution that **(1)** harms and prematurely kills large numbers of people, **(2)** lowers food and timber productivity, and **(3)** decreases biodiversity.

Those who favor doing nothing or waiting before acting point out that there is a 50% chance that we are *overestimating* the impact of rising greenhouse gases. However, those urging action now point out that there is also a 50% chance that we are *underestimating* such effects.

How Can We Reduce the Threat of Climate Change from Human Activities? Figure 13-14 presents a variety of prevention and cleanup solutions analysts have suggested for slowing climate change from increased greenhouse gas emissions. Basically it boils down to **(1)** wasting less energy by improving energy efficiency (Section 20-2, p. 516), **(2)** using less oil and coal, which produce greenhouse gases, and **(3)** relying more on cleaner energy sources such as natural gas (p. 499), wind (p. 531), solar (p. 523), and hydrogen (p. 534). Gradually implementing such solutions over the next 20–30 years could simultaneously reduce the threats from global warming, air pollution, deforestation, and biodiversity loss.

How Can We Use Government Regulation to Reduce Greenhouse Gas Emissions? Governments could significantly reduce CO_2 emissions over a decade by

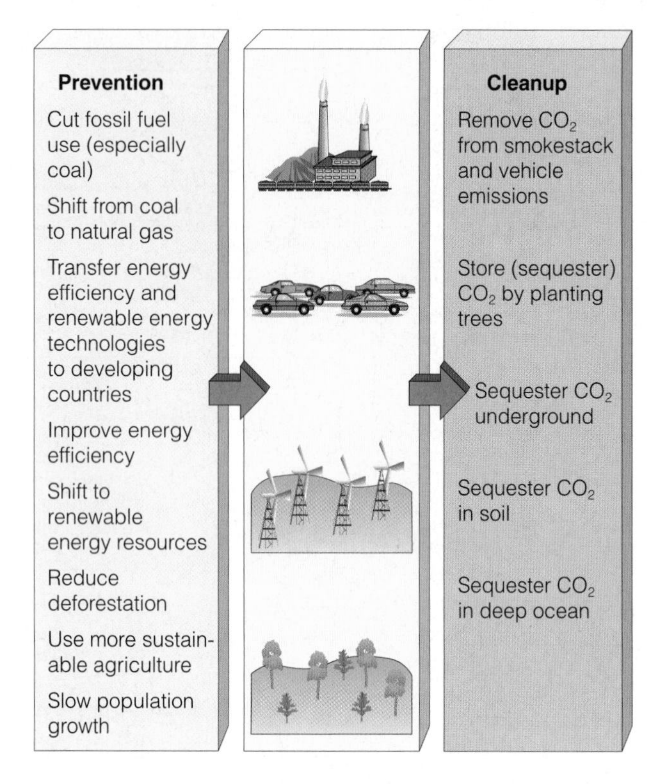

Figure 13-14 Solutions: methods for slowing atmospheric warming during the 21st century.

- Phasing out government subsidies for coal and oil (about $300 billion per year globally and $20 billion annually in the United States).
- Retaining government subsidies for natural gas, which could help us make the 40- to 50-year transition to an age of energy efficiency and renewable energy. When burned, natural gas emits only half as much CO_2 per unit of energy as coal (Figure 19-11, p. 498), and it emits far smaller amounts of most other air pollutants.
- Increasing emphasis on reducing emissions of methane, which absorbs and releases about 20 times as much heat per molecule as CO_2 (Table 13-1). Ways to reduce methane inputs include **(1)** capturing and burning methane gas released by landfills (Figure 15-9, p. 384) and **(2)** reducing leaks from tanks, pipelines, and other natural gas handling facilities.
- Agreeing to global and national limits on greenhouse gas emissions and encouraging industries and countries to meet these limits by selling and trading greenhouse gas emission permits in the marketplace. For example, a coal-burning power plant in Ohio could earn credits to allow some CO_2 emissions by financing a CO_2-removing tree planting project in Oregon or Costa Rica. Some analysts oppose this approach

Energy Efficiency to the Rescue

SOLUTIONS

According to energy expert Amory Lovins (Guest Essay, p. 517), *the major remedies for slowing global warming are things we should be doing already even if there were no threat of global warming.* Lovins argues that we should **(1)** waste less energy, **(2)** reduce air pollution by cutting down on our use of fossil fuels (especially coal and oil) and switching to renewable forms of energy, and **(3)** harvest trees more sustainably.

According to Lovins, improving energy efficiency (Section 20-2, p. 516) would be the fastest, cheapest, and surest way to slash emissions of CO_2 and most other air pollutants within two decades, using existing technology. Lovins estimates that increased energy efficiency would also save **(1)** $1 trillion per year in reduced energy costs globally (as much as the annual global military budget) and **(2)** $300 billion per year in the United States.

Lovins also warns that using fossil fuels such as natural gas (CH_4) as a source of hydrogen to power fuel cells (Figure 20-8, p. 520) instead of hydrogen derived from H_2O (Figure 20-27, p. 535) could increase the threat of global warming. The reason is that stripping hydrogen from carbon-based fossil fuels leaves behind carbon dioxide, which would probably be vented into the atmosphere.

Using energy more efficiently would also **(1)** reduce pollution, **(2)** help protect biodiversity, **(3)** help prevent arguments between governments about how CO_2 reductions should be divided up and enforced, **(4)** make the world's supplies of fossil fuel last longer, **(5)** reduce international tensions over who gets the dwindling oil supplies, and **(6)** allow more time to phase in renewable energy sources. A growing number of companies have gotten the message and are **(1)** saving money and energy while reducing their carbon dioxide emissions and **(2)** selling their emission reductions in the global marketplace.

Critical Thinking

1. Do you agree that improving energy efficiency should be done regardless of its impact on the threat of global warming? Explain.

2. Why do you think there has been little emphasis on improving energy efficiency?

because **(1)** carbon fuels are burned in so many homes, vehicles, factories, and cropfields that it is almost impossible to monitor such a system to see whether trading partners are living up to their agreements, and **(2)** it leads to disagreement between developed countries and developing countries by allowing developed countries to gain credits toward their CO_2 reduction goals by planting trees instead of adopting measures to reduce fossil fuel use, improve energy efficiency, and shift to noncarbon renewable energy resources.

- Allocating emissions equally on a per capita basis throughout the world. For example, each American on average would be allowed the same emissions as the average Indian. (Currently, the average American produces 25 times more CO_2 than the average Indian.) Many developing countries contend that it is fairer and much more democratic to allocate greenhouse gas emissions. Developed countries strongly oppose this approach.

- Phasing in **(1)** output-based *carbon taxes* on each unit of CO_2 emitted by fossil fuels (especially coal and gasoline) or **(2)** input-based *energy taxes* on each unit of fossil fuel (especially coal and gasoline) that is burned. Costa Rica has enacted a 15% carbon tax and uses a third of the revenues to finance tree-planting projects by farmers.

- Decreasing taxes on income, labor, and profits to match any increases in consumption taxes on carbon emissions or fossil fuel use (Solutions, p. 33).

- Greatly increasing government subsidies for energy-efficiency and renewable-energy technologies to help speed up the shift to these alternatives (Chapter 20, p. 513).

- Having each country agree to improve energy efficiency by a specified amount (such as 2–5%) every year until global emissions of greenhouse gases have been cut by at least 50%. Each country's progress could be measured by calculating the annual change in its use of carbon fuel divided by its gross domestic product.

- Funding the transfer of energy efficiency and renewable energy technologies from developed countries to developing countries. Increasing the current tax on each international currency transaction by a quarter of a penny could finance this technology transfer, which would then generate wealth for developing countries. This would generate revenues of $200–300 billion a year for projects such as fuel-cell factories in South Africa, wind farms in India and China, and vast solar-powered hydrogen farms in the sunny Middle East.

- Increasing use of nuclear power (Section 19-5, p. 502) because it produces only about one-sixth as much CO_2 per unit of electricity as coal (Figure 19-11, p. 498). Other analysts oppose this because of **(1)** the danger of large-scale releases of highly radioactive materials from nuclear power-plant accidents (p. 490), **(2)** its high cost (p. 509) even with large government subsidies (at least $10 billion a year in the United States), and **(3)** the problem of storing radioactive wastes safely for thousands of years (p. 506).

- Using government regulations and subsidies to **(1)** reduce deforestation (p. 445) and **(2)** encourage increased use of sustainable agriculture (Figure 16-29, p. 425, and Solutions, right).
- Establishing national policies and funding international efforts to slow population growth (p. 260). If we cut per capita greenhouse gas emissions in half but world population doubles, we're back where we started.

Organic Farming to the Rescue

SOLUTIONS

Recent studies by the Rodale Institute in Pennsylvania and other scientists suggest that a wholesale switch to organic farming could help slow global warming. They have shown that crops grown with organic fertilizer produce equivalent yields but much less CO_2 through respiration than crops grown with commercial fertilizer.

CO_2 outputs also are reduced because organic farming uses 50% less energy than conventional farming methods. Farmers can also reduce greenhouse emissions by adopting well-known soil conservation methods such as **(1)** using conservation tillage (p. 218) to reduce or eliminate plowing, **(2)** using cover crops in winter, and **(3)** preserving buffer strips of trees along riverbanks.

Critical Thinking

List three things governments could do to encourage farmers to switch to more sustainable organic farming and soil conservation methods.

How Can We Remove CO_2 from the Atmosphere? Scientists are evaluating several ways to remove CO_2 from the atmosphere or from smokestacks and store (sequester) it in **(1)** immature trees, **(2)** plants that store it in the soil, **(3)** deep underground reservoirs, and **(4)** the deep ocean.

One way to remove CO_2 from the atmosphere temporarily would be to *plant trees* over an area equivalent to the size of Australia in a huge global reforestation program. Such a program would also help restore degraded lands. However, studies suggest that a global reforestation program (requiring each person in the world to plant and tend to an average of 1,000 trees every year) would offset only about 3 years of our current CO_2 emissions from burning fossil fuels.

Recent scientific studies call into question the effectiveness of trees (p. 310) and soils (p. 310) as long-term carbon sinks. There are also ecological concerns. To get a short-term increase in carbon uptake, managers of carbon sequestration projects might be tempted to replace naturally diverse forest areas with tree plantations and thus reduce some of the earth's biodiversity.

Other sequestering approaches include collecting CO_2 from smokestacks (and natural gas wells) and **(1)** *pumping it deep underground* into unminable coal seams and abandoned oil fields or **(2)** *injecting it into the deep ocean.* However, any method of underground or deep-sea sequestration would take a costly investment in materials, transportation of the CO_2 (presumably by pipeline) to storage sites, and infrastructure. In addition, injecting large quantities of CO_2 into the ocean could upset the global carbon cycle and some forms of deep-sea life in unpredictable ways. Another problem is that current methods can remove only about 30% of the CO_2 from smokestack emissions, and using them would probably double the cost of electricity.

Can Technofixes Save Us? Some scientists have suggested various technofixes for reducing the threat of global warming, including **(1)** adding iron to the oceans to stimulate the growth of marine algae (which could remove more CO_2 through photosynthesis but would return it to the atmosphere a short time later unless the carbon was somehow deposited in the deep ocean), **(2)** unfurling gigantic foil-surfaced sun mirrors in space or placing such mirrors on about 50,000 orbiting satellites to reduce solar input, **(3)** releasing trillions of reflective balloons filled with helium into the atmosphere, and **(4)** injecting sunlight-reflecting sulfate particulates or firing sulfur dioxide cannonballs into the stratosphere to cool the earth's surface (this would turn the sky white and increase depletion of stratospheric ozone).

Many of these costly schemes might not work, and most probably would have unpredictable short- and long-term harmful environmental effects. Moreover, once started, those that work could never be stopped without a renewed rise in CO_2 levels. Instead of spending huge sums of money on such schemes, many scientists believe it would be more effective and less expensive to **(1)** improve energy efficiency (Solutions, p. 315) and **(2)** shift to renewable forms of energy that do not produce carbon dioxide (Chapter 20, p. 513).

What Has Been Done to Reduce Greenhouse Gas Emissions? At the 1992 Earth Summit in Rio de Janeiro, Brazil (p. 43), 106 nations approved a Convention on Climate Change in which developed countries committed themselves to reducing their emissions of CO_2 and other greenhouse gases to 1990 levels by the year 2000. However, the convention did not *require* countries to reach this goal, and most countries did not achieve this goal.

In December 1997, more than 2,200 delegates from 161 nations met in Kyoto, Japan, to negotiate a new treaty to help slow global warming. The resulting treaty would **(1)** require 38 developed countries to cut greenhouse emissions to an average of about 5.2% below 1990 levels by 2012, **(2)** not require developing countries to make any cuts in their greenhouse gas emissions until the second stage of the treaty, and **(3)** allow emissions trading.

Some analysts praise the Kyoto agreement as a small but important step in dealing with the problem of global warming and hope that the conditions of the treaty will be strengthened in future negotiating sessions. However, according to computer models, the 5.1% reduction goal of the Kyoto Protocol would shave only about 0.06°C (0.1°F) off the 0.7–1.7°C (1–3°F) temperature rise projected by 2060.

The U.S. Congress has not ratified the treaty, mostly because of **(1)** its failure to require emission reductions from developing countries and **(2)** intensive lobbying by a coalition of coal, oil, steel, chemical, and automobile companies opposed to the treaty who argued that it would have a devastating impact on the U.S. economy and workers. In 2001, President George Bush said that he was not interested in participating in the Kyoto treaty. He declared the treaty dead because it did not include developing countries such as China and would hurt the U.S. economy. This decision set off strong protests by many scientists, citizens, and leaders throughout most of the world.

What Must Be Done to Reduce Greenhouse Gas Emissions? Politicians in developed countries continue to argue about how to decrease CO_2 emissions by about 5% over 1990 levels between 2008 and 2012. However, *according to latest global climate models, the world needs to reduce emissions of greenhouse gases (not just CO_2) by at least 50% by 2018 to stabilize concentrations of CO_2 in the air at their present levels.*

The *bad news* is that such a change is extremely unlikely for political and economic reasons because it would take rapid, widespread changes in **(1)** industrial processes, **(2)** energy sources, **(3)** transportation options, and **(4)** individual lifestyles. It also takes vision, courage, and bold leadership by the leaders of the United States (by far the greatest contributor of greenhouse gases) and other nations throughout the world.

The *good news* is that some progress is being made.

- The United Kingdom is committed to a 12.5% reduction in CO_2 emissions by 2010, and a royal commission has called for 60% cuts by 2050.

- Germany is considering cuts of 50%.

- The low-lying Netherlands, which faces a serious threat from rising sea levels, plans to cut its CO_2 emissions by 80% in the next 40 years.

- A number of developing countries, including China (the world's third largest emitter of CO_2, after the United States and the European Union), are voluntarily improving energy efficiency and installing wind, solar, and small hydropower projects even though they are exempt from cuts under the first round of the Kyoto treaty. For example, a 2001 report by the Natural Resources Defense Council found that China reduced its CO_2 emissions by 17% between 1997 and 2000 while its economy grew by 36%. (By comparison CO_2 emissions in the United States rose by 14% between 1997 and 2000.) China accomplished this by **(1)** switching from coal to cleaner energy resources by phasing out all coal subsidies, closing coal mines, and shutting down inefficient coal-fired electric plants, **(2)** stepping up its 20-year commitment to promoting energy efficiency, and **(3)** restructuring its economy to reduce dependence on fossil fuels and increase use of renewable energy resources.

- Since 2000 some of the major automobile companies and oil companies (with the exception of ExxonMobil) have dropped out of the Global Climate Coalition, which opposed the Kyoto treaty and most other action on reducing the threat of global warming. Their CEOs **(1)** indicated that global warming was a potential risk that should be addressed and **(2)** agreed with most economists that dealing with this problem would stimulate the economy and create many new jobs.

- DuPont has committed to reducing its greenhouse gas emissions by 65% between 1990 and 2010. British Petroleum (BP) **(1)** is investing large sums in solar power and anticipates doing $1 billion a year in sales of solar technologies by 2010 and **(2)** aims to cut its greenhouse gas emissions to 20% below 1990 levels by 2010. Shell Oil has created a $500-million renewable energy company and plans to cut its greenhouse emissions to 10% below 1990 levels by 2002.

- DaimlerChrysler and Ford have invested $1 billion in a joint venture to begin selling fuel-cell-powered cars (Figure 20-8, p. 520) by 2004. Toyota, Honda, and General Motors also plan to sell such cars within a few years.

How Can We Prepare for Global Warming? Without a much greater sense of urgency and bold leadership by the United States, many (perhaps most) of the actions climate experts have recommended for slowing atmospheric warming (Figure 13-14) either will not be done or will be done too slowly. As a result, a growing number of analysts suggest that we should also

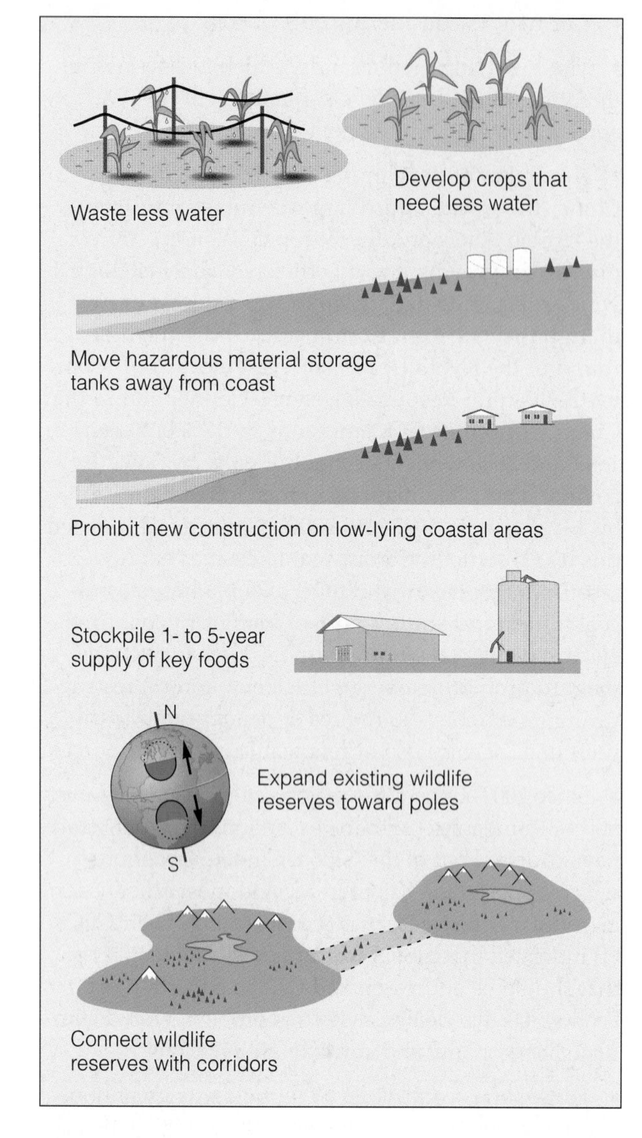

Figure 13-15 Solutions: ways to prepare for the possible long-term effects of climate caused by increased atmospheric temperatures.

begin to prepare for the possible harmful effects of long-term atmospheric warming and climate change (Figure 13-11). Figure 13-15 shows some ways to do this.

See the website material for this chapter for some actions you can take to reduce the threat of global warming.

13-6 OZONE DEPLETION IN THE STRATOSPHERE

What Is the Threat from Ozone Depletion? A layer of ozone in the lower stratosphere (Figure 12-2, p. 278, and Figure 12-3, p. 279) keeps about 95% of the sun's harmful ultraviolet (UV) radiation from reaching the earth's surface. Measuring instruments on balloons, aircraft, and satellites show seasonal depletion (thinning) of ozone concentrations in the stratosphere above Antarctica and the Arctic. Similar measurements reveal a lower overall thinning everywhere except over the tropics.

Based on these measurements and chemical models, the overwhelming consensus of researchers in this field is that ozone depletion (thinning) in the stratosphere is a serious long-term threat to **(1)** humans, **(2)** many other animals, and **(3)** the sunlight-driven primary producers (mostly plants) that support the earth's food chains and webs.

What Causes Ozone Depletion? From Dream Chemicals to Nightmare Chemicals This situation started when Thomas Midgley, Jr., a General Motors chemist, discovered the first **chlorofluorocarbon (CFC)** in 1930, and chemists made similar compounds to create a family of highly useful CFCs. The two most widely used were CFC-11 (trichlorofluoromethane, CCl_3F) and CFC-12 (dichlorodifluoromethane, CCl_2F_2), known by their trade name as Freons.

These chemically stable (nonreactive), odorless, nonflammable, nontoxic, and noncorrosive compounds seemed to be dream chemicals. Cheap to make, they became popular as **(1)** coolants in air conditioners and refrigerators (replacing toxic sulfur dioxide and ammonia), **(2)** propellants in aerosol spray cans, **(3)** cleaners for electronic parts such as computer chips, **(4)** sterilants for hospital instruments, **(5)** fumigants for granaries and ship cargo holds, and **(6)** bubbles in plastic foam used for insulation and packaging. Between 1960 and the early 1990s, CFC production rose sharply.

However, CFCs were too good to be true. In 1974, calculations by two University of California–Irvine chemists, Sherwood Rowland and Mario Molina, indicated that CFCs were lowering the average concentration of ozone in the stratosphere. They shocked both the scientific community and the $28-billion-per-year CFC industry by calling for an immediate ban of CFCs in spray cans (for which substitutes were available).

According to Rowland and Molina,

- Large quantities of CFCs were being released into the troposphere mostly from **(1)** the use of CFCs as propellants in spray cans, **(2)** leaks from refrigeration and air conditioning equipment, and **(3)** the production and burning of plastic foam products.
- CFCs remain in the troposphere because they are insoluble in water and are chemically unreactive.
- Over 11–20 years they rise into the stratosphere mostly through convection, random drift, and the turbulent mixing of air in the troposphere.
- In the stratosphere, the CFC molecules break down under the influence of high-energy UV radiation. This

releases highly reactive chlorine atoms, which speed up the breakdown of very reactive ozone (O_3) into O_2 and O in a cyclic chain of chemical reactions (Figure 13-16). This causes ozone in various parts of the stratosphere to be destroyed faster than it is formed.

- Each CFC molecule can last in the stratosphere for 65–385 years (depending on its type), with the most widely used CFCs lasting 75–111 years. During that time, each chlorine atom released from these molecules can convert up to 100,000 molecules of O_3 to O_2.

According to Rowland and Molina's calculations and later models and atmospheric measurements of CFCs in the stratosphere, these dream molecules have turned into a nightmare of global ozone destroyers.

The CFC industry (led by the DuPont Company), a powerful, well-funded adversary with a lot of profits and jobs at stake, attacked Rowland and Molina's calculations and conclusions. However, they held their ground, expanded their research, and explained the meaning of their calculations to other scientists, elected officials, and the media. It was not until 1988—14 years after Rowland and Molina's study—that DuPont officials acknowledged that CFCs were depleting the ozone layer and agreed to stop producing them once they found substitutes. In 1995, Rowland and Molina received the Nobel prize in chemistry for their work.

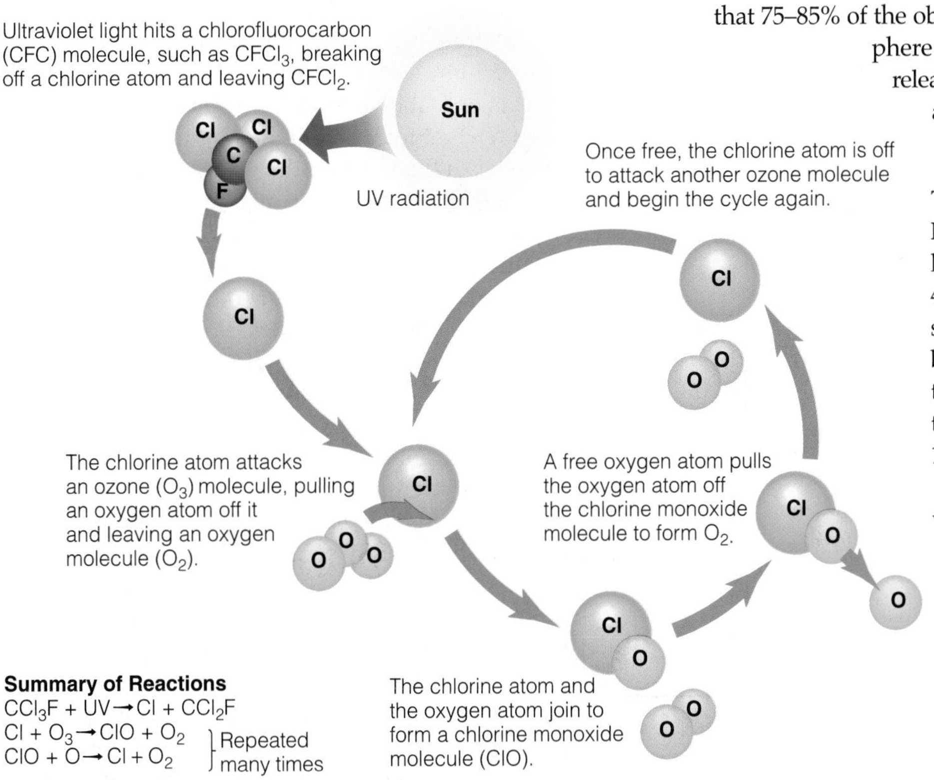

What Other Chemicals Deplete Stratospheric Ozone? Other ozone-depleting compounds (ODCs) include

- *Halons* and *HBFCs*, used in fire extinguishers
- *Methyl bromide* (CH_3Br), a widely used fumigant
- *Carbon tetrachloride* (CCl_4), a cheap, highly toxic solvent
- *Methyl chloroform*, or 1,1,1-trichloroethane ($C_2H_3Cl_3$), used as a cleaning solvent for clothes and metals and as a propellant in more than 160 consumer products such as correction fluid, dry-cleaning sprays, spray adhesives, and other aerosols
- *Hydrogen chloride* (HCl), emitted into the stratosphere by U.S. space shuttles

The oceans and occasional volcanic eruptions also release chlorine and bromine compounds into the troposphere. However, most of these chlorine compounds do not make it to the stratosphere because they easily dissolve in water and are washed out of the troposphere in rain. Bromine compounds may be less likely to be washed out of the troposphere, but further study is needed to confirm this possibility. Measurements and models indicate that 75–85% of the observed ozone losses in the stratosphere since 1976 are the result of ODCs released into the atmo-sphere by human activities beginning in the 1950s.

Why Is There Seasonal Thinning of Ozone over the Poles? In 1984, researchers analyzing satellite data discovered that 40–50% of the ozone in the upper stratosphere over Antarctica was being destroyed during the antarctic spring and early summer (September–December), especially since 1976 (Figure 13-17).

Figure 13-18 shows the seasonal variation of ozone over Antarctica during 1997. The observed seasonal loss during the antarctic summer has been incorrectly called an *ozone hole*. A more accurate term is *ozone thinning* because the ozone depletion varies with altitude (Figure 13-18) and location.

The total area of the atmosphere above Antarctica that suffers from ozone thinning during the peak season varies from year

Ultraviolet light hits a chlorofluorocarbon (CFC) molecule, such as $CFCl_3$, breaking off a chlorine atom and leaving $CFCl_2$.
Sun
UV radiation
Cl
C
F
Once free, the chlorine atom is off to attack another ozone molecule and begin the cycle again.
The chlorine atom attacks an ozone (O_3) molecule, pulling an oxygen atom off it and leaving an oxygen molecule (O_2).
A free oxygen atom pulls the oxygen atom off the chlorine monoxide molecule to form O_2.
The chlorine atom and the oxygen atom join to form a chlorine monoxide molecule (ClO).
O
Summary of Reactions
$CCl_3F + UV \rightarrow Cl + CCl_2F$
$Cl + O_3 \rightarrow ClO + O_2$
$ClO + O \rightarrow Cl + O_2$
Repeated many times

Figure 13-16 A simplified summary of how chlorofluorocarbons (CFCs) and other chlorine-containing compounds destroy ozone in the stratosphere. Note that chlorine atoms are continuously regenerated as they react with ozone. Thus, they act as *catalysts*, chemicals that speed up chemical reactions without being used up by the reaction. Bromine atoms released from bromine-containing compounds that reach the stratosphere also destroy ozone by a similar mechanism.

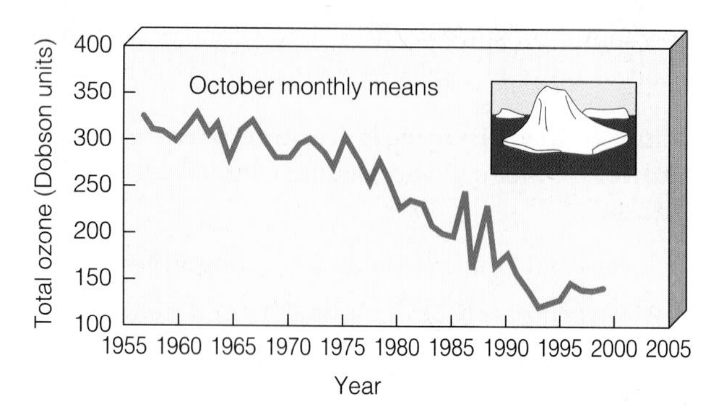

Figure 13-17 Mean total level of ozone for October over the Halley Bay measuring station in Antarctica, 1956–1999. (Data from British Antarctic Survey and World Meteorological Organization)

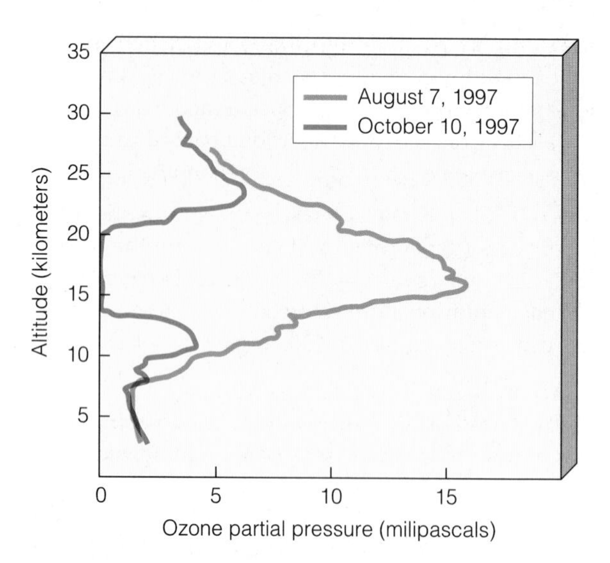

Figure 13-18 Variation of ozone level with altitude over Antarctica during 1997. Note the severe depletion of ozone during October (during the antarctic summer) and its return to more normal levels in August (during the antarctic winter). (Data from National Oceanic and Atmospheric Administration)

to year (Figure 13-19). In 2000, seasonal ozone thinning above Antarctica was the largest ever and covered an area three times the size of the continental United States.

Measurements indicate that CFCs are the primary culprits. Each sunless winter, steady winds blow in a circular pattern over the earth's poles. This creates a *polar vortex*: a huge swirling mass of very cold air that is isolated from the rest of the atmosphere until the sun returns a few months later.

When water droplets in clouds enter this circling stream of extremely frigid air, they form tiny ice crystals. The surfaces of these ice crystals collect CFCs and other ODCs in the stratosphere and speed up (catalyze) the chemical reactions that release Cl atoms and ClO. Instead of entering a chain reaction of ozone destruction (Figure 13-16), the ClO atoms combine with one another to form Cl_2O_2 molecules. In the dark of winter, the Cl_2O_2 molecules cannot react with ozone, so they accumulate in the polar vortex.

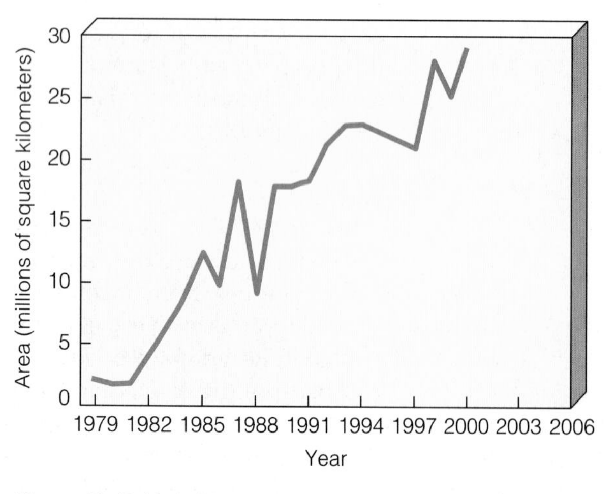

Figure 13-19 Variation in the average area of ozone thinning above Antarctica between September 7 and October 10 from 1979 to 1999. (Data from National Oceanic and Atmospheric Administration)

When sunlight and the antarctic spring return 2–3 months later (in October), the light breaks up the stored Cl_2O_2 molecules, releasing large numbers of Cl atoms and initiating the catalyzed chlorine cycle (Figure 13-16). Within weeks, this typically destroys 40–50% of the ozone above Antarctica (and 100% in some places). The returning sunlight **(1)** gradually melts the ice crystals, **(2)** breaks up the vortex of trapped polar air, and **(3)** allows it to begin mixing again with the rest of the atmosphere. Then new ozone forms over Antarctica until the next dark winter (Figure 13-18).

When the vortex breaks up, huge masses of ozone-depleted air above Antarctica flow northward and linger for a few weeks over parts of Australia, New Zealand, South America, and South Africa. This raises biologically damaging UV-B levels in these areas by 3–10% and in some years by as much as 20%.

In 1988, scientists discovered that similar but less severe ozone thinning occurs over the Arctic during the arctic spring and early summer (February–June), with a seasonal ozone loss of 11–38% (compared with a typical 50% loss above Antarctica). However, in 1997 and 1999, the average seasonal loss over the Arctic was about 60%. When this mass of air above the Arctic breaks up each spring, large masses of ozone-depleted air flow south to linger over parts of Europe, North America, and Asia. According to a 1998 model developed by National Aeronautics and Space Agency (NASA) scientists at the Goddard Institute for Space Studies, ozone depletion over the Antarctic and Arctic will be at its worst between 2010 and 2019.

The Cancer You Are Most Likely to Get

Research indicates that years of exposure to UV-B ionizing radiation in sunlight is the primary cause of *squamous cell* (Figure 13-20, left) and *basal cell* (Figure 13-20, center) *skin cancers*. Together these two types make up 96% of all skin cancers. Typically there is a 15- to 40-year lag between excessive exposure to UV-B and development of these cancers.

Caucasian children and adolescents who get only a single severe sunburn double their chances of getting these two types of cancers. Some 90–95% of these types of skin cancer can be cured if detected early enough, although their removal may leave disfiguring scars. These cancers kill only 1–2% of their victims, but this still amounts to about 2,300 deaths in the United States each year.

A third type of skin cancer, *malignant melanoma* (Figure 13-20, right), occurs in pigmented areas such as moles anywhere on the body's surface. Within a few months, this type of cancer can spread to other organs. It kills about one-fourth of its victims (most under age 40) within 5 years, despite surgery, chemotherapy, and radiation treatments. Each year it kills about 100,000 people (including more than 7,700 Americans), mostly Caucasians. It can be cured if detected early enough, but recent studies show that some melanoma survivors have a recurrence more than 15 years later.

Recent evidence suggests that about 90% of sunlight's melanoma-causing effect may come from exposure to UV-A and 10% from UV-B. Some sunscreens do not protect from UV-A, and tanning booth lights emit mostly UV-A.

Evidence indicates that people (especially Caucasians) who get three or more blistering sunburns before age 20 are five times more likely to develop malignant melanoma than those who have never had severe sunburns. About 10% of those who get malignant melanoma have an inherited gene that makes them especially susceptible to the disease.

To protect yourself, the safest course is to **(1)** stay out of the sun (especially between 10 A.M. and 3 P.M., when UV levels are highest) and **(2)** avoid tanning parlors. When you are in the sun, wear **(1)** tightly woven protective clothing, **(2)** a wide-brimmed hat, and **(3)** sunglasses that protect against UV-A and UV-B radiation (ordinary sunglasses may actually harm your eyes by dilating your pupils so that more UV radiation strikes the retina).

Because UV rays can penetrate clouds, overcast skies do not protect you; neither does shade, because UV rays can reflect off sand, snow, water, or patio floors. People who take antibiotics and women who take birth control pills are more susceptible to UV damage.

Use a sunscreen that offers protection against both UV-A and UV-B and has a protection factor of 15 or more (25 if you have light skin). Apply to all exposed skin and reapply it after swimming or excessive perspiration. Most people do not realize that the protection factors for sunscreens are based on using one full ounce of the product—far more than most people apply. Some people also increase their risk of skin cancer by falsely assuming that sunscreens allow them to spend more time in the sun.

Children who use a sunscreen with a protection factor of 15 every time they are in the sun from age 1 to age 18 decrease their chance of getting skin cancer by 80%. Babies under a year old should not be exposed to the sun at all and should not have sunscreens applied until they are at least 6 months old.

Become familiar with your moles and examine your skin at least once a month. The warning signs of skin cancer are **(1)** a change in the size, shape, or color of a mole or wart (the major sign of malignant melanoma, which must be treated quickly), **(2)** sudden appearance of dark spots on the skin, or **(3)** a sore that keeps oozing, bleeding, and crusting over but does not heal. Be alert for precancerous growths (reddish-brown spots with a scaly crust). If you observe any of these signs, consult a doctor immediately.

Critical Thinking

What precautions, if any, do you take to reduce your chances of getting skin cancer from exposure to sunlight? Explain why you do or do not take such precautions.

Why Should We Be Worried About Ozone Depletion? Life in the Ultraviolet Zone Why should we care about ozone loss? From a human standpoint the answer is that with less ozone in the stratosphere, more biologically damaging UV-A and UV-B radiation will reach the earth's surface. This will give humans **(1)** worse sunburns, **(2)** more eye cataracts (a clouding of the eye's lens that reduces vision and can cause blindness if not corrected), and **(3)** more skin cancers (Connections, above).

According to UNEP estimates, the additional UV-B radiation reaching the earth's surface resulting from an annual 10% loss of global ozone (already a strong possibility within a few years) could lead to **(1)** 300,000 additional cases of squamous cell cancer (Figure 13-20, left) and basal cell cancer (Figure 13-20, center) worldwide each year, **(2)** 4,500–9,000 additional cases of potentially fatal malignant melanoma (Figure 13-20, right) each year, and **(3)** 1.5 million new cases of cataracts (which account for more than half of the world's 25–35 million cases of blindness) each year.

Other effects of increased UV exposure are

- Immune system suppression, which makes the body more susceptible to infectious diseases and some forms of cancer.
- An increase in damaging acid deposition (Figure 12-8, p. 286) and eye-burning photochemical smog in the troposphere (Figure 12-5, p. 282).
- Lower yields of key crops such as corn, rice, cotton, soybeans, beans, peas, sorghum, and wheat, with estimated losses totaling $2.5 billion per year in the United States before the mid-21st century.
- Decline in forest productivity of the many tree species sensitive to UV-B radiation. This could reduce CO_2 uptake and enhance global warming.
- Increased breakdown and degradation of materials such as various types of paints, plastics, and outdoor materials. Such damage could cost billions of dollars per year.
- Reduction in the productivity of surface-dwelling phytoplankton, which could **(1)** upset aquatic food

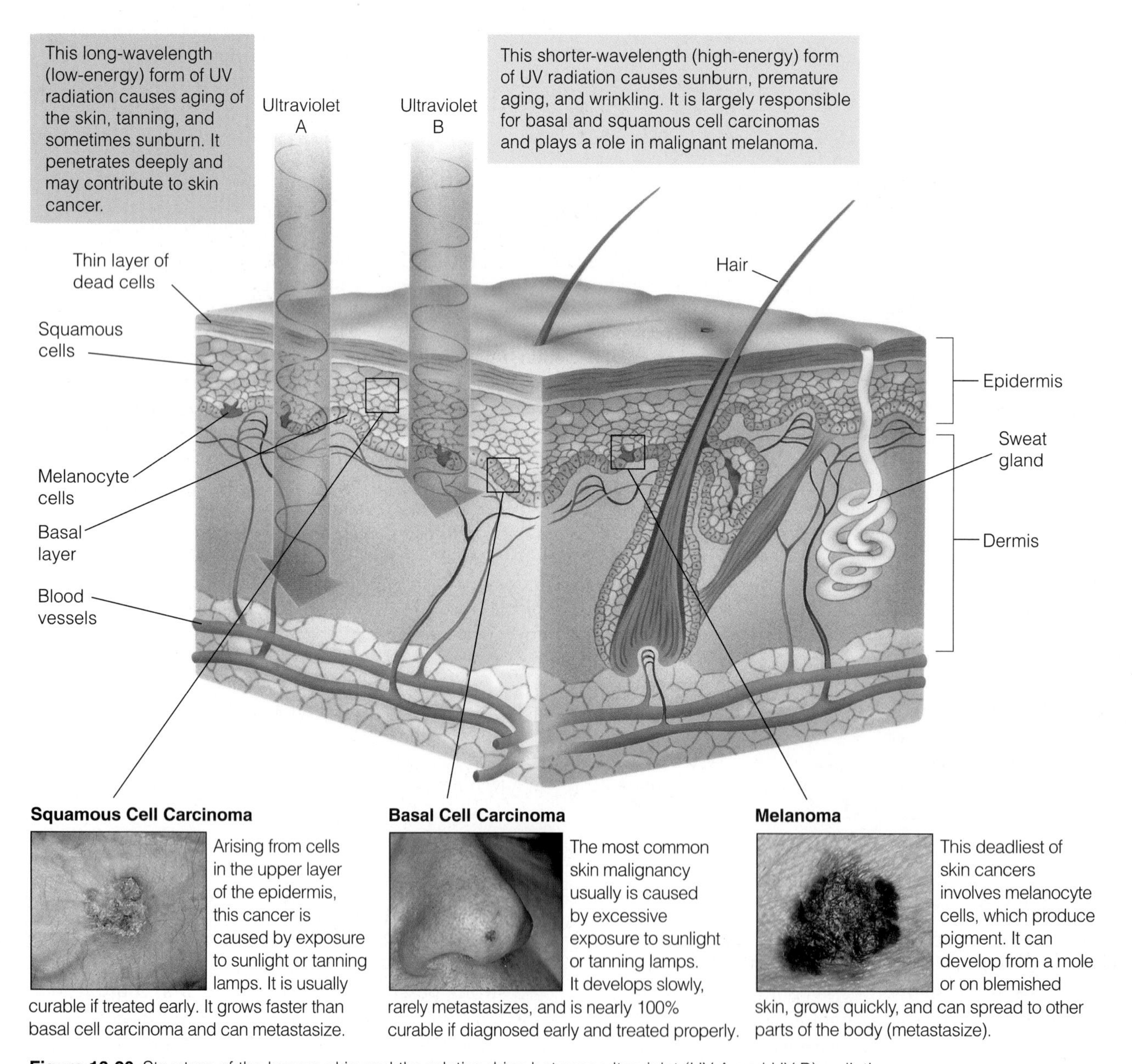

This long-wavelength (low-energy) form of UV radiation causes aging of the skin, tanning, and sometimes sunburn. It penetrates deeply and may contribute to skin cancer.
Ultraviolet A
Ultraviolet B
This shorter-wavelength (high-energy) form of UV radiation causes sunburn, premature aging, and wrinkling. It is largely responsible for basal and squamous cell carcinomas and plays a role in malignant melanoma.
Thin layer of dead cells
Squamous cells
Melanocyte cells
Basal layer
Blood vessels
Hair
Epidermis
Sweat gland
Dermis
Squamous Cell Carcinoma
Arising from cells in the upper layer of the epidermis, this cancer is caused by exposure to sunlight or tanning lamps. It is usually curable if treated early. It grows faster than basal cell carcinoma and can metastasize.
Basal Cell Carcinoma
The most common skin malignancy usually is caused by excessive exposure to sunlight or tanning lamps. It develops slowly, rarely metastasizes, and is nearly 100% curable if diagnosed early and treated properly.
Melanoma
This deadliest of skin cancers involves melanocyte cells, which produce pigment. It can develop from a mole or on blemished skin, grows quickly, and can spread to other parts of the body (metastasize).

Figure 13-20 Structure of the human skin and the relationships between ultraviolet (UV-A and UV-B) radiation and the three types of skin cancer. The incidence of these types of cancer is rising, mostly because more fair-skinned people have increased their exposure to sunlight by moving to areas with sunnier climates and spending more of their leisure time exposed to sunlight. If ozone-destroying chemicals continue to reduce stratospheric ozone levels, incidence of these types of cancers is expected to rise. (The Skin Cancer Foundation)

webs, **(2)** decrease yields of seafood eaten by people, and **(3)** possibly accelerate global warming by decreasing oceanic CO_2 uptake.

Humans can make cultural adaptations to increased UV-B radiation by **(1)** staying out of the sun, **(2)** protecting their skin with clothing, and **(3)** applying sunscreens (Connections, p. 321). However, UV-sensitive plants and animals that help support us and other forms of life cannot make such changes except through the long process of biological evolution.

13-7 SOLUTIONS: PROTECTING THE OZONE LAYER

How Can We Protect the Ozone Layer? *The scientific consensus of researchers in this field is that we should immediately stop producing all ozone-depleting chemicals.* Some *good news* is that substitutes are available for most uses of CFCs, and others are being developed (Individuals Matter, right).

Can Technofixes Save Us? What about a quick fix from technology, so that we can keep on using CFCs? One proposed scheme for removing ozone from the atmosphere is to launch a fleet of 20–30 football-field-long, radio-controlled blimps into the stratosphere above Antarctica. A huge curtain of electrical wires hanging from the blimps would inject negatively charged electrons into the stratosphere to react with and remove ozone-destroying chlorine atoms (Cl). However, atmospheric chemist Ralph Ciecerone believes that this plan will not work because other chemical species in the stratosphere snatch electrons more readily than does chlorine. This scheme could also have unpredictable side effects on atmospheric chemistry.

Others have suggested using tens of thousands of lasers to blast CFCs out of the atmosphere before they can reach the stratosphere. However, enormous amounts of energy would be needed to do this, and decades of research would be needed to perfect the types of lasers needed. Moreover, we cannot predict the possible effects of such powerful laser blasts on climate, birds, or planes.

What Is Being Done to Reduce Ozone Depletion? Some Hopeful Progress In 1987, 36 nations meeting in Montreal developed a treaty, commonly known as the *Montreal Protocol,* to cut emissions of CFCs (but not other ozone depleters) into the atmosphere by about 35% between 1989 and 2000. After hearing more bad news about seasonal ozone thinning above Antarctica in 1989, representatives of 93 countries met in London in 1990 and in Copenhagen in 1992 and adopted a new protocol accelerating the phaseout of key ozone-depleting chemicals.

The landmark international agreements reached so far and signed by 175 countries are important examples of global cooperation in response to a serious global environmental problem. However, according to a 1998 World Meteorological Organization (WMO) study by 350 scientists, the ozone layer will

- Continue to be depleted for several decades because of **(1)** the 11- to 20-year time lag between when ODCs are released into the stratosphere and when they reach the stratosphere and **(2)** their persistence for decades in the stratosphere.
- Return to 1980 levels by about 2050 and to 1950 levels by about 2100 (Figure 13-21), assuming that **(1)** the international agreements are followed and **(2)** there are no major volcanic eruptions (which can temporarily deplete stratospheric ozone). Without the 1992 international agreement, ozone depletion would be a much more serious threat (Figure 13-21).

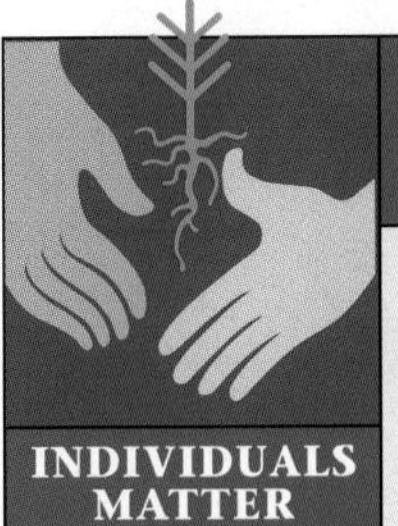

Ray Turner and His Refrigerator

Ray Turner, an aerospace manager at Hughes Aircraft in California, made an important low-tech, ozone-saving discovery by using his head—and his refrigerator. His concern for the environment led him to look for a cheap and simple substitute for the CFCs used as cleaning agents to remove films of oxidation from the electronic circuit boards manufactured at his plant.

He started by looking in his refrigerator. He decided to put drops of various substances on a corroded penny to see whether any of them removed the film of oxidation. Then he used his soldering gun to see whether solder would stick to the surface of the penny, indicating that the film had been cleaned off.

First, he tried vinegar. No luck. Then he tried some ground-up lemon peel, also a failure. Next he tried a drop of lemon juice and watched as the solder took hold. The rest, as they say, is history.

Today, Hughes Aircraft uses inexpensive citrus-based solvents that are CFC-free to clean circuit boards. This new cleaning technique has reduced circuit board defects by about 75% at Hughes. And Turner got a hefty bonus. Now other companies, such as AT&T, clean computer boards and chips using acidic chemicals extracted from cantaloupes, peaches, and plums. Maybe you can find a solution to an environmental problem in your refrigerator, grocery, drugstore, or backyard.

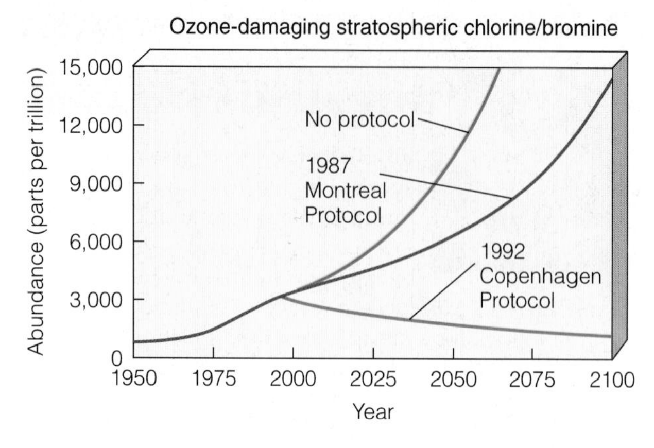

Figure 13-21 Projected concentrations of ozone-depleting chemicals (ODCs) in the stratosphere under three scenarios: **(1)** no action, **(2)** the 1987 Montreal Protocol, and **(3)** the 1992 Copenhagen Protocol.

One piece of disturbing news in the 1998 WMO study is that ozone depletion in the stratosphere has been cooling the troposphere. It is estimated that this has helped offset or disguise as much as 30% of global warming caused by our greenhouse gas emissions. Thus, restoring the ozone layer could lead to an increase in global warming.

The ozone treaty set an important precedent for global cooperation and action to avert potential global disaster. Nations and companies agreed to work together to solve this problem because

- There was convincing and dramatic scientific evidence of a serious problem.
- CFCs were produced by a small number of international companies.
- The certainty that CFC sales would decline unleashed the economic and creative resources of the private sector to find even more profitable substitute chemicals.

International cooperation in dealing with projected atmospheric warming is much more difficult because

- We lack clear-cut and dramatic evidence that there is a serious problem. Because the offending gases cannot be smelled or seen, we must rely on scientific experts, sophisticated instruments, and computer models to describe and project the possible consequences of global warming throughout the world.
- Greenhouse gas emissions result from the actions of hundreds of different large and politically powerful industries (such as coal, oil, chemicals, automobiles, and steel) and billions of consumers.
- Reducing greenhouse gas emissions will take far-reaching changes within most industries and lifestyle changes for billions of consumers.

Thus, reducing and adapting to the threat of global warming is a difficult political and scientific challenge. However, it can be done over the next few decades (Figures 13-14 and 13-15). Moreover, numerous economic studies show that meeting this challenge will **(1)** save huge amounts of money, **(2)** create many jobs, and **(3)** be much more profitable for many of the world's major businesses than doing nothing or waiting to act.

The atmosphere is the key symbol of global interdependence. If we can't solve some of our problems in the face of threats to this global commons, then I can't be very optimistic about the future of the world.

MARGARET MEAD

REVIEW QUESTIONS

1. Define the boldfaced terms in this chapter.

2. Summarize briefly how the earth's climate has changed over the past 900,000 years and over the past 140 years. Distinguish between *glacial* and *interglacial* periods.

3. How do scientists get information about past changes in the earth's climate?

4. What three factors determine the average temperature of the atmosphere near the earth's surface?

5. What is the earth's *natural greenhouse effect*? How widely is this theory accepted? What are the two major greenhouse gases?

6. What is *global warming*? List three human activities that increase the input of greenhouse gases into the troposphere and could enhance the earth's natural greenhouse effect. Describe changes at the earth's poles and in Greenland that indicate that the troposphere has warmed in recent decades.

7. Why does rapid climate change over a few decades to 100 years pose a serious threat to human life, wildlife, and the world's economies?

8. Describe how scientists develop mathematical models to make projections about future climate change. According to the latest models, between 2000 and 2100 about how much increase is projected for **(a)** the global average temperature and **(b)** global sea levels?

9. Explain how each of the following factors might enhance or dampen global warming: **(a)** changes in solar output, **(b)** changes in the earth's reflectivity (albedo), **(c)** the oceans, **(d)** water vapor content and clouds, **(e)** air pollution, and **(f)** effects of increased carbon dioxide levels on photosynthesis and methane emissions. How rapidly might climates shift? What is the range of error in current projections of changes in the earth's average atmospheric temperature?

10. Why should we worry about a possible rise of only one to a few degrees in the average temperature at the earth's surface?

11. Explain how atmospheric warming might affect each of the following: **(a)** food production, **(b)** water supplies, **(c)** forests, **(d)** biodiversity, **(e)** sea levels, **(f)** weather extremes, **(g)** human health, and **(h)** developing countries.

12. What are the four schools of thought about what should be done about global warming?

13. According to climate scientists, by how much will we need to reduce the current level of fossil fuel use to stabilize carbon dioxide emissions at their current level?

14. List eight prevention methods and five cleanup methods for slowing climate change from increased greenhouse gas emissions.

15. Explain how improving energy efficiency and relying more on organic farming could help reduce greenhouse gas emissions.

16. Summarize progress made in developing an international treaty to help reduce greenhouse gas emissions.

17. List seven ways in which we might prepare for and adjust to the harmful effects of global warming.

18. What is stratospheric *ozone depletion,* and how serious is this problem? What types of chemicals cause ozone depletion? How do these chemicals cause such depletion?

19. Explain how seasonal ozone thinning occurs each year in the troposphere over the earth's poles.

20. What are the major harmful effects of ozone depletion on **(a)** human health, **(b)** crop yields, **(c)** forest productivity, **(d)** materials such as plastics and paints, and **(e)** plankton productivity?

21. Distinguish between *squamous cell skin cancer, basal cell skin cancer,* and *malignant melanoma.* List ways in which you can reduce your chances of getting skin cancer.

22. If all ozone-depleting chemicals were banned now, about how long would it take for average concentrations of ozone in the stratosphere to return to **(a)** 1980 levels and **(b)** 1950 levels?

23. Summarize the progress that has been made in reducing the threat of ozone depletion and explain the importance of such efforts.

24. List three factors that helped countries agree to an international treaty to phase out ozone-depleting chemicals. List three reasons why getting countries to develop an international treaty to reduce greenhouse gas emissions is much more difficult.

CRITICAL THINKING

1. In preparation for the 1992 UN Conference on the Human Environment in Rio de Janeiro, president George Bush Sr.'s top economic adviser gave an address in Williamsburg, Virginia, to representatives of governments from a number of countries. He told his audience not to worry about global warming because the average temperature increases scientists are predicting were much less than the temperature increase he experienced in coming from Washington, D.C., to Williamsburg. What is the fundamental flaw in this reasoning?

2. What changes might occur in the global hydrologic cycle (Figure 4-26, p. 95) if the atmosphere were to experience significant **(a)** warming or **(b)** cooling? Explain.

3. What effect would clearing forests and converting them to grasslands and crops have on the earth's **(a)** reflectivity (albedo) and **(b)** average surface temperature? Explain.

4. Which of the four schools of thought about what should be done about possible global warming (pp. 313–314) do you favor? Explain.

5. Explain why you agree or disagree with each of the proposals listed in **(a)** Figure 13-14 for slowing down emissions of greenhouse gases into the atmosphere and **(b)** Figure 13-15 for preparing for the effects of global warming. What might be the harmful effects on your life of *not* taking these actions?

6. What consumption patterns and other features of your lifestyle directly add greenhouse gases to the atmosphere? Which, if any, of these things would you be willing to give up to slow global warming and reduce other forms of air pollution?

7. You have been diagnosed with treatable basal cell skin cancer (Figure 13-20, center). Explain why you could increase your chances of getting more of such cancers within 15–40 years by moving to Australia or Florida.

PROJECTS

1. As a class, conduct a poll of students at your school to determine **(a)** whether they understand the difference between global warming of the troposphere and ozone depletion in the stratosphere and **(b)** whether they believe that global warming from an enhanced greenhouse effect is a very serious problem, a moderately serious problem, or of little concern. Tally the results to see whether there are differences related to year in school, political leaning (liberal, conservative, moderate), or sex of poll participants.

2. As a class, conduct a poll of students at your school to determine whether they believe that stratospheric ozone depletion is a very serious problem, a moderately serious problem, or of little concern. Tally the results to see whether there are differences related to year in school, political leaning (liberal, conservative, moderate), or sex of poll participants.

3. Use the library or the internet to determine how the current government policy on global warming in the country where you live compares with the policy suggestions made by climate scientists in Figure 13-14.

4. Make a concept map of this chapter's major ideas, using the section heads and subheads and the key terms (in boldface). Look on the website for this book for information about making concept maps.

INTERNET STUDY RESOURCES AND RESOURCES FOR FURTHER READING AND RESEARCH

The website for this book contains helpful study aids and many ideas for further reading and research. Log on to

http://www.brookscole.com/product/0534389872s

and click on the Chapter-by-Chapter area. Choose Chapter 13 and select a resource:

- "Flash Cards" allows you to test your mastery of the Terms and Concepts to Remember for this chapter.
- "Tutorial Quizzes" provides a multiple-choice practice quiz.
- "Student Guide to InfoTrac" will lead you to Critical Thinking Projects that use InfoTrac College Edition as a research tool.
- "References" lists the major books and articles consulted in writing this chapter.
- "Hypercontents" takes you to an extensive list of sites with news, research, and images related to individual sections of the chapter.

INFOTRAC COLLEGE EDITION

Improve your skills with InfoTrac College Edition, a searchable online database of articles from more than 700 periodicals. Log on to

http://www.infotrac-college.com

or access InfoTrac through the website for this book. Try to find the following articles:

1. The Climbing Cost of Climate Change (effect of natural disasters on insurance costs) (Brief Article). *Earth Island Journal*, Summer 2001 v16 i2 p19. According to a United Nations report, global warming may cost the world several billion dollars a year unless urgent efforts are made to curb emissions of carbon dioxide and the other greenhouse gases. *Hint*: Enter the search term "climate change" using the keywords "climate change."

2. The Climate Domino (melting permafrost affects climate). *The Ecologist*, April 2001 v31 i3 p10. According to this report, permafrost is melting, and organic material contained in it is releasing carbon. This carbon would add to climate change, producing a feedback loop. *Hint*: Enter the search term "domino" using the keywords "climate change."

14 WATER RESOURCES AND WATER POLLUTION

Water Wars in the Middle East

If there is another war between countries in the Middle East, it could be fought over water, not oil. Most water in this dry region comes from three shared river basins: the Nile, Jordan, and Tigris–Euphrates (Figure 14-1). Water in much of this arid region is already in short supply.

Ethiopia, which controls the headwaters that feed 86% of the Nile's flow, plans to divert more of this water; so does Sudan. This could reduce the amount of water available to water-short Egypt, whose terrain is desert except for a green area of irrigated cropland running down its middle along the Nile and its delta. Between 2001 and 2025, Egypt's population is expected to increase from 70 million to 96 million, greatly increasing the demand for already scarce water.

Egypt's options are to **(1)** go to war with Sudan and Ethiopia to obtain more water, **(2)** cut population growth, **(3)** improve irrigation efficiency, **(4)** spend $2 billion to build the world's longest concrete canal and pump water out of Lake Nasser (the reservoir created from the Nile by the Aswan High Dam) to create more irrigated farmland in the middle of the desert, **(5)** import more grain to reduce the need for irrigation water, **(6)** work out water-sharing agreements with other countries, or **(7)** suffer the harsh human and economic consequences.

The Jordan Basin is by far the most water-short region, with fierce competition for its water between Jordan, Syria, Palestine (Gaza and the West Bank), and Israel (Figure 14-1). The combined populations of these already water-short countries are projected to increase from 32 million to 52 million between 2001 and 2025. Some good news is that in 1994 Israel and Jordan signed a peace treaty that addressed their disputes over water from the Jordan River basin.

Syria plans to build dams and withdraw more water from the Jordan River, decreasing the downstream water supply for Jordan and Israel. Israel warns that it will consider destroying the largest dam that Syria plans to build.

Turkey, located at the headwaters of the Tigris and Euphrates rivers, controls how much water flows downstream to Syria and Iraq before emptying into the Persian Gulf (Figure 14-1). Turkey is building 24 dams along the upper Tigris and Euphrates rivers to **(1)** generate huge quantities of electricity, **(2)** irrigate a large area of land, and **(3)** create about 3.5 million jobs for its 65 million people.

If completed, these dams will reduce the flow of water downstream to Syria and Iraq by up to 35% in normal years and much more in dry years. Syria also plans to build a large dam along the Euphrates River to divert water arriving from Turkey. This will leave little water for Iraq and could lead to war between Syria and Iraq.

Clearly, water distribution will be a key issue in any peace talks in this region. Resolving these problems will require a combination of **(1)** regional cooperation in allocating water supplies, **(2)** slowed population growth, **(3)** improved efficiency in water use, and **(4)** increased water prices to encourage water conservation and improve irrigation efficiency.

According to Lester Brown (Guest Essay, p. 20) and Christopher Flavin of the Worldwatch Institute, "The spreading scarcity of fresh water may be the most underestimated resource issue facing the world in this new millennium."

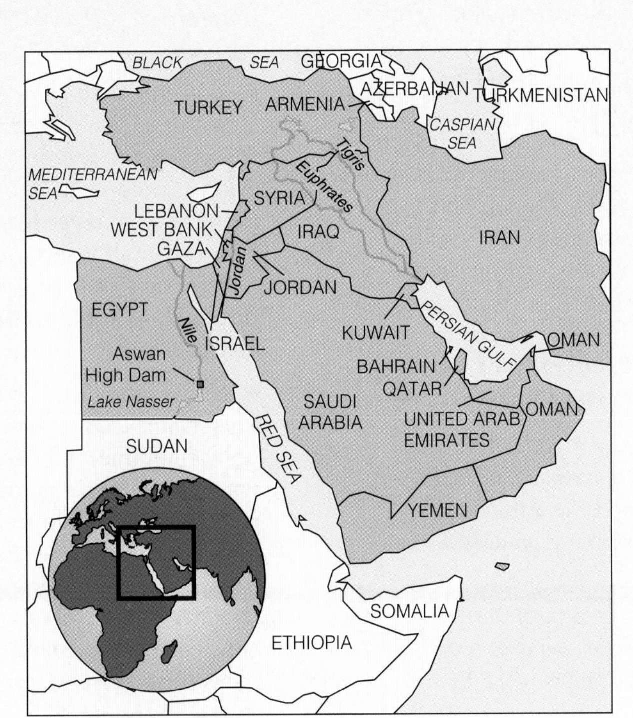

Figure 14-1 The Middle East, whose countries have some of the highest population growth rates in the world. Because of the dry climate, food production depends heavily on irrigation. Existing conflicts between countries in this region over access to water may soon overshadow both long-standing religious and ethnic clashes and attempts to take over valuable oil supplies.

Our liquid planet glows like a soft blue sapphire in the hard-edged darkness of space. There is nothing else like it in the solar system. It is because of water.

JOHN TODD

This chapter addresses the following questions:

- What are water's unique physical properties?
- How much fresh water is available to us, and how much of it are we using?
- What causes freshwater shortages, and what can be done about this problem?
- What are the causes of flooding, and what can be done to reduce the risk of flooding and flood damage?
- What pollutes water, where do the pollutants come from, and what effects do they have?
- What are the major water pollution problems of streams, lakes, and groundwater?
- What are the major water pollution problems of oceans?
- How can we prevent and reduce water pollution?
- How can we use the earth's water more sustainably?

14-1 WATER'S IMPORTANCE AND UNIQUE PROPERTIES

Why Is Water So Important? We live on the water planet, with a precious film of water—most of it salt water—covering about 71% of the earth's surface (Figure 6-29, p. 144). All organisms are made up mostly of water; a tree is about 60% water by weight, and you and most animals are about 50–65% water.

Each of us needs only about a dozen cupfuls of water per day to survive, but huge amounts of water are needed to supply us with food, shelter, and our other needs and wants. Water also plays a key role in **(1)** sculpting the earth's surface, **(2)** moderating climate, and **(3)** diluting pollutants.

What Are Some Important Properties of Water? Water is a remarkable substance with a unique combination of properties:

- *There are strong forces of attraction (called hydrogen bonds) between molecules of water.* These attractive forces are the major factor determining water's unique properties.
- *Water exists as a liquid over a wide temperature range because of the strong forces of attraction between water molecules.* Its high boiling point of 100°C (212°F) and low freezing point of 0°C (32°F) mean that water remains a liquid in most climates on the earth.
- *Liquid water changes temperature very slowly because it can store a large amount of heat without a large change in temperature.* This high heat capacity **(1)** helps protect living organisms from temperature fluctuations, **(2)** moderates the earth's climate, and **(3)** makes water an excellent coolant for car engines, power plants, and heat-producing industrial processes.
- *It takes a lot of heat to evaporate liquid water because of the strong forces of attraction between its molecules.* Water absorbs large amounts of heat as it changes into water vapor and releases this heat as the vapor condenses back to liquid water. This is a primary factor in distributing heat throughout the world and thus plays an important role in determining the climates of various areas (Figure 6-6, p. 125). This property also makes water evaporation an effective cooling process, which is why you feel cooler when perspiration or bathwater evaporates from your skin.
- *Liquid water can dissolve a variety of compounds.* This enables it to **(1)** carry dissolved nutrients into the tissues of living organisms, **(2)** flush waste products out of those tissues, **(3)** serve as an all-purpose cleanser, and **(4)** help remove and dilute the water-soluble wastes of civilization. Water's superiority as a solvent also means that water-soluble wastes pollute it easily.
- *Water molecules can break down (ionize) into hydrogen ions (H^+) and hydroxide ions (OH^-), which help maintain a balance between acids and bases in cells, as measured by the pH of water solutions* (Figure 3-5, p. 63).
- *Water filters out wavelengths of ultraviolet radiation* (Figure 3-8, p. 66) *that would harm some aquatic organisms.*
- *The strong attractive forces between the molecules of liquid water cause its surface to contract (high surface tension) and to adhere to and coat a solid (high wetting ability).* These cohesive forces pull water molecules at the surface layer together so strongly that it can support small insects. The combination of high surface tension and wetting ability allow water to rise through a plant from the roots to the leaves (capillary action).
- *Unlike most liquids, water expands when it freezes.* This means that ice has a lower density (mass per unit of volume) than liquid water. Thus ice floats on water. Without this property, lakes and streams in cold climates would freeze solid and lose most of their current forms of aquatic life. Because water expands upon freezing, it can also **(1)** break pipes, **(2)** crack engine blocks (which is why we use antifreeze), **(3)** break up streets, and **(4)** fracture rocks (thus helping form soil).

Water is the lifeblood of the biosphere. It connects us to one another, to other forms of life, and to the entire planet. Despite its importance, water is one of our most poorly managed resources. We waste it and pollute it. We also charge too little for making it available. This encourages still greater waste and pollution of this resource, for which there is no substitute.

14-2 SUPPLY, RENEWAL, AND USE OF WATER RESOURCES

How Much Fresh Water Is Available? Only a tiny fraction of the planet's abundant water is available to us as fresh water (Figure 14-2). About 97.4% by volume is found in the oceans and is too salty for drinking, irrigation, or industry (except as a coolant).

Most of the remaining 2.6% that is fresh water is **(1)** locked up in ice caps or glaciers or **(2)** in groundwater too deep or salty to be used (Figure 14-2).

Thus, only about 0.014% of the earth's total volume of water is easily available to us as soil moisture, usable groundwater, water vapor, and lakes and streams (Figure 14-2). If the world's water supply were only 100 liters (26 gallons), our usable supply of fresh water would be only about 0.014 liter (2.5 teaspoons).

Fortunately, the available fresh water amounts to a generous supply. Moreover, this water is continuously collected, purified, recycled, and distributed in the solar-powered *hydrologic cycle* (Figure 4-26, p. 95) as long as we do not **(1)** overload it with slowly degradable and non-degradable wastes or **(2)** withdraw it from underground supplies faster than it is replenished. Unfortunately, in some parts of the world we are doing both.

Differences in average annual precipitation divide the world's countries and people into water haves and have-nots. For example, Canada, with only 0.5% of the world's population, has 20% of the world's fresh water, whereas China, with 21% of the world's people, has only 7% of the supply.

As population, irrigation, and industrialization increase, water shortages in already water-short regions will intensify and heighten tensions between and within countries (p. 327). Global warming (Figure 13-8, p. 306) can **(1)** increase global rates of evaporation, **(2)** shift precipitation patterns, and **(3)** disrupt water supplies and thus food supplies. Some areas will get more precipitation and some less. River flows will change. Monsoons and hurricanes are likely to intensify. The average sea level will rise from thermal expansion of the oceans and partial melting of ice caps and mountain glaciers (Connections, p. 308).

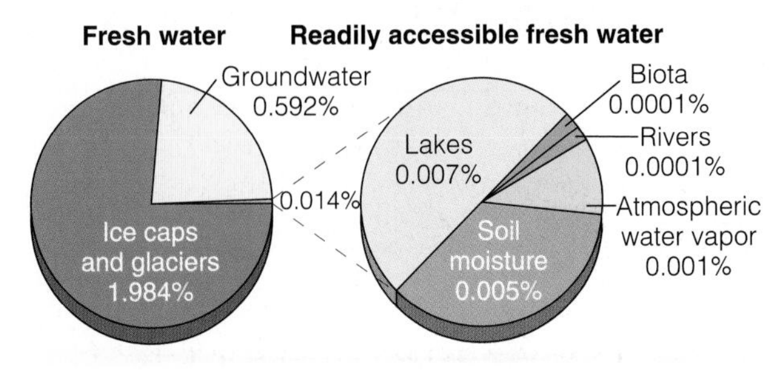

Figure 14-2 The planet's water budget. Only a tiny fraction by volume of the world's water supply is fresh water available for human use.

What Is Surface Water? The fresh water we use first arrives as the result of precipitation. Precipitation that does not infiltrate the ground or return to the atmosphere by evaporation (including transpiration) is called **surface runoff** that flows into streams, lakes, wetlands, and reservoirs.

About two-thirds of the world's annual runoff is lost in seasonal floods and is not available for human use. The remaining one-third is **reliable runoff** that generally can be counted on as a stable source of water from year to year if water flows are not disrupted or shifted by climate change.

A **watershed**, also called a **drainage basin**, is a region from which water drains into a stream, lake, reservoir, wetland, or other body of water.

What Is Groundwater? Some precipitation infiltrates the ground and percolates downward through voids (pores, fractures, crevices, and other spaces) in soil and rock (Figure 14-3). The water in these voids is called **groundwater**.

Close to the surface, the voids have little moisture in them. However, below some depth, in what is called the **zone of saturation,** the voids are completely filled with water. The **water table** is located at the top of the zone of saturation. It falls in dry weather and rises in wet weather.

Porous, water-saturated layers of sand, gravel, or bedrock through which groundwater flows are called **aquifers** (Figure 14-3). Aquifers are like large, elongated sponges through which groundwater seeps. Any area of land through which water passes downward or laterally into an aquifer is called a **recharge area.** Aquifers are replenished naturally by precipitation that percolates downward through soil and rock in what is called **natural recharge**, but some are recharged from the side by *lateral recharge*.

Groundwater moves from the *recharge area* through an aquifer and out to a *discharge area* (well, spring, lake, geyser, stream, or ocean) as part of the hydrologic cycle (Figure 4-26, p. 95). Groundwater normally moves from points of high elevation and pressure to points of lower elevation and pressure. This movement is quite slow, typically only a meter or so (about 3 feet) per year and rarely more than 0.3 meter (1 foot) per day.

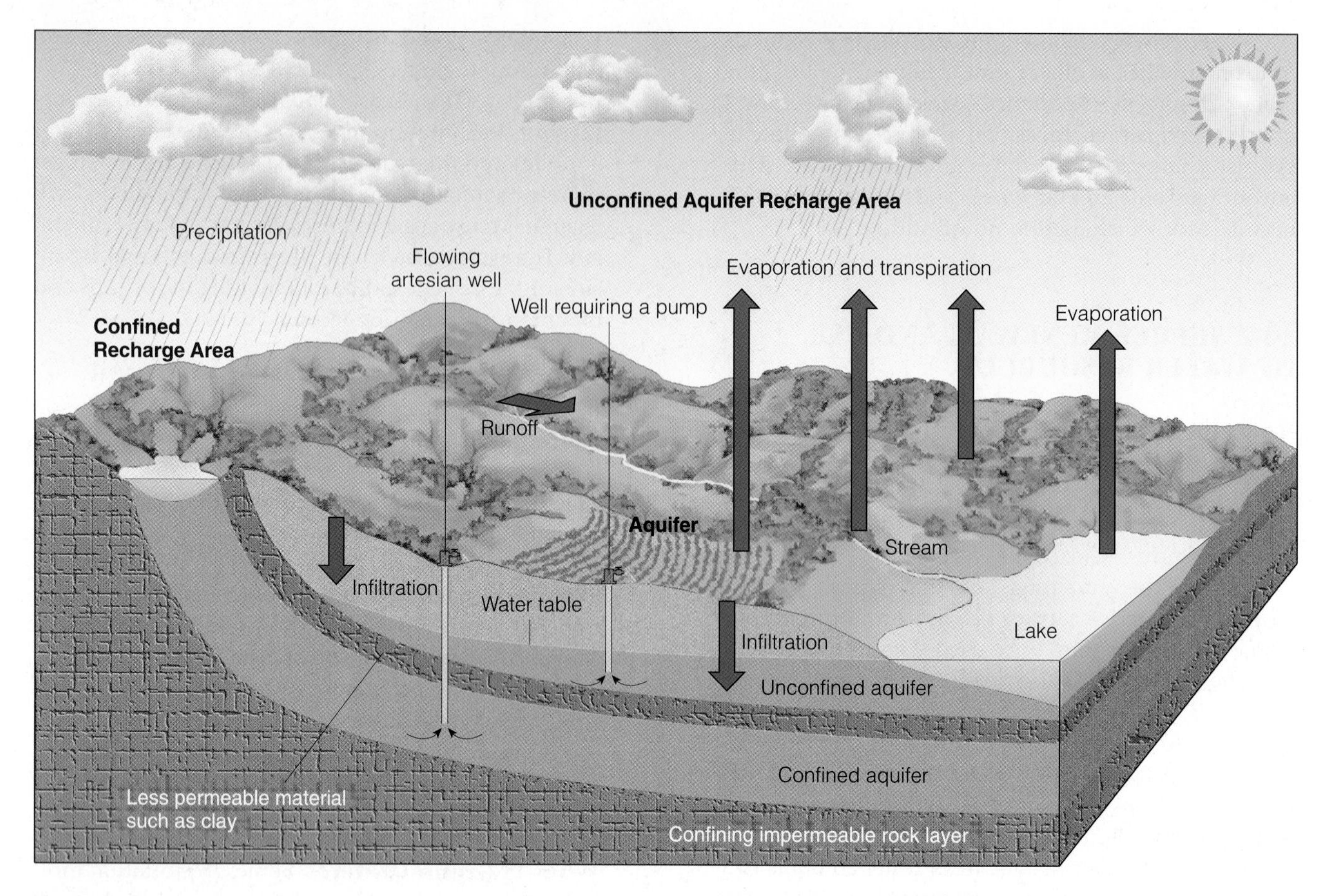

Figure 14-3 The groundwater system. An *unconfined aquifer* is an aquifer with a water table. A *confined aquifer* is bounded above and below by less permeable beds of rock. Groundwater in this type of aquifer is confined under pressure.

Some aquifers get very little (if any) recharge and on a human time scale are nonrenewable resources. They are often found fairly deep underground and were formed tens of thousands of years ago. Withdrawals from such aquifers amount to *water mining* that, if kept up, will deplete these ancient deposits of water.

How Much of the World's Reliable Water Supply Are We Using? Since 1900, global water use has increased about ninefold and per capita use has quadrupled, with irrigation accounting for the largest increase in water use (Figure 14-4). As a result, humans now withdraw about 35% of the world's reliable runoff. At least another 20% of this runoff is left in streams to transport goods by boats, dilute pollution, and sustain fisheries.

Thus, we are directly or indirectly using more than half of the world's reliable runoff. Because of increased population growth and economic development, global withdrawal rates of surface water are projected to **(1)** at least double in the next two decades and **(2)** exceed the reliable surface runoff in a growing number of areas.

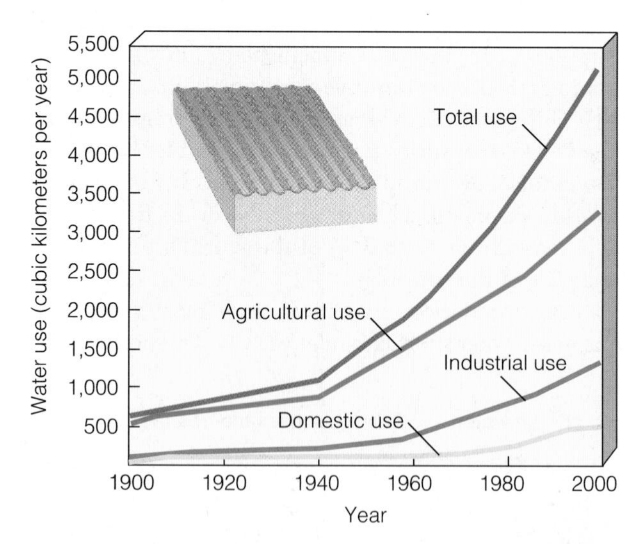

Figure 14-4 Global water use, 1900–2000. Between 2000 and 2054, the world's population is expected to increase by about 3 billion people. (Data from World Commission on Water Use in the 21st Century)

How Do We Use the World's Fresh Water? Uses of withdrawn water vary from one region to another and from one country to another (Figure 14-5). Worldwide, about 70% of all water withdrawn each year from rivers, lakes, and aquifers is used to **(1)** irrigate

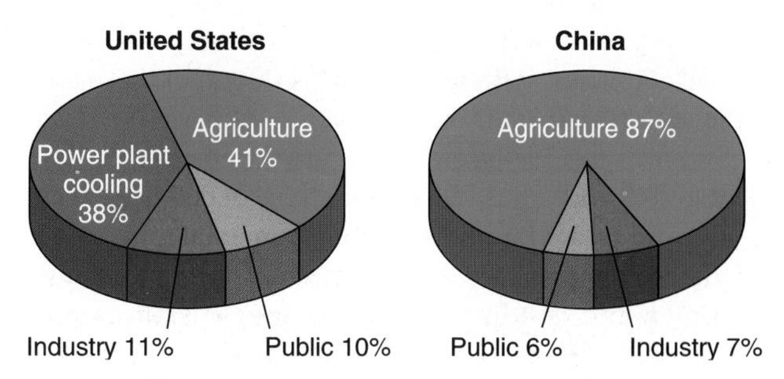

Figure 14-5 Water use in the United States and China. The United States has the world's highest per capita use of water, amounting to an average of 4,800 liters (1,280 gallons) per person per day in 1999. Between 1980 and 1999, total water use in the United States decreased by 10% despite a 17% increase in population, mostly because of more efficient irrigation. (Data from Worldwatch Institute and World Resources Institute)

17% of the world's cropland and **(2)** produce about 40% of the world's food.

Industry uses about 20% of the water withdrawn each year, and cities and residences use the remaining 10%. Agriculture and manufacturing use large amounts of water to produce common products (Figure 14-6).

Case Study: Freshwater Resources in the United States Although the United States has plenty of fresh water, much of it is **(1)** in the wrong place at the wrong time or **(2)** contaminated by agricultural and industrial practices. The eastern states usually have ample precipitation, whereas many western states have too little (Figure 14-7, top).

In the East, the largest uses for water are for energy production, cooling, and manufacturing. The largest use by far in the West is for irrigation (which accounts for about 85% of all water use).

In many parts of the eastern United States the most serious water problems are **(1)** flooding, **(2)** occasional urban shortages, and **(3)** pollution. For example, the 3 million residents of Long Island, New York, get most of their water from an aquifer that is becoming severely contaminated.

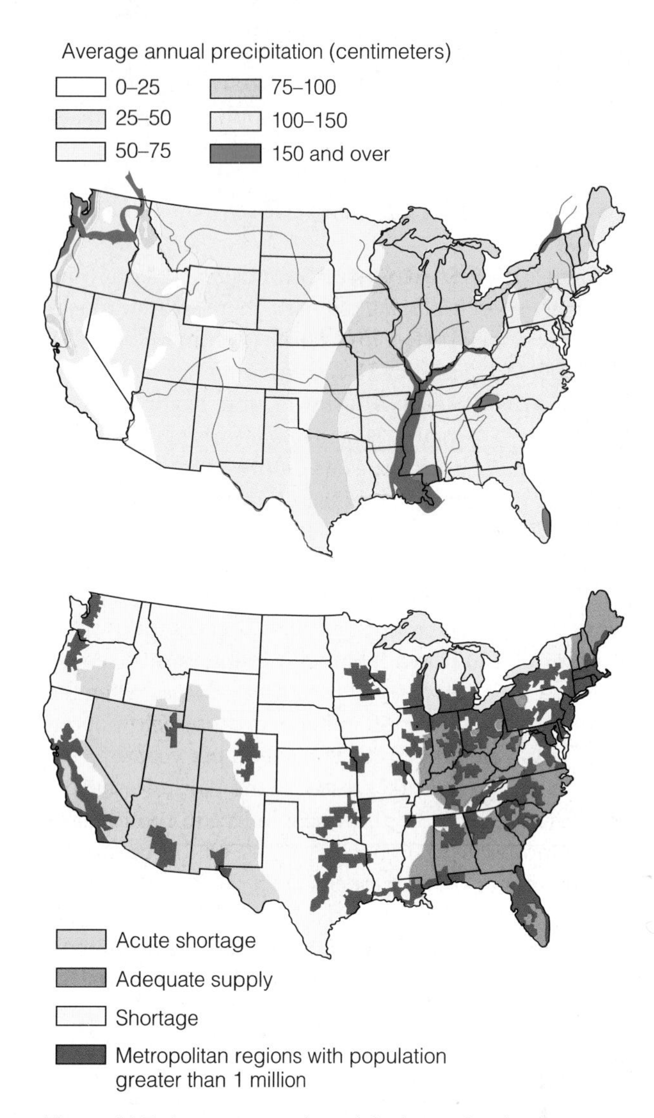

Figure 14-7 Average annual precipitation and major rivers (top) and water deficit regions in the continental United States and their proximity to metropolitan areas with populations greater than 1 million (bottom). (Data from U.S. Water Resources Council and U.S. Geological Survey)

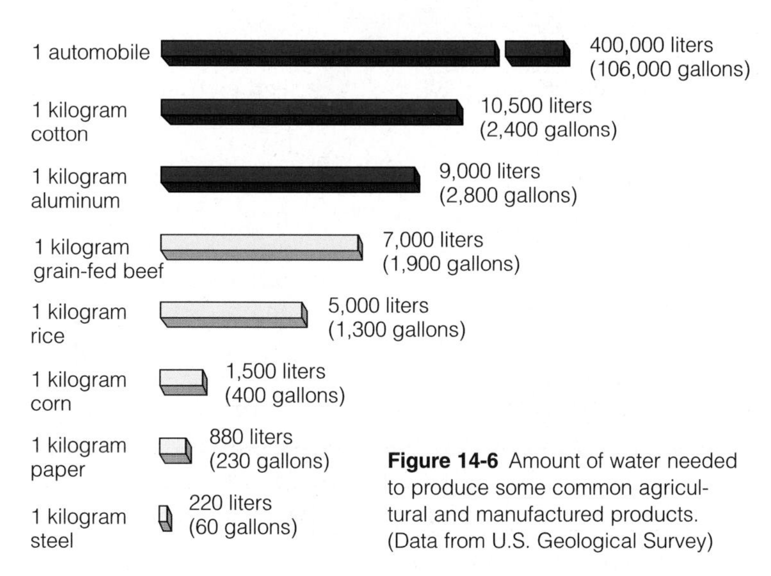

Figure 14-6 Amount of water needed to produce some common agricultural and manufactured products. (Data from U.S. Geological Survey)

The major water problem in the arid and semiarid areas of the western half of the country is a shortage of runoff, caused by **(1)** low precipitation (Figure 14-7, top), **(2)** high evaporation, and **(3)** recurring prolonged drought. Water tables in many areas are dropping rapidly as farmers and cities deplete aquifers faster than they are recharged.

In the United States, many major urban centers (especially those in the West and Midwest) are located in areas that do not have enough water (Figure 14-7, bottom). Water experts project that conflicts over water supplies within and between states will intensify as more industries and people migrate west and compete with farmers for scarce water. These shortages could worsen as global warming causes climate changes and shifts in water supplies in some areas.

14-3 TOO LITTLE WATER

What Causes Freshwater Shortages? According to water expert Malin Falkenmark, there are four causes of water scarcity: **(1)** a *dry climate* (Figure 6-3, p. 123), **(2)** *drought* (a period of 21 days or longer in which precipitation is at least 70% lower and evaporation is higher than normal), **(3)** *desiccation* (drying of the soil because of such activities as deforestation and overgrazing by livestock), and **(4)** *water stress* (low per capita availability of water caused by increasing numbers of people relying on limited runoff levels).

Figure 14-8 shows the degree of stress on the world's major river systems, based on comparing the amount of water available with the amount used by humans. A country is said to be *water stressed* when the volume of reliable runoff per person drops below about 1,700 cubic meters (60,000 cubic feet) per year.

According to a 2000 study by the World Resources Institute (WRI), 2.3 billion people live in river basins under moderate to high water stress. Of this group, 1.7 million live in areas of *water scarcity*, where annual per capita water availability falls below 1,000 cubic meters (35,000 cubic feet). If current water consumption patterns continue, the WRI projects that by 2025 at least 3.4 billion people will live in water-stressed river basins in 50 countries, with more than 2.4 billion of these people suffering from more dire water scarcity.

Since the 1970s, water scarcity intensified by prolonged drought has killed more than 24,000 people per year and created millions of environmental refugees. In water-short rural areas in developing countries, many women and children must walk long distances each day, carrying heavy jars or cans, to get a meager and sometimes contaminated supply of water.

A number of environmental, political, and economic analysts believe that *access to water resources, already a key foreign policy and environmental security issue for water-short countries, will become even more important over the next 10–20* years. Two countries share almost 150 of the world's 214 major river systems (57 of them in Africa), and another 50 are shared by 3 to 10 countries. Some 40% of the world's population already clashes over water, especially in the Middle East (p. 327).

Some areas have lots of water, but the largest rivers (which carry most of the runoff) are far from agricultural and population centers where the water is needed. For example, South America has the largest annual water runoff of any continent, but 60% of the runoff flows through the Amazon River in remote areas where few people live.

In some areas, overall precipitation may be plentiful, but most arrives during short periods, or it cannot be collected and stored because of a lack of water storage capacity. For example, only a few hours of rain pro-

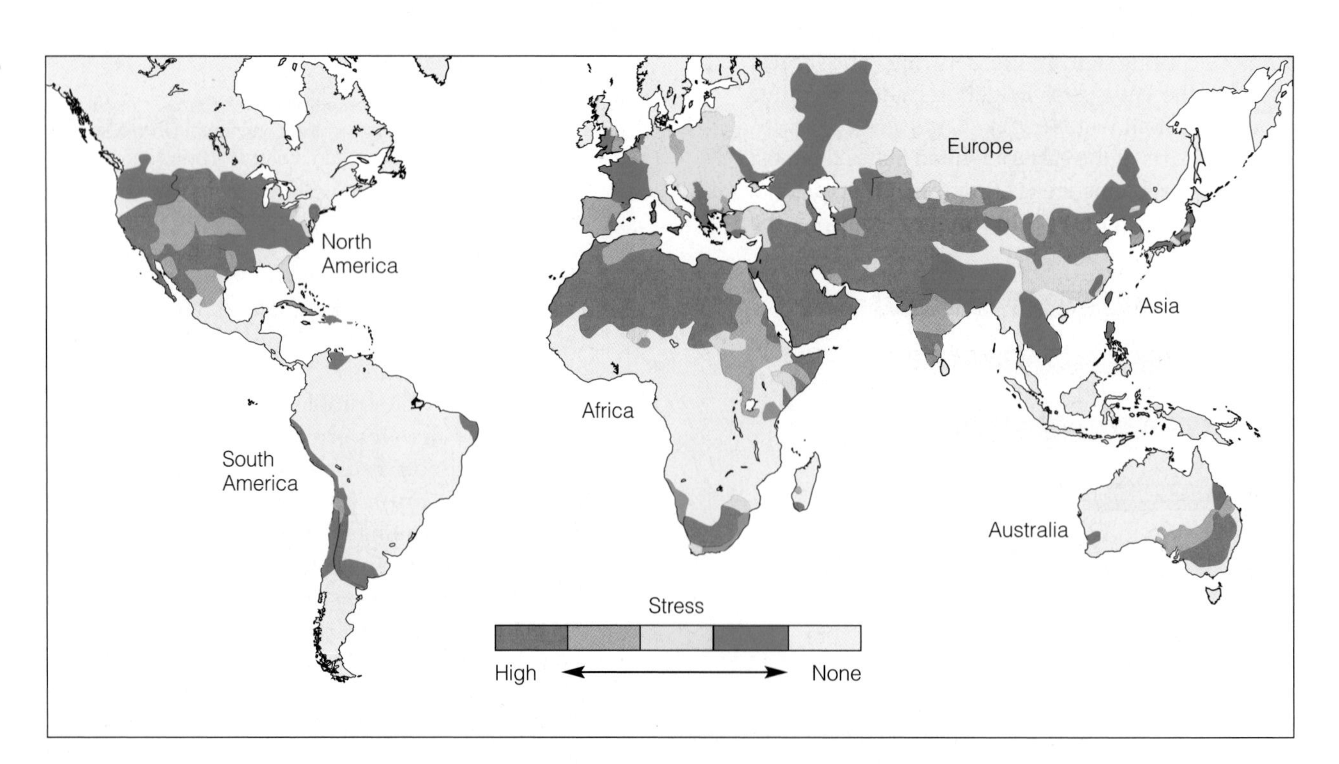

Figure 14-8 Stress on the world's major river basins, based on a comparison of the amount of water available with the amount used by humans. (Data from World Commission on Water Use in the 21st Century)

vide more than half of India's rainfall during a four-month monsoon season.

Even when there is a plentiful supply of water, most of the 1.2 billion poor people living on less than $1 a day cannot afford a safe supply of drinking water. Most are cut off from municipal water supplies and must **(1)** collect water from unsafe sources or **(2)** buy water (often coming from polluted rivers) from private vendors at high prices. In developing countries, people not connected to municipal water supplies on average pay 12 times more per liter of water than people connected to such systems and in some areas pay up to 100 times as much.

How Can We Increase Freshwater Supplies? There are five ways to increase the supply of fresh water in a particular area: **(1)** Build dams and reservoirs to store runoff, **(2)** bring in surface water from another area, **(3)** withdraw groundwater, **(4)** convert salt water to fresh water (desalination), and **(5)** improve the efficiency of water use.

In developed countries, people tend to live where the climate is favorable and then bring in water from another watershed. In developing countries, most people (especially the rural poor) must settle where the water is and try to capture and use the precipitation they need.

What Are the Pros and Cons of Large Dams and Reservoirs? Large dams and reservoirs have benefits and drawbacks (Figure 14-9). The main purpose of dams and large reservoirs is to capture and store runoff and release it as needed for **(1)** controlling floods, **(2)** producing hydroelectric power, and **(3)** supplying water for irrigation and for towns and cities. Reservoirs also provide recreational activities such as swimming, fishing, and boating.

So far, the world's dams have increased the annual runoff available for human use by nearly one-third. However, a series of dams on a river, especially in arid areas, can reduce downstream flow to a trickle and prevent it from reaching the sea as a part of the hydrologic cycle. According to the World Commission on Water in the 21st Century, half of the world's major rivers are going dry part of the year or are seriously polluted. In addition to threatening the water supplies for 500 million people, this engineering approach to river management often impairs the important ecological services rivers provide (Figure 14-10).

Case Study: China's Three Gorges Dam When completed, China's Three Gorges project on the mountainous upper reaches of the Yangtze River will be the world's largest hydroelectric dam and reservoir. According to Chinese officials, this superdam, with the electric output of 20 large coal-burning or nuclear power plants, will

- Supply power to industries and to 150 million people.

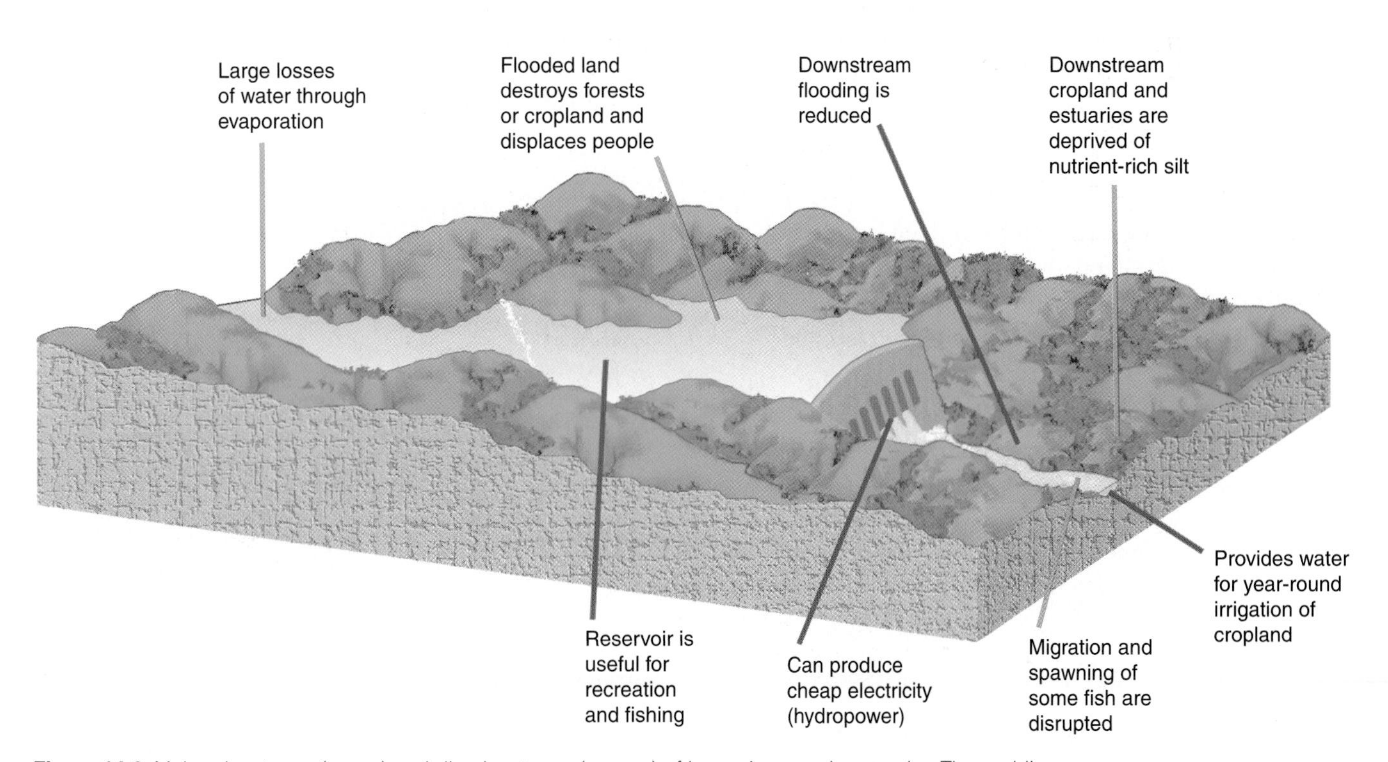

Figure 14-9 Main advantages (green) and disadvantages (orange) of large dams and reservoirs. The world's 45,000 large dams now impound about 14% of the world's runoff.

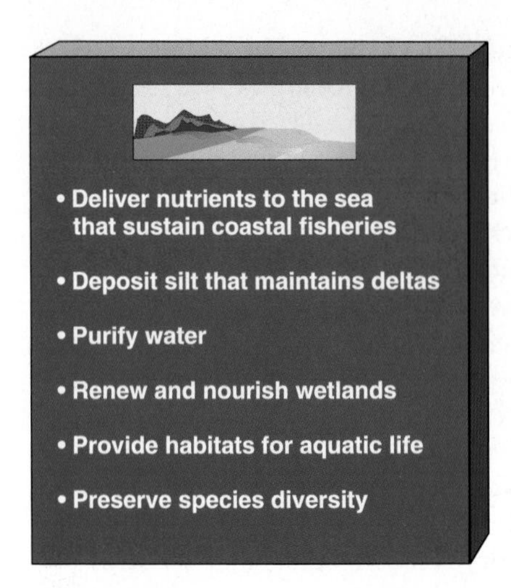

• Deliver nutrients to the sea that sustain coastal fisheries
• Deposit silt that maintains deltas
• Purify water
• Renew and nourish wetlands
• Provide habitats for aquatic life
• Preserve species diversity

Figure 14-10 Some ecological services provided by rivers. Currently, the services are given little or no monetary value when the costs and benefits of dam and reservoir projects are assessed. According to environmental economists, attaching even crudely estimated monetary values to these ecosystem services would help sustain rivers.

- Help China reduce its dependence on coal, which causes severe air pollution and releases enormous amounts of the greenhouse gas carbon dioxide into the atmosphere.
- Hold back the Yangtze River's floodwaters, which have killed more than 500,000 people during the past 100 years, including 4,000 people in 1998. According to Chinese officials, 400 million people living in the Yangtze River Valley will benefit from the dam. This greatly exceeds the 1.9 million people who will be relocated from the area to be flooded to form a gigantic 596-kilometer-long (370-mile-long) reservoir behind the dam.
- Reduce flooding and silting of the river by eroded soil. In 1998 logging was banned in the upper Yangtze watershed; $2 billion will be spent to reforest this watershed.

Critics point to a number of drawbacks for the Yangtze dam and reservoir project:

- Forming the huge reservoir will flood large areas of productive farmland and forests and displace about 1.9 million people from their homes.
- The region's entire ecosystem will be drastically changed.
- Water pollution will increase because of the river's reduced water flow.
- If the reservoir fills up with sediment and overflows (especially if the reservoir is kept filled at a high level, as planned, to provide maximum hydroelectric power), half a million people will be exposed to severe flooding.
- Annual deposits of nutrient-rich sediments below the dam will be reduced.
- The reduced downstream water flow will promote saltwater intrusion into drinking water supplies near the mouth of the river.

What Are the Pros and Cons of Large-Scale Water Transfers? The California Experience Tunnels, aqueducts, and underground pipes can transfer stream runoff collected by dams and reservoirs from water-rich areas to water-poor areas. Although such transfers have benefits, they also create environmental problems (Case Study, right). Indeed, most of the world's dam projects and large-scale water transfers illustrate the important ecological principle that *you cannot do just one thing.*

One of the world's largest watershed transfer projects is the *California Water Project*. In California, the basic water problem is that 75% of the population lives south of Sacramento but 75% of the state's rain occurs north of Sacramento.

The California Water Project uses a maze of giant dams, pumps, and aqueducts to transport water from water-rich northern California to heavily populated areas and to arid and semiarid agricultural regions, mostly in southern California (Figure 14-11).

For decades, northern and southern Californians have been feuding over how the state's water should be allocated under this project. Southern Californians say they need more water from the north to support Los Angeles, San Diego, and other growing urban areas and to grow more crops. Agriculture uses 74% of the water withdrawn in California, much of it for water-thirsty crops.

Opponents in the north say that sending more water south would **(1)** degrade the Sacramento River, **(2)** threaten fisheries, and **(3)** reduce the flushing action that helps clean San Francisco Bay of pollutants. They also argue that much of the water sent south is wasted unnecessarily and that making irrigation just 10% more efficient would provide enough water for domestic and industrial uses in southern California. However, if

Figure 14-11 The California Water Project and the Central Arizona Project involve large-scale water transfers from one watershed to another. Arrows show the general direction of water flow.

CASE STUDY

The Aral Sea Water Transfer Disaster

The shrinking of the Aral Sea (Figure 14-12) is a result of a large-scale water transfer project in an area of the former Soviet Union with the driest climate in central Asia. Since 1960, enormous amounts of irrigation water have been diverted from the inland Aral Sea and its two feeder rivers to irrigate cotton, vegetable, fruit, and rice crops and create one of the world's largest irrigated areas. The irrigation canal, the world's longest, stretches over 1,300 kilometers (800 miles).

This water diversion project (coupled with droughts) has caused a regional ecological, economic, and health disaster, described by one former Soviet official as "ten times worse than the 1986 Chernobyl nuclear power-plant accident" (p. 490).

Since 1960, as more and more water from the feeder rivers has been diverted to irrigate crops, **(1)** the sea's salinity has tripled, **(2)** its surface area has shrunk by 54% (Figure 14-12), **(3)** its volume has decreased by 75%, **(4)** its two supply rivers have become mere trickles, and **(5)** about 36,000 square kilometers (14,000 square miles) of former lake bottom has become a human-made desert covered with glistening white salt. The process continues, and within 10–20 years the once enormous Aral Sea may break up into three small brine lakes.

Twenty of the area's 24 native fish species have become extinct. This has devastated the area's fishing industry, which once provided work for more than 60,000 people. Fishing villages and boats once on the sea's coastline now are in the middle of a salt desert and have been abandoned.

Wetlands have shrunk by 85% and, combined with high levels of pollution from agricultural chemicals, have greatly reduced waterfowl populations. Roughly half the area's bird and mammal species have disappeared.

Winds pick up the salty dust that encrusts the lake's now-exposed bed and blow it onto fields as far as 300 kilometers (190 miles) away. As the salt spreads, it kills wildlife, crops, and other vegetation and pollutes water.

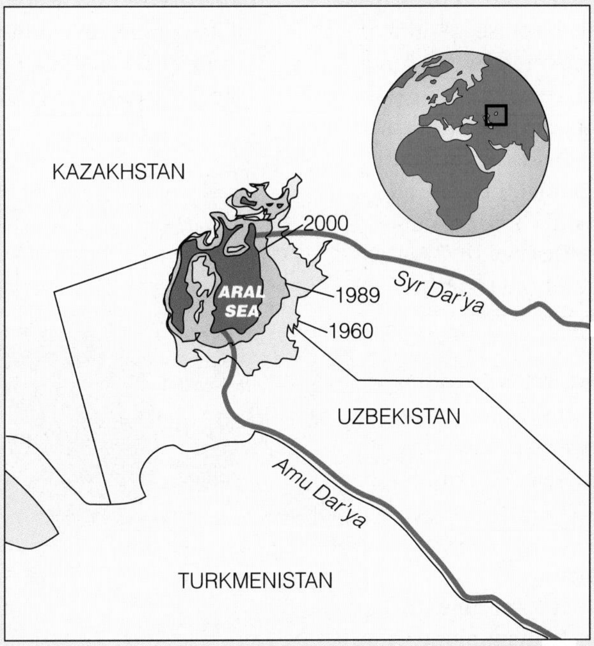

Figure 14-12 Once the world's fourth-largest freshwater lake, the Aral Sea has been shrinking and getting saltier since 1960 because most of the water from the rivers that replenish it has been diverted to grow cotton and food crops. As the lake shrinks, it leaves behind a salty desert, economic ruin, increasing health problems, and severe ecological disruption.

To raise yields, farmers have increased inputs of herbicides, insecticides, fertilizers, and irrigation water on some crops. Many of these chemicals have percolated downward and accumulated to dangerous levels in the groundwater, from which most of the region's drinking water comes. The lower river flows have also concentrated salts, pesticides, and other toxic chemicals, making surface water supplies hazardous to drink.

The Aral Sea basin may have one of the world's worst soil salinization (Figure 9-25, p. 217) problems. This situation is getting worse as irrigators apply more water to their fields during the season when crops are not grown to flush the accumulated salts out of the root zone before planting the next crop. This extra input of irrigation water **(1)** increases water use, **(2)** further shrinks the Aral Sea and increases the salt blowing onto cropland, and **(3)** adds more salt from irrigation water to the soil.

Conversion of much of the Aral Sea to a salt desert has also affected the area's climate. The once-huge sea acted as a thermal buffer that moderated the heat of summer and the extreme cold of winter. Now there is less rain, summers are hotter and drier, winters are colder, and the growing season is shorter. This, coupled with severe salinization of almost a third of the area's cropland, has caused crop yields to drop 20–50%.

More water could be withdrawn to flush out and lessen the area's acute salt problem. However, Russian scientists estimate that freeing up this much water would mean retiring about half of the area's irrigated cropland, an unthinkable solution considering the region's already dire economic conditions.

Winds whip up fertilizer and pesticide residues from the poisoned agricultural land and salt from the bare floor of the shriveled Aral Sea. The combination of toxic dust, salt, and contaminated water has caused serious health problems for a growing number of the 58 million

(continued)

CASE STUDY (Continued)

people living in the Aral Sea's watershed, and the area's population is expected to increase to 83 million by 2025. Such problems include abnormally high rates of **(1)** infant mortality, **(2)** tuberculosis, **(3)** anemia, **(4)** respiratory illness (one of the world's highest), **(5)** eye diseases (from salt dust), **(6)** throat cancer, **(7)** kidney and liver diseases (especially cancers), **(8)** arthritic diseases, **(9)** typhoid fever, and **(10)** hepatitis.

Can the Aral Sea be saved and can the area's serious ecological and human health problems be reduced? Since 1999 the United Nations and the World Bank have funded a $600-million program to **(1)** purify drinking water, **(2)** upgrade irrigation and drainage systems to improve irrigation efficiency, flush salts from croplands, and boost crop productivity, and **(3)** construct wetlands and artificial lakes to help restore aquatic vegetation, wildlife, and fisheries. However, this process will take decades and will not prevent the shrinkage of the Aral Sea into a few brine lakes.

Critical Thinking

What ecological and economic lessons can we learn from the Aral Sea tragedy?

water supplies in northern California and in the Colorado River basin drop sharply because of global warming, the amount of water delivered by the huge distribution system will plummet.

Pumping out more groundwater is not the answer because groundwater is already being withdrawn faster than it is replenished throughout much of California. To most analysts, quicker and cheaper solutions are **(1)** improving irrigation efficiency (p. 339) and **(2)** allowing farmers to sell their legal rights to withdraw certain amounts of water from rivers.

What Are the Pros and Cons of Withdrawing Groundwater? Pumping groundwater from aquifers has several advantages over tapping more erratic flows from streams. Groundwater **(1)** can be removed as needed year-round, **(2)** is not lost by evaporation, and **(3)** usually is less expensive to develop than surface water systems.

Aquifers provide drinking water for almost one-third of the planet's people. In Asia alone, more than 1 billion people depend on groundwater for drinking. In the United States, about 51% of the drinking water (96% in rural areas and 20% in urban areas) and 43% of irrigation water is pumped from aquifers.

However, overuse of groundwater can cause or intensify several problems: **(1)** *water table lowering,* **(2)** *aquifer depletion* (Figure 14-13, top), **(3)** *aquifer subsidence* (sinking of land when groundwater is withdrawn, Figure 14-13, bottom), **(4)** *intrusion of salt water into aquifers,* **(5)** *drawing of chemical contamination in groundwater toward wells,* and **(6)** *reduced stream flow* because of diminished flows of groundwater into streams. Groundwater can also be contaminated by industrial and agricultural activities, septic tanks, and other sources, as discussed on p. 352.

In the United States groundwater is being withdrawn at four times its replacement rate. The most serious overdrafts are occurring **(1)** in parts of the huge Ogallala Aquifer, extending from southern South Dakota to central Texas (Case Study, right) and **(2)** in parts of the arid southwestern United States (Figure 14-13, top),

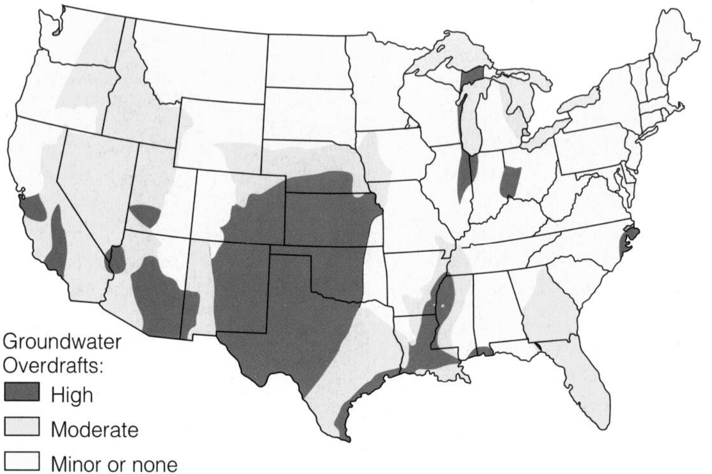

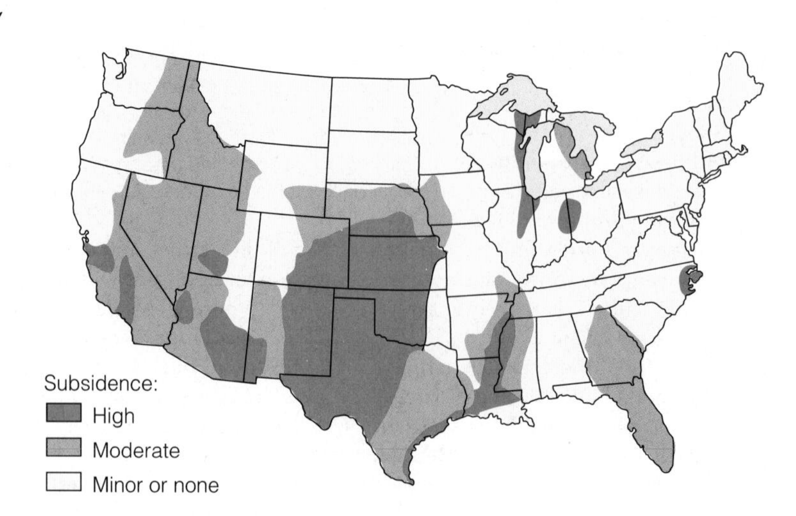

Figure 14-13 Areas of greatest aquifer depletion and groundwater contamination (top) and ground subsidence (bottom) in the continental United States. Aquifer depletion is also high in Hawaii and Puerto Rico (not shown on map). (Data from U.S. Water Resources Council and U.S. Geological Survey)

Mining Groundwater: The Shrinking Ogallala Aquifer

CASE STUDY

Large amounts of water have been pumped from the Ogallala, the world's largest known aquifer (Figure 14-14). This water has helped transform vast areas of arid high plains prairie land into one of the largest and most productive agricultural regions in the United States.

Mostly because of irrigated farming, this region produces 20% of U.S. agricultural output (including 40% of its feedlot beef), valued at $32 billion per year. This has brought prosperity to many farmers and merchants in this region, but the hidden environmental and economic cost has been increasing aquifer depletion in some areas.

Although this aquifer is gigantic, it is essentially a nonrenewable aquifer (stored during the retreat of the last ice age about 15,000–30,000 years ago) with an extremely slow recharge rate. About 200,000 wells pump water out of this aquifer. In some areas, water is being pumped out of the aquifer 8–10 times faster than the aquifer's natural recharge rate.

The northernmost states (Wyoming, North Dakota, South Dakota, and parts of Colorado) still have ample supplies. However, supplies in parts of the southern states, where the aquifer is thinner (Figure 14-14) are being depleted rapidly, with about two-thirds of the aquifer's depletion taking place in the Texas High Plains.

Water experts project that at the current rate of withdrawal, one-fourth of the aquifer's original supply will be depleted by 2020 and much sooner in areas where it is shallow. It will take thousands of years to replenish the aquifer.

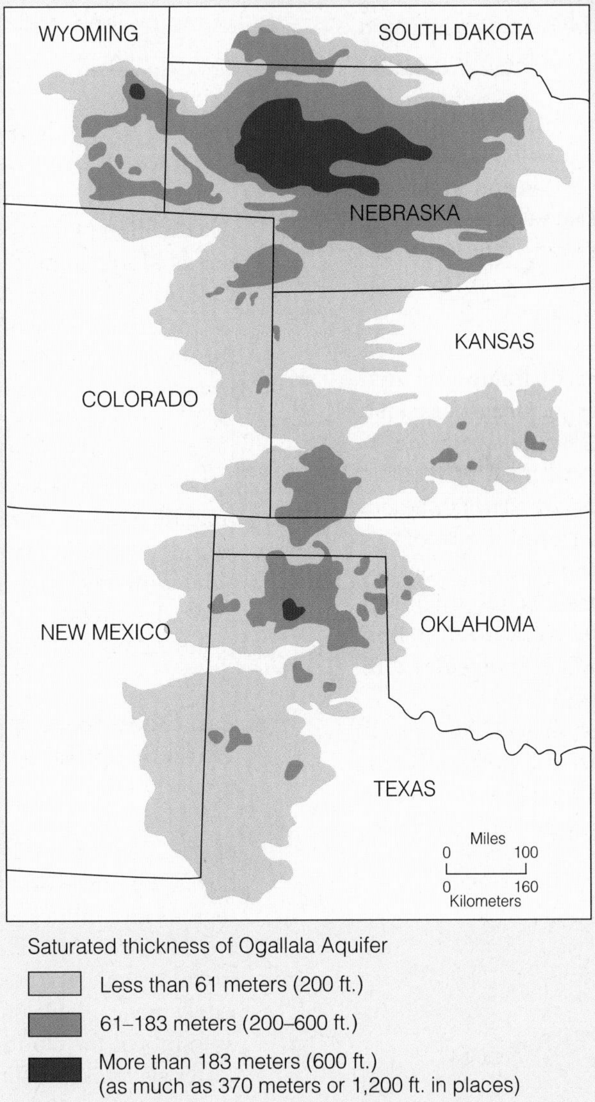

Saturated thickness of Ogallala Aquifer

- Less than 61 meters (200 ft.)
- 61–183 meters (200–600 ft.)
- More than 183 meters (600 ft.) (as much as 370 meters or 1,200 ft. in places)

Figure 14-14 The Ogallala, the world's largest known aquifer. If the water in this aquifer were above ground, it could cover all 50 states with 0.5 meter (1.5 feet) of water. Water withdrawn from this aquifer is used to grow crops, raise cattle, and provide cities and industries with water. As a result, this aquifer, which is renewed very slowly, is being depleted (especially at its thin southern end in parts of Texas, New Mexico, Oklahoma, and Kansas). (Data from U.S. Geological Survey)

Government subsidies designed to increase crop production also increase depletion of the Ogallala by **(1)** encouraging the growth of water-thirsty cotton in the lower basin, **(2)** providing crop disaster payments, and **(3)** providing tax breaks in the form of groundwater depletion allowances, with larger breaks for heavier groundwater use.

Depletion of this essentially nonrenewable water resource can be delayed if farmers **(1)** use more efficient forms of irrigation (Figure 14-16, p. 339), **(2)** switch to crops that need less water, or **(3)** irrigate less land.

Cities using this groundwater can also implement policies and technologies to reduce their water use and waste. People enjoying the benefits of this aquifer can help by installing water-saving toilets and showerheads and converting their lawns to plants that can survive in an arid climate with little watering.

Critical Thinking

1. What are the pros and cons of giving government subsidies to farmers and ranchers using water withdrawn from the Ogallala to grow crops and raise livestock that need large amounts of irrigation water? How do you benefit from such subsidies?

2. Should these subsidies be reduced or eliminated and replaced with subsidies that encourage farmers to use more efficient forms of irrigation and switch to crops that need less water? Explain.

especially California's Central Valley, which supplies about half the country's vegetables and fruits.

Aquifer depletion also is a problem in **(1)** Saudi Arabia, **(2)** central and northern China, **(3)** northwest and southern India (where one-fourth of the country's grain is being produced by unsustainable groundwater withdrawal), **(4)** northern Africa (especially Libya and Tunisia), **(5)** southern Europe, **(6)** the Middle East, and **(7)** parts of Mexico, Thailand, and Pakistan.

When fresh water from an aquifer near a coast is withdrawn faster than it is recharged, salt water intrudes into the aquifer (Figure 14-15). Such intrusion can contaminate the drinking water of many towns and cities along coastal areas.

Ways to prevent or slow groundwater depletion include **(1)** controlling population growth, **(2)** not planting water-intensive crops such as cotton and sugarcane in dry areas, **(3)** shifting to crops that need less water in dry areas, **(4)** developing crop strains that need less water and are more resistant to heat stress, and **(5)** wasting less irrigation water (pp. 339–340).

How Useful Is Desalination? Removing dissolved salts from ocean water or from brackish (slightly salty) groundwater, called **desalination**, is another way to increase supplies of fresh water. The two most widely used methods are **(1)** *distillation,* which involves heating salt water until it evaporates (and leaves behind salts in solid form) and condenses as fresh water, and **(2)** *reverse osmosis,* in which salt water is pumped at high pressure through a thin membrane whose pores allow water molecules, but not dissolved salts, to pass through.

About 13,300 desalination plants in 120 countries (especially in the arid Middle East and parts of North Africa) meet less than 0.2% of the world's water needs. Desalination would have to increase 25-fold just to supply 5% of current world water use.

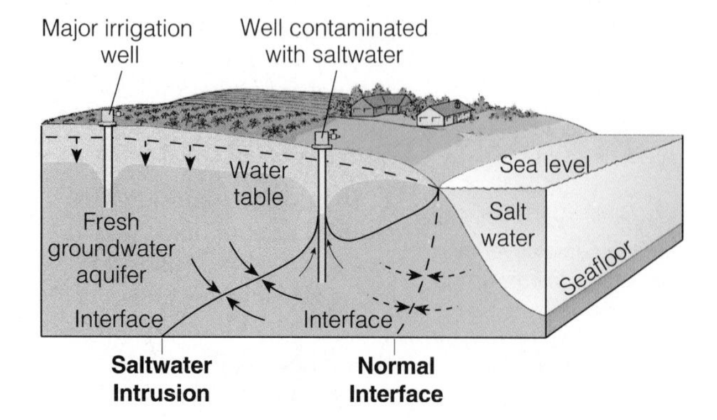

Figure 14-15 Saltwater intrusion along a coastal region. When the water table is lowered, the normal interface (dotted line) between fresh and saline groundwater moves inland (solid line), making coastal groundwater drinking supplies unusable.

This is unlikely because desalination has two major disadvantages:

- *It is expensive because it takes large amounts of energy.* Desalinating water costs 2–3 times as much as the conventional purification of fresh water.
- *It produces large quantities of wastewater (brine) containing high levels of salt and other minerals.* Dumping the concentrated brine into the ocean near the plants increases the local salt concentration and threatens food resources in estuary waters, and dumping it on land could contaminate groundwater and surface water.

Desalination can provide fresh water for coastal cities in arid countries (such as sparsely populated Saudi Arabia and Israel), where the cost of getting fresh water by any method is high. In the United States, desalination plants are used to meet some of the water needs along some coastal areas of Florida, southern California, Virginia, North Carolina, and Texas.

Scientists are working to develop new membranes for reverse osmosis that can separate water from salt more efficiently and under less pressure. If successful, this could help bring down the cost of using desalinization to produce drinking water. However, desalinated water probably will never be cheap enough to irrigate conventional crops or meet much of the world's demand for fresh water unless **(1)** affordable solar-powered distillation plants can be developed and **(2)** someone can figure out what to do with the resulting mountains of salt.

Can Cloud Seeding and Towing Icebergs Improve Water Supplies? For decades several countries, particularly the United States, have been experimenting with seeding clouds with tiny particles of chemicals (such as silver iodide). The particles form water condensation nuclei and thus produce more rain over dry regions and more snow over mountains.

However, cloud seeding **(1)** is not useful in very dry areas, where it is most needed, because rain clouds rarely are available there and **(2)** would introduce large amounts of the cloud-seeding chemicals into soil and water systems, possibly harming people, wildlife, and agricultural productivity.

Another obstacle to cloud seeding is legal disputes over the ownership of water in clouds. During the 1977 drought in the western United States, the attorney general of Idaho accused officials in neighboring Washington of "cloud rustling" and threatened to file suit in federal court.

There also have been proposals to tow huge icebergs to arid coastal areas (such as Saudi Arabia and southern California) and then to pump the fresh water from the melting bergs ashore. However, the technology for doing this is not available and the costs may be too high, especially for water-short developing countries.

What Are the Benefits of Reducing Water Waste? Mohamed El-Ashry of the World Resources Institute estimates that *65–70% of the water people use throughout the world is lost through evaporation, leaks, and other losses*. The United States, the world's largest user of water, does slightly better but still loses about 50% of the water it withdraws. El-Ashry believes that it is economically and technically feasible to reduce such water losses to 15%, thereby meeting most of the world's water needs for the foreseeable future.

This will involve greatly increased use of water-saving technologies and practices that do more with less water. This will **(1)** decrease the burden on wastewater plants, **(2)** reduce the need for expensive dams and water transfer projects that destroy wildlife habitats and displace people, **(3)** slow depletion of groundwater aquifers, and **(4)** save energy and money.

Why Do We Waste So Much Water? According to water resource experts, two major causes of water waste are

- *Government subsidies of water supply projects that create artificially low water prices and lack of subsidies for improving water efficiency.* According to Sandra Postel, "By heavily subsidizing water, governments give out the false message that it is abundant and can afford to be wasted—even as rivers are drying up, aquifers are being depleted, fisheries are collapsing, and species are going extinct." However, farmers, industries, and others benefiting from water subsidies argue that they **(1)** promote settlement and agricultural production in arid and semiarid areas, **(2)** stimulate local economies, and **(3)** help lower prices of food and manufactured goods for consumers.
- *Fragmented watershed management.* The Chicago, Illinois, metropolitan area, for example, has 349 water supply systems divided among some 2,000 local units of government over a six-county area.

Solutions: How Can We Waste Less Water in Irrigation? Much of the irrigation water applied throughout the world does not reach the targeted crops. Most irrigation systems distribute water from a groundwater well or a surface water source and allow it to flow by gravity through unlined ditches in cropfields so that the water can be absorbed by crops (Figure 14-16, left). This flood irrigation method **(1)** delivers far more water than needed for crop growth and **(2)** typically allows only 60% of the water to reach crops because of evaporation, seepage, and runoff.

Water waste can be reduced by using more efficient irrigation systems such as

- *Center-pivot low-pressure sprinklers* (Figure 14-16, right), which typically allow 80% of the water input to reach crops and reduce water use over conventional gravity flow systems by 25%.

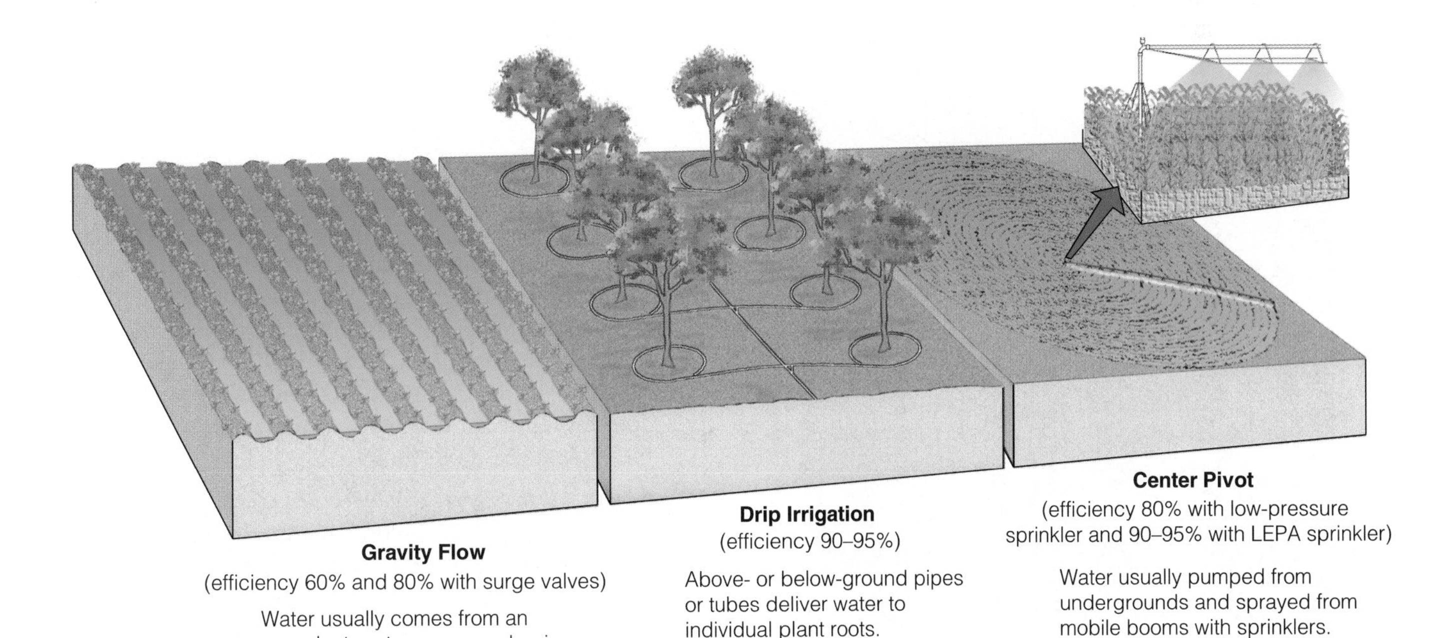

Gravity Flow
(efficiency 60% and 80% with surge valves)
Water usually comes from an aqueduct system or a nearby river.
Drip Irrigation
(efficiency 90–95%)
Above- or below-ground pipes or tubes deliver water to individual plant roots.
Center Pivot
(efficiency 80% with low-pressure sprinkler and 90–95% with LEPA sprinkler)
Water usually pumped from undergrounds and sprayed from mobile booms with sprinklers.

Figure 14-16 Major irrigation systems. Because of high initial costs, center-pivot irrigation and drip irrigation are used on only about 1% of the world's irrigated cropland each. However, this may change because of the development of new low-cost drip irrigation systems (Solutions, p. 341).

- *Low-energy precision application (LEPA) sprinklers.* This form of center-pivot irrigation allows 90–95% of the water input to reach crops by spraying it closer to the ground and in larger droplets than the center-pivot, low-pressure system. LEPA sprinklers use 20–30% less energy than low-pressure sprinklers and typically use 37% less water than conventional gravity flow systems.

- *Using surge or time-controlled valves on conventional gravity flow irrigation systems* (Figure 14-16, left). These valves send water down irrigation ditches in pulses instead of in a continuous stream. This can raise irrigation efficiency to 80% and reduce water use by 25%.

- *Using soil moisture detectors to water crops only when they need it.* For example, some farmers in Texas bury a $1 cube of gypsum, the size of a lump of sugar, at the root zone of crops. Wires embedded in the gypsum are run back to a small, portable meter that indicates soil moisture. Farmers using this technique can use 33–66% less irrigation water.

- *Drip irrigation systems* (Figure 14-16, center, and Solutions, right), which can raise water efficiency to 90–95% and reduce water use by 37–70%.

Other ways to reduce irrigation water waste are listed in Figure 14-17. Since 1950 water-short Israel has used many of these techniques to slash irrigation water waste by about 84% while irrigating 44% more land. Israel now treats and reuses 65% of its municipal sewage water for crop production and plans to increase this to 80% by 2025. The government also **(1)** gradually removed most water subsidies to raise the price of irrigation water to one of the highest in the world, **(2)** imports most of its water-intensive wheat and meat, and **(3)** concentrates on growing fruits, vegetables, and flowers that need less water.

- Lining canals bringing water to irrigation ditches
- Leveling fields with lasers
- Irrigating at night to reduce evaporation
- Using soil and satellite sensors and computer systems to monitor soil moisture and add water only when necessary
- Polyculture
- Organic farming
- Growing water efficient crops using drought-resistant and salt-tolerant crop varieties
- Irrigating with treated urban waste water
- Importing water-intensive crops and meat

Figure 14-17 Methods for reducing water waste in irrigation.

Many of the world's poor farmers cannot afford to use most modern technological methods for increasing irrigation and irrigation efficiency. Such farmers increase irrigation by using small-scale and low-cost traditional technologies such as **(1)** pedal-powered treadle pumps to move water through irrigation ditches (widely used in Bangladesh), **(2)** animal-powered irrigation pumps, **(3)** buckets with holes for drip irrigation, **(4)** check dams, ponds, and tanks to collect rainwater for irrigation, **(5)** terracing (Figure 9-26a, p. 219) to reduce water loss on crops grown on steep terrain, and **(6)** cultivating seasonally waterlogged wetlands, delta lands, and valley bottoms.

Solutions: How Can We Waste Less Water in Industry, Homes, and Businesses? Ways to use water more efficiently in industries, homes, and businesses include the following:

- *Redesigning manufacturing processes.* A paper mill in Hadera, Israel, uses one-tenth as much water as most of the world's other paper mills, and a German paper plant nearly eliminated water use by completely recycling and purifying its water. Manufacturing aluminum from recycled scrap rather than virgin ore can reduce water needs by 97%.

- *Replacing green lawns in arid and semiarid regions with vegetation adapted to a dry climate.* This form of landscaping, called *xeriscaping* (pronounced "ZER-i-scaping"), reduces water use by 30–85% and sharply reduces inputs of labor, fertilizer, and fuel and the production of polluted runoff, air pollution, and yard wastes.

- *Using drip irrigation to water gardens and other vegetation around homes and businesses.*

- *Fixing leaks in water mains, pipes, toilets, and faucets.* Leaks waste about half of the water supply in many cities in developing countries and 20–35% of water withdrawn from public supplies in the United States and the United Kingdom. Leaks from toilet valves, dripping faucets, and aging pipes account for about one-tenth of the water used in the typical U.S. household.

- *Using water meters to monitor and charge for municipal water use.* In Boulder, Colorado, introducing water meters reduced water use by more than one-third. About one-fifth of all U.S. public water systems do not have water meters and charge a single low rate for almost unlimited use of high-quality water. Many apartment dwellers have little incentive to conserve water because their water use is included in their rent.

The Promise of Drip Irrigation

SOLUTIONS

The development of inexpensive, weather-resistant, and flexible plastic tubing after World War II paved the way for use of a new form of microirrigation called *drip irrigation* (Figure 14-16, middle). It consists of a network of perforated plastic tubing, installed at or below the ground surface. The small holes or emitters in the tubing deliver drops of water at a slow and steady rate close to plant roots.

This technique, developed in Israel in the 1960s and now used by half the country's farmers, has a number of advantages, including the following:

- *Adaptability*. The tubing system can easily be fitted to match the patterns of crops in a field and left in place or moved to different locations.
- *Efficiency*, with 90–95% of the water input reaching crops.
- *Lower operating costs* because 37–70% less energy is needed to pump this water at low pressure and less labor is needed to move sprinkler systems.
- *Ability to apply fertilizer solutions in precise amounts*, which reduces fertilizer use and waste, salinization, and water pollution from fertilizer runoff.
- *An increase in crop yields of 20–90% by getting more crop growth per drop.*
- *Healthier plants and higher yields* because plants are neither underwatered nor overwatered.

Despite these advantages, drip irrigation is used on less than 1% of the world's irrigated area. The main reason is that the capital cost of conventional drip irrigation systems is too high for most poor farmers and for use on low-value row crops. However, drip irrigation is economically feasible for high-profit fruit, vegetable, and orchard crops and for home gardens.

Some *good news* is that the capital cost of a newly developed drip irrigation system is one-tenth as much per hectare as conventional drip systems. This and other low-cost drip irrigation systems could bring about a revolution in more sustainable irrigated agriculture that would **(1)** increase food yields, **(2)** reduce water use and waste, and **(3)** lessen some of the environmental problems associated with agriculture (Figure 16-9, p. 405).

Critical Thinking

Should governments provide subsidies to farmers who use drip irrigation based on how much water they save? Explain.

- *Having ordinances requiring water conservation in water-short cities.* Because of such ordinances, the desert city of Tucson, Arizona, consumes half as much water per person as Las Vegas, a desert city with even less rainfall and less emphasis on water conservation (Spotlight, p. 342).
- *Requiring or encouraging use of water-saving toilets and showerheads.* Since 1994 all new toilets sold in the United States must use no more than 6 liters (1.6 gallons) per flush, and similar laws have been passed in Mexico and in Ontario, Canada. A low-flow showerhead costing about $20 saves about $34–56 per year in water heating costs. Audits conducted by students in Brown University's environmental studies program showed that the school could save $44,000 a year by using low-flow showerheads in dormitories.
- *Using washing machines that load from the front instead of the top.* Such machines **(1)** use 40–75% less water, **(2)** make clothes last longer because they are not agitated, **(3)** use less energy, and **(4)** save money.
- *Reusing gray water from bathtubs, showers, bathroom sinks, and clothes washers for irrigating lawns and nonedible plants and raising fish.* About 50–75% of the water used by a typical house could be reused as gray water. In the United States, California has become the first state to legalize reuse of gray water to irrigate landscapes. About 65% of the wastewater in Israel is reused.
- *Installing or leasing systems that purify and completely recycle wastewater from houses, apartments, or office buildings.* In Tokyo, Japan, all the water used in Mitsubishi's 60-story office building is purified for reuse by an automated recycling system.
- *Collecting and using rainwater for flushing toilets, irrigating gardens, watering lawns, and putting out fires.* In Tokyo, Japan, large tanks on top of 579 city buildings capture rainwater.
- *Reducing personal water use and waste* by actions such as those listed on the website for this chapter.

Raising the price of water for domestic and industrial consumers (as Israel has done) is one way to reduce wasteful water use. You might think that charging more for water supplied by public water systems would hurt the poor. Instead, this usually lowers the cost of water for the poor because most are paying 10 to 12 times more per liter of water to buy it from private water vendors than citizens receiving often purer water from public systems.

Running Out of Water in Las Vegas, Nevada

SPOTLIGHT

Las Vegas, Nevada, located in the Mojave Desert, is an artificial aquatic wonderland of large trees, green lawns and golf courses, waterfalls, and swimming pools. The city is also one of the fastest-growing cities in the United States, with its population doubling from 550,000 in 1985 to 1.2 million in 1998. It is estimated that the city uses more water per person than any other city in the world.

Tucson, Arizona, in the Sonora Desert, gets only 30 centimeters (12 inches) of rainfall a year, and Las Vegas averages only 10 centimeters (4 inches). Tucson, now a model of water conservation, began a strict water conservation program in 1976, including raising water rates 500% for some residents.

In contrast, Las Vegas only recently started to encourage water conservation by **(1)** raising water rates sharply (but they are still less than half those in Tucson) and **(2)** encouraging replacement of lawns with rocks and native plants that survive on little water.

Water experts project that even if these recent water conservation efforts are successful, the city will run short of water by 2007.

Critical Thinking

1. If you were an elected official in charge of Las Vegas, what three things would you do to improve water conservation? What might be the political implications of instituting such a program?

2. If water shortages by 2007 limit the growth of the area's population, would you consider this good or bad? Explain.

14-4 TOO MUCH WATER

What Are the Causes and Effects of Flooding? Heavy rain or rapid melting of snow is the major cause of natural flooding by streams. This causes water in a stream to overflow its normal channel and flood the adjacent area, called a **floodplain** (Figure 14-18). Floodplains, which include highly productive wetlands, help **(1)** provide natural flood and erosion control, **(2)** maintain high water quality, and **(3)** recharge groundwater.

People have settled on floodplains since the beginnings of agriculture. They have many advantages, including **(1)** fertile soil, **(2)** ample water for irrigation, **(3)** flat land suitable for crops, buildings, highways, and railroads, and **(4)** availability of nearby rivers for transportation and recreation. In the United States, 10 million households and businesses with property valued at $1 trillion exist in flood-prone areas.

Floods are a natural phenomenon and have several benefits. They **(1)** provide the world's most productive

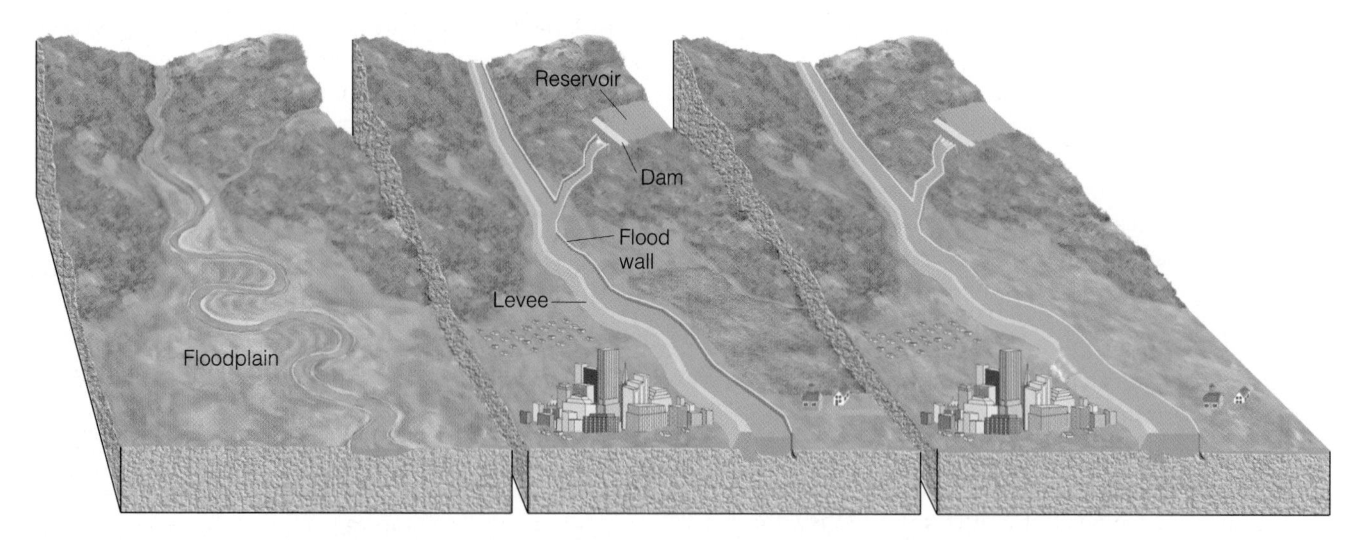

Figure 14-18 Land in a natural floodplain (left) often is flooded after prolonged rains. When the floodwaters recede, silt deposits are left behind, creating a nutrient-rich soil. To reduce the threat of flooding (and thus allow people to live in floodplains), rivers have been **(1)** dammed to create reservoirs that store and release water as needed, **(2)** narrowed and straightened (channelization), and **(3)** equipped with protective levees and walls (middle). These alterations can give a false sense of security to floodplain dwellers living in high-risk areas. In the long run, such measures can greatly increase flood damage because they can be overwhelmed by prolonged rains (right), as happened in the midwestern United States during the summer of 1993.

farmland because they are regularly covered with nutrient-rich silt left after floodwaters recede, **(2)** recharge groundwater, and **(3)** refill wetlands.

However, each year floods kill thousands of people and causes tens of billions of dollars in property damage. Between 1985 and 1999, floods killed about 275,000 people, 96% of them in developing countries.

Floods, like droughts, usually are considered natural disasters, but since the 1960s human activities have contributed to the sharp rise in flood deaths and damages. Three ways humans increase the severity of flood damage are by **(1)** removing water-absorbing vegetation, especially on hillsides (Figure 14-19), **(2)** draining wetlands that absorb floodwaters and reduce the severity of flooding, and **(3)** living on floodplains (Connections, p. 347).

In developed countries, people deliberately settle on floodplains and then expect dams, levees, and other devices to protect them from floodwaters. In many developing countries, the poor have little choice but to try to survive in flood-prone areas (Connections, p. 347).

Urbanization also increases flooding by replacing water-absorbing vegetation, soil, and wetlands with highways, parking lots, and buildings that cannot absorb rainwater. If sea levels rise during the next century, as projected, many low-lying croplands, cities, and islands will be under water (Figure 13-13, p. 313).

Solutions: How Can We Reduce Flood Risks? Ways humans can reduce the risk from flooding include

- *Straightening and deepening streams* (channelization, Figure 14-18, middle). Channelization can reduce upstream flooding, but the increased flow of water can also **(1)** increase upstream bank erosion and downstream flooding and sediment deposition and **(2)** reduce habitats for aquatic wildlife by removing bank vegetation and increasing stream velocity.
- *Building levees* (Figure 14-18, middle). Levees contain and speed up stream flow but **(1)** increase the water's capacity for doing damage downstream and **(2)** do not protect against unusually high and powerful floodwaters, as occurred in 1993 when two-thirds of the levees built along the Mississippi River were damaged or destroyed.
- *Building dams.* A flood control dam built across a stream can reduce flooding by storing water in a reservoir and releasing it gradually. Dams have a number of advantages and disadvantages (Figure 14-9).
- *Restoring wetlands* to take advantage of the natural flood control provided by floodplains.
- *Managing floodplains* to get people out of flood-prone areas (Figure 14-20). This prevention approach is based on thousands of years of experience that can

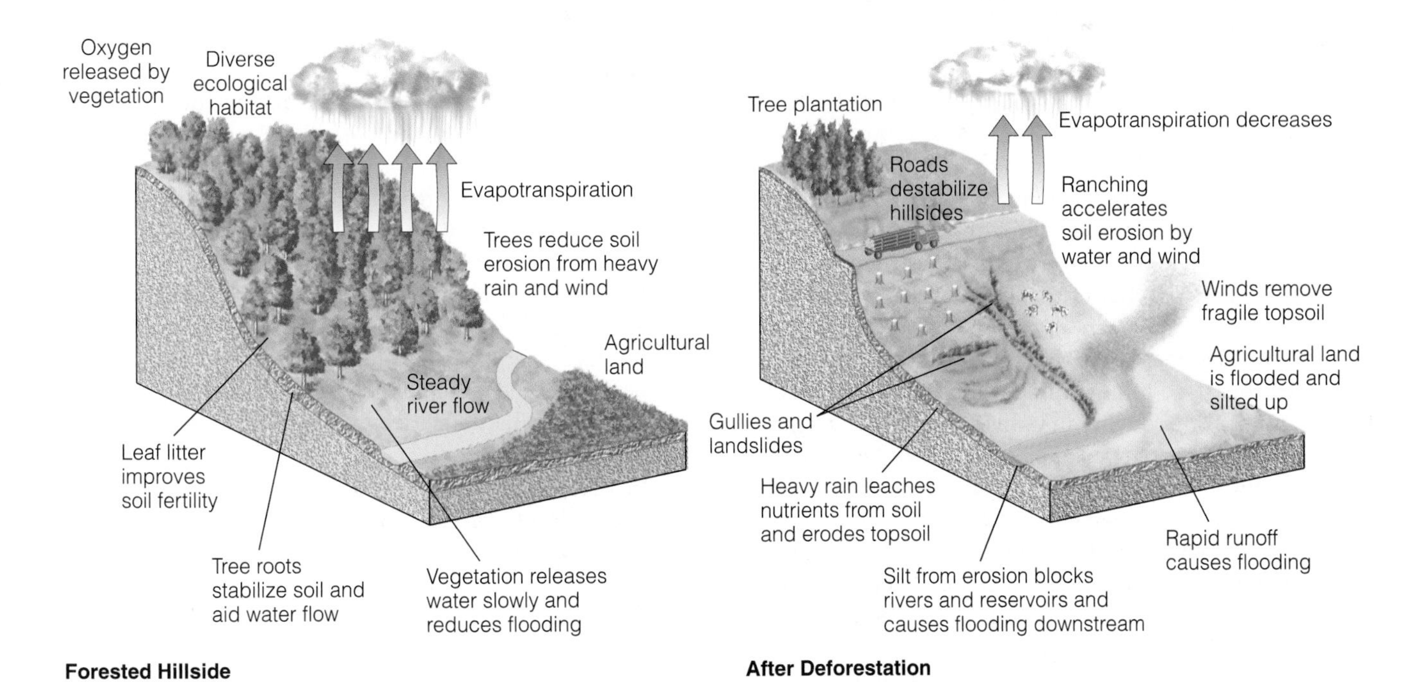

Figure 14-19 A hillside before and after deforestation. Once a hillside has been deforested for timber and fuelwood, livestock grazing, or unsustainable farming, water from precipitation **(1)** rushes down the denuded slopes, **(2)** erodes precious topsoil, and **(3)** floods downstream areas. A 3,000-year-old Chinese proverb says, "To protect your rivers, protect your mountains."

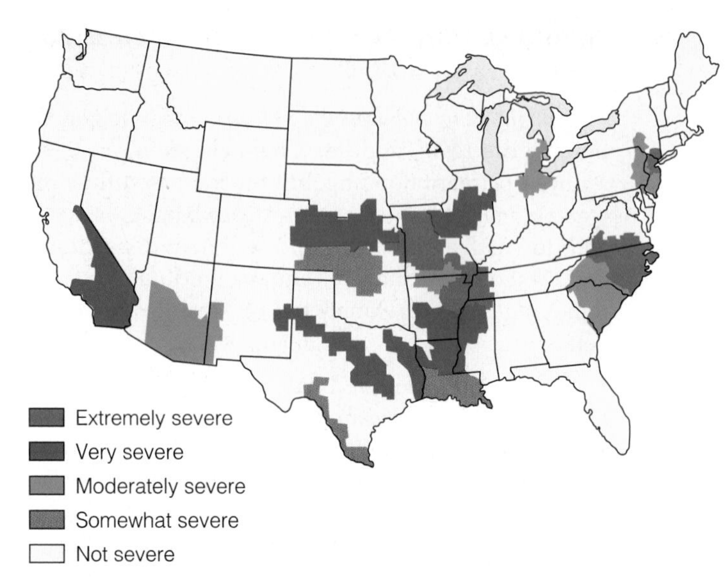

Figure 14-20 Generalized map of *flood-prone areas* in the United States. More detailed state and local maps are used to show the likelihood and severity of floods within these general areas. Flood frequency data and maps of flood-prone areas do not tell us when floods will occur, but they give a general idea of how often and where floods might occur, based on an area's history. (Data from U.S. Geological Survey)

be summed up in one idea: *Sooner or later the river (or the ocean) always wins.*

14-5 TYPES, EFFECTS, AND SOURCES OF WATER POLLUTANTS

What Are the Major Types and Effects of Water Pollutants? **Water pollution** is any chemical, biological, or physical change in water quality that has a harmful effect on living organisms or makes water unsuitable for desired uses. Table 14-1 lists the major classes of water pollutants along with their major human sources and harmful effects.

How Do We Measure Water Quality? Various methods are used to determine water quality. A good indicator of water quality in terms of infectious agents is the number of colonies of *coliform bacteria* present in a 100-milliliter (0.1-quart) sample of water. The World Health Organization recommends a coliform bacteria count of 0 colonies per 100 milliliters for drinking water, and the U.S. Environmental Protection Agency (EPA) recommends a maximum level for swimming water of 200 colonies per 100 milliliters.

Water pollution from oxygen-demanding wastes and plant nutrients can be determined by measuring the level of dissolved oxygen. The quantity of oxygen-demanding wastes in water can be determined by measuring the **biological oxygen demand (BOD)**: the amount of dissolved oxygen needed by aerobic decomposers to break down the organic materials in a certain volume of water over a 5-day incubation period at 20°C (68°F).

Chemical analysis can determine the presence and concentrations of most inorganic and organic chemicals that pollute water. Living organisms can also be used as *indicator species* to monitor water pollution. For example, the tissues of *filter-feeding mussels*, harvested from the sediments of coastal waters, can be analyzed for the presence of various industrial chemicals, toxic metals (such as mercury and lead), and pesticides. Aquatic plants such as cattails can be removed and analyzed to determine pollution in areas contaminated with fuels, solvents, and other organic chemicals.

What Are Point and Nonpoint Sources of Water Pollution? **Point sources** discharge pollutants at specific locations through pipes, ditches, or sewers into bodies of surface water. Examples include **(1)** factories, **(2)** sewage treatment plants (which remove some but not all pollutants), **(3)** active and abandoned underground mines, and **(4)** oil tankers. Because point sources are at specific places, they are fairly easy to identify, monitor, and regulate. In developed countries many industrial discharges are strictly controlled, whereas in most developing countries such discharges are largely uncontrolled.

Nonpoint sources are sources that cannot be traced to any single discharge site. They are usually large land areas or airsheds that pollute water by runoff, subsurface flow, or deposition from the atmosphere. Examples include **(1)** acid deposition (Figure 12-8, p. 286) and **(2)** runoff of chemicals into surface water from croplands, livestock feedlots, logged forests, urban streets, lawns, golf courses, and parking lots.

Nonpoint pollution from agriculture includes **(1)** sediments, **(2)** inorganic fertilizers, **(3)** manure, **(4)** salts dissolved in irrigation water, and **(5)** pesticides. In the United States, such pollution is responsible for an estimated 64% of the total mass of pollutants entering streams and 57% of those entering lakes. Little progress has been made in controlling nonpoint water pollution because of the difficulty and expense of identifying and controlling discharges from so many diffuse sources.

Is the Water Safe to Drink? About one-fourth of people in developing countries do not have access to clean drinking water. In China, an estimated 700 million people drink contaminated water, and only 6 of China's 27 largest cities provide drinking water that meets government standards. In Russia, half of all tap water is unfit to drink, and a third of the aquifers are too contaminated for drinking purposes. In India an estimated 300 million people lack access to safe water. About 290 million Africans—more than the entire U.S. population—do not have access to safe drinking water.

Table 14-1 Major Categories of Water Pollutants States

Infectious Agents
Examples: Bacteria, viruses, protozoa, and parasitic worms
Major Human Sources: Human and animal wastes
Harmful Effects: Disease

Oxygen-Demanding Wastes
Examples: Organic waste such as animal manure and plant debris that can be decomposed by aerobic (oxygen-requiring) bacteria
Major Human Sources: Sewage, animal feedlots, paper mills, and food processing facilities
Harmful Effects: Large populations of bacteria decomposing these wastes can degrade water quality by depleting water of dissolved oxygen. This causes fish and other forms of oxygen-consuming aquatic life to die.

Inorganic Chemicals
Examples: Water-soluble **(1)** acids, **(2)** compounds of toxic metals such as lead (Pb), arsenic (As), and selenium (Se), and **(3)** salts such as NaCl in ocean water and fluorides (F^-) found in some soils
Major Human Sources: Surface runoff, industrial effluents, and household cleansers
Harmful Effects: Can **(1)** make freshwater unusable for drinking or irrigation, **(2)** cause skin cancers and crippling spinal and neck damage (F^-), **(3)** damage the nervous system, liver, and kidneys (Pb and As), **(4)** harm fish and other aquatic life, **(5)** lower crop yields, and **(6)** accelerate corrosion of metals exposed to such water.

Organic Chemicals
Examples: Oil, gasoline, plastics, pesticides, cleaning solvents, detergents
Major Human Sources: Industrial effluents, household cleansers, surface runoff from farms and yards
Harmful Effects: can **(1)** threaten human health by causing nervous system damage (some pesticides), reproductive disorders (some solvents), and some cancers (gasoline, oil, and some solvents) and **(2)** harm fish and wildlife

Plant Nutrients
Examples: Water-soluble compounds containing nitrate (NO_3^-), phosphate (PO_4^{3-}), and ammonium (NH_4^+) ions
Major Human Sources: Sewage, manure, and runoff of agricultural and urban fertilizers
Harmful Effects: Can cause excessive growth of algae and other aquatic plants, which die, decay, deplete water of dissolved oxygen, and kill fish. Drinking water with excessive levels of nitrates lowers the oxygen-carrying capacity of the blood and can kill unborn children and infants ("blue-baby syndrome").

Sediment
Examples: Soil, silt
Major Human Sources: Land erosion
Harmful Effects: Can **(1)** cloud water and reduce photosynthesis, **(2)** disrupt aquatic food webs, **(3)** carry pesticides, bacteria, and other harmful substances, **(4)** settle out and destroy feeding and spawning grounds of fish, and **(5)** clog and fill lakes, artificial reservoirs, stream channels, and harbors.

Radioactive Materials
Examples: Radioactive isotopes of iodine, radon, uranium, cesium, and thorium
Major Human Sources: Nuclear power plants, mining and processing of uranium and other ores, nuclear weapons production, natural sources
Harmful Effects: Genetic mutations, miscarriages, birth defects, and certain cancers

Heat (Thermal Pollution)
Examples: Excessive heat
Major Human Sources: Water cooling of electric power plants (Figure 19–20, p. 503) and some types of industrial plants. Almost half of all water withdrawn in the United States each year is for cooling electric power plants.
Harmful Effects: Lowers dissolved oxygen levels and makes aquatic organisms more vulnerable to disease, parasites, and toxic chemicals. When a power plant first opens or shuts down for repair, fish and other organisms adapted to a particular temperature range (Figure 4-13, p. 85) can be killed by the abrupt change in water temperature—known as *thermal shock*.

The United Nations estimates that it would cost about $23 billion a year over 8–10 years to bring low-cost safe water and sanitation to the 1.4 billion people who do not have access to clean drinking water. These expenditures could prevent many of the 5 million deaths (including 2 million children under age 5) and 3.4 billion cases of illness caused each year by unsafe water. Currently, the world is spending about $16 billion a year on clean water efforts. The $7-billion shortfall is about equal to what the world spends every 4 days for military purposes.

14-6 POLLUTION OF FRESHWATER STREAMS, LAKES, AND AQUIFERS

What Are the Water Pollution Problems of Streams? Flowing streams, including large ones called *rivers,* can recover rapidly from degradable, oxygen-demanding wastes and excess heat through a combination of dilution and bacterial decay. This natural recovery process works as long as **(1)** streams are not overloaded with these pollutants and **(2)** their flow is not reduced by drought, damming, or diversion for agriculture and industry. However, these natural dilution and biodegradation processes do not eliminate slowly degradable and nondegradable pollutants.

In a flowing stream, the breakdown of degradable wastes by bacteria depletes dissolved oxygen, which reduces or eliminates populations of organisms with high oxygen needs until the stream is cleansed of wastes. The depth and width of the resulting *oxygen sag curve* (Figure 14-21), and thus the time and distance needed for a stream to recover, depend on the volume of incoming degradable wastes and the

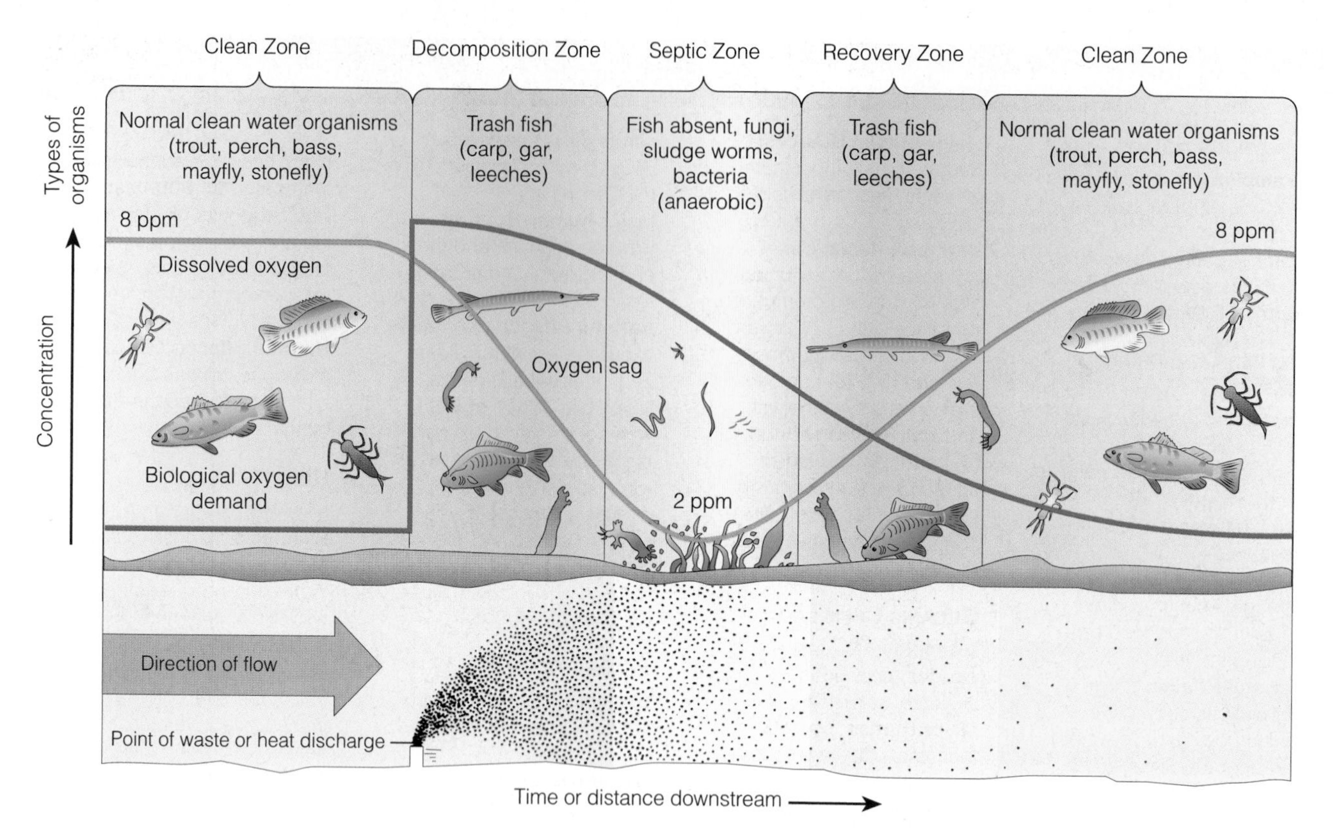

Figure 14-21 Dilution and decay of degradable, oxygen-demanding wastes and heat, showing the oxygen sag curve (orange) and the curve of oxygen demand (blue). Depending on flow rates and the amount of pollutants, streams recover from oxygen-demanding wastes and heat if they are given enough time and are not overloaded.

stream's **(1)** volume, **(2)** flow rate, **(3)** temperature, and **(4)** pH level (Figure 3-5, p. 63). Similar oxygen sag curves can be plotted when heated water from industrial and power plants is discharged into streams.

Although rivers and streams contain only about 1.3% of the world's nonfrozen fresh water, they provide important ecological services (Figure 14-10). Major threats to the ecological services and the biodiversity of rivers are **(1)** pollution, **(2)** disruption of water flows and species composition by dams (Figure 14-9), channelization (Figure 14-18, middle), and the diversion of water for irrigation and urban areas, and **(3)** overfishing.

What Progress Have We Made in Reducing Stream Pollution? Requiring cities to withdraw their drinking water downstream rather than upstream (as is done now) would improve water quality dramatically. Then each city would have to clean up its own waste outputs rather than passing them downstream. However, upstream users, who already have the use of fairly clean water without high cleanup costs, fight this pollution prevention approach.

Here is some *good news*. Water pollution control laws enacted in the 1970s have greatly increased the number and quality of wastewater treatment plants in the United States and many other developed countries. Laws have also required industries to reduce or eliminate point-source discharges into surface waters. These efforts, spurred by individuals (Individuals Matter, p. 348), have enabled the United States to hold the line against increased pollution of most of its streams by disease-causing agents and oxygen-demanding wastes. This is an impressive accomplishment given the rise in economic activity and population since the laws were passed.

One success story is the cleanup of Ohio's Cuyahoga River, which was so polluted that in 1959 and again in 1969 it caught fire and burned for several days as it flowed through Cleveland, Ohio. The highly publicized image of this burning river prompted city, state, and federal officials to **(1)** enact laws limiting the discharge of industrial wastes into the river and sewage systems and **(2)** appropriate funds to upgrade sewage treatment facilities. Today the river has made a comeback and is widely used by boaters and anglers.

A spectacular cleanup has also occurred in Great Britain. In the 1950s, the Thames River was little more than a flowing anaerobic sewer. However, after more than 40 years of effort and hundreds of millions of dollars spent

Living Dangerously on Floodplains in Bangladesh

Bangladesh is one of the world's **(1)** most densely populated countries, with 134 million people packed into an area roughly the size of Wisconsin, and **(2)** poorest countries, with an average per capita GNP of about $350, or 96¢ per day.

The people of Bangladesh depend on moderate annual flooding during the summer monsoon season to **(1)** grow rice and **(2)** help maintain soil fertility in the delta basin by receiving an annual deposit of eroded Himalayan soil.

However, excessive flooding can be disastrous. In the past, great floods occurred every 50 years or so, but since the 1970s they have come about every 4 years.

Bangladesh's increased flood problems begin in the Himalayan watershed. A combination of rapid population growth, deforestation, overgrazing, and unsustainable farming on steep, easily erodible mountain slopes has greatly diminished the soil's ability to absorb water. Instead of being absorbed and released slowly, water from the monsoon rains runs off the denuded Himalayan foothills, carrying vital topsoil with it (Figure 14-19).

This runoff, combined with heavier-than-normal monsoon rains, has increased the severity of flooding along Himalayan rivers and downstream in Bangladesh. For example, a disastrous flood in 1998 **(1)** covered two-thirds of Bangladesh's land area for 9 months, **(2)** leveled 2 million homes, **(3)** drowned at least 2,000 people, **(4)** left 30 million people homeless, **(5)** destroyed more than one-fourth of the country's crops, which caused thousands of people to die of starvation, and **(6)** caused at least $3.4 billion in damages.

Living on Bangladesh's coastal floodplain also carries dangers from storm surges and cyclones. Since 1961, 17 devastating cyclones have slammed into Bangladesh. In 1970, as many as 1 million people drowned in one storm, and another surge killed an estimated 139,000 people in 1991.

In their struggle to survive, the poor in Bangladesh have cleared many of the country's coastal mangrove forests for fuelwood, farming, and aquaculture ponds for raising shrimp. This has led to more severe flooding because these coastal wetlands shelter Bangladesh's low-lying coastal areas from storm surges and cyclones. Damages and deaths from cyclones in areas of Bangladesh still protected by mangrove forests have been much lower than in areas where the forests have been cleared.

Critical Thinking

1. Bangladesh's population is growing rapidly and is expected to increase from 134 million to 180 million between 2001 and 2025. How could slowing its rate of population growth help reduce poverty and the harmful impacts of excessive flooding?

2. How could reforestation in the upstream countries of Bhutan, China, India, and Nepal reduce flooding in those countries and in Bangladesh?

by British taxpayers and private industry, the Thames has made a remarkable recovery. Commercial fishing is thriving, and many species of waterfowl and wading birds have returned to their former feeding grounds.

What Is the Bad News About Stream Pollution? Despite progress in improving stream quality in most developed countries, large fish kills and drinking water contamination still occur. Most of these disasters are caused by **(1)** accidental or deliberate releases of toxic inorganic and organic chemicals by industries or mines, **(2)** malfunctioning sewage treatment plants, and **(3)** nonpoint runoff of pesticides and nutrients (eroded soil, fertilizer, and animal waste) from cropland or animal feedlots (Individuals Matter, p. 351).

Available data indicate that stream pollution from discharges of sewage and industrial wastes is a serious and growing problem in most developing countries, where waste treatment is practically nonexistent. Numerous streams in the former Soviet Union and in eastern European countries are severely polluted. Currently, more than two-thirds of India's water resources are polluted with industrial wastes and sewage. Of the 78 streams monitored in China, 54 are seriously polluted with untreated sewage and industrial wastes. In Latin America and Africa, most streams passing through urban or industrial areas are severely polluted.

What Are the Pollution Problems of Lakes? In lakes, reservoirs, and ponds, dilution often is less effective than in streams because

- Lakes and reservoirs often contain stratified layers (Figure 6-42, p. 154) that undergo little vertical mixing.
- They have little flow. For example, the flushing and changing of water in lakes and large artificial reservoirs can take from 1 to 100 years, compared with several days to several weeks for streams.
- Ponds contain small volumes of water.

As a result, lakes, reservoirs, and ponds are more vulnerable than streams to contamination by **(1)** plant

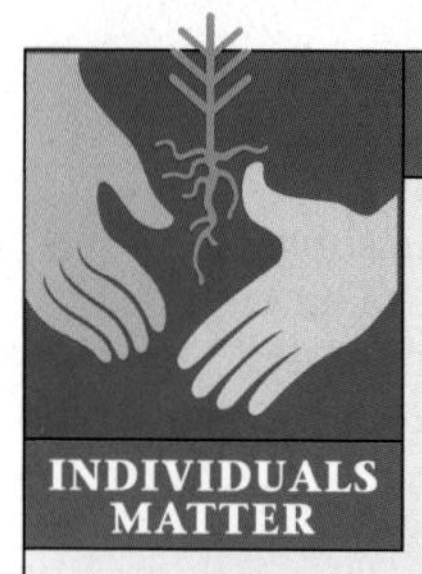

Rescuing a River

In the 1960s, Marion Stoddart moved to Groton, Massachusetts, on the Nashua River, then considered one of the nation's filthiest rivers. Dead fish bobbed on its waves, and at times the water was red, green, or blue from pigments discharged by paper mills.

Instead of thinking nothing could be done, she committed herself to restoring the Nashua and establishing public parklands along its banks.

She did not start by filing lawsuits or organizing demonstrations. Instead she created a careful cleanup plan and approached state officials with it in 1962. They laughed, but she was not deterred and began practicing the most time-honored skill of politics: one-on-one persuasion. She identified the power brokers in the riverside communities and began to educate them, win them over, and get them to cooperate in cleaning up the river.

She got the state to ban open dumping in the river. When promised federal matching funds for building the treatment plant failed to materialize, Stoddart gathered 13,000 signatures on a petition sent to President Richard Nixon. The funds arrived in a hurry.

Stoddart's next success was getting a federal grant to beautify the river. She hired high school dropouts to clear away mounds of debris. When the river cleanup was completed, she persuaded communities along the river to create some 2,400 hectares (6,000 acres) of riverside park and woodlands along both banks.

Now, almost four decades later, the Nashua is still clean. Several new water treatment plants have been built, and a citizens' group founded by Stoddart keeps watch on water quality. The river supports many kinds of fish and other wildlife, and its waters are used for canoeing and other kinds of recreation.

For her efforts, the UN Environment Programme has named Stoddart as an outstanding worldwide worker for the environment. However, she might say that the blue and canoeable Nashua itself is her best reward.

nutrients, **(2)** oil, **(3)** pesticides, and **(4)** toxic substances such as lead, mercury, and selenium. These contaminants can destroy both bottom life and fish and birds that feed on contaminated aquatic organisms.

Many toxic chemicals and acids also enter lakes and reservoirs from the atmosphere (Figure 12-8, p. 286). Concentrations of some chemicals, such as DDT (Figure 10-4, p. 227), PCBs (Figure 14-22), some radioactive isotopes, and some mercury compounds, can be biologically magnified as they pass through food webs in lakes.

Lakes also receive inputs of nutrients and silt from the surrounding land basin as a result of natural erosion and runoff. This natural nutrient enrichment of lakes is called **eutrophication**. Over time, some lakes become more eutrophic (Figure 6-43, bottom, p. 155), but others do not because of differences in the surrounding drainage basin.

Near urban or agricultural areas, human activities can greatly accelerate the input of plant nutrients to a lake, which results in a process known as **cultural eutrophication**. Such a change is caused mostly by nitrate- and phosphate-containing effluents from **(1)** sewage treatment plants, **(2)** runoff of fertilizers and animal wastes, and **(3)** accelerated erosion of nutrient-rich topsoil (Figure 14-23).

During hot weather or drought, this nutrient overload produces dense growths of organisms such as algae, cyanobacteria, water hyacinths, and duckweed. Large masses of algae die, fall to the bottom, and are decomposed by aerobic bacteria, which depletes dissolved oxygen (in both the surface layer of water near the shore and in the bottom layer). This oxygen depletion can kill fish and other aerobic aquatic animals. If excess nutrients continue to flow into a lake, anaerobic bacteria take over and produce gaseous decomposition products such as smelly, highly toxic hydrogen sulfide and flammable methane.

About one-third of the 100,000 medium to large lakes and about 85% of the large lakes near major population centers in the United States suffer from some degree of cultural eutrophication. One-fourth of China's lakes are classified as eutrophic.

Ways to *prevent* or reduce cultural eutrophication include **(1)** advanced waste treatment to remove nitrates and phosphates, **(2)** bans or limits on phosphates in household detergents and other cleaning agents, and **(3)** soil conservation and land-use control to reduce nutrient runoff.

Major *cleanup methods* are **(1)** removing excess weeds, **(2)** controlling undesirable plant growth with herbicides and algicides, and **(3)** pumping air through lakes and reservoirs to avoid oxygen depletion (an expensive and energy-intensive method).

As usual, pollution prevention is more effective and usually cheaper in the long run than cleanup. If excessive inputs of limiting plant nutrients stop, a lake usually can return to its previous state.

Seattle's Lake Washington is a success story of recovery from severe eutrophication caused by decades of sewage inputs. The recovery took place within about

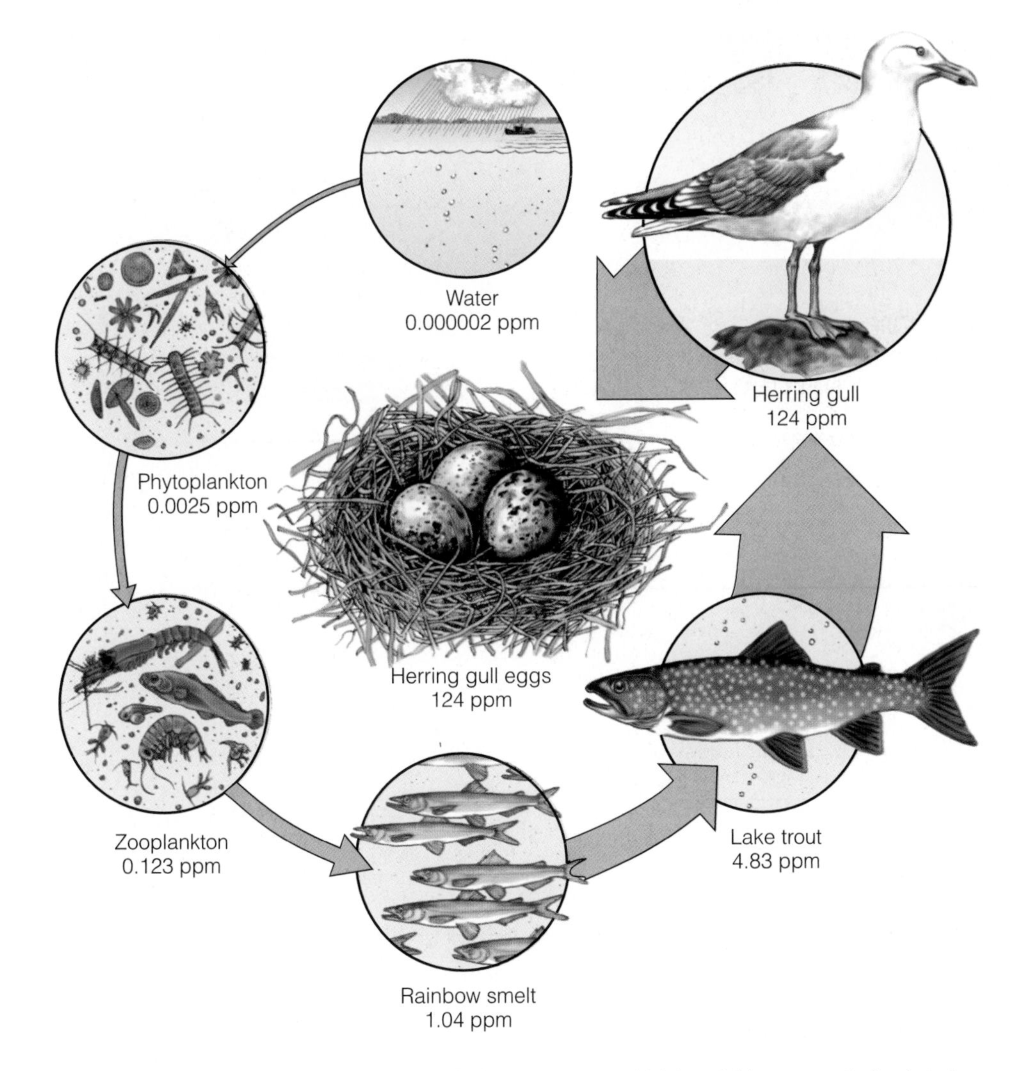

Figure 14-22 *Biological magnification* of PCBs (polychlorinated biphenyls) in an aquatic food chain in the Great Lakes. Most of the 209 different PCBs are **(1)** insoluble in water, **(2)** soluble in fats, and **(3)** resistant to biological and chemical degradation—properties that result in their accumulation in the tissues of organisms and their biological amplification in food chains and webs. Although the long-term health effects on people exposed to low levels of PCBs are unknown, high doses of PCBs in laboratory animals produce **(1)** liver and kidney damage, **(2)** gastric disorders, **(3)** birth defects, **(4)** skin lesions, **(5)** hormonal changes, **(6)** smaller penis size, and **(7)** tumors. Boys in Taiwan exposed to PCBs while in their mothers' wombs developed abnormally small penises. In the United States, manufacture and use of PCBs have been banned since 1976; before then, millions of metric tons of these long-lived chemicals were released into the environment, and many of them are still found in bottom sediments of lakes, streams, and oceans.

4 years after the sewage was diverted into Puget Sound. This worked for three reasons: **(1)** A large body of water (Puget Sound) was available to receive the sewage wastes, **(2)** the lake had not yet filled with weeds and sediment because of its large size and depth, and **(3)** corrective action was taken before the lake had become a shallow, highly eutrophic lake (Figure 6-43, bottom, p. 155). Today, the lake's water quality is good, but there is concern about increased urban runoff caused by the area's rapidly growing population.

Case Study: Chemical and Biological Disruption in the Great Lakes The five interconnected Great Lakes (Figure 14-24, p. 352) contain at least 95% of the fresh surface water in the United States and 20% of the world's fresh surface water. The Great Lakes basin is home for about 38 million people—about 30% of the Canadian population and 14% of the U.S. population.

Despite their enormous size, these lakes are vulnerable to pollution from point and nonpoint sources because less than 1% of the water entering the Great Lakes flows out to the St. Lawrence River each year. In addition to land runoff, these lakes receive large quantities of acids, pesticides, and other toxic chemicals by deposition from the atmosphere (often blown in from hundreds or thousands of kilometers away).

By the 1960s, many areas of the Great Lakes were suffering from **(1)** severe cultural eutrophication, **(2)** huge fish kills, and **(3)** contamination from bacteria and a variety of toxic industrial wastes. The impact on Lake Erie was particularly intense because it **(1)** is the shallowest of the Great Lakes and **(2)** has the highest concentrations of people and industrial activity along its shores. Many bathing beaches had to be closed, and by 1970 the lake had lost nearly all its native fish.

Here is some *good news*. Since 1972, a $20-billion Great Lakes pollution control program has been carried out jointly by Canada and the United States. This joint program has **(1)** significantly decreased levels of phosphates, coliform bacteria, and many toxic industrial chemicals, **(2)** decreased algae blooms, **(3)** increased dissolved oxygen levels and sport and commercial fish catches, and **(4)** allowed most swimming beaches to reopen.

These improvements occurred mainly because of **(1)** new or upgraded sewage treatment plants, **(2)** better treatment of industrial wastes, and **(3)** banning of phosphate detergents, household cleaners, and water conditioners.

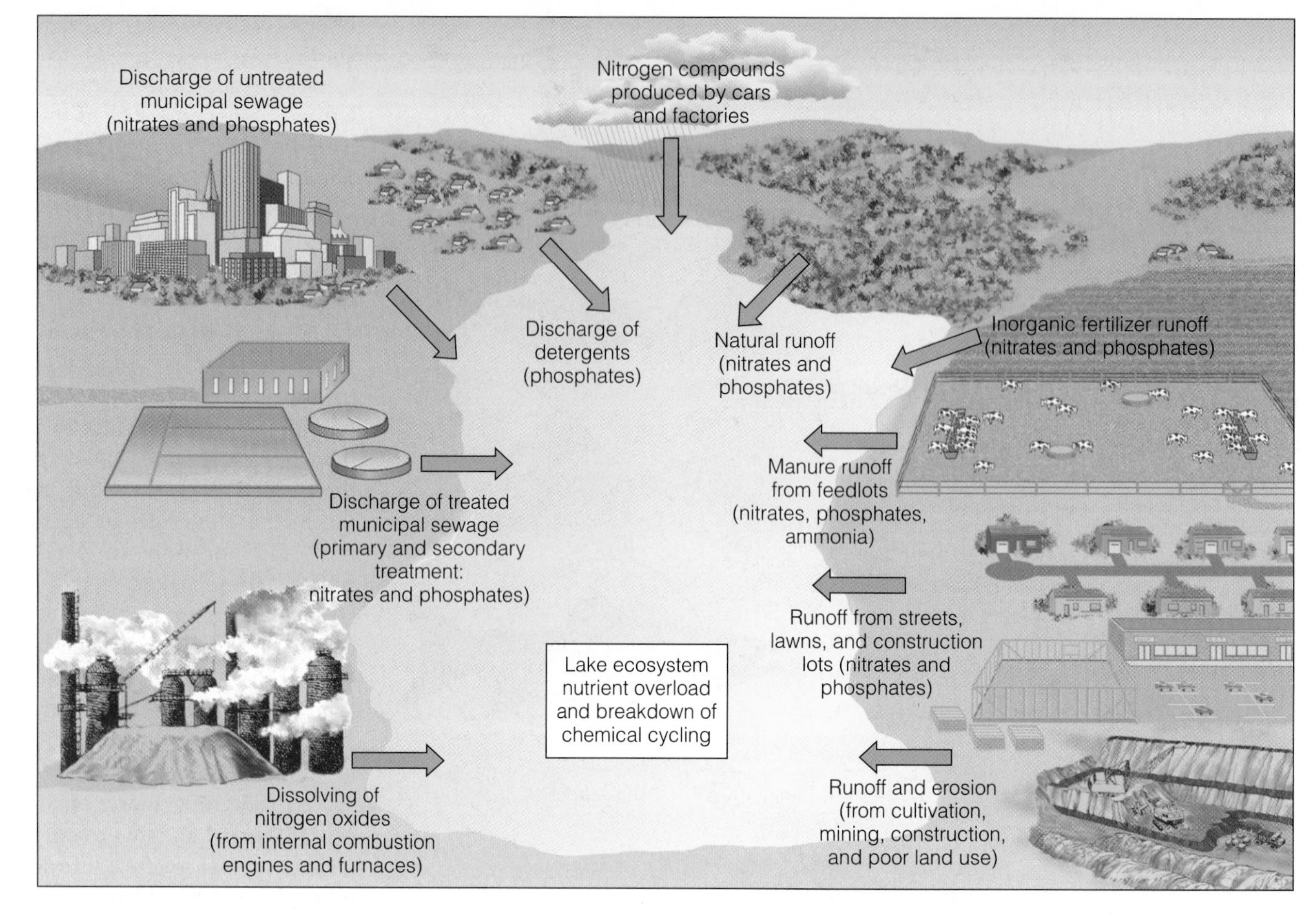

Figure 14-23 Principal sources of nutrient overload causing *cultural eutrophication* in lakes. The amount of nutrients from each source varies according to the types and amounts of human activities occurring in each airshed and watershed. Levels of dissolved oxygen drop when enlarged populations of algae and plants (stimulated by increased nutrient input) die and are decomposed by aerobic bacteria. Lowered oxygen levels can **(1)** kill fish and other aquatic life and **(2)** reduce biodiversity and the aesthetic and recreational value of the lake.

Here is some *bad news*:

- Less than 3% of the lakes' shoreline is clean enough for swimming or for supplying drinking water.
- Nonpoint land runoff of pesticides and fertilizers from urban sprawl has surpassed industrial pollution as the greatest threat to the lakes.
- Forty-three toxic hot spots (Figure 14-24) are still heavily polluted.
- About 50% of the input of toxic compounds comes from **(1)** atmospheric deposition of pesticides, **(2)** mercury from coal-burning plants, and **(3)** other toxic chemicals from as far as Mexico and Russia.
- Toxic chemicals such as PCBs have **(1)** built up in food chains and webs (Figure 14-22), **(2)** contaminated many types of sport fish, and **(3)** depleted populations of birds, river otters, and other animals feeding on contaminated fish.
- A survey by Wisconsin biologists revealed that one fish in four taken from the Great Lakes is unsafe for human consumption.

In 1991, the U.S. government passed a law requiring **(1)** accelerated cleanup of the lakes (especially 43 toxic hot spots) and **(2)** an immediate reduction in air pollutant emissions in the region. However, a lack of federal and state funds has delayed progress toward these goals.

Some environmentalists call for a ban on **(1)** the use of chlorine as a bleach in the pulp and paper industry around the Great Lakes, **(2)** all new incinerators in the area, and **(3)** discharge into the lakes of 70 toxic chemicals that threaten human health and wildlife. Officials of these industries strongly oppose such bans.

Great Lakes fisheries also face threats from biological disruption caused by invasion of nonnative species. In 1986, larvae of a nonnative species, the *zebra mussel*, arrived in ballast water discharged from a European ship near Detroit, Michigan. With no known natural enemies, these thumbnail-sized mussels have **(1)** displaced other

Tracking Down the *Pfiesteria* Cell from Hell

JoAnn M. Burkholder is professor of aquatic biology and marine science at North Carolina State University. She knows what it is to **(1)** be sickened by a newly identified fish-killing microbe and **(2)** experience the political heat when you go public with your research to alert people about a potentially serious health threat.

In 1986, she investigated why a colleague's laboratory research fish were dying mysteriously and discovered that the culprit was a new microbe so tiny that dozens could fit on the head of a pin. She and her codiscoverer named it *Pfiesteria piscida* (pronounced "fee-STEER-e-ah pis-kuh-SEED-uh"), but some biologists call it the cell from hell.

She and her colleagues discovered that this complex microscopic organism could assume at least 24 guises in its lifetime. Without suitable prey, the microbe can masquerade as a plant or lie dormant for years. Then under certain conditions these microbes can change from alga eaters into fish-killing dinoflagellates that release neurotoxins that **(1)** stun fish in rivers and coastal estuaries and **(2)** usually kill them within 10 minutes to several hours.

The neurotoxin can also form an aerosol above the water. In 1993, Burkholder and her chief research aide breathed toxic fumes released in tanks of fish dying from *Pfiesteria* attacks. They experienced **(1)** nausea, **(2)** burning eyes and cramps, **(3)** weakness, **(4)** slow-healing sores, **(5)** difficulty breathing, and **(6)** severe loss of memory and mental powers. Eventually they recovered but still cannot exercise strenuously without severe shortness of breath and the onset of respiratory illness.

Since then more than 100 researchers, anglers, and water-skiers in North Carolina, Virginia, and Maryland have experienced one or more of these symptoms when exposed to water or air contaminated by *Pfiesteria* toxins.

In 1995, Burkholder and her colleagues detected a second species of fish-eating *Pfiesteria* in North Carolina's New River after a major spill from a hog-waste lagoon. These two species of single-cell organisms live in waters from the Chesapeake Bay to the Gulf Coast of Florida and Alabama and each year cause more than $60 million in losses to U.S. fisheries and tourism.

Through lab and field research, Burkholder developed evidence that connected outbreaks or blooms of *Pfiesteria* with excessive levels of nitrogen (as nitrates) and phosphorus (as phosphates) in rivers and estuaries. High levels of such nutrients are found in runoff from fertilized croplands, industrial development, and animal feedlots (especially those used to raise hogs and chickens) into rivers flowing into coastal estuaries.

In 1991, Burkholder went public with her findings and urged North Carolina state legislators to put curbs on hog farming and enact much tougher laws to reduce the flow of nutrients and other pollutants into the state's rivers.* Hog farmers, developers, farming interests, fishing industry officials (worried about whether it is safe to eat fish and shellfish from affected rivers and estuaries), tourist industry officials (alarmed about a negative image of the state's huge coastal recreational industry), and some state officials reacted negatively to her political activism.

Some challenged her character and competence and accused her of using the results of preliminary research to push for questionable policies. She also received some anonymous death threats.

Burkholder has not backed down and continues to criticize state health officials and legislators for not taking her concerns about public health seriously enough. Under the glare of state and national publicity, the state now supports research on the problem and since 1997 **(1)** has had a moratorium on construction of new hog farms and **(2)** is looking for better ways to deal with hog waste.

Research by other scientists has confirmed the link between *Pfiesteria* outbreaks and nutrient overloading of rivers. Since 1997, a number of federal and state environmental, health, and agricultural agencies have set up a coordinated research effort to learn more about what triggers outbreaks of these organisms and how they affect humans and other organisms.

*For a popularized description of her research and political battle to alert the public and elected officials to the dangers posed by this microbe, see Rodney Barker's *And the Waters Turned to Blood: The Ultimate Biological Threat* (New York: Simon & Schuster, 1997).

mussel species, **(2)** depleted the food supply for other Great Lakes species, **(3)** clogged irrigation pipes, **(4)** shut down water intake systems for power plants and city water supplies, **(5)** fouled beaches, **(6)** grown in huge masses on boat hulls, piers, pipes, rocks, and almost any exposed aquatic surface, and **(7)** cost the Great Lakes basin at least $700 million per year (annual costs could reach $5 billion within a few years).

However, zebra mussels may be *good news* for a number of aquatic plants. By consuming algae and other microorganisms, the mussels increase water clarity, which permits deeper penetration of sunlight and more photosynthesis. This allows some native plants to thrive and return the plant composition of Lake Erie (and presumably other lakes) closer to what it was 100 years ago. Because the plants provide food and increase

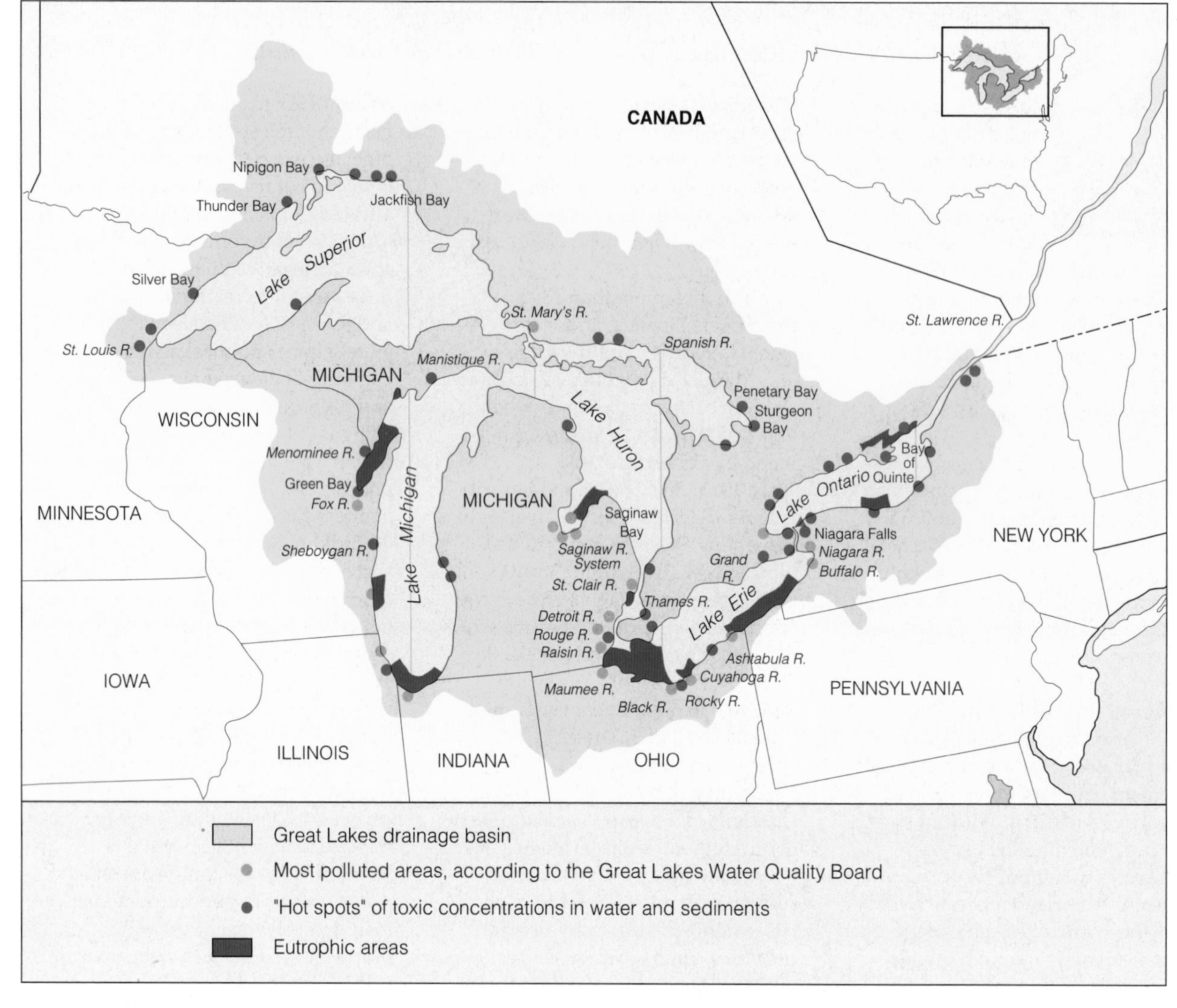

Figure 14-24 The Great Lakes basin and the locations of some of its water quality problems. The Great Lakes region is dotted with several hundred abandoned toxic waste sites that are listed by the EPA as Superfund sites to receive cleanup priority (p. 389). (Data from Environmental Protection Agency)

dissolved oxygen, their comeback may benefit certain aquatic animals (including the mussels).

There is more *bad news*, however. In 1991 a larger and potentially more destructive species, the *quagga mussel*, invaded the Great Lakes, probably discharged in the ballast water of a Russian freighter. It can survive at greater depths and tolerate more extreme temperatures than the zebra mussel. There is concern that it may eventually colonize areas such as the Chesapeake Bay and waterways in parts of Florida.

By 1999, a European fish, the *round goby*, had invaded all of the Great Lakes. The *good news* is that these tiny predators have a voracious appetite for zebra mussels. The *bad news* is that these bottom-dwelling fish also devour eggs and the young of any fish sharing their habitat. This includes prized recreational fish such as perch, walleye, and smallmouth bass.

Why Is Groundwater Pollution Such a Serious Problem? According to many scientists, a serious threat to human health is the out-of-sight pollution of groundwater, a prime source of water for drinking and irrigation. Groundwater supplies about 75% of the drinking water in Europe, 51% in the United States, 32% in Asia, and 29% in Latin America.

Studies indicate that groundwater pollution comes from numerous sources as we dump more of our wastes into **(1)** storage lagoons, **(2)** septic tanks, **(3)** landfills, **(4)** hazardous waste dumps, and **(5)** deep injection wells, and **(6)** store gasoline, oil, solvents, and hazardous wastes in metal underground tanks that after 20–40 years can corrode and leak (Figure 14-25). Groundwater is also contaminated by people who dump or spill gasoline, oil, and paint thinners and other organic solvents onto the ground.

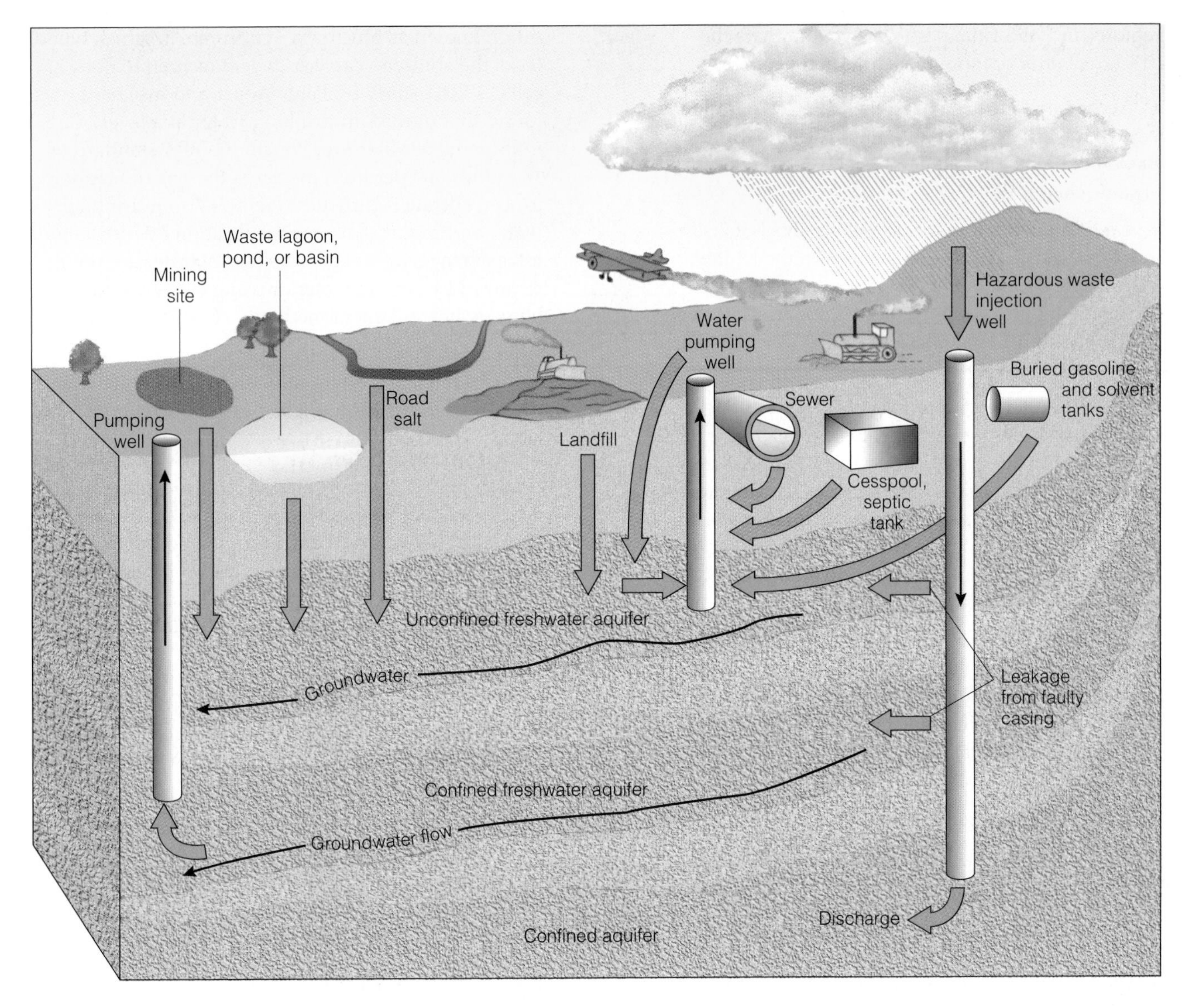

Figure 14-25 Principal sources of groundwater contamination in the United States.

Although experts rate groundwater pollution as a low-risk ecological problem, they consider pollutants in drinking water (much of it from groundwater) a high-risk health problem (Figure 10-12, left, p. 238). Human health risks come mostly from groundwater contaminated with **(1)** petrochemicals (such as gasoline and oil), **(2)** organic solvents (such as trichloroethylene, or TCE), **(3)** pesticides, **(4)** arsenic (As), **(5)** lead (Pb), and **(6)** fluoride (F^-; Table 14-1).

When groundwater becomes contaminated, it cannot cleanse itself of *degradable wastes* as flowing surface water does (Figure 14-21) because **(1)** it flows so slowly (usually less than 0.3 meters or 1 foot per day) that contaminants are not diluted and dispersed effectively, **(2)** it has much smaller populations of decomposing bacteria, and **(3)** its cold temperature slows down the chemical reactions that decompose wastes.

This means that it can take hundreds to thousands of years for contaminated groundwater to cleanse itself of *degradable* wastes, and on a human time scale *nondegradable wastes* (such as toxic lead, arsenic, and fluoride) are there permanently.

What Is the Extent of Groundwater Pollution? The answer is that *we do not know* because few countries go to the great expense of locating, tracking, and testing aquifers. However, studies made by scientists in scattered parts of the world indicate that

- Up to 25% of usable groundwater in the United States is contaminated (and in some areas as much as 75%). In New Jersey, for example, every major aquifer is contaminated.
- The EPA has documented groundwater contamination by 74 pesticides in 38 states, mostly in farming

regions. In California, pesticides contaminate the drinking water of more than 1 million people.

- According to the EPA and the U.S. Geological Survey, about 45% of *municipal* groundwater supplies in the United States are contaminated with one or more organic chemicals.

- An EPA survey found that **(1)** one-third of 26,000 industrial waste ponds and lagoons in the United States have no liners to prevent toxic liquid wastes from seeping into aquifers, and **(2)** one-third of these sites are within 1.6 kilometers (1 mile) of a drinking water well.

- The EPA estimates that at least 100,000 underground tanks storing gasoline, diesel fuel, home heating oil, and toxic solvents are leaking their contents into groundwater in the United States. During this century, scientists expect many of the millions of such tanks installed around the world in recent decades to corrode, leak, contaminate groundwater, and become a major global health problem.

- Determining the extent of a leak from an underground tank can cost $25,000–250,000, and cleanup costs range from $10,000 to more than $250,000. If the chemical reaches an aquifer, effective cleanup is rarely possible. Replacing a leaking tank costs an additional $10,000–60,000.

- About 60% of the liquid hazardous waste in the United States is disposed of by injection into deep underground wells (Figure 14-25). Although these wastes are injected below the deepest sources of drinking water, **(1)** some of the injection pipes can leak, and **(2)** some of the wastes have entered aquifers used for drinking water in parts of Texas, Florida, Oklahoma, and Ohio.

- Groundwater contamination by toxic *arsenic (As)* can occur when tubewells are drilled in areas where deep soils and underground rocks are naturally rich in arsenic. In India's state of West Bengal and parts of Bangladesh, 28–57 million people are drinking water with arsenic levels 5–100 times the World Health Organization limit. In the United States, there is controversy over standards for arsenic in drinking water (Spotlight, p. 356).

- Groundwater used as a source of drinking water can also be contaminated with nitrate ions (NO_3^-), especially in agricultural areas where nitrates in fertilizer can be leached into groundwater. Some of the nitrate in drinking water is transformed in the body to nitrite ions (NO_2^-). In the stomach, colon, and bladder, these ions can be converted to N-nitroso compounds, which can cause cancer in a variety of organs in more than 40 test animal species. The conversion of nitrates in tap water to nitrites in infants under 6 months old can cause a potentially fatal condition known as blue baby syndrome, in which blood lacks the ability to carry sufficient oxygen to body cells. A 2001 study by Peter Weyer and his colleagues found that nitrates in drinking water at a level of 2.5 milligrams per liter (one-fourth the EPA standard of 10 milligrams per liter) increases the risk of bladder cancer. Nitrate is difficult to remove from drinking water and requires expensive filtration by distillation, ion exchange, or reverse osmosis. Nitrates are not removed by boiling water or using conventional home water treatment methods.

- In coastal areas, excessive pumping of water from aquifers can lead to contamination of drinking water by saltwater intrusion (Figure 14-15).

Solutions: How Can We Protect Groundwater? Contaminated aquifers are almost impossible to clean because of their **(1)** enormous volume, **(2)** inaccessibility, and **(3)** slow movement. Pumping polluted groundwater to the surface, cleaning it up, and returning it to the aquifer is extremely expensive. Thus, *preventing contamination is the only effective way to protect groundwater resources*. Ways to do this include

- *Monitoring aquifers near landfills and underground tanks*
- *Requiring leak detection systems for underground tanks used to store hazardous liquids*
- *Banning or more strictly regulating disposal of hazardous wastes in deep injection wells and landfills*
- *Storing hazardous liquids above ground in tanks with systems that detect and collect leaking liquids*

14-7 OCEAN POLLUTION

How Much Pollution Can the Oceans Tolerate? The oceans are the ultimate sink for much of the waste matter we produce. Oceans can dilute, disperse, and degrade large amounts of raw sewage, sewage sludge, oil, and some types of degradable industrial waste, especially in deep-water areas. Some forms of marine life have proved to be more resilient than originally expected. This has led some analysts to suggest that it is safer to dump sewage sludge and most other hazardous wastes into the deep ocean than to bury them on land or burn them in incinerators.

Other scientists disagree, pointing out that we know less about the deep ocean than we do about outer space. They add that dumping waste in the ocean would delay urgently needed pollution prevention and promote further degradation of this vital part of the earth's life-support system.

How Do Pollutants Affect Coastal Areas? Coastal areas—especially wetlands and estuaries, coral reefs, and mangrove swamps (Figure 6-28, p. 143)—

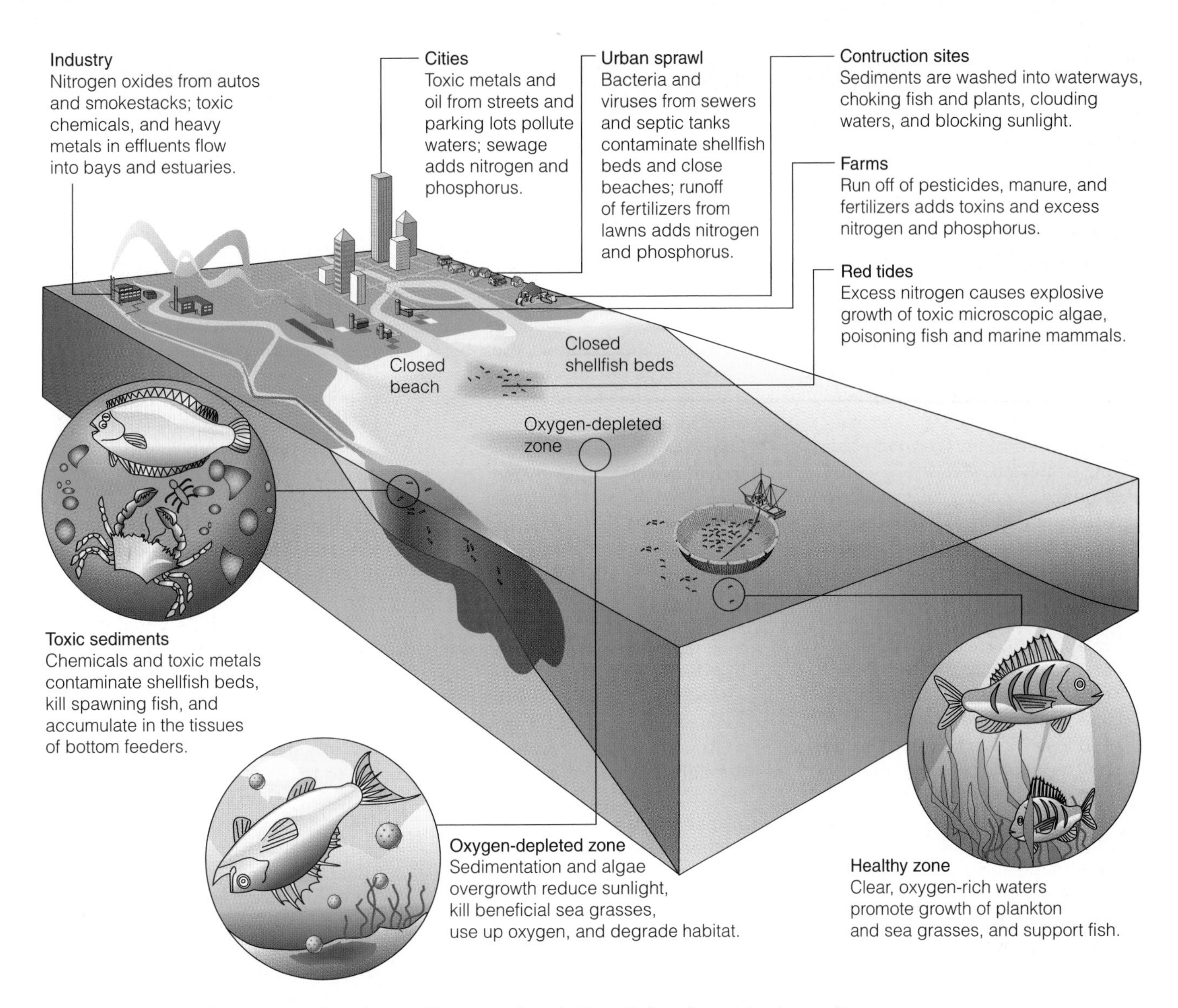
Industry
Nitrogen oxides from autos and smokestacks; toxic chemicals, and heavy metals in effluents flow into bays and estuaries.
Cities
Toxic metals and oil from streets and parking lots pollute waters; sewage adds nitrogen and phosphorus.
Urban sprawl
Bacteria and viruses from sewers and septic tanks contaminate shellfish beds and close beaches; runoff of fertilizers from lawns adds nitrogen and phosphorus.
Contruction sites
Sediments are washed into waterways, choking fish and plants, clouding waters, and blocking sunlight.
Farms
Run off of pesticides, manure, and fertilizers adds toxins and excess nitrogen and phosphorus.
Red tides
Excess nitrogen causes explosive growth of toxic microscopic algae, poisoning fish and marine mammals.
Closed shellfish beds
Closed beach
Oxygen-depleted zone
Toxic sediments
Chemicals and toxic metals contaminate shellfish beds, kill spawning fish, and accumulate in the tissues of bottom feeders.
Oxygen-depleted zone
Sedimentation and algae overgrowth reduce sunlight, kill beneficial sea grasses, use up oxygen, and degrade habitat.
Healthy zone
Clear, oxygen-rich waters promote growth of plankton and sea grasses, and support fish.

Figure 14-26 How residential areas, factories, and farms contribute to the pollution of coastal waters and bays.

bear the brunt of our enormous inputs of wastes into the ocean (Figure 14-26). This is not surprising because **(1)** about 40% the world's population lives on or within 100 kilometers (62 miles) of the coast, **(2)** 14 of the world's 15 largest metropolitan areas, each with 10 million people or more, are near coastal waters (Figure 11-20, p. 261), and **(3)** coastal populations are growing more rapidly than the global population.

In most coastal developing countries (and in some coastal developed countries), municipal sewage and industrial wastes are dumped into the sea without treatment. About 85% of the sewage from large cities along the Mediterranean Sea, which has a coastal population of 200 million people during tourist season, is discharged into the sea untreated. This causes widespread beach pollution and shellfish contamination. A recent study found that about one-fourth of the people using coastal beaches develop ear infections, sore throats and eyes, respiratory disease, or gastrointestinal disease.

Runoff of sewage and agricultural wastes into coastal waters and acid deposition from the atmosphere (Figure 12-8, p. 286) introduce large quantities of nitrate (NO_3^-) and phosphate (PO_4^{3-}) plant nutrients, which can cause explosive growth of harmful algae.

These harmful algal blooms (HABs)—called red, brown, or green tides, depending on their color—can **(1)** release waterborne and airborne toxins that damage fisheries, **(2)** kill some fish-eating birds, **(3)** reduce tourism, and **(4)** poison seafood. When the algae die and decompose, coastal waters are depleted of oxygen, and a variety of marine species die.

Each year some 61 large *oxygen-depleted zones* (sometimes inaccurately called dead zones) form in the world's coastal waters and in landlocked seas such as the Baltic and Black Seas because of excessive fertilizer inputs. In these zones, much of the aquatic life dies or moves elsewhere. The biggest such zone in U.S. waters and the third largest in the world forms

SPOTLIGHT

How Much Arsenic Should Be Allowed in Drinking Water?

Arsenic is a naturally occurring element in various rocks and can be released into groundwater and surface water from the natural weathering of minerals in deep soils and rocks. In addition, arsenic can be released into the air and water from **(1)** coal burning, **(2)** copper and lead smelting, **(3)** municipal trash incinerators, **(4)** wood-preserving treatments, **(5)** leaching from landfills containing arsenic-laden ash produced by coal-burning power plants, and **(6)** use of certain arsenic-containing pesticides.

According to a 1999 report by the U.S. National Academy of Sciences, long-term exposure to arsenic in drinking water **(1)** can cause cancer of the skin, bladder, and lungs, **(2)** may cause kidney and liver cancer, **(3)** can cause skin lesions and hardening of the skin (keratosis), and **(4)** has been linked to adult-onset diabetes, cardiovascular disease, anemia, and disorders of the immune, nervous, and reproductive systems. In 2001, researchers at the Dartmouth Medical School reported that arsenic is also a potent endocrine disrupter (Connections, p. 231).

The acceptable level of arsenic in U.S. drinking water has been 50 parts per billion (50 ppb) since 1942. This is five times the international standard of 10 ppb adopted in 1993 by the World Health Organization and in 1998 by the 15-member European Union.

In 1962, the U.S. Public Health Service proposed lowering the U.S. drinking water standard for arsenic from 50 ppb to 10 ppb. After more than 25 years of scientific reviews, including a 1999 study by the National Academy of Sciences, the EPA proposed that the U.S. drinking water standard for arsenic be reduced to the international standard of 10 ppb.

According to WHO scientists, even the 10 ppb standard is not safe. Based on health concerns, many scientists believe that this standard should be lowered to 3–5 ppb.

Running drinking water through activated alumina can absorb arsenic. Reverse osmosis also removes arsenic, but this method is expensive.

According to the EPA, implementing the 10-ppb standard would **(1)** require about 3,000 communities with high levels of arsenic in groundwater (Figure 14-27) to improve their drinking water treatment at an average annual cost of about $12 per person, **(2)** cost about $181 million per year, **(3)** make drinking water safer for at least 11 million Americans (Figure 14-27), and **(4)** provide annual benefits of $140–198 million from reduction of arsenic-related bladder and liver cancers. Reductions in the incidence of other arsenic-related health problems would provide larger financial benefits, which so far have not been estimated by the EPA.

Mining, coal, and lumber products and other arsenic-producing companies (who could face tougher regulations) oppose the 10-ppb standard and say that the estimated costs are too low. Also, some small communities contend that they cannot afford to implement the new standard. For example, for affected communities with 500 or fewer homes, the annual water bill could increase by $162–327 per household. However, the federal government could subsidize some or all treatment costs for small systems.

In March 2001, President George Bush (under lobbying pressure from mining interests, coal companies, the wood products

every summer in a narrow stretch of the Gulf of Mexico (Figure 14-28).

Case Study: The Chesapeake Bay The Chesapeake Bay, the largest estuary in the United States, is in trouble because of human activities. Between 1940 and 2001, the number of people living in the Chesapeake Bay area grew from 3.7 million to 17 million, and within a few years its population may reach 18 million.

The estuary receives wastes from point and nonpoint sources scattered throughout a huge drainage basin that includes 9 large rivers and 141 smaller streams and creeks in parts of six states (Figure 14-29). The bay has become a huge pollution sink because **(1)** it is quite shallow, and **(2)** only 1% of the waste entering it is flushed into the Atlantic Ocean.

Phosphate and nitrate levels have risen sharply in many parts of the bay, causing algae blooms and oxygen depletion (Figure 14-29). Studies have shown that point sources, primarily sewage treatment plants, contribute about 60% by weight of the phosphates. Nonpoint sources—mostly runoff of fertilizer and animal wastes from urban, suburban, and agricultural land and deposition from the atmosphere—account for about 60% by weight of the nitrates.

Large quantities of pesticides also run off cropland and urban lawns, and industries discharge large amounts of toxic wastes, often in violation of their discharge permits. Commercial harvests of oysters, crabs, and several important fish have fallen sharply since 1960 because of a combination of overfishing, pollution, and disease.

In the 1980s, the Chesapeake Bay Program, the country's most ambitious attempt at *integrated coastal management*, was implemented. Results have been impressive. Between 1985 and 2000, phosphorus levels declined 27% and nitrogen levels dropped 16%, a

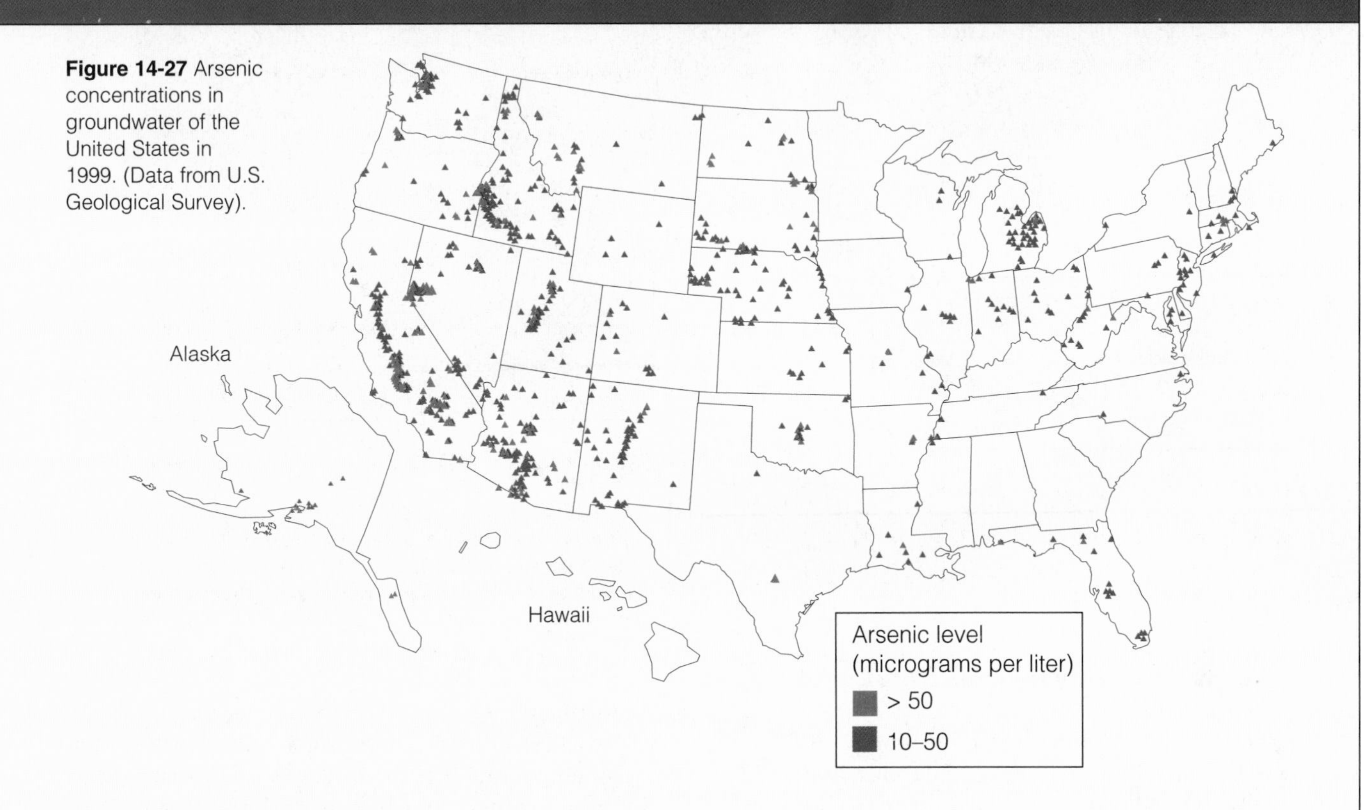

Figure 14-27 Arsenic concentrations in groundwater of the United States in 1999. (Data from U.S. Geological Survey).

industry, and other arsenic producers) withdrew implementation of the new EPA standard, arguing that it would cost too much to implement. Health scientists and environmentalists were outraged and accused the Bush administration of caving to the demands of industries whose activities release toxic levels of arsenic into drinking water supplies.

Critical Thinking

Do you believe that the drinking standard for arsenic in U.S. drinking water should be lowered from 50 ppb to 10 ppb? Explain. If you lived in a community affected by such a standard, how much more per year would you be willing to pay on your water bill to implement the standard?

significant achievement given the increasing population in the watershed and the fact that nearly 40% of the nitrogen inputs come from the atmosphere.

Reaching the declared goal of a 40% reduction in nutrient levels and a significant improvement in habitat water quality throughout the bay will be difficult because the area's population is expected to grow by 25% between 1995 and 2020. So far, however, the Chesapeake Bay Program shows what can be done when diverse interested parties work together to achieve goals that benefit both wildlife and people.

What Pollutants Do We Dump into the Ocean?

Industrial waste dumping off U.S. coasts has stopped, although it still occurs in a number of other developed countries and some developing countries. However, barges and ships still legally dump large quantities of **dredge spoils** (materials, often laden with toxic metals, scraped from the bottoms of harbors and rivers to maintain shipping channels) at 110 sites off the Atlantic, Pacific, and Gulf coasts.

In addition, many countries dump into the ocean large quantities of sewage **sludge**, a gooey mixture of toxic chemicals, infectious agents, and settled solids removed from wastewater at sewage treatment plants. Since 1992, this practice has been banned in the United States.

Fifty countries with at least 80% of the world's merchant fleet have agreed not to dump sewage and garbage at sea, but this agreement is difficult to enforce and often is violated. Most ship owners save money by dumping wastes at sea and risk only small fines if they are caught. Each year, as many as 2 million seabirds and more than 100,000 marine mammals (including whales, seals, dolphins, and sea lions) die when they ingest or become entangled in fishing nets,

Figure 14-28 A large zone of oxygen-depleted water forms for half of the year in the Gulf of Mexico as a result of oxygen-depleting algal blooms. It is created by huge inputs of nitrate (NO_3^-) and phosphate (PO_4^{3-}) plant nutrients from the massive Mississippi River Basin.

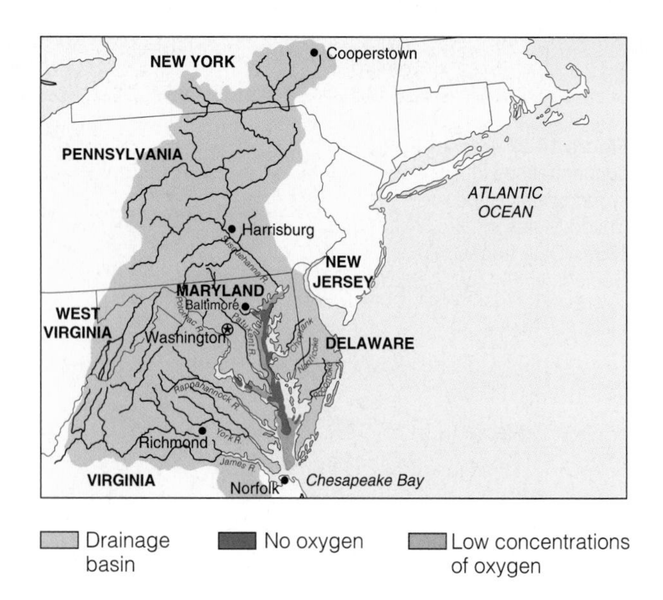

Figure 14-29 Chesapeake Bay, the largest estuary in the United States, is severely degraded as a result of water pollution from point and nonpoint sources in six states and from deposition of air pollutants.

ropes, and other debris dumped into the sea or discarded on beaches.

Under the London Dumping Convention of 1972, 100 countries agreed not to dump highly toxic pollutants and high-level radioactive wastes in the open sea beyond the boundaries of their national jurisdictions. Since 1983, these same nations have observed a moratorium on the dumping of low-level radioactive wastes at sea, which in 1994 became a permanent ban. However, France, Great Britain, Russia, China, and Belgium may legally exempt themselves from this ban. In 1992, it was learned that for decades the former Soviet Union had been dumping large quantities of high- and low-level radioactive wastes into the Arctic Ocean and its tributaries.

What Are the Effects of Oil on Ocean Life? *Crude petroleum* (oil as it comes out of the ground) and *refined petroleum* (fuel oil, gasoline, and other processed petroleum products, Figure 19-7, p. 495) are accidentally or deliberately released into the environment from a number of sources.

Tanker accidents and blowouts at offshore drilling rigs (when oil escapes under high pressure from a borehole in the ocean floor) get most of the publicity because of their high visibility. However, more oil is released **(1)** during normal operation of offshore wells, **(2)** from washing tankers and releasing the oily water, and **(3)** from pipeline and storage tank leaks.

Natural oil seeps also release large amounts of oil into the ocean at some sites, but most ocean oil pollution comes from activities on land. Almost half (some experts estimate 90%) of the oil reaching the oceans is waste oil dumped, spilled, or leaked onto the land or into sewers by cities, industries, and people changing their own motor oil. Worldwide, about 10% of the oil that reaches the ocean comes from the atmosphere, mostly from smoke emitted by oil fires.

The effects of oil on ocean systems depend on a number of factors: **(1)** type of oil (crude or refined), **(2)** amount released, **(3)** distance of release from shore, **(4)** time of year, **(5)** weather conditions, **(6)** average water temperature, and **(7)** ocean currents.

Volatile organic hydrocarbons in oil immediately kill a number of aquatic organisms, especially in their vulnerable larval forms. Some other chemicals form tarlike globs that float on the surface and coat the feathers of birds (especially diving birds) and the fur of marine mammals. This destroys their natural insulation and buoyancy, causing many of them to drown or die of exposure from loss of body heat. Heavy oil components that sink to the ocean floor or wash into estuaries can smother bottom-dwelling organisms such as crabs, oysters, mussels, and clams or make them unfit for human consumption. Some oil spills have killed reef corals.

Research shows that most (but not all) forms of marine life recover from exposure to large amounts of *crude oil* within 3 years. However, recovery from exposure to *refined oil*, especially in estuaries, can take 10 years or longer. The effects of spills in cold waters and in shallow enclosed gulfs and bays generally last longer.

Oil slicks that wash onto beaches can have a serious economic impact on coastal residents, who lose income from fishing and tourist activities. Oil-polluted beaches washed by strong waves or currents become clean after about a year, but beaches in sheltered areas remain contaminated for several years. Estuaries and salt marshes suffer the most and longest-lasting damage. Despite their localized harmful effects, experts rate oil spills as a low-risk ecological problem (Figure 10-12, left, p. 238).

How Can Oil Spills Be Cleaned Up? If they are not too large, oils spills can be partially cleaned up by mechanical, chemical, fire, and natural methods. *Mechanical methods* include using **(1)** floating booms to contain the oil spill or keep it from reaching sensitive areas, **(2)** skimmer boats to vacuum up some of the oil into collection barges, and **(3)** absorbent pads or large mesh pillows filled with feathers or hair to soak up oil on beaches or in waters too shallow for skimmer boats.

Chemical methods include using **(1)** coagulating agents to cause floating oil to clump together for easier pickup or to sink to the bottom, where it usually does less harm, and **(2)** dispersing agents to break up oil slicks. However, these agents can also damage some types of organisms. Fire can burn off floating oil, but crude oil is hard to ignite, and this approach produces air pollution. In time, the natural action of wind and waves mixes or emulsifies oil with water (like emulsified salad dressing), and bacteria biodegrade some of the oil.

These methods remove only part of the oil, and none work well on a large spill. Scientists estimate that no more than 15% of the oil from a major spill can be recovered. This explains why preventing oil pollution is the most effective and in the long run the least costly approach.

Solutions: How Can We Protect Coastal Waters? The key to protecting oceans is to reduce the flow of pollution from the land and from streams emptying into the ocean. Such efforts must be integrated with efforts to prevent and control air pollution because an estimated 33% of all pollutants entering the ocean worldwide comes from air emissions from land-based sources.

Analysts have suggested the following measures to prevent and reduce excessive pollution of coastal waters:

Prevention

- *Use separate sewage and storm runoff lines in coastal urban areas.* Otherwise, excessive rainfall causes lines carrying sewage and storm runoff to overflow and release raw sewage into coastal waters.
- *Discourage or ban ocean dumping of sludge and hazardous dredged materials.*
- *Protect sensitive and ecologically valuable coastal areas from development, oil drilling, and oil shipping.*
- *Regulate coastal development.*
- *Require double hulls for all oil tankers.*
- *Recycle used oil.*

Cleanup

- *Improve oil spill cleanup capabilities.*
- *Require at least secondary treatment of coastal sewage, or use wetlands* (Solutions, p. 364), *solar aquatic* (p. 362), *or other treatment methods.*

14-8 SOLUTIONS: PREVENTING AND REDUCING SURFACE WATER POLLUTION

What Can We Do About Water Pollution from Nonpoint Sources? Ways to help control nonpoint water pollution, most of it from agriculture, include the following:

- Reduce fertilizer runoff into surface waters and leaching into aquifers by using slow-release fertilizer and using none on steeply sloped land.
- Reduce the need for fertilizer by alternating plantings between row crops and soybeans or other nitrogen-fixing plants.
- Plant buffer zones of vegetation between cultivated fields and nearby surface water.
- Reduce pesticide runoff by applying pesticides only when needed and using biological control or integrated pest management (p. 424).
- Control runoff and infiltration of manure from animal feedlots by **(1)** improving manure control, **(2)** planting buffers, and **(3)** not locating feedlots and animal waste on steeply sloped land near surface water and in flood zones.
- Reduce soil erosion and flooding by reforesting critical watersheds.

What Can We Do About Water Pollution from Point Sources? The Legal Approach According to Sandra Postel, director of the Global Water Policy Project, most cities in developing countries discharge 80–90% of their untreated sewage directly into rivers and streams, which are used for drinking water, bathing, and washing clothes.

In developed countries, most wastes from point sources are purified to varying degrees. The Federal Water Pollution Control Act of 1972 (renamed the Clean Water Act when it was amended in 1977) and the 1987 Water Quality Act form the basis of U.S. efforts to control pollution of the country's surface waters.

In 1995, the EPA developed a *discharge trading policy* designed to use market forces to reduce water pollution (as has been done with sulfur dioxide for air pollution control, p. 295). The policy would allow a water pollution source, such as an industrial plant or a sewage treatment plant, to sell credits for its excess reductions to another facility that cannot reduce its discharges as cheaply.

Here is some *good news*. The Clean Water Act of 1972 led to the following improvements in U.S. water quality between 1972 and 1998: **(1)** The percentage of U.S. rivers and lakes tested that are fishable and swimmable increased from 36% to 62%, **(2)** the amount of topsoil lost through agricultural runoff was cut by about 1.1 billion metric tons (1 billion tons) annually, **(3)** the proportion of the U.S. population served by sewage treatment plants increased from 32% to 74%, and **(4)** annual wetland losses decreased by 83%.

Here is some *bad news*: **(1)** About 44% of lakes, 38% of rivers (up from 26% in 1984), and 32% of tested estuaries in the United States still are unsafe for fishing, swimming, and other recreational uses, **(2)** hog, poultry, and cattle farm runoff pollutes 70% of U.S. rivers, **(3)** fish caught in more than 1,400 different waterways are unsafe to eat because of high levels of pesticides and other toxic substances, **(4)** less than 2% of the country's 5.8 million kilometers (3.6 million miles) of tested streams are healthy enough to be considered high quality, **(5)** the number of polluted streams could be 10–20 times higher because as many as 400,000 streams have not been tested for water quality, **(6)** 40% of the country's surface and groundwater is unsafe for human use, and **(7)** a 2001 study by the National Academy of Sciences found that the area of wetlands in the United States is continuing to fall, despite a government goal of "no net loss" in the area and function of wetlands.

Some environmentalists and a 2001 report by the EPA's inspector general call for the Clean Water Act to be strengthened by **(1)** increasing funding and authority to control nonpoint sources of pollution, **(2)** increasing monitoring of state programs to see that pollution permits are not allowed to expire, **(3)** strengthening programs to prevent and control toxic water pollution, **(4)** providing more funding and authority for integrated watershed and airshed planning to protect groundwater and surface water from contamination, **(5)** requiring states to do a better job of monitoring and enforcing water pollution laws, and **(6)** expanding the rights of citizens to bring lawsuits to ensure that water pollution laws are enforced. The National Academy of Sciences also calls for **(1)** halting the loss of wetlands, **(2)** higher standards for restoration, and **(3)** creating new wetlands before any natural wetlands are allowed to be filled.

Many people oppose these proposals, contending that the Clean Water Act's regulations and government wetland regulations already are too restrictive and costly. Farmers and developers **(1)** see the law as a curb on their rights as property owners to fill in wetlands and **(2)** believe that they should be compensated for property value losses because of federal wetland protection regulations. State and local officials want more discretion in testing for and meeting water quality standards. They argue that in many communities it is unnecessary and too expensive to test for all the water pollutants required by federal law.

What Can We Do About Water Pollution from Point Sources? The Technological Approach As population, urbanization, and industrialization grow, the volume of wastewater needing treatment will increase rapidly. In rural and suburban areas with suitable soils, sewage from each house usually is discharged into a **septic tank** (Figure 14-30). About 25% of all homes in the United States are served by septic tanks, which should be cleaned out every 3–5 years by a reputable contractor so that they will not contribute to groundwater pollution.

In U.S. urban areas, most waterborne wastes from homes, businesses, factories, and storm runoff flow through a network of sewer pipes to wastewater treatment plants. Some cities have separate lines for stormwater runoff, but in 1,200 U.S. cities the lines for these two systems are combined because it is cheaper. When rains cause combined sewer systems to overflow, they discharge untreated sewage directly into surface waters.

When sewage reaches a treatment plant, it can undergo up to three levels of purification. **Primary sewage treatment** is a *mechanical* process that uses

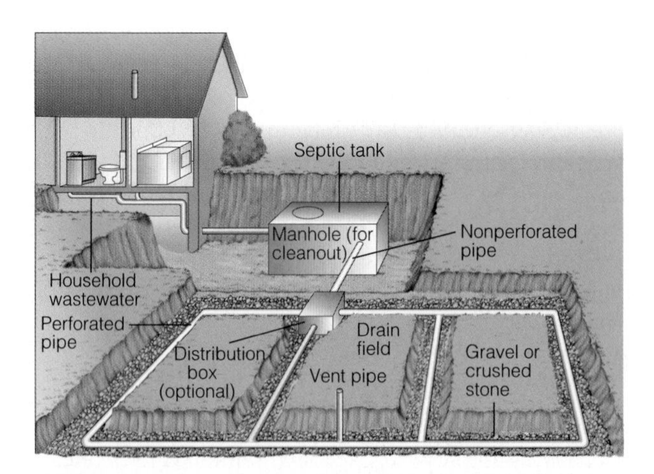

Figure 14-30 *Septic tank system* used for disposal of domestic sewage and wastewater in rural and suburban areas. This system traps greases and large solids and discharges the remaining wastes over a large drainage field. As these wastes percolate downward, the soil filters out some potential pollutants, and soil bacteria decompose biodegradable materials. To be effective, septic tank systems must be **(1)** properly installed in soils with adequate drainage, **(2)** not placed too close together or too near well sites, and **(3)** pumped out when the settling tank becomes full.

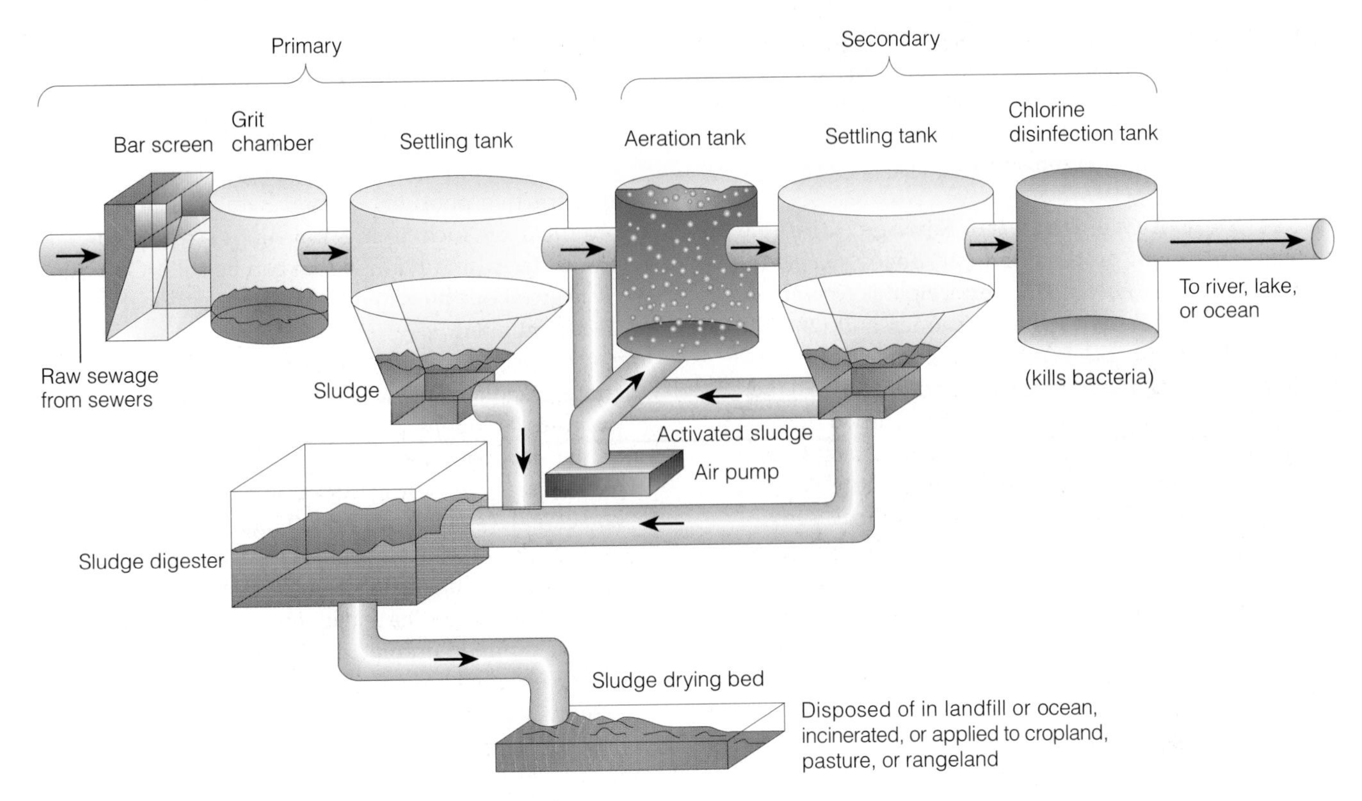

Figure 14-31 Primary and secondary sewage treatment.

screens to filter out debris such as sticks, stones, and rags and allows suspended solids to settle out as sludge in a settling tank (Figure 14-31). Improved primary treatment uses chemically treated polymers to remove suspended solids more thoroughly. By itself, primary treatment removes about 60% of the suspended solids and 30% of the oxygen-demanding organic wastes from sewage but removes no phosphates, nitrates, salts, radioisotopes, or pesticides.

Secondary sewage treatment is a *biological* process in which aerobic bacteria are used to remove up to 90% of biodegradable, oxygen-demanding organic wastes (Figure 14-31). Some treatment plants use *trickling filters*, in which aerobic bacteria degrade sewage as it seeps through a bed of crushed stones covered with bacteria and protozoa. Others use an *activated sludge process*, in which the sewage is pumped into a large tank and mixed for several hours with bacteria-rich sludge and air bubbles to facilitate degradation by microorganisms. The water then goes to a sedimentation tank, where most of the suspended solids and microorganisms settle out as sludge. The sludge produced by primary or secondary treatment is broken down in an anaerobic digester and **(1)** incinerated, **(2)** dumped into the ocean or a landfill, or **(3)** applied to land as fertilizer.

A combination of primary and secondary treatment (Figure 14-31) removes about **(1)** 97% by weight of the suspended solids, **(2)** 95–97% of the oxygen-demanding organic wastes, **(3)** 70% of most toxic metal compounds and nonpersistent synthetic organic chemicals, **(4)** 70% of the phosphorus (mostly as phosphates), **(5)** 50% of the nitrogen (mostly as nitrates), and **(6)** 5% of dissolved salts. Almost no long-lived radioactive isotopes or persistent organic substances such as pesticides are removed.

As a result of the Clean Water Act, most U.S. cities have combined primary and secondary sewage treatment plants (Figure 14-31). However, government studies have found that **(1)** at least two-thirds of these plants have violated water pollution regulations, **(2)** 500 cities have failed to meet federal standards for sewage treatment plants, and **(3)** 34 East Coast cities simply screen out large floating objects from their sewage before discharging it into coastal waters.

Advanced sewage treatment is a series of specialized chemical and physical processes that remove specific pollutants left in the water after primary and secondary treatment. Advanced treatment is rarely used because such plants typically cost twice as much to build and four times as much to operate as secondary plants.

Before water is discharged after primary, secondary, or advanced treatment, it is **(1)** bleached to remove water coloration and **(2)** disinfected to kill disease-carrying bacteria and some but not all viruses. The usual method for doing this is *chlorination*. However, chlorine can react with organic materials in water to form small amounts of chlorinated hydrocarbons, some of which cause cancers in test animals and may damage the human nervous, immune, and endocrine systems (Connections, p. 231). Use of other disinfectants, such as ozone and ultraviolet light, is increasing, but they cost more than chlorination and are not as long lasting.

What Should We Do with Sewage Sludge? Sewage treatment produces a toxic, gooey *sludge*. In the United States, about **(1)** 9% by weight is converted to compost for use as a soil conditioner, and **(2)** 36% is applied to farmland, forests, golf courses, cemeteries, parkland, highway medians, and degraded land as fertilizer. The remaining 55% is **(1)** dumped in conventional landfills (where it can contaminate groundwater) or **(2)** incinerated (which can pollute the air with traces of toxic chemicals and produces a toxic ash, which usually is buried in landfills that EPA experts say will leak eventually).

From an environmental standpoint, it is desirable to recycle the plant nutrients in sewage sludge to the soil on land not used to grow food crops. As long as harmful bacteria and toxic chemicals are not present or are removed, sludge can also be used to fertilize land used for food crops or livestock. However, removing bacteria (usually by heating), toxic metals, and organic chemicals is expensive and is rarely done in the United States.

How Can We Treat Sewage by Working with Nature? Some communities and individuals are seeking better ways to purify contaminated water by working with nature. Ecologist John Todd designs, builds, and operates innovative ecological wastewater treatment systems called *living machines*. They look like aquatic botanical gardens and are powered by the sun in greenhouses or outdoors, depending on the climate.

This ecological purification process begins when sewage flows into a passive solar greenhouse or outdoor site containing rows of large open tanks populated by an increasingly complex series of organisms. In the first set of tanks, algae and microorganisms decompose organic wastes into nutrients that are taken up by aquatic plants such as water hyacinths, cattails, and bulrushes.

After flowing though several of these natural purification tanks, algae and organic waste are filtered out as the water passes through an artificial marsh of sand, gravel, and bulrush plants. Some of the plants also absorb (sequester) toxic metals such as lead and mercury and secrete natural antibiotic compounds that kill pathogens.

Next the water flows into engineered ecosystems in aquarium tanks, where snails and zooplankton consume microorganisms and are in turn consumed by crayfish, tilapia, and other fish that can be eaten or sold as bait. After 10 days, the clear water flows into a second artificial marsh for final filtering and cleansing.

The water can be made pure enough to drink by using ultraviolet light or passing the water through an ozone generator, usually immersed out of sight in an attractive pond or wetland habitat. The chief byproducts of such living machines are ornamental plants, trees, and baitfish that can be sold. Operating costs are about the same as for a conventional sewage treatment plant.

Some communities use nearby natural wetlands to treat sewage, and others create artificial wetlands for such purposes (Solutions, p. 364). Mark Nelson has developed a small, low-tech, and inexpensive artificial wetland system to treat raw sewage from hotels, restaurants, and homes in developing countries (Figure 14-32). This *wastewater garden* system removes 99.9% of fecal coliform bacteria and more than 80% of the nitrates and phosphates from incoming sewage that in most developing countries often is dumped untreated into the ocean or into shallow holes in the ground. The water flowing out of such systems can be used to irrigate gardens or fields or for flushing toilets and thus helps save water.

How Is Drinking Water Purified? Treatment of water for drinking by city dwellers is much like wastewater treatment. Areas that depend on surface water usually store it in a reservoir for several days to improve clarity and taste by allowing the dissolved oxygen content to increase and suspended matter to settle out. The water is then pumped to a purification plant, where it is treated to meet government drinking water standards. Usually the water is run through sand filters and activated charcoal before it is disinfected. In areas with very pure groundwater sources, little treatment is necessary.

In tropical countries without centralized water treatment systems, the WHO is urging people to purify their own drinking water by exposing a plastic bottle filled with contaminated water to the sun.

How Is the Quality of Drinking Water Protected? About 54 countries, most of them in North America and Europe, have safe drinking water standards. The U.S. Safe Drinking Water Act of 1974 requires the EPA to establish national drinking water standards, called *maximum contaminant levels*, for any pollutants that may have adverse effects on human health.

Privately owned wells are not required to meet federal drinking water standards, primarily because of the costs of testing each well regularly (at least $1,000) and opposition to mandatory testing and compliance by some homeowners.

It is difficult to estimate how many people in the United States get sick or die each year from drinking

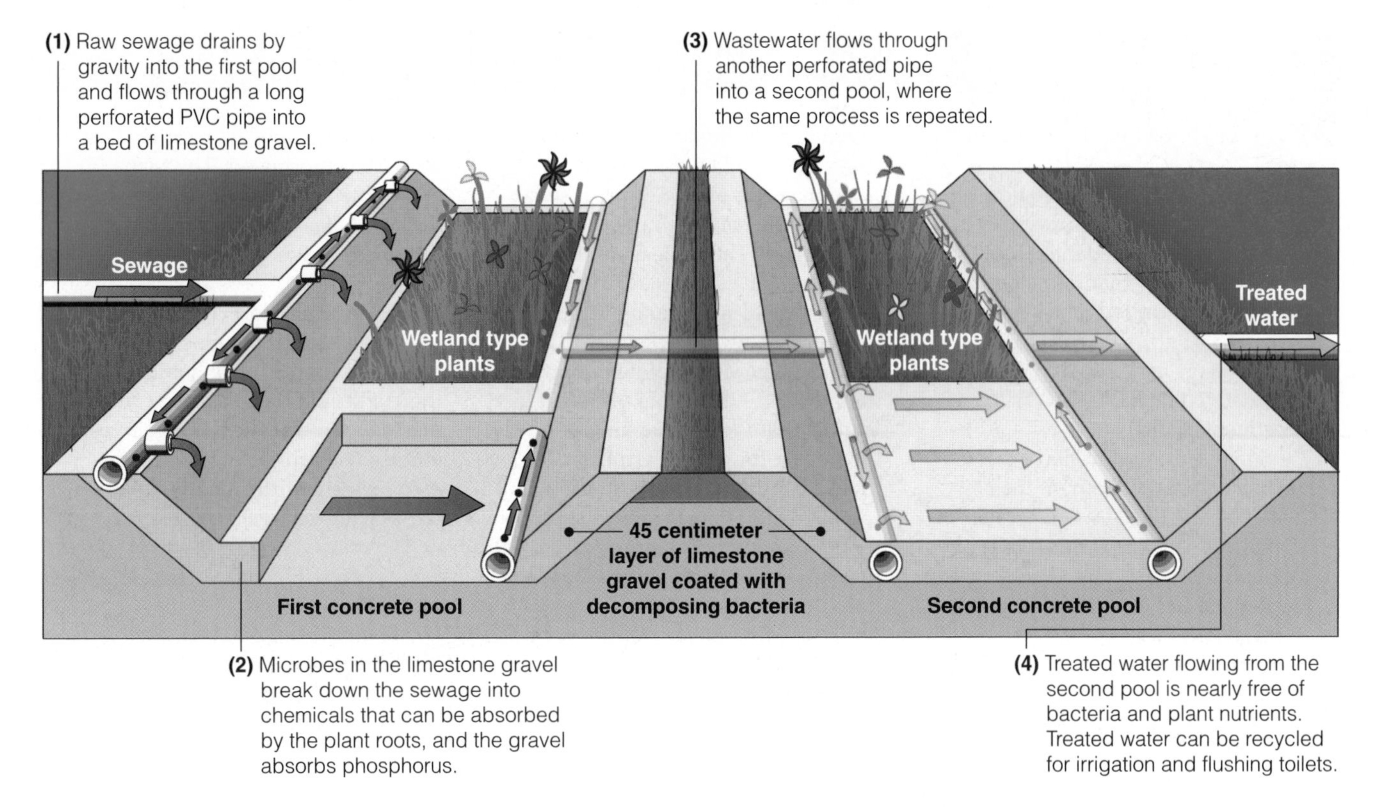
(1) Raw sewage drains by gravity into the first pool and flows through a long perforated PVC pipe into a bed of limestone gravel.
(3) Wastewater flows through another perforated pipe into a second pool, where the same process is repeated.
Sewage
Wetland type plants
Wetland type plants
Treated water
45 centimeter layer of limestone gravel coated with decomposing bacteria
First concrete pool
Second concrete pool
(2) Microbes in the limestone gravel break down the sewage into chemicals that can be absorbed by the plant roots, and the gravel absorbs phosphorus.
(4) Treated water flowing from the second pool is nearly free of bacteria and plant nutrients. Treated water can be recycled for irrigation and flushing toilets.

Figure 14-32 Wastewater garden. This small, artificial, gravity-fed wetland system for treating sewage from hotels, restaurants, and homes in developing countries was developed by the creators of the Biosphere 2 facility in Arizona (p. 23). This wastewater treatment system uses only 1.9–3.8 square meters (20–30 square feet) of space per person.

contaminated water. Estimates include **(1)** 7 million illnesses and 1,200 deaths per year according to a 1994 study by the Natural Resources Defense Council, **(2)** 1 million illnesses and about 900 deaths according to a 1997 report by the Centers for Disease Control and Prevention, and **(3)** 239,000 illnesses and 50 deaths according to a 1999 EPA study.

Environmentalists call for the U.S. Safe Drinking Water Act to be strengthened by **(1)** improving treatment by combining at least half of the 50,000 water systems that serve fewer than 3,300 people each with larger ones nearby, **(2)** strengthening and enforcing public notification requirements about violations of drinking water standards, and **(3)** banning all lead in new plumbing pipes, faucets, and fixtures (current law allows fixtures with up to 10% lead to be sold as lead-free). According to the Natural Resources Defense Council (NRDC), such improvements would cost about $30 a year per U.S. household.

However, Congress is being pressured by water-polluting industries to weaken the Safe Drinking Water Act by **(1)** eliminating national drinking water tests, **(2)** suspending the requirement that the media be advised of emergency water health violations and that water system officials notify their customers of such violations, **(3)** allowing states to give drinking water systems a permanent right to violate the standard for a given contaminant if the provider claims it cannot afford to comply, and **(4)** removing the requirement that water systems use affordable, feasible technology to remove cancer-causing contaminants.

Is Bottled Water the Answer? Despite some problems, experts say that the United States has some of the world's cleanest drinking water. Yet about half of all Americans worry about getting sick from tap water contaminants, and many drink bottled water or install expensive water purification systems. Studies indicate that many of these consumers are being ripped off and in some cases may end up drinking water that is dirtier than they can get from their taps.

An estimated one-third of the bottled water purchased in the United States is contaminated with bacteria. To be safe, consumers purchasing bottled water should determine whether the bottler belongs to the International Bottled Water Association (IBWA)* and

*Check for the IBWA seal of approval on the bottle or contact the International Bottled Water Association (113 North Henry Street, Alexandria, VA 22314; phone: 703-683-5213) for a member list.

SOLUTIONS

Using Wetlands to Treat Sewage

Waste treatment is one of the important ecological services provided by wetlands. More than 150 cities and towns in the United States now use natural and artificial wetlands to treat sewage as a low-tech, low-cost alternative to expensive waste treatment plants.

Some communities have created artificial wetlands to treat their water, as the residents of Arcata, California, did, under the leadership of Humboldt State University professors Robert Gearheart and George Allen.

In this coastal town of 17,000, some 63 hectares (155 acres) of wetlands has been created between the town and the adjacent Humboldt Bay. The marshes, developed on land that was once a dump, act as an inexpensive, natural waste treatment plant. The project cost less than half the estimated cost of a conventional treatment plant.

Here's how it works: First, sewage is held in sedimentation tanks, where the solids settle out as sludge that is removed and processed for use as fertilizer. The liquid is pumped into oxidation ponds, where remaining wastes are broken down by bacteria. After a month or so, the water is released into the artificial marshes, where it is further filtered and cleansed by plants and bacteria.

Although the water is clean enough to be discharged directly into the bay, state law requires that it first be chlorinated. The town chlorinates the water and then dechlorinates it before sending it into the bay, where oyster beds thrive.

The marshes and lagoons also serve as an Audubon Society bird sanctuary and provide habitats for thousands of otters, seabirds, and marine animals. The town even celebrates its natural sewage treatment system with an annual "Flush with Pride" festival.

Critical Thinking

List some possible drawbacks to creating artificial wetlands to treat sewage. Do these drawbacks outweigh the benefits? Explain.

adheres to its testing requirements. Some companies pay $2,500 annually to obtain more stringent certification by the National Sanitation Foundation, an independent agency that tests for 200 chemical and biological contaminants.

Before drinking expensive bottled water and buying costly home water purifiers, health officials suggest that consumers have their water tested by local health authorities or private labs (not companies trying to sell water purification equipment) to **(1)** identify what contaminants, if any, must be removed and **(2)** recommend the type of purification needed to remove such contaminants. Independent experts contend that unless tests show otherwise, for most urban and suburban Americans served by large municipal drinking water systems, home water treatment systems are not worth the expense and maintenance hassles.

Buyers should carefully check out companies selling water purification equipment and be wary of claims that the EPA has approved a treatment device. Although the EPA does *register* such devices, it neither tests nor approves them.

14-9 SOLUTIONS: ACHIEVING A MORE SUSTAINABLE WATER FUTURE

How Can We Use and Manage Water More Sustainably? Sustainable water use is based on the commonsense principle stated in an old Inca proverb: "The frog does not drink up the pond in which it lives." Figure 14-33 lists ways to implement this principle.

The challenge in developing such a *blue revolution* is to implement a mix of strategies built around **(1)** irrigating crops more efficiently, **(2)** using water-saving technologies in industries and homes, and **(3)** improving and integrating management of water basins and groundwater supplies.

Accomplishing such a revolution in water use and management will be difficult and controversial. However, water experts contend that not developing such strategies will eventually lead to **(1)** economic and health problems, **(2)** increased environmental degradation and loss of biodiversity, **(3)** heightened tensions and perhaps armed conflicts over water supplies (p. 327), **(4)** larger numbers of environmental refugees, and **(5)** threats to national and global military, economic, and environmental security.

How Can We Reduce Water Pollution? An Integrated Approach According to environmentalists, a more sustainable approach to dealing with water pollution requires that we shift our emphasis from pollution cleanup to pollution prevention by **(1)** *reducing* the toxicity or volume of pollutants (for example, replacing organic solvent-based inks and paints with water-based materials), **(2)** *reusing* wastewater instead of discharging it (for example, reusing treated wastewater for irrigation), and **(3)** *recycling* pollutants (for example, cleaning up and recycling contaminated solvents for reuse) instead of discharging them.

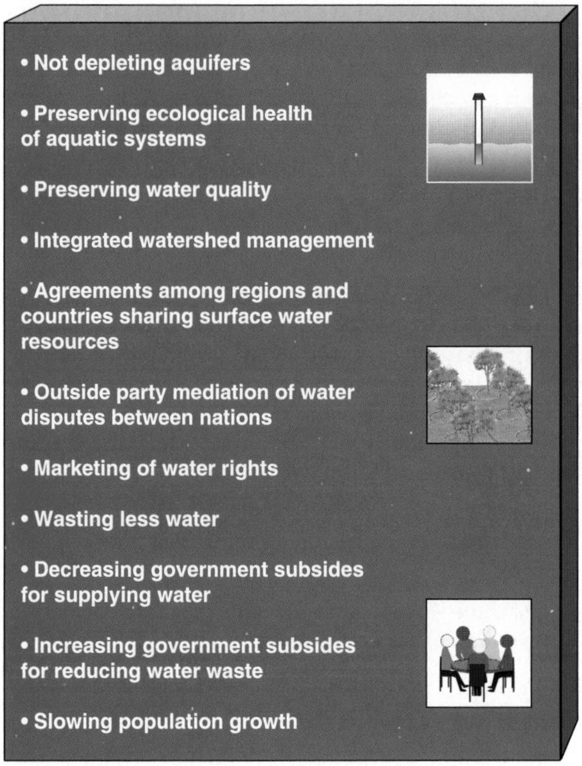

- Not depleting aquifers
- Preserving ecological health of aquatic systems
- Preserving water quality
- Integrated watershed management
- Agreements among regions and countries sharing surface water resources
- Outside party mediation of water disputes between nations
- Marketing of water rights
- Wasting less water
- Decreasing government subsides for supplying water
- Increasing government subsides for reducing water waste
- Slowing population growth

Figure 14-33 Methods for achieving more sustainable use of the earth's water resources.

To make such a shift, we need to accept that the environment—air, water, soil, and life—is an interconnected whole. Without an integrated approach to all forms of pollution, environmentalists argue that we will continue to shift environmental problems from one part of the environment to another. See the website material for this chapter for some actions you can take to help reduce water pollution.

It is not until the well runs dry that we know the worth of water.

BENJAMIN FRANKLIN

REVIEW QUESTIONS

1. Define the boldfaced terms in this chapter.

2. Explain why there is a danger of water wars in the Middle East.

3. List nine unique properties of water and explain the importance of each property.

4. What percentage of the earth's total volume of water is available for use by us?

5. Distinguish between *surface runoff, reliable runoff, watershed, groundwater, zone of saturation, water table, aquifer, recharge area,* and *natural recharge*. Explain how the water in some aquifers can be depleted.

6. Since 1900 how much has the total use and per capita use of water by humans increased? About what percentage of the world's reliable surface runoff is used by humanity? About what percentage of the water we withdraw each year is used for **(a)** irrigation, **(b)** industry, and **(c)** residences and cities?

7. What are the major water uses and problems of **(a)** the eastern United States and **(b)** the western United States?

8. List four causes of water scarcity. About how many people in the world live in countries suffering from water stress, and how many people are expected to face such shortages by 2025?

9. List five ways to increase the supply of fresh water in a particular area.

10. List the major pros and cons of building large dams and reservoirs to supply fresh water. List the pros and cons of China's Three Gorges dam project. List six ecological services provided by rivers.

11. List the major pros and cons of supplying water by transferring it from one watershed to another. List the pros and cons of **(a)** the Aral Sea water transfer project in Central Asia and **(b)** the California Water Project.

12. List the pros and cons of supplying more water by withdrawing groundwater. Summarize the problems of withdrawing groundwater from the Ogallala Aquifer in the United States. List five ways to prevent or slow groundwater depletion.

13. What is *saltwater intrusion*, and what harm does it cause?

14. List the pros and cons of increasing supplies of fresh water by **(a)** desalination of salt water, **(b)** cloud seeding, and **(c)** towing icebergs to water-short areas.

15. What percentage of the water used by people throughout the world is wasted? List four benefits of conserving water. List two major causes of water waste.

16. List ways to reduce water waste in **(a)** irrigation and **(b)** industry, homes, and businesses. List six advantages of *drip irrigation* and explain how it could help bring about a revolution in improving water efficiency. Explain why raising water prices can economically benefit the urban poor in many areas.

17. What is a *floodplain*? What three major services do floodplains provide? List four reasons why so many people live on floodplains. List the major benefits and disadvantages of floods. List three ways in which humans increase the damages from floods. Describe the nature and causes of the flooding problems in Bangladesh.

18. List the pros and cons of trying to reduce flood risks by **(a)** straightening and deepening streams (channelization), **(b)** building levees, **(c)** building dams, **(d)** restoring wetlands, and **(e)** managing floodplains.

19. What is *water pollution*? Describe how a *coliform bacteria count*, measurement of *biological oxygen demand*, and *biological indicators* can be used to determine water quality. What are eight types of water pollutants, and what are the major sources and effects of each type?

20. Distinguish between *point* and *nonpoint sources of water pollution* and give two examples of each type. Which type is easier to control? Why?

21. What percentage of the world's people in developing countries does not have access to clean drinking water?

22. What are the major water pollution problems of streams? Explain how streams can handle some loads of biodegradable wastes, and explain the limitations of this approach.

23. Summarize the good and bad news about attempts to prevent or control stream pollution.

24. What are the major water pollution problems of lakes? List three reasons why dilution of pollution often is less effective in lakes than in streams.

25. Distinguish between *eutrophication* and *cultural eutrophication*. What are the major causes of cultural eutrophication? List three methods for **(a)** preventing cultural eutrophication and **(b)** cleaning up cultural eutrophication.

26. Summarize the good and bad news about attempts to reduce water pollution in the Great Lakes. Give three examples of harmful invasions of the Great Lakes by nonnative species.

27. List five major sources of groundwater contamination. List three reasons why groundwater pollution is such a serious problem.

28. List three examples indicating the seriousness of groundwater pollution. List four ways to prevent groundwater contamination.

29. List the major pollution problems of the oceans. Why are most of these problems found in coastal areas?

30. Summarize the major pollution problems of the Chesapeake Bay in the United States and the progress made in dealing with these problems.

31. Distinguish between ocean pollution from *dredge spoils* and from *sewage sludge*.

32. Distinguish between *crude petroleum* and *refined petroleum* and summarize the major effects of oil pollution on ocean systems. What are the three major sources of oil pollution in the world's oceans? List **(a)** six ways to prevent oil pollution and **(b)** two ways to help clean up oil pollution.

33. List six ways to help prevent and reduce pollution of coastal waters.

34. List six ways to help prevent water pollution from nonpoint sources.

35. Explain how the United States and most developed countries have reduced water pollution from point sources by enacting laws, and summarize the good and bad news about such efforts. List six ways in which environmentalists believe water pollution control laws in the United States should be strengthened, and list two reasons why there is oppositionto such changes.

36. Distinguish between *septic tanks*, *primary sewage treatment*, *secondary sewage treatment*, and *advanced sewage treatment* as ways to reduce water pollution. List ways to deal with the sludge produced by waste treatment methods. What are the pros and cons of each approach?

37. Describe how sewage can be purified by working with nature using **(a)** John Todd's living machines and **(b)** artificial wetlands.

38. How is drinking water purified? How is the quality of drinking water protected in the United States? How successful have these efforts been? List three ways in which environmentalists believe the U.S. Safe Drinking Water Act should be strengthened and four ways in which opponents believe it should be weakened. List the pros and cons of drinking bottled water.

39. List 11 ways to use the world's water more sustainably and 5 disadvantages of not implementing such strategies.

40. List three ways to shift the emphasis from cleanup to prevention of water pollution.

CRITICAL THINKING

1. How do human activities increase the harmful effects of prolonged drought? How can we reduce these effects?

2. Do you believe that the projected benefits of China's Three Gorges dam and reservoir project on the Yangtze River will outweigh its potential drawbacks? Explain. What are the alternatives?

3. What role does population growth play in water supply problems?

4. Should the prices of water for all uses be raised sharply to include more of its environmental costs and to encourage water conservation? Explain. What harmful and beneficial effects might this have on **(a)** business and jobs, **(b)** your lifestyle and the lifestyles of any children or grandchildren you might have, **(c)** the poor, and **(d)** the environment?

5. What are the pros and cons of **(a)** gradually phasing out government subsidies of irrigation projects in the western United States (or in the country where you live) to increase water conservation and **(b)** providing government subsidies to farmers for improving irrigation efficiency?

6. List five major ways to conserve water for personal use (see website material for this chapter). Which, if any, of these practices do you now use or intend to use?

7. How do human activities contribute to flooding and flood damage? How can these effects be reduced?

8. Why is dilution not always the solution to water pollution? Give examples and conditions for which this solution is or is not applicable.

9. Which of the eight categories of pollutants listed in Table 14-1 is most likely to originate from **(a)** point sources and **(b)** nonpoint sources?

10. Should all dumping of wastes and untreated sewage in the ocean be banned? Explain. If so, where would you put the wastes instead? What exceptions would you permit, and why? How would you enforce such regulations?

11. Congratulations. You have been placed in charge of managing the world's water resources. What are the three most important things you would do?

12. Congratulations. You have been placed in charge of sharply reducing water pollution throughout the world. What are the three most important things you would do?

PROJECTS

1. In your community,

- **a.** What are the major sources of the water supply?
- **b.** How is water use divided between agricultural, industrial, power plant cooling, and public uses? Who are the biggest consumers of water?
- **c.** What has happened to water prices during the past 20 years? Are they too low to encourage water conservation and reuse?
- **d.** What water supply problems are projected?
- **e.** How is water being wasted?

2. Develop a water conservation plan for your school and submit it to school officials.

3. Consult with local officials to identify any floodplain areas in your community. Develop a map showing these areas and the types of activities (such as housing, manufacturing, roads, and recreational use) found on these lands.

4. In your community,

- **a.** What are the principal nonpoint sources of surface water and groundwater contamination?
- **b.** What is the source of drinking water?
- **c.** How is drinking water treated?
- **d.** How many times during each of the past 5 years have levels of tested contaminants violated federal standards? Was the public notified about the violations?
- **e.** Is fishing prohibited in any lakes or rivers in your region because of pollution? Are people warned about this?
- **f.** Is groundwater contamination a problem? If so, where, and what has been done about the problem?
- **g.** Is there a vulnerable aquifer or critical recharge zone that should be protected to ensure the quality of groundwater? Is your local government aware of this? What action (if any) has it taken?

5. Arrange a class or individual tour of a sewage treatment plant in your community. Compare the processes it uses with those shown in Figure 14-31. What happens to the sludge produced by this plant? What improvements, if any, would you suggest for this plant?

6. Make a concept map of this chapter's major ideas, using the section heads and subheads and the key terms (in boldface type). See material on the website for this book about how to prepare concept maps.

INTERNET STUDY RESOURCES AND RESOURCES FOR FURTHER READING AND RESEARCH

The website for this book contains helpful study aids and many ideas for further reading and research. Log on to

http://www.brookscole.com/product/0534389872s

and click on the Chapter-by-Chapter area. Choose Chapter 14 and select a resource:

- "Flash Cards" allows you to test your mastery of the Terms and Concepts to Remember for this chapter.
- "Tutorial Quizzes" provides a multiple-choice practice quiz.
- "Student Guide to InfoTrac" will lead you to Critical Thinking Projects that use InfoTrac College Edition as a research tool.
- "References" lists the major books and articles consulted in writing this chapter.
- "Hypercontents" takes you to an extensive list of sites with news, research, and images related to individual sections of the chapter.

INFOTRAC COLLEGE EDITION

Improve your skills with InfoTrac College Edition, a searchable online database of articles from more than 700 periodicals. Log on to

http://www.infotrac-college.com

or access InfoTrac through the website for this book. Try to find the following articles:

1. China Journal I. Henry Petroski. *American Scientist,* May 2001 v89 i3 p198. This firsthand report on the Yangtze River and China's Three Gorges dam project discusses the choices involved in connection with the project. *Hint:* Enter the search term "china journal" using the keyword "dams."

2. Runoff from Farms and Cities Threatening Coastal Ecosystems. *Clean Water Report,* March 12, 2001 v39 i6 pNA. Coastal fisheries and ecosystems are still in peril, despite 30 years of reductions in ocean dumping, pollution from waste treatment facilities, and toxics, such as DDT, according to the Pew Oceans Commission. *Hint:* Enter the search term "runoff ecosystems" using the keywords "ocean pollution."

15 SOLID AND HAZARDOUS WASTE

There Is No "Away": Love Canal

Between 1942 and 1953, Hooker Chemicals and Plastics (owned by OxyChem since 1968) **(1)** sealed chemical wastes containing at least 200 different chemicals into steel drums and **(2)** dumped them into an old canal excavation (called Love Canal after its builder, William Love) near Niagara Falls, New York.

In 1953, Hooker Chemicals filled the canal, covered it with clay and topsoil, and sold it to the Niagara Falls school board for $1. In the deed the company inserted a disclaimer denying legal liability for any injury caused by the wastes. In 1957, Hooker warned the school board not to disturb the clay cap because of the possible danger from toxic wastes.

By 1959 an elementary school, playing fields, and 949 homes had been built in the 10-square-block Love Canal area (Figure 15-1). Roads and sewer lines crisscrossed the dump site, some of them disrupting the clay cap covering the wastes. In the 1960s, an expressway was built at one end of the dump. It blocked groundwater from migrating to the Niagara River and allowed contaminated groundwater and rainwater to build up and overflow the disrupted cap.

Residents began complaining to city officials in 1976 about chemical smells and chemical burns their children received playing in the canal area, but these complaints were ignored. In 1977, chemicals began leaking from the badly corroded steel drums into storm sewers, gardens, basements of homes next to the canal, and the school playground.

In 1978, after much media publicity and pressure from residents led by Lois Gibbs (a mother galvanized into action as she watched her children come down with one illness after another; Guest Essay, p. 372), the state acted. It closed the school and arranged for the 239 homes closest to the dump to be evacuated, purchased, and destroyed.

Two years later, after protests from families still living fairly close to the landfill, President Jimmy Carter **(1)** declared Love Canal a federal disaster area, **(2)** had the remaining families relocated, and **(3)** offered federal funds to buy 564 more homes. Residents of all but 72 of the homes moved out. Some residents who remained claim that the entire problem was exaggerated by other residents, environmentalists, and the media.

After more than 15 years of court cases, OxyChem agreed in 1994 to **(1)** pay a $98-million settlement to New York State and **(2)** be responsible for all future treatment of wastes and wastewater at the Love Canal site. In 1999, the company also agreed to reimburse the federal government and New York State $7.1 million for the Love Canal cleanup.

Because of the difficulty in linking exposure to a variety of chemicals to specific health effects (Section 10-2, p. 226), the long-term health effects of exposure to hazardous chemicals on Love Canal residents remain unknown and controversial. However, for the rest of their lives the evacuated families will worry about the possible effects of the chemicals on themselves and their children and grandchildren.

The dumpsite has been covered with a new clay cap and surrounded by a drainage system that pumps leaking wastes to a new treatment plant. In June 1990, the U.S. Environmental Protection Agency (EPA) allowed state officials to begin selling 234 of the remaining houses in the area (renamed Black Creek Village) at 10–20% below market value. Most of the houses have been sold. Buyers must sign an agreement stating that New York State and the federal government make no guarantees or representations about the safety of living in these homes.

The Love Canal incident is a vivid reminder that **(1)** we can never really throw anything away, **(2)** wastes do not stay put, and **(3)** preventing pollution is much safer and cheaper than trying to clean it up.

Figure 15-1 The Love Canal housing development near Niagara Falls, New York, was built near a hazardous-waste dump site. The photo shows the area when it was abandoned in 1980. In 1990, the EPA allowed people to buy some of the remaining houses and move back into the area. (New York State Department of Environmental Conservation)

Solid wastes are only raw materials we're too stupid to use.
ARTHUR C. CLARKE

This chapter addresses the following questions:

- What are solid waste and hazardous waste, and how much of each type do we produce?
- What can we do to reduce, reuse, and recycle solid waste and hazardous waste?
- What are we doing to recycle aluminum, paper, and plastics?
- What are the advantages and disadvantages of burning or burying wastes?
- What can we do to reduce exposure to lead, hazardous chlorine compounds, and dioxins?
- How is hazardous waste regulated in the United States?
- How can we make the transition to a more sustainable low-waste society?

15-1 WASTING RESOURCES

What Is Solid Waste, and How Much Is Produced? The United States, with only 4.7% of the world's population, produces about 33% of the world's **solid waste**: any unwanted or discarded material that is not a liquid or a gas. About 97.5% of this solid waste comes from **(1)** mining (Section 9-4, p. 200), **(2)** oil (p. 494) and natural gas production (p. 499), **(3)** agriculture (Figure 16-9, p. 405), and **(4)** industrial activities used to produce goods and services for consumers (Figure 15-2).

Another 1.5% of solid waste produced in the United States is **municipal solid waste** (MSW) from homes and businesses in or near urban areas. The amount of MSW, often called *garbage*, currently produced in the United States each year amounts to about 200 million metric tons (506 billion pounds)—almost twice as much as in 1970. This is enough waste to fill a bumper-to-bumper convoy of garbage trucks encircling the globe eight times. This amounts to an average of 820 kilograms (1,800 pounds) per person in the United States. This is the world's highest per capita solid waste production and many times the rate in developing countries.

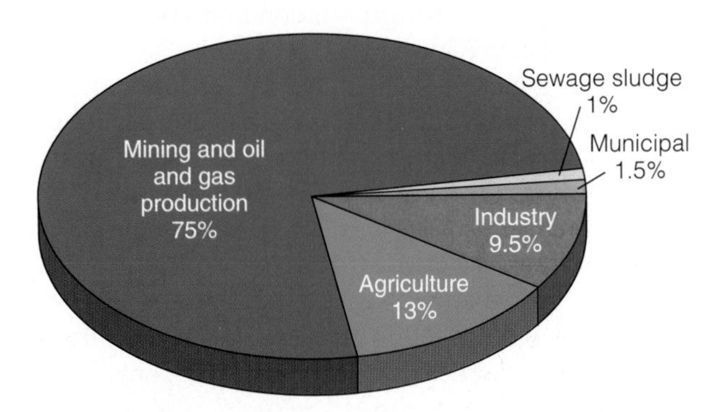

Figure 15-2 Sources of the estimated 11 billion metric tons (12 billion tons) of solid waste produced each year in the United States. Mining, agricultural, and industrial activities produce 55 times as much solid waste as household activities. (Data from U.S. Environmental Protection Agency and U.S. Bureau of Mines)

What Does It Mean to Live in a High-Waste Society? Here are a few of the solid wastes U.S. consumers throw away:

- Enough aluminum to rebuild the country's entire commercial airline fleet every 3 months
- Enough tires each year to encircle the planet almost three times
- About 18 billion disposable diapers per year, which if linked end to end would reach to the moon and back seven times
- About 2 billion disposable razors, 19 million computers, and 8 million television sets each year
- Some 8.6 million metric tons (17 billion pounds) of polystyrene peanuts used to protect items during shipping each year
- Used carpet each year that would cover 7,800 square kilometers (3,000 square miles)
- About 1.7 billion metric tons (3.7 trillion pounds) of construction waste per year—an average of 6 metric tons (13,200 pounds) per person
- About 2.5 million nonreturnable plastic bottles every hour
- About 670,000 metric tons (1.5 billion pounds) of edible food per year
- Enough office paper each year to build a 3.5-meter (11-foot) high wall from New York City to San Francisco, California
- Some 186 billion pieces of junk mail (an average of 662 per American) each year, about 45% of which are thrown in the trash unopened

This is only part of the 1.5% of all solid waste labeled "municipal" in Figure 15-2.

What Is Hazardous Waste, and How Much Is Produced? In the United States, **hazardous waste** is legally defined as any discarded solid or liquid material that **(1)** contains one or more of 39 toxic, carcinogenic, mutagenic, or teratogenic compounds (Section 10-3, p. 229) at levels that exceed established limits (including many solvents, pesticides, and paint strippers), **(2)** catches fire easily (gasoline, paints, and

Table 15-1 Common Household Toxic and Hazardous Materials

Cleaning Products	Disinfectants; Drain, toilet, and window cleaners; Oven cleaners; Bleach and ammonia; Cleaning solvents and spot removers; Septic tank cleaners
Paint and Building Products	Latex and oil-based paints; Paint thinners, solvents, and strippers; Stains, varnishes, and lacquers; Wood preservatives; Acids for etching and rust removal; Asphalt and roof tar
Gardening and Pest Control Products	Pesticide sprays and dusts; Weed killers; Ant and rodent killers; Flea powder
Automotive Products	Gasoline; Used motor oil; Antifreeze; Battery acid; Solvents; Brake and transmission fluid; Rust inhibitor and rust remover
General Products	Dry cell batteries (mercury and cadmium); Artists' paints and inks; Glues and cements

solvents), **(3)** is reactive or unstable enough to explode or release toxic fumes (acids, bases, ammonia, and chlorine bleach), or **(4)** is capable of corroding metal containers such as tanks, drums, and barrels (industrial cleaning agents and oven and drain cleaners).

This official definition of hazardous wastes (mandated by Congress) does *not* include the following materials: **(1)** radioactive wastes (p. 506), **(2)** hazardous and toxic materials discarded by households (Table 15-1), **(3)** mining wastes, **(4)** oil- and gas-drilling wastes (routinely discharged into surface waters or dumped into unlined pits and landfills), **(5)** liquid waste containing organic hydrocarbon compounds (80% of all liquid hazardous waste), **(6)** cement kiln dust, produced when liquid hazardous wastes are burned in a cement kiln, and **(7)** wastes from the thousands of small businesses and factories that generate less than 100 kilograms (220 pounds) of hazardous waste per month.

As a result, *hazardous-waste laws do not regulate 95% of the country's hazardous waste.* In most other countries, especially developing countries, little, if any, of the hazardous waste is regulated.

Including all categories, the EPA estimates that at least 5.5 billion metric tons (12 trillion pounds) of hazardous waste are produced each year in the United States—an average of 20 metric tons (43,000 pounds) per person. This amounts to about 75% of the world's hazardous waste. According to U.S. government toxic release data, in 1999 the two industries releasing the largest quantities of toxic chemicals into the environment were **(1)** mining (accounting for almost half of the nation's total, including large quantities of arsenic; Spotlight, p. 356) and **(2)** electric power plants (accounting for about 15% of the toxic releases).

15-2 PRODUCING LESS WASTE AND POLLUTION

What Are Our Options? There are two ways to deal with the solid and hazardous waste we create: **(1)** *waste management* and **(2)** *pollution (waste) prevention.* Waste management is a *high-waste approach* (Figure 3-15, p. 73) that views waste production as an unavoidable product of economic growth. It attempts to manage the resulting wastes in ways that reduce environmental harm, mostly by **(1)** burying them, **(2)** burning them, or **(3)** shipping them off to another state or country. In effect, it transfers solid and hazardous waste from one part of the environment to another.

Preventing pollution and waste is a *low-waste approach* that recognizes that there is no "away" and views most solid and hazardous waste either **(1)** as potential resources (that we should be recycling, composting, or reusing) or **(2)** as harmful substances that we should not be using in the first place (Figure 3-16, p. 74, and Figures 15-3 and 15-4). This approach focuses on discouraging waste production and encouraging waste prevention (Guest Essay, p. 372).

According to the U.S. National Academy of Sciences (Figures 15-3 and 15-4), the low-waste approach has the following hierarchy of goals: **(1)** *reduce* waste and pollution, **(2)** *reuse* as many things as possible, **(3)** *recycle and compost* as much waste as possible, **(4)** *chemically or biologically treat or incinerate* waste that cannot be reduced, reused, recycled, or composted, and **(5)** *bury* what is left in state-of-the-art landfills or aboveground vaults after the first four goals have been met.

Scientists estimate that in a low-waste society 60–80% of the solid and hazardous waste produced could be eliminated through *reduction, reuse, recycling* (including composting), and *redesign* of manufactur-

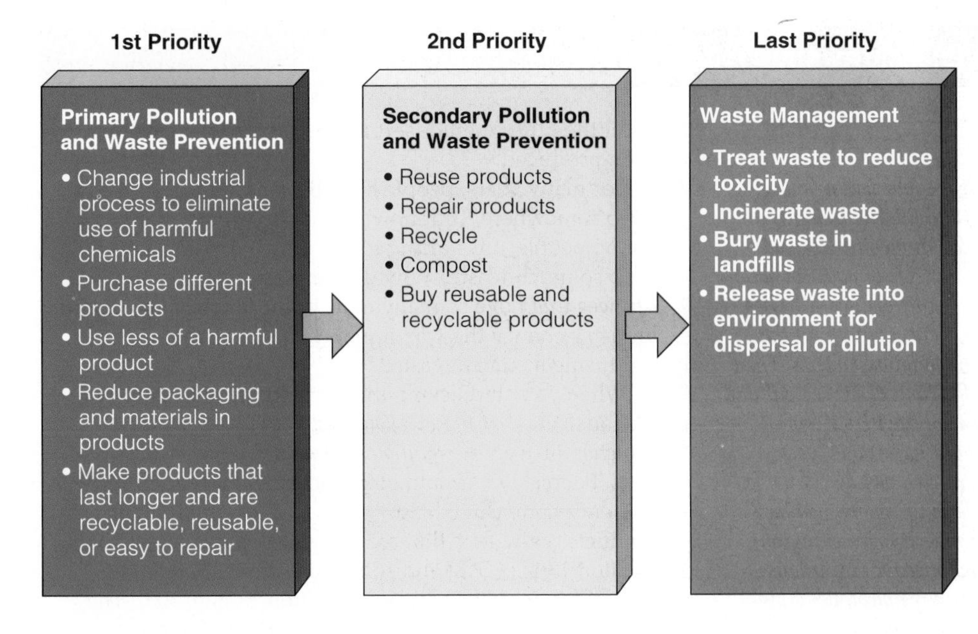

1st Priority
Primary Pollution and Waste Prevention
• Change industrial process to eliminate use of harmful chemicals
• Purchase different products
• Use less of a harmful product
• Reduce packaging and materials in products
• Make products that last longer and are recyclable, reusable, or easy to repair
2nd Priority
Secondary Pollution and Waste Prevention
• Reuse products
• Repair products
• Recycle
• Compost
• Buy reusable and recyclable products
Last Priority
Waste Management
• Treat waste to reduce toxicity
• Incinerate waste
• Bury waste in landfills
• Release waste into environment for dispersal or dilution

Figure 15-3 *Solutions*: priorities suggested by prominent scientists for dealing with material use and solid waste. To date, these waste-reduction priorities have not been followed in the United States (and in most other countries). Instead, most efforts are devoted to waste management (bury it or burn it). (U.S. Environmental Protection Agency and U.S. National Academy of Sciences)

ing processes and buildings. Currently, the order of priorities shown in Figures 15-3 and 15-4 for dealing with solid and hazardous wastes is reversed in the United States (and in most other countries).

Solutions: How Can We Reduce Waste and Pollution? Ways to reduce resource use, waste, and pollution (Figure 3-16, p. 74) include

▪ *Decreasing consumption*. Before any purchase this involves asking questions such as **(1)** Do I really need this? **(2)** Can I buy it secondhand? and **(3)** Can I borrow or lease it?

▪ *Doing more with less by redesigning manufacturing processes and products to use less material and energy per unit of output of goods and services* (Solutions, p. 374, and Guest Essay, p. 375).

▪ *Redesigning manufacturing processes to produce less waste and pollution*. Most toxic organic solvents can be recycled within plants or replaced with water-based or citrus-based solvents (Individuals Matter, p. 323). Hydrogen peroxide can be used instead of toxic chlorine to bleach paper and other materials. A CO_2-based process can replace dry cleaning with toxic organic solvents such as perchloroethylene (PERC).

▪ *Developing products that are easy to repair, reuse, remanufacture, compost, or recycle*. Several European auto manufacturers design their cars for easy disassembly and for reuse and recycling of up to 80% of their parts (75% in the United States). Xerox's latest photocopier, with every part reusable or recyclable for easy remanufacturing, should eventually save the company $1 billion in manufacturing costs.

▪ *Designing products to last longer*. Tires are now being produced that have an average life of 97,000 kilometers (60,000 miles), but researchers believe this could be extended to at least 160,000 kilometers (100,000 miles).

▪ *Eliminating or reducing unnecessary packaging*. Here are some key questions environmentalists believe designers, manufacturers, and consumers should ask about packaging: **(1)** Is it necessary? **(2)** Can it use fewer materials? **(3)** Can it be reused? **(4)** Are the resources that went into it

Produce Less Waste

Manipulate processes to eliminate or reduce production

Recycle and reuse

Convert to Less Hazardous or Nonhazardous Substances

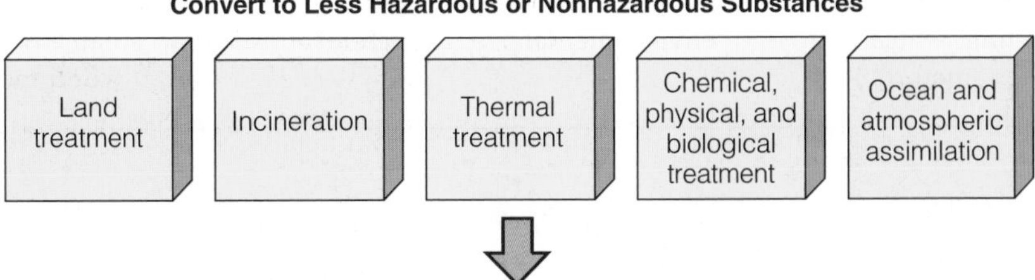

Put in Perpetual Storage

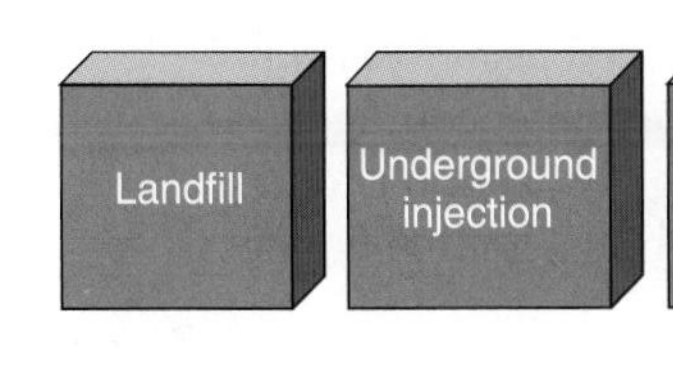

Surface impoundments

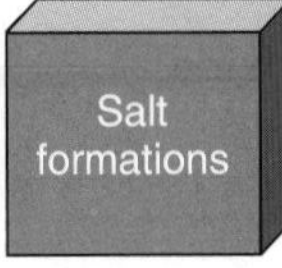

Arid region unsaturated zone

Figure 15-4 *Solutions*: priorities suggested by prominent scientists for dealing with hazardous waste. To date, these priorities have not been followed in the United States (and in most other countries). (U.S. National Academy of Sciences)

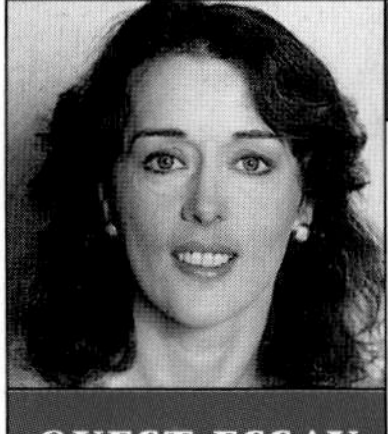

GUEST ESSAY

We Have Been Asking the Wrong Questions About Wastes

Lois Marie Gibbs

In 1977, Lois Marie Gibbs was a young homemaker with two children living near the Love Canal toxic dumpsite. She had never engaged in any sort of political action until her children began experiencing unexplained illnesses and she learned that toxic chemicals were oozing from the dumpsite into many of the area's yards and basements. Then she organized her neighborhood and became the president and major strategist for the Love Canal Homeowners Association. This grassroots political action brought hazardous-waste issues to national prominence and spurred passage of the federal Superfund legislation to help clean up abandoned hazardous-waste sites. Lois Gibbs then moved to Washington, D.C., and formed Citizens' Clearinghouse for Hazardous Wastes (renamed the Center for Health, Environment, and Justice), an organization that has helped more than 7,000 community organizations protect themselves from hazardous wastes. Her story is told in her autobiography, Love Canal: My Story *(State University of New York Press, 1982). She was also the subject of a CBS movie,* Lois Gibbs: The Love Canal, *which aired in 1982. Her latest book is* Dying from Dioxin *(Boston: South End Press, 1995).*

Just about everyone knows our environment is in danger. One of the most serious threats is the massive amount of waste we put into the air, water, and ground every year. All across the United States and around the world are thousands of places that have been, and continue to be, polluted by toxic chemicals, radioactive waste, and just plain garbage.

For generations, the main question people have asked is "Where do we put all this waste? It's got to go somewhere." That is the wrong question, as has been shown by a series of experiments in waste disposal and by the simple fact that there is no "away" in "throwaway."

We tried dumping our waste in the oceans. That does not work We tried injecting it into deep, underground wells. That does not work. We've been trying to build landfills that do not leak, but according to the EPA all landfills eventually leak. We've been trying to get rid of waste by burning it in high-tech incinerators; that only produces different types of pollution, such as air pollution and toxic ash. Even recycling, which is a very good thing to do, suffers from the same problem as all the other methods: It addresses waste *after* it has been produced.

For many years, people have been assuming, "It's got to go somewhere," but now many people, especially young people, are starting to ask why. Why do we produce so much waste? Why do we need products and services that have so many toxic by-products? Why can't industry change the way it makes things so that it stops producing so much waste?

When you start asking these questions, you start getting answers that lead to *pollution prevention* and *waste reduction* instead of *pollution control* and *waste management*. People, young and old, who care about pollution prevention are challenging companies to stop making products with gases that reduce ozone in the ozone layer [Section 13-6, p. 318] and contribute to the threatening possibility of global warming [Section 13-2, p. 303]. They are asking why so many goods are wrapped in excessive, throwaway packaging. They are challenging companies that sell pesticides, cleaning fluids, batteries, and other hazardous products to either **(1)** remove the toxins from those products or **(2)** take them back for recovery or recycling rather than disposing of them in the environment. They are demanding alternatives to throwaway materials in general.

Since 1988, hundreds of student groups have contacted my organization to get help and advice in taking these effective types of actions. Oregon students took legal action to get rid of cups and plates made from bleached paper because the paper contains the deadly poison dioxin. As a result the school systems switched to reusable cups, plates, and utensils.

Dozens of student groups have joined with local grassroots organizations to get toxic-waste sites cleaned up or to stop construction of new toxic-waste sites, radioactive-waste sites, or waste incinerators.

Waste issues are not simply environmental issues; they are also tied up with our economy, which is geared to producing and then disposing of waste. *Somebody* is making money from every scrap of waste and has a vested interest in keeping things the way they are.

Waste issues are also issues of *justice* and *fairness*. There is a lot of debate between industry officials and environmentalists, especially those in federal and state environmental agencies, about so-called acceptable risk

(continued)

nonrenewable or renewable? **(5)** Does it contain the highest feasible amount of consumer-discarded (postconsumer) recycled material? **(6)** Is it designed to be recycled easily? **(7)** Can it be incinerated without producing harmful air pollutants or a toxic ash? **(8)** Can it be buried and decomposed in a landfill without producing chemicals that can contaminate groundwater?

- *Using trash taxes* to reduce waste and using the revenues to reduce taxes on income and wealth (Solutions, p. 33), as a number of European countries have done. A related *pay-as-you-throw* system reduces solid waste and encourages recycling by basing garbage collection charges on the amount of waste a household generates for disposal.

GUEST ESSAY (continued)

[Section 10-5, p. 236]. These industry officials and environmentalists decide the degree of people's exposure to toxic chemicals but do not ask the people who will actually be exposed how they feel about it.

Risk analysts often say, "There's only a one in a million chance of increased death from this toxic chemical." That may be true, but suppose I took a pistol and went to the edge of your neighborhood and began shooting into it. There's probably only a one in a million chance that I'd hit somebody, but would you issue me a license to do that?

As long as we do not stand up for our rights and demand that "bullets" in the form of hazardous chemicals not be "fired" in our neighborhoods, we are giving environmental regulators and waste producers a license to kill a certain number of us without our even being consulted.

From my personal experience, I know that decisions made to dump wastes at Love Canal [p. 368] and in thousands of other places were not made purely on the basis of the best available scientific knowledge. The same holds true for decisions about how to manage the wastes we produce today and how to produce less waste.

We live in a world that is shaped by decisions based on money and power. If you really want to understand what's behind any given environmental issue, the first question you should ask is, "Who stands to profit from this?" Then ask, "Who is going to pay the price?" You can then identify both sides of the issue and decide whether you want to be part of the problem or part of the solution.

Critical Thinking

1. Do you believe that we should put primary emphasis on pollution prevention and waste reduction? Explain.

2. What changes would you be willing to make in your own lifestyle to prevent pollution and reduce waste?

15-3 SOLUTIONS: CLEANER PRODUCTION AND SELLING SERVICES INSTEAD OF THINGS

What Is the Ecoindustrial Revolution? Some analysts urge us to bring about an *ecoindustrial revolution* over the next 50 years as a way to help achieve industrial, economic, and environmental sustainability. The goals of this emerging concept of *cleaner production* (Guest Essay, p. 375), or *industrial ecology*, are to redesign all industrial products and processes integrate them into **(1)** an essentially closed system of cyclical material flows (Figure 3-16, p. 74) or **(2)** a network in which the wastes of one manufacturer become raw materials for another, and companies take back packaging and used products from consumers for reuse, recycling, repair, or remanufacturing.

In effect, companies would mimic natural chemical cycles (Section 4-6, p. 94) and interact in complex *resource exchange webs* similar to food webs in natural ecosystems. A prototype of this industrial ecosystem concept exists in Kalundborg, Denmark, where **(1)** a coal-fired power plant, **(2)** an oil refinery, **(3)** a sulfuric acid producer, **(4)** a sheetrock plant, **(5)** a pharmaceutical plant, **(6)** a cement manufacturer, **(7)** local farms, **(8)** horticulture greenhouses, **(9)** a fish farm, and **(10)** nearby homes are working together to save money by exchanging and converting their wastes into resources for one another.

Halifax (Nova Scotia) and more than 20 U.S. cities—including Baltimore (Maryland), Rochester (New York), Chattanooga (Tennessee), and the Brownsville/Matamoros region along the Texas–Mexico border—have announced plans to build ecoindustrial parks similar to the one in Kalundborg.

These important forms of *biomimicry*, which eliminate most waste, also provide economic benefits to businesses by

- Reducing the costs of controlling pollution and complying with pollution regulations
- Improving the health and safety of workers by reducing exposure to toxic and hazardous material (and thus reducing company health-care insurance costs)
- Reducing future legal liability for toxic and hazardous wastes
- Stimulating companies to come up with new, environmentally beneficial chemicals, processes, and products that can be sold worldwide
- Giving companies a better image among consumers based on results rather than on public relations campaigns

In 1975, the Minnesota Mining and Manufacturing Company (3M), which makes 60,000 different products in 100 manufacturing plants, began a Pollution Prevention Pays (3P) program. It **(1)** redesigned equipment and processes, **(2)** used fewer hazardous raw materials, **(3)** identified hazardous chemical outputs (and recycled or sold them as raw materials to other companies), and **(4)** began making more nonpolluting products.

By 1998, **(1)** 3M's overall waste production was down by one-third, **(2)** its air pollutant emissions per unit

Doing More with Less: Increasing Resource Productivity

SOLUTIONS

During the last few decades a *design revolution* has allowed businesses to use less material and energy per unit of goods and services, mostly by **(1)** finding substitutes for products that use less material and **(2)** redesigning or improving products so that they take less material and energy to produce.

Today, a single fiber-optic cable containing about 65 grams (3 ounces) of silica (SiO_2) can carry many times more electronic messages than a comparable length of cable containing 0.9 metric tons (1 ton) of copper. Increased use of wireless communication will greatly reduce the use of wires of any type.

Paper documents such as product catalogs, phone directories, technical reference manuals, and parts directories can be accessed on CD-ROMs or at various internet sites, saving millions of dollars and tons of paper. All the phone books in the United States can be put on about three CD-ROMs, and a single DVD-ROM could hold all the world's phone numbers.

A skyscraper built today includes about 35% less steel than the same building built in the 1960s because of the use of lighter-weight but higher-strength steel. Use of such steel and replacement of many steel parts with lightweight plastics and composite materials has **(1)** reduced the weight of cars by about 25% without compromising performance and safety, **(2)** increased fuel efficiency, and **(3)** reduced the average weight per unit of appliances such as stoves, washers, dryers, air conditioners, TV sets, and computers.

Conventional lumber is being replaced by engineered structural beams and joists and wall framing (studs) made by compressing and gluing wood wastes or by gluing many layers of wood together. Such products make roof and floor supports in houses so rigid that no internal load-bearing walls are needed. This **(1)** allows more flexible and useful living space, **(2)** reduces the wood needed for internal walls by at least 70%, and **(3)** doubles the amount of space for insulation (paid for by saved wood and lumber), which saves energy and allows smaller and less costly heating and cooling systems.

Since the mid-1970s **(1)** the thickness of plastic grocery bags has been reduced by 70% without sacrificing strength, **(2)** plastic milk jugs weigh 40% less, **(3)** aluminum drink cans contain one-third less aluminum, **(4)** steel cans are 60% lighter, **(5)** disposable diapers contain 50% less paper pulp, and **(6)** plastic frozen food bags weigh 89% less.

Because of increased population and per capita consumption, the total amount of municipal solid waste in the United States continues to grow. However, improvements in material efficiency (dematerialization) and increased recycling and composting have helped cut the rate of growth of such waste in half since 1990.

These improvements in resource productivity are important, but according to some analysts they can be greatly increased through a new *resource productivity revolution*. In their 1999 book *Natural Capitalism*, Paul Hawken (Guest Essay, p. 6), Amory Lovins (Guest Essay, p. 517), and Hunter Lovins contend that we already have the knowledge and technology to greatly increase *resource productivity* by getting 75–90% more work or service from each unit of material resources we use. According to these analysts, such a revolution in *resource productivity* by doing more with less would

- Sharply decrease the depletion and degradation of the earth's natural capital (Figure 2-3, p. 26, and Figure 4-33, p. 103) that supports all economies (Guest Essay, p. 6).
- Give companies and countries making such improvements a huge competitive advantage in the global marketplace.
- Help reduce unemployment and poverty by making it more profitable to employ people. This is especially important in developing countries, where human labor is an abundant and underused economic resource.
- Not be as difficult as it might seem because the current use of matter and energy resources to fuel economic growth is extremely inefficient and amounts to throwing money away. According to a 1989 report by the National Academy of Sciences, only about 6% of the vast flows of materials through the U.S. economy end up in products. In addition, about 43% of the energy flowing through the U.S. economy is unnecessarily wasted (Figure 20-2, p. 514).

To these analysts, the only major impediments to such an economic and ecological revolution are laws, policies, taxes, and subsidies that continue to reward inefficient resource use.

Critical Thinking

Do you believe that it is possible to decrease resource waste by 75–90% within the next 20 years? Explain. What might be some disadvantages of making such a shift? Do you believe that such disadvantages outweigh the advantages? Explain.

of production were reduced by 70%, and **(3)** the company had saved more than $750 million in waste disposal and material costs. Since 1990 a growing number of companies have adopted similar pollution prevention programs.

What Is a Service Flow Economy, and What Are Its Advantages? In the mid-1980s, German chemist Michael Braungart and Swiss industry analyst Walter Stahel independently proposed a new economic model

GUEST ESSAY

Cleaner Production: A New Environmentalism for the 21st Century

Peter Montague

Peter Montague is director of the Environmental Research Foundation in Washington, D.C., which studies and informs the public about environmental problems and the technologies and policies that might help solve them. He has served as project administrator of a hazardous-waste research program at Princeton University and has taught courses in environmental impact analysis at the University of New Mexico. He is the coauthor of two books on toxic heavy metals in the natural environment and is editor of Rachel's Environment and Health Biweekly, *an informative newsletter on environmental problems and solutions.*

Environmentalism as we have known it for the last 31 years is dead. The environmentalism of the 1970s advocated strict numerical controls on releases into the environment of *dangerous wastes* (any unwanted or uncontrolled materials that can harm living things or disrupt ecosystems).

However, after several decades of effort by government regulatory agencies and concerned citizens (the environmental movement), most dangerous chemicals are not regulated in any way. Even the few that are covered by regulations have not been controlled adequately.

In short, the *pollution management* approach to environmental protection has failed; *pollution prevention* is our only hope. An ounce of prevention really is worth a pound of cure.

Here is the situation facing environmentalists today:

- *All waste disposal—landfilling, incineration, deep-well injection—is polluting because "disposal" means dispersal into the environment.* Once wastes are created, they cannot be contained or controlled because of the scientific laws of matter conservation [p. 66] and energy conservation [p. 71]. The old environmentalism failed to recognize this important truth and thus squandered enormous resources trying to achieve the impossible.
- *The inevitable result of our reliance on waste treatment and disposal systems has been an unrelenting buildup of toxic synthetic materials in humans and other forms of life worldwide.* For example, breast milk of women in industrialized countries is so contaminated with pesticides and industrial hydrocarbons that if human milk were bottled and sold commercially, it could be banned by the U.S. Food and Drug Administration as unsafe for human consumption. If a whale beaches itself on U.S. shores and dies, its body must be treated as "hazardous waste" because whales contain concentrations of PCBs [polychlorinated biphenyls; Figure 14-22, p. 349] legally defined as hazardous.
- *The ability of humans and other life-forms to adapt to changes in their environment is strictly limited by the genetic code each form of life inherits.* Continued contamination at a rate hundreds of times faster than we can adapt will subject humans to increasingly widespread sickness and degradation of the species and could ultimately lead to extinction.
- *Damage to humans (and other life-forms) is abundantly documented.* Birds, fish, and humans in industrialized countries are enduring steadily rising levels of cancer, genetic mutations, and damage to their nervous, immune, and hormonal systems [Connections, p. 231] as a result of pollution.

If we will but look, the handwriting is on the wall everywhere.

To deal with these problems, industrial societies must abandon their reliance **(1)** on waste treatment and disposal and **(2)** on the regulatory system of numerical standards created to manage the damage that results from relying on waste disposal instead of waste prevention. We must quickly move the industrialized and industrializing countries to new technical approaches accompanied by new industrial goals: *clean production* or zero-discharge systems.

The concept of clean production involves industrial systems that avoid or eliminate dangerous wastes and dangerous products and minimize the use and waste of raw materials, water, and energy. Goods manufactured in a clean production process must not damage natural ecosystems throughout their entire life cycle including **(1)** raw material selection, extraction, and processing, **(2)** product conceptualization, design, manufacture, and assembly [Guess Essay, p. 272], **(3)** material transport during all phases, **(4)** industrial and household usage, and **(5)** reintroduction of the product into industrial systems or into the environment when it no longer serves a useful function.

Clean production does not rely on *end-of-pipe* pollution controls such as filters or scrubbers [Figure 12-18, p. 296] or chemical, physical, or biological treatment [Figure 14-31, p. 361]. Measures that **(1)** pretend to reduce the volume of waste by incineration or concentration, **(2)** mask the hazard by dilution, or **(3)** transfer pollutants from one environmental medium to another are also excluded from the concept of clean production.

A new industrial pattern, and thus a new environmentalism, is emerging. Human survival and life quality depend on our willingness to make and pay for the changes needed to shift to this cleaner form of industrial production.

Critical Thinking

1. Some environmentalists point to the successes of the *pollution management* approach to environmental protection practiced during the past 31 years and do not agree that it has failed. What is your position? Explain.

2. List three undesirable economic, health, consumption, and lifestyle changes you might experience as a consequence of putting much greater emphasis on pollution prevention.

that would provide profits while greatly reducing resource use and waste. Their idea involves shifting from our current *material flow economy* (Figure 3-15, p. 73) to a *service flow economy* over the next few decades. Instead of buying most goods outright, customers would lease or rent the *services* such goods provide.

With such a service flow or product stewardship economy, a product produced by a manufacturer remains as an asset that yields more profit if it **(1)** uses the minimum amount of materials, **(2)** lasts as long as possible, **(3)** is easy to maintain, repair, remanufacture, reuse, or recycle, and **(4)** provides customers with the services they want instead of trying to keep selling them newer models of outmoded products. This economic shift is under way:

- Since 1992 the Xerox Corporation has been leasing most of its copy machines as part of its mission to provide *document services* instead of selling photocopiers. The company replaces or upgrades copier cartridges and other parts in its copiers at no extra cost to the customer. When the service contract expires, Xerox takes the machine back for reuse or remanufacture and has a goal of sending no material to landfills or incinerators. To save money, machines are designed to **(1)** use recycled paper, **(2)** have few parts, **(3)** be energy efficient, and **(4)** emit as little noise, heat, ozone, and copier chemicals as possible.

- Ray Anderson, CEO of a large carpet tile company, plans to lease rather than sell carpet (Individuals Matter, right).

- For years, 160 firms, called *chauffagistes,* have been providing 10 million buildings in metropolitan France with heat. These firms provide *warmth services* by contracting to keep a client's space within a specified temperature during certain hours at a designated cost.

- Carrier, the world's leading maker of air-conditioning equipment, now sells leases to provide its customers with *cooling services.* Carrier also teams up with other service providers to install superwindows and more efficient lighting and make other energy-efficiency upgrades that reduce the cooling needs of its customers. Carrier makes money doing this by having to install less or even no air conditioning equipment.

- Dow and several other chemical companies are doing a booming business in leasing organic solvents (mostly used to remove grease from surfaces), photographic developing chemicals, and dyes and pigments. In this *chemical service* business, the company **(1)** delivers the chemicals, **(2)** helps the client set up a recovery system, **(3)** takes away the recovered chemicals, and **(4)** delivers new chemicals as needed.

15-4 REUSE

What Are the Advantages of Refillable Containers? *Reuse* is a form of waste reduction that **(1)** extends resource supplies, **(2)** keeps high-quality matter resources from being reduced to low-matter-quality waste (Figure 3-15, p. 73), and **(3)** reduces energy use and pollution even more than recycling.

Two examples of reuse are refillable glass beverage bottles and refillable soft drink bottles made of polyethylene terephthalate (PET) plastic. Unlike throwaway and recyclable cans and bottles, refillable beverage bottles create local jobs related to their collection and refilling. Moreover, studies by Coca-Cola and PepsiCo of Canada show that their soft drinks in 0.5-liter (16-ounce) bottles cost one-third less in refillable bottles than in throwaway bottles.

Denmark has led the way by banning all beverage containers that cannot be reused. To encourage use of refillable glass bottles, Ecuador has a refundable beverage container deposit fee that is 50% of the cost of the drink. In Finland, 95% of the soft drink, beer, wine, and spirits containers are refillable, and in Germany, 73% are refillable.

In 1964, 89% of all soft drinks and 50% of all beer in the United States was sold in refillable glass bottles. Today such bottles make up only about 7% of the beer and soft drink market, and only 10 states have refillable glass bottles. The disappearance of most local bottling companies has led to a loss of local jobs, income, and tax revenues. Some people call for reinstatement of this bottling reuse system in the United States; others say it is not practical because the system of collections and returns has been dismantled. What do you think?

Other examples of reusable items are

- Metal or plastic lunchboxes.

- Plastic containers for storing lunchbox items and refrigerator leftovers instead of using throwaway plastic wrap and aluminum foil.

- Cloth shopping bags (Solutions, p. 378).

- Shipping pallets made of recycled plastic waste instead of throwaway wood pallets. In 1991, Toyota shifted entirely to reusable shipping containers. A similar move by the Xerox Corporation saves the company \$2–5 million per year.

- *Tool libraries* (such as those in Berkeley, California, and Takoma Park, Maryland) where people can check out a variety of power and hand tools.

- E-paper, a flexible and cordless computer screen being developed by Xerox that **(1)** looks like a sheet of paper, **(2)** uses no energy for storing or viewing writing or images, and **(3)** can be elec-

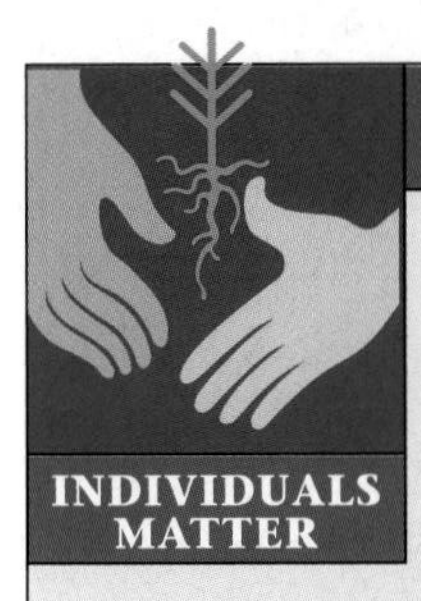

Ray Anderson

Ray Anderson is CEO of Interface, an Atlanta, Georgia–based company that makes carpet tiles. The company has 26 factories in 6 countries, customers in 110 countries, and more than $1 billion in annual sales.

Ray changed the way he viewed the world and his business after reading Paul Hawken's book *The Ecology of Commerce* (Guest Essay, p. 6). In 1994, he announced plans to develop the nation's first totally sustainable green corporation.

He has implemented hundreds of projects with the goals of **(1)** zero waste, **(2)** greatly reduced energy use, and **(3)** eventually zero use of fossil fuels by relying on renewable solar energy. By 1999, the company had reduced resource waste by almost 30% and reduced energy waste enough to save $100 million. One of Interface's factories in California is being run on solar cells to produce the world's first solar-made carpet.

To achieve the goal of zero waste, Anderson plans to stop selling carpet and lease it as a way to control recycling. For a monthly fee, the company will **(1)** install, clean, and inspect the carpet on a monthly basis, **(2)** repair worn carpet tiles overnight, and **(3)** recycle worn-out tiles into new carpeting. As Anderson puts it, "We want to harvest yesterday's carpets and recycle them with zero scrap going to the landfill and zero emissions into the ecosystem—and run the whole thing on sunlight."

DuPont and several other chemical companies have developed processes to remove the nylon and plastic PVC fibers in carpet and recycle it into other lower-quality products (open-loop recycling).

Interface has gone further and developed a new polymer material, called Solenium, that **(1)** when worn out can be completely recycled back into new carpet tiles (more desirable closed-loop recycling), **(2)** does not mildew, **(3)** is highly stain resistant, and **(4)** is easily cleaned with water. Making this material takes fewer steps and produces 99.7% less waste than making normal carpet, and the material lasts about four times longer than conventional carpet.

The company also has plans to install and lease a raised-floor system that goes beneath its carpet tiles and integrate this with cooling and heating services provided by other service companies.

Anderson is one of a growing number of business leaders committed to a finding a more economically and ecologically sustainable way to do business while still making a profit for stockholders. Since he instituted his environmentally friendly policies, the company's share price and earnings have increased, and he says he is having a blast.

tronically written and rewritten at least a million times, making it equivalent to more than a million sheets of paper.

15-5 RECYCLING

What Are the Two Types of Recycling? Recycling has a number of benefits to people and the environment (Figure 15-5). There are two types of recycling for materials such as glass, metals, paper, and plastics:

- *Primary*, or *closed-loop, recycling*, in which wastes discarded by consumers (*postconsumer wastes*) are recycled to produce new products of the same type (such as newspaper into newspaper, and aluminum cans into aluminum cans). This reduces pollution and use of virgin resources and saves energy.
- *Secondary*, or *open-loop, recycling*, in which waste materials are converted into different and usually lower-quality products.

Primary recycling reduces the amount of virgin materials in a product by 20–90%, whereas secondary recycling reduces virgin material by 25% at most. A material that is a good candidate for recycling is **(1)** easily isolated from other wastes, **(2)** available in large quantities in a fairly uniform form, and **(3)** valuable.

Environmentalists urge us not to be misled by labels claiming that paper and plastic bags or other items are recyclable. Just about anything is recyclable. What counts is **(1)** whether an item is actually recycled and **(2)** whether we complete the recycling loop by buying products using the maximum feasible content of postconsumer recycled materials.

Case Study: Recycling Municipal Solid Waste in the United States In 1999, about 28% of U.S. MSW was recycled or composted (Solutions, p. 379)—the highest rate of any industrialized country. The United States has more than 8,800 municipal curbside recycling programs serving 51% of the population.

These programs recycle **(1)** 98% of the steel used in cars, **(2)** 96% of car batteries, **(3)** 74% of aluminum cans, **(4)** 70% of lead, **(5)** 49% of wastepaper and paperboard, **(6)** 40% of yard waste, and **(7)** 27% of glass containers. Pilot studies in several U.S. communities show that 60–80% recycling and composting rates are possible.

What Kind of Grocery Bags Should We Use?

SOLUTIONS

When you're offered a choice between plastic or paper bags for your groceries, which should you choose? The answer is *neither*. Both are environmentally harmful, and the question of which is the more damaging has no clear-cut answer.

On one hand, plastic bags degrade slowly in landfills and can harm wildlife if swallowed, and producing them pollutes the environment. On the other hand, producing the brown paper bags used in most supermarkets uses trees and pollutes the air and water.

Overall, white or clear polyethylene plastic bags take less energy for manufacture and cause less damage to the environment than do paper bags not made mostly from recycled paper.

Instead of having to choose between paper and plastic bags, you can bring your own *reusable* canvas or string containers to the store, and save and reuse any paper or plastic bags you get. Using a reusable bag just five times displaces the pollution caused by the manufacture of the bag. To encourage people to bring their own reusable bags, stores in the Netherlands charge for paper or plastic bags.

Critical Thinking

1. Apply similar reasoning to determine what kind of cup (plastic, paper, or reusable) you should use whenever possible. How could you solve the problem of getting coffee or other beverages at fast-food places and at workplaces?

2. Do you believe that grocery stores should charge for paper or plastic bags and sell reusable bags to encourage reuse? Explain. How would you implement such a policy in all major grocery stores to provide an even economic playing field for all consumers?

Studies show that one of the best ways to encourage recycling is a *pay-as-you-throw* program that bases garbage collection charges on the amount of waste a household generates for disposal; materials sorted out for recycling are hauled away free. Currently, more than 2,800 communities in North America have such systems.

Is Centralized Recycling of Mixed Solid Waste the Answer? Large-scale recycling can be accomplished by collecting mixed urban waste and transporting it to centralized *materials-recovery facilities (MRFs)*. There, machines shred and automatically separate the mixed waste to recover valuable materials for

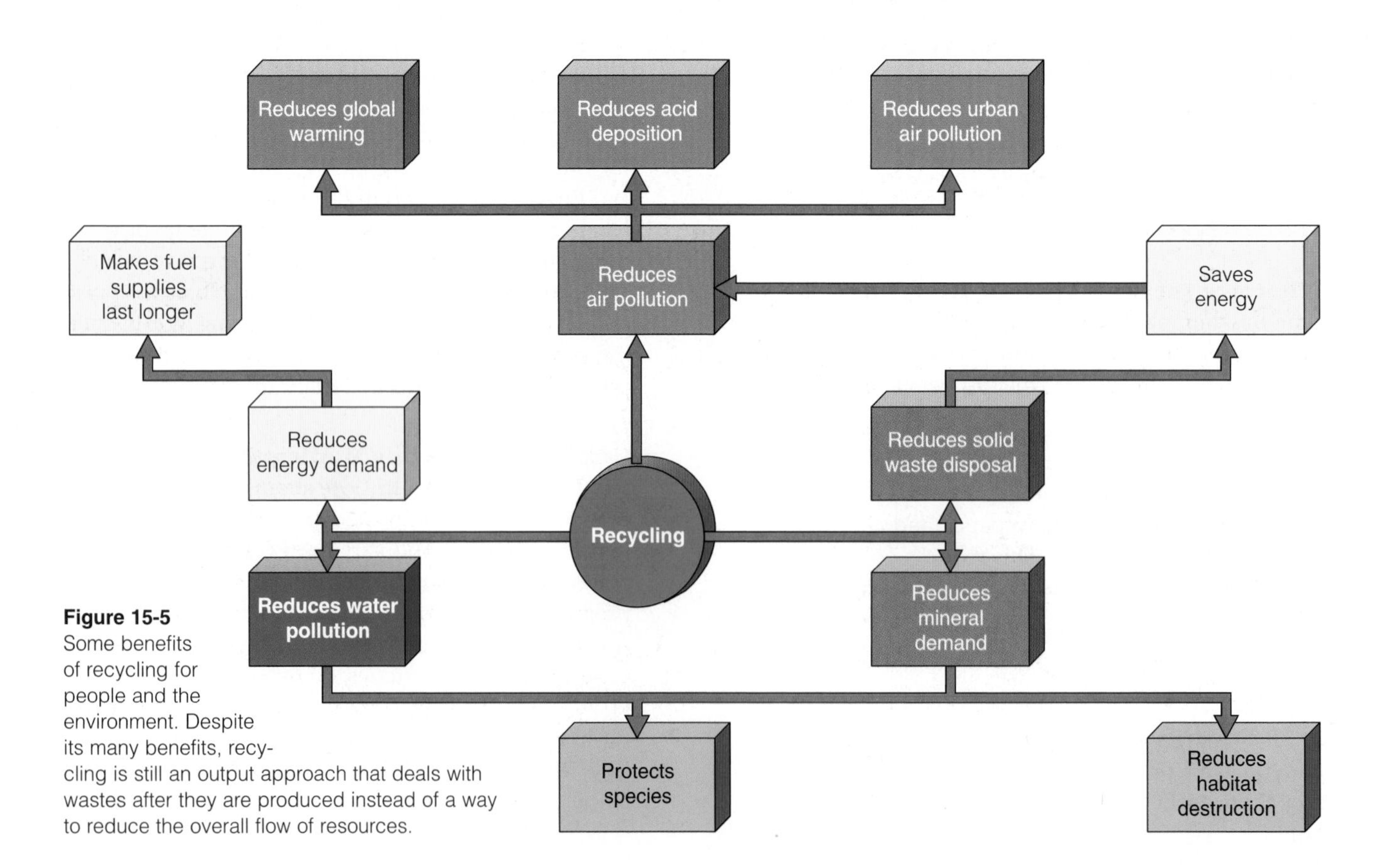

Figure 15-5 Some benefits of recycling for people and the environment. Despite its many benefits, recycling is still an output approach that deals with wastes after they are produced instead of a way to reduce the overall flow of resources.

Using Composting to Recycle Biodegradable Wastes

SOLUTIONS

Biodegradable organic waste such as paper, food scraps, and lawn waste is a valuable resource that can composted to produce plant nutrients that can be recycled to the soil. Biodegradable wastes make up about 35% by weight of the MSW output in the United States and could be converted to compost (p. 220).

However, only about 5% of the MSW in the United States is composted (compared to 17% in France and 10% in Switzerland). Some cities in Austria, Belgium, Denmark, Germany, Luxembourg, and Switzerland recover and compost more than 85% of their biodegradable wastes.

Individuals can compost such wastes in backyard bins or indoor containers. They can also be collected and composted in centralized community facilities, as is done in many western European countries.

The resulting compost can be used as **(1)** an organic soil fertilizer or conditioner, **(2)** topsoil, or **(3)** landfill cover. Compost also can be used to help restore eroded soil on hillsides and along highways, strip-mined land, overgrazed areas, and eroded cropland.

To be successful, a large-scale composting program must **(1)** overcome siting problems (few people want to live near a giant compost pile or plant), **(2)** control odors, and **(3)** exclude toxic materials that can contaminate the compost and make it unsuitable for use as a fertilizer on crops and lawns.

Three ways to control or reduce odors for large-scale composting operations are **(1)** enclosing the facilities and filtering the air inside (but residents near large composting plants still complain of unacceptable odors), **(2)** creating municipal compost operations near existing landfills or at other isolated sites, and **(3)** decomposing biodegradable wastes in a closed metal container in which air is recirculated to give precise control of available oxygen and temperature (a technique that has been used successfully in the Netherlands for 20 years).

Critical Thinking

How would you increase the rate of composting of biodegradable wastes in your community?

sale to manufacturers as raw materials (Figure 15-6). The remaining paper, plastics, and other combustible wastes are recycled or burned to produce steam or electricity to run the recovery plant or to sell to nearby industries or homes. Ash from the incinerator is buried in a landfill.

There are more than 225 MRFs in the United States. However, such plants **(1)** are expensive to build, operate, and maintain (which is why some have been shut down), **(2)** can emit toxic air pollutants if not operated properly, and **(3)** produce a toxic ash that must be disposed of safely. MRFs also must have a large input of garbage to make them financially successful. Thus, their owners have a vested interest in increasing *throughput* of matter and energy resources to produce more trash, the reverse of what prominent scientists believe we should be doing (Figure 15-3).

Is Separating Solid Wastes for Recycling the Answer? Many solid-waste experts argue that it makes more sense economically and environmentally for households and businesses to keep trash separate in recyclable and reusable categories (such as glass, paper, metals, certain types of plastics, and compostable materials). Then compartmentalized city collection trucks, private haulers, or volunteer recycling organizations pick up the segregated wastes and sell them to scrap dealers, compost plants, and manufacturers. Another alternative (especially in less populated areas) is to establish a network of drop-off centers, buy-back centers, and deposit-refund programs in which people deliver and sell or donate their separated recyclable materials.

This *source separation* approach **(1)** produces little air and water pollution, **(2)** has low startup costs and moderate operating costs, **(3)** saves more energy and provides more jobs per unit of material than MRFs, landfills, and incinerators, **(4)** yields cleaner and usually more valuable recyclables, and **(5)** educates people about the need for waste reduction, reuse, and recycling.

Aluminum and paper separated out for recycling are worth a lot of money. As a result, in a growing number of cities people steal these materials from curbside containers set out by residents and from unprotected recycling drop-off centers and sell them. This undermines municipal recycling programs by lowering the income available from selling these high-value materials.

Does Recycling Make Economic Sense? The answer is yes and no, depending on different ways of looking at the economic and environmental benefits and costs of recycling. Critics contend that recycling

- Has become almost a religion that is above criticism regardless of how much it costs communities
- Does not make sense if it costs more to recycle materials than to send them to a landfill or incinerator (as is the case in some areas)

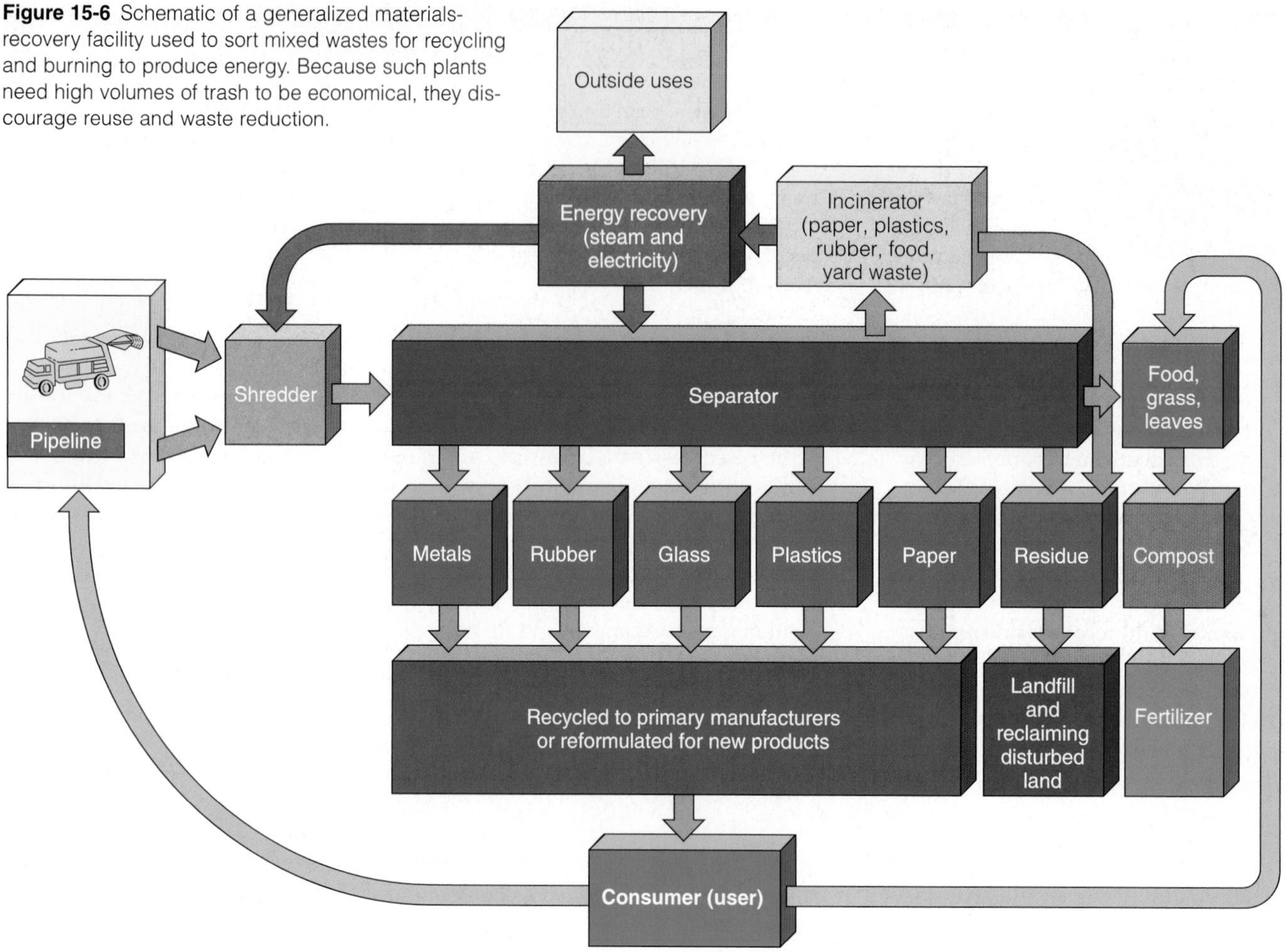

Figure 15-6 Schematic of a generalized materials-recovery facility used to sort mixed wastes for recycling and burning to produce energy. Because such plants need high volumes of trash to be economical, they discourage reuse and waste reduction.

- Often is not needed to save landfill space because many areas in the United States are not running out of landfill space
- May make economic sense for valuable and easy-to-recycle materials (such as aluminum, paper, and steel) but not for cheap or plentiful resources (such as glass from silica) and most plastics (which are expensive to recycle)

However, recycling proponents argue that

- Recycling programs should not be judged on whether they pay for themselves any more than are conventional garbage disposal systems based on land burial or incineration.
- The primary benefit of recycling is not a reduction in the use of landfills and incinerators but the other important benefits it provides for people and the environment (Figure 15-5).
- Studies show that the net economic, health, and environmental benefits of recycling (Figure 15-5) far outweigh the costs.
- Programs with **(1)** a single pickup system (for both materials to be recycled and garbage that cannot be recycled) instead of a more expensive dual collection system and **(2)** a pay-as-you-throw system tend to make money and have higher recycling rates.

Why Don't We Have More Reuse and Recycling? Three factors that hinder recycling (and reuse) are

- Failure to include the environmental and health costs of raw materials in the market prices of consumer items (p. 28)
- More tax breaks and subsidies for resource-extracting industries than for recycling and reuse industries
- Lack of large, steady markets for recycled materials

Analysts have suggested that these obstacles can be overcome by

- Taxing virgin resources and phasing out subsidies and tax breaks for extracting virgin resources
- Lowering or eliminating taxes on recycled materials based on postconsumer waste content
- Providing subsidies for reuse and postconsumer waste recycling
- Greatly increasing use of the *pay-as-you-throw* system

- Encouraging or requiring government purchases of recycled and reused products to help increase demand and lower prices
- Viewing landfilling and incineration of solid wastes as last resorts to be used only for wastes that cannot be reused, composted, or recycled (Figure 15-3)
- Requiring labels on all products listing preconsumer and postconsumer recycled content

15-6 CASE STUDIES: RECYCLING ALUMINUM, WASTEPAPER, AND PLASTICS

How Much Aluminum Is Being Recycled? The United States gets 60% of its aluminum from virgin ore in a process that uses enormous amounts of energy. Recycling the remaining 40% **(1)** produces 95% less air pollution and 97% less water pollution and **(2)** uses 95% less energy than mining and processing aluminum ore. In 2000, 74% of aluminum beverage cans produced in the United States were recycled. Despite this progress, if laid end-to-end the 26% of aluminum cans not recycled would wrap around the planet more than 120 times.

Recycling aluminum cans is great, but many environmentalists believe that these cans are an example of an unnecessary item that could be replaced by more energy-efficient and less polluting refillable glass or PET plastic bottles. One way to encourage this change would be to place a heavy tax on nonrefillable containers and no tax on reusable beverage containers, as is done in at least nine countries.

How Much Paper Is Being Recycled? Paper (especially newspaper and cardboard) is one of the easiest materials to recycle. Recycling paper involves removing its ink, glue, and coating and reconverting it to pulp that is pressed again into new paper. The perception that recycled paper is weak, coarse, flecked, and too expensive is no longer true. A variety of high-quality recycled papers (including the paper used in this book) are available to meet all types of printing demands at competitive prices.

In 2000, the United States recycled about 49% of its wastepaper (up from 25% in 1989) and 70% of its corrugated cardboard containers. At least 10 other countries recycle 50–96% of their wastepaper and paperboard, with a global recycling rate of 43%. Despite a 49% recycling rate, the amount of paper thrown away each year in the United States is more than all the paper consumed in China (where the recycling rate is only 27%).

Recycling paper **(1)** does not involve cutting new trees, **(2)** saves energy because it takes 30–64% less energy to produce the same weight of recycled paper as to make the paper from trees, **(3)** reduces air pollution from pulp mills by 74–95%, **(4)** lowers water pollution by 35%, **(5)** helps prevent groundwater contamination by toxic ink left after paper rots in landfills over a 30- to 60-year period, **(6)** conserves large quantities of water, **(7)** takes little or no bleaching because the fibers recycled from white paper have already been bleached, **(8)** can save landfill space, **(9)** creates five times more jobs than harvesting trees for pulp, and **(10)** can save money.

Chlorine (Cl_2) and chlorine compounds (such as chlorine dioxide, ClO_2) used to bleach paper **(1)** corrode processing equipment, **(2)** are hazardous for workers, **(3)** are hard to recover and reuse, and **(4)** are extremely harmful when released into the environment. A growing number of paper mills (mostly in Europe) are replacing chlorine-based bleaching chemicals with oxygen-based chemicals such as hydrogen peroxide (H_2O_2) or ozone (O_3). Such processes **(1)** nearly eliminate the release of air pollutants (including highly toxic chlorine-containing dioxin, p. 388), **(2)** use less water and energy, **(3)** allow reuse of the water many times, **(4)** reduce the treatment needed for water that is discharged, and **(5)** save money.

Buying recycled paper products can save trees and energy and reduce pollution, but it does not necessarily reduce solid waste. Only products made from *postconsumer waste*—waste intercepted on its way from consumer to the landfill or incinerator—does that.

Most recycled paper is made from *preconsumer waste*: scraps and cuttings recovered from paper and printing plants. Because paper manufacturers have always recycled this waste, it has never contributed to landfill problems. Now this paper is labeled "recycled" as a marketing ploy, giving the false impression that people who buy such products (often at higher prices) are helping to reduce solid waste. Most "recycled" paper has no more than 50% recycled fibers, with only 10% from postconsumer waste. Environmentalists propose that governments require companies to use labels giving postconsumer recycled content and indicating whether the paper was bleached with chlorine or by a chlorine-free process.

Is It Feasible to Recycle Plastics? Plastics are various types of polymer molecules made by chemically linking monomer molecules (petrochemicals) produced mostly from oil and natural gas.

Currently, only about 7% by weight of all plastic wastes and 10% of plastic containers in the United States are recycled, mostly because

- Plastics are very difficult to isolate from other wastes because they **(1)** occur in so many different and often difficult-to-identify forms of resins, **(2)** sometimes consist of composites or laminated layers of different plastics, and **(3)** contain stabilizers and other chemicals that must be removed before recycling.

- Recovering individual plastic resins does not yield much material because only small amounts of any given resin are used per product.
- The price of oil is so low (Figure 19-9, p. 497) that the cost of virgin plastic resins (except for PET, used mostly in plastic drink bottles) is about 40% lower than that of recycled resins.

Thus, mandating that plastic products contain a certain amount of recycled plastic resins is **(1)** unlikely to work and **(2)** could hinder the use of plastics in reducing the resource content and weight (Solutions, p. 374) of many widely used items such as plastic bags, bottles, and other containers.

Environmentalists recognize the beneficial qualities of plastics: **(1)** durability (in products such as car and machine parts, carpeting, toys, furniture, reusable tubs and containers, and refillable bottles), **(2)** light weight, **(3)** unbreakability (compared to glass), and **(4)** in some cases reusability as containers.

On the other hand,

- The plastics industry is among the leading producers of hazardous waste.
- In landfills, toxic cadmium and lead compounds used as binders, colorants, and heat stabilizers can leach out of plastics into groundwater and surface water.
- Most plastics are nondegradable or take 200–400 years to degrade.
- Some widespread uses of plastics—especially excessive and often unnecessary single-use packaging and throwaway beverage and food containers—could be sharply reduced and replaced with less harmful and less wasteful alternatives.

15-7 DETOXIFYING, BURNING, BURYING, AND EXPORTING WASTES

How Can Hazardous Waste Be Detoxified? Denmark has the world's most comprehensive and effective hazardous-waste detoxification program. Hazardous and toxic waste from industries and households is delivered to 21 transfer stations throughout the country. All waste is then transferred to a large treatment facility. There about 75% of the waste is detoxified, and the rest is buried in a carefully designed and monitored landfill.

Some consider biological treatment of hazardous waste, or *bioremediation*, to be the wave of the future for cleaning up some types of toxic and hazardous waste. In this process, microorganisms (usually natural or genetically engineered bacteria) and enzymes **(1)** destroy toxic or hazardous substances or **(2)** convert them to harmless compounds.

Bioremediation is effective for a number of organic wastes, including pesticides, gasoline, diesel fuel, polychlorinated biphenyls (PCBs), and organic solvents. However, it does not appear to work very well for **(1)** toxic metals, **(2)** highly concentrated chemical wastes, or **(3)** complete digestion of some complex mixtures of toxic chemicals.

Another biological way to treat hazardous wastes is *phytoremediation*, which involves using natural or genetically engineered plants to filter and remove contaminants. Selected plants can be used to clean up soil and water contaminated with chemicals such as pesticides, organic solvents, radioactive metals, and toxic metals such as lead and mercury. Plants are also used to remove contaminants in sewage treatment systems that work with nature (p. 362 and Figure 14-32, p. 363).

Phytoremediation **(1)** is inexpensive, **(2)** does not involve heavy machinery that produces air pollution, and **(3)** can reduce the amount of material dumped in landfills. On the other hand, **(1)** it is often slow (it can take several growing seasons to clean a site), **(2)** it is effective only at depths that plant roots can reach, and **(3)** in some cases animals may feed on pollutant-containing leaves.

Is Burning Solid and Hazardous Waste the Answer? In the United States, about 16% of the mixed trash in municipal solid waste is combusted in about 170 *mass-burn incinerators* (Figure 15-7). About 80% of the hazardous waste is burned in 172 commercial incinerators, cement kilns, and lightweight aggregate kilns. The other 20% is combusted in industrial boilers and other types of industrial furnaces. Figure 15-8 lists the pros and cons of using incinerators to burn solid and hazardous waste.

Since 1985, there has been a decrease in the use of incineration for treating wastes in some parts of the world because of high costs, health threats from air pollution, and intense citizen opposition. For example, **(1)** Sweden banned the construction of new incinerators in 1985, **(2)** Rhode Island and West Virginia banned solid-waste incineration in 1992, **(3)** several solid-waste incinerators in the United States have been shut down because of excessive costs and pollution, **(4)** more than 280 new incinerator projects have been blocked, delayed, or canceled in the United States since 1985, **(5)** in 1999, the Philippines became the first country to ban all waste incineration, followed by Costa Rica, and **(6)** a growing number of hospitals are destroying infectious material by using steam heat and pressure in

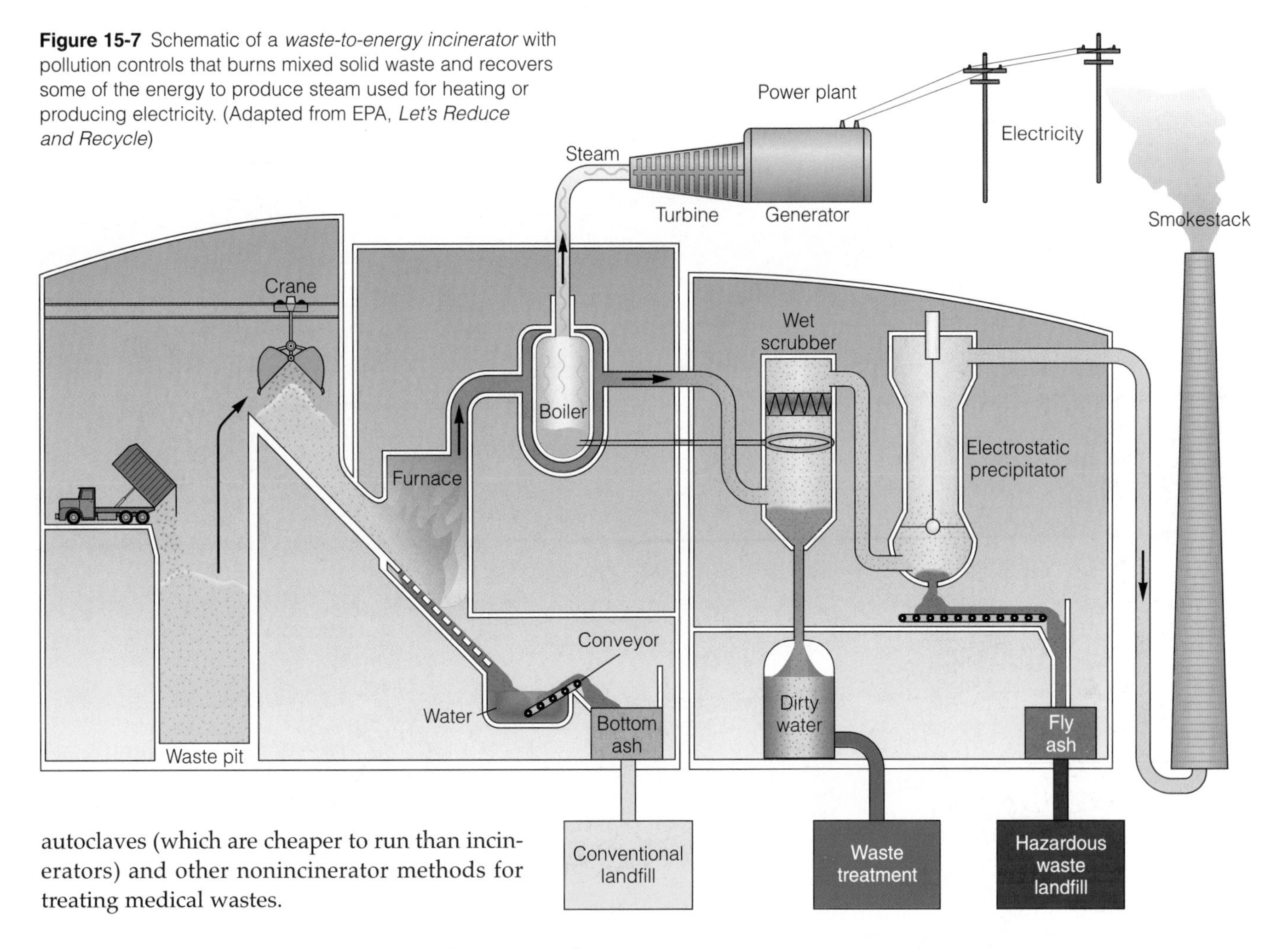

Figure 15-7 Schematic of a *waste-to-energy incinerator* with pollution controls that burns mixed solid waste and recovers some of the energy to produce steam used for heating or producing electricity. (Adapted from EPA, *Let's Reduce and Recycle*)

autoclaves (which are cheaper to run than incinerators) and other nonincinerator methods for treating medical wastes.

Is Land Disposal of Solid Waste the Answer? About 54% by weight of the MSW in the United States is buried in sanitary landfills (compared to 90% in the United Kingdom, 80% in Canada, 15% in Japan, and 12% in Switzerland). In a **sanitary landfill**, solid wastes are **(1)** spread out in thin layers, **(2)** compacted, and **(3)** covered daily with a fresh layer of clay or plastic foam.

Modern state-of-the-art landfills on geologically suitable sites are lined with clay and plastic before being filled with garbage (Figure 15-9). The bottom is covered with a second impermeable liner, usually made of several layers of clay, thick plastic, and sand. This liner collects *leachate* (rainwater contaminated as it percolates through the solid waste) and is intended to prevent its leakage into groundwater. Collected leachate is pumped from the bottom of the landfill, stored in tanks, and sent to a regular sewage treatment plant or an on-site treatment plant. When full, the landfill is covered with clay, sand, gravel, and topsoil to prevent water from seeping in. Several wells are drilled around the landfill to monitor any leakage of leachate into nearby groundwater. Figure 15-10 lists the pros and cons of using sanitary landfills for solid-waste disposal.

Since 1997, only state-of-the art landfills (Figure 15-9) are allowed to operate in the United States. As a result, many older and small landfills have been closed and replaced with larger local and regional modern landfills. Some cities in the eastern United States, notably Philadelphia and New York, are having trouble

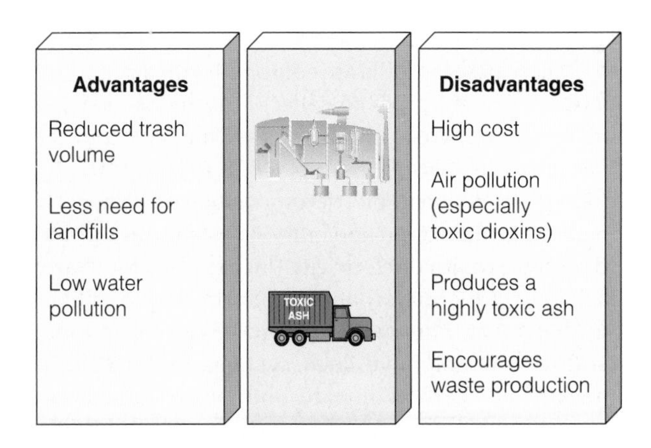

Advantages
Reduced trash volume
Less need for landfills
Low water pollution
TOXIC ASH
Disadvantages
High cost
Air pollution (especially toxic dioxins)
Produces a highly toxic ash
Encourages waste production

Figure 15-8 Advantages and disadvantages of incinerating solid and hazardous waste.

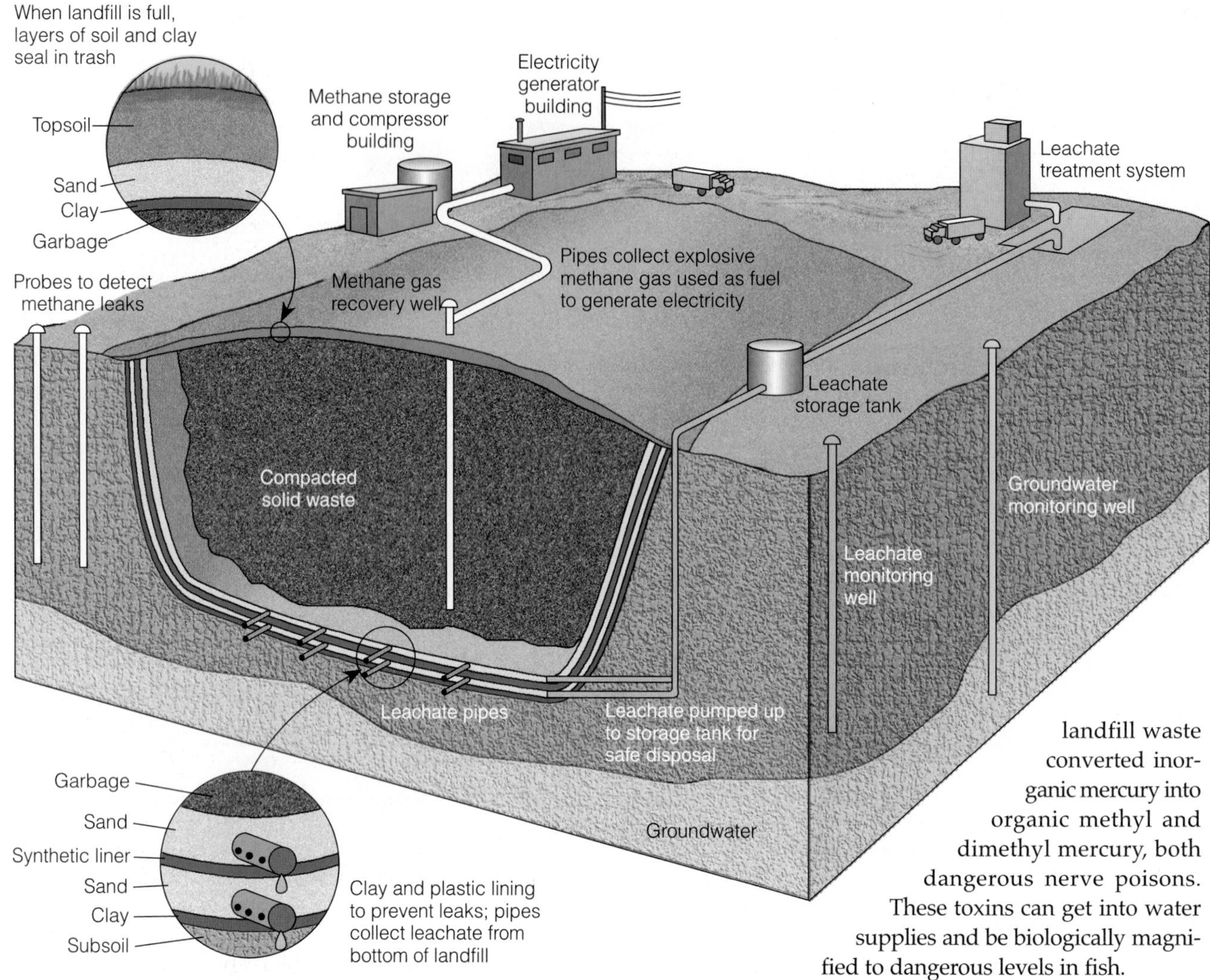

Figure 15-9 State-of-the-art sanitary landfills are designed to eliminate or minimize environmental problems that plague older landfills. Even such state-of-the-art landfills are expected to leak eventually, passing both the effects of contamination and cleanup costs on to future generations.

finding nearby landfills. However, there is no shortage of landfill space in most parts of the United States.

These new landfills are equipped with a connected network of vent pipes to collect landfill gas (consisting mostly of two greenhouse gases, methane and carbon dioxide; Table 13-1, p. 305) released by the underground (anaerobic) decomposition of wastes. The methane is filtered out and burned in small gas turbines to produce steam or electricity (Figure 15-9) for nearby facilities (Individuals Matter, p. 387) or sold to utilities.

However, thousands of older and abandoned landfills do not have such systems and will emit methane and carbon dioxide, both potent greenhouse gases, for decades. A 2001 study by Steve Linberg and his associates found very high levels of highly toxic methyl mercury and dimethyl mercury in gases emitted by landfills. Apparently bacteria in the landfill waste converted inorganic mercury into organic methyl and dimethyl mercury, both dangerous nerve poisons. These toxins can get into water supplies and be biologically magnified to dangerous levels in fish.

Contamination of groundwater and nearby surface water by leachate from unlined and lined older landfills is a serious problem. Some 86% of older U.S. landfills studied have contaminated groundwater, and a fifth of all Superfund hazardous-waste sites (p. 389) are former municipal landfills that will cost billions of dollars to clean up.

Modern double-lined landfills (Figure 15-9) delay the release of toxic leachate into groundwater below landfills but do not prevent it. These landfills are designed to accept waste for 10–40 years, and current EPA regulations require owners to maintain and monitor landfills for at least 30 years after they are closed. However, they could begin to leak after this period, passing the health risks and costs of contamination to future generations.

According to G. Fred Lee (an experienced landfill consultant) and Ann Christy (a researcher at Ohio State University), the best solution to the leachate problem is to **(1)** apply clean water to landfills continuously and **(2)** collect and treat the resulting leachate in carefully designed and monitored facilities. They contend that after 10–20 years of such washing, little potential for groundwater pollution should remain. This wetting

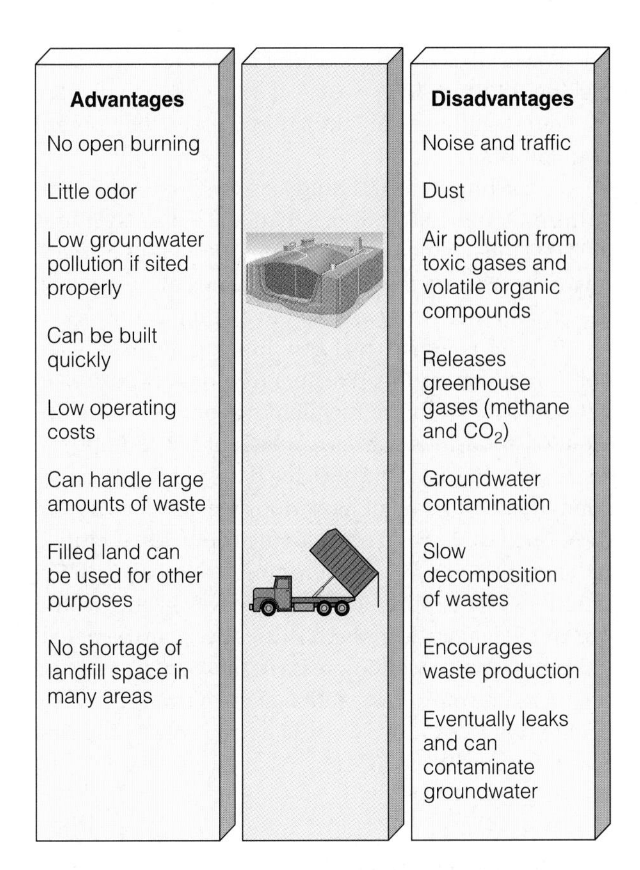

Advantages
No open burning
Little odor
Low groundwater pollution if sited properly
Can be built quickly
Low operating costs
Can handle large amounts of waste
Filled land can be used for other purposes
No shortage of landfill space in many areas
Disadvantages
Noise and traffic
Dust
Air pollution from toxic gases and volatile organic compounds
Releases greenhouse gases (methane and CO_2)
Groundwater contamination
Slow decomposition of wastes
Encourages waste production
Eventually leaks and can contaminate groundwater

Figure 15-10 Advantages and disadvantages of using sanitary landfills to dispose of solid waste.

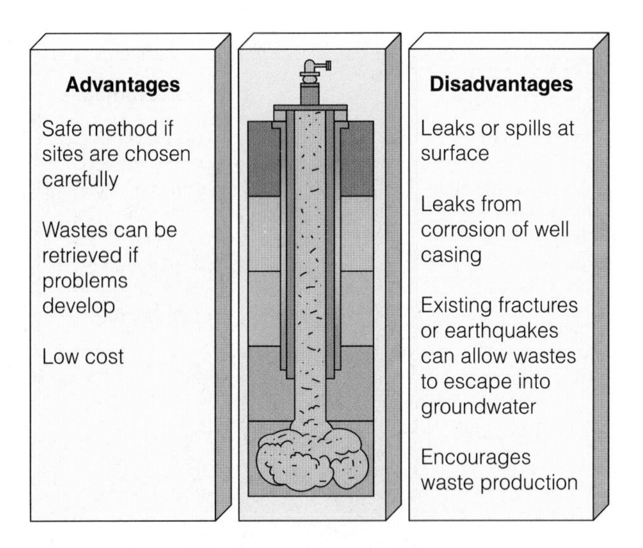

Advantages
Safe method if sites are chosen carefully
Wastes can be retrieved if problems develop
Low cost
Disadvantages
Leaks or spills at surface
Leaks from corrosion of well casing
Existing fractures or earthquakes can allow wastes to escape into groundwater
Encourages waste production

Figure 15-11 Advantages and disadvantages of injecting liquid hazardous wastes into deep underground wells.

would also **(1)** hasten the breakdown of wastes to 5–10 years instead of 100 years or longer, **(2)** provide more room for more trash in the same landfill, and **(3)** allow old landfills to be dug out and used again.

Is Land Disposal of Hazardous Waste the Answer? Hazardous waste in the United States is disposed of on land in **(1)** deep underground wells, **(2)** surface impoundments such as ponds, pits, or lagoons, **(3)** state-of-the-art landfills, and **(4)** aboveground storage facilities.

In *deep-well disposal,* liquid hazardous wastes (such as cleaning solutions, metals, cyanides, and corrosive solutions) are pumped under pressure through a pipe into dry, porous geologic formations or zones of rock far beneath aquifers tapped for drinking and irrigation water (Figure 14-25, p. 353). In theory, these liquids soak into the porous rock material and are isolated from overlying groundwater by essentially impermeable layers of rock. Figure 15-11 lists the pros and cons of deep-well disposal of liquid hazardous wastes. Many scientists believe that current regulations for deep-well disposal are inadequate and should be improved.

Surface impoundments are excavated depressions such as ponds, pits, or lagoons into which liquid hazardous wastes are drained and stored (Figure 14-25, p. 353). As water evaporates, the waste settles and becomes more concentrated. Figure 15-12 lists the pros and cons of this method. EPA studies found that 70% of these storage basins in the United States have no liners, and as many as 90% may threaten groundwater. According to the EPA, all liners are likely to leak eventually and can contaminate groundwater.

Liquid and solid hazardous waste also can be put into drums or other containers and buried in carefully designed and monitored *secure hazardous-waste landfills* (Figure 15-13). Sweden goes further and buries its concentrated hazardous wastes in underground vaults made of reinforced concrete. By contrast, in the United Kingdom, most hazardous wastes are mixed with household garbage and stored in hundreds of conventional landfills throughout the country.

With proper design and maintenance, secure landfills can effectively store hazardous wastes. However, **(1)** this method is expensive, **(2)** burrowing animals may make holes in the clay cap, **(3)** the liners are likely to leak eventually, and **(4)** this method discourages hazardous waste reduction.

Hazardous wastes also can be stored in carefully designed *aboveground buildings.* These two-story buildings **(1)** are built of reinforced concrete to prevent damage by storms and hurricanes and to help contain any leakage and **(2)** use fans and filters to create a negative air pressure to prevent the release of toxic gases.

The first floor contains no wastes but has inspection walkways so people can easily check for leaks from the upper story. Any leakage is collected, treated, solidified, and returned to the storage building. Earthquakes could damage such structures, but this is also a potential problem for all other methods for storing hazardous wastes.

Each year there are more than 500,000 shipments of hazardous wastes (mostly to landfills and incinerators in trucks or by train) in the United States. On average,

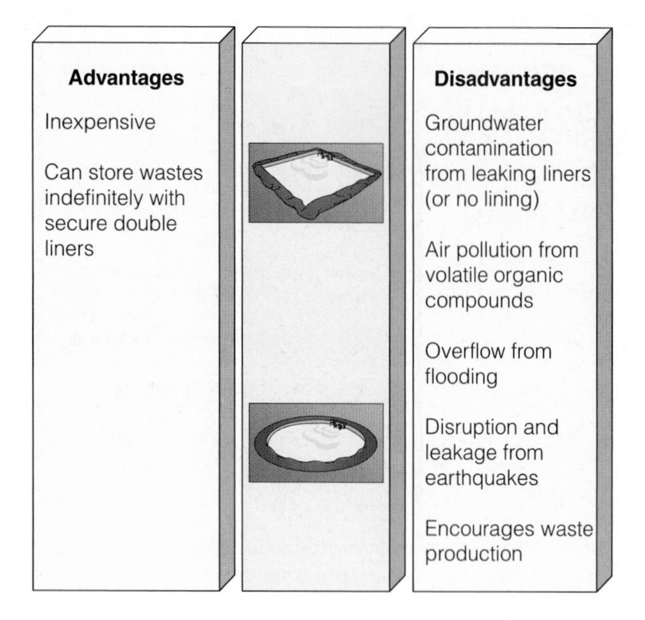

Advantages
Inexpensive
Can store wastes indefinitely with secure double liners
Disadvantages
Groundwater contamination from leaking liners (or no lining)
Air pollution from volatile organic compounds
Overflow from flooding
Disruption and leakage from earthquakes
Encourages waste production

Figure 15-12 Advantages and disadvantages of storing liquid hazardous wastes in surface impoundments.

trucks and trains carrying hazardous materials are involved in about 13,000 accidents per year in the United States. In a typical year, these accidents **(1)** kill about 100 people, **(2)** cause more than 10,000 injuries, and **(3)** necessitate evacuation of more than 500,000 people. Most communities do not have the equipment and trained personnel to deal with hazardous-waste spills.

Is Exporting Hazardous Waste the Answer? Some hazardous-waste producers in the United States and several other industrialized countries have been getting rid of some of these wastes by legally (or illegally) shipping them to other countries, especially developing countries.

Figure 15-13 Secure hazardous-waste landfill.

Gas vent
Topsoil
Earth
Sand
Plastic cover
Impervious clay cap
Clay cap
Leak detection system
Bulk waste
Reactive wastes in drums
Impervious clay
Earth
Groundwater monitering well
Water table
Groundwater
Double leachate collection system
Plastic double liner

Waste disposal firms can charge high prices for picking up hazardous wastes. If they can then dispose of them (legally or illegally) at low costs, they pocket huge profits.

According to a 2001 study by the Food and Agriculture Organization, more than 450,000 metric tons (500,000 tons) of banned or expired pesticides are seriously threatening the health of millions of people and the environment in nearly all developing countries.

In 1989, countries met and drew up the Basel Convention on Hazardous Waste. It requires exporters to get approval from the recipient nation before a shipment of hazardous wastes can be sent but did not ban trade in such wastes. In 1995, the Basel Convention was strengthened to ban all hazardous waste exports from developed countries to developing countries. If ratified by enough countries and enforced, this ban on hazardous-waste exports will help. However, this would not end illegal trade in these wastes because the potential profits are much too great. To most environmental scientists, the only real solution to the hazardous-waste problem is to produce as little as possible in the first place (Figure 15-4, p. 371).

15-8 CASE STUDIES: LEAD, CHLORINE, AND DIOXINS

How Can We Reduce Exposure to Lead? Each year, 12,000–16,000 American children under age 9 are treated for acute lead poisoning, and about 200 die. About 30% of the survivors suffer from palsy, partial paralysis, blindness, and mental retardation.

Lead can also cause damage at levels far below those that cause acute lead poisoning, especially in children and unborn fetuses. Research indicates that children under age 6 and unborn fetuses with fairly low blood levels of lead are especially vulnerable to **(1)** nervous system impairment, **(2)** a lowered IQ (by 4–7 points), **(3)** a shortened attention span, **(4)** hyperactivity, **(5)** hearing damage, and **(6)** various behavior disorders.

It is *good news* that between 1976 and 1999 the percentage of U.S. children ages 1–5 with lead levels above the current blood level standard of5 micrograms per deciliter (5 μg/dl) dropped from 85% to 4%, preventing at least 9 million childhood lead poisonings. The primary reason is that goverment regulations banned **(1)** leaded gasoline in 1976 (with a gradual phaseout by 1986) and **(2)** lead-based paints in 1970 (but illegal use continued until about 1978).

GUEST ESSAY

Environmental Justice for All

Robert D. Bullard

Robert D. Bullard is professor of sociology and director of the Environmental Justice Resource Center at Clark Atlanta University. For more than a decade, he has conducted research in the areas of urban land use, housing, community development, industrial facility siting, and environmental justice. He is the author of four books and more than three dozen articles, monographs, and scholarly papers that address concerns about environmental justice. His book Dumping in Dixie: Race, Class, and Environmental Quality, 2d ed. *(Westview Press, 1994), has become a standard text in the field. Other books are* Confronting Environmental Racism *(South End Press, 1993) and* Unequal Protection: Environmental Justice and Communities of Color *(Sierra Club Books, 1994).*

Despite widespread media coverage and volumes written on the U.S. environmental movement, environmentalism and social justice have seldom been linked. Nevertheless, an environmental revolution has been taking shape in the United States that combines the environmental and social justice movements into one framework.

People of color (African-Americans, Latinos, Asians, Pacific Islanders, and Native Americans), working-class people, and poor people in the United States suffer disproportionately from industrial toxins, dirty air and drinking water, unsafe work conditions, and the location of noxious facilities such as municipal landfills, incinerators, and toxic-waste dumps.

The *environmental justice* movement attempts to dismantle **(1)** exclusionary zoning ordinances, **(2)** discriminatory land-use practices, **(3)** differential enforcement of environmental regulations, **(4)** disparate siting of risky technologies, and **(5)** the dumping of toxic waste on the poor and people of color in the United States and in developing countries.

Despite the government's attempts to level the playing field, all communities are not created equal when it comes to resolving environmental and public health concerns. More than 300,000 farm workers (more than 90% of whom are people of color) and their children are poisoned by pesticides sprayed on crops in the United States. Some 3–4 million children (many of them African-Americans or Latinos living in the inner city) are poisoned by lead-based paint in old buildings, lead-soldered pipes and water mains, lead-tainted soil contaminated by industry, and air pollutants from smelters.

All communities do not bear the same burden or reap the same benefits from industrial expansion. This is true in the case of the mostly African-American Emelle, Alabama (home of the nation's largest hazardous-waste landfill), Navajo lands in Arizona where uranium is mined, and the 2,000 factories, known as *maquiladores*, located just across the U.S. border in Mexico.

Nationally, 60% of African-Americans and 50% of Latinos live in communities with at least one uncontrolled toxic-waste site. Three of the five largest hazardous-waste landfills are located in communities that are predominantly African-American or Latino.

Environmental justice does not stop at the U.S. border. Environmental injustices exist from the *favelas* of Rio de Janeiro, Brazil, to the shantytowns of Johannesburg, South Africa. Members of the environmental justice movement are also questioning the wasteful and nonsustainable development models being exported to the developing world.

Grassroots leaders are demanding justice. Residents of communities such as West Dallas and Texarkana (Texas), West Harlem (New York), Rosebud (South Dakota), Kettleman City (California), and Sunrise, Lions, and Wallace (Louisiana) see their struggle for environmental justice as a life-and-death matter. Unfortunately, their stories of environmental racism are not broadcast into the nation's living rooms during the nightly news, nor are they splashed across the front pages of national newspapers and magazines. To a large extent, the communities that are the victims of environmental injustice remain invisible to the larger society.

The environmental justice movement is led, planned, and to a large extent funded by people who are not part of the established environmental community or the "Big 10" environmental organizations. Most environmental justice groups are small and operate with resources generated from the local community.

For too long these groups and their leaders have been invisible and their stories muted. This is changing as these grassroots groups are forcing their issues onto the nation's environmental agenda.

The United States has a long way to go in achieving environmental justice for all its citizens. The membership of decision-making boards and commissions still does not reflect the racial, ethnic, and cultural diversity of the country, and token inclusion of people of color on boards and commissions does not necessarily mean that their voices will be heard or their cultures respected. The ultimate goal of any inclusion strategy should be to democratize the decision-making process and empower disenfranchised people to speak and do for themselves.

Critical Thinking

1. How would you define environmental injustice? Can you identify any examples of environmental injustice in your community?

2. Have you been a victim of environmental injustice? Compare your answers with those of other members of your class.

Most members of such groups recognize that health risks from incinerators and landfills, when averaged over the entire country, are quite low. However, they also know that the risks for the people near these facilities are much higher. These people, not the rest of the population, are the ones whose health, lives, and property values are being threatened.

Manufacturers and waste industry officials point out that something must be done with the toxic and hazardous wastes produced to provide people with certain goods and services. They contend that if local citizens adopt a "not in my backyard" (NIMBY) approach, the waste still ends up in someone's backyard.

Many citizens do not accept this argument. To them, the best way to deal with most toxic or hazardous wastes is to produce much less of them, as suggested by the National Academy of Sciences (Figure 15-4, p. 371, and Guest Essay, p. 372). For such materials, their goal is "not in anyone's backyard" (NIABY) or "not on planet Earth" (NOPE) by emphasizing pollution prevention and use of the precautionary principle.

What Can Be Done at the International Level? Between 1989 and 1994, an international treaty to limit transfer of hazardous waste from one country to another was developed (p. 386). In 2000, delegates from 122 countries completed a global treaty to control 12 *persistent organic pollutants* (POPs), which will go into effect when ratified by 50 countries.

These widely used toxic chemicals are insoluble in water and soluble in fatty tissues. This allows them to be concentrated in the fatty tissues of humans and other organisms feeding at high trophic levels in food webs to levels hundreds of thousand times higher than in the general environment (Figure 10-4, p. 227, and Figure 14-22, p. 349). They can also be transported long distances by wind and water.

The list of 12 chemicals, called the *dirty dozen*, includes DDT and 8 other chlorine-containing persistent pesticides, PCBs, dioxins, and furans. The goals of the treaty are to **(1)** ban or phase out use of these chemicals and **(2)** detoxify or isolate stockpiles of such chemicals in warehouses and dumps. About 25 countries will still be allowed to use DDT to combat malaria until safer alternatives are available. Developed nations will provide developing nations about $150 million per year to help them switch to safer alternatives for the 12 POPs.

How Can We Make the Transition to a Low-Waste Society? According to physicist Albert Einstein, "A clever person solves a problem, a wise person avoids it." To prevent pollution and reduce waste, many environmental scientists urge us to understand and live by four key principles: **(1)** Everything is connected, **(2)** there is no "away" for the wastes we produce, **(3)** dilution is not always the solution to pollution, and **(4)** the best and cheapest way to deal with waste and pollution is to produce less of them and then reuse and recycle most of the materials we use (Figures 15-3 and 15-4).

Visible signs of **(1)** *cleaner production* (p. 373 and Guest Essay, p. 375), **(2)** increased *resource productivity* (Solutions, p. 374), and **(3)** *service flow* businesses (p. 374) are emerging. Ecoindustrial parks (p. 373) are being built and planned.

Such revolutions start off slowly but can accelerate rapidly as their economic, ecological, and health advantages become more apparent to investors, business leaders, elected officials, and citizens. See the website material for this chapter for some actions you can take to reduce your production of solid waste and hazardous waste.

Nearly all environmental and social harm is caused by the uneconomically wasteful use of human and natural resources.

PAUL HAWKEN

REVIEW QUESTIONS

1. Define the boldfaced terms in this chapter.

2. Summarize what happened at Love Canal and the lessons learned from this environmental problem.

3. Distinguish between *solid waste* and *municipal solid waste*. What are the major sources of solid waste in the United States? Give five examples of solid waste thrown away in the United States.

4. What is *hazardous waste*? Hazardous-waste laws regulate what percentage of the overall hazardous waste produced in the United States?

5. Distinguish between the *high-waste* and *low-waste* approaches to solid and hazardous waste management. According to the U.S. National Academy of Sciences, what should be the five goals of solid and hazardous waste management in order of their importance? According to scientists, what percentage of solid and hazardous waste produced could be eliminated through a combination of waste reduction, reuse, and recycling (including composting)?

6. What does Lois Gibbs (p. 372) think we do with solid and hazardous waste, and what major questions does she believe we should ask about such wastes?

7. List seven ways to reduce waste and pollution. What is *resource productivity*? List five ways in which resource productivity has been improved. By how much do some experts think we can improve resource productivity? List four advantages of greatly increasing resource productivity.

8. What is *cleaner production*? Describe the resource exchange system used in Denmark and the pollution prevention program implemented by the Minnesota Mining and Manufacturing (3M) company in the United States. List five economic benefits of cleaner production.

9. Explain why Peter Montague (p. 375) believes that the regulation and standards approach used to control pol-

lution during the past 31 years has failed. Describe what he thinks should replace this approach.

10. What is a *service flow economy*? What are four economic advantages of such an economy for businesses and consumers? List four examples of how a service economy is being implemented. Describe how Ray Anderson (p. 377) is developing a carpet tile service economy business.

11. What is *reuse*, and what are three advantages of using this approach for waste reduction? List five examples of reuse.

12. Distinguish between *primary (closed-loop)* and *secondary (open-loop) recycling*. What is *compost*, and how is it used as a way to deal with solid waste? About what percentage of the municipal solid waste produced in the United States is recycled and composted, and what percentage do experts believe could be recycled and composted? What is a *pay-as-you-throw* garbage collection program, and how does it encourage recycling?

13. Distinguish between the *centralized recycling* of mixed solid waste and *consumer separation* of solid waste, and list the pros and cons of each approach to recycling.

14. List the pros and cons of recycling. What three factors hinder recycling and reuse? List seven ways to encourage more recycling and reuse.

15. Summarize the recycling of **(a)** aluminum, **(b)** wastepaper, and **(c)** plastics.

16. Describe how paper is recycled. List nine benefits from recycling paper. Distinguish between *preconsumer* and *postconsumer paper waste*.

17. List three reasons why so few plastics are recycled. List four advantages and four disadvantages of using plastics.

18. Describe Denmark's hazardous-waste detoxification program. What are *bioremediation* and *phytoremediation*?

19. Describe the major components of a *mass-burn incinerator* and a *sanitary landfill*. List the pros and cons of dealing with solid and hazardous waste by **(a)** burning it in incinerators and **(b)** burying it in sanitary landfills.

20. List the pros and cons of storing hazardous wastes in **(a)** deep underground wells, **(b)** surface impoundments, **(c)** secure landfills, and **(d)** aboveground buildings. Describe what is being done at the international level about the exporting of hazardous wastes from one country to another.

21. Describe the hazards of lead exposure and list nine ways to reduce such exposure.

22. What are three major problems with many chlorine-containing compounds? What are the three major uses of chlorine? How can we reduce exposure to harmful chlorine-containing compounds?

23. What are *dioxins*? How are they produced, what harm can they cause, and how can we reduce exposure to these hazardous chemicals?

24. How is the Resource Conservation and Recovery Act used to deal with the problem of hazardous wastes in the United States?

25. What is the Superfund Act, and what are its strengths and weaknesses?

26. What are *brownfields*, and what has been done to help redevelop such sites in the United States?

27. Describe international efforts to control use of 12 persistent organic pollutants (POPs).

28. Summarize Robert Bullard's views about environmental justice (p. 391).

29. List four principles that can be used as guidelines for making the transition to a low-waste society.

CRITICAL THINKING

1. Explain why you support or oppose requiring that **(a)** all beverage containers be reusable, **(b)** all households and businesses put recyclable materials into separate containers for curbside pickup, **(c)** garbage-collecting systems implement the pay-as-you-throw approach, and **(d)** consumers pay for plastic or paper bags at grocery and other stores to encourage the use of reusable shopping bags.

2. Use the second law of energy (p. 71) to explain why a *properly designed* source-separation recycling program takes less energy and produces less pollution than a centralized program that collects mixed waste over a large area and hauls it to a centralized facility where machinery or workers separate the wastes for recycling.

3. What short- and long-term disadvantages (if any) might an ecoindustrial revolution based on cleaner production (p. 373) bring? Do you believe that it will be possible to phase in such a revolution in the country where you live over the next two to three decades? Explain. What are the three most important strategies for doing this?

4. What short- and long-term disadvantages (if any) might there be in shifting to a service flow economy (p. 374)? Do you believe that it will be possible to phase in such a shift in the country where you live over the next two to three decades? Explain. What are the three most important strategies for doing this?

5. Would you oppose having a hazardous-waste landfill, waste treatment plant, deep-injection well, or incinerator in your community? Explain. If you oppose these disposal facilities, how do you believe the hazardous waste generated in your community and your state should be managed?

6. Give your reasons for agreeing or disagreeing with each of the following proposals for dealing with hazardous waste:

a. Reducing the production of hazardous waste and encouraging recycling and reuse of hazardous materials by charging producers a tax or fee for each unit of waste generated
b. Banning all land disposal and incineration of hazardous waste to encourage recycling, reuse, and treatment and to protect air, water, and soil from contamination
c. Providing low-interest loans, tax breaks, and other financial incentives to encourage industries producing hazardous waste to reduce, recycle, reuse, treat, and decompose such waste

d. Banning the shipment of hazardous waste from one country to another

7. Congratulations! You have just been put in charge of bringing about a cleaner production, resource productivity, and service flow economic revolution throughout the world over the next 20 years. List the three most important components of your strategy.

PROJECTS

1. For 1 week, keep a list of the solid waste you throw away. What percentage of this waste consists of materials that could be recycled, reused, or burned for energy? What percentage of the items could you have done without in the first place? Tally and compare the results for your entire class.

2. What percentage of the municipal solid waste in your community is **(a)** landfilled, **(b)** incinerated, **(c)** composted, and **(d)** recycled? What technology is used in local landfills and incinerators? What leakage and pollution problems have local landfills or incinerators had? Does your community have a recycling program? Is it voluntary or mandatory? Does it have curbside collection? Drop-off centers? Buyback centers?

3. What hazardous wastes are produced **(a)** at your school and **(b)** in your community? What happens to these wastes?

4. Make a concept map of this chapter's major ideas, using the section heads and subheads and the key terms (in boldface). Look on the website for this book for information about making concept maps.

INTERNET STUDY RESOURCES AND RESOURCES FOR FURTHER READING AND RESEARCH

The website for this book contains helpful study aids and many ideas for further reading and research. Log on to

http://www.brookscole.com/product/0534389872s

and click on the Chapter-by-Chapter area. Choose Chapter 15 and select a resource:

- "Flash Cards" allows you to test your mastery of the Terms and Concepts to Remember for this chapter.
- "Tutorial Quizzes" provides a multiple-choice practice quiz.
- "Student Guide to InfoTrac" will lead you to Critical Thinking Projects that use InfoTrac College Edition as a research tool.
- "References" lists the major books and articles consulted in writing this chapter.
- "Hypercontents" takes you to an extensive list of sites with news, research, and images related to individual sections of the chapter.

INFOTRAC COLLEGE EDITION

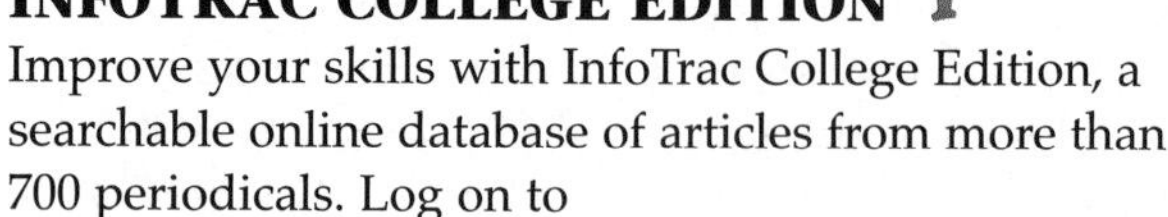

Improve your skills with InfoTrac College Edition, a searchable online database of articles from more than 700 periodicals. Log on to

http://www.infotrac-college.com

or access InfoTrac through the website for this book. Try to find the following articles:

1. Acute Respiratory Situation from Landfill Emissions and Air Pollution. *Life Sciences & Biotechnology Update*, May 2001 v2001 i05 pNA. Discussion of the air pollution effects of landfill emissions. *Hint*: Enter the search term "landfill emissions" using the keyword "pollution."

2. In the Future, People Like Me Will Go to Jail. Eileen Gunn. *Fortune* May 24, 1999 v139 i10 p190+ (1). Ray Anderson, co-chair of the President's Council on Sustainable Development and a "born-again green industrialist" from Georgia, is trying to sell sustainability to his fellow CEOs. *Hint*: Enter the search term "future people" using the keyword "sustainability."

Part IV

SUSTAINING BIODIVERSITY

It is the responsibility of all who are alive today to accept the trusteeship of wildlife and to hand on to posterity, as a source of wonder and interest, knowledge, and enjoyment, the entire wealth of diverse animals and plants. This generation has no right by selfishness, wanton or intentional destruction, or neglect, to rob future generations of this rich heritage. Extermination of other creatures is a disgrace to humankind.

World Wildlife Charter

16 FOOD RESOURCES

Perennial Crops on the Kansas Prairie

When you think about farms in Kansas, you probably picture seemingly endless fields of wheat or corn plowed up and planted each year. By 2040 the picture might change, thanks to pioneering research at the nonprofit Land Institute near Salina, Kansas.

The institute, headed by plant geneticist Wes Jackson, is experimenting with an ecological approach to agriculture on the midwestern prairie. This approach relies on planting a mixture of different crops planted in the same area, a technique called *polyculture*. The goal is to grow food crops by planting a mix of **(1)** perennial grasses (Figure 16-1, right), **(2)** legumes (a source of nitrogen fertilizer; Figure 16-1, left), **(3)** sunflowers, **(4)** grain crops, and **(5)** plants that provide natural insecticides in the same field.

The institute's goal is to raise food by mimicking many of the natural conditions of the prairie without losing fertile grassland soil (Figure 9-20, top right, p. 213). Institute researchers believe that *perennial polyculture* can be blended with *modern monoculture* to reduce its harmful environmental effects.

Because these plants are perennials, the soil does not have to be plowed up and prepared each year to replant them. This takes much less labor than conventional monoculture or diversified organic farms that grow annual crops. It also reduces **(1)** soil erosion because the unplowed soil is not exposed to wind and rain, **(2)** pollution caused by chemical fertilizers and pesticides, and **(3)** the need for irrigation because the deep roots of such perennials retain more water than annuals.

Thirty years of research by the institute have shown that various mixtures of perennials grown in parts of the midwestern prairie could be used as important food sources. One such mix of perennial crops includes **(1)** *eastern gamma grass* (a warm season grass that is a relative of corn with three times as much protein as corn and twice as much as wheat; Figure 16-1, right), **(2)** *mammoth wildrye* (a cool season grass that is distantly related to rye, wheat, and barley), **(3)** *Illinois bundleflower* (a wild nitrogen-producing legume that can enrich the soil and whose seeds can serve as livestock feed; Figure 16-1, left), and **(4)** *Maximilian sunflower* (which produces seeds with as much protein as soybeans).

These discoveries may help the growing human population produce and distribute enough food to meet everyone's basic nutritional needs without degrading the soil, water, air, and biodiversity that support all food production.

Figure 16-1 The Land Institute in Salina, Kansas, is a farm, a prairie laboratory, and a school dedicated to changing the way we grow food. It advocates growing a diverse mixture of edible perennial plants to supplement traditional annual monoculture crops. Two of these perennial crops are **(1)** eastern gamma grass (right) and **(2)** the Illinois bundleflower (left).

There are two spiritual dangers in not owning a farm. One is the danger of supposing that breakfast comes from the grocery, and the other that heat comes from the furnace.

ALDO LEOPOLD

This chapter addresses the following questions:

- How is the world's food produced?
- How much has food production increased, how serious is malnutrition, and what are the environmental effects of producing food?
- How can we increase production of **(a)** crops, **(b)** meat, and **(c)** fish and shellfish?
- How do government policies affect food production?
- What are pesticides, and what are the pros and cons of using chemicals to kill insects and weeds?
- What alternatives are there to using conventional pesticides, and what are the advantages and disadvantages of each alternative?
- How can we design and shift to more sustainable agricultural systems?

16-1 HOW IS FOOD PRODUCED?

What Three Systems Provide Us with Food? Some Good and Bad News Historically, humans have depended on three systems for their food supply: **(1)** *croplands* (mostly for producing grain), **(2)** *rangelands* (for producing meat mostly from grazing livestock), and **(3)** *oceanic fisheries*.

Some *good news* is that

- Since 1950 there has been a staggering increase in global food production from all three systems.
- This phenomenal growth in food productivity occurred because of technological advances such as **(1)** increased use of tractors and farm machinery and high-tech fishing boats and gear, **(2)** inorganic chemical fertilizers, **(3)** irrigation, **(4)** pesticides, **(5)** high-yield varieties of wheat, rice, and corn, **(6)** densely populated feedlots and enclosed pens for raising cattle, pigs, and chickens, **(7)** intensive ocean fishing, and **(8)** aquaculture ponds for raising some types of fish and shellfish.

To feed the world's 9.3 billion people projected by 2050, we must **(1)** produce and equitably distribute more food than has been produced since agriculture began about 10,000 years ago and **(2)** do this in an environmentally sustainable manner. Some analysts believe that we can continue expanding the use of industrialized agriculture to produce the necessary food.

Some *bad news* is that other analysts contend that future food production may be limited by **(1)** environmental degradation, **(2)** pollution, **(3)** lack of water for irrigation, **(4)** overgrazing by livestock, **(5)** overfishing, and **(6)** loss of vital ecological services (Figure 4-33, p. 103) as human activities continue to take over or degrade more of the planet's *net primary productivity* (Figure 4-24, p. 94), which supports all life. The rest of this chapter is devoted to analyzing the pros and cons of the world's crop, meat, and fish production systems and how these systems can be made more sustainable.

What Plants and Animals Feed the World? Although the earth has perhaps 30,000 plant species with parts that people can eat, only 15 plant and 8 terrestrial animal species supply 90% of our food. Just three grain crops—*wheat, rice,* and *corn*—provide more than half the calories people consume. These three grains, and most other food crops, are *annuals,* whose seeds must be replanted each year.

Two-thirds of the world's people survive primarily on traditional grains (mainly rice, wheat, and corn), mostly because they cannot afford meat. As incomes rise, people consume more grain, but indirectly in the form of meat (mostly beef, pork, and chicken), eggs, milk, cheese, and other products of grain-eating domesticated livestock.

Fish and shellfish are an important source of food for about 1 billion people, mostly in Asia and in coastal areas of developing countries. However, on a global scale fish and shellfish supply less than 1% of the energy and less than 6% of the protein in the human diet.

What Are the Major Types of Food Production? All crop production involves replacing species-rich late successional communities such as mature grasslands (Figure 6-19, p. 135) and forests (Figure 6-24, p. 140) with an early successional community (Figure 7-14, p. 173) consisting of a single crop (*monoculture,* Figure 6-20, p. 136) or a mixture of crops (*polyculture*).

There are two major types of agricultural systems: industrialized and traditional. **Industrialized agriculture**, or **high-input agriculture**, uses large amounts of fossil fuel energy, water, commercial fertilizers, and pesticides to produce huge quantities of single crops (monocultures) or livestock animals for sale. Practiced on about 25% of all cropland, mostly in developed countries (Figure 16-2), high-input industrialized agriculture has spread since the mid-1960s to some developing countries.

Plantation agriculture is a form of industrialized agriculture practiced primarily in tropical developing countries. It involves growing cash crops (such as bananas, coffee, soybeans, and sugarcane, cocoa, and vegetables) on large monoculture plantations, mostly for sale in developed countries.

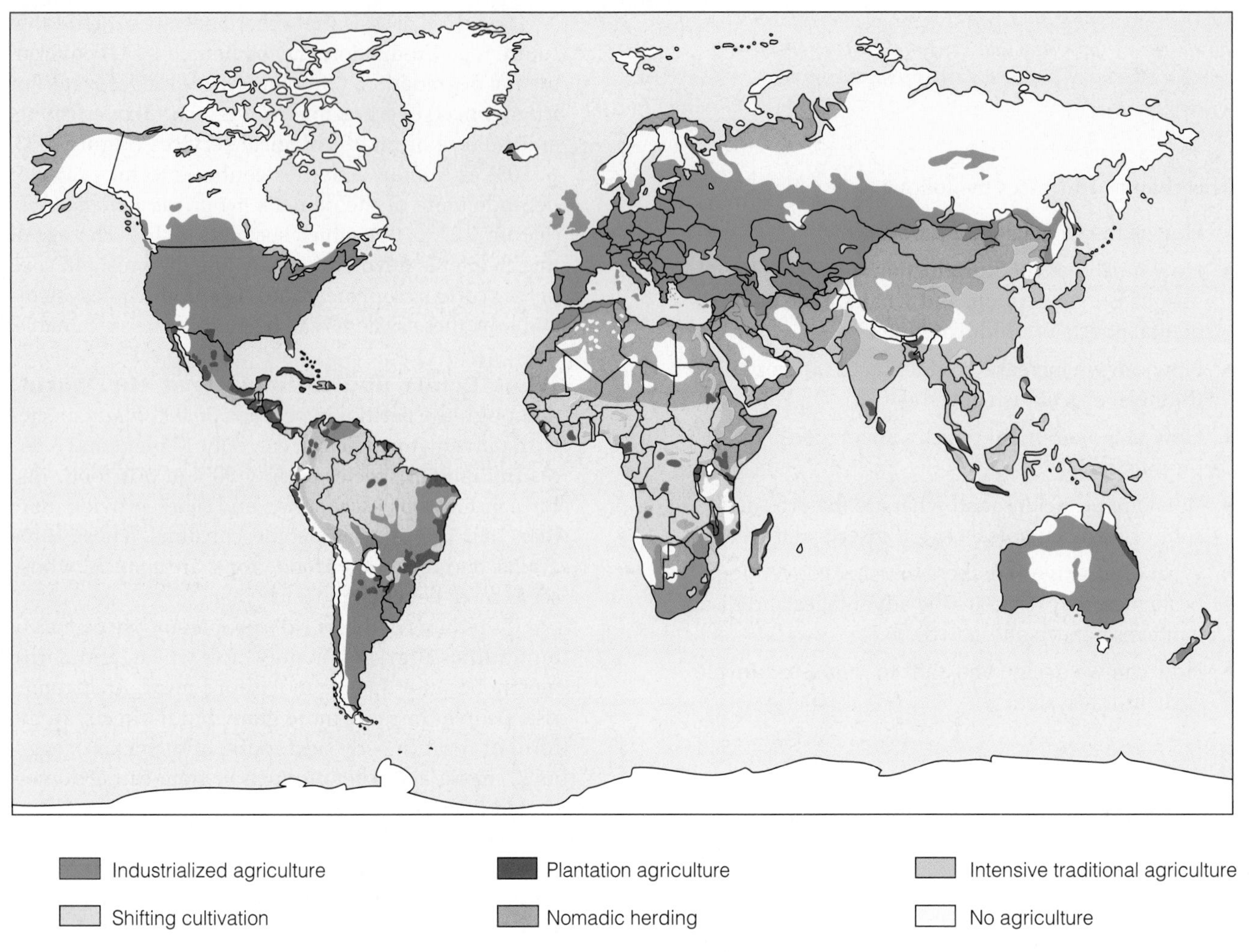

Figure 16-2 Locations of the world's principal types of food production. Excluding Antarctica and Greenland, agricultural systems cover almost one-third of the earth's land surface and account for an annual output of food worth about $1.3 trillion.

An increasing amount of livestock production in developed countries is industrialized. Large numbers of cattle are brought to densely populated feedlots, where they are fattened up for about 4 months before slaughter. Most pigs and chickens in developed countries spend their entire lives in densely populated pens and cages and are fed mostly grain grown on cropland.

Traditional agriculture consists of two main types, which together are practiced by about 2.7 billion people (44% of the world's people) in developing countries and provide about 20% of the world's food supply. **Traditional subsistence agriculture** typically uses mostly human labor and draft animals to produce only enough crops or livestock for a farm family's survival. Examples of this very low-input type of agriculture include numerous forms of shifting cultivation in tropical forests (Figure 1-15, p. 15) and nomadic livestock herding (Figure 16-2).

In **traditional intensive agriculture**, farmers increase their inputs of human and draft labor, fertilizer, and water to get a higher yield per area of cultivated land to produce enough food to feed their families and to sell for income. Croplands, like natural ecosystems, provide ecological and economic services (Figure 16-3).

How Have Green Revolutions Increased Food Production? High-Input Monocultures in Action Farmers can produce more food by **(1)** farming more land or **(2)** getting higher yields per unit of area from existing cropland. Since 1950 most of the increase in global food production has come from increased yields per unit of area of cropland in a process called the **green revolution**.

This process involves three steps:

- Developing and planting monocultures (Figure 6-20, p. 136) of selectively bred or genetically engineered high-yield varieties of key crops such as rice, wheat, and corn
- Producing high yields by using large inputs of fertilizer, pesticides, and water on crops
- Increasing the intensity and frequency of cropping

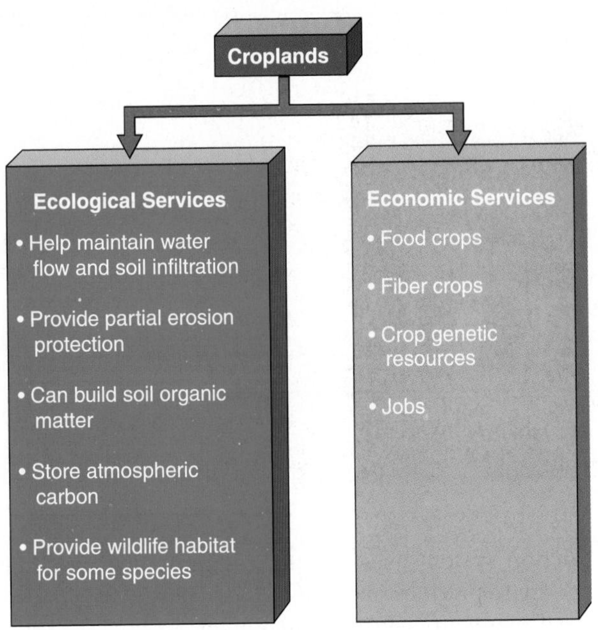

Figure 16-3 Ecological and economic services provided by croplands.

This high-input approach dramatically increased crop yields in most developed countries between 1950 and 1970 in what is called the *first green revolution* (Figure 16-4).

A *second green revolution* has been taking place since 1967 (Figure 16-4), introducing fast-growing dwarf varieties of rice and wheat, specially bred for tropical and subtropical climates, into several developing countries. With enough fertile soil and fertilizer, water, and pesticides, yields of these new plants (Figure 16-5) can be two to five times those of traditional wheat and rice varieties. The fast growth also allows farmers to grow two or even three crops a year (multiple cropping) on the same land. Producing more food on less land is also an important way to protect biodiversity by saving large areas of forests, grasslands, wetlands, and easily eroded mountain terrain from being used to grow food.

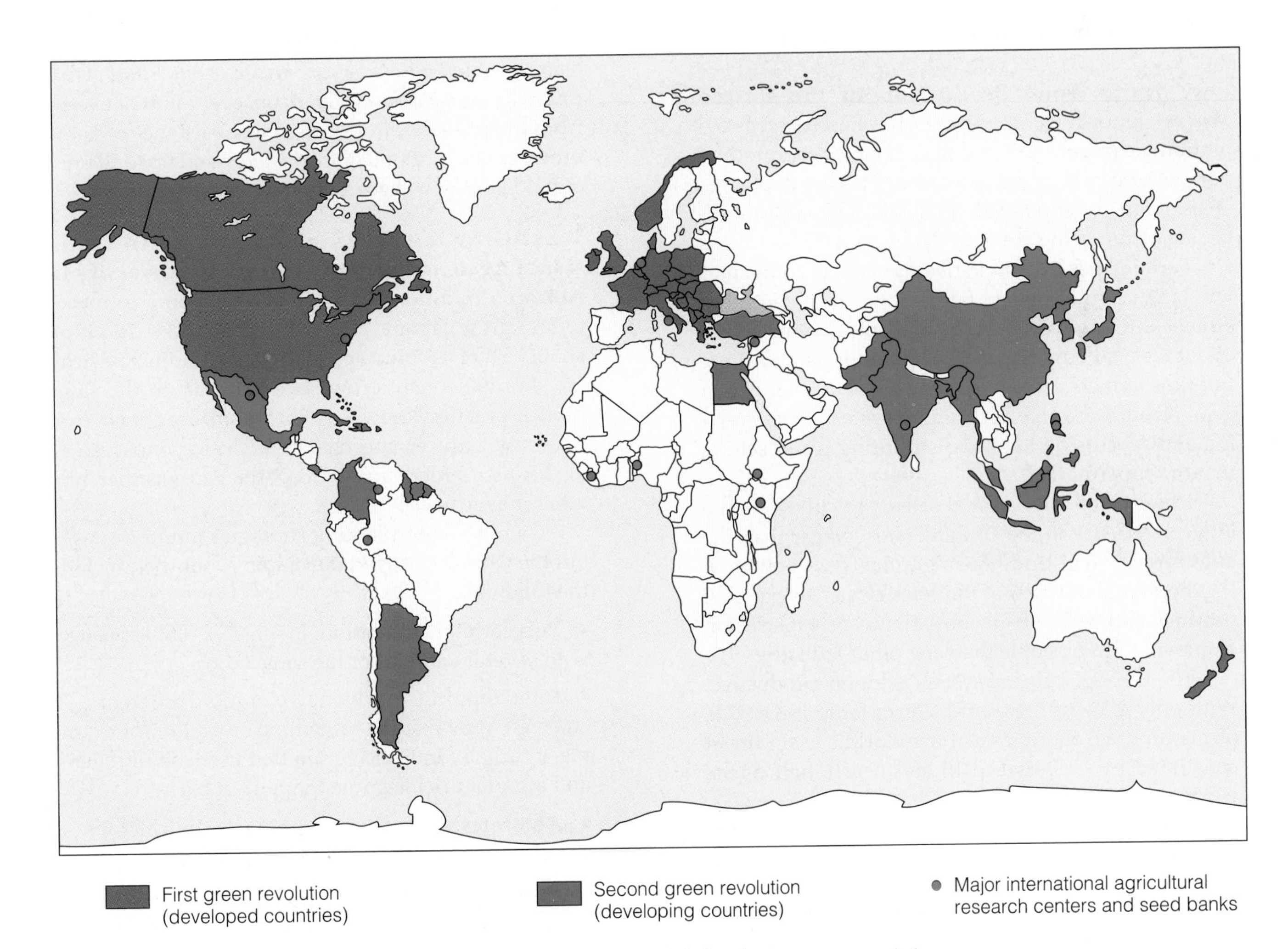

Figure 16-4 Countries whose crop yields per unit of land area increased during the two green revolutions. The first took place in developed countries between 1950 and 1970; the second has occurred since 1967 in developing countries with enough rainfall or irrigation capacity. Several agricultural research centers and gene or seed banks play a key role in developing high-yield crop varieties.

Figure 16-5 A high-yield, semi-dwarf variety of rice called IR-8 (left), a part of the second green revolution, was produced by crossbreeding two parent strains of rice: PETA from Indonesia (center) and DGWG from China (right). The shorter and stiffer stalks of the new variety allow the plants to support larger heads of grain without toppling over and increase the benefit of applying more fertilizer. (International Rice Research Institute, Manila)

These yield increases depend not only on fertile soil and ample water but also on high inputs of fossil fuels to **(1)** run machinery, **(2)** produce and apply inorganic fertilizers and pesticides, and **(3)** pump water for irrigation. All told, high-input, green revolution agriculture uses about 8% of the world's oil output.

Case Study: Food Production in the United States Since 1940, U.S. farmers have used green-revolution techniques to more than double crop production without cultivating more land. This has kept large areas of forests, grasslands, wetlands, and easily erodible land from being converted to farmland.

Farming has become *agribusiness* as big companies and larger family-owned farms have taken control of almost three-fourths of U.S. food production. Only about 650,000 Americans (2% of the population) are full-time farmers. However, about 9% of the population is involved in the U.S. agricultural system, from growing and processing food to distributing it and selling it at the supermarket.

In terms of total annual sales, agriculture is the biggest industry in the United States, bigger than the automotive, steel, and housing industries combined. It generates about 18% of the country's gross national product and 19% of all jobs in the private sector, employing more people than any other industry.

The U.S. agricultural system is highly productive. With only 0.3% of the world's farm labor force, U.S. farms produce about 17% of the world's grain (most consumed by U.S. livestock) and nearly half of the world's grain exports.

This industrialization of agriculture has been made possible by the availability of cheap energy, most of it from oil. Agriculture consumes about 17% of all commercial energy in the United States each year (Figure 16-6). Most plant crops in the United States provide more food energy than the energy used to grow them. However, if we include livestock, the U.S. food production system uses about three units of fossil fuel energy to produce one unit of food energy.

Energy efficiency is much lower if we look at the whole U.S. food system. Considering the energy used to grow, store, process, package, transport, refrigerate, and cook all plant and animal food, *about 10 units of nonrenewable fossil fuel energy are needed to put 1 unit of food energy on the table.* By comparison, every unit of energy from human labor in traditional subsistence farming provides at least 1 unit of food energy and up to 10 units of food energy using traditional intensive farming.

What Growing Techniques Are Used in Traditional Agriculture? Low-Input Agrodiversity in Action Traditional farmers in developing countries today grow about 20% of the world's food on about 75% of its cultivated land. Many traditional farmers simultaneously grow several crops on the same plot, a practice known as **interplanting**. Such crop diversity reduces the chance of losing most or all of the year's food supply to pests, bad weather, and other misfortunes.

Common interplanting strategies found throughout the world, mostly in developing countries, include the following:

- **Polyvarietal cultivation**, in which a plot is planted with several varieties of the same crop.
- **Intercropping**, in which two or more different crops are grown at the same time on a plot (for example, a carbohydrate-rich grain that uses soil nitrogen and a protein-rich legume that puts it back).
- **Agroforestry**, or **alley cropping**, in which crops and trees are planted together (Figure 9-26c, p. 219).
- **Polyculture**, a more complex form of intercropping in which many different plants maturing at various times are planted together. Advantages of this low-input approach include **(1)** less need for fertilizer and water because root systems at different

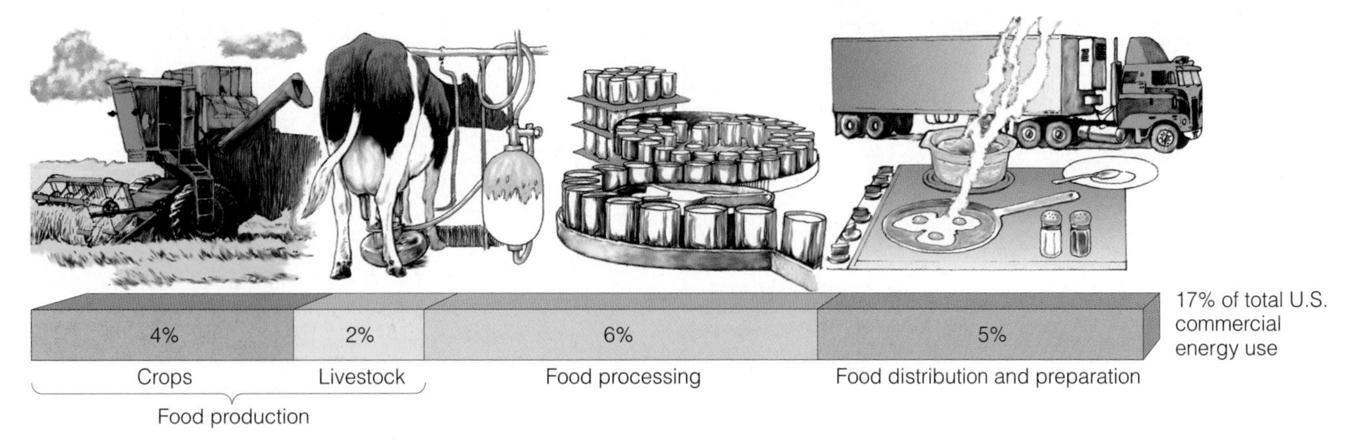

Figure 16-6 In the United States, industrialized agriculture uses about 17% of all commercial energy. On average, a piece of food eaten in the United States has traveled 2,100 kilometers (1,300 miles).

depths in the soil capture nutrients and moisture efficiently, **(2)** protection from wind and water erosion because the soil is covered with crops year-round, **(3)** little or no need for insecticides because multiple habitats are created for natural predators of crop-eating insects, **(4)** little or no need for herbicides because weeds have trouble competing with the multitude of crop plants, and **(5)** insurance against bad weather because of the diversity of crops raised. Wes Jackson is using this technique to grow perennial crops on prairie land in the United States (p. 396).

Recent ecological research on crop yields of 14 artificial ecosystems found that on average low-input polyculture (with four or five different crop species) produces higher yields per hectare of land than high-input monoculture. This finding has important implications for developing high-yield sustainable agriculture in developing countries by combining the techniques of traditional high-yield interplanting with increased inputs of organic fertilizer and irrigation.

16-2 FOOD PRODUCTION, NUTRITION, AND ENVIRONMENTAL EFFECTS

How Much Has Food Production Increased? Figure 16-7 illustrates the success of using high-input monoculture farming to produce food and ward off sharp rises in hunger and malnutrition. Here is some *good news*. Between 1950 and 1990,

- World grain production nearly tripled (Figure 16-7, left).
- Per capita production rose by about 36% (Figure 16-7, right).
- Average food prices adjusted for inflation dropped by 25%.
- The amount of food traded in the world market quadrupled.

Despite these impressive achievements in food production, there is some *disturbing news*:

- Population growth is outstripping food production and distribution in areas that support about 2 billion people, especially in sub-Saharan Africa,* with extreme poverty and one of the world's highest population growth rates.
- Since 1985 global grain production has leveled off (Figure 16-7, left) and per capita grain production has declined (Figure 16-7, right). According to some analysts, this happened because more efficient food production has lowered the price farmers get for grain and reduced their incentive to grow more food. Other analysts attribute much of this leveling off of grain production to **(1)** limits on the amounts of water, fertilizer, and pesticides that green-revolution crops can tolerate and **(2)** loss of productivity from erosion and salinization of soil and lack of irrigation water.
- Two traditional sources of animal protein in the human diet—*rangelands* (which account for much of the world's beef and mutton production) and *fisheries*—appear to be approaching their productive limits.
- If everyone ate the diet typical of a person in a developed country, with 30–40% of the calories coming from animal products, the world's current agricultural system would support only an estimated 2.5 billion people.

How Serious Are Undernutrition and Malnutrition? According to the United Nations and the World Bank, **(1)** one out of six people in developing countries does not get enough food, and **(2)** one

*Sub-Saharan Africa includes all of Africa's countries except South Africa and the six countries north of the Sahara desert.

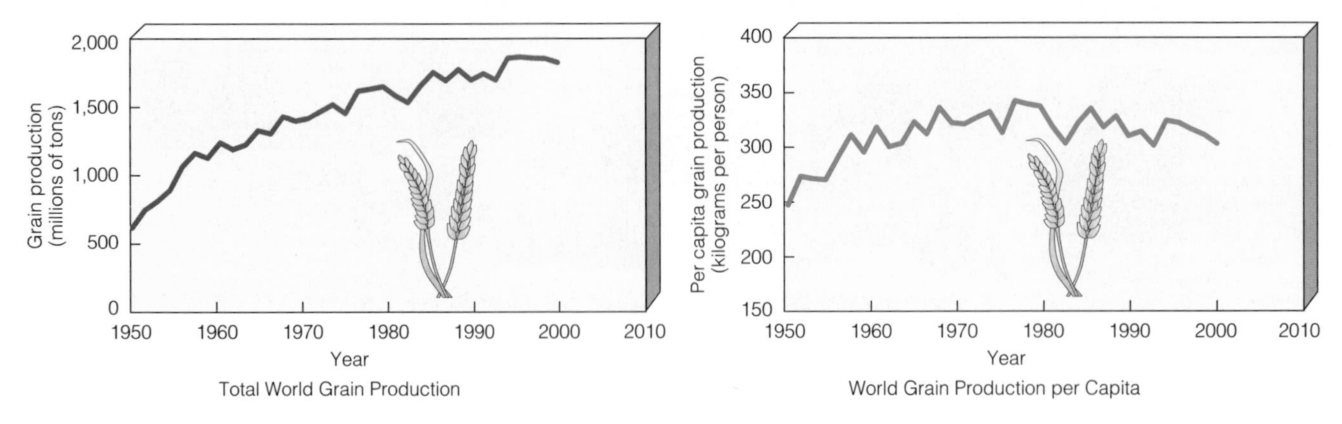

Figure 16-7 Total worldwide grain production of wheat, corn, and rice, and per capita grain production, 1950–2000. In order, the world's three largest grain-producing countries in 2000 were China, the United States, and India. (Data from U.S. Department of Agriculture, Worldwatch Institute, and UN Food and Agriculture Organization)

out of seven in developed countries eats too much. People who are underfed and underweight and those who are overfed and overweight face similar health problems: **(1)** lower life expectancy, **(2)** greater susceptibility to disease and illness, and **(3)** lower productivity and life quality.

To maintain good health and resist disease, people need **(1)** fairly large amounts of *macronutrients* such as protein, carbohydrates, and fats and **(2)** smaller amounts of *micronutrients* consisting of various vitamins (such as A, C, and E) and minerals (such as iron, iodine, and calcium).

People who cannot grow or buy enough food to meet their basic energy needs suffer from **undernutrition**. *Seriously undernourished* people get less than 80% of the minimum calorie intake needed. Children in this category are likely to **(1)** suffer from mental retardation and stunted growth and **(2)** be much more susceptible to infectious diseases (such as measles and diarrhea), which kill one child in four in developing countries.

People who are forced to live on a low-protein, high-carbohydrate diet consisting only of grains such as wheat, rice, or corn often suffer from **malnutrition**: deficiencies of protein and other key nutrients. Many of the world's desperately poor people, especially children, suffer from both undernutrition and malnutrition.

The two most common nutritional deficiency diseases are marasmus and kwashiorkor. *Marasmus* (from the Greek word *marasmos*, "to waste away") occurs when a diet is low in both calories and protein (Figure 1-7, p. 9). Most victims are either nursing infants of malnourished mothers or children who do not get enough food after being weaned from breast-feeding. If the child is treated in time with a balanced diet, most of these effects can be reversed.

Kwashiorkor (meaning "displaced child" in a West African dialect) is a severe protein deficiency occurring in infants and children ages 1–3, usually after the arrival of a new baby deprives them of breast milk. The displaced child's diet changes to grain or sweet potatoes, which provide enough calories but not enough protein. If it is caught soon enough, most of the harmful effects can be cured with a balanced diet. Otherwise, children who survive their first year or two suffer from stunted growth and mental retardation.

Here is some *good news*. Despite population growth, the estimated number of chronically malnourished people fell from 918 million in 1970 to 826 million in 2000 (96% of them in developing countries). However, according to the UN the *bad news* is that about one of every six people in developing countries (including one of every three children below age 5) is chronically undernourished or malnourished.

Such people are disease prone, and adults are too weak to work productively or think clearly. As a result, their children also tend to be underfed, malnourished, and susceptible to disease. If these children survive to adulthood, many are locked in a malnutrition–poverty cycle (Figure 16-8) that can continue for generations.

According to the World Health Organization (WHO), each year at least 10 million people, half of them children under age 5, die prematurely from **(1)** undernutrition, **(2)** malnutrition, or **(3)** normally nonfatal diseases such as measles and diarrhea worsened by malnutrition. Such undernutrition and malnutrition have been described by the UN as a "silent and invisible global emergency with a massive impact on children" that could be prevented (Solutions, p. 404).

The most widespread micronutrient deficiencies in developing countries involve *vitamin A*, *iron*, and *iodine*. According to WHO, an estimated 124 million children in developing countries are deficient in vitamin A. This puts them at risk for blindness (about 500,000 cases per year) and premature death because even mild vitamin A deficiency reduces children's resistance to infectious diseases such as diarrhea and measles. According

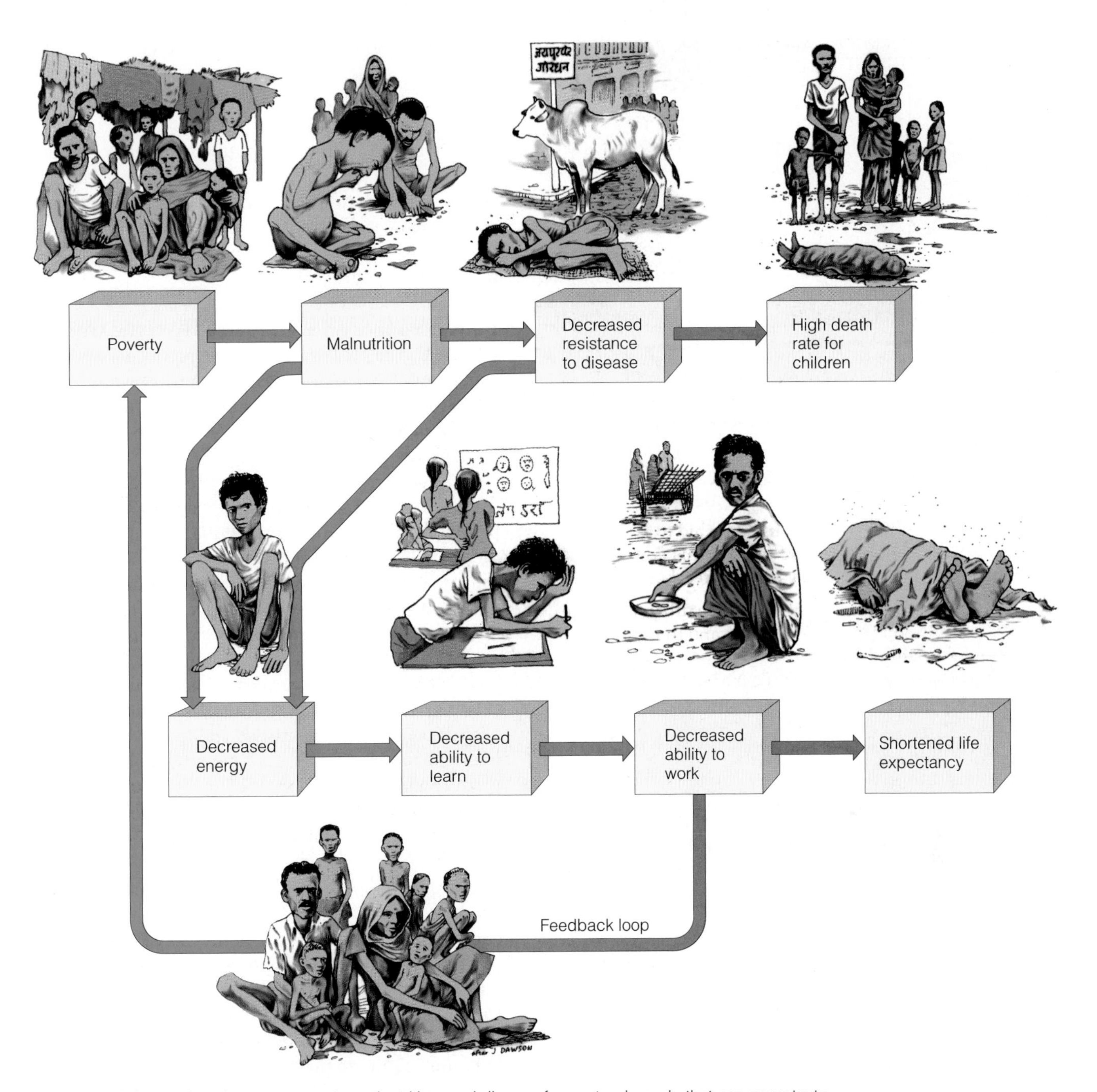

Figure 16-8 Interactions between poverty, malnutrition, and disease form a tragic cycle that can perpetuate such conditions in succeeding generations of families.

to the United Nations Children's Fund (UNICEF), improved vitamin A nutrition through use of vitamin-fortified rice could prevent 1–2 million premature deaths per year among children under age 5.

Other nutritional deficiency diseases are caused by the lack of minerals. Too little *iron* (a component of hemoglobin that transports oxygen in the blood) causes anemia. This mineral deficiency **(1)** causes fatigue, **(2)** makes infection more likely, **(3)** increases a woman's chances of dying in childbirth, **(4)** increases an infant's chances of dying of infection during its first year of life, and **(5)** cripples efforts to improve primary school education because developing brains need adequate iron to learn. According to a 1999 survey by WHO, one of every three people, mostly women and children in tropical developing countries, suffers from iron deficiency.

Elemental *iodine*, found in seafood and crops grown in iodine-rich soils, is essential for the functioning of the thyroid gland, which produces a hormone that controls the body's rate of metabolism. Lack of iodine can cause **(1)** stunted growth, **(2)** mental

Saving Children

SOLUTIONS

According to the United Nations Children's Fund (UNICEF), one-half to two-thirds of childhood deaths from nutrition-related causes could be prevented at an average annual cost of only $5–10 per child, or only 10–19¢ per week. This life-saving program would involve the following measures:

- Immunizing children against childhood diseases such as measles
- Encouraging breast-feeding (except by mothers with AIDS)
- Preventing dehydration from diarrhea by giving infants a mixture of sugar and salt in a glass of water
- Preventing blindness by giving people a vitamin A capsule twice a year at a cost of about 75¢ per person or fortifying common foods with vitamin A and other micronutrients at a cost of about 10¢ per person annually
- Providing family-planning services to help mothers space births at least 2 years apart
- Increasing education for women, with emphasis on nutrition, drinking water sterilization, and child care

Critical Thinking

How much money (if any) would you be willing to spend each year to help implement such a program for saving children? Why has insufficient money been allocated for such a program?

retardation, and **(3)** goiter (an abnormal enlargement of the thyroid gland that can lead to deafness).

Some *good news* is that programs by WHO and UNICEF to get countries to establish programs to add iodine to salt slashed the percentage of the world's people with iodine deficiency from 29% to 13% between 1994 and 1997. The *bad news* is that this still leaves at least 740 million people in developing countries with too little iodine in their diet.

How Serious Is Overnutrition? About one out of seven adults in developed countries suffers from **overnutrition**, a condition in which food energy intake exceeds energy use and causes excess body fat (obesity). Too much food, too little exercise, or both can cause overnutrition.

Overnutrition is the second leading cause of preventable deaths after smoking (p. 224), mostly from heart disease, cancer, stroke, and diabetes. A study of thousands of Chinese villagers indicates that the healthiest diet for humans is largely vegetarian, with only 10–15% of calories coming from fat. This is in contrast to the typical meat-based diet, in which 40% of the calories come from fat.

About 61% of the U.S. adult population is overweight and 27% is obese—the highest percentage of any developed country. The $36 billion Americans spend each year trying to lose weight is almost twice the $19 billion needed to eliminate undernutrition and malnutrition in the world.

Do We Produce Enough Food to Feed the World's People? The *good news* is that we produce more than enough food to meet the basic nutritional needs of every person on the earth today. If distributed equally, the grain currently produced worldwide is enough to give everyone a meatless subsistence diet.

The *bad news* for those not getting enough to eat is that food is not distributed equally among the world's people because of differences in **(1)** soil, **(2)** climate, **(3)** political and economic power, and **(4)** average per capita income throughout the world (Figure 1-4, p. 8).

Most agricultural experts agree that *the principal cause of hunger and malnutrition is and will continue to be poverty*, which prevents poor people from growing or buying enough food regardless of how much is available. For example, according to the UN, in the 1990s nearly 80% of all malnourished children lived in countries with food surpluses. Thus poverty, not lack of food production, is the real food problem for about one out of six people. Ways to decrease hunger and malnutrition by reducing poverty are discussed in Section 2-4 (p. 34).

What Are the Environmental Effects of Producing Food? Agriculture has significant harmful effects on air, soil, water, and biodiversity (Figure 16-9). David Pimentel has estimated that the harmful environmental costs not included in the prices of food in the United States are $150–200 billion per year.

According to Norman Myers (Guest Essay, p. 118), the future ability to produce more food will be limited by a combination of **(1)** soil erosion (Figure 9-22, p. 214), **(2)** desertification (Figure 9-23, p. 215), **(3)** salinization and waterlogging of irrigated lands (Figure 9-25, p. 217), **(4)** water deficits and droughts (Section 14-3, p. 332), **(5)** loss of wild species that provide the genetic resources for new foods and improved forms of existing foods, and **(6)** the effects of global warming (Section 13-4, p. 311, and Figure 13-11, p. 311).

According to a 2000 study by the UN-affiliated International Food Policy Research Institute, nearly 40% of the world's cropland is seriously degraded (including 75% in Central America, 20% in Africa, and 11% in Asia). There is growing concern that such environmental factors may limit food production in India and China, the world's two most populous countries.

Biodiversity Loss

Loss and degradation of habitat from clearing grasslands and forests and draining wetlands

Fish kills from pesticide runoff

Killing of wild predators to protect livestock

Loss of genetic diversity from replacing thousands of wild crop strains with a few monoculture strains

Soil

Erosion

Loss of fertility

Salinization

Waterlogging

Desertification

Air Pollution

Greenhouse gas emissions from fossil fuel use

Other air pollutants from fossil fuel use

Pollution from pesticide sprays

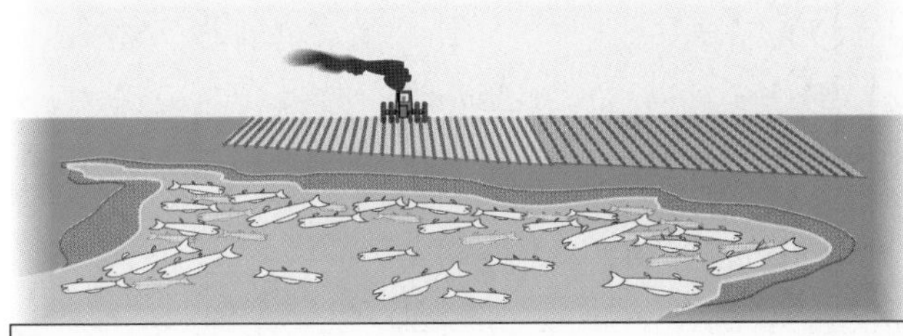

Water

Aquifer depletion

Increased runoff and flooding from land cleared to grow crops

Sediment pollution from erosion

Fish kills from pesticide runoff

Surface and groundwater pollution from pesticides and fertilizers

Overfertilization of lakes and slow-moving rivers from runoff of nitrates and phosphates from fertilizers, livestock wastes, and food processing wastes

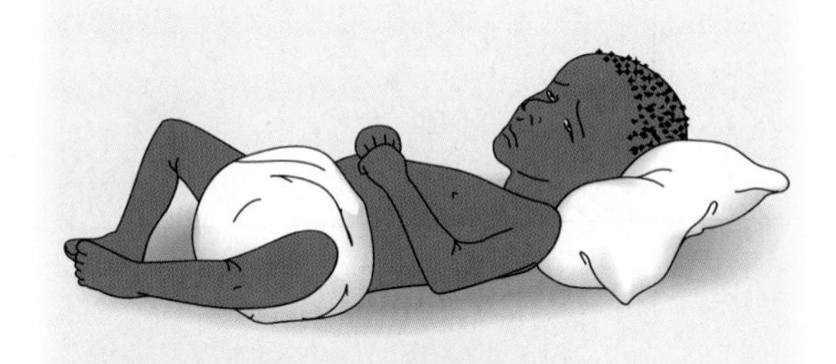

Human Health

Nitrates in drinking water

Pesticide residues in drinking water, food, and air

Contamination of drinking and swimming water with disease organisms from livestock wastes

Bacterial contamination of meat

Figure 16-9 Major environmental effects of food production.

16-3 INCREASING WORLD CROP PRODUCTION

How Can Crossbreeding Be Used to Develop Genetically Improved Crop Strains? For centuries farmers and scientists have used traditional methods of *crossbreeding* through artificial selection to develop genetically improved varieties of crop strains and livestock. Agricultural experts expect most future increases in food yields per hectare on existing cropland to result from **(1)** improved crossbred plant strains and **(2)** expansion of green-revolution technology to new parts of the world. For example, new corn and rice strains are expected to raise yields of these two crops.

However, traditional crossbreeding **(1)** is a slow process, typically taking 15 years or more to produce a commercially valuable new variety, **(2)** can combine traits only from species that are close to one another genetically, and **(3)** produces varieties that are useful for only about 5–10 years before their effectiveness is reduced by pests and diseases.

Is Genetic Engineering the Answer? Scientists are working to create new green revolutions—actually *gene revolutions*—by using genetic engineering and other forms of biotechnology to develop new genetically improved strains of crops. **Genetic engineering** or **gene splicing** of food crops is the insertion of an alien gene

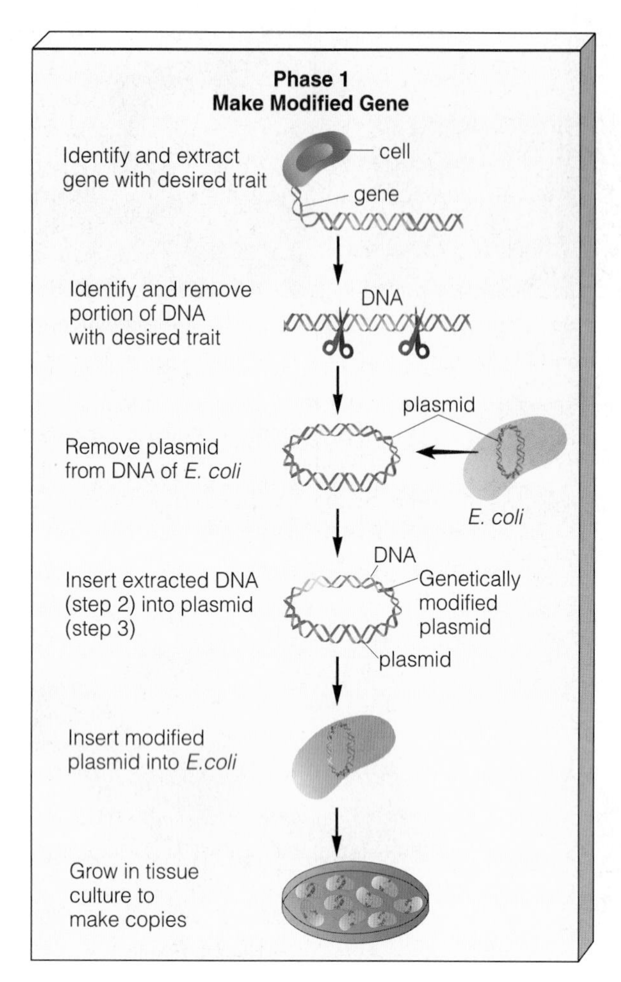
Phase 1
Make Modified Gene
Identify and extract gene with desired trait
cell
gene
Identify and remove portion of DNA with desired trait
DNA
Remove plasmid from DNA of E. coli
plasmid
E. coli
Insert extracted DNA (step 2) into plasmid (step 3)
DNA
Genetically modified plasmid
plasmid
Insert modified plasmid into E.coli
Grow in tissue culture to make copies

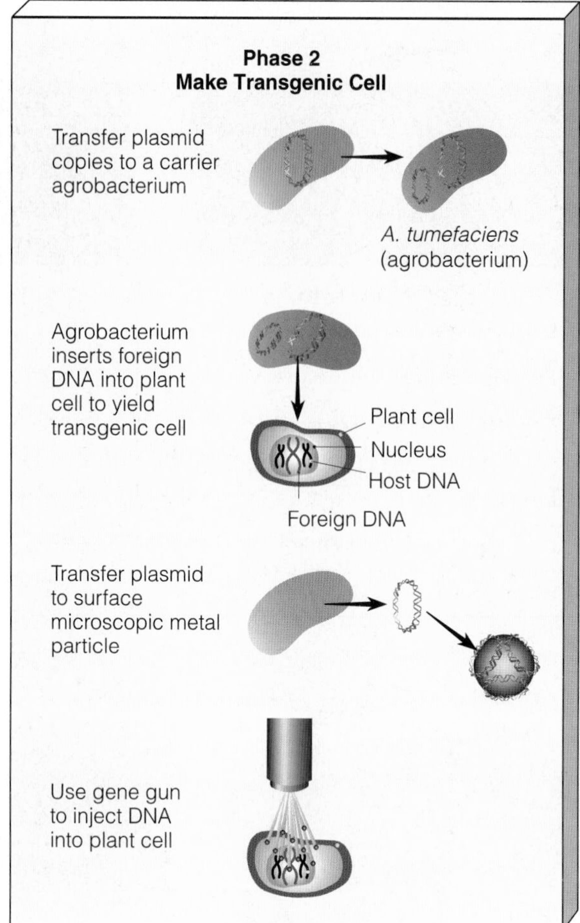
Phase 2
Make Transgenic Cell
Transfer plasmid copies to a carrier agrobacterium
A. tumefaciens (agrobacterium)
Agrobacterium inserts foreign DNA into plant cell to yield transgenic cell
Plant cell
Nucleus
Host DNA
Foreign DNA
Transfer plasmid to surface microscopic metal particle
Use gene gun to inject DNA into plant cell

into a commercially valuable plant (or animal) to give it new beneficial genetic traits. Such organisms are called **genetically modified organisms (GMOs)**.

Compared with traditional crossbreeding, gene splicing **(1)** takes about half as much time to develop a new crop or animal variety, **(2)** cuts costs, and **(3)** allows the insertion of genes from almost any other organism into crop or animal cells. Figure 16-10 outlines the steps involved in developing a genetically modified or transgenic plant.

Scientists also use *advanced tissue culture techniques* to produce only the desired parts of a plant such as its oils or fruits. Instead of moving a valuable gene trait from one species to another, genetic engineers are also investigating how to achieve the same result by goading a plant into a mutation that activates genes that provide the desired trait. If this can be done, **(1)** scientists will be able to speed up the rate of beneficial mutations, which is a major driving force in biological evolution (p. 109), and **(2)** sidestep some of the current critics of biotechnology.

Ready or not, the world is entering the *age of genetic engineering*. Nearly two-thirds of the food products on U.S. supermarket shelves contain genetically engineered crops, and the proportion is increasing rapidly.

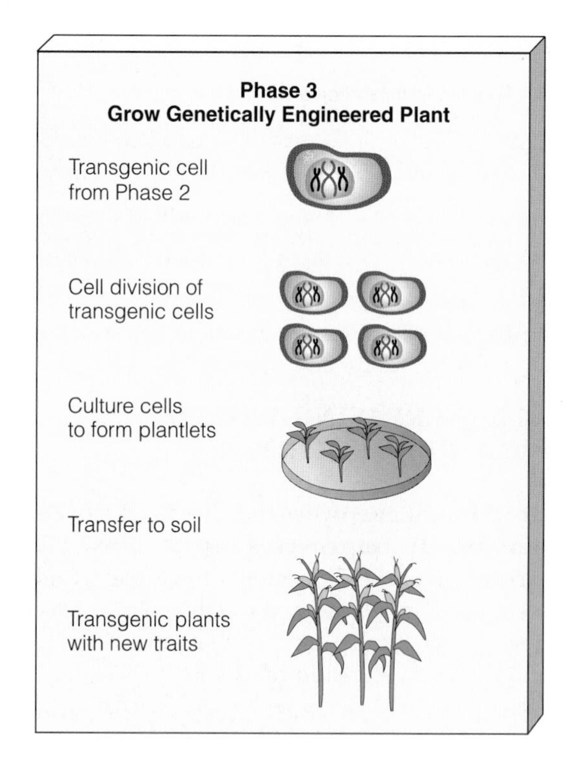
Phase 3
Grow Genetically Engineered Plant
Transgenic cell from Phase 2
Cell division of transgenic cells
Culture cells to form plantlets
Transfer to soil
Transgenic plants with new traits

Figure 16-10 Steps in genetically modifying a plant.

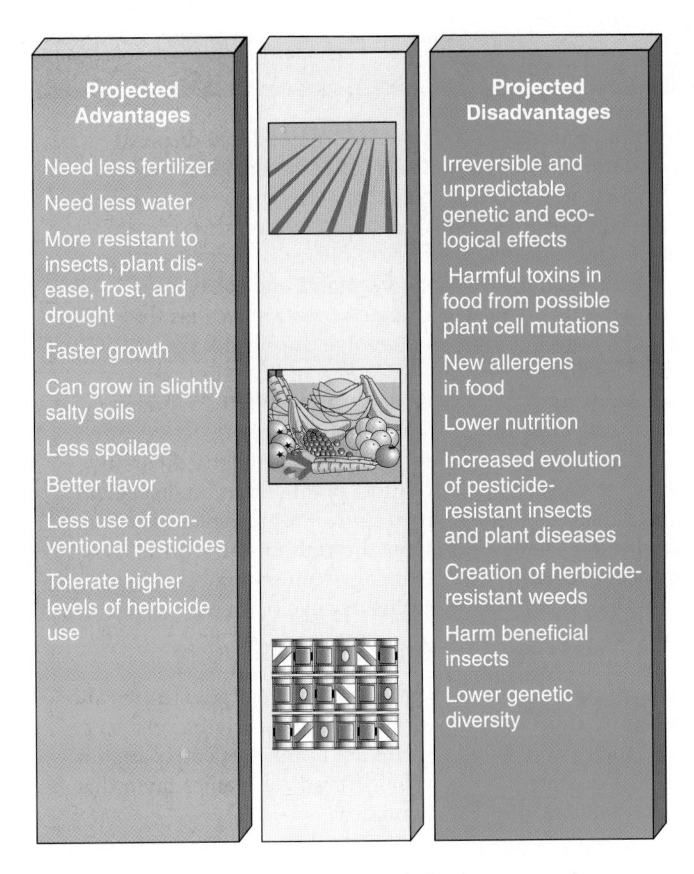

Projected Advantages
Need less fertilizer
Need less water
More resistant to insects, plant disease, frost, and drought
Faster growth
Can grow in slightly salty soils
Less spoilage
Better flavor
Less use of conventional pesticides
Tolerate higher levels of herbicide use
Projected Disadvantages
Irreversible and unpredictable genetic and ecological effects
Harmful toxins in food from possible plant cell mutations
New allergens in food
Lower nutrition
Increased evolution of pesticide-resistant insects and plant diseases
Creation of herbicide-resistant weeds
Harm beneficial insects
Lower genetic diversity

Figure 16-11 Projected advantages and disadvantages of genetically modified crops and foods.

There is growing controversy over the use of *genetically modified food* (GMF). Such food is seen as a potentially sustainable way to solve world food problems by its producers and investors and is called potentially dangerous "Frankenfood" by its critics. Figure 16-11 summarizes the pros and cons of this new technology.

Critics recognize the potential benefits of genetically modified crops but say that **(1)** we know far too little about the potential harm from widespread use of such crops to human health and ecosystems, and **(2)** genetically modified organisms cannot be recalled if they cause harmful genetic and ecological effects. They call for **(1)** more controlled field experiments, **(2)** more research and long-term safety testing to better understand the risks, **(3)** stricter regulation of this rapidly developing technology, and **(4)** mandatory labeling of such foods to provide consumers with a choice (as is now required in Japan, Europe, South Korea, Canada, Australia, and New Zealand).

Growing numbers of food manufacturers and retailers, first in Europe and now in the United States, have stopped selling GMFs. As a result, some major food biotech companies have put their biotech crop operations on a back burner or are getting out of the business.

Can We Continue Expanding the Green Revolution? Many analysts believe that it is possible to produce enough food to feed the 8 billion people projected by 2025 through new advances in gene-splicing technology and increased use of new and existing high-yield green-revolution techniques. Other analysts point to the following factors that have limited the success of the green and gene revolutions to date and may continue to do so:

- Without huge amounts of fertilizer and water, most green-revolution crop varieties produce yields that are no higher (and are sometimes lower) than those from traditional strains; this is why the second green revolution has not spread to many arid and semiarid areas such as much of Africa and Australia (Figure 16-4).
- Green-revolution and genetically engineered crop strains and their needed inputs of water, fertilizer, and pesticides cost too much for most subsistence farmers in developing countries.
- Continuing to increase fertilizer, water, and pesticide inputs eventually produces no additional increase in crop yields.
- Grain yields per hectare are still increasing in many parts of the world, but at a much slower rate. For example, grain yields rose about 2.1% a year between 1950 and 1990 but dropped to about 1% per year between 1990 and 2000.
- According to Indian economist Vandana Shiva, overall gains in crop yields from new green- and gene-revolution varieties may be much lower than claimed. The reason is that the yields are based on comparisons between the output per hectare of old and new *monoculture* varieties rather than between the even higher yields per hectare for *polyculture* cropping systems and the new monoculture varieties that often replace them.
- Yields may start dropping for a number of reasons: **(1)** The soil erodes, loses fertility, and becomes salty and waterlogged (Figure 9-25, p. 217), **(2)** underground and surface water supplies become depleted and polluted with pesticides and nitrates from fertilizers, and **(3)** populations of rapidly breeding pests develop genetic immunity to widely used pesticides (p. 420).
- Increased loss of biodiversity can limit the genetic raw material needed for future green and gene revolutions (Spotlight, p. 408).

Will People Try New Foods? Some analysts recommend greatly increased cultivation of less widely known plants to supplement or replace such staples as wheat, rice, and corn. One of many possibilities

SPOTLIGHT

Shrinking the World's Genetic Plant Library

The UN Food and Agriculture Organization estimates that by 2000 two-thirds of all seeds planted in developing countries were of uniform strains. Such genetic uniformity increases the vulnerability of food crops to pests, diseases, and harsh weather. Many biologists argue that this decreased variability, plus growing extinction rates of plant species, can limit the genetic raw material available to support future green and gene revolutions.

For example, in the mid-1970s, a valuable wild corn species, the only known perennial strain of corn, was barely saved from extinction. Crossbreeding this perennial strain with commercial varieties could reduce the need for yearly plowing and sowing, which would reduce soil erosion and save water and energy. Even more important, this wild corn **(1)** has a built-in genetic resistance to four of the eight major corn viruses, and **(2)** grows better in cooler and damper habitats than established commercial strains. Overall, the economic benefits of cultivating this wild plant could total several billion dollars per year.

Wild varieties of the world's most important plants can be collected and stored in **(1)** gene or seed banks, **(2)** agricultural research centers, and **(3)** botanical gardens. However, space and money severely limit the number of species that can be preserved.

Other limitations include **(1)** inability to successfully store seeds of many plants (such as potatoes), **(2)** irreversible loss of stored seeds because of power failures, fires, or unintentional disposal, **(3)** death of stored seeds unless they are periodically planted (germinated) and then stored again, and **(4)** difficulty in reintroducing stored plants and seeds into changed habitats because they do not evolve during storage.

Because of these limitations, ecologists and plant scientists warn that the only effective way to preserve the genetic diversity of most plant and animal species is to protect representative ecosystems throughout the world from agriculture and other forms of development.

Critical Thinking

What are the major advantages and disadvantages of relying on a shrinking number of crop varieties? Why do seed companies favor this approach?

is the *winged bean*, a protein-rich legume now common only in New Guinea and Southeast Asia. This fast-growing plant produces so many different edible parts that it has been called a supermarket on a stalk. It also needs little fertilizer because of nitrogen-fixing nodules in its roots.

Insects—called *microlivestock*—are also important potential sources of protein, vitamins, and minerals in many parts of the world. There are about 1,500 edible insect species. Some are important food items in many parts of the world. Examples include **(1)** black ant larvae (served in tacos in Mexico), **(2)** giant waterbugs (crushed into vegetable dip in Thailand), **(3)** *mopani* or emperor moth caterpillars (eaten in South Africa), **(4)** cockroaches (eaten by Kalahari desert dwellers), **(5)** lightly toasted butterflies (a favorite food in Bali), and **(6)** French-fried ants (sold on the streets of Bogotá, Colombia). Most of these insects are 58–78% protein by weight—three to four times as protein-rich as beef, fish, or eggs. Two problems are **(1)** getting farmers to take the financial risk of cultivating new types of food crops and **(2)** convincing consumers to try new foods.

Some plant scientists believe we should rely more on polycultures of perennial crops, which are better adapted to regional soil and climate conditions than most annual crops (p. 396). Using perennials would also **(1)** eliminate the need to till soil and replant seeds each year, **(2)** greatly reduce energy use, **(3)** save water, and **(4)** reduce soil erosion and water pollution from eroded sediment. However, large seed companies that make their money selling farmers seeds each year for annual crops generally oppose this.

Is Irrigating More Land the Answer? About 40% of the world's food production comes from irrigated land. This includes 70% of the grain harvest in China and 50% in India (compared to 15% in the United States).

Between 1950 and 2000, the world's irrigated area tripled, with most of the growth occurring from 1950 to 1978. Since 1978 the amount of irrigated land per person has been falling and is projected to fall much more between 2000 and 2050 (Figure 16-12). Reasons for this downward trend include the following:

- World population has grown faster than irrigated agriculture since 1978.
- Chronic water shortages could affect 2.8 billion people by 2025 (Section 14-3, p. 332).
- Water is being pumped too rapidly from aquifers in many of the world's food-growing areas (Case Study, p. 337).
- Irrigation water is used inefficiently (p. 339).
- Crop productivity is decreased by soil salinization (Figure 9-25, p. 217) on irrigated cropland.

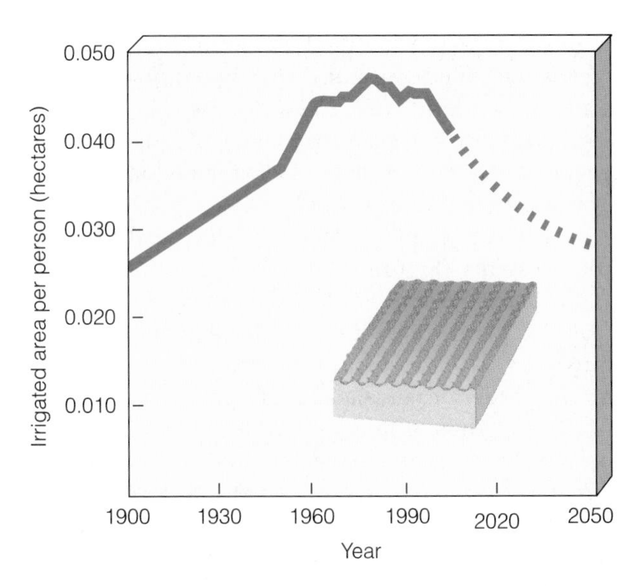

Figure 16-12 World irrigated area of cropland per person, 1900–2000, with projections to 2050. (Data from United Nations Food and Agriculture Organization, U.S. Census Bureau, and the Worldwatch Institute)

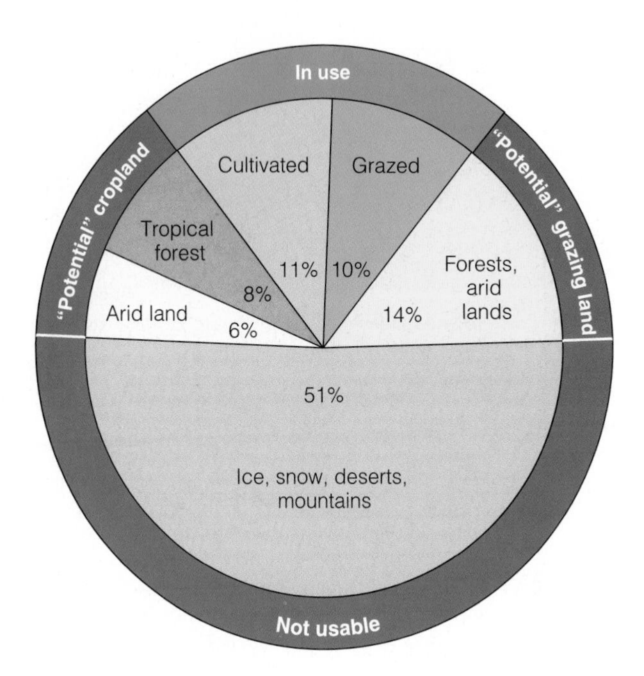

Figure 16-13 Classification of the earth's land. Theoretically, we could double the amount of cropland by clearing tropical forests and irrigating arid lands. However, converting these lands into cropland would **(1)** destroy valuable forest resources, **(2)** reduce the earth's biodiversity, **(3)** affect water quality and quantity, and **(4)** cause other serious environmental problems, usually without being cost-effective.

- Increasing urbanization puts city dwellers and farmers in competition for limited water supplies.
- Water supplies in many food-growing areas may be disrupted by global warming (p. 312).
- The majority of the world's farmers don't have enough money to irrigate their crops. Thus, rainfall is the source of water for 83% of the world's cropland.

Key methods for using water more sustainably in crop production are **(1)** increasing irrigation efficiency (p. 339 and Solutions, p. 341), **(2)** shifting to crops that need less water (for example, depending less on water-thirsty rice and sugarcane and more on wheat and sorghum), and **(3)** withdrawing water from aquifers no faster than they are replenished.

Is Cultivating More Land the Answer? Theoretically, the world's cropland could be more than doubled by clearing tropical forests and irrigating arid land (Figure 16-13). However, much of this is marginal land where cultivation is unlikely to be sustainable.

Most of the land cleared in rain forests has nutrient-poor soils (Figure 9-20, lower left, p. 213) and cannot support crop growth for more than a couple of years. In addition, potential cropland in savanna and other semiarid land in Africa cannot be used for farming or livestock grazing. The reason is the presence of 22 species of the tsetse fly, which transmits a protozoan parasite that causes incurable sleeping sickness in humans and a fatal disease in livestock.

Some researchers hope to develop new methods of intensive cultivation in tropical areas. However, other scientists argue that it makes more ecological and economic sense to combine ancient methods of shifting cultivation (Figure 1-15, p. 15) with various forms of polyculture.

Much of the world's potentially cultivable land lies in dry areas, especially in Australia and Africa. Large-scale irrigation in these areas would **(1)** require large, expensive dam projects with a mixture of beneficial and harmful impacts (Figure 14-9, p. 333), **(2)** use large inputs of fossil fuel to pump water long distances, **(3)** deplete groundwater supplies by removing water faster than it is replenished, and **(4)** require expensive efforts to prevent erosion, groundwater contamination, salinization, and waterlogging, all of which reduce crop productivity.

Thus, much of the world's new cropland that could be developed would be on land that is marginal for raising crops and requires expensive inputs of fertilizer, water, and energy. Furthermore, these potential increases in cropland would not offset the projected loss of almost one-third of today's cultivated cropland caused by erosion, overgrazing, waterlogging, salinization, and urbanization.

Such expansion in cropland would also reduce wildlife habitats and thus the world's biodiversity. According to the UN Food and Agriculture Organization (FAO), cultivating all potential cropland in developing countries would reduce forests, woodlands, and permanent pasture by 47%. Clearing these forests

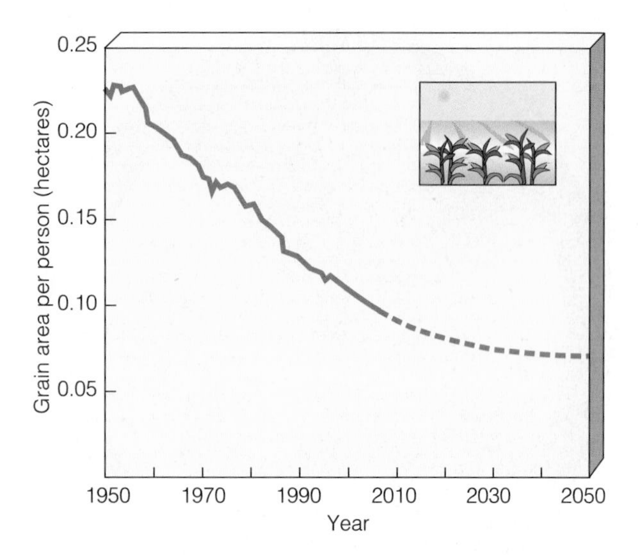

Figure 16-14 Average grain area per person worldwide, 1950–2000, with projections to 2050. (Data from U.S. Department of Agriculture and Worldwatch Institute)

would also release a huge amount of carbon dioxide into the atmosphere and accelerate global warming (Figure 13-8, p. 306), which would cause shifts in the areas where crops could be grown.

Thus, *many analysts believe that significant expansion of cropland is unlikely over the next few decades*. If this assessment is correct, the world's grainland area per person, which dropped by half between 1950 and 1998 (because the human population grew seven times faster than grainland expansion), is expected to decline further (Figure 16-14).

Can We Grow More Food in Urban Areas? Food experts believe that people in urban areas could live more sustainably and save money by growing food in empty lots, on rooftops and balconies, and in their own backyards and by raising fish in tanks and sewage lagoons. Currently, about 800 million urban gardens provide about 15% of the world's food. More food could be grown in urban areas. A study by the UN Center for Human Settlements estimated that up to 50% of the total area in many cities in developing countries is vacant public land that could be used to produce food.

16-4 PRODUCING MORE MEAT

What Are Rangeland and Pasture? About 40% of the earth's ice-free land is **rangeland** that supplies forage or vegetation for grazing (grass-eating) and browsing (shrub-eating) animals. About 1.3 billion cattle and 1 billion sheep graze on about 42% of the world's rangeland. Much of the rest is too dry, cold, or remote from population centers to support large numbers of livestock. About 29% of the total U.S. land area is rangeland, most of it short-grass prairie in the arid and semiarid western half of the country. Livestock also graze in **pastures**: managed grasslands or enclosed meadows usually planted with domesticated grasses or other forage.

Most rangeland grasses have deep and complex root systems (Figure 9-20, top right, p. 213) that **(1)** help anchor the plants, **(2)** extract underground water so plants can withstand drought, and **(3)** store nutrients so plants can grow again after a drought or fire. Blades of rangeland grass grow from the base, not the tip. Thus, as long as only its upper half—called its *metabolic reserve*—is eaten and its lower half remains, rangeland grass is a renewable resource that can be grazed again and again. The exposed metabolic reserve of a grass plant is where photosynthesis takes place to provide food for the deep roots of rangeland grasses. If all or most of the lower half (metabolic reserve) of the plant is eaten, the plant is weakened and can die.

How Is Meat Produced, and What Are Its Environmental Consequences? Globally, animal products provide about 15% of the energy and 30% of the protein in the human diet. In 2000, meat and meat products accounted for 30–40% of agriculture's global economic output. According to Worldwatch Institute estimates, the world's meat production is likely to more than double between 2000 and 2050 as affluence rises in middle-income developing countries and people begin consuming more meat.

Most of the world's 1.3 billion cattle and 1 billion sheep and other grazing livestock are raised on rangeland by open grazing or nomadic herding (Figure 16-2). Although these animals graze about 25% of the world's land, they produce only about 9% of the world's beef and 30% of its mutton.

Analysts expect most future increases in meat production to come from densely populated *feedlots*, where animals are fattened for slaughter by feeding on grain grown on cropland or meal produced from fish. Feedlots account for about 40% of the world's meat production and more than half of the world's poultry and pork.

This industrialized approach increases meat productivity. However, it also

- Concentrates pollution problems such as **(1)** foul odors, **(2)** water pollution when lagoons storing animal wastes collapse or are flooded, **(3)** and contamination of drinking water wells by nitrates from animal wastes

- Increases pressure on the world's **(1)** grain supply because feedlot livestock consume grain produced on cropland instead of feeding on natural grasses and

Some Environmental Consequences of Meat Production

The meat-based diet of affluent people in developed and developing countries has the following environmental effects:

- More than half of the world's cropland (19% in the United States) is used to produce livestock feed grain (mostly field corn, sorghum, and soybeans).
- Livestock and fish raised for food consume about 37% of the world's grain production (70% of U.S. production).
- Livestock use more than half the water withdrawn from rivers and aquifers each year, mostly to **(1)** irrigate crops fed to livestock and **(2)** wash away manure from crowded livestock pens and feedlots.
- Manure washing off the land or leaking from lagoons used to store animal wastes is a significant source of water pollution that kills fish by depleting dissolved oxygen.
- About 14% of U.S. topsoil loss is directly associated with livestock grazing.
- Overgrazing of sparse vegetation and trampling of the soil by too many livestock are the major causes of desertification in arid and semiarid areas (Figure 9-23, p. 215).
- Cattle belch out about 16% of the methane (a greenhouse gas that is about 25 times more potent than carbon dioxide) released into the atmosphere (Figure 13-4, p. 304).
- Some of the nitrogen in commercial inorganic fertilizer used to grow livestock feed is converted to nitrous oxide, another greenhouse gas (Figure 13-4, p. 304).
- Livestock in the United States produce 13 times more waste (manure) than is produced by the country's human population. Only about half of this nutrient-rich livestock waste is recycled to the soil.

Some environmentalists have called for reducing livestock production (especially cattle) to decrease its environmental effects and to feed more people. This would decrease the environmental impact of livestock production, but it would not free up much land or grain to feed more of the world's hungry people.

Cattle and sheep that graze on rangeland use a resource (grass) that humans cannot eat, and most of this land is not suitable for growing crops. Moreover, because of poverty, insufficient economic aid, and the nature of global economic and food distribution systems, very little if any additional grain grown on land used to raise livestock or livestock feed would reach the world's hungry people.

Critical Thinking

Would you be willing to eat less meat or not eat any meat? Explain.

(2) fish supplies because fish is diverted to feed livestock instead of being consumed directly by people

- Increases the spread of infectious livestock diseases such as **(1)** *mad cow disease*, which since 1985 has infected cows in 12 European nations and killed more than 80 people in Great Britain and several people in France and Ireland, and **(2)** *highly infectious hoof-and-mouth disease*, which since 2000 has infected large numbers of cattle, pigs, and sheep in Europe and Brazil

Livestock and fish vary in the efficiency with which they convert grain into animal protein (Figure 16-15). A more sustainable form of agriculture would involve shifting to more grain-efficient sources of animal protein, such as poultry. Livestock production also has an enormous environmental impact (Connections, above).

What Are the Effects of Overgrazing and Undergrazing? Overgrazing can also limit livestock production. **Overgrazing** occurs when too many animals graze for too long and exceed the carrying capacity of a grassland area. Excessive numbers of domestic livestock feeding for too long in a particular area cause most overgrazing.

Such overgrazing **(1)** lowers the net primary productivity of grassland vegetation (Figure 4-24, p. 93), **(2)** reduces grass cover and exposes the soil to erosion by water and wind (Figure 16-16, left), **(3)** compacts the soil (which reduces its capacity to hold water), **(4)** promotes invasion of exposed land by woody shrubs such as mesquite and prickly pear cactus, and **(5)** is a major cause of desertification (Figure 9-23, p. 215).

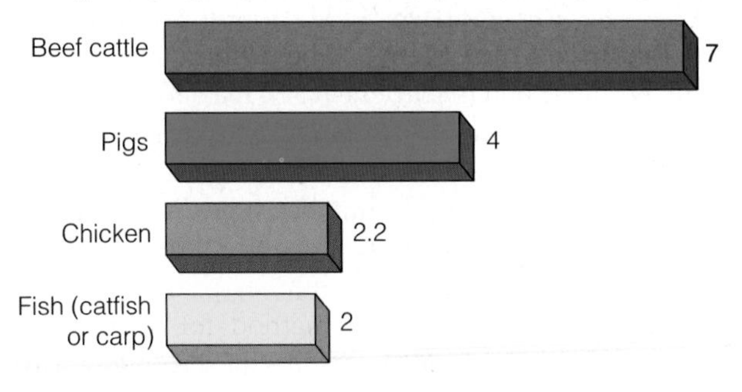

Figure 16-15 Efficiency of converting grain into animal protein, in kilograms of grain per kilogram of body weight added. (Data from U.S. Department of Agriculture)

Figure 16-16 Rangeland: overgrazed (left) and lightly grazed (right). (USDA, Natural Resources Conservation Service)

Some grassland can suffer from **undergrazing**, where absence of grazing for long periods (at least 5 years) can reduce the net primary productivity of grassland vegetation and grass cover. Moderate grazing of such areas removes accumulation of standing dead material and stimulates new biomass production.

What Is the Condition of the World's Rangelands? *Range condition* usually is classified as **(1)** *excellent* (containing more than 75% of its potential forage production), **(2)** *good* (51–75%), **(3)** *fair* (26–50%), or **(4)** *poor* (0–25%). Limited data from surveys in various countries indicate that overgrazing by livestock has caused as much as 20% of the world's rangeland to lose productivity, mostly by desertification (Figure 9-23, p. 215).

Most of the rangeland in the United States is in the West. About 60% is privately owned, and the rest is public land managed by the Bureau of Land Management (BLM) and the U.S. Forest Service. Only about 2% of the 120 million cattle and 10% of the 20 million sheep raised in the United States graze on public rangelands.

In 1990 (the latest data available), the BLM and the General Accounting Office rated 68% of nonarctic U.S. public rangeland as being in unsatisfactory (fair or poor) condition, compared with 84% in 1936. This is a great improvement, but there is still a long way to go. Conservation biologists and some range experts also point out that surveys of U.S. rangeland condition neglect the damage livestock inflict on vital riparian zones (Spotlight, right).

How Can Rangelands Be Managed Sustainably to Produce More Meat? The primary goal of sustainable rangeland management is to maximize livestock productivity without overgrazing or undergrazing rangeland vegetation. *Rangeland management* methods include **(1)** controlling the number, types, and distribution of livestock grazing on land and **(2)** restoring and improving rangeland.

The most widely used method for sustainable rangeland management is controlling the number of grazing animals and the duration of their grazing in a given area so that the carrying capacity of the area is not exceeded. However, determining the carrying capacity of a range site is difficult and costly. In addition, rangeland carrying capacity varies with factors such as **(1)** climatic conditions (especially drought), **(2)** past grazing use, **(3)** soil type, **(4)** invasions by new species, **(5)** kinds of grazing animals, and **(6)** intensity of grazing.

Livestock tend to cluster around natural water sources and stock ponds. As a result, areas around water sources tend to be overgrazed and other areas can be undergrazed. Managers can prevent this and help promote more uniform use of rangeland by **(1)** fencing off damaged rangeland and riparian zones, **(2)** moving livestock from one grazing area to another, **(3)** providing supplemental feed at selected sites, and **(4)** situating water holes and tanks and salt blocks in strategic places.

A more expensive and less widely used method of rangeland management involves suppressing the growth of unwanted plants by herbicide spraying, mechanical removal, or controlled burning. A cheaper way to discourage unwanted vegetation is controlled, short-term trampling by large numbers of livestock.

16-5 CATCHING AND RAISING MORE FISH

How Are Fish and Shellfish Harvested? The world's third major food-producing system consists of **fisheries**: concentrations of particular aquatic species suitable for commercial harvesting in a given ocean

Endangered Riparian Zones

According to some wildlife and rangeland experts, estimates of rangeland condition do not take into account severe damage to heavily grazed thin strips of lush vegetation along streams called **riparian zones**. Because cattle need lots of water, they **(1)** congregate near riparian zones, **(2)** feed there, and **(3)** trample and overgraze riparian vegetation.

These ecologically important zones **(1)** help prevent floods, **(2)** help keep streams from drying out during droughts by storing and releasing water slowly from spring runoff and summer storms, and **(3)** provide habitats, food, water, and shade for wildlife in the arid and semiarid Western lands.

Studies indicate that 65–75% of the wildlife in the western United States is dependent on riparian habitats. According to a 1999 study in the *Journal of Soil and Water Conservation*, livestock grazing has damaged approximately 80% of stream and riparian ecosystems in the United States.

Riparian areas can be restored by **(1)** using fencing to restrict access to degraded areas and **(2)** developing off-stream watering sites for livestock. Sometimes protected areas can recover in a few years.

Critical Thinking

Do you believe that riparian zones on public rangelands in the United States should receive stronger protection? Explain. If so, how would you see that such protection is provided?

area or inland body of water. Some commercially important marine species of fish and shellfish are shown in Figure 16-17.

The world's commercial fishing industry is dominated by industrial fishing fleets using satellite positioning equipment, sonar, huge nets, spotter planes, and factory ships that can process and freeze their catches. About 55% of the annual commercial catch of fish and shellfish comes from the ocean using harvesting methods shown in Figure 16-18. About 99% of this catch is taken from plankton-rich coastal waters, but this vital coastal zone is being disrupted and polluted (p. 354).

The remainder of the annual catch comes from using **(1)** aquaculture to raise marine and freshwater fish in ponds and underwater cages (33%) and **(2)** inland freshwater fishing from lakes, rivers, reservoirs, and ponds (12%). About one-third of the world fish harvest is used as animal feed, fish meal, and oils.

The *good news* is that between 1950 and 1999, **(1)** the annual commercial fish catch (marine plus freshwater harvest) increased almost fivefold (Figure 16-19, left),

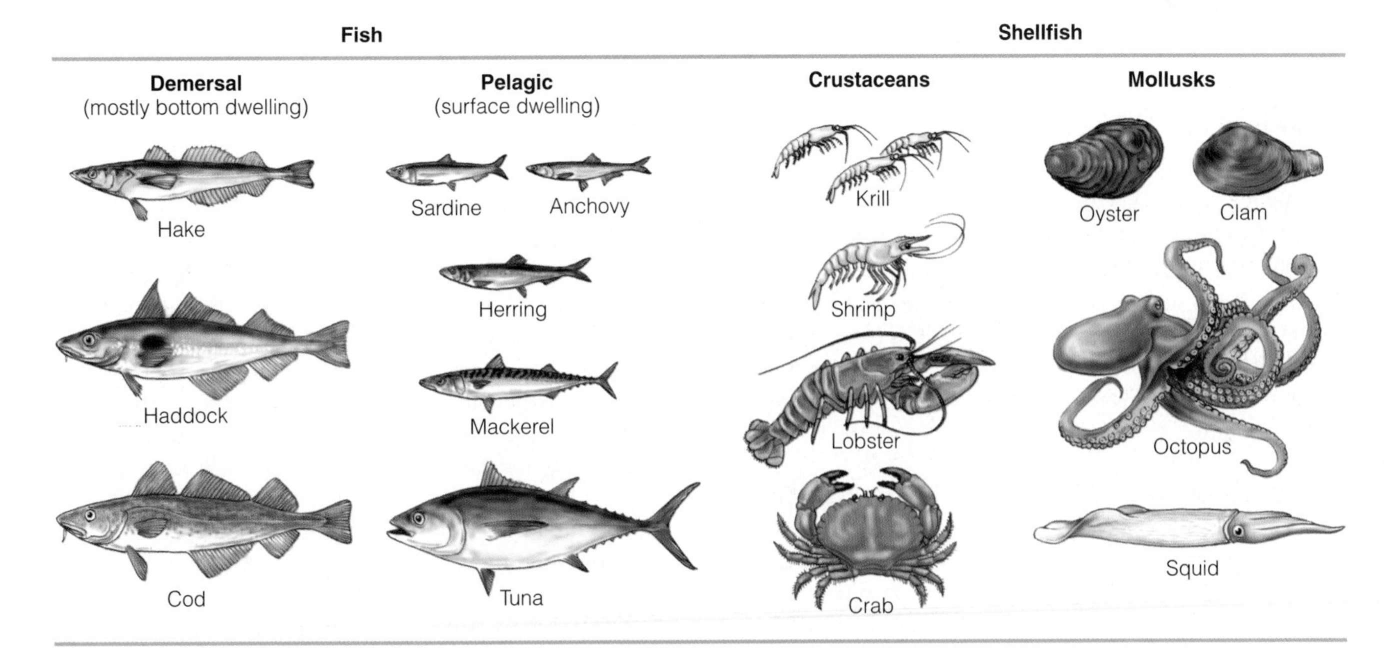

Figure 16-17 Some major types of commercially harvested marine fish and shellfish.

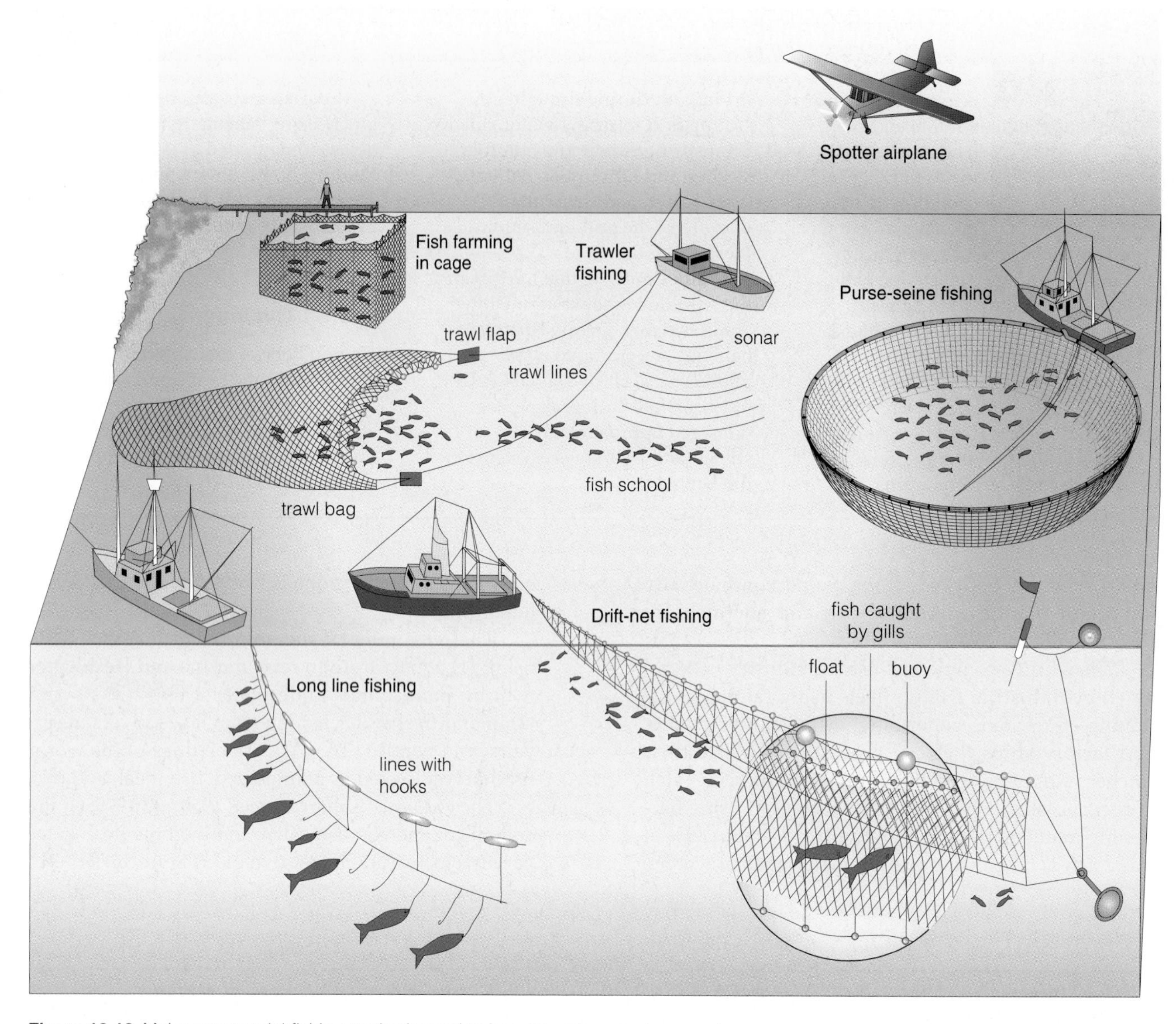

Figure 16-18 Major commercial fishing methods used to harvest various marine species.

and **(2)** the per capita seafood catch more than doubled (Figure 16-19, right).

The *bad news* is that **(1)** the commercial fish catch has increased little since 1982 (Figure 16-19, left) and **(2)** the per capita commercial fish catch has been falling since 1992 (Figure 16-19, right) and may continue to decline because of overfishing, pollution, habitat loss, and population growth.

Connections: How Are Overfishing and Habitat Degradation Affecting Fish Harvests? Fish are potentially renewable resources as long as the annual harvest leaves enough breeding stock to renew the species for the next year. Ideally, an annual **sustainable yield**—the size of the annual catch that could be harvested indefinitely without a decline in the population of a species—should be established for each species to avoid depleting the stock.

However, determining sustainable yields is not easy because **(1)** it is difficult to estimate mobile aquatic populations, and **(2)** sustainable yields shift from year to year because of changes in climate, pollution, and other factors. Furthermore, sustainably harvesting the entire annual surplus of one species may severely reduce the population of other species that rely on it for food.

Overfishing is the taking of so many fish that too little breeding stock is left to maintain numbers; that is, overfishing is a harvest that exceeds a species' sustainable yield. Prolonged overfishing leads to **commercial extinction**, when the population of a species declines to the point at which it is no longer profitable to hunt for them. Fishing fleets then move to a new

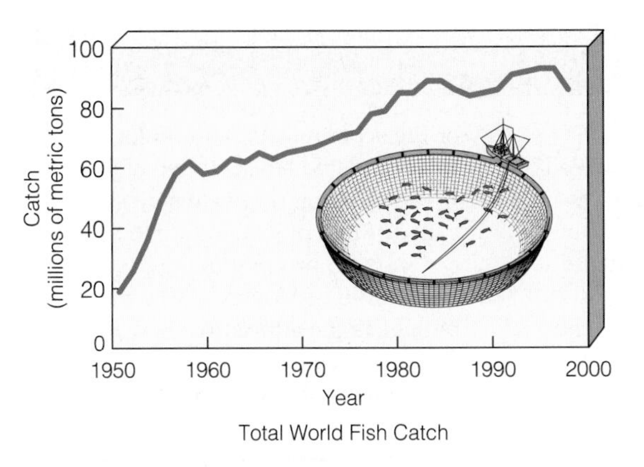

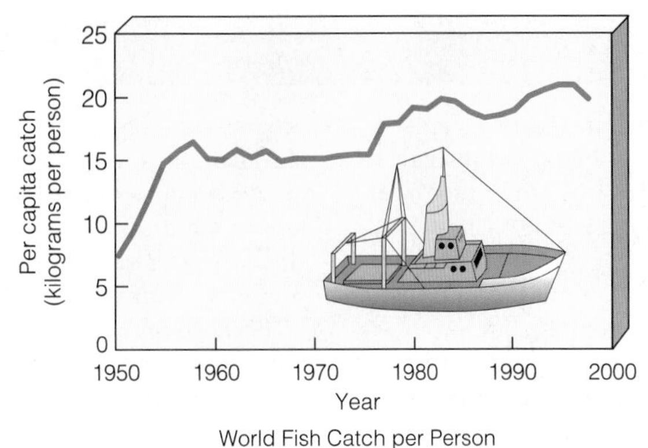

Figure 16-19 World fish catch and world fish catch per person, 1950–99. Worldwide per capita fish catch did not rise much between 1968 and 1989 and has dropped since then. (Data from UN Food and Agriculture Organization and Worldwatch Institute)

species or a new region, hoping that the overfished species will recover eventually.

Fisheries are also depleted by high levels of *bycatch*, the nontarget fish that are caught in nets and then thrown back into the sea, usually dead or dying. Nearly one-fourth of the annual global fish catch is bycatch that depletes marine biodiversity and does not provide food for people.

According to a 2001 report by the UN Food and Agriculture Organization, about 75% of the world's 200 commercially valuable marine fish species are either overfished or fished to their biological limit. The primary cause of this depletion is too many fishing boats pursuing too few fish—another example of the tragedy of the commons (Connections, p. 416). This decline in available fish in many areas helps explain why there are more than 100 disputes over rights to marine fisheries between countries.

According to the U.S. National Fish and Wildlife Foundation, 14 major commercial fish species in U.S. waters (accounting for one-fifth of the world's annual catch and half of all U.S. stocks) are so depleted that even if all fishing stopped immediately it would take up to 20 years for stocks to recover (Figure 16-20). A 2001 report by the U.S. Department of Commerce found that 107 out of 127 species taken in U.S. marine waters were in jeopardy of being overfished.

Degradation, destruction, and pollution of wetlands, estuaries, coral reefs, salt marshes, and mangroves also threaten populations of fish and shellfish. An estimated 80–90% of the global commercial marine catch comes from coastal waters within 320 kilometers (200 miles) of the shoreline.

Projected global warming over the next 50–100 years (Figure 13-8, p. 306) is also a serious threat to the global fish catch. Warmer ocean water can degrade or destroy highly productive coral reefs (p. 150) and enhance the harmful effects of habitat degradation and pollution on fish populations. The thinning of the ozone layer (Section 13-17, p. 320) can also damage surface-dwelling marine species by leading to increased penetration of harmful UV radiation into ocean waters.

Is Aquaculture the Answer? **Aquaculture**, in which fish and shellfish are raised for food, supplies about 33% of the world's commercial fish harvest. China is the world leader in aquaculture (producing about 68% of the world's output), followed by India and Japan.

There are two basic types of aquaculture. **Fish farming** involves cultivating fish in a controlled environment, often a pond or tank, and harvesting them when they reach the desired size. **Fish ranching**

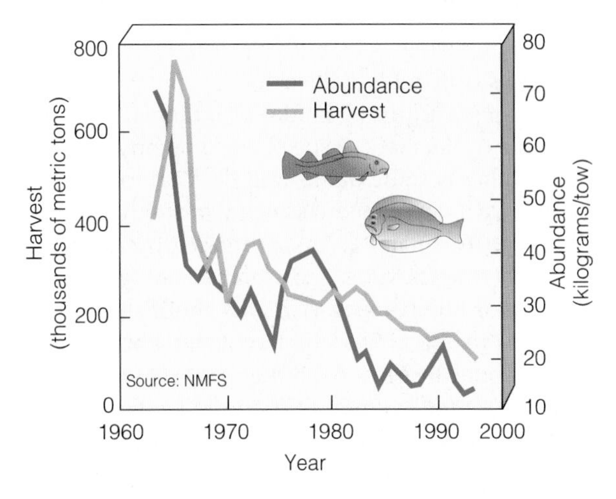

Figure 16-20 The harvest of groundfishes (yellowtail flounder, haddock, and cod) in the Georges Bank off the coast of New England in the North Atlantic, once one of the world's most productive fishing grounds, has declined sharply since 1965. Stocks dropped to such low levels that since December 1994 the National Marine Fisheries Services has banned fishing of these species in the Georges Bank. (Data from U.S. National Marine Fisheries Service)

CONNECTIONS

Commercial Fishing and the Tragedy of the Commons

In the 1970s and 1980s, extensive investment in fishing fleets, aided by government and international development agency subsidies, helped to significantly boost the fish catch (Figure 16-19, left). However, since 1975 the size of the industrial fishing fleet has expanded twice as fast as the rise in catches.

Thus, there are now too many boats fishing for a declining number of fish. This leads to overfishing, an example of the tragedy of the commons (Connections, p. 11).

A 1998 study by Daniel Pauly warned that the current harvesting of species at increasingly lower trophic levels in ocean food webs can lead to **(1)** decreased chances for recovery of species at the top of ocean food webs by reducing stocks of the smaller fish they feed on, **(2)** collapse of marine ecosystems, **(3)** a drop in aquatic biodiversity, and **(4)** a loss of high-quality protein for humans.

Because of overfishing and the overcapacity of the fishing fleet, it costs the global fishing industry about $125 billion a year to catch $70 billion worth of fish. Most of the $55-billion annual deficit of the industry is made up by government subsidies such as fuel tax exemptions, price controls, low-interest loans, and grants for fishing gear.

Critics contend that such subsidies promote overfishing. They argue that eliminating or significantly lowering these subsidies would reduce the size of the fishing fleet by encouraging free-market competition and would allow some of the economically and biologically depleted stocks to recover.

Eliminating these subsidies will cause of loss of jobs for some fishers and fish processors in coastal communities. However, to fishery biologists the alternative is worse. Continuing to subsidize excess fishing allows fishers to keep their jobs a little longer while making less and less money until the fishery collapses. Then all jobs are gone, and fishing communities suffer even more.

Critical Thinking

Do you believe that government subsidies for the fishing industry should be eliminated or sharply reduced? Explain. How would you feel about eliminating such subsidies if your livelihood depended on fishing?

involves **(1)** holding anadromous species such as salmon (which live part of their lives in fresh water and part in salt water) in captivity for the first few years of their lives (usually in fenced-in areas or floating cages in coastal lagoons and estuaries, Figure 16-18), **(2)** releasing them, and then **(3)** harvesting the adults when they return to spawn (Figure 16-21).

Species cultivated in developing countries (mostly by inland aquaculture) include carp, tilapia, milkfish, clams, and oysters. These species feed on phytoplankton and other aquatic plants and thus eat low on the food chain. In developed countries and some rapidly developing countries in Asia, aquaculture is used mostly to **(1)** stock lakes and streams with game fish or **(2)** to raise expensive fish and shellfish such as oysters, catfish, crayfish, rainbow trout, shrimp, and salmon. Aquaculture now produces **(1)** 90% of all oysters, **(2)** 40% of all salmon (75% in the United States), **(3)** 50% of internationally traded shrimp and prawns, and **(4)** 65% of freshwater fish sold in the global marketplace. Many of these species are fed grain or fish meal.

Figure 16-22 lists the major advantages and disadvantages of aquaculture. Some analysts project that freshwater and saltwater aquaculture production could double during the next 10 years. Other analysts warn that the harmful environmental effects of aquaculture (Figure 16-22) could limit future production.

Aquaculture has been promoted as a way to boost the global seafood harvest while taking the pressure off of the world's overharvested marine fisheries. However, a 2000 study by Stanford University researcher Rosamond Taylor found that increased use of aquaculture has stressed marine fisheries by **(1)** raising the demand for some ocean fish such as anchovies that are ground into fish meal that is fed to some aquaculture species and **(2)** creating vast amounts of animal waste that have fouled coastal areas, which are important sources of fish and shellfish.

Traditional aquaculture in which farmed fish eat plants and detritus is being replaced by intensive farming of large carnivorous fish whose wild populations have been decimated by overfishing. This increases overfishing of smaller marine species used to feed farmed carnivorous species. Depleting the seas to feed aquaculture farms could cause the collapse of both marine fisheries and carnivorous aquaculture.

Even under the most optimistic projections, increasing both the wild catch and aquaculture will not increase world food supplies significantly. The reason is that fish and shellfish supply only about 1% of the energy and 6% of the protein in the human diet.

How Can We Develop a More Sustainable Approach to Fishery Management? Analysts sug-

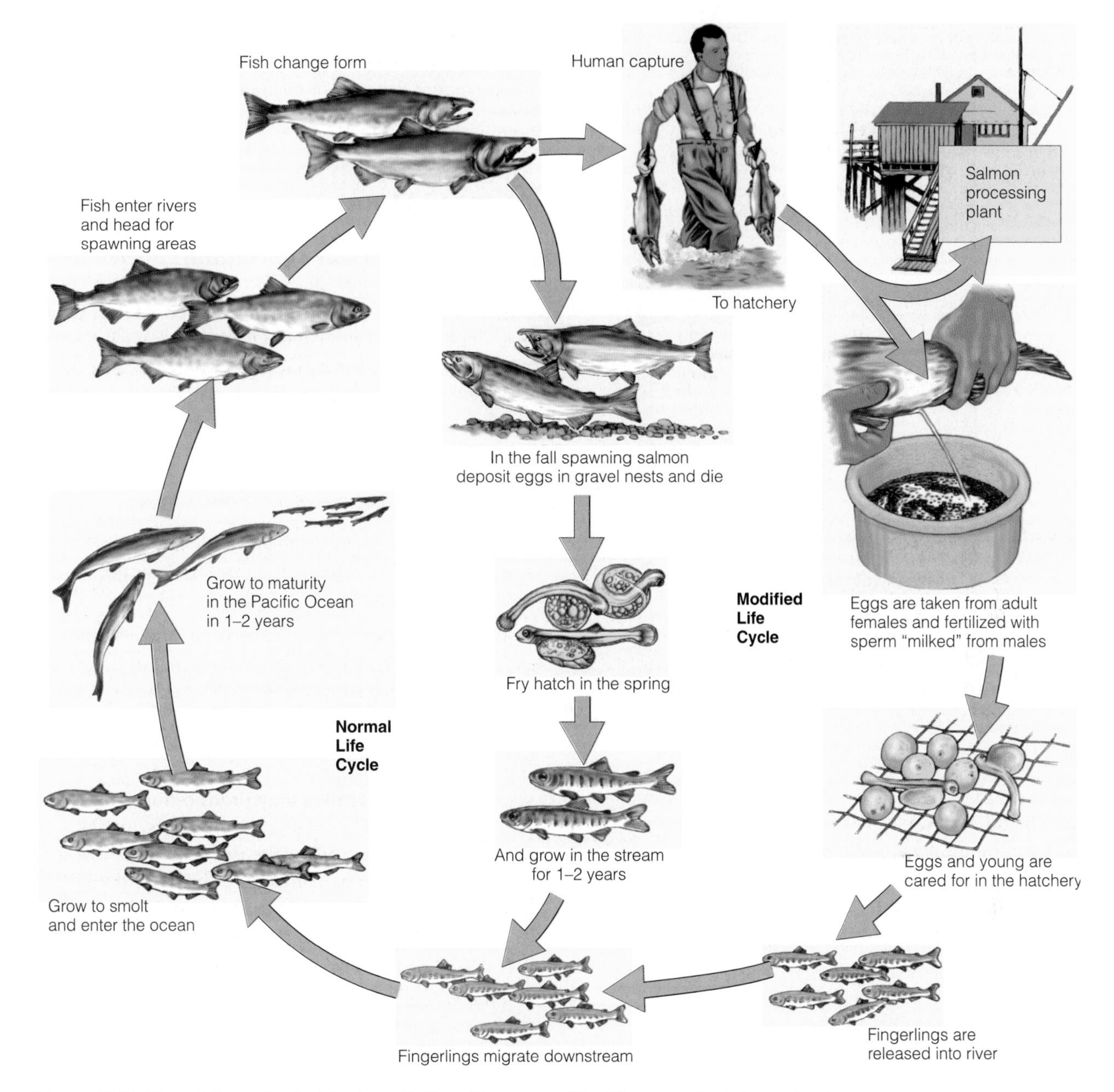
Fish change form
Human capture
Salmon processing plant
Fish enter rivers and head for spawning areas
To hatchery
In the fall spawning salmon deposit eggs in gravel nests and die
Grow to maturity in the Pacific Ocean in 1–2 years
Modified Life Cycle
Eggs are taken from adult females and fertilized with sperm "milked" from males
Fry hatch in the spring
Normal Life Cycle
And grow in the stream for 1–2 years
Eggs and young are cared for in the hatchery
Grow to smolt and enter the ocean
Fingerlings migrate downstream
Fingerlings are released into river

Figure 16-21 Normal life cycle of wild salmon (left) and human-modified life cycle of hatchery-raised salmon (right). Salmon spend part of their lives in fresh water and part in salt water.

gest the following measures for managing global fisheries more sustainably and protecting marine biodiversity:

Fishery Regulations

- *Set, monitor, and enforce fishery quotas well below their estimated maximum sustained yields.*
- *Improve monitoring and enforcement of fishing regulations.*

Economic Approaches

- *Sharply reduce or eliminate fishing subsidies* (Connections, left).
- *Impose fees for harvesting fish and shellfish from publicly owned and managed offshore waters and use the money for government fishery management, as Australia does.*

Bycatch

- *Reduce bycatch levels* by **(1)** using wider-mesh nets to allow smaller species and smaller individuals of the targeted species to escape, **(2)** having observers on fishing vessels, **(3)** licensing boats to catch several species instead of only one target species, and **(4)** enacting laws that prohibit throwing edible and marketable fish back to sea

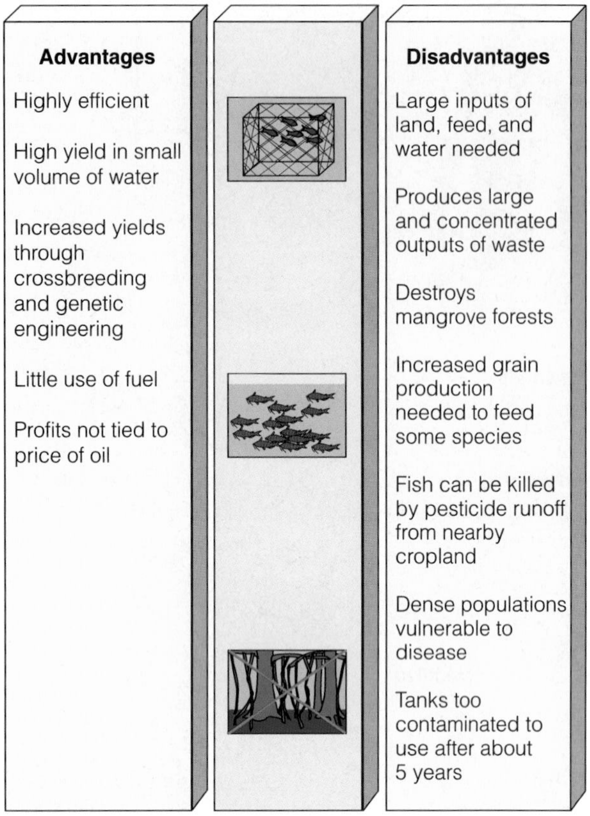

Advantages	Disadvantages
Highly efficient	Large inputs of land, feed, and water needed
High yield in small volume of water	Produces large and concentrated outputs of waste
Increased yields through crossbreeding and genetic engineering	Destroys mangrove forests
Little use of fuel	Increased grain production needed to feed some species
Profits not tied to price of oil	Fish can be killed by pesticide runoff from nearby cropland
	Dense populations vulnerable to disease
	Tanks too contaminated to use after about 5 years

Figure 16-22 Advantages and disadvantages of aquaculture.

(as Namibia and Norway have done, with the law enforced by onboard observers).

Protected Areas

- *Establish no-fishing marine areas, seasonal fishery closures, and marine protected areas to allow depleted fish species to recover.*
- *Strengthen commitment to marine biodiversity protection and integrated coastal management programs that promote both sustainable fishing and the ecological health of marine ecosystems.*

Consumer Information

- *Use labels that allow consumers to identify fish that have been harvested sustainably.* See the website material for this chapter for lists of seafood species that consumers should consider not buying and those that are safe for now, as compiled in 2000 by the National Audubon Society and the Monterey Bay Aquarium.

Aquaculture

- *Restrict location of fish farms to reduce loss of mangrove forests and other threatened coastal environments.*
- *Enact and enforce stricter pollution regulations for aquaculture operations.*
- *Increase production of herbivorous aquaculture fish species (such as carp, tilapia, and shellfish) that need little or no grain or fish meal in their diets.*

16-6 GOVERNMENT AGRICULTURAL POLICY

How Do Government Agricultural Policies Affect Food Production? Agriculture is a financially risky business. Whether farmers have a good year or a bad year depends on factors over which they have little control: weather, crop prices, crop pests and diseases, interest rates, and the global market. Because of the need for reliable food supplies despite fluctuations in these factors, most governments provide various forms of assistance to farmers and consumers.

Three approaches are to

- *Keep food prices artificially low.* This makes consumers happy but means that farmers may not be able to make a living.
- *Give farmers subsidies to keep them in business and encourage them to increase food production.* Globally, government price supports and other subsidies for agriculture total more than $500 billion per year (including $100 billion per year in the United States). If government subsidies are too generous and the weather is good, farmers may produce more food than can be sold. The resulting surplus depresses food prices, which reduces the financial incentive for farmers in developing countries to increase domestic food production. Moreover, the taxes paid by citizens in developed countries to provide agricultural subsidies can offset the lower food prices they enjoy.
- *Eliminate most or all price controls and subsidies and let farmers and fishers respond to market demand without government interference.* However, some analysts urge that any phaseout of farm and fishery subsidies should be coupled with increased aid for the poor and the lower middle class, who would suffer the most from any increase in food prices.

Many environmentalists believe that instead of eliminating all subsidies, we should use them to reward farmers and ranchers who **(1)** protect the soil, **(2)** conserve water, **(3)** reforest degraded land, **(4)** protect and restore wetlands, **(5)** conserve wildlife, and **(6)** practice more sustainable agriculture and fishing.

16-7 PROTECTING FOOD RESOURCES: PESTICIDES AND PEST CONTROL

What Are Pesticides, and How Are They Used? A **pest** is any species that **(1)** competes with us for food,

(2) invades lawns and gardens, (3) destroys wood in houses, (4) spreads disease, or (5) is simply a nuisance. In natural ecosystems and many polyculture agroecosystems (p. 396), *natural enemies* (predators, parasites, and disease organisms) control the populations of 50–90% of pest species as part of the earth's ecological services (Figure 4-33, p. 103) and help keep any one species from taking over for very long.

When we replace polyculture agriculture with monoculture agriculture and spray fields with huge amounts of pesticides, we upset many of these natural population checks and balances. Then we must devise ways to protect our monoculture crops, tree farms, and lawns from insects and other pests that nature once controlled at no charge.

To help control pest organisms, we have developed a variety of **pesticides** (or *biocides*): chemicals to kill organisms we consider undesirable. Common types of pesticides include **(1)** *insecticides* (insect-killers), **(2)** *herbicides* (weed-killers), **(3)** *fungicides* (fungus-killers), **(4)** *nematocides* (roundworm-killers), and **(5)** *rodenticides* (rat- and mouse-killers).

Since 1950 pesticide use has increased more than 50-fold, and most of today's pesticides are more than 10 times as toxic as those used in the 1950s. Worldwide, about 2.3 million metric tons (2.5 million tons) of these pesticides are used yearly. About 75% of these chemicals are used in developed countries, but use in developing countries is soaring.

According to the U.S. Environmental Protection Agency (EPA), the average lawn in the United States is doused with 10 times more synthetic pesticides per hectare than U.S. cropland. Each year, more than 250,000 people in the United States become ill because of household pesticide use, and such pesticides are a major source of accidental poisonings and deaths for children under age 5.

Some pesticides, called *broad-spectrum agents*, are toxic to many species; others, called *selective* or *narrow-spectrum agents*, are effective against a narrowly defined group of organisms. Pesticides vary in their *persistence*, the length of time they remain deadly in the environment. In 1962, biologist Rachel Carson warned against relying on synthetic organic chemicals to kill insects and other species we deem pests (Individuals Matter, p. 420).

What Is the Case for Pesticides? Proponents of conventional chemical pesticides contend that their benefits outweigh their harmful effects. Here are some of the major benefits of conventional pesticides:

- *They save human lives.* Since 1945, DDT and other chlorinated hydrocarbon and organophosphate insecticides probably have prevented the premature deaths of at least 7 million people from insect-transmitted diseases such as **(1)** malaria (carried by the *Anopheles* mosquito, p. 235), **(2)** bubonic plague (rat fleas), **(3)** typhus (body lice and fleas), and **(4)** sleeping sickness (tsetse fly).

- *They increase food supplies and lower food costs.* About 55% of the world's potential human food supply is lost to pests before (35%) or after (20%) harvest. Pests before and after harvest destroy an estimated 37% of the potential U.S. food supply; insects cause 13% of these losses, plant pathogens 12%, and weeds 12%. Without pesticides, these losses would be worse, and food prices would rise.

- *They increase profits for farmers.* Pesticide companies estimate that every $1 spent on pesticides leads to an increase in U.S. crop yields worth approximately $4 (but studies have shown that this benefit drops to about $2 if the harmful effects of pesticides are included).

- *They work faster and better than alternatives.* Pesticides can **(1)** control most pests quickly and at a reasonable cost, **(2)** have a long shelf life, **(3)** are easily shipped and applied, and **(4)** are safe when handled properly. When genetic resistance occurs, farmers can use stronger doses or switch to other pesticides.

- *When used properly, their health risks are very low compared with their benefits.* According to Elizabeth Whelan, director of the American Council on Science and Health (ACSH), which presents the position of the pesticide industry, "The reality is that pesticides, when used in the approved regulatory manner, pose no risk to either farm workers or consumers."

- *Newer pesticides are safer and more effective than many older pesticides.* Greater use is being made of botanicals and microbotanicals, derived originally from plants, that are safer to users and less damaging to the environment. Genetic engineering is also being used to develop pest-resistant crop strains and genetically altered crops that produce pesticides (Figure 16-11).

- *Many new pesticides are used at very low rates per unit area compared with those of older products.* For example, application amounts per hectare for many new herbicides are 1/100 the rates for older ones, and genetically engineered crops could reduce the use of toxic insecticides (Figures 16-10 and 16-11).

Scientists continue to search for the ideal pest-killing chemical, which would

- *Kill only the target pest*
- *Harm no other species*
- *Disappear or break down into something harmless after doing its job*
- *Not cause genetic resistance in target organisms*

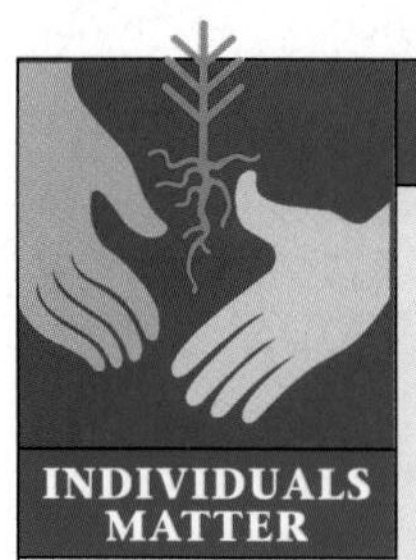

Rachel Carson

Rachel Carson (Figure 16-23) began her professional career as a biologist for the Bureau of U.S. Fisheries (later to become the U.S. Fish and Wildlife Service). In that capacity, she **(1)** carried out research on oceanography and marine biology, **(2)** wrote articles about the oceans and topics related to the environment, and **(3)** became editor-in-chief of the bureau's publications in 1949.

In 1951, she wrote *The Sea Around Us*, which described in easily understandable terms the natural history of oceans and the harm that humans were doing them. The book **(1)** was on the best-seller list for 86 weeks, **(2)** sold more than 2 million copies, **(3)** was translated into 32 languages, and **(4)** won a National Book Award.

During the late 1940s and throughout the 1950s, the use of DDT and related compounds to kill insects that ate food crops, attacked trees, bothered people, and transmitted diseases such as malaria expanded rapidly.

In 1958, DDT was sprayed to control mosquitoes near the home and private bird sanctuary of Olga Huckins, a good friend of Carson. After the spraying, Huckins witnessed the agonizing deaths of several of her birds, and in distress she asked Carson whether she could find someone to investigate the effects of pesticides on birds and other wildlife.

Carson decided to look into the issue herself and quickly found that almost no independent research on the environmental effects of pesticides existed. As a well-trained scientist, Carson **(1)** surveyed the scientific literature, **(2)** became convinced that pesticides could harm wildlife and humans, and **(3)** built a case against the widespread use of pesticides.

In 1962 she published her findings in popular form in *Silent Spring*, an allusion to the silencing of "robins, catbirds, doves, jays, wrens, and scores of other bird voices" because of their exposure to pesticides. She pointed out that "for the first time in the history of the world, every human being is now subjected to dangerous chemicals, from the moment of conception until death."

Carson's book was read by many scientists, politicians, and policy makers and was embraced by the public. However, the chemical industry viewed the book as a serious threat to booming pesticide sales and mounted a campaign to discredit Carson. A parade of critical reviewers and industry scientists claimed that her book **(1)** was full of inaccuracies, **(2)** made selective use of research findings, and **(3)** failed to give a balanced account of the benefits of pesticides.

Some critics even claimed that, as a woman, she was incapable of understanding the highly scientific and technical subject of pesticides. Others charged that she was a hysterical woman and a radical nature lover trying to scare the American public in order to sell books.

During this period of intense controversy Carson was suffering from terminal cancer, but she was able to defend her research and strongly counter her critics. She died in 1964, about 18 months after the publication of *Silent Spring*, without knowing that many historians considered her work a key element in the birth the modern environmental movement in the United States.

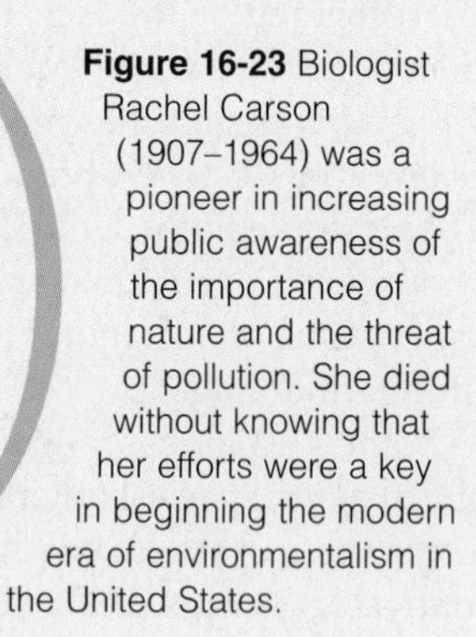

Figure 16-23 Biologist Rachel Carson (1907–1964) was a pioneer in increasing public awareness of the importance of nature and the threat of pollution. She died without knowing that her efforts were a key in beginning the modern era of environmentalism in the United States.

- *Be more cost-effective than doing nothing*

The search continues, but so far no known natural or synthetic pesticide chemical meets all or even most of these criteria.

What Is the Case Against Pesticides? Opponents of widespread pesticide use believe that their harmful effects outweigh their benefits. Here are some of the major problems with conventional pesticides:

- *They accelerate the development of genetic resistance to pesticides.* Insects breed rapidly, and within 5–10 years (much sooner in tropical areas) they can develop immunity to pesticides through directional natural selection (Figure 5-5, left, p. 111) and come back stronger than before. Weeds and plant disease organisms also develop genetic resistance, but more slowly. Since 1945, at least 520 insect and mite species, 273 weed species, 150 plant diseases, and 10 rodent species (mostly rats) have developed genetic resistance to one or more pesticides (Figure 16-24). Because of genetic resistance, many widely used insecticides (such as DDT) no longer protect people from insect-transmitted diseases in some parts of the

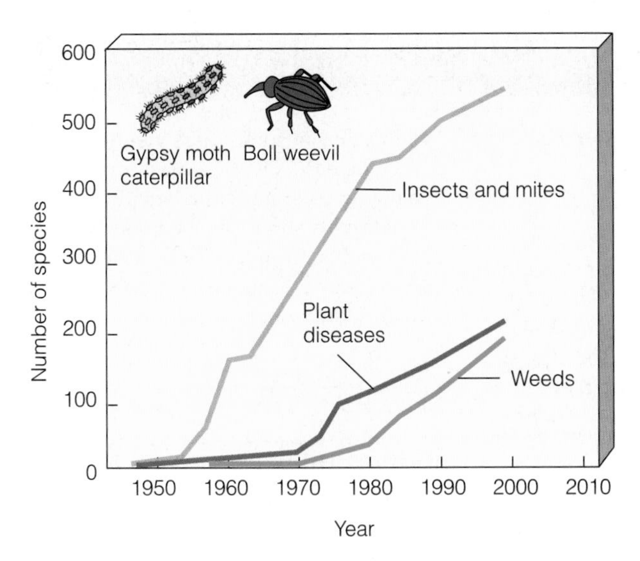

Figure 16-24 Rise of genetic resistance to pesticides, 1945–1998. (Data from U.S. Department of Agriculture and the Worldwatch Institute)

world. This has led to resurgences of diseases such as malaria (Figure 10-9, p. 235). Genetic resistance can also put farmers on a *pesticide treadmill,* whereby they pay more and more for a pest control program that often becomes less and less effective.

- *Broad-spectrum insecticides kill natural predators and parasites that help control the populations of pest species.* Wiping out natural predators can also unleash new pests whose populations the predators had previously held in check, causing other unexpected effects (Connections, p. 190). Currently 100 of the 300 most destructive insect pests in the United States were secondary pests that became major pests after widespread use of insecticides. Mostly because of genetic resistance and reduction of natural predators, pesticide use has not reduced crop losses to pests in the United States (Spotlight, p. 422).

- *Pesticides do not stay put.* According to the U.S. Department of Agriculture (USDA), **(1)** no more than 2% (and often less than 0.1%) of the insecticide applied to crops by aerial spraying or ground spraying reaches the target pests, and **(2)** less than 5% of herbicide applied to crops reaches the target weeds. Pesticides that miss their target pests can end up in the **(1)** air, **(2)** surface water, **(3)** groundwater, **(4)** bottom sediments, **(5)** food, and **(6)** nontarget organisms, including humans and wildlife. Crops that have been genetically altered to release small amounts of pesticides directly to pests can help overcome this problem but can also promote genetic resistance to such pesticides.

- *Some pesticides harm wildlife.* According to the USDA and the U.S. Fish and Wildlife Service, each year in the United States pesticides applied to cropland **(1)** wipe about 20% of U.S. honeybee colonies and damage another 15%, costing farmers at least $200 million per year from reduced pollination of vital crops, **(2)** kill more than 67 million birds and 6–14 million fish, and **(3)** menace about 20% of the endangered and threatened species in the United States.

- *They can threaten human health.* According to the WHO and the UN Environment Programme (UNEP), an estimated 3 million agricultural workers in developing countries (at least 300,000 in the United States) are seriously poisoned by pesticides each year. This results in an estimated 18,000 deaths (about 25 in the United States)—an average of 490 premature deaths each day. Health officials believe that the actual number of pesticide-related illnesses and deaths among the world's farm workers probably is greatly underestimated because of **(1)** poor records, **(2)** lack of doctors and disease reporting in rural areas, and **(3)** inaccurate diagnoses. Each year 110,000 Americans, mostly children, get sick from misuse or unsafe storage of pesticides in the home, and about 20 die. According to the EPA, approximately 165 of the active ingredients approved for use in U.S. pesticide products are known or suspected human carcinogens. Some scientists are becoming increasingly concerned about possible **(1)** genetic mutations, **(2)** birth defects, **(3)** nervous system disorders (especially behavioral disorders), and **(4)** effects on the immune and endocrine systems from long-term exposure to low levels of various pesticides (Connections, p. 231).

Is the U.S. Public Adequately Protected from Exposure to Pesticides? There is controversy over how well the public in the United States is protected from the harmful effects of pesticides. According to studies by National Academy of Sciences, **(1)** federal laws regulating pesticide use in the United States are inadequate and poorly enforced by the EPA, Food and Drug Administration (FDA), and USDA, and **(2)** exposure to pesticide residues in food causes 4,000–20,000 cases of cancer per year, which lead to 2,000–10,000 premature deaths per year in the United States.

Representatives from pesticide companies dispute these findings. Indeed, the food industry denies that eating food grown using pesticides for the past 50 years has ever harmed anyone in the United States.

Between 1972 and 2000, the EPA banned or severely restricted the use of 56 active pesticide ingredients. However, banned or unregistered pesticides may be manufactured in the United States and exported to other countries (Connections, p. 423).

What Are Other Ways to Control Pests? Many scientists believe that we should greatly increase the

SPOTLIGHT

How Effective Have Synthetic Pesticides Been in Reducing Crop Losses?

Studies indicate that pesticides have not been as effective in reducing crop losses in the United States as agricultural experts had hoped. David Pimentel, an expert in insect ecology, has evaluated data from more than 300 agricultural scientists and economists and come to the following conclusions:

- Although the use of synthetic pesticides has increased 33-fold since 1942, it is estimated that more of the U.S. food supply is lost to pests today (37%) than in the 1940s (31%). Losses attributed to insects almost doubled (from 7% to 13%) despite a 10-fold increase in the use of synthetic insecticides.

- The estimated environmental, health, and social costs of pesticide use in the United States range from $4 to $10 billion per year. The International Food Policy Research Institute puts the estimate much higher, at $100–200 billion per year, or $5–10 in damages for every dollar spent on pesticides.

- Alternative pest control practices (pp. 422–424) could halve the use of chemical pesticides on 40 major U.S. crops without reducing crop yields.

Numerous studies and experience show that pesticide use can be reduced sharply without reducing yields, and in some cases yields even increase. Over the past few years, Sweden has cut pesticide use in half with almost no decrease in crop yields. Campbell Soup uses no pesticides on tomatoes it grows in Mexico, and yields have not dropped. After a 65% cut in pesticide use on rice in Indonesia, yields increased by 15%.

Critical Thinking

Pesticide proponents argue that although crop losses to pests are higher today than in the past, without the widespread use of pesticides losses would be even higher. Explain why you agree or disagree with this argument.

use of biological, ecological, and other alternative methods for controlling pests and diseases that affect crops and human health. Such methods include using

- *Cultivation practices* such as **(1)** rotating the types of crops planted in a field each year, **(2)** adjusting planting times so that major insect pests either starve or get eaten by their natural predators, **(3)** growing crops in areas where their major pests do not exist, **(4)** planting trap crops to lure pests away from the main crop, and **(5)** increasing the use of polyculture (p. 396), which uses plant diversity to reduce losses to pests.

- *Genetic engineering* (Figure 16-25) *to speed up the development of pest- and disease-resistant crop strains.* However, there is controversy over whether the projected advantages of the increasing use of genetically modified plants and foods outweigh their projected disadvantages (Figure 16-11).

- *Biological pest control, in which natural predators* (Figure 16-26), *parasites, and disease-causing bacteria and viruses can be imported to regulate pest populations.* This approach **(1)** focuses on selected target species, **(2)** is nontoxic to other species, **(3)** can save large amounts of money ($25 for every $1 invested in controlling 70 pests in the United States and $178 for each $1 invested by Nigerian farmers in using parasitic wasps to fight the cassava mealybug), and **(4)** minimizes genetic resistance. However, biological control agents **(1)** cannot always be mass-produced, **(2)** often are slower acting and more difficult to apply than conventional pesticides, **(3)** must be protected from pesticides sprayed in nearby fields, and **(4)** can sometimes multiply and become pests themselves.

- *Insect birth control,* in which laboratory-raised males of insect pests are sterilized by radiation or chemicals and released into an infested area to mate unsuccessfully with fertile wild females. This has been used to control **(1)** the screwworm fly, a major livestock pest from the southeastern United States (Figure 16-27), and **(2)** the Mediterranean fruit fly (medfly) during a 1990 outbreak in California. Problems include **(1)** high costs,

Figure 16-25 The results of one example of using genetic engineering to reduce pest damage. Both tomato plants were exposed to destructive caterpillars. The normal plant's leaves are almost gone (left), whereas the genetically altered plant (right) shows little damage. (Monsanto)

CONNECTIONS

What Goes Around Can Come Around

U.S. pesticide companies can make and export to other countries pesticides that have been banned or severely restricted—or never even approved—in the United States. Between 1992 and 2000, U.S. exports of such pesticides (most to developing countries) averaged more than 27 metric tons (30 tons) per day. Other industrial countries also export banned and unapproved pesticides.

However, what goes around can come around. In what environmentalists call a *circle of poison*, residues of some of these banned or unapproved chemicals exported to other countries can return to the exporting countries on imported food. Persistent pesticides such as DDT can also be carried by winds from other countries to the United States.

Environmentalists have urged Congress—without success—to ban such exports. Supporters of pesticide exports argue that **(1)** such sales increase economic growth and provide jobs, **(2)** if the United States did not export pesticides, other countries would, and **(3)** banned pesticides are exported only with the consent of the importing countries.

In 1998, more than 50 countries met to finalize an international treaty that requires exporting countries to have informed consent from importing countries for exports of 22 pesticides and 5 industrial chemicals. In 2000, more than 100 countries developed an international agreement to ban or phase out the use of 12 especially hazardous persistent organic pollutants (9 of them persistent chlorinated hydrocarbon pesticides).

Critical Thinking

Should U.S. companies be allowed to export pesticides that have been banned, severely restricted, or not approved for use in the United States? Explain.

(2) difficulties in knowing the mating times and behaviors of each target insect, **(3)** the large number of sterile males needed, **(4)** the few species for which this strategy works, and **(5)** the need to release sterile males continually to prevent resurgence.

- *Sex attractants.* Sex attractants (called pheromones) can lure pests into traps or attract their natural predators into cropfields (usually the more effective approach). These chemicals **(1)** attract only one species, **(2)** work in trace amounts, **(3)** have little chance of causing genetic resistance, and **(4)** are not harmful to nontarget species. However, it is costly and time-consuming to identify, isolate, and produce the specific sex attractant for each pest or predator.

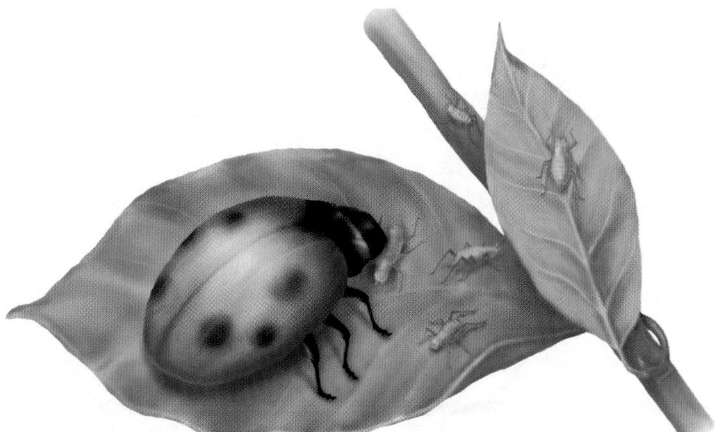

Figure 16-26 Biological pest control: An adult convergent ladybug (left) is consuming an aphid (right).

- *Hormones that disrupt an insect's normal life cycle, causing the insect to fail to reach maturity and reproduce* (Figure 16-28). Insect hormones have the same advantages as sex attractants. However, they **(1)** take weeks to kill an insect, **(2)** often are ineffective with large infestations of insects, **(3)** sometimes break down before they can act, **(4)** must be applied at exactly the right time in the target insect's life cycle, **(5)** can sometimes affect the target's predators and other nonpest species, and **(6)** are difficult and costly to produce.

- *Spraying insects with hot water.* So far, the system has worked well on cotton, alfalfa, and potato fields and in citrus groves in Florida, where the spray machine was invented. The cost is roughly equal to that of using chemical pesticides.

- *Exposing foods to high-energy gamma radiation.* Such *food irradiation* extends food shelf life and kills **(1)** insects, **(2)** parasitic worms (such as trichinae in pork), and **(3)** bacteria (such as salmonella, which infects at least 51,000 Americans and kills 2,000 each year, and *E. coli*, which infects more than 20,000 Americans and kills about 250 each year). According to the U.S. FDA and the WHO, more than 2,000 studies over three decades show that foods exposed to low doses of ionizing radiation are safe for human consumption. Critics of food irradiation argue that **(1)** irradiating food forms trace amounts of certain chemicals called free radicals, some of which have caused

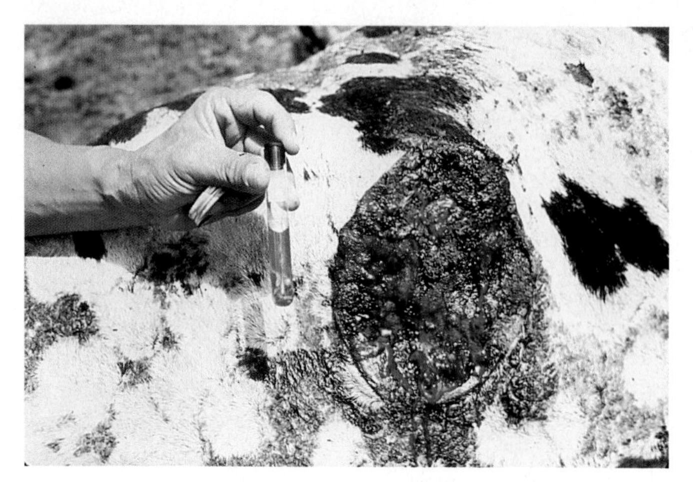

Figure 16-27 Infestation of a steer by screwworm fly larvae in Texas. An adult steer can be killed in 10 days by thousands of maggots feeding on a single wound. (U.S. Department of Agriculture)

cancer in laboratory animals, **(2)** we do not know the long-term health effects of eating irradiated food, **(3)** current levels of irradiation do not destroy botulinum spores, but they do destroy the bacteria that give off the rotten odor warning us of their presence, and **(4)** consumers want fresh, wholesome food, not old, possibly less nutritious food made to appear fresh by irradiation, and want clear labels on all irradiated food so they can make informed choices.

Is Integrated Pest Management the Answer? An increasing number of pest control experts and farmers believe that the best way to control crop pests is a carefully designed **integrated pest management (IPM)** program. In this approach, each crop and its pests are evaluated as parts of an ecological system. Then a control program is developed that includes cultivation, biological, and chemical methods applied in proper sequence and with the proper timing.

The overall aim of IPM is not to eradicate pest populations but to reduce crop damage to an economically tolerable level. Fields are monitored carefully, and when an economically damaging level of pests is reached, farmers first use biological methods (natural predators, parasites, and disease organisms) and cultivation controls, including vacuuming up harmful bugs. Small amounts of insecticides (mostly botanicals or microbotanicals) are applied only as a last resort, and different chemicals are used to slow development of genetic resistance and to avoid killing predators of pest species.

In 1986, the Indonesian government **(1)** banned the use of 57 of the 66 pesticides used on rice, **(2)** phased out pesticide subsidies over a 2-year period, and **(3)** used some of the money to help launch a nationwide program to switch to IPM, including a major farmer education program. The results were dramatic: Between 1987 and 1992, **(1)** pesticide use dropped by 65%, **(2)** rice production rose by 15%, and **(3)** more than 250,000 farmers were trained in IPM techniques. By 1993, the program had saved the Indonesian government more than $1.2 billion (most from elimination of $120-million-per-year pesticide subsidies), more than enough to fund its IPM program.

The experiences of countries such as China, Brazil, Indonesia, Australia, Bangladesh, the Philippines, Mexico, and the United States show that a well-designed IPM program can **(1)** reduce pesticide use and pest control costs by 50–90%, **(2)** reduce preharvest pest-induced crop losses by 50%, **(3)** improve crop yields, **(4)** reduce inputs of fertilizer and irrigation water, and **(5)** slow the development of genetic resistance because pests are assaulted less often and with lower doses of pesticides.

Thus, IPM is an important form of *pollution prevention* that reduces risks to wildlife and human health. Consumer's Union estimates that if all U.S. farmers practiced IPM by 2020, public health risks from pesticides would drop by 75%.

Despite its promise, IPM, like any other form of pest control, has some disadvantages. It requires expert knowledge about each pest situation, it is slower acting than conventional pesticides, and methods developed for a crop in one area might not apply to areas with even slightly different growing conditions. Although long-term costs typically are lower than those of using conventional pesticides, initial costs may be higher.

Widespread use of IPM is hindered by government subsidies of conventional chemical pesticides and by opposition from

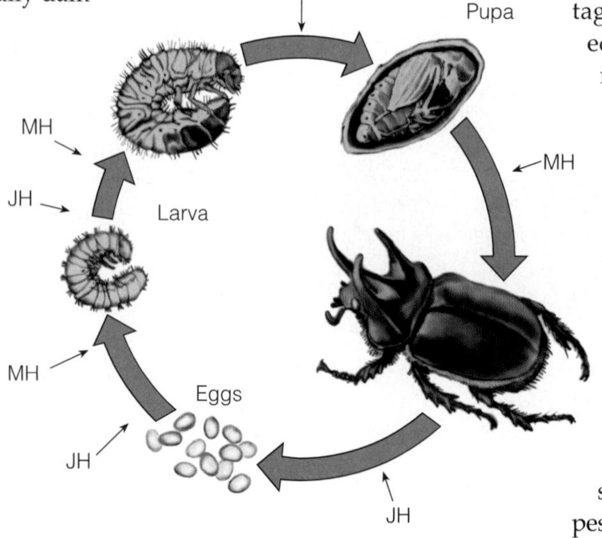

Figure 16-28 For normal insect growth, development, and reproduction to occur, certain juvenile hormones (JH) and molting hormones (MH) must be present at genetically determined stages in the insect's life cycle. If applied at the proper time, synthetic hormones disrupt the life cycles of insect pests and help control their populations.

agricultural chemical companies, whose pesticide sales would drop sharply. In addition, farmers get most of their information about pest control from pesticide salespeople (and in the United States from USDA county farm agents). Most of these advisers do not have adequate training in IPM.

A 1996 study by the National Academy of Sciences recommended that the United States shift from chemically based approaches to ecologically based pest management approaches. A growing number of scientists urge Congress to promote IPM in the United States by **(1)** adding a 2% sales tax on pesticides and using the revenue to fund IPM research and education, **(2)** setting up a federally supported IPM demonstration project on at least one farm in every county, **(3)** training USDA field personnel and county farm agents in IPM so that they can help farmers use this alternative, **(4)** providing federal and state subsidies, and perhaps government-backed crop insurance, to farmers who use IPM or other approved alternatives to pesticides, and **(5)** gradually phasing out subsidies to farmers who depend almost entirely on convential pesticides as effective IPM methods are developed for major pest species. See the website material for this chapter for some actions you can take to reduce your use of and exposure to pesticides.

16-8 SOLUTIONS: SUSTAINABLE AGRICULTURE

What Is Sustainable Agriculture? Evidence suggests that two of the world's major food production systems, rangelands and oceanic fisheries, have reached or are near their limits. If these trends continue, most future growth in food production will have to come from **(1)** increasing the amount of land used to grow crops, **(2)** increasing crop yields, and **(3)** using less water for irrigation (p. 339).

However, the total area of cropland is unlikely to expand because of a lack of affordable and environmentally sustainable land (p. 409). In addition, increasing the yields per area of existing cropland may be limited because of **(1)** a lack of water for irrigation (Figure 16-12), **(2)** reduced genetic diversity (Spotlight, p. 408), **(3)** a leveling off of yields per hectare, and **(4)** the environmental effects of food production, which degrade existing cropland (Figure 16-9).

If these projections are correct, the two main tools in trying to reduce hunger, malnutrition, poverty, and the harmful environmental effects of agriculture will be **(1)** slowing population growth (Section 11-3, p. 253) and **(2)** developing and phasing in systems of **sustainable agriculture** or **low-input agriculture** (also called **organic farming**) over the next three decades.

Gordon Conway, president of the Rockefeller Foundation, calls this a new *doubly green revolution* that will **(1)** increase crop yields in an environmentally sustainable manner and **(2)** benefit the poor more directly than the first two green revolutions. Currently, organic farming is used on less than 1% of the world's cropland (0.2% in the United States) but on 6–10% of the cropland in many European countries (2.8% in the 15-member European Union). However, this type of farming is growing rapidly and in 2000 was a $25-billion global market.

Figure 16-29 lists the major components of more sustainable, low-input agriculture. Studies have shown that low-input organic farming **(1)** produces equivalent yields with lower carbon dioxide emissions (Solutions, p. 316), **(2)** uses about 50% less energy than conventional farming, **(3)** improves soil fertility, and **(4)** generally is more profitable for the farmer than high-input farming.

Most proponents of more sustainable agriculture are not opposed to high-yield agriculture. Instead, they see it as vital for protecting the earth's biodiversity by reducing the need to cultivate new and often marginal land. They call for using more environmentally sustainable forms of both high-yield polyculture and high-yield monoculture for growing crops.

Can We Make the Transition to More Sustainable Agriculture? A growing number agricultural analysts believe that over the next 30 years we must

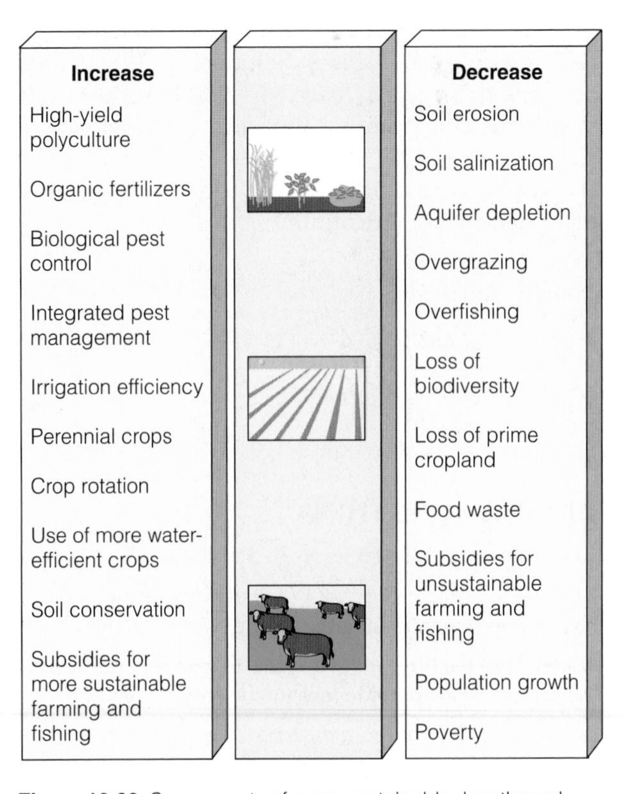

Increase
High-yield polyculture
Organic fertilizers
Biological pest control
Integrated pest management
Irrigation efficiency
Perennial crops
Crop rotation
Use of more water-efficient crops
Soil conservation
Subsidies for more sustainable farming and fishing
Decrease
Soil erosion
Soil salinization
Aquifer depletion
Overgrazing
Overfishing
Loss of biodiversity
Loss of prime cropland
Food waste
Subsidies for unsustainable farming and fishing
Population growth
Poverty

Figure 16-29 Components of more sustainable, low-throughput agriculture.

make a transition from unsustainable and environmentally harmful agriculture (Figure 16-9) to more sustainable forms of agriculture (Figure 16-29).

In developed countries, even a partial shift to more environmentally sustainable food production will not be easy. It will be opposed by **(1)** agribusiness, **(2)** successful farmers with large investments in unsustainable forms of industrialized agriculture, **(3)** specialized farmers unwilling to learn the more demanding art of farming sustainably, and **(4)** many consumers unwilling or unable to pay higher prices for food when more of agriculture's harmful environmental and health costs (Figure 16-9) are included in the market prices of food.

To help farmers make the transition to more sustainable agriculture, analysts call for

- Greatly increasing research on sustainable agriculture and improving human nutrition
- Setting up demonstration projects throughout each country so that farmers can see how sustainable agricultural systems work
- Increasing agricultural aid to developing countries (which, when adjusted for inflation, declined by about 50% between 1982 and 2000), with emphasis on developing more sustainable, low-input agriculture
- Establishing training programs in sustainable agriculture for farmers and government agricultural officials and encouraging the creation of college curricula in sustainable agriculture and human nutrition

Sustainable agriculture involves applying the four principles of sustainability (Solutions, p. 191) to food production. The goal is to feed the world's people while sustaining and restoring the earth's natural capital (Figure 4-33, p. 103). See the website material for this chapter for some actions you can take to help promote more sustainable agriculture.

The need to bring birth rates well below death rates, increase food production while protecting the environment, and distribute food to all who need it is the greatest challenge our species has ever faced.

PAUL AND ANNE EHRLICH

REVIEW QUESTIONS

1. Define the boldfaced terms in this chapter.

2. What are *perennial crops*? What advantages do they have over conventional annual crops?

3. What are the three main systems that supply our food? What three crops provide most of the world's food?

4. Distinguish between *industrialized agriculture, plantation agriculture, traditional subsistence agriculture,* and *traditional intensive agriculture.*

5. What is a *green revolution,* and what three steps does it involve? Distinguish between the first and second green revolutions.

6. Explain how producing more food on less land can help protect biodiversity.

7. Describe the nature and importance of the agricultural industry in the United States. How energy-efficient is industrialized agriculture in the United States?

8. Distinguish between *interplanting, polyvarietal cultivation, intercropping, agroforestry,* and *polyculture.* What are the major advantages and disadvantages of polyculture?

9. List four pieces of good news about world food production since 1950. List four pieces of disturbing news about present and future food production.

10. Distinguish between *undernutrition, malnutrition,* and *overnutrition.* What are the two most common nutritional deficiency diseases, and how do they differ? What are the effects of deficiencies of **(a)** vitamin A, **(b)** iron, and **(c)** iodine?

11. About how many chronically malnourished people are there in the world? About how many people die each year from undernutrition, malnutrition, or diseases worsened by malnutrition? List six major ways to reduce sickness and premature death of children from malnutrition. What is the primary cause of hunger in the world?

12. List the major harmful environmental effects of producing food.

13. Describe how genetically improved crop strains are developed by **(a)** *crossbreeding* and **(b)** *genetic engineering.*

14. List the pros and cons of growing more food by **(a)** increasing crop yields through green revolutions, **(b)** genetically modifying crops and foods, **(c)** getting people to switch to new types of foods, **(d)** irrigating more cropland, **(e)** cultivating more land, and **(f)** growing more food in urban areas.

15. List seven factors that could limit food production increases through green revolution and genetic engineering techniques. Why should we be concerned about loss of genetic diversity in agricultural crop strains?

16. What is *rangeland*? Explain how rangeland grass can be a renewable resource for livestock and wild herbivores. How is meat produced in developed and developing countries? What are the advantages and disadvantages of raising livestock in factorylike production facilities?

17. Distinguish between *overgrazing* and *undergrazing.* What are three major effects of overgrazing? Describe the general condition of rangelands throughout the world and in the United States. Describe three ways to manage rangelands more sustainably for meat production. What are *riparian areas,* and why are they important?

18. What are some of the harmful environmental effects of meat production?

19. What are *fisheries*? Distinguish between *trawling, purse seine, longlining,* and *drift-net* methods for harvesting fish. Describe trends in the world's total fish catch and per capita fish catch since 1950 and explain why the per capita catch is expected to decline.

20. What is the *sustainable yield* of a fishery? Distinguish between *overfishing* and *commercial extinction* of a fish species. About what percentage of the world's fish stocks are in decline because of overfishing and pollution?

21. Distinguish between *fish farming* and *fish ranching*. What are the pros and cons of aquaculture?

22. List three policy types governments use to affect food production. List the pros and cons of each approach.

23. Distinguish between *broad-spectrum* and *narrow-spectrum pesticides.* Summarize the role of Rachel Carson in alerting people to the potential dangers of widespread pesticide use.

24. List **(a)** seven reasons for relying on chemical pesticides, **(b)** five traits of the ideal pest-killing chemical, and **(c)** five reasons for relying less on chemical pesticides.

25. List four cultivation practices that can be used to help control pests. List the pros and cons of using the following pest control methods: **(a)** genetic engineering to speed up the development of pest- and disease-resistant crop strains, **(b)** biological control, **(c)** insect birth control through sterilization, **(d)** insect sex attractants, **(e)** insect hormones (pheromones), **(f)** food irradiation, and **(g)** integrated pest management (IPM).

26. What is *sustainable agriculture*, why is it important, and what are its major components?

CRITICAL THINKING

1. Summarize the major economic and ecological advantages and limitations of each of the following proposals for increasing world food supplies and reducing hunger over the next 30 years: **(a)** cultivating more land by clearing tropical forests and irrigating arid lands, **(b)** catching more fish in the open sea, **(c)** producing more fish and shellfish with aquaculture, and **(d)** increasing the yield per area of cropland.

2. Why should it matter to people in developed countries that many people in developing countries are malnourished and hungry? What are the three most important things you believe should be done to reduce hunger **(a)** in the country where you live and **(b)** in the world?

3. Some people argue that starving people could get enough food by eating nonconventional plants and insects; others point out that most starving people do not know what plants and insects are safe to eat and cannot take a chance on experimenting when even the slightest illness could kill them. If you had no money to grow or buy food, would you collect and eat protein-rich grasshoppers, moths, beetles, or other insects?

4. What could happen to energy-intensive agriculture in the United States and other industrialized countries if world oil prices rose sharply?

5. Should **(a)** all price supports and other government subsidies paid to farmers be eliminated and **(b)** governments phase in agricultural tax breaks and subsidies to encourage farmers to switch to more sustainable farming? Explain your answers. Try to consult one or more farmers in answering these questions.

6. What are the economic and ecological advantages and disadvantages of relying more on **(a)** a small number of genetic varieties of major crops and livestock, **(b)** genetically modified food, and **(c)** perennial food crops? Explain why you support or oppose each of these approaches.

7. Suppose you live near a coastal area and a company wants to use a fairly large area of coastal marshland for an aquaculture operation. If you were an elected local official, would you support or oppose such a project? Explain. What safeguards or regulations would you impose on the operation?

8. If increased mosquito populations threatened you with malaria, would you spray DDT in your yard and inside your home to reduce the risk? Explain. What are the alternatives?

9. Do you believe that farmers should be given economic incentives for switching to IPM? Explain your position.

10. Congratulations. You have just been put in charge of the world. List the three most important features of your agricultural policy.

PROJECTS

1. If possible, visit both a conventional industrialized farm and an organic or low-input farm. Compare **(a)** soil erosion and other forms of land degradation, **(b)** use and costs of energy, **(c)** use and costs of pesticides and inorganic fertilizer, **(d)** use and costs of natural pest control and organic fertilizer, **(e)** yields per hectare for the same crops, and **(f)** overall profit per hectare for the same crops.

2. Try to gather data evaluating the harmful environmental effects of nearby agriculture on your local community. What things are being done to reduce these effects?

3. Use health and other local government records to estimate how many people in your community suffer from undernutrition or malnutrition. Has this problem increased or decreased since 1980? What are the basic causes of this hunger problem, and what is being done to alleviate it? Share the results of your study with local officials and then present your own plan for improving efforts to reduce hunger in your community.

4. How are bugs and weeds controlled in **(a)** your yard and garden, **(b)** the grounds of your school, and

(c) public school grounds, parks, and playgrounds in your community?

5. Make a survey of all pesticides used in or around your home. Compare the results for your entire class.

6. Make a concept map of this chapter's major ideas, using the section heads and subheads and the key terms (in boldface). Look on the website for this book for information about making concept maps.

INTERNET STUDY RESOURCES AND RESOURCES FOR FURTHER READING AND RESEARCH

The website for this book contains helpful study aids and many ideas for further reading and research. Log on to

http://www.brookscole.com/product/0534389872s

and click on the Chapter-by-Chapter area. Choose Chapter 16 and select a resource:

- "Flash Cards" allows you to test your mastery of the Terms and Concepts to Remember for this chapter.
- "Tutorial Quizzes" provides a multiple-choice practice quiz.
- "Student Guide to InfoTrac" will lead you to Critical Thinking Projects that use InfoTrac College Edition as a research tool.
- "References" lists the major books and articles consulted in writing this chapter.
- "Hypercontents" takes you to an extensive list of sites with news, research, and images related to individual sections of the chapter.

INFOTRAC COLLEGE EDITION

Improve your skills with InfoTrac College Edition, a searchable online database of articles from more than 700 periodicals. Log on to

http://www.infotrac-college.com

or access InfoTrac through the website for this book. Try to find the following articles:

1. The True Cost of Food (environmental sustainability). Michael F. Jacobson. *Nutrition Action Healthletter*, May 2001 v28 i4 p2. Brief discussion of reasons to eat lower on the food chain. *Hint*: Enter the search term "food cost" using the keywords "living sustainably."

2. Grains of Hope: Genetically Engineered Crops Could Revolutionize Farming. Protesters Fear They Could Also Destroy the Ecosystem. You Decide (Science) (Cover Story). *Time*, July 31, 2000 v156 i5 p38+. Discussion of the benefits and drawbacks of genetically engineered crops. *Hint*: Enter the search term "grains hope" using the keywords "increasing world crop production."

17 SUSTAINING TERRESTRIAL AND AQUATIC BIODIVERSITY: THE ECOSYSTEM APPROACH

Who's Afraid of the Big Gray Wolf?

At one time, the gray wolf (Figure 17-1) ranged over most of North America. Between 1850 and 1900, however, an estimated 2 million wolves were shot, trapped, and poisoned by ranchers, hunters, and government employees. The idea was to make the West and the Great Plains safe for livestock and for big game animals prized by hunters.

This strategy worked. When the U.S. Endangered Species Act (p. 477) was passed in 1973, there were only about 400–500 gray wolves in the lower 48 states, primarily in Minnesota and Michigan. In 1974, the gray wolf was listed as endangered in all 48 lower states except Minnesota. Alaska was not included because it had 6,000–8,000 gray wolves.

With such protection, the gray wolf population has increased to about 3,000 in seven states (Arizona, Idaho, Michigan, Minnesota, New Mexico, Wisconsin, and Wyoming). Therefore, in 2000 the U.S. Fish and Wildlife Service (USFWS) proposed that the gray wolf be moved from the endangered to the threatened category in these states.

Figure 17-1 The gray wolf is an endangered species in the lower 48 states (except Minnesota, where it is listed as threatened). Ranchers, hunters, miners, and loggers have vigorously opposed efforts to return this keystone species to its former habitat in the Yellowstone National Park area. However, wolves were reintroduced beginning in 1995 and now number around 350.

Ecologists now recognize the important role this keystone predator species once played in parts of the West and the Great Plains. These wolves **(1)** culled herds of bison, elk, caribou, and mule deer, **(2)** kept down coyote populations, and **(3)** provided uneaten meat for scavengers such as ravens, bald eagles, ermines, and foxes.

In recent years, herds of elk, moose, and antelope have proliferated, devastating some of the area's vegetation, increasing erosion, and threatening the niches of other wildlife species. Reintroducing a keystone species such as the gray wolf into a terrestrial ecosystem is one way to help sustain its biodiversity and prevent environmental degradation.

In 1987, the USFWS proposed reintroducing gray wolves into the Yellowstone ecosystem, an idea that brought outraged protests. Some objections came from ranchers who feared the wolves would attack their cattle and sheep; one enraged rancher said that the idea was "like reintroducing smallpox." Other protests came from **(1)** hunters who feared that the wolves would kill too many big game animals and **(2)** miners and loggers who worried that the government would force them to cease operations on wolf-populated federal lands.

Since 1995, federal wildlife officials have caught gray wolves in Canada and relocated them in Yellowstone National Park and northern Idaho. By 2001, the population of these wolves had grown by natural reproduction to around 350.

National Park Service officials trap or shoot wolves that kill livestock outside park areas. In addition, a private fund established by Defenders of Wildlife pays ranchers for sheep or cattle verified as having been killed by gray wolves that have wandered out of the Yellowstone ecosystem.

Some ranchers and hunters say that they will take care of the wolves quietly—what they call the "shoot, shovel, and shut up" solution. Meanwhile, there are continuing efforts in Congress to eliminate the program and its funding.

Forests, grasslands, parks, wilderness, and other storehouses of terrestrial biodiversity and oceans, rivers, and wetlands are coming under increasing pressure from population growth and economic development. Biodiversity researchers and protectors urge us to use these renewable natural resources more sustainably, as discussed in this chapter.

Forests precede civilizations, deserts follow them.
FRANÇOIS-AUGUSTE-RENÉ DE CHATEAUBRIAND

This chapter addresses the following questions:

- How have human activities affected the earth's biodiversity?
- What are the major types of public lands in the United States, and how are they used?
- Why are forest resources important, and how are they used and managed in the world and in the United States?
- Are tropical deforestation and fuelwood shortages serious problems? If so, what can we do about them?
- What problems do parks face, and how should we manage them?
- How should we establish, design, protect, and manage terrestrial nature reserves?
- How can we protect and sustain aquatic systems?
- What is ecological restoration, and why is it important?

17-1 HUMAN IMPACTS ON BIODIVERSITY

How Have Human Activities Affected Global Biodiversity? Figure 17-2 summarizes the major connections between human activities and the earth's biodiversity. Here are some examples of how human activities have decreased and degraded the earth's biodiversity:

- Humans have taken over, disturbed, or degraded 40-50% of the earth's land surface (Figure 1-3, p. 5), especially by filing in wetlands and converting grasslands and forests to cropfields and urban areas. Over the next 100 years, such activities may degrade another third of the planet's land surface.
- Humans use, waste, or destroy about **(1)** 27% of the earth's total potential net primary productivity and **(2)** 40% of the net primary productivity of the planet's terrestrial ecosystems (Figure 4-25, p. 94), and this resource use is expected to increase.
- About half of the world's wetlands were lost during the last century.
- Logging and land conversion have reduced global forest cover by at least 20% and possibly as much as 50%.
- An estimated 27% of the world's coral reefs (Figure 6-28, p. 143) have been severely damaged, and another 70% could be severely damaged or eliminated by 2050.
- Biologists estimate that the current global extinction rate of species is 100 to 1,000 times what it would be without human-induced changes.

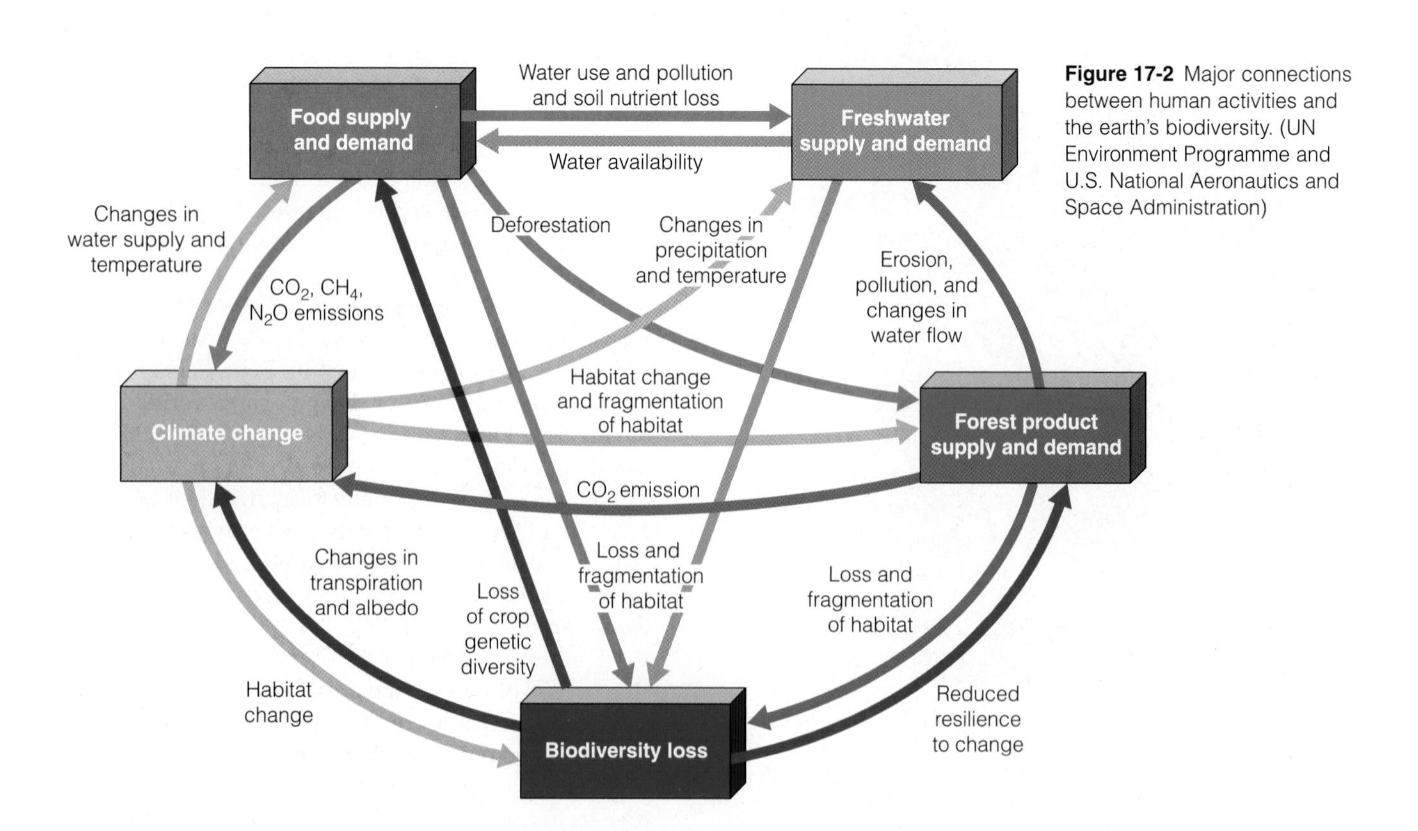

Figure 17-2 Major connections between human activities and the earth's biodiversity. (UN Environment Programme and U.S. National Aeronautics and Space Administration)

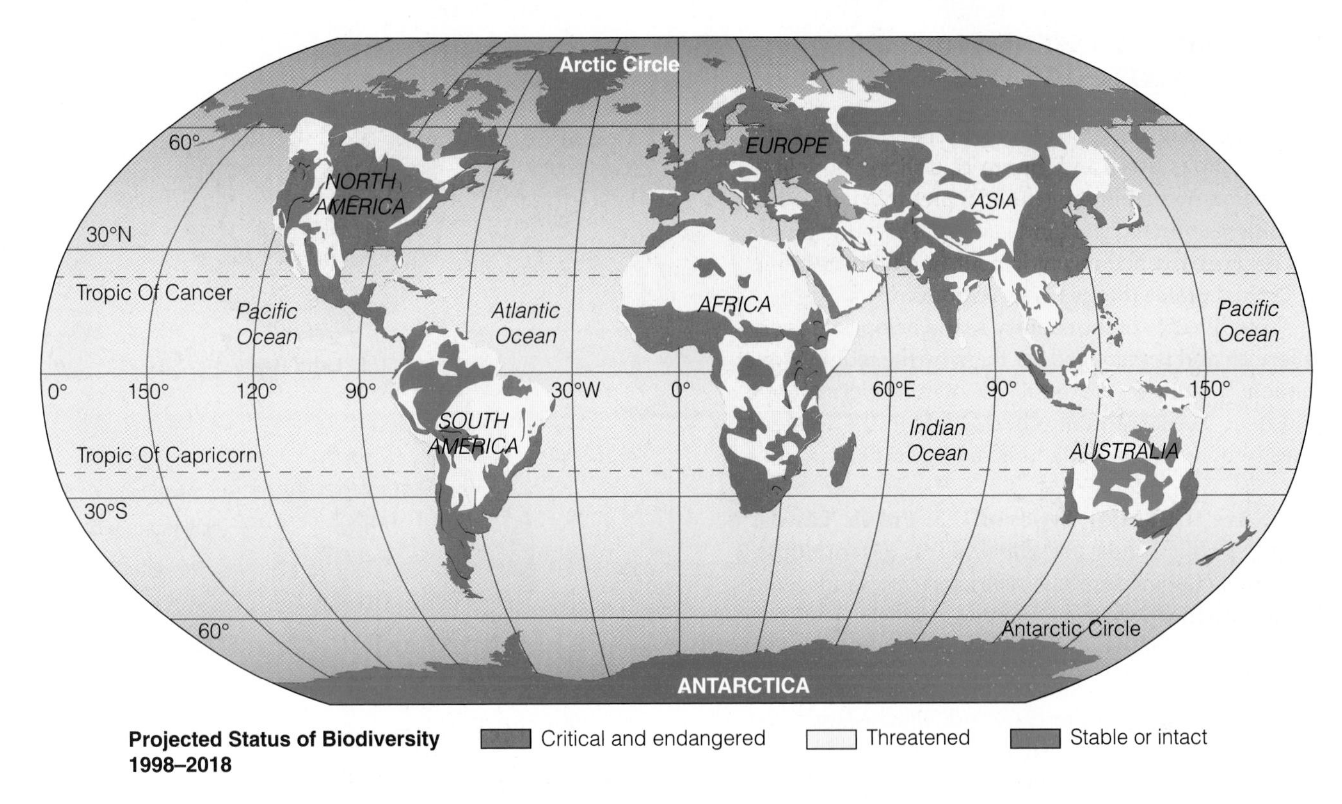

Figure 17-3 Projected status of the earth's biodiversity between 1998 and 2018. (Data from World Resources Institute, World Conservation Monitoring Center, and Conservation International)

- About 75% of the world's major marine fish stocks are overfished or are being fished at their biological limit (p. 414).
- A 1999 study by the U.S. Geological Survey found that **(1)** 95–98% of the virgin forests in the lower 48 states have been destroyed since 1620, **(2)** 96% of the virgin southeastern coastal plain longleaf pine forests are gone, **(3)** 98% of tallgrass prairie in the Midwest and Great Plains has disappeared, **(4)** 99% of California's native grassland, 91% of its wetlands, and 85% of its redwood forests are gone, **(5)** 81% of the nation's fish communities have been disturbed by human activities, and **(7)** 53% of the country's original wetlands have been destroyed.

These threats to the world's biodiversity are projected to increase sharply by 2018 (Figure 17-3).

Should Efforts to Conserve Biodiversity Focus on Sustaining Species or Ecosystems? Figure 17-4 outlines the goals, strategies, and tactics for **(1)** preserving and restoring the ecosystems and aquatic systems that provide habitats and resources for the world's species, as discussed in this chapter, and **(2)** preventing the premature extinction of species, as discussed in Chapter 18.

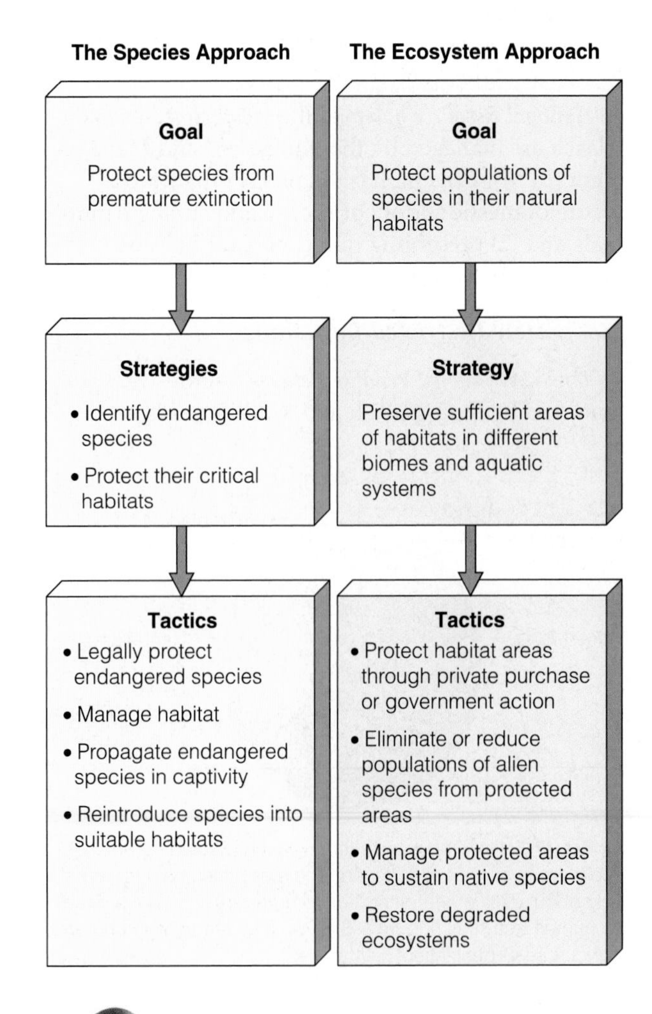

Figure 17-4 Goals, strategies, and tactics for protecting biodiversity.

17-2 LAND USE IN THE WORLD AND THE UNITED STATES

How Is Land Used? Figure 17-5 shows how the world's land is used, and Figure 17-6 shows how land is used and owned in the United States. No nation has set aside as much of its land—about 42%—for public use, resource extraction, enjoyment, and wildlife as has the United States (Figure 17-6, bottom).

Roughly 35% of the country's land belongs to every American and is managed for them by the federal government (Figure 17-6). About 73% of this federal public land is in Alaska, and another 22% is in the western states (where 60% of all land is public land).

What Are the Major Types of U.S. Public Lands? Federal public lands are classified as **(1)** multiple-use lands, **(2)** moderately restricted-use lands, and **(3)** restricted-use lands.

Multiple-Use Lands

- The 156 forests (Figure 17-7) and 20 grasslands of the *National Forest System* are managed by the U.S. Forest Service. These forests are used for logging, mining, livestock grazing, farming, oil and gas extraction, recreation, sport hunting, sport and commercial fishing, and conservation of watershed, soil, and wildlife resources. Off-road vehicles usually are restricted to designated routes.
- *National Resource Lands* in the western states and Alaska are managed by the Bureau of Land Management (BLM). The emphasis is on **(1)** providing a secure domestic supply of energy and strategic minerals and **(2)** preserving rangelands for livestock grazing under a permit system.

Moderately Restricted-Use Lands

- The 524 *National Wildlife Refuges* (Figure 17-7) are managed by the U.S. Fish and Wildlife Service (USFWS). Most refuges protect habitats and breeding areas for waterfowl and big game to provide a harvestable supply for hunters; a few protect endangered species from extinction. Permitted activities include sport hunting, trapping, sport and commercial fishing, oil and gas development, mining, logging, grazing, some military activities, and farming as long as the Department of the Interior finds such uses compatible with the purposes of each unit.

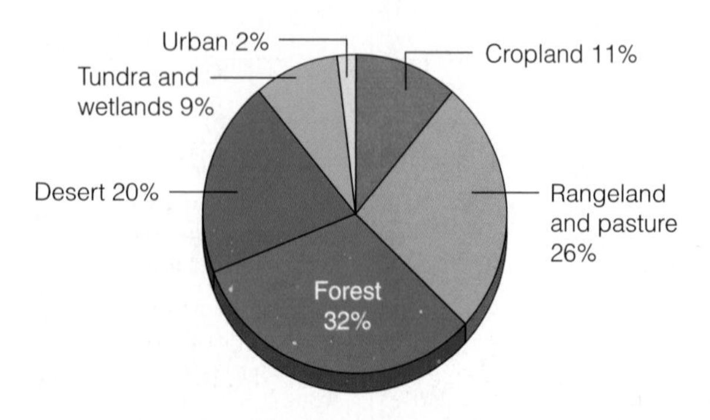

Figure 17-5 How the world's land is used. Excluding uninhabitable areas of rock, ice, desert, and steep mountain terrain, only about 27% of the planet's land area remains undisturbed by human activities (Figure 1-3, p. 5). (Data from UN Food and Agriculture Organization)

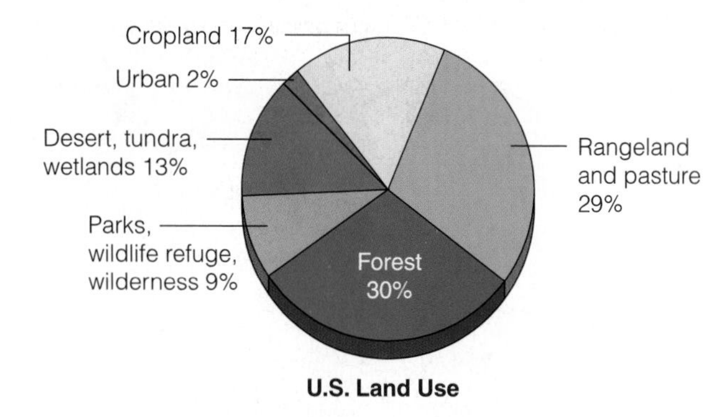

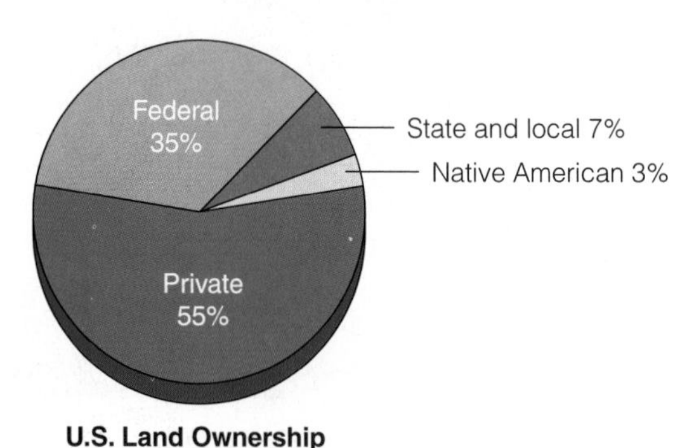

Figure 17-6 How land is used (top) and owned (bottom) in the United States. (Data from U.S. Department of Agriculture)

Restricted-Use Lands

- The 379 units of the *National Park System* include 55 major parks (mostly in the West; Figure 17-7) and 324 national recreation areas, monuments, memorials, battlefields, historic sites, parkways, trails, rivers, seashores, and lakeshores managed by the National Park Service. National parks may be used only for camping, hiking, sport fishing, and boating. Motor vehicles are permitted only on roads, although off-road vehicles are permitted in some parks. In national recreation areas, these same activities, plus sport hunting, mining, and oil and gas drilling, are allowed.
- The 630 roadless areas of the *National Wilderness Preservation System* lie within the national parks, national wildlife refuges, and national forests. They are managed by the National Park Service (42%), Forest Service (33%), USFWS (20%), and BLM (5%). These areas are open only for recreational activities such as hiking, sport fishing, camping, nonmotorized

Figure 17-7 National forests, national parks, and wildlife refuges managed by the U.S. federal government. U.S. citizens jointly own these and other public lands. (Data from U.S. Geological Survey)

boating, and, in some areas, sport hunting and horseback riding. Roads, motorized vehicles, logging, livestock grazing, mining, commercial activities, and buildings are banned, except when they predate the wilderness designation.

How Should U.S. Public Lands Be Managed? Because of the forest, oil, natural gas, coal, and other mineral resources they contain, there has been intense controversy over how public lands should be used and managed. Most conservation biologists and ecological economists and many free-market economists believe that the following four principles should govern use of public land:

- Protecting biodiversity, wildlife habitats, and the ecological functioning of public land ecosystems should be the primary goal.

- No one should be given subsidies or tax breaks for using or extracting resources on public lands. In recent years, the government has given more than $1 billion a year in subsidies to mining (Pro/Con, p. 206), logging, and grazing interests using U.S. public lands.

- The American people deserve fair compensation for the use of their property.

- All users or extractors of resources on public lands should be fully responsible for any environmental damage they cause.

Most of these proposals are based on Aldo Leopold's land-use ethic (Individuals Matter, p. 47).

There is strong opposition to these ideas. Economists, developers, and resource extractors tend to view areas of the earth's surface in terms of their **(1)** usefulness in providing mineral, timber, and other resources, **(2)** potential for urban development, and **(3)** ability to increase short-term economic growth. Many people with these views have lobbied members of the U.S. Congress to pass laws that would

- Sell public lands or their resources to corporations or individuals, usually at less than market value

- Slash federal funding for regulatory administration of public lands

- Cut all old-growth forests in the national forests and replace them with tree farms

- Open all national parks, national wildlife refuges, and wilderness areas to oil drilling, mining, off-road vehicles, and commercial development

- Do away with the National Park Service and launch a 20-year construction program of new concessions and theme parks run by private firms in the national parks

- Continue mining on public lands under the provisions of the 1872 Mining Law (Pro/Con, p. 206), which allows mining interests to **(1)** pay no royalties to taxpayers for hard-rock minerals they remove and **(2)** not be responsible for any environmental damage they cause

- Redefine protected wetlands so that about half of them would no longer be protected

In 2001, the Bush administration began pushing hard to open large tracts of public land to logging, drilling for oil and natural gas, mining, off-road vehicles, and nuclear-waste storage.

Case Study: Livestock and U.S. Public Rangeland About 23,000 U.S. ranchers hold permits to graze about 2% of the country's cattle and 10% of the sheep on BLM and Forest Service rangelands in 16 western states. Permit holders pay the federal government a grazing fee for this privilege.

Since 1981, Congress has set grazing fees on public rangeland at one-fourth to one-tenth the going rate for comparable private land. This means that taxpayers give the roughly 1% of U.S. ranchers with federal grazing permits subsidies amounting to $20–150 million a year, depending on how the costs of federal rangeland management are calculated.

The public subsidy does not end with low grazing fees. Other government costs include water pipelines, stock ponds, weed control, livestock predator control, clearing of undesirable vegetation, grass planting, erosion, and biodiversity loss. When these costs are added, taxpayers are giving 23,000 ranchers an annual subsidy of about $2 billion (an average of $69,000 per rancher) to produce 2% of the country's beef and 10% of its mutton. It is not surprising that politically influential permit holders have fought hard to block any change in this system.

Ranchers with permits say that grazing fees on public lands should be low because most private rangeland is more productive than public rangeland. Environmentalists contend that **(1)** this varies with the land involved, and **(2)** even when this is true current grazing fees on public rangelands are much too low.

Here are three positions on the use of public rangelands (mostly in the west):

- Phase out commercial grazing of livestock on western public lands over the next 10–15 years. Proponents contend that water-poor western rangeland (Figure 14-7, p. 331) is not a good place to raise cattle and sheep, which **(1)** need a lot of water and can overgraze rangeland (Figure 16-16, left, p. 412) and **(2)** degrade riparian areas (Spotlight, p. 413).

- Get government out of the management of public rangeland and allow it to be **(1)** managed by grazing boards made up of ranchers exempt from most environmental laws or **(2)** sold to private ranching or development interests.

- Get progressive ranchers (Individuals Matter, right), federal land managers, and environmentalists to work together to **(1)** develop more sustainable ways to manage public rangelands and **(2)** help keep ranchers using public and private lands in business so that rangeland is not destroyed by being converted to houses, condos, small "ranchettes," and tourist attractions.

Although their numbers are small, some coalitions of environmentalists, federal range managers, and ranchers are working together to develop more sustainable ways of using public rangeland such as

- *Banning or strictly limiting grazing on riparian areas* (Spotlight, p. 413).

The Eco-Rancher

Wyoming rancher Jack Turnell is one of a new breed of cowpuncher who gets along with environmentalists. He talks about riparian ecology and biodiversity as fluently as he talks about cattle: "I guess I have learned how to bridge the gap between the environmentalists, the bureaucracies, and the ranching industry."

Turnell grazes cattle on his 32,000-hectare (80,000-acre) ranch south of Cody, Wyoming, and on 16,000 hectares (40,000 acres) of Forest Service land on which he has grazing rights. For the first decade after he took over the ranch, he raised cows the conventional way. Since then, he has made some changes.

Turnell disagrees with the proposals by some environmentalists to raise grazing fees and remove sheep and cattle from public rangeland. He believes that if ranchers are kicked off the public range, ranches like his will be sold to developers and chopped up into vacation sites and homes ("ranchettes"), destroying the range for wildlife and livestock alike.

At the same time, he believes that ranches can be operated in more ecologically sustainable ways. To demonstrate this, Turnell **(1)** began rotating his cows away from the riparian areas (Spotlight, p. 413), **(2)** gave up most uses of fertilizers and pesticides, and **(3)** crossed his Hereford and Angus cows with a French breed that tends to congregate less around water. He makes most of his ranching decisions in consultation with range and wildlife scientists, and uses photographs to monitor changes in range condition.

The results have been impressive. Willows and other plant life line riparian areas on the ranch and Forest Service land. This has provided lush habitat for an expanding population of wildlife, including pronghorn antelope, deer, moose, elk, bear, and mountain lions. In addition, this eco-rancher makes more money because the higher-quality grass puts more meat on his cattle.

- *Banning or strictly limiting grazing on rangeland where it* **(1)** *endangers habitats of threatened and endangered species,* **(2)** *causes desertification* (Figure 9-23, p. 215), *or* **(3)** *is otherwise ecologically unsustainable.*
- *Allowing individuals and environmental groups to purchase grazing permits and not use the land for grazing.* In 2000, the U.S. Supreme Court upheld this practice.

So far, western ranchers with grazing permits have wielded enough political power to see that most such measures are not implemented by elected officials and government land management agencies.

17-3 MANAGING AND SUSTAINING FORESTS

What Are the Major Types of Forests? Forests with 50% or more tree cover occupy about 32% of the earth's land surface and provide numerous ecological and economic services (Figure 17-8). There are three general types of forests, based primarily on climate: *tropical, temperate,* and *polar (boreal)* (Figure 6-14, p. 130), described in detail in Section 6-4 (p. 137). These forests can be further classified as

- **Old-growth forests** (sometimes called *frontier forests*): uncut forests or regenerated forests that have not been seriously disturbed by human activities or natural disasters for at least several hundred years. Old-growth forests provide ecological niches for a multitude of wildlife species (Figure 6-23, p. 139).
- **Second-growth forests:** stands of trees resulting from secondary ecological succession (Figure 7-14, p. 173) after the trees in an area have been removed by **(1)** human activities such as clear-cutting for

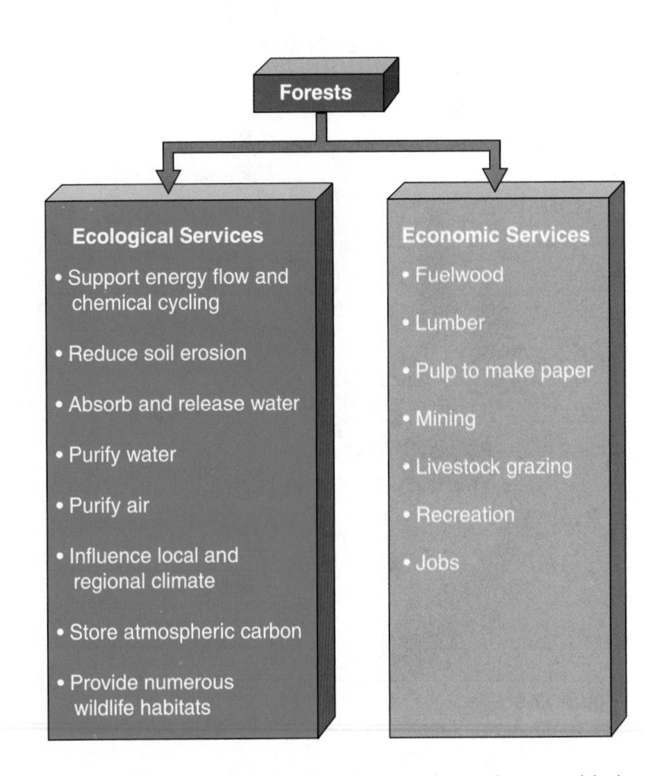

Figure 17-8 Major ecological and economic services provided by forests.

timber or conversion to cropland or **(2)** natural forces such as fire, hurricanes, or volcanic eruption.

- **Tree farms** or **plantations**: managed tracts with uniformly aged trees of one species (see photo on p. xvi) that are harvested by clear-cutting as soon as they become commercially valuable. Then they are replanted and clear-cut again on regular cycles (Figure 17-9). These farms are created mostly by clearing old-growth and second-growth forests and now occupy about 3% of the world's forest area.

What Are the Major Types of Forest Management? The total volume of wood produced by a particular stand of forest varies as it goes through different stages of growth and ecological succession (Figure 17-10). If the goal is to produce fuelwood or fiber for paper production in the shortest time, the forest usually is harvested on a short rotation cycle (point A in Figure 17-10), well before the volume of wood produced peaks. Typically, pulpwood farms are harvested in 6- to 10-year rotations in the tropics and 20- to 30-year rotations in temperate regions (Figure 17-9).

Harvesting at point B in Figure 17-10 gives the maximum yield of wood per unit of time. If the goal is high-quality wood for fine furniture or veneer, managers use longer rotations to develop larger, older-growth trees (point C in Figure 17-10), whose rate of growth has leveled off and is much lower than that of young trees.

There are two forest management systems:

- **Even-aged management**, which involves maintaining trees in a given stand at about the same age and size. In this approach, sometimes called *industrial forestry*, a biologically diverse old-growth or second-growth forest is replaced with a simplified *tree farm* of one or two fast-growing and economically desirable species that can be harvested every 10–100 years, depending on the species (Figure 17-9).

- **Uneven-aged management**, which involves maintaining a variety of tree species in a stand at many ages and sizes to foster natural regeneration. Here the goals are **(1)** biological diversity, **(2)** long-term sustainable production of high-quality timber (Figure 17-10, point C), **(3)** selective cutting of mature or intermediate-aged trees, and **(4)** multiple use of the forest for timber, wildlife, watershed protection, and recreation.

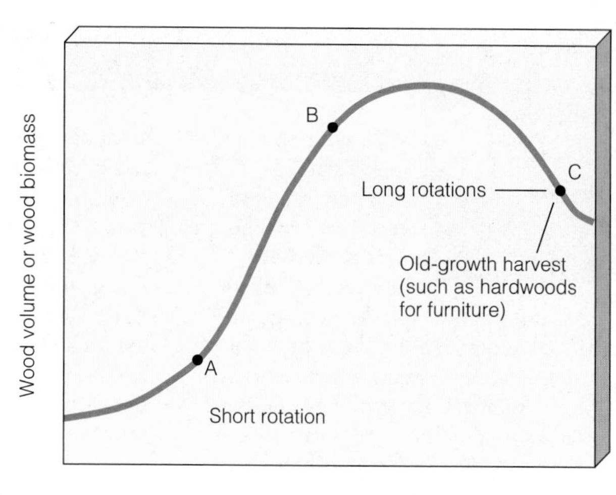

Figure 17-10 Changes in wood volume over various growth and harvest cycles in forest management. Forest management occurs over a cycle of decisions and events called a *rotation* (Figure 17-9). The most important steps in a rotation include **(1)** taking an inventory of the site, **(2)** developing a forest management plan, **(3)** building roads into the site, **(4)** preparing the site for harvest, **(5)** harvesting timber, and **(6)** regenerating and managing the site until the next harvest.

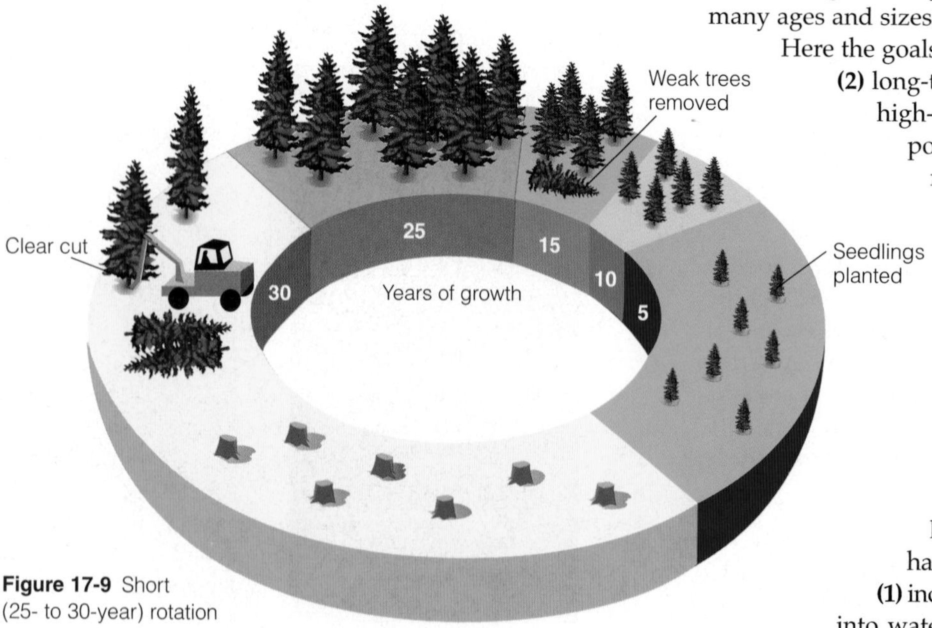

Figure 17-9 Short (25- to 30-year) rotation cycle of cutting and regrowth of a monoculture tree plantation in modern industrial forestry. In tropical countries, where trees can grow more rapidly year-round, the rotation cycle can be 6–10 years.

How Are Trees Harvested? Before large amounts of timber can be harvested, roads must be built for access and timber removal. Even with careful design, logging roads have a number of harmful effects (Figure 17-11), such as **(1)** increased erosion and sediment runoff into waterways, **(2)** habitat fragmentation and biodiversity loss, **(3)** exposure of forests to invasion by nonnative pests, diseases, and wildlife species, and **(4)** opening of once-inaccessible forests to

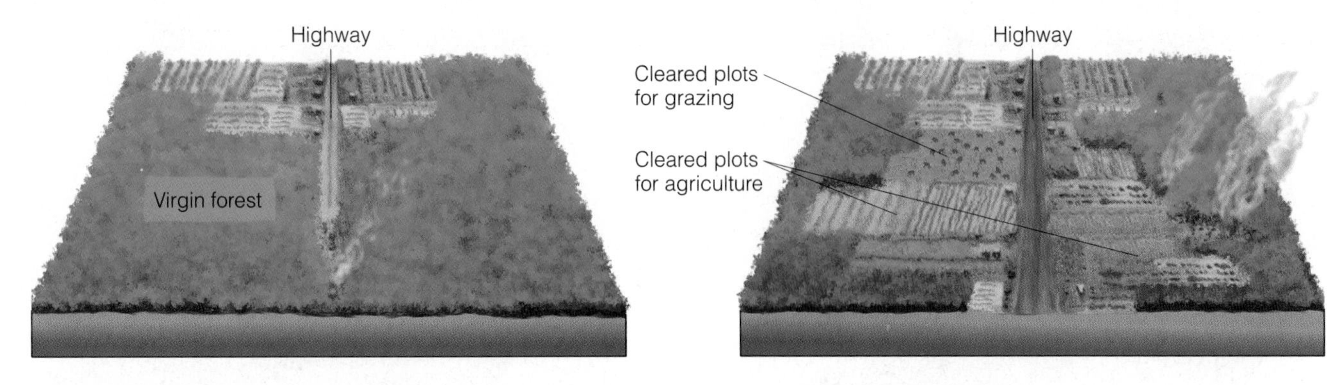

Figure 17-11 Building roads into previously inaccessible forests paves the way to fragmentation, degradation, and destruction.

farmers, miners, ranchers, hunters, and off-road vehicle users. In addition, logging roads on public lands in the United States disqualify the land for protection as wilderness.

Once loggers can reach a forest, they use various methods to harvest the trees (Figure 17-12). With **selective cutting**, intermediate-aged or mature trees in an uneven-aged forest are cut singly or in small groups (Figure 17-12a). Selective cutting **(1)** reduces crowding, **(2)** encourages growth of younger trees, **(3)** maintains an uneven-aged stand of trees of different species, **(4)** allows natural regeneration from the surrounding trees, **(5)** can help protect the site from soil erosion and wind damage, **(6)** can be used to remove diseased trees, and **(7)** allows a forest to be used for multiple purposes.

However, much of the logging in tropical forests involves a form of selective cutting called *high grading*, which involves cutting and removing only the largest and best specimens of the most desirable species. Studies have shown that for every large tree that is felled, 16 or 17 other trees are damaged or pulled down because the tree canopies in tropical forests usually are connected by a network of vines. This reduction of the forest canopy (Figure 6-23, p. 139) **(1)** causes the forest floor to become warmer, drier, and more flammable and **(2)** increases erosion of the forest's thin and usually nutrient-poor soil (Figure 9-20, bottom left, p. 213).

Some tree species grow best in full or moderate sunlight in moderate to large clearings. Methods for harvesting such species include

- **Shelterwood cutting**, which removes all mature trees in two or typically three cuttings over a period of about 10 years (Figure 17-12b)
- **Seed-tree cutting**, which harvests nearly all a stand's trees in one cutting, leaving a few uniformly distributed seed-producing trees to regenerate the stand (Figure 17-12c)
- **Clear-cutting**, which removes of all trees from an area in a single cutting (Figure 17-12d)

According to timber companies, clear-cutting **(1)** increases timber yield per hectare, **(2)** permits reforesting with genetically improved stocks of fast-growing trees, **(3)** shortens the time needed to establish a new stand of trees (Figure 17-9), **(4)** takes less skill and planning than other harvesting methods, **(5)** usually provides the maximum economic return in the shortest time, and, **(6)** if done carefully and responsibly, often is the best way to harvest tree farms and stands of some tree species that need full or moderate sunlight for growth.

On the negative side, clear-cutting **(1)** leaves moderate to large forest openings, **(2)** eliminates most recreational value for several decades, **(3)** reduces biodiversity, disrupts ecosystem processes, and destroys and fragments some wildlife habitats, **(4)** makes nearby trees more vulnerable to being blown down by windstorms, and **(5)** leads to severe soil erosion, sediment water pollution, and flooding when done on steep slopes (Figure 14-19, p. 343).

A clear-cutting variation that can allow a sustainable timber yield without widespread destruction is **strip cutting**, in which a strip of trees is clear-cut along the contour of the land, with the corridor narrow enough to allow natural regeneration within a few years (Figure 17-12e). After regeneration, another strip is cut above the first, and so on. This allows clear-cutting of a forest in narrow strips over several decades with minimal damage.

What Is Happening to the World's Forests? Forests are renewable resources as long as the rate of cutting and degradation does not exceed the rate of regrowth. However, according to the World Resources Institute, during the past 8,000 years human activities have reduced the original forest cover of the earth by at least 20% and possibly by as much as 50%.

Some good news is that the area of many temperate forests in North America and Europe is increasing slightly because of reforestation from secondary ecological succession (Figure 7-14, p. 173) on cleared forest

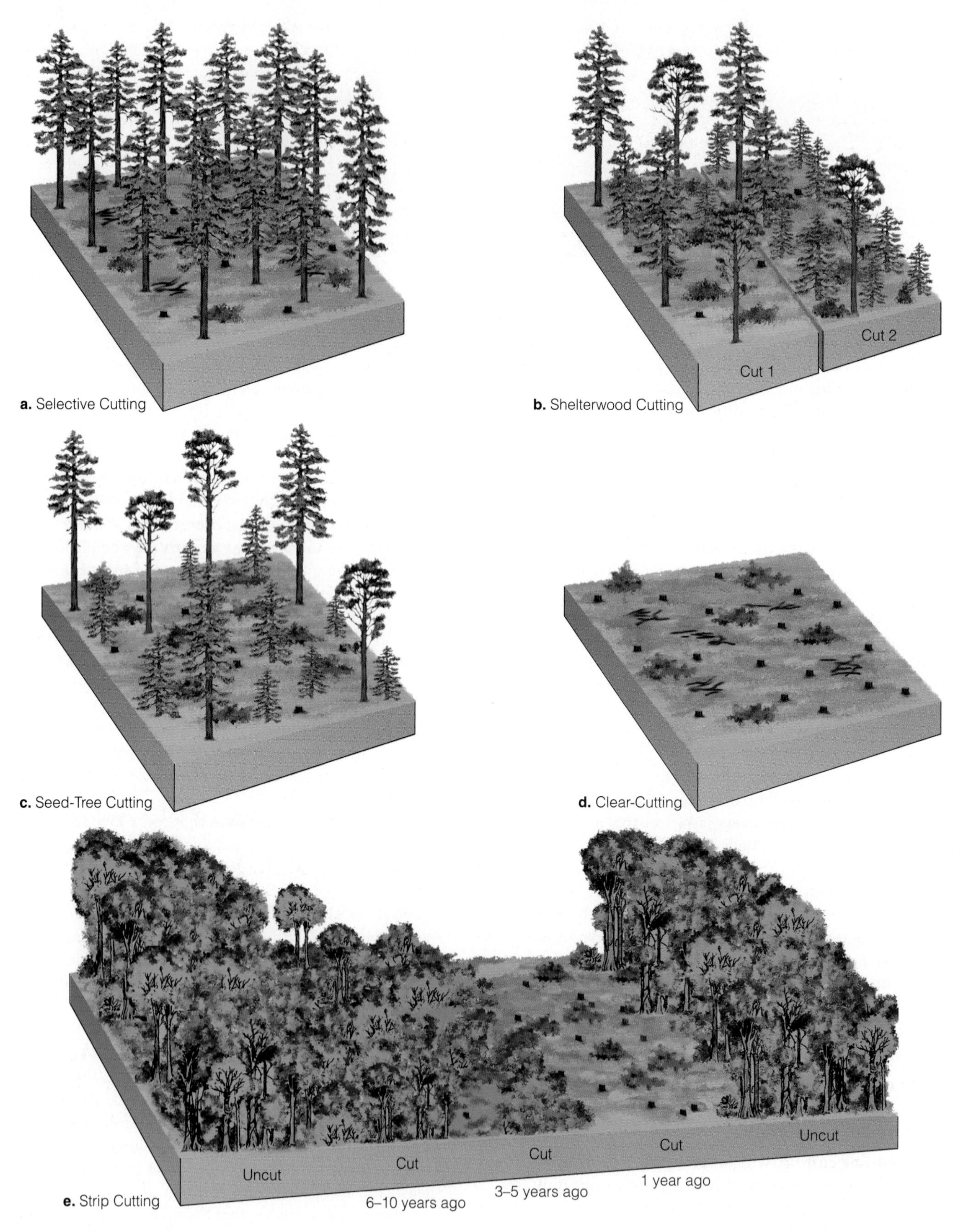

Figure 17-12 Tree harvesting methods.

areas and abandoned croplands. However, forests are being cut or fragmented and degraded faster than they can regenerate in many parts the world, especially in tropical areas of Latin America, Africa, and Asia.

Deforestation can provide land for crops, livestock, or urban growth. However, it also **(1)** decreases the overall net primary productivity (Figure 4-24, p. 93) of the cleared area, **(2)** reduces the stock of nutrients once stored in the trees and leaf litter, **(3)** diminishes biodiversity, **(4)** makes the soil more prone to erosion and drying, **(5)** increases the rate of runoff of water and soil nutrients from the land, **(6)** reduces the uptake of CO_2 from the atmosphere, and **(7)** adds CO_2 to the atmosphere if cleared trees are burned or allowed to decay. According to a 2001 study by the World Resources Institute, over the past 140 years deforestation has contributed an estimated 30% of the atmospheric buildup of CO_2 (Figure 13-4, top, p. 304).

How Can Forests Be Managed More Sustainably? Instead of increased use of conventional short-rotation industrial forestry (Figure 17-9), biodiversity researchers and a growing number of foresters call for more sustainable forest management that

- *Grows more timber on long rotations,* generally about 100–200 years (point C, Figure 17-10), depending on the species and soil quality
- *Emphasizes* **(1)** *selective cutting of individual trees or small groups of most tree species* (Figure 17-12a), **(2)** *strip cutting* (Figure 17-12e) *instead of conventional clear-cutting, and* **(3)** *not clear-cutting (including seed-tree and shelterwood cutting) on land that slopes more than 15°*
- *Minimizes fragmentation of remaining larger blocks of forest*
- *Stops or sharply reduces road building into uncut forested areas*
- *Uses road-building and logging methods that minimize soil erosion and compaction*
- *Leaves most standing dead trees (snags) and fallen timber to maintain diverse wildlife habitats and to be recycled as nutrients* (Figure 4-14, p. 87)
- *Has timber grown by sustainable methods certified and labeled by outside certifying groups* (Solutions, p. 440)
- *Includes the estimated ecological services provided by trees and a forest* (Spotlight, p. 441) *in estimates of their economic value*

Some foresters have suggested that the United States and the world meet most of the demand for wood and wood products by growing genetically improved trees on carefully managed tree farms (Figure 17-9 and photo on p. xvi). According to one estimate, *tree farms* occupying only about 5% of the world's forested land could supply the entire world demand for forest products (except fuelwood). This would **(1)** reduce pressures to cut timber in old-growth and second-growth forests and **(2)** help protect wildlife habitats.

Conservation biologists and environmentalists support using this approach only if

- Harvesting of remaining old-growth forests is banned or sharply reduced.
- Harvesting of second-growth forests is phased out over a 10- to 20-year period as the tree farms are phased in.
- Tree farms are established only on lands that are truly degraded, not on any newly cleared land or existing cropland.
- All government subsidies and tax breaks for harvesting timber from old-growth and second-growth forests are phased out and replaced with subsidies for establishing tree farms in truly degraded areas.

How Do Fires Affect Forest Ecosystems? Several types of fires can affect forest ecosystems. Some, called *surface fires* (Figure 17-13, left), usually burn only undergrowth and leaf litter on the forest floor. These fires can kill seedlings and small trees but spare most mature trees and allow most wild animals to escape.

Occasional surface fires set by lightning or under controlled conditions by forest managers **(1)** burn away flammable ground material and help prevent more destructive fires, **(2)** release valuable mineral nutrients tied up in slowly decomposing litter and undergrowth,

Surface fire

Crown fire

Figure 17-13 Surface fires (left) usually burn undergrowth and leaf litter on a forest floor and can help prevent more destructive crown fires (right) by removing flammable ground material. Sometimes forest managers deliberately set carefully controlled surface fires to prevent buildup of flammable ground material in forests.

Certifying Sustainably Grown Timber

SOLUTIONS

Collins Pine owns and manages a large area of productive timberland in northeastern California. The company's goals are to maintain ecological, economic, and social sustainability by using selective cutting (since the 1940s). The timber the company harvests each year is certified as being sustainably produced by Scientific Certification Systems (SCS). It is part of the nonprofit Forest Stewardship Council (FSC), formed in 1993 to develop a list of environmentally sound practices for use in certifying timber and products made from such timber.

Each year, SCS evaluates all of Collins's landholdings to ensure that **(1)** cutting has not exceeded long-term forest regeneration, **(2)** roads and harvesting systems have not caused unreasonable ecological damage, **(3)** management of soils, downed wood, and standing snags provides adequate nutrient cycling and wildlife habitat, and **(4)** the company is a good employer and a good steward of its land and water resources.

Another successful example of forest certification involves the Menominee nation. Since 1890, they have been selectively harvesting (Figure 17-12a) trees of mixed species and ages from their tribal reservation land near Green Bay, Wisconsin, which is the state's single largest tract of virgin forest. Lumber from the tribal forest has been certified by the Rainforest Alliance's Smart Wood Program as being harvested in an environmentally and socially responsible manner. Each year thousands of people flock to the reservation to look at a sustainable forest operation that honors wildlife and biodiversity while providing jobs and income for the local economy.

In 2001, the World Wildlife Fund called on the world's five largest wood harvesting and processing companies and the five largest companies buying wood products to adopt the responsible and sustainable management processes of the FSC. If done, these 10 companies could halt logging of old-growth forests and still meet the world's industrial wood and wood fiber needs on one-fifth of the world's forests—an area twice the size of India.

Critical Thinking

Suppose you inherited a sustainable timber company operation similar to the one described in this box. If you switched to clear-cutting most of the land, you could **(1)** make millions in a short time, **(2)** sell the clear-cut land, and **(3)** invest the money elsewhere at a much higher rate of return than long-term sustainable timber harvesting. What would you do with your inheritance? Explain.

(3) increase the activity of underground nitrogen-fixing bacteria, **(4)** stimulate the germination of certain tree seeds (such as those of the giant sequoia, lodgepole pine, and jack pine), and **(5)** help control pathogens and insects. In addition, some wildlife species such as deer, moose, elk, muskrat, woodcock, and quail depend on occasional surface fires **(1)** to maintain their habitats and **(2)** provide food in the form of vegetation that sprouts after fires.

Some extremely hot fires, called *crown fires* (Figure 17-13, right), may start on the ground but eventually burn whole trees and leap from treetop to treetop. They usually occur in forests where the absence of surface fires for several decades has allowed dead wood, leaves, and other flammable ground litter to build up. These rapidly burning fires can **(1)** destroy most vegetation, **(2)** kill wildlife, and **(3)** increase soil erosion.

Sometimes surface fires go underground and burn partially decayed leaves or peat. Such *ground fires* are most common in northern peat bogs. They may smolder for days or weeks before being detected and are difficult to extinguish.

Four approaches used to protect forest resources from fire are **(1)** *prevention,* **(2)** *prescribed burning* (setting controlled ground fires to prevent buildup of flammable material), **(3)** *presuppression* (early detection and control of fires), and **(4)** *suppression* (fighting fires once they have started). Ways to prevent forest fires include **(1)** requiring burning permits, **(2)** closing all or parts of a forest to travel and camping during periods of drought and high fire danger, and **(3)** educating the public. The Smokey Bear educational campaign of the Forest Service and the National Advertising Council, for example, has **(1)** prevented countless forest fires in the United States, **(2)** saved many lives, and **(3)** prevented billions of dollars in losses.

This educational program has convinced most members of the public that all forest fires are bad and should be put out. According to ecologists, however, preventing all fires can increase the likelihood of highly destructive crown fires (Figure 17-13, right) because it allows large quantities of highly flammable underbrush and undergrowth and smaller trees to accumulate in some forests. This can convert what would have been harmless surface fires into fires intense enough to destroy the larger fire-resistant species needed for forest regeneration.

How Do Air Pollution and Climate Change Threaten Forests? Forests at high elevations and

SPOTLIGHT

What Is the Economic Value of the Earth's Ecological Services?

In 1997, a team of 13 ecologists, economists, and geographers attempted to estimate how much the earth's natural ecological services (Figure 4-33, p. 103) are worth in monetary terms.

According to this crude appraisal led by ecological economist Robert Costanza of the University of Maryland, the economic value of income from the earth's natural capital is at least $36 trillion per year. This is close to the annual gross world product of about $40 trillion. To provide an annual natural income of $36 trillion per year, the world's natural capital would have a value of at least $500 trillion—an average of about $82,000 for each person on earth.

To make these estimates, the researchers divided the earth's surface into 16 biomes (Figure 6-13, p. 129) and aquatic life zones (they omitted deserts and tundra because of a lack of data). Then they **(1)** agreed on a list of 17 goods and services provided by nature (Figure 4-33, p. 103) and **(2)** sifted through more than 100 studies that attempted to put a dollar value on such services in the 16 different types of ecosystems.

Some analysts believe that such estimates are misleading and dangerous because they put a dollar value on ecosystem services that have an infinite value because they are irreplaceable. In the 1970s, economist E. F. Schumacher warned that "to undertake to measure the immeasurable is absurd" and is a "pretense that everything has a price."

The researchers admit that their estimates rely on many assumptions and omissions and could easily be too low by a factor of 10 to 1 million or more. For example, their calculations **(1)** included only estimates of the ecosystem services themselves, not the natural capital that generates them, and **(2)** omitted the value of nonrenewable minerals and fuels. They also recognize that **(1)** as the supply of ecosystem services declines their value will rise sharply and **(2)** such services can be viewed as having an infinite value.

However, they contend that their estimates are much more accurate than the *very low* or *zero* value the market usually assigns to these ecosystem services. They hope such estimates will call people's attention to the facts that **(1)** the earth's ecosystem services are essential for all humans and their economies (Figure 2-3, p. 26) and **(2)** their economic value is huge.

Critical Thinking

Should we put a price on nature's services? Explain. What are the alternatives?

those downwind from urban and industrial centers are exposed to a variety of air pollutants that can harm trees (especially conifers) and make them more vulnerable to drought, diseases, and insects (Figure 12-11, p. 288). The solution is to reduce emissions of the offending pollutants from coal-burning power plants, industrial plants, and motor vehicles (Section 12-7, p. 293).

In coming decades, an even greater threat to forests (especially temperate and boreal forests) may come from regional climate changes brought about by global warming that can **(1)** increase the threat of forest fires in some areas and **(2)** cause some forest types to die out in some areas (Figure 13-12, p. 312). Ways to deal with global warming are discussed in Section 13-5, p. 313.

17-4 FOREST RESOURCES AND MANAGEMENT IN THE UNITED STATES

What Is the Status of Forests in the United States? Forests **(1)** cover about 30% of U.S. land area (Figure 17-6, top), **(2)** provide habitats for more than 80% of the country's wildlife species, and **(3)** supply about two-thirds of the nation's total water runoff.

Some *good news* is that forests (including tree farms) in the United States today cover more area and often are healthier than they were in 1920, when the country's population was around 100 million. Many of the old-growth forests that were cleared or partially cleared between 1620 (Figure 17-14, left) and 1960 have grown back naturally through secondary ecological succession as fairly diverse second-growth (and in some cases third-growth) forest.

However, some *bad news* is that

- Between 1620 and 1998, most of the existing old-growth forests in the lower 48 states were cut (Figure 17-14).
- Since the mid-1960s, an increasing area of the nation's remaining old-growth and fairly diverse second-growth forests has been clear-cut and replaced with tree farms, cropfields, or urban development. According to biodiversity researchers, this **(1)** reduces overall forest biodiversity and **(2)** disrupts ecosystem processes such as energy flow and chemical cycling.

Others argue that tree farms help preserve overall forest biodiversity by reducing the pressure to clear-cut more diverse old-growth and second-growth forests.

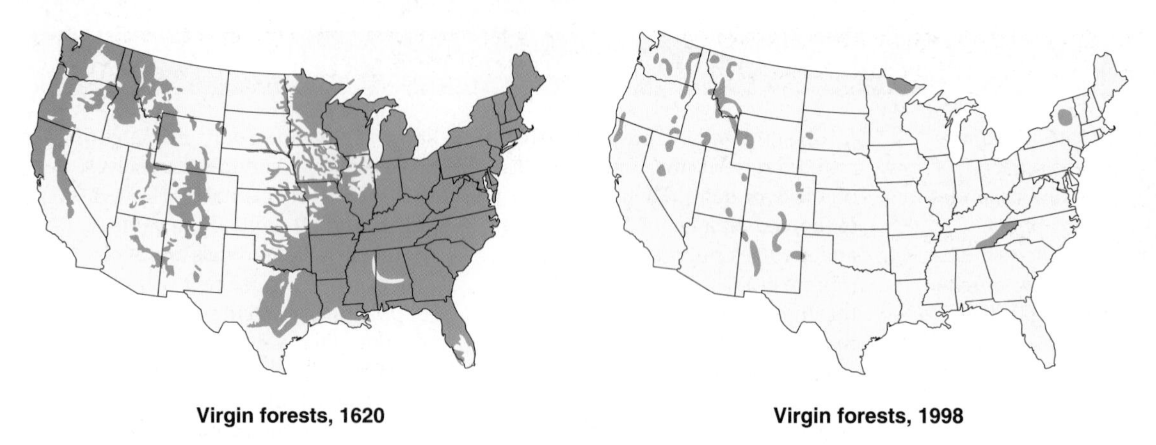

Figure 17-14 Vanishing old-growth forests in the United States, 1620 and 1998. Since 1960, some of these forests have been replaced with second-growth forests and tree plantations. (Data from the Wilderness Society and the U.S. Forest Service)

The basic question is, "What percentage of the nation's forests should be converted to tree farms to meet the wood and pulp needs of consumers and still preserve the overall biodiversity of the nation's forests?"

Case Study: How Should U.S. National Forests Be Managed? The 156 national forests (Figure 17-7) managed by the U.S. Forest Service

- Contain about 19% of the country's forest area and supply about 3% of the nation's softwood timber (down from 15% in the 1980s).
- Serve as grazing lands for more than 3 million cattle and sheep each year.
- Provide about $4 billion worth of minerals, oil, and natural gas per year.
- Contain a network of more than 612,000 kilometers (380,000 miles) of roads, equal in area to the entire U.S. interstate highway system. Most are narrow dirt roads built at taxpayers' expense for logging by private companies.
- Provide habitats for almost 200 threatened and endangered species and hundreds of other wild species and fish.
- Are the principal habitats for thousands of pollinator species that contribute $4–7 billion per year to U.S. agriculture.
- Provide some of the country's cleanest drinking water for more than 60 million Americans in some 3,400 communities. The estimated value of this water is $3.7 billion per year.
- Contain about one-third of the country's protected wilderness area.
- Receive more visits for recreation, hunting, and fishing than any other federal public lands.

The Forest Service is required by law to manage national forests according to the principles of **(1)** *sustained yield* (which states that potentially renewable tree resources should not be harvested or used faster than they are replenished) and **(2)** *multiple use* (which says that each of these forests should be managed for a variety of simultaneous uses such as sustainable timber harvesting, recreation, livestock grazing, watershed protection, and wildlife).

There is much controversy over the use of the national forests. Timber companies push to cut as much of this timber as possible at low prices. Biodiversity experts and environmentalists believe that **(1)** tree harvesting in public forests should be reduced or eliminated, **(2)** timber companies should pay more for trees they harvest from national forests, and **(3)** national forests should be managed primarily to provide recreation and to sustain their biodiversity, water resources, and other ecological services.

By law, the Forest Service must sell timber for no less than the cost of reforesting the cleared land. However, the cost of access roads is not included in this price and is provided as a subsidy to logging companies. Usually, the companies also get the timber itself for less than they would pay a private landowner for an equivalent amount of timber. As a result, the Forest Service's timber-cutting program loses money because revenue from timber sales does not cover the costs of **(1)** road building, **(2)** timber sale preparation, **(3)** administration, and **(4)** other overhead costs.

Because of such subsidies, timber sales from U.S. federal lands have turned a profit for taxpayers in only three of the last 100 years. According to a 2000 study by the National Forest Protection Alliance, logging in national forests **(1)** causes more economic harm than good to local communities near such forests, **(2)** costs the nation's taxpayers about $1.2 billion per year in logging subsidies (an estimate confirmed by the U.S. Congressional Research Service), and **(3)** costs taxpayers and nearby communities many billions of dollars more each year from rivers polluted and fisheries damaged by sediments from logged lands, increased flooding, lost recreational opportunities, and degraded scenery.

Timber company officials argue that logging in national forests

- Helps satisfy the country's demand for wood
- Provides cheap timber that benefits consumers by keeping lumber and paper prices down
- Improves forest health by removing diseased trees and helping prevent forest fires
- Provides jobs and stimulates economic growth in nearby communities

Environmentalists respond that

- Timber cutting in national forests provides only about 3% of the country's wood, and there is ample private forestland to meet the country's demand for wood.
- Below-cost timber prices in national forests increase pressure for timber cutting in national and state forests and have little effect on the consumer costs of lumber and paper.
- Communities relying heavily on 25% of the proceeds from national forest timber sales experience severe economic slumps as the timber is depleted and timber companies move to other areas.
- Recreation in national forests provides many more jobs and much more income for local communities than does logging. According to a 2000 study by the economic accounting firm Econorthwest, recreation, hunting, and fishing in national forests generate about 2.9 million jobs and add about $234 billion to the national economy each year. By contrast, logging, mining, grazing, and other extractive uses in these forests add about 407,000 jobs and $23 billion to the economy per year.
- According to the Congressional Budget Office, eliminating money-losing timber sales in the national forest would save taxpayers about $1.6 billion over the next 10 years.

In 2001, timber company officials were pleased that the Bush administration **(1)** proposed increasing the timber sale program in the national forests, **(2)** appointed a new Forest Service Chief expected to be far more favorable to timber interests, and **(3)** moved to reverse decisions by the Clinton administration to block road construction in roadless areas of national forests (which makes them ineligible for protection in the National Wilderness System; Case Study, p. 455).

17-5 TROPICAL DEFORESTATION AND THE FUELWOOD CRISIS

What Are Tropical Forests, and How Fast Are They Being Cleared and Degraded? Tropical forests cover about 6% of the earth's land area and include **(1)** rain forests (which receive rainfall almost daily), **(2)** deciduous forests with one or two dry seasons each year, **(3)** dry deciduous forests, and **(4)** forests on hills and mountains.

Climatic and biological data suggest that mature tropical forests once covered at least twice as much area as they do today, with most of the destruction occurring since 1950. Satellite scans and ground-level surveys used to estimate forest destruction indicate that large areas of tropical forests are being cut rapidly in parts of South America (especially Brazil), Africa, and Asia. Haiti has lost 99% of its original forest cover, the Philippines 97%, and Madagascar 84%.

Brazil's vast Amazon Basin is the world's largest remaining tropical forest and is home for up to 30% of the world's plant and animal life. In 1970, only 1% of this forest area had been deforested. By 2000 almost 15% had been destroyed—a loss of forest area the size of France in 30 years. According to a 2001 study by Penn State researcher James Alcock, without immediate and aggressive action to change current forest destruction and degradation practices, Brazil's Amazon rain forests **(1)** will not be able to sustain themselves within 10–15 years and **(2)** will largely disappear within 40–50 years.

There is debate over the current rates of tropical deforestation and degradation because of **(1)** difficulties in interpreting satellite images, **(2)** different ways of defining deforestation and forest degradation, and **(3)** political and economic factors that cause countries to hide or exaggerate deforestation. Consequently, estimates of total global tropical forest loss per year vary from 50,000 square kilometers (19,300 square miles) to 170,000 square kilometers (65,600 square miles). Scientists estimate that each year an equivalent area of these forests is seriously degraded and fragmented.

Why Should We Care About Tropical Forests? To most biologists, cutting and degrading most remaining old-growth tropical forests is a serious global environmental problem because of the important ecological and economic services they provide, with some species providing a variety of services (Solutions, p. 444). Here are some of the important services provided by tropical forests:

Ecological Services

- They provide habitats for 50–90% of the earth's terrestrial species (Figure 4-16, p. 88). Many of these

The Incredible Neem Tree

SOLUTIONS

Wouldn't it be nice if there were a single plant that could **(1)** quickly reforest bare land, **(2)** provide fuelwood and lumber in dry areas, **(3)** provide alternatives to toxic pesticides, **(4)** be used to treat numerous diseases, and **(5)** help control population growth? There is: the *neem tree*, a broadleaf evergreen member of the mahogany family.

This remarkable tropical species, native to India and Burma, is ideal for reforestation because it can grow to maturity in only 5–7 years. It grows well in poor soil in semiarid lands such as those in Africa, providing abundant fuelwood, lumber, and lamp oil.

It also contains various natural pesticides. Chemicals from its leaves and seeds can repel or kill more than 200 insect species, including termites, gypsy moths, locusts, boll weevils, and cockroaches.

Extracts from neem seeds and leaves (Figure 17-15) can be used to fight bacterial, viral, and fungal infections. Indeed, the tree's chemicals have allegedly relieved so many different afflictions that the tree has been called a "village pharmacy." Its twigs are used as an antiseptic toothbrush, and oil from its seeds is used to make toothpaste and soap.

That is not all. Neem-seed oil evidently acts as a strong spermicide that might be used in producing a much-needed male birth control pill. According to a study by the U.S. National Academy of Sciences, the neem tree "may eventually benefit every person on the planet."

Despite its numerous advantages, ecologists caution against widespread planting of neem trees outside its native range. If introduced as a nonnative species to other ecosystems, it could take over and displace other species because of its rapid growth and resistance to pests.

Critical Thinking

Assume that you are an elected county official. Someone proposes widespread planting of neem trees on abandoned cropland in the area where you live. Would you favor this proposal? Explain. What tests might you want to run before approving this project?

species have highly specialized niches (Figure 6-23, p. 139) that make them vulnerable to extinction.

- They play an important role in removing some of the excess CO_2 we are putting into the atmosphere (Figure 13-5, p. 305).
- They provide an array of other important ecosystem services (Figure 4-33, p. 103, and Figure 17-8).

Economic Services

- They supply **(1)** half of the world's annual harvest of hardwood, **(2)** hundreds of food products including coffee, tea, cocoa, spices, nuts, and tropical fruits, and **(3)** materials such as natural latex rubber, resins, dyes, and essential oils.
- About 90% of today's food crops were domesticated from wild tropical plants (Figure 16-5, p. 400). The genetic diversity of existing wild plant species is needed to develop future crop strains using traditional crossbreeding and genetic engineering (Figure 16-10, p. 406).
- At least 40% of all medicines (worth at least $200 billion per year) and 80% of the top 150 prescription drugs used in the United States were originally derived from living organisms, mostly plants from tropical developing countries (Figure 17-15).

What Causes Tropical Deforestation? Tropical deforestation results from a number of interconnected causes (Figure 17-16). All these factors are related to increasing use of the net primary productivity (Figure 4-24, p. 93) and mineral resources of the earth's forests for human use as a result of **(1)** population growth (Figure 11-7, p. 248), **(2)** poverty (Section 2-4, p. 34), and **(3)** government policies that encourage deforestation. Population growth and poverty combine to drive subsistence farmers and the landless poor to tropical forests, where they try to grow enough food to survive.

Government subsidies can accelerate deforestation by **(1)** making timber or other resources cheap relative to their ecological value and **(2)** encouraging the poor to colonize tropical forests by giving them title to land they clear, as is done in Indonesia, Mexico, and Brazil.

The process of degrading a tropical or other type of forest often begins with a road (Figure 17-11), usually cut by logging companies. Once the forest becomes accessible, it can be cleared and degraded by a number of factors, including

- *Unsustainable forms of small-scale farming.* Instead of practicing various methods of traditional and potentially sustainable shifting cultivation (Figure 1-15, p. 15), many poor people migrating to tropical forests practice unsustainable farming that depletes soils and destroys large tracts of forests.
- *Cattle ranching.* Cattle ranches, sometimes supported by government subsidies, often are established on cropland that has been exhausted by small-scale farmers and then abandoned or sold to ranchers. When torrential rains and overgrazing turn the usually thin and nutrient-poor tropical forest soils

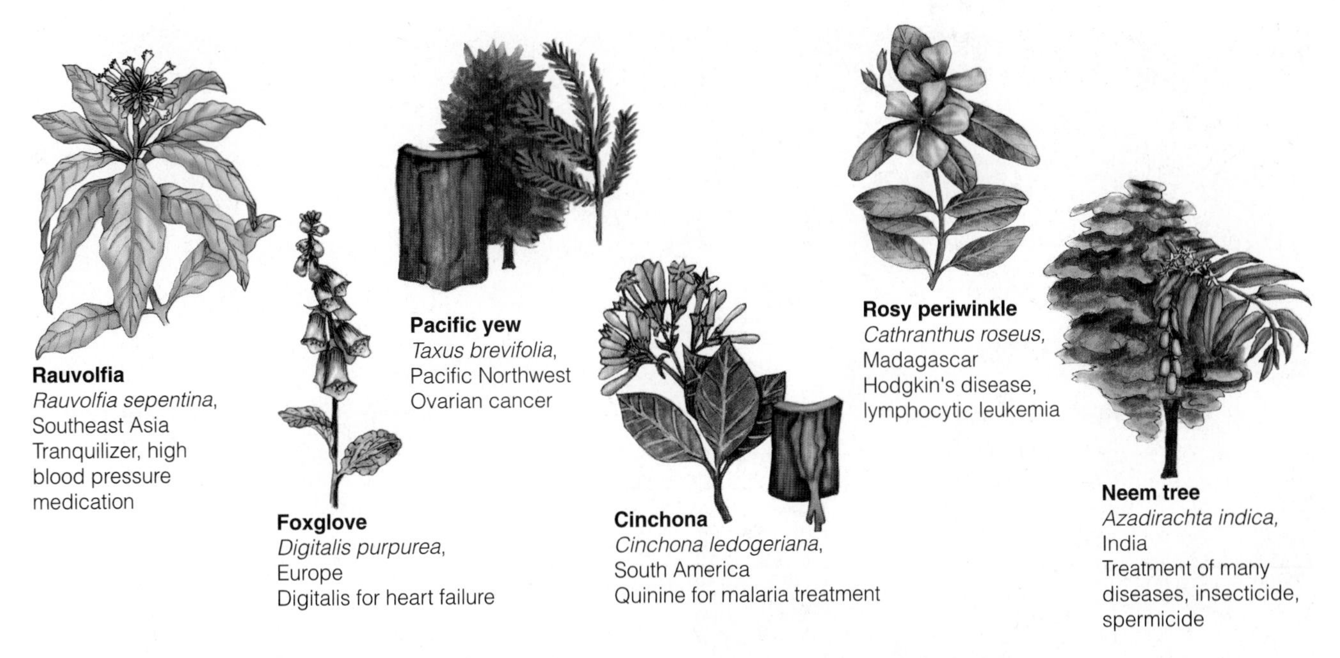

Figure 17-15 Nature's pharmacy. Parts of these and a number of other plants and animals (many of them found in tropical forests) are used to treat a variety of human ailments and diseases. About 70% of the 3,000 plants identified by the National Cancer Institute as sources of cancer-fighting chemicals come from tropical forests. Despite their economic and health potential, fewer than 1% of the estimated 125,000 flowering plant species in tropical forests (and a mere 1,100 of the world's 260,000 known plant species) have been examined for their medicinal properties. Many of these tropical plant species are likely to become extinct before we can learn about their ecological and economic values.

(Figure 9-20, bottom left, p. 213) into eroded wastelands, ranchers move to another area and repeat this destructive process.

- *Clearing large areas of tropical forest for raising cash crops.* Large plantations grow crops such as sugarcane, bananas, pineapples, peppers, strawberries, cotton, tea, and coffee, mostly for export to developed countries.
- *Mining and oil drilling.*
- *Building dams on rivers that flood large areas of tropical forests.*
- *Commercial logging.* Japan alone accounts for 53% of the world's tropical timber imports, followed by Europe (32%) and the United States (15%). After depleting tropical timber in much of Asia, cutting is now shifting to Latin America and Africa. Timber exports to developed countries contribute significantly to tropical forest depletion and degradation, but domestic use accounts for more than 80% of the trees cut in developing countries.
- *Increasing forest fires.* Between 1997 and 1999, huge areas of forest were burned in Indonesia, Malaysia, Brazil, Guatemala, Nicaragua, and Mexico, mostly by **(1)** farmers using fire to prepare fields for planting or cattle-grazing and **(2)** corporations clearing forests to establish pulp, palm oil, and rubber plantations. The resulting highly polluted air **(1)** sickened tens of millions of people, **(2)** killed hundreds, **(3)** caused billions of dollars in damage, and **(4)** released large amounts of carbon dioxide into the atmosphere.

Solutions: How Can We Reduce Tropical Deforestation and Degradation? A number of analysts have suggested the following ways to protect tropical forests and use them more sustainably:

- Identify and move rapidly to protect areas of tropical forests that are rich in unique animal and plant species and in imminent danger.
- Mount global efforts to reduce the poverty (Section 2-4, p. 34) that leads the poor to use forests (and other resources) unsustainably for short-term survival.
- Slow population growth to help reduce the flow of the landless poor to tropical forests.
- Establish programs to help new settlers in tropical forests learn how to practice small-scale sustainable agriculture and forestry (Solutions, p. 447).
- Phase out government subsidies that encourage unsustainable forest use and phase in tariffs, taxes, user fees, and subsidies that favor more sustainable forestry and biodiversity protection.
- Encourage governments to protect large areas of tropical forests. Currently less than 5% of the world's

Figure 17-16 Major interconnected causes of the destruction and degradation of tropical forests. These factors ultimately are related to **(1)** population growth, **(2)** poverty, and **(3)** government policies that encourage deforestation.

tropical forests are part of parks and preserves, and many exist in name only.

- Use debt-for-nature swaps and conservation easements to encourage countries to protect tropical forests. In a *debt-for-nature swap,* participating countries act as custodians for protected forest reserves in return for foreign aid or debt relief; in *conservation easements,* a private organization, country, or group of

Sustainable Agriculture and Forestry in Tropical Forests

SOLUTIONS

A combination of the knowledge of indigenous peoples, ecological research, and modern technology can show people how to grow crops and harvest timber in tropical forests more sustainably.

One approach would be to show people migrating to forests how to grow crops using forms of more sustainable agroforestry developed by indigenous peoples throughout the world. The Lacandon Maya Indians of Chiapas, Mexico, for example, have developed a multilayered system of agroforestry that allows them to cultivate up to 75 crop species on 1-hectare (2.5-acre) plots for up to seven consecutive years. After that, they plant a new plot to allow regeneration of the soil in the original plot.

In the lush rain forests of Peru's Palcazú Valley, Yaneshá Indians use strip cutting (Figure 17-12e) to harvest tropical trees for lumber. Widely spaced thin trips are harvested in an alternating pattern (for example, strips 1, 3, 5, and then 2, 4, and 6), and any given strip is allowed to regenerate 30–40 years before being harvested again. This type of harvesting **(1)** reduces environmental degradation and **(2)** promotes rapid natural regeneration of native tree species from stumps or from seeds of nearby trees. Tribe members also act as consultants to help other forest dwellers set up similar systems.

Another way to maintain biodiversity is to grow coffee in shade tree plantations under a diverse canopy of trees instead of in unshaded plantations. However, shade tree coffee plantations in the tropics are being converted rapidly to higher-yield unshaded coffee plantations that need more pesticides and fertilizers.

Some coffee drinkers are fighting this loss of biodiversity by buying only coffee certified (by organizations such as the Rainforest Alliance) as being grown under shade trees. In the United States, such coffee is available through a number of coffee distributors and mail-order retailers.

Critical Thinking

What applications (if any) might these practices have for growing food and harvesting timber in the United States and other developed countries? How would you encourage use of such approaches for more sustainable use of forests?

countries compensates other countries for protecting selected forest areas.

- Establish an international system for **(1)** evaluating and labeling timber produced by sustainable methods and **(2)** encouraging wood suppliers and buyers to use this system (Solutions, p. 440).
- Enact and enforce laws to reduce illegal timber cutting. In Cambodia, for example, the estimated amount of money lost to the national treasury because of illegal logging is equal to the entire national budget. In Brazil, the government acknowledges that illegal harvesting accounts for nearly 80% of the timber extracted from the Amazon rain forest.
- Use gentler methods for harvesting trees. For example, **(1)** cutting canopy vines (lianas) before felling a tree can reduce damage to neighboring trees by 20–40%, and **(2)** using the least obstructed paths to remove the logs can halve the damage to other trees.
- Concentrate peasant farming, tree farms, crop plantations, and ranching activities on already-cleared tropical forest areas.
- Mount global efforts to reforest and rehabilitate degraded tropical forests and watersheds (Individuals Matter, p. 448).
- Encourage citizens to work together to protect tropical forests. In Brazil, 500 conservation organizations have formed a coalition to preserve the country's remaining tropical forests.
- Reduce the waste and overconsumption of industrial timber, paper, and other resources by consumers, especially in developed countries.

17-6 MANAGING AND SUSTAINING NATIONAL PARKS

How Popular Is the Idea of National Parks? Today, there are more than 1,100 national parks larger than 10 square kilometers (4 square miles) in more than 120 countries. These parks cover a total area equal to that of Alaska, Texas, and California combined.

The U.S. National Park System, established in 1912, is dominated by 55 national parks, most of them in the West (Figure 17-7), sometimes called *America's crown jewels*. These national parks are supplemented by state, county, and city parks. Most state parks are located near urban areas and have about twice as many visitors per year as the national parks.

How Are Parks Being Threatened? Parks everywhere are under pressure from external and internal threats. A 1999 study by the World Bank and the World Wildlife Fund found that most parks and wildlife

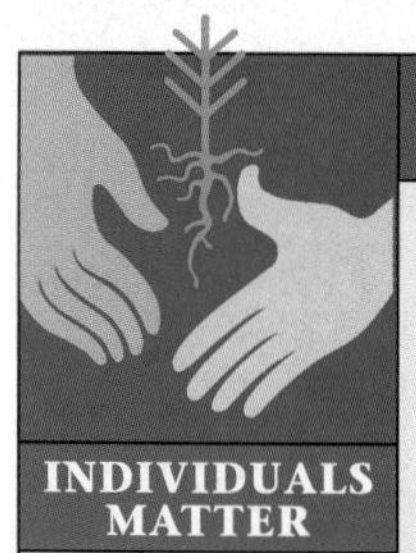

Kenya's Green Belt Movement

In Kenya, Wangari Maathai (Figure 17-17) started the Green Belt Movement. The goals of this highly regarded women's self-help group are to establish tree nurseries, raise seedlings, and plant a tree for each of Kenya's almost 30 million people. By 2000, the 100,000 members of this grassroots group had planted and protected more than 10 million trees.

This project's success has sparked the creation of similar programs in more than 30 other African countries. This inspiring leader has said,

I don't really know why I care so much. I just have something inside me that tells me that there is a problem and I have to do something about it. And I'm sure it's the same voice that is speaking to everyone on this planet, at least everybody who seems to be concerned about the fate of the world, the fate of this planet.

Figure 17-17 Wangari Maathai, the first Kenyan woman to earn a Ph.D. (in anatomy) and to head an academic department (veterinary medicine) at the University of Nairobi, organized the internationally acclaimed Green Belt Movement in 1977.

reserves in developing countries are poorly managed, and only 1% are protected. The other 99% are mostly *paper parks* that exist in name only and have no protection. Most of these parks are invaded by **(1)** local people who need wood, cropland, game animals, and other natural products for their daily survival and **(2)** loggers, miners, and wildlife poachers who kill animals to obtain and sell items such as rhino horns, elephant tusks, and furs.

In addition, most of the world's national parks are too small to sustain many large animal species. Many have been invaded by nonnative species that can reduce the populations of native species and cause ecological disruption.

Popularity is one of the biggest problems of national and state parks in the United States (and other developed countries). Because of increased numbers of roads, cars, and affluent people, annual recreational visits to major national parks increased more than fourfold and visits to state parks sevenfold between 1950 and 2000.

Many parks provide spectacular scenery and enjoyable recreational experiences for visitors. However, because of underfunding, visitors to some of the most heavily used national and state parks find **(1)** closed campgrounds, **(2)** uncollected garbage, **(3)** debris on trails and beaches, **(4)** dirty toilets, and **(5)** fewer nature lectures by park rangers.

During the summer, users of the most popular U.S. national and state parks often face hour-long entrance backups. Many visitors to heavily used national and state parks leave the city to commune with nature, only to find the parks as noisy, congested, and stressful as the places they came from.

U.S. Park Service rangers now spend an increasing amount of their time on law enforcement instead of conservation, management, and education. Currently, there is one ranger for every 84,200 visitors to the major national parks. Many overworked and underpaid rangers are leaving for better-paying jobs.

Nonnative species have moved into or been introduced into many parks. Wild boars (imported to North Carolina in 1912 for hunting) threaten vegetation in part of the Great Smoky Mountains National Park. The Brazilian pepper tree has invaded Florida's Everglades National Park. Mountain goats in Washington's Olympic National Park trample native vegetation and accelerate soil erosion. While some nonnative species have moved into parks, some native species of animals and plants (including many threatened or endangered species) are being killed or removed illegally in almost half of U.S. national parks.

Nearby human activities that threaten wildlife and recreational values in many national parks include **(1)** mining, **(2)** logging, **(3)** livestock grazing, **(4)** coal-burning power plants, **(5)** water diversion, and **(6)** urban development. Polluted air drifts hundreds of kilometers, killing ancient trees in California's Sequoia National Park and blurring the awesome vistas at Arizona's Grand Canyon. Car smog is damaging scores of plant species in the heavily used Great Smoky Mountains National Park. According to the National Park Service, air pollution affects scenic views in national parks more than 90% of the time.

That is not all. Mountains of trash wash ashore daily at Padre Island National Seashore in Texas. Water use in Las Vegas (Spotlight, p. 342) threatens to shut down geysers in the Death Valley National Monument.

Visitors to Sequoia National Park and Kentucky's Mammoth Cave complain of raw sewage flowing through a parking lot. Unless a massive ecological restoration project is successful, Florida's Everglades National Park may dry up because of water diversion for urban areas and agriculture.

Solutions: How Can Management of U.S. National Parks Be Improved? The U.S. National Park Service has two goals that increasingly conflict: **(1)** to preserve nature in parks and **(2)** to make nature more available to the public. In 1988, the Wilderness Society and the National Parks and Conservation Association made a number of suggestions for sustaining and expanding the national park system, including the following:

- *Require integrated management plans for parks and other nearby federal lands.*
- *Increase the budget for* **(1)** *adding new parkland near the most threatened parks and* **(2)** *buying private lands inside parks.*
- *Locate all new and some existing commercial facilities and visitor parking areas outside parks and provide shuttle buses for entering and touring most parks,* as was done in 2000 in Zion National Park, Utah. Shuttle bus systems are available at the Yosemite, Grand Canyon, and Denali National Parks.
- *Require private concessionaires who provide campgrounds, restaurants, hotels, and other services for park visitors to* **(1)** *compete for contracts,* **(2)** *pay franchise fees equal to 22% of their gross (not net) receipts, and* **(3)** *pour some of their receipts back into parks.* Private concessionaires in national parks pay an average of only about 6–7% of their gross receipts in franchise fees to the government, and many large concessionaires with long-term contracts pay as little as 0.75% of their gross receipts.
- *Allow concessionaires to lease but not own facilities inside parks.*
- *Provide more funds for park system maintenance and repairs.* Currently, there is a $4-billion backlog of maintenance, repairs, and high-priority construction projects at a time when park usage and external threats to the parks are increasing.
- *Survey the condition and types of wildlife species in parks.*
- *Limit the number of visitors to crowded park areas.*
- *Increase the number and pay of park rangers.*
- *Encourage volunteers to give tours and lectures to visitors.*
- *Encourage individuals and corporations to donate money for park maintenance and repair.*

17-7 ESTABLISHING, DESIGNING, AND MANAGING NATURE RESERVES

How Much of the Earth's Land Should We Protect from Human Exploitation? Most ecologists and conservation biologists believe that the best way to preserve biodiversity is through a worldwide network of reserves, parks, wildlife sanctuaries, wilderness, and other protected areas. Currently, more than 17,000 nature reserves, parks, wildlife refuges, and other areas provide strict or partial protection for about 8% of the world's land area. However, many existing reserves **(1)** are too small to provide any real protection for the wild species that live on them and **(2)** receive so little protection that their resources often are extracted illegally and unsustainably.

According to conservation biologists, at least 10% of the globe's land area should be *strictly* protected. Establishing and managing an expanded global system of biodiversity reserves that includes multiple examples of all the earth's biomes will take action and funding by **(1)** national governments (Solutions, p. 450), **(2)** private groups (Solutions, p. 451), and **(3)** cooperative ventures involving governments, businesses, and private conservation groups.

Most developers and resource extractors and some economists oppose protecting even the current 8% of the earth's remaining undisturbed ecosystems. They contend that such areas contain valuable resources that would add to economic growth and should be developed now. Ecologists and conservation biologists disagree and view protected areas as islands of biodiversity that are **(1)** vital parts of the earth's natural resources that sustain all life and economies (Figure 4-33, p. 103) and **(2)** centers of evolution.

What Principles Should Be Used to Establish and Manage Nature Reserves? According to most ecologists and conservation biologists, the selection, design, and management of biodiversity reserves should be guided by the following three ecological principles:

- Ecosystems are rarely at a stable point (the "balance of nature" concept) and thus cannot be locked up and protected from human disturbances. Instead, they are mostly in an ever-changing *nonequilibrium state* because of internal and external disturbances caused by natural processes and human activity (Section 7-6, p. 175).
- Ecosystems and communities that experience fairly frequent but moderate disturbances have the greatest diversity of species, known as the *intermediate disturbance hypothesis* (Figure 7-16, p. 175).
- We should view most reserves as *habitat islands* surrounded by a sea of developed and fragmented

Parks in Costa Rica

SOLUTIONS

Once tropical forests completely covered Costa Rica, which is smaller in area than West Virginia. Between 1963 and 1983, however, politically powerful ranching families **(1)** cleared much of the country's forests to graze cattle and **(2)** exported most of the beef to the United States and western Europe.

Despite such widespread forest loss (which continues today), tiny Costa Rica is a superpower of biodiversity, with an estimated 500,000 plant and animal species. A single park in Costa Rica is home for more bird species than all of North America.

In the mid-1970s, Costa Rica established a system of national parks and conservation reserves that by 2000 included 12% of its land (6% of it in reserves for indigenous peoples). Costa Rica now has a larger proportion of land devoted to biodiversity conservation than any other country.

The country's parks and reserves are consolidated into eight *megareserves* designed to sustain about 80% of the country's biodiversity (Figure 17-18). Each reserve contains a protected inner core surrounded by buffer zones that local and indigenous people use for sustainable logging, food growing, cattle grazing, hunting, fishing, and ecotourism.

One reason for this accomplishment in biodiversity protection was the establishment of the Organization of Tropical Studies in 1963. It is a consortium of more than 50 U.S. and Costa Rican universities with the goal of promoting research and education in tropical ecology. The resulting infusion of several thousand scientists **(1)** has helped Costa Ricans appreciate their country's great biodiversity and **(2)** led to the establishment in 1989 of the National Biodiversity Institute (INBio), a private nonprofit organization set up to survey and catalog the country's biodiversity. This biodiversity conservation strategy has paid off. Today, the $1 billion a year tourism business (almost two-thirds of it from ecotourists) is the country's largest source of income.

However, legal and illegal deforestation threatens this plan because of the country's **(1)** population growth, **(2)** poverty (which affects 10% of its people), and **(3)** lack of government regulations. Currently, this small country still loses roughly 400 square kilometers (150 square miles) of primary forest per year—four times the rate of loss in Brazil.

In addition, without careful government control the 1 million tourists visiting Costa Rica each year can **(1)** degrade some of the protected areas and **(2)** lead to too many hotels, resorts, and other potentially harmful forms of development.

Critical Thinking

Why do you think Costa Rica has been able to set aside a much larger percentage of its land for national parks and nature reserves than any other country? Do you agree or disagree with the implementation of such a plan in the country where you live?

Figure 17-18 Costa Rica has consolidated its parks and reserves into eight *megareserves* designed to sustain about 80% of the country's rich biodiversity.

land. As a result, their species diversity and ecological functioning can be evaluated to some degree by using the *theory of island biogeography* (Figure 7-5, p. 162).

Experience shows that two important principles in establishing, managing, and protecting reserves are to

- Include local people in the planning and design of a reserve.
- Create "user-friendly" reserves that allow local people to use parts of a reserve or a buffer zone surrounding a reserve for sustainable timber cutting, livestock grazing, growing crops, hunting, and fishing (Solutions, p. 452). This gives local people a vested interest in protecting a reserve from unsustainable uses.

Should Reserves Be as Large as Possible? There is general agreement that large reserves **(1)** are the only way to maintain viable populations of large, wide-ranging species (such as panthers, elephants, and grizzly bears), **(2)** sustain more species, **(3)** minimize the area of outside edges exposed to natural disturbances (such as fires and hurricanes), invading species, and human disturbances from nearby developed areas, and **(4)** provide greater habitat diversity than small preserves.

The Nature Conservancy

SOLUTIONS

Private groups play an important role in establishing wildlife refuges and other protected areas to protect biological diversity. For example, since 1951 the Nature Conservancy has preserved more than **(1)** 44,500 square kilometers (17,200 square miles) of vital wildlife habitats in the United States and **(2)** 243,000 square kilometers (94,000 square miles) throughout Canada, Latin America, the Caribbean, Asia, and the Pacific Ocean.

The Nature Conservancy has more than 1 million members throughout the world. It has one of the lowest overhead rates of any nonprofit organization, and 85% of all contributions go directly to its conservation programs.

The Nature Conservancy began in 1951 when an association of professional ecologists wanted to use their scientific knowledge to conserve natural areas. Since then, this science-based organization has used the most sophisticated scientific knowledge available to identify and rank sites **(1)** that are unique and ecologically significant and **(2)** whose biodiversity or existence is threatened by development or other human activities.

Once sites are identified, **(1)** a variety of techniques are used to see that they receive legal protection and **(2)** science-based management plans are used to maintain or restore the ecological health of each site and to provide long-term stewardship.

This conservation organization uses private and corporate donations to maintain a fund for buying ecologically important pieces of land or wetlands threatened by development when no other option is available. If it cannot buy land for habitat protection, the conservancy helps landowners obtain tax benefits in exchange for accepting legal restrictions or conservation easements preventing development. Other techniques include long-term management agreements and debt-for-nature swaps. Landowners have also received sizable tax deductions by donating their land to the Nature Conservancy in exchange for lifetime occupancy rights.

Through such efforts, this organization has created the world's largest system of private natural areas and wildlife sanctuaries, using the guiding principle of land conservation through private action.

Critical Thinking

Do you favor this private approach to protecting biodiversity over the government approach to protecting public lands and endangered species? Explain. Would you be in favor of selling or giving some public lands to private groups such as the Nature Conservancy? Explain.

However, research indicates that in some locales several well-placed, medium-sized reserves may better protect a greater variety of habitat types and more populations of rare species than a single large one. If fire, epidemic, or some other disaster wipes out the population of a species in one reserve, the species might survive in the other reserves. When establishing a comprehensive system of reserves over a large region, a mixture of both large and small reserves (Figure 17-18) may be the best way to protect a greater variety of species and communities.

Should Reserves Be Connected by Corridors? Some conservation biologists suggest that establishing protected habitat corridors between reserves can **(1)** help support more species, **(2)** allow migration of vertebrates that need large ranges, **(3)** permit migration of individuals and populations when environmental conditions in a reserve deteriorate, **(4)** help preserve animals that must make seasonal migrations to obtain food, and **(5)** enable some species to shift their ranges if global warming makes their current ranges uninhabitable.

On the other hand, corridors can **(1)** threaten once-isolated populations by allowing the movement of pest species, disease, fire, and exotic species between reserves, **(2)** subject migrating species to increased exposure to natural predators, human hunters, and pollution, and **(3)** be costly to acquire, protect, and manage.

What Areas Should Receive Top Priority for Establishing Reserves? In reality, few countries are physically, politically, or financially able to set aside and protect large biodiversity reserves. Because there will not be enough money or political support to protect most of the world's terrestrial biodiversity, ecologists suggest using two approaches:

- Focusing international efforts on establishing a variety of large and small reserves in the world's most biodiverse countries (Figure 17-20) and threatened species-rich areas within such countries. This is a *prevention strategy* designed to reduce the future loss of biodiversity.
- An *emergency action* strategy that identifies and quickly protects *biodiversity hot spots* (Figure 17-21). They are areas that are especially rich in plant and animal species that are found nowhere else and are in great danger of extinction or serious ecological disruption. Currently, extreme malnutrition is pervasive among the 1.1 billion people who live in or near the world's 25 biodiversity hot spots. In 19 of these 25 hot spots,

Biosphere Reserves

SOLUTIONS

In 1971, the UN Educational, Scientific, and Cultural Organization (UNESCO) created the Man and the Biosphere (MAB) Programme to improve the relationship between people and the environment. The program proposed setting up at least one (and ideally five or more) *biosphere reserves* in each of the earth's 193 biogeographical zones.

Ideally, each reserve should **(1)** be large enough to prevent gradual species loss and **(2)** combine both conservation and sustainable use of natural resources. Today there are more than 350 biosphere reserves in 90 countries.

A reserve must **(1)** be nominated by its national government, **(2)** meet certain size requirements, and **(3)** contain three zones (Figure 17-19):

- A *core area* containing an important ecosystem that the government legally protects from all human activities except nondestructive research and monitoring.
- A *buffer zone* that surrounds and protects the core area. In this zone, emphasis is on **(1)** nondestructive research, education, and recreation and **(2)** sustainable logging, agriculture, livestock grazing, hunting, and fishing by local people as long as such activities do not harm the core.
- A second *buffer* or *transition zone*, which combines conservation and more intensive but sustainable forestry, grazing, hunting, fishing, agriculture, and recreation by local people and ecotourists.

Biosphere Reserve

Core area

Buffer zone 1

Buffer zone 2

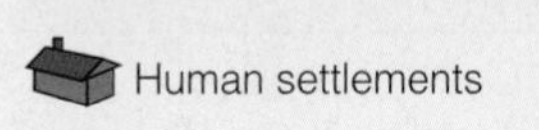

Tourism and education center

Figure 17-19 Design of a model biosphere reserve. In traditional parks and wildlife reserves, well-defined boundaries keep people out and wildlife in. By contrast, biosphere reserves recognize people's needs for access to sustainable use of various resources in parts of the reserve.

So far, most biosphere reserves fall short of the ideal (Figure 17-19) and receive too little funding for their protection and management. An international fund to help countries protect and manage biosphere reserves would cost about $100 million per year—about what the world's nations spend on weapons every 90 minutes.

Critical Thinking

Would you be willing to spend $10 a year to help establish and maintain a global network of biosphere reserves? Can you think of any disadvantages of such a system?

population is growing more rapidly than in the world as a whole.

According to the World Conservation Union, ways to help protect biodiversity in the world's biodiversity hot spots include

- Helping these countries decrease their rate of population growth
- Helping farmers living in or near hot spots learn how to practice *ecoagriculture* by growing more food while conserving critical wildlife habitats (Solutions, p. 447)
- Increasing agricultural productivity on current farmlands to reduce the need to convert more land to agriculture
- Establishing protected area networks near farming plots, ranchland, and fisheries
- Establishing networks of wildlife habitat in nonfarmed areas and connecting them with larger protected areas

What Is Wilderness? One way to protect undeveloped lands from human exploitation is by legally setting them aside as wilderness. According to the U.S. Wilderness Act of 1964, wilderness consists of areas "of undeveloped land affected primarily by the forces of nature, where man is a visitor who does not remain." U.S. president Theodore Roosevelt summarized what we should do with wilderness: "Leave it as it is. You cannot improve it."

The U.S. Wilderness Society estimates that a wilderness area should contain at least 4,000 square kilometers (1,500 square miles); otherwise, it can be

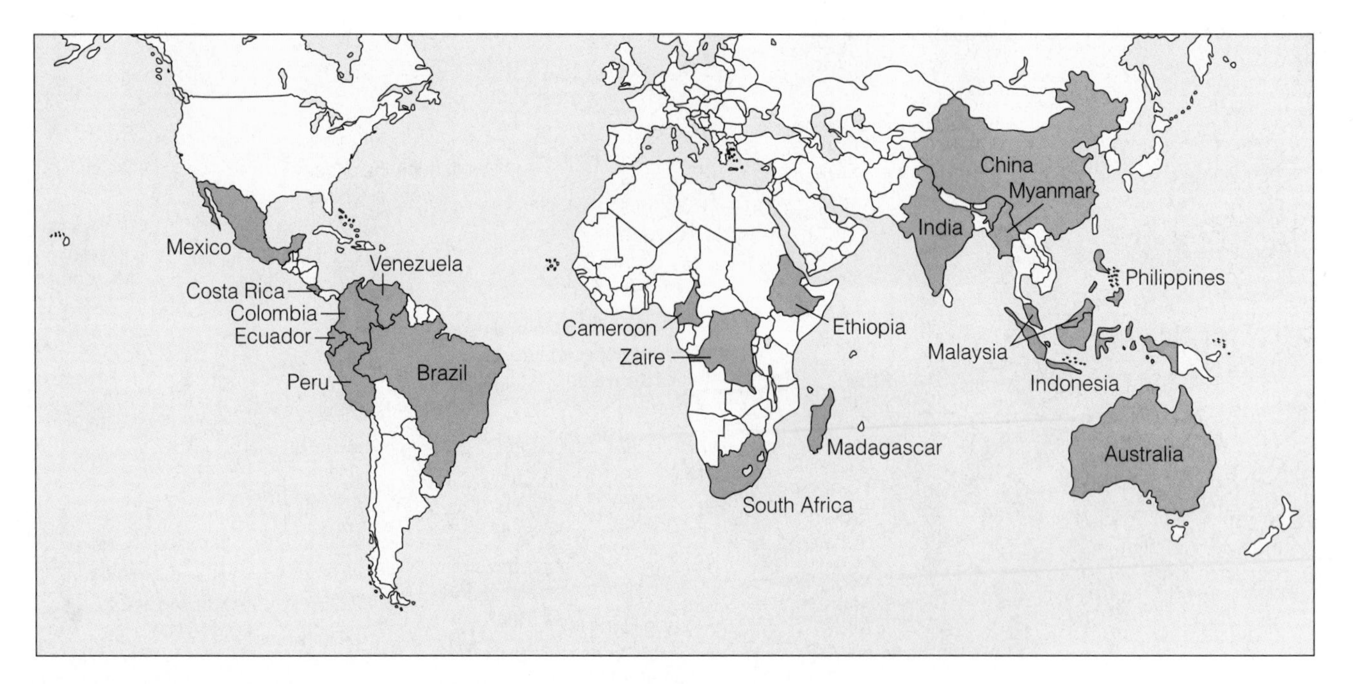

Figure 17-20 The earth's 19 most biodiverse countries. Conservation biologists believe that efforts to preserve repositories of biodiversity as protected wilderness areas and other nature reserves should be concentrated in these 19 countries, which contain at least 60% of the world's known terrestrial species. Protecting the world's biodiversity will take financial and scientific help from developed countries. (Data from Conservation International and World Wildlife Fund)

affected by air, water, and noise pollution from nearby human activities.

Why Preserve Wilderness? According to wilderness supporters, we need wild places where people can **(1)** experience the beauty of nature and observe natural biological diversity and **(2)** enhance their mental and physical health by getting away from noise, stress, development, and large numbers of people. Wilderness preservationist John Muir advised us,

> *Climb the mountains and get their good tidings. Nature's peace will flow into you as the sunshine into the trees. The winds will blow their freshness into you, and the storms their energy, while cares will drop off like autumn leaves.*

Even those who never use the wilderness may want to know it is there, a feeling expressed by novelist Wallace Stegner:

> *Save a piece of country . . . and it does not matter in the slightest that only a few people every year will go into it. This is precisely its value. . . . We simply need that wild country available to us, even if we never do more than drive to its edge and look in. For it can be a means of reassuring ourselves of our sanity as creatures, a part of the geography of hope.*

Some critics have argued that protecting wilderness for its scenic and recreational value for a small number of people is an outmoded concept that keeps some areas of the planet from being economically useful to humans. To most biologists, however, *the most important reasons for protecting wilderness and other areas from exploitation and degradation are* **(1)** *to preserve the biodiversity they contribute as a vital part of the earth's natural capital* (Figure 4-33, p. 103) *and* **(2)** *to protect them as centers for evolution in response to mostly unpredictable changes in environmental conditions* (Guest Essay, p. 118). In other words, wilderness is a biodiversity and wildness bank and an ecoinsurance policy. As a result, wilderness advocates call for more wilderness areas to be protected in **(1)** the United States (Case Study, p. 455) and other species-rich countries (Figure 17-20) and **(2)** the world's most endangered hot spots (Figure 17-21).

How Should Wilderness Be Managed? To protect the most popular areas from damage, wilderness managers **(1)** designate sites where camping is allowed, **(2)** limit the number of people using these sites at any one time, **(3)** use wilderness rangers to patrol vulnerable areas, and **(4)** enlist volunteers to pick up trash discarded by thoughtless users who do not follow the *Leave No Trace* (LNT) wilderness ethic.

Environmental historian and wilderness expert Roderick Nash suggests dividing wilderness areas into three categories:

- Easily accessible, popular areas that would be intensively managed and have trails, bridges, hiker's

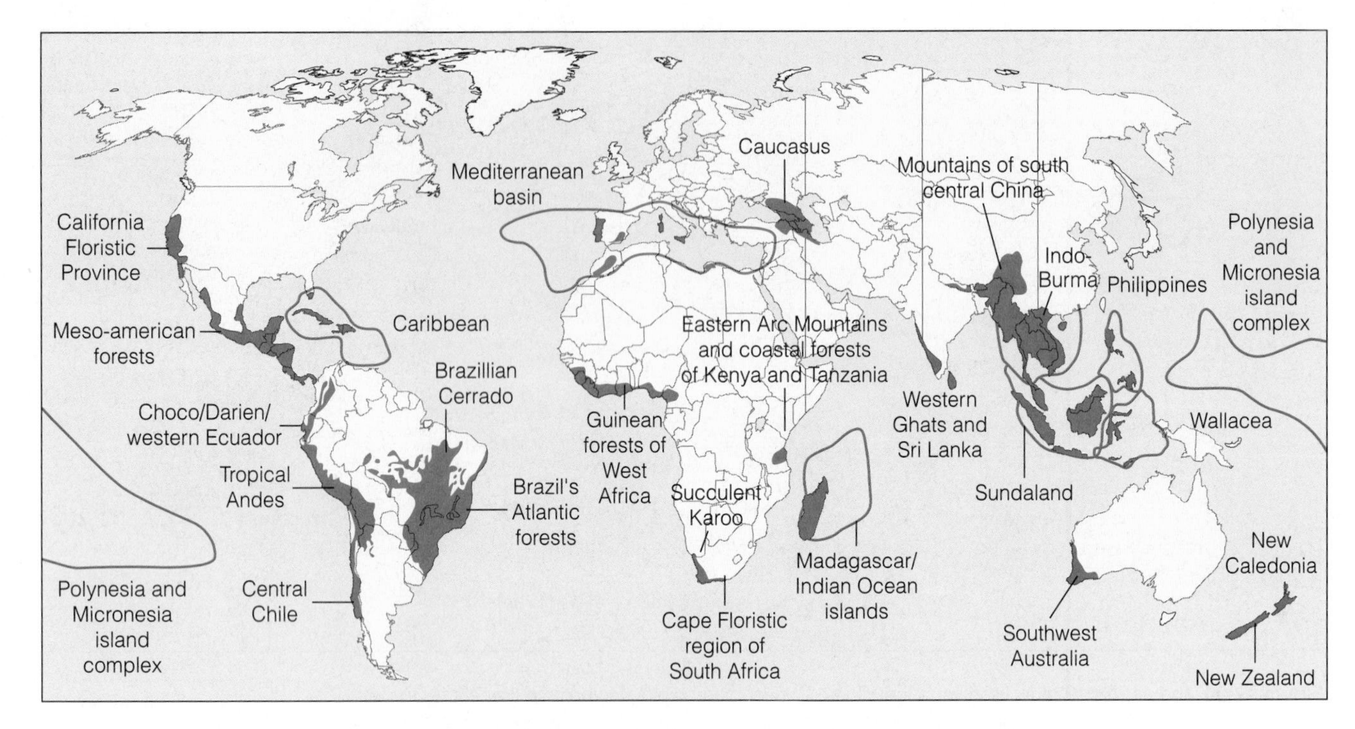

Figure 17-21 Twenty-five hot spots identified by ecologists as being important but endangered centers of biodiversity that contain a large number of endemic plant and animal species found nowhere else. (Data from Conservation International)

huts, outhouses, assigned campsites, and extensive ranger patrols.

- Large, remote wilderness areas that would be used only by people who get a permit by demonstrating their wilderness skills.
- Biologically unique areas that would be left undisturbed as gene pools of plant and animal species, with no human entry allowed.

17-8 PROTECTING AND SUSTAINING AQUATIC SYSTEMS

How Can We Protect and Sustain Marine Systems? Ways to sustain the biodiversity of marine systems include

- Establishing protected areas
- Using integrated coastal management
- Regulating and preventing ocean pollution, as discussed in Section 14-7, p. 354.

What Is the Role of Protected Marine Sanctuaries? Individual governments have used several international treaties, agreements, and actions to protect living marine resources in parts of the world. Under the United Nations Law of the Sea, all coastal nations have **(1)** sovereignty over the waters and seabed up to 19 kilometers (12 miles) offshore and **(2)** almost total jurisdiction over their Exclusive Economic Zone (EEZ), which extends 320 kilometers (200 miles) offshore. Taken together, the nations of the world have jurisdiction over 36% of the ocean surface and 90% of the world's fish stocks.

In 2001, the National Center for Ecological Analysis and Synthesis reported that scientific studies show that within fully protected marine reserves **(1)** fish populations double, **(2)** fish size grows by 30%, **(3)** fish reproduction triples, and **(4)** species diversity is 23% higher. Furthermore, this improvement happens within 2–4 years after protection from fishing and other exploitive activities is implemented and lasts for decades.

Some *good news* is that

- The United Nations Environment Programme has spearheaded efforts to develop 12 regional agreements to protect large marine areas shared by several countries.
- About 90 of the world's 350 biosphere reserves (Solutions, p. 452) include coastal or marine habitats that receive partial protection.
- The United States has designated 12 marine sanctuaries.
- Since 1986, the World Conservation Union (IUCN) has helped establish a global system of *marine protected areas* (MPAs), mostly at the national level. The 1,300 existing MPAs help protect about 0.2% of the earth's total ocean area.

The U.S. National Wilderness Preservation System

CASE STUDY

In the United States, preservationists have been trying to save wild areas from development since 1900. Overall, they have fought a losing battle. Not until 1964 did Congress pass the Wilderness Act, which allowed the government to protect undeveloped tracts of public land from development as part of the National Wilderness Preservation System.

To wilderness supporters, the *good news* is that there was a 10-fold increase in the area of protected wilderness in the United States between 1970 and 2000. The *bad news* is that **(1)** only about 4.6% of U.S. land is protected as wilderness, with almost three-fourths of it in Alaska, and **(2)** only 1.8% of the land area of the lower 48 states is protected, most of it in the West. According to a 1999 study by the World Conservation Union (IUCN), the United States ranks 42nd among nations in terms of terrestrial area protected as wilderness, and Canada is in 36th place.

More *bad news* is that **(1)** of the 413 wilderness areas in the lower 48 states, only four are larger than 4,000 square kilometers, **(2)** the system includes only 81 of the country's 233 distinct ecosystems, and **(3)** most wilderness areas in the lower 48 states are habitat islands in a sea of development.

Almost 400,000 square kilometers (150,000 square miles) in scattered blocks of public lands could qualify for designation as wilderness, about 60% of it in the national forests. Conservation biologists urge wilderness protection for all of this land.

Some wilderness advocates also urge that *wilderness recovery areas* be created by **(1)** closing and obliterating nonessential roads in large areas of public lands, **(2)** restoring wildlife habitats, **(3)** allowing natural fires to burn, and **(4)** reintroducing species that have been driven from such areas.

Some ecologists and conservation biologists have proposed *The Wildlands Project* (TWP) to establish a network of protected wildlands and ecosystems throughout as much of the United States as possible. This science-based approach has the goal of reconnecting many of the existing fragmented and threatened pieces of wild nature for use by wildlife and people in regional networks of protected core wilderness areas, buffer zones, wildlife corridors, wilderness recovery areas, and areas undergoing ecological restoration. This would be accomplished through cooperative efforts between government agencies, scientists, conservation groups, and private owners and users.

Expansion of wilderness areas, wilderness recovery areas, and TWP are strongly opposed by timber, mining, ranching, energy, and other interests who want to **(1)** extract resources from these and most other public lands or **(2)** convert them to private ownership.

Critical Thinking

1. Should a large portion of U.S. Forest Service and Bureau of Land Management land be reclassified as wilderness recovery areas, closed off for many uses, and allowed to undergo natural restoration as wildlife habitat? Explain.

2. Do you support or oppose TWP? Explain.

Some *bad news* is that

- Currently, less than 0.01% of the world's ocean area consists of fully protected marine reserves. In the United States, the total area of fully protected marine habitat is only about 130 square kilometers (50 square miles).
- Many existing marine sanctuaries and MPAs allow extractive activities that are prohibited in marine reserves.
- Stresses from nearby coastal areas can disrupt marine reserves unless they are also protected as part of an integrated coastal management plan.

In 1997, a group of international marine scientists called for governments to increase fully protected marine reserves to be 20% of the ocean's surface by 2020.

What Is the Role of Integrated Coastal Management? *Integrated coastal management* is a community-based attempt to develop and use coastal resources in a sustainable manner. The overall aim is for groups competing for the use of coastal resources to **(1)** identify shared problems and goals and **(2)** agree to workable and cost-effective solutions that preserve biodiversity and environmental quality while meeting economic and social needs.

Ideally, the overall goal is to zone the land and sea portions of an entire coastal area. Such zoning would include **(1)** some fully protected marine reserves where no exploitive human activities are allowed and **(2)** other zones where different kinds and levels of human activities are permitted. Australia's huge Great Barrier Reef Marine Park is managed in this way. Currently, more than 100 integrated coastal management programs are being developed throughout the world.

In the United States,

- Ninety coastal counties are working to establish coastal management systems, but fewer than 20 of these plans have been implemented.

- Since the early 1980s, people have worked together with some success to develop an integrated coastal management plan for the Chesapeake Bay (Case Study, p. 356, and Figure 14-29, p. 358).

How Can Wetlands Be Sustained and Protected? Coastal wetlands (p. 145 and Figure 6-34, p. 148) and inland wetlands (p. 155) are reservoirs of aquatic biodiversity that provide many important ecological and economic services (Figure 6-31, p. 145). These systems are under intense pressure from development and pollution (p. 147 and p. 155).

Ecologists and environmentalists call for several strategies to protect and sustain wetlands. They include

- Enacting and enforcing laws to protect existing wetlands from destruction and degradation
- Using comprehensive land-use planning to steer developers, farmers, and resource extractors away from existing wetlands
- Requiring creation and evaluation of a new wetland before destroying any existing wetland
- Restoring degraded wetlands

Many developers, farmers, and resource extractors oppose such wetland protection.

How Are Wetlands Protected in the United States? In the United States, a federal permit is required to fill or to deposit dredged or fill material into wetlands occupying more than 1.2 hectares (3 acres). Some *good news* is that according to the U.S. Fish and Wildlife Service, this law has helped cut the average annual wetland loss by 80% since 1969.

Some *bad news* is that **(1)** there are continuing attempts to weaken this law by using unscientific criteria to classify areas as wetlands, **(2)** only about 8% of remaining inland wetlands are under federal protection, and **(3)** federal, state, and local wetland protection is weak.

17-9 ECOLOGICAL RESTORATION

How Can We Rehabilitate and Restore Damaged Ecosystems? Some *bad news* is that almost every place on the earth has been degraded at least to some degree by human activities. The *good news* is that much of the environmental damage we have inflicted on nature is at least partially reversible through ecological restoration.

By studying how natural ecosystems recover, scientists are learning how to speed up repair operations by using several approaches.

- *Restoration* involves trying to return a particular degraded habitat or ecosystem to a condition as similar as possible to its predegraded state. Three difficulties are **(1)** lack of knowledge about the previous composition of a system, **(2)** changes in climate, soil, and species composition that may make it impossible to restore an area to an earlier state, and **(3)** the need to deal with a moving target because ecosystems are always undergoing change.
- *Rehabilitation* involves any attempt to restore at least some of a degraded system's natural species and ecosystem functions. Examples include **(1)** removing pollutants and **(2)** replanting areas such as mining sites, landfills, and clear-cut forests to reduce soil erosion.
- *Replacement* involves replacing a degraded ecosystem with another type of ecosystem. For example, a productive pasture or tree farm may replace a degraded forest.
- *Creating artificial ecosystems* involves using ecological principles to create human-designed ecosystems for specific purposes. Examples are the ecological wastewater treatment systems developed by John Todd (p. 362) and the artificial wetlands developed to treat sewage in Arcata, California (Solutions, p. 364).

The basic steps in all forms of ecological restoration and rehabilitation are as follows:

- Identify what caused the degradation (such as pollution, farming, overgrazing, mining, or invading species).
- Eliminate or sharply reduce these factors. Examples include **(1)** removing toxic soil pollutants, **(2)** adding nutrients to depleted soil, **(3)** adding new topsoil, and **(4)** eliminating disruptive nonnative species.
- Protect the area from further degradation and from the disruptive effects of fires (Individuals Matter, right).
- Monitor restoration efforts, assess success, and use adaptive ecosystem management to modify strategies as needed.

Is Ecological Restoration the Best Approach? Some analysts worry that environmental restoration could encourage continuing environmental destruction and degradation by suggesting that any ecological harm we do can be undone. However, ecologists point out that preventing ecosystem damage in the first place is cheaper and more effective than any form of ecological restoration.

Another concern is government policies that allow developers to destroy one ecosystem or wetland if they

Ecological Restoration of a Tropical Dry Forest in Costa Rica

INDIVIDUALS MATTER

Costa Rica is the site of one of the world's largest *ecological restoration* projects. In the lowlands of the country's Guanacastle National Park (Figure 17-18), a small tropical dry deciduous forest has been burned, degraded, and fragmented by large-scale conversion to cattle ranches and farms. Now it is being restored and relinked to the rain forest on adjacent mountain slopes. The goal is to eliminate damaging nonnative grass and cattle and reestablish a tropical dry forest ecosystem over the next 100–300 years.

The keys to restoring the resulting grasslands to tropical dry forest are to **(1)** exclude fires and cattle grazing from much of the park and **(2)** enhance seed dispersal from remaining small patches of native woodland.

The strategy seems to be working. Areas of the park that were covered with monoculture expanses of nonnative grasses 10–15 years ago are now healthy second-growth dry forests of native species.

Daniel Janzen, professor of biology at the University of Pennsylvania and a leader in the field of restoration ecology, has helped galvanize international support and has raised more than $10 million for this restoration project.

He recognizes that ecological restoration and protection of the park will fail unless the people in the surrounding area believe that they will benefit from such efforts. Janzen's vision is to make the nearly 40,000 people who live near the park an essential part of the restoration of the degraded forest, a concept he calls *biocultural restoration.*

By actively participating in the project, local residents reap educational, economic, and environmental benefits. Local farmers make money by sowing large areas with tree seeds and planting seedlings started in Janzen's lab. Local grade school, high school, and university students and citizens' groups study the ecology of the park and go on field trips to the park. The park's location near the Pan American Highway makes it an ideal area for ecotourism, which stimulates the local economy.

The project also serves as a training ground in tropical forest restoration for scientists from all over the world. Research scientists working on the project give guest classroom lectures and lead some of the field trips.

Janzen recognizes that in a few decades today's children will be running the park and the local political system. If they understand the importance of their environment, they are more likely to protect and sustain its biological resources. He believes that education, awareness, and involvement—not guards and fences—are the best ways to restore degraded ecosystems and to protect largely intact ecosystems from unsustainable use.

protect, restore, or "create" a similar one of roughly the same size. This trade-off or *mitigation* approach is preferable to wanton destruction of ecosystems. However, according to ecological restoration expert John Berger, "The purpose of ecological restoration is to repair previous damage, not legitimize further destruction."

According to a 2001 study by the U.S. National Academy of Sciences, the current approach, designed to ensure "no net loss" of U.S. wetlands, is a failure, and human-made wetlands often are a poor substitute for the real thing. The report called for major changes in the system for designing and more carefully regulating replacement wetlands. Ecologists also believe that no natural wetland or ecosystem should be destroyed until its replacement has been successfully established—the opposite of current policy in the United States. See the website material for this chapter for some actions you can take to help sustain the earth's biodiversity.

We abuse land because we regard it as a commodity belonging to us. When we see land as a community to which we belong, we may begin to use it with love and respect.

ALDO LEOPOLD

REVIEW QUESTIONS

1. Define the boldfaced terms in this chapter.

2. List the pros and cons of reintroducing the gray wolf as a keystone predator species in the Yellowstone ecosystem.

3. List five ways in which human activities have decreased and degraded the earth's biodiversity.

4. List the major goals, strategies, and tactics for protecting the earth's biodiversity by using **(a)** the species approach and **(b)** the ecosystem approach.

5. How is land used in the world and in the United States? What percentage of the land in the United States does the federal government own and manage as public lands? What are the five major types of public lands in the United States? What are the major uses allowed on each type?

6. List four principles that most biologists and some economists believe should govern the use of public land in the United States. Compare these principles with the ways in which users of mineral and other resources believe these lands should be used, owned, and managed.

7. List the pros and cons of providing government subsidies to ranchers holding permits to graze livestock on

public lands in the United States. List three major positions on use of public rangelands. List three ways suggested by some ranchers, environmentalists, and federal range managers for using public rangeland more sustainably.

8. List **(a)** eight important ecological services provided by forests and **(b)** seven important economic benefits of forests.

9. Distinguish between *old-growth forests, second-growth forests*, and *tree farms*.

10. Describe the *rotation cycle* for harvesting and managing a forest. Distinguish between *even-aged* and *uneven-aged management* of a forest and list the advantages and disadvantages of each type.

11. Describe five major ways for harvesting trees. List the advantages and disadvantages of **(a)** selective cutting and **(b)** clear-cutting.

12. Summarize the state of forest growth and loss in the world's temperate and tropical forests. What are the pros and cons of deforestation?

13. List eight ways to use forests more sustainably. Describe **(a)** how industrial forestry based on tree farms could lead to more sustainable use of forests and **(b)** four conditions necessary for this approach to work.

14. Distinguish between *surface, crown*, and *ground* forest fires. What are the benefits of fire for some plant and animal species?

15. Summarize the threats to forests from **(a)** air pollution and **(b)** global warming.

16. Describe two pieces of good news and two pieces of bad news about the general extent and condition of forests in the United States.

17. How are national forests in the United States used? List nine services provided by U.S. national forests. What are the principles of *sustained yield* and *multiple use* that are supposed to guide the use of national forests in the United States?

18. What are the pros and cons of giving government subsidies to companies cutting timber in national forests? List the pros and cons of making **(a)** timber cutting and **(b)** recreation the primary use of national forests.

19. Describe and list the advantages of programs for certifying that timber has been grown sustainably.

20. Summarize the estimated economic values of the earth's ecological services.

21. How rapidly are tropical forests being cleared and degraded? Describe the many uses of the tropical *neem tree* and list the pros and cons of widespread planting of neem trees.

22. List three major ecological services and three economic services provided by tropical forests.

23. List three underlying causes and seven direct causes of tropical deforestation and degradation.

24. List 13 ways to reduce tropical deforestation and degradation. Describe **(a)** ways to achieve more sustainable farming and logging in tropical forests and **(b)** Kenya's Green Belt movement.

25. What are the major threats to national parks in the United States and in other countries? List 11 ways to improve national park management in the United States. Describe efforts by Costa Rica to establish reserves to protect its biodiversity.

26. List the pros and cons of strictly protecting at least 10% of the earth's land surface as nature reserves. Describe the role of the Nature Conservancy in establishing nature reserves in the United States and other parts of the world.

27. List three scientific principles and two social principles for guiding the selection, design, and management of nature reserves. Describe the need to establish nature reserves in *megadiversity countries* and *biodiversity hot spots*.

28. List the pros and cons of establishing large reserves or several small reserves with the same total area. List the pros and cons of establishing corridors between nature reserves.

29. What is a *biosphere reserve*?

30. List two strategies conservation biologists believe should be used to protect most of the earth's terrestrial biodiversity.

31. What is *wilderness*, and why is it important? List the pros and cons of protecting more wilderness. List three ways to manage wilderness.

32. Describe protecting and sustaining marine biodiversity by **(1)** establishing fully protected marine reserves and **(2)** using integrated coastal management.

33. List four ways to protect and sustain wetlands. Describe wetland protection in the United States.

34. Distinguish between ecosystem *restoration, rehabilitation*, and *replacement*. List four basic steps in carrying out ecological restoration.

35. Describe efforts to restore a degraded tropical dry forest in Costa Rica.

36. What are two concerns some people have about ecological restoration?

CRITICAL THINKING

1. Do you agree or disagree with the program to reintroduce populations of the gray wolf in the Yellowstone ecosystem? Explain. Could the money be better spent on other wildlife programs? If so, what programs would you suggest? Do you favor reintroducing grizzly bears to the Yellowstone or other ecosystems in the western United States? Explain.

2. Explain why you agree or disagree with the four principles biologists and some economists have suggested for using public land in the United States (pp. 433–434).

3. Explain why you agree or disagree with each of the proposals listed on pp. 434–435 for providing more sustainable use of public rangeland in the United States. What might be some drawbacks of implementing these proposals?

4. Should private companies that harvest timber from U.S. national forests continue to be subsidized by federal payments for reforestation and for building and maintaining access roads? Explain.

5. In the early 1990s, Miguel Sanchez, a subsistence farmer in Costa Rica, was offered $600,000 by a hotel developer for a piece of land that he and his family had been using sustainably for many years. The land contained an old-growth rain forest and a black sand beach in an area under rapid development. Sanchez refused the offer. What would you have done if you were a poor subsistence farmer in Miguel Sanchez's position? Explain your decision.

6. Should the United States (or the country where you live) meet most of its demand for wood and wood products by growing genetically improved trees on carefully managed tree farms? Explain. If you agree, what regulations and safeguards would you impose to help ensure that such an approach would be sustainable and have a low environmental impact?

7. Explain why you agree or disagree with each of the proposals listed on pp. 445–447 for protecting the world's tropical forests. Should developed countries provide most of the money to preserve the remaining tropical forests in developing countries? Explain.

8. List five actions you could take to help preserve some of the world's tropical forests. Which, if any, of these actions do you plan to carry out?

9. Explain why you agree or disagree with each of the proposals listed on p. 449 concerning the U.S. national park system.

10. Should more wilderness areas be preserved in the United States, especially in the lower 48 states (or in the country where you live)? Explain. What might be some drawbacks of doing this?

11. Congratulations. You have been put in charge of protecting and sustaining the world's biodiversity. List the three most important features of your policies for using and managing **(a)** forests, **(b)** parks, **(c)** wilderness and other protected biodiversity reserves, and **(d)** aquatic systems.

PROJECTS

1. Obtain a topographic map of the region where you live and use it to identify local, state, and federally owned lands in the form of parks, rangeland, forests, and wilderness areas. Identify the government agency responsible for managing each of these areas and evaluate how well these agencies are preserving the natural resources on this land on your behalf.

2. If possible, try to visit **(a)** a diverse old-growth forest, **(b)** an area that has been recently clear-cut, and **(c)** an area that was clear-cut 5–10 years ago. Compare the biodiversity, soil erosion, and signs of rapid water runoff in each of the three areas.

3. Survey the condition of a nearby wetland, coastal area, river, or stream. Has its condition improved or deteriorated during the last 10 years? What local, state, or national efforts are being made to protect this aquatic system? Develop a plan for protecting this system.

4. Use the library and internet to find one example of a successful ecological restoration project not discussed in this chapter and one that failed. For each example, describe **(a)** the strategy used, **(b)** the ecological principles involved, and **(c)** why the project succeeded or failed.

5. Make a concept map of this chapter's major ideas, using the section heads and subheads and the key terms (in boldface). Look on the website for this book for information about making concept maps.

INTERNET STUDY RESOURCES AND RESOURCES FOR FURTHER READING AND RESEARCH

The website for this book contains helpful study aids and many ideas for further reading and research. Log on to

http://www.brookscole.com/product/0534389872s

and click on the Chapter-by-Chapter area. Choose Chapter 17 and select a resource:

- "Flash Cards" allows you to test your mastery of the Terms and Concepts to Remember for this chapter.
- "Tutorial Quizzes" provides a multiple-choice practice quiz.
- "Student Guide to InfoTrac" will lead you to Critical Thinking Projects that use InfoTrac College Edition as a research tool.
- "References" lists the major books and articles consulted in writing this chapter.
- "Hypercontents" takes you to an extensive list of sites with news, research, and images related to individual sections of the chapter.

INFOTRAC COLLEGE EDITION

Improve your skills with InfoTrac College Edition, a searchable online database of articles from more than 700 periodicals. Log on to

http://www.infotrac-college.com

or access InfoTrac through the website for this book. Try to find the following articles:

1. Can We Afford to Conserve Biodiversity? Alexander James, Kevin J. Gaston, Andrew Balmford. *BioScience*, Jan 2001 v51 i1 p43. This article examines the cost of maintaining biodiversity. *Hint*: Enter the search term "conserve biodiversity" using the keyword "biodiversity."

2. Down on the Fish Farm: Developing Effluent Standards for Aquaculture. Karen Kreeger. *BioScience*, Nov 2000 v50 i11 p949. Discussion of environmental impacts of wastewater from using aquaculture to raise a variety of fish and other types of seafood. *Hint*: Enter the search term "fish farm" using the keywords "stream pollution."

18 SUSTAINING WILD SPECIES

The Passenger Pigeon

In the early 1800s, bird expert Alexander Wilson watched a single migrating flock of an estimated 2 million passenger pigeons darken the sky for more than 4 hours. By 1914 the passenger pigeon (Figure 18-1) had disappeared forever.

How could a species that was once the most common bird in North America become extinct in only a few decades? The answer is humans. The main reasons for the extinction of this species were **(1)** uncontrolled commercial hunting and **(2)** loss of the bird's habitat and food supply as forests were cleared to make room for farms and cities.

Passenger pigeons were good to eat, their feathers made good pillows, and their bones were widely used for fertilizer. They were easy to kill because they flew in gigantic flocks and nested in long, narrow colonies.

Commercial hunters would capture one pigeon alive, sew its eyes shut, and tie it to a perch called a stool. Soon a curious flock would land beside this "stool pigeon." Then the birds would be shot or ensnared by nets that might trap more than 1,000 birds at once.

Beginning in 1858, passenger pigeon hunting became a big business. Shotguns, traps, artillery, and even dynamite were used. Burning grass or sulfur below their roosts sometimes suffocated birds. Live birds were used as targets in shooting galleries. In 1878, one professional pigeon trapper made $60,000 by killing 3 million birds at their nesting grounds near Petoskey, Michigan.

By the early 1880s, only a few thousand birds remained. At that point, recovery of the species was doomed because the females laid only one egg per nest. On March 24, 1900, a young boy in Ohio shot the last known wild passenger pigeon. The last passenger pigeon on earth, a hen named Martha after Martha Washington, died in the Cincinnati Zoo in 1914. Her stuffed body is now on view at the National Museum of Natural History in Washington, D.C.

Eventually, all species become extinct or evolve into new species. However, biologists estimate that every day 3–200 species become *prematurely extinct* primarily because of human activities. Studies indicate that this rate of loss of biodiversity is increasing as the human population **(1)** grows, **(2)** consumes more resources, **(3)** disturbs more of the earth's land, and **(4)** uses more of the earth's net plant productivity that supports all species.

Figure 18-1 Passenger pigeons, extinct in the wild since 1900. The last known passenger pigeon died in the Cincinnati Zoo in 1914. (John James Audubon/The New York Historical Society)

The last word in ignorance is the person who says of an animal or plant: "What good is it?" . . . If the land mechanism as a whole is good, then every part of it is good, whether we understand it or not. . . . Harmony with land is like harmony with a friend; you cannot cherish his right hand and chop off his left.

ALDO LEOPOLD

This chapter addresses the following questions:

- Are human activities causing a new mass extinction?
- Why should we care about species extinction?
- What human activities endanger wildlife?
- How can we use laws, treaties, and sanctuaries to prevent premature extinction of species?
- How can we manage game animals more sustainably?

18-1 SPECIES EXTINCTION

What Are Three Types of Species Extinction? Biologists distinguish between three levels of species extinction:

- *Local extinction* occurs when a species is no longer found in an area it once inhabited but is still found elsewhere in the world. Most local extinctions involve losses of one or more populations of species.
- *Ecological extinction* occurs when there are so few members of a species left that it can no longer play its ecological roles in the biological communities where it is found.
- *Biological extinction* occurs when a species is no longer found anywhere on the earth (Figures 18-1 and 18-2).

How Serious Is the Problem of Premature Extinction? Here are some recent estimates of extinction, caused or hastened mostly by human activities:

- A 2000 joint study by the World Conservation Union and Conservation International and a 1999 study by the World Wildlife Fund found that **(1)** 34% of the world's fish species (51% of freshwater species), **(2)** 25% of amphibians (Connections, p. 164), **(3)** 24% of mammals, **(4)** 20% of reptiles, **(5)** 14% of plants, and **(6)** 12% of bird species are threatened with premature extinction.
- The International Union for Conservation (IUCN) estimates that 29% of the known plant species in the United States are threatened with extinction.
- In 1999, Peter Raven, president of the International Botanical Congress, reported, "We are predicting the extinction of about two-thirds of all mammal, butterfly, and plant species by the end of the 21st century."
- According to Norman Myers (Guest Essay, p. 118), "Within just a few human generations, we shall—in the absence of greatly expanded conservation efforts—impoverish the biosphere to an extent that will persist for at least 200,000 human generations or twenty times longer than the period since humans emerged as a species."

What Are Endangered and Threatened Species? Biologists classify species heading toward biological extinction as either *endangered* or *threatened* (Figure 18-3). An **endangered species** has so few individual survivors that the species could soon become extinct over all or most of its natural range. A **threatened** or **vulnerable species** is still abundant in its natural range but is declining in numbers and is likely to become endangered in the near future. Endangered and threatened species are ecological smoke alarms. Some species have characteristics that make them more vulnerable

Figure 18-2 Some animal species that have become prematurely extinct largely because of human activities, mostly habitat destruction and overhunting.

Figure 18-3 Species that are endangered or threatened with premature extinction largely because of human activities. Almost 30,000 of the world's species and 1,200 of those in the United States are officially listed as being in danger of becoming extinct. Most biologists believe that the actual number of species at risk is much larger.

than others to biological extinction (Figure 18-4 and Case Study, p. 465).

How Does Background Extinction Differ from Mass Extinction? Evolutionary biologists estimate that more than 99.9% of all the species that have ever existed are now extinct because of a combination of background and mass extinctions.

Each year, a small number of species become extinct naturally at a low rate, a phenomenon called the **background extinction** or **natural rate of extinction**. According to the fossil record, the earth's estimated background extinction rate is about **(1)** 3–14 species per year if the earth has 14 million species and **(2)** 1–5 species per year if there are 5 million species.

In contrast, **mass extinction** is a rise in extinction rates above the background level. It is a catastrophic, widespread (often global) event in which large groups of existing species (perhaps 25–70%) are wiped out. Most mass extinctions are believed to result from one or a combination of global climate changes that kill many species and leave behind those able to adapt to the new conditions.

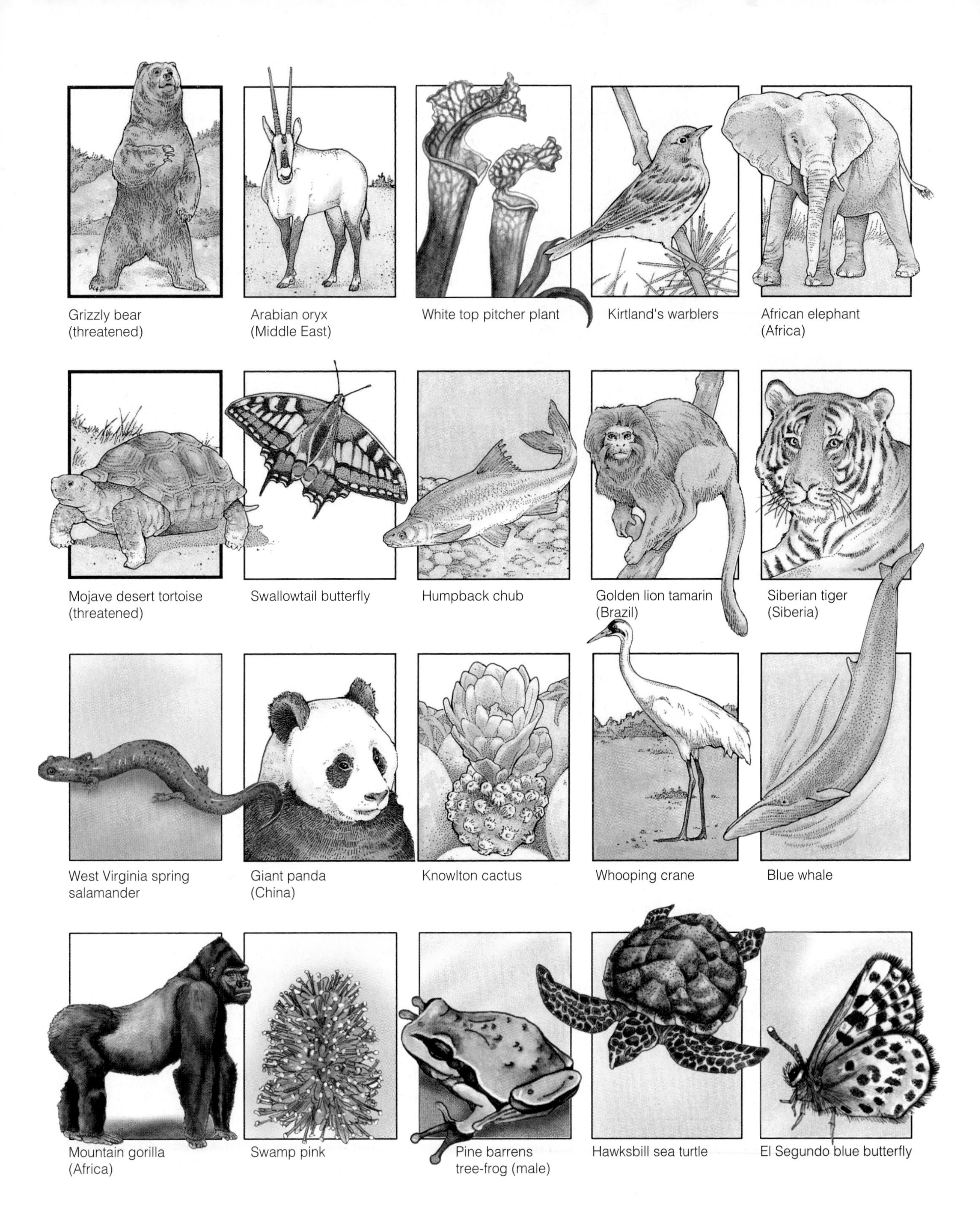
Grizzly bear
(threatened)
Arabian oryx
(Middle East)
White top pitcher plant
Kirtland's warblers
African elephant
(Africa)
Mojave desert tortoise
(threatened)
Swallowtail butterfly
Humpback chub
Golden lion tamarin
(Brazil)
Siberian tiger
(Siberia)
West Virginia spring
salamander
Giant panda
(China)
Knowlton cactus
Whooping crane
Blue whale
Mountain gorilla
(Africa)
Swamp pink
Pine barrens
tree-frog (male)
Hawksbill sea turtle
El Segundo blue butterfly

Characteristic	Examples
Low reproductive rate (K-strategist)	Blue whale, giant panda, rhinoceros
Specialized niche	Blue whale, giant panda, Everglades kite
Narrow distribution	Many island species, elephant seal, desert pupfish
Feeds at high trophic level	Bengal tiger, bald eagle, grizzly bear
Fixed migratory patterns	Blue whale, whooping crane, sea turtles
Rare	Many island species, African violet, some orchids
Commercially valuable	Snow leopard, tiger, elephant, rhinoceros, rare plants and birds
Large territories	California condor, grizzly bear, Florida panther

Figure 18-4 Characteristics of species that are prone to extinction.

Fossils and geological evidence indicate that the earth's species have experienced five great mass extinctions (20–60 million years apart) during the past 500 million years in which large numbers of species became extinct each year for tens of thousands to millions of years (Figure 5-9, p. 116). Evidence also shows that these mass extinctions were followed by other periods, called *adaptive radiations*, when the diversity of life increased and spread for 10 million years or more (Figure 5-9, p. 116, and Figure 5-10, p. 117).

How Do Scientists Estimate Extinction Rates? Estimates of species extinction rates are based on

- Observations of how the number of species present increases with the size of an area
- Models for estimating the risk that a particular species will become endangered or extinct within a certain time based on factors such as trends in population size and changes in habitat availability
- Models such as the theory of island biogeography (Figure 7-5, p. 162, and Figure 7-6, p. 162)
- Changes in species diversity at different latitudes (Figure 7-3, p. 161)
- Statistical sampling mostly in tropical forests
- Differing assumptions about the **(1)** earth's total number of species (5–100 million, with a best estimate of 12–14 million), **(2)** proportion of these species found in tropical forests (50–80%), and **(3)** rate of clearing of tropical forests (0.5–2% per year) (Spotlight, p. 466)

Because of the uncertainties involved, biodiversity researchers estimate that extinction rates range from 1,000 to 75,000 species per year (a loss of about 3–200 species per day).

Some biologists believe that loss of *local populations* of key species may be a better indicator of biodiversity loss than species extinction because these local populations provide most of the ecological services of an area. According to a 1997 study by biologist Jennifer Hughes and her colleagues at Stanford University, the estimated loss of populations is tens of thousands of times greater than the estimated rate of species loss.

Are Human Activities Causing a New Mass Extinction? According to a 1998 survey, 70% of the biologists polled believe that **(1)** we are in the midst of a new mass extinction caused mostly by human activities and **(2)** this loss of species will pose a major threat to human health and economies in the 21st century. To these biologists, we are not heeding Aldo Leopold's (Individuals Matter, p. 47) warning about preserving species as we tinker with the earth: "To keep every cog and wheel is the first precaution of intelligent tinkering."

Mass extinctions occurred long before humans evolved. However, biologists point out two important differences between the current mass extinction most biologists believe we are bringing about and those of the past:

- It is taking place in only a few decades rather than over thousands to millions of years (Figure 5-9, p. 116).
- We are eliminating or degrading many biologically diverse environments (such as tropical forests, tropical coral reefs, wetlands, and estuaries) that in the past have served as evolutionary centers for the 5- to 10-million-year recovery of biodiversity after a mass extinction (Figure 5-10, p. 117, and Guest Essay, p. 118).

Some analysts say that global and national estimates of extinction are exaggerated. They point out that

- We do not know how many species there are, with estimates of 5–100 million and a best guess of 12–14 million.
- We have identified and named only about 1.4–1.8 million species and know a fair amount about only one-third of these species.
- We do not know how important various species are to the functioning and sustainability of their ecosystems and to humans because we understand the detailed roles and interactions of only a small number of the world's known species.

CASE STUDY

Bats Are Getting a Bad Rap

Despite their variety (950 known species) and worldwide distribution, bats—the only mammals that can fly—have certain traits that make them vulnerable to extinction: **(1)** They reproduce slowly, and **(2)** many bat species that live in huge colonies in caves and abandoned mines become vulnerable to destruction when people block the passageways or disturb their hibernation.

Bats play important ecological roles. About 70% of all bat species feed on crop-damaging nocturnal insects and other pest species such as mosquitoes, making them the primary control agents for such insects.

In some tropical forests and on many tropical islands, pollen-eating bats pollinate flowers, and fruit-eating bats distribute plants throughout tropical forests by excreting undigested seeds (p. 159). As *keystone species*, they are vital for maintaining plant biodiversity and for regenerating large areas of tropical forest cleared by human activities. If you enjoy bananas, cashews, dates, figs, avocados, or mangos, you can thank bats.

Many people mistakenly view bats as fearsome, filthy, aggressive, rabies-carrying bloodsuckers. However, most bat species are harmless to people, livestock, and crops. In the United States, only 10 people have died of bat-transmitted disease in four decades of recordkeeping; more Americans die each year from falling coconuts.

Because of unwarranted fears of bats and lack of knowledge about their vital ecological roles, several bat species have been driven to extinction. Currently, 26% of the world's bat species, including the ghost bat (Figure 18-3), are listed as endangered or threatened. Conservation biologists urge us to view bats as valuable allies, not as enemies.

Critical Thinking

How do you feel about bats? What would you do if you found bats flying around your yard at night?

- Documenting the extinction of even a widely studied species is very difficult.

Scientists agree that their estimates of extinction rates are based on inadequate data and sampling, and they are continuously striving to get better data. However, they point out that

- This is all the information we have, and such estimates can give us an idea of general trends in biodiversity loss.
- Even if the best available estimates of extinction rates are 100–1,000 times higher than the actual rates, there has been a serious loss and degradation of global biodiversity.
- Because of projected increases in population and resource consumption per person, the rate and extent of biodiversity loss is very likely to increase.
- Arguing over the numbers and waiting to get better data should not be used as excuses for not implementing a *precautionary strategy* to reduce the threats to human health and economies and to the world's ecosystems and wildlife from this potentially serious global ecological problem.

18-2 WHY SHOULD WE CARE ABOUT SPECIES EXTINCTION?

Why Preserve Wild Species and Ecosystems?

If all species eventually become extinct, why should we worry about losing a few more because of our activities? Does it matter that **(1)** the passenger pigeon (Figure 18-1), **(2)** the great auk (Figure 18-2), **(3)** the green sea turtle (photo on p. 53), **(4)** the 30–50 remaining Florida panthers (photo on p. 395), or **(5)** some unknown plant or insect in a tropical forest becomes prematurely extinct because of human activities?

Because ecosystems are constantly changing in response to changing environmental conditions (Section 7-6, p. 175), why should we try to preserve ecosystems? Does it matter that tropical forests, grasslands, wetlands, coral reefs, and other systems making up the earth's ecological diversity are being destroyed or degraded by human activities?

Conservation biologists and ecologists contend that the answer to these questions is *yes* because wild species, natural ecosystems, and the earth's overall biodiversity have two types of value:

- **Instrumental value** because of their economic and ecological usefulness to us
- **Intrinsic value** because they exist, regardless of whether they have any usefulness to us

Examples of the instrumental values of the genes, species, and ecosystems that make up the earth's biodiversity include

- Ecological services (Figure 4-33, p. 103).
- Economic services such as **(1)** development of food crops by crossbreeding and genetic engineering

SPOTLIGHT

Is There Really an Extinction Crisis?

Some social scientists and a few biologists question the existence of a human-caused extinction crisis. Some of these critics also point out that biologists contending that we are in the midst of a human-caused extinction crisis are making the questionable assumption that any loss of habitat also means a net loss in species, usually in some proportion to the amount of habitat lost.

For example, when an old-growth forest is cleared, a number of species (many of them with specialized niches) are lost. However, a cleared area that returns by secondary ecological succession to a second-growth forest can still support some of the original species. There may be a net loss of species, but not as high as projected by some estimates.

Biologists using models and field data estimate the annual loss of tropical forest habitat at about 1.8% per year. Edward O. Wilson and several biologists who have counted species in patches of tropical forest before and after destruction or degradation estimate that this 1.8% loss in habitat results in roughly a 0.5% loss of species.

Do such estimates add up to an extinction crisis? Let us assume, as Wilson and many other biologists do, that a loss of 1 million species over 50–150 years (a very short time in evolutionary history) represents an extinction crisis, with an extinction rate comparable to that during the last mass extinction 65 million years ago (Figure 5-9, p. 116).

If we assume the global decline in species to be 0.5% per year, then we will lose 25,000 species per year if there are 5 million species and 70,000 per year if there are 14 million species. If these assumptions are correct, we will lose 1 million species in 40 years if there are 5 million species and in 14 years with 100 million species.

Assume, however, that the estimate of 0.5% species loss per year is too high for the earth as a whole because of replacement of some species by ecological succession or other factors. If it is 0.25% per year, then we will lose 1 million species in 80 years with 5 million species and in 29 years with 14 million species. Even if we halve the estimated species loss again, to 0.125% per year, we can still lose 1 million species within 57 to 160 years, enough to qualify the situation as an extinction crisis. According to biodiversity expert Edward O. Wilson, "Clearly, we are in the midst of one of the great extinction spasms of geological history."

Mathematical models recently developed by ecologists indicate a time lag of several generations between habitat loss and extinction, primarily because habitat loss also removes potential colonization sites for the emergence of new species. If these models are correct, biologists may be greatly underestimating the magnitude of current and projected biotic impoverishment.

Biologists do not contend that their estimates of extinction rates are precise enough to make firm predictions. Instead, they argue that there is ample evidence that we are destroying and degrading wildlife habitats at an increasing rate and that our actions certainly are leading to a significant loss of species, even though the number and rate vary in different parts of the world.

Critical Thinking

Do you believe that we are in the midst of a sixth mass extinction caused mostly by human activities? Explain. If so, list three ways in which you contribute to this loss of biodiversity.

(Figure 16-10, p. 406), **(2)** fuelwood and lumber, **(3)** paper, and **(4)** medicine (Figure 17-15, p. 445).

- Recreational pleasure provided by wild plants and animals and natural ecosystems. For example, each year Americans spend more than three times more to watch wildlife than they do to watch movies or professional sporting events. Wildlife tourism, or *ecotourism*, generates at least $500 billion per year worldwide. Conservation biologist Michael Soulé estimates that over a lifetime of 60 years, a Kenyan elephant is worth about $1 million in ecotourist revenue—many times more than its tusks are worth when sold illegally for their ivory. See the website material for this chapter for ways to evaluate ecotours.

Is It Ethically Important to Preserve Wild Species? Some people believe that all wild species and ecosystems have an inherent right to exist that is unrelated to their usefulness to humans. According to this intrinsic view, we have an ethical responsibility to **(1)** protect species from becoming prematurely extinct as a result of human activities and **(2)** prevent the degradation of the world's ecosystems and its overall biodiversity.

Some people distinguish between the survival rights of plants and those of animals, mostly for practical reasons. Poet Alan Watts once said that he was a vegetarian "because cows scream louder than carrots." Other people distinguish between various types of species. For example, they might think little about killing a mosquito, cockroach (Spotlight, p. 113), or rat or ridding the world of disease-causing bacteria.

Some proponents go further and assert that each individual organism, not just each species, has a right to survive without human interference. Others apply this to individuals of some species but not to those of other species. Unless they are strict vegetarians, for example, some people might see no harm in having others kill domesticated animals in slaughterhouses to provide them with meat, leather, and other products. However, these same people might deplore the killing of wild animals such as deer, squirrels, or rabbits.

Others emphasize the importance of preserving the whole spectrum of biodiversity by protecting entire ecosystems rather than individual species or organisms, as discussed in Chapter 17.

18-3 EXTINCTION THREATS FROM HABITAT LOSS AND DEGRADATION

What Is the Role of Habitat Loss and Degradation? Figure 18-5 shows the underlying and direct causes of the endangerment and premature extinction of wild species. Biologists agree that the greatest threat to wild species is habitat loss (Figure 18-6), degradation, and fragmentation. According to biodiversity researchers, tropical deforestation (Section 17-5, p. 443) is the greatest eliminator of species, followed by **(1)** destruction of coral reefs and wetlands (pp. 145 and 150), **(2)** plowing of grasslands (Figure 6-20, p. 136), and **(3)** pollution of freshwater (Section 14-6, p. 345) and marine habitats (Section 14-7, p. 354). Globally, temperate biomes have been affected more by habitat disturbance, degradation, and fragmentation than have tropical biomes because of widespread development in temperate developed countries over the past 200 years.

According to the Nature Conservancy, the major types of habitat disturbance threatening endangered species in the United States, in order of importance, are **(1)** agriculture, **(2)** commercial development, **(3)** water development, **(4)** outdoor recreation including off-road vehicles, **(5)** livestock grazing, and **(6)** pollution.

Island species, many of them *endemic species* found nowhere else on earth, are especially vulnerable to extinction. Scientists have used the theory of island biogeography (Figure 7-5, p. 162) to predict the number and percentage of species that would become extinct when habitats on islands are destroyed, degraded, or fragmented. They have also applied the model to national parks, tropical rain forests, lakes, and nature reserves, which often can be viewed as *habitat islands* in an inhospitable sea of human-altered habitat.

What Is the Role of Habitat Fragmentation? *Habitat fragmentation* occurs when a large, continuous area of habitat is reduced in area and divided into a patchwork of isolated areas or fragments. The three main problems caused by habitat fragmentation are

- A decrease in the sustainable population size for many species when an existing population is divided into two or more isolated subpopulations.
- Increased surface area or edge, which makes some species more vulnerable to **(1)** predators, **(2)** competition from nonnative and pest species, **(3)** wind, and **(4)** fire.
- Creation of barriers that limit the ability of some species to **(1)** disperse and colonize new areas, **(2)** find enough to eat, and **(3)** find mates. Some species of migrating birds face loss, degradation, or fragmentation of their seasonal habitats (Case Study, p. 469).

Types of species that are vulnerable to local and regional extinction because of habitat fragmentation are those that **(1)** are rare, **(2)** need to roam unhindered over a large area, **(3)** cannot rebuild their population because of a low reproductive capacity, **(4)** have specialized niches (habitat or resource needs), and **(5)** are sought by people for furs, food, medicines, or other uses.

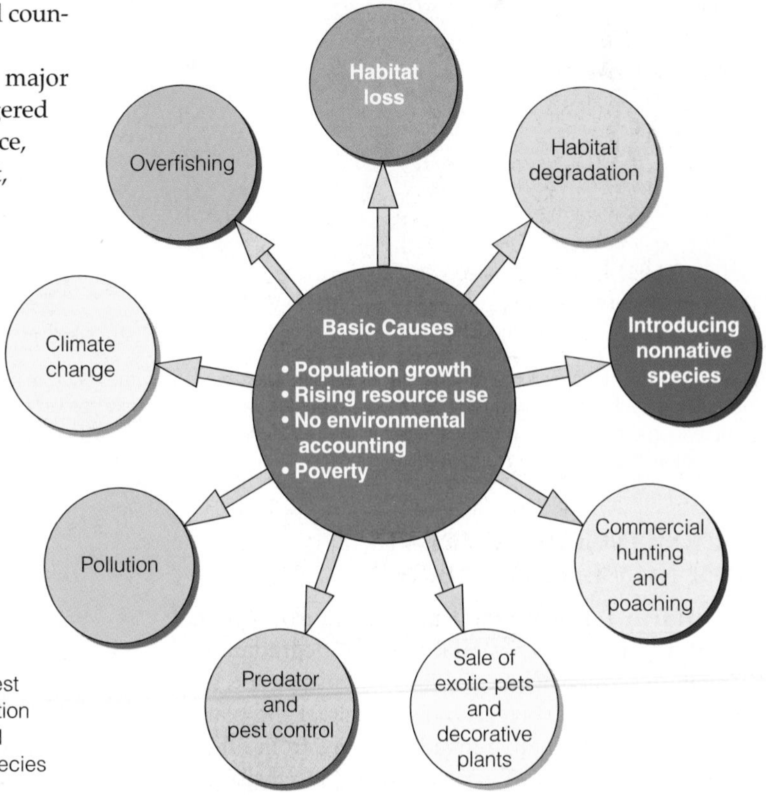

Figure 18-5 Underlying and direct causes of depletion and premature extinction of wild species. The two biggest direct causes of wildlife depletion and premature extinction are **(1)** habitat loss, fragmentation, and degradation and **(2)** deliberate or accidental introduction of nonnative species into ecosystems.

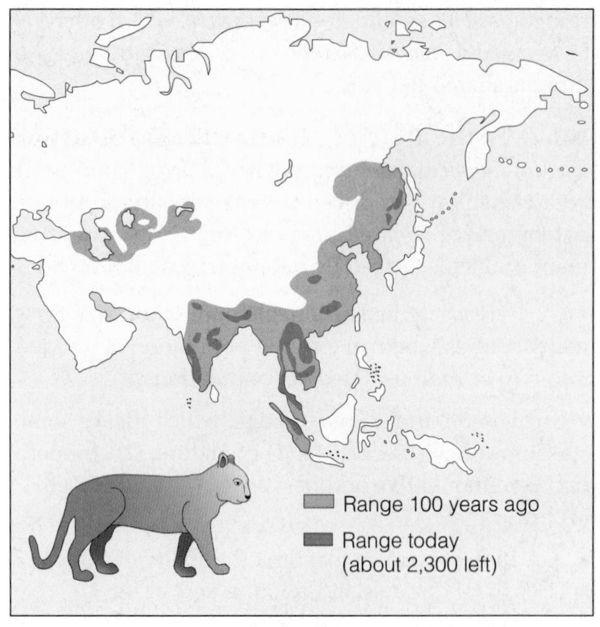

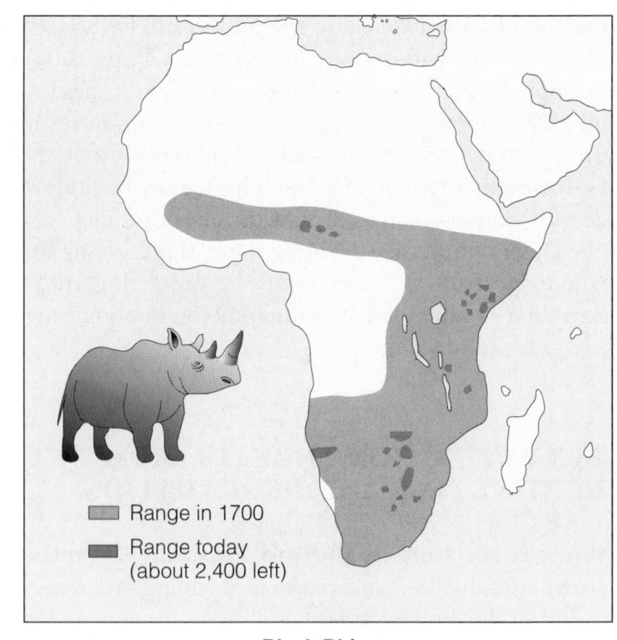

Black Rhino

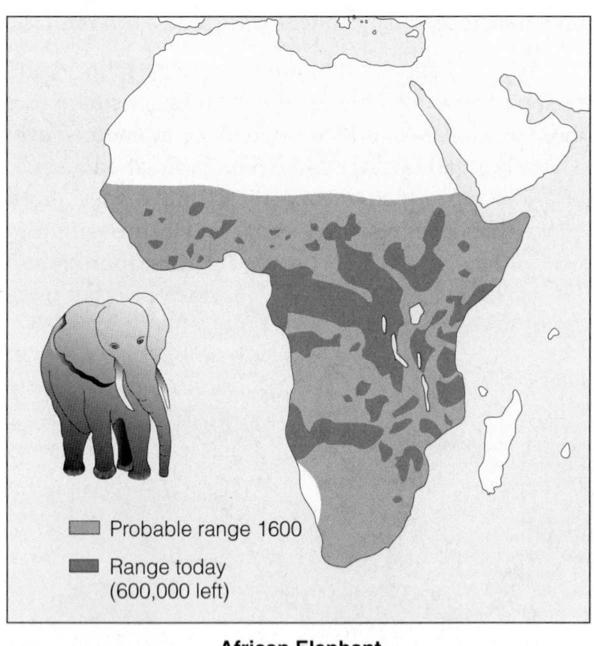

African Elephant

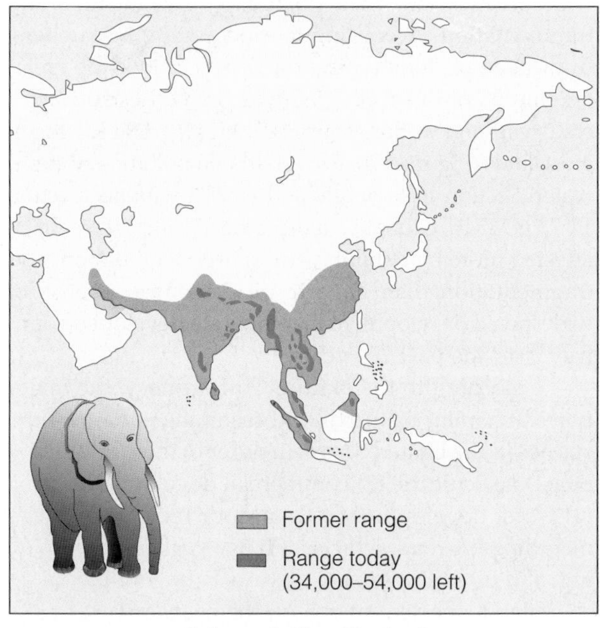

Asian or Indian Elephant

Figure 18-6 Reductions in the ranges of four wildlife species, mostly the result of habitat loss and hunting. What will happen to these and millions of other species when the world's human population doubles and per capita resource consumption rises sharply in the next few decades? (Data from International Union for the Conservation of Nature and World Wildlife Fund)

18-4 EXTINCTION THREATS FROM NONNATIVE SPECIES

What Harm Do Nonnative Species Cause in the United States? After habitat loss and degradation, deliberate or accidental introduction of nonnative species into ecosystems is the biggest cause of animal and plant extinctions. The nonnative invaders arrive from other continents as **(1)** stowaways on aircraft, **(2)** in the bilge water of tankers (p. 349), and **(3)** as hitchhikers on imported products such as wooden packing crates.

The estimated 50,000 nonnative species in the United States

- Result in at least $137 billion per year in damages and pest control costs (an average loss of $16 million per hour) according to a 2000 study by David Pimentel.

- Threaten 49% of the more than 1,200 endangered and threatened species in the United States (and 95% of those in Hawaii) according to the U.S. Fish and Wildlife Service.

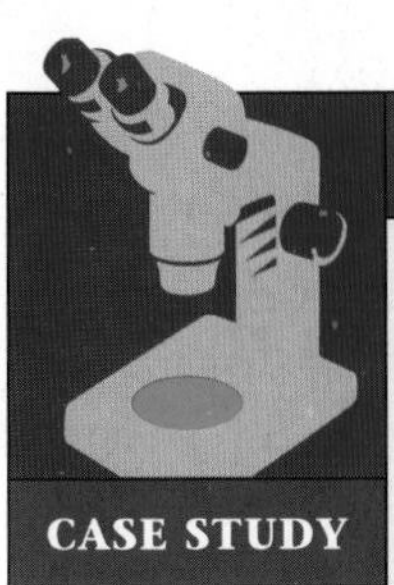

The Plight of Migrating Birds

CASE STUDY

Migrating bird species face a double habitat problem. Nearly half of the 700 U.S. bird species spend two-thirds of the year in the tropical forests of Central or South America or on Caribbean islands. During the summer they return to North America to breed.

A U.S. Fish and Wildlife study showed that between 1978 and 1987, populations of 44 of the 62 surveyed species of insect-eating, migratory songbirds in North America declined; 20 species experienced drops of 25–45%.

Researchers have identified several possible culprits:

- Logging of tropical forests in the birds' winter habitats.
- Fragmentation of their summer forest and grassland habitats in North America. The intrusion of farms, freeways, and suburbs breaks forests into patches. This makes it easier for **(1)** predators to feast on the eggs and the young of migrant songbirds and **(2)** parasitic cowbirds to lay their eggs in the nests of various songbird species and have those birds raise their young for them.
- Deaths of at least 4 million migrating songbirds each year when they fly into TV, radio, and phone towers.

However, there is no universal trend in North American songbird populations. Evidence indicates that some species are generally declining, whereas populations of other species are declining in some areas and increasing in other areas.

Approximately 70% of the world's 9,900 known bird species are declining in numbers (58%) or are threatened with extinction (12%), mostly because of habitat loss and fragmentation. Conservation biologists view this decline of bird species as an early warning of the greater loss of biodiversity to come. Birds are excellent environmental indicators because they **(1)** live in every climate and biome, **(2)** respond quickly to environmental changes in their habitats, and **(3)** are easy to track and count.

In addition to serving as indicator species, birds play important ecological roles such as **(1)** helping control populations of insects (including the spruce budworm, gypsy moth, and tent caterpillar, which decimate many tree species) and rodents, **(2)** pollinating a wide variety of flowering plants, and **(3)** spreading plants throughout their habitats by consuming plant seeds and excreting them in their droppings.

Critical Thinking

What three things would you do to help prevent the population decline of migrating birds?

- Are blamed for about 68% of fish extinctions in the United States between 1900 and 2000.

What Is the Role of Deliberately Introduced Species? Deliberate introduction of nonnative species can be beneficial or harmful depending on the species and where they are introduced. We depend heavily on nonnative organisms for ecosystem services, food (Figure 16-5, p. 400), shelter, medicine (Figure 17-15, p. 445), and aesthetic enjoyment.

However, some introduced species have no natural predators, competitors, parasites, or pathogens to control their numbers in their new habitats. This can allow them to reduce or wipe out the populations of many native species and trigger ecological disruptions (Table 18-1). One example of a deliberately introduced species is the kudzu ("CUD-zoo") vine, which grows rampant in the southeastern United States (Connections, p. 470).

Here are a few other examples of harmful nonnative species that were deliberately imported into the United States:

- An estimated 1 million *wild (feral) pigs* are roaming the state of Florida, hogging food from endangered animals, rooting up farm fields, and causing traffic accidents. Game and wildlife officials have had little success in controlling their numbers with hunting and trapping and say there is no way to stop them.
- The Australian *melaleuca tree,* introduced in south Florida in 1906, was planted as windbreaks and fence rows. Now it has invaded 217,000 hectares (536,000 acres) of the forest and grassland ecosystems of the Florida Everglades. Every day it takes over another 30 hectares (74 acres).
- The estimated 30 million *feral cats* and 41 million *outdoor pet cats* introduced into the United States kill about 568 million birds per year.

What Is the Role of Accidentally Introduced Species? In 1918, extremely aggressive fire ants were introduced accidentally into the United States in Mobile, Alabama. They may have arrived on shiploads of lumber or coffee imported from South America or by hitching a ride in the soil-containing ballast of cargo ships.

Without natural predators, these ants have spread rapidly by land and water (they can float) throughout the South, from Texas to Florida and as far north as Tennessee and Virginia (Figure 18-8) and are also found in Puerto Rico. Recently they have invaded California and New Mexico.

Wherever the fire ant has gone, up to 90% of native ant populations have been sharply reduced or wiped out. Their extremely painful stings have also killed deer fawn, birds, livestock, pets, and at least 80 people allergic to

Deliberate Introduction of the Kudzu Vine

In the 1930s, the *kudzu vine* was imported from Japan and planted in the southeastern United States to help control soil erosion. It does control erosion, but it is so prolific and difficult to kill that it engulfs hillsides, trees, abandoned houses and cars, stream banks, patches of forest, and anything else in its path (Figure 18-7).

This vine, sometimes called "the vine that ate the South," has spread throughout much of the southern United States and could spread as far north as the Great Lakes by 2040 if projected global warming occurs.

Although kudzu is considered a menace in the United States, Asians use a powdered kudzu starch in beverages, gourmet confections, and herbal remedies for a range of diseases. A Japanese firm has built a large kudzu farm and processing plant in Alabama and ships the extracted starch to Japan.

Critical Thinking

On balance, do you think the potential beneficial uses of the kudzu vine outweigh its harmful ecological effects? Explain.

Figure 18-7 Kudzu taking over a house and a truck. This vine can grow 0.3 meter (1 foot) per day and is now found from East Texas to Florida and as far north as southeastern Pennsylvania and Illinois. Kudzu was deliberately introduced into the United States for erosion control, but it cannot be stopped by being dug up or burned. Grazing by goats and repeated doses of herbicides can destroy it, but goats and herbicides also destroy other plants, and herbicides can contaminate water supplies. Recently, scientists have found a common fungus (*Myrithecium verrucaria*) that can kill kudzu within a few hours, apparently without harming other plants.

their venom. They have also **(1)** invaded cars and caused accidents by attacking drivers, **(2)** made cropfields unplowable, **(3)** disrupted phone service and electrical power, **(4)** caused some fires by chewing through underground cables, and **(5)** cost the United States an estimated $600 million per year.

Widespread pesticide spraying in the 1950s and 1960s temporarily reduced fire ant populations. In the end, however, this chemical warfare hastened the advance of the rapidly multiplying fire ant by **(1)** reducing populations of many native ant species and **(2)** promoting development of genetic resistance to heavily used pesticides in the rapidly multiplying fire ants.

Researchers at the U.S. Department of Agriculture are experimenting with use of biological control to reduce fire ant plantations. Before widespread use of these agents, however, researchers must be sure they will not cause problems for native ant species or become pests themselves.

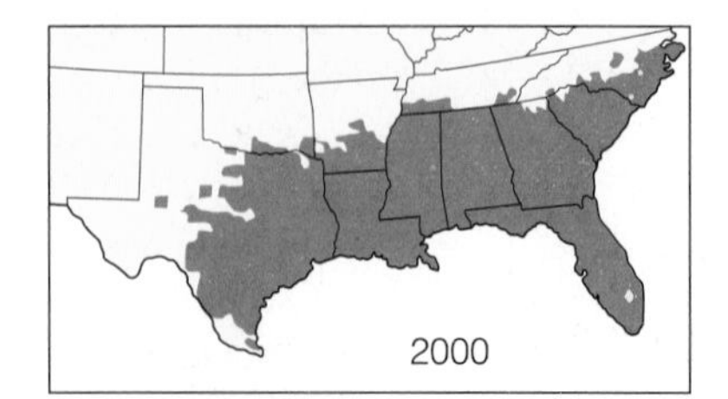

Figure 18-8 Expansion of the fire ant in southern states, 1918–2000. This invader is also found in Puerto Rico, New Mexico, and California. (Data from U.S. Department of Agriculture)

In 1985, *tiger mosquitoes,* which breed in scrap tires, arrived in the United States in a Japanese ship carrying tires to a Houston, Texas, recapping plant. Since then they have spread to 25 states. Aggressive biters, these mosquitoes can transmit 17 potentially fatal tropical viruses, including dengue fever, yellow fever, and forms of

Table 18-1 Damage Caused by Species Imported into the United States

Name	Origin	Mode of Transport	Type of Damage
Mammals			
European wild boar	Russia	Intentionally imported (1912), escaped captivity	Destroys habitat by rooting; damages crops
Nutria (cat-sized rodent)	Argentina	Intentionally imported (1940), escaped captivity	Alters marsh ecology; damages levees and earth dams; destroys crops
Birds			
European starling	Europe	Intentionally released (1890)	Competes with native songbirds; damages crops; transmits swine diseases; causes airport nuisance
House sparrow	England	Intentionally released by Brooklyn Institute (1853)	Damages crops; displaces native songbirds
Fish			
Carp	Germany	Intentionally released (1877)	Displaces native fish; uproots water plants; lowers waterfowl populations
Sea lamprey	North Atlantic Ocean	Entered Great Lakes via Welland Canal (1829)	Wiped out lake trout, lake whitefish, and sturgeon in Great Lakes
Walking catfish	Thailand	Imported into Florida	Destroys bass, bluegill, and other fish
Insects			
Argentine fire ant	Argentina	Probably entered via coffee shipments from Brazil (1918)	Damages crops; destroys native ant species
Camphor scale insect	Japan	Accidentally imported on nursery stock (1920s)	Damaged nearly 200 plant species in Louisiana, Texas, and Alabama
Japanese beetle	Japan	Accidentally imported on irises or azaleas (1911)	Defoliates more than 250 species of trees and other plants, including many of commercial importance
Plants			
Water hyacinth	Central America	Accidentally introduced (1882)	Clogs waterways; shades out other aquatic vegetation
Chestnut blight (fungus)	Asia	Accidentally imported on nursery plants (1900)	Killed nearly all eastern U.S. chestnut trees; disturbed forest ecology
Dutch elm disease (fungus)	Europe	Accidentally imported on infected elm timber used for veneers (1930)	Killed millions of elms; disturbed forest ecology

Source: Modified from *Biological Conservation* by David W. Ehrenfeld, 1970. Holt, Rinehart & Winston, Inc.

encephalitis. Another harmful invader is the Formosan termite (Case Study, p. 476).

What Can Be Done to Reduce the Threat from Nonnative Species? Once a nonnative species has become established in an ecosystem, its wholesale removal is almost impossible—somewhat like trying to get smoke back into a chimney or trying to unscramble an egg. Thus, the best way to limit the harmful impacts of nonnative species is to prevent them from being introduced and becoming established. This can be done by

- Identifying major characteristics that allow species to become successful invaders and the types of communities that are vulnerable to invaders and using this information to screen out potentially harmful invaders (Figure 18-9)
- Stepping up inspections of goods coming into a country
- Identifying major harmful invader species and passing international laws banning their transfer from one country to another (as is now done for endangered species)
- Requiring ships to **(1)** discharge their ballast (bilge) water and replace it with saltwater at sea before entering ports and **(2)** sterilize such water

Characteristics of Successful Invader Species

- High reproductive rate, short generation time (r-selected species)
- Pioneer species
- Long lived
- High dispersal rate
- Release growth-inhibiting chemicals into soil
- Generalists
- High genetic variability

Characteristics of Ecosystems Vulnerable to Invader Species

- Similar climate to habitat of invader
- Absence of predators on invading species
- Early successional species
- Low diversity of native species
- Absence of fire
- Disturbed by human activities

Figure 18-9 Some general characteristics of successful invader species and ecosystems vulnerable to invading species.

18-5 EXTINCTION THREATS FROM HUNTING AND POACHING

What Is the Role of Commercial Hunting and Poaching? The international trade in wild plants and animals brings in an estimated $20 billion a year. At least one-fourth of this trade involves the illegal sale of endangered and threatened species or their parts. Organized crime has moved into illegal wildlife smuggling because of the huge profits involved (second only to drug smuggling). At least two-thirds of all live animals smuggled around the world die in transit. The demand for illegal wildlife comes mostly from wealthy collectors and other consumers in Asia, the Middle East, North America, and Europe.

Worldwide, some 622 animal and plant species face premature extinction, mostly because of illegal trade. To poachers **(1)** a live mountain gorilla is worth $150,000, **(2)** a gyrfalcon $120,000, **(3)** a panda pelt $100,000 (with only about 1,000 pandas thought to remain in the wild), **(4)** a chimpanzee $50,000, **(5)** an Imperial Amazon macaw $30,000, and **(6)** rhinoceros horn as much as $28,600 per kilogram ($13,000 per pound) mostly because of its use in dagger handles in the Middle East and as a fever reducer and alleged aphrodisiac in China and other parts of Asia.

In much of West Africa, wildlife in the form of bushmeat is an important source of protein for many local people (Figure 18-10). The growing bushmeat trade also generates revenues of more than $150 million per year and can endanger species such as the gorilla.

Figure 18-10 Bushmeat, such as this gorilla head, is consumed as a source of protein by local people in parts of West Africa and sold in the national and international marketplace. You can find bushmeat on the menu in Cameroon and the Congo in West Africa as well as in Paris, France, and Brussels, Belgium. (Karl Ammann, BioSynergy Institute)

In 1950, there were an estimated 100,000 tigers in the world. Despite international protection, today only about 5,000–7,500 tigers are left (about 4,000 of them in India), mostly because of habitat loss and poaching for their furs and bones. Bengal tigers are at risk because a tiger fur sells for $100,000 in Tokyo. With the body parts of a single tiger worth more than $10,000, it is not surprising that illegal hunting has skyrocketed, especially in India. Without emergency action, there may be few or no tigers left in the wild within 20 years.

As more species become endangered, the demand for them on the black market soars, hastening their chances of premature extinction from poaching. Most poachers are not caught, and the money to be made far outweighs the risk of fines and the much smaller risk of imprisonment.

Case Study: Should Commercial Whaling Be Resumed? *Cetaceans* are an order of mostly marine mammals ranging in size from the 0.9-meter (3-foot) porpoise to the giant 15- to 30-meter (50- to 100-foot) blue whale. They are divided into two major groups: toothed whales and baleen whales (Figure 18-11).

Toothed whales, such as the porpoise, sperm whale, and killer whale (orca), bite and chew their food and feed mostly on squid, octopus, and other marine animals. *Baleen whales*, such as the blue, gray, humpback, and finback, are filter feeders. Instead of teeth, they have several hundred horny plates made of baleen, or whalebone, that hang down from the upper jaw. These plates filter plankton from the seawater, especially tiny shrimplike krill (Figure 4-18, p. 90). Baleen whales are the most abundant group of cetaceans.

Whales are fairly easy to kill because of their large size and their need to come to the surface to breathe. Mass slaughter has become very efficient with the use of fast ships, harpoon guns, and inflation lances (which pump dead whales full of air and make them float).

Whale harvesting, mostly in international waters, has followed the classic pattern of a tragedy of the commons (p. 11), with whalers killing an estimated 1.5 million whales between 1925 and 1975. This overharvesting **(1)** drove the populations of 8 of the 11 major species to the point of commercial extinction at which it no longer paid to hunt and kill them, and **(2)** drove some commercially prized species such as the giant blue whale to the brink of biological extinction (Case Study, p. 477).

In 1946, the International Convention for the Regulation of Whaling established the International Whaling Commission (IWC) to regulate the whaling industry by setting annual quotas to prevent overharvesting and commercial extinction. However, IWC quotas often were based on inadequate data or were ignored by whaling countries. Without any powers of enforcement, the IWC has been unable to stop the decline of most commercially hunted whale species to the point at which they were commercially extinct.

In 1970, the United States stopped all commercial whaling and banned all imports of whale products. Under intense pressure from environmentalists, the U.S. government, and governments of many nonwhaling countries in the IWC, the IWC has imposed a moratorium on commercial whaling since 1986. As a result, the estimated number of whales killed commercially worldwide dropped from 42,480 in 1970 to about 1,200 in 2001. Despite the ban, IWC members Japan and Norway have continued to hunt certain whale species, and Iceland resumed hunting whales in 2002. Japan, Norway, Iceland, Russia, and a growing number of small tropical island countries (which Japan has brought into the IWC to support its position) are working to reverse the IWC ban on commercial whaling.

18-6 OTHER EXTINCTION THREATS

What Is the Role of Predators and Pest Control? People try to exterminate species that compete with them for food and game animals. For example, U.S. fruit farmers exterminated the Carolina parakeet around 1914 because it fed on fruit crops. The species was easy prey because when one member of a flock was shot, the rest of the birds hovered over its body, making themselves easy targets.

African farmers kill large numbers of elephants to keep them from trampling and eating food crops. Many ranchers, farmers, and hunters in the United States support the killing of coyotes, wolves, and other species that can prey on livestock and on species prized by game hunters.

Since 1929, U.S. ranchers and government agencies have poisoned 99% of North America's prairie dogs because horses and cattle sometimes step into the burrows and break their legs. This has also nearly wiped out the endangered black-footed ferret (Figure 18-3, about 600 left in the wild), which preyed on the prairie dog.

What Is the Role of the Market for Exotic Pets and Decorative Plants? The global legal and illegal trade in wild species for use as pets is a huge and very profitable business. However, for every live animal captured and sold in the pet market, an estimated 50 other animals are killed.

About 25 million U.S. households have exotic birds as pets, 85% of them imported. More than 60 bird species, mostly parrots, are endangered or threatened because of this wild bird trade. According to the U.S. Fish and Wildlife Service, collectors of exotic birds may pay $10,000 for a threatened hyacinth macaw smuggled out of Brazil; however, during its lifetime, a single macaw left in the wild might yield as much as $165,000

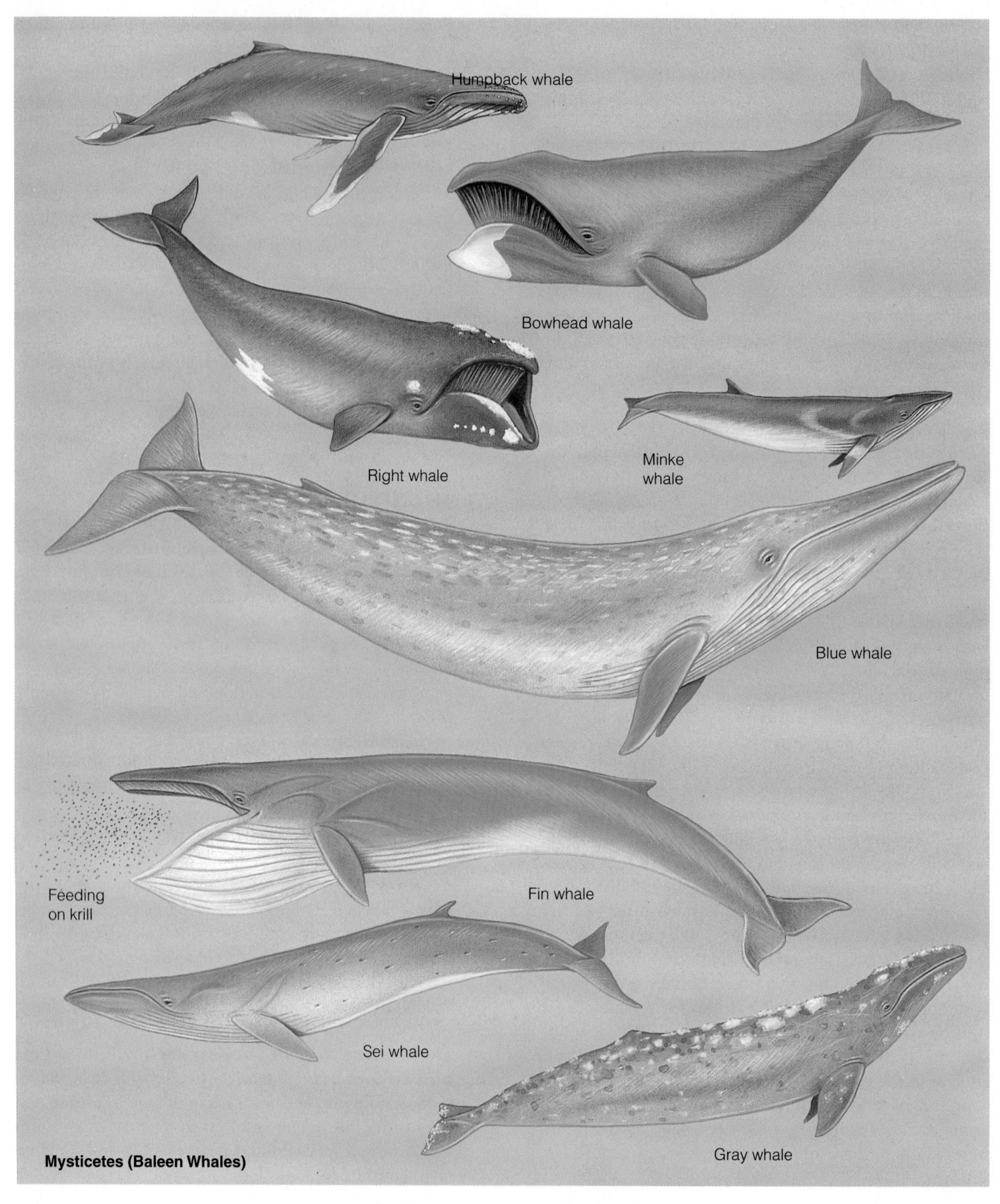

Figure 18-11 Examples of cetaceans, which can be classified as baleen whales and toothed whales.

in tourist income. A 1992 study suggested that keeping a pet bird indoors for more than 10 years doubles a person's chances of getting lung cancer from inhaling tiny particles of bird dander.

Other wild species whose populations are depleted because of the pet trade include amphibians, reptiles, mammals, and tropical fish (taken mostly from the coral reefs of Indonesia and the

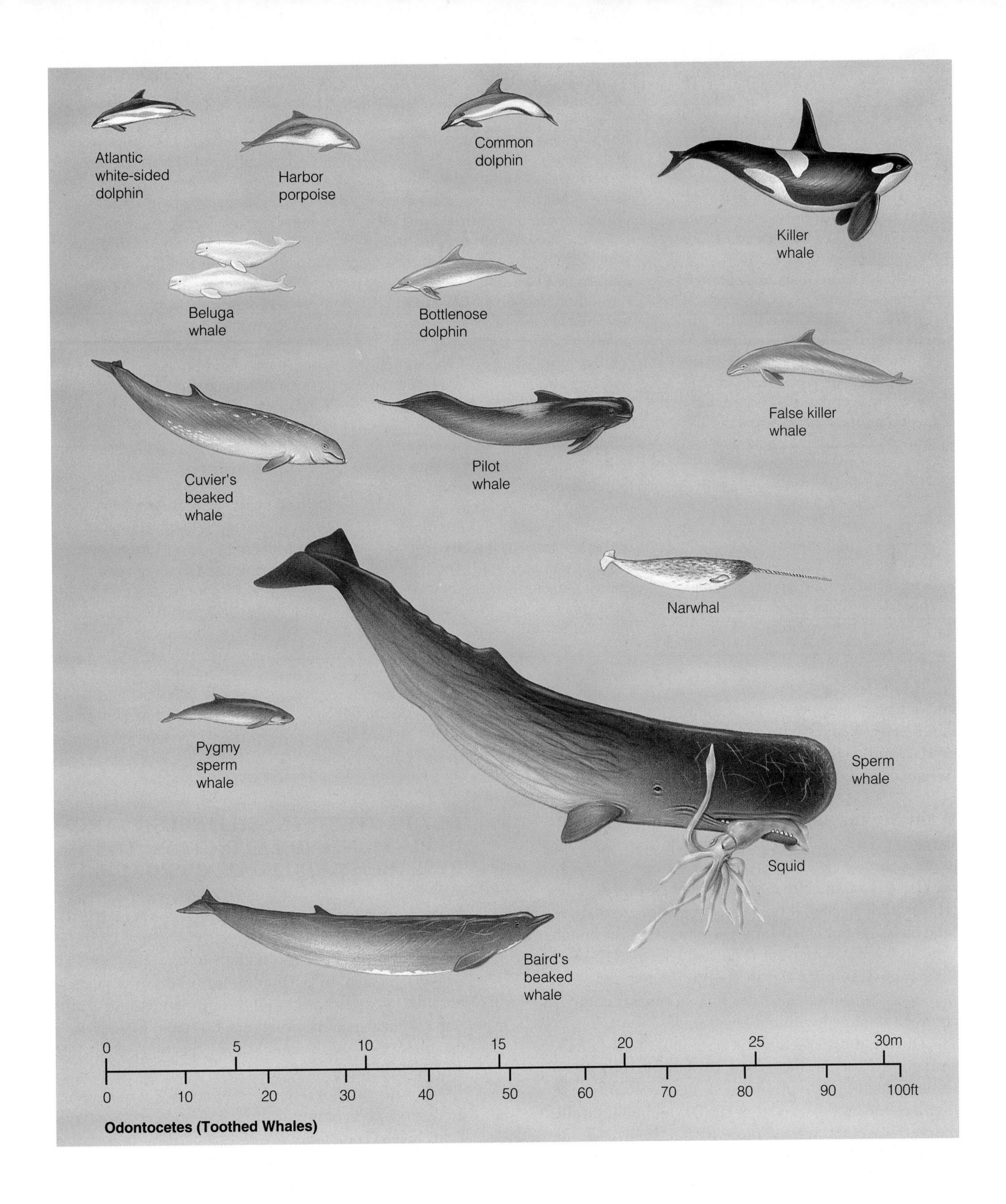

Philippines). Divers catch tropical fish by using plastic squeeze bottles of cyanide to stun them. For each fish caught alive, many more die. In addition, the cyanide solution kills the coral animals that create the reef, which is a center for marine biodiversity (Figure 6-38, p. 151).

Some exotic plants, especially orchids and cacti, are endangered because they are gathered (often illegally)

The Termite from Hell

CASE STUDY

Forget killer bees and fire ants. The homeowner's nightmare is the Formosan termite. It is the most voracious, aggressive, and prolific of more than 2,000 known termite species.

These termites invaded the Hawaiian Islands by 1900 and probably arrived on the U.S. mainland during or soon after World War II in wooden packing materials on military cargo ships that docked in southern ports such as New Orleans, Louisiana, and Houston, Texas.

They can quickly munch through wooden beams and plywood and can even chew through brick. Their huge colonies, which can contain up to 73 million insects (73 times as many as a nest of native termites), can be found under houses or in their attics.

Formosan termites also consume wood nine times faster than domestic termites. Domestic termites must be in contact with soil, which can be treated around the outside of a building to reduce infestation. However, Formosan termites can establish a colony in an attic. This makes barrier treatments around the perimeter of a building almost worthless in fighting these pests.

Over the past decade, the Formosan termite has caused more damage in New Orleans than hurricanes, floods, and tornadoes combined. As many as 90% of the houses and one-third of the oak trees in the city are thought to be infested. The famous French Quarter has one of the most concentrated infestations in the world.

Once confined to Louisiana, they have invaded at least a dozen other states, including Alabama, Florida, Mississippi, North and South Carolina, Texas, and California. They cause at least $1 billion in damage each year.

Most pesticides do not work on these termites. In New Orleans, the U.S. Department of Agriculture is using a variety of techniques all at once in an attempt to control the species in a heavily infested 15-block area of the French Quarter. They hope to develop a mix of techniques for dealing with these invaders elsewhere.

Critical Thinking

What important ecological roles do termites play in nature? If the Formosan termite and other termite species could be eradicated (a highly unlikely possibility), would you favor doing this? Explain.

and sold to collectors to decorate houses, offices, and landscapes. A collector may pay $5,000 for a single rare orchid, and a single rare mature crested saguaro cactus can earn cactus rustlers as much as $15,000.

What Are the Roles of Climate Change and Pollution? There is concern that human activities such as greenhouse gas emissions and deforestation can bring about rapid climate change over several decades (Figure 13-8, p. 306). If these projected climate changes take place, **(1)** many wild species may not have enough time to adapt or migrate (Figure 13-12, p. 312), and **(2)** some wildlife species in well-protected and well-managed terrestrial reserves and ocean sanctuaries could be depleted within a few decades.

During the 1950s and 1960s, populations of fish-eating birds such as the osprey, cormorant, brown pelican, and bald eagle plummeted. Research indicated that a chemical derived from DDT, when biologically magnified in food webs (Figure 10-4, p. 227), made the birds' eggshells so fragile that they could not reproduce successfully. Also hard-hit were such predatory birds as the prairie falcon, sparrow hawk, and peregrine falcon, which help control populations of rabbits, ground squirrels, and other crop-eaters. Since the U.S. ban on DDT in 1972, most of these species have made a comeback. However, according to the U.S. Fish and Wildlife Service, pesticides in use today menace about 20% of the endangered and threatened species in the United States.

18-7 PROTECTING WILD SPECIES FROM DEPLETION AND EXTINCTION: THE RESEARCH AND LEGAL APPROACH

How Can Bioinformatics Help Protect Biodiversity? To protect biodiversity, we need fundamental information about the biology and ecology of wild species. In particular, information is needed about **(1)** species names, **(2)** descriptions, **(3)** distributions, **(4)** status of populations, **(5)** habitat needs, and **(6)** interactions with other species (Section 7-3, p. 165, and Section 7-4, p. 169).

Bioinformatics is the applied science of managing, analyzing, and communicating biological information. It involves **(1)** *building computer databases* to organize and store useful biological information, **(2)** *developing computer tools to find, visualize, and analyze* the information, and **(3)** *communicating* the information, especially using the internet. Bioinformatics is being applied to many aspects of biology, ranging from storing DNA sequences to storing names, descriptions, and locations of collections of biological organisms in museums.

Near Extinction of the Blue Whale

CASE STUDY

The biologically endangered blue whale (Figure 18-11) is the world's largest animal. Fully grown, it is more than 30 meters (100 feet) long—longer than three train boxcars—and weighs more than 25 elephants. The adult has a heart the size of a Volkswagen Beetle car, and some of its arteries are so big that a child could swim through them.

Blue whales spend about 8 months of the year in Antarctic waters. There they find an abundant supply of krill (Figure 4-18, p. 90), which they filter daily by the trillions from seawater. During the winter, they migrate to warmer waters, where their young are born.

Before commercial whaling began, an estimated 200,000 blue whales roamed the Antarctic Ocean. Today, the species has been hunted to near biological extinction for its oil, meat, and bone.

A combination of prolonged overharvesting and certain natural characteristics of blue whales caused its decline. Their huge size made them easy to spot. They were caught in large numbers because they grouped together in their Antarctic feeding grounds. They also take 25 years to mature sexually and have only one offspring every 2–5 years. This low reproductive rate makes it difficult for the species to recover once its population falls beneath a certain threshold.

Blue whales have not been hunted commercially since 1964 and have been classified as an endangered species since 1975. Despite this protection, some marine biologists believe that too few blue whales—an estimated 1,000–3,000—remain for the species to recover and avoid extinction.

Their extinction could also be hastened by melting polar ice (Connections, p. 308) because of climate change. This melting reduces populations of krill, which are a main food source for blue whales (Figure 4-18, p. 90).

Critical Thinking

Opponents of commercial whaling contend that resuming commercial whaling for some whale species such as minke, pilot, and gray could lead to illegal harvesting of blue whales. Japan contends that excess population of minkes in Antarctic waters is threatening the blue whale population by consuming much of the krill that they eat (Figure 4-18, p. 90). What scientific evidence would you want to have to resolve this issue?

How Can International Treaties Help Protect Endangered Species? Several international treaties and conventions help protect endangered or threatened wild species. One of the most far-reaching is the 1975 *Convention on International Trade in Endangered Species* (CITES). This treaty, now signed by 152 countries, **(1)** lists some 900 species that cannot be commercially traded as live specimens or wildlife products because they are in danger of extinction and **(2)** restricts international trade of 29,000 other species because they are at risk of becoming threatened.

CITES has helped reduce international trade in many threatened animals, including elephants, crocodiles, and chimpanzees. However, the effects of this treaty are limited because **(1)** enforcement is difficult and spotty, **(2)** convicted violators often pay only small fines, **(3)** member countries can exempt themselves from protecting any listed species, and **(4)** much of the highly profitable illegal trade in wildlife and wildlife products goes on in countries that have not signed the treaty.

The *Convention on Biological Diversity* (CBD), which has been ratified by 172 countries, legally binds signatory governments to reversing the global decline of biological diversity. However, its implementation has proceeded slowly because **(1)** some key countries, such as the United States, have not ratified the treaty, and **(2)** it contains no severe penalties or other enforcement mechanisms.

How Can National Laws Help Protect Endangered Species? The United States controls imports and exports of endangered wildlife and wildlife products through two important laws:

- The *Lacey Act of 1900,* which prohibits transporting live or dead wild animals or their parts across state borders without a federal permit.
- The *Endangered Species Act of 1973* (ESA, amended in 1982 and 1988), which makes it illegal for Americans to import or trade in any product made from an endangered or threatened species unless it is used **(1)** for an approved scientific purpose or **(2)** to enhance the survival of the species.

The ESA authorizes the National Marine Fisheries Service (NMFS) to identify and list endangered and threatened ocean species and the U.S. Fish and Wildlife Service (USFWS) to identify and list all other endangered and threatened species. These species cannot be hunted, killed, collected, or injured in the United States.

Any decision by either agency to add or remove a species from the list must be based on biology only, not on economic or political considerations. However, economic factors can be used in **(1)** deciding whether and how to protect endangered habitat and **(2)** developing recovery plans for listed species. The act also forbids federal agencies to carry out, fund, or authorize projects that would jeopardize an endangered or threatened species or destroy or modify the critical habitat it needs to survive. On private lands, fines and even jail sentences can be imposed to ensure protection of the habitats of endangered species.

Between 1973 and 2001, the number of U.S. species on the official endangered and threatened list increased from 92 to more than 1,200 species (about 60% of them plants and 40% animals). According to a 2000 study by the Nature Conservancy, about 33% of the country's species are at risk of extinction, and 15% of them are at high risk. This amounts to about 30,000 species, compared to the roughly 1,200 species currently protected under the ESA. The study found that many of the country's rarest and most imperiled species are concentrated in a few hot spots (Figure 18-12).

The ESA generally requires the secretary of the interior to designate and protect the *critical habitat* needed for the survival and recovery of each listed species. By June 2001, however, only 124 designated critical habitats had been established.

Getting listed is only half the battle. Next, the USFWS or the NMFS is supposed to prepare a plan to help the species recover. By 2001, final recovery plans had been developed and approved for about 75% of the endangered or threatened U.S. species, but about half of those plans exist only on paper.

The ESA requires that all commercial shipments of wildlife and wildlife products enter or leave the country through one of nine designated ports. Few illegal shipments are confiscated (Figure 18-13) because the 60 USFWS inspectors can examine only about one-fourth of at least 90,000 shipments that enter and leave the United States each year. Even if caught, many violators are not prosecuted, and convicted violators often pay only a small fine.

How Can Private Landowners Be Encouraged to Protect Endangered Species? One problem is that the ESA has encouraged some developers, timber companies, and other private landowners to avoid government regulation by managing their land to reduce its use by endangered species. The National Association for Homebuilders, for example, has published practical tips for developers and other landowners to avoid ESA issues. Suggestions include **(1)** planting crops, **(2)** plowing fields between crops to prevent native vegetation and endangered species from occupying the fields, **(3)** clearing forests, and **(4)** burning or managing vegetation to make it unsuitable for local endangered species.

In 1982, Congress amended the ESA to allow the secretary of the interior to use *habitat conservation plans* (HCPs) to strike a compromise between the interests of private landowners and the interests of endangered and threatened species without reducing the recovery chances of a protected species. With an HCP, landowners, developers, or loggers are allowed to destroy some

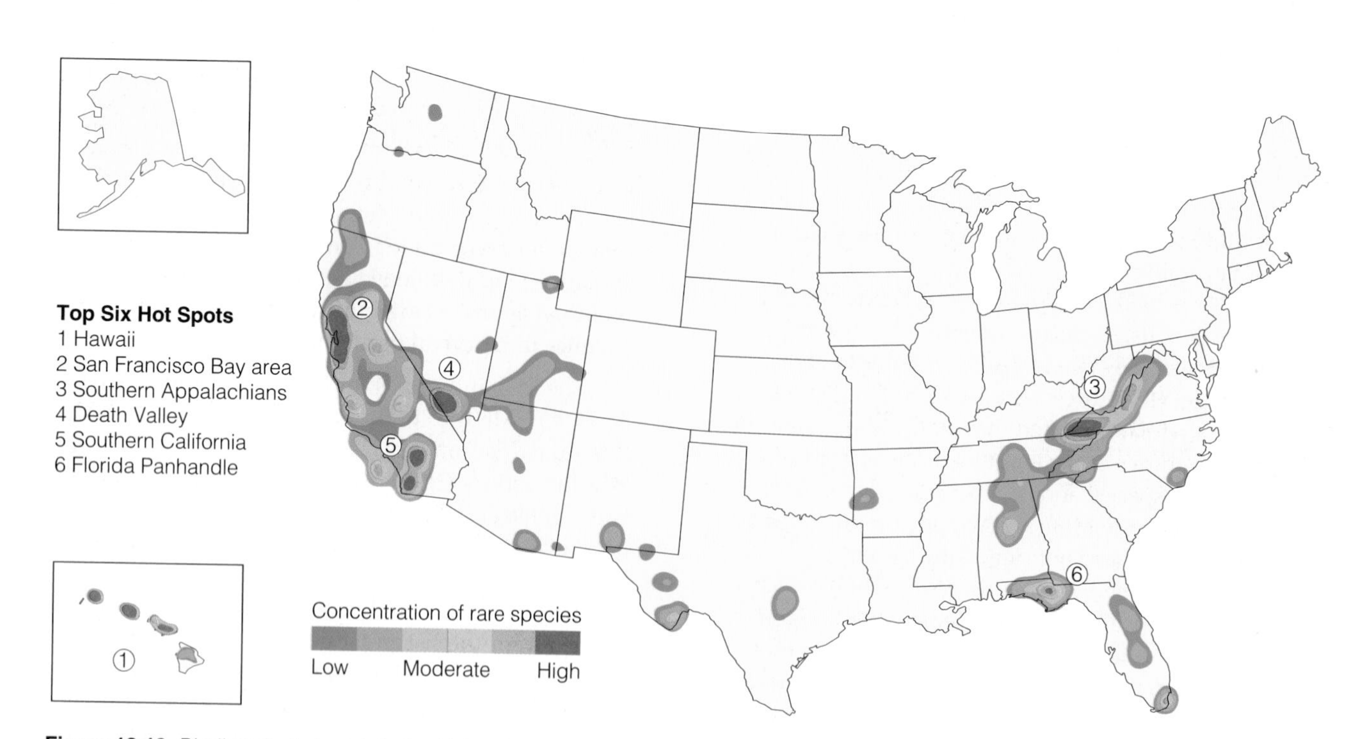

Figure 18-12 *Biodiversity hot spots* in the United States. This map shows areas that contain the largest concentrations of rare and potentially endangered species. (Data from State Natural Heritage Programs, the Nature Conservancy, and Association for Biodiversity Information)

Figure 18-13 Confiscated products made from endangered species. Because of a scarcity of funds and inspectors, probably no more than one-tenth of the illegal wildlife trade in the United States is discovered. The situation is even worse in most other countries. (Steve Hillebrand/U.S. Fish and Wildlife Service)

critical habitat or kill all or a certain number of endangered or threatened species on private land in exchange for taking steps to protect that species.

Such protective measures might include **(1)** setting aside a part of the species' habitat as a preserve, **(2)** paying to relocate the species to another suitable habitat, or **(3)** paying money to have the government buy suitable habitat elsewhere. Once the plan is approved it cannot be changed, even if new data show that the plan is inadequate to protect a species and help it recover.

Some wildlife conservationists support this approach because it can head off use of evasive techniques and reduce political pressure to seriously weaken or eliminate the ESA. However, there is growing concern that such plans are being developed without enough scientific evaluation of their effects on a species' recovery. Suggestions for improving HCPs include **(1)** developing scientific standards for such plans, **(2)** having plans reviewed by a scientific advisory committee, and **(3)** requiring more compensating efforts by private landowners as a buffer when important data such as population trends or the impacts on the affected species are not known.

In 1999, the USFWS approved two new approaches for encouraging private landowners to protect threatened or endangered species:

- *Safe harbor agreements* in which landowners voluntarily agree to take specified steps to restore, improve, or maintain habitat for threatened or endangered species located on their land. In return, landowners get **(1)** technical help from local conservation agencies, **(2)** government assurances that the land, water, or other natural resources involved will not face future restrictions once the agreement is over, and **(3)** assurances that after the agreement has expired landowners can return the property to its original condition without penalty.

- Voluntary *candidate conservation agreements* in which landowners agree to take specific steps to help conserve a species whose population is declining but is not yet listed as endangered or threatened. Participating landowners receive technical help and assurances that no additional resource use restrictions will be imposed on the land covered by the agreement if the species is listed as endangered or threatened in the future.

How Can We Protect Endangered and Threatened Marine Species? The Endangered Species Act has also been used to protect a number of endangered and threatened marine reptiles (turtles, Figure 18-14) and mammals (especially whales, seals, and sea lions; Figure 18-15, p. 482).

Three of eight major sea turtle species (Figure 18-14) are endangered (Kemp's ridley, leatherbacks, and hawksbills), and the rest are threatened. The world's sea turtle species are endangered or threatened because of **(1)** loss or degradation of beach habitat (where they come ashore to lay their eggs), **(2)** legal and illegal taking of their eggs, **(3)** their increased use as sources of food, medicinal ingredients, tortoiseshell (for jewelry), and leather from their flippers (with some turtles selling for up to $1,500 in China), and **(4)** unintentional capture and drowning by commercial fishing boats (especially shrimp trawlers).

Two major problems with protecting marine biodiversity by protecting endangered species are **(1)** lack of knowledge about marine species and **(2)** difficulty in monitoring and enforcing treaties to protect marine species, especially in the open ocean.

Should the Endangered Species Act Be Weakened? Opponents of the ESA contend that it has **(1)** not worked and **(2)** caused severe economic losses by hindering development on private land. Since 1995 there have been efforts to weaken the ESA by

- Making protection of endangered species on private land voluntary

- Having the government pay landowners if it forces them to stop using part of their land to protect endangered species

- Making it harder and more expensive to list newly endangered species by requiring government wildlife

Figure 18-14 Major species of sea turtles that have roamed the seas for 150 million years, showing their relative adult sizes. Three of these species (Kemp's ridley, leatherbacks, and hawksbills) are endangered, and the rest are threatened as a result of human activities.

officials to navigate through a series of hearings and peer-review panels

- Giving the secretary of the interior **(1)** the power to permit a listed species to become extinct without trying to save it and **(2)** power over the listing of species (proposed by the Bush administration in 2001)
- Allowing the secretary of the interior to give any state, county, or landowner permanent exemption from the law, with no requirement for public notification or comment
- Prohibiting the public from commenting on or bringing lawsuits to change poorly designed HCPs
- Limiting the ability of environmental and other citizen groups to sue the Interior Department to get rare plants and animals listed as endangered by providing insufficient funds for the department to respond to such suits (proposed to Congress in 2001 by the Bush administration)

Should the Endangered Species Act Be Strengthened? Most conservation biologists and wildlife scientists contend that the ESA has not been a failure (Spotlight, p. 484). They also refute the charge that the ESA has caused severe economic losses.

- Since 1979 only about 0.05% of the almost 200,000 projects evaluated by the USFWS have been blocked or canceled as a result of the ESA.
- The act does allow for economic concerns. By law, a decision to list a species must be based solely on science. However, once a species is listed, economic considerations can be weighed against species protection in protecting critical habitat and designing and implementing recovery plans.
- The act allows a special Cabinet-level panel, called the "God Squad," to exempt any federal project from having to comply with the act if the economic costs are too high.
- The act allows the government to **(1)** issue permits and exemptions to landowners with listed species living on their property and **(2)** use habitat conservation plans, safe harbor agreements, and candidate conservation agreements to bargain with private landowners.

A study by the U.S. National Academy of Sciences recommended changes to make the ESA more scien-

tifically sound and effective by **(1)** greatly increasing the meager funding for implementing the act, **(2)** developing recovery plans more quickly, **(3)** developing guidelines to avoid provisions that are scientifically or economically unsound and that spell out which actions are likely to harm recovery, and **(4)** establishing a core "survival habitat" as a temporary emergency measure when a species is first listed that could support the species for 25–50 years.

Most biologists and wildlife conservationists believe that the United States should develop a new system to protect and sustain biological diversity and ecosystem function based on three principles: **(1)** Find out what species and ecosystems we have, **(2)** locate and protect the most endangered ecosystems and species, and **(3)** give private landowners financial incentives (tax breaks and write-offs), technical help, and assurances of no future requirements (safe harbor and candidate conservation agreements) for helping protect endangered species and ecosystems.

Should We Try to Protect All Endangered and Threatened Species? Because of limited funds, information, and trained personnel, only a few endangered and threatened species can be saved. Some analysts suggest that the limited funds available for preserving threatened and endangered wildlife be concentrated on species that **(1)** have the best chance for survival, **(2)** have the most ecological value to an

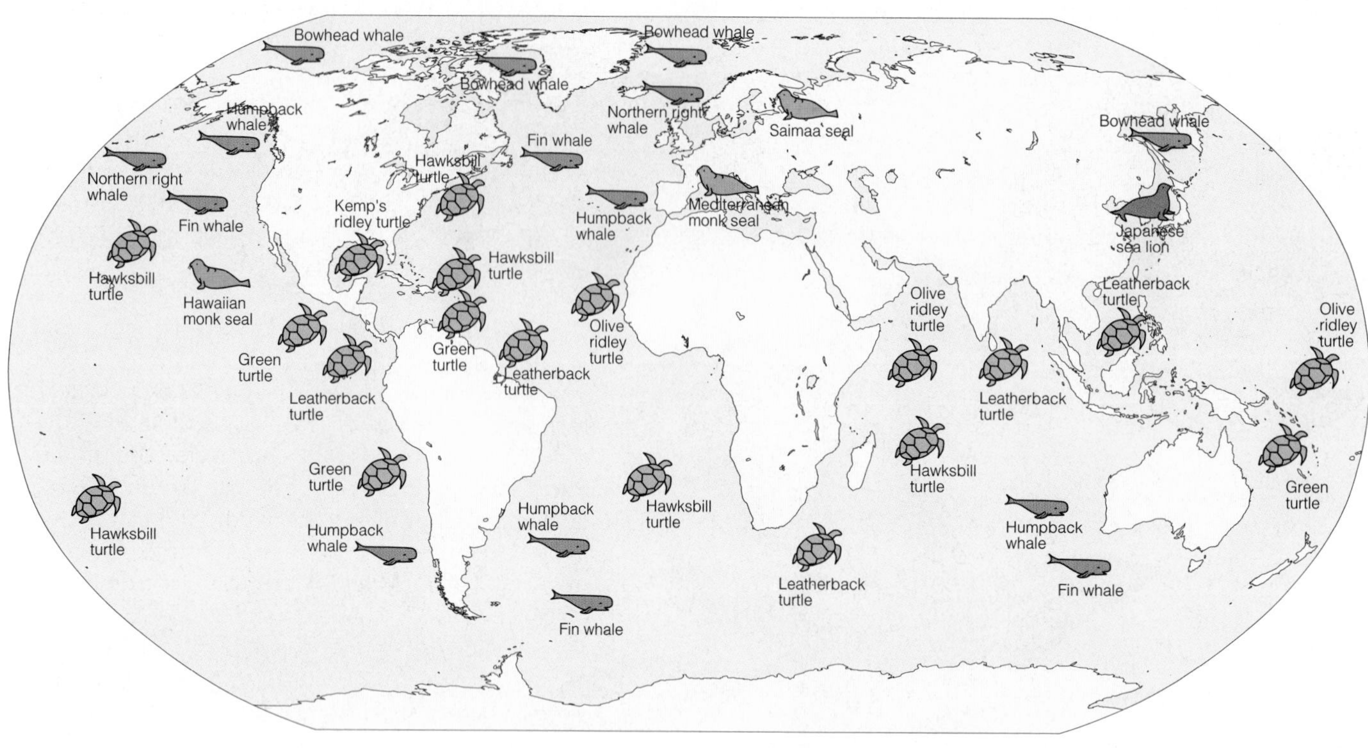

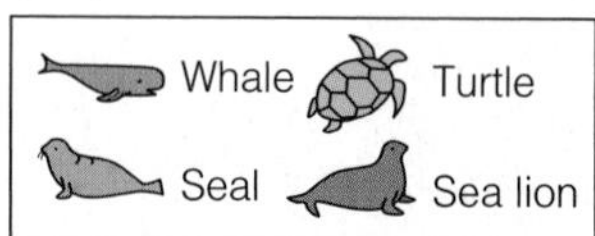

Figure 18-15 Endangered marine mammals (whales, seals, and sea lions) and reptiles (turtles). Many marine fish, seabirds (Figure 7-7, p. 166), and invertebrate species are also threatened.

ecosystem, and **(3)** are potentially useful for agriculture, medicine, or industry.

New research suggests that the health of an ecosystem should be judged not on the basis of sheer numbers of species but on which species **(1)** play keystone roles and **(2)** are tolerant of environmental change such as acid deposition, climate change, and toxins. Researchers say preserving 10 key and adaptable species in an ecosystem is likely to be better than preserving 10,000 weak ones. In addition to protecting keystone species, biologists call for protection of *keystone resources* such as **(1)** salt licks and mineral pools that provide essential minerals for wildlife, **(2)** deep pools in streams and springs that serve as refuges for fish and other aquatic species during dry periods, and **(3)** hollow tree trunks used as breeding sites and homes for many bird and mammal species.

Others oppose selective protection of species on ethical grounds or contend that we don't have enough biological information to make such evaluations. Proponents argue that **(1)** in effect we are already deciding by default which species to save and that **(2)** despite limited knowledge, the selective approach is more effective and a better use of limited funds than the current one. What do you think?

18-8 PROTECTING WILD SPECIES FROM DEPLETION AND EXTINCTION: THE SANCTUARY APPROACH

How Can Wildlife Refuges and Other Protected Areas Help Protect Endangered Species? Since 1903, when President Theodore Roosevelt established the first U.S. federal wildlife refuge at Pelican Island, Florida, the National Wildlife Refuge System has grown to 522 refuges. Some 34 million Americans visit these refuges each year to hunt, fish, hike, or watch birds and other wildlife.

More than three-fourths of the refuges are wetlands for protecting migratory waterfowl. About 20% of the U.S. endangered and threatened species have habitats in the refuge system, and some refuges have been set aside for specific endangered species. These have helped Florida's key deer, the brown pelican, and the trumpeter swan to recover.

Conservation biologists urge the establishment of more refuges for endangered plants. They also urge Congress and state legislatures to allow abandoned military lands that contain significant wildlife habitat to become national or state wildlife refuges.

According to a General Accounting Office study, activities considered harmful to wildlife occur in nearly 60% of the nation's wildlife refuges. There is much controversy over whether to allow oil and gas development in Alaska's Arctic National Wildlife Refuge (Pro/Con, p. 500). According to a 2001 study by the Audubon Society, the U.S. Refuge System is in a state of crisis. Pollutants and invasive species are degrading many refuges, and the system has a $1.6-billion backlog of unmet operations and maintenance needs.

Research shows that fully protected marine reserves can be effective in building up populations of depleted fish and other aquatic species (p. 454). However, so far only a tiny fraction of the world's oceans have been set aside as fully protected marine reserves.

Can Gene Banks, Botanical Gardens, and Farms Help Save Most Endangered Species? Gene or seed banks are used to preserve genetic information and endangered plant species by storing their seeds in refrigerated, low-humidity environments (Spotlight, p. 408). Most of the world's 50 seed banks have focused on storing seeds of the approximately 100 plant species that provide about 90% of the food consumed by humans. However, some banks are devoting more attention to storing seeds for a wider range of species that may be threatened with extinction or a loss of genetic diversity.

Scientists urge that many more such banks be established, especially in developing countries. However, some species cannot be preserved in gene banks, and maintaining the banks is very expensive.

The world's 1,600 botanical gardens and arboreta contain about 4 million living plants, representing about 80,000 species or approximately 30% of the world's known plant species. The world's largest botanical garden is the Royal Botanical Gardens of England at Kew. It contains an estimated 25,000 species of living plants, or about 10% of the world's total. About 2,700 of these species are listed as threatened.

Botanical gardens are increasingly focusing on cultivation of rare and endangered plant species. In the United States, the Center for Plant Conservation has coordinated efforts by 28 botanical gardens to store nearly 600 endangered U.S. plant species and to propagate and reintroduce some of them into the wild. In addition, botanical gardens help educate an estimated 150 million visitors a year about the need for plant conservation. However, these sanctuaries have too little storage capacity and too little funding to preserve most of the world's rare and threatened plants.

We can take pressure off some endangered or threatened species by raising individuals on farms for commercial sale. Two examples are using **(1)** farms in Florida to raise alligators (Connections, p. 175) for their meat and hides and **(2)** butterfly farms in Papua New Guinea, where many butterfly species are threatened by habitat destruction and fragmentation, commercial overexploitation, and environmental degradation.

Can Zoos and Aquariums Help Protect Most Endangered Species? Zoos, aquariums, game parks, and animal research centers are being used to preserve some individuals of critically endangered animal species, with the long-term goal of reintroducing the species into protected wild habitats.

Two techniques for preserving endangered terrestrial species are egg pulling and captive breeding. *Egg pulling* involves collecting wild eggs laid by critically endangered bird species and hatching them in zoos or research centers. In *captive breeding*, some or all of the wild individuals of a critically endangered species are captured for breeding in captivity, with the aim of reintroducing the offspring into the wild.

Other techniques for increasing the populations of captive species include **(1)** artificial insemination, **(2)** surgical implantation of eggs of one species into a surrogate mother of another species (embryo transfer), **(3)** use of incubators, **(4)** cross-fostering, in which the young of a rare species are raised by parents of a similar species, **(5)** using computer databases of the family lineages of species in zoos and DNA analysis to match individuals for mating and prevent genetic erosion through inbreeding, and **(6)** genetic cloning.

Captive breeding programs at zoos in Phoenix, Arizona, and San Diego and Los Angeles, California, temporarily saved the nearly extinct Arabian oryx (Figure 18-3), a large antelope that once lived throughout the Middle East. By 1960, it had been hunted nearly to extinction in the wild (with only about 48 left) by people riding in jeeps and helicopters and wielding rifles and machine guns. Between 1980 and 1996, the return of oryx bred in captivity to protected habitats in the Middle East allowed the population to reach about 400. However, since 1996 poachers have been capturing these animals for sale to private zoos. As a result, since 1996 the wild population has been reduced to about 100, and the species is again facing extinction.

Endangered species now being bred in captivity in the United States and returned to the wild include the **(1)** peregrine falcon (removed from the endangered species list in 1999) and **(2)** black-footed ferret (Figure 18-3). However, most reintroductions fail because of **(1)** lack of suitable habitat, **(2)** inability of individuals bred in captivity to survive in the wild, or **(3)** renewed overhunting or capture of some returned species (such as the Arabian oryx).

Lack of space and money limit efforts to maintain populations of endangered species in zoos and research centers. The captive population of each species must number 100 to 500 individuals to avoid extinction

Accomplishments of the Endangered Species Act

SPOTLIGHT

Critics of the ESA call it an expensive failure because only a few species have been removed from the endangered list. Most biologists strongly disagree that the act has been a failure, for several reasons. *First*, species are listed only when they are already in serious danger of extinction. This is like setting up a poorly funded hospital emergency room that takes only the most desperate cases, often with little hope for recovery, and then saying it should be shut down because it has not saved enough patients.

Second, it takes decades for most species to become endangered or threatened. Thus, it should not be surprising that it usually takes decades to bring a species in critical condition back to the point where it can be removed from the list. Expecting the ESA (which has been in existence only since 1973) to quickly repair the biological depletion of many decades is unrealistic.

The most important measure of the law's success is that the conditions of almost 40% of the listed species are stable or improving. A hospital emergency room taking only the most desperate cases yet stabilizing or improving the condition of 40% of its patients would be considered an astounding success.

Third, the federal endangered species budget was only $110 million in 2000 (up from $23 million in 1993)—about one-third the cost of one C-17 transport plane. The amount spent by the federal government to help protect endangered species amounts to only about 34¢ a year per U.S. citizen, and that represents a threefold increase since 1993.

To most biologists, it is amazing that so much has been accomplished in stabilizing or improving the condition of almost 40% of the listed species on a shoestring budget.

Critical Thinking

Explain why you agree or disagree each of the proposals made on pp. 479–480 to **(a)** weaken the ESA and **(b)** strengthen the act.

through accident, disease, or loss of genetic diversity through inbreeding. Recent genetic research indicates that 10,000 or more individuals are needed for an endangered species to maintain its capacity for biological evolution.

Today's zoos and research centers have space to preserve healthy and sustainable populations of only about 925 of the 2,000 large vertebrate species that could vanish from the planet. According to one estimate, if all the space in U.S. zoos were used for captive breeding, only about 100 large animal species could be sustained on a long-term basis.

Public aquariums that exhibit unusual and attractive fish and some marine animals such as seals and dolphins have been successful in educating the public about need to protect such species. In the United States, more than 35 million people visit aquariums each year. However, unlike zoos, public aquariums have not yet served as effective gene banks for endangered marine species, especially marine mammals that need large volumes of water.

Instead of seeing zoos and aquariums as Noah's Arks, some critics see them as prisons for once-wild animals. They also contend that zoos and aquariums foster the notion that we do not need to preserve large numbers of wild species in their natural habitats.

Whether one agrees or disagrees with this position, conservation biologists point out that zoos, aquariums, and botanical gardens are not a biologically or economically feasible solution for **(1)** most of the world's current endangered species and **(2)** the much larger number expected to become endangered over the next few decades.

18-9 WILDLIFE MANAGEMENT

How Can Wildlife Populations Be Managed? **Wildlife management** involves manipulating wildlife populations (especially game species) and their habitats for their welfare and for human benefit. The *wildlife management approach* manages game species for sustained yield by **(1)** using laws to regulate hunting and fishing, **(2)** establishing harvest quotas, **(3)** developing population management plans, **(4)** improving wildlife habitat, and **(5)** using international treaties to protect migrating game species such as waterfowl.

In the United States, funds for state game management programs come from the sale of hunting and fishing licenses and from federal taxes on hunting and fishing equipment. Two-thirds of the states also have provisions on state income tax returns that allow individuals to contribute money to state wildlife programs.

Only 10% of all government wildlife dollars are spent to study or benefit nongame species, which make up nearly 90% of the country's wildlife species. Since the passage of the Wildlife Restoration Act in 1937 there has been a large increase in the number of game animals (such as white-tailed deer, wild turkeys, Rocky Mountain elk, and pronghorn antelope) sought by many sport hunters.

The first step in wildlife management is to decide which species are to be managed in a particular area. Ecologists and conservation biologists emphasize preserving biodiversity, wildlife conservationists are concerned about endangered species, birdwatchers want the greatest diversity of bird species, and hunters want large populations of game species. In the United States and other developed countries, most wildlife management is devoted to producing surpluses of game animals and birds for hunters.

After goals have been set, the wildlife manager must develop a management plan. Ideally, this is based on the principles of **(1)** ecological succession (Figure 7-14, p. 173), **(2)** wildlife population dynamics (Figure 8-3, p. 183, and Figure 8-7, p. 185), **(3)** an understanding of the cover, food, water, space, and other habitat needs of each species to be managed, and **(4)** the *maximum sustained yield (MSY)* of a population, in which harvested individuals are not removed faster than they can be replaced through reproduction. The manager must also consider **(1)** the number of potential hunters, **(2)** their likely success rates, and **(3)** the regulations for preventing excessive hunting.

This information is difficult, expensive, and time-consuming to obtain. It involves much educated guesswork and trial and error, which is why wildlife management is as much an art as a science. Management plans must also be sensitive to political pressures from conflicting groups and to budget constraints.

How Useful Is Sport Hunting in Managing Wildlife Populations? Most developed countries use sport hunting laws to manage populations of game animals. Licensed hunters are allowed to hunt only during certain portions of the year to protect animals during their mating season. Limits are set on the size, number, and sex of animals that can be killed and on the number of hunters allowed in a given area.

Close control of sport hunting is difficult. Accurate data on game populations may not exist and may cost too much to get. People in communities near hunting areas, who benefit from money spent by hunters, may seek to have hunting quotas raised.

There is controversy over sport hunting. Proponents argue that without carefully regulated sport hunting, deer and other large game animals exceed the carrying capacity of their habitats and destroy vegetation they and other species need. For example, between 1900 and 2001, the estimated population of white-tailed deer in the United States increased from 500,000 to 25–30 million. Many of these deer **(1)** invade subdivisions and eat shrubs and home gardens, **(2)** raid farmers' fields and orchards, **(3)** threaten rare plants and animals in some areas, **(4)** help spread Lyme disease, carried by deer ticks to humans, and **(5)** are involved in more than 500,000 deer–vehicle collisions each year that kill more Americans than any other wild animal, injure thousands, and cost at least $1.1 billion. Wildlife biologist William Porter estimates that without hunting, the U.S. deer population would be five times its current level and cause extensive ecological damage.

According to its proponents, sport hunting also **(1)** provides recreational pleasure for millions of people (15 million in the United States), **(2)** stimulates local economies, and **(3)** provides money through sales of hunting licenses and taxes on firearms and ammunition (more than $1.7 billion since 1937) that is used to buy, restore, and maintain wildlife habitats and to support wildlife research in the United States. Environmental groups such as the Sierra Club and Defenders of Wildlife support carefully controlled sport hunting as a way to preserve biological diversity by helping to prevent depletion of native species of plants and animals.

However some individuals and groups, including the Humane Society, oppose sport hunting. They argue that **(1)** it inflicts unnecessary pain and suffering on wild animals (most of which are not killed to supply food humans need for survival), and **(2)** game managers create a surplus of game animals by deliberately eliminating their natural predators (such as wolves, p. 429) and then claim that the surplus must be harvested by hunters to prevent habitat degradation or starvation of the game. Instead of eliminating natural predators, say opponents, wildlife managers should reintroduce them to reduce the need for sport hunting.

Supporters of hunting point out that **(1)** populations of many game species (such as deer) are so large that predators such as wolves cannot control them, and **(2)** because most wildlife habitats are fragmented, introducing predators can lead to the loss of nearby livestock. However, critics of hunting contend that deer (which make up only 2% of the 200 million animals hunters kill in the United States each year) are being used as a smokescreen argument to allow killing of many other game species that do not threaten vegetation. What do you think?

How Can Populations of Migratory Waterfowl Be Managed? Migratory birds—including ducks, geese, swans, and many songbirds (Case Study, p. 469)—make north–south journeys from one habitat to another each year, usually to find food, suitable climate, and other conditions necessary for reproduction. Such bird species use many different north–south routes called **flyways**, but only about 15 are considered major routes (Figure 18-16).

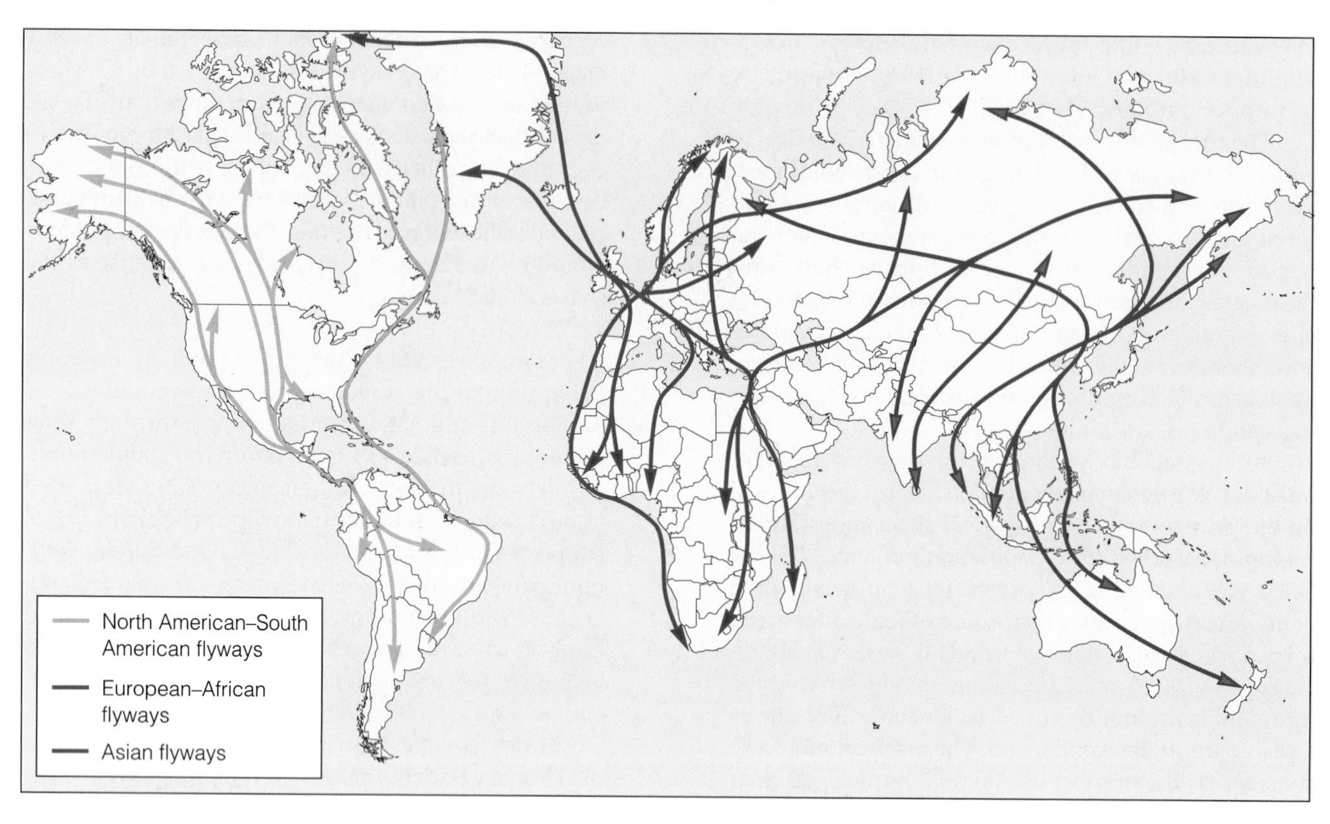

Figure 18-16 Major flyways used by migratory birds, mostly waterfowl. Each route has a number of subroutes.

Some countries along such flyways have entered into agreements and treaties to protect crucial habitats needed by such species, both along their migration routes and at each end of their journeys. Wildlife officials manage waterfowl by **(1)** regulating hunting, **(2)** protecting existing habitats, and **(3)** developing new habitats, including artificial nesting sites, ponds, and nesting islands.

More than 75% of the federal wildlife refuges in the United States are wetlands used by migratory birds. Local and state agencies and private conservation groups such as Ducks Unlimited, the Audubon Society, and the Nature Conservancy (Solutions, p. 451) have also established waterfowl refuges.

Since 1934 the Migratory Bird Hunting and Conservation Stamp Act has required waterfowl hunters to buy a duck stamp each season they hunt. Revenue from these sales goes into a fund to buy land and easements for the benefit of waterfowl.

In this chapter, we have seen that protecting the species that make up part of the earth's biodiversity from premature extinction is a difficult, controversial, and challenging responsibility. See the website for this chapter for some actions you can take to help protect wildlife and preserve biodiversity.

A greening of the human mind must precede the greening of the earth. A green mind is one that cares, saves, and shares. These are the qualities essential for conserving biological diversity now and forever.
M. S. SWAMINATHAN

REVIEW QUESTIONS

1. Define the boldfaced terms in this chapter.

2. What factors led to the extinction of the passenger pigeon in the United States?

3. Distinguish between *local, ecological,* and *biological* extinction of a species.

4. Describe recent estimates of extinction rates in the world and the United States.

5. Distinguish between *endangered* and *threatened* species. List characteristics that make species vulnerable to biological extinction.

6. Distinguish between *background* and *mass extinction.* List six methods scientists use to estimate extinction rates. List two reasons why most biologists believe that we are in the midst of sixth mass extinction caused by human activities.

7. List reasons why some analysts believe that the global and national estimates of biodiversity loss and degradation are inaccurate and exaggerated. Describe the response of biologists to these claims.

8. Distinguish between *instrumental* and *intrinsic* values of wildlife and give three examples of each type of value.

9. What are four underlying causes of the population reduction and extinction of wild species? Describe how each of the following factors contributes to the premature extinction of species and give an example of a species affected by each factor: **(a)** habitat loss and degradation, **(b)** habitat fragmentation, **(c)** deliberately introduced nonnative species, **(d)** accidentally introduced nonnative species, **(e)** commercial hunting and illegal hunting (poaching), **(f)** predator and pest control, **(g)** the legal and illegal market for exotic pets and decorative plants, and **(h)** climate change and pollution.

10. Give two examples of **(a)** deliberately introduced harmful species and **(b)** accidentally introduced harmful species.

11. What are major characteristics of **(a)** successful invader species and **(b)** ecosystems vulnerable to invader species?

12. List four ways to reduce the threat from nonnative species.

13. Why are many island species especially vulnerable to extinction? Why are some migrating birds vulnerable to premature extinction? What are some important ecological roles of bats, and why are they vulnerable to extinction?

14. Describe the near extinction of the blue whale.

15. What is *bioinformatics*, and how can it be used to help protect biodiversity?

16. List the benefits and limitations of protecting species using **(a)** the Convention on International Trade in Endangered Species (CITES) and **(b)** the Endangered Species Act (ESA) in the United States.

17. List measures that would strengthen and weaken the ESA in the United States.

18. Distinguish between *habitat conservation plans*, *safe harbor agreements*, and *candidate conservation plans* used as ways to help implement the Endangered Species Act.

19. What are the pros and cons of trying to protect all endangered and threatened species? List three guidelines that could be used to decide which species to protect.

20. Summarize the pros and cons of using the following to help protect endangered species: **(a)** wildlife refuges, **(b)** gene banks and botanical gardens, and **(c)** zoos, aquariums, and animal research centers. Distinguish between *egg pulling* and *captive breeding*.

21. What is *wildlife management*, and what are the major steps in managing wildlife? What are the major limitations of wildlife management?

22. List the pros and cons of sport hunting.

23. List ways to help protect migratory waterfowl.

CRITICAL THINKING

1. Discuss your gut-level reaction to the following statement: "Eventually all species become extinct. Thus, it does not really matter that the passenger pigeon is extinct and that the blue whale, the whooping crane, the California condor, and the world's remaining rhinoceros and tiger species are endangered mostly because of human activities." Be honest about your reaction, and give arguments for your position.

2. **(a)** Do you accept the ethical position that each *species* has the inherent right to survive without human interference, regardless of whether it serves any useful purpose for humans? Explain. Would you extend this right to the *Anopheles* mosquito, which transmits malaria? What about infectious bacteria? **(b)** Do you believe that each *individual* of an animal species has an inherent right to survive? Explain. Would you extend such rights to individual plants and microorganisms? What about tigers that kill people? Explain.

3. Explain why you agree or disagree with **(a)** using animals for research, **(b)** keeping animals captive in a zoo, and **(c)** killing surplus animals produced by a captive breeding program in a zoo when no suitable habitat is available for their release.

4. Your lawn and house are invaded by fire ants, which can cause painful bites. What would you do?

5. Which of the following statements best describes your feelings toward wildlife: **(a)** As long as it stays in its space, wildlife is OK, **(b)** as long as I do not need its space, wildlife is OK, **(c)** I have the right to use wildlife habitat to meet my own needs, **(d)** when you've seen one redwood tree, fox, elephant, or some other form of wildlife you've seen them all, so lock up a few of each species in a zoo or wildlife park and do not worry about protecting the rest, and **(e)** wildlife should be protected.

6. List your three favorite wild species. Examine why they are your favorites. Are they cute and cuddly-looking, like the giant panda and the koala? Do they have human-like qualities, like apes or penguins that walk upright? Are they large, like elephants or blue whales? Are they beautiful, like tigers and monarch butterflies? Are any of them plants? Are any of them species such as bats, sharks, snakes, or spiders that most people are afraid of? Are any of them microorganisms that help keep you alive? Reflect on what your choice of favorite species tells you about your attitudes toward most wildlife.

7. Environmental groups in a heavily forested state want to restrict logging in some areas to save the habitat of an endangered squirrel. Timber company officials argue that the well-being of one type of squirrel is not as important as the well-being of the many families affected if the restriction causes them to lay off hundreds of workers. If you had the power to decide this issue, what would you do and why? Can you come up with a compromise?

8. If you moved to the suburbs and deer invaded your yard and ate your shrubs and vegetables, what would you do?

9. Recently scientists have begun using gene-transfer techniques to clone some endangered species as a way to help prevent their extinction. However, some conservationists are concerned that this could divert money and attention from what endangered species need most: protected habitats and protection from poaching. What is your opinion on this issue? Explain.

10. Congratulations. You have been put in charge of preventing the premature extinction of the world's existing species from human activities. What would be the three major components of your program to accomplish this goal?

PROJECTS

1. Make a log of your own consumption of all products for a single day. Relate your level and types of consumption to the **(a)** decline of wildlife species and **(b)** increased destruction, degradation, and fragmentation of wildlife habitats in the United States (or the country where you live) and in tropical forests.

2. Identify examples of habitat destruction or degradation in your community that have had harmful effects on the populations of various wild plant and animal species. Develop a management plan for rehabilitating these habitats and species.

3. Choose a particular endangered animal or plant species and use the library or the internet to find out what is being done to protect it from extinction. Develop a protection plan for this species.

4. Make a concept map of this chapter's major ideas, using the section heads and subheads and the key terms (in boldface). Look on the website for this book for information about making concept maps.

INTERNET STUDY RESOURCES AND RESOURCES FOR FURTHER READING AND RESEARCH

The website for this book contains helpful study aids and many ideas for further reading and research. Log on to

http://www.brookscole.com/product/0534389872s

and click on the Chapter-by-Chapter area. Choose Chapter 18 and select a resource:

- "Flash Cards" allows you to test your mastery of the Terms and Concepts to Remember for this chapter.
- "Tutorial Quizzes" provides a multiple-choice practice quiz.
- "Student Guide to InfoTrac" will lead you to Critical Thinking Projects that use InfoTrac College Edition as a research tool.
- "References" lists the major books and articles consulted in writing this chapter.
- "Hypercontents" takes you to an extensive list of sites with news, research, and images related to individual sections of the chapter.

INFOTRAC COLLEGE EDITION

Improve your skills with InfoTrac College Edition, a searchable online database of articles from more than 700 periodicals. Log on to

http://www.infotrac-college.com

or access InfoTrac through the website for this book. Try to find the following articles:

1. Going Down in History (passenger pigeon and other birds on brink of extinction). Simon Reeve. *Geographical*, March 2001 v73 i3 p60. The story of the extinction of the passenger pigeon is used as an illustration of what may be happening now to other birds. *Hint*: Enter the search term "passenger pigeon" using the keyword "extinction."

2. Pet Trade Blues (the efforts and moral problems involved in attempting to save Brazil's Lear's macaws from extinction). Richard Hartley. *International Wildlife*, March–April 2000 pNA. This article addresses the issue of the pet industry and animal smuggling. *Hint*: Enter the search term "pet trade" using the keywords "using regulations to improve environmental quality."

PART V

ENERGY RESOURCES

A country that runs on energy cannot afford to waste it.

BRUCE HANNON

19 NONRENEWABLE ENERGY RESOURCES

Bitter Lessons from Chernobyl

Chernobyl is a chilling word recognized around the globe as the site of a major nuclear disaster (Figure 19-1). On April 26, 1986, a series of explosions in one of the reactors in a nuclear power plant in Ukraine (then part of the Soviet Union) blew the massive roof off the reactor building and flung radioactive debris and dust high into the atmosphere. A huge radioactive cloud spread over much of Belarus, Russia, the Ukraine, and other parts of Europe and eventually encircled the planet.

Here are some consequences of this disaster, caused by poor reactor design and human error:

- In 1998, the Ukrainian Health Ministry put the official death toll from the accident at 3,576. However, Greenpeace Ukraine estimates that by 1995 the total death toll from the accident was about 32,000.
- According to the United Nations, almost 400,000 people were forced to leave their homes, probably never to return. Most were not evacuated until at least 10 days after the accident.
- According to a United Nations study, some 160,000 square kilometers (62,000 square miles)—about the size of the state of Florida—of the former Soviet Union remains contaminated with radioactivity.
- More than half a million people were exposed to dangerous levels of radioactivity. In Belarus, where 70% of the radiation was deposited, the World Health Organization says that thyroid cancer rates among children are 100 times the levels before the accident.
- Government officials say that the total cost of the accident will reach at least $358 billion, many times more than the value of all the nuclear electricity ever generated in the former Soviet Union.

The environmental refugees evacuated from the Chernobyl region had to leave their possessions behind and say good-bye to **(1)** lush, green wheat fields and blossoming apple trees, **(2)** land their families had farmed for generations, **(3)** cows and goats that would be shot because the grass they ate was radioactive, and **(4)** their radioactivity-poisoned cats and dogs. They will not be able to return. Today the Chernobyl power plant remains one of the most dangerous places on the earth.

Chernobyl taught us that *a major nuclear accident anywhere is a nuclear accident everywhere.*

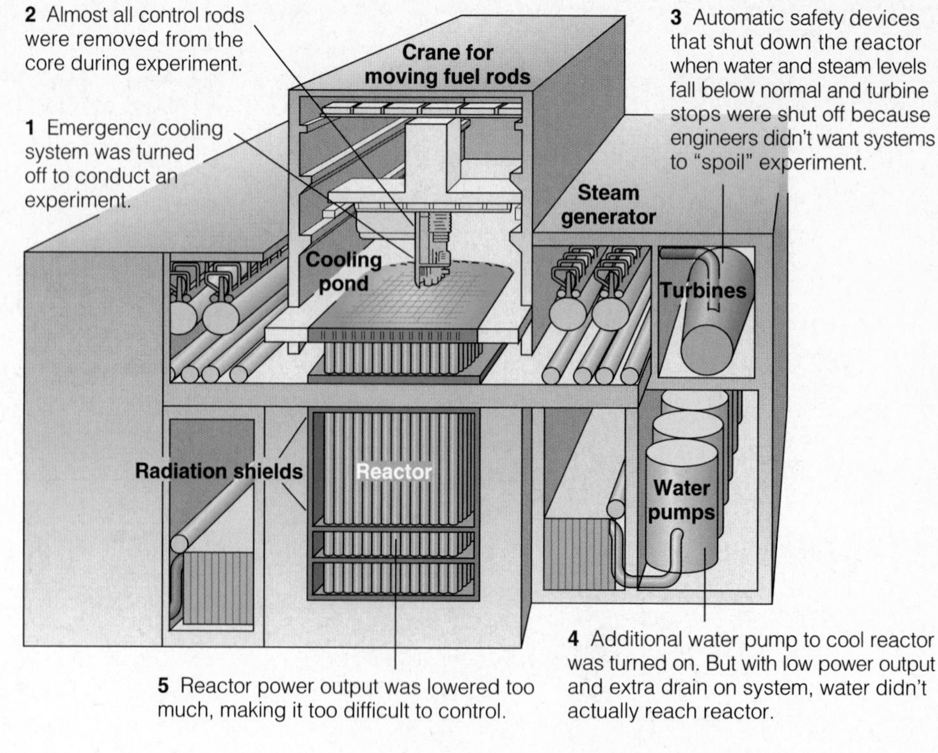

2 Almost all control rods were removed from the core during experiment.
1 Emergency cooling system was turned off to conduct an experiment.
3 Automatic safety devices that shut down the reactor when water and steam levels fall below normal and turbine stops were shut off because engineers didn't want systems to "spoil" experiment.
Crane for moving fuel rods
Steam generator
Cooling pond
Turbines
Radiation shields
Reactor
Water pumps
4 Additional water pump to cool reactor was turned on. But with low power output and extra drain on system, water didn't actually reach reactor.
5 Reactor power output was lowered too much, making it too difficult to control.

Figure 19-1 Major events leading to the Chernobyl nuclear power-plant accident on April 26, 1986, in the former Soviet Union. The accident happened because **(1)** engineers turned off most of the reactor's automatic safety and warning systems (to keep them from interfering with an unauthorized safety experiment), **(2)** the safety design of the reactor was inadequate (there was no secondary containment shell, as in Western-style reactors, and **(3)** a design flaw led to unstable operation at low power. After the reactor exploded, crews exposed themselves to lethal levels of radiation to put out fires and encase the shattered reactor in a hastily constructed concrete tomb. This 19-story concrete tomb is sagging and is full of holes that allow water to seep in and radioactive dust to drift out. Building a new tomb for the reactor will cost at least $1.5 billion—money the Ukrainian government does not have.

Typical citizens of advanced industrialized nations each consume as much energy in six months as typical citizens in developing countries consume during their entire life.

MAURICE STRONG

This chapter addresses the following questions:

- How should we evaluate energy alternatives?
- What are the advantages and disadvantages of oil?
- What are the advantages and disadvantages of natural gas?
- What are the advantages and disadvantages of coal?
- What are the advantages and disadvantages of conventional nuclear fission, breeder nuclear fission, and nuclear fusion?

19-1 EVALUATING ENERGY RESOURCES

What Types of Energy Do We Use? *Some 99% of the energy used to heat the earth and all of our buildings comes directly from the sun* (photo on p. 489). Without this direct input of essentially inexhaustible solar energy, the earth's average temperature would be −240°C (−400°F), and life as we know it would not exist.

This direct input of solar energy also produces several other *indirect forms of renewable solar energy*: **(1)** wind, **(2)** falling and flowing water (hydropower), and **(3)** biomass (solar energy converted to chemical energy stored in chemical bonds of organic compounds in trees and other plants).

Commercial energy sold in the marketplace makes up the remaining 1% of the energy we use to supplement the earth's direct input of solar energy. Most commercial energy comes from extracting and burning nonrenewable mineral resources obtained from the earth's crust, primarily carbon-containing fossil fuels (petroleum, natural gas, and coal; Figure 19-2).

Over the past 60,000 years, major cultural changes (Section 1-6, p. 13) and technological advances have greatly increased energy use per person (Figure 19-3). As a result of such advances, about 81% of the commercial energy consumed in the world comes from *nonrenewable* energy resources (75% from fossil fuels and 6% from nuclear power; Figure 19-4, left).

In developing countries, the most important supplemental source of renewable energy is from *biomass* (mostly fuelwood and charcoal made from fuelwood). It is the main source of energy for heating and cooking for roughly half the world's population. Within a few decades, one-fourth of the world's population in developed countries may face an oil shortage. However, half the world's people in developing countries already face a fuelwood shortage and cannot afford to use fossil fuels.

Energy resources come and go as a result of shortages, changing costs, and new technologies. For example, during the 20th century there were major shifts in the world's major sources of energy. Specifically, the use of **(1)** *coal* dropped from 55% to 21%, **(2)** *oil* increased from 2% to 32%, **(3)** *natural gas* rose from 1%

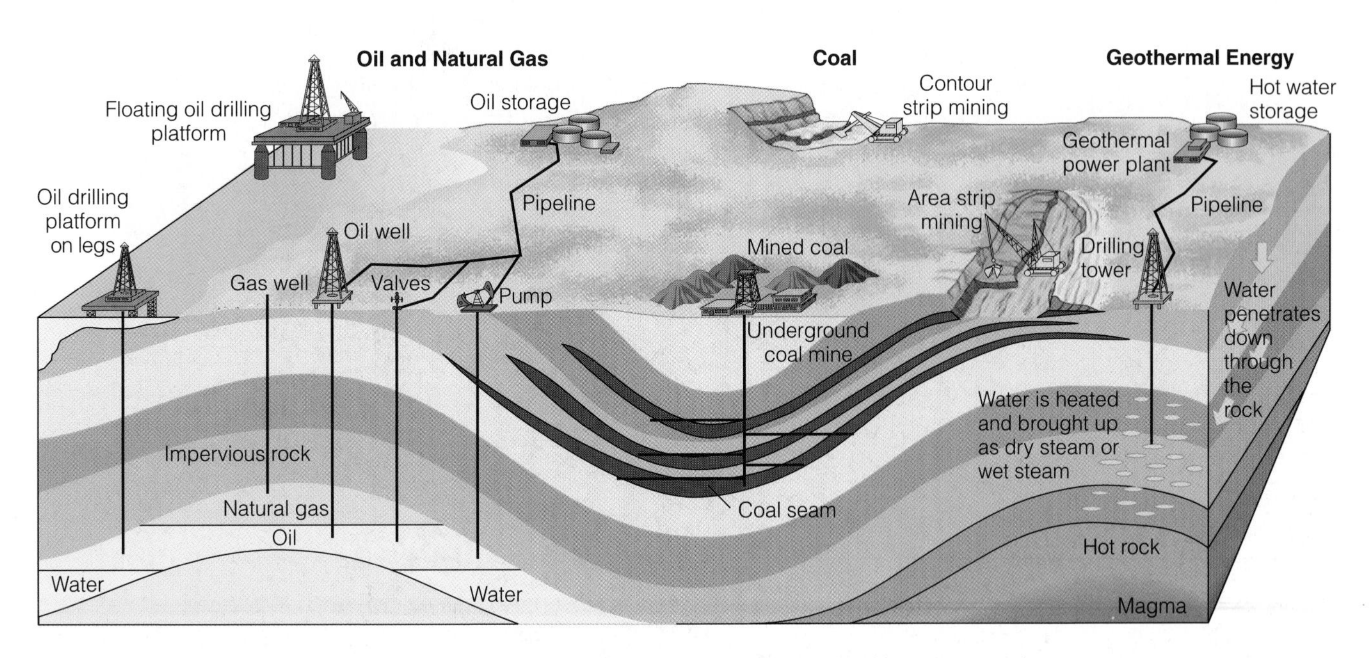

Figure 19-2 Important nonrenewable energy resources that can be removed from the earth's crust are coal, oil, natural gas, and some forms of geothermal energy. Nonrenewable uranium ore is also extracted from the earth's crust and then processed to increase its concentration of uranium-235, which can be used as a fuel in nuclear reactors to produce electricity.

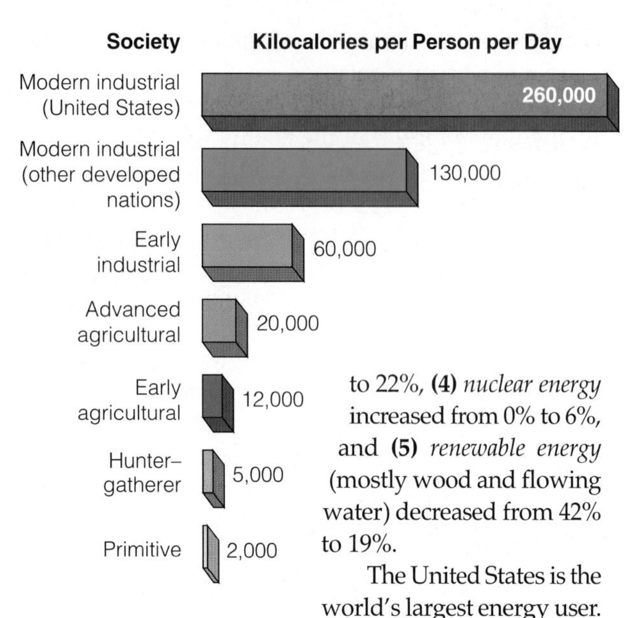

Figure 19-3 Average direct and indirect daily energy use per person at various stages of human cultural development. A typical citizen in a modern industrial society uses 65 to 130 times as much energy per day as a hunter–gatherer.

to 22%, **(4)** *nuclear energy* increased from 0% to 6%, and **(5)** *renewable energy* (mostly wood and flowing water) decreased from 42% to 19%.

The United States is the world's largest energy user. With only 4.7% of the population, it uses 24% of the world's commercial energy. In contrast, India, with 17% of the world's people, uses about 3% of the world's commercial energy.

About 91% of the commercial energy used in the United States come from *nonrenewable* energy resources (84% from fossil fuels and 7% from nuclear power, Figure 19-4, right). Energy use per person in the United States and Canada is about twice as high as in Japan, Germany, France, and the United Kingdom and at least 100 times as high as that in China and India.

How Should We Evaluate Energy Resources? The world's current dependence on nonrenewable fossil fuels (Figure 19-4) is the primary cause of **(1)** air and water pollution, **(2)** land disruption, and **(3)** greenhouse gas emissions. Moreover, affordable oil, the most widely used energy resource in developed countries, may be **(1)** *economically depleted* within 42–93 years and **(2)** gradually replaced by other energy resources.

For many analysts, the key problem is not running out of oil or other fossil fuels. Instead, it is reducing the disruptive economic and ecological effects of projected global warming (Figure 13-11, p. 311) by **(1)** reducing energy waste (Solutions, p. 315) and **(2)** shifting to a mix of new energy resources that produce little or no carbon dioxide and other greenhouse gases.

What is our best immediate energy option? The general consensus is to cut out unnecessary energy

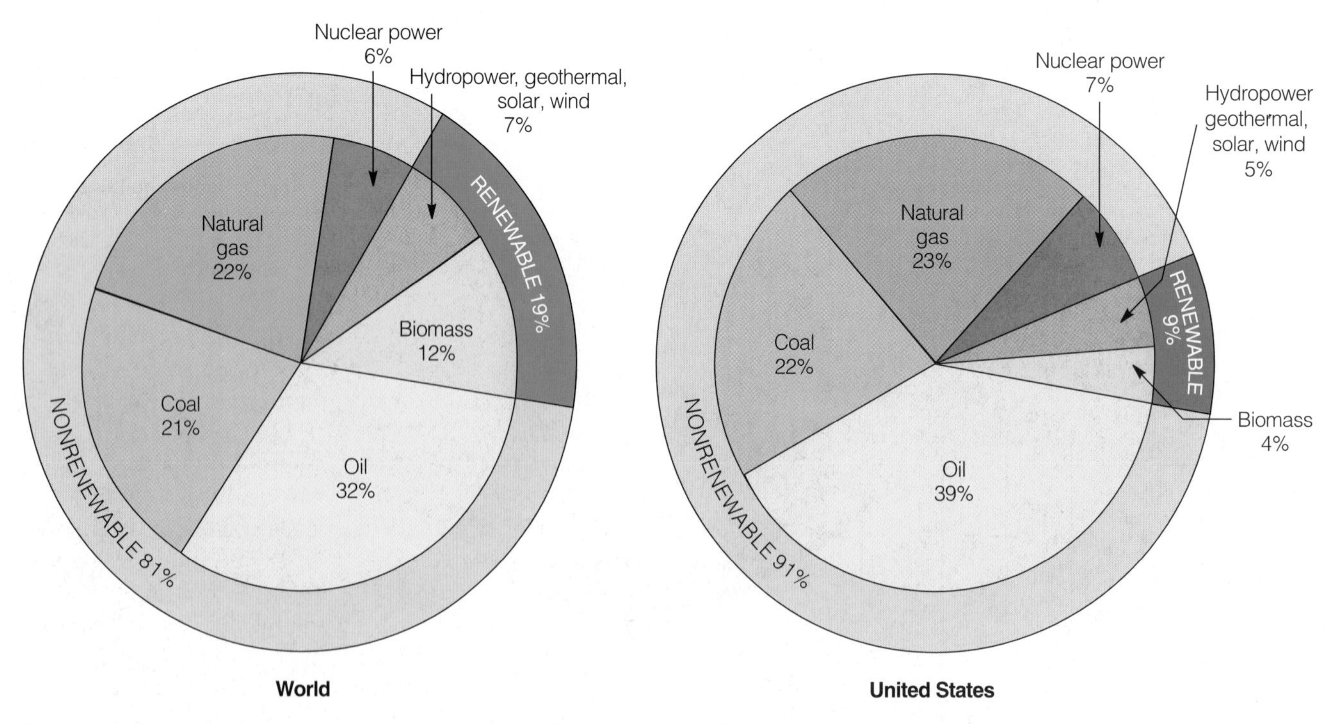

Figure 19-4 Commercial energy use by source for the world and the United States in 1999. Commercial energy amounts to only 1% of the energy used in the world; the other 99% is direct solar energy received from the sun and is not sold in the marketplace. (Data from U.S. Department of Energy, British Petroleum, and Worldwatch Institute)

waste by improving energy efficiency. This means using energy technologies such as **(1)** insulation, **(2)** high-mileage vehicles, and **(3)** more energy-efficient light bulbs and appliances, as discussed in Section 20-2, p. 516.

There is disagreement about our next best energy option. Some say we should get much more of the energy we need from a mix of renewable energy sources such as **(1)** the sun, **(2)** wind, **(3)** flowing water (hydropower), **(4)** biomass, **(5)** hydrogen gas, and **(6)** heat stored in the earth's interior (geothermal energy), as discussed in Chapter 20.

Some think we can find more oil. Others say we should burn more coal and synthetic liquid and gaseous fuels made from coal because it is our most plentiful energy resource. Many believe that fairly abundant and clean-burning natural gas is the answer, especially as a transition fuel to a new solar–hydrogen age built around improved energy efficiency and increased use of renewable energy (Figure 19-5). Some think nuclear power is the answer. These nonrenewable energy options are evaluated in this chapter.

Experience shows that it usually takes at least 50 years and huge investments to phase in new energy alternatives (Figure 19-5) to the point where they provide 10–20% of total energy use. What energy resources will we be using over the next 20 to 50 years? Making such projections involves answering the following questions for *each* energy alternative:

- How much of the energy source will be available in the **(1)** near future (the next 15–25 years) and **(2)** long term (the next 25–50 years)?
- What is this source's net energy yield?

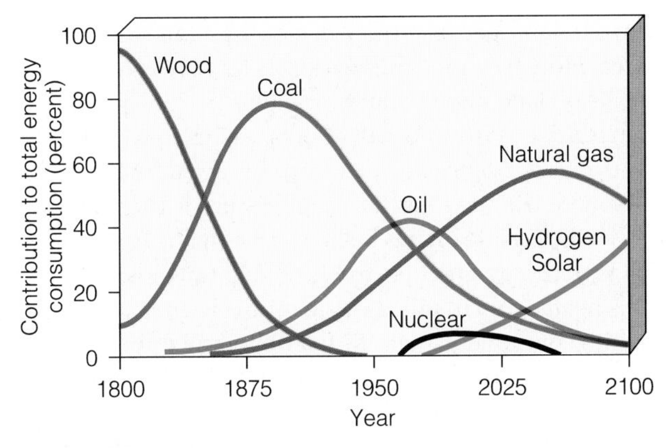

Figure 19-5 Shifts in the use of commercial energy resources in the United States since 1800, with projected changes to 2100. Shifts from wood to coal and then from coal to oil and natural gas have each taken about 50 years. The projected shift to 2100 is only one of many possible scenarios depending on a variety of assumptions. (Data from U.S. Department of Energy)

- How much will it cost to develop, phase in, and use this energy resource?
- How will extracting, transporting, and using the energy resource affect the environment, human health, and the earth's climate?

What Is Net Energy? The Only Energy That Really Counts It takes energy to get energy. For example, oil must be **(1)** found, **(2)** pumped up from beneath the ground or ocean floor, **(3)** transferred to a refinery and converted to useful fuels (such as gasoline, diesel fuel, and heating oil), **(4)** transported to users, and **(5)** burned in furnaces and cars before it is useful to us. Each of these steps uses energy, and the second law of energy (p. 71) tells us that each time we use energy to perform a task, some of it is always wasted and is degraded to low-quality energy.

The usable amount of *high-quality energy* (Figure 3-9, p. 67) available from a given quantity of an energy resource is its **net energy**: the total amount of energy available from an energy resource minus the energy needed to find, extract, process, and get that energy to consumers. It is calculated by estimating the total energy available from the resource over its lifetime minus the amount of energy **(1)** used (the first law of energy), **(2)** automatically wasted (the second law of energy), and **(3)** unnecessarily wasted in finding, processing, concentrating, and transporting the useful energy to users.

Net energy is like your net spendable income (your wages minus taxes and other deductions). For example, suppose that for every 10 units of energy in oil in the ground we have to use and waste 8 units of energy to find, extract, process, and transport the oil to users. Then we have only 2 units of *useful energy* available from every 10 units of energy in the oil.

We can express net energy as the ratio of useful energy produced to the useful energy used to produce it. In the example just given, the *net energy ratio* would be 10/8, or 1.25. The higher the ratio, the greater the net energy. When the ratio is less than 1, there is a net energy loss.

Figure 19-6 shows estimated net energy ratios for various types of space heating, high-temperature heat for industrial processes, and transportation. Currently, oil has a high net energy ratio because much of it comes from large, accessible deposits such as those in the Middle East. When those sources are depleted, the net energy ratio of oil will decline and prices will rise. Then more money and high-quality fossil-fuel energy will be needed to find, process, and deliver new oil from **(1)** widely dispersed small deposits, **(2)** deposits buried deep in the earth's crust, or **(3)** deposits located in remote areas.

Conventional nuclear energy has a low net energy ratio because large amounts of energy are needed to

Space Heating
Passive solar
Natural gas
Oil
Active solar
Coal gasification
Electric resistance heating (coal-fired plant)
Electric resistance heating (natural-gas-fired plant)
Electric resistance heating (nuclear plant)

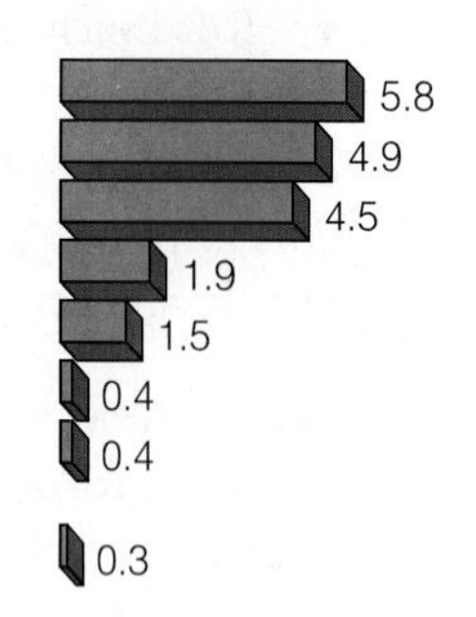

High-Temperature Industrial Heat
Surface-mined coal
Underground-mined coal
Natural gas
Oil
Coal gasification
Direct solar (highly concentrated by mirrors, heliostats, or other devices)

Solar Electricity
Photovoltaic (solar) cells

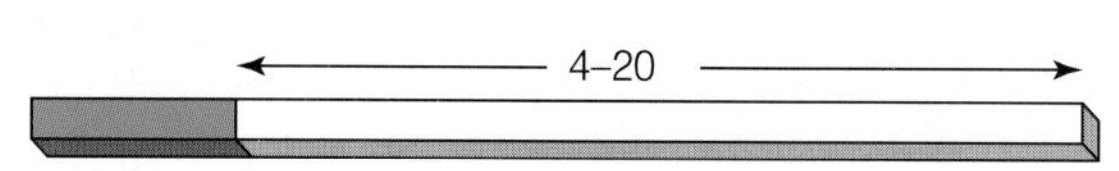

Transportation
Natural gas
Gasoline (refined crude oil)
Biofuel (ethyl alcohol)
Coal liquefaction
Oil shale

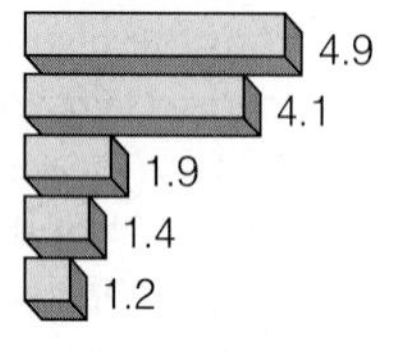

Figure 19-6 Net energy ratios for various energy systems over their estimated lifetimes. The higher the net energy ratio, the greater the net energy available. (Data from U.S. Department of Energy and Colorado Energy Research Institute, *Net Energy Analysis*, 1976; and Howard T. Odum and Elisabeth C. Odum, *Energy Basis for Man and Nature*, 3d ed., New York: McGraw-Hill, 1981)

(1) extract and process uranium ore, **(2)** convert it into a usable nuclear fuel, **(3)** build and operate nuclear power plants, **(4)** dismantle the highly radioactive plants after their 15–60 years of useful life, and **(5)** store the resulting highly radioactive wastes for thousands of years.

19-2 OIL

What Is Crude Oil, and How Is It Extracted and Processed? **Petroleum**, or **crude oil** (oil as it comes out of the ground), is a thick liquid consisting of hundreds of combustible hydrocarbons along with small amounts of sulfur, oxygen, and nitrogen impurities. This fossil fuel was produced by the decomposition of dead organic matter from plants and animals that were **(1)** buried under lake and ocean sediments hundreds of millions of years ago and **(2)** subjected to high temperatures and pressures over millions of years as part of the carbon cycle (Figure 4-27, p. 96).

Deposits of crude oil and natural gas often are trapped together under a dome deep within the earth's crust on land or under the seafloor (Figure 19-2). The crude oil is dispersed in pores and cracks in underground rock formations, somewhat like water saturating a sponge. A well can be drilled, and the crude oil that is drawn by gravity out of the rock pores and into the bottom of the well can be pumped out.

On average, producers get only about 35% of the oil out of an oil deposit. The well is then abandoned because the *heavy crude oil* that remains is too difficult or expensive to recover. As oil prices rise, it can become economical to remove about 10–25% of this remaining heavy oil. However, the net energy yield for such recovered oil is lower because it takes the energy in one-third of a barrel of refined oil to retrieve each barrel of heavy crude oil.

Drilling for oil causes moderate damage to the earth's land because the wells occupy fairly little land area. However, oil drilling always involves **(1)** some oil spills on land and in aquatic systems (p. 358) and **(2)** the harmful environmental effects associated with the extraction, processing, and use of any nonrenewable resource from the earth's crust (Figure 9-10, p. 203).

Once crude oil has been extracted, it is transported to a *refinery* by pipeline, truck, or ship (oil tanker). There it is heated and distilled in gigantic columns to separate it into components with different boiling points (Figure 19-7). Some of the products of oil distillation, called **petrochemicals**, are used as raw materials in industrial organic chemicals, pesticides, plastics, synthetic fibers, paints, medicines, and many other products.

Who Has the World's Oil Supplies? Oil *reserves* are identified deposits from which oil can be extracted profitably at current prices with current technology. The 11 countries that make up the Organization of Petro-

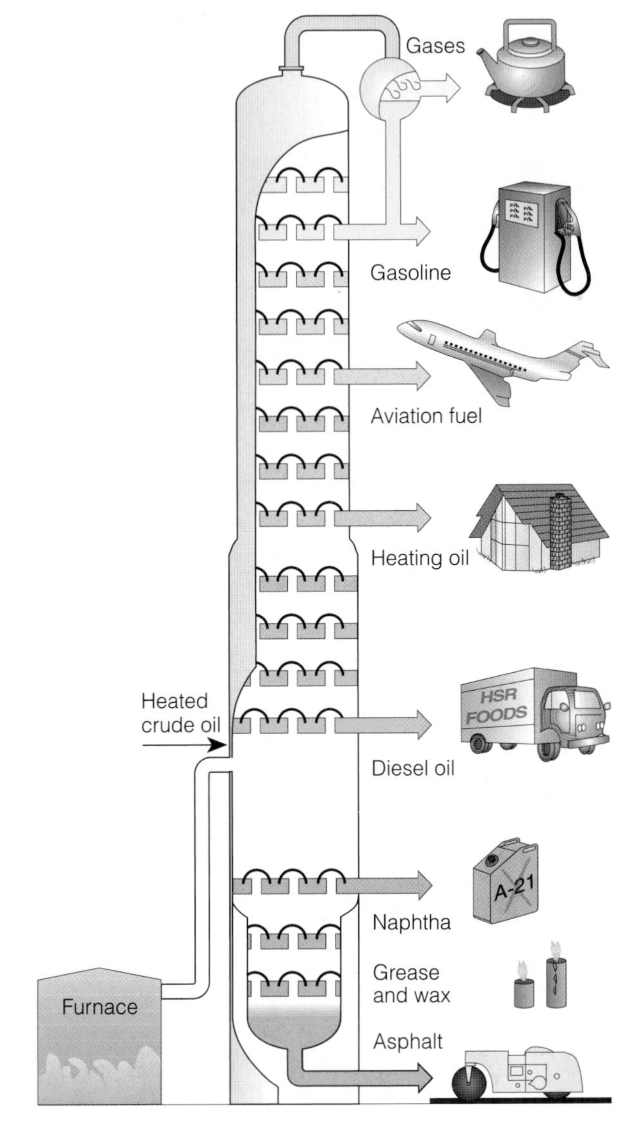

Figure 19-7 Refining crude oil. Based on their boiling points, components are removed at various levels in a giant distillation column. The most volatile components with the lowest boiling points are removed at the top of the column.

leum Exporting Countries (OPEC)* have 67% of the world's crude oil reserves, which explains why OPEC is expected to have long-term control over world oil supplies and prices. According to a 2001 study by U.S. Department of Energy, OPEC produced 40% of the world's oil in 2000 and is projected to produce 50% of the global supply by 2020.

Saudi Arabia, with 26%, has by far the largest proportion of the world's crude oil reserves, followed by Iraq, Kuwait, Iran, and United Arab Emigrates, each with 9–10%.

The remaining global crude oil reserves are found in **(1)** Latin America (14%, with 7% in Venezuela and 5% in Mexico), **(2)** Africa (7%), **(3)** the former Soviet Union (6%), **(4)** Asia (4%, with 3% in China), **(5)** the United States (3%), and **(6)** Europe (2%).

Figure 19-8 shows the locations of the major known deposits of fossil fuels (oil, natural gas, and coal) in **(1)** the United States and Canada and **(2)** ocean areas where more crude oil and natural gas might be found.

The United States has only 3% of the world's oil reserves. However, it uses nearly 30% of the crude oil extracted worldwide each year (68% of it for transportation), mostly because oil is an abundant, convenient, and cheap fuel (Figure 19-9). Despite an upsurge in exploration and test drilling, U.S. oil extraction has declined since 1985, and domestic supplies are not expected to increase significantly.

In 2001, the United States imported about 55% of the oil it used (up from 36% in 1973 during the OPEC oil embargo), mostly because of declining domestic oil reserves and increased oil use. In 2000, the U.S. bill for oil imports was about $100 billion—an average of $11 million per hour. According to the Department of Energy (DOE), the United States could be importing 61% or more of the oil it uses by 2010.

This dependence on imported oil (about half of it from OPEC countries) and the possibility of much higher oil prices within 10–20 years could drain the United States, Japan, and other major oil-importing nations of vast amounts of money. Economists warn that this could lead to severe inflation and widespread economic recession, perhaps even a major depression.

How Long Will Oil Supplies Last, and What Are the Pros and Cons of Oil? Production of the world's estimated oil reserves is expected to peak between 2010 and 2030 (Figure 19-10, top), and production of estimated U.S. reserves peaked in 1975 (Figure 19-10, bottom). *Identified* global reserves of oil should last about 53 years at the current usage rate, and 42 years if usage increases as projected by about 2% per year. Undiscovered oil that is thought to exist might add another 20–40 years to global oil supplies, probably at higher prices. Thus, *known and projected supplies of oil are projected to be 80% depleted within 42–93 years depending on the annual rate of use.*

U.S. oil reserves should last about **(1)** 15–24 years (Figure 19-10, bottom) at current consumption rates and **(2)** 10–15 years if consumption increases as projected. However, potential reserves might yield an additional 24 years of production. Thus, *U.S. oil supplies are projected to be 80% depleted within 10–48 years, depending on the annual rate of use.*

*OPEC was formed in 1960 so that developing countries with much of the world's known and projected oil supplies could get a higher price for this resource. Today its members are Algeria, Indonesia, Iran, Iraq, Kuwait, Libya, Nigeria, Qatar, Saudi Arabia, United Arab Emirates, and Venezuela.

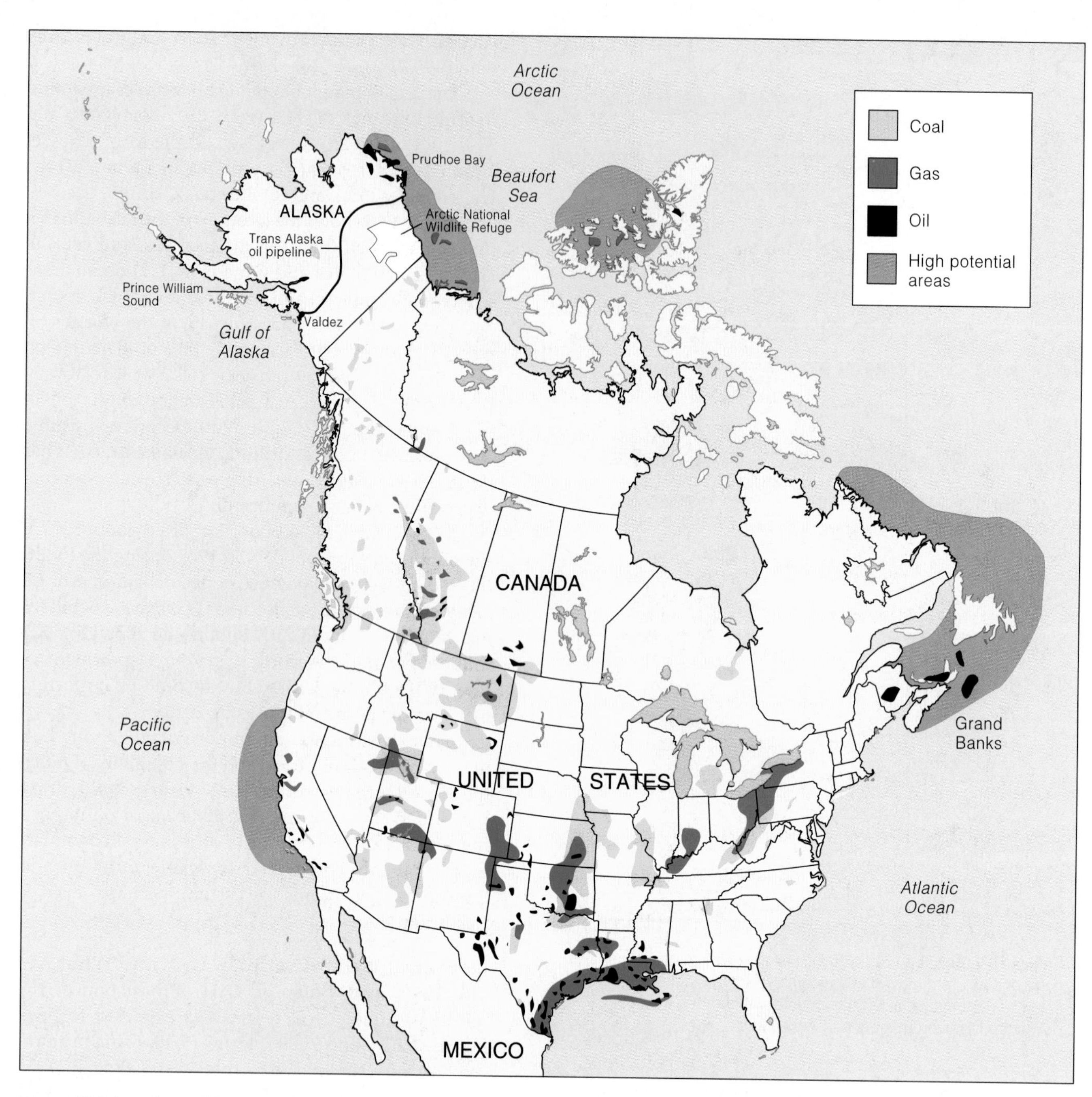

Figure 19-8 Locations of the major known deposits of oil, natural gas, and coal in North America and offshore areas where more crude oil and natural gas might be found. Geologists do not expect to find very much new oil and natural gas in North America. (Data from Council on Environmental Quality and U.S. Geological Survey)

Some analysts contend that rising oil prices (when oil consumption exceeds oil production) will stimulate exploration and lead to enough new reserves to meet future demand through the next century or longer. Other analysts argue that such projections about future oil supplies ignore the consequences of the high (2–5% per year) exponential growth in oil consumption.

Even assuming that we continue to use crude oil at the current rate,

- Saudi Arabia, with the largest known crude oil reserves, could supply all the world's oil needs for only about 10 years.
- The estimated reserves under Alaska's North Slope (the largest ever found in North America) would meet current world demand for only 6 months or U.S. demand for 3 years.
- The estimated reserves in Alaska's Arctic National Wildlife Refuge would meet the current global oil

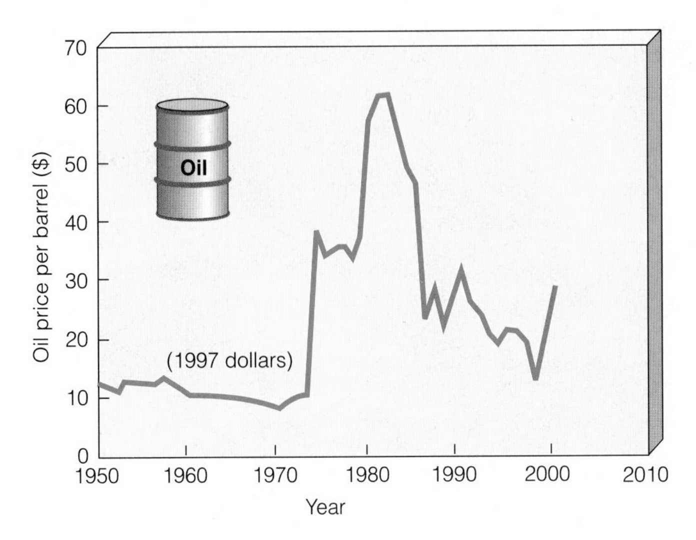

Figure 19-9 Inflation-adjusted price of oil, 1950–2000. When adjusted for inflation, oil costs about the same as it did in 1975. Although low oil prices have stimulated economic growth, they have **(1)** discouraged improvements in energy efficiency and **(2)** increased use of renewable energy resources. (Data from U.S. Department of Energy and Department of Commerce)

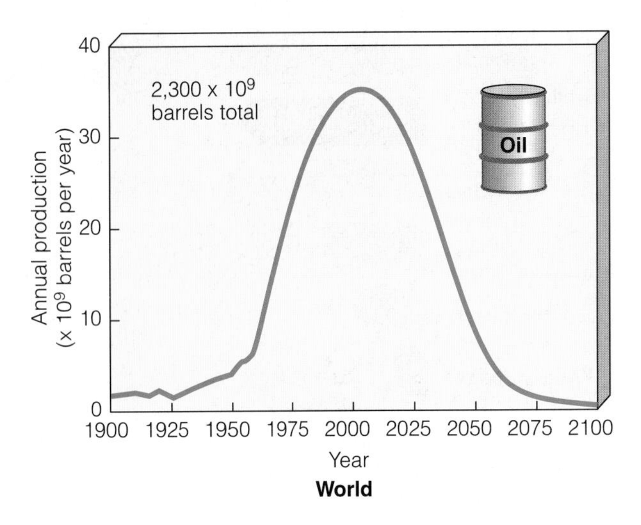

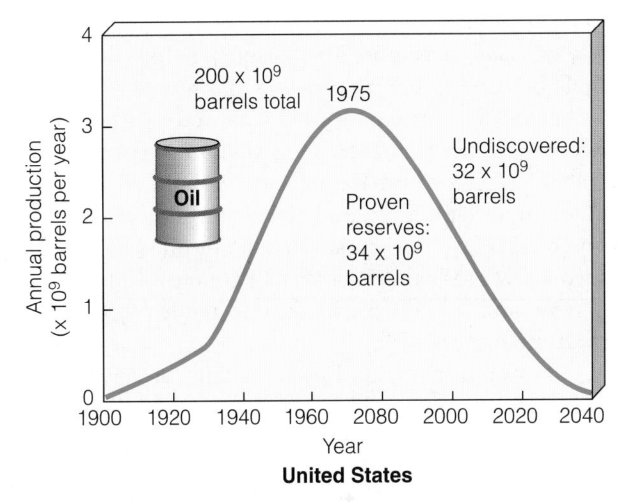

Figure 19-10 Petroleum production curves for the world (top) and the United States (bottom). (Data from U.S. Geological Survey)

demand for only 1–5 months and U.S. oil demand for 7–24 months (Pro/Con, p. 500).

In short, just to keep using oil at the *current rate*, we must discover and add to global oil reserves the equivalent of a new Saudi Arabian supply *every 10 years*.

There is talk of the United States reducing its dependence on imported oil and having more control over global oil prices by increasing domestic oil supplies. Most analysts consider this unrealistic because the United States **(1)** has only 3% of the world's oil reserves, **(2)** uses 30% of the world's annual oil production, and **(3)** produces most of its oil at a high cost of $5–7.50 per barrel, compared with production costs of about $1.50 per barrel in Persian Gulf countries. Consequently, opening all the U.S. coastal waters, forests, and wild places to drilling would hardly put a dent in world oil prices or meet much of the U.S. demand for oil. Critics also point out that some of the oil from Alaska's North Slope—which supplies 17% of U.S. oil—is shipped to Asia instead of being used to reduce U.S. oil imports.

Critics also urge the government to force the oil companies to **(1)** stop shorting taxpayers by about $100 million a year on oil royalty payments and **(2)** pay the several billion dollars they owe in back bills for underpayment of oil royalties. According to sworn government and court evidence, the oil industry shorts taxpayers in annual oil royalty payments by **(1)** falsifying prices, **(2)** using phony bills of sale, and **(3)** deliberately misclassifying high-quality oil as low-quality oil.

The problem is that the current system allows the industry great leeway in deciding **(1)** how to measure the amount of oil removed from public lands and **(2)** how much it should pay the government for the oil. This is somewhat like going to a gas station and being allowed to **(1)** bring your own gauge to measure how much gasoline you pump into your car and **(2)** decide how much it is worth.

Burning any fossil fuel releases carbon dioxide into the atmosphere and thus can promote global warming. Figure 19-11 compares the relative amounts of CO_2 emitted per unit of energy by the major fossil fuels and nuclear power. Currently, burning oil mostly as gasoline and diesel fuel for transportation accounts for 43% of global carbon dioxide emissions. Figure 19-12 lists the pros and cons of using conventional crude oil as an energy resource.

How Useful Are Heavy Oils from Oil Shale and Tar Sands? **Oil shale** is a fine-grained sedimentary rock (Figure 19-13, left) containing solid combustible organic material called *kerogen*. This material can be distilled from oil shale by heating it in a large container

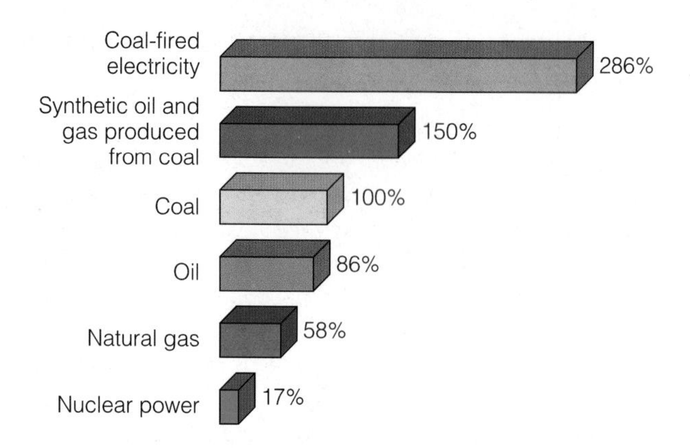

Figure 19-11 Carbon dioxide emissions per unit of energy produced by use of various fuels, expressed as percentages of emissions produced by coal.

Figure 19-13 Oil shale (left) and the shale oil extracted from it (right). Big U.S. oil shale projects have been canceled because of excessive cost. (U.S. Department of Energy)

to yield *shale oil* (Figure 19-13, right). Before the thick shale oil can be sent by pipeline to a refinery, it must be heated to increase its flow rate and processed to remove sulfur, nitrogen, and other impurities. Estimated potential global supplies of shale oil are 200 times larger than estimated global supplies of conventional oil. However, most deposits of oil shale are of such a low grade that it takes more energy and money to mine and convert the kerogen to crude oil than the resulting fuel is worth.

Tar sand (or oil sand) is a mixture of clay, sand, water, and a combustible organic material called *bitumen* (a thick, high-sulfur heavy oil). The bitumen is removed, purified, and chemically upgraded into a synthetic crude oil suitable for refining.

The world's largest known deposits of tar sands, the Athabasca Tar Sands, lie in northern Alberta, Canada. About 10% of these deposits lie close enough to the surface to be surface mined. Currently, these deposits supply about 21% of Canada's oil needs. These deposits could supply all of Canada's projected oil needs for about 33 years at its current consumption rate, but they would last the world only about 2 years. Other large deposits of tar sands are in Utah, Venezuela, Colombia, and Russia.

Figure 19-14 lists the pros and cons of using heavy oil from oil shale and tar sand as energy resources. Because of low net energy yields and the high costs needed to develop and process them, neither of these

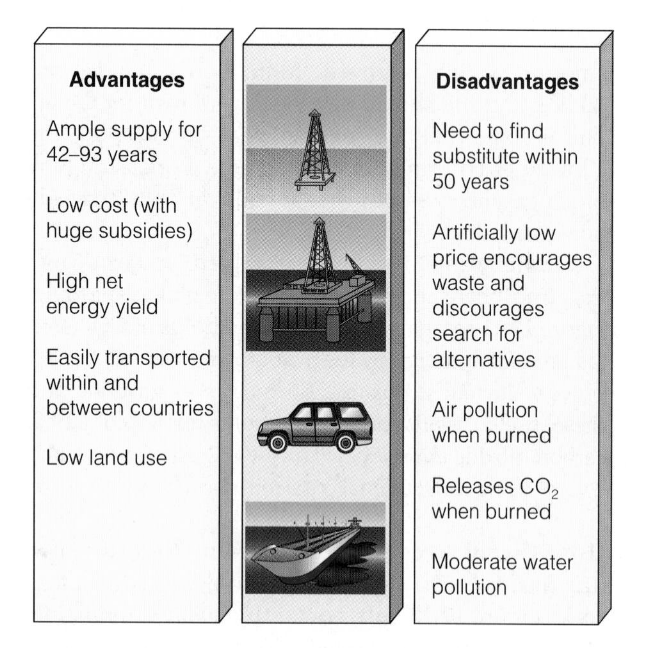

Figure 19-12 Advantages and disadvantages of using conventional oil as an energy resource.

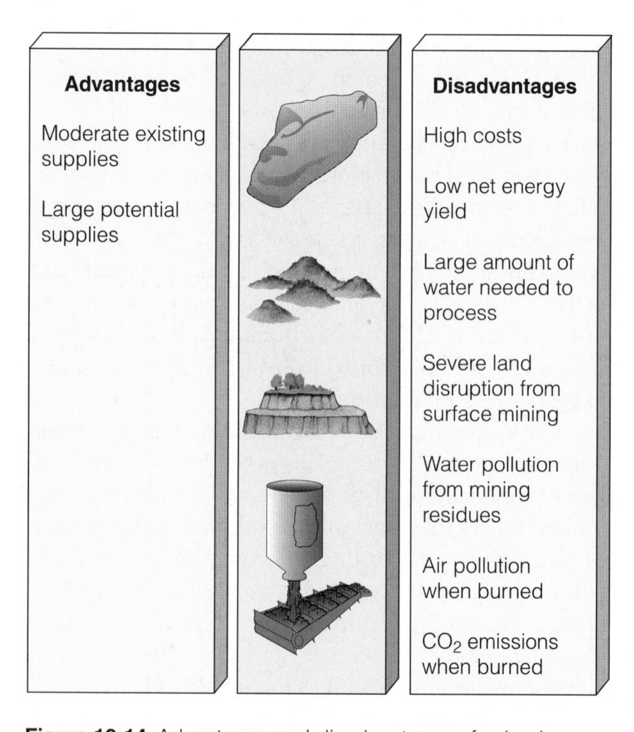

Figure 19-14 Advantages and disadvantages of using heavy oils from oil shale and tar sand as energy resources.

resources is expected to provide much of the world's energy in the foreseeable future.

19-3 NATURAL GAS

What Is Natural Gas? In its underground gaseous state, **natural gas** is a mixture of **(1)** 50–90% by volume of methane (CH_4), the simplest hydrocarbon, **(2)** smaller amounts of heavier gaseous hydrocarbons such as ethane (C_2H_6), propane (C_3H_8), and butane (C_4H_{10}), and **(3)** small amounts of highly toxic hydrogen sulfide (H_2S), a by-product of naturally occurring sulfur in the earth.

Conventional natural gas lies above most reservoirs of crude oil (Figure 19-2). *Unconventional natural gas* is found by itself in other underground sources. One such source is *methane hydrate,* which is composed of small bubbles of natural gas trapped in ice crystals deep under the arctic permafrost and beneath deep ocean sediments. So far it costs too much to get natural gas from such unconventional sources, but the extraction technology is being developed rapidly.

When a natural gas field is tapped, propane and butane gases are liquefied and removed as **liquefied petroleum gas (LPG)**. LPG is stored in pressurized tanks for use mostly in rural areas not served by natural gas pipelines. The rest of the gas (mostly methane) is **(1)** dried to remove water vapor, **(2)** cleansed of poisonous hydrogen sulfide and other impurities, and **(3)** pumped into pressurized pipelines for distribution. At a very low temperature of −184°C (−300°F), natural gas can be converted to **liquefied natural gas (LNG)**. This highly flammable liquid can then be shipped to other countries in refrigerated tanker ships.

Who Has the World's Natural Gas Supplies? Russia and Kazakhstan have about 42% of the world's natural gas reserves. Other countries with large known natural gas reserves are Iran (15%), Qatar (5%), Saudi Arabia (4%), Algeria (4%), the United States (3%), Nigeria (3%), and Venezuela (3%).

Geologists expect to find more natural gas, especially in unexplored developing countries. Most U.S. natural gas reserves are located in the same places as crude oil (Figures 19-2 and 19-8).

How Long Will Natural Gas Supplies Last? The outlook for natural gas supplies is much better than for oil. *At the current consumption rate, known reserves and undiscovered, potential reserves of conventional natural gas are expected to last the world for 125 years and the United States for 65–80 years.*

It is estimated that *conventional* supplies of natural gas, plus *unconventional* supplies available at higher prices, will last at least **(1)** 200 years at the current consumption rate and **(2)** 80 years if usage rates rise 2% per year. Thus, *global supplies of conventional and unconventional natural gas should last 205–325 years, depending on how rapidly natural gas is used.*

What Is the Future of Natural Gas? Figure 19-15 lists the pros and cons of using natural gas as an energy resource. In *combined-cycle natural gas systems,* natural gas is burned in combustion turbines, which are essentially giant jet engines bolted to the ground. This system can **(1)** produce electricity more efficiently than burning coal or oil or using nuclear power, **(2)** produce much less carbon dioxide (Figure 19-11) and smog-causing nitrogen oxides than coal-burning power plants, and **(3)** provide backup power for solar energy and wind power systems. Smaller combined-cycle natural gas units being developed could supply all the heat and electricity needs of an apartment building or office building.

Much natural gas is stranded at wells and costs too much to transport to users. Currently this stranded gas is burned off (flared) at natural gas wells. In addition to wasting natural gas, this adds carbon dioxide to the atmosphere. This practice may change. In 1999, DaimlerChrysler and Syntropetroleum of Oklahoma announced development of a chemical process to convert stranded natural gas at well sites to a clean-burning fuel for use in diesel, fuel cell, and electric hybrid vehicles.

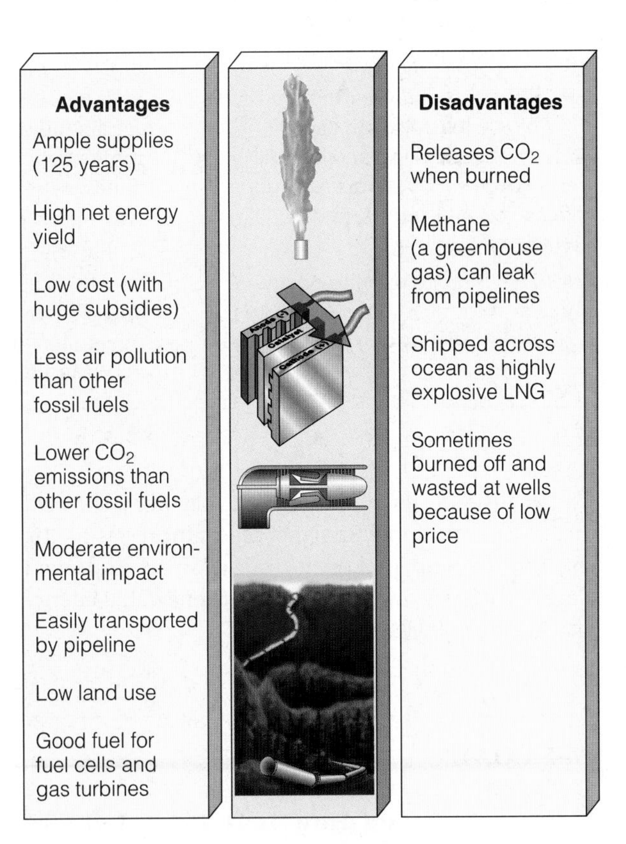

Advantages
Ample supplies (125 years)
High net energy yield
Low cost (with huge subsidies)
Less air pollution than other fossil fuels
Lower CO_2 emissions than other fossil fuels
Moderate environmental impact
Easily transported by pipeline
Low land use
Good fuel for fuel cells and gas turbines
Disadvantages
Releases CO_2 when burned
Methane (a greenhouse gas) can leak from pipelines
Shipped across ocean as highly explosive LNG
Sometimes burned off and wasted at wells because of low price

Figure 19-15 Advantages and disadvantages of using conventional natural gas as an energy resource.

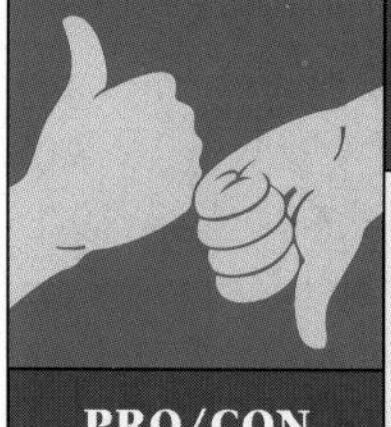

PRO/CON

Should Oil and Gas Development Be Allowed in the Arctic National Wildlife Refuge?

The Arctic National Wildlife Refuge (ANWR) on Alaska's North Slope (Figure 19-8, p. 496) contains more than one-fifth of all the land in the U.S. National Wildlife Refuge System and has been called the crown jewel of the system. The refuge's coastal plain, its most biologically productive part, is the only stretch of Alaska's arctic coastline not open to oil and gas development.

The Alaskan National Interest Lands Conservation Act of 1989 requires specific authorization from Congress before drilling or other development can take place on this coastal plain. For years, U.S. oil companies have been lobbying Congress to grant them permission to carry out exploratory drilling in the coastal plain because they believe that this area might contain oil and natural gas deposits.

Oil company officials, with the support of the Bush administration, argue that

- Possible oil and natural gas in ANWR's coastal plain could **(1)** increase U.S. oil and natural gas supplies, **(2)** reduce U.S. dependence on oil imports, and **(3)** lower energy prices.
- They seek to open to oil and gas development only about 800 hectares (2,000 acres) of the coastal plain region.
- They have developed Alaska's Prudhoe Bay oil fields without significant harm to wildlife.
- They can develop this area in an environmentally responsible manner with little lasting environmental impact by using new oil-drilling technology. These advances **(1)** greatly reduce the area covered by gravel drilling pads, buildings, and equipment and **(2)** inject drilling wastes deep into the round instead of into huge surface pits.

Environmentalists and many biologists oppose this proposal and urge Congress to designate the entire coastal plain as protected wilderness. They contend that

- The Department of Interior estimates that there is only a 19% chance of finding as much economically recoverable oil there as the United States consumes every 7–24 months. This will have no effect on oil prices or oil imports because the potential supply is too little and Persian Gulf oil is much cheaper to produce.
- It would take at least 10 years for any oil or natural gas from the refuge to become available and another 15 years for the field to reach peak production level.
- Improving fuel efficiency (p. 518) is a much faster, cheaper, and cleaner way to save far more oil. For example, requiring new SUVs and light trucks to get the same mileage per gallon as new cars would save more oil in 10 years than would ever be produced from the ANWR.
- The 800 hectares (2,000 acres) to be developed would be spread across 35 separate and far-flung drilling sites that would require construction of a network of roads and pipelines, spanning a much larger area.
- Between 1996 and 1999, there have been 400 oil spills or oil-related pollution incidents per year at Alaska's Prudhoe Bay.
- Potential degradation of any portion of this irreplaceable wildlife area is not worth the risk. A 1995 study by the Department of the Interior concluded that long-lasting ecological harm would be caused by oil drilling in the refuge's fragile tundra ecosystem.
- Carrying out any sort of drilling or exploration there will disqualify the refuge from being added to the U.S. wildreness system.
- Improvements in slant drilling technology may enable oil companies someday to drill the refuge from outside its boundaries.

Critical Thinking

Do you believe that oil companies should be allowed to explore and remove oil and natural gas from this wildlife refuge? Why or why not?

Because of its advantages over oil, coal, and nuclear energy, some analysts see natural gas as the best fuel to help make the transition to improved energy efficiency and greater use of renewable energy over the next 50 years.

19-4 COAL

What Is Coal, and How Is It Extracted and Processed? **Coal** is a solid fossil fuel formed in several stages as buried plant remains are subjected to intense heat and pressure over many millions of years (Figure 19-16). Coal contains **(1)** small amounts of sulfur (which is released into the atmosphere as SO_2 when coal is burned) and **(2)** trace amounts of mercury and radioactive materials (which are also released into the atmosphere when coal is burned). Anthracite, which is about 98% carbon, is the most desirable type of coal because of its high heat content and low sulfur content (Figure 19-16). However, because it takes much longer to form, it is less common and therefore more expensive.

Figure 19-16 Stages in coal formation over millions of years. Peat is a soil material made of moist, partially decomposed organic matter. Lignite and bituminous coal are sedimentary rocks, whereas anthracite is a metamorphic rock (Figure 9-6, p. 199).

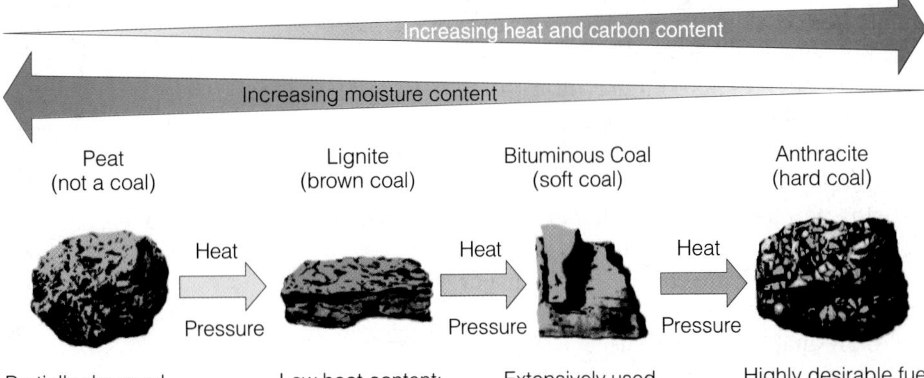

Some coal is extracted underground (Figure 9-9, p. 202) by miners working in tunnels and shafts. Such mining is one of the world's most dangerous occupations because of accidents and black lung disease (caused by prolonged inhalation of coal dust particles). When coal lies close to the earth's surface it is extracted by **(1)** *area strip mining* (Figure 9-8c, p. 201) on flat terrain and **(2)** *contour strip mining* (Figure 9-8d, p. 201) on hilly or mountainous terrain.

After coal is removed, it is transported (usually by train) to a processing plant, where it is broken up, crushed, and then washed to remove impurities. The coal is then dried and shipped (again usually by train) to users, mostly power plants and industrial plants.

How Is Coal Used, and Where Are the Largest Supplies? Coal provides about 21% of the world's commercial energy (and 22% in the United States). It is used to generate 62% of the world's electricity (53% in the United States) and make 75% of its steel.

About 66% of the world's proven coal reserves and 85% of the estimated undiscovered coal deposits are located in the United States (with 24% of global reserves), Russia, China, and India. Half of global coal consumption takes place in the United States (26%) and China (24%). In the United States, coal is burned to generate about 52% of the country's electricity.

How Long Will Coal Supplies Last? Coal is the world's most abundant fossil fuel. *Identified* world reserves of coal should last at least **(1)** 225 years at the current usage rate and **(2)** 65 years if usage rises 2% per year. The world's *unidentified* coal reserves are projected to last about **(1)** 900 years at the current consumption rate and **(2)** 149 years if the usage rate increases 2% per year. Thus, *identified and unidentified supplies of coal could last the world for 214–1,125 years, depending on the rate of usage.*

China, with 11% of the world's reserves, has enough coal to last 300 years at its current rate of consumption. Identified U.S. coal reserves should last about 300 years at the current consumption rate, and unidentified U.S. coal resources could extend those supplies for perhaps 100 years, at a higher cost.

What Is the Future of Coal? Figure 19-17 lists the pros and cons of using coal as an energy resource. Coal is very abundant, but it has the highest environmental impact of any fossil fuel from **(1)** land disturbance (Figure 9-8, p. 201), **(2)** air pollution, **(3)** CO_2 emissions (Figure 19-11, accounting for about 36% of the world's annual emissions), **(4)** release of particles of toxic mercury when burned, **(5)** release of thousands of times more radioactive particles into the atmosphere per unit of energy produced than does a normally operating nuclear power plant, and **(6)** water pollution (Figure

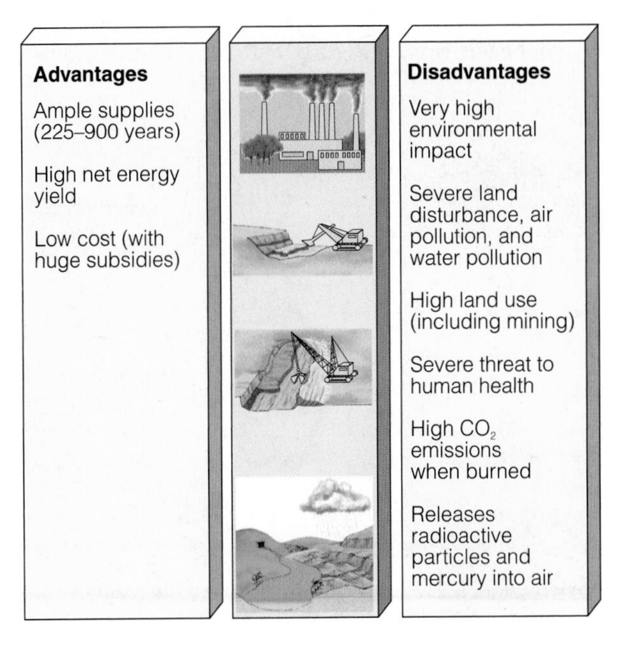

Advantages
Ample supplies (225–900 years)
High net energy yield
Low cost (with huge subsidies)
Disadvantages
Very high environmental impact
Severe land disturbance, air pollution, and water pollution
High land use (including mining)
Severe threat to human health
High CO_2 emissions when burned
Releases radioactive particles and mercury into air

Figure 19-17 Advantages and disadvantages of using coal as an energy resource.

9-11, p. 203). Each year in the United States alone, air pollutants from coal burning **(1)** kill thousands of people (with estimates ranging from 65,000 to 200,000), **(2)** cause at least 50,000 cases of respiratory disease, and **(3)** result in several billion dollars of property damage.

New ways, such as *fluidized-bed combustion*, have been developed to burn coal more cleanly and efficiently and may be phased in over the next several decades. However, this approach does little to reduce carbon dioxide emissions that are considered the major culprit in global warming (Figure 13-8, p. 306).

Many analysts project a decline in coal use over the next 40–50 years because of **(1)** its high CO_2 emissions (Figure 19-11) and harmful health effects and **(2)** the availability of less environmentally harmful ways to produce electricity such as burning natural gas in combined-cycle gas turbines, wind energy (p. 531), solar cells (p. 526), and hydrogen (p. 534).

However, in 2001 the Bush administration **(1)** scuttled the Kyoto treaty, designed to reduce carbon dioxide emissions (p. 317), **(2)** proposed a $2-billion government subsidy program for research into cleaner coal technologies, **(3)** pro-posed exempting older and highly polluting U.S. coal-burning plants from the latest Clean Air Act standards for another 10 years, **(4)** would allow coal-fired power plants to qualify for government renewable energy subsidies merely by mixing a small amounts of biomass such as wood chips or agriculture waste with the coal they burn, and **(5)** asked Congress to reduce government research subsidies for energy efficiency by 30% and renewable energy by 40%.

What Are the Pros and Cons of Converting Solid Coal into Gaseous and Liquid Fuels? Solid coal can be converted into **synthetic natural gas (SNG)** by **coal gasification** (Figure 19-18) or into a liquid fuel

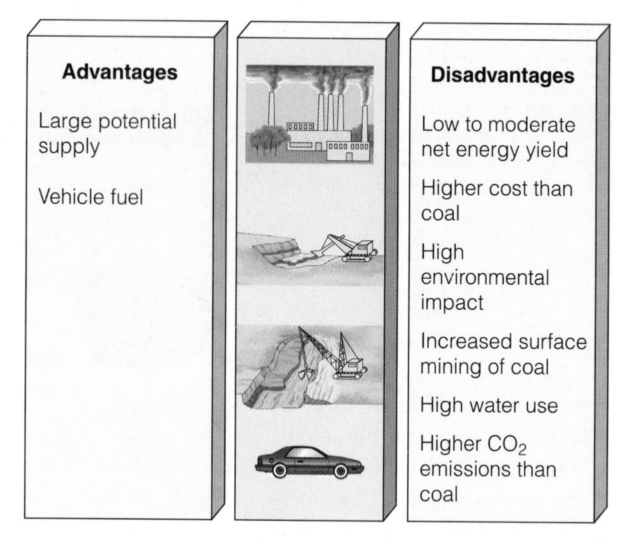

Figure 19-19 Advantages and disadvantages of using synthetic natural gas (SNG) and liquid synfuels produced from coal.

such as methanol or synthetic gasoline by **coal liquefaction**. Figure 19-19 lists the pros and cons of using these *synfuels* produced from coal. Most analysts expect synfuels to play only a minor role as an energy resource in the next 30–50 years.

19-5 NUCLEAR ENERGY

How Does a Nuclear Fission Reactor Work? To evaluate the pros and cons of nuclear power, we must know how a conventional nuclear power plant and its accompanying nuclear fuel cycle work. In a nuclear fission chain reaction, neutrons split the nuclei of atoms

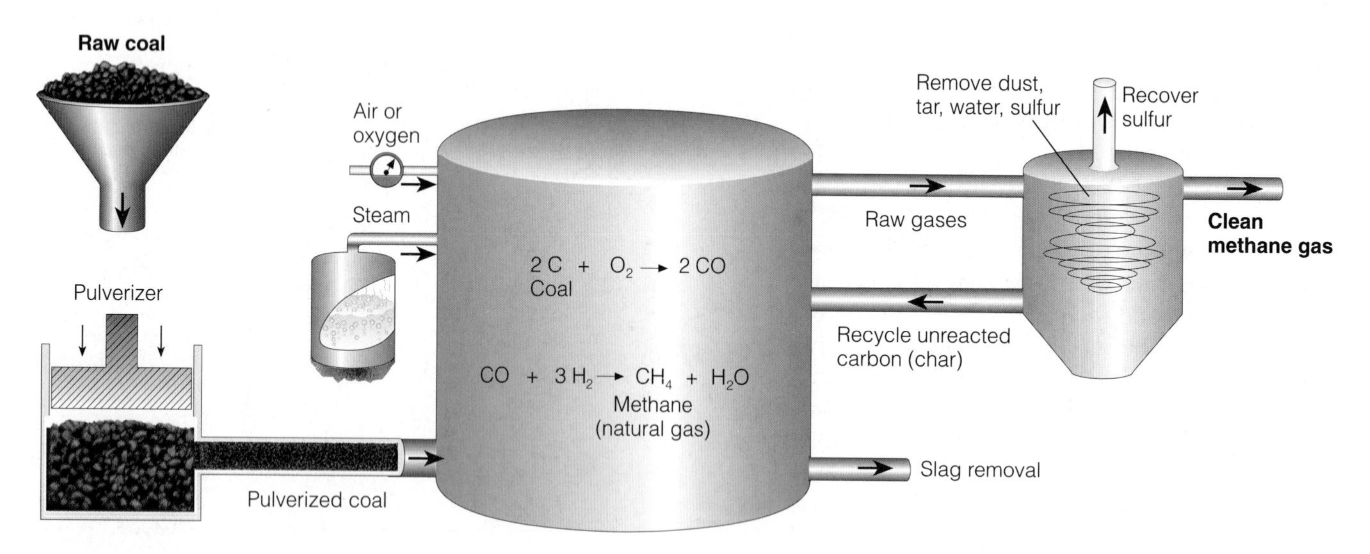

Figure 19-18 *Coal gasification.* Generalized view of one method for converting solid coal into synthetic natural gas (methane).

such as uranium-235 and plutonium-239 and release energy mostly as high-temperature heat (Figure 3-12, p. 70). In the reactor of a nuclear power plant, the rate of fission is controlled and the heat generated is used to produce high-pressure steam, which spins turbines that generate electricity.

Light-water reactors (LWRs) like the one diagrammed in Figure 19-20 produce about 85% of the world's nuclear-generated electricity (100% in the United States). An LWR has the following key parts:

- *Core* containing 35,000–70,000 long, thin fuel rods, each packed with fuel pellets (each about one-third the size of a cigarette). Each pellet contains the energy equivalent of 0.9 metric tons (1 ton) of coal.
- *Uranium oxide fuel* consisting of about 97% nonfissionable uranium-238 and 3% fissionable uranium-235. To create a suitable fuel, the concentration of uranium-235 in the ore is increased (enriched) from 0.7% (its natural concentration in uranium ore) to 3% by removing some of the uranium-238.
- *Control rods*, which are moved in and out of the reactor core to absorb neutrons and thus regulate the rate of fission and amount of power the reactor produces.
- *Moderator*, which slows down the neutrons emitted by the fission process so that the chain reaction can be kept going. This is a material such as **(1)** liquid water (75% of the world's reactors, called *pressurized water reactors*, Figure 19-20), **(2)** solid graphite (20% of reactors), or **(3)** heavy water (D_2O, 5% of reactors). Graphite-moderated reactors can also produce fissionable plutonium-239 for nuclear weapons.
- Coolant, usually water, which circulates through the reactor's core to remove heat (to keep fuel rods and other materials from melting) and produce steam for generating electricity.

Nuclear power plants, each with one or more reactors, are only one part of the nuclear fuel cycle involved in producing nuclear power (Figure 19-21). *In evaluating*

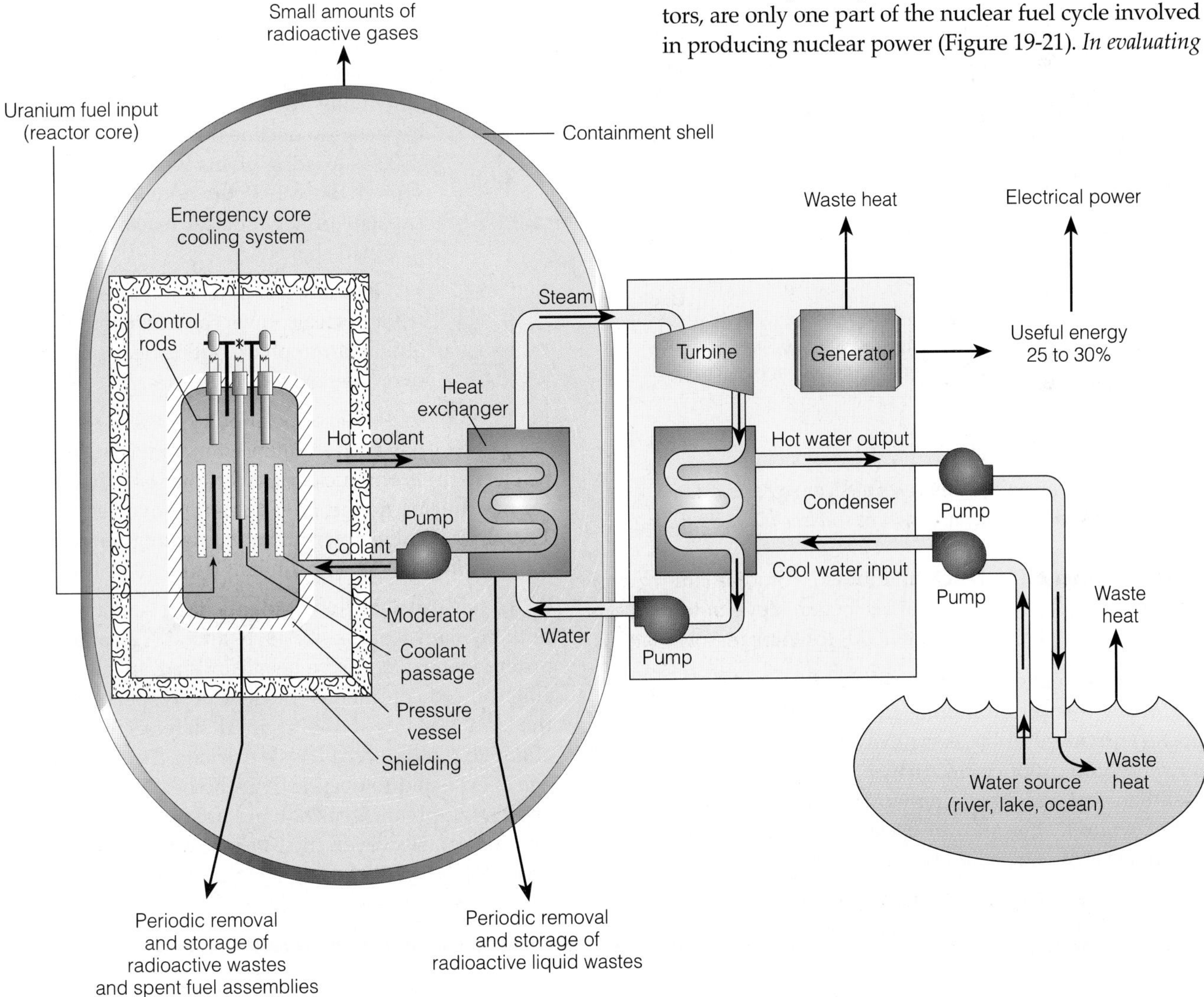

Figure 19-20 Light-water-moderated and -cooled nuclear power plant with a pressurized water reactor.

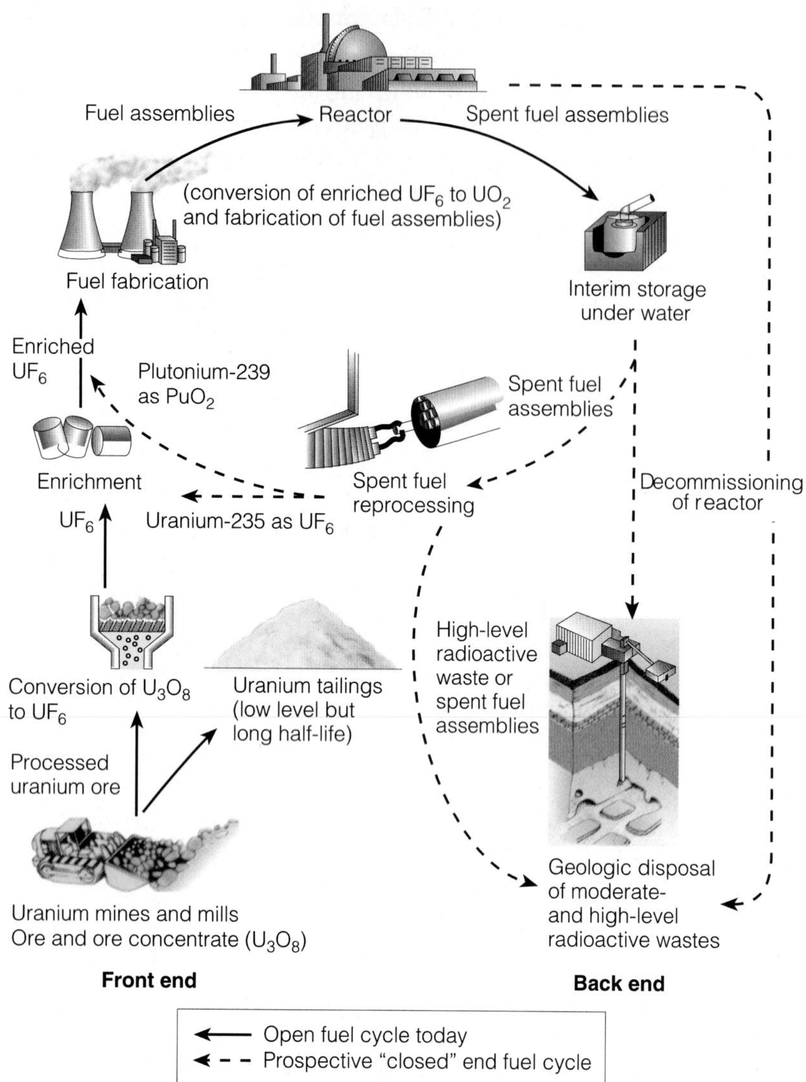

Figure 19-21 The nuclear fuel cycle.

the safety and economic feasibility of nuclear power, we need to look at this entire cycle, not just the nuclear plant itself.

What Happened to Nuclear Power? Studies indicate that U.S. utility companies began developing nuclear power plants in the late 1950s for three reasons:

- The Atomic Energy Commission (which had the conflicting roles of promoting and regulating nuclear power) promised utility executives that nuclear power would produce electricity at a much lower cost than coal and other alternatives. Indeed, President Dwight D. Eisenhower declared in a 1953 speech that nuclear power would be "too cheap to meter."
- The government paid about one-fourth of the cost of building the first group of commercial reactors and guaranteed that there would be no cost overruns.
- After insurance companies refused to insure nuclear power, Congress passed the Price–Anderson Act to protect the U.S. nuclear industry and utilities from significant liability to the general public in case of accidents.* This act is up for renewal by Congress in 2002.

In the 1950s, researchers predicted that by the year 2000, at least 1,800 nuclear power plants would supply 21% of the world's commercial energy (25% in the United States), and most of the world's electricity.

However, after almost 50 years of development, enormous government subsidies, and an investment of $2 trillion, these goals have not been met. Instead,

- By the end of 2000, 438 commercial nuclear reactors in 32 countries were producing 6% of the world's commercial energy and 16% of its electricity.
- Since 1989 the growth in electricity production from nuclear power has essentially leveled off, and its capacity is expected to decline between 2003 and 2020 as existing plants wear out and are retired. Since 1975 the number and capacity of new nuclear reactors has declined sharply.
- Germany (with 31% of its electricity from nuclear power) and Sweden (with 39%) plan to phase out nuclear power over the next 20 to 30 years.
- Public opposition to nuclear power has been so intense in Japan (with 36% of its electricity from nuclear power) that only two sites for new nuclear plants have been approved since 1979. In 2000, the government cut its long-term target for building new reactors in half (from 20 to 10).
- No new nuclear power plants have been ordered in the United States since 1978, and all 120 plants ordered since 1973 have been canceled. In 2000, the 103 licensed commercial nuclear power reactors in the United States at 64 sites in 31 states generated about 20% of the country's electricity. This percentage is expected to decline over the next two decades as existing plants wear out and are retired (decommissioned). However, the Bush administration proposes to slow this decline by **(1)** allowing addition

*This act limits the nuclear industry's liability in case of an accident to $9.5 billion, with the government (taxpayers) paying most of this. According to the U.S. Nuclear Regulatory Commission, a worst-case nuclear accident would cause more than $300 billion in damages.

of new reactors to some existing power plants, **(2)** allowing plants to use new technologies to increase their output, **(3)** streamlining the permitting process for building new nuclear power with less evaluation and input from citizens, and **(4)** increasing the 40-year operating licenses for many existing plants to 60 years.

China is not following this global trend in the declining use of nuclear power. It plans to build more than 50 new plants by 2020 as a way to reduce its dependence on highly polluting coal. France, which gets 76% of its electricity from nuclear power, is also a strong advocate of nuclear power. However, in 1999 it placed a moratorium on further construction of nuclear power plants.

The major reasons for the failure of nuclear power to grow as projected are **(1)** multi-billion-dollar construction cost overruns, **(2)** stricter government safety regulations, **(3)** higher operating costs and more malfunctions than expected, **(4)** poor management, **(5)** public concerns about safety after the 1986 Chernobyl (p. 490) and 1979 Three Mile Island (Pennsylvania) accidents, and **(6)** investor concerns about the economic feasibility of nuclear power.

What Are the Pros and Cons of Nuclear Power? Figure 19-22 lists the major advantages and disadvantages of nuclear power. Using nuclear power to produce electricity has some important advantages over coal-burning power plants (Figure 19-23).

How Safe Are Nuclear Power Plants and Other Nuclear Facilities? Because of the built-in safety features, the risk of exposure to radioactivity from nuclear power plants in the United States and most other developed countries is *extremely low*. However, a partial or complete meltdown or explosion is possible, as accidents at the Chernobyl (p. 490) nuclear power plant in Ukraine and the Three Mile Island plant in Pennsylvania have taught us.

The U.S. Nuclear Regulatory Commission (NRC) estimates that there is a 15–45% chance of a complete core meltdown at a U.S. reactor during the next 20 years. The NRC also found that 39 U.S. reactors have an 80% chance of containment shell failure from a meltdown or an explosion of gases inside the containment structures.

In addition, nearly half of the 103 operating nuclear reactors in the United States have failed mock terrorist attacks against them. The 2001 destruction of New York City's World Trade Center towers has suggested that a similar attack by a large commercial plane could **(1)** break open a reactor's containment shell (Figure 19-20) and **(2)** set off a reactor meltdown that could create a Chernobyl-like radioactive disaster (p. 490).

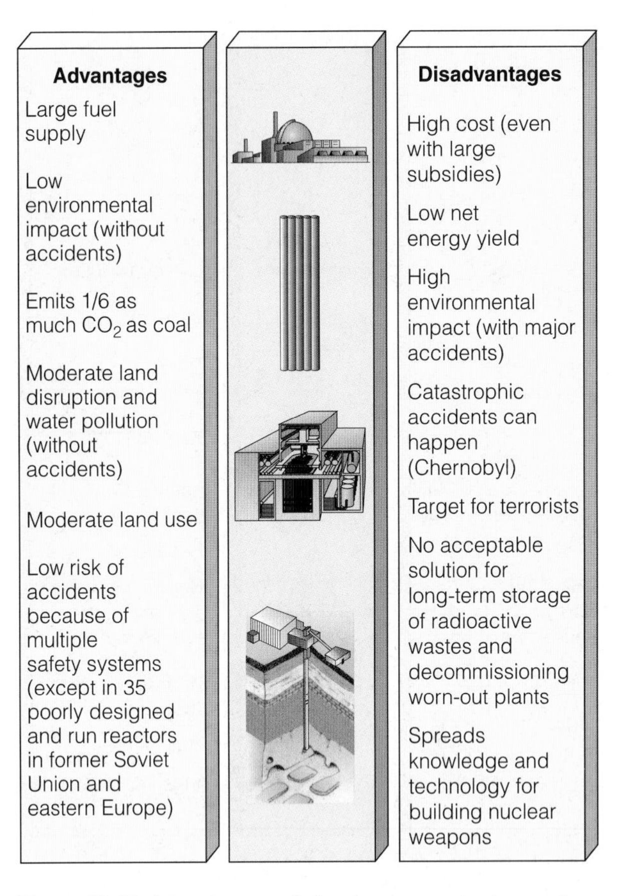

Figure 19-22 Advantages and disadvantages of using nuclear power to produce electricity. This evaluation includes the entire nuclear fuel cycle (Figure 19-21).

Nuclear scientists and government officials throughout the world urge the shutdown of 35 poorly designed and poorly operated nuclear reactors in some republics of the former Soviet Union and in eastern Europe. However, without economic aid from the world's developed countries, it is unlikely that these potentially dangerous plants will be closed and replaced with safer nuclear or nonnuclear alternatives.

In the United States, there is widespread public distrust in the ability of the NRC and the Department of Energy (DOE) to enforce nuclear safety in commercial (NRC) and military (DOE) nuclear facilities. Congressional hearings in 1987 uncovered evidence that high-level NRC staff members **(1)** destroyed documents and obstructed investigations of criminal wrongdoing by utilities, **(2)** suggested ways utilities could evade commission regulations, and **(3)** provided utilities and their contractors with advance notice of "surprise" inspections. In 1996, George Galatis, a respected senior nuclear engineer, said, "I believe in nuclear power but after seeing the NRC in action I'm convinced a serious accident is not just likely, but inevitable. . . . They're asleep at the wheel."

The nuclear power industry claims that nuclear power plants in the United States have not killed anyone and cause far less environmental harm than coal-burning

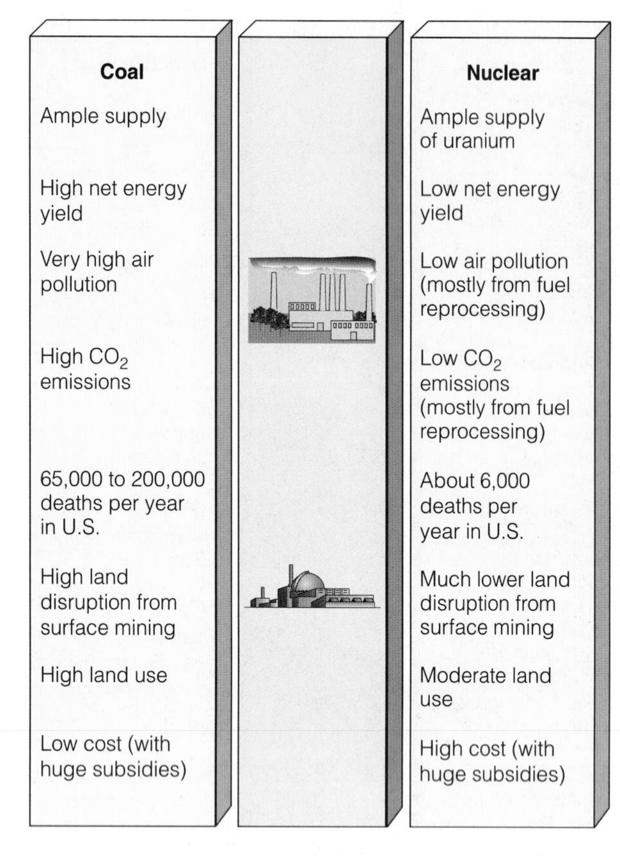

Coal
Ample supply
High net energy yield
Very high air pollution
High CO_2 emissions
65,000 to 200,000 deaths per year in U.S.
High land disruption from surface mining
High land use
Low cost (with huge subsidies)
Nuclear
Ample supply of uranium
Low net energy yield
Low air pollution (mostly from fuel reprocessing)
Low CO_2 emissions (mostly from fuel reprocessing)
About 6,000 deaths per year in U.S.
Much lower land disruption from surface mining
Moderate land use
High cost (with huge subsidies)

Figure 19-23 Comparison of the risks of using nuclear power and coal-burning plants to produce electricity.

plants (Figure 19-23). However, according to U.S. National Academy of Sciences estimates, U.S. nuclear plants cause 6,000 premature deaths and 3,700 serious genetic defects each year. If correct, this annual death toll is much smaller than the 65,000 to 200,000 deaths per year caused by coal-burning plants in the United States. However, critics point out that the estimated annual deaths from both of these types of plants are unacceptable, given the availability of much less harmful alternatives such as combined-cycle natural gas turbines, wind power (p. 531), solar cells (p. 526), and hydrogen (p. 534).

What Do We Do with Low-Level Radioactive Waste? Each part of the nuclear fuel cycle (Figure 19-21) produces low-level and high-level solid, liquid, and gaseous radioactive wastes with various half-lives (Table 3-1, p. 69). Wastes classified as *low-level radioactive wastes* give off small amounts of ionizing radiation and must be stored safely for 100–500 years before decaying to safe levels.

From the 1940s to 1970, most low-level radioactive waste produced in the United States (and most other countries) was put into steel drums and dumped into the ocean; the United Kingdom and Pakistan still dispose of their low-level radioactive wastes in this way.

Today, low-level waste materials from commercial nuclear power plants, hospitals, universities, industries, and other producers in the United States are put in steel drums and shipped to the two remaining regional landfills run by federal and state governments. Attempts to build new regional dumps for low-level radioactive waste using improved technology have met with fierce public opposition.

What Should We Do with High-Level Radioactive Waste? *High-level radioactive wastes* give off large amounts of ionizing radiation for a short time and small amounts for a long time. Such wastes must be stored safely for at least 10,000 years and about 240,000 years if long-lived plutonium-239 (Table 3-1, p. 69) is not removed by reprocessing. Most high-level radioactive wastes are **(1)** spent fuel rods from commercial nuclear power plants (now being stored in pools of water or dry storage casks at plant sites) and **(2)** an assortment of wastes from plants that produce plutonium and tritium for nuclear weapons.

After 50 years of research, scientists still do not agree on whether there is any safe method of storing these wastes. Some scientists believe that the long-term safe storage or disposal of high-level radioactive wastes is technically possible. Others disagree, pointing out that it is impossible to demonstrate that any method will work for the 10,000–240,000 years of fail-safe storage needed for such wastes.

Here are some of the proposed methods and their possible drawbacks:

- *Bury it deep underground* (Figure 19-24, and Case Study, p. 508). This favored strategy is under study by all countries producing nuclear waste. In 2001, the U.S. National Academy of Sciences concluded that the geological repository option is the only scientifically feasible long-term solution for safely isolating such wastes. However, according to an earlier report by the U.S. National Academy of Sciences, "Use of geological information to pretend to be able to make very accurate predictions of long-term site behavior is scientifically unsound."

- *Shoot it into space or into the sun.* Costs would be very high, and a launch accident, such as the explosion of the space shuttle *Challenger,* could disperse high-level radioactive wastes over large areas of the earth's surface. This strategy has been abandoned for now.

- *Bury it under the antarctic ice sheet or the Greenland ice cap.* The long-term stability of the ice sheets is not known. They could be destabilized by heat from the wastes, and retrieving the wastes would be difficult or impossible if the method failed. This strategy is prohibited by international law.

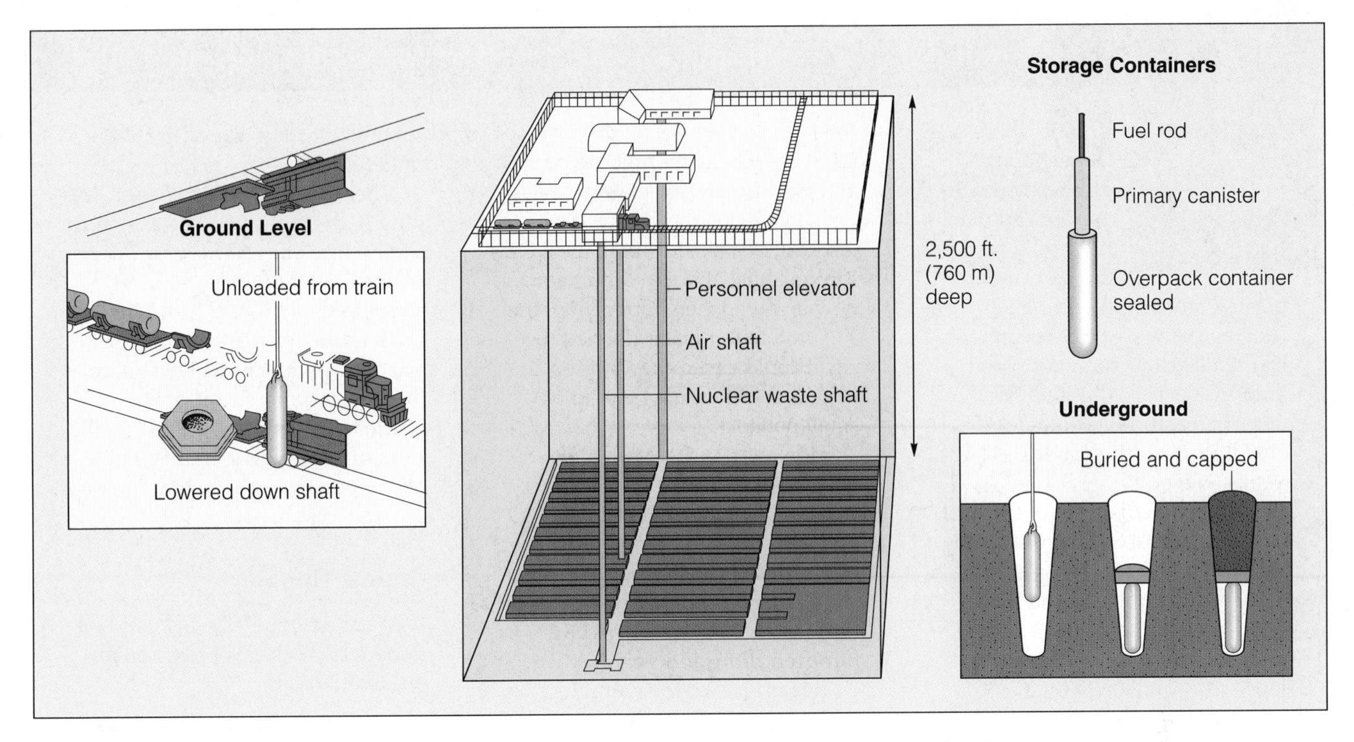

Figure 19-24 Proposed general design for deep-underground permanent storage of high-level radioactive wastes from commercial nuclear power plants in the United States. (U.S. Department of Energy)

- *Dump it into descending subduction zones in the deep ocean* (Figure 9-4, middle, p. 197). However, **(1)** wastes eventually might be spewed out somewhere else by volcanic activity, **(2)** containers might leak and contaminate the ocean before being carried downward, and **(3)** retrieval would be impossible if the method did not work. This strategy is prohibited by international law.

- *Bury it in thick deposits of mud on the deep ocean floor in areas that tests show have been geologically stable for 65 million years.* The waste containers eventually would corrode and release their radioactive contents. This approach is prohibited by international law.

- *Change it into harmless, or less harmful, isotopes.* Currently there is no way to do this. Even if a method was developed, **(1)** costs probably would be very high, and **(2)** the resulting toxic materials and low-level (but very long-lived) radioactive wastes would still need to be disposed of safely.

How Widespread Are Contaminated Radioactive Sites? In 1992, the EPA estimated that as many as 45,000 sites in the United States may be contaminated with radioactive materials, 20,000 of them belonging to the DOE and the Department of Defense. According to the DOE, it will cost taxpayers at least $230 billion over the next 75 years to clean up these facilities (with some analysts estimating that the price will range from $400 to $900 billion). According to a 2000 DOE report, more than two-thirds of 144 highly contaminated sites used to produce nuclear weapons will never be completely cleaned up and will have to be protected and monitored for centuries. Some critics question spending so much money on a problem that is ranked by scientific advisers to the EPA as a low-risk ecological problem and not among the top high-risk human health problems (Figure 10-12, p. 238).

The radioactive contamination situation in the United States pales in comparison to the post–Cold War legacy of nuclear waste and contamination in the republics of the former Soviet Union. Land in various parts of the former Soviet Union is dotted with **(1)** areas severely contaminated by nuclear accidents, **(2)** 25 operating nuclear power plants with flawed and unsafe designs, **(3)** nuclear-waste dump sites, **(4)** radioactive waste-processing plants, **(5)** contaminated nuclear test sites, and **(6)** coastal waters where nuclear wastes and retired nuclear-powered submarines were dumped.

A 1957 explosion of a nuclear-waste storage tank at Mayak, a plutonium production facility in southern Russia, spewed 2.5 times as much radiation into the atmosphere as the Chernobyl accident. Because of radioactive contamination, **(1)** no one can live within about 2,600 square kilometers (1,000 square miles) of the surrounding area, and **(2)** nearby Lake Karachay (which was a dump site for the facility's radioactive wastes between 1949 and 1967) is so radioactive that standing on its shores for about an hour would be fatal.

Underground Storage of High-Level Radioactive Wastes in the United States

CASE STUDY

In 1985, the U.S. Department of Energy (DOE) announced plans to build a repository for underground storage of high-level radioactive wastes from commercial nuclear reactors on federal land in the Yucca Mountain desert region, 160 kilometers (100 miles) northwest of Las Vegas, Nevada.

By 2001, the federal government had spent about $7 billion studying and testing the potential of the site as a permanent underground repository. The proposed facility (Figure 19-24), which is expected to cost as much as $60 billion to build (financed by a tax on nuclear power), is scheduled to open by 2010.

It may never open, mostly because **(1)** rock fractures may allow water to leak into the site and corrode radioactive waste storage casks, **(2)** there is an active volcano nearby, and **(3)** there are 36 active earthquake faults on the site. In 1998, Jerry Szymanski, formerly the DOE's top geologist at Yucca Mountain and now an outspoken opponent of the site, said that if water flooded the site it could cause an explosion so large that "Chernobyl would be small potatoes."

However, in 1999 scientists working on the site released a preliminary report **(1)** finding nothing so far to disqualify the site and **(2)** contending that it could safely store high-level radioactive wastes for several hundred thousand years.

Regardless of the storage method, most citizens strongly oppose the location of low-level or high-level nuclear-waste disposal facilities anywhere near them. Even if the problem is technically solvable, it may be politically unacceptable.

There is also concern about shipping high-level nuclear waste to the site by truck or rail from nuclear plants scattered throughout 31 states in the United States. The DOE estimates that a shipment accident could cost $19–50 billion in damages and cleanup costs. Nuclear power proponents say the risks of an accident are negligible, but opponents believe that the risks are underestimated.

Critical Thinking

1. What do you believe should be done with high-level radioactive wastes? Explain.

2. Would you favor having high-level nuclear waste transported by truck or train through the area where you live to the Yucca Mountain site? Explain.

What Can We Do with Worn-Out Nuclear Plants? After approximately 15–40 years of operation, a nuclear reactor becomes dangerously contaminated with radioactive materials, and many of its parts are worn out. Unless the plant's life can be extended by expensive renovation, it must be *decommissioned* or retired (the last step in the nuclear fuel cycle, Figure 19-21) by **(1)** dismantling it and storing its large volume of highly radioactive materials in high-level nuclear-waste storage facilities (Figure 19-24), **(2)** putting up a physical barrier and setting up full-time security for 30–100 years before the plant is dismantled, or **(3)** enclosing the entire plant in a tomb that must last for several thousand years.

At least 228 large commercial reactors worldwide (20 in the United States) are scheduled for retirement between 2000 and 2012.

By 2030 all U.S. reactors will have to be retired, based on the life of their current operating licenses. Some utility companies and the Bush administration have proposed extending the life of current reactors to 50–60 years. However, opponents contend that this could increase the risk of nuclear accidents in aging reactors.

What Is the Connection Between Nuclear Reactors and the Spread of Nuclear Weapons? Currently, 60 countries—1 of every 3 in the world—have nuclear weapons or the knowledge and ability to build them. Information and fuel needed to build these nuclear weapons have come mostly from the research and commercial nuclear reactors that the United States and 14 other countries have been giving away and selling in the international marketplace for decades.

Between 1986 and 2000, the number of nuclear warheads held by the world's five largest nuclear powers declined by 55% from 70,000 to 31,535. However, this **(1)** still leaves enough nuclear weapons to kill everyone in the world at least 30 times and **(2)** greatly increases the amount of bomb-grade plutonium-239 removed from these weapons that must be kept from use in nuclear weapons.

Can Nuclear Power Reduce Dependence on Oil and Global Warming? Proponents of nuclear power in the United States claim that it will help reduce dependence on imported oil. However, environmentalists point out that the use of nuclear power has little effect on U.S. oil use because **(1)** oil produces only about 2–3% of the electricity in the United States and **(2)** the major use for oil is in transportation, which would not be affected by increasing nuclear power production.

Nuclear power advocates also contend that increased use of nuclear power would reduce the threat of global warming by eliminating the carbon dioxide emissions created by burning coal to produce electricity. Environmentalists point out that this is only partially correct. Nuclear plants themselves are not emitters of CO_2. However, the nuclear fuel cycle (Figure 19-21) does produce some CO_2, although it is much less than that produced by burning coal or natural gas to produce electricity (Figure 19-11). Environmentalists and energy experts (Guest Essay, p. 517) argue that reducing energy waste (Section 20-1, p. 514) and increasing the use of wind turbines (p. 531), solar cells (p. 526), and hydrogen (p. 534) to produce electricity are better ways to reduce CO_2 emissions.

Can We Afford Nuclear Power? Experience has shown that nuclear power is an expensive way to boil water to produce electricity, even when huge government subsidies partially shield it from free-market competition with other energy sources.

Costs rose dramatically in the 1970s and 1980s because of unanticipated safety problems and stricter regulations after the Three Mile Island and Chernobyl accidents. In 1995, the World Bank said that nuclear power is too costly and risky. *Forbes* business magazine has called the failure of the U.S. nuclear power program "the largest managerial disaster in U.S. business history, involving $1 trillion in wasted investment and $10 billion in direct losses to stockholders."

In recent years the operating cost of many U.S. nuclear power plants has dropped, mostly because of less down time. However, environmentalists and economists point out that the cost of nuclear power must be based on the entire nuclear fuel cycle, not merely the operating cost of individual plants. When these costs, including nuclear-waste disposal and decommissioning of worn-out plants—are included, the overall cost—of nuclear power is very high, even with huge government subsidies that help shield it from competing in the marketplace with other energy alternatives. Other energy alternatives do not have the high costs of long-term waste storage and decommissioning.

The U.S. nuclear industry hopes to persuade the federal government and utility companies to build hundreds of smaller second-generation plants using standardized designs, which they claim are safer and can be built more quickly (in 3–6 years). These *advanced light-water reactors (ALWRs)* have built-in *passive safety features* designed to make explosions or the release of radioactive emissions almost impossible. However, according to *Nucleonics Week*, an important nuclear industry publication, "Experts are flatly unconvinced that safety has been achieved—or even substantially increased—by the new designs."

One proposed new design is called a *pebble bed modular reactor (PBMR)*. About 10,000 uranium oxide fuel particles, each the size of a pencil point, are encapsulated into a microsphere ball. The ball is coated with several layers of graphite and a silicon dioxide outer shell to prevent leakage of radiation. The core of a commercial 110-megawatt pebble bed reactor would be filled with about 400,000 pebbles. Such small reactors could be added to existing reactors as needed.

Designers say that the balls are meltdown-proof. To make this reactor affordable, they contend that there is no need for an emergency core cooling system and an airtight containment dome used in light-water reactors (Figure 19-20).

Edwin Lyman and other nuclear physicists, oppose this type of reactor. They contend that

- A crack in the reactor could expose the graphite protective coatings to air. At a high temperature the graphite could burn and release massive amounts of radioactivity—similar to what happened at Chernobyl (Figure 19-1).
- This is an old design that has been rejected in England and Germany for safety reasons.
- This technology would create about 10 times the volume of high-level radioactive waste per unit of electricity as a conventional nuclear reactor.
- A lack of a containment shell would make it easier for terrorists to enter such reactor facilities and **(1)** steal nuclear fuel material that could be used to make nuclear weapons or **(2)** blow it up to release large amounts of radioactivity.
- This approach does not eliminate the expense and hazards of long-term radioactive waste storage and power-plant decommissioning.

Is Breeder Nuclear Fission a Feasible Alternative? Some nuclear power proponents urge the development and widespread use of **breeder nuclear fission reactors**, which generate more nuclear fuel than they consume by converting nonfissionable uranium-238 into fissionable plutonium-239. Because breeders would use more than 99% of the uranium in ore deposits, the world's known uranium reserves would last at least 1,000 years, and perhaps several thousand years.

However, if the safety system of a breeder reactor fails, the reactor could lose some of its liquid sodium coolant, which ignites when exposed to air and reacts explosively if it comes into contact with water. This could cause a runaway fission chain reaction and perhaps a nuclear explosion powerful enough to blast open the containment building and release a cloud of highly radioactive gases and particulate matter. Leaks of flammable liquid sodium can also cause fires, as has happened with all experimental breeder reactors built so far.

In addition, existing experimental breeder reactors produce plutonium so slowly that it would take 100–200 years for them to produce enough plutonium to fuel a significant number of other breeder reactors. In 1994, the United States ended government-supported research for breeder technology after providing about $9 billion in research and development funding.

In December 1986, France opened a commercial-size breeder reactor. It was so expensive to build and operate that after spending $13 billion the government closed it permanently in 1998. Because of this experience, other countries have abandoned their plans to build full-size commercial breeder reactors.

Is Nuclear Fusion a Feasible Alternative? Scientists hope that controlled nuclear fusion will provide an almost limitless source of high-temperature heat and electricity. Research has focused on the D–T nuclear fusion reaction, in which two isotopes of hydrogen—deuterium (D) and tritium (T)—fuse at about 100 million degrees (Figure 3-13, p. 71).

After 50 years of research and huge expenditures of mostly government funds, controlled nuclear fusion is still in the laboratory stage. None of the approaches tested so far have produced more energy than they use. In 1989, two chemists claimed to have achieved deuterium–deuterium (D–D) nuclear fusion at room temperature using a simple apparatus, but subsequent experiments have not substantiated their claim.

If researchers can eventually get more energy out of nuclear fusion than they put in, the next step would be to build a small fusion reactor and then scale it up to commercial size, an extremely difficult engineering problem. The estimated cost of a commercial fusion reactor is several times that of a comparable conventional fission reactor.

Proponents contend that with greatly increased federal funding, a commercial nuclear fusion power plant might be built by 2030. However, many energy experts do not expect nuclear fusion to be a significant energy source until 2100, if then.

What Should Be the Future of Nuclear Power in the United States? Since 1948 nuclear energy has received about 65% of all federal energy research and development funds in the United States. Some analysts call for phasing out all or most government subsides and tax breaks for nuclear power and using such funds to subsidize and accelerate the development of promising other energy technologies such as **(1)** natural gas turbines, **(2)** improving energy efficiency, **(3)** forms of renewable energy such as wind, solar cells, and hydrogen (Chapter 20), and **(4)** smaller and more flexible *micropower* turbines and fuel cells (Section 20-9, p. 538).

To these analysts, nuclear power is a complex, expensive, inflexible, and centralized way to produce electricity that is too vulnerable to terrorist attack. They believe that it is a technology whose time has passed in a world where electricity will increasingly be provided by small, decentralized, easily expandable power plants such as natural gas turbines, wind turbines, arrays of solar cells, and fuel cells.

Proponents of nuclear power argue that governments should continue funding research and development and pilot plant testing of potentially safer and cheaper reactor designs along with breeder fission and nuclear fusion. They argue that we need to keep these nuclear options available for use in the future if natural gas turbines, improved energy efficiency, and various renewable energy options fail to **(1)** keep up with electricity demands and **(2)** reduce CO_2 emissions to acceptable levels.

We are embarked on the beginning of the last days of the Age of Oil.

MIKE BOWLIN (CEO, ARCO OIL)

REVIEW QUESTIONS

1. Define the boldfaced terms in this chapter.

2. What supplies 99% of the energy we use? What percentage of the remaining 1% of the energy we use comes from nonrenewable and from renewable energy in **(a)** the world, **(b)** the United States, and **(c)** developing countries?

3. How long does it usually take to phase in a new energy alternative to the point where it accounts for 10–20% of total energy use? What four questions should we try to answer about each energy resource?

4. What is *net energy*, and why is it important in evaluating an energy resource? Why is the net energy for oil from the Middle East high and that for nuclear power low?

5. What is *petroleum* or *crude oil*? What happens to crude oil at a refinery? What are *petrochemicals*?

6. Who has most of the world's oil reserves? What percentage of the world's oil reserves is found in the United States? What percentage of the world's oil does the United States use? What percentage of the oil used in the United States is imported? How long are known and projected supplies of conventional oil expected to last in **(a)** the world and **(b)** the United States? List the pros and cons of drilling for oil and natural gas in Alaska's Arctic National Wildlife Refuge.

7. What are the pros and cons of using oil as an energy resource?

8. What are the pros and cons of using heavy oil from shale oil and tar sand as energy resources?

9. What is *natural gas*? Who has most of the world's reserves of natural gas? Distinguish between *liquefied petroleum gas (LPG)* and *liquefied natural gas (LNG)*.

10. How long are known and projected supplies of natural gas expected to last in **(a)** the world and **(b)** the United States? What is a *combined-cycle natural gas turbine system*, and why is its use rising rapidly? What are the pros and cons of using natural gas as an energy resource?

11. What is *coal*, and how is it formed? Distinguish between *lignite*, *bituminous*, and *anthracite* coal. How is coal extracted from the earth's crust? How is coal used? What four countries have the largest coal reserves?

12. How long are known and projected supplies of coal expected to last **(a)** in the world and **(b)** in the United States?

13. What are the pros and cons of using coal as an energy resource? What are the pros and cons of converting solid coal into gaseous and liquid fuels?

14. Describe how a *nuclear fission reactor* works. What are the five major components of a *light-water nuclear reactor*, and what role does each play? What is the *nuclear fuel cycle*?

15. List three reasons why commercial nuclear power plants were developed in the United States after World War II. List five factors that have contributed to the leveling off of the use of nuclear power plants to produce electricity.

16. List the major advantages and disadvantages of using conventional nuclear fission to produce electricity. Compare the advantages and disadvantages of using nuclear power and burning coal to produce electricity.

17. How safe are nuclear power plants? Describe the Chernobyl nuclear power-plant accident in the former Soviet Union.

18. What is being done with *low-level radioactive waste* produced by the nuclear fuel cycle? What are the options for dealing with *high-level radioactive waste*? List the pros and cons of the proposed site for storing high-level nuclear wastes at Yucca Mountain in Nevada. How widespread are contaminated radioactive waste sites in **(a)** the United States and **(b)** the former Soviet Union?

19. What are the three options for retiring (decommissioning) nuclear power plants?

20. What is the relationship between the development of commercial nuclear power and the spread of nuclear weapons throughout much of the world?

21. How useful is nuclear power in reducing **(a)** dependence on oil and **(b)** global warming?

22. What is the likely financial future of nuclear power?

23. What are the pros and cons of using **(a)** breeder nuclear fission and **(b)** nuclear fusion as an energy resource? What are the pros and cons of continuing large-scale government subsidies for research and development of conventional nuclear power, breeder fission, and nuclear fusion?

CRITICAL THINKING

1. Just to continue using oil at the current rate (not the projected higher exponential increase in its annual use) we must discover and add to global oil reserves the equivalent of a new Saudi Arabian supply (the world's largest) *every 10 years*. Do you believe this is possible? If not, what effects might this have on your life and on the life of a child or grandchild you might have?

2. List five things you can do to reduce your dependence on oil and resources such as gasoline and most plastics derived from oil (see website material for Chapter 19). Which of these things do you actually plan to do?

3. The United States now imports about 55% of the oil it uses and could be importing 64% of its oil by 2020. Explain why you are for or against continuing to increase oil imports. What do you believe are the three best ways to reduce dependence on oil imports?

4. Explain why you agree or disagree with the following proposals by various energy analysts to solve U.S. energy problems: **(a)** Find and develop more domestic supplies of oil, **(b)** place a heavy federal tax on gasoline and imported oil to help reduce the waste of oil resources, **(c)** increase dependence on nuclear power, and **(d)** phase out all nuclear power plants by 2020.

5. Explain why you agree or disagree with each of the following proposals made by the U.S. nuclear power industry: **(a)** Provide at least $100 billion in government subsidies to build a large number of better-designed nuclear fission power plants to reduce dependence on imported oil and slow global warming, **(b)** prevent the public from participating in hearings for licensing nuclear power plants and the Yucca Mountain high-level waste depository (Figure 19-24) and on safety issues at the nation's nuclear reactors, **(c)** restore government subsidies to develop a breeder nuclear fission reactor program, and **(d)** greatly increase federal subsidies for developing nuclear fusion.

6. If you had to choose, would you rather live next door to a coal-fired power plant or a nuclear plant? Explain.

7. Should the United States and other developed countries provide economic and technical aid for closing 35 poorly designed and poorly operated nuclear reactors in some republics of the former Soviet Union and in eastern Europe? Explain.

PROJECTS

1. Write a two-page scenario of what your life might be like without oil. Compare and discuss the scenarios developed by members of your class.

2. Make a concept map of this chapter's major ideas, using the section heads and subheads and the key terms (in boldface). Look on the website for this book for information about making concept maps.

INTERNET STUDY RESOURCES AND RESOURCES FOR FURTHER READING AND RESEARCH

The website for this book contains helpful study aids and many ideas for further reading and research. Log on to

http://www.brookscole.com/product/0534389872s

and click on the Chapter-by-Chapter area. Choose Chapter 19 and select a resource:

- "Flash Cards" allows you to test your mastery of the Terms and Concepts to Remember for this chapter.
- "Tutorial Quizzes" provides a multiple-choice practice quiz.
- "Student Guide to InfoTrac" will lead you to Critical Thinking Projects that use InfoTrac College Edition as a research tool.
- "References" lists the major books and articles consulted in writing this chapter.
- "Hypercontents" takes you to an extensive list of sites with news, research, and images related to individual sections of the chapter.

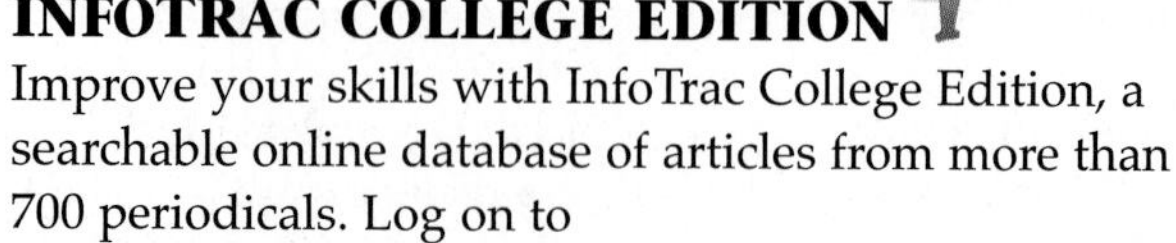

INFOTRAC COLLEGE EDITION

Improve your skills with InfoTrac College Edition, a searchable online database of articles from more than 700 periodicals. Log on to

http://www.infotrac-college.com

or access InfoTrac through the website for this book. Try to find the following articles:

1. Depletion and Denial: The Final Years of Oil Supplies. Colin J. Campbell. *USA Today (Magazine)*, Nov 2000 v129 i2666 p18. Oil production is approaching its peak, if it hasn't been reached already, and we need to plan for future energy sources. Do we realize this? *Hint*: Enter the search term "oil supplies" using the keyword "oil."

2. Stop the Nuclear Greenwash (clean development mechanism could become subsidy for the nuclear industry). *The Ecologist*, June 2000 v30 i4 p12. Enthusiasm for expanded use of nuclear power to counteract global warming is growing. There is opposition from environmentalists. *Hint*: Enter the search term "greenwash" using the keywords "nuclear energy."

20 ENERGY EFFICIENCY AND RENEWABLE ENERGY

The Coming Energy-Efficiency and Renewable-Energy Revolution

Energy experts Hunter and Amory Lovins (Guest Essay, p. 517) have built a large, passively heated, superinsulated, partially earth-sheltered home and office in Snowmass, Colorado (Figure 20-1), where winter temperatures can drop to $-40°C$ ($-40°F$).

This structure, which also houses the research center for the Rocky Mountain Institute, an office used by 40 people, **(1)** gets 99% of its space and water heating and 95% of its daytime lighting from the sun and **(2)** uses one-tenth the usual amount of electricity for a structure of its size.

Today's superinsulating windows mean that a house can have large numbers of windows without much heat loss in cold weather or heat gain in hot weather. Thinner insulation material now being developed will allow roofs and walls to be insulated far better than in today's best superinsulated houses.

A small but growing number of people in developed and developing countries are getting their electricity from *solar cells* that convert sunlight directly into electricity. They can be attached like shingles to a roof or applied to window glass as a coating. Solar-cell prices are high but are falling rapidly.

Many scientists and executives of oil and automobile companies believe we are in the beginning stages of a *solar–hydrogen revolution*. Electricity produced by large banks of solar cells or farms of wind turbines could be passed through water to make hydrogen gas (H_2), which could be used to fuel vehicles, industries, and buildings. Another solution is to burn hydrogen in energy-efficient *fuel cells* that produce electricity to **(1)** run cars and appliances, **(2)** heat water, and **(3)** heat and cool buildings.

Burning hydrogen produces **(1)** water vapor, **(2)** small amounts of controllable nitrogen oxides, and **(3)** no carbon dioxide. Thus, shifting to hydrogen as our primary energy resource during the 21st century would eliminate most of the world's air pollution and greatly slow global warming.

These are only a few of the components of the *energy-efficiency and renewable-energy revolution* that many analysts believe will help us make the transition to more sustainable societies over the next 40–50 years.

Figure 20-1 The Rocky Mountain Institute in Colorado. This facility is a home and a center for the study of energy efficiency and sustainable use of energy and other resources. It is also an example of energy-efficient passive solar design. (Robert Millman/Rocky Mountain Institute)

If the United States wants to save a lot of oil and money and increase national security, there are two simple ways to do it: Stop driving Petropigs and stop living in energy sieves.

AMORY B. LOVINS

This chapter addresses the following questions:

- What are the advantages and disadvantages of improving energy efficiency?
- What are the advantages and disadvantages of using solar energy to **(1)** heat buildings and water and **(2)** produce electricity?
- What are the advantages and disadvantages of using flowing water and solar energy stored as heat in water to produce electricity?
- What are the advantages and disadvantages of using wind to produce electricity?
- What are the advantages and disadvantages of burning plant material (biomass) to **(1)** heat buildings and water, **(2)** produce electricity, and **(3)** propel vehicles (biofuels)?
- What are the advantages and disadvantages of producing hydrogen gas and using it to **(1)** produce electricity, **(2)** heat buildings and water, and **(3)** propel vehicles?
- What are the advantages and disadvantages of extracting heat from the earth's interior (geothermal energy)?
- What are the advantages and disadvantages of using smaller, decentralized micropower sources to **(1)** heat buildings and water, **(2)** produce electricity, and **(3)** propel vehicles?
- How can we make a transition to a more sustainable energy future?

20-1 THE IMPORTANCE OF IMPROVING ENERGY EFFICIENCY

What Is Energy Efficiency? Doing More with Less **Energy efficiency** is the percentage of total energy input into an energy conversion device or system that **(1)** does useful work and **(2)** is not converted to low-quality, essentially useless heat. Improving the energy efficiency of a car motor, home heating system, or other energy conversion device involves using less energy to do more useful work.

You may be surprised to learn that *84% of all commercial energy used in the United States is wasted* (Figure 20-2). About 41% of this energy is wasted automatically because of the degradation of energy quality imposed by the second law of energy (p. 71). However, about 43% is wasted unnecessarily, mostly by **(1)** using fuel-

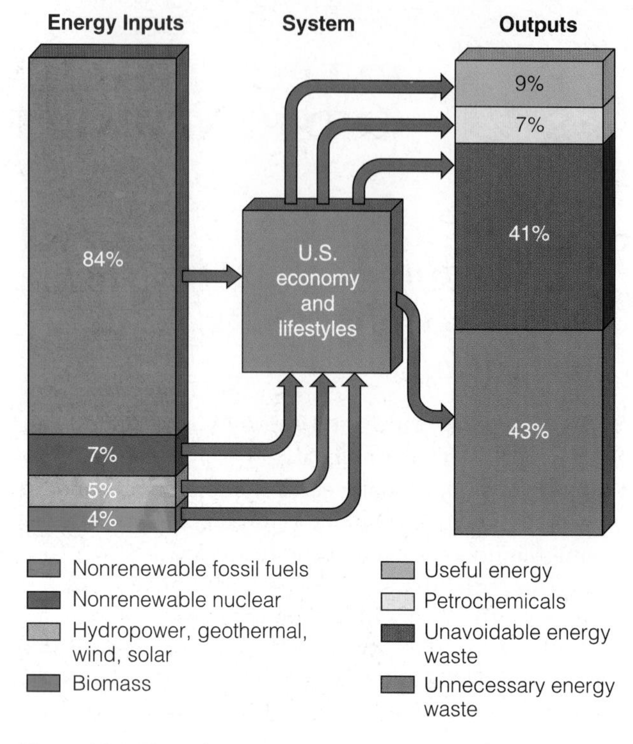

Figure 20-2 Flow of commercial energy through the U.S. economy. Note that only 16% of all commercial energy used in the United States ends up performing useful tasks or being converted to petrochemicals; the rest is either automatically and unavoidably wasted because of the second law of energy (41%) or wasted unnecessarily (43%).

wasting motor vehicles, furnaces, and other devices and **(2)** living and working in leaky, poorly insulated, poorly designed buildings (Guest Essay, p.517).

According to the U.S. Department of Energy (DOE), the United States unnecessarily wastes as much energy as two-thirds of the world's population consumes. Improvements in energy efficiency since the Organization of Petroleum Exporting Countries (OPEC) oil embargo in 1973 have cut U.S. energy bills by $275 billion a year. However, unnecessary energy waste still costs the United States about $300 billion per year (an average of $570,000 per minute)—more than the country's annual military budget. Reducing energy waste has a number of economic, military, and environmental advantages (Figure 20-3).

The energy conversion devices we use vary in their energy efficiencies (Figure 20-4). We can save energy and money by buying more energy-efficient cars, lighting, heating systems, water heaters, air conditioners, and appliances. Some energy-efficient models may cost more initially, but in the long run they usually save money by having a lower **life-cycle cost**: initial cost plus lifetime operating costs.

The net energy efficiency of the entire energy delivery process for a space heater, water heater, or car is determined by the efficiency of each step in the energy

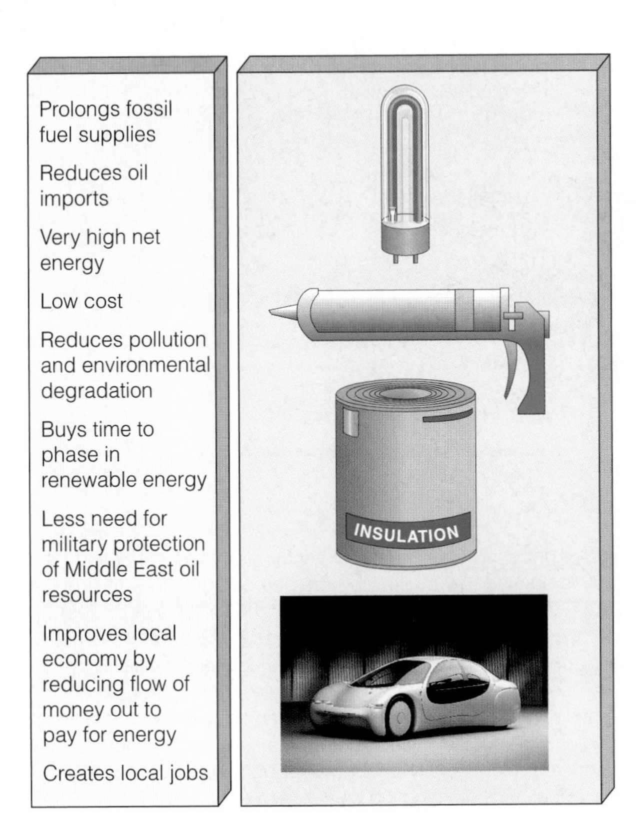

Figure 20-3 Advantages of reducing energy waste. Global improvements in energy efficiency could save the world about $1 trillion per year

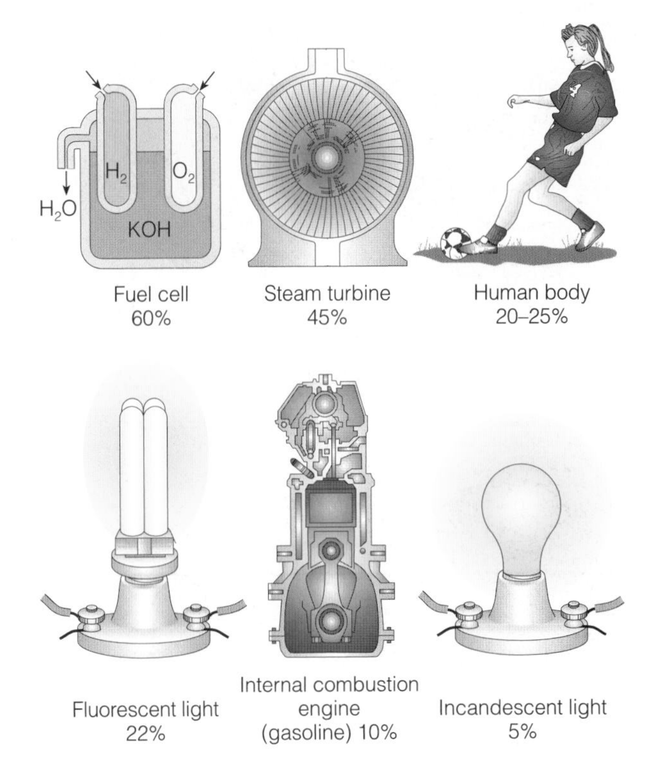

Figure 20-4 Energy efficiency of some common energy conversion devices.

conversion process. For example, the sequence of energy-using (and energy-wasting) steps involved in using electricity produced from fossil or nuclear fuels is

Extraction ⟶ Transportation ⟶ Processing ⟶
Transportation to power plant ⟶ Electric generation ⟶
Transmission ⟶ End use

Figure 20-5 shows the net energy efficiency for heating two well-insulated homes, **(1)** one with electricity produced at a nuclear power plant, transported by wire to the home, and converted to heat (electric resistance heating) and **(2)** the other heated passively, with an input of direct solar energy through high-efficiency windows facing the sun, with heat stored in heat-absorbing materials for slow release.

This analysis shows that the process of **(1)** converting the high-quality energy in nuclear fuel to high-quality heat at several thousand degrees in the power plant, **(2)** converting this heat to high-quality electricity, **(3)** transmitting the electricity to users, and **(4)** using the electricity to provide low-quality heat for warming a house to only about 20°C (68°F) is very wasteful of high-quality energy (Figure 3-9, p. 67). Burning coal or any fossil fuel at a power plant to supply electricity for heating water or space is also inefficient. It is much less wasteful to **(1)** collect solar energy from the environment, **(2)** store the resulting heat in heat-absorbing materials, and, **(3)** if necessary, use a small backup heating system to raise its temperature slightly to provide space heating or household hot water.

Figure 20-5 shows that one way to waste less energy (and money) is not using high-quality energy to do a job that can be done with lower-quality energy (Figure 3-9, p. 67). This helps explain why using electricity to heat a house (resistance heating) costs about **(1)** three times more per unit of energy than using a heat pump (useful in warm to moderate climates only) and **(2)** twice as much as heating with oil or with an efficient natural gas furnace.

Perhaps the three least efficient energy-using devices in widespread use today are **(1)** *incandescent light bulbs* (which waste 95% of the energy input), **(2)** *vehicles with internal combustion engines* (which waste 86–90% of the energy in their fuel), and **(3)** *nuclear power plants* producing electricity for space heating or water heating (which waste 86% of the energy in their nuclear fuel and probably 92% when the energy needed to deal with radioactive wastes and retired nuclear plants is included). Energy experts call for us to replace these devices or greatly improve their energy efficiency over the next few decades.

Coal-burning power plants also are big energy wasters. About 34% of the energy in coal burned in a typical electric power plant is used to produce

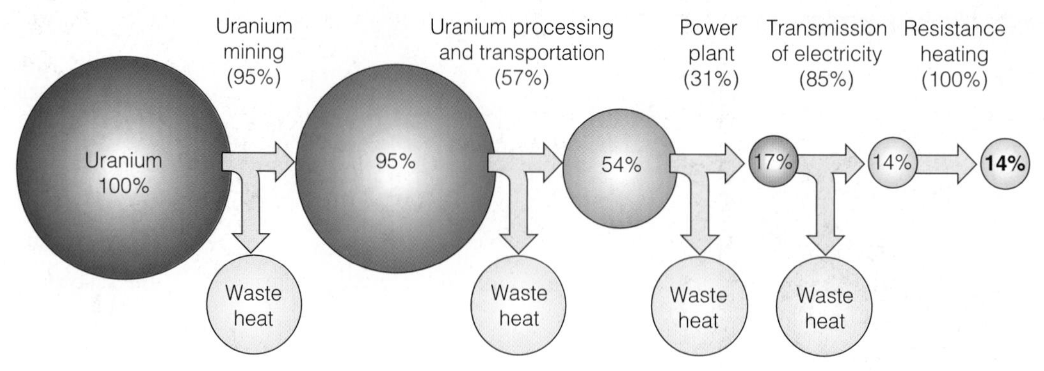

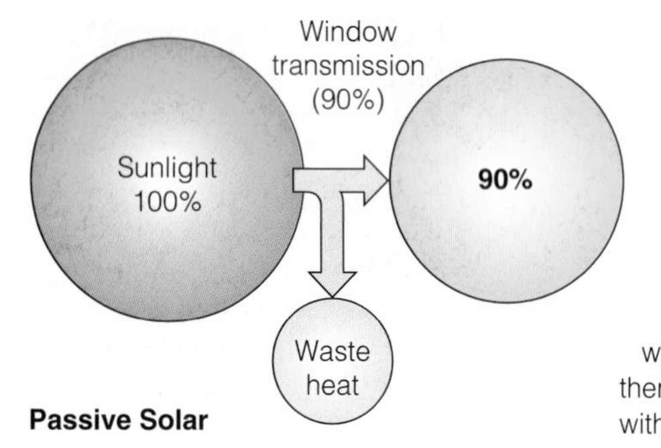

Figure 20-5 Comparison of net energy efficiency for two types of space heating. The cumulative net efficiency is obtained by multiplying the percentage shown inside the circle for each step by the energy efficiency for that step (shown in parentheses). Because of the second law of thermodynamics, in most cases the greater the number of steps in an energy conversion process, the lower its net energy efficiency. About 86% of the energy used to provide space heating by electricity produced at a nuclear power plant is wasted. If the additional energy needed to deal with nuclear wastes and to retire highly radioactive nuclear plants after their useful life is included, then the net energy yield for a nuclear plant is only about 8% (or 92% waste). By contrast, with passive solar heating, only about 10% of incoming solar energy is wasted.

electricity, and the remaining 66% ends up as waste heat that flows into the environment. As a result, U.S. coal-burning power plants throw away as much heat as all the energy used by Japan, the world's second largest economy.

20-2 WAYS TO IMPROVE ENERGY EFFICIENCY

How Can We Use Waste Heat? Could we save energy by recycling energy? No. The second law of energy tells us that we cannot recycle energy. However, we can slow the rate at which waste heat flows into the environment when high-quality energy is degraded.

For a house, the best way to do this is to **(1)** insulate it thoroughly, **(2)** eliminate air leaks (Figure 3-1, p. 54), and **(3)** equip it with an air-to-air heat exchanger to prevent buildup of indoor air pollutants.

In office buildings and stores, waste heat from lights, computers, and other machines can be collected and distributed to reduce heating bills during cold weather. During hot weather, the collected heat can be vented outdoors to reduce cooling bills.

How Can We Save Energy in Industry? Three important ways to save energy and money in industry are

- **Cogeneration**, or *combined heat and power (CHP)* systems, in which two useful forms of energy (such as steam and electricity) are produced from the same fuel source. These systems **(1)** have an efficiency of up to 80% (compared to about 30—40% for coal-fired boilers and nuclear power plants) and **(2)** emit two-thirds less carbon dioxide per unit of energy produced than conventional coal-fired boilers. Cogeneration has been widely used in Western Europe for years, and its use in the United States (now producing only 9% of the country's electricity) and China is growing. In Germany, small cogeneration units that run on natural gas or liquefied petroleum gas (LPG) supply restaurants, apartment buildings, and houses with all their energy. In 6–8 years, they pay for themselves in saved fuel and electricity.

- *Replacing energy-wasting electric motors.* Running electric motors (mostly in industry) consumes about half of all electricity produced in the United States. Most of these motors are inefficient because they run only at full speed, with their output throttled to match the task. Each year a heavily used electric motor consumes 10 times its purchase cost in electricity—equivalent to using $200,000 worth of gasoline each year to fuel a $20,000 car. The costs of replacing such motors with new adjustable-speed drive motors would **(1)** be paid back in about 1 year and **(2)** save

GUEST ESSAY

Technology Is the Answer (But What Was the Question?)

Amory B. Lovins

Physicist and energy consultant Amory B. Lovins is one of the world's most respected experts on energy strategy. In 1989, he received the Delphi Prize for environmental work; in 1990, the Wall Street Journal *named him one of the 39 people most likely to change the course of business in the 1990s. He is research director at Rocky Mountain Institute, a nonprofit resource policy center that he and Hunter Lovins, founded in Snowmass, Colorado, in 1982. He has served as a consultant to more than 200 utilities, private industries, and international organizations and to many national, state, and local governments. He is active in energy affairs in more than 35 countries and has published several hundred papers and a dozen books on energy strategies and policies.*

It is fashionable to suppose that we're running out of energy and ask how can we get more of it. However, the more important questions are **(1)** How much energy do we need? and **(2)** What are the cheapest and least environmentally harmful ways to meet these needs?

How much energy it takes to make steel, run a car, or keep ourselves comfortable in our houses depends on how cleverly we use energy. For example, it is now cheaper to double the efficiency of most industrial electric motor drive systems than to fuel existing power plants to make electricity. Just this one saving can more than replace the entire U.S. nuclear power program. We know how to make lights five times as efficient as those currently in use and how to make household appliances that give us the same work as now but use one-fifth as much energy (saving money in the process).

Ten automakers have made good-sized, peppy, safe prototype cars averaging 30–60 kilometers per liter (70–140 miles per gallon), and within a decade automakers could have cars getting 64–128 kpl (150–300 mpg) on the road if consumers demanded such cars. We know today how to make new buildings (and many old ones) so heat-tight (but still well ventilated) that they need essentially no outside energy to maintain comfort year-round, even in severe climates. In fact, I live and work in one [Figure 20-1].

These energy-saving measures are all cheaper than going out and getting more energy. However, the old view of the energy problem included a worse mistake than forgetting to ask how much energy we needed: It sought more energy, in any form, from any source, at any price, as if all kinds of energy were alike.

Just as there are different kinds of food, so there are many different forms of energy whose different prices and qualities suit them to different uses [Figure 3-9, p. 67]. After all, there is no demand for energy as such; nobody wants raw kilowatt-hours or barrels of sticky black goo. People instead want energy services: comfort, light, mobility, hot showers, cold beverages, and the ability to cook food and make cement. In developing energy resources we should start by asking, "What tasks do we want energy for, and what amount, type, and source of energy will do each task most cheaply?"

Electricity is a high-quality, expensive form of energy. An average kilowatt-hour delivered in the United States in 1999 was priced at about 6.7¢, equivalent to buying the heat content of oil costing $111 per barrel—almost four times the average world price of crude oil in 2000. No new nuclear plants are being built in the United States (and only a few are being built in other parts of the world). One reason is that the average cost of electricity from the last nuclear power plant built in the United States is about 13.5¢ per kilowatt-hour, equivalent on a heat basis to buying oil at about $216 per barrel.

Such costly energy might be worthwhile if it were used only for the premium tasks that need it, such as lights, motors, electronics, and smelters. However, those special uses—only 8% of all delivered U.S. energy needs—are already met twice over by today's power stations. Two-fifths of the electricity used in the United States is for uneconomic, low-grade uses such as water heating, space heating, and air conditioning. Yet no matter how efficiently we use electricity (even with heat pumps), we can never get our money's worth on these applications.

Thus, *supplying more electricity is irrelevant to the energy problem we have.* Even though electricity accounts for almost all the federal energy research-and-development budget and at least half the national energy investment, it is the wrong kind of energy to meet the nation's needs economically. Arguing about what kind of new power station to build—coal, nuclear, or solar—is like shopping for the best buy in antique Chippendale chairs to burn in your stove or for expensive brandy to put in your car's gas tank.

The real question is, "What is the cheapest way to do low-temperature heating and cooling?" The answer is weather-stripping, insulation, heat exchangers, greenhouses, superwindows (which have as much insulating value as the outside wall of a typical house), roof overhangs, trees, and so on. These measures generally cost about 0.5–2¢ per kilowatt-hour, the lowest-cost way by far to supply energy.

If we need more electricity, we should get it from the cheapest sources first. In approximate order of increasing price, these include

- Converting to efficient lighting equipment. This would save the United States electricity equal to the output of 120 large power plants, plus $30 billion a year in fuel and maintenance costs.
- Using more efficient electric motors to save up to half the energy used by motor systems. This would save

(continued)

GUEST ESSAY (continued)

electricity equal to the output of another 150 large power plants and repay the cost in about a year.

- Displacing the electricity now used for water heating and for space heating and cooling with good architecture, weatherization, insulation, and mostly passive solar techniques.
- Improving the energy efficiency of appliances, smelters, and the like.

Just these four measures can quadruple U.S. electrical efficiency, making it possible to run today's economy with no changes in lifestyles and using no power plants, whether old or new or fueled with oil, gas, coal, uranium, or solar energy. We would need only the present hydroelectric capacity, readily available small-scale hydroelectric projects, and a modest amount of wind power.

If we still wanted more electricity, the next cheapest sources would include **(1)** cogenerating electricity and heat in industrial plants and power plants, **(2)** using low-temperature heat engines run by industrial waste heat or by solar ponds, **(3)** filling empty turbine bays and upgrading equipment in existing big dams, **(4)** using modern wind machines or small-scale hydroelectric turbines in good sites, **(5)** using combined-cycle natural gas turbines, and perhaps **(6)** using recently developed more efficient solar cells when their price is reduced by mass production.

It is only after we have exhausted all these cheaper opportunities that we would even consider building a new central power station of any kind—the slowest and costliest known way to get more electricity (or to save oil).

To emphasize the importance of starting with energy end uses rather than energy sources, consider a story from France. In the mid-1970s, energy conservation planners in the French government found that their biggest need for energy was to heat buildings and that even with good heat pumps, electricity would be the costliest way to do this. So they had a fight with their government-owned and -run utility company; they won, and electric heating was supposed to be discouraged or even phased out because it was so wasteful of money and fuel.

Meanwhile, down the street, the energy supply planners (who were far more numerous and influential in the French government) said, "Look at all that nasty imported oil coming into our country. We must replace that oil with some other source of energy. Voilà! Nuclear reactors can give us energy, so we'll build them all over the country." However, they paid little attention to who would use that extra energy and no attention to relative prices.

Thus, these two groups of the French energy establishment went on with their respective solutions to two different, indeed contradictory, French energy problems: *more energy of any kind* versus *the right kind to do each task in the most inexpensive way.* It was only in 1979 that these conflicting perceptions collided. The supply-side planners suddenly realized that the only thing they would be able to sell all that nuclear electricity for would be electric heating, which they had just agreed not to do.

Every industrial country is in this embarrassing position. Supply-oriented planners think the problem boils down to whether to build coal or nuclear power stations (or both). Energy-use planners realize that *no* kind of new power station can be an economic way to meet the needs for using electricity to provide low- and high-temperature heat and for the vehicular liquid fuels that are 92% of our energy problem.

So if we want to provide energy services at the lowest cost, we need to begin by determining what we need the energy for!

Critical Thinking

1. The author argues that building more nuclear, coal, or other electrical power plants to supply electricity for the United States is unnecessary and wasteful of energy and money. List your reasons for agreeing or disagreeing with this viewpoint.

2. Explain why you agree or disagree that increasing the supply of energy, instead of improving energy efficiency, is the wrong answer to our energy problems.

---

an amount of energy equal to that generated by 150 large (1,000-megawatt) power plants.

- *Switching to high-efficiency lighting* (Guest Essay, p. 517).

How Can We Save Energy in Transportation? According to most energy analysts, the best way to save energy (especially oil) and money in transportation is to *increase the fuel efficiency of motor vehicles.*

Between 1973 and 1985, the average fuel efficiency doubled for new American cars and rose 37% for all passenger cars on the road because of government-mandated standards, called the Corporate Average Fuel Economy (CAFE) standards. However, between 1985 and 2000 the average fuel efficiency of new vehicles in the United States fell by about 8% because of the popularity of **(1)** sport utility vehicles (SUVs), minivans, and light trucks (subject to much lower mileage standards than cars) and **(2)** larger, less efficient autos. According to the EPA, increasing average fuel economy by 1.3 kpl (3 mpg) would **(1)** save $25 billion a year in fuel costs, **(2)** reduce CO_2 emissions, and **(3)** save 1 million barrels of oil per day.

According to a 2001 study by the American Council for an Energy-Efficient Economy, increasing the fuel economy of new cars, light trucks, SUVs, and minivans by just 5% a year for 10 years would **(1)** save 10–20 times more oil than the projected supply from the Arctic National Wildlife Refuge (Pro/Con, p. 500) and **(2)** more than three times the oil in the nation's current proven oil reserves.

However, automakers have successfully opposed any increase in the CAFE standards, and they are lobbying Congress to eliminate them.

Since 1985, at least 10 automobile companies have developed prototype cars with fuel efficiencies of 30–60 kilometers per liter (70–140 miles per gallon). These cars **(1)** are manufactured with light and strong materials, **(2)** meet current safety and air pollution standards, **(3)** carry four or five passengers, and **(4)** accelerate as rapidly as most current models.

If such cars were mass-produced, their slightly higher costs would be more than offset by their fuel savings. The problem is that there is little consumer interest in fuel-efficient cars mostly because **(1)** the inflation-adjusted price of gasoline today in the United States is low despite increases in gasoline prices in 2000 (Figure 20-6), and **(2)** two-thirds of consumers prefer SUVs and other large, inefficient vehicles.

Are Hybrid and Fuel-Cell Cars the Answer? There is growing interest in developing *superefficient cars* that could eventually get 34–128 kilometers per liter (80–300 miles per gallon). One type of highly efficient car uses **(1)** a small *hybrid electric internal combustion engine* that runs on gasoline or some other liquid fuel and **(2)** a small battery (recharged by the internal combustion engine) to provide the energy needed for acceleration and hill climbing (Figure 20-7). In 1999, Toyota and Honda began selling the first generation of fuel-efficient hybrid engine cars. Toyota has developed a 17-kilometer-per-liter (40-mile-per-gallon) hybrid minivan. General Motors plans to offer hybrid electric engines on many of its sedans, hatchbacks, and SUVs by 2004.

Another type of superefficient car is an electric vehicle that uses *fuel cells* (Figure 20-8). Such cars are 30–50% efficient (compared to 10–14% efficiency for gasoline-powered vehicles) and emit about 20% less CO_2 per kilometer than conventional cars. As a result,

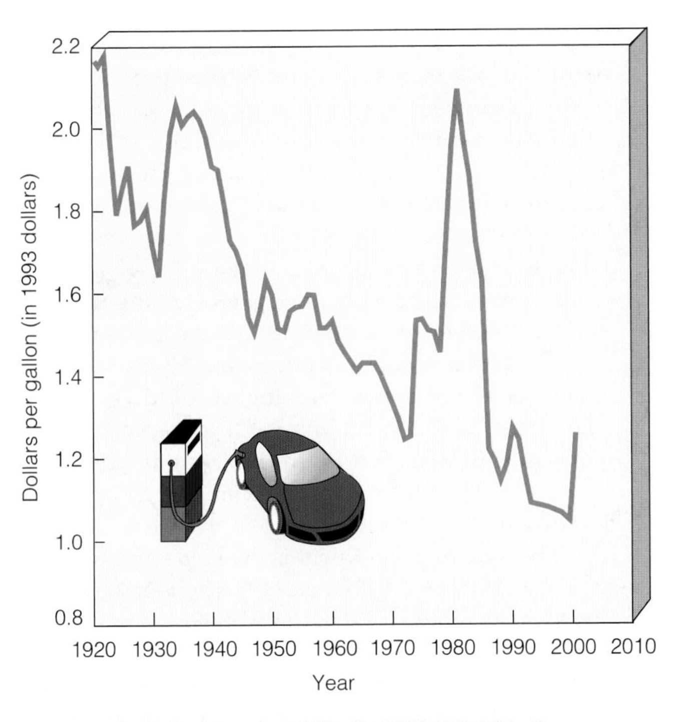

2.2
2.0
1.8
1.6
1.4
1.2
1.0
0.8
Dollars per gallon (in 1993 dollars)
1920
1930
1940
1950
1960
1970
1980
1990
2000
2010
Year

Figure 20-6 Real price of gasoline (in 1993 dollars) in the United States, 1920–2000. The 212 million motor vehicles in the United States use about 40% of the world's gasoline. (U.S. Department of Energy)

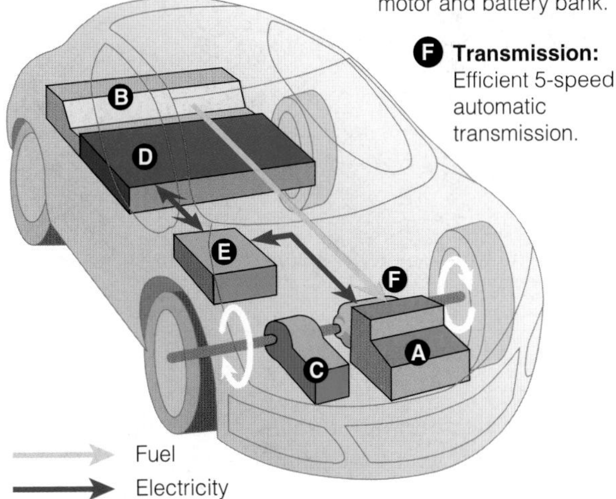

A Combustion engine: Small, efficient internal combustion engine powers vehicle with low emissions.
B Fuel tank: Liquid fuel such as gasoline, diesel, or ethanol runs small combustion engine.
C Electric motor: Traction drive provides additional power, recovers braking energy to recharge battery.
D Battery bank: High-density batteries power electric motor for increased power.
E Regulator: Controls flow of power between electric motor and battery bank.
F Transmission: Efficient 5-speed automatic transmission.
Fuel
Electricity

Figure 20-7 General features of a car powered by a *hybrid gas–electric engine.* A small internal combustion engine recharges the batteries, thus reducing the need for heavy banks of batteries and solving the problem of the limited range of conventional electric cars. The bodies of future models of such cars probably will be made of lightweight composite plastics that **(1)** offer more protection in crashes, **(2)** do not need to be painted, **(3)** do not rust, **(4)** can be recycled, and **(5)** have fewer parts than conventional cars. (Concept information from DaimlerChrysler, Ford, Honda, and Toyota)

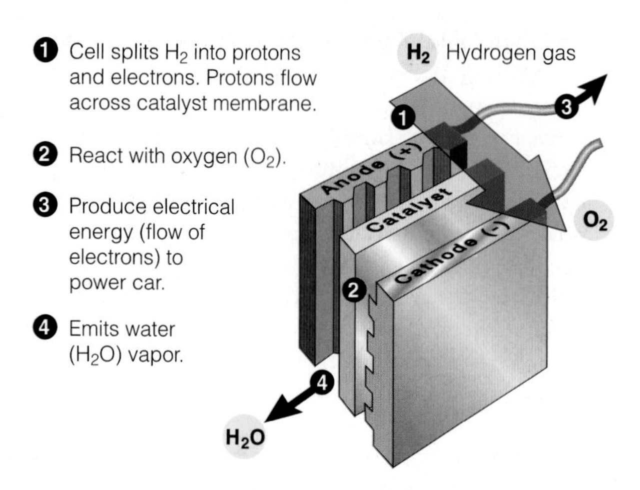

Figure 20-8 General features of an electric car powered by a *fuel cell* running on hydrogen gas. Fuel cells are similar to a battery with a hydrogen fuel tank. As long as there is some form of hydrogen available as a fuel, the cell can produce electricity. Such cars **(1)** will be almost pollution-free, emitting only water vapor, small amounts of nitrogen oxides, and with some cells CO_2 and **(2)** should get at least twice the mileage of comparable gasoline-powered cars. Several automobile companies have developed prototypes and are working to get costs down and improve hydrogen fuel storage systems. Early models could be on the road by 2004. (Concept information from DaimlerChrysler, Ford, Ballard, Toyota, and Honda)

A Fuel cell stack: Hydrogen and oxygen combine chemically to produce electricity.

B Fuel tank: Hydrogen gas or liquid or solid metal hydride stored on board or made from gasoline or methanol.

C Turbo compressor: Sends pressurized air to fuel cell.

D Traction inverter: Module converts DC electricity from fuel cell to AC for use in electric motor.

E Electric motor / transaxle: Converts electrical energy to mechanical energy to turn wheels.

Fuel

Electricity

fuel-cell cars running on hydrogen should get 37–47 kilometers per liter (87–108 miles per gallon).

Most major automobile companies have developed prototype fuel-cell cars and hope to begin marketing them by 2004 using two different approaches. Some will be fueled by methanol (CH_3OH), a liquid usually produced from natural gas. A *liquid reformer system* in the engine extracts the hydrogen from methanol for use in the fuel cell. Drivers would fill up with methanol at pumps in conventional filing stations. A new process developed in 2000 allows fuel cells to produce hydrogen directly from hydrocarbons such as natural gas, gasoline, or diesel fuel. Instead of using complicated liquid fuel reformers, General Motors and Honda use a chemical process to store hydrogen fuel as a *solid metal hydride* compound that can be heated as needed to provide H_2. Toyota plans to begin selling a commercial fuel-cell car in Japan by 2003.

How Can Electric Bicycles Reduce Energy Use and Waste? For urban trips, some people may begin using *electric bicycles,* now being sold by several companies. These bicycles are powered by a small electric motor and cost about $1,100. They **(1)** travel at up to 30 kilometers per hour (18 miles per hour), **(2)** go about 48 kilometers (30 miles) without pedaling on a full electric charge, and **(3)** produce no pollution during operation (and only a small amount for the electricity used in recharging them).

How Can We Save Energy in Buildings? Atlanta's 13-story Georgia Power Company building uses 60% less energy than conventional office buildings of the same size. The largest surface of the building faces south to capture solar energy. Each floor extends out over the one below it, blocking out the higher summer sun to reduce air conditioning costs but allowing warming by the lower winter sun. Energy-efficient lights focus on desks rather than illuminating entire rooms. The Georgia Power model and other existing cost-effective commercial building technologies could **(1)** reduce energy use by 75% in buildings, **(2)** cut carbon dioxide emissions in half, and **(3)** in the United States save more than $130 billion per year in energy bills—an average of $15 million an hour.

The federal government is the largest energy user in the United States, with some 50,000 buildings. According to a 2001 study by five national energy labs, the government could save $1 billion a year by installing currently available energy-efficient technologies such as compact fluorescent lights and energy-efficient appliances. The estimated $5.2-billion investment needed to do this would be paid back in savings within about 5 years, an attractive rate of return.

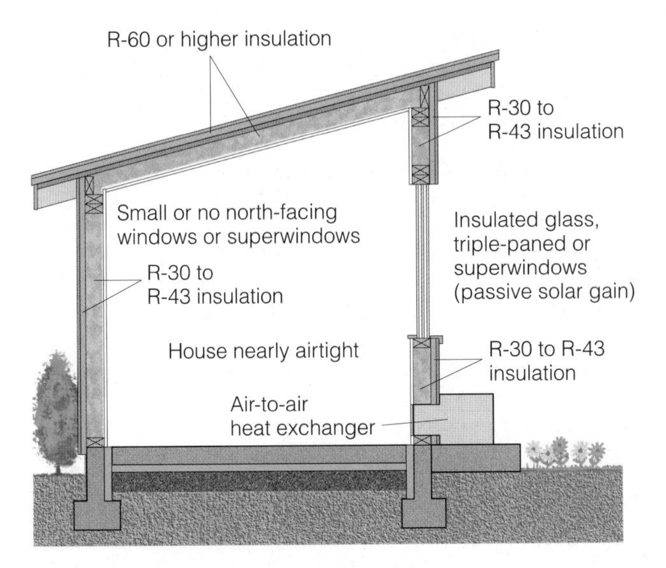

Figure 20-9 Major features of a superinsulated house. Such a house is so heavily insulated and so airtight that it can be warmed by heat from direct sunlight, appliances, and human bodies, with little or no need for a backup heating system. An air-to-air heat exchanger prevents buildup of indoor air pollution.

There are a number of ways to improve the energy efficiency of buildings, some of them discussed in the opening of this chapter (p. 513). One is to build more *superinsulated houses* (Figure 20-9). Such houses typically cost 5% more to build than conventional houses of the same size. However, this extra cost is paid back by energy savings within about 5 years and can save a homeowner $50,000–100,000 over a 40-year period.

Since the mid-1980s there has been growing interest in building superinsulated houses called *strawbale houses* with walls consisting of compacted bales of certain types of straw (available at a low cost almost everywhere) covered with plaster or adobe. By 2000, there were more than 1,200 such homes built or under construction in the United States (Guest Essay, p. 531). Using straw, an *annually* renewable agricultural residue often burned as a waste product, for the walls reduces the need for wood and thus slows deforestation. The main problem is getting banks and other moneylenders to recognize the potential of this and other unconventional types of housing and provide homeowners with construction loans.

Another way to save energy is to *use the most energy-efficient ways to heat houses* (Figure 20-10). The most energy-efficient ways to heat space are **(1)** a superinsulated house, **(2)** passive solar heating, **(3)** heat pumps in warm climates (but not in cold climates because at low temperatures they automatically switch to costly electric resistance heating), and **(4)** a high-efficiency (85–98%) natural gas furnace. The most wasteful and expensive way is to use electric resistance heating with the electricity produced by a coal-fired or nuclear power plant.

The energy efficiency of existing houses and buildings can be improved significantly by **(1)** *adding insulation,* **(2)** *plugging leaks, and* **(3)** *installing energy-saving windows and lighting.* About one-third of heated air in U.S. homes and buildings escapes through closed windows and holes and cracks (Figure 3-1, p. 54)—equal to the energy in all the oil flowing through the Alaska pipeline every year. During hot weather these windows and cracks also let heat in, increasing the use of air conditioning.

Replacing all windows in the United States with low-E (low-emissivity) windows would cut these expensive losses by two-thirds and reduce CO_2 emissions. Widely available superinsulating windows insulate as well as 8–12 sheets of glass. Although they cost 10–15% more than double-glazed windows, this cost is paid back rapidly by the energy they save. Even better windows will reach the market soon.

Simply wrapping a water heater in a $20 insulating jacket can save a homeowner $45 a year and reduce CO_2 emissions. Leaky ducts allow 20–30% of a home's heating and cooling energy to escape and draw unwanted moisture and heat into the home. Careful sealing can reduce this loss. Some designs for new homes keep the ducts inside the home's thermal envelope so that escaping hot or cool air leaks into the living space.

An energy-efficient way to heat hot water for washing and bathing is to use tankless instant water heaters (about the size of bookcase loudspeakers) fired by natural gas or LPG (Guest Essay, p. 531). These devices, widely used in many parts of Europe, heat the water instantly as it flows through a small burner chamber and provide hot water only when it is needed.* A well-insulated, conventional natural gas or LPG water heater is fairly efficient. However, all conventional natural gas and electric resistance heaters waste energy by keeping a large tank of water hot all day and night and can run out after a long shower or two.

Using electricity produced by any type of power plant is the most inefficient and expensive way to heat water for washing and bathing. A $425 electric water heater can cost $5,900 in energy over its 20-year life, compared to about $4,000 for a comparable natural gas water heater over the same period.

Setting higher energy-efficiency standards for new buildings would also save energy. Building codes could require that all new houses use 60–80% less energy than conventional houses of the same size, as has been done in Davis, California. Because of tough energy-efficiency standards, the average Swedish home consumes about

*They work great. I used them in a passively heated and cooled office and home for 15 years.

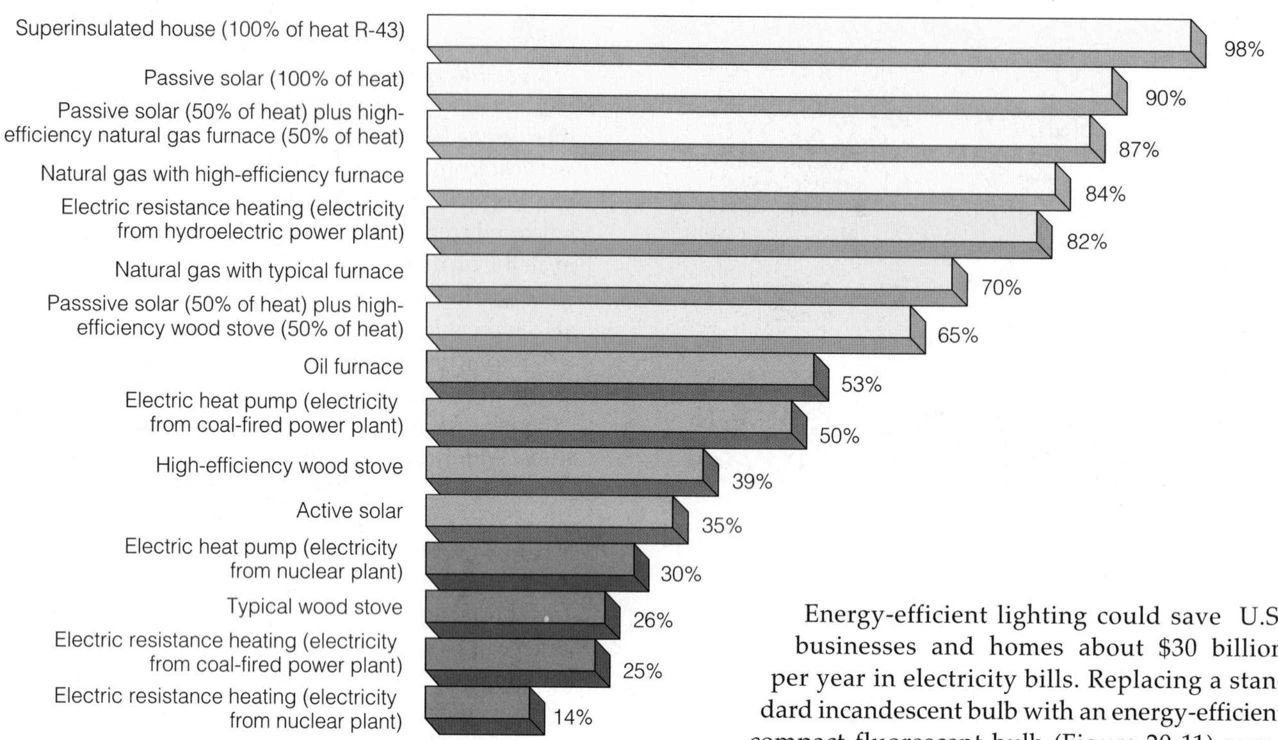

Figure 20-10 Net energy efficiencies for various ways to heat an enclosed space such as a house. (Data from Howard T. Odum)

one-third as much energy as the average American home of the same size.

Another way to save energy is to *buy the most energy-efficient appliances and lights.** Federal energy efficiency standards set for more than 20 appliances-including a tripling of refrigerator efficiency—since 1978 **(1)** have saved more than $25 billion in energy costs in the United States and **(2)** are expected to save U.S. consumers another $46 billion between 2000 and 2020.

In 1992, the EPA and DOE introduced the Energy Star program as a voluntary effort to identify and promote energy-efficient appliances, homes and buildings, office equipment, lighting, computers, and consumer electronics. Between 1994 and 2000 this program saved enough energy to light 213 million homes year round. If all consumers and business followed the voluntary Energy Star standards over the next decade, the national annual energy bill would be reduced by $200 billion.

*Each year the American Council for an Energy-Efficient Economy (ACEEE) publishes a list of the most energy-efficient major appliances mass-produced for the U.S. market. A copy can be obtained from the council at 1001 Connecticut Ave. NW, Suite 801, Washington, DC 20036. Each year they also publish *A Consumer Guide to Home Energy Savings*, available in bookstores or from the ACEEE.

Energy-efficient lighting could save U.S. businesses and homes about $30 billion per year in electricity bills. Replacing a standard incandescent bulb with an energy-efficient compact fluorescent bulb (Figure 20-11) saves about $48–70 per bulb over its 10-year life. Thus, replacing 25 incandescent bulbs in a house or building with energy-efficient fluorescent bulbs saves $1,250–1,750. Students in Brown University's environmental studies program showed that the school could save more than $40,000 per year just by replacing the incandescent light bulbs in exit signs with compact fluorescent bulbs.

Despite their advantages, less than 10% of U.S. homes use these energy- and money-saving bulbs, which pay for themselves in 2–4 years. Studies show that such low use has resulted from **(1)** the high initial cost of the bulbs, **(2)** lack of information and education about life-cycle costing, **(3)** initial lack of suitable light fixtures for these larger bulbs (light fixtures and smaller bulbs are now available), and **(4)** dissatisfaction with the color and intensity of the light produced compared with incandescent bulbs (corrected with newer bulbs).

If all U.S. households used the most efficient frost-free refrigerator now available, 18 large (1,000-megawatt) power plants could close. Microwave ovens can cut electricity use for cooking by 25–50% (but not if used for defrosting food). Clothes dryers with moisture sensors cut energy use by 15%, and front-loading washers use 50% less energy than top-loading models but cost about the same. Increased use of the internet for business and shopping transactions reduces energy use and decreases emissions of carbon dioxide and other air pollutants (Connections, p. 530).

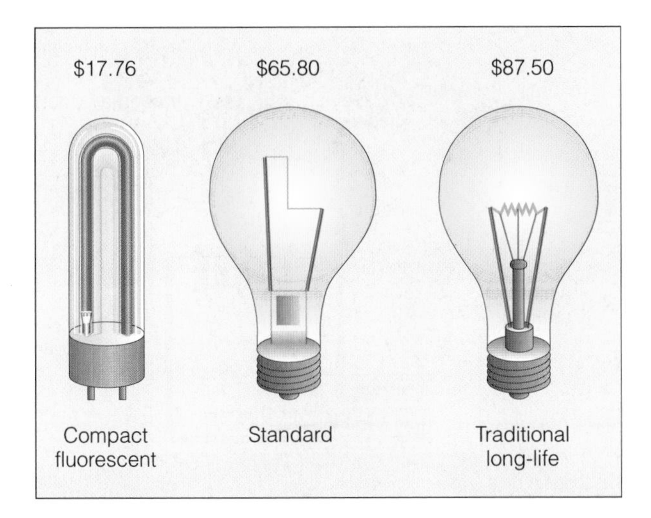

Figure 20-11 Cost of electricity for comparable light bulbs used for 10,000 hours. Because conventional incandescent bulbs are only 5% efficient and last only 1,500 hours, they waste enormous amounts of energy and money and add to the heat load of houses during hot weather. Socket-type fluorescent lights use one-fourth as much electricity as conventional bulbs. Although these bulbs cost $6–15 per bulb, they last up to 100,000 hours (60–70 times longer than conventional incandescent bulbs and 25 times longer than halogen bulbs), saving a lot of money (compared with less efficient incandescent and halogen bulbs) over their long life. Recently, smaller compact fluorescent bulbs (costing about $6) that fit into ordinary light fixtures and ones that can be dimmed have been developed. (Data from Electric Power Research Institute)

Why Aren't We Doing More to Reduce Energy Waste? With such an impressive array of benefits (Figure 20-3), why isn't there more emphasis on improving energy efficiency? The major reasons are as follows:

- A glut of low-cost fossil fuels (Figure 19-9, p. 497, and Figure 20-6). As long as energy is cheap, people are more likely to waste it and not make investments in improving energy efficiency.
- Lack of sufficient tax breaks and other economic incentives for consumers and businesses to invest in improving energy efficiency.
- Lack of information about **(1)** the availability of energy-saving devices and **(2)** the amount of money such items can save consumers by using *life-cycle cost* analysis.

20-3 USING SOLAR ENERGY TO PROVIDE HEAT AND ELECTRICITY

What Are the Major Advantages and Disadvantages of Solar Energy? According to Royal Dutch/Shell International Petroleum, **(1)** 50% of the energy needs of industrialized countries could be met by natural gas and renewable energy by 2020, and **(2)** renewables alone could account for 50% of world energy production by 2050.

Figure 20-12 lists some of the advantages and disadvantages of making a shift to greatly increased use of direct solar energy and indirect forms of solar energy such as wind. Like fossil fuels and nuclear power (Chapter 19), each renewable energy alternative has a mix of advantages and disadvantages, as discussed in the remainder of this chapter.

How Can We Use Solar Energy to Heat Houses and Water? Buildings and water can be heated by solar energy using two methods: passive and active (Figure 20-13). A **passive solar heating system** absorbs and stores heat from the sun directly within a structure (Figures 20-1, 20-13 (left), and 20-14 and Guest Essay, p. 517). Energy-efficient windows, greenhouses, and

Advantages

- Save money (wind)
- Reduce air pollution (99% less than coal)
- Greatly reduce CO_2 emissions
- Reduce dependence on imported oil
- Last as long as coal and nuclear plants (30–40 years)
- Land use less than for coal
- Low land use with new solar cell and window glass system
- Backup and storage devices available (such as gas turbines, batteries, and flywheels)
- Backup need reduced by distributing and storing solar-produced hydrogen gas

Disadvantages

- Making solar cells produces toxic chemicals
- Solar systems last only 30–40 years
- Take large amounts of land because of diffuse nature of sunlight
- Can damage fragile desert ecosystems used to collect solar energy
- Need backup systems at night and during cloudy and rainy weather

Figure 20-12 Major advantages and disadvantages of using direct and indirect solar energy systems to produce heat and electricity. Specific advantages and disadvantages of different direct and indirect solar and other renewable energy systems are discussed in this chapter.

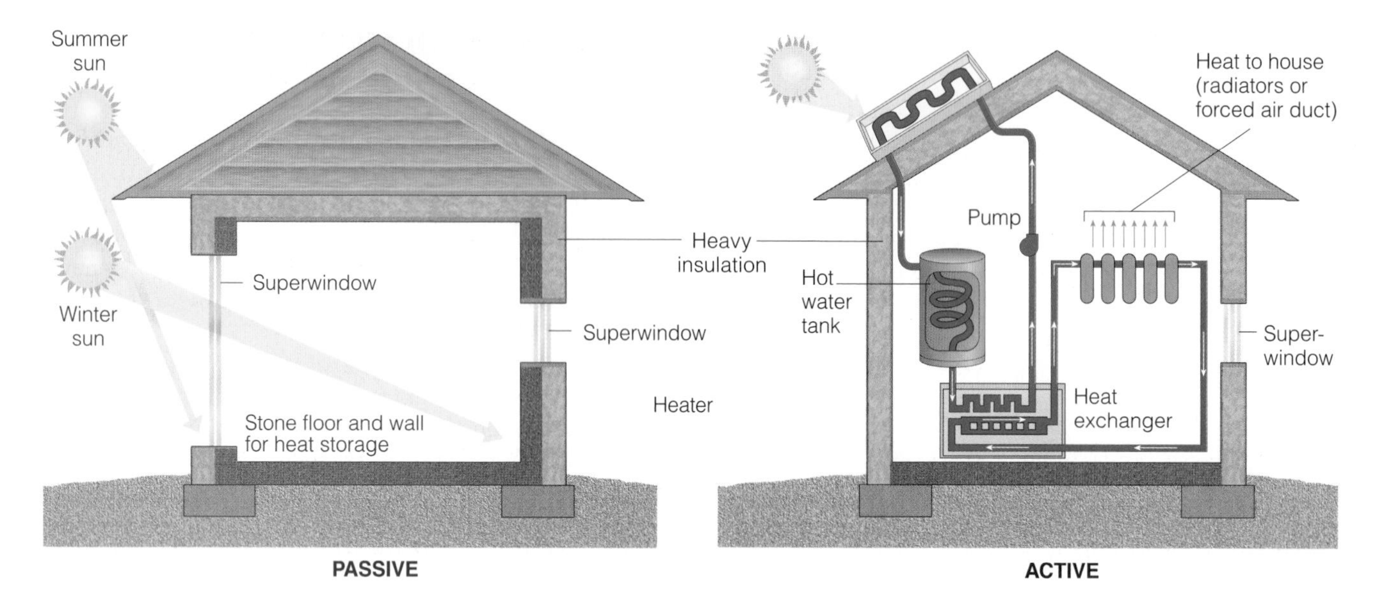

Figure 20-13 Passive and active solar heating for a home.

sunspaces face the sun to collect solar energy by direct gain. Walls and floors of concrete, adobe, brick, stone, salt-treated timber, and water in 55-gallon drums store much of the collected solar energy as heat and release it slowly throughout the day and night. A small backup heating system such as a vented natural gas or propane heater may be used but is not necessary in many climates.

On a life-cycle cost basis, good passive solar and superinsulated design is the cheapest way to heat a home or small building in regions where ample sunlight is available during daytime. Such a system usually adds 5–10% to the construction cost, but the life cycle cost of operating such a house is 30–40% lower. The typical payback time for passive solar features is 3–7 years.

In an **active solar heating system**, collectors absorb solar energy, and a fan or a pump supplies part of a building's space-heating or water-heating needs (Figure 20-13, right). Several connected collectors usually are mounted on the roof with an unobstructed exposure to the sun. Some of the heat can be used directly, and the rest can be stored in insulated tanks containing rocks, water, or a heat-absorbing chemical for release as needed. Active solar collectors can also supply hot water. Most analysts do not expect widespread use of active solar collectors for heating houses because of high costs, maintenance, and unappealing appearance.

Figure 20-15 lists the major advantages and disadvantages of using passive or active solar energy for heating buildings. Passive solar cannot be used to heat existing homes and buildings **(1)** not oriented to receive sunlight and **(2)** whose access to sunlight is blocked by other buildings and structures.

How Can We Cool Houses Naturally? Ways to make a building cooler include

- Using superinsulation and superinsulating windows (Figure 20-9, p. 521)
- Blocking the high summer sun with deciduous trees, window overhangs, or awnings (Figure 20-14, top left)
- Using windows and fans to take advantage of breezes and keep air moving
- Suspending reflective insulating foil in an attic to block heat from radiating down into the house
- Placing plastic *earth tubes* 3–6 meters (10–20 feet) underground where the earth is cool year-round and using a tiny fan to pipe cool and partially dehumidified air into an energy-efficient house (Figure 20-14, top left)*
- Using solar-powered evaporative air conditioners (which work well only in dry climates and cost too much for residential use)

How Can We Use Solar Energy to Generate High-Temperature Heat and Electricity? Several so-called *solar thermal systems* collect and transform radiant energy from the sun into high-temperature thermal energy (heat), which can be used directly or converted to electricity (Figure 20-16). In one such *central receiver system*, called a *power tower*, huge arrays of computer-controlled mirrors called *heliostats* track the sun and focus sunlight on a central heat collection tower (Figure 20-16a).

*They work. I used them in a passively heated and cooled office and home for 15 years. People allergic to pollen and molds should add an air purification system, but this is also necessary with a conventional cooling system.

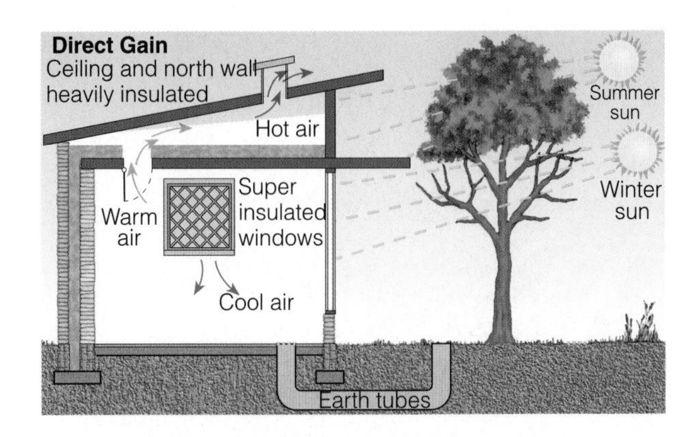

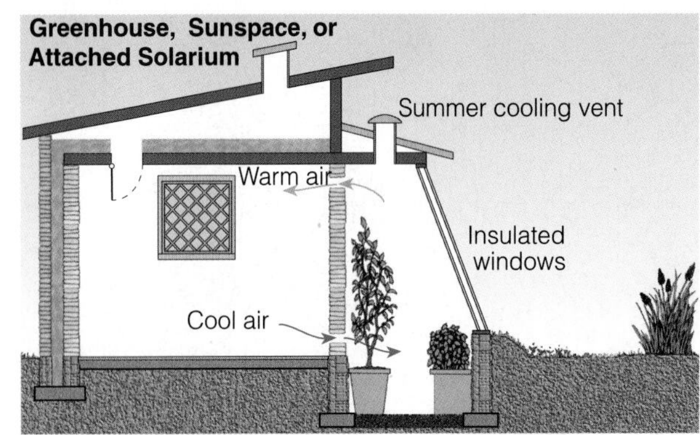

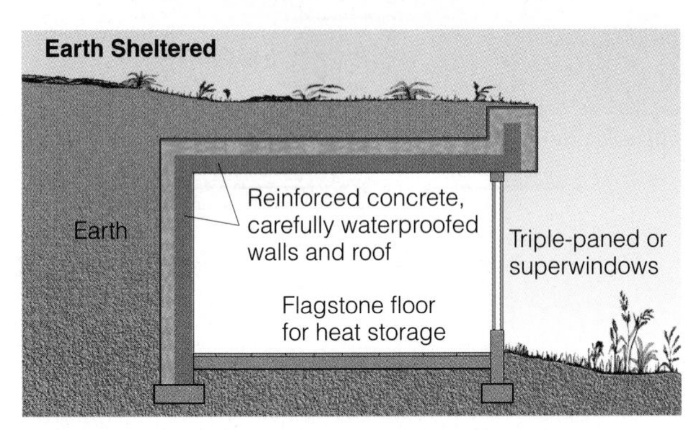

Figure 20-14 Three examples of passive solar design for houses.

A government-subsidized power tower system, called Solar Two, began operating in the California desert in 1996. However, this experimental plant **(1)** cost about eight times more to build than a coal-fired plant, **(2)** produced electricity at about twice the cost of a coal-fired plant, and **(3)** was shut down in 1999.

In a *solar thermal plant* or *distributed receiver system*, sunlight is collected and focused on oil-filled pipes running through the middle of curved solar collectors (Figure 20-16b). This concentrated sunlight can generate temperatures high enough for industrial processes or for producing steam to run turbines and generate electricity. At night or on cloudy days, high-efficiency combined-cycle natural gas turbines can supply backup electricity as needed. In California's Mojave Desert, such a solar thermal system with a natural gas turbine backup system produced power much more cheaply than nuclear power plants. However, the company went bankrupt, partly because of a lack of tax breaks equal to those for fossil fuel and nuclear power plants.

Another type of distributed receiver system uses *parabolic dish collectors* (which look somewhat like TV satellite dishes) instead of parabolic troughs. These collectors can track the sun along two axes and generally are more efficient than troughs. A pilot plant is being built in northern Australia. The DOE projects that within 10–20 years parabolic dishes with a natural gas turbine backup should be able to produce electrical power costing about the same as that from coal-burning plants.

Another approach for intensifying incoming solar energy about 80,000 times is a *nonimaging optical solar concentrator*. With this technology, the sun's rays are allowed to scramble instead of being focused on a particular point (Figure 20-16c). Because of their high efficiency and ability to generate extremely high temperatures, nonimaging concentrators may make solar energy practical for widespread industrial and commercial use within 10–20 years.

Inexpensive solar cookers can focus and concentrate sunlight and cook food, especially in rural villages in sunny developing countries. They can be made by fitting an insulated box big enough to hold three or four pots with a transparent, removable top (Figure 20-16d). Solar cookers reduce **(1)** deforestation for fuelwood, **(2)** the time and labor needed to collect firewood, and **(3)** indoor air pollution from smoky fires.

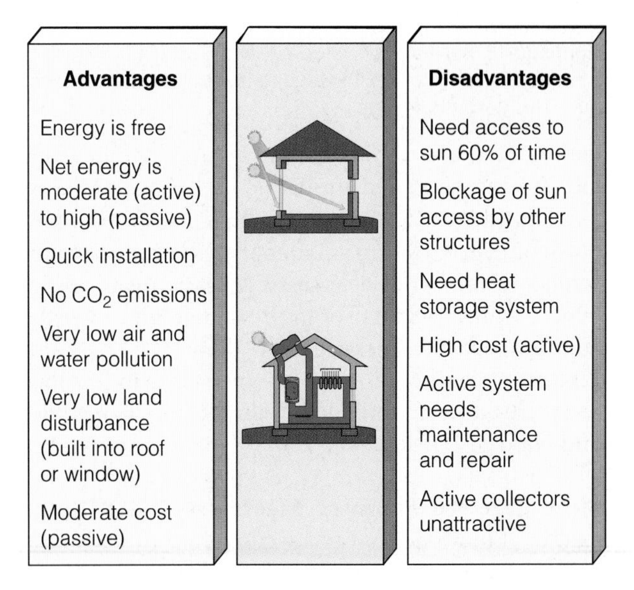

Figure 20-15 Advantages and disadvantages of heating a house with passive or active solar energy.

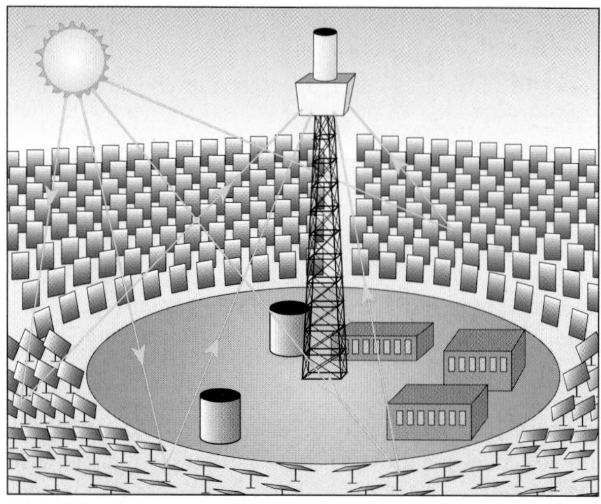

(a) Solar Power Tower

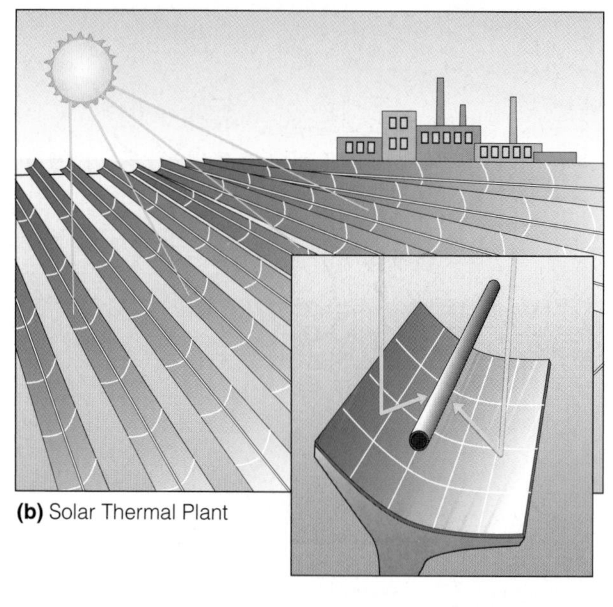

(b) Solar Thermal Plant

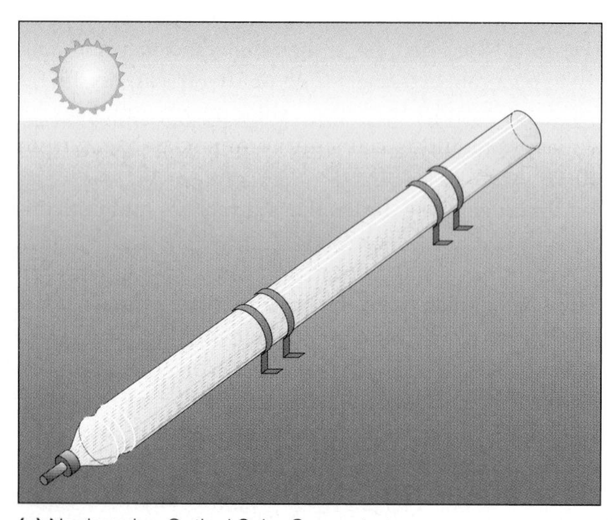

(c) Nonimaging Optical Solar Concentrator

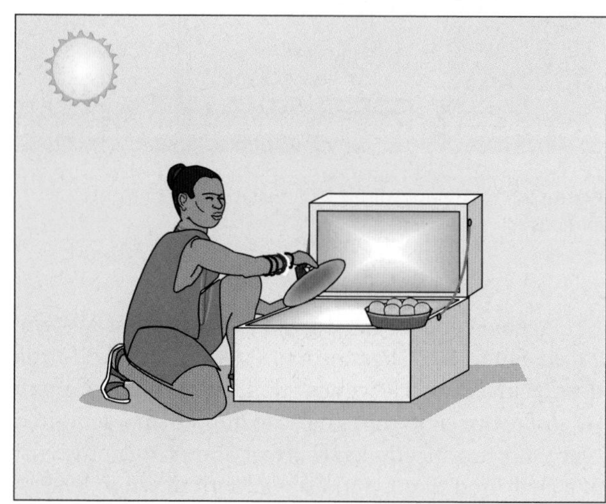

(d) Solar Cooker

Figure 20-16 Several ways to collect and concentrate solar energy to produce high-temperature heat and electricity. Because of their high costs (except for solar ovens), such systems are not expected to provide much of the world's energy.

Figure 20-17 lists the advantages and disadvantages of concentrating solar energy to produce high-temperature heat or electricity. Most analysts do not expect widespread use of such technologies over the next few decades because of **(1)** their high costs and **(2)** the availability of much cheaper ways to produce electricity such as combined-cycle natural gas turbines and wind turbines.

How Can We Produce Electricity with Solar Cells? Solar energy can be converted directly into electrical energy by **photovoltaic (PV) cells**, commonly called **solar cells** (Figure 20-18). A solar cell is a transparent wafer that contains a *semiconductor* material with a thickness ranging from less than that of a human hair to that of a sheet of paper. Sunlight energizes and causes electrons in the semiconductor to flow, creating an electrical current.

Because a single solar cell produces only a tiny amount of electricity, many cells are wired together in modular panels to produce the amount of electricity needed. The direct current (DC) electricity produced can be **(1)** stored in batteries and used directly or **(2)** converted to conventional alternating-current (AC) electricity by a separate inverter or an inverter built into the cells (Guest Essay, p. 531).

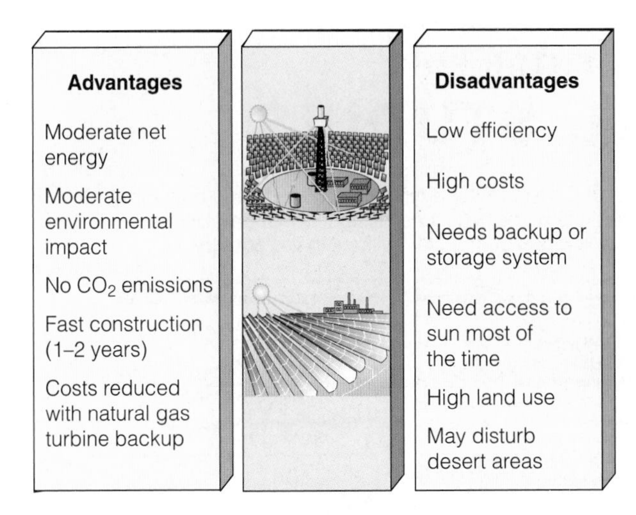

Advantages
Moderate net energy
Moderate environmental impact
No CO_2 emissions
Fast construction (1–2 years)
Costs reduced with natural gas turbine backup
Disadvantages
Low efficiency
High costs
Needs backup or storage system
Need access to sun most of the time
High land use
May disturb desert areas

Figure 20-17 Advantages and disadvantages of using solar energy to generate high-temperature heat and electricity.

Traditional-looking solar-cell roof shingles and photovoltaic panels that resemble metal roofs (developed in Japan) reduce the cost of solar-cell installations by saving on roof costs (Figure 20-18). With this technology, the roof becomes a building's power plant. A German company is testing a *solar–electric window* that incorporates solar cells into a semitransparent glazing that simultaneously generates electricity and provides filtered light during daylight hours.

Easily expandable banks of solar cells can be used to **(1)** provide electricity for rural villages in developing countries, **(2)** produce electricity at a small power plant, using combined-cycle natural gas turbines to provide backup power when the sun is not shining, and **(3)** convert water to hydrogen gas that can be distributed to energy users by pipeline, as natural gas is. Researchers are developing **(1)** flywheels, **(2)** improved deep-cycle batteries, and **(3)** supercapacitors to store solar (or wind) power for later use as needed. One promising system is to **(1)** use rooftop solar cells to produce hydrogen when the sun is shining, **(2)** store the hydrogen, and **(3)** use it in a fuel cell to provide electricity and heat as needed.

Oberlin College in Ohio recently built a new environmental studies center incorporating major elements of *green design*, including use of passive solar energy and solar cells (Figure 20-19). David Orr (head of Oberlin's environmental studies program, Guest Essay, p. 272) designed the building with the help of more than 250 students, faculty, and townspeople.

Solar cells are an ideal technology for providing electricity to 2 billion people in rural areas in most developing countries who have no electricity. With financing from the World Bank, India (the world's number-one market for solar cells) is installing solar-cell systems in 38,000 villages, and Zimbabwe is bringing solar electricity to 2,500 villages.

Figure 20-20 lists the advantages and disadvantages of solar cells. With an aggressive program, analysts project that solar cells could supply 17% of the world's electricity by 2020—as much as nuclear power does today after more than 50 years of development—at a lower

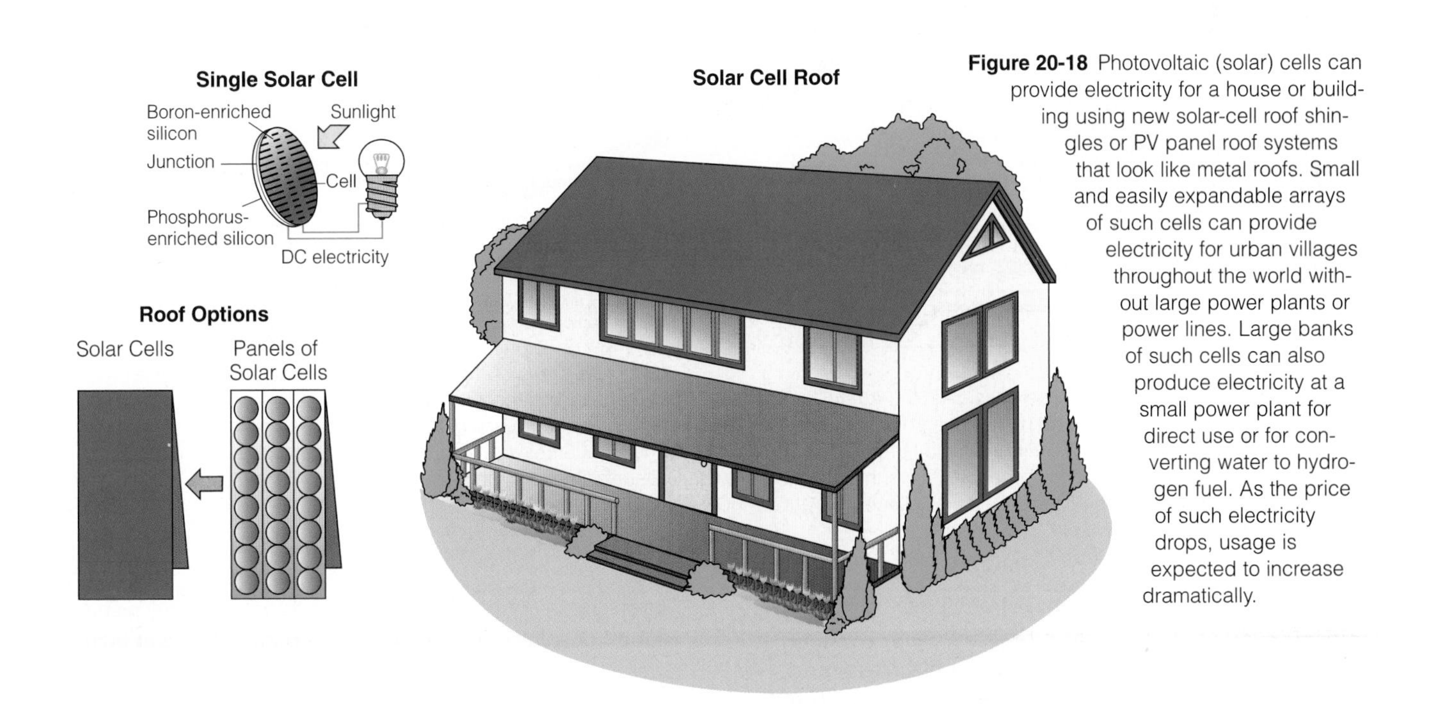

Figure 20-18 Photovoltaic (solar) cells can provide electricity for a house or building using new solar-cell roof shingles or PV panel roof systems that look like metal roofs. Small and easily expandable arrays of such cells can provide electricity for urban villages throughout the world without large power plants or power lines. Large banks of such cells can also produce electricity at a small power plant for direct use or for converting water to hydrogen fuel. As the price of such electricity drops, usage is expected to increase dramatically.

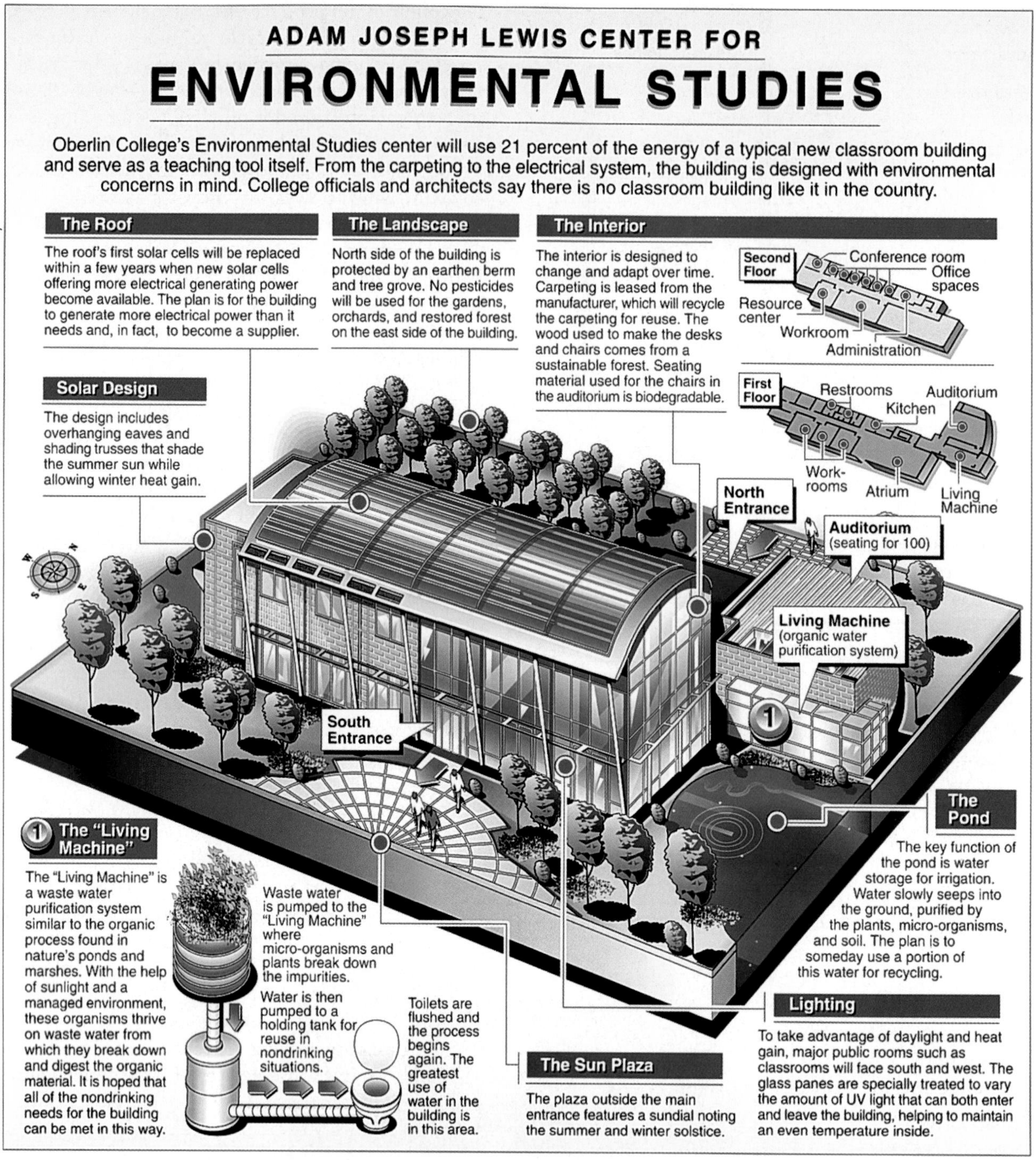

ADAM JOSEPH LEWIS CENTER FOR

ENVIRONMENTAL STUDIES

Oberlin College's Environmental Studies center will use 21 percent of the energy of a typical new classroom building and serve as a teaching tool itself. From the carpeting to the electrical system, the building is designed with environmental concerns in mind. College officials and architects say there is no classroom building like it in the country.

The Roof

The roof's first solar cells will be replaced within a few years when new solar cells offering more electrical generating power become available. The plan is for the building to generate more electrical power than it needs and, in fact, to become a supplier.

The Landscape

North side of the building is protected by an earthen berm and tree grove. No pesticides will be used for the gardens, orchards, and restored forest on the east side of the building.

The Interior

The interior is designed to change and adapt over time. Carpeting is leased from the manufacturer, which will recycle the carpeting for reuse. The wood used to make the desks and chairs comes from a sustainable forest. Seating material used for the chairs in the auditorium is biodegradable.

Second Floor
Conference room
Office spaces
Resource center
Workroom
Administration

First Floor
Restrooms
Auditorium
Kitchen
Work-rooms
Atrium
Living Machine

Solar Design

The design includes overhanging eaves and shading trusses that shade the summer sun while allowing winter heat gain.

North Entrance

Auditorium (seating for 100)

Living Machine (organic water purification system)

South Entrance

The Pond

The key function of the pond is water storage for irrigation. Water slowly seeps into the ground, purified by the plants, micro-organisms, and soil. The plan is to someday use a portion of this water for recycling.

1 The "Living Machine"

The "Living Machine" is a waste water purification system similar to the organic process found in nature's ponds and marshes. With the help of sunlight and a managed environment, these organisms thrive on waste water from which they break down and digest the organic material. It is hoped that all of the nondrinking needs for the building can be met in this way.

Waste water is pumped to the "Living Machine" where micro-organisms and plants break down the impurities.

Water is then pumped to a holding tank for reuse in nondrinking situations.

Toilets are flushed and the process begins again. The greatest use of water in the building is in this area.

The Sun Plaza

The plaza outside the main entrance features a sundial noting the summer and winter solstice.

Lighting

To take advantage of daylight and heat gain, major public rooms such as classrooms will face south and west. The glass panes are specially treated to vary the amount of UV light that can both enter and leave the building, helping to maintain an even temperature inside.

Figure 20-19 Major elements of green design in the Adam Joseph Lewis Center for Environmental Studies at Oberlin College in Ohio. There were no waste products as a result of the construction and no use of toxic materials. The building is the brainchild of David Orr (head of Oberlin's Environmental Studies program) and a number of students, faculty, and townspeople. In North Carolina, Catawba College's Center for the Environment is also a model of green design. (Oberlin College)

cost and much lower risk. With a strong push from governments and private investors, by 2050 solar cells could provide as much as 25% of the world's electricity and at least 35% of the electricity in the United States. If such projections are correct, the production, sale, and installation of solar cells (already a $2.5-billion-per-year business) could become one of the world's largest and fastest-growing businesses.

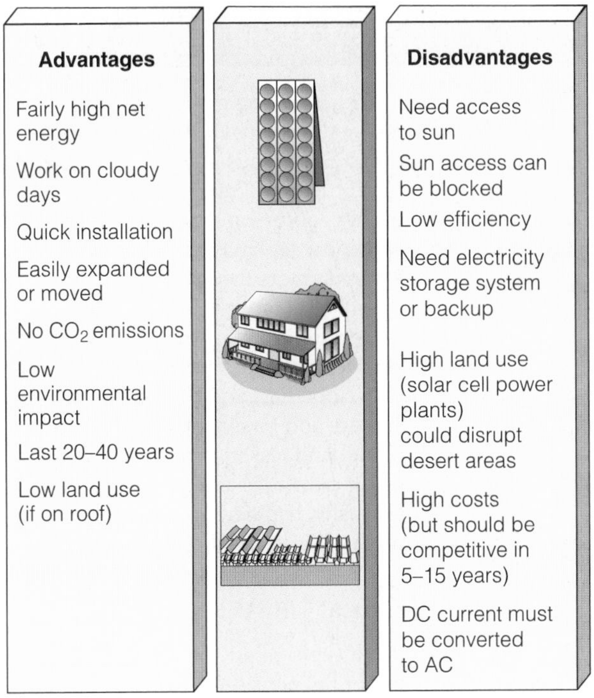

Advantages

Fairly high net energy

Work on cloudy days

Quick installation

Easily expanded or moved

No CO_2 emissions

Low environmental impact

Last 20–40 years

Low land use (if on roof)

Disadvantages

Need access to sun

Sun access can be blocked

Low efficiency

Need electricity storage system or backup

High land use (solar cell power plants) could disrupt desert areas

High costs (but should be competitive in 5–15 years)

DC current must be converted to AC

Figure 20-20 Advantages and disadvantages of using solar cells to produce electricity.

Advantages

Moderate to high net energy

High efficiency (80%)

Low-cost electricity

Long life span

No CO_2 emissions during operation

May provide flood control below dam

Provides water for year-round irrigation

Disadvantages

High construction costs

High environmental impact

High CO_2 emissions from biomass decay in shallow tropical reservoirs

Floods natural areas

Converts land habitat to lake habitat

Danger of collapse

Uproots people

Decreases fish harvest below dam

Decreases flow of natural fertilizer (silt) to land below dam

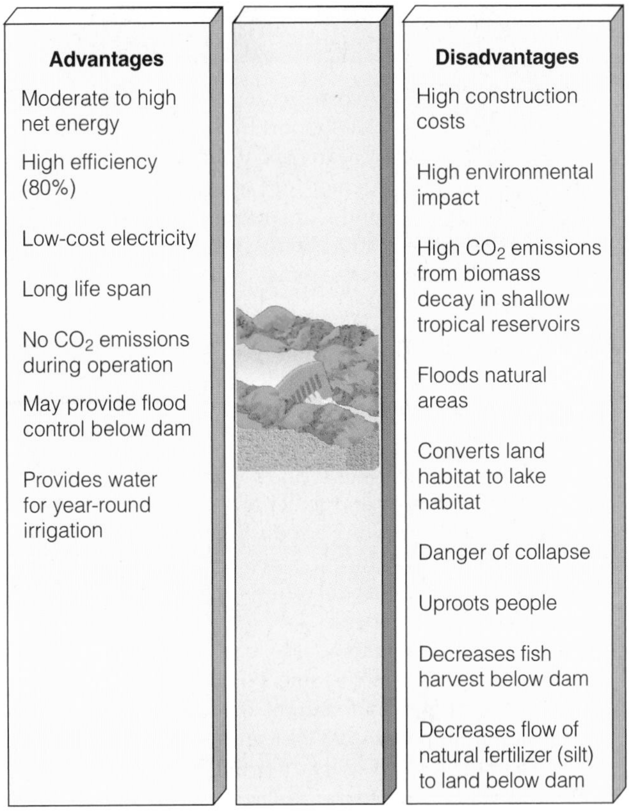

Figure 20-21 Advantages and disadvantages of using large dams and reservoirs to produce electricity.

20-4 PRODUCING ELECTRICITY FROM MOVING WATER AND FROM HEAT STORED IN WATER

How Can We Produce Electricity Using Hydropower Plants? Electricity can be produced from flowing water by

- *Large-scale hydropower*, in which a high dam is built across a large river to create a reservoir (Figure 14-9, p. 333). Some of the water stored in the reservoir is allowed to flow through huge pipes at controlled rates, spinning turbines and producing electricity.
- *Small-scale hydropower*, in which a low dam with no reservoir (or only a small one) is built across a small stream and the stream's flow of water is used to spin turbines and produce electricity.
- *Pumped-storage hydropower*, in which pumps using surplus electricity from a conventional power plant pump water from a lake or a reservoir to another reservoir at a higher elevation. When more electricity is needed, water in the upper reservoir is released, flows through turbines, and generates electricity on its return to the lower reservoir.

Hydropower supplies about **(1)** 6% of the world's total commercial energy (4% in the United States) and **(2)** 20% of the world's electricity (10% in the United States but about 63% of the power used along the West Coast). Hydropower supplies about 99% of the electricity in Norway, 75% in New Zealand, 50% in developing countries, and 25% in China.

Figure 20-21 lists the advantages and disadvantages of using large-scale hydropower plants to produce electricity. According to the United Nations, only about 13% of the world's technically exploitable potential for hydropower has been developed, with much of this untapped potential in South Asia (especially China, p. 333), South America, and parts of the former Soviet Union.

Because of increasing concern about the environmental and social consequences of large dams (Figure 20-21 and Figure 14-9, p. 333), there has been growing pressure on the World Bank and other development agencies to stop funding new large-scale hydropower projects. In 2000, the World Commission on Dams published a study indicating that hydropower is a major emitter of greenhouse gases. This occurs because reservoirs that power the dams can trap rotting vegetation, which can emit greenhouse gases such as CO_2 and CH_4.

Small-scale hydropower projects eliminate most of the harmful environmental effects of large-scale projects, but they can **(1)** threaten recreational activities and aquatic life, **(2)** disrupt the flow of wild and scenic rivers, and **(3)** destroy wetlands. In addition, their electrical output can vary with seasonal changes in stream flow.

Using the Internet to Save Energy and Paper and Reduce Global Warming

CONNECTIONS

According to a 2000 report by researchers at the Center for Energy and Climate Solutions, increasing use of the internet is saving energy by

- Allowing **(1)** more employees to work at home and **(2)** mobile workers to be assigned flexible office space only when they are in the office.
- Saving energy and material resources by allowing a reduction in retail, manufacturing, warehouse, and commercial office space. Some online companies keep no merchandise in warehouses and have it shipped to customers directly from manufacturers. For example, Amazon.com uses one-sixteenth as much energy per unit of floor space to sell a book as a regular store.
- Using less energy to get products from sellers to consumers. Sending a package by overnight air uses about 40% less fuel than driving round-trip to the mall to get the product. Shipping by rail saves even more energy. However, shipping a larger number of packages to individuals instead of bulk shipments to businesses causes a huge increase in packaging.
- Reducing the energy, paper, and materials used to produce, package, and market consumer items such as computer software and music CDs by allowing them to be downloaded.
- Reducing energy used in transporting goods by using internet-based systems to auction off empty shipping space on trucks, aircraft, and trains.

The resulting energy savings also reduce air pollution and carbon dioxide emissions by reducing fossil fuel use. Paper production (a highly energy- and resource-intensive industry) may also decrease as **(1)** consumers use the internet to download software and view magazines, newspapers, research articles, telephone directories, encyclopedias, and books, **(2)** consumers and businesses send more e-mail and less mail and reports by using envelopes and packages, and **(3)** easily updated online catalogs replace paper catalogs.

Critical Thinking

What types of environmental harm might be increased by greatly expanded use of the internet to sell more and more goods in the global marketplace?

Is Producing Electricity from Tides and Waves a Useful Option? Twice a day in high and low tides, water that flows into and out of coastal bays and estuaries can spin turbines to produce electricity. Two large tidal energy facilities are currently operating, one at La Rance in France and the other in Canada's Bay of Fundy. However, most analysts expect tidal power to make only a tiny contribution to world electricity supplies. There are few suitable sites, and construction costs are high.

The kinetic energy in ocean waves, created primarily by wind, is another potential source of electricity. Most analysts expect wave power to make little contribution to world electricity production, except in a few coastal areas with the right conditions (such as western England). Construction costs are moderate to high and the net energy yield is moderate, but equipment can be damaged or destroyed by saltwater corrosion and severe storms.

How Can We Produce Electricity from Heat Stored in Water? Japan and the United States have been evaluating the use of the large temperature differences (between the cold, deep waters and the sun-warmed surface waters) of tropical oceans for producing electricity. If economically feasible, this would be done in *ocean thermal energy conversion* (OTEC) plants anchored to the bottom of tropical oceans in suitable sites. However, most energy analysts believe that the large-scale extraction of energy from ocean thermal gradients may never compete economically with other energy alternatives.

Saline solar ponds, usually located near inland saline seas or lakes in areas with ample sunlight, can be used to produce electricity. Heat accumulated during the day in the denser bottom layer can be used to produce steam that spins turbines, generating electricity. A small experimental saline solar pond power plant on the shore of the Israeli side of the Dead Sea operated for several years but was closed in 1989 because of high operating costs.

Freshwater solar ponds can be used to heat water and space. A shallow hole is dug and lined with concrete. A number of large, black plastic bags, each filled with several centimeters of water, are placed in the hole and then covered with fiberglass insulation panels. The panels let sunlight in but keep most of the heat stored in the water during the daytime from being lost to the atmosphere. When the water in the bags has reached its peak temperature in the afternoon, a computer turns on pumps to transfer hot water from the bags to large, insulated tanks for distribution.

Saline and freshwater solar ponds **(1)** use no energy storage and backup systems, **(2)** emit no air pollution, and **(3)** have a moderate net energy yield. Freshwater solar ponds can be built in almost any sunny area and have moderate construction and operating costs.

Living Lightly on the Earth at Round Mountain Organics

GUEST ESSAY

Nancy Wicks

Nancy Wicks is an ecopioneer trying to live her ideals. She grew up in a small town in Iowa and did undergraduate studies in a village in Nepal. Both of these life experiences inspired her to live more sustainably by creating Round Mountain Organics, a high altitude organic garden in the Rocky Mountains near Crested Butte, Colorado. Nancy lives in a passive solar strawbale house, which is powered by the wind and sun. She is a fanatic reuser, recycler, and composter. She received the "Sustainable Business of the Year 2000" award from the High Country Citizens' Alliance.

After studying in Nepal, where sustainability is a do or die situation, I have tried incorporate as many sustainable practices into my life as possible at my house and organic garden business called Round Mountain Organics. This includes being a member of a buying co-op (where buying in bulk not only saves resources but also saves money), reusing everything from plastic and paper bags to trays and pots for garden plants, and composting food waste (which saves money on trash bills and fertilizes the soil).

After moving onto the land that is now home to Round Mountain Organics, I spent 4 years planning and building an octagonal strawbale house with a stucco exterior—the first such house to be built in the county. I chose to build with straw because **(1)** straw is a natural building material and renewable resource, **(2)** there is a surplus of straw after harvesting grains such as wheat (which I used), oats, barley, and rice, **(3)** its insulation value of R-54 comes in handy when you live in an area where winter temperatures can dip to 40°F below zero, and **(4)** they are easy to build (with only 1 week needed to put up the strawbale walls). Since the 1980s strawbale houses have also been built in Arizona and New Mexico to beat the heat.

I used a passive solar design by orienting the house to the south to take advantage of Colorado's abundant sunshine. During the day the insulated window covers are drawn up and the sun shines onto flagstone tiles covering a cement slab that stores and releases heat slowly to keep the house comfortably warm or cool regardless of outside conditions. At night the window covers are let down to hold the heat in.

Because I live in one of the world's sunniest places, I decided not to get hooked up to the electrical grid and instead get my electricity from a small wind turbine and panels of solar cells. The electricity is stored in a bank of 12 batteries, and an inverter converts the stored direct current (DC) electricity to ordinary 120-volt alternating current (AC).

If it's cloudy and not windy for a couple of days (which is rare), I fire up a small gas generator to charge the batteries. I also use some propane to provide hot water with a small on-demand water heater. Only a small pilot light stays lit until the hot water faucet is turned on. Then a large flame is ignited that the water is piped through. This way, I do not use energy to keep a tank of water hot around the clock.

I use many energy-saving devices. They include compact florescent light bulbs, an oversized pressure tank so the well pump does not have to kick on every time the faucet turns on, and a superinsulated energy-efficient DC refrigerator.

I use organic gardening to grow flowers, herbs, and vegetables for my own use and for sale to local residents and restaurants. I incorporate some pioneering organic gardening techniques such as Rudolph Steiner's biodynamics (developed in 1924) and Bill Mollison's permaculture (developed in 1978).

Insect pests are picked off by hand, and beneficial insects such as ladybugs are used to eat harmful insects such as aphids. Compost, aged animal manure, and cover crops that are plowed in as green manure are used to add nutrients to the soil. Crop rotation is used so as not to deplete the soil of nutrients.

Cold frames (a type of mini-greenhouse) are used to extend the growing season to 150 days in the cold climate where there are only about 90 days without a killing frost. My latest endeavor is building a passively heated strawbale greenhouse to provide the community with fresh salad greens, herbs, and flowers all winter long. The chicken coop is in the northeast corner of the greenhouse, with the heat given off by the chickens helping to warm the greenhouse.

My next venture is to start a nonprofit, Round Mountain Sustainable Living Institute, to educate people on how they can live in harmony with the earth.

Critical Thinking

Would you like to live a lifestyle similar to that of Nancy Wicks? Explain. Why do you think more people do not try to live more sustainably, as she does? List three ways to help encourage people to adopt such a lifestyle.

20-5 PRODUCING ELECTRICITY FROM WIND

How Rapidly Has the Use of Wind Power Grown?

Wind power is the world's fastest-growing energy resource, at 27% per year and more than $4 billion in sales of wind turbines in 2000. In 2000, wind turbines (Figure 20-22) worldwide produced almost 18,000 megawatts of electricity, enough to meet the needs of 5.5 million homes. If its current growth rate continues,

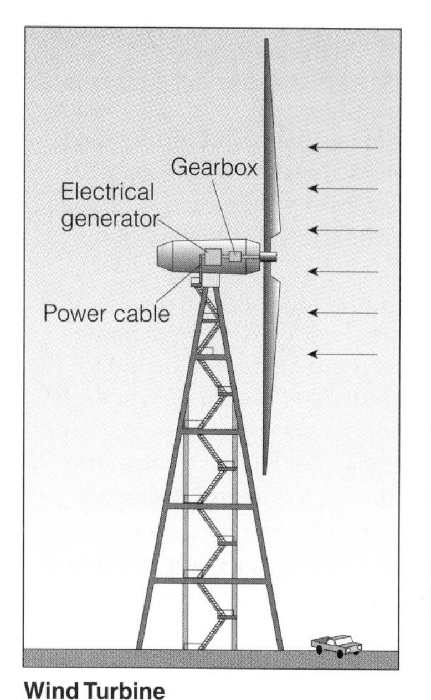

Wind Turbine

Wind Farm

Figure 20-22 Wind turbines can be used to produce electricity individually or in clusters called wind farms.

wind power could produce 10% of the world's electricity by 2020.

Despite its rapid growth, wind power produced only about 1% of the energy used in the United States in 2000 because it is still in its infancy. However, the DOE has launched a program designed to have 5% of the country's energy and 10–25% of its electricity produced by the wind by 2020. In 2000, the price of electricity produced by wind in the United States was about the same as that produced by new gas- and coal-fired power plants. Within a few years, wind power could be the country's cheapest way to produce electricity. Despite the potential for wind power, the Bush administration proposed cutting the federal wind research budget in 2002 in half while greatly increasing research funds for coal and nuclear power.

What Areas Have the Greatest Potential for Wind Power? Figure 20-23 shows the potential areas for use of wind power in the United States. The DOE calls the midwestern United States the "Saudi Arabia of wind." The Dakotas and Texas alone have enough wind resources to meet all the nation's electricity needs. Sizable wind farm projects are being developed in 12 states. Individuals can also use small wind turbines to supply some or all of their electricity (Guest Essay, p. 531). Denmark (with wind generating about 13% of its electricity) is the world's largest user of wind and producer of wind turbines. Wind power also is being developed rapidly in Germany (the world's third largest user of wind power), Spain, and India (the world's number-two market for wind energy).

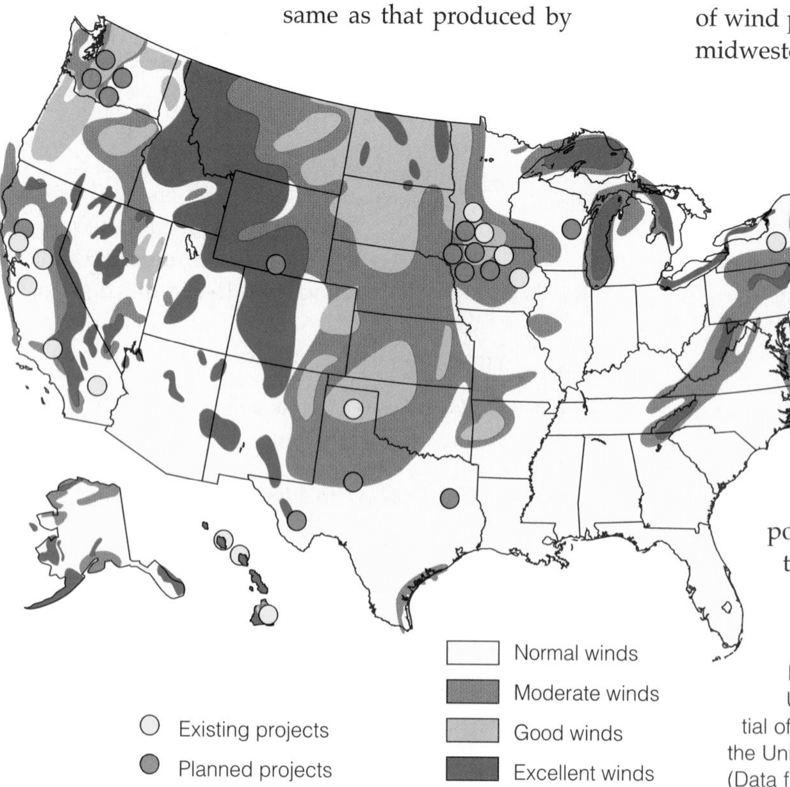

Figure 20-23 Potential for use of wind power in the United States. In principle, exploiting the wind potential of just three states could provide all the power needs of the United States: North Dakota, South Dakota, and Texas. (Data from U.S. Department of Energy)

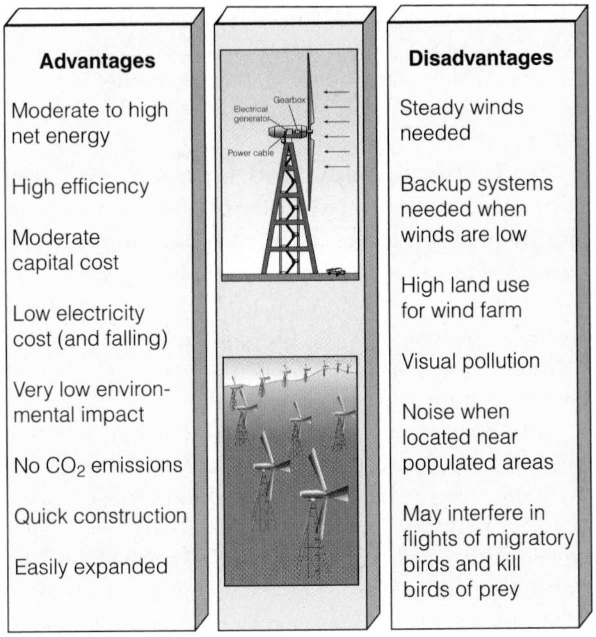

Advantages	Disadvantages
Moderate to high net energy	Steady winds needed
High efficiency	Backup systems needed when winds are low
Moderate capital cost	High land use for wind farm
Low electricity cost (and falling)	Visual pollution
Very low environmental impact	Noise when located near populated areas
No CO_2 emissions	May interfere in flights of migratory birds and kill birds of prey
Quick construction	
Easily expanded	

Electrical generator
Gearbox
Power cable

Figure 20-24 Advantages and disadvantages of using wind to produce electricity. Wind power experts project that by 2050 wind power could supply more than 10% of the world's electricity and 10–25% of the electricity used in the United States.

In the long run, electricity from large wind farms in remote areas might be used to make hydrogen gas from water during off-peak periods. The hydrogen could then be fed into a pipeline and storage system.

What Are the Major Advantages and Disadvantages of Wind Power? Figure 20-24 lists the advantages and disadvantages of using wind to produce electricity. Some environmentalists and other critics have pointed out that wind turbines have been responsible for the death of approximately 10,000 predatory birds (such as some types of hawks, kestrels, vultures, and eagles) in the United States over the past 20 years. The problem is that wind turbine towers attract bird prey, which attracts predatory birds that sometimes get caught in the blades of the wind turbines. Environmentalists and wind turbine manufacturers are working on this problem.

Oil spills, air pollution, water pollution, and release of toxic wastes from use of fossil fuels such as coal and oil have also killed enormous numbers of birds, fish, and other forms of wildlife. The key questions are **(1)** which types of energy resources lead to the lowest loss of wildlife and **(2)** how loss of wildlife from use of any energy resource can be minimized.

European scientists are working on the problem of reduced wind turbine efficiency from the buildup of dead bugs on the blades. This is similar to bugs splatting and sticking to the windshield of a car.

Farmers can boost their income by leasing some of their cropland for wind turbines while growing crops around the turbines. For example, a farmer in Iowa who leases 0.10 hectare (0.25 acre) of cropland to the local utility as a site for a wind turbine typically gets $2,000 a year in royalties from the electricity produced. In a good year the site occupied by the turbine could produce only $100 of corn.

20-6 PRODUCING ENERGY FROM BIOMASS

How Useful Is Burning Solid Biomass? *Biomass* is plant materials and animal wastes used as sources of energy. Biomass comes in many forms and can be burned directly as a solid fuel or converted into gaseous or liquid **biofuels** (Figure 20-25).

Most biomass is burned **(1)** directly for heating, cooking and industrial processes or **(2)** indirectly to drive turbines and produce electricity. Burning wood and manure for heating and cooking supplies about 12% of the world's energy and about 30% of the energy used in developing countries. Almost 70% of the people living in developing countries heat their homes and cook their food by burning wood or charcoal. However, about 2.7 billion people in these countries cannot find (or are too poor to buy) enough fuelwood to meet their needs.

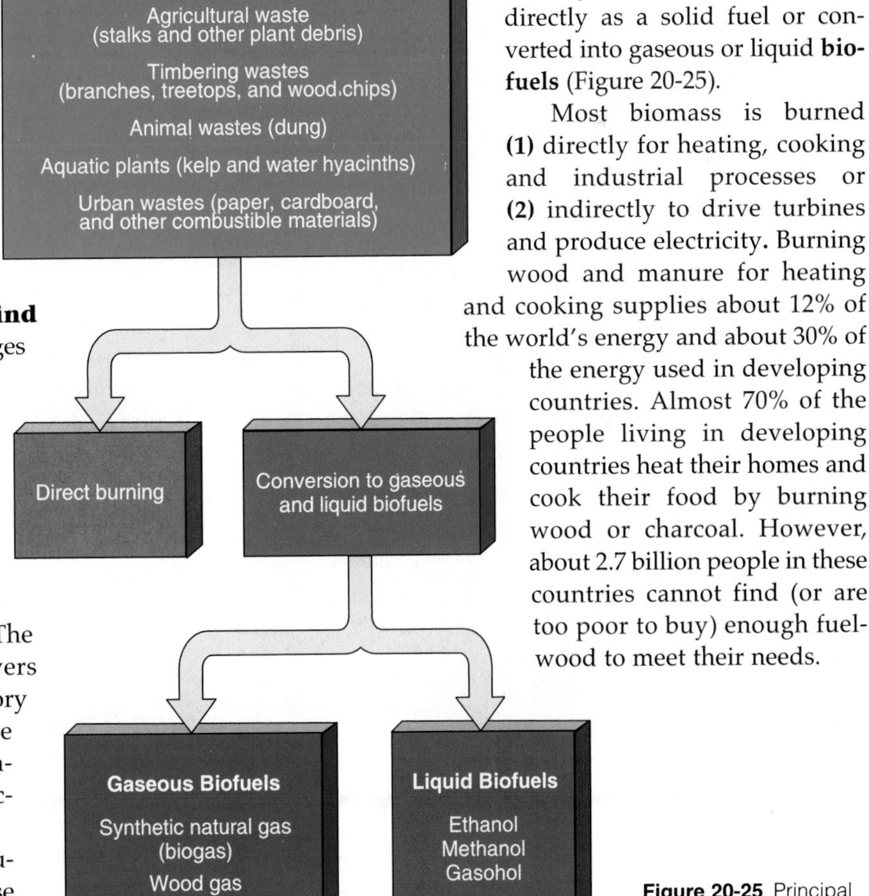

Figure 20-25 Principal types of biomass fuel.

In the United States, biomass is used to supply about 4% of the country's commercial energy and 2% of its electricity (produced by about 350 biomass power plants). The U.S. government has a goal of increasing the use of biomass energy to 9% of the country's total commercial energy by 2010.

One way to produce biomass fuel is to plant, harvest, and burn large numbers of **(1)** fast-growing trees (especially cottonwoods, poplars, sycamores, willows, and leucaenas), **(2)** shrubs, **(3)** perennial grasses (such as switchgrass), and **(4)** water hyacinths in *biomass plantations.*

In agricultural areas, *crop residues* (such as sugarcane residues, rice husks, cotton stalks, and coconut shells) and *animal manure* can be collected and burned or converted into biofuels. According to a 1999 study by the Union of Concerned Scientists, energy crops and crop wastes from the Midwest alone could theoretically provide about 16% of the electricity used in the United States, without irrigation and without competing with food crops for land. Some ecologists argue that it makes more sense to use animal manure as a fertilizer and crop residues to feed livestock, retard soil erosion, and fertilize the soil.

Figure 20-26 lists the general advantages and disadvantages of burning solid biomass as a fuel. One problem is that burning biomass produces carbon dioxide. However, if the rate of use of biomass does not exceed the rate at which it is replenished by new plant growth (which takes up CO_2), there is no net increase in CO_2 emissions.

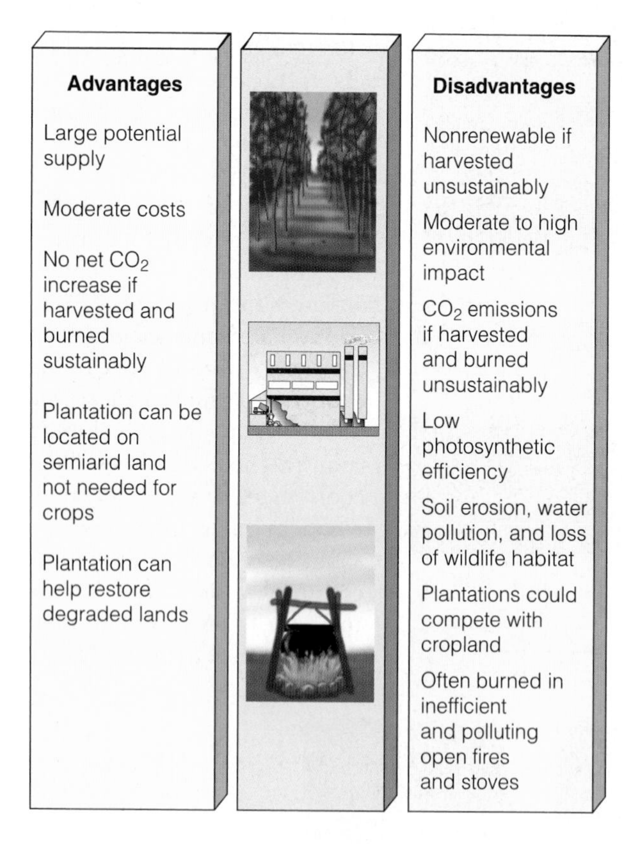

Advantages
Large potential supply
Moderate costs
No net CO2 increase if harvested and burned sustainably
Plantation can be located on semiarid land not needed for crops
Plantation can help restore degraded lands
Disadvantages
Nonrenewable if harvested unsustainably
Moderate to high environmental impact
CO2 emissions if harvested and burned unsustainably
Low photosynthetic efficiency
Soil erosion, water pollution, and loss of wildlife habitat
Plantations could compete with cropland
Often burned in inefficient and polluting open fires and stoves

Figure 20-26 General advantages and disadvantages of burning solid biomass as a fuel.

Is Producing Gaseous and Liquid Fuels from Solid Biomass a Useful Option? Bacteria and various chemical processes can convert some forms of biomass into gaseous and liquid biofuels (Figure 20-25). Examples include **(1)** *biogas,* a mixture of 60% methane and 40% carbon dioxide, **(2)** *liquid ethanol* (ethyl, or grain alcohol), and **(3)** and *liquid methanol* (methyl, or wood alcohol).

In China, anaerobic bacteria in more than 6 million *biogas digesters* convert plant and animal wastes into methane fuel for heating and cooking. These simple devices can be built for about $50 including labor. After the biogas has been separated, the solid residue is used as fertilizer on food crops or, if contaminated, on trees. When they work, biogas digesters are very efficient. However, they are also slow and unpredictable, a problem that could be corrected with development of more reliable models.

Some analysts believe that liquid ethanol and methanol produced from biomass could replace gasoline and diesel fuel when oil becomes too scarce and expensive. *Ethanol* (C_2H_5OH) can be made from sugar and grain crops (sugarcane, sugar beets, sorghum, sunflowers, and corn) by fermentation and distillation. Gasoline mixed with 10–23% pure ethanol makes *gasohol,* which can be burned in conventional gasoline engines and is sold as super unleaded or ethanol-enriched gasoline.

Another alcohol, *methanol* (CH_3OH), is made mostly from natural gas but also can be produced at a higher cost from wood, wood wastes, agricultural wastes (such as corncobs), sewage sludge, garbage, and coal. Some of the first generation of cars using hydrogen-powered fuel cells (Figure 20-8) will use reformers to convert methanol to hydrogen. The advantages and disadvantages of using ethanol, methanol, and several other fuels as alternatives to gasoline are summarized in Table 12-4, p. 297. According to a 1997 analysis by David Pimentel and two other researchers, "Large-scale biofuel production is not an alternative to the current use of oil and is not even an advisable option to cover a significant fraction of it."

20-7 THE SOLAR–HYDROGEN REVOLUTION

What Can We Use to Replace Oil? Good-Bye Oil and Smog, Hello Hydrogen When oil is gone (or when what's left costs too much to use), what will we use to fuel vehicles, industry, and buildings? Many scientists and executives of major oil companies and automobile companies say the fuel of the future is hydrogen gas (H_2) (Table 12-4, p. 297).

When hydrogen gas burns in air, it combines with oxygen gas in the air and produces nonpolluting water vapor and some nitrogen oxides (produced because air is 78% nitrogen).* This eliminates most of the air pollution problems we face today and greatly reduces the threats from global warming by emitting no carbon dioxide.

There is very little hydrogen gas (H_2) around, but we can get it from something we have plenty of: *water*. Water can be split by electricity or high temperatures into gaseous hydrogen and oxygen (Figure 20-27). Hydrogen can also be produced **(1)** by *reforming*, in which chemical processes are used to separate hydrogen from carbon atoms in organic compounds such as methane (CH_4) or methanol (CH_3OH), **(2)** by gasification of coal or biomass, and **(3)** by some types of algae and bacteria (Spotlight, p. 536). Producing hydrogen from hydrocarbons and gasification of coal or biomass would add some CO_2 to the atmosphere. However, producing hydrogen by using electricity from solar cells to decompose water (Figure 20-27) would produce no CO_2 and thus be a way to slow global warming.

What Is the Catch? If you think using hydrogen as an energy source sounds too good to be true, you're right. Several problems must be solved to make hydrogen one of our primary energy resources, but scientists are making rapid progress in finding solutions to these problems.

One problem is that it takes energy (and thus money) to produce this fuel. We could burn coal to produce high-temperature heat or use electricity from coal-burning and nuclear power plants to split water and produce hydrogen. However, this subjects us to the harmful environmental effects associated with using these fuels (Figure 19-17, p. 501, and Figure 19-22, p. 505), and it costs more than the hydrogen fuel is worth.

Most proponents of using hydrogen gas believe that if we are to get its very low pollution benefits, the energy to produce the gas from water must come from the sun, probably **(1)** in the form of electricity generated by sources such as hydropower, solar cells, solar thermal power plants (Figure 20-16b, p. 526), and wind farms and **(2)** perhaps eventually from bacteria and algae (Spotlight, p. 536).

If scientists and engineers can learn how to use sunlight to decompose water cheaply enough, they will set in motion a *solar–hydrogen revolution* over the next 50 years and change the world as much as the agricultural and industrial revolutions did. Currently, using solar energy to produce hydrogen gas is too costly, but the costs of using solar energy to produce electricity are coming down. The goal of the DOE is to have the cost of hydrogen gas equal to that of natural gas by 2030.

Hydrogen can be stored

- In *compressed gas storage tanks*. The technology is available, but the costs of tanks and compression are high, and tanks are too heavy for use in motor vehicles.
- *As liquid hydrogen*. Condensing hydrogen gas into more dense liquid form allows a larger quantity of hydrogen to be stored and transported. However, this conversion takes a large input of energy and is costly.
- *As solid metal hydride compounds*, which when heated decompose and release hydrogen gas. This is a safe and efficient way to store hydrogen, but an input of energy is needed to release the hydrogen.
- *By absorption on activated charcoal*, which when heated releases hydrogen gas. Like hydrides, this is a

*Water vapor is a potent greenhouse gas. However, because there is already so much of it in the atmosphere, human additions of this gas are insignificant.

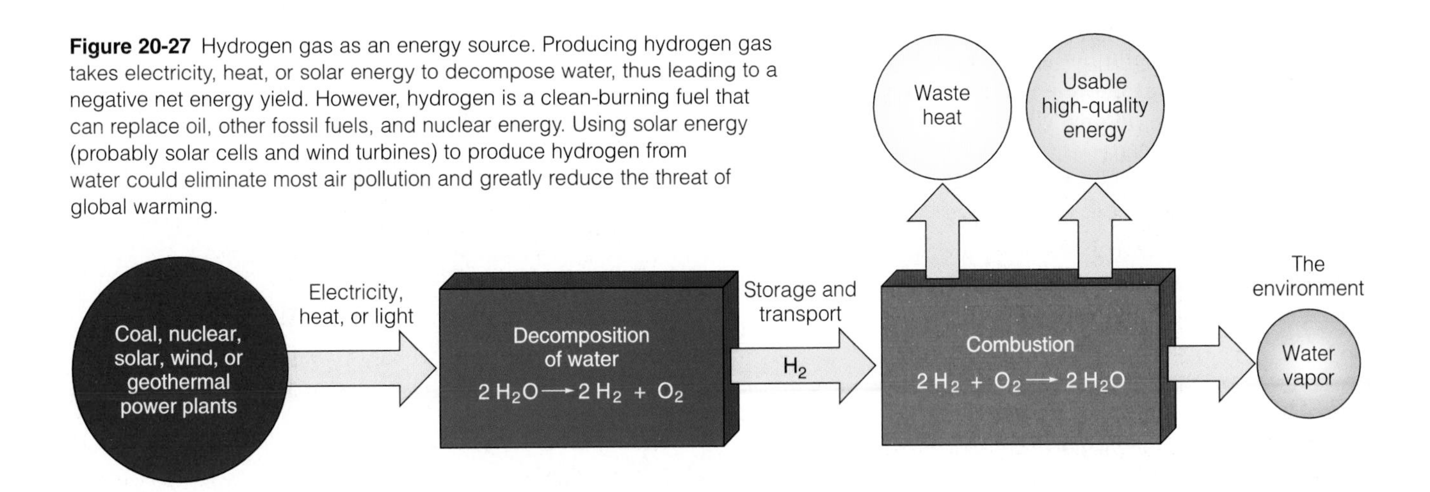

Figure 20-27 Hydrogen gas as an energy source. Producing hydrogen gas takes electricity, heat, or solar energy to decompose water, thus leading to a negative net energy yield. However, hydrogen is a clean-burning fuel that can replace oil, other fossil fuels, and nuclear energy. Using solar energy (probably solar cells and wind turbines) to produce hydrogen from water could eliminate most air pollution and greatly reduce the threat of global warming.

Producing Hydrogen from Green Algae Found in Pond Scum

SPOTLIGHT

In a few decades we may be able to use large-scale cultures of green algae to produce hydrogen gas. This simple plant grows all over the world and is commonly found in pond scum.

When living in ordinary air and sunlight, green algae carry out photosynthesis like other plants and produce carbohydrates and oxygen gas. However, in 2000, Tasios Melis, a researcher at the University of California, Berkeley, found a way to make these algae produce bubbles of hydrogen rather than oxygen.

First, he grew cultures of hundreds of billions of the algae in the normal way with plenty of sunlight, nutrients, and water. Then he cut off their supply of two key nutrients: sulfur and oxygen. Within 20 hours, the plant cells underwent a metabolic change and switched from an oxygen-producing to a hydrogen-producing metabolism, allowing the researcher to collect hydrogen gas bubbling from the culture.

Melis believes that he can increase the efficiency of this hydrogen-producing process tenfold. If so, sometime in the future a biological hydrogen plant might cycle an algae–water mixture through a system of clear tubes exposed to sunlight to produce hydrogen. The gene responsible for producing the hydrogen might even be transferred to other plants to produce hydrogen.

Critical Thinking

What might be some ecological problems related to the widespread use of this method for producing hydrogen?

safe and efficient way to store hydrogen, but an input of energy is needed to release the hydrogen.

- *Inside glass microspheres.* Currently, tiny glass spheres are being developed for this purpose.

Unlike gasoline, metal hydrides, charcoal powders, and glass microspheres containing hydrogen will not explode or burn if a vehicle's tank is ruptured in an accident. However, it's difficult to store enough hydrogen gas in a car as a compressed gas, liquid, or a solid for it to run very far, a problem similar to the one the electric car faces. Scientists and engineers are seeking solutions to this problem.

Another possibility is to power a car with a *fuel cell* (Figures 20-4 and 20-8) in which hydrogen and oxygen gas combine to produce electrical current. Fuel cells produce no air pollution and have energy efficiencies of 65–95%, several times the efficiency of conventional gasoline-powered engines and electric cars.

A number of prototype fuel-cell systems for cars, buses, homes, and buildings are being tested and evaluated. All major automobile companies have developed fuel-cell cars (Figure 20-8) and hope to begin marketing them by 2004.

In 1999, DaimlerChrysler, Royal Dutch Shell, and Norsk Hydro announced plans to turn the tiny country of Iceland into the world's first "hydrogen economy" by 2030–2040. The country's abundant renewable geothermal energy and hydropower will be used to produce hydrogen from seawater, with the H_2 used to run its buses, passenger cars, fishing vessels, and factories.

Figure 20-28 lists the pros and cons of using hydrogen as an energy resource. The DOE has a goal of hydrogen energy providing 10% of all U.S. energy consumption by 2025. Even if this is only partially accomplished, it could greatly reduce emissions of CO_2 and other air pollutants and decrease U.S. dependence on oil imports.

Advantages	Disadvantages
Can be produced from water	Not found in nature
Low environmental impact	Energy is needed to produce fuel
No CO_2 emissions	Negative net energy
Good substitute for oil	High costs (but expected to come down)
Competitive price if environmental and social costs are included in cost comparisons	Short driving range for current fuel-cell cars
Easier to store than electricity	
Safer than gasoline and natural gas	
High efficiency (65–95%) in fuel cells	

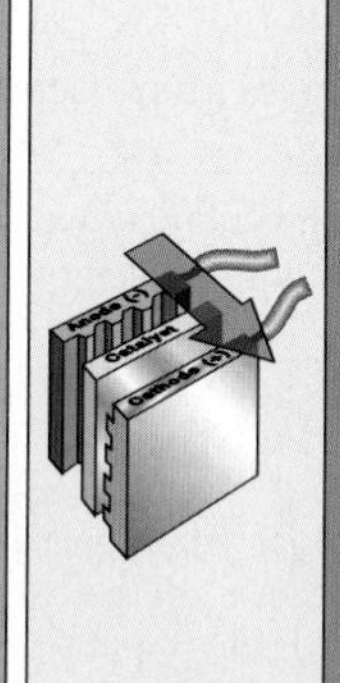

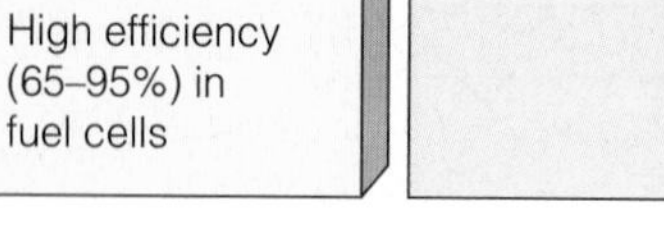

Figure 20-28 Advantages and disadvantages of using hydrogen as a fuel for vehicles and for providing heat and electricity.

20-8 GEOTHERMAL ENERGY

How Can We Tap the Earth's Internal Heat? Going Underground Heat contained in underground rocks and fluids is an important source of energy. Over millions of years, this **geothermal energy** from the earth's mantle (Figure 9-2, p. 195) has been transferred to underground reservoirs of **(1)** *dry steam* (steam with no water droplets), **(2)** *wet steam* (a mixture of steam and water droplets), and **(3)** *hot water* trapped in fractured or porous rock at various places in earth's crust.

If such geothermal sites are close to the surface, wells can be drilled to extract the dry steam, wet steam (Figure 20-29), or hot water. This thermal energy can be used to heat homes and buildings and to produce electricity. Currently, about 22 countries (most of them in the developing world) are extracting energy from geothermal sites to produce about 1% of the world's electricity.

However, these geothermal reservoirs can be depleted if heat is removed faster than natural processes renew it. Thus, geothermal resources can be nonrenewable on a human time scale, but the potential supply is so vast that it is usually classified as a renewable energy resource.

Figure 20-29 Tapping the earth's heat or geothermal energy in the form of wet steam to produce electricity.

Geothermal electricity meets the electricity needs of 6 million Americans and supplies 6% of California's electricity. The world's largest operating geothermal system, called *The Geysers*, extracts energy from a dry steam reservoir north of San Francisco, California. However, heat is being withdrawn from this geothermal site about 80 times faster than it is being replenished, converting this potentially renewable resource to a nonrenewable source of energy. In 1999, Santa Monica, California, became the first city in the world to get all its electricity from geothermal energy. In Iceland, every building is heated by hot spring water.

Three other nearly nondepletable sources of geothermal energy are **(1)** *molten rock* (magma), **(2)** *hot dry-rock zones*, where molten rock that has penetrated the earth's crust heats subsurface rock to high temperatures, and **(3)** low- to moderate-temperature *warm-rock reservoir deposits*, which could be used to preheat water and run heat pumps for space heating and air conditioning. Research is being carried out in several countries to see whether hot dry-rock zones, which can be found almost anywhere about 8–10 kilometers (5–6 miles) below the earth's surface, can provide affordable geothermal energy.

Figure 20-30 lists the pros and cons of using geothermal energy. Currently, the cost of tapping geothermal energy is too high for all but the most concentrated and accessible sources. According to the U.S. Geothermal Energy Association, at best geothermal energy could meet 5% of all U.S. energy needs within several decades.

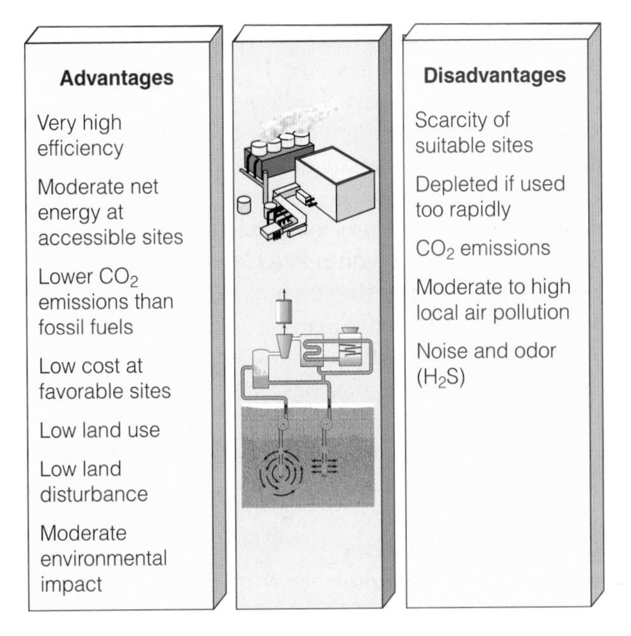

Advantages
Very high efficiency
Moderate net energy at accessible sites
Lower CO_2 emissions than fossil fuels
Low cost at favorable sites
Low land use
Low land disturbance
Moderate environmental impact
Disadvantages
Scarcity of suitable sites
Depleted if used too rapidly
CO_2 emissions
Moderate to high local air pollution
Noise and odor (H_2S)

Figure 20-30 Advantages and disadvantages of using geothermal energy for space heating and to produce electricity or high-temperature heat for industrial processes.

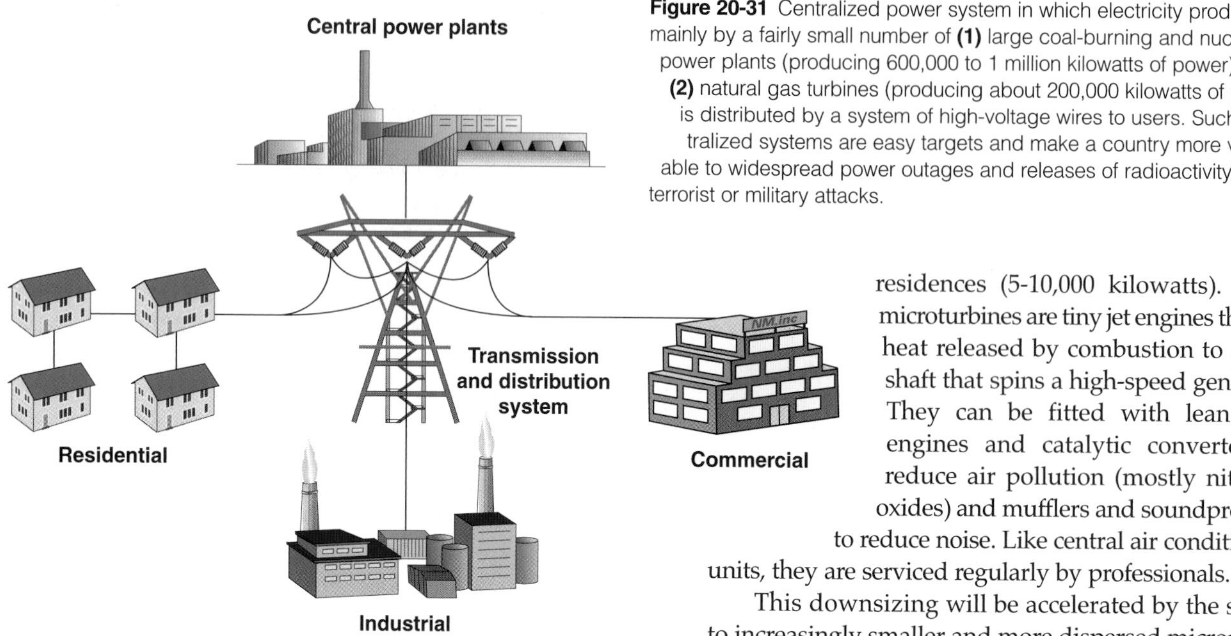

Figure 20-31 Centralized power system in which electricity produced mainly by a fairly small number of **(1)** large coal-burning and nuclear power plants (producing 600,000 to 1 million kilowatts of power) and **(2)** natural gas turbines (producing about 200,000 kilowatts of power) is distributed by a system of high-voltage wires to users. Such centralized systems are easy targets and make a country more vulnerable to widespread power outages and releases of radioactivity from terrorist or military attacks.

20-9 ENTERING THE AGE OF DECENTRALIZED MICROPOWER

What Is Micropower? According to Chuck Linderman, director of energy supply policy for the Edison Electric Institute, the era of big central power-plant systems (Figure 20-31) is over. Most energy analysts believe that the chief feature of electricity production over the next few decades is significant *decentralization* to dispersed, small-scale power systems (Figure 20-32). These **micropower systems** generate 1–10,000 kilowatts of power. In a special August 1999 issue of *Business Week* titled "21 Ideas for the 21st Century," the drastic downsizing of power-producing systems to micropower plants headed the list. This shift from centralized *macropower* to dispersed *micropower* is analogous to the computer industry's shift from large, centralized mainframes to increasingly smaller, widely dispersed PCs, laptops, and handheld computers.

This trend of power-plant downsizing involves increased use of **(1)** moderate-size industrial cogeneration plants (50,000 kilowatts) and **(2)** energy-efficient, natural gas-burning generators (micro-turbines) for commercial buildings and residences (5-10,000 kilowatts). These microturbines are tiny jet engines that use heat released by combustion to spin a shaft that spins a high-speed generator. They can be fitted with lean burn engines and catalytic converters to reduce air pollution (mostly nitrogen oxides) and mufflers and soundproofing to reduce noise. Like central air conditioning units, they are serviced regularly by professionals.

This downsizing will be accelerated by the switch to increasingly smaller and more dispersed micropower systems such as **(1)** *wind turbines* (1–3,000 kilowatts), **(2)** low-cost *microturbines* for businesses (25–300 kilowatts), **(3)** energy-efficient *Stirling engines* (0.1–100

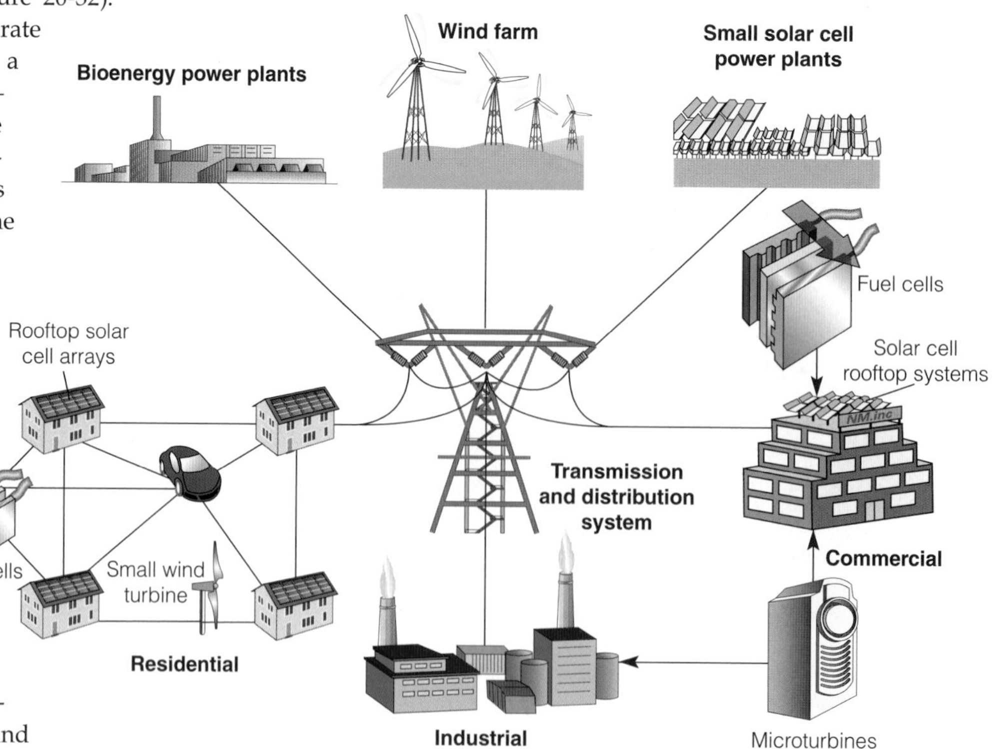

Figure 20-32 Decentralized power system in which electricity is produced by a large number of dispersed, small-scale *micropower systems* (producing 1–10,000 kilowatts of power). Some would produce power on site and others would feed the power they produce into a conventional electrical distribution system. Over the next few decades, many energy and financial analysts expect a shift to this type of power system.

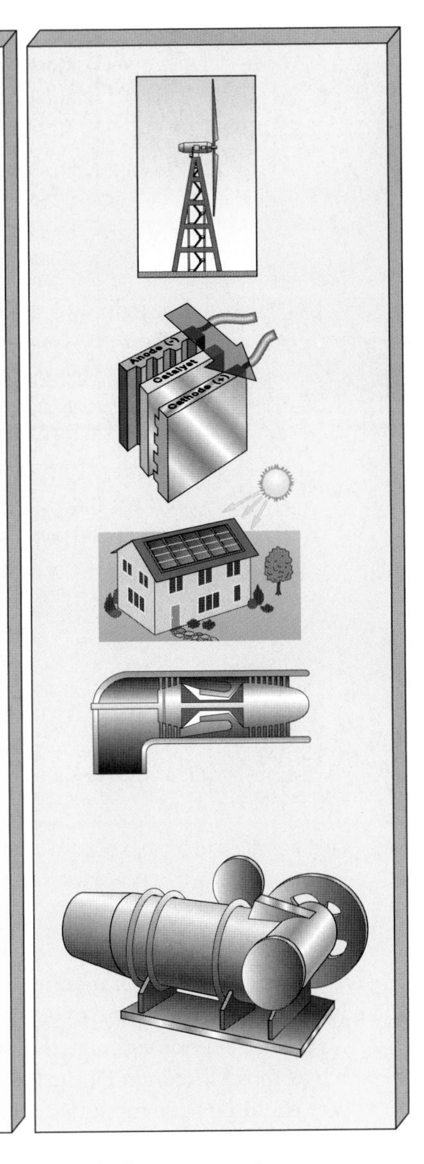

Small modular units
Fast factory production
Fast installation (hours to days)
Can add or remove modules as needed
High energy efficiency (60–80%)
Low or no CO_2 emissions
Low air pollution emissions
Reliable
Easy to repair
Much less vulnerable to power outages and terrorist attack
Useful anywhere
Especially useful in rural areas in developing countries with no power
Can use locally available renewable energy resources
Easily financed (costs included in mortgage and commercial loan)
Anode
Catalyst
Cathode

Figure 20-33 Some advantages of micropower systems.

kilowatts), **(4)** efficient, quiet, reliable, low-maintenance *fuel cells* (1–10,000 kilowatts), and **(5)** quiet, reliable, low-maintenance household *solar panels and solar roofs* (1–1,000 kilowatts, Figure 20-18). Figure 20-33 lists some of the advantages of decentralized micropower systems (Figure 20-32) over traditional macropower systems (Figure 20-31).

20-10 SOLUTIONS: A SUSTAINABLE ENERGY STRATEGY

What Are the Best Energy Alternatives? We have a variety of nonrenewable and renewable energy resources, each with certain advantages and disadvantages. Many scientists and energy experts who have evaluated these energy alternatives have come to the following general conclusions:

- *There will be a shift from centralized macropower systems* (Figure 20-31) *to smaller, decentralized micropower systems* (Figures 20-32 and 20-33).
- *The best alternatives are a combination of improved energy efficiency and using natural gas as a fuel to make the transition to increased use of a variety of small-scale, decentralized, locally available renewable energy resources.*
- *Because there is not enough money to develop all energy alternatives, governments and private companies must choose carefully which alternatives to support.*
- *Over the next 50 years the choice is not between using nonrenewable fossil fuels and various types of renewable energy.* Because of their supplies and low prices, fossil fuels will continue to be used in large quantities. The key questions are **(1)** how we can reduce the harmful environmental impacts of widespread fossil fuel use (especially to reduce air pollution and slow global warming) and **(2)** what roles improving energy efficiency and depending more on some forms of renewable energy can play in achieving these goals.

What Role Does Economics Play in Energy Resource Use? To most analysts the key to making a shift to a more sustainable energy resources and societies is not technology but economics and politics. Governments use three basic economic and political strategies to help stimulate or dampen the short-term and long-term use of a particular energy resource:

- *Allowing all energy resources to compete in a free market* (p. 24) *without any government interference.* This is rarely politically feasible because of well-entrenched government intervention into the marketplace in the form of subsidies, taxes, and regulations. Furthermore, the free-market approach, with its emphasis on short-term profit, can inhibit development of new energy resources, which can rarely compete economically in their early stages without government support.
- *Trying to keep energy prices artificially low to encourage use of selected energy resources.* This is done mostly by **(1)** providing research and development subsidies and tax breaks and **(2)** enacting regulations that help stimulate the development and use of energy resources receiving such support. For decades, this approach has been used to help stimulate the development and use of fossil fuels and nuclear power in most developed countries. This has created an uneven economic playing field that **(1)** encourages energy waste and rapid depletion of a nonrenewable energy resource and **(2)** discourages the development of renewable energy alternatives that are not getting at least the same level of subsidies and tax breaks. For example, fairly meager U.S. government research and development funding for renewable energy

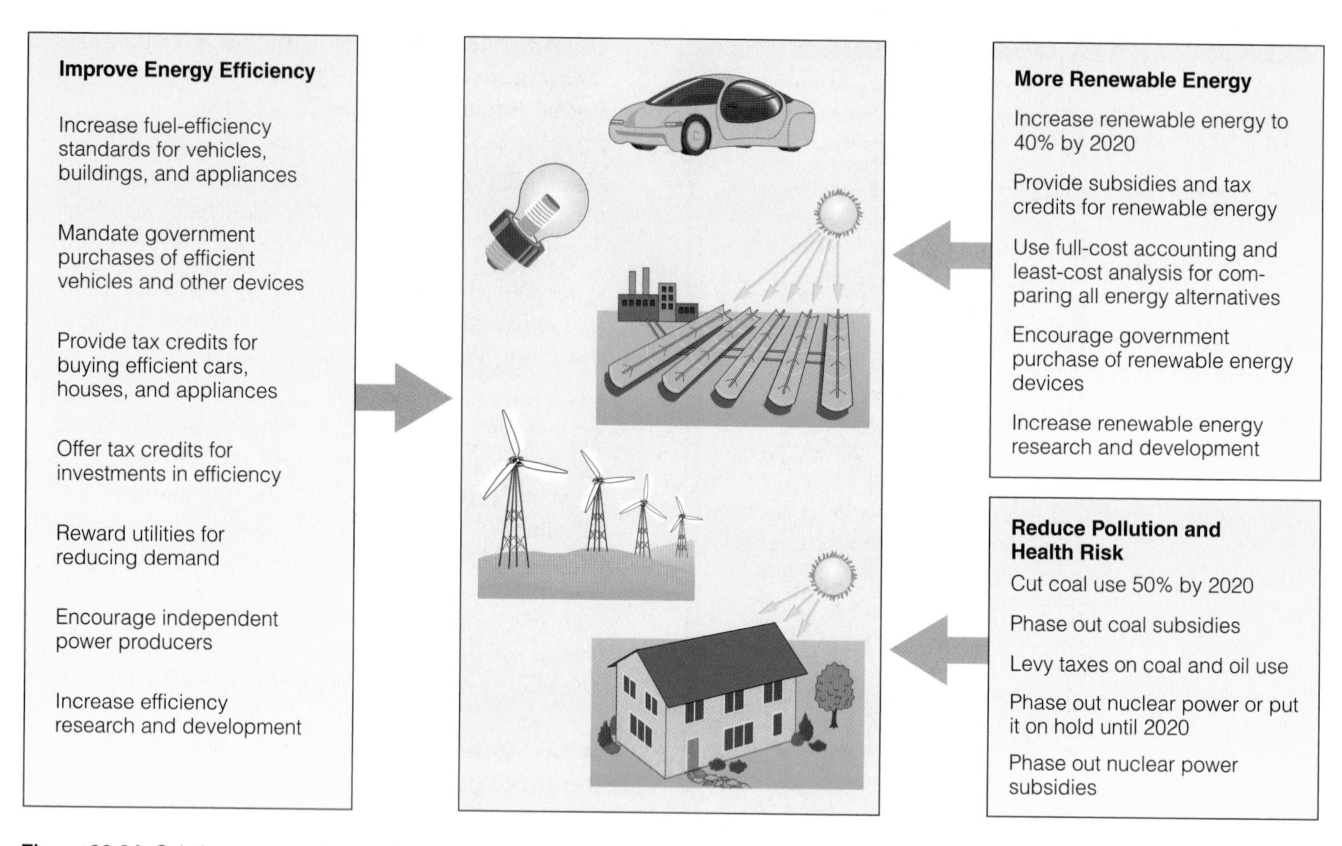

Improve Energy Efficiency
Increase fuel-efficiency standards for vehicles, buildings, and appliances
Mandate government purchases of efficient vehicles and other devices
Provide tax credits for buying efficient cars, houses, and appliances
Offer tax credits for investments in efficiency
Reward utilities for reducing demand
Encourage independent power producers
Increase efficiency research and development
More Renewable Energy
Increase renewable energy to 40% by 2020
Provide subsidies and tax credits for renewable energy
Use full-cost accounting and least-cost analysis for comparing all energy alternatives
Encourage government purchase of renewable energy devices
Increase renewable energy research and development
Reduce Pollution and Health Risk
Cut coal use 50% by 2020
Phase out coal subsidies
Levy taxes on coal and oil use
Phase out nuclear power or put it on hold until 2020
Phase out nuclear power subsidies

Figure 20-34 *Solutions*: suggestions various analysts have made to help make the transition to a more sustainable energy future.

declined by 87% between 1980 and 1990 and by 2002 was still 63% below 1980 levels.

- *Keeping energy prices artificially high to discourage use of a resource.* Governments can raise the price of an energy resource by **(1)** withdrawing existing tax breaks and other subsidies, **(2)** enacting restrictive regulations, or **(3)** adding taxes on its use. This **(1)** increases government revenues, **(2)** encourages improvements in energy efficiency, **(3)** reduces dependence on imported energy, and **(4)** decreases use of an energy resource that has a limited future supply.

Many economists favor *increasing taxes on fossil fuels* as a way to reduce air and water pollution and slow global warming. The tax revenues would be used to **(1)** reduce income taxes on wages and profits, **(2)** improve energy efficiency, **(3)** encourage use of renewable energy resources, and **(4)** provide energy assistance to the poor and lower middle class. Some economists believe that the public might accept these higher taxes if income and payroll taxes were lowered as gasoline or other fossil fuel taxes were raised (Solutions, p. 33).

How Can We Develop a More Sustainable Energy Future? Figure 20-34 lists a variety of strategies analysts have suggested for making the transition to a more sustainable energy future over the next few decades.

Energy experts estimate that implementing policies such as those shown in Figure 20-34 over the next 20–30 years could **(1)** save money, **(2)** create a net gain in jobs, **(3)** reduce greenhouse gas emissions, **(4)** sharply reduce air and water pollution, and **(5)** increase national security. See the website for this chapter for some actions that you can take to promote a more sustainable energy future.

A transition to renewable energy is inevitable, not because fossil fuel supplies will run out—large reserves of oil, coal, and gas remain in the world—but because the costs and risks of using these supplies will continue to increase relative to renewable energy.

MOHAMED EL-ASHRY

REVIEW QUESTIONS

1. Define the boldfaced terms in this chapter.

2. What is *energy efficiency*? How much of the energy used in the United States is wasted? What percentage of this is wasted because of the second law of energy, and what percentage is wasted unnecessarily? What is *life-cycle cost*? What are three of the least efficient energy-using devices?

3. Explain why we cannot recycle energy. List three ways to slow down the flow of heat from **(a)** a house and **(b)** an office building.

4. What are the advantages of saving energy?

5. What is *cogeneration*, and how efficient is it compared with producing electricity by a conventional coal-burning or nuclear power plant? List two other ways to save energy in industry.

6. What do most experts believe is the best way to save energy in transportation?

7. List the pros and cons of using **(a)** hybrid cars, **(b)** fuel-cell cars, and **(c)** electric bicycles.

8. Describe how we can save energy in homes by using **(a)** superinsulated houses and **(b)** strawbale houses. What are the four most efficient ways to heat a house? Describe ways to make an existing house more energy efficient. What are the most efficient and least efficient ways to heat water for washing and bathing? List the pros and cons of switching from inefficient incandescent and halogen light bulbs to efficient compact fluorescent light bulbs.

9. Describe how using the internet can save energy and help reduce carbon dioxide emissions.

10. List three reasons why there is little emphasis on saving energy in the United States, despite its important benefits.

11. What are the major advantages and disadvantages of relying more on direct and indirect renewable energy from the sun?

12. Distinguish between a *passive solar heating system* and an *active solar heating system* and list the pros and cons of each system.

13. Describe three ways to cool houses naturally.

14. Distinguish between the following solar systems used to generate high-temperature heat and electricity: **(a)** power tower, **(b)** solar thermal plant, **(c)** parabolic dish collection system, **(d)** nonimaging optical solar concentrator, and **(e)** solar stoves. List the advantages and disadvantages of concentrating solar energy to produce high-temperature heat or electricity.

15. What is a *solar cell*? List the advantages and disadvantages of using solar cells to produce electricity.

16. Distinguish between *large-scale hydropower*, *small-scale-hydropower*, and *pumped-storage hydropower* systems. List the advantages and disadvantages of using hydropower to produce electricity.

17. List the advantages and disadvantages of using the following systems for storing heat in water to produce electricity: **(a)** ocean thermal energy conversion (OTEC), **(b)** saline solar ponds, and **(c)** freshwater solar ponds.

18. List the advantages and disadvantages of using wind to produce electricity.

19. List the advantages and disadvantages of **(a)** burning solid biomass as a source of energy and **(b)** producing gaseous and liquid fuels from solid biomass.

20. What is the *solar–hydrogen revolution*? List the advantages and disadvantages of using hydrogen as a source of energy.

21. What is *geothermal energy*? Describe three types of geothermal reservoirs. List the advantages and disadvantages of using geothermal energy to produce heat and electricity.

22. What is *micropower*, and what are its advantages over macropower electricity systems? Describe five types of micropower systems.

23. What four conclusions have energy experts reached about possible future energy alternatives?

24. Summarize the three different economic approaches that can be used to stimulate or dampen the use of a particular energy resource. List the pros and cons of each approach.

25. What are major ways to help make the transition to a more sustainable energy future?

CRITICAL THINKING

1. A homebuilder installs electric baseboard heat and claims that "it's the cheapest and cleanest way to go." Apply your understanding of the second law of energy (thermodynamics) to evaluate this claim.

2. Someone tells you that we can save energy by recycling it. How would you respond?

3. Should the Corporate Average Fuel Economy (CAFE) standards for motor vehicles used in the United States be increased, left at 1985 levels (the current situation), or eliminated? Explain. Should the CAFE standards for light trucks, vans, and sport utility vehicles be increased to the same level as for cars? Explain. List the positive and negative effects on your health and lifestyle if CAFE standards are **(a)** increased or **(b)** eliminated.

4. What are the five most important things an individual can do to save energy at home and in transportation (see website for this chapter)? Which, if any, of these do you currently do? Which, if any, do you plan to do?

5. Explain why you agree or disagree with the following proposals by various energy analysts: **(a)** Federal subsidies for all energy alternatives should be eliminated so that all energy choices can compete in a pure free-market system (p. 24), **(b)** all government tax breaks and other subsidies for conventional fuels (oil, natural gas, coal), synthetic natural gas and oil, and nuclear power (fission and fusion) should be removed and replaced with subsidies and tax breaks for improving energy efficiency and developing solar, wind, geothermal, and biomass energy alternatives, and **(c)** development of solar and wind energy should be left to private enterprise and receive little or no help from the federal government, but nuclear energy and fossil fuels should continue to receive large federal subsidies.

6. Explain why you agree or disagree with the proposals suggested in Figure 20-34 as ways to promote a more sustainable energy future.

7. Congratulations. You have just been put in charge of the world. List the five most important features of your energy policy.

PROJECTS

1. Make a study of energy use in your school and use the findings to develop an energy-efficiency improvement program. Present your plan to school officials.

2. Learn how easy it is to produce hydrogen gas from water using a battery, some wire for two electrodes, and a dish of water. Hook a wire to each of the poles of the battery, immerse the electrodes in the water, and observe bubbles of hydrogen gas being produced at the negative electrode and bubbles of oxygen at the positive electrode. Carefully add a small amount of battery acid to the water and notice that this increases the rate of hydrogen production.

3. Make a concept map of this chapter's major ideas, using the section heads and subheads and the key terms (in boldface). Look on the website for this book for information about making concept maps.

INTERNET STUDY RESOURCES AND RESOURCES FOR FURTHER READING AND RESEARCH

The website for this book contains helpful study aids and many ideas for further reading and research. Log on to

http://www.brookscole.com/product/0534389872s

and click on the Chapter-by-Chapter area. Choose Chapter 20 and select a resource:

- "Flash Cards" allows you to test your mastery of the Terms and Concepts to Remember for this chapter.
- "Tutorial Quizzes" provides a multiple-choice practice quiz.
- "Student Guide to InfoTrac" will lead you to Critical Thinking Projects that use InfoTrac College Edition as a research tool.
- "References" lists the major books and articles consulted in writing this chapter.
- "Hypercontents" takes you to an extensive list of sites with news, research, and images related to individual sections of the chapter.

INFOTRAC COLLEGE EDITION

Improve your skills with InfoTrac College Edition, a searchable online database of articles from more than 700 periodicals. Log on to

http://www.infotrac-college.com

or access InfoTrac through the website for this book. Try to find the following articles:

1. The First H-Powered Country (hydrogen fuel). *Earth Island Journal*, Summer 2001 v16 i2 p14. Iceland currently uses geothermal energy for most of its space heating and electrical generation. Now it is planning to replace internal combustion engines with hydrogen-powered fuel cells. *Hint*: Enter the search term "iceland" using the keyword "hydrogen."

2. Oil, Profits, and the Question of Alternative Energy. Richard Rosentreter. *The Humanist*, Sept 2000 v60 i5 p8. Oil companies own many of the patents on alternative energy devices. This article suggests that they have no incentive to pursue alternative energy while the profits from oil are high, and they would rather spend money on exploring for oil. *Hint*: Enter the search term "oil profits" using the keywords "oil" and "alternative energy."

APPENDIX 1

UNITS OF MEASURE

LENGTH

Metric
1 kilometer (km) = 1,000 meters (m)
1 meter (m) = 100 centimeters (cm)
1 meter (m) = 1,000 millimeters (mm)
1 centimeter (cm) = 0.01 meter (m)
1 millimeter (mm) = 0.001 meter (m)

English
1 foot (ft) = 12 inches (in)
1 yard (yd) = 3 feet (ft)
1 mile (mi) = 5,280 feet (ft)
1 nautical mile = 1.15 miles

Metric–English
1 kilometer (km) = 0.621 mile (mi)
1 meter (m) = 39.4 inches (in)
1 inch (in) = 2.54 centimeters (cm)
1 foot (ft) = 0.305 meter (m)
1 yard (yd) = 0.914 meter (m)
1 nautical mile = 1.85 kilometers (km)

AREA

Metric
1 square kilometer (km^2) = 1,000,000 square meters (m^2)
1 square meter (m^2) = 1,000,000 square millimeters (mm^2)
1 hectare (ha) = 10,000 square meters (m^2)
1 hectare (ha) = 0.01 square kilometer (km^2)

English
1 square foot (ft^2) = 144 square inches (in^2)
1 square yard (yd^2) = 9 square feet (ft^2)
1 square mile (mi^2) = 27,880,000 square feet (ft^2)
1 acre (ac) = 43,560 square feet (ft^2)

Metric–English
1 hectare (ha) = 2.471 acres (ac)
1 square kilometer (km^2) = 0.386 square mile (mi^2)
1 square meter (m^2) = 1.196 square yards (yd^2)
1 square meter (m^2) = 10.76 square feet (ft^2)
1 square centimeter (cm^2) = 0.155 square inch (in^2)

VOLUME

Metric
1 cubic kilometer (km^3) = 1,000,000,000 cubic meters (m^3)
1 cubic meter (m^3) = 1,000,000 cubic centimeters (cm^3)
1 liter (L) = 1,000 milliliters (mL) = 1,000 cubic centimeters (cm^3)
1 milliliter (mL) = 0.001 liter (L)
1 milliliter (mL) = 1 cubic centimeter (cm^3)

English
1 gallon (gal) = 4 quarts (qt)
1 quart (qt) = 2 pints (pt)

Metric–English
1 liter (L) = 0.265 gallon (gal)
1 liter (L) = 1.06 quarts (qt)
1 liter (L) = 0.0353 cubic foot (ft^3)
1 cubic meter (m^3) = 35.3 cubic feet (ft^3)
1 cubic meter (m^3) = 1.30 cubic yards (yd^3)
1 cubic kilometer (km^3) = 0.24 cubic mile (mi^3)
1 barrel (bbl) = 159 liters (L)
1 barrel (bbl) = 42 U.S. gallons (gal)

MASS

Metric
1 kilogram (kg) = 1,000 grams (g)
1 gram (g) = 1,000 milligrams (mg)
1 gram (g) = 1,000,000 micrograms (μg)
1 milligram (mg) = 0.001 gram (g)
1 microgram (μg) = 0.000001 gram (g)
1 metric ton (mt) = 1,000 kilograms (kg)

English
1 ton (t) = 2,000 pounds (lb)
1 pound (lb) = 16 ounces (oz)

Metric–English
1 metric ton (mt) = 2,200 pounds (lb) = 1.1 tons (t)
1 kilogram (kg) = 2.20 pounds (lb)
1 pound (lb) = 454 grams (g)
1 gram (g) = 0.035 ounce (oz)

ENERGY AND POWER

Metric
1 kilojoule (kJ) = 1,000 joules (J)
1 kilocalorie (kcal) = 1,000 calories (cal)
1 calorie (cal) = 4,184 joules (J)

Metric–English
1 kilojoule (kJ) = 0.949 British thermal unit (Btu)
1 kilojoule (kJ) = 0.000278 kilowatt-hour (kW-h)
1 kilocalorie (kcal) = 3.97 British thermal units (Btu)
1 kilocalorie (kcal) = 0.00116 kilowatt-hour (kW-h)
1 kilowatt-hour (kW-h) = 860 kilocalories (kcal)
1 kilowatt-hour (kW-h) = 3,400 British thermal units (Btu)
1 quad (Q) = 1,050,000,000,000,000 kilojoules (kJ)
1 quad (Q) = 2,930,000,000,000 kilowatt-hours (kW-h)

TEMPERATURE CONVERSIONS

Fahrenheit (°F) to Celsius (°C): °C = (°F − 32.0)/1.80
Celsius (°C) to Fahrenheit (°F): °F = (°C × 1.80) + 32.0

Appendix 2

MAJOR EVENTS IN U.S. ENVIRONMENTAL HISTORY

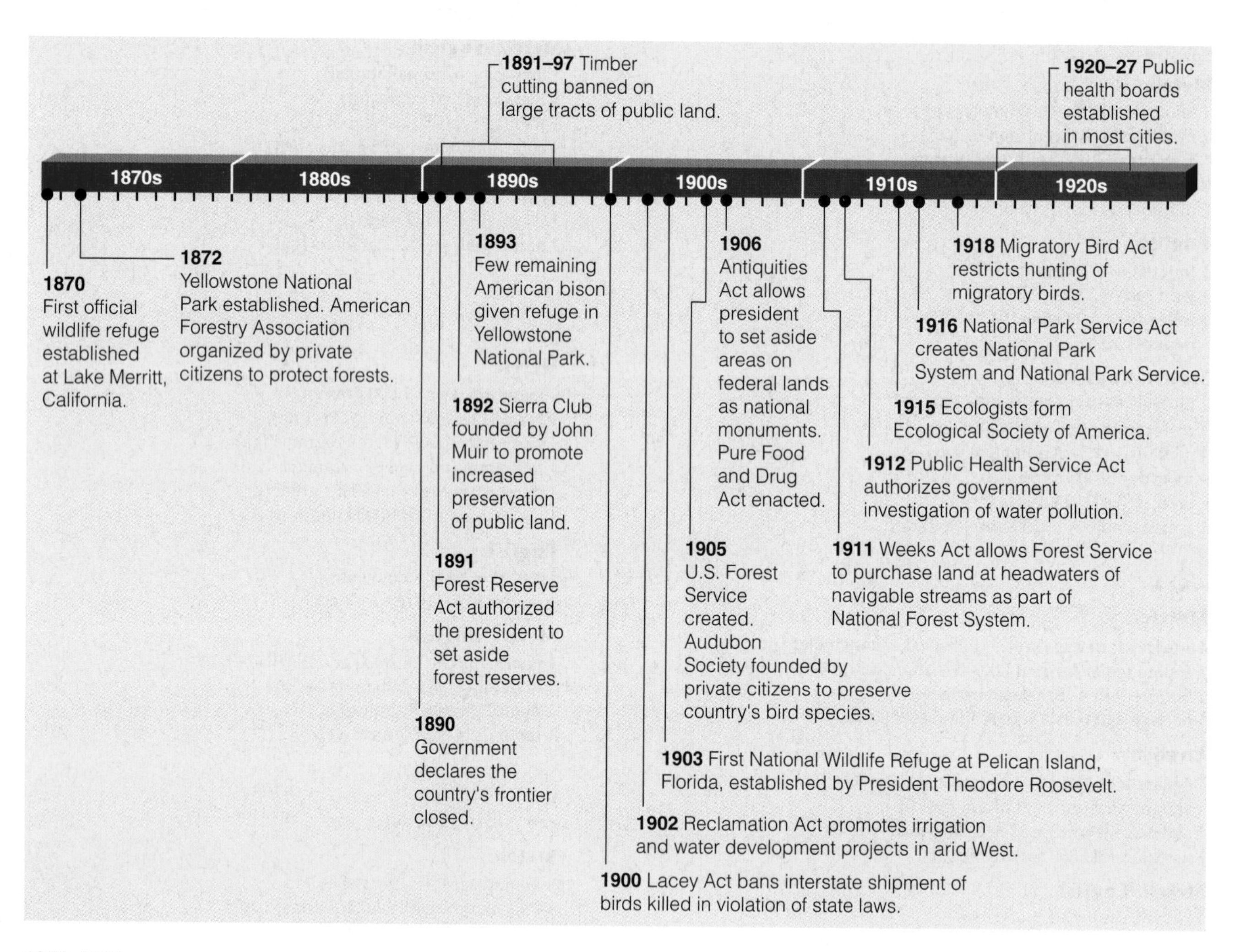

1870s
1880s
1890s
1900s
1910s
1920s
1870 First official wildlife refuge established at Lake Merritt, California.
1872 Yellowstone National Park established. American Forestry Association organized by private citizens to protect forests.
1890 Government declares the country's frontier closed.
1891 Forest Reserve Act authorized the president to set aside forest reserves.
1891–97 Timber cutting banned on large tracts of public land.
1892 Sierra Club founded by John Muir to promote increased preservation of public land.
1893 Few remaining American bison given refuge in Yellowstone National Park.
1900 Lacey Act bans interstate shipment of birds killed in violation of state laws.
1902 Reclamation Act promotes irrigation and water development projects in arid West.
1903 First National Wildlife Refuge at Pelican Island, Florida, established by President Theodore Roosevelt.
1905 U.S. Forest Service created. Audubon Society founded by private citizens to preserve country's bird species.
1906 Antiquities Act allows president to set aside areas on federal lands as national monuments. Pure Food and Drug Act enacted.
1911 Weeks Act allows Forest Service to purchase land at headwaters of navigable streams as part of National Forest System.
1912 Public Health Service Act authorizes government investigation of water pollution.
1915 Ecologists form Ecological Society of America.
1916 National Park Service Act creates National Park System and National Park Service.
1918 Migratory Bird Act restricts hunting of migratory birds.
1920–27 Public health boards established in most cities.

1870-1930

Figure 1 Examples of the increased role of the federal government in resource conservation and public health and establishment of key private environmental groups, 1870-1930.

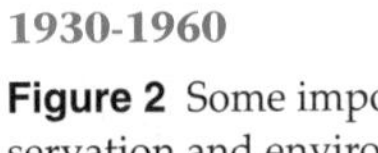

1930s
1940s
1950s
1938 Federal Food, Drug, and Cosmetic Act regulates consumer foods, drugs, and cosmetics.
1937 Federal Aid in Wildlife Restoration Act levies federal tax on gun and ammunition sales, with funds used for wildlife research and protection.
1935 Soil Conservation Act creates Soil Erosion Service. Wilderness Society founded.
1934 Taylor Grazing Act regulates livestock grazing on public lands. Migratory Bird Hunting Stamp Act requires federal license for duck hunters, with funds used for waterfowl refuges.
1933 Civilian Conservation Service established.
1948 Air pollution disaster at Donora, Pennsylvania, kills 22 and sickens 7,000 people.
1947 Federal Insecticide, Fungicide, and Rodenticide Act regulates use of pesticides.
1940 U.S. Fish and Wildlife Service created to manage National Wildlife Refuge system and protect endangered species.
1957 Price–Anderson Act greatly limits liability of power plant owners and the government in cases of a major nuclear power plant accident.
1956 Water Pollution Control Act provides grants to states for water treatment plants.
1954 Atomic Energy Act promotes development of nuclear power plants.

1930-1960

Figure 2 Some important conservation and environmental events between 1930 and 1960.

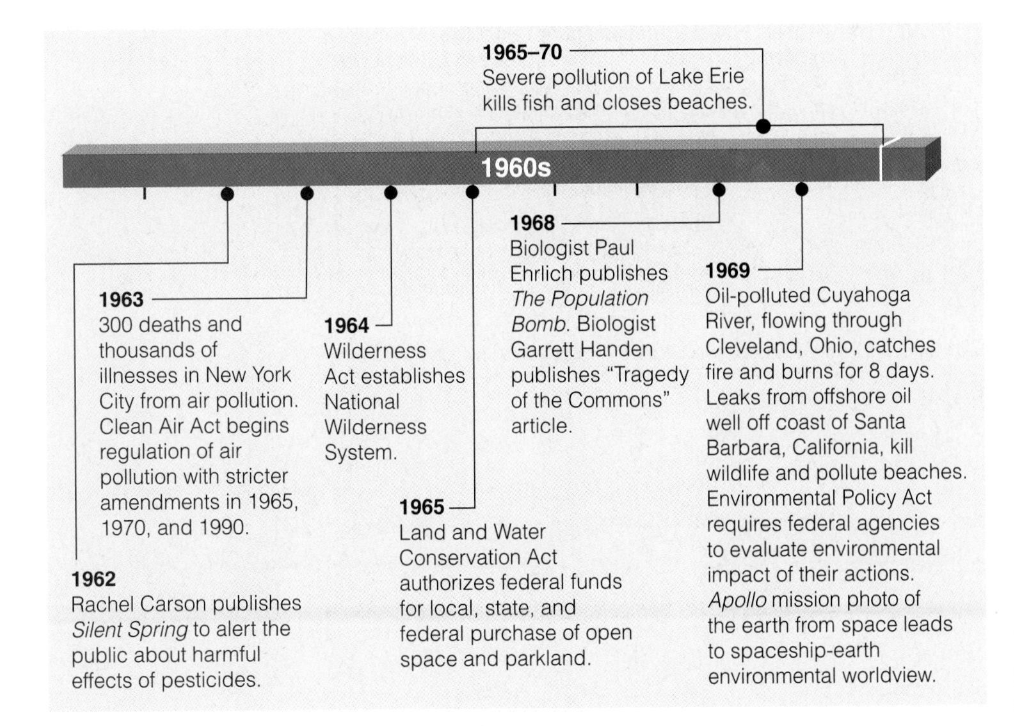

1965–70
Severe pollution of Lake Erie kills fish and closes beaches.
1960s
1963
300 deaths and thousands of illnesses in New York City from air pollution. Clean Air Act begins regulation of air pollution with stricter amendments in 1965, 1970, and 1990.
1962
Rachel Carson publishes Silent Spring to alert the public about harmful effects of pesticides.
1964
Wilderness Act establishes National Wilderness System.
1965
Land and Water Conservation Act authorizes federal funds for local, state, and federal purchase of open space and parkland.
1968
Biologist Paul Ehrlich publishes The Population Bomb. Biologist Garrett Handen publishes "Tragedy of the Commons" article.
1969
Oil-polluted Cuyahoga River, flowing through Cleveland, Ohio, catches fire and burns for 8 days. Leaks from offshore oil well off coast of Santa Barbara, California, kill wildlife and pollute beaches. Environmental Policy Act requires federal agencies to evaluate environmental impact of their actions. Apollo mission photo of the earth from space leads to spaceship-earth environmental worldview.

1960s

Figure 3 Some important environmental events during the 1960s.

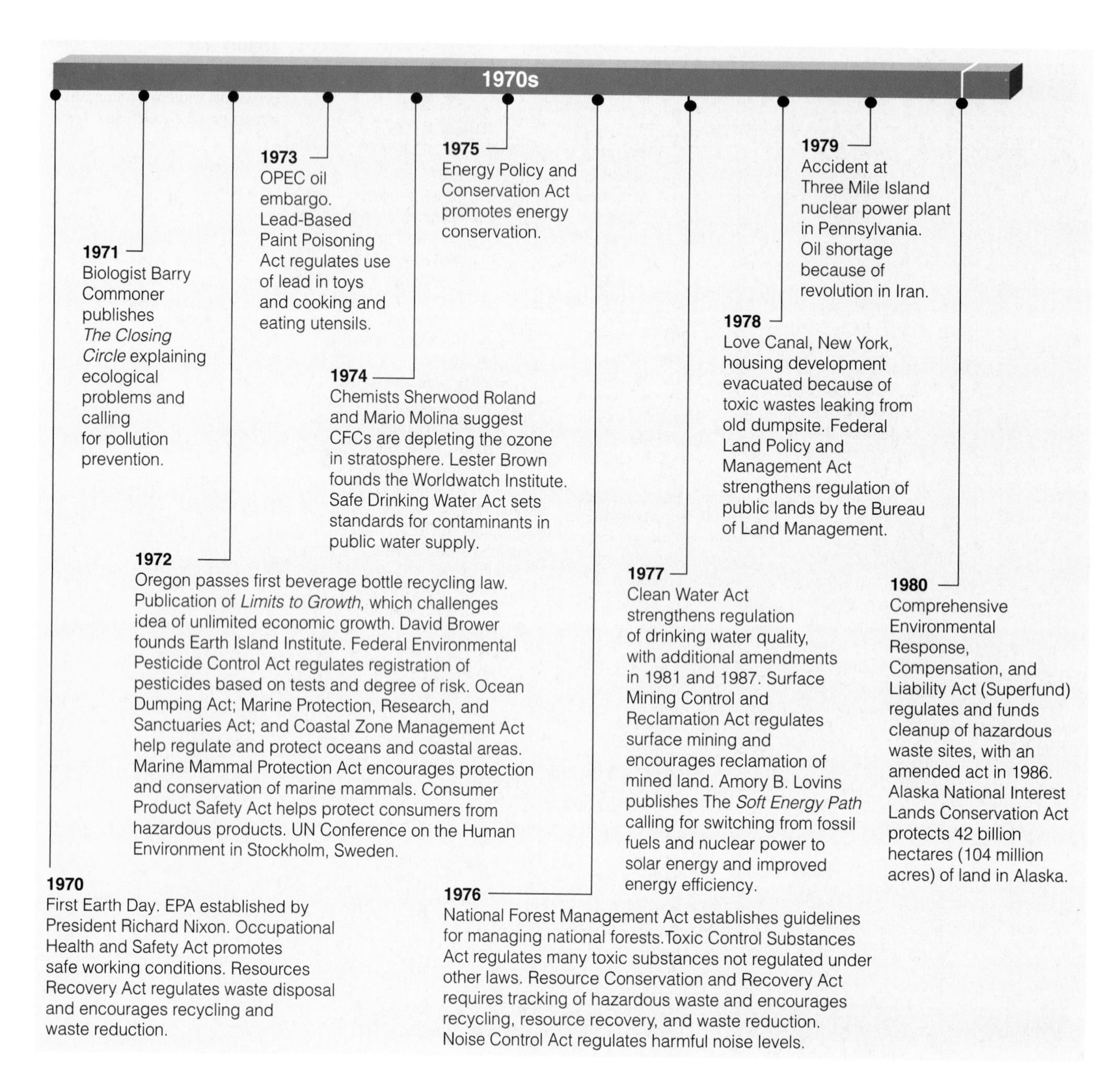
1970s
1971
Biologist Barry Commoner publishes The Closing Circle explaining ecological problems and calling for pollution prevention.
1973
OPEC oil embargo. Lead-Based Paint Poisoning Act regulates use of lead in toys and cooking and eating utensils.
1975
Energy Policy and Conservation Act promotes energy conservation.
1979
Accident at Three Mile Island nuclear power plant in Pennsylvania. Oil shortage because of revolution in Iran.
1978
Love Canal, New York, housing development evacuated because of toxic wastes leaking from old dumpsite. Federal Land Policy and Management Act strengthens regulation of public lands by the Bureau of Land Management.
1974
Chemists Sherwood Roland and Mario Molina suggest CFCs are depleting the ozone in stratosphere. Lester Brown founds the Worldwatch Institute. Safe Drinking Water Act sets standards for contaminants in public water supply.
1972
Oregon passes first beverage bottle recycling law. Publication of Limits to Growth, which challenges idea of unlimited economic growth. David Brower founds Earth Island Institute. Federal Environmental Pesticide Control Act regulates registration of pesticides based on tests and degree of risk. Ocean Dumping Act; Marine Protection, Research, and Sanctuaries Act; and Coastal Zone Management Act help regulate and protect oceans and coastal areas. Marine Mammal Protection Act encourages protection and conservation of marine mammals. Consumer Product Safety Act helps protect consumers from hazardous products. UN Conference on the Human Environment in Stockholm, Sweden.
1977
Clean Water Act strengthens regulation of drinking water quality, with additional amendments in 1981 and 1987. Surface Mining Control and Reclamation Act regulates surface mining and encourages reclamation of mined land. Amory B. Lovins publishes The Soft Energy Path calling for switching from fossil fuels and nuclear power to solar energy and improved energy efficiency.
1980
Comprehensive Environmental Response, Compensation, and Liability Act (Superfund) regulates and funds cleanup of hazardous waste sites, with an amended act in 1986. Alaska National Interest Lands Conservation Act protects 42 billion hectares (104 million acres) of land in Alaska.
1970
First Earth Day. EPA established by President Richard Nixon. Occupational Health and Safety Act promotes safe working conditions. Resources Recovery Act regulates waste disposal and encourages recycling and waste reduction.
1976
National Forest Management Act establishes guidelines for managing national forests.Toxic Control Substances Act regulates many toxic substances not regulated under other laws. Resource Conservation and Recovery Act requires tracking of hazardous waste and encourages recycling, resource recovery, and waste reduction. Noise Control Act regulates harmful noise levels.

1970s

Figure 4 Some important environmental events during the 1970s, sometimes called the *environmental decade.*

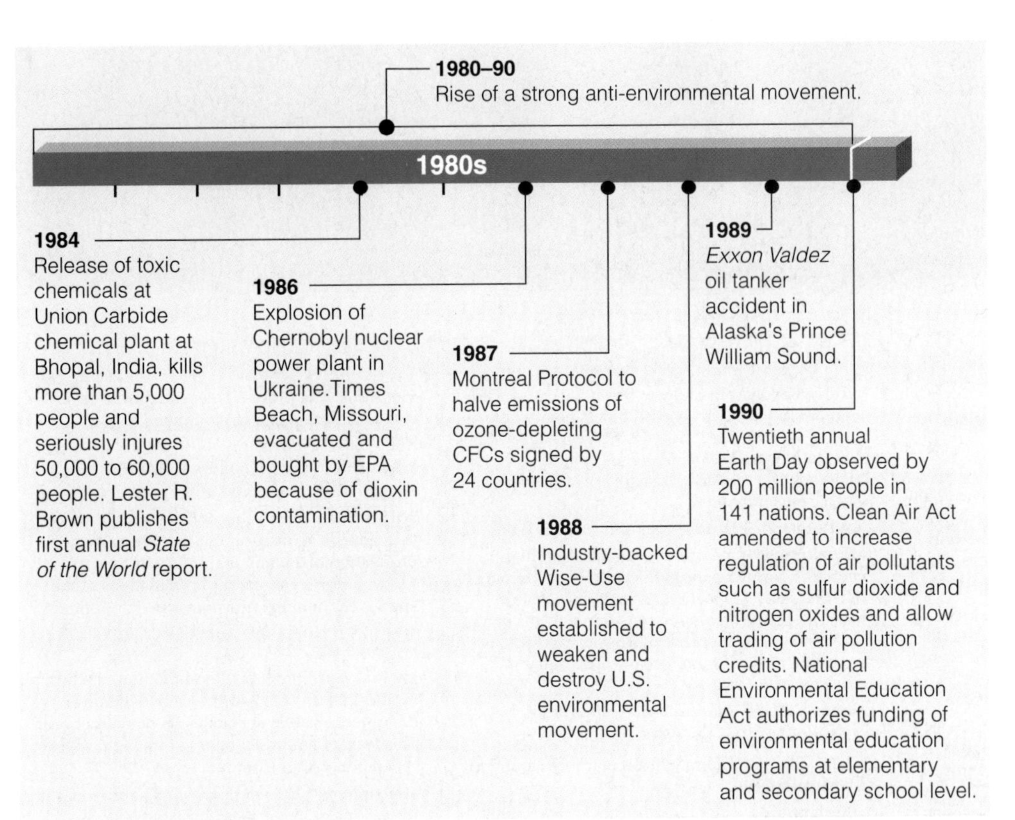

1980–90
Rise of a strong anti-environmental movement.
1980s
1984
Release of toxic chemicals at Union Carbide chemical plant at Bhopal, India, kills more than 5,000 people and seriously injures 50,000 to 60,000 people. Lester R. Brown publishes first annual *State of the World* report.
1986
Explosion of Chernobyl nuclear power plant in Ukraine.Times Beach, Missouri, evacuated and bought by EPA because of dioxin contamination.
1987
Montreal Protocol to halve emissions of ozone-depleting CFCs signed by 24 countries.
1988
Industry-backed Wise-Use movement established to weaken and destroy U.S. environmental movement.
1989
Exxon Valdez oil tanker accident in Alaska's Prince William Sound.
1990
Twentieth annual Earth Day observed by 200 million people in 141 nations. Clean Air Act amended to increase regulation of air pollutants such as sulfur dioxide and nitrogen oxides and allow trading of air pollution credits. National Environmental Education Act authorizes funding of environmental education programs at elementary and secondary school level.

1980s

Figure 5 Some important environmental events during the 1980s.

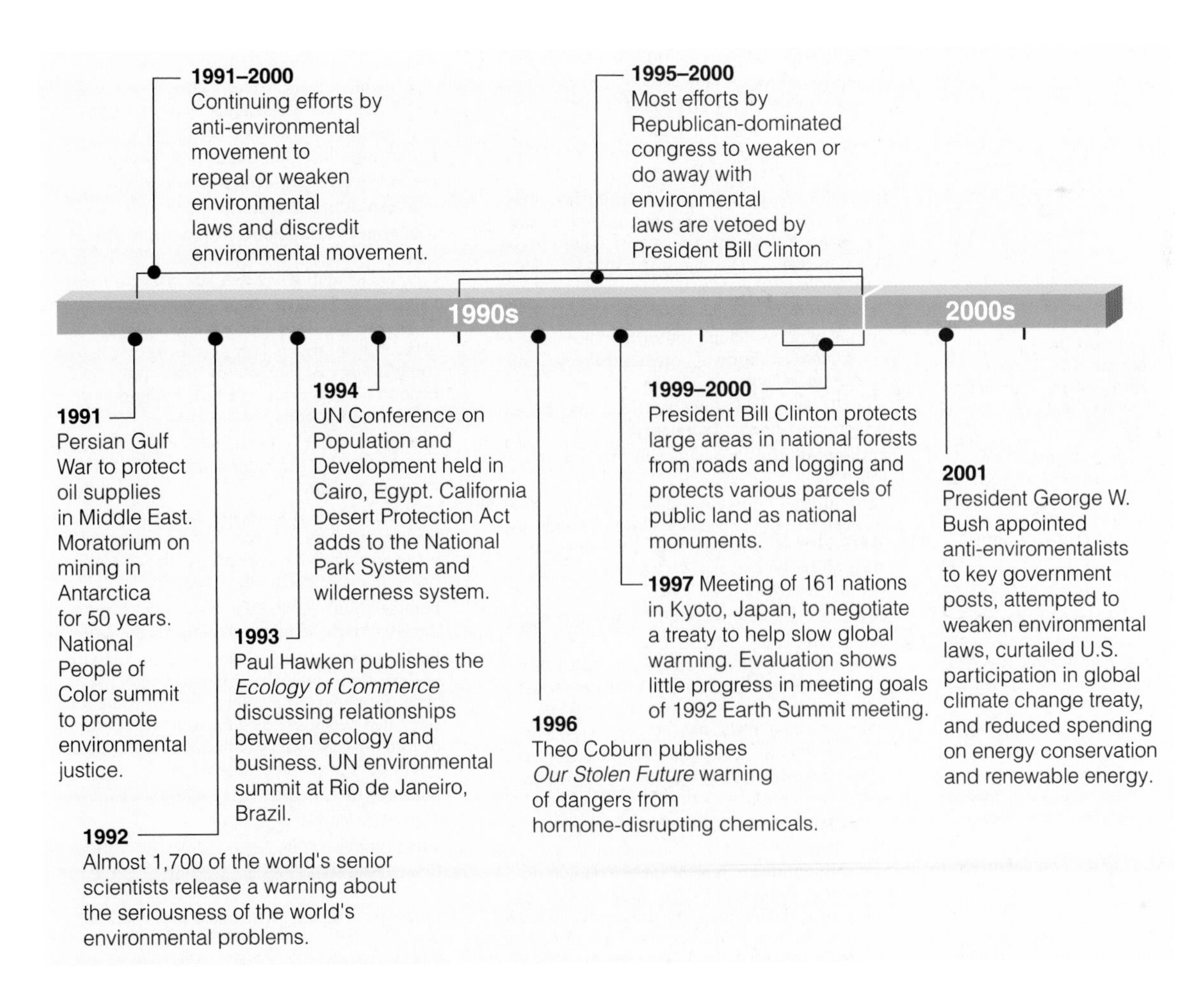

1990–2001

Figure 6 Some important environmental events, 1990–2001.

GLOSSARY

abiotic Nonliving. Compare *biotic.*

acclimation Adjustment to slowly changing new conditions. Compare *threshold effect.*

acid See *acid solution.*

acid deposition The falling of acids and acid-forming compounds from the atmosphere to the earth's surface. Acid deposition is commonly known as *acid rain*, a term that refers only to wet deposition of droplets of acids and acid-forming compounds.

acid rain See *acid deposition.*

acid solution Any water solution that has more hydrogen ions (H^+) than hydroxide ions (OH^-); any water solution with a pH less than 7. Compare *basic solution, neutral solution.*

active solar heating system System that uses solar collectors to capture energy from the sun and store it as heat for space heating and water heating. Liquid or air pumped through the collectors transfers the captured heat to a storage system such as an insulated water tank or rock bed. Pumps or fans then distribute the stored heat or hot water throughout a dwelling as needed. Compare *passive solar heating system.*

adaptation Any genetically controlled structural, physiological, or behavioral characteristic that helps an organism survive and reproduce under a given set of environmental conditions. It usually results from a beneficial mutation. See *biological evolution, differential reproduction, mutation, natural selection.*

adaptive management Flexible management that views attempts to solve problems as experiments, analyzes failures to see what went wrong, and tries to modify and improve an approach before abandoning it. Because of the inherent unpredictability of complex systems, it often uses the precautionary principle as a management tool. See *precautionary principle.*

adaptive radiation Process in which numerous new species evolve to fill vacant and new ecological niches in changed environments, usually after a mass extinction. Typically, this takes millions of years.

adaptive trait See *adaptation.*

advanced sewage treatment Specialized chemical and physical processes that reduce the amount of specific pollutants left in wastewater after primary and secondary sewage treatment. This type of treatment usually is expensive. See also *primary sewage treatment, secondary sewage treatment.*

aerobic respiration Complex process that occurs in the cells of most living organisms, in which nutrient organic molecules such as glucose ($C_6H_{12}O_6$) combine with oxygen (O_2) and produce carbon dioxide (CO_2), water (H_2O), and energy. Compare *photosynthesis.*

age structure Percentage of the population (or number of people of each sex) at each age level in a population.

agricultural revolution Gradual shift from small, mobile hunting and gathering bands to settled agricultural communities in which people survived by learning how to breed and raise wild animals and to cultivate wild plants near where they lived. It began 10,000–12,000 years ago. Compare *environmental revolution, hunter–gatherers, industrial revolution, information and globalization revolution.*

agroforestry Planting trees and crops together.

air pollution One or more chemicals in high enough concentrations in the air to harm humans, other animals, vegetation, or materials. Excess heat and noise can also be considered forms of air pollution. Such chemicals or physical conditions are called air pollutants. See *primary pollutant, secondary pollutant.*

albedo Ability of a surface to reflect light.

alien species See *nonnative species.*

allele Slightly different molecular form found in a particular gene.

alley cropping Planting of crops in strips with rows of trees or shrubs on each side.

alpha particle Positively charged matter, consisting of two neutrons and two protons, that is emitted as a form of radioactivity from the nuclei of some radioisotopes. See also *beta particle, gamma rays.*

altitude Height above sea level. Compare *latitude.*

anaerobic respiration Form of cellular respiration in which some decomposers get the energy they need through the breakdown of glucose (or other nutrients) in the absence of oxygen. Compare *aerobic respiration.*

ancient forest See *old-growth forest.*

animal manure Dung and urine of animals that can be used as a form of organic fertilizer. Compare *green manure.*

annual Plant that grows, sets seed, and dies in one growing season. Compare *perennial.*

anthropocentric Human-centered. Compare *biocentric.*

aquaculture Growing and harvesting of fish and shellfish for human use in freshwater ponds, irrigation ditches, and lakes, or in cages or fenced-in areas of coastal lagoons and estuaries. See *fish farming, fish ranching.*

aquatic Pertaining to water. Compare *terrestrial.*

aquatic life zone Marine and freshwater portions of the biosphere. Examples include freshwater life zones (such as lakes and streams) and ocean or marine life zones (such as estuaries, coastlines, coral reefs, and the deep ocean).

aquifer Porous, water-saturated layers of sand, gravel, or bedrock that can yield an economically significant amount of water.

arable land Land that can be cultivated to grow crops.

area strip mining Type of surface mining used where the terrain is fairly flat. An earthmover strips away the overburden, and a power shovel digs a cut to remove the mineral deposit. After the mineral is removed, the trench is filled with overburden, and a new cut is made parallel to the previous one. The process is repeated over the entire site. Compare *dredging, open-pit mining, subsurface mining.*

arid Dry. A desert or other area with an arid climate has little precipitation.

artificial selection Process by which humans select one or more desirable genetic traits in the population of a plant or animal and then use *selective breeding* to end up with populations of the species containing large numbers of individuals with the desired traits. Compare *genetic engineering, natural selection.*

asexual reproduction Reproduction in which a mother cell divides to produce two identical daughter cells that are clones of the mother cell. This type of reproduction is common in single-celled organisms. Compare *sexual reproduction.*

atmosphere The whole mass of air surrounding the earth. See *stratosphere, troposphere.*

atom Minute unit made of subatomic particles that is the basic building block of all chemical elements and thus all matter; the smallest unit of an element that can exist and still have the unique characteristics of that element. Compare *ion, molecule.*

atomic number Number of protons in the nucleus of an atom. Compare *mass number.*

autotroph See *producer.*

background extinction Normal extinction of various species as a result of changes in local environmental conditions. Compare *mass extinction.*

bacteria Prokaryotic, one-celled organisms. Some transmit diseases. Most act as decomposers and get the nutrients they need by breaking down complex organic compounds in the tissues of living or dead organisms into simpler inorganic nutrient compounds.

barrier islands Long, thin, low offshore islands of sediment that generally run parallel to the shore along some coasts.

basic solution Water solution with more hydroxide ions (OH^-) than hydrogen ions (H^+); water solution with a pH greater than 7. Compare *acid solution, neutral solution.*

beneficiation Separation of an ore mineral from the waste mineral material (gangue). See *tailings.*

benefit–cost analysis Estimates and comparison of short-term and long-term costs (losses) and benefits (gains) from an economic decision. If the estimated benefits exceed the estimated costs, the decision to buy an economic good or provide a public good is considered worthwhile.

benthos Bottom-dwelling organisms. Compare *decomposer, nekton, plankton.*

beta particle Swiftly moving electron emitted by the nucleus of a radioactive isotope. See also *alpha particle, gamma rays.*

bioaccumulation An increase in the concentration of a chemical in specific organs or tissues at a level higher than would normally be expected. Compare *biomagnification.*

biocentric Life-Centered. Compare *anthropocentric.*

biodegradable Capable of being broken down by decomposers.

biodegradable pollutant Material that can be broken down into simpler substances (elements and compounds) by bacteria or other decomposers. Paper and most organic wastes such as animal manure are biodegradable but can take decades to biodegrade in modern landfills. Compare *degradable pollutant, nondegradable pollutant, slowly degradable pollutant.*

biodiversity Variety of different species (*species diversity*), genetic variability among individuals within each species (*genetic diversity*), variety of ecosystems (*ecological diversity*), and functions such as energy flow and matter cycling needed for the survival of species and biological communities (*functional diversity*).

biofuel Gas or liquid fuel (such as ethyl alcohol) made from plant material (biomass).

biogeochemical cycle Natural processes that recycle nutrients in various chemical forms from the nonliving environment to living organisms and then back to the nonliving environment. Examples are the carbon, oxygen, nitrogen, phosphorus, sulfur, and hydrologic cycles.

bioinformatics Applied science of managing, analyzing, and communicating biological information.

biological amplification See *biomagnification.*

biological community See *community.*

biological diversity See *biodiversity.*

biological evolution Change in the genetic makeup of a population of a species in successive generations. If continued long enough, it can lead to the formation of a new species. Note that populations—not individuals—evolve. See also *adaptation, differential reproduction, natural selection, theory of evolution.*

biological oxygen demand (BOD) Amount of dissolved oxygen needed by aerobic decomposers to break down the organic materials in a given volume of water at a certain temperature over a specified time period.

biological pest control Control of pest populations by natural predators, parasites, or disease-causing bacteria and viruses (pathogens).

biomagnification Increase in concentration of DDT, PCBs, and other slowly degradable, fat-soluble chemicals in organisms at successively higher trophic levels of a food chain or web. Compare *bioaccumulation.*

biomass Organic matter produced by plants and other photosynthetic producers; total dry weight of all living organisms that can be supported at each trophic level in a food chain or web; dry weight of all organic matter in plants and animals in an ecosystem; plant materials and animal wastes used as fuel.

biome Terrestrial regions inhabited by certain types of life, especially vegetation. Examples are various types of deserts, grasslands, and forests.

biosphere Zone of earth where life is found. It consists of parts of the atmosphere (the troposphere), hydrosphere (mostly surface water and groundwater), and lithosphere (mostly soil and surface rocks and sediments on the bottoms of oceans and other bodies of water) where life is found. Sometimes called the *ecosphere.*

biotic Living organisms. Compare *abiotic.*

biotic potential Maximum rate at which the population of a given species can increase when there are no limits on its rate of growth. See *environmental resistance.*

birth rate See *crude birth rate.*

breeder nuclear fission reactor Nuclear fission reactor that produces more nuclear fuel than it consumes by converting nonfissionable uranium-238 into fissionable plutonium-239.

broadleaf deciduous plants Plants such as oak and maple trees that survive drought and cold by shedding their leaves and becoming dormant. Compare *broadleaf evergreen plants, coniferous evergreen plants.*

broadleaf evergreen plants Plants that keep most of their broad leaves year-round. Examples are the trees found in the canopies of tropical rain forests. Compare *broadleaf deciduous plants, coniferous evergreen plants.*

buffer Substance that can react with hydrogen ions in a solution and thus hold the acidity or pH of a solution fairly constant. See *pH.*

cancer Group of more than 120 different diseases, one for each type of cell in the human body. Each type of cancer produces a tumor in which cells multiply uncontrollably and invade surrounding tissue.

capitalism See *capitalist market economic system.* Compare *pure command economic system, pure free-market economic system.*

capitalist market economic system Economic system built around controlling market prices of goods and services, global free trade, and maximizing profits for the owners or stockholders whose financial capital the company is using to do business. Compare *pure command economic system, pure free-market economic system.*

carbon cycle Cyclic movement of carbon in different chemical forms from the environment to organisms and then back to the environment.

carcinogen Chemicals, ionizing radiation, and viruses that cause or promote the development of cancer. See *cancer.* Compare *mutagen, teratogen.*

carnivore Animal that feeds on other animals. Compare *herbivore, omnivore.*

carrying capacity (K) Maximum population of a particular species that a given habitat can support over a given period of time.

cell Smallest living unit of an organism. Each cell is encased in an outer membrane or wall and contains genetic material (DNA) and other parts to perform its life function. Organisms such as bacteria consist of only one cell, but most of the organisms we are familiar with contain many cells. See *eukaryotic cell, prokaryotic cell.*

centrally planned economy See *pure command economic system.*

CFCs See *chlorofluorocarbons.*

chain reaction Multiple nuclear fissions, taking place within a certain mass of a fissionable isotope, that release an enormous amount of energy in a short time.

chemical One of the millions of different elements and compounds found naturally and synthesized by humans. See *compound, element.*

chemical change Interaction between chemicals in which there is a change in the chemical composition of the elements or compounds involved. Compare *nuclear change, physical change.*

chemical evolution Formation of the earth and its early crust and atmosphere, evolution of the biological molecules necessary for life, and evolution of systems of chemical reactions needed to produce the first living cells. These processes are believed to have occurred about 1 billion years before biological evolution. Compare *biological evolution.*

chemical formula Shorthand way to show the number of atoms (or ions) in the basic structural unit of a compound. Examples are H_2O, NaCl, and $C_6H_{12}O_6$.

chemical reaction See *chemical change.*

chemosynthesis Process in which certain organisms (mostly specialized bacteria) extract inorganic compounds from their environment and convert them into organic nutrient compounds without the presence of sunlight. Compare *photosynthesis.*

chlorinated hydrocarbon Organic compound made up of atoms of carbon, hydrogen, and chlorine. Examples are DDT and PCBs.

chlorofluorocarbons (CFCs) Organic compounds made up of atoms of carbon, chlorine, and fluorine. An example is Freon-12 (CCl_2F_2), used as a refrigerant in refrigerators and air conditioners and in making plastics such as Styrofoam. Gaseous CFCs can deplete the ozone layer when they slowly rise into the stratosphere and their chlorine atoms react with ozone molecules. Use of these molecules is being phased out.

chromosome A grouping of various genes and associated proteins in plant and animal cells that carry certain types of genetic information. See *genes.*

clear-cutting Method of timber harvesting in which all trees in a forested area are removed in a single cutting. Compare, *seed-tree cutting, selective cutting, shelterwood cutting, strip cutting.*

climate Physical properties of the troposphere of an area based on analysis of its weather records over a long period (at least 30 years). The two main factors determining an area's climate are *temperature*, with its seasonal variations, and the amount and distribution of *precipitation.* Compare *weather.*

climax community See *mature community.*

coal Solid, combustible mixture of organic compounds with 30–98% carbon by weight, mixed with various amounts of water and small amounts of sulfur and nitrogen compounds. It is formed in several stages as the remains of plants are subjected to heat and pressure over millions of years.

coal gasification Conversion of solid coal to synthetic natural gas (SNG).

coal liquefaction Conversion of solid coal to a liquid hydrocarbon fuel such as synthetic gasoline or methanol.

coastal wetland Land along a coastline, extending inland from an estuary, that is covered with salt water all or part of the year. Examples are marshes, bays, lagoons, tidal flats, and mangrove swamps. Compare *inland wetland.*

coastal zone Warm, nutrient-rich, shallow part of the ocean that extends from the high-tide mark on land to the edge of a shelflike extension of continental land masses known as the continental shelf. Compare *open sea.*

coevolution Evolution in which two or more species interact and exert selective pressures on each other that can lead each species to undergo various adaptations. See *evolution, natural selection.*

cogeneration Production of two useful forms of energy, such as high-temperature heat or steam and electricity, from the same fuel source.

commensalism An interaction between organisms of different species in which one type of organism benefits and the other type is neither helped nor harmed to any great degree. Compare *mutualism.*

commercial extinction Depletion of the population of a wild species used as a resource to a level at which it is no longer profitable to harvest the species.

commercial inorganic fertilizer Commercially prepared mixture of plant nutrients such as nitrates, phosphates, and potassium applied to the soil to restore fertility and increase crop yields. Compare *organic fertilizer.*

common-property resource Resource that people normally are free to use; each user can deplete or degrade the available supply. Most are renewable and are owned by no one. Examples are clean air, fish in parts of the ocean not under the control of a coastal country, migratory birds, gases of the lower atmosphere, and the ozone content of the upper atmosphere (strato-sphere). See *tragedy of the commons.*

community Populations of all species living and interacting in an area at a particular time.

community development See *ecological succession.*

competition Two or more individual organisms of a single species (*intraspecific competition*) or two or more individuals of different species (*interspecific competition*) attempting to use the same scarce resources in the same ecosystem.

competitive exclusion principle No two species can occupy exactly the same fundamental niche indefinitely in a habitat where there is not enough of a particular resource to meet the needs of both species. See *ecological niche, fundamental niche, realized niche.*

compost Partially decomposed organic plant and animal matter that can be used as a soil conditioner or fertilizer.

compound Combination of atoms, or oppositely charged ions, of two or more different elements held together by attractive forces called chemical bonds. Compare *element.*

concentration Amount of a chemical in a particular volume or weight of air, water, soil, or other medium.

condensation nuclei Tiny particles on which droplets of water vapor can collect.

coniferous evergreen plants Cone-bearing plants (such as spruces, pines, and firs) that keep some of their narrow, pointed leaves (needles) all year. Compare *broadleaf deciduous plants, broadleaf evergreen plants.*

coniferous trees Cone-bearing trees, mostly evergreens, that have needle-shaped or scalelike leaves. They produce wood known commercially as softwood. Compare *deciduous plants.*

consensus science Scientific data, models, theories, and laws that are widely accepted by scientists considered experts in the area of study. These results of science are very reliable. Compare *frontier science.*

conservation Sensible and careful use of natural resources by humans. People with this view are called *conservationists.*

conservation biologist Biologist who investigates human impacts on the diversity of life found on the earth (biodiversity) and develops practical plans for preserving such biodiversity. Compare *conservationist, ecologist, environmentalist, environmental scientist, preservationist, restorationist.*

conservation biology Multidisciplinary science created to deal with the crisis of maintaining the genes, species, communities, and ecosystems that make up earth's biological diversity. Its goals are to investigate human impacts on biodiversity and to develop practical approaches to preserving biodiversity.

conservationist Person concerned with using natural areas and wildlife in ways that sustain them for current and future generations of humans and other forms of life. Compare *conservation biologist, ecologist, environmentalist, environmental scientist, preservationist, restorationist.*

conservation-tillage farming Crop cultivation in which the soil is disturbed little (minimum-tillage farming) or not at all (no-till farming) to reduce soil erosion, lower labor costs, and save energy. Compare *conventional-tillage farming.*

constancy Ability of a living system, such as a population, to maintain a certain size. Compare *inertia, resilience.* See *homeostasis.*

consumer Organism that cannot synthesize the organic nutrients it needs and gets its organic nutrients by feeding on the tissues of producers or of other consumers; generally divided into *primary consumers* (herbivores), *secondary consumers* (carnivores), *tertiary (higher-level) consumers, omnivores,* and *detritivores* (decomposers and detritus feeders). In economics, one who uses economic goods.

contour farming Plowing and planting across the changing slope of land, rather than in straight lines, to help retain water and reduce soil erosion.

contour strip mining Form of surface mining used on hilly or mountainous terrain. A power shovel cuts a series of terraces into the side of a hill. An earthmover removes the overburden, and a power shovel extracts the coal or ore, with the overburden from each new terrace dumped onto the one below. Compare *area strip mining, dredging, subsurface mining.*

controlled burning Deliberately set, carefully controlled surface fires that reduce flammable litter and decrease the chances of damaging crown fires. See *ground fire, surface fire.*

conventional-tillage farming Crop cultivation method in which a planting surface is made by plowing land, breaking up the exposed soil, and then smoothing the surface. Compare *conservation-tillage farming.*

convergent plate boundary Area where earth's lithospheric plates are pushed together. See *subduction zone.* Compare *divergent plate boundary, transform fault.*

coral reef Formation produced by massive colonies containing billions of tiny coral animals, called polyps, that secrete a stony substance (calcium carbonate) around themselves for protection. When the corals die, their empty outer skeletons form layers that cause the reef to grow. They are found in the coastal zones of warm tropical and subtropical oceans.

core Inner zone of the earth. It consists of a solid inner core and a liquid outer core. Compare *crust, mantle.*

cost–benefit analysis See *benefit–cost analysis.*

critical mass Amount of fissionable nuclei needed to sustain a nuclear fission chain reaction.

crop rotation Planting a field, or an area of a field, with different crops from year to year to reduce soil nutrient depletion. A plant such as corn, tobacco, or cotton, which removes large amounts of nitrogen from the soil, is planted one year. The next year a legume such as soybeans, which adds nitrogen to the soil, is planted.

crown fire Extremely hot forest fire that burns ground vegetation and treetops. Compare *controlled burning, ground fire, surface fire.*

crude birth rate Annual number of live births per 1,000 people in the population of a geographic area at the midpoint of a given year. Compare *crude death rate.*

crude death rate Annual number of deaths per 1,000 people in the population of a geographic area at the midpoint of a given year. Compare *crude birth rate.*

crude oil Gooey liquid consisting mostly of hydrocarbon compounds and small amounts of compounds containing oxygen, sulfur, and nitrogen. Extracted from underground accumulations, it is sent to oil refineries, where it is converted to heating oil, diesel fuel, gasoline, tar, and other materials.

crust Solid outer zone of the earth. It consists of oceanic crust and continental crust. Compare *core, mantle.*

cultural eutrophication Overnourishment of aquatic ecosystems with plant nutrients (mostly nitrates and phosphates) because of human activities such as agriculture, urbanization, and discharges from industrial plants and sewage treatment plants. See *eutrophication.*

DDT Dichlorodiphenyltrichloroethane, a chlorinated hydrocarbon that has been widely used as a pesticide but is now banned in some countries.

death rate See *crude death rate.*

debt-for-nature swap Agreement in which a certain amount of foreign debt is canceled in exchange for local currency investments that will improve natural resource management or protect certain areas in the debtor country from harmful development.

deciduous plants Trees, such as oaks and maples, and other plants that survive during dry seasons or cold seasons by shedding their leaves. Compare *coniferous trees, succulent plants.*

decomposer Organism that digests parts of dead organisms and cast-off fragments and wastes of living organisms by breaking down the complex organic molecules in those materials into simpler inorganic compounds and then absorbing the soluble nutrients. Producers return most of these chemicals to the soil and water for reuse. Decomposers consist of various bacteria and fungi. Compare *consumer, detritivore, producer.*

deforestation Removal of trees from a forested area without adequate replanting.

degradable pollutant Potentially polluting chemical that is broken down completely or reduced to acceptable levels by natural physical, chemical, and biological processes. Compare *biodegradable pollutant, nondegradable pollutant, slowly degradable pollutant.*

degree of urbanization Percentage of the population in the world, or a country, living in areas with a population of more than 2,500 people (higher in some countries). Compare *urban growth.*

democracy Government by the people through their elected officials and appointed representatives. In a *constitutional democracy,* a constitution provides the basis of government authority and puts restraints on government power through free elections and freely expressed public opinion.

demographic transition Hypothesis that countries, as they become industrialized, have declines in death rates followed by declines in birth rates.

depletion time The time it takes to use a certain fraction, usually 80%, of the known or estimated supply of a nonrenewable resource at an assumed rate of use. Finding and extracting the remaining 20% usually costs more than it is worth.

desalination Purification of salt water or brackish (slightly salty) water by removal of dissolved salts.

desert Biome in which evaporation exceeds precipitation and the average amount of precipitation is less than 25 centimeters (10 inches) a year. Such areas have little vegetation or have widely spaced, mostly low vegetation. Compare *forest, grassland.*

desertification Conversion of rangeland, rain-fed cropland, or irrigated cropland to desertlike land, with a drop in agricultural productivity of 10% or more. It usually is caused by a combination of overgrazing, soil erosion, prolonged drought, and climate change.

detritivore Consumer organism that feeds on detritus, parts of dead organisms, and cast-off fragments and wastes of living organisms. The two principal types are *detritus feeders* and *decomposers.*

detritus Parts of dead organisms and cast-off fragments and wastes of living organisms.

detritus feeder Organism that extracts nutrients from fragments of dead organisms and their cast-off parts and organic wastes. Examples are earthworms, termites, and crabs. Compare *decomposer.*

deuterium (D; hydrogen-2) Isotope of the element hydrogen, with a nucleus containing one proton and one neutron and a mass number of 2.

developed country Country that is highly industrialized and has a high per capita GNP. Compare *developing country*.

developing country Country that has low to moderate industrialization and low to moderate per capita GNP. Most are located in Africa, Asia, and Latin America. Compare *developed country*.

dew point Temperature at which condensation occurs for a given amount of water vapor.

dieback Sharp reduction in the population of a species when its numbers exceed the carrying capacity of its habitat. See *carrying capacity*.

differential reproduction Phenomenon in which individuals with adaptive genetic traits produce more living offspring than do individuals without such traits. See *natural selection*.

dioxins Family of 75 different chlorinated hydrocarbon compounds formed as unwanted by-products in chemical reactions involving chlorine and hydrocarbons, usually at high temperatures.

discount rate The economic value a resource will have in the future compared with its present value.

dissolved oxygen (DO) content Amount of oxygen gas (O_2) dissolved in a given volume of water at a particular temperature and pressure, often expressed as a concentration in parts of oxygen per million parts of water.

disturbance A discrete event that disrupts an ecosystem or community. Examples of *natural disturbances* include fires, hurricanes, tornadoes, droughts, and floods. Examples of *human-caused disturbances* include deforestation, overgrazing, and plowing.

divergent plate boundary Area where earth's lithospheric plates move apart in opposite directions. Compare *convergent plate boundary, transform fault*.

DNA (deoxyribonucleic acid) Large molecules in the cells of organisms that carry genetic information in living organisms. Compare *wild species*.

domesticated species Wild species tamed or genetically altered by crossbreeding for use by humans for food (cattle, sheep, and food crops), pets (dogs and cats), or enjoyment (animals in zoos and plants in gardens).

dose The amount of a potentially harmful substance an individual ingests, inhales, or absorbs through the skin. Compare *response*. See *dose–response curve, median lethal dose*.

dose–response curve Plot of data showing effects of various doses of a toxic agent on a group of test organisms. See *dose, median lethal dose, response*.

doubling time The time it takes (usually in years) for the quantity of something growing exponentially to double. It can be calculated by dividing the annual percentage growth rate into 70.

drainage basin See *watershed*.

dredge spoils Materials scraped from the bottoms of harbors and streams to maintain shipping channels. They are often contaminated with high levels of toxic substances that have settled out of the water. See *dredging*.

dredging Type of surface mining in which chain buckets and draglines scrape up sand, gravel, and other surface deposits covered with water. It is also used to remove sediment from streams and harbors to maintain shipping channels. See *dredge spoils*.

drift-net fishing Catching fish in huge nets that drift in the water.

drought Condition in which an area does not get enough water because of **(1)** lower-than-normal precipitation or **(2)** higher-than-normal temperatures that increase evaporation.

early successional plant species Plant species found in the early stages of succession that **(1)** grow close to the ground, **(2)** can establish large populations quickly under harsh conditions, and **(3)** have short lives. Compare *late successional plant species, midsuccessional plant species*.

earth capital See *natural resources*.

earthquake Shaking of the ground resulting from the fracturing and displacement of rock, which produces a fault, or from subsequent movement along the fault.

earth resources See *natural resources*.

ecological diversity The variety of forests, deserts, grasslands, oceans, streams, lakes, and other biological communities interacting with one another and with their nonliving environment. See *biodiversity*. Compare *functional diversity genetic diversity, species diversity*.

ecological efficiency Percentage of energy transferred from one trophic level to another in a food chain or web.

ecological footprint A measure of the ecological impact of the **(1)** consumption of food, wood products, and other resources, **(2)** use of buildings, roads, garbage dumps, and other things that consume land space, and **(3)** destruction of the forests needed to absorb the CO_2 produced by burning fossil fuels.

ecological niche Total way of life or role of a species in an ecosystem. It includes all physical, chemical, and biological conditions a species needs to live and reproduce in an ecosystem. See *fundamental niche, realized niche*.

ecological restoration Deliberate alteration of a degraded habitat or ecosystem to restore as much of its ecological structure and function as possible.

ecological succession Process in which communities of plant and animal species in a particular area are replaced over time by a series of different and often more complex communities. See *primary succession, secondary succession*.

ecologist Biological scientist who studies relationships between living organisms and their environment. Compare *conservation biologist, conservationist, environmentalist, environmental scientist, preservationist, restorationist*.

ecology Study of the interactions of living organisms with one another and with their nonliving environment of matter and energy; study of the structure and functions of nature.

economic decision Deciding **(1)** what goods and services to produce, **(2)** how to produce them, **(3)** how much to produce, and **(4)** how to distribute them to people.

economic depletion Exhaustion of 80% of the estimated supply of a nonrenewable resource. Finding, extracting, and processing the remaining 20% usually costs more than it is worth; may also apply to the depletion of a renewable resource, such as a fish or tree species.

economic development Improvement of living standards by economic growth. Compare *economic growth, environmentally sustainable economic development*.

economic growth Increase in the capaity to provide people with goods and services produced by an economy; an increase in GNP. Compare *economic development, environmentally sustainable economic development*.

economic resources Natural resources, capital goods, and labor used in an economy to produce material goods and services. See *natural resources*.

economic system Method that a group of people uses to choose **(1)** what goods and services to produce, **(2)** how to produce them, **(3)** how much to produce, and **(4)** how to distribute them to people. See *capitalist market economic system, pure command economic system, pure free-market economic system*.

economy System of production, distribution, and consumption of economic goods.

ecosphere See *biosphere*.

ecosystem Community of different species interacting with one another and with the chemical and physical factors making up its nonliving environment.

ecosystem services Natural services or natural capital that support life on the earth and are essential to the quality of human life and the functioning of the world's economies. See *natural resources*.

edge effect The existence of a greater number of species and a higher population density in a transition zone (ecotone) between two ecosystems than in either adjacent ecosystem.

electromagnetic radiation Forms of kinetic energy traveling as electromagnetic waves. Examples are radio waves, TV waves, microwaves, infrared radiation, visible light, ultraviolet radiation, X-rays, and gamma rays. Compare *ionizing radiation, nonionizing radiation*.

electron (*e*) Tiny particle moving around outside the nucleus of an atom. Each electron has one unit of negative charge and almost no mass. Compare *neutron, proton*.

element Chemical, such as hydrogen (H), iron (Fe), sodium (Na), carbon (C), nitrogen (N), or oxygen (O), whose distinctly different atoms serve as the basic building blocks of all matter. There are 92 naturally occurring elements. Another 23 have been made in laboratories. Two or more elements combine to form compounds that make up most of the world's matter. Compare *compound*.

endangered species Wild species with so few individual survivors that the species could soon become extinct in all or most of its natural range. Compare *threatened species*.

endemic species Species that is found in only one area. Such species are especially vulnerable to extinction.

energy Capacity to do work by performing mechanical, physical, chemical, or electrical tasks or to cause a heat transfer between two objects at different temperatures.

energy efficiency Percentage of the total energy input that does useful work and is not converted into low-quality, usually useless heat in an energy conversion system or process. See *energy quality, net energy*. Compare *material efficiency*.

energy productivity See *energy efficiency*.

energy quality Ability of a form of energy to do useful work. High-temperature heat and the chemical energy in fossil fuels and nuclear fuels are concentrated high-quality energy. Low-quality energy such as low-temperature heat is dispersed or diluted and cannot do much useful work. See *high-quality energy, low-quality energy*.

environment All external conditions and factors, living and nonliving (chemicals and energy), that affect an organism or other specified system during its lifetime.

environmental degradation Depletion or destruction of a renewable resource such as soil, grassland, forest, or wildlife that is used faster than it is naturally replenished. If such use continues, the resource can become nonrenewable (on a human time scale) or nonexistent (extinct). See also *sustainable yield*.

environmental ethics Our beliefs about what is right or wrong environmental behavior.

environmentalist Person who is concerned about the impact of people on environmental quality and believe that some human actions are degrading parts of the earth's life-support systems for humans and many other forms of life. Compare *conservation biologist, conservationist, ecologist, environmental scientist, preservationist, restorationist*.

environmental justice Fair treatment and meaningful involvement of all people regardless of race, color, sex, national origin, or income with respect to the development, implementation, and enforcement of environmental laws, regulations, and policies.

environmentally sustainable economic development Development that **(1)** *encourages* environmentally sustainable forms of economic growth that meet the basic needs of the current generations of humans and other species without preventing future generations of humans and other species from meeting their basic needs and **(2)** *discourages* environmentally harmful and unsustainable forms of economic growth. It is the economic component of an *environmentally sustainable society.* Compare *economic development, economic growth.*

environmentally sustainable society Society that satisfies the basic needs of its people without depleting or degrading its natural resources and thereby preventing current and future generations of humans and other species from meeting their basic needs.

environmental movement Efforts by citizens at the grassroots level to demand that political leaders enact laws and develop policies to **(1)** curtail pollution, **(2)** clean up polluted environments, and **(3)** protect pristine areas and species from environmental degradation.

environmental resistance All the limiting factors that act together to limit the growth of a population. See *biotic potential, limiting factor.*

environmental revolution Cultural change involving halting population growth and altering lifestyles, political and economic systems, and the way we treat the environment so that we can help sustain the earth for ourselves and other species. This involves working with the rest of nature by learning more about how nature sustains itself. See *environmental wisdom worldview.* Compare *agricultural revolution, hunter–gatherers, industrial revolution, information and globalization revolution.*

environmental science Study of how we and other species interact with one another and with the nonliving environment (matter and energy). It is a physical and social science that integrates knowledge from a wide range of disciplines, including physics, chemistry, biology (especially ecology), geology, geography, resource technology and engineering, resource conservation and management, demography (the study of population dynamics), economics, politics, sociology, psychology, and ethics.

environmental scientist Scientist who uses information from the physical sciences and social sciences to **(1)** understand how the earth works, **(2)** learn how humans interact with the earth, and **(3)** develop solutions to environmental problems. Compare *conservation biologist, conservationist, ecologist, preservationist, restorationist.*

environmental wisdom worldview Beliefs that **(1)** nature exists for all the earth's species, not just for us, and we are not in charge of the rest of nature; **(2)** there is not always more, and it's not all for us; **(3)** some forms of economic growth are beneficial and some are harmful, and our goals should be to design economic and political systems that encourage earth-sustaining forms of growth and discourage or prohibit earth-degrading forms; and **(4)** our success depends on learning to cooperate with one another and with the rest of nature instead of trying to dominate and manage earth's life-support systems primarily for our own use. Compare *frontier environmental worldview, planetary management worldview, spaceship-earth worldview.*

environmental worldview How people think the world works, what they think their role in the world should be, and what they believe is right and wrong environmental behavior (environmental ethics).

EPA U.S. Environmental Protection Agency; responsible for managing federal efforts to control air and water pollution, radiation and pesticide hazards, environmental research, hazardous waste, and solid-waste disposal.

epidemiology Study of the patterns of disease or other harmful effects from toxic exposure within defined groups of people to find out why some people get sick and some do not.

epiphyte Plant that uses its roots to attach itself to branches high in trees, especially in tropical forests.

erosion Process or group of processes by which loose or consolidated earth materials are dissolved, loosened, or worn away and removed from one place and deposited in another. See *weathering.*

estuary Partially enclosed coastal area at the mouth of a river where its fresh water, carrying fertile silt and runoff from the land, mixes with salty seawater.

eukaryotic cell Cell containing a *nucleus,* a region of genetic material surrounded by a membrane. Membranes also enclose several of the other internal parts found in a eukaryotic cell. Compare *prokaryotic cell.*

euphotic zone Upper layer of a body of water through which sunlight can penetrate and support photosynthesis.

eutrophication Physical, chemical, and biological changes that take place after a lake, estuary, or slow-flowing stream receives inputs of plant nutrients—mostly nitrates and phosphates—from natural erosion and runoff from the surrounding land basin. See *cultural eutrophication.*

eutrophic lake Lake with a large or excessive supply of plant nutrients, mostly nitrates and phosphates. Compare *mesotrophic lake, oligotrophic lake.*

evaporation Conversion of a liquid into a gas.

even-aged management Method of forest management in which trees, sometimes of a single species in a given stand, are maintained at about the same age and size and are harvested all at once. Compare *uneven-aged management.*

evergreen plants Plants that keep some of their leaves or needles throughout the year. Examples are ferns and cone-bearing trees (conifers) such as firs, spruces, pines, redwoods, and sequoias. Compare *deciduous plants, succulent plants.*

evolution See *biological evolution.*

exhaustible resource See *nonrenewable resource.*

exotic species See *nonnative species.*

experiment Procedure a scientist uses to study some phenomenon under known conditions. Some experiments are conducted in the laboratory, but others are conducted in nature. The resulting scientific data or facts must be verified or confirmed by repeated observations and measurements, ideally by several different investigators.

exponential growth Growth in which some quantity, such as population size or economic output, increases by a fixed percentage of the whole in a given time period; when the increase in quantity over time is plotted, this type of growth yields a curve shaped like the letter *J.* Compare *linear growth.*

external benefit Beneficial social effect of producing and using an economic good that is not included in the market price of the good. Compare *external cost, full cost.*

external cost Harmful social effect of producing and using an economic good that is not included in the market price of the good. Compare *external benefit, full cost, internal cost.*

externalities Social benefits ("goods") and social costs ("bads") not included in the market price of an economic good. See *external benefit, external cost.* Compare *full cost, internal cost.*

extinction Complete disappearance of a species from the earth. This happens when a species cannot adapt and successfully reproduce under new environmental conditions or when it evolves into one or more new species. Compare *speciation.* See also *endangered species, threatened species.*

family planning Providing information, clinical services, and contraceptives to help people choose the number and spacing of children they want to have.

famine Widespread malnutrition and starvation in a particular area because of a shortage of food, usually caused by drought, war, flood, earthquake, or other catastrophic events that disrupt food production and distribution.

feedback loop Circuit of sensing, evaluating, and reacting to changes in environmental conditions as a result of information fed back into a system; it occurs when one change leads to some other change, which eventually reinforces or slows the original change. See *negative feedback loop, positive feedback loop.*

feedlot Confined outdoor or indoor space used to raise hundreds to thousands of domesticated livestock. Compare *rangeland.*

fermentation See *anaerobic respiration.*

fertilizer Substance that adds inorganic or organic plant nutrients to soil and improves its ability to grow crops, trees, or other vegetation. See *commercial inorganic fertilizer, organic fertilizer.*

financial resources Cash, investments, and monetary institutions used to support the use of natural resources and human resources to provide economic goods and services. Compare *human resources, manufactured resources, natural resources.*

first law of energy See *first law of thermodynamics.*

first law of thermodynamics In any physical or chemical change, no detectable amount of energy is created or destroyed, but in these processes energy can be changed from one form to another; you can't get more energy out of something than you put in; in terms of energy quantity, you can't get something for nothing (there is no free lunch). This law does not apply to nuclear changes, in which energy can be produced from small amounts of matter. See also *second law of thermodynamics.*

fishery Concentrations of particular aquatic species suitable for commercial harvesting in a given ocean area or inland body of water.

fish farming Form of aquaculture in which fish are cultivated in a controlled pond or other environment and harvested when they reach the desired size. See also *fish ranching.*

fish ranching Form of aquaculture in which members of a fish species such as salmon are held in captivity for the first few years of their lives, released, and then harvested as adults when they return from the ocean to their freshwater birthplace to spawn. See also *fish farming.*

fissionable isotope Isotope that can split apart when hit by a neutron at the right speed and thus undergo nuclear fission. Examples are uranium-235 and plutonium-239.

floodplain Flat valley floor next to a stream channel. For legal purposes the term often is applied to any low area that has the potential for flooding, including certain coastal areas.

flows See *throughputs.*

flyway Generally fixed route along which waterfowl migrate from one area to another at certain seasons of the year.

food chain Series of organisms in which each eats or decomposes the preceding one. Compare *food web*.

food web Complex network of many interconnected food chains and feeding relationships. Compare *food chain*.

forest Biome with enough average annual precipitation (at least 76 centimeters, or 30 inches) to support growth of various tree species and smaller forms of vegetation. Compare *desert, grassland*.

fossil fuel Products of partial or complete decomposition of plants and animals that occur as crude oil, coal, natural gas, or heavy oils as a result of exposure to heat and pressure in the earth's crust over millions of years. See *coal, crude oil, natural gas*.

fossils Skeletons, bones, shells, body parts, leaves, seeds, or impressions of such items that provide recognizable evidence of organisms that lived long ago.

free-access resource See *common-property resource*.

Freons See *chlorofluorocarbons*.

frontier environmental worldview Viewing undeveloped land as a hostile wilderness to be conquered (cleared, planted) and exploited for its resources as quickly as possible. Compare *environmental wisdom worldview, planetary management worldview, spaceship-earth worldview*.

frontier science Preliminary scientific data, hypotheses, and models that have not been widely tested and accepted. Compare *consensus science*.

full cost Cost of a good when its internal costs and its estimated short- and long-term external costs are included in its market price. Compare *external cost, internal cost*.

functional diversity Biological and chemical processes or functions such as energy flow and matter cycling needed for the survival of species and biological communities. See *biodiversity, ecological diversity, genetic diversity, species diversity*.

fundamental niche The full potential range of the physical, chemical, and biological factors a species can use if there is no competition from other species. See *ecological niche*. Compare *realized niche*.

fungicide Chemical that kills fungi.

game species Type of wild animal that people hunt or fish for, for sport and recreation and sometimes for food.

gamma rays A form of ionizing electromagnetic radiation with a high energy content emitted by some radioisotopes. They readily penetrate body tissues. See also *alpha particle, beta particle*.

gangue Waste or undesired material in an ore. See *ore*.

GDP See *gross domestic product*.

gene flow Movement of genes between populations, which can lead to changes in the genetic composition of local populations.

gene mutation See *mutation*.

gene pool The sum total of all genes found in the individuals of the population of a particular species.

generalist species Species with a broad ecological niche. They can live in many different places, eat a variety of foods, and tolerate a wide range of environmental conditions. Examples are flies, cockroaches, mice, rats, and human beings. Compare *specialist species*.

genes Coded units of information about specific traits that are passed on from parents to offspring during reproduction. They consist of segments of DNA molecules found in chromosomes.

gene splicing See *genetic engineering*.

genetic adaptation Changes in the genetic makeup of organisms of a species that allow the species to reproduce and gain a competitive advantage under changed environmental conditions. See *differential reproduction, evolution, mutation, natural selection*.

genetically modified organism (GMO) Organism whose genetic makeup has been modified by genetic engineering.

genetic diversity Variability in the genetic makeup among individuals within a single species. See *biodiversity*. Compare *ecological diversity, functional diversity, species diversity*.

genetic drift Change in the genetic composition of a population by chance. It is especially important for small populations.

genetic engineering Insertion of an alien gene into an organism to give it a beneficial genetic trait. Compare *artificial selection, natural selection*.

genome Complete set of genetic information for an organism.

geographic isolation Separation of populations of a species for fairly long times into different areas.

geology Study of the earth's dynamic history. Geologists study and analyze rocks and the features and processes of the earth's interior and surface.

geothermal energy Heat transferred from the earth's underground concentrations of **(1)** dry steam (steam with no water droplets), **(2)** wet steam (a mixture of steam and water droplets), or **(3)** hot water trapped in fractured or porous rock.

globalization Broad process of global social, economic, and environmental change that leads to an increasingly integrated world. See *information and globalization revolution*.

global warming Warming of the earth's atmosphere as a result of increases in the concentrations of one or more greenhouse gases primarily as a result of human activities. See *greenhouse effect, greenhouse gases*.

GNP See *gross national product*.

grassland Biome found in regions where moderate annual average precipitation (25 to 76 centimeters, or 10 to 30 inches) is enough to support the growth of grass and small plants but not enough to support large stands of trees. Compare *desert, forest*.

greenhouse effect A natural effect that releases heat in the atmosphere (troposphere) near the earth's surface. Water vapor, carbon dioxide, ozone, and several other gases in the lower atmosphere (troposphere) absorb some of the infrared radiation (heat) radiated by the earth's surface. This causes their molecules to vibrate and transform the absorbed energy into longer-wavelength infrared radiation (heat) in the troposphere. If the atmospheric concentrations of these greenhouse gases rise and they are not removed by other natural processes, the average temperature of the lower atmosphere will increase gradually. Compare *global warming*.

greenhouse gases Gases in the earth's lower atmosphere (troposphere) that cause the greenhouse effect. Examples are carbon dioxide, chlorofluorocarbons, ozone, methane, water vapor, and nitrous oxide.

green manure Freshly cut or still-growing green vegetation that is plowed into the soil to increase the organic matter and humus available to support crop growth. Compare *animal manure*.

green revolution Popular term for introduction of scientifically bred or selected varieties of grain (rice, wheat, maize) that, with high enough inputs of fertilizer and water, can greatly increase crop yields.

gross domestic product (GDP) Total market value in current dollars of all goods and services produced *within a country*, usually during a year. Compare *gross national product, gross world product*.

gross national product (GNP) Total market value in current dollars of all goods and services produced by an economy, usually during a year. Compare *gross domestic product, gross world product*.

gross primary productivity (GPP) The rate at which an ecosystem's producers capture and store a given amount of chemical energy as biomass in a given length of time. Compare *net primary productivity*.

gross world product (GWP) Market value in current dollars of all goods and services produced in the world each year. Compare *gross domestic product, gross national product*.

ground fire Fire that burns decayed leaves or peat deep below the ground surface. Compare *crown fire, surface fire*.

groundwater Water that sinks into the soil and is stored in slowly flowing and slowly renewed underground reservoirs called aquifers; underground water in the zone of saturation, below the water table. Compare *runoff, surface water*.

gully reclamation Restoring land suffering from gully erosion by seeding gullies with quick-growing plants, building small dams to collect silt and gradually fill in the channels, and building channels to divert water away from the gully.

habitat Place or type of place where an organism or population of organisms lives. Compare *ecological niche*.

habitat fragmentation Breakup of a habitat into smaller pieces, usually as a result of human activities.

half-life Time needed for one-half of the nuclei in a radioisotope to emit its radiation. Each radioisotope has a characteristic half-life, which may range from a few millionths of a second to several billion years. See *radioactivity*.

hazard Something that can cause injury, disease, economic loss, or environmental damage. See also *risk*.

hazardous chemical Chemical that can cause harm because it **(1)** is flammable or explosive, **(2)** can irritate or damage the skin or lungs (such as strong acidic or alkaline substances), or **(3)** can cause allergic reactions of the immune system (allergens). See also *toxic chemical*.

hazardous waste Any solid, liquid, or containerized gas that **(1)** can catch fire easily, **(2)** is corrosive to skin tissue or metals, **(3)** is unstable and can explode or release toxic fumes, or **(4)** has harmful concentrations of one or more toxic materials that can leach out. See also *toxic waste*.

heat Total kinetic energy of all the randomly moving atoms, ions, or molecules within a given substance, excluding the overall motion of the whole object. Heat always flows spontaneously from a hot sample of matter to a colder sample of matter. This is one way to state the second law of thermodynamics. Compare *temperature*.

herbicide Chemical that kills a plant or inhibits its growth.

herbivore Plant-eating organism. Examples are deer, sheep, grasshoppers, and zooplankton. Compare *carnivore, omnivore*.

heterotroph See *consumer*.

high-input agriculture See *industrialized agriculture*.

high-quality energy Energy that is concentrated and has great ability to perform useful work. Examples are high-temperature heat and the energy in electricity, coal, oil, gasoline, sunlight, and nuclei of uranium-235. Compare *low-quality energy*.

high-quality matter Matter that is concentrated and contains a high concentration of a useful resource. Compare *low-quality matter*.

high-throughput economy The situation in most advanced industrialized countries, in which ever-increasing economic growth is sustained by maximizing the rate at which matter and energy resources are used, with little emphasis on pollution prevention, recycling, reuse, reduction of unnecessary waste, and other forms of resource conservation. Compare *low-throughput economy, matter-recycling economy*.

homeostasis Maintenance of favorable internal conditions in a system despite fluctuations in external conditions. See *constancy, inertia, resilience*.

host Plant or animal on which a parasite feeds.

human capital See *human resources*.

human resources Physical and mental talents of people used to produce, distribute, and sell an economic good. Compare *financial resources, manufactured resources, natural resources*.

humus Slightly soluble residue of undigested or partially decomposed organic material in topsoil. This material helps retain water and water-soluble nutrients, which can be taken up by plant roots.

hunter–gatherers People who get their food by gathering edible wild plants and other materials and by hunting wild animals and fish. Compare *agricultural revolution, environmental revolution, industrial revolution, information and globalization revolution*.

hydrocarbon Organic compound of hydrogen and carbon atoms. The simplest hydrocarbon is methane (CH_4), the major component of natural gas.

hydroelectric power plant Structure in which the energy of falling or flowing water spins a turbine generator to produce electricity.

hydrologic cycle Biogeochemical cycle that collects, purifies, and distributes the earth's fixed supply of water from the environment to living organisms and then back to the environment.

hydropower Electrical energy produced by falling or flowing water. See *hydroelectric power plant*.

hydrosphere The earth's **(1)** liquid water (oceans, lakes, other bodies of surface water, and underground water), **(2)** frozen water (polar ice caps, floating ice caps, and ice in soil, known as permafrost), and **(3)** small amounts of water vapor in the atmosphere. See also *hydrologic cycle*.

identified resources Deposits of a particular mineral-bearing material of which the location, quantity, and quality are known or have been estimated from direct geological evidence and measurements. Compare *undiscovered resources*.

igneous rock Rock formed when molten rock material (magma) wells up from the earth's interior, cools, and solidifies into rock masses. Compare *metamorphic rock, sedimentary rock*. See *rock cycle*.

immature community Community at an early stage of ecological succession. It usually has a low number of species and ecological niches and cannot capture and use energy and cycle critical nutrients as efficiently as more complex, mature communities. Compare *mature community*.

immigrant species See *nonative species*.

immigration Migration of people into a country or area to take up permanent residence.

indicator species Species that serve as early warnings that a community or ecosystem is being degraded. Compare *keystone species, native species, nonative species*.

industrialized agriculture Using large inputs of energy from fossil fuels (especially oil and natural gas), water, fertilizer, and pesticides to produce large quantities of crops and livestock for domestic and foreign sale. Compare *subsistence farming*.

industrial revolution Use of new sources of energy from fossil fuels and later from nuclear fuels, and use of new technologies, to grow food and manufacture products. Compare *agricultural revolution, environmental revolution, hunter–gatherers, information and globalization revolution*.

industrial smog Type of air pollution consisting mostly of a mixture of sulfur dioxide, suspended droplets of sulfuric acid formed from some of the sulfur dioxide, and a variety of suspended solid particles. Compare *photochemical smog*.

inertia Ability of a living system to resist being disturbed or altered. Compare *constancy, resilience*.

infant mortality rate Number of babies out of every 1,000 born each year that die before their first birthday.

infiltration Downward movement of water through soil.

information and globalization revolution Use of new technologies such as the telephone, radio, television, computers, the internet, automated databases, and remote sensing satellites to enable people to have increasingly rapid access to much more information on a global scale. Compare *agricultural revolution, environmental revolution, hunter–gatherers, industrial revolution*.

inherent value See *intrinsic value*.

inland wetland Land away from the coast, such as a swamp, marsh, or bog, that is covered all or part of the time with fresh water. Compare *coastal wetland*.

inorganic compounds All compounds not classified as organic compounds. See *organic compounds*.

inorganic fertilizer See *commercial inorganic fertilizer*.

input Matter, energy, or information entering a system. Compare *output, throughput*.

input pollution control See *pollution prevention*.

insecticide Chemical that kills insects.

instrumental value Value of an organism, species, ecosystem, or the earth's biodiversity based on its usefulness to us. Compare *intrinsic value*.

integrated pest management (IPM) Combined use of biological, chemical, and cultivation methods in proper sequence and timing to keep the size of a pest population below the size that causes economically unacceptable loss of a crop or livestock animal.

intercropping Growing two or more different crops at the same time on a plot. For example, a carbohydrate-rich grain that depletes soil nitrogen and a protein-rich legume that adds nitrogen to the soil may be intercropped. Compare *monoculture, polyculture, polyvarietal cultivation*.

intermediate goods See *manufactured resources*.

internal cost Direct cost paid by the producer and the buyer of an economic good. Compare *external benefit, external cost, full cost*.

interplanting Simultaneously growing a variety of crops on the same plot. See *agroforestry, intercropping, polyculture, polyvarietal cultivation*.

interspecific competition Members of two or more species trying to use the same limited resources in an ecosystem. See *competition, competitive exclusion principle, intraspecific competition*.

intertidal zone The area of shoreline between low and high tides.

intraspecific competition Two or more organisms of a single species trying to use the same limited resources in an ecosystem. See *competition, interspecific competition*.

intrinsic rate of increase (r) Rate at which a population could grow if it had unlimited resources. Compare *environmental resistance*.

intrinsic value Value of an organism, species, ecosystem, or the earth's biodiversity based on its existence, regardless of whether it has any usefulness to us. Compare *instrumental value*.

inversion See *temperature inversion*.

invertebrates Animals that have no backbones. Compare *vertebrates*.

ion Atom or group of atoms with one or more positive (+) or negative (−) electrical charges. Compare *atom, molecule*.

isotopes Two or more forms of a chemical element that have the same number of protons but different mass numbers because they have different numbers of neutrons in their nuclei.

***J*-shaped curve** Curve with a shape similar to that of the letter *J*; can represent prolonged exponential growth.

keystone species Species that play roles affecting many other organisms in an ecosystem. Compare *indicator species, native species, nonative species*.

kinetic energy Energy that matter has because of its mass and speed or velocity. Compare *potential energy*.

K-selected species Species that produce a few, often fairly large offspring but invest a great deal of time and energy to ensure that most of those offspring reach reproductive age. Compare *r-selected species*.

K-strategists See *K-selected species*.

lake Large natural body of standing fresh water formed when water from precipitation, land runoff, or groundwater flow fills a depression in the earth created by **(1)** glaciation, **(2)** earth movement, **(3)** volcanic activity, or **(4)** a giant meteorite. See *eutrophic lake, mesotrophic lake, oligotrophic lake*.

landfill See *sanitary landfill*.

land-use planning Process for deciding the best present and future use of each parcel of land in an area.

late successional plant species Mostly trees that can tolerate shade and form a fairly stable complex forest community. Compare *early successional plant species, midsuccessional plant species*.

latitude Distance from the equator. Compare *altitude*.

law of conservation of energy See *first law of thermodynamics*.

law of conservation of matter In any physical or chemical change, matter is neither created nor destroyed but merely changed from one form to another; in physical and chemical changes, existing atoms are rearranged into different spatial patterns (physical changes) or different combinations (chemical changes).

law of tolerance The existence, abundance, and distribution of a species in an ecosystem are determined by whether the levels of one or more physical or chemical factors fall within the range tolerated by the species. See *threshold effect*.

LD_{50} See *median lethal dose*.

LDC See *developing country*.

leaching Process in which various chemicals in upper layers of soil are dissolved and carried to lower layers and, in some cases, to groundwater.

less developed country (LDC) See *developing country*.

life-cycle cost Initial cost plus lifetime operating costs of an economic good. Compare *full cost*.

life expectancy Average number of years a newborn infant can be expected to live.

limiting factor Single factor that limits the growth, abundance, or distribution of the population of a species in an ecosystem. See *limiting factor principle.*

limiting factor principle Too much or too little of any abiotic factor can limit or prevent growth of a population of a species in an ecosystem, even if all other factors are at or near the optimum range of tolerance for the species.

linear growth Growth in which a quantity increases by some fixed amount during each unit of time. Compare *exponential growth.*

liquefied natural gas (LNG) Natural gas converted to liquid form by cooling to a very low temperature.

liquefied petroleum gas (LPG) Mixture of liquefied propane (C_3H_8) and butane (C_4H_{10}) gas removed from natural gas and used as a fuel.

lithosphere Outer shell of the earth, composed of the crust and the rigid, outermost part of the mantle outside the asthenosphere; material found in earth's plates. See *crust, mantle.*

loams Soils containing a mixture of clay, sand, silt, and humus. Good for growing most crops.

logistic growth Pattern in which exponential population growth occurs when the population is small, and population growth decreases steadily with time as the population approaches the carrying capacity. See *S-shaped curve.*

low-input agriculture See *sustainable agriculture.*

low-quality energy Energy that is dispersed and has little ability to do useful work. An example is low-temperature heat. Compare *high-quality energy.*

low-quality matter Matter that is dilute or dispersed or contains a low concentration of a useful resource. Compare *high-quality matter.*

low-throughput economy Economy based on working with nature by **(1)** recycling and reusing discarded matter, **(2)** preventing pollution, **(3)** conserving matter and energy resources by reducing unnecessary waste and use, **(4)** not degrading renewable resources, **(5)** building things that are easy to recycle, reuse, and repair, **(6)** not allowing population size to exceed the carrying capacity of the environment, and **(7)** preserving biodiversity. See *environmental worldview.* Compare *high-throughput economy, matter-recycling economy.*

low-waste society See *low-throughput economy.*

LPG See *liquefied petroleum gas.*

macroevolution Long-term, large-scale evolutionary changes among groups of species. Compare *microevolution.*

magma Molten rock below the earth's surface.

malnutrition Faulty nutrition, caused by a diet that does not supply an individual with enough protein, essential fats, vitamins, minerals, and other nutrients needed for good health. Compare *overnutrition, undernutrition.*

mangrove swamps Swamps found on the coastlines in warm tropical climates. They are dominated by mangrove trees, any of about 55 species of trees and shrubs that can live partly submerged in the salty environment of coastal swamps.

mantle Zone of the earth's interior between its core and its crust. Compare *core, crust.* See *lithosphere.*

manufactured capital See *manufactured resources.*

manufactured resources Manufactured items made from natural resources and used to produce and distribute economic goods and services bought by consumers. These include tools, machinery, equipment, factory buildings, and transportation and distribution facilities. Compare *financial resources, human resources, natural resources.*

manure See *animal manure, green manure.*

mass The amount of material in an object.

mass extinction A catastrophic, widespread, often global event in which major groups of species are wiped out over a short time compared with normal (background) extinctions. Compare *background extinction.*

mass number Sum of the number of neutrons (n) and the number of protons (p) in the nucleus of an atom. It gives the approximate mass of that atom. Compare *atomic number.*

mass transit Buses, trains, trolleys, and other forms of transportation that carry large numbers of people.

material efficiency Total amount of material needed to produce each unit of goods or services. Also called *resource productivity.* Compare *energy efficiency.*

matter Anything that has mass (the amount of material in an object) and takes up space. On the earth, where gravity is present, we weigh an object to determine its mass.

matter quality Measure of how useful a matter resource is, based on its availability and concentration. See *high-quality matter, low-quality matter.*

matter-recycling economy Economy that emphasizes recycling the maximum amount of all resources that can be recycled. The goal is to allow economic growth to continue without depleting matter resources and without producing excessive pollution and environmental degradation. Compare *high-throughput economy, low-throughput economy.*

mature community Fairly stable community in an advanced stage of ecological succession; usually has a diverse array of species and ecological niches; captures and uses energy and cycles critical chemicals more efficiently than simpler, immature communities. Compare *immature community.*

maximum sustainable yield See *sustainable yield.*

MDC See *developed country.*

median lethal dose (LD_{50}) Amount of a toxic material per unit of body weight of test animals that kills half the test population in a certain time.

megacity City with 10 million or more people.

mesotrophic lake Lake with a moderate supply of plant nutrients. Compare *eutrophic lake, oligotrophic lake.*

metabolism Ability of a living cell or organism to capture and transform matter and energy from its environment to supply its needs for survival, growth, and reproduction.

metamorphic rock Rock produced when a preexisting rock is subjected to high temperatures (which may cause it to melt partially), high pressures, chemically active fluids, or a combination of these agents. Compare *igneous rock, sedimentary rock.* See *rock cycle.*

metastasis Spread of malignant (cancerous) cells from a tumor to other parts of the body.

metropolitan area See *urban area.*

microclimates Local climatic conditions that differ from the general climate of a region. Various topographic features of the earth's surface such as mountains and cities typically create them.

microevolution The small genetic changes a population undergoes. Compare *macroevolution.*

microorganisms Organisms such as bacteria that are so small that they can be seen only by using a microscope.

midsuccessional plant species Grasses and low shrubs that are less hardy than early successional plant species. Compare *early successional plant species, late successional plant species.*

mineral Any naturally occurring inorganic substance found in the earth's crust as a crystalline solid. See *mineral resource.*

mineral resource Concentration of naturally occurring solid, liquid, or gaseous material in or on the earth's crust in a form and amount such that extracting and converting it into useful materials or items is currently or potentially profitable. Mineral resources are classified as *metallic* (such as iron and tin ores) or *nonmetallic* (such as fossil fuels, sand, and salt).

minimum-tillage farming See *conservation-tillage farming.*

minimum viable population (MVP) Estimate of the smallest number of individuals necessary to ensure the survival of a population in a region for a specified time period, typically ranging from decades to 100 years.

mixture Combination of one or more elements and compounds.

model An approximate representation or simulation of a system being studied.

molecule Combination of two or more atoms of the same chemical element (such as O_2) or different chemical elements (such as H_2O) held together by chemical bonds. Compare *atom, ion.*

monoculture Cultivation of a single crop, usually on a large area of land. Compare *polyculture, polyvarietal cultivation.*

more developed country (MDC) See *developed country.*

multiple use Use of an ecosystem such as a forest for a variety of purposes such as timber harvesting, wildlife habitat, watershed protection, and recreation. Compare *sustainable yield.*

municipal solid waste Solid materials discarded by homes and businesses in or near urban areas. See *solid waste.*

mutagen Chemical or form of radiation that causes inheritable changes (mutations) in the DNA molecules in the genes found in chromosomes. See *carcinogen, mutation, teratogen.*

mutation A random change in DNA molecules making up genes that can yield changes in anatomy, physiology, or behavior in offspring. See *mutagen.*

mutualism Type of species interaction in which both participating species generally benefit. Compare *commensalism.*

native species Species that normally live and thrive in a particular ecosystem. Compare *indicator species, keystone species, nonative species.*

natural capital See *natural resources.*

natural gas Underground deposits of gases consisting of 50-90% by weight methane gas (CH_4) and small amounts of heavier gaseous hydrocarbon compounds such as propane (C_3H_8) and butane (C_4H_{10}).

natural greenhouse effect Heat buildup in the troposphere because of the presence of certain gases, called greenhouse gases. Without this effect, the earth would be nearly as cold as Mars, and life as we know it could not exist. Compare *global warming.*

natural law See *scientific law.*

natural radioactive decay Nuclear change in which unstable nuclei of atoms spontaneously shoot out particles (usually alpha or beta particles) or energy (gamma rays) at a fixed rate.

natural rate of extinction See *background extinction.*

natural recharge Natural replenishment of an aquifer by precipitation, which percolates downward through soil and rock. See *recharge area.*

natural resources The earth's natural materials and processes that sustain other species and us. Compare *financial resources, human resources, manufactured resources.*

natural selection Process by which a particular beneficial gene (or set of genes) is reproduced in succeeding generations more than other genes. The result of natural selection is a population that contains a greater proportion of organisms better adapted to certain environmental conditions. See *adaptation, biological evolution, differential reproduction, mutation.*

negative feedback loop Situation in which a change in a certain direction provides information that causes a system to change less in that direction. Compare *positive feedback loop.*

nekton Strongly swimming organisms found in aquatic systems. Compare *benthos, plankton.*

net energy Total amount of useful energy available from an energy resource or energy system over its lifetime, minus the amount of energy **(1)** used (the first energy law), **(2)** automatically wasted (the second energy law), and **(3)** unnecessarily wasted in finding, processing, concentrating, and transporting it to users.

net primary productivity (NPP) Rate at which all the plants in an ecosystem produce net useful chemical energy; equal to the difference between the rate at which the plants in an ecosystem produce useful chemical energy (gross primary productivity) and the rate at which they use some of that energy through cellular respiration. Compare *gross primary productivity.*

neutral solution Water solution containing an equal number of hydrogen ions (H^+) and hydroxide ions (OH^-); water solution with a pH of 7. Compare *acid solution, basic solution.*

neutron (*n*) Elementary particle in the nuclei of all atoms (except hydrogen-1). It has a relative mass of 1 and no electric charge. Compare *electron, proton.*

niche See *ecological niche.*

nitrogen cycle Cyclic movement of nitrogen in different chemical forms from the environment to organisms and then back to the environment.

nitrogen fixation Conversion of atmospheric nitrogen gas into forms useful to plants by lightning, bacteria, and cyanobacteria; it is part of the nitrogen cycle.

noise pollution Any unwanted, disturbing, or harmful sound that **(1)** impairs or interferes with hearing, **(2)** causes stress, **(3)** hampers concentration and work efficiency, or **(4)** causes accidents.

nondegradable pollutant Material that is not broken down by natural processes. Examples are the toxic elements lead and mercury. Compare *biodegradable pollutant, degradable pollutant, slowly degradable pollutant.*

nonnative species Species that migrate into an ecosystem or are deliberately or accidentally introduced into an ecosystem by humans. Compare *native species.*

nonpersistent pollutant See *degradable pollutant.*

nonpoint source Large or dispersed land areas such as cropfields, streets, and lawns that discharge pollutants into the environment over a large area. Compare *point source.*

nonrenewable resource Resource that exists in a fixed amount (stock) in various places in the earth's crust and has the potential for renewal by geological, physical, and chemical processes taking place over hundreds of millions to billions of years. Examples are copper, aluminum, coal, and oil. We classify these resources as exhaustible because we are extracting and using them at a much faster rate than they were formed. Compare *renewable resource.*

nontransmissible disease A disease that is not caused by living organisms and does not spread from one person to another. Examples are most cancers, diabetes, cardiovascular disease, and malnutrition. Compare *transmissible disease.*

no-till farming See *conservation-tillage farming.*

nuclear change Process in which nuclei of certain isotopes spontaneously change, or are forced to change, into one or more different isotopes. The three principal types of nuclear change are natural radioactivity, nuclear fission, and nuclear fusion. Compare *chemical change, physical change.*

nuclear energy Energy released when atomic nuclei undergo a nuclear reaction such as the spontaneous emission of radioactivity, nuclear fission, or nuclear fusion.

nuclear fission Nuclear change in which the nuclei of certain isotopes with large mass numbers (such as uranium-235 and plutonium-239) are split apart into lighter nuclei when struck by a neutron. This process releases more neutrons and a large amount of energy. Compare *nuclear fusion.*

nuclear fusion Nuclear change in which two nuclei of isotopes of elements with a low mass number (such as hydrogen-2 and hydrogen-3) are forced together at extremely high temperatures until they fuse to form a heavier nucleus (such as helium-4). This process releases a large amount of energy. Compare *nuclear fission.*

nucleus Extremely tiny center of an atom, making up most of the atom's mass. It contains one or more positively charged protons and one or more neutrons with no electrical charge (except for a hydrogen-1 atom, which has one proton and no neutrons in its nucleus).

nutrient Any food or element an organism must take in to live, grow, or reproduce.

nutrient cycle See *biogeochemical cycle.*

oil See *crude oil.*

old-growth forest Virgin and old, second-growth forests containing trees that are often hundreds, sometimes thousands of years old. Examples include forests of Douglas fir, western hemlock, giant sequoia, and coastal redwoods in the western United States. Compare *second-growth forest, tree farm.*

oligotrophic lake Lake with a low supply of plant nutrients. Compare *eutrophic lake, mesotrophic lake.*

omnivore Animal that can use both plants and other animals as food sources. Examples are pigs, rats, cockroaches, and people. Compare *carnivore, herbivore.*

open-pit mining Removing minerals such as gravel, sand, and metal ores by digging them out of the earth's surface and leaving an open pit. Compare *area strip mining, contour strip mining.*

open sea The part of an ocean that is beyond the continental shelf. Compare *coastal zone.*

ore Part of a metal-yielding material that can be economically and legally extracted at a given time. An ore typically contains two parts: the ore mineral, which contains the desired metal, and waste mineral material (gangue).

organic compounds Compounds containing carbon atoms combined with each other and with atoms of one or more other elements such as hydrogen, oxygen, nitrogen, sulfur, phosphorus, chlorine, and fluorine. All other compounds are called *inorganic compounds.*

organic farming Producing crops and livestock naturally by using organic fertilizer (manure, legumes, compost) and natural pest control (bugs that eat harmful bugs, plants that repel bugs, and environmental controls such as crop rotation) instead of using commercial inorganic fertilizers and synthetic pesticides and herbicides. See *sustainable agriculture.*

organic fertilizer Organic material such as animal manure, green manure, and compost, applied to cropland as a source of plant nutrients. Compare *commercial inorganic fertilizer.*

organism Any form of life.

other resources Identified and undiscovered resources not classified as reserves. Compare *identified resources, reserves, undiscovered resources.*

output Matter, energy, or information leaving a system. Compare *input, throughput.*

output pollution control See *pollution cleanup.*

overburden Layer of soil and rock overlying a mineral deposit. It is removed during surface mining.

overfishing Harvesting so many fish of a species, especially immature fish, that there is not enough breeding stock left to replenish the species.

overgrazing Destruction of vegetation when too many grazing animals feed too long and exceed the carrying capacity of a rangeland or pasture area.

overnutrition Diet so high in calories, saturated (animal) fats, salt, sugar, and processed foods and so low in vegetables and fruits that the consumer runs high risks of diabetes, hypertension, heart disease, and other health hazards. Compare *malnutrition, undernutrition.*

oxygen-demanding wastes Organic materials that are usually biodegraded by aerobic (oxygen-consuming) bacteria if there is enough dissolved oxygen in the water. See also *biological oxygen demand.*

ozone depletion Decrease in concentration of ozone (O_3) in the stratosphere. See *ozone layer.*

ozone layer Layer of gaseous ozone (O_3) in the stratosphere that protects life on earth by filtering out most harmful ultraviolet radiation from the sun.

PANs Peroxyacyl nitrates. Group of chemicals found in photochemical smog.

parasite Consumer organism that lives on or in and feeds on a living plant or animal, known as the host, over an extended period of time. The parasite draws nourishment from and gradually weakens its host; it may or may not kill the host. See *parasitism.*

parasitism Interaction between species in which one organism, called the parasite, preys on another organism, called the host, by living on or in the host. See *host, parasite.*

parts per billion (ppb) Number of parts of a chemical found in 1 billion parts of a particular gas, liquid, or solid.

parts per million (ppm) Number of parts of a chemical found in 1 million parts of a particular gas, liquid, or solid.

parts per trillion (ppt) Number of parts of a chemical found in 1 trillion parts of a particular gas, liquid, or solid.

passive solar heating system System that captures sunlight directly within a structure and converts it into low-temperature heat for space heating or for heating water for domestic use without the use of mechanical devices. Compare *active solar heating system.*

pasture Managed grassland or enclosed meadow that usually is planted with domesticated grasses or other forage to be grazed by livestock. Compare *feedlot, rangeland.*

pathogen Organism that produces disease.

PCBs See *polychlorinated biphenyls.*

per capita GNP Annual gross national product (GNP) of a country divided by its total population. See *gross national product.*

percolation Passage of a liquid through the spaces of a porous material such as soil.

perennial Plant that can live for more than 2 years. Compare *annual.*

permafrost Perennially frozen layer of the soil that forms when the water there freezes. It is found in arctic tundra.

permeability The degree to which underground rock and soil pores are interconnected and thus a measure of the degree to which water can flow freely from one pore to another. Compare *porosity*.

perpetual resource An essentially inexhaustible resource on a human time scale. Solar energy is an example. Compare *nonrenewable resource, renewable resource*.

persistence How long a pollutant stays in the air, water, soil, or body. See also *inertia*.

persistent pollutant See *slowly degradable pollutant*.

pest Unwanted organism that directly or indirectly interferes with human activities.

pesticide Any chemical designed to kill or inhibit the growth of an organism that people consider undesirable. See *fungicide, herbicide, insecticide*.

petrochemicals Chemicals obtained by refining (distilling) crude oil. They are used as raw materials in manufacturing most industrial chemicals, fertilizers, pesticides, plastics, synthetic fibers, paints, medicines, and many other products.

petroleum See *crude oil*.

pH Numeric value that indicates the relative acidity or alkalinity of a substance on a scale of 0 to 14, with the neutral point at 7. Acid solutions have pH values lower than 7, and basic or alkaline solutions have pH values greater than 7.

phosphorus cycle Cyclic movement of phosphorus in different chemical forms from the environment to organisms and then back to the environment.

photochemical smog Complex mixture of air pollutants produced in the lower atmosphere by the reaction of hydrocarbons and nitrogen oxides under the influence of sunlight. Especially harmful components include ozone, peroxyacyl nitrates (PANs), and various aldehydes. Compare *industrial smog*.

photosynthesis Complex process that takes place in cells of green plants. Radiant energy from the sun is used to combine carbon dioxide (CO_2) and water (H_2O) to produce oxygen (O_2) and carbohydrates (such as glucose, $C_6H_{12}O_6$) and other nutrient molecules. Compare *aerobic respiration, chemosynthesis*.

photovoltaic cell (solar cell) Device in which radiant (solar) energy is converted directly into electrical energy.

physical change Process that alters one or more physical properties of an element or a compound without altering its chemical composition. Examples are changing the size and shape of a sample of matter (crushing ice and cutting aluminum foil) and changing a sample of matter from one physical state to another (boiling and freezing water). Compare *chemical change, nuclear change*.

phytoplankton Small, drifting plants, mostly algae and bacteria, found in aquatic ecosystems. Compare *plankton, zooplankton*.

pioneer community First integrated set of plants, animals, and decomposers found in an area undergoing primary ecological succession. See *immature community, mature community*.

pioneer species First hardy species, often microbes, mosses, and lichens, that begin colonizing a site as the first stage of ecological succession. See *ecological succession, pioneer community*.

planetary management worldview Beliefs that **(1)** we are the planet's most important species; **(2)** there is always more, and it's all for us; **(3)** all economic growth is good, more economic growth is better, and the potential for economic growth is limitless; and **(4)** our success depends on how well we can understand, control, and manage the earth's life-support systems for our own benefit. See *spaceship-earth worldview*. Compare *environmental wisdom worldview*.

plankton Small plant organisms (phytoplankton) and animal organisms (zooplankton) that float in aquatic ecosystems.

plantation agriculture Growing specialized crops such as bananas, coffee, and cacao in tropical developing countries, primarily for sale to developed countries.

plates Various-sized areas of the earth's lithosphere that move slowly around with the mantle's flowing asthenosphere. Most earthquakes and volcanoes occur around the boundaries of these plates. See *lithosphere, plate tectonics*.

plate tectonics Theory of geophysical processes that explains the movements of lithospheric plates and the processes that occur at their boundaries. See *lithosphere, plates*.

point source Single identifiable source that discharges pollutants into the environment. Examples are the **(1)** smokestack of a power plant or an industrial plant, **(2)** drainpipe of a meatpacking plant, **(3)** chimney of a house, or **(4)** exhaust pipe of an automobile. Compare *nonpoint source*.

poison A chemical that in one dose kills exactly 50% of the animals (usually rats and mice) in a test population (usually 60 to 200 animals) within a 14-day period. See *median lethal dose*.

politics Process through which individuals and groups try to influence or control government policies and actions that affect the local, state, national, and international communities.

pollutant A particular chemical or form of energy that can adversely affect the health, survival, or activities of humans or other living organisms. See *pollution*.

pollution An undesirable change in the physical, chemical, or biological characteristics of air, water, soil, or food that can adversely affect the health, survival, or activities of humans or other living organisms.

pollution cleanup Device or process that removes or reduces the level of a pollutant after it has been produced or has entered the environment. Examples are automobile emission control devices and sewage treatment plants. Compare *pollution prevention*.

pollution prevention Device or process that **(1)** prevents a potential pollutant from forming or entering the environment or **(2)** sharply reduces the amount entering the environment. Compare *pollution cleanup*.

polychlorinated biphenyls (PCBs) Group of 209 different toxic, oily, synthetic chlorinated hydrocarbon compounds that can be biologically amplified in food chains and webs.

polyculture Complex form of intercropping in which a large number of different plants maturing at different times are planted together. See also *intercropping*. Compare *monoculture, polyvarietal cultivation*.

polyvarietal cultivation Planting a plot of land with several varieties of the same crop. Compare *intercropping, monoculture, polyculture*.

population Group of individual organisms of the same species living in a particular area.

population change An increase or decrease in the size of a population. It is equal to (Births + Immigration) – (Deaths + Emigration).

population density Number of organisms in a particular population found in a specified area or volume.

population dispersion General pattern in which the members of a population are arranged throughout its habitat.

population distribution Variation of population density over a particular geographic area. For example, a country has a high population density in its urban areas and a much lower population density in rural areas.

population dynamics Major abiotic and biotic factors that tend to increase or decrease the population size and age and sex composition of a species.

population size Number of individuals making up a population's gene pool.

porosity Percentage of space in rock or soil occupied by voids, whether the voids are isolated or connected. Compare *permeability*.

positive feedback loop Situation in which a change in a certain direction provides information that causes a system to change further in the same direction. Compare *negative feedback loop*.

potential energy Energy stored in an object because of its position or the position of its parts. Compare *kinetic energy*.

poverty Inability to meet basic needs for food, clothing, and shelter.

ppb See *parts per billion*.

ppm See *parts per million*.

ppt See *parts per trillion*.

precautionary principle When there is scientific uncertainty about potentially serious harm from chemicals or technologies, decision makers should act to prevent harm to humans and the environment. See *pollution prevention*.

precipitation Water in the form of rain, sleet, hail, and snow that falls from the atmosphere onto the land and bodies of water.

predation Situation in which an organism of one species (the predator) captures and feeds on parts or all of an organism of another species (the prey).

predator Organism that captures and feeds on parts or all of an organism of another species (the prey).

predator–prey relationship Interaction between two organisms of different species in which one organism, called the *predator*, captures and feeds on parts or all of another organism, called the *prey*.

preservationist Person concerned primarily with setting aside or protecting undisturbed natural areas from harmful human activities. Compare *conservation biologist, conservationist, ecologist, environmentalist, environmental scientist, restorationist*.

prey Organism that is captured and serves as a source of food for an organism of another species (the predator).

primary consumer Organism that feeds on all or part of plants (herbivore) or on other producers. Compare *detritivore, omnivore, secondary consumer*.

primary pollutant Chemical that has been added directly to the air by natural events or human activities and occurs in a harmful concentration. Compare *secondary pollutant*.

primary productivity See *gross primary productivity, net primary productivity*.

primary sewage treatment Mechanical sewage treatment in which large solids are filtered out by screens and suspended solids settle out as sludge in a sedimentation tank. Compare *advanced sewage treatment, secondary sewage treatment*.

primary succession Sequential development of communities in a bare area that has never been occupied by a community of organisms. Compare *secondary succession*.

probability A mathematical statement about how likely it is that something will happen.

producer Organism that uses solar energy (green plant) or chemical energy (some bacteria) to manufacture the organic compounds it needs as nutrients from simple inorganic compounds obtained from its environment. Compare *consumer, decomposer*.

prokaryotic cell Cell that doesn't have a distinct nucleus. Other internal parts are also not enclosed by membranes. Compare *eukaryotic cell.*

proton (*p*) Positively charged particle in the nuclei of all atoms. Each proton has a relative mass of 1 and a single positive charge. Compare *electron, neutron.*

pure capitalism See *pure free-market economic system.*

pure command economic system System in which all economic decisions are made by the government or some other central authority. Compare *capitalist market economic system, pure free-market economic system.*

pure free-market economic system System in which all economic decisions are made in the market, where buyers and sellers of economic goods interact freely, with no government or other interference. Compare *capitalist market economic system, pure command economic system.*

pyramid of biomass Diagram representing the biomass, or total dry weight of all living organisms, that can be supported at each trophic level in a food chain or food web. See *pyramid of energy flow, pyramid of numbers.*

pyramid of energy flow Diagram representing the flow of energy through each trophic level in a food chain or food web. With each energy transfer, only a small part (typically 10%) of the usable energy entering one trophic level is transferred to the organisms at the next trophic level. Compare *pyramid of biomass, pyramid of numbers.*

pyramid of numbers Diagram representing the number of organisms of a particular type that can be supported at each trophic level from a given input of solar energy at the producer trophic level in a food chain or food web. Compare *pyramid of biomass, pyramid of energy flow.*

radiation Fast-moving particles (particulate radiation) or waves of energy (electromagnetic radiation). See *alpha particle, beta particle, gamma rays.*

radiation temperature inversion Temperature inversion that typically occurs at night in which a layer of warm air lies atop a layer of cooler air nearer the ground as the air near the ground cools faster than the air above it. As the sun rises and warms the earth's surface, the inversion normally disappears by noon and disperses the pollutants built up during the night. See *temperature inversion.* Compare *subsidence temperature inversion.*

radioactive decay Change of a radioisotope to a different isotope by the emission of radioactivity.

radioactive isotope See *radioisotope.*

radioactive waste Waste products of nuclear power plants, research, medicine, weapon production, or other processes involving nuclear reactions. See *radioactivity.*

radioactivity Nuclear change in which unstable nuclei of atoms spontaneously shoot out "chunks" of mass, energy, or both at a fixed rate. The three principal types of radioactivity are gamma rays and fast-moving alpha particles and beta particles.

radioisotope Isotope of an atom that spontaneously emits one or more types of radioactivity (alpha particles, beta particles, gamma rays).

rain shadow effect Low precipitation on the far side (leeward side) of a mountain when prevailing winds flow up and over a high mountain or range of high mountains. This creates semiarid and arid conditions on the leeward side of a high mountain range.

rangeland Land that supplies forage or vegetation (grasses, grasslike plants, and shrubs) for grazing and browsing animals and is not intensively managed. Compare *feedlot, pasture.*

range of tolerance Range of chemical and physical conditions that must be maintained for populations of a particular species to stay alive and grow, develop, and function normally. See *law of tolerance.*

realized niche Parts of the fundamental niche of a species that are actually used by that species. See *ecological niche, fundamental niche.*

recharge area Any area of land allowing water to pass through it and into an aquifer. See *aquifer, natural recharge.*

recycling Collecting and reprocessing a resource so that it can be made into new products. An example is collecting aluminum cans, melting them down, and using the aluminum to make new cans or other aluminum products. Compare *reuse.*

reforestation Renewal of trees and other types of vegetation on land where trees have been removed; can be done naturally by seeds from nearby trees or artificially by planting seeds or seedlings.

reliable runoff Surface runoff of water that generally can be counted on as a stable source of water from year to year.

renewable resource Resource that can be replenished fairly rapidly (hours to several decades) through natural processes. Examples are trees in forests, grasses in grasslands, wild animals, fresh surface water in lakes and streams, most groundwater, fresh air, and fertile soil. If such a resource is used faster than it is replenished, it can be depleted and converted into a nonrenewable resource. Compare *nonrenewable resource* and *perpetual resource.* See also *environmental degradation.*

replacement-level fertility Number of children a couple must have to replace them. The average for a country or the world usually is slightly higher than 2 children per couple (2.1 in the United States and 2.5 in some developing countries) because some children die before reaching their reproductive years. See also *total fertility rate.*

reproduction Production of offspring by one or more parents.

reproductive isolation Long-term geographic separation of members of a particular sexually reproducing species.

reproductive potential See *biotic potential.*

reserves Resources that have been identified and from which a usable mineral can be extracted profitably at present prices with current mining technology. See *identified resources, undiscovered resources.*

reserve-to-production ratio Number of years reserves of a particular nonrenewable mineral will last at current annual production rates. See *reserves.*

resilience Ability of a living system to restore itself to original condition after being exposed to an outside disturbance that is not too drastic. See *constancy, inertia.*

resource Anything obtained from the living and nonliving environment to meet human needs and wants. It can also be applied to other species.

resource partitioning Process of dividing up resources in an ecosystem so that species with similar needs (overlapping ecological niches) use the same scarce resources at different times, in different ways, or in different places. See *ecological niche, fundamental niche, realized niche.*

resource productivity See *material efficiency.*

respiration See *aerobic respiration.*

response The amount of health damage caused by exposure to a certain dose of a harmful substance or form of radiation. See *dose, dose–response curve, median lethal dose.*

restoration ecology Research and scientific study devoted to restoring, repairing, and reconstructing damaged ecosystems.

restorationist Scientist or other person devoted to the partial or complete restoration of natural areas that have been degraded by human activities. Compare *conservation biologist, conservationist, ecologist, environmental scientist, preservationist.*

reuse Using a product over and over again in the same form. An example is collecting, washing, and refilling glass beverage bottles. Compare *recycling.*

riparian zones Thin strips and patches of vegetation that surround streams. They are very important habitats and resources for wildlife.

risk The probability that something undesirable will result from deliberate or accidental exposure to a hazard. See *risk analysis, risk assessment, risk–benefit analysis, risk management.*

risk analysis Identifying hazards, evaluating the nature and severity of risks (*risk assessment*), using this and other information to determine options and make decisions about reducing or eliminating risks (*risk management*), and communicating information about risks to decision makers and the public (*risk communication*).

risk assessment Process of gathering data and making assumptions to estimate short- and long-term harmful effects on human health or the environment from exposure to hazards associated with the use of a particular product or technology. See *risk–benefit analysis.*

risk–benefit analysis Estimate of the short- and long-term risks and benefits of using a particular product or technology. See *risk, risk assessment.*

risk communication Communicating information about risks to decision makers and the public. See *risk, risk analysis, risk–benefit analysis.*

risk management Using risk assessment and other information to determine options and make decisions about reducing or eliminating risks. See *risk, risk analysis, risk-benefit analysis, risk communication.*

rock Any material that makes up a large, natural, continuous part of earth's crust. See *mineral.*

rock cycle Largest and slowest of the earth's cycles, consisting of geologic, physical, and chemical processes that form and modify rocks and soil in the earth's crust over millions of years.

r-selected species Species that reproduce early in their life span and produce large numbers of usually small and short-lived offspring in a short period of time. Compare *K-selected species.*

r-strategists See *r-selected species.*

rule of 70 Doubling time (in years) = 70 / percentage growth rate. See *doubling time, exponential growth.*

runoff Fresh water from precipitation and melting ice that flows on the earth's surface into nearby streams, lakes, wetlands, and reservoirs. See *reliable runoff, surface runoff, surface water.* Compare *groundwater.*

rural area Geographic area in the United States with a population of less than 2,500. The number of people used in this definition may vary in different countries. Compare *urban area.*

salinity Amount of various salts dissolved in a given volume of water.

salinization Accumulation of salts in soil that can eventually make the soil unable to support plant growth.

saltwater intrusion Movement of salt water into freshwater aquifers in coastal and inland areas as groundwater is withdrawn faster than it is recharged by precipitation.

sanitary landfill Waste disposal site on land in which waste is spread in thin layers, compacted, and covered with a fresh layer of clay or plastic foam each day.

scavenger Organism that feeds on dead organisms that were killed by other organisms or died naturally. Examples are vultures, flies, and crows. Compare *detritivore.*

science Attempts to discover order in nature and use that knowledge to make predictions about what should happen in nature. See *consensus science, frontier science, scientific data, scientific hypothesis, scientific law, scientific methods, scientific model, scientific theory.*

scientific data Facts obtained by making observations and measurements. Compare *model, scientific hypothesis, scientific law, scientific methods, scientific model, scientific theory.*

scientific hypothesis An educated guess that attempts to explain a scientific law or certain scientific observations. Compare *scientific data, scientific law, scientific methods, scientific model, scientific theory.*

scientific law Description of what scientists find happening in nature over and over in the same way, without known exception. See *first law of thermodynamics, law of conservation of matter, second law of thermodynamics.* Compare *scientific data, scientific hypothesis, scientific methods, scientific model, scientific theory.*

scientific methods The ways scientists gather data and formulate and test scientific hypotheses, models, theories, and laws. See *scientific data, scientific hypothesis, scientific law, scientific model, scientific theory.*

scientific model A simulation of complex processes and systems. Many are mathematical models that are run and tested using computers.

scientific theory A well-tested and widely accepted scientific hypothesis. Compare *scientific data, scientific hypothesis, scientific law, scientific methods, scientific model.*

secondary consumer Organism that feeds only on primary consumers. Compare *detritivore, omnivore, primary consumer.*

secondary pollutant Harmful chemical formed in the atmosphere when a primary air pollutant reacts with normal air components or other air pollutants. Compare *primary pollutant.*

secondary sewage treatment Second step in most waste treatment systems in which aerobic bacteria break down up to 90% of degradable, oxygen-demanding organic wastes in wastewater. This usually is done by bringing sewage and bacteria together in trickling filters or in the activated sludge process. Compare *advanced sewage treatment, primary sewage treatment.*

secondary succession Sequential development of communities in an area in which natural vegetation has been removed or destroyed but the soil is not destroyed. Compare *primary succession.*

second-growth forest Stands of trees resulting from secondary ecological succession. Compare *old-growth forest, tree farm.*

second law of energy See *second law of thermodynamics.*

second law of thermodynamics In any conversion of heat energy to useful work, some of the initial energy input is always degraded to a lower-quality, more dispersed, less useful energy, usually low-temperature heat that flows into the environment; you can't break even in terms of energy quality. See *first law of thermodynamics.*

sedimentary rock Rock that forms from the accumulated products of erosion and in some cases from the compacted shells, skeletons, and other remains of dead organisms. Compare *igneous rock, metamorphic rock.* See *rock cycle.*

seed-tree cutting Removal of nearly all trees on a site in one cutting, with a few seed-producing trees left uniformly distributed to regenerate the forest. Compare *clear-cutting, selective cutting, shelterwood cutting, strip cutting.*

selective cutting Cutting of intermediate-aged, mature, or diseased trees in an uneven-aged forest stand, either singly or in small groups. This encourages the growth of younger trees and maintains an uneven-aged stand. Compare *clear-cutting, seed-tree cutting, shelterwood cutting, strip cutting.*

septic tank Underground tank for treating wastewater from a home in rural and suburban areas. Bacteria in the tank decompose organic wastes, and the sludge settles to the bottom of the tank. The effluent flows out of the tank into the ground through a field of drainpipes.

sexual reproduction Reproduction in organisms that produce offspring by combining sex cells or *gametes* (such as ovum and sperm) from both parents. This produces offspring that have combinations of traits from their parents. Compare *asexual reproduction.*

shelterbelt See *windbreak.*

shelterwood cutting Removal of mature, marketable trees in an area in a series of partial cuttings to allow regeneration of a new stand under the partial shade of older trees, which are later removed. Typically, this is done by making two or three cuts over a decade. Compare *clear-cutting, seed-tree cutting, selective cutting, strip cutting.*

shifting cultivation Clearing a plot of ground in a forest, especially in tropical areas, and planting crops on it for a few years (typically 2–5 years) until the soil is depleted of nutrients or the plot has been invaded by a dense growth of vegetation from the surrounding forest. Then a new plot is cleared and the process is repeated. The abandoned plot cannot successfully grow crops for 10–30 years. See also *slash-and-burn cultivation.*

slash-and-burn cultivation Cutting down trees and other vegetation in a patch of forest, leaving the cut vegetation on the ground to dry, and then burning it. The ashes that are left add nutrients to the nutrient-poor soils found in most tropical forest areas. Crops are planted between tree stumps. Plots must be abandoned after a few years (typically 2–5 years) because of loss of soil fertility or invasion of vegetation from the surrounding forest. See also *shifting cultivation.*

slowly degradable pollutant Material that is slowly broken down into simpler chemicals or reduced to acceptable levels by natural physical, chemical, and biological processes. Compare *biodegradable pollutant, degradable pollutant, nondegradable pollutant.*

sludge Gooey mixture of toxic chemicals, infectious agents, and settled solids removed from wastewater at a sewage treatment plant.

smart growth Form of urban planning that recognizes that urban growth will occur but uses zoning laws and an array of other tools to **(1)** prevent sprawl, **(2)** direct growth to certain areas, **(3)** protect ecologically sensitive and important lands and waterways, and **(4)** develop urban areas that are more environmentally sustainable and more enjoyable places to live.

smelting Process in which a desired metal is separated from the other elements in an ore mineral.

smog Originally a combination of smoke and fog but now used to describe other mixtures of pollutants in the atmosphere. See *industrial smog, photochemical smog.*

soil Complex mixture of inorganic minerals (clay, silt, pebbles, and sand), decaying organic matter, water, air, and living organisms.

soil conservation Methods used to reduce soil erosion, prevent depletion of soil nutrients, and restore nutrients already lost by erosion, leaching, and excessive crop harvesting.

soil erosion Movement of soil components, especially topsoil, from one place to another, usually by wind, flowing water, or both. This natural process can be greatly accelerated by human activities that remove vegetation from soil.

soil horizons Horizontal zones that make up a particular mature soil. Each horizon has a distinct texture and composition that vary with different types of soils. See *soil profile.*

soil permeability Rate at which water and air move from upper to lower soil layers. Compare *porosity.*

soil porosity See *porosity.*

soil profile Cross-sectional view of the horizons in a soil.

soil structure How the particles that make up a soil are organized and clumped together. See also *soil permeability, soil texture.*

soil texture Relative amounts of the different types and sizes of mineral particles in a sample of soil.

solar capital Solar energy from the sun reaching the earth. Compare *natural resources.*

solar cell See *photovoltaic cell.*

solar collector Device for collecting radiant energy from the sun and converting it into heat. See *active solar heating system, passive solar heating system.*

solar energy Direct radiant energy from the sun and a number of indirect forms of energy produced by the direct input. Principal indirect forms of solar energy include wind, falling and flowing water (hydropower), and biomass (solar energy converted into chemical energy stored in the chemical bonds of organic compounds in trees and other plants).

solid waste Any unwanted or discarded material that is not a liquid or a gas. See *municipal solid waste.*

spaceship-earth worldview View of the earth as a spaceship: a machine that we can understand, control, and change at will by using advanced technology. See *planetary management worldview.* Compare *environmental wisdom worldview.*

specialist species Species with a narrow ecological niche. They may be able to **(1)** live in only one type of habitat, **(2)** tolerate only a narrow range of climatic and other environmental conditions, or **(3)** use only one type or a few types of food. Compare *generalist species.*

speciation Formation of two species from one species as a result of divergent natural selection in response to changes in environmental conditions; usually takes thousands of years. Compare *extinction.*

species Group of organisms that resemble one another in appearance, behavior, chemical makeup and processes, and genetic structure. Organisms that reproduce sexually are classified as members of the same species only if they can actually or potentially interbreed with one another and produce fertile offspring.

species diversity Number of different species and their relative abundances in a given area. See *biodiversity.* Compare *ecological diversity, genetic diversity.*

species equilibrium model See *theory of island biogeography.*

spoils Unwanted rock and other waste materials produced when a material is removed from the earth's surface or subsurface by mining, dredging, quarrying, and excavation.

***S*-shaped curve** Leveling off of an exponential, *J*-shaped curve when a rapidly growing population exceeds the carrying capacity of its environment and ceases to grow.

stability Ability of a living system to withstand or recover from externally imposed changes or stresses. See *constancy, inertia, resilience.*

stewardship View that because of our superior intellect and power or because of our religious beliefs, we have an ethical responsibility to manage and care for domesticated plants and animals and the rest of nature. Compare *environmental wisdom worldview, planetary management worldview.*

stratosphere Second layer of the atmosphere, extending about 17–48 kilometers (11–30 miles) above the earth's surface. It contains small amounts of gaseous ozone (O_3), which filters out about 95% of the incoming harmful ultraviolet (UV) radiation emitted by the sun. Compare *troposphere.*

stream Flowing body of surface water. Examples are creeks and rivers.

strip cropping Planting regular crops and close-growing plants, such as hay or nitrogen-fixing legumes, in alternating rows or bands to help reduce depletion of soil nutrients.

strip cutting A variation of clear-cutting in which a strip of trees is clear-cut along the contour of the land, with the corridor narrow enough to allow natural regeneration within a few years. After regeneration, another strip is cut above the first, and so on. Compare *clear-cutting, seed-tree cutting, selective cutting, shelterwood cutting.*

strip mining Form of surface mining in which bulldozers, power shovels, or stripping wheels remove large chunks of the earth's surface in strips. See *area strip mining, contour mining, surface mining.* Compare *subsurface mining.*

subatomic particles Extremely small particles—electrons, protons, and neutrons—that make up the internal structure of atoms.

subduction zone Area in which oceanic lithosphere is carried downward (subducted) under the island arc or continent at a convergent plate boundary. A trench ordinarily forms at the boundary between the two converging plates. See *convergent plate boundary.*

subsidence Slow or rapid sinking of part of the earth's crust that is not slope-related.

subsidence temperature inversion Inversion of normal air temperature layers when a large mass of warm air moves into a region at a high altitude and floats over a mass of colder air near the ground. This keeps the air over a city stagnant and prevents vertical mixing and dispersion of air pollutants. See *temperature inversion.* Compare *radiation temperature inversion.*

subsurface mining Extraction of a metal ore or fuel resource such as coal from a deep underground deposit. Compare *surface mining.*

succession See *ecological succession, primary succssion, secondary succession.*

succulent plants Plants, such as desert cacti, that survive in dry climates by having no leaves, thus reducing the loss of scarce water. They store water and use sunlight to produce the food they need in the thick, fleshy tissue of their green stems and branches. Compare *deciduous plants, evergreen plants.*

sulfur cycle Cyclic movement of sulfur in different chemical forms from the environment to organisms and then back to the environment.

superinsulated house House that is heavily insulated and extremely airtight. Typically, active or passive solar collectors are used to heat water, and an air-to-air heat exchanger is used to prevent buildup of excessive moisture and indoor air pollutants.

surface fire Forest fire that burns only undergrowth and leaf litter on the forest floor. Compare *crown fire, ground fire.* See controlled burning.

surface mining Removing soil, subsoil, and other strata and then extracting a mineral deposit found fairly close to the earth's surface. See *area strip mining, contour strip mining.* Compare *subsurface mining.*

surface runoff Water flowing off the land into bodies of surface water. See *reliable runoff.*

surface water Precipitation that does not infiltrate the ground or return to the atmosphere by evaporation or transpiration. See *runoff.* Compare *groundwater.*

survivorship curve Graph showing the number of survivors in different age groups for a particular species.

sustainability Ability of a system to survive for some specified (finite) time.

sustainable agriculture Method of growing crops and raising livestock based on organic fertilizers, soil conservation, water conservation, biological pest control, and minimal use of nonrenewable fossil-fuel energy.

sustainable development See *environmentally sustainable economic development.*

sustainable living Taking no more potentially renewable resources from the natural world than can be replenished naturally and not overloading the capacity of the environment to cleanse and renew itself by natural processes.

sustainable society A society that manages its economy and population size without doing irreparable environmental harm by overloading the planet's ability to absorb environmental insults, replenish its resources, and sustain human and other forms of life over a specified period, usually hundreds to thousands of years. During this period, it satisfies the needs of its people without depleting natural resources and thereby jeopardizing the prospects of current and future generations of humans and other species.

sustainable yield (sustained yield) Highest rate at which a potentially renewable resource can be used without reducing its available supply throughout the world or in a particular area. See also *environmental degradation.*

synergistic interaction Interaction of two or more factors or processes so that the combined effect is greater than the sum of their separate effects.

synergy See *synergistic interaction.*

synfuels Synthetic gaseous and liquid fuels produced from solid coal or sources other than natural gas or crude oil.

synthetic natural gas (SNG) Gaseous fuel containing mostly methane produced from solid coal.

system A set of components that function and interact in some regular and theoretically predictable manner.

tailings Rock and other waste materials removed as impurities when waste mineral material is separated from the metal in an ore.

temperature Measure of the average speed of motion of the atoms, ions, or molecules in a substance or combination of substances at a given moment. Compare *heat.*

temperature inversion Layer of dense, cool air trapped under a layer of less dense, warm air. This prevents upward-flowing air currents from developing. In a prolonged inversion, air pollution in the trapped layer may build up to harmful levels. See *radiation temperature inversion, subsidence temperature inversion.*

teratogen Chemical, ionizing agent, or virus that causes birth defects. Compare *carcinogen, mutagen.*

terracing Planting crops on a long, steep slope that has been converted into a series of broad, nearly level terraces with short vertical drops from one to another that run along the contour of the land to retain water and reduce soil erosion.

terrestrial Pertaining to land. Compare *aquatic.*

territoriality Process in which organisms patrol or mark an area around their home, nesting, or major feeding site and defend it against members of their own species.

tertiary (higher-level) consumers Animals that feed on animal-eating animals. They feed at high trophic levels in food chains and webs. Examples are hawks, lions, bass, and sharks. Compare *detritivore, primary consumer, secondary consumer.*

tertiary sewage treatment See *advanced sewage treatment.*

theory of evolution Widely accepted scientific idea that all life-forms developed from earlier life-forms. Although this theory conflicts with the creation stories of many religions, it is the way biologists explain how life has changed over the past 3.6–3.8 billion years and why it is so diverse today.

theory of island biogeography The number of species found on an island is determined by a balance between two factors: the **(1)** *immigration rate* (of species new to the island) from other inhabited areas and **(2)** *extinction rate* (of species established on the island). The model predicts that at some point the rates of immigration and extinction will reach an equilibrium point that determines the island's average number of different species (species diversity).

thermal inversion See *temperature inversion.*

thermocline Zone of gradual temperature decrease between warm surface water and colder deep water in a lake, reservoir, or ocean.

threatened species Wild species that is still abundant in its natural range but is likely to become endangered because of a decline in numbers. Compare *endangered species.*

threshold effect The harmful or fatal effect of a small change in environmental conditions that exceeds the limit of tolerance of an organism or population of a species. See *law of tolerance.*

throughput Rate of flow of matter, energy, or information through a system. Compare *input, output.*

throwaway society See *high-throughput economy.*

time delay Time lag between the input of a stimulus into a system and the response to the stimulus.

tolerance limits Minimum and maximum limits for physical conditions (such as temperature) and concentrations of chemical substances beyond which no members of a particular species can survive. See *law of tolerance.*

total fertility rate (TFR) Estimate of the average number of children who will be born alive to a woman during her lifetime if she passes through all her childbearing years (ages 15–44) conforming to age-specific fertility rates of a given year. In simpler terms, it is an estimate of the average number of children a woman will have during her childbearing years.

totally planned economy See *pure command economic system.*

toxic chemical Chemical that is fatal to humans in low doses or fatal to more than 50% of test animals at stated concentrations. Most are neurotoxins, which attack nerve cells. See *carcinogen, hazardous chemical, mutagen, teratogen.*

toxicity Measure of how harmful a substance is.

toxicology Study of the adverse effects of chemicals on health.

toxic waste Form of hazardous waste that causes death or serious injury (such as burns, respiratory diseases, cancers, or genetic mutations). See *hazardous waste.*

traditional intensive agriculture Producing enough food for a farm family's survival and perhaps a surplus that can be sold. This type of agriculture uses higher inputs of labor, fertilizer, and water than traditional subsistence agriculture.

See *traditional subsistence agriculture*. Compare *industrial agriculture*.

traditional subsistence agriculture Production of enough crops or livestock for a farm family's survival and, in good years, a surplus to sell or put aside for hard times. Compare *industrial agriculture, traditional intensive agriculture*.

tragedy of the commons Depletion or degradation of a renewable resource to which people have free and unmanaged access. An example is the depletion of commercially desirable fish species in the open ocean beyond areas controlled by coastal countries. See *common-property resource*.

transform fault Area where the earth's lithospheric plates move in opposite but parallel directions along a fracture (fault) in the lithosphere. Compare *convergent plate boundary, divergent plate boundary*.

transmissible disease A disease that is caused by living organisms (such as bacteria, viruses, and parasitic worms) and can spread from one person to another by air, water, food, or body fluids (or in some cases by insects or other organisms). Compare *nontransmissible disease*.

transpiration Process in which water **(1)** is absorbed by the root systems of plants, **(2)** moves up through the plants, **(3)** passes through pores (stomata) in their leaves or other parts, and **(4)** evaporates into the atmosphere as water vapor.

tree farm Site planted with one or only a few tree species in an even-aged stand. When the stand matures it is usually harvested by clear-cutting and then replanted. These farms normally are used to grow rapidly growing tree species for fuelwood, timber, or pulpwood. See *even-aged management*. Compare *old-growth forest, second-growth forest, uneven-aged management*.

tree plantation See *tree farm*.

trophic level All organisms that are the same number of energy transfers away from the original source of energy (for example, sunlight) that enters an ecosystem. For example, all producers belong to the first trophic level, and all herbivores belong to the second trophic level in a food chain or a food web.

troposphere Innermost layer of the atmosphere. It contains about 75% of the mass of earth's air and extends about 17 kilometers (11 miles) above sea level. Compare *stratosphere*.

true cost See *full cost*.

undergrazing Reduction of the net primary productivity of grassland vegetation and grass cover from absence of grazing for long periods (at least 5 years). Compare *overgrazing*.

undernutrition Consuming insufficient food to meet one's minimum daily energy needs for a long enough time to cause harmful effects. Compare *malnutrition, overnutrition*.

undiscovered resources Potential supplies of a particular mineral resource, believed to exist because of geologic knowledge and theory, although specific locations, quality, and amounts are unknown. Compare *identified resources, reserves*.

uneven-aged management Method of forest management in which trees of different species in a given stand are maintained at many ages and sizes to permit continuous natural regeneration. Compare *even-aged management*.

upwelling Movement of nutrient-rich bottom water to the ocean's surface. This can occur far from shore but usually occurs along certain steep coastal areas where the surface layer of ocean water is pushed away from shore and replaced by cold, nutrient-rich bottom water.

urban area Geographic area with a population of 2,500 or more. The number of people used in this definition may vary, with some countries setting the minimum number of people at 10,000–50,000.

urban growth Rate of growth of an urban population. Compare *degree of urbanization*.

urban heat island Buildup of heat in the atmosphere above an urban area. The large concentration of cars, buildings, factories, and other heat-producing activities produces this heat.

urbanization See *degree of urbanization*.

urban sprawl Growth of low-density development on the edges of cities and towns. See *smart growth*.

utilitarian value See *instrumental value*.

vertebrates Animals that have backbones. Compare *invertebrates*.

volcano Vent or fissure in the earth's surface through which magma, liquid lava, and gases are released into the environment.

water cycle See *hydrologic cycle*.

waterlogging Saturation of soil with irrigation water or excessive precipitation so that the water table rises close to the surface.

water pollution Any physical or chemical change in surface water or groundwater that can harm living organisms or make water unfit for certain uses.

watershed Land area that delivers water, sediment, and dissolved substances via small streams to a major stream (river).

water table Upper surface of the zone of saturation, in which all available pores in the soil and rock in the earth's crust are filled with water.

weather Short-term changes in the temperature, barometric pressure, humidity, precipitation, sunshine, cloud cover, wind direction and speed, and other conditions in the troposphere at a given place and time. Compare *climate*.

weathering Physical and chemical processes in which solid rock exposed at earth's surface is changed to separate solid particles and dissolved material, which can then be moved to another place as sediment. See *erosion*.

wetland Land that is covered all or part of the time with salt water or fresh water, excluding streams, lakes, and the open ocean. See *coastal wetland, inland wetland*.

wilderness Area where the earth and its community of life have not been seriously disturbed by humans and where humans are only temporary visitors.

wildlife management Manipulation of populations of wild species (especially game species) and their habitats for **(1)** human benefit, **(2)** the welfare of other species, and **(3)** the preservation of threatened and endangered wildlife species.

wild species Species found in the natural environment. Compare *domesticated species*.

windbreak Row of trees or hedges planted to partially block wind flow and reduce soil erosion on cultivated land.

wind farm Cluster of small to medium-sized wind turbines in a windy area to capture wind energy and convert it into electrical energy.

worldview How people think the world works and what they think their role in the world should be. See *environmental wisdom worldview, planetary management worldview, spaceship-earth worldview*.

zero population growth (ZPG) State in which the birth rate (plus immigration) equals the death rate (plus emigration) so that the population of a geographic area is no longer increasing.

zone of aeration Zone in soil that is not saturated with water and that lies above the water table. See *water table, zone of saturation*.

zone of saturation Area where all available pores in soil and rock in the earth's crust are filled by water. See *water table, zone of aeration*.

zoning Regulating how various parcels of land can be used.

zooplankton Animal plankton. Small floating herbivores that feed on plant plankton (phytoplankton). Compare *phytoplankton*.

INDEX

Note: Page numbers in **boldface type** indicate definitions of key terms. Page numbers followed by *f* and *t* indicate figures and tables, respectively.